a LANGE medical book

D0743664

basic histology

Text & Atlas

tenth edition

Luiz Carlos Junqueira, MD, PhD
Professor Emeritus
Medical School
University of São Paulo
São Paulo, Brazil
Honorary Research Associate in Biology
Harvard University
Boston, Massachusetts
Formerly Research Associate
University of Chicago Medical School
Chicago, Illinois
Honorary Member
American Association of Anatomists
Emeritus Member
American Society of Cell Biology

José Carneiro, MD, PhD
Professor Emeritus
Institute of Biomedical Sciences
University of São Paulo
São Paulo, Brazil
Formerly Research Associate
Department of Anatomy
McGill University Medical School
Montreal, Canada
Formerly Visiting Associate Professor
Department of Anatomy
University of Virginia Medical School
Charlottesville, Virginia

Lange Medical Books McGraw-Hill
Medical Publishing Division
New York Chicago San Francisco Lisbon London
Madrid Mexico City Milan New Delhi San Juan Seoul
Singapore Sydney Toronto

The McGraw-Hill Companies

Basic Histology, Tenth Edition

Copyright © 2003 by The **McGraw-Hill Companies**, Inc. All rights reserved. Printed in the United States of America. Except as permitted under the United States Copyright Act of 1976, no part of this publication may be reproduced or distributed in any form or by any means, or stored in a data base or retrieval system, without the prior written permission of the publisher.

Previous editions copyright © 1998, 1995, 1989, 1986 by Appleton & Lange

Original title: Histologia Basica, 8th ed. © 1995 by Editora Guanabara Koogan S.A., Rio de Janeiro, Brazil

Many of the illustrations in this book were initially prepared with financial aid from the Fundação de Amparo à Pesquisa do Estado de São Paulo, and National Research Council (CNPq).

3 4 5 6 7 8 9 0 KGP/KGP 0 9 8 7 6 5 4

ISSN: 0891-2106
Set ISBN: 0-07-137829-4
Book ISBN: 0-07-141365-0
CD-ROM ISBN: 0-07-140834-7

Notice

Medicine is an ever-changing science. As new research and clinical experience broaden our knowledge, changes in treatment and drug therapy are required. The authors and the publisher of this work have checked with sources believed to be reliable in their efforts to provide information that is complete and generally in accord with the standards accepted at the time of publication. However, in view of the possibility of human error or changes in medical sciences, neither the authors nor the publisher nor any other party who has been involved in the preparation or publication of this work warrants that the information contained herein is in every respect accurate or complete, and they disclaim all responsibility for any errors or omissions or for the results obtained from use of the information contained in this work. Readers are encouraged to confirm the information contained herein with other sources. For example and in particular, readers are advised to check the product information sheet included in the package of each drug they plan to administer to be certain that the information contained in this work is accurate and that changes have not been made in the recommended dose or in the contraindications for administration. This recommendation is of particular importance in connection with new or infrequently used drugs.

This book was set in Times Roman by TechBooks, Inc.
The editors were Janet Foltin, Harriet Lebowitz, and Peter J. Boyle; the production supervisor was Philip Galea; the art manager was Charissa Baker; the designer was Eve Siegel; the index was prepared by Katherine Pitcoff.
Quebecor Kingsport was printer and binder.

This book is printed on acid-free paper.

INTERNATIONAL EDITION
Set ISBN: 0-07-121565-4
Book ISBN: 0-07-121222-1
CD-ROM ISBN: 0-07-140834-7

Copyright © 2003. Exclusive rights by The McGraw-Hill Companies, Inc. for manufacture and export. This book cannot be re-exported from the country to which it is consigned by McGraw-Hill. The International Edition is not available in North America.

Contents

Contributors

Paulo Alexandre Abrahamsohn, MD, PhD
Professor
Department of Histology and Embryology
Institute of Biomedical Sciences
University of São Paulo
São Paulo, Brazil
Chapter 1: Histology & Its Methods of Study; Chapter 20: Hypophysis; Chapter 21: Adrenals, Islets of Langerhans, Thyroid, Parathyroids, & Pineal Gland; Chapter 22: The Male Reproductive System; and Chapter 23: The Female Reproductive System

Marinilce Fagundes dos Santos, DDS, PhD
Assistant Professor
Department of Histology and Embryology
Institute of Biomedical Sciences
University of São Paulo
São Paulo, Brazil
Chapter 15: Digestive Tract and Chapter 16: Organs Associated with the Digestive Tract

Telma Maria Tenório Zorn, MD, PhD
Professor
Department of Histology and Embryology
Institute of Biomedical Sciences
University of São Paulo
São Paulo, Brazil
Chapter 5: Connective Tissue and Chapter 11: The Circulatory System

Preface

The tenth edition of *Basic Histology* continues to be a concise, well-illustrated exposition of the basic facts and interpretation of microscopic anatomy. The authors of this text recognize that students of biologic structure share a common goal—namely, to better understand how structure and function are integrated in the molecules, cells, tissues, and organs of a living creature. Histology is the branch of science that centers on the biology of cells and tissues within an organism and, as such, serves as the foundation on which pathology and pathophysiology are built. In this edition, we continue to emphasize the relationships and concepts that inextricably link cell and tissue structure with their functions as they constitute the fabric of a living organism.

In revising *Basic Histology,* our intent is to provide our readers with the most contemporary and useful text possible. We do this in two ways: by describing the most important recent developments in the sciences basic to histology and by recognizing that our readers are faced with the task of learning an ever-increasing number of facts in an ever-decreasing period of time. Because of this, every attempt has been made to present the information as concisely as possible and organize it in a way that facilitates learning.

INTENDED AUDIENCE

This text is designed for students in professional schools of medicine, veterinary medicine, dentistry, nursing, and allied health sciences. It is also a useful, ready reference for both undergraduate students of microscopic anatomy and others in the structural biosciences.

ORGANIZATION

Because the study of histology requires a firm foundation in cell biology, *Basic Histology* begins with an accurate, up-to-date description of the structure and function of cells and their products and a brief introduction into the molecular biology of the cell. This foundation is followed by a description of the four basic tissues of the body, emphasizing how cells become specialized to perform the specific functions of these tissues. Finally, we devote individual chapters to each of the organs and organ systems of the human body. Here, the emphasis on spatial arrangements of the basic tissues provides the key to understanding the functions of each organ. Again, we emphasize cell biology as the most fundamental approach to the study of structure and function.

As a further aid to learning, color photomicrographs and electron micrographs amplify the text and remind the reader of the laboratory basis of the study of histology. In addition, we place particular emphasis on full-color diagrams, three-dimensional illustrations, and charts to summarize morphologic and functional features of cells, tissues, and organs.

NEW TO THIS EDITION

- All chapters have been revised to reflect new findings and interpretations, and the emphasis on human histology has been further strengthened.
- The chapter on microscopy and techniques includes new information on methods that permit analysis of molecules, cells, and tissues.
- New information on the molecular biology of the genome and its regulation is included in the chapter on the nucleus.
- New information on the organization and molecular composition of the extracellular matrix has been included in the chapter on connective tissue.
- A discussion of the mechanisms of signal transduction in intercellular communication that adds to the student's understanding of tissue organization has been included in the chapter on the cell.
- The chapter on nerve tissue and the nervous system has been extensively rewritten to include contemporary concepts and information regarding neurons and glial cells and their interactions.
- The chapter on the immune system has been further revised to include current information and to organize that material into a readily assimilated body of knowledge.
- The 600+ illustrations that appear throughout the book include numerous color photomicrographs prepared from new tissue samples with distinctive labeling that clearly pinpoints the elements of interest in each figure. These new micrographs of resin-embedded specimens provide clearer detail of cell and tissue organization.
- All existing diagrams have been revised in full color and new color figures have been added to enhance the usefulness of the text.

ACKNOWLEDGMENTS

We wish to thank the following professors who critically read several parts of this book: Edna T. Kimura (thyroid gland), Nancy Amaral Rebouças (in situ hybridization), Sirlei Daffre (protein separation) and Wolfgang G. W. Zorn (blood vessels).The chapter on the immune system was prepared with the help of Professor Flávio Alvim Braga. We also extend our appreciation to the staff of McGraw-Hill, Janet Foltin, Harriet Lebowitz, Charissa Baker, Phil Galea, Peter Boyle, and Jim Halston as well as to Mary McKenney for her editorial expertise.

We are pleased to announce that Italian, Spanish, Dutch, Indonesian, Japanese, Turkish, Korean, German, Serbo-Croatian, French, Portuguese, Greek, and Chinese translations of *Basic Histology* are now available.

Luiz Carlos Junqueira, MD
José Carneiro, MD
November, 2002

Histology & Its Methods of Study 1

Histology (Gr. *histo,* web or tissue, + *logos,* study) is the study of the tissues of the body and of how these tissues are arranged to constitute organs. Four fundamental tissues are recognized: epithelial tissue, connective tissue, muscular tissue, and nervous tissue.

Tissues are made of cells and extracellular matrix, two components that were formerly considered separate entities. The extracellular matrix consists of many kinds of molecules, some of which are highly organized and form complex structures, such as collagen fibrils and basement membranes. The main functions formerly attributed to the extracellular matrix were to furnish mechanical support for the cells, to transport nutrients to the cells, and to carry away catabolites and secretory products. Recent work has shown that, although the cells produce the extracellular matrix they are influenced and sometimes controlled by molecules of the matrix. There is thus an intense interaction between cells and matrix. Moreover, many molecules of the matrix are recognized by and attach to receptors present on cell surfaces. Most of these receptors are molecules that cross the cell membranes and connect to molecules within the cytoplasm. Thus, cells and extracellular matrix form a continuum that functions together and reacts to stimuli and inhibitors together.

Each of the fundamental tissues is formed by several kinds of cells and typically by specific associations of cells and extracellular matrix. These characteristic associations facilitate the recognition of the many subtypes of tissues by students. Most organs are formed by an orderly combination of several tissues, except the central nervous system, which is formed almost solely by nervous tissue. The precise combination of these tissues allows the functioning of each organ and of the organism as a whole.

The small size of cells and matrix components makes histology dependent on the use of microscopes. Advances in chemistry, physiology, immunology, and pathology—and the interactions among these fields—are essential for a better knowledge of tissue biology. Familiarity with the tools and methods of any branch of science is essential for a proper understanding of the subject. This chapter reviews some of the more common methods used to study cells and tissues and the principles involved in these methods.

PREPARATION OF TISSUES FOR MICROSCOPIC EXAMINATION

The most common procedure used in the study of tissues is the preparation of histologic sections that can be studied with the aid of the light microscope. Under the light microscope, tissues are examined via a light beam that is transmitted through the tissue. Since tissues and organs are usually too thick for light to pass through them, they must be sectioned to obtain thin, translucent sections. However, living cells, very thin layers of tissues, or transparent membranes of living animals (eg, the mesentery, the tail of a tadpole, the wall of a hamster's cheek pouch) can be observed directly in the microscope without first sectioning the tissue. It is then possible to study these structures for long periods and under varying physiologic or experimental conditions. In most cases, however, tissues must be sliced into thin sections and attached on glass slides before they can be examined. These sections are precisely cut from tissues previously prepared for sectioning, using fine cutting instruments called **microtomes.**

The ideal microscope tissue preparation should be preserved so that the tissue on the slide has the same structure and molecular composition as it had in the body. This is sometimes possible but—as a practical matter—seldom feasible, and artifacts, distortions, and loss of components due to the preparation process are almost always present.

Fixation

If a permanent section is desired, tissues must be fixed. To avoid tissue digestion by enzymes present within the cells (autolysis) or by bacteria and to preserve the structure and molecular composition, pieces of organs should be promptly and adequately treated before or as soon as possible after removal from the animal's body. This treatment—**fixation**—can be done by chemical or, less frequently, physical methods. In chemical fixation, the tissues are usually immersed in solutions of stabilizing or cross-linking agents called **fixatives.** Because the fixative needs some time to fully diffuse into the tissues, the tissues are usually cut into small fragments before fixation to facilitate the penetration of the fixative and to guarantee preservation of the tissue. Intravascular perfusion of fixatives can be used. Because the fixative in this case rapidly reaches the tissues through the blood vessels, fixation is greatly improved.

One of the best fixatives for routine light microscopy is a buffered isotonic solution of 4% formaldehyde. The chemistry of the process involved in fixation is complex and not always well understood. Formaldehyde and glutaraldehyde, another widely used fixative, are known to react with the amine groups (NH_2) of tissue proteins. In the case of glutaraldehyde, the fixing action is reinforced by virtue of its being a dialdehyde, which can cross-link proteins.

In view of the high resolution afforded by the electron microscope, greater care in fixation is necessary to preserve

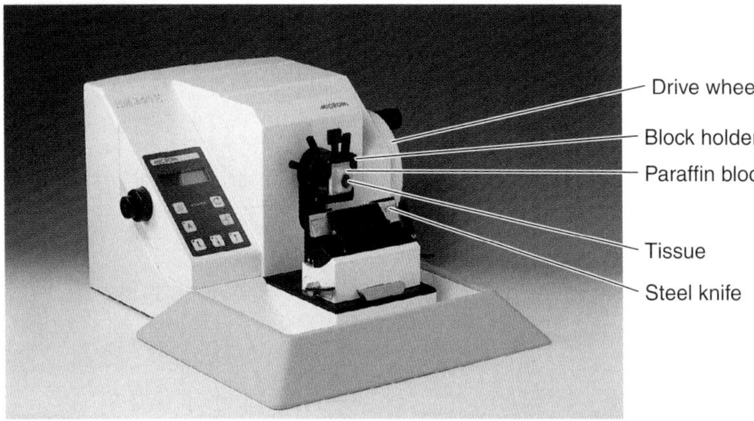

Figure 1–1. Microtome for sectioning resin- and paraffin-embedded tissues for light microscopy. Rotation of the drive wheel moves the tissue-block holder up and down. Each turn of the drive wheel advances the specimen holder a controlled distance, generally between 1 and 10 μm. After each forward move, the tissue block passes over the knife edge, which cuts the sections. (Courtesy of Microm.)

ultrastructural detail. Toward that end, a double fixation procedure, using a buffered glutaraldehyde solution followed by a second fixation in buffered osmium tetroxide, has become a standard procedure in preparations for fine structural studies. The effect of osmium tetroxide is to preserve and stain lipids and proteins.

Embedding

Tissues are usually embedded in a solid medium to facilitate sectioning. To obtain thin sections with the microtome, tissues must be infiltrated after fixation with embedding substances that impart a rigid consistency to the tissue. Embedding materials include paraffin and plastic resins. Paraffin is used routinely for light microscopy; resins are used for both light and electron microscopy.

The process of paraffin embedding, or tissue impregnation, is ordinarily preceded by two main steps: **dehydration** and **clearing.** The water is first extracted from the fragments to be embedded by bathing them successively in a graded series of mixtures of ethanol and water (usually from 70% to 100% ethanol). The ethanol is then replaced with a solvent miscible with the embedding medium. In paraffin embedding, the solvent used is usually xylene. As the tissues are infiltrated with the solvent, they generally become transparent (clearing). Once the tissue is impregnated with the solvent, it is placed in melted paraffin in the oven, typically at 58–60 °C. The heat causes the solvent to evaporate, and the spaces within the tissues become filled with paraffin. The tissue together with its impregnating paraffin gets hard after being taken out of the oven. Tissues to be embedded with plastic resin are also dehydrated in ethanol and—depending on the kind of resin used—subsequently infiltrated with plastic solvents. The ethanol or the solvents are later replaced by plastic solutions that are hardened by means of cross-linking polymerizers. Plastic embedding prevents the shrinking effect of the high temperatures needed for paraffin embedding and gives much better results.

The hard blocks containing the tissues are then taken to a microtome (Figure 1–1) and are sectioned by the microtome's steel or glass blade to a thickness of 1–10 μm. Remember that 1 micrometer (1 μm) = 0.001 mm = 10^{-6} m; 1 nanometer (1 nm) = 0.001 μm = 10^{-6} mm = 10^{-9} m. The sections are floated on water and transferred to glass slides to be stained.

A completely different way to prepare tissue sections is to submit the tissues to rapid freezing. In this process, the tissues are fixed by freezing (physically, not chemically) and at the same time become hard and thus ready to be sectioned. A freezing microtome—the **cryostat** (Gr. *kryos,* cold, + *statos,* standing)—has been devised to section the frozen tissues. Because this method allows the rapid preparation of sections without going through the long embedding procedure described above, it is routinely used in hospitals to study specimens during surgical procedures. Freezing of tissues is also effective in the histochemical study of very sensitive enzymes or small molecules, since freezing does not inactivate most enzymes. Because immersion of tissues in solvents such as xylene dissolves the tissue lipids, the use of frozen sections is advised when these compounds are to be studied.

Staining

To be studied in the microscope most sections must be stained. With few exceptions, most tissues are colorless, so observing them unstained in the light microscope is useless. Methods of staining tissues have therefore been devised that not only make the various tissue components conspicuous but also permit distinctions to be made between them. The dyes stain tissue components more or less selectively. Most of these dyes behave like acidic or basic compounds and have a tendency to form electrostatic (salt) linkages with ionizable radicals of the tissues. Tissue components that stain more readily with basic dyes are termed **basophilic** (Gr. *basis,* base, + *phileo,* to love); those with an affinity for acid dyes are termed **acidophilic.**

Examples of basic dyes are toluidine blue and methylene blue. Hematoxylin behaves like a basic dye, that is, it stains the basophilic tissue components. The main tissue components that ionize and react with basic dyes do so because of acids in their composition (nucleic acids, glycosaminoglycans, and acid glycoproteins). Acid dyes (eg, orange G, eosin, acid fuchsin) stain the acidophilic components of tissues such as mitochondria, secretory granules, and collagen.

Of all dyes, the combination of **hematoxylin and eosin (H&E)** is the most commonly used. Hematoxylin stains the cell nucleus and other acidic structures (such as RNA-rich portions of the cytoplasm and the matrix of hyaline cartilage) blue. In contrast, eosin stains the cytoplasm and collagen pink. Many other dyes, such as the **trichromes** (eg, Mallory's stain, Masson's stain), are used in different histologic procedures. The trichromes, besides showing the nuclei and cytoplasm very well, help to differentiate collagen from smooth muscle. A good technique for differentiating collagen is the use of picrosirius, especially when associated with polarized light (see Polarizing Microscopy).

In many procedures (see Immunocytochemistry), the sections become labeled by a precipitate, but cells and cell limits are often not visible. In this case a **counterstain** is used. A counterstain is usually a single stain that is applied to a section to allow the recognition of nuclei or cytoplasm.

Although most stains are useful in visualizing the various tissue components, they usually provide no insight into the chemical nature of the tissue being studied. In addition to tissue staining with dyes, impregnation with metals such as silver and gold is a common method, especially in studies of the nervous system.

The whole procedure, from fixation to observing a tissue in a light microscope, may take from 12 hours to 2½ days, depending on the size of the tissue, the fixative, and the embedding medium.

LIGHT MICROSCOPY

Conventional light, phase contrast, differential interference, polarizing, confocal, and fluorescence microscopy are all based on the interaction of light and tissue components. With the light microscope, stained preparations are usually examined by means of light that passes through the specimen. The microscope is composed of mechanical and optical parts (Figure 1–2). The optical components consist of 3 systems of lenses: condenser, objective, and eyepiece. The **condenser** collects and focuses light, producing a cone of light that illuminates the object to be observed. The **objective** lenses enlarge and project the illuminated image of the object in the direction of the eyepiece. The **eyepiece** further magnifies this image and projects it onto the viewer's retina, a photographic plate, or (to obtain a digital image) a detector such as a charged coupled device camera. The total magnification is obtained by multiplying the magnifying power of the objective and ocular lenses.

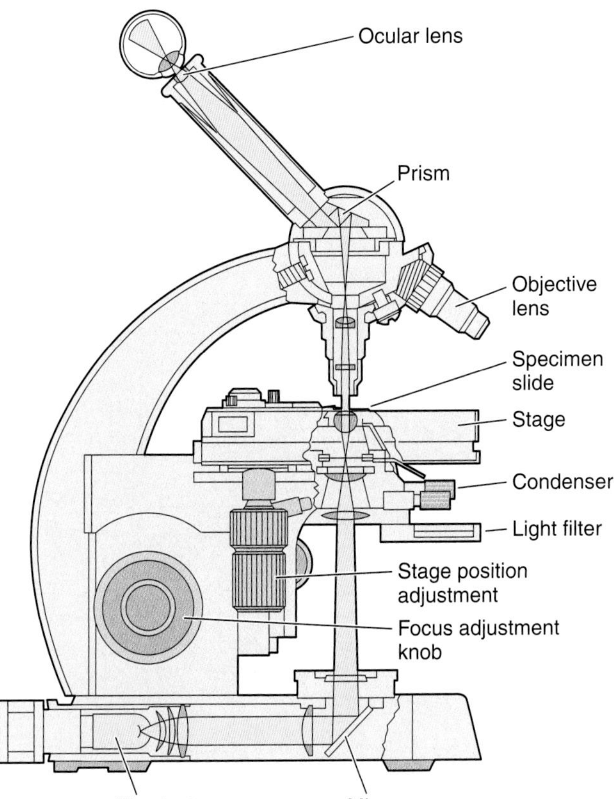

Figure 1–2. Schematic drawing of a light microscope showing its main components and the pathway of light from the substage lamp to the eye of the observer. (Courtesy of Carl Zeiss Co.)

Resolution

The critical factor in obtaining a crisp, detailed image with the microscope is its **resolving power,** that is, the smallest distance between two particles at which they can be seen as separate objects. The maximal resolving power of the light microscope is approximately 0.2 μm; this power permits good images magnified 1000–1500 times. Objects smaller than 0.2 μm (such as a membrane or a filament of actin) cannot be distinguished with this instrument. Likewise, two objects, such as two mitochondria or two lysosomes, will be seen as only one object if they are separated by less than 0.2 μm. The quality of the image—its clarity and richness of detail—depends on the microscope's resolving power. The magnification is of value only when accompanied by high resolution. The resolving power of a microscope depends mainly on the quality of its objective lens. The eyepiece lens only enlarges the image obtained by the objective; it does not improve resolution. For this reason, when one compares objectives of different magnifications, one sees that those that provide higher magnification also have higher resolving power.

Highly sensitive video cameras enhance the power of the light microscope and allow the capture of digitized images that can be fed into computers for quantitative image analysis and printing.

The frontiers of light microscopy have been redefined by the use of video cameras highly sensitive to light. With cameras and image-enhancement programs, objects that may not be visible when viewed directly through the ocular may be made visible in the video screen. These video systems are also useful for studying living cells for long periods of time, because they use low-intensity light and thus avoid the cellular damage that can result from intense illumination.

The electronic images from video cameras can be easily digitized and adapted to the specific requirements of an experiment through computer programming. For example, contrast enhancement is an important computer-assisted technique that may reveal to the investigator a structural image not immediately seen when the specimen is observed directly in the microscope. Software developed for image analysis allows the measurement of microscopic structures.

PHASE CONTRAST MICROSCOPY & DIFFERENTIAL INTERFERENCE MICROSCOPY

Some optical arrangements allow the observation of unstained cells and tissue sections. Unstained biologic specimens are usually transparent and difficult to view in detail, since all parts of the specimen have almost the same optical density. Phase contrast microscopy, however, uses a lens system that produces visible images from transparent objects (Figure 1–3).

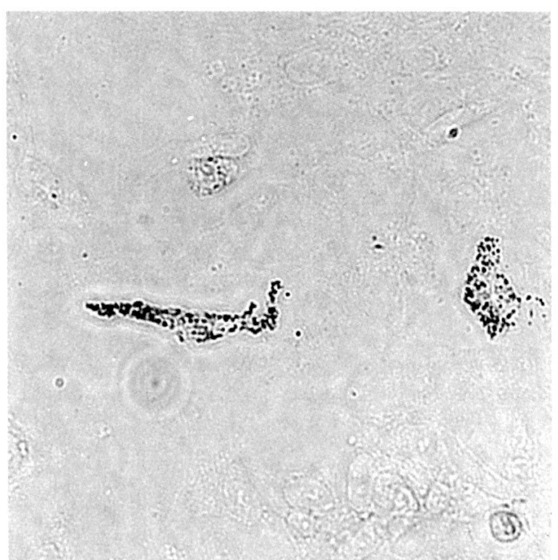

A

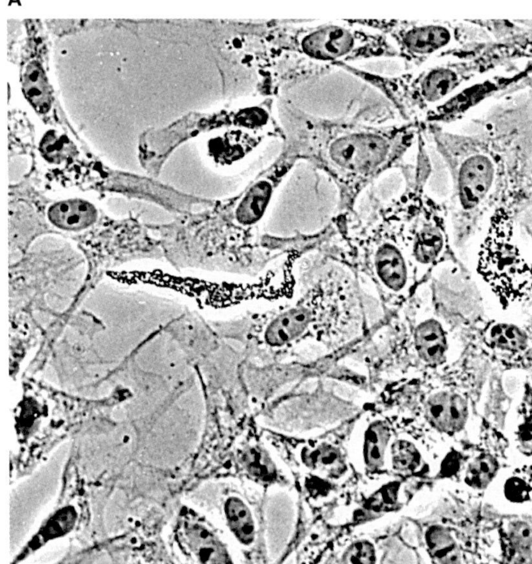

B

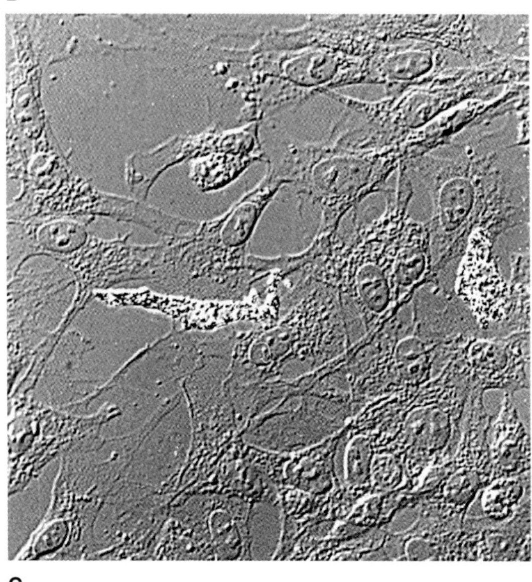

C

Figure 1–3. Cultured neural crest cells seen with different optical techniques. The cells are unstained, and the same cells appear in all photographs. Use the two pigmented cells for orientation in each image. **A:** Conventional light microscopy. **B:** Phase contrast microscopy. **C:** Nomarski differential interference microscopy. High magnification. (Courtesy of S Rogers.)

Phase contrast microscopy is based on the principle that light changes its speed when passing through cellular and extracellular structures with different refractive indices. These changes are used by the phase contrast system to cause the structures to appear lighter or darker relative to each other, which makes this kind of microscopy a powerful tool to observe living cells. Another way to observe unstained cells or tissue sections is Nomarski differential interference microscopy, which produces an apparently 3-dimensional image (Figure 1–3).

POLARIZING MICROSCOPY

Polarizing microscopy allows the recognition of structures made of highly organized molecules. When normal light passes through a **polarizing filter** (such as a Polaroid), it exits vibrating in only one direction. If a second filter is placed in the microscope above the first one, with

Figure 1–4. Polarized light microscopy. A small piece of rat mesentery was stained with the picrosirius method, which stains collagen fibers. The mesentery was then placed on the slide and observed by transparency. Under polarized light, collagen fibers exhibit intense birefringence and appear brilliant or yellow. Medium magnification.

its main axis perpendicular to the first filter, no light passes through. If, however, tissue structures containing oriented molecules (such as cellulose, collagen, microtubules, and microfilaments) are located between the two polarizing filters, their repetitive, oriented molecular structure rotates the axis of the light emerging from the polarizer. Consequently, they appear as bright structures against a dark background (Figure 1–4). The ability to rotate the direction of vibration of polarized light is called **birefringence** and is a feature of crystalline substances or substances containing highly oriented molecules.

CONFOCAL MICROSCOPY

Confocal microscopy allows the precise focusing of a very thin plane of a cell or section. The depth of focus in the light microscope is relatively long, especially when small magnification objectives are used. This means that a rather wide extent of the specimen is seen in focus simultaneously, causing superimposition of the image of a 3-dimensional object. One of the most important features of the confocal microscope is that only a very thin plane of the specimen is seen in focus at a time. The principles on which this is based are the following: (1) the specimen is illuminated by a very small beam of light (whereas in the common light microscope, a large beam of light floods the specimen); (2) the image collected from the specimen must pass through a small pinhole. The result is that only the image originating from the focused plane reaches the detector whereas the images from in front of and behind this plane are blocked (Figure 1–5). The harmful glare of the out-of-focus objects is lost, and the definition of the focused object becomes better and allows the localization of any specimen component with much greater precision than in the common light microscope.

For practical reasons, the following arrangement is used in most confocal microscopes (Figure 1–6): (1) the illumination is provided by a laser source; (2) because it is a very small point, it must be moved over the specimen (scanned) to allow the observation of a larger area of the specimen; (3) the component of the specimen that is of interest must be labeled with a fluorescent molecule (meaning that a routine section cannot be studied); (4) the light that is reflected by the specimen is used to form an image; (5) the reflected light is captured by a detector, so that the signal can be electronically enhanced to be seen in a monitor.

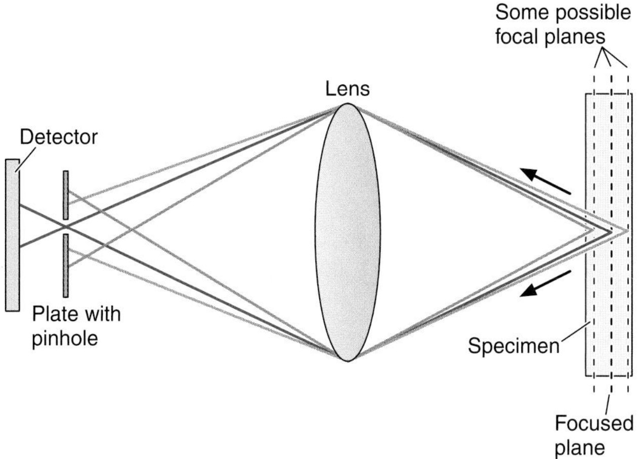

Figure 1–5. Principle of confocal microscopy. While a very small spot of light originating from one plane of the section crosses the pinhole and reaches the detector, rays originating from other planes are blocked by the blind. Thus, only one very thin plane of the specimen is focused at a time.

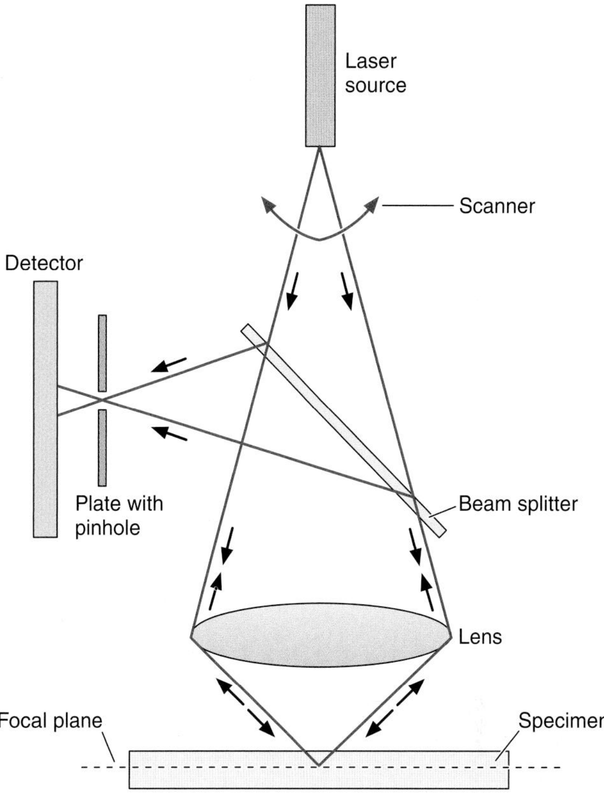

Figure 1–6. Practical arrangement of a confocal microscope. Light from a laser source hits the specimen and is reflected. A beam splitter directs the reflected light to a pinhole and a detector. Light from components of the specimen that are above or below the focused plane are blocked by the blind. The laser scans the specimen so that a larger area of the specimen can be observed.

Because only a very thin focal plane (also called an optical section) is focused at a time, it is possible to reunite several focused planes of one specimen and reconstruct them into a 3-dimensional image. To accomplish the reconstruction and many of its other features, the confocal microscope depends on heavy computing capacity.

FLUORESCENCE MICROSCOPY

When certain substances are irradiated by light of a proper wavelength, they emit light with a longer wavelength. This phenomenon is called fluorescence. In fluorescence microscopy, tissue sections are usually irradiated with UV light, and the emission is in the visible portion of the spectrum. The fluorescent substances appear brilliant on a dark background. For this method, the microscope has a strong UV light source and special filters that select rays of different wavelengths emitted by the substances.

Fluorescent compounds that have an affinity for cell macromolecules may be used as fluorescent stains. Acridine orange,

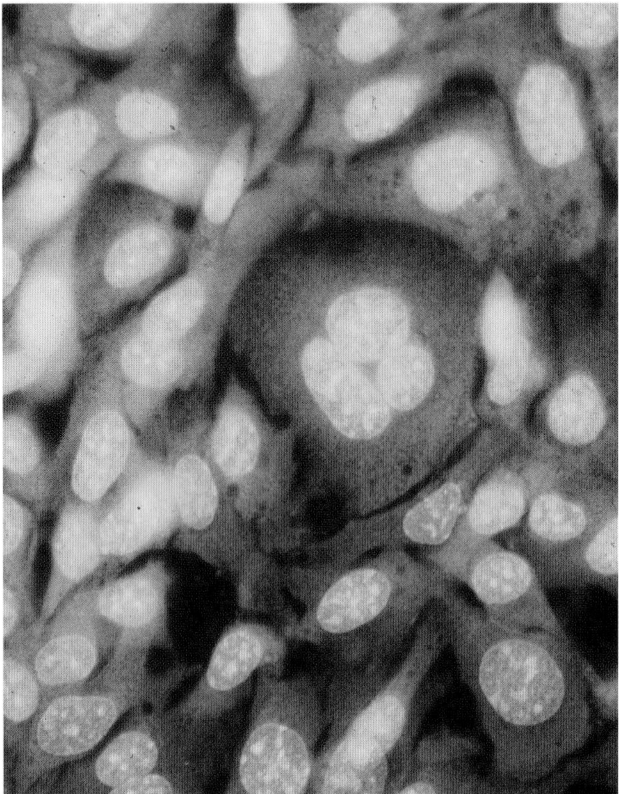

Figure 1–7. Photomicrograph of kidney cells in culture, stained with acridine orange. Under a fluorescence microscope, DNA (within the nuclei) emits yellow light, and the RNA-rich cytoplasm appears reddish or orange. (Courtesy of A Geraldes and JMV Costa.)

which can combine with DNA and RNA, is an example. When observed in the fluorescence microscope, the DNA–acridine orange complex emits a yellowish-green light, and the RNA–acridine orange complex emits a reddish-orange light. It is thus possible to identify and localize nucleic acids in the cells (Figure 1–7). Another important application of fluorescence microscopy is achieved by coupling fluorescent substances (such as fluorescein isothiocyanate) to marker molecules that will specifically bind to components of the tissues and will thus allow the identification of these components under the microscope (see Detection Methods Using High-Affinity Interactions between Molecules).

ELECTRON MICROSCOPY

Transmission and scanning electron microscopes are based on the interaction of electrons and tissue components.

Transmission Electron Microscopy

The transmission electron microscope is an imaging system that theoretically permits a very high resolution (0.1 nm) (Figure 1–8). In practice, however, the resolution obtained by most good instruments is around 3 nm. This high resolution allows magnifications of up to 400,000 times to be viewed with details. Unfortunately, this level of magnification applies only to isolated molecules or particles. Very thin tissue sections can be observed with details at magnifications of up to about 120,000 times.

The transmission electron microscope functions on the principle that a beam of electrons can be deflected by electromagnetic fields in a manner similar to light deflection in glass lenses. Electrons are released by heating a very thin metallic (usually tungsten) filament (cathode) in a vacuum. The released electrons are then submitted to a voltage difference of 60–120 kV between the cathode and the anode, which is a metallic plate with a hole in its center (Figure 1–9). Electrons are thus attracted to the anode and accelerated to high speeds. They pass through the central opening in the anode, forming a constant stream (or beam) of electrons that penetrate the tube of the microscope. The beam passes inside electric coils and is deflected in a way roughly analogous to that which occurs in optical lenses, because electrons change their path when submitted to electromagnetic fields. For this reason, the electric coils of the electron microscopes are called electromagnetic lenses.

The configuration of the electron microscope is very similar to that of the optical microscope, although the optics of the electron microscope are usually placed upside down (Figure 1–9). The first lens is a condenser that focuses the beam of electrons on the section. Some electrons interact with atoms of the section and continue their course, while others simply cross the specimen without interacting. Most electrons reach the objective lens, which forms a magnified image that is then projected through other magnifying lenses. Because the human eye is not sensitive to electrons, the image is finally projected on a fluorescent screen or is registered by photographic plates or a charged coupled device camera. Because most of the image in the transmission electron microscope is produced by the balance between

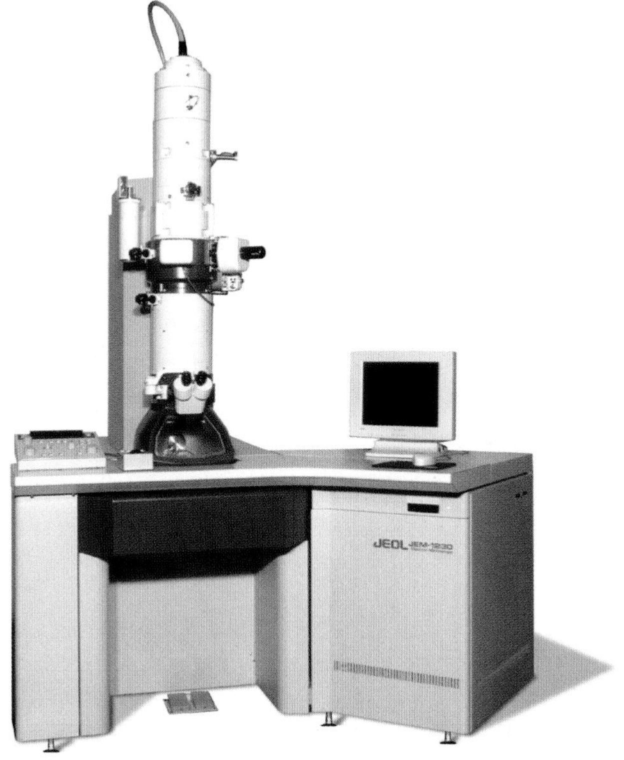

Figure 1–8. Photograph of the JEM-1230 transmission electron microscope. (Courtesy of JEOL USA, Inc., Peabody, MA.)

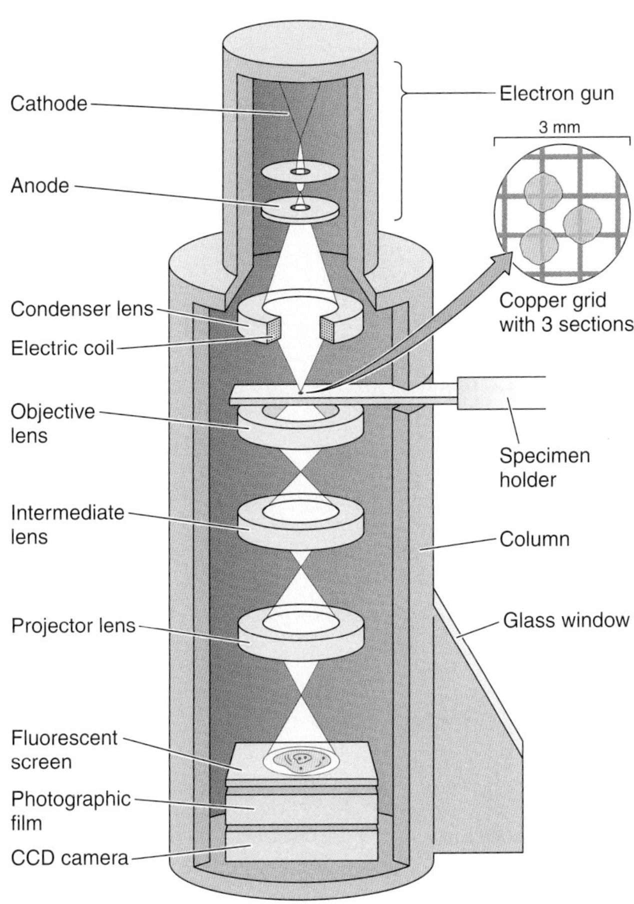

Figure 1–9. Schematic view of a transmission electron microscope with its lenses and the pathway of the electrons. CCD, charged coupled device.

the electrons that hit the fluorescent screen (or the photographic plate) and the electrons that are retained in the tube of the microscope, the resulting image is always in black and white. Dark areas of an electron micrograph are usually called electron dense, whereas light areas are called electron lucent.

To provide a good interaction between the specimen and the electrons, electron microscopy requires very thin sections (40–90 nm); therefore, embedding is performed with a hard epoxy plastic. The blocks thus obtained are so hard that glass or diamond knives are usually necessary to section them. The extremely thin sections are collected on small metal grids and transferred to the interior of the microscope to be analyzed.

Freezing techniques allow the examination of tissues by electron microscopy without the need for fixation and embedding. There are fewer artifacts than with conventional tissue preparation, although the technique is usually arduous. Frozen tissues may be sectioned and submitted to cytochemistry or immunocytochemistry or may be fractured (cryofracture, freeze fracture) to reveal details of the internal structure of the membranes.

Scanning Electron Microscopy

Scanning electron microscopy permits pseudo–3-dimensional views of the surfaces of cells, tissues, and organs. This electron microscope produces a very narrow electron beam that is moved sequentially (scanned) from point to point across the specimen. Unlike the electrons in the transmission electron microscope, those in the scanning microscope do not pass through the specimen (Figure 1–10). The electron beam interacts with a very thin metal coating previously applied to the specimen and produces reflected or emitted electrons. These electrons are captured by a detector that transmits them to amplifiers and other devices so that in the end the signal is projected into a cathode ray tube (a monitor), resulting in a black-and-white image. The resulting photographs are easily understood, since they present a view that appears to be illuminated from above, just as our ordinary macroscopic world is filled with highlights and shadows caused by illumination from above. The scanning electron microscope shows only surface views. The inside of organs can be analyzed by freezing the organs and fracturing them to expose their internal surfaces.

AUTORADIOGRAPHY OF TISSUE SECTIONS

Autoradiography is the study of biological events in tissue sections using radioactivity. Autoradiography permits the localization of radioactive substances in tissues by means of emitted radiation effects on photographic emulsions. Silver bromide crystals present in the emulsion act as microdetectors of radioactivity in the same way that they respond to light in common photography. The first step of autoradiog-

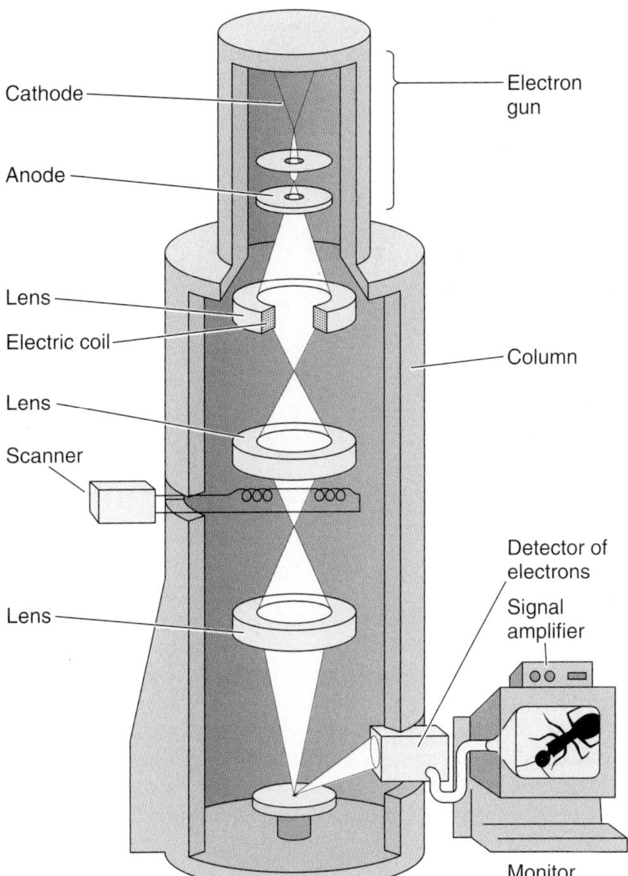

Figure 1–10. Schematic view of a scanning electron microscope.

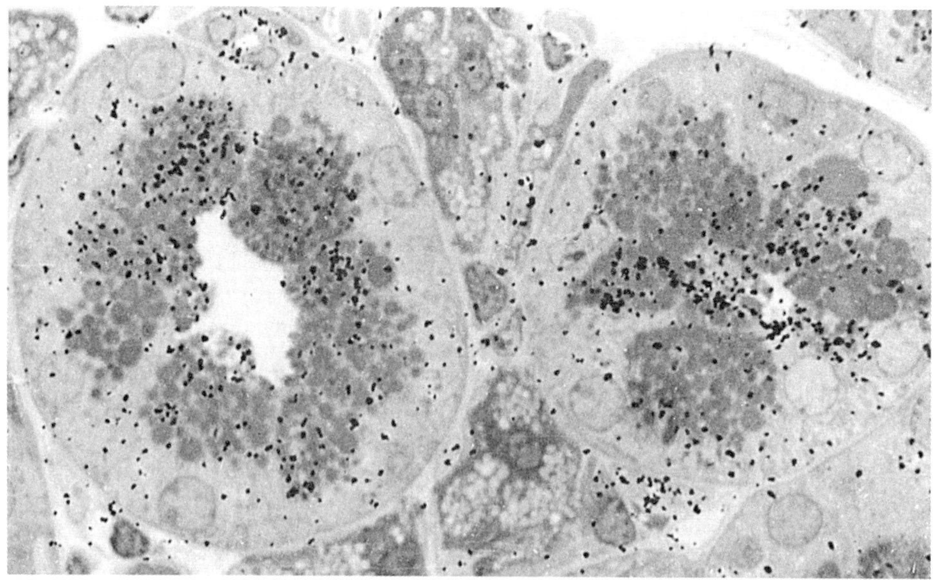

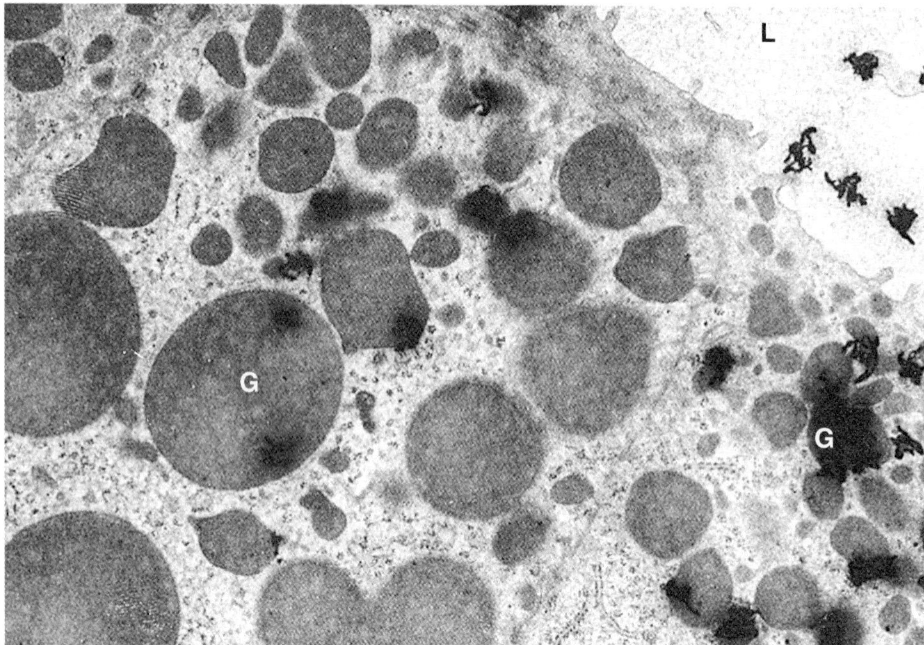

Figure 1–11. Autoradiographs from the submandibular gland of a mouse injected with ^{3}H-fucose 8 h before being killed. **Top:** With a light microscope it is possible to observe black silver grains indicating radioactive regions in the cells. Most radioactivity is in the granules of the cells of the granular ducts of the gland. High magnification. **Bottom:** The same tissue prepared for electron-microscope autoradiography. The silver grains in this enlargement appear as coiled structures localized mainly over the granules (G) and in the gland lumen (L). High magnification. (Courtesy of TG Lima and A Haddad.)

raphy is to deliver a radioactive molecule to the cells. A variety of molecules, including radioactive amino acids, radioactive nucleotides, and radioactive sugars, can be used, depending on the purpose of the study. These molecules are called precursors, because they may be used by the cells to synthesize larger molecules, such as proteins, nucleic acids, or polysaccharides and glycoproteins. The tissue sections are prepared and are covered with photographic emulsion. The slides are kept in light-proof boxes; after an adequate exposure time they are developed photographically and examined. The silver bromide crystals hit by radiation are reduced to small black granules of metallic silver, which reveal the existence of radioactivity in the tissue. The structures that contain radioactive molecules become covered by these granules. This procedure can be used in both light and electron microscopy (Figure 1–11).

Much information becomes available by localizing radioactivity in tissue components. Thus, if a radioactive amino acid is used, it is possible to know which cells in a tissue produce more protein and which cells produce less, because the number of silver granules formed over the cells is proportional to the intensity of protein synthesis. If a radioactive precursor of DNA (such as radioactive thymidine) is used, it is possible to know which cells in a tissue (and how many) are preparing to divide. Dynamic events may also be analyzed. For instance, if one wishes to know where in the cell a protein is produced, if it is secreted, and which path it follows in the cell before being secreted, sev-

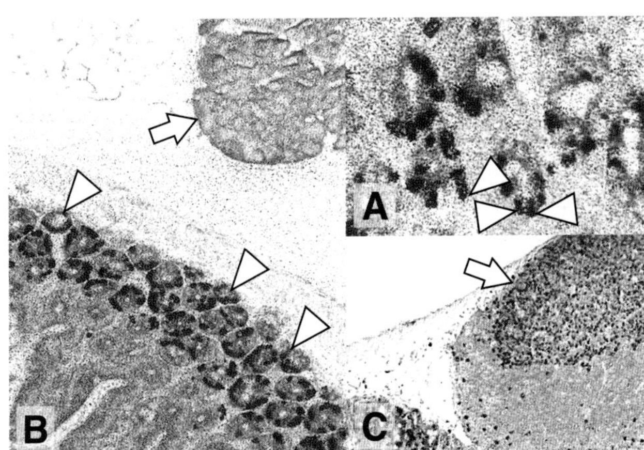

Figure 1–12. Autoradiographs of tissue sections from a mouse that was injected with ^{3}H-thymidine. **A:** Because the autoradiographs were exposed for a very long time, the radioactive nuclei became heavily labeled and appear covered by clouds of dark granules (arrowheads). High magnification. **B:** Many cells were dividing at the base of the intestinal glands (arrowheads), but no cells were dividing in the pancreas (long arrow). Low magnification. **C:** A section of a lymph node shows that cell division occurs mostly at the germinal centers of this structure (arrow). Low magnification. (Courtesy of TMT Zorn, M Soto-Suazo, CMR Pellegrini, and WE Stumpf.)

eral animals are injected with a radioactive amino acid and are killed at different times after the injection. Autoradiographs of the sections, taken at various times throughout the experiment, will show the migration of the radioactive proteins. If one wishes to know where new cells are produced in an organ and where they migrate, several animals are injected with radioactive thymidine and are killed at different times after the injection. Autoradiographs of the sections will show where the cells divide and where (or if) they migrate (Figure 1–12).

CELL AND TISSUE CULTURE

Live cells and tissues can be maintained and studied outside the body. In a complex organism, tissues and organs are formed by several kinds of cells. These cells are bathed in blood plasma, which contains hundreds of different molecules. Cell and tissue culture has been very helpful in isolating the effect of a single molecule on one type of cell or tissue. It also allows the direct observation of the behavior of living cells under a microscope. Several experiments that cannot be performed in the living animal can be reproduced **in vitro.**

The cells and tissues are grown in complex solutions of known composition (salts, amino acids, vitamins) to which serum components are frequently added. In preparing cultures from a tissue or organ, cells must be initially dispersed mechanically or by treating the tissue with enzymes. Once isolated, the cells can be cultivated in a suspension or spread out on a Petri dish or glass slide, to which they adhere, usually as a single layer of cells (Figure 1–3). Cultures of cells that are isolated in this way are called **primary cell cultures.** Many cell types were once isolated in this way from normal or pathological tissue and have been maintained in vitro ever since because they have been immortalized and now constitute a permanent **cell line.** Most cells obtained from normal tissues have a finite, genetically programmed life span. Certain changes, however (mainly related to oncogenes; see Chapter 3), can promote cell immortality, a process called **transformation,** which may be a first step in a normal cell's becoming a cancer cell. Because of transformation and other improvements in culture technology, most cell types can now be maintained in the laboratory indefinitely. All procedures with living cells and tissues must be performed in a sterile area, using sterile solutions and equipment.

MEDICAL APPLICATION

Cell culture has been widely used for the study of the metabolism of normal and cancerous cells and for the development of new drugs. This technique is also useful in the study of parasites that grow only within cells, such as viruses, mycoplasma, and some protozoa (Figure 1–13). In cytogenetic research, determination of human karyotypes (the number and morphology of an individual's chromosomes) is accomplished by the short-term cultivation of blood lymphocytes or of skin fibroblasts. By examining cells during mitotic division in tissue cultures, one can detect anomalies in the number and morphology of the chromosomes that have been shown to be related and are diagnostic of numerous diseases collectively called genetic disorders. In addition, cell culture is central to contemporary techniques of molecular biology and recombinant DNA technology.

CELL FRACTIONATION

Organelles and other components of cells and tissues can be isolated by cell fractionation. This is the physical process by which centrifugal force is used to separate organelles and cellular components as a function of their sedimentation coefficients. The sedimentation coefficient of a particle depends on its size, form, and density and on the viscosity of the medium (Figure 1–14). The organelles obtained with these techniques can be analyzed for purity in the electron microscope (Figure 1–15), and their chemical composition and functions can be studied in vitro.

HISTOCHEMISTRY & CYTOCHEMISTRY

The terms **histochemistry** and **cytochemistry** are used mainly to indicate methods for localizing different substances in tissue sections. Several procedures are used to obtain this type of information, most of them based on specific chemical reactions or on high-affinity interactions between macromolecules. These methods usually produce insoluble colored or electron-dense compounds that enable the localization of specific substances by means of light or electron microscopy.

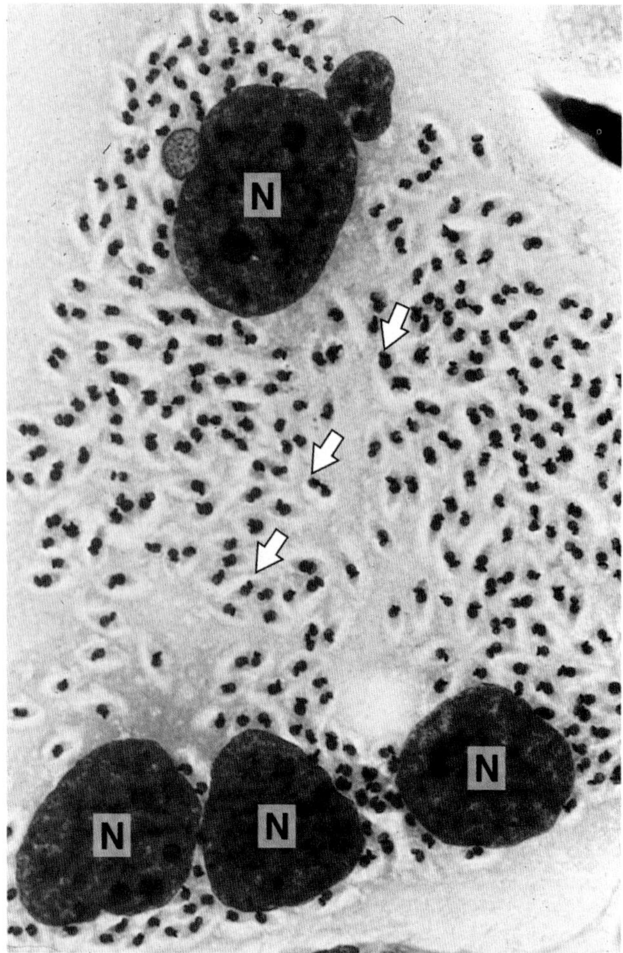

Figure 1–13. Photomicrograph of chicken fibroblasts that were grown in tissue culture and infected by the protozoan *Trypanosoma cruzi.* Although the borders of the cells are not readily visible, their nuclei (N) can be easily seen. Many trypanosomes are present within each cell (arrows). High magnification. (Courtesy of S Yoneda.)

Ions

Several ions (eg, calcium, iron, phosphate) have been localized in tissues with these methods, using chemical reactions that produce a dark insoluble product (Figure 1–16).

Nucleic Acids

DNA can be identified and quantified in cell nuclei using the Feulgen reaction, which produces a red color in DNA. DNA and RNA can also be analyzed by staining cells or tissue sections with a basic stain. The basophilia due to either nucleic acid can be abolished by previous digestion with DNase or RNase.

Proteins

Although there are general methods to detect proteins in tissue sections, the histochemical methods usually do not permit localization of specific proteins in cells and tissues. **Immunocytochemistry,** presented later in this chapter, can do so.

Several histochemical methods, however, can be used to reveal, more or less specifically, a large group of proteins, the **enzymes.** These methods usually employ the capacity of the enzymes to react with specific chemical bonds. Most histoenzymatic methods work in the following way: (1) tissue sections are immersed in a solution that contains the substrate of the enzyme intended for study; (2) the enzyme is allowed to act on its substrate; (3) at this stage or at a later stage in the method, the section is put in contact with a marker compound; (4) this compound reacts with a molecule that results from the degradation or transformation of the substrate; (5) the final reaction product, which must be insoluble and which is visible by light or electron microscopy only if it is colored or electron dense, precipitates over the site that contains the enzyme. When examining such a section in the microscope, one can see the cells (or organelles) covered with a colored or electron-dense material.

Some examples of enzymes that can be detected are the following:

Phosphatases are enzymes widely found in the body. They split the bond between a phosphate group and an alcohol residue of phosphorylated molecules. The colored insoluble reaction product of phosphatases is usually lead phosphate or lead sulfide. Alkaline phosphatases, which have their maximum activity at an alkaline pH, can be detected (Figure 1–17). Acid phosphatases are frequently used to demonstrate **lysosomes,** cytoplasmic organelles that contain acid phosphatase (Figure 1–18).

Dehydrogenases remove hydrogen from one substrate and transfer it to another. There are many dehydrogenases in the body, and they play an important role in several metabolic processes. Dehydrogenases are detected histochemically by incubating nonfixed tissue sections in a substrate solution containing a molecule that receives hydrogen and precipitates as an insoluble colored compound. By this method, succinate dehydrogenase—a key enzyme in the citric acid (Krebs) cycle—can be localized in mitochondria.

Peroxidase, which is present in several types of cells, is an enzyme that promotes the oxidation of certain substrates with the transfer of hydrogen ions to hydrogen peroxide, forming molecules of water.

In this method, sections of adequately fixed tissue are incubated in a solution containing hydrogen peroxide and 3,3-diaminoazobenzidine. The latter compound is oxidized in the presence of peroxidase, resulting in an insoluble, brown, electron-dense precipitate that permits the localization of peroxidase activity by light and electron microscopy. Peroxidase activity in blood cells, which is important in the diagnosis of leukemias, can be detected by this method.

Since peroxidase is extremely active and produces an appreciable amount of insoluble precipitate in a short time, it has also been used for an important practical application: tagging other compounds. Molecules of peroxidase can be purified, isolated, and coupled with another molecule. Later in this chapter, applications of tagging molecules with peroxidase are presented.

Polysaccharides & Oligosaccharides

Polysaccharides in the body occur either in a free state or combined with proteins and lipids. In the combined state, they constitute an extremely complex heterogeneous group. They can be demonstrated by the periodic acid–Schiff (PAS) reaction, which is based on the transformation of 1,2-glycol groups present in the sugars into aldehyde residues. These residues are then revealed by Schiff's reagent, which produces a purple or

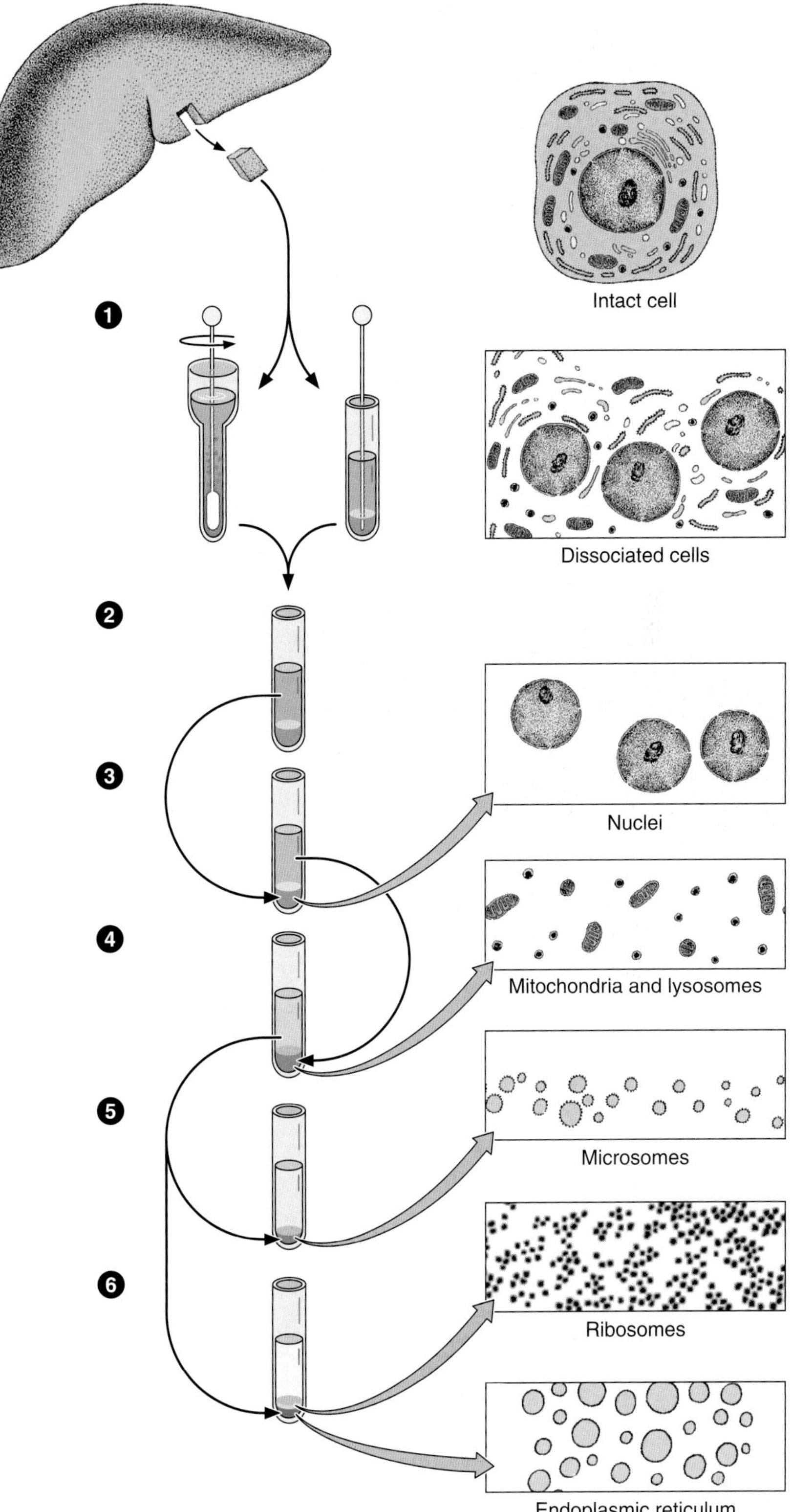

Figure 1–14. Cell fractionation allows the isolation of cell constituents by differential centrifugation. The drawings at right show the cellular organelles at the bottom of each tube after centrifugation. Centrifugal force is expressed by *g,* which is equivalent to the force of gravity. (**1**) A fragment of tissue is minced with razor blades or scissors and dissociated with a homogenizer or by ultrasound. (**2**) The dissociated tissue is left standing for about 20 min. Clumps of cells and fibers of extracellular matrix precipitate to the bottom. (**3**) The supernatant is centrifuged at 1000 *g* for 20 min. Nuclei precipitate. (**4**) The supernatant is centrifuged at 10,000 *g* for 20 min. Mitochondria and lysosomes precipitate. (**5**) The supernatant is centrifuged at 105,000 *g* for 120 min. Microsomes precipitate. (**6**) If the supernatant is first treated with sodium deoxycholate and then centrifuged at 105,000 *g* for 120 min, the microsomes dissociate and precipitate separately as endoplasmic reticulum membranes and ribosomes. (Redrawn and reproduced, with permission, from Bloom W, Fawcett DW: *A Textbook of Histology,* 9th ed. Saunders, 1968.)

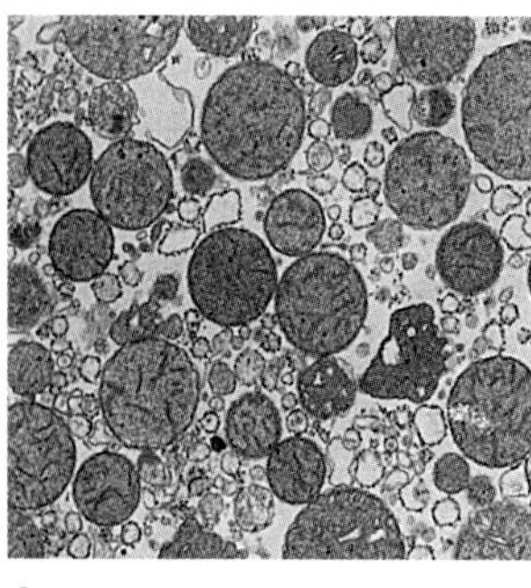

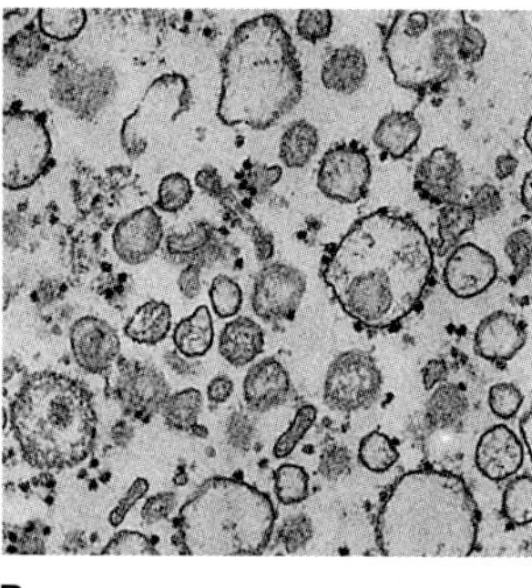

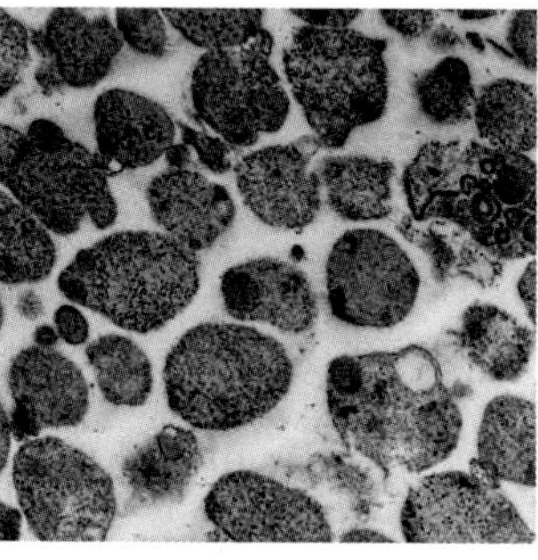

A B C

Figure 1–15. Electron micrographs of 3 cell fractions isolated by density gradient centrifugation. **A:** Mitochondrial fraction, contaminated with microsomes. **B:** Microsomal fraction. **C:** Lysosomal fraction. High magnifications. (Courtesy of P Baudhuin.)

magenta color in areas of the section with an accumulation of polysaccharides.

A ubiquitous free polysaccharide in the body is **glycogen,** which can be demonstrated by the PAS reaction in liver, striated muscle, and other tissues where it accumulates.

Glycoproteins are protein molecules associated with small, branched chains of sugars (oligosaccharides). The protein chain predominates in weight and volume over the oligosaccharide chain. While some glycoproteins contain no acidic groups (neutral glycoproteins) and are PAS-positive, others have limited amounts of carboxyl or sulfate radicals. Since both glycogen and neutral glycoproteins are PAS-positive, the specificity of the PAS reaction can be improved by comparing the staining of regular sections with that of sections pretreated with an enzyme that breaks glycogen (eg, salivary amylase). Structures that stain intensely with the PAS reaction but do not stain when pretreated with amylase contain glycogen. Figure 1–19 shows examples of structures stained by the PAS reaction.

Glycosaminoglycans are strongly anionic, unbranched long-chain polysaccharides containing aminated monosaccharides (amino sugars). A large number of glycosaminoglycan chains inserted at regular intervals along a protein core constitute the **proteoglycans.** Some of the significant constituents of connective

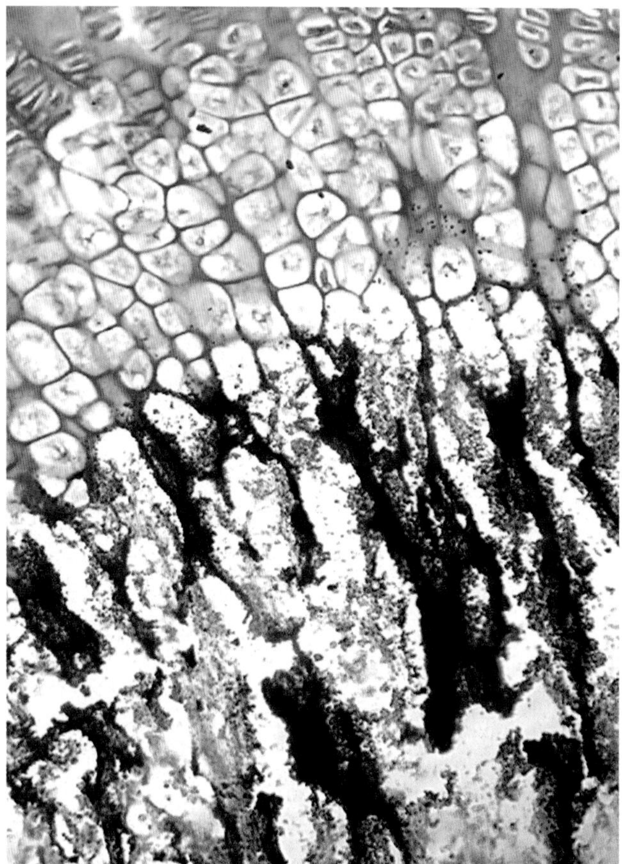

Figure 1–16. Photomicrograph of a bone section treated with a histochemical technique to demonstrate calcium ions. The dark precipitate indicates the presence of calcium phosphate in calcified bone and cartilage. Noncalcified cartilage tissue (stained in pink) is in the upper portion of the figure. Medium magnification.

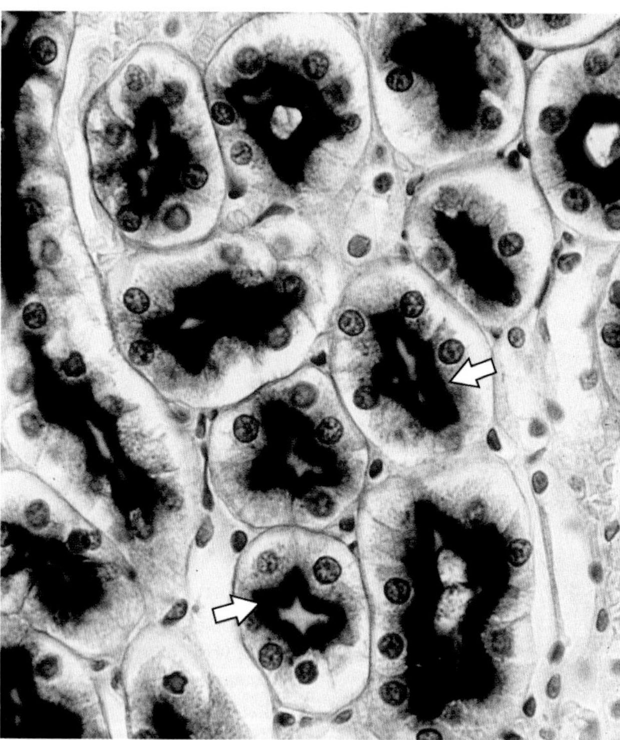

Figure 1–17. Photomicrograph of a rat kidney section treated by the Gomori method to demonstrate the enzyme alkaline phosphatase. The sites where this enzyme is present (cell surface) stain intensely with black (arrows). Medium magnification.

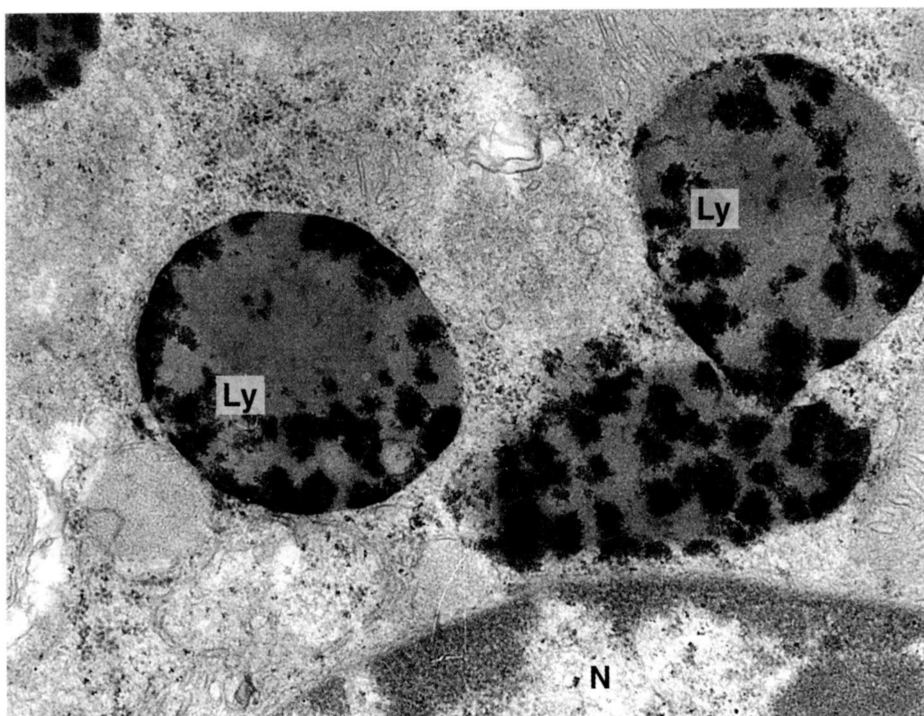

Figure 1–18. Detection of acid phosphatase. Electron micrograph of a rat kidney cell showing 3 lysosomes (ly) above the nucleus (N). The dark precipitate within these structures is lead phosphate that precipitated on places where acid phosphatase was present. × 25,000. (Courtesy of E Katchburian.)

tissue matrices are proteoglycans (see Chapters 5 and 7). Unlike in the glycoproteins, the carbohydrate chains in proteoglycans constitute the major component of the molecule. Glycosaminoglycans and acidic glycoproteins are strongly anionic because of their high content of carboxyl and sulfate groups. For this reason, they react strongly with the alcian blue dye.

Lipids

Lipids are best revealed with dyes that are soluble in lipids. Frozen sections are immersed in alcohol solutions saturated with the dye. Sudan IV and Sudan black are the most commonly used dyes. The dye dissolves in the cellular lipid droplets, which become stained in red or black. Additional methods used for the localization of cholesterol and its esters, phospholipids, and glycolipids are useful in diagnosing metabolic diseases in which there are intracellular accumulations of different kinds of lipids.

MEDICAL APPLICATION

Several histochemical procedures are frequently used in laboratory diagnosis of diseases that result in the storage of iron, glycogen, glycosaminoglycans and other substances. Examples are the Perls' reaction for iron (eg, hemochromatosis, hemosiderosis), the PAS-amylase reaction for glycogen (glycogenosis), the alcian blue staining for glycosaminoglycans (mucopolysaccharidosis), and the lipid staining (sphingolipidosis).

DETECTION METHODS USING HIGH-AFFINITY INTERACTIONS BETWEEN MOLECULES

A specific molecule present in a tissue section may be identified by using compounds that specifically interact with the molecule. The compounds that will interact with the molecule must be

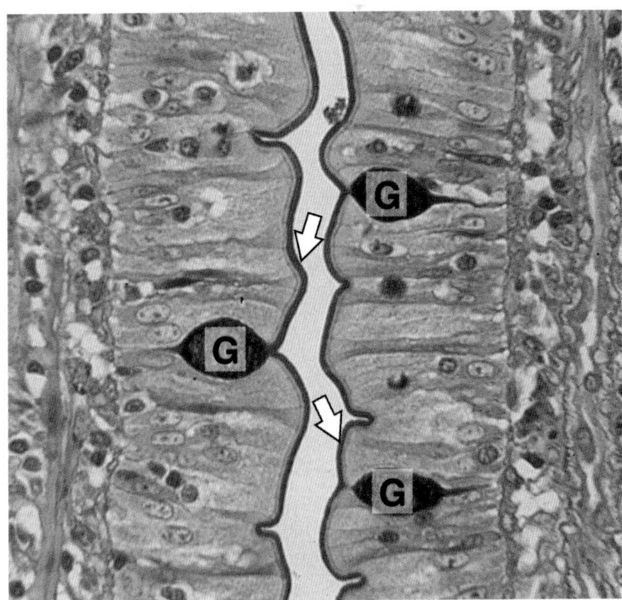

Figure 1–19. Photomicrograph of an intestinal villus stained by PAS. Staining is intense in the cell surface brush border (arrows) and in the secretory product of goblet cells (G) because of their high content of polysaccharides. The counterstain was hematoxylin. High magnification.

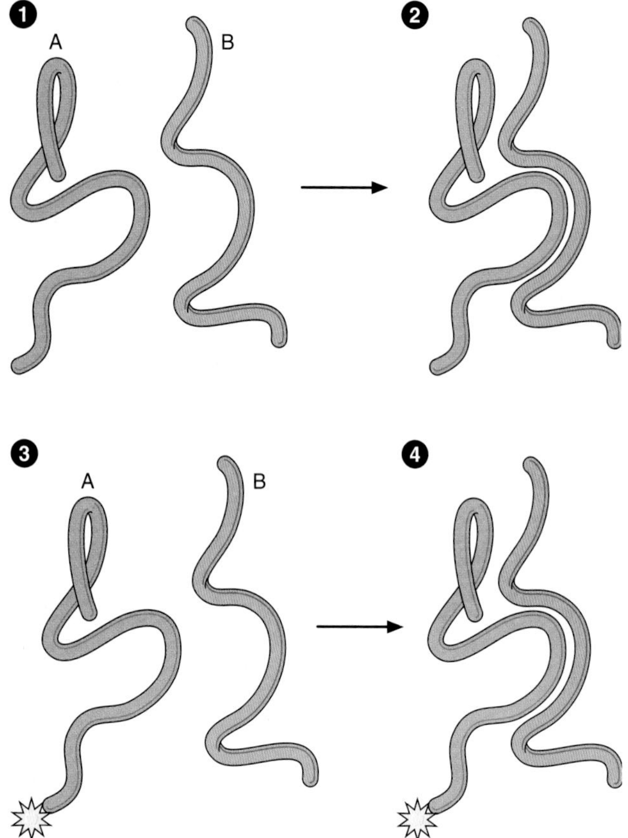

Figure 1–20. Compounds that have affinity toward another molecule can be tagged with a label and used to identify that molecule. (**1**) Molecule A has a high and specific affinity toward a portion of molecule B. (**2**) When A and B are mixed, A binds to the portion of B it recognizes. (**3**) Molecule A may be tagged with a label that can be visualized with a light or electron microscope. The label can be a fluorescent compound, an enzyme such as peroxidase, a gold particle, or a radioactive atom. (**4**) If molecule B is present in a cell or extracellular matrix that is incubated with labeled molecule A, molecule B can be detected.

tagged with a label that can be detected under the light or electron microscope (Figure 1–20). The most commonly used labels are fluorescent compounds (which can be seen with a fluorescence or laser microscope), radioactive atoms (which can be detected with autoradiography), molecules of peroxidase (which can be detected after demonstration of the enzyme with hydrogen peroxide and 3,3-diaminoazobenzidine) or other enzymes (which can be detected with their respective substrates), and metal (usually gold) particles that can be observed with light and electron microscopy. These methods are mainly used for detecting sugars, proteins, and nucleic acids.

Phalloidin, protein A, lectins, and antibodies are examples of compounds that interact specifically with other molecules.

Phalloidin, which is extracted from a mushroom (*Amanita phalloides*), interacts strongly with actin and is usually labeled with fluorescent dyes to demonstrate microfilaments.

Protein A is a protein obtained from *Staphylococcus aureus* that binds to the Fc region of immunoglobulin (antibody) molecules. When protein A is tagged with a label, immunoglobulins can be detected (see Immunocytochemistry).

Lectins are proteins or glycoproteins that are derived mainly from plant seeds and that bind to carbohydrates with high affinity and specificity. Different lectins bind to specific sugars or to sequences of sugar residues. They bind to glycoproteins, proteoglycans, and glycolipids and are widely used to characterize membrane molecules containing specific sequences of sugar residues.

Immunocytochemistry

A highly specific interaction between molecules is that between an antigen and its antibody. For this reason, methods using labeled antibodies have proved most useful in identifying and localizing specific proteins.

The body has cells that are able to distinguish its own molecules (self) from foreign ones. When exposed to foreign molecules—called **antigens**—the body may respond by producing proteins—**antibodies**—that react specifically and bind to the antigen, thus helping to eliminate the foreign substance. Antibodies are proteins of a large family, the **immunoglobulin** family.

In immunocytochemistry, a tissue section (or cells in culture) that one believes may contain a certain protein is incubated in a solution containing an antibody to this protein. The antibody binds specifically to the protein, whose location can then be seen with either the light or electron microscope, depending on the type of compound used to label the antibody.

One of the most important requirements for immunocytochemistry is the availability of an antibody against the protein that is to be detected. This means that the protein must have been previously purified and isolated so that antibodies can be produced. Some methods for protein isolation can be seen in Figures 1–21 and 1–22.

Polyclonal and Monoclonal Antibodies

Let us suppose that one's objective is to produce antibodies against protein *x* of a certain animal species (eg, a rat or a human). If protein *x* is already isolated, it is injected into an animal of another species (eg, a rabbit or a goat). If the protein is sufficiently different for this animal to recognize it as foreign—that is, as an antigen—the animal will produce antibodies against the protein (eg, rabbit antibody against rat *x* or goat antibody against human *x*). These antibodies are collected from the animal's plasma and used for immunocytochemistry.

Several groups (clones) of lymphocytes of the animal that was injected with protein *x* may recognize different parts of protein *x* and produce a different antibody against each part. These antibodies constitute a mixture of **polyclonal antibodies.**

It is possible, however, to provide protein *x* for lymphocytes maintained in cell culture (actually, lymphocytes fused with tumor cells). The different clones of lymphocytes will produce different antibodies against the several parts of protein *x*. Each clone can be isolated and cultured separately so that the different antibodies against protein *x* can be collected separately. Each of these antibodies is a **monoclonal antibody.** There are several advantages to using a monoclonal antibody rather than a polyclonal antibody: for instance, a monoclonal antibody can be selected to be highly specific and to bind strongly to the protein to be detected. Therefore, there will be less nonspecific binding to the other proteins that could make the detection of protein *x* difficult.

In the **direct method of immunocytochemistry,** the antibody (either monoclonal or polyclonal) must be tagged with an appropriate label. A tissue section is incubated with the antibody for some time so that the antibody interacts with and binds to protein *x*. The section is then washed to remove the antibody (Figure 1–23). Depending on the label that was used (fluorescent compound, enzyme, gold particles), the section can be observed with a light or electron microscope. If peroxidase or another enzyme was used as a label, the enzyme must be detected before the tissue section is observed in the microscope (see Histochemistry & Cytochemistry). The areas of the tissue section that contain protein *x* will become fluorescent or will be covered by gold particles or by a dark precipitate because of the presence of the labeling enzyme.

The **indirect method of immunocytochemistry** is more sensitive but requires more steps. Let us suppose that one's objective is to detect protein *x*, present in rats. Before proceeding to the immunochemical reaction, two procedures are needed: (1) antibodies to rat protein *x* must first be produced in an animal of another species (eg, a rabbit); (2) in a parallel procedure, immunoglobulin from a normal (noninjected) rabbit must be injected into an animal of a third species (eg, a sheep or a goat). Rabbit immunoglobulins are considered foreign by a sheep or a goat and are thus capable of inducing the production of an antibody (an anti-antibody or anti-immunoglobulin) in that animal.

The indirect immunocytochemical detection is performed by initially incubating a section of a rat tissue believed to contain protein *x* with rabbit anti-*x* antibody. After washing, the tissue sections are incubated with labeled sheep or goat antibody against rabbit antibodies. The anti-antibodies will recognize the rabbit antibody that had recognized protein *x* (Figure 1–24). Protein *x* can then be detected by using a microscopic technique appropriate for the label used in the secondary antibody. There are other indirect methods that involve the use of other intermediate molecules, such as the biotin-avidin technique.

MEDICAL APPLICATION

Immunocytochemistry has contributed significantly to research in cell biology and to the improvement of medical diagnostic procedures. Figures 1–25 to 1–28 show examples of immunocytochemical detection of molecules. Table 1–1 shows some of the routine applications of immunocytochemical procedures in clinical practice.

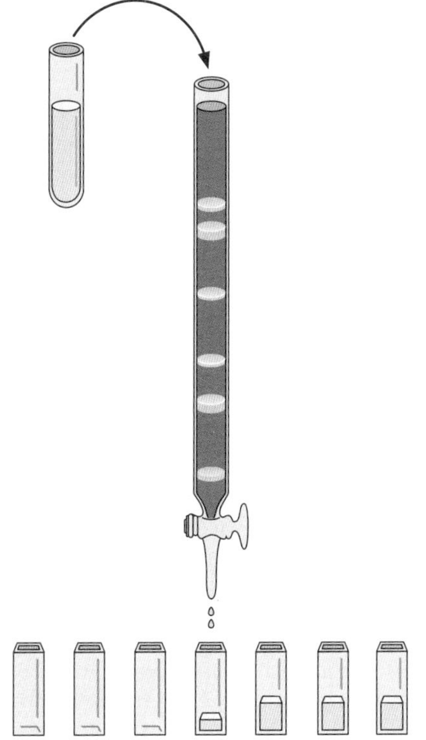

Figure 1–21. Ultracentrifugation (**A**) and chromatography (**B**): methods of protein isolation. **A:** A mixture of proteins obtained from homogenized cells or tissues is submitted to centrifugation at high speed for several hours. The proteins separate into several bands, depending on the size and density of the protein molecules. The ultracentrifugation medium is drained and collected in several fractions that contain different proteins, which can be analyzed further. **B:** A solution containing a mixture of proteins obtained from homogenized cells or tissues is added to a column filled with particles that have different chemical properties. For instance, the particles may have different electrostatic charges (attracting proteins according to their charge) or different sizes of pores (acting as sieves for different-sized molecules). As the proteins migrate through the column, their movement is slowed according to their interaction with the particles. When the effluent is recovered, the different groups of proteins may be collected separately.

Hybridization Techniques

The central challenge in modern cell biology is to understand the workings of the cell in molecular detail. This goal requires techniques that permit analysis of the molecules involved in the process of information flow from DNA to protein. Many techniques are based on **hybridization.** Hybridization is the binding between two single strands of nucleic acids (DNA with DNA, RNA with RNA, or RNA with DNA) that recognize each other if the strands are complementary. The greater the similarities of the sequences, the more readily do complementary strands form "hybrid" double-strand molecules. Hybridization thus allows the specific identification of sequences of DNA or RNA.

In Situ Hybridization

When applied directly to cells and tissue sections, smears, or chromosomes of squashed mitotic cells, the technique is called in situ hybridization. This technique is ideal for determining if

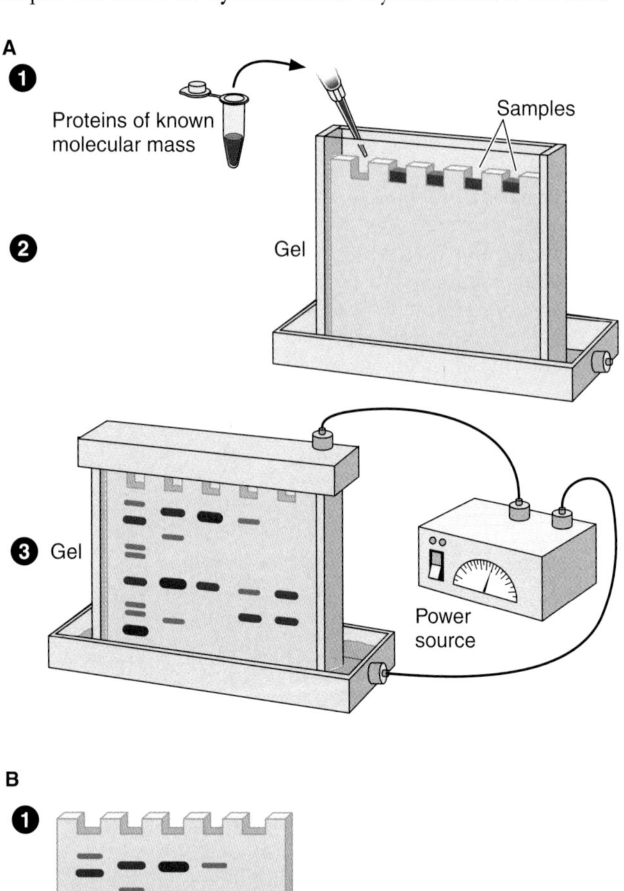

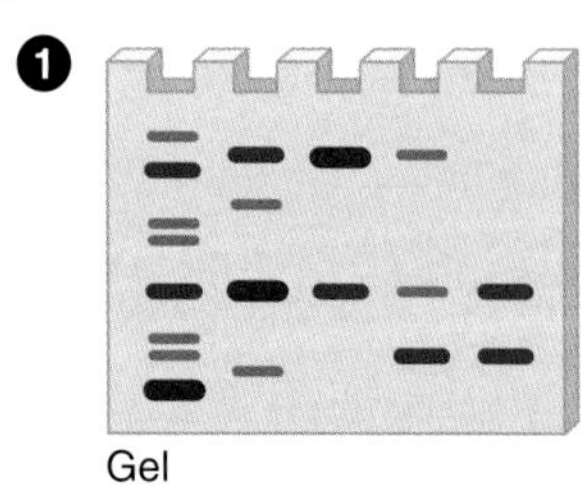

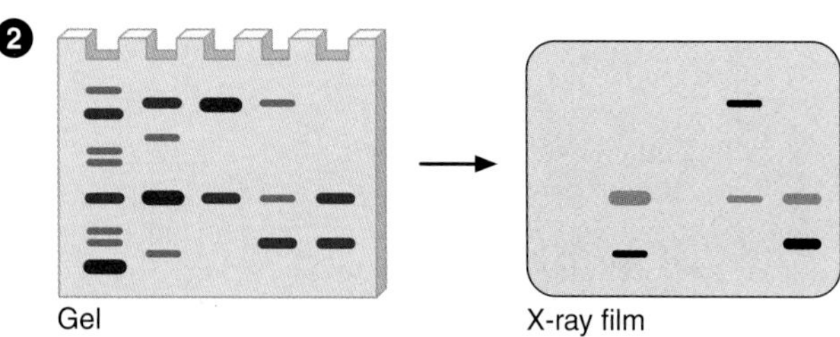

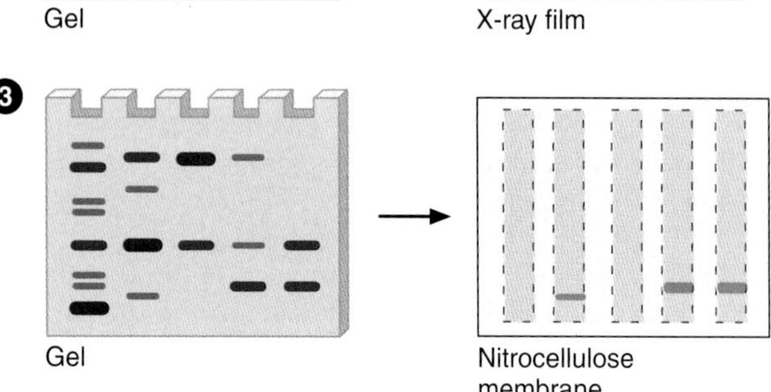

Table 1–1. Partial list of commonly used proteins (antigens) important for the immunocytochemical diagnosis and treatment of disease.

Antigens	Diagnosis
Intermediate filament proteins	
Cytokeratins	Tumors of epithelial origin
Glial fibrillary acid protein	Tumors of some glial cells
Vimentin	Tumors of connective tissue
Desmin	Tumors of muscle
Other proteins	
Protein and polypeptide hormones	Protein or polypeptide hormone–producing tumors
Carcinoembryonic antigen (CEA)	Glandular tumors, mainly of the digestive tract and breast
Steroid hormone receptors	Breast duct–cell tumors
Antigens produced by viruses	Specific virus infections

Figure 1–22. Gel electrophoresis: a method of protein isolation. **A:** Isolation of proteins. (**1**) Mixtures of proteins are obtained from homogenized cells or tissues. They are usually treated with a strong detergent (sodium dodecyl sulfate) and with mercaptoethanol to unfold and separate the protein subunits. (**2**) The samples are put on top of a slab of polyacrylamide gel, which is submitted to an electrical field. The proteins migrate along the gel according to their size and shape. (**3**) A mixture of proteins of known molecular mass is added to the gel as a reference to identify the molecular mass of the other proteins. **B:** Detection and identification of the proteins. (**1**) Staining. All proteins will stain the same color. The color intensity is proportional to the protein concentration. (**2**) Autoradiography. Radioactive proteins can be detected by autoradiography. An x-ray film is apposed to the gel for a certain time and then developed. Radioactive proteins will appear as dark bands in the film. (**3**) Immunoblotting. The proteins can be transferred from the gel to a nitrocellulose membrane. The membrane is incubated with an antibody made against proteins that may be present in the sample.

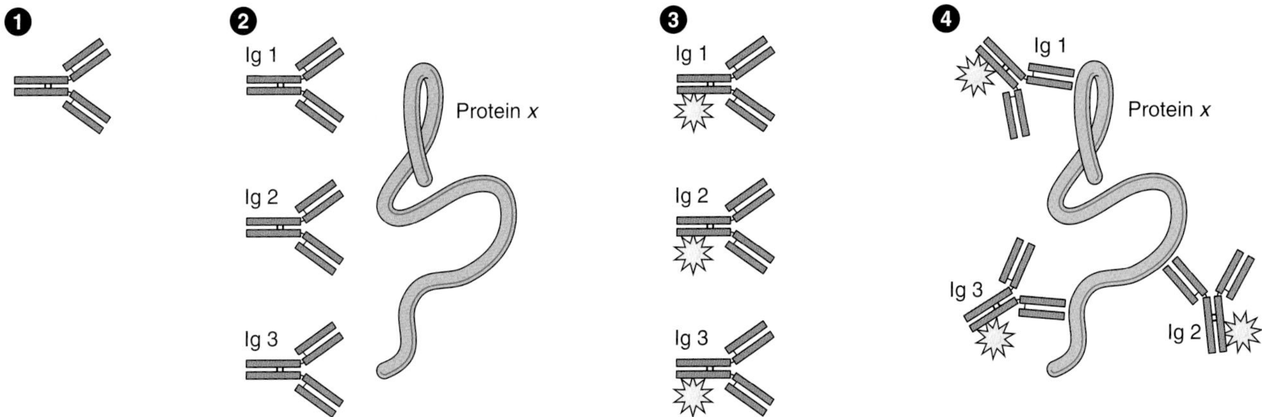

Figure 1–23. Direct method of immunocytochemistry. (**1**) Immunoglobulin molecule (Ig). (**2**) Production of a polyclonal antibody. Protein *x* from a rat is injected into a rabbit. Several rabbit Igs are produced against protein *x*. (**3**) Labeling the antibody. The rabbit Igs are tagged with a label. (**4**) Immunocytochemical reaction. The rabbit Igs recognize and bind to different parts of protein *x*.

a cell has a specific sequence of DNA (such as a gene or part of a gene), for identifying the cells in which a specific gene is being transcribed, or for determining the localization of a gene in a specific chromosome. The DNA inside the cell must be initially denatured by heat or by denaturing agents so that both strands of the DNA separate. They are then ready to be hybridized with a segment of single-stranded DNA or RNA that is complementary to the sequence one wishes to detect. This sequence is called a **probe.** The probe may be obtained by cloning, by PCR amplification of the target sequence, or by synthesis if the desired sequence is short. The probe must be tagged with a label, usually a radioactive isotope (which can be localized by autoradiography) or a modified nucleotide (digoxygenin), which can be identified by immunocytochemistry.

In in situ hybridization, the tissue section, cultured cells, smears, or chromosomes of squashed mitotic cells must first be heated to separate the double strands of their DNA. Then a solution containing the probe is placed over the specimen for a period of time necessary for hybridization. After washing off the excess probe, the localization of the bound probe is revealed through its label (Figure 1–29).

Hybridization can also be performed with purified DNA or RNA in solid supports. Mixtures of nucleic acids are separated by electrophoresis in an agarose gel or a polyacrylamide gel. The polyacrylamide gel allows higher resolution. After electrophoresis, the molecules of nucleic acids of different sizes are transferred to a nylon or nitrocellulose sheet by solvent drag: a buffer flows through the gel and membrane by capillarity, carrying the nu-

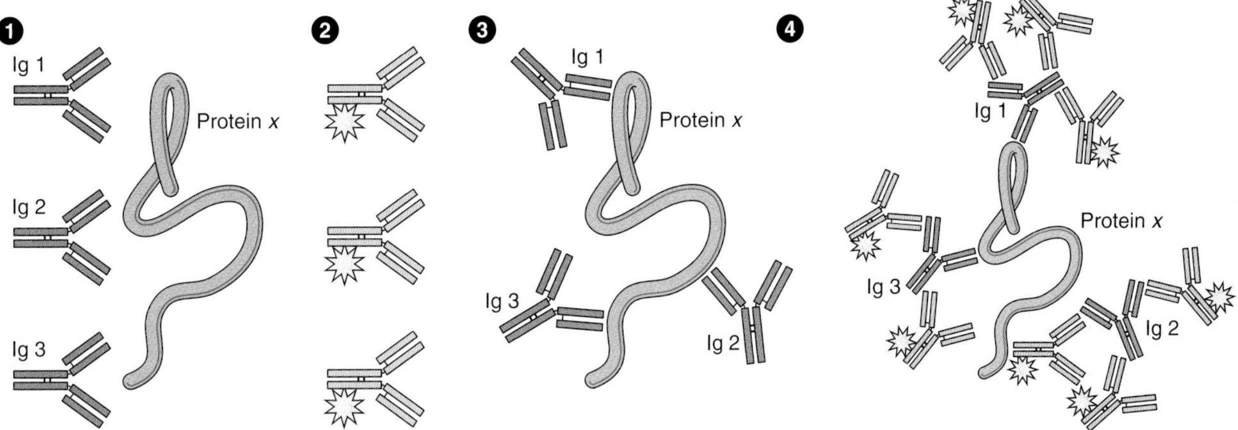

Figure 1–24. Indirect method of immunocytochemistry. (**1**) Production of primary polyclonal antibody. Protein *x* from a rat is injected into a rabbit. Several rabbit immunoglobulins (Ig) are produced against protein *x*. (**2**) Production of secondary antibody. Ig from a nonimmune rabbit is injected into a goat. Goat Igs against rabbit Ig are produced. The goat Igs are then isolated and tagged with a label. (**3**) First step of immunocytochemical reaction. The rabbit Igs recognize and bind to different parts of protein *x*. (**4**) Second step of immunocytochemical reaction. Labelled goat Igs recognize and bind to different parts of rabbit immunoglobulin molecules, therefore labeling protein *x*.

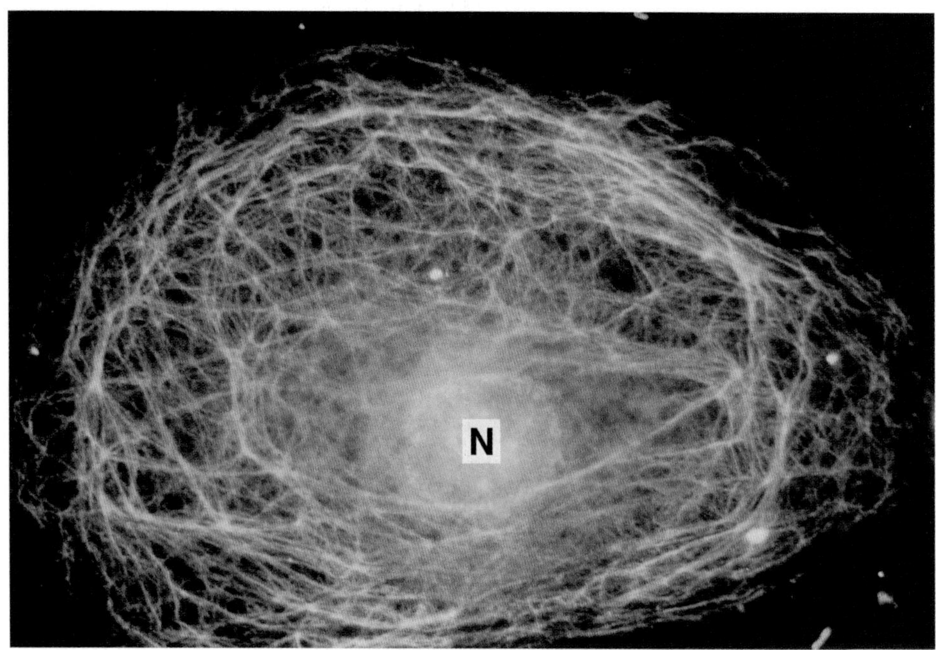

Figure 1–25. Photomicrograph of a mouse decidual cell grown in vitro. The protein desmin, which forms intermediate filaments, was detected with an indirect immunofluorescence technique. A mesh of fluorescent intermediate filaments occupies most of the cytoplasm. The nucleus (N) is stained blue. High magnification. (Courtesy of Fabiano G Costa.)

cleic acid molecules that bind strongly to the nylon or nitrocellulose sheet, where the nucleic acids can be further analyzed. This technique of DNA identification is called Southern blotting. When electrophoresis of RNA is performed, the technique is called Northern blotting.

Hybridization techniques are highly specific and are routinely used in research, clinical diagnosis, and forensic medicine.

PROBLEMS IN THE INTERPRETATION OF TISSUE SECTIONS

Distortions and Artifacts Caused by Tissue Processing

A key point to be remembered in studying and interpreting stained tissue sections in microscope preparations is that the observed product is the end result of a series of processes that begin with fixation and finish with the staining of the section. Several steps of this procedure may distort the tissues, delivering an image that may differ from the structures as they were when alive. One cause of distortion is the shrinkage produced by the fixative, by the ethanol, and by the heat needed for paraffin embedding. Shrinkage is lessened when specimens are embedded in resin.

A consequence of shrinkage is the appearance of artificial spaces between cells and other tissue components. Another source of artificial spaces is the loss of molecules that were not properly kept in the tissues by the fixative or that were removed by the dehydrating and clearing fluids.

All these artificial spaces and other distortions caused by the section preparation procedure are called **artifacts.** Other artifacts may include wrinkles of the section (which may be confused with a blood capillary), precipitates of stain (which may be confused with cytoplasmic granules), and many more. Students must be aware of the existence of artifacts and try to recognize them so as not to be confused by these distortions.

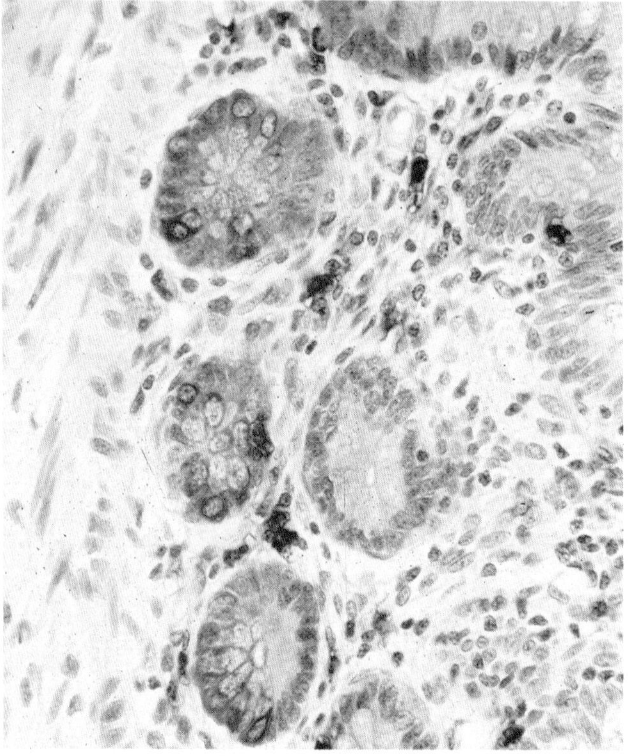

Figure 1–26. Photomicrograph of a section of small intestine in which an antibody against the enzyme lysozyme was applied to demonstrate lysosomes in macrophages and Paneth cells. The brown color results from the reaction done to show peroxidase, which was linked to the secondary antibody. Nuclei counterstained with hematoxylin. Medium magnification.

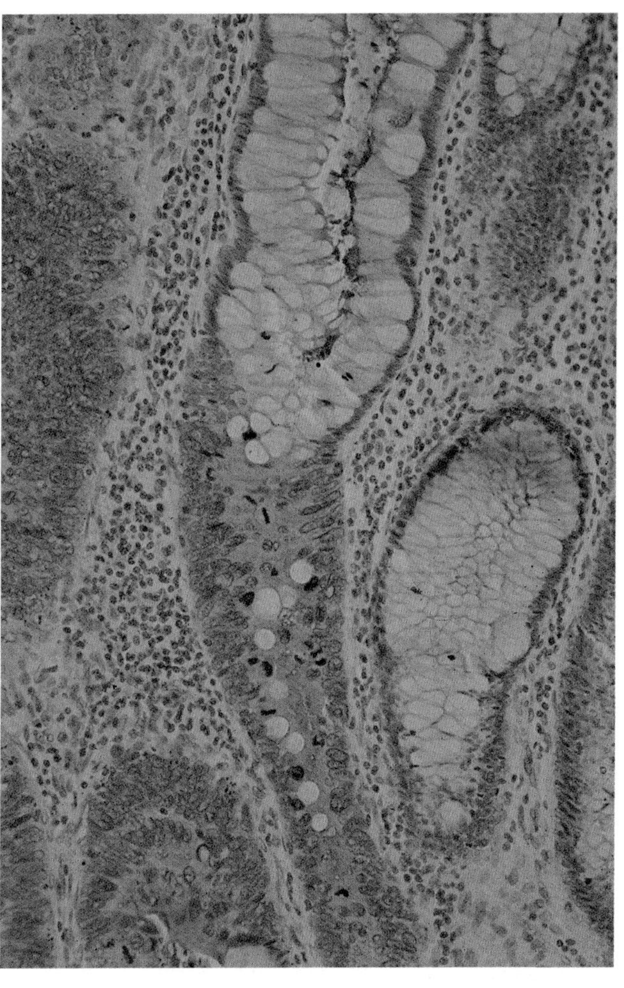

Figure 1–27. Carcinoembryonic antigen is a protein present in several malignant tumors mainly of the breast and intestines. This photomicrograph is an immunocytochemical demonstration of carcinoembryonic antigen in a section of large intestine adenocarcinoma. The antibody was labeled with peroxidase, and the counterstain was hematoxylin. Medium magnification.

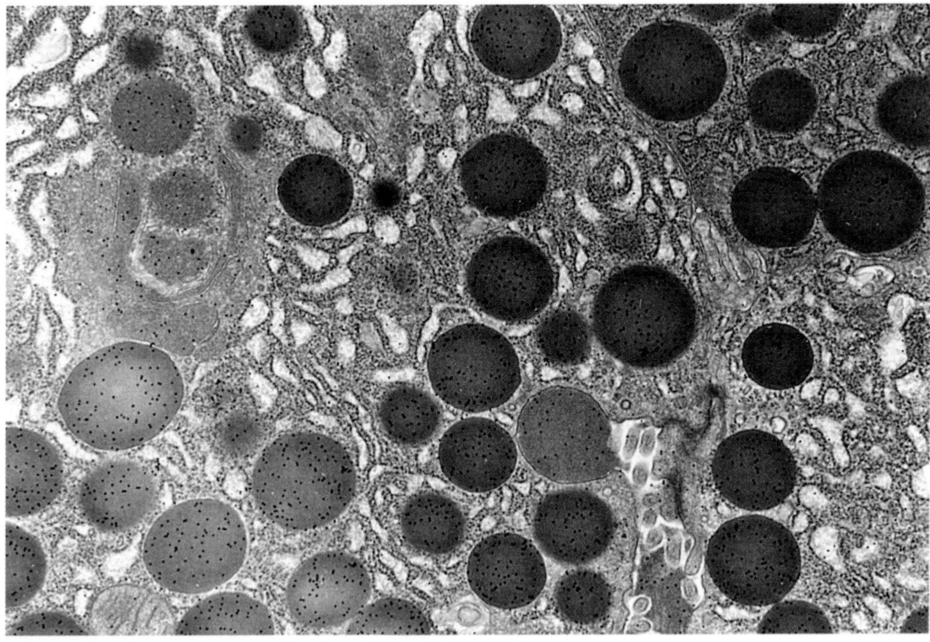

Figure 1–28. Electron micrograph showing a section of a pancreatic acinar cell that was incubated with anti-amylase antibody and stained by protein A coupled with gold particles. Protein A has high affinity toward antibody molecules. The gold particles appear as very small black dots over the mature secretory granules and the forming granules in the Golgi complex. (Courtesy of M Bendayan.)

Totality of the Tissue

Another difficulty in the study of histological sections is the impossibility of differentially staining all tissue components on only one slide. Thus, when observing cells under a light microscope, it is almost impossible to see the nuclei, mitochondria, lysosomes, and peroxisomes, surrounded by a basement membrane as well as by collagen and elastic and reticular fibers. It is necessary to examine several preparations, each one stained by a different method, before an idea of the whole composition and structure of a tissue can be obtained. The transmission electron microscope, on the other hand, allows the observation of a cell with all its organelles and inclusions surrounded by the components of the extracellular matrix.

Two Dimensions and Three Dimensions

When a 3-dimensional volume is cut into very thin sections, the sections seem to have only 2 dimensions: length and width. This often leads the observer to err if he or she does not realize that a sectioned ball looks like a circle and that a sectioned tube looks like a ring (Figure 1–30). When a section is observed under the microscope, the student must always imagine that something may be missing in front of or behind that section, because many structures are thicker than the section. It must also be remembered that the structures within a tissue are usually sectioned randomly.

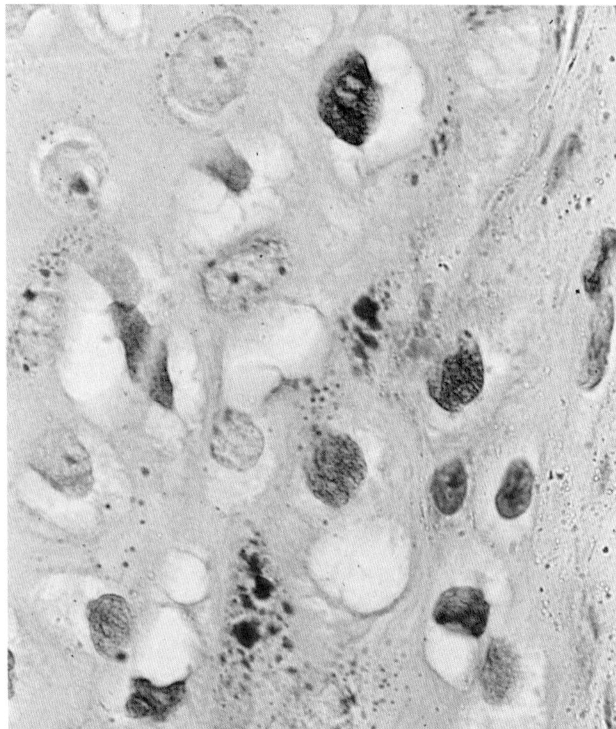

Figure 1–29. Tissue section of a benign epithelial tumor (condyloma) submitted to in situ hybridization. The brown areas are places where DNA of human papillomavirus type 2 is present. The counterstain was hematoxylin. Medium magnification. (Courtesy of JE Levi.)

To understand the architecture of an organ, one must study sections made in different planes. Sometimes only the study of serial sections and their reconstruction into a 3-dimensional volume make it possible to understand a complex organ.

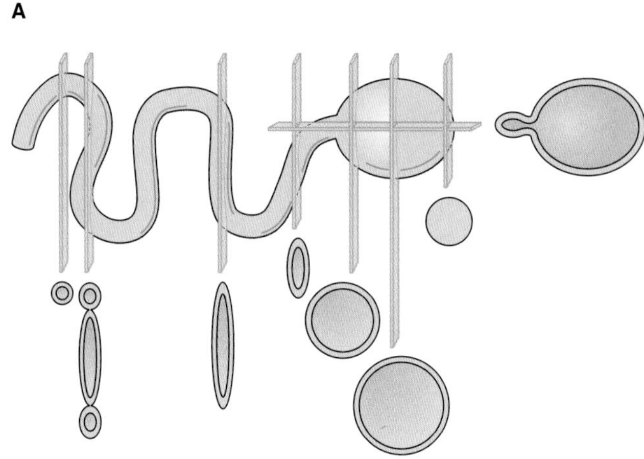

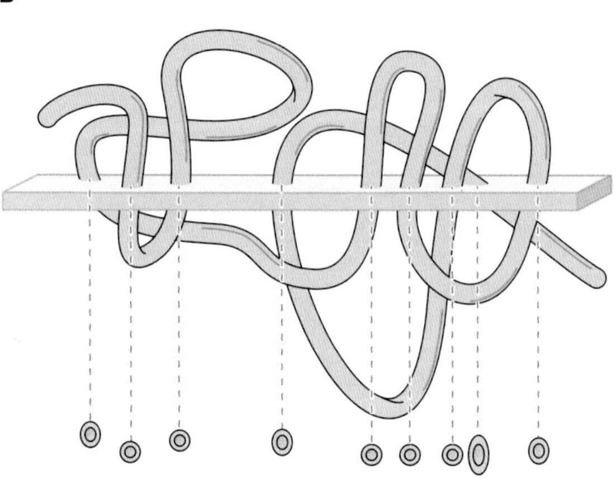

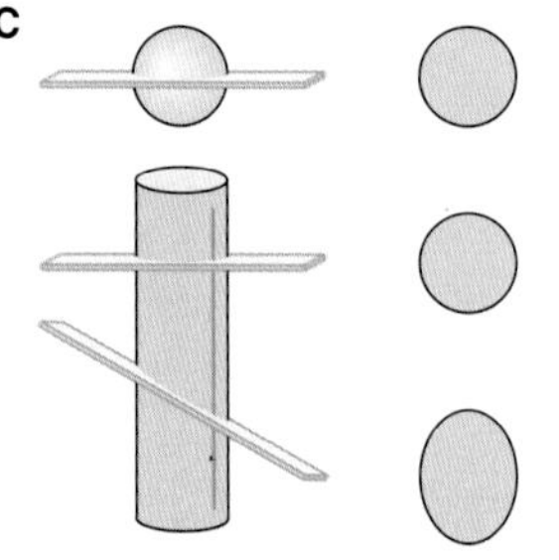

Figure 1–30. How different 3-dimensional structures may appear when thin-sectioned. **A:** Different sections through a hollow ball and a hollow tube. **B:** A section through a single coiled tube may appear as sections of many separate tubes. **C:** Sections through a solid ball (above) and sections through a solid cylinder (below).

REFERENCES

Alberts B et al: *Molecular Biology of the Cell,* 3rd ed. Garland, 1994.

Bancroft JD, Stevens A: *Theory and Practice of Histological Techniques,* 2nd ed. Churchill Livingstone, 1990.

Cuello ACC: *Immunocytochemistry.* Wiley, 1983.

Darnell J, Lodish H, Baltimore D: *Molecular Cell Biology,* 2nd ed. Scientific American Books, 1990.

Hayat MA: *Stains and Cytochemical Methods.* Plenum, 1993.

James J: *Light Microscopic Techniques in Biology and Medicine.* Martinus Nijhoff, 1976.

Junqueira LCU et al: Differential staining of collagen types I, II and III by Sirius Red and polarization microscopy. Arch Histol Jpn 1978;41:267.

Meek GA: *Practical Electron Microscopy for Biologists.* Wiley, 1976.

Pease AGE: *Histochemistry: Theoretical and Applied,* 4th ed. Churchill Livingstone, 1980.

Rochow TG, Tucker PA: *Introduction to Microscopy by Means of Light, Electrons, X Rays, or Acoustics.* Plenum Press, 1994.

Rogers AW: *Techniques of Autoradiography,* 3rd ed. Elsevier, 1979.

Rubbi CP: *Light Microscopy. Essential Data.* Wiley, 1994.

Spencer M: *Fundamentals of Light Microscopy.* Cambridge Univ Press, 1982.

Stoward PJ, Polak JM (editors): *Histochemistry: The Widening Horizons of Its Applications in Biological Sciences.* Wiley, 1981.

The Cytoplasm

2

Cells are the structural units of all living organisms. There are two fundamentally different types of cells, but so many biochemical similarities exist between them that some investigators have postulated that one group evolved from the other.

The **prokaryotic** (Gr. *pro,* before, + *karyon,* nucleus) cell is found only in bacteria. These cells are small (1–5 μm long), typically have a cell wall outside the plasmalemma, and lack a nuclear envelope separating the genetic material (DNA) from other cellular constituents. In addition, prokaryotes have no histones (specific basic proteins) bound to their DNA and usually no membranous organelles.

In contrast, **eukaryotic** (Gr. *eu,* good, + *karyon*) cells are larger and have a distinct nucleus surrounded by a nuclear envelope (Figure 2–1). Histones are associated with the genetic material, and numerous membrane-limited organelles are found in the cytoplasm. This book is concerned exclusively with eukaryotic cells.

CELLULAR DIFFERENTIATION

The human organism presents about 200 different cell types, all derived from the zygote, a single cell formed by fertilization of an oocyte by a spermatozoon. The first cellular divisions of the zygote originate cells called **blastomeres,** which are able to form all cell types of the adult. Through this process, called **cell differentiation,** the cells synthesize specific proteins, change their shape, and become very efficient in specialized functions. For example, muscle cell precursors elongate into spindle-shaped cells that synthesize and accumulate myofibrillar proteins (actin, myosin). The resulting cell efficiently converts chemical energy into contractile force.

The main cellular functions performed by specialized cells in the body are listed in Table 2–1.

CELL ECOLOGY

Since the body experiences considerable environmental diversity (eg, normal and pathological conditions), the same cell type can exhibit different characteristics and behaviors in different regions and circumstances. Thus, macrophages and neutrophils (both of which are phagocytic defense cells) will shift from oxidative metabolism to glycolysis in an anoxic, inflammatory environment. Cells that appear to be structurally similar may react in different ways because they have different families of receptors for signaling molecules (such as hormones and extracellular matrix macromolecules). For example, because of their diverse library of receptors, breast fibroblasts and uterine smooth muscle cells are exceptionally sensitive to female sex hormones.

CELL COMPONENTS

The cell is composed of two basic parts: **cytoplasm** (Gr. *kytos,* cell, + *plasma,* thing formed) and **nucleus** (L. *nux,* nut). Individual cytoplasmic components are usually not clearly distinguishable in common hematoxylin-and-eosin–stained preparations; the nucleus, however, appears intensely stained dark blue or black.

Cytoplasm

The outermost component of the cell, separating the cytoplasm from its extracellular environment, is the **plasma membrane**

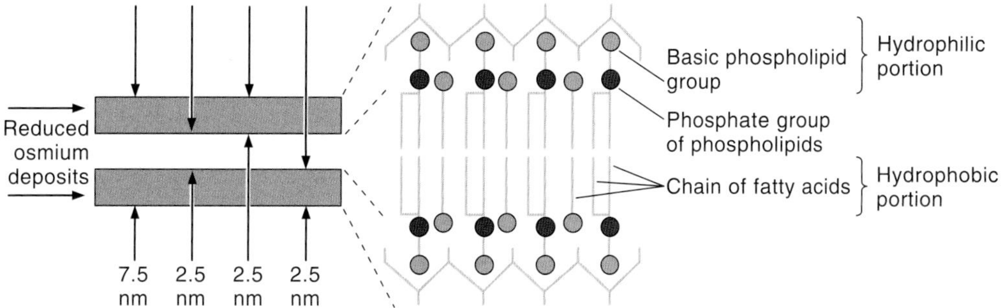

Figure 2–1. The ultrastructure and molecular organization (**right**) of the cell membrane. The dark lines at left represent the two dense layers observed in the electron microscope; these are caused by the deposit of osmium in the hydrophilic portions of the phospholipid molecules.

Table 2–1. Cellular functions in some specialized cells.

Function	Specialized Cell(s)
Movement	Muscle cell
Synthesis and secretion of enzymes	Pancreatic acinar cells
Synthesis and secretion of mucous substances	Mucous-gland cells
Synthesis and secretion of steroids	Some adrenal gland, testis, and ovary cells
Ion transport	Cells of the kidney and salivary gland ducts
Intracellular digestion	Macrophages and some white blood cells
Transformation of physical and chemical stimuli into nervous impulses	Sensory cells
Metabolite absorption	Cells of the intestine

(**plasmalemma**). However, even if the plasma membrane is the external limit of the cell, there is a continuum between the interior of the cell and extracellular macromolecules. The plasma membrane contains proteins called **integrins** that are linked to cytoplasmic cytoskeletal filaments and to extracellular molecules. Through these linkages there is a constant exchange of influences, in both ways, between the extracellular matrix and the cytoplasm. The cytoplasm itself is composed of a matrix, or **cytosol,** in which are embedded the **organelles,** the **cytoskeleton,** and **deposits** of carbohydrates, lipids, and pigments.

The cytoplasm of eukaryotic cells is divided into several distinct compartments by membranes that regulate the intracellular traffic of ions and molecules. These compartments concentrate enzymes and the respective substrates, thus increasing the efficiency of the cell.

Plasma Membrane

All eukaryotic cells are enveloped by a limiting membrane composed of phospholipids, cholesterol, proteins, and chains of oligosaccharides covalently linked to phospholipids and protein molecules. The cell, or plasma, membrane functions as a selective barrier that regulates the passage of certain materials into and out of the cell and facilitates the transport of specific molecules. One important role of the cell membrane is to keep constant the intracellular milieu, which is different from the extracellular fluid. Membranes also carry out a number of specific recognition and regulatory functions (to be discussed later), playing an important role in the interactions of the cell with its environment.

Membranes range from 7.5 to 10 nm in thickness and consequently are visible only in the electron microscope. Electron micrographs reveal that the plasmalemma—and, for that matter, all other organellar membranes—exhibit a trilaminar structure after fixation in osmium tetroxide (Figure 2–1). Because all membranes have this appearance, the 3-layered structure has been designated the **unit membrane** (Figure 2–2). The 3 layers seen in the electron microscope are apparently produced by the deposit of reduced osmium on the hydrophilic groups present on each side of the lipid bilayer.

Membrane phospholipids, such as phosphatidylcholine (lecithin) and phosphatidylethanolamine (cephalin), consist of two long, nonpolar (hydrophobic) hydrocarbon chains linked to a charged (hydrophilic) head group. Cholesterol is also a constituent of cell membranes. Within the membrane, phospholipids are most stable when organized into a double layer with their hydrophobic (nonpolar) chains directed toward the center of the membrane and their hydrophilic (charged) heads directed outward (Figure 2–1). Cholesterol breaks up the close packing of the phospholipid long chains, and this disruption makes the membrane more fluid. The cell controls the fluidity of the membranes through the amount of cholesterol. The lipid composition of each half of the bilayer is different. For example, in red blood cells (erythrocytes), phosphatidylcholine and sphingomyelin are more abundant in the outer half of the membrane, whereas phosphatidylserine and phosphatidylethanolamine are more concentrated in the inner half. Some of the lipids, known as glycolipids, possess oligosaccharide chains that extend outward from the surface of the cell membrane and thus contribute to the lipid asymmetry (Figures 2–3A and 2–4).

Proteins, which are a major molecular constituent of membranes (about 50% w/w in the plasma membrane), can be divided into two groups. **Integral proteins** are directly incorporated within the lipid bilayer, whereas **peripheral proteins** exhibit a looser association with membrane surfaces. The loosely bound peripheral proteins can be easily extracted from cell membranes with salt solutions, whereas integral proteins can be extracted only by drastic methods that use detergents. Some integral proteins span the membrane one or more times, from one side to the other. Accordingly, they are called **one-pass** or **multipass transmembrane proteins** (Figure 2–4).

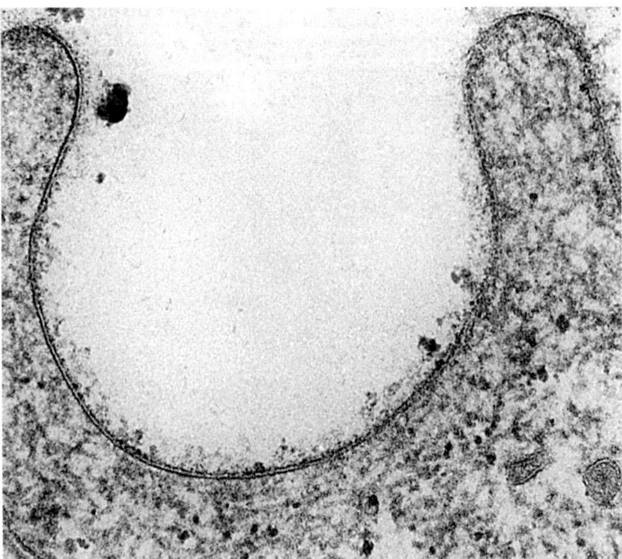

Figure 2–2. Electron micrograph of a section of the surface of an epithelial cell, showing the unit membrane with its two dark lines enclosing a clear band. The granular material on the surface of the membrane is the glycocalyx. ×100,000.

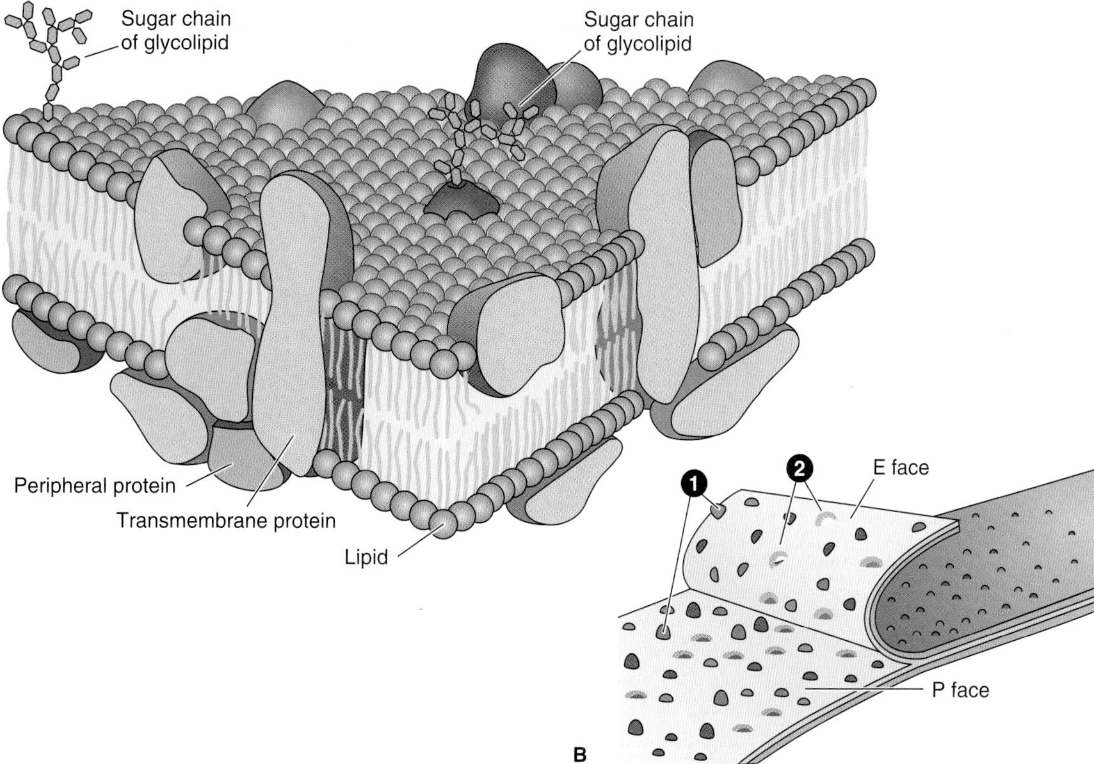

Figure 2–3. **A:** The fluid mosaic model of membrane structure. The membrane consists of a phospholipid double layer with proteins inserted in it (integral proteins) or bound to the cytoplasmic surface (peripheral proteins). Integral membrane proteins are firmly embedded in the lipid layers. Some of these proteins completely span the bilayer and are called transmembrane proteins, whereas others are embedded in either the outer or inner leaflet of the lipid bilayer. The dotted line in the integral membrane protein is the region where hydrophobic amino acids interact with the hydrophobic portions of the membrane. Many of the proteins and lipids have externally exposed oligosaccharide chains. **B:** Membrane cleavage occurs when a cell is frozen and fractured (cryofracture). Most of the membrane particles (**1**) are proteins or aggregates of proteins that remain attached to the half of the membrane adjacent to the cytoplasm (P, or protoplasmic, face of the membrane). Fewer particles are found attached to the outer half of the membrane (E, or extracellular, face). For every protein particle that bulges on one surface, a corresponding depression (**2**) appears in the opposite surface. Membrane splitting occurs along the line of weakness formed by the fatty acid tails of membrane phospholipids, since only weak hydrophobic interactions bind the halves of the membrane along this line. (Modified and reproduced, with permission, from Krstíc RV: *Ultrastructure of the Mammalian Cell.* Springer-Verlag, 1979.)

Freeze-fracture electron-microscope studies indicate that many integral proteins are distributed as globular molecules intercalated among the lipid molecules (Figure 2–3B). Some of these proteins are only partially embedded in the lipid bilayer, so that they may protrude from either the outer or inner surface. Other proteins are large enough to extend across the two lipid layers and protrude from both membrane surfaces (transmembrane proteins). The carbohydrate moieties of glycoproteins and glycolipids project from the external surface of the plasma membrane; they are important components of specific molecules called **receptors** that participate in important interactions such as cell adhesion, recognition, and response to protein hormones. As with lipids, the distribution of membrane proteins is different in the two surfaces of the cell membranes. Therefore, all membranes in the cell are asymmetric.

Integration of the proteins within the lipid bilayer is mainly the result of hydrophobic interactions between the lipids and nonpolar amino acids present on the outer shell of the integral proteins. Some integral proteins are not bound rigidly in place and are able to move within the plane of the cell membrane (Figure 2–5). However, unlike lipids, most membrane proteins are restricted in their lateral diffusion by attachment to the cytoskeletal components. In most epithelial cells, the tight junctions (see Chapter 4) prevent lateral diffusion of transmembrane proteins and even the diffusion of membrane lipids of the outer leaflet.

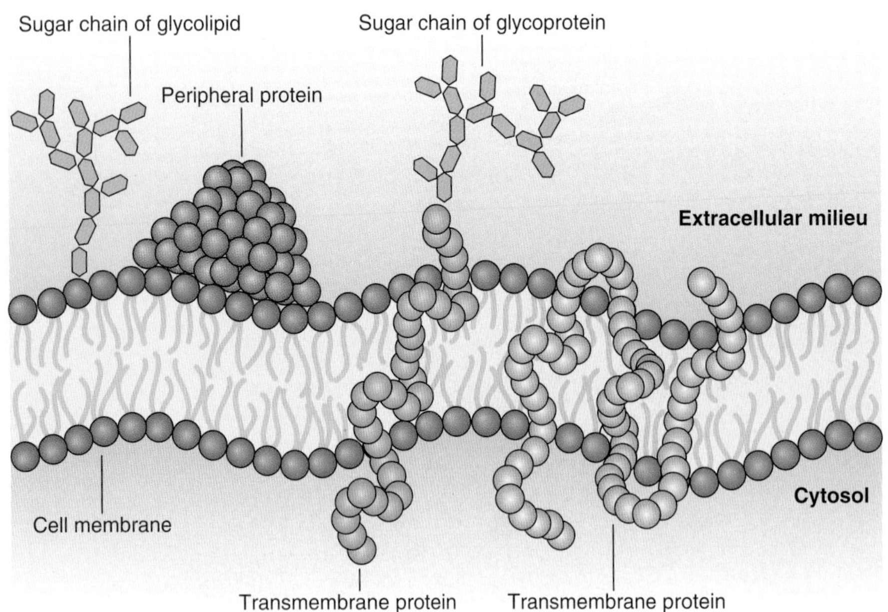

Figure 2–4. Schematic drawing of the molecular structure of the plasma membrane. Note the one-pass and multipass transmembrane proteins. The drawing shows a peripheral protein in the external face of the membrane, but the proteins are present mainly in the cytoplasmic face, as shown in Figure 2–3. (Redrawn and reproduced, with permission, from Junqueira LC, Carneiro J: *Biologia Celular e Molecular,* 6th ed. Editora Guanabara, 1997.)

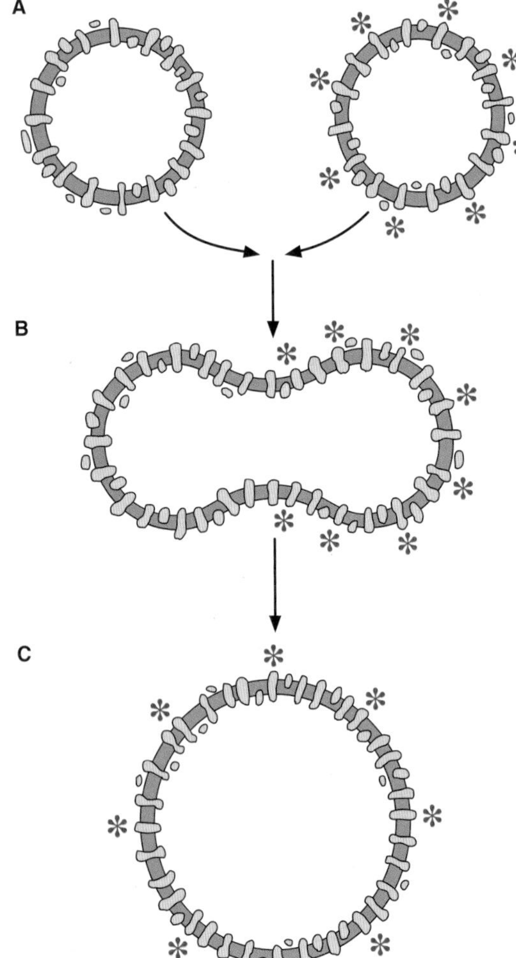

The mosaic disposition of membrane proteins, in conjunction with the fluid nature of the lipid bilayer, constitutes the basis of the **fluid mosaic model** for membrane structure shown in Figure 2–3A. Membrane proteins are synthesized in the rough endoplasm reticulum; their molecules are completed in the Golgi apparatus; and they are transported in vesicles to the cell surface (Figure 2–6).

In the electron microscope the external surface of the cell shows a fuzzy carbohydrate-rich region called **glycocalyx** (Figure 2–2). This layer is made of carbohydrate chains linked to membrane proteins and lipids and of cell-secreted glycoproteins and proteoglycans. The glycocalyx has a role in cell recognition and attachment to other cells and to extracellular molecules. The plasma membrane is the site where materials are exchanged between the cell and its environment. Some ions, such as Na^+, K^+, and Ca^{2+}, are transported across the cell membrane through integral membrane proteins, using energy from the breakdown of adenosine triphosphate (ATP). Mass transfer of material also occurs through the plasma membrane. This bulk uptake of material is known as **endocytosis** (Gr. *endon,* within, + *kytos*). The corresponding name for release of material in bulk is **exocytosis.** However, at the molecular level, exocytosis and endocytosis are different processes that utilize different protein molecules.

Figure 2–5. Experiment demonstrating the fluid nature of proteins within the cell membrane. The plasmalemma is shown as 2 parallel lines (representing the lipid portion) in which proteins are embedded. In this experiment, 2 types of cells derived from tissue cultures (one with a fluorescent marker **[right]** and one without) are fused (**A** → **B**) through the action of the Sendai virus. Minutes after the fusion of the membranes, the fluorescent marker of the labeled cell spreads to the entire surface of the fused cells (**C**). However, in many cells, most transmembrane proteins are stabilized in place by anchoring to the cytoskeleton.

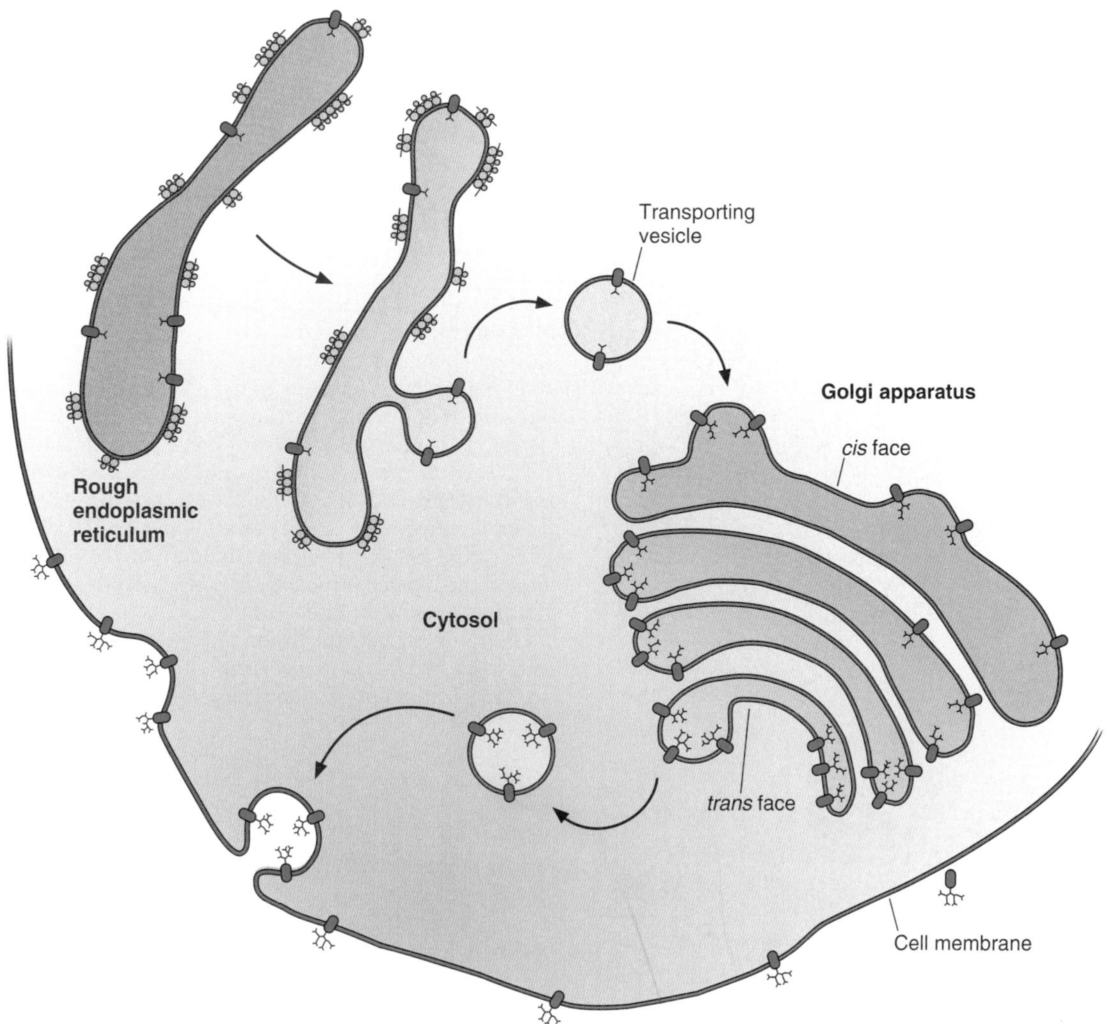

Figure 2–6. The proteins of the plasmalemma are synthesized in the rough endoplasmic reticulum and then transported in vesicles to the Golgi complex, where they may be modified and transferred to the cell membrane. This example shows the synthesis and transport of a glycoprotein, which is an integral protein of the membrane. (Redrawn and reproduced, with permission, from Junqueira LC, Carneiro J: *Biologia Celular e Molecular,* 6th ed. Editora Guanabara, 1997.)

Fluid-Phase Pinocytosis

In fluid-phase pinocytosis, small invaginations of the cell membrane form and entrap extracellular fluid and anything in solution in the fluid. **Pinocytotic vesicles** (about 80 nm in diameter) pinch off from the cell surface (Figure 4–24), and most eventually fuse with lysosomes (see the section on lysosomes later in this chapter). In the lining cells of capillaries (endothelial cells), however, pinocytotic vesicles may move to the surface opposite their origin. There they fuse with the plasma membrane and release their contents onto the cell surface, thus accomplishing bulk transfer of material across the cell (Figure 11–4).

Receptor-Mediated Endocytosis

Receptors for many substances, such as low-density lipoproteins and protein hormones, are located at the cell surface. The receptors are either originally widely dispersed over the surface or aggregated in special regions called **coated pits.** Binding of the ligand (a molecule with high affinity for a receptor) to its receptor causes widely dispersed receptors to accumulate in coated pits (Figure 2–7). The coating on the cytoplasmic surface of the membrane is composed of several polypeptides, the major one being clathrin. These proteins form a lattice composed of pentagons and hexagons very similar in arrangement to the struts in a geodesic dome. The coated pit invaginates and pinches off from the cell membrane, forming a coated vesicle that carries the ligand and its receptor into the cell.

The coated vesicles soon lose their clathrin coat and fuse with **endosomes,** a system of vesicles (Figure 2–7) and tubules located in the cytosol near the cell surface (early endosomes) or deeper in the cytoplasm (late endosomes). Together they constitute the **endosomal compartment.** Whether early and late endosomes are separate compartments or one is a precursor of the other is still an open question. The membrane of all endosomes contains ATP-driven H^+ pumps that acidify their interior. The

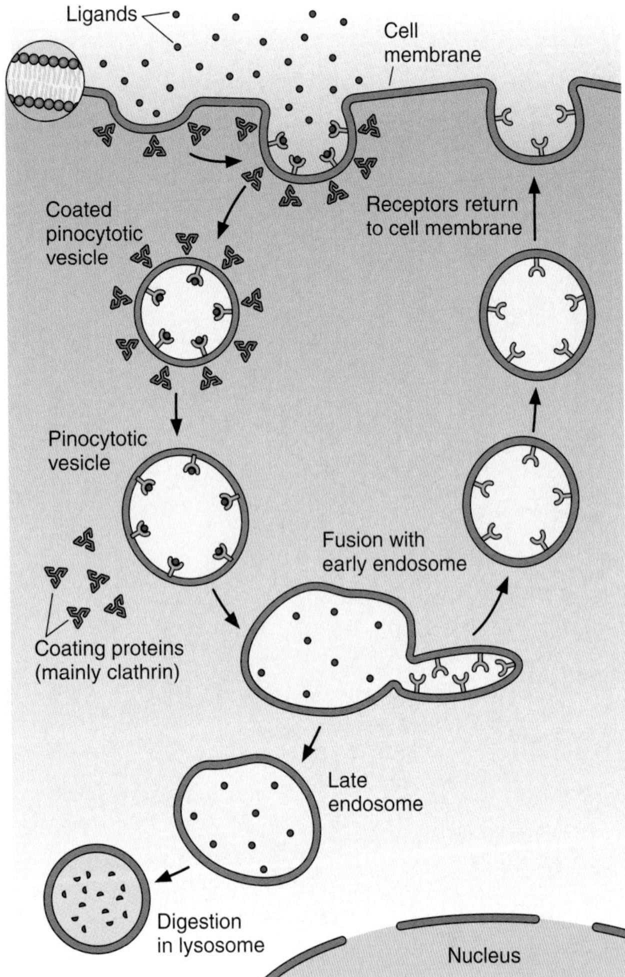

Figure 2–7. Schematic representation of the endocytic pathway and membrane trafficking. Ligands, such as hormones and growth factors, bind to specific surface receptors and are internalized in pinocytotic vesicles coated with clathrin and other proteins. After the liberation of the coating molecules, the pinocytotic vesicles fuse with the endosomal compartment, where the low pH causes the separation of the ligands from their receptors. Membrane with receptors is returned to the cell surface to be reused. The ligands typically are transferred to lysosomes. The cytoskeleton with motor proteins is responsible for all vesicle movements described.

clathrin molecules separated from the coated vesicles are moved back to the cell membrane to participate in the formation of new coated pits.

Molecules penetrating the endosomes may take more than one pathway (Figure 2–7). Receptors that are separated from their ligand by the acidic pH of the endosomes may return to the cell membrane to be reused. For example, low-density lipoprotein receptors (Figure 2–8) are recycled several times. The ligands typically are transferred to late endosomes. However, some ligands are returned to the extracellular milieu to be used again. An example of this activity is the iron-transporting protein transferrin.

Phagocytosis

Phagocytosis literally means "cell eating" and can be compared to pinocytosis, which means "cell drinking." Certain cell types, such as macrophages and polymorphonuclear leukocytes, are specialized for incorporating and removing foreign bacteria, protozoa, fungi, damaged cells, and unneeded extracellular constituents. For example, after a bacterium becomes bound to the surface of a macrophage, cytoplasmic processes of the macrophage are extended and ultimately surround the bacterium. The edges of these processes fuse, enclosing the bacterium in an intracellular **phagosome.**

Exocytosis is the term used to describe the fusion of a membrane-limited structure with the plasma membrane, resulting in the release of its contents into the extracellular space without compromising the integrity of the plasma membrane. A typical example is the release of stored products from secretory cells, such as those of the exocrine pancreas and the salivary glands (Figure 4–26). The fusion of membranes in exocytosis is a complex process. Because cell membranes exhibit a high density of negative charges (phosphate residues of the phospholipids), membrane-covered structures coming close to each other will not fuse but will rather repel each other, unless specific interactions facilitate the fusion process. Consequently, exocytosis is mediated by a number of specific proteins. Usually, Ca^{2+} regulates the process. For example, an increase in cytosolic Ca^{2+} often triggers exocytosis.

During endocytosis, portions of the cell membrane become an endocytotic vesicle; during exocytosis, the membrane is returned to the cell surface. This phenomenon is called **membrane trafficking** (Figures 2–7 and 2–8). In several systems, membranes are conserved and reused several times during repeated cycles of endocytosis.

Signal Reception

Cells in a multicellular organism need to communicate with one another to regulate their development into tissues, to control their growth and division, and to coordinate their functions. Many cells form communicating junctions that couple adjacent cells, allowing the exchange of ions and small molecules (see Chapter 4). Through these channels, also called gap junctions, signals pass directly from cell to cell without reaching the extracellular fluid. In other cases, cells display membrane-bound signaling molecules that influence other cells in direct physical contact.

Extracellular signaling molecules, or messengers, mediate 3 kinds of communication between cells. In **endocrine signaling,** hormones are carried in the blood to target cells throughout the body; in **paracrine signaling,** chemical mediators are rapidly metabolized so that they act on local cells only; and in **synaptic signaling,** neurotransmitters act only on adjacent nerve cells through special contact areas called **synapses** (see Chapter 9). In some cases, paracrine signals act on the same cell type that produced the messenger molecule, a phenomenon called **autocrine signaling.** Each cell type in the body contains a distinctive set of receptor proteins that enable it to respond to a complementary set of signaling molecules in a specific, programmed way (Figure 2–9).

Signaling molecules differ in their water solubility. Small **hydrophobic signaling molecules,** such as steroid and thyroid hormones, diffuse through the plasma membrane of the target cell and activate receptor proteins inside the cell. In contrast, **hydrophilic signaling molecules,** including neurotransmitters,

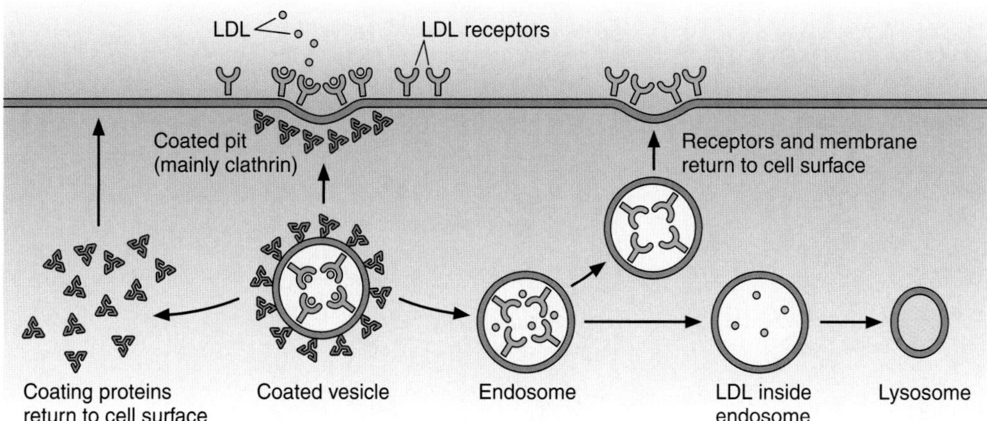

Figure 2–8. Internalization of low-density lipoproteins (LDL) is important to keep the concentration of LDL in body fluids low. LDL, which is rich in cholesterol, binds with high affinity to its receptors in the cell membranes. This binding activates the formation of pinocytotic vesicles from coated pits. The vesicles soon lose their coating, which is returned to the inner surface of the plasmalemma: the uncoated vesicles fuse with endosomes. In the next step, the LDL is transferred to lysosomes for digestion and separation of their components to be utilized by the cell.

most hormones, and local chemical mediators (paracrine signals), activate receptor proteins on the surface of target cells. These receptors, which span the cell membrane, relay information to a series of intracellular intermediaries that ultimately pass the signal to its final destination in either the cytoplasm or the nucleus. The numerous intercellular hydrophilic messengers rely on membrane proteins that direct the flow of information from the receptor to the rest of the cell. The best studied of these proteins are the **G proteins,** so named because they bind to guanine nucleotides. Once a **first messenger** (hormone, neurotransmitter, paracrine signal) binds to a receptor, conformational changes occur in the receptor; this, in turn, activates the G protein–guanosine diphosphate complex (Figure 2–10). A guanosine diphosphate–guanosine triphosphate exchange releases the α subunit of the G protein, which acts on other membrane-bound intermediaries called **effectors.** Often, the effector is an enzyme that converts an inactive precursor molecule into an active **second messenger,** which can diffuse through the cytoplasm and carry the signal beyond the cell membrane. Second messengers trigger a cascade of molecular reactions that lead to changes in cell behavior. The examples listed in Table 2–2 illustrate the diversity of G proteins present in various tissues and their roles in regulating important cell functions.

MEDICAL APPLICATION

Several diseases have been shown to be due to defective receptors. For example, pseudohypoparathyroidism and a type of dwarfism are due to nonfunctioning parathyroid and growth hormone receptors. In these two conditions the glands produce the respective hormones, but the target cells do not respond, because they lack normal receptors.

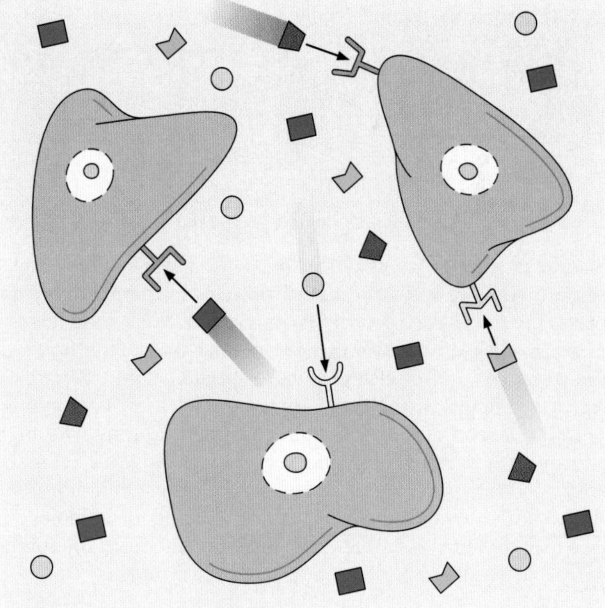

Figure 2–9. Cells respond to chemical signals according to the library of receptors they have. In this schematic representation, 3 cells appear with different receptors, and the extracellular environment contains several ligands which will interact with the appropriate receptors. Considering that the extracellular environment contains a multitude of molecules, it is important that ligands and the respective receptors exhibit complementary morphology and great affinity.

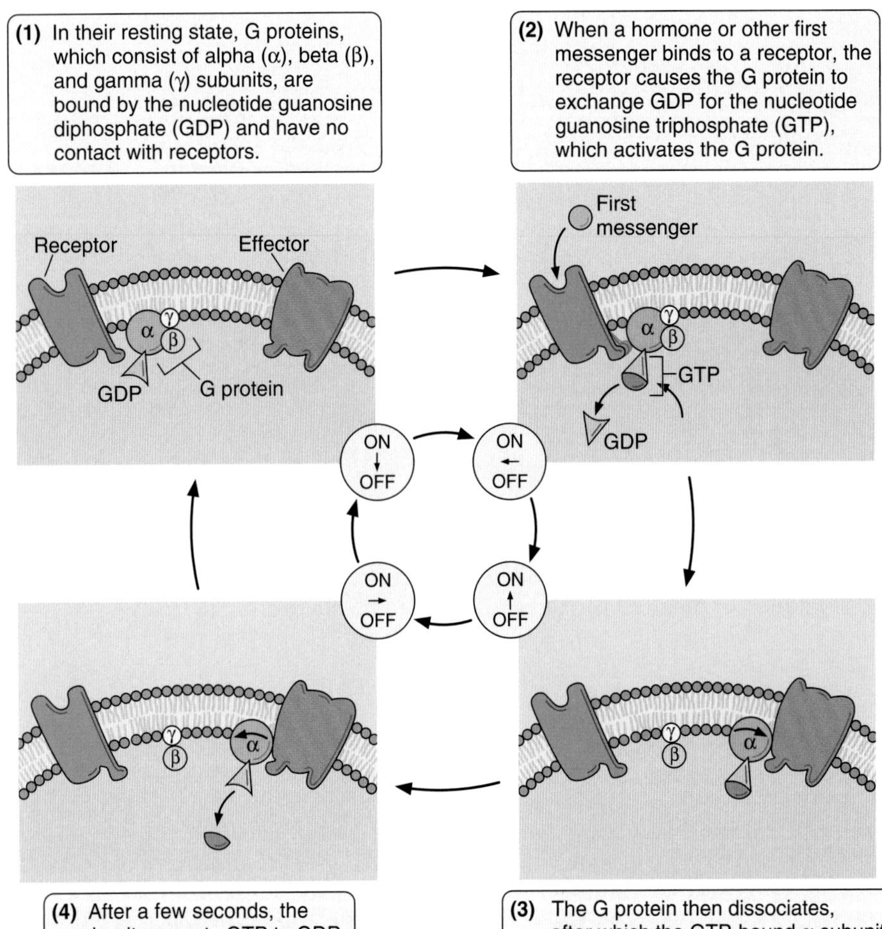

Figure 2–10. **Diagram illustrating how G proteins switch effectors on and off.** (Modified and reprinted, with permission, from Linder M, Gilman AG: G proteins. Sci Am 1992;267:56.)

Signaling Mediated by Intracellular Receptors

Steroid hormones are small hydrophobic (lipid-soluble) molecules; binding reversibly to carrier proteins in the plasma transports them in the blood. Once released from their carrier proteins, they diffuse through the plasma membrane lipids of the target cell and bind reversibly to specific steroid hormone–receptor proteins in the cytoplasm or the nucleus. The binding of hormone activates the receptor, enabling it to bind with high affinity to specific DNA sequences; this generally increases the level of transcription from specific genes. Each steroid hormone is recognized by a different member of a family of homologous receptor proteins. Thyroid hormones are modified lipophilic amino acids that also act on intracellular receptors.

Mitochondria

Mitochondria (Gr. *mitos,* thread, + *chondros,* granule) are spherical or filamentous organelles 0.5–1 μm wide that can attain a length of up to 10 μm (Figure 2–11). They tend to accumulate in parts of the cytoplasm where the utilization of energy is more intense, such as the apical ends of ciliated cells (Figure 17–3), in the middle piece of spermatozoa (Figure 22–9), or at the base of ion-transferring cells (Figure 4–24).

These organelles transform the chemical energy of the metabolites present in cytoplasm into energy that is easily accessible to the cell. About 50% of this energy is stored as high-energy phosphate bonds in ATP molecules, and the remaining 50% is dissipated as heat used to maintain body temperature. Through the activity of the enzyme ATPase, ATP promptly releases energy when required by the cell to perform any type of work, whether it is osmotic, mechanical, electrical, or chemical.

Mitochondria have a characteristic structure under the electron microscope (Figures 2–12 and 2–13A). They are composed of an **outer** and an **inner mitochondrial membrane;** the inner membrane projects folds, termed **cristae,** into the interior of the mitochondrion. These membranes enclose two compartments. The compartment located between the two membranes is termed the **intermembrane space.** The inner membrane encloses the other compartment—the intercristae, or matrix, space. Mitochondrial membranes contain a large number of protein molecules, compared with other cell membranes. Most mitochondria have flat, shelflike cristae in their interiors (Figures 2–12 and 2–13A), whereas cells that secrete steroids (eg, adrenal gland; see Chapter 4) frequently contain tubular cristae (Figure 4–37). The cristae increase the internal surface area of mitochondria and contain enzymes and other components of oxidative phosphorylation and electron transport systems. The adenosine diphos-

Table 2–2. A sampling of physiologic effects mediated by G proteins.

Stimulus	Affected Cell Type	G Protein	Effector	Effect
Epinephrine, glucagon	Liver cells	G_s	Adenylyl cyclase	Breakdown of glycogen
Epinephrine, glucagon	Fat cells	G_s	Adenylyl cyclase	Breakdown of fat
Luteinizing hormone	Ovarian follicles	G_s	Adenylyl cyclase	Increased synthesis of estrogen and progesterone
Antidiuretic hormone	Kidney cells	G_s	Adenylyl cyclase	Conservation of water by kidney
Acetylcholine	Heart muscle cells	G_i	Potassium channel	Slowed heart rate and decreased pumping force
Enkephalins, endorphins, opioids	Brain neurons	G_i/G_o	Calcium and potassium channels, adenylyl cyclase	Changed electrical activity of neurons
Angiotensin	Smooth muscle cells in blood vessels	G_q	Phospholipase C	Muscle contraction; elevation of blood pressure
Odorants	Neuroepithelial cells in nose	G_{olf}	Adenylyl cyclase	Detection of odorants
Light	Rod and cone cells in retina	G_t	Cyclic GMP phosphodiesterase	Detection of visual signals
Pheromone	Baker's yeast	GPA1	Unknown	Mating of cells

Reproduced, with permission, from Linder M, Gilman AG: G proteins. Sci Am 1992;267:56.

phate (ADP) to ATP phosphorylating system is localized in globular structures connected to the inner membrane by cylindrical stalks (Figure 2–12). The globular structures are a complex of proteins with ATP synthetase activity that, in the presence of ADP plus inorganic phosphate and energy, form ATP. The chemiosmotic theory suggests that ATP synthesis occurs at the expense of a flow of protons across this globular unit (Figure 2–14).

The number of mitochondria and the number of cristae in each mitochondrion are related to the energetic activity of the cells in which they reside. Thus, cells with a high-energy metabolism (eg, cardiac muscle, cells of some kidney tubules) have abundant mitochondria with a large number of closely packed cristae, whereas cells with a low-energy metabolism have few mitochondria with short cristae.

Between the cristae is an amorphous **matrix,** rich in protein and containing circular molecules of DNA and the 3 varieties of RNA. In a great number of cell types, the mitochondrial matrix also exhibits rounded electron-dense granules rich in Ca^{2+}. Although the function of this cation in mitochondria is not completely understood, it may be important in regulating the activity of some mitochondrial enzymes; another functional role is related to the necessity of keeping the cytosolic concentration of Ca^{2+} low. Mitochondria will pump in Ca^{2+} when its concentration in the cytosol is high. Enzymes for the citric acid (Krebs) cycle and fatty acid β-oxidation are found to reside within the matrix space.

The DNA isolated from the mitochondrial matrix is double stranded and has a circular structure, very similar to that of bacterial chromosomes. These strands are synthesized within the mitochondrion; their duplication is independent of nuclear DNA replication. Mitochondria contain the 3 types of RNA: ribosomal RNA (rRNA), messenger RNA (mRNA), and transfer RNA. Mitochondrial ribosomes are smaller than cytosolic ribosomes and are comparable to bacterial ribosomes. Protein synthesis occurs in mitochondria, but because of the reduced amount of mitochondrial DNA, only a small proportion of the mitochondrial proteins is produced locally. Most are encoded by nuclear DNA and synthesized in polyribosomes located in the cytosol. These proteins have a small amino acid sequence that is a signal for their mitochondrial destination, and they are transported into mitochondria by an energy-requiring mechanism.

The initial degradation of carbohydrates and fats is carried out in the cytoplasmic matrix. The metabolic end product of these extramitochondrial metabolic pathways is acetyl coenzyme A, which then enters mitochondria. Within mitochondria, acetyl coenzyme A combines with oxaloacetate to form citric acid. Within the citric acid cycle, several reactions of decarboxylation produce CO_2, and specific reactions catalyzed by dehydrogenase result in the removal of 4 pairs of H^+ ions. The H^+ ions ultimately react with oxygen to form H_2O. Through the action of cytochromes *a, b,* and *c,* coenzyme Q, and cytochrome oxidase, the **electron transport system,** located in the inner mitochondrial membrane, releases energy that is captured at 3 points of this system through the formation of ATP from ADP and inorganic phosphate. Under aerobic conditions, the combined activity of extramitochondrial glycolysis and the citric acid cycle as well as the electron transport system gives rise to 36 molecules of ATP per molecule of glucose. This is 18 times the energy obtainable under anaerobic circumstances, when only the glycolytic pathway can be used.

In the process of mitosis, each daughter cell receives approximately half the mitochondria originally present in the parent cell. New mitochondria originate from preexisting mitochondria by growth and subsequent division (fission) of the organelle itself.

The fact that mitochondria have some characteristics in common with bacteria has led to the hypothesis that mitochondria originated from an ancestral aerobic prokaryote that adapted to an endosymbiotic life within an eukaryotic host cell.

MEDICAL APPLICATION

Several mitochondrial deficiency diseases have been described, and most of them are characterized by muscular dysfunction. Because of their high-energy metabolism, skeletal muscle fibers are very sensitive to mitochondrial defects. These diseases typically begin with drooping of the upper eyelid and progress to difficulties in swallowing and limb weakness. DNA mutations or defects that can occur in the mitochondria or the cell nucleus cause them. Mitochondrial inheritance is maternal, because few, if any, mitochondria from the sperm nucleus remain in the cytoplasm of the zygote. In the case of nuclear DNA defects, inheritance may be from either parent or both parents. Generally, in these diseases the mitochondria show morphological changes (Figure 2–13B).

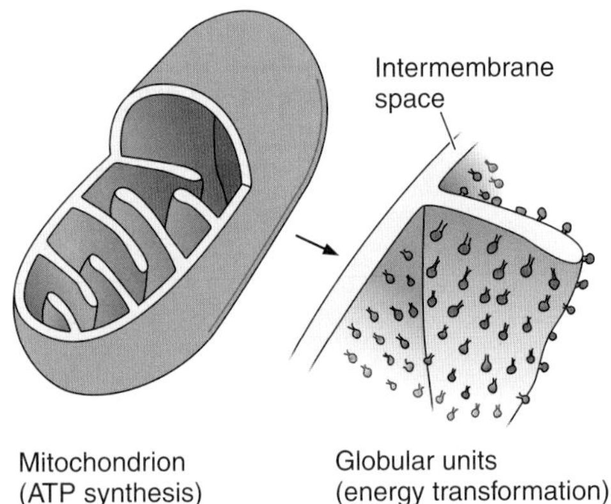

Figure 2–12. Three-dimensional representation of a mitochondrion with its cristae penetrating the matrix space. Note that 2 membranes delimiting an intermembrane space form the wall of the mitochondrion. The cristae are covered with globular units that participate in the formation of ATP.

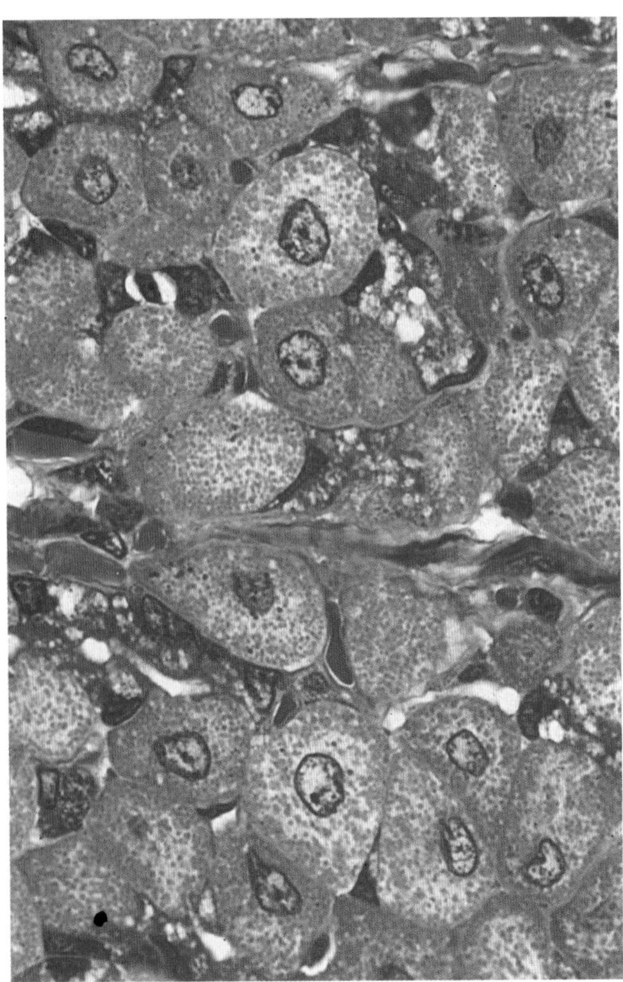

Figure 2–11. Photomicrograph of the stomach inner covering. The large cells show many round and elongated mitochondria in the cytoplasm. The central nuclei are also clearly seen. High magnification.

Ribosomes

Ribosomes are small electron-dense particles, about 20 × 30 nm in size. They are composed of 4 types of rRNA and almost 80 different proteins.

There are two classes of ribosomes: One class is found in prokaryotes, chloroplasts, and mitochondria; the other is found in eukaryotic cells. Both classes of ribosomes are composed of two different-sized subunits.

In eukaryotic cells, the RNA molecules of both subunits are synthesized within the nucleus. Their numerous proteins are synthesized in the cytoplasm and then enter the nucleus and associate with rRNAs. Subunits then leave the nucleus, via nuclear pores, to enter the cytoplasm and participate in protein synthesis.

Ribosomes are intensely basophilic because of the presence of numerous phosphate groups of the constituent rRNA that act as polyanions. Thus, sites in the cytoplasm that are rich in ribosomes stain intensely with basic dyes, such as methylene and toluidine blue. These basophilic sites also stain with hematoxylin.

The individual ribosomes (Figure 2–15A) are held together by a strand of mRNA to form **polyribosomes** (**polysomes**). The message carried by mRNA is a code for the amino acid sequence of proteins being synthesized by the cell, and the ribosomes play a crucial role in decoding, or translating, this message during protein synthesis. Proteins synthesized for use within the cell and destined to remain in the cytosol (eg, hemoglobin in immature erythrocytes) are synthesized on polyribosomes existing as isolated clusters within the cytoplasm. Polyribosomes that are attached to the membranes of the endoplasmic reticulum (via their large subunits) translate mRNAs that code for proteins that are segregated into the cisternae of the reticulum (Figure 2–15B). These proteins can be secreted (eg, pancreatic and salivary enzymes) or stored in the cell (eg, enzymes of lysosomes, proteins within granules of white blood cells [leukocytes]). In addition, integral proteins of the plasma membrane are synthesized on polyribosomes attached to membranes of the endoplasmic reticulum (Figure 2–6).

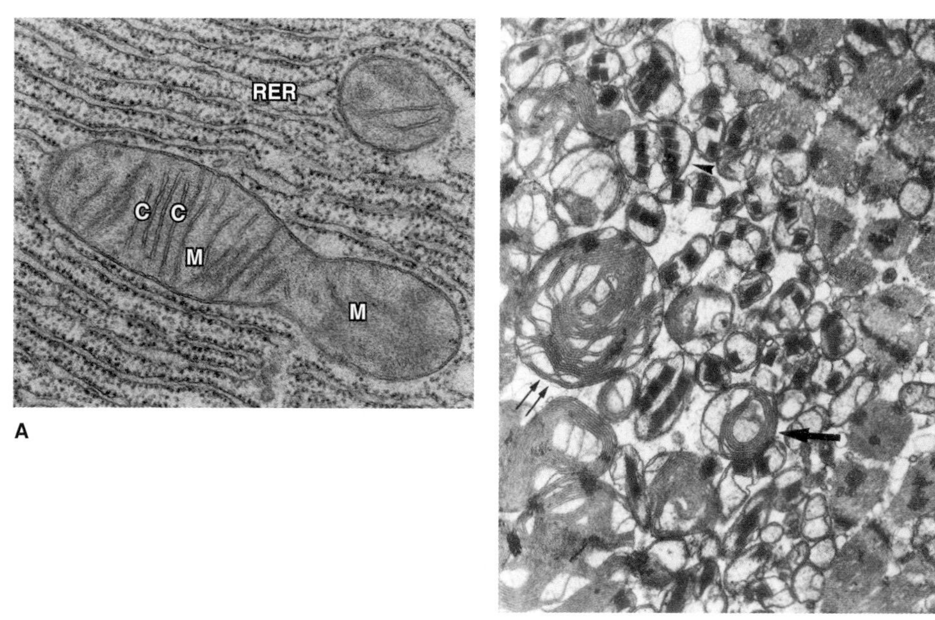

Figure 2–13. Structural lability of mitochondria. **A:** Electron micrograph of a section of rat pancreas. A mitochondrion with its membranes, cristae (C), and matrix (M) is seen in the center. Numerous flattened cisternae of rough endoplasmic reticulum (RER) with ribosomes on their cytoplasmic surfaces are also visible. ×50,000. **B:** Electron micrograph of striated muscle from a patient with mitochondrial myopathy. The mitochondria are profoundly modified, showing marked swelling of the matrix.

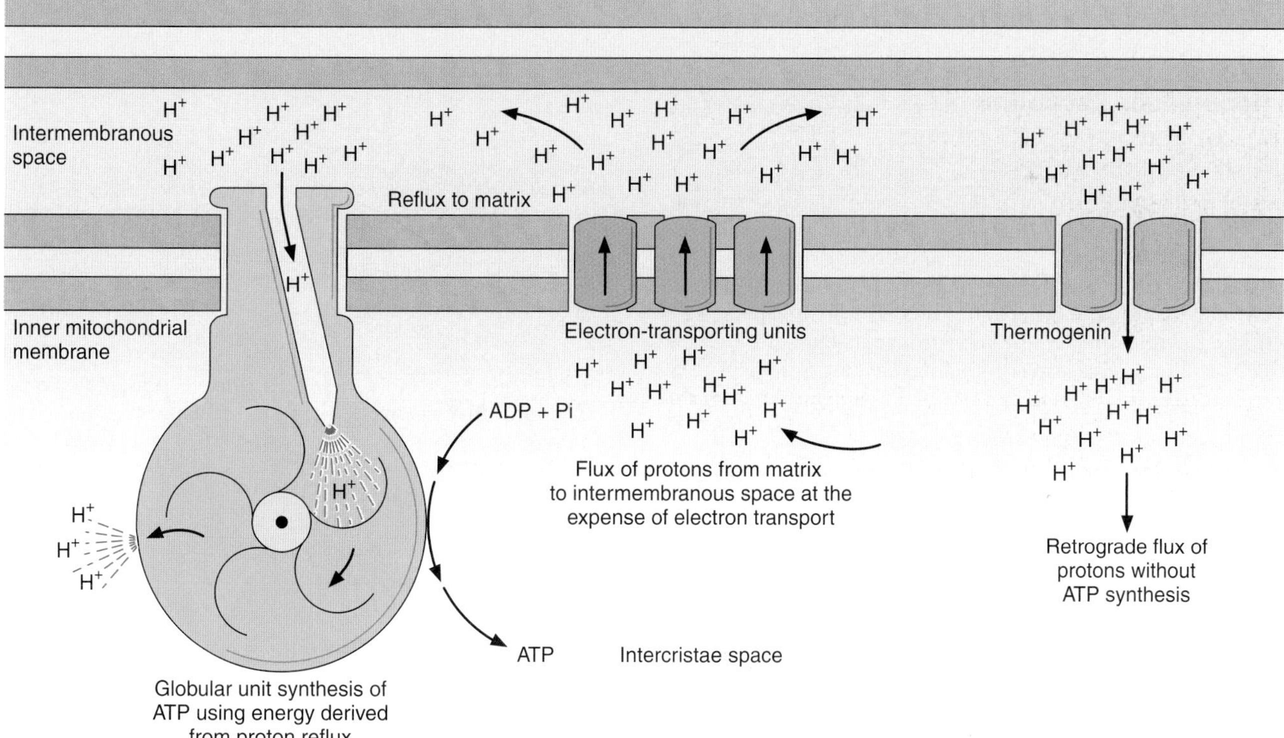

Figure 2–14. The chemiosmotic theory of mitochondrial energy transduction. **Middle:** The flux of protons is directed from the matrix to the intermembranous space promoted at the expense of energy derived from the electron transport system in the inner membrane. **Left:** Half the energy derived from proton reflux produces ATP; the remaining energy produces heat. **Right:** The protein thermogenin, present in multilocular adipose tissue, forms a shunt for reflux of protons. This reflux, which dissipates energy as heat, does not produce ATP (see Chapter 6).

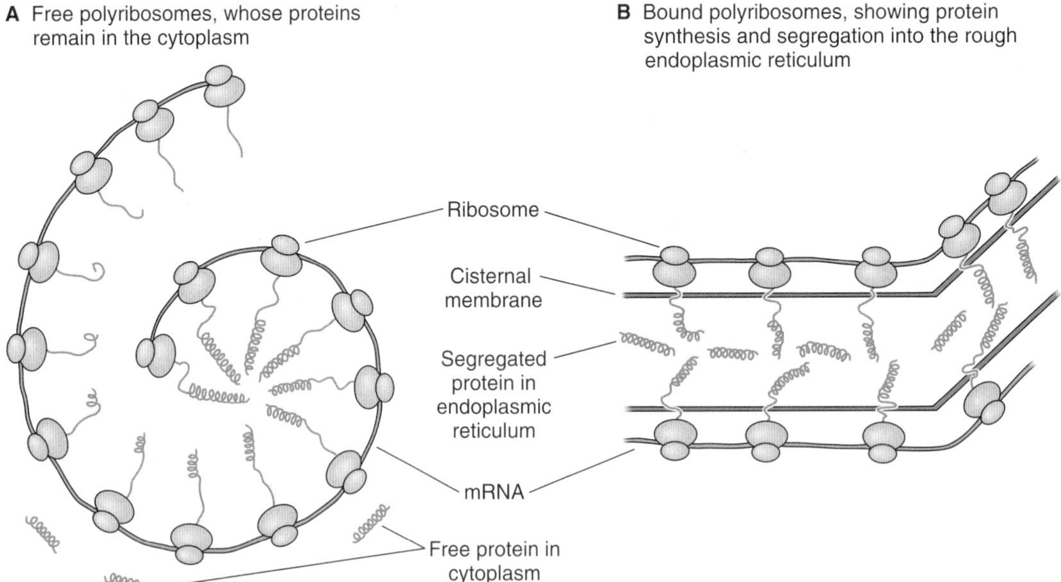

Figure 2–15. Diagram illustrating (**A**) the concept that cells synthesizing proteins (represented here by spirals) that are to remain within the cytoplasm possess (free) polyribosomes (ie, nonadherent to the endoplasmic reticulum). In **B,** where the proteins are segregated in the endoplasmic reticulum and may eventually be extruded from the cytoplasm (export proteins), not only do the polyribosomes adhere to the membranes of rough endoplasmic reticulum, but the proteins produced by them are injected into the interior of the organelle across its membrane. In this way, the proteins, especially enzymes such as ribonucleases and proteases, which could have undesirable effects on the cytoplasm, are separated from it.

Endoplasmic Reticulum

The cytoplasm of eukaryotic cells contains an anastomosing network of intercommunicating channels and sacs formed by a continuous membrane, which encloses a space called a **cisterna.** In sections, cisternae appear separated, but high-resolution microscopy of whole cells reveals that they are continuous. This membrane system is called the endoplasmic reticulum (Figure 2–16). In many places the cytosolic side of the membrane is covered by polyribosomes synthesizing protein molecules, which are injected into the cisternae. This permits the distinction between the two types of endoplasmic reticulum: **rough** and **smooth.**

Rough Endoplasmic Reticulum

Rough endoplasmic reticulum (RER) is prominent in cells specialized for protein secretion, such as pancreatic acinar cells (digestive enzymes), fibroblasts (collagen), and plasma cells (immunoglobulins). The RER consists of saclike as well as parallel stacks of flattened cisternae (Figure 2–13), limited by membranes that are continuous with the outer membrane of the nuclear envelope. The name "rough endoplasmic reticulum" alludes to the presence of polyribosomes on the cytosolic surface of this structure's membrane (Figures 2–16 and 2–17). The presence of polyribosomes also confers basophilic staining properties on this organelle when viewed with the light microscope.

The principal function of the RER is to segregate proteins not destined for the cytosol. Additional functions include the initial (core) glycosylation of glycoproteins, the synthesis of phospholipids, the assembly of multichain proteins, and certain post-translational modifications of newly formed polypeptides.

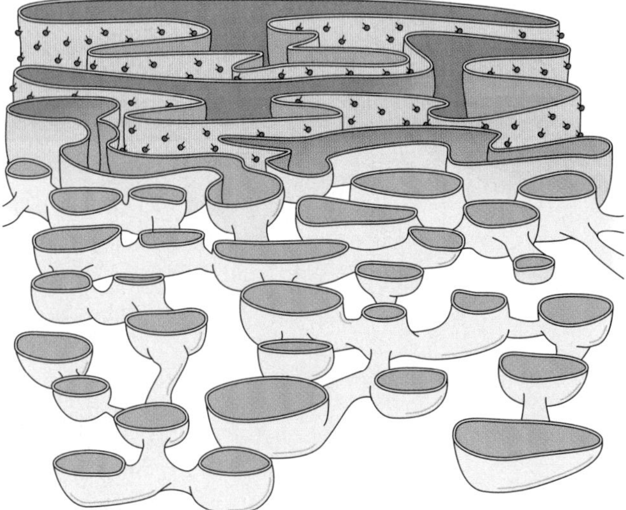

Figure 2–16. The endoplasmic reticulum is an anastomosing network of intercommunicating channels and sacs formed by a continuous membrane. Note that the smooth endoplasmic reticulum (foreground) is devoid of ribosomes, the small dark dots that are present in the rough endoplasmic reticulum (background). The cisternae of the smooth reticulum are tubular, whereas in the rough reticulum they are flat sacs.

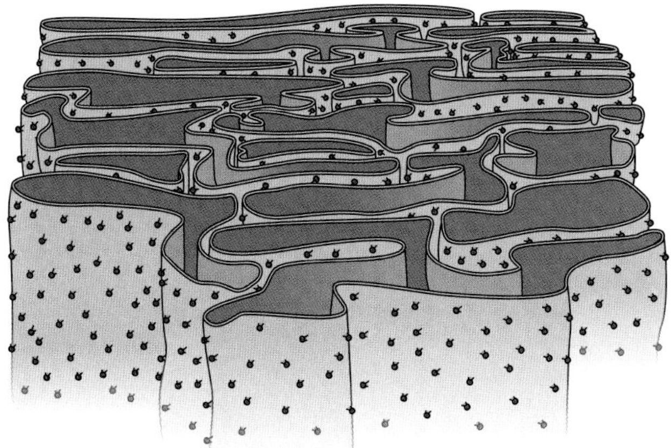

Figure 2–17. Schematic representation of a small portion of the rough endoplasmic reticulum to show the shape of its cisternae and the presence of numerous ribosomes which are part of polyribosomes. It should be kept in mind that the cisternae appear separated in sections made for electron microscopy, but they form a continuous tunnel in the cytoplasm.

All protein synthesis begins on polyribosomes that are not attached to the endoplasmic reticulum. mRNAs of proteins destined to be segregated in the endoplasmic reticulum contain an additional sequence of bases at their 5′ end that code for approximately 20–25 mainly hydrophobic amino acids called the **signal sequence.** Upon translation, the signal sequence interacts with a complex of 6 nonidentical polypeptides plus a 7S RNA molecule that is referred to as the **signal-recognition particle (SRP)**. SRP inhibits further polypeptide elongation until the SRP-polyribosome complex binds to a receptor in the membrane of the RER, the **docking protein.** Upon binding to the docking protein, SRP is released from the polyribosomes, allowing the translation to continue (Figure 2–18).

Once inside the lumen of the RER, a specific enzyme, signal peptidase, located at the inner surface of the RER removes the signal sequence. Translation of the protein continues, accompanied by intracisternal secondary and tertiary structural changes as well as certain post-translational modifications such as hydroxylation, glycosylation, sulfating, and phosphorylation.

Proteins synthesized in the RER can have several destinations: intracellular storage (eg, in lysosomes and specific granules of leukocytes), provisional intracellular storage of proteins for export (eg, in the pancreas, some endocrine cells), and as a component of other membranes (eg, integral proteins). Figure 2–19 shows several cell types with clear differences in the destination of the proteins they synthesize.

Smooth Endoplasmic Reticulum

Smooth endoplasmic reticulum (SER) also takes the form of a membranous network within the cell; however, its ultrastructure differs from that of RER in two ways. First, SER lacks the associated polyribosomes that characterize RER. SER membranes therefore appear smooth rather than granular. Second, its cisternae are more tubular and more likely to appear as a profusion of interconnected channels of various shapes and sizes than as stacks of flattened cisternae (Figures 2–16 and 4–37). SER is continuous with the RER (Figure 2–16).

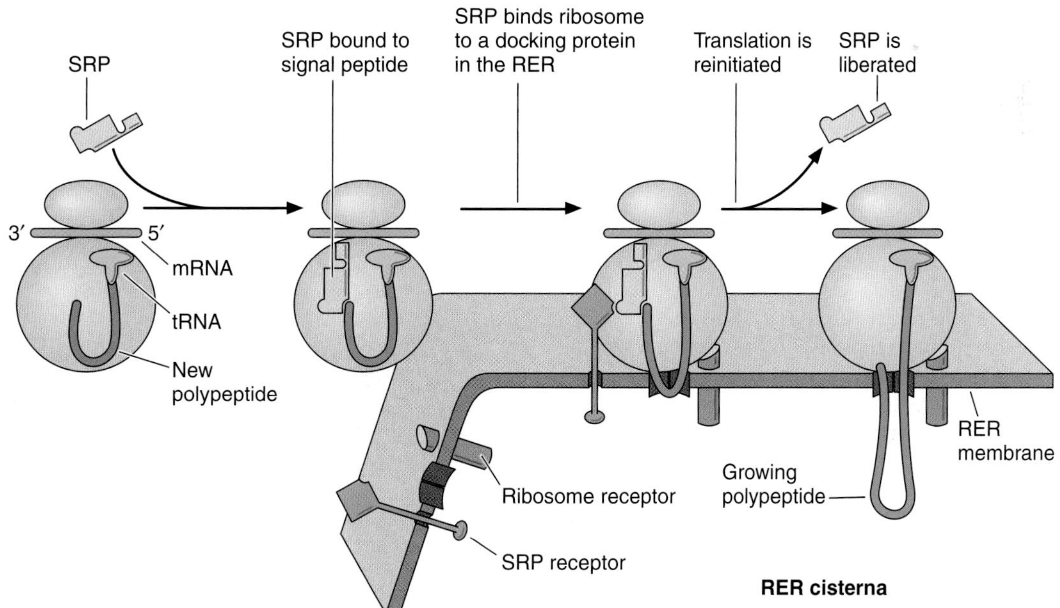

Figure 2–18. The transport of proteins across the membrane of the rough endoplasmic reticulum (RER). The ribosomes bind to mRNA, and the signal peptide is initially bound to a signal-recognition particle (SRP). Ribosomes bind to the RER by interacting with the SRP and a ribosomal receptor. The signal peptide is then removed by a signal peptidase (not shown). These interactions cause the opening of a pore through which the protein is extruded into the RER.

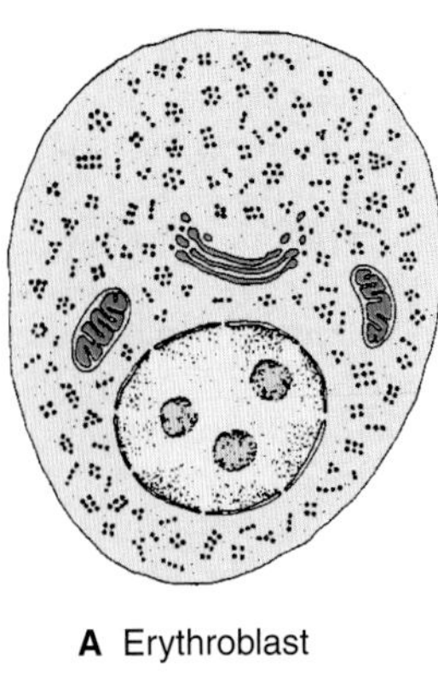

A Erythroblast

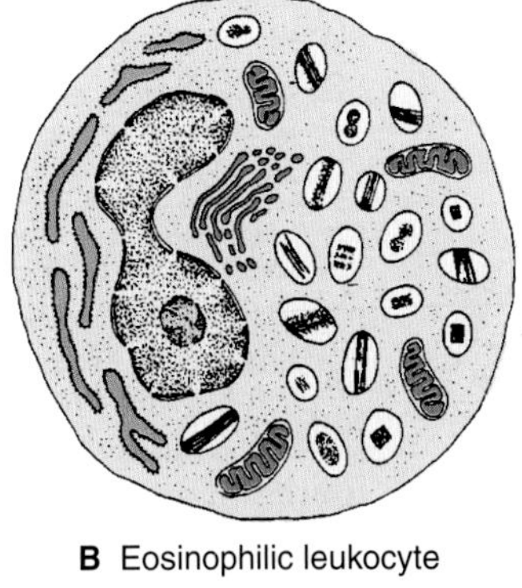

B Eosinophilic leukocyte

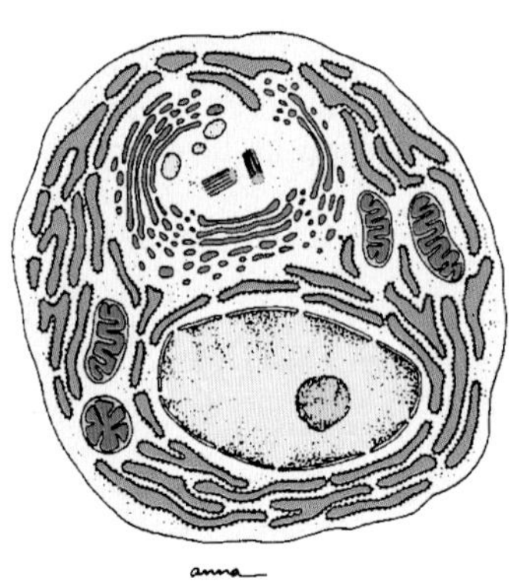

C Plasma cell

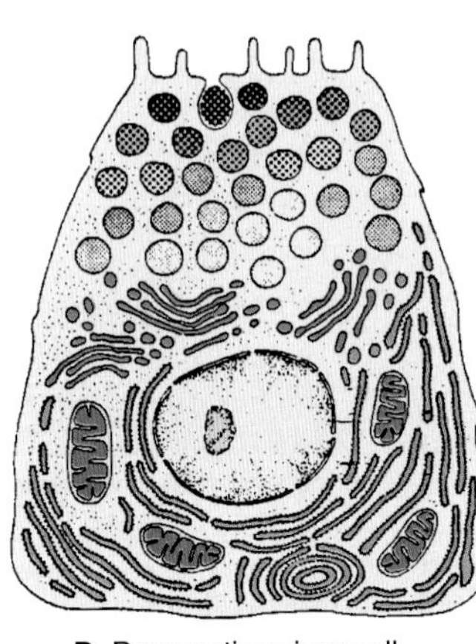

D Pancreatic acinar cell

Figure 2–19. The ultrastructure of a cell that synthesizes (but does not secrete) proteins on free polyribosomes (**A**); a cell that synthesizes, segregates, and stores proteins in organelles (**B**); a cell that synthesizes, segregates, and directly exports proteins (**C**); and a cell that synthesizes, segregates, stores in supranuclear granules, and exports proteins (**D**).

SER is associated with a variety of specialized functional capabilities. In cells that synthesize steroid hormones (eg, cells of the adrenal cortex), SER occupies a large portion of the cytoplasm and contains some of the enzymes required for steroid synthesis (Figure 4–37). SER is abundant in liver cells, where it is responsible for the oxidation, conjugation, and methylation processes employed by the liver to degrade certain hormones and neutralize noxious substances such as barbiturates. Another important function of SER is the synthesis of phospholipids for all cell membranes. The phospholipid molecules are transferred from the SER to other membranes (1) by vesicles that detach and are moved along cytoskeletal elements by the action of motor proteins, (2) through direct communication with the RER, or (3) by transfer proteins (Figure 2–20). SER contains the enzyme glucose-6-phosphatase, which is involved in the utilization of glucose originating from glycogen in liver cells. This enzyme is also found in RER, an example of the lack of absolute partitioning of functions between these organelles. SER participates in the contraction process in muscle cells, where it appears in a specialized form, called the **sarcoplasmic reticulum,** that is involved in the sequestration and release of the calcium ions that regulate muscular contraction (see Chapter 10).

Golgi Complex (Golgi Apparatus)

The Golgi complex completes post-translational modifications and packages and places an address on products that have been synthesized by the cell. This organelle is composed of smooth membrane–limited **cisternae** (Figures 2–21, 2–22, and 2–23). In highly polarized cells, such as mucus-secreting goblet cells (Figure 4–30), the Golgi complex occupies a characteristic position in the cytoplasm between the nucleus and the apical plasma membrane.

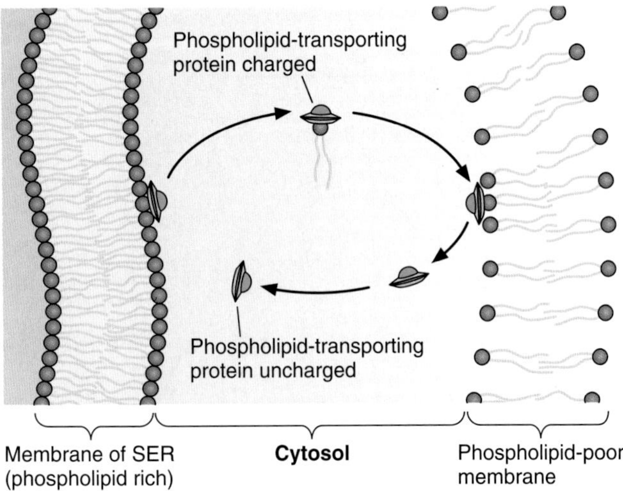

Figure 2–20. Schematic representation of a phospholipid-transporting amphipathic protein. Phospholipid molecules are transported from lipid-rich (SER) to lipid-poor membranes. (Redrawn and reproduced, with permission, from Junqueira LC, Carneiro J: *Biologia Celular e Molecular,* 6th ed. Editora Guanabara, 1997.)

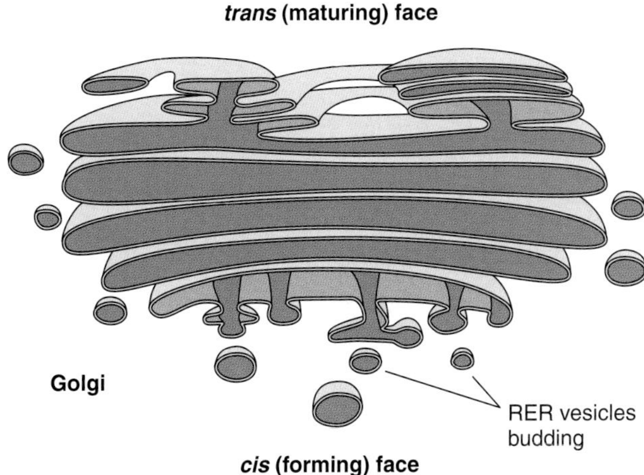

Figure 2–21. Three-dimensional representation of a Golgi complex. Through transport vesicles that fuse with the Golgi *cis* face, the complex receives several types of molecules produced in the rough endoplasmic reticulum (RER). After Golgi processing, these molecules are released from the Golgi *trans* face in larger vesicles to constitute secretory vesicles, lysosomes, or other cytoplasmic components.

In most cells, there is also polarity in Golgi structure and function. Near the Golgi complex, the RER can sometimes be seen budding off small vesicles (transport vesicles) that shuttle newly synthesized proteins to the Golgi complex for further processing. The Golgi cisterna nearest this point is called the forming, convex, or *cis* face. On the opposite side of the Golgi complex, which is the maturing, concave, or *trans* face, large Golgi vacuoles accumulate (Figure 2–21). These are sometimes called **condensing vacuoles.**

These structures bud from the Golgi cisternae, generating vesicles that will transport proteins to various sites. Cytochemical methods and the electron microscope have shown that the Golgi cisternae present different enzymes at different *cis-trans* levels and that the Golgi complex is important in the glycosylation, sulfating, phosphorylation, and limited proteolysis of proteins. Furthermore, the Golgi complex initiates packing, concentration, and storage of secretory products. Figure 2–23 gives an overall view of the currently accepted concepts regarding transit of material through the Golgi complex.

Lysosomes

Lysosomes are sites of intracellular digestion and turnover of cellular components. Lysosomes (Gr. *lysis,* solution, + *soma,* body) are membrane-limited vesicles that contain a large variety of hydrolytic enzymes (more than 40) whose main function is intracytoplasmic digestion (Figures 2–24, 2–25, and 2–26). Lysosomes are particularly abundant in cells exhibiting phagocytic activity (eg, macrophages, neutrophilic leukocytes). Although the nature and activity of lysosomal enzymes vary depending on the cell type, the most common enzymes are acid phosphatase, ribonuclease, deoxyribonuclease, proteases, sulfatases, lipases, and β-glucuronidase. As

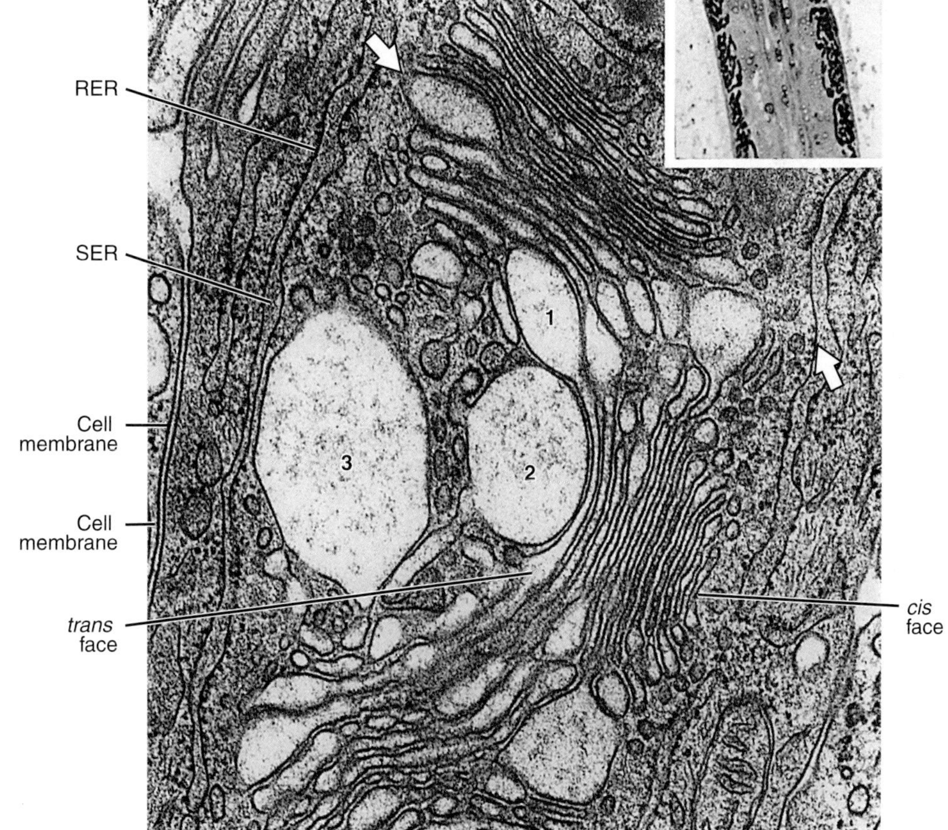

Figure 2–22. Electron micrograph of a Golgi complex of a mucous cell. To the right is a cisterna (arrow) of the rough endoplasmic reticulum containing granular material. Close to it are small vesicles containing this material. This is the *cis* face of the complex. In the center are flattened and stacked cisternae of the Golgi complex. Dilatations can be observed extending from the ends of the cisternae. These dilatations gradually detach themselves from the cisternae and fuse, forming the secretory granules (**1, 2,** and **3**). This is the *trans* face. Near the plasma membrane of two neighboring cells is endoplasmic reticulum with a smooth section (SER) and a rough section (RER). ×30,000. **Inset:** The Golgi complex as seen in 1-μm sections of epididymis cells impregnated with silver. ×1200.

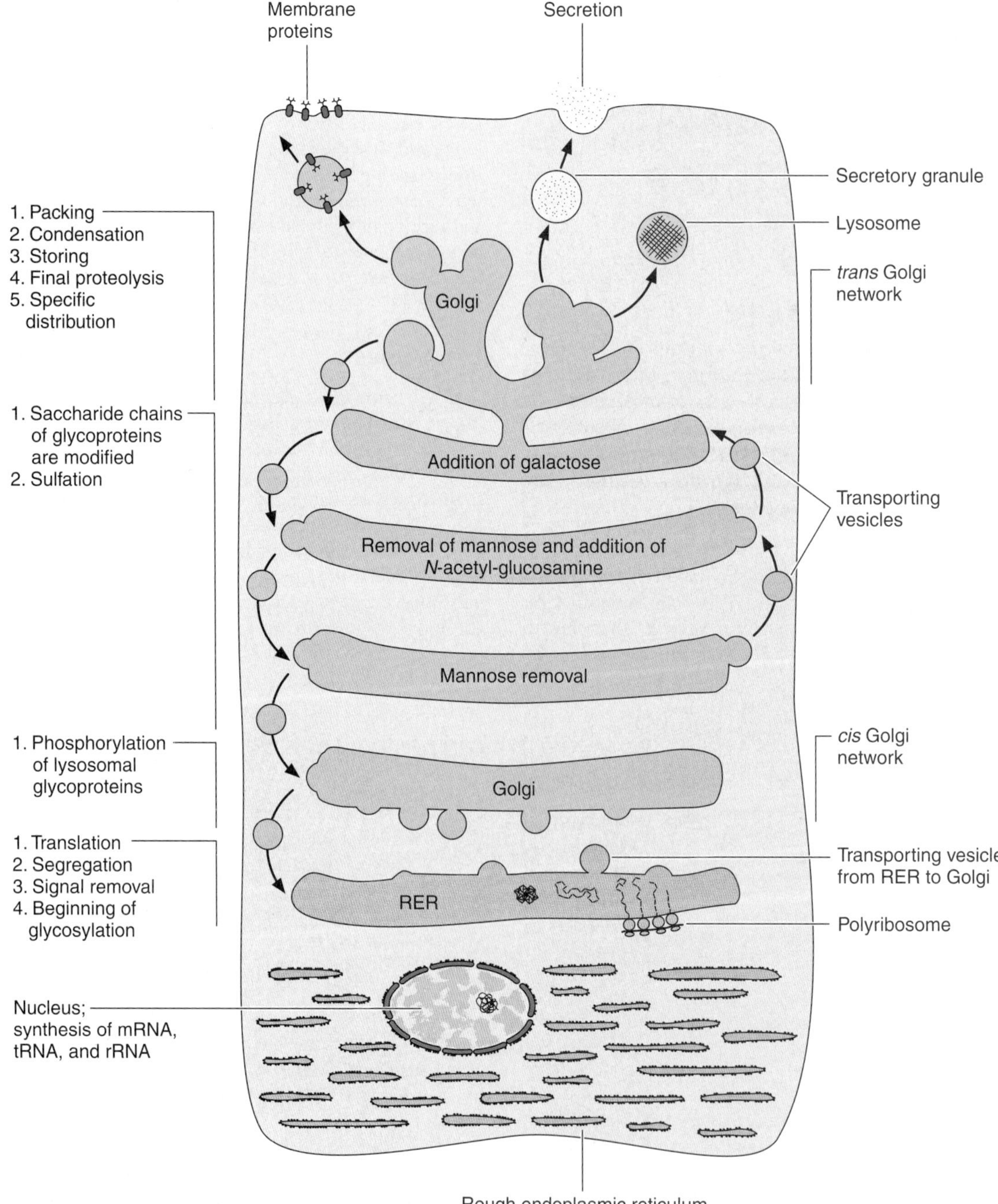

Figure 2–23. Main events occurring during trafficking and sorting of proteins through the Golgi complex. Numbered at the left are the main molecular processes that take place in the compartments indicated. Note that the labeling of lysosomal enzymes starts early in the *cis* Golgi network. In the *trans* Golgi network, the glycoproteins combine with specific receptors that guide them to their destination. On the left side of the drawing is the returning flux of membrane, from the Golgi to the endoplasmic reticulum. (Redrawn and reproduced, with permission, from Junqueira LC, Carneiro J: *Biologia Celular e Molecular,* 6th ed. Editora Guanabara, 1997.)

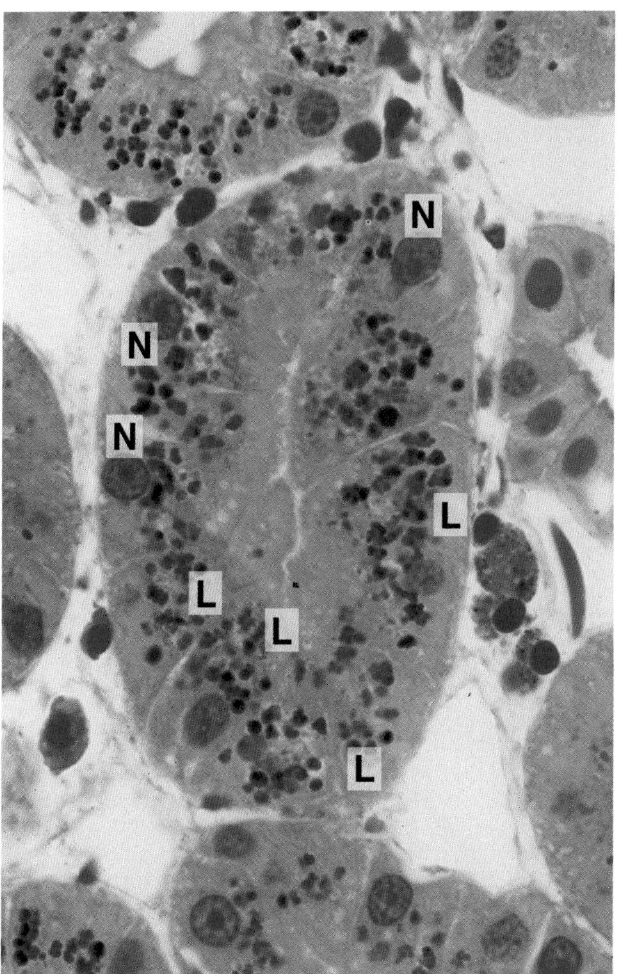

Figure 2–24. Photomicrograph of a kidney tubule whose lumen appears in the center as a long slit. The numerous dark-stained cytoplasmic granules are lysosomes (L), organelles abundant in these kidney cells. The cell nuclei (N), some showing a nucleolus, are also seen in the photograph as dark-stained corpuscles. Toluidine blue stain. High magnification.

can be seen from this list, lysosomal enzymes are capable of breaking down most biologic macromolecules. Lysosomal enzymes have optimal activity at an acidic pH.

Lysosomes, which are usually spherical, range in diameter from 0.05 to 0.5 μm and present a uniformly granular, electron-dense appearance in electron micrographs. In a few cells, such as macrophages and neutrophilic leukocytes, primary lysosomes are larger, up to 0.5 μm in diameter, and thus just visible with the light microscope.

The enveloping membrane separates the lytic enzymes from the cytoplasm, preventing the lysosomal enzymes from attacking and digesting cytoplasmic components. The fact that the lysosomal enzymes are practically inactive at the pH of the cytosol (~7.2) is an additional protection of the cell against leakage of lysosomal enzymes.

Lysosomal enzymes are synthesized and segregated in the RER and subsequently transferred to the Golgi complex, where the enzymes are modified and packaged as lysosomes. These enzymes have oligosaccharides attached to them with one or more of the mannose residues phosphorylated at the 6′ position by a phosphotransferase. There are receptors for mannose 6-phosphate-containing proteins in the RER and Golgi complex that allow these proteins to be diverted from the main secretory pathway and segregated in lysosomes.

Lysosomes that have not entered into a digestive event are identified as **primary lysosomes.**

Lysosomes can digest materials taken into the cell from its environment. The material is taken into a phagosome or phagocytic vacuole (Figure 2–27); primary lysosomes then fuse with the membrane of the phagosome and empty their hydrolytic enzymes into the vacuole. Digestion follows, and the composite structure is now termed a **secondary lysosome.**

Secondary lysosomes are generally 0.2–2 μm in diameter and present a heterogeneous appearance in electron microscopes because of the wide variety of materials they may be digesting.

After digestion of the contents of the secondary lysosome, nutrients diffuse through the lysosomal limiting membrane and enter the cytosol. Indigestible compounds are retained within the vacuoles, which are now called **residual bodies** (Figures 2–27 and 2–28). In some long-lived cells (eg, neurons, heart muscle), large quantities of residual bodies accumulate and are referred to as **lipofuscin,** or **age pigment.**

Another function of lysosomes concerns the turnover of cytoplasmic organelles. Under certain conditions, a membrane may enclose organelles or portions of cytoplasm. Primary lysosomes fuse with this structure and initiate the lysis of the enclosed cytoplasm. The resulting secondary lysosomes are known as **autophagosomes** (Gr. *autos,* self, + *phagein,* to eat, + *soma*), indicating that their contents are intracellular in origin. Cytoplasmic digestion by autophagosomes is enhanced in secretory cells that have accumulated excess secretory product. The digested products of lysosomal hydrolysis are recycled by the cell to be reutilized by the cytoplasm.

MEDICAL APPLICATION

In some cases, primary lysosomes release their contents extracellularly, and their enzymes act in the extracellular milieu. An example is the destruction of bone matrix by the collagenases synthesized and released by osteoclasts during normal bone tissue formation (see Chapter 8). Lysosomal enzymes acting in the extracellular milieu also play a significant role in the response to inflammation or injury. Several possible pathways relating to lysosome activities are schematically illustrated in Figure 2–27.

Lysosomes play an important role in the metabolism of several substances in the human body, and consequently many diseases have been ascribed to deficiencies of lysosomal enzymes. In metachromatic leukodystrophy, there is an intracellular accu-

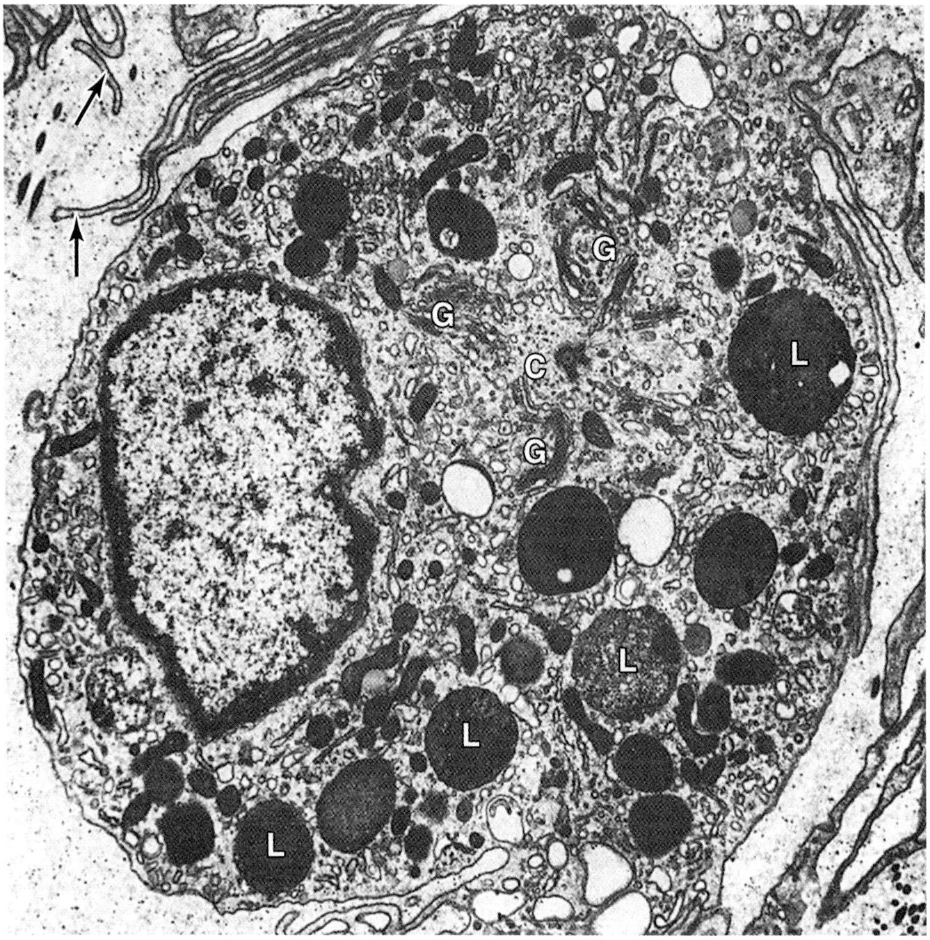

Figure 2–25. Electron micrograph of a macrophage. Note the abundant cytoplasmic extensions (arrows). In the center is a centriole (C) surrounded by Golgi cisternae (G). Secondary lysosomes (L) are abundant. ×15,000.

mulation of sulfated cerebrosides caused by lack of lysosomal sulfatases. In most of these diseases, a specific lysosomal enzyme is absent or inactive, and certain molecules (eg, glycogen, cerebrosides, gangliosides, sphingomyelin, glycosaminoglycans) are not digested. As a result, these substances accumulate in the cells, interfering with their normal functions. This diversity of affected cell types explains the variety of clinical symptoms observed in lysosomal diseases (Table 2–3).

Table 2–3. Examples of diseases caused by lysosomal enzyme failure and accumulation of undigested material in different cell types.

Disease	Faulty Enzyme	Main Organs Affected
Hurler	α L-iduronidase	Skeleton and nervous system
Sanfilippo Syndrome A	Heparan sulfate sulfamidase	Skeleton and nervous system
Tay-Sachs	Hexosaminidase-A	Nervous system
Gaucher	β D-glycosidase	Liver and spleen
I-cell disease	Phosphotransferase	Skeleton and nervous system

I-cell disease (inclusion cell disease) is a rare inherited condition clinically characterized by defective physical growth and mental retardation and is due to a deficiency in a phosphorylating enzyme normally present in the Golgi complex. Lysosomal enzymes coming from the RER are not phosphorylated in the Golgi complex. Nonphosphorylated protein molecules are not separated to form lysosomes, instead following the main secretory pathway. The secreted lysosomal enzymes are present in the blood of patients with I-cell disease, whereas their lysosomes are empty. Cells of these patients show large inclusion granules that interfere with normal cellular metabolism.

Proteasomes

Proteasomes are multiple-protease complexes that digest proteins targeted for destruction by attachment to ubiquitin. Protein degradation is essential to remove excess enzyme and other proteins that become unnecessary to the cell after they perform their normal functions, and also to remove proteins that were incorrectly folded. Protein encoded

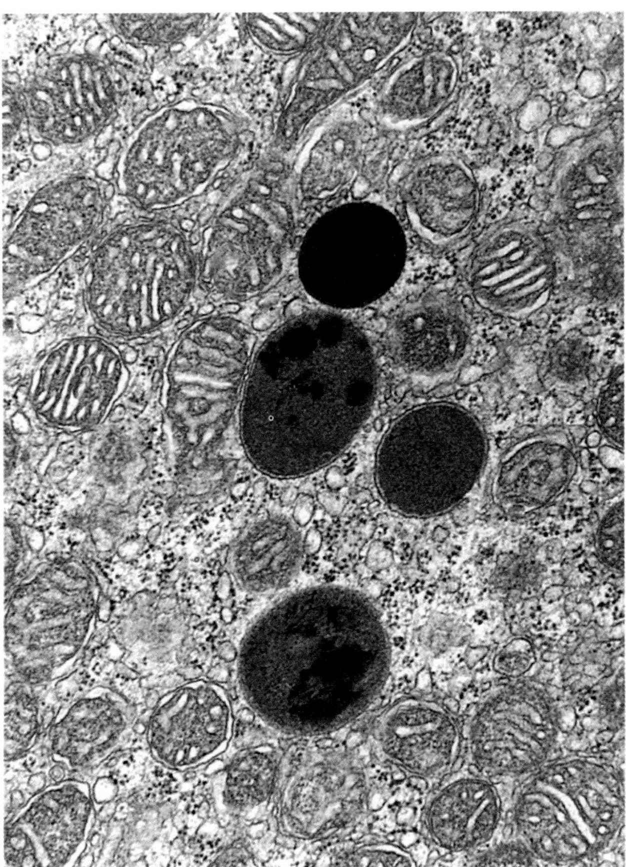

Figure 2–26. Electron micrograph showing 4 dark secondary lysosomes surrounded by numerous mitochondria.

Peroxisomes, or Microbodies

Peroxisomes (peroxide + *soma*) are spherical membrane-limited organelles whose diameter ranges from 0.5 to 1.2 μm (see Figure 2–39). Like the mitochondria, they utilize oxygen but do not produce ATP and do not participate directly in cellular metabolism. Peroxisomes oxidize specific organic substrates by removing hydrogen atoms that are transferred to molecular oxygen (O_2). This activity produces hydrogen peroxide (H_2O_2), a substance that is very damaging to the cell. However, H_2O_2 is eliminated by the enzyme **catalase,** which is present in peroxisomes. Catalase transfers oxygen atoms from H_2O_2 to several compounds and also decomposes H_2O_2 to H_2O and O_2 ($2\ H_2O_2 \rightarrow 2\ H_2O + O_2$). Catalase activity also has clinical implications: It degrades several toxic molecules and prescription drugs, particularly in liver and kidney peroxisomes. For example, 50% of ingested ethyl alcohol is degraded to acetic aldehyde in liver and kidney peroxisomes. Liver and kidney peroxisomes show a higher variation in their enzyme complement than do other peroxisomes. Their homogeneous matrix contains D- and L-amino oxidases, catalase, and hydroxyacid oxidase. In some species, but not humans, a crystalline nucleoid is present that is composed of urate oxidase.

Peroxisomes contain enzymes involved in lipid metabolism. Thus, the β-oxidation of long-chain fatty acids (18 carbons and longer) is preferentially accomplished by peroxisomal enzymes that differ from their mitochondrial counterparts. Certain reactions leading to the formation of bile acids and cholesterol also have been localized in highly purified peroxisomal fractions.

Peroxisomal enzymes are synthesized on free cytosolic polyribosomes, with a small sequence of amino acids located near the carboxyl terminus that functions as an import signal. Proteins with this signal are recognized by receptors located in the membrane of peroxisomes and internalized into the organelle. The peroxisome grows in size and is divided into two smaller peroxisomes, by a mechanism not completely understood.

by virus should also be destroyed. Proteasomes deal primarily with proteins as individual molecules, whereas lysosomes digest bulk material introduced into the cell or whole organelles and vesicles.

The proteasome has a core particle with the shape of a barrel made of 4 rings stacked on each other. At each end of the core particle is a regulatory particle that contains ATPase and recognizes proteins with ubiquitin molecules attached. Ubiquitin is a small protein (76 amino acids) found in all cells and is highly conserved during evolution—it has virtually the same structure from bacteria to humans. Ubiquitin targets proteins for destruction as follows. A molecule of ubiquitin binds to a lysine residue in the protein to be degraded. Then other ubiquitin molecules attach to the first one; the complex is recognized by the regulatory particle; the protein is unfolded by the ATPases using energy from ATP; and the protein is translocated into the core particle, where it is broken into peptides of about 8 amino acids each. These peptides are transferred to the cytosol by a process yet unknown. The ubiquitin molecules are released by the regulatory particles for reuse.

The 8-amino-acid peptides may be broken down to amino acids by cytosol enzymes, or they may have other destinations (eg, in some cells they participate in the immune response).

MEDICAL APPLICATION

A large number of disorders arise from defective peroxisomal proteins, because this organelle is involved in several metabolic pathways. Probably the most common peroxisomal disorder is X-chromosome-linked adrenoleukodystrophy, caused by a defective integral membrane protein that participates in transporting very long-chain fatty acids into the peroxisome for β-oxidation. Accumulation of these fatty acids in body fluids destroys the myelin sheaths in nerve tissue, causing severe neurological symptoms. Deficiency in peroxisomal enzymes causes the fatal Zellweger syndrome, with severe muscular impairment, liver and kidney lesions, and disorganization of the central and peripheral nervous systems. Electron microscopy reveals empty peroxisomes in liver and kidney cells of these patients.

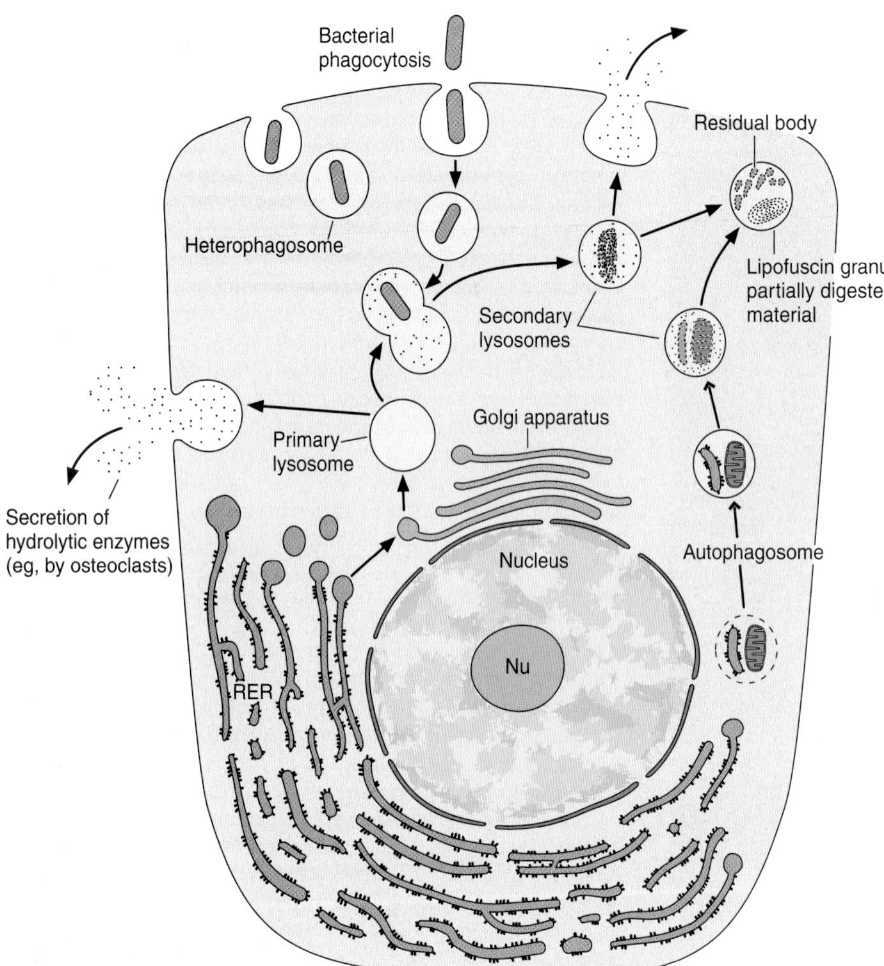

Figure 2–27. Current concepts of the functions of lysosomes. Synthesis occurs in the rough endoplasmic reticulum (RER), and the enzymes are packaged in the Golgi complex. Note the heterophagosomes, in which bacteria are being destroyed, and the autophagosomes, with RER and mitochondria in the process of digestion. Heterophagosomes and autophagosomes are secondary lysosomes. The result of their digestion can be excreted, but sometimes the secondary lysosome creates a residual body, containing remnants of undigested molecules. In some cells, such as osteoclasts, the lysosomal enzymes are secreted to the extracellular environment. Nu, nucleolus.

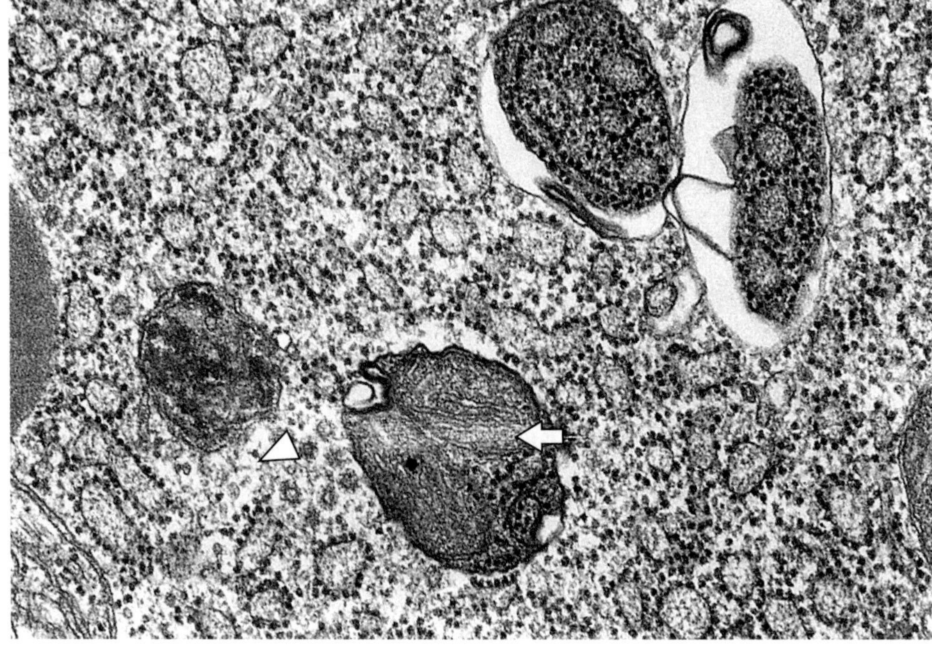

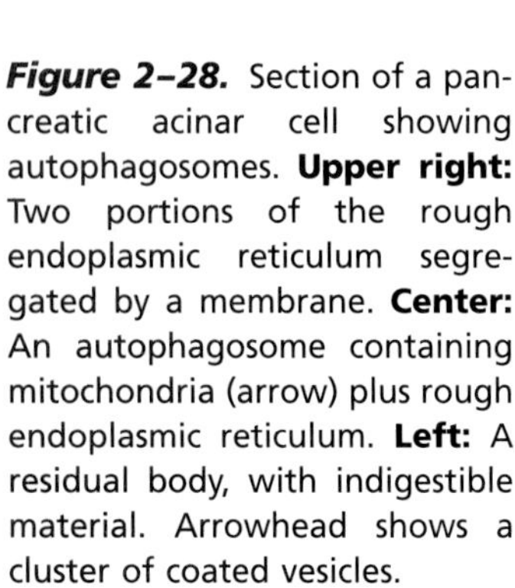

Figure 2–28. Section of a pancreatic acinar cell showing autophagosomes. **Upper right:** Two portions of the rough endoplasmic reticulum segregated by a membrane. **Center:** An autophagosome containing mitochondria (arrow) plus rough endoplasmic reticulum. **Left:** A residual body, with indigestible material. Arrowhead shows a cluster of coated vesicles.

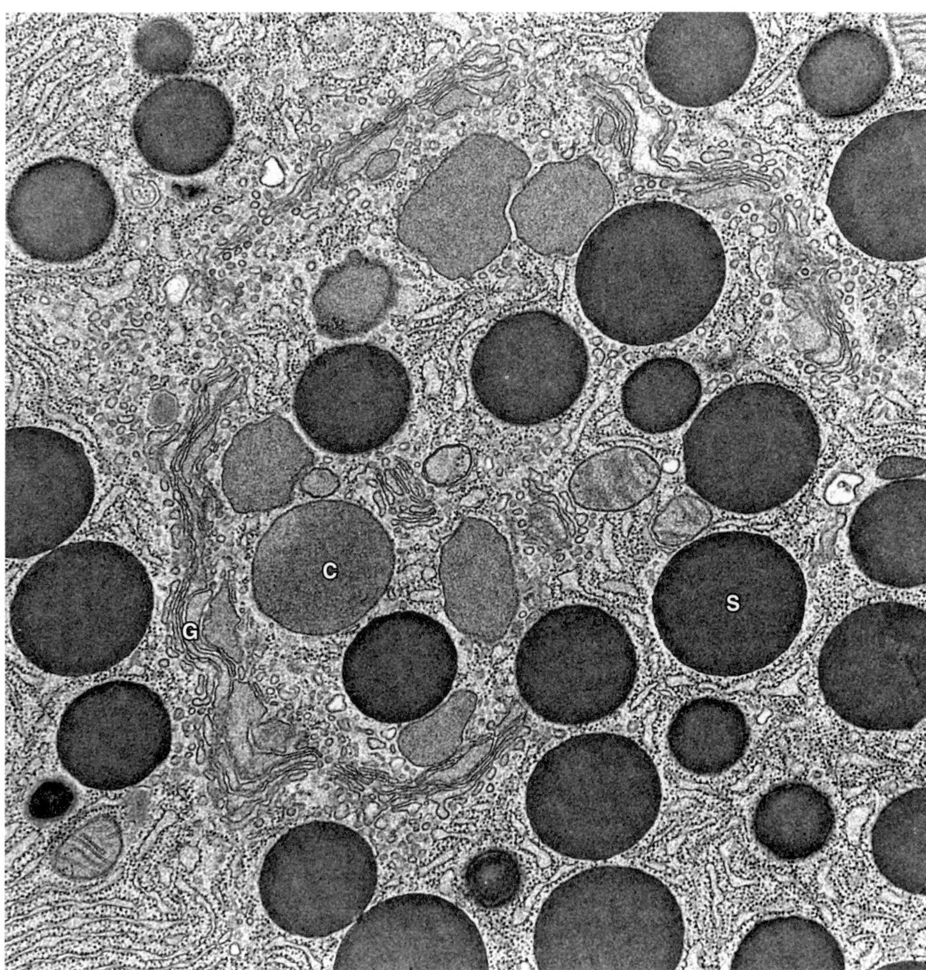

Figure 2–29. Electron micrograph of a pancreatic acinar cell from the rat. Numerous mature secretory granules (S) are seen in association with condensing vacuoles (C) and the Golgi complex (G). ×18,900.

Secretory Vesicles, or Granules

Secretory vesicles are found in those cells that store a product until its release is signaled by a metabolic, hormonal, or neural message (regulated secretion). These vesicles are surrounded by a membrane and contain a concentrated form of the secretory product (Figure 2–29). The contents of some secretory vesicles may be up to 200 times more concentrated than those in the cisternae of the RER. Secretory vesicles containing digestive enzymes are referred to as **zymogen granules.**

THE CYTOSKELETON

The cytoplasmic cytoskeleton is a complex network of microtubules, actin filaments (microfilaments), and intermediate filaments. These structural proteins provide for the shaping of cells and also play an important role in the movements of organelles and intracytoplasmic vesicles. The cytoskeleton also participates in the movement of entire cells.

Microtubules

Within the cytoplasmic matrix of eukaryotic cells are tubular structures known as microtubules (Figures 2–30, 2–31, and 2–32). Microtubules are also found in cytoplasmic processes called cilia (Figure 2–33) and flagella. They have an outer diameter of 24 nm, consisting of a dense wall 5 nm thick and a hollow core 14 nm wide. Microtubules are variable in length, and individual tubules can attain lengths of several micrometers. Occasionally, arms or bridges are found linking two or more tubules (Figure 2–34).

The subunit of a microtubule is a heterodimer composed of α and β **tubulin** molecules of closely related amino acid composition, each with a molecular mass of about 50 kDa.

Under appropriate conditions (in vivo or in vitro), tubulin subunits polymerize to form microtubules. With special staining procedures, tubulin can be seen as heterodimers organized into a spiral. A total of 13 units are present in one complete turn of the spiral (Figure 2–34).

Polymerization of tubulins to form microtubules in vivo is directed by a variety of structures collectively known as **microtubule organizing centers.** These structures include cilia, basal bodies, and centrosomes. Microtubule growth, via subunit polymerization, occurs more rapidly at one end of existing microtubules. This end is referred to as the plus (+) end, and the other extremity is the minus (−) end. Tubulin polymerization is under control of the concentration of Ca^{2+} and of the microtubule associated proteins, or **MAPs.** Microtubule stability is variable; for example, microtubules of cilia are stable, whereas microtubules of the mitotic spindle have a short duration. The antimitotic alkaloid colchicine binds specifically to tubulin, and when the complex tubulin-colchicine binds to microtubules, it prevents the addition of more tubulin in the plus (+) extremity. Mitotic microtubules are broken down because the depolymerization continues, mainly at the minus (−) end, and the lost tubulin units are not replaced. Another alkaloid that interferes with the mitotic microtubule is taxol, which accelerates the formation of microtubules but at the same time stabilizes them. All cytosolic tubulin is used in stable microtubules, and no tubulin is left for the formation of the mitotic spindle. Another alkaloid, vinblastine, acts by depolymerizing formed microtubules and, in a second step, aggregating to form paracrystalline arrays of tubulin.

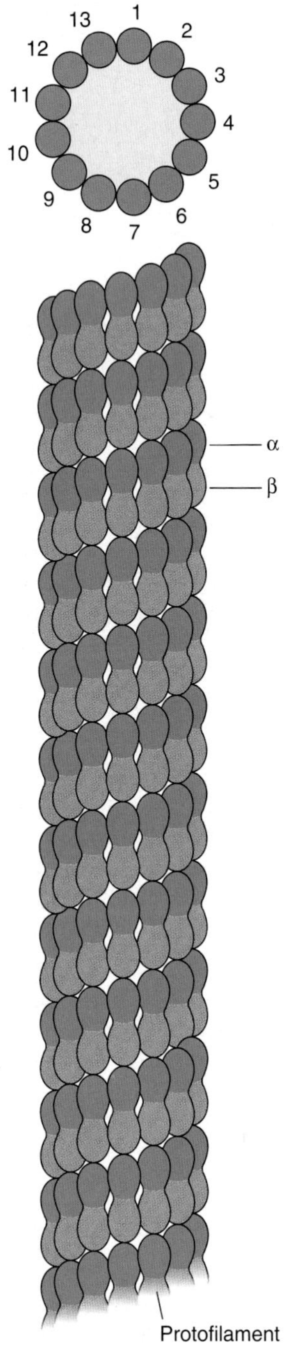

Figure 2–30. Molecular organization of a microtubule. In this polarized structure there is an alternation of the two subunits (α and β) of the tubulin molecule. Tubulin molecules are arranged to form 13 protofilaments, as seen in the cross section in the upper part of the drawing.

MEDICAL APPLICATION

The antimitotic alkaloids are useful tools in cell biology (eg, colchicine is used to arrest chromosomes in metaphase and to prepare karyotypes) and in cancer chemotherapy (eg, vinblastine, vincristine, and taxol are used to arrest cell proliferation in tumors). Because tumor cells proliferate rapidly, they are more affected by antimitotic drugs than are normal cells. However, chemotherapy has many undesirable consequences. For example, some normal blood-forming cells and the epithelial cells that cover the digestive tract also show a high rate of proliferation and are adversely affected by chemotherapy.

Cytoplasmic microtubules are stiff structures that play a significant role in the development and maintenance of cell shape. They are usually present in a proper orientation, either to effect development of a given cellular asymmetry or to maintain it. Procedures that disrupt microtubules result in the loss of this cellular asymmetry.

Microtubules also participate in the intracellular transport of organelles and vesicles. Examples include axoplasmic transport in neurons, melanin transport in pigment cells, chromosome movements by the mitotic spindle, and vesicle movements among different cell compartments. In each of these examples, movement is related to the presence of complex microtubule networks, and such activities are suspended if microtubules are disrupted. The transport guided by microtubules is under the control of special proteins called **motor proteins,** which use energy to move molecules and vesicles.

Microtubules provide the basis for several complex cytoplasmic components, including centrioles, basal bodies, cilia, and flagella. **Centrioles** are cylindrical structures (0.15 μm in diameter and 0.3–0.5 μm in length) composed primarily of short, highly organized microtubules (Figure 2–34). Each centriole shows 9 sets of microtubules arranged in triplets. The microtubules are so close together that adjacent microtubules of a triplet share a common wall. Close to the nucleus of nondividing cells is a **centrosome** (Figure 2–35) made of a pair of centrioles surrounded by a granular material. In each pair, the long axes of the centrioles are at right angles to each other. Before cell division, more specifically during the S period of the interphase, each centrosome duplicates itself so that now each centrosome

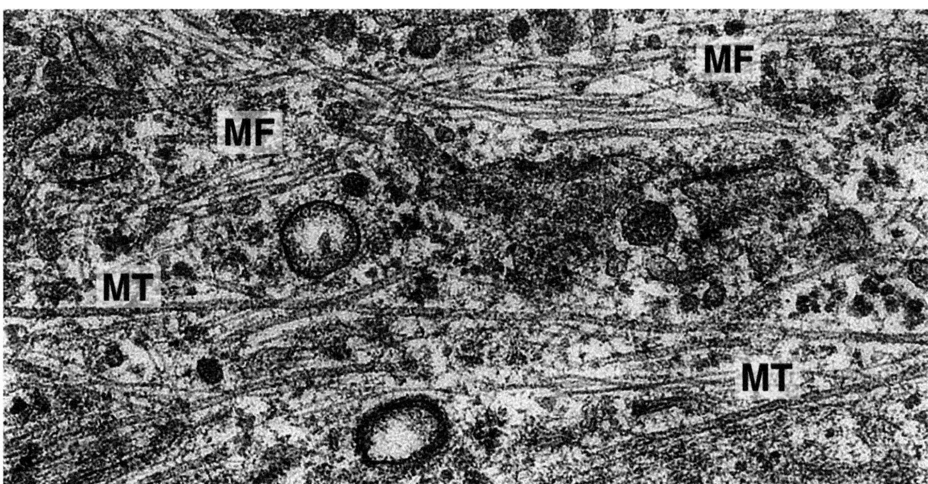

Figure 2–31. Electron micrograph of fibroblast cytoplasm. Note the microfilaments (MF) and microtubules (MT). ×60,000. (Courtesy of E Katchburian.)

has two pairs of centrioles. During mitosis, the centrosomes divide in two, move to opposite poles of the cell, and become organizing centers for the microtubules of the mitotic spindle.

Cilia and **flagella** (singular, cilium, flagellum) are motile processes, covered by cell membrane, with a highly organized microtubule core. Ciliated cells typically possess a large number of cilia, each about 2–3 μm in length. Flagellated cells have only one flagellum, with a length close to 100 μm. In humans, the spermatozoa are the only cell type with a flagellum. The main function of cilia is to sweep fluid from the surface of cell sheets. Both cilia and flagella possess the same core organization.

This core consists of 9 pairs of microtubules surrounding 2 central microtubules. This sheaf of microtubules, possessing a **9 + 2 pattern,** is called an axoneme (Gr. *axon,* axis, + *nema,* thread). Each of the 9 peripheral pairs shares a common wall (Figure 2–34). The microtubules in the central pair are enclosed within a **central sheath.** Adjacent peripheral pairs are linked to each other by protein bridges called **nexins** and to the central sheath by **radial spokes.** The microtubules of each pair are identified as A and B. A is a complete microtubule with 13 heterodimers, whereas B has only 10 heterodimers (in a cross section). Extending from the surface of microtubule A are pairs of arms formed by the protein **dynein,** which has ATPase activity.

At the base of each cilium or flagellum is a **basal body,** essentially similar to a centriole, that controls the assembly of the axoneme.

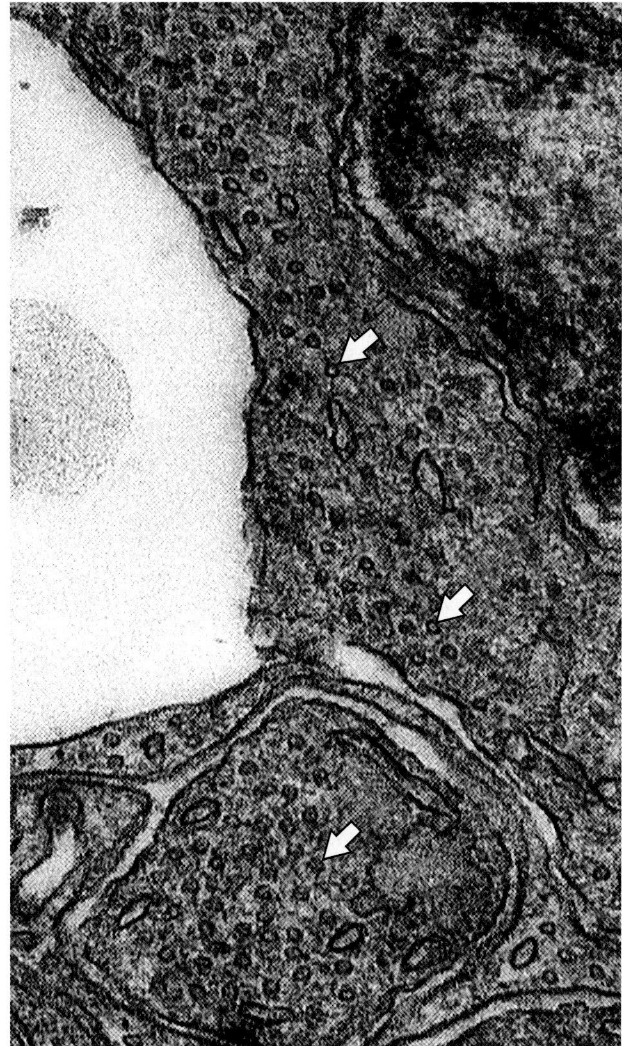

Figure 2–32. Electron micrograph of a section of a photosensitive retinal cell. Note the accumulation of transversely sectioned microtubules (arrows). Reduced slightly from ×80,000.

MEDICAL APPLICATION

*Several mutations have been described in the proteins of the cilia and flagella. They are responsible for the **immotile cilia syndrome,** the symptoms of which are immotile spermatozoa, male infertility, and chronic respiratory infections caused by the lack of the cleansing action of cilia in the respiratory tract.*

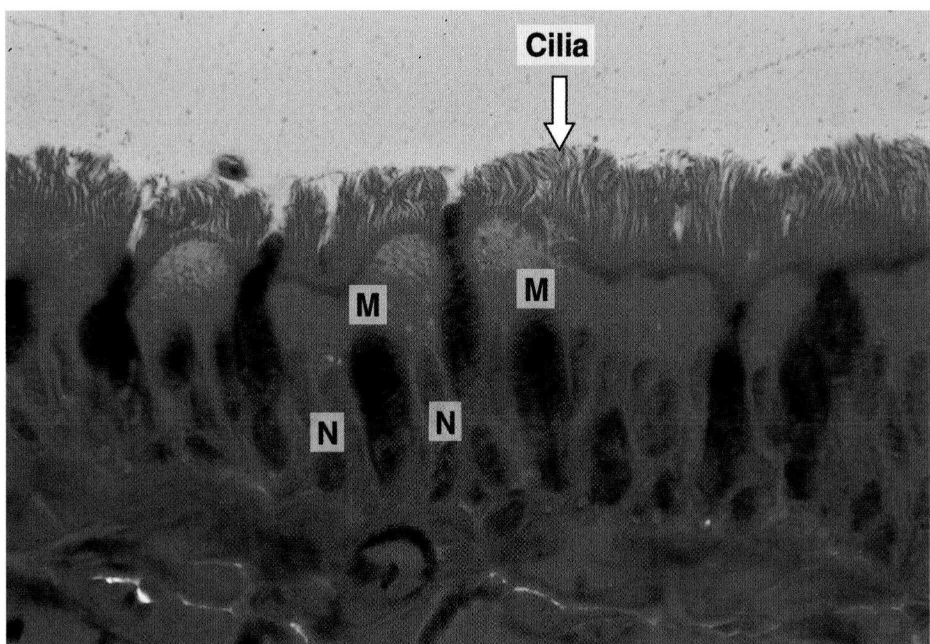

Figure 2–33. Photomicrograph of the epithelium covering the inner surface of the respiratory airways. Most cells in this epithelium contain numerous cilia in their apices (free upper extremities). N, cell nuclei; M, cytoplasmic mucus secretion, which appears dark in this preparation. H&E stain. High magnification.

Actin Filaments

Contractile activity in muscle cells results primarily from an interaction between two proteins: **actin** and **myosin.** Actin is present in muscle as a thin (5–7 nm in diameter) filament composed of globular subunits organized into a double-stranded helix (Figure 2–36). Structural and biochemical studies reveal that there are several types of actin and that this protein is present in all cells.

Within cells, microfilaments can be organized in many forms. (1) In skeletal muscle, they assume a paracrystalline array integrated with thick (16-nm) myosin filaments. (2) In most cells, microfilaments form a thin sheath just beneath the plasmalemma, called the **cell cortex.** These filaments appear to be associated with membrane activities such as endocytosis, exocytosis, and cell migratory activity. (3) Microfilaments are intimately associated with several cytoplasmic organelles, vesicles, and granules. The filaments are believed to play a role in moving and shifting cytoplasmic components (cytoplasmic streaming). (4) Microfilaments are associated with myosin and form a "purse-string" ring of filaments whose constriction results in the cleavage of mitotic cells. (5) In most cells, microfilaments are found scattered in what appears to be an unorganized fashion within the cytoplasm (Figure 2–31).

Although actin filaments in muscle cells are structurally stable, in nonmuscle cells they readily dissociate and reassemble. Actin filament polymerization appears to be under the direct control of minute changes in Ca^{2+} and cyclic AMP levels. A large number of actin-binding proteins have been demonstrated in a wide variety of cells, and much current research is focused on how these proteins regulate the state of polymerization and lateral aggregation of actin filaments. Their importance can be deduced from the fact that only about half the cell's actin is in the form of microfilaments.

Presumably, most actin filament–related activities depend upon the interaction of **myosin** with actin. (The structure and activity of the thick myosin filaments are described in the section on muscle tissues.)

Intermediate Filaments

Ultrastructural and immunocytochemical investigations reveal that a third major filamentous structure is present in eukaryotic cells. In addition to the thin (actin) and thick (myosin) filaments, cells contain a class of intermediate-sized filaments with an average diameter of 10–12 nm (Figure 2–37 and Table 2–4). Several proteins that form intermediate filaments have been isolated and localized by immunocytochemical means.

Keratins (Gr. *keras,* horn) are a family of approximately 20 proteins found in epithelia. They are encoded by a family of

Table 2–4. Examples of intermediate filaments found in eukaryotic cells.

Filament Type	Cell Type	Examples
Keratins	Epithelium	Both keratinizing and nonkeratinizing epithelia
Vimentin	Mesenchymal cells	Fibroblasts, chondroblasts, macrophages, endothelial cells, vascular smooth muscle
Desmin	Muscle	Striated and smooth muscle (except vascular smooth muscle)
Gilial fibrillary acidic proteins	Glial cells	Astrocytes
Neurofilaments	Neurons	Nerve cell body and processes

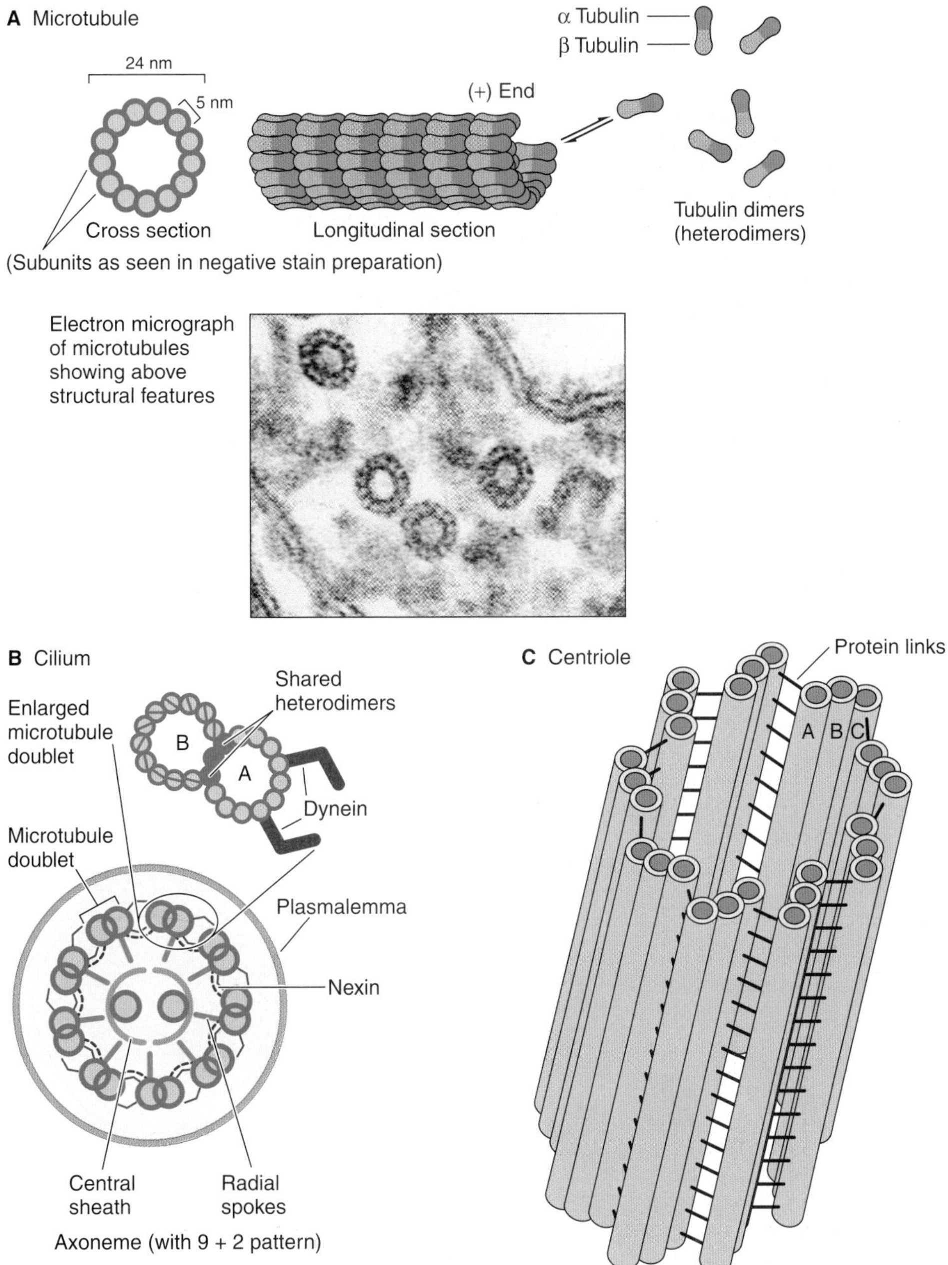

Figure 2–34. Schematic representation of microtubules, cilia, and centrioles. **A:** Microtubules as seen in the electron microscope after fixation with tannic acid in glutaraldehyde. The unstained tubulin subunits are delineated by the dense tannic acid. Cross sections of tubules reveal a ring of 13 subunits of dimers arranged in a spiral. Changes in microtubule length are due to the addition or loss of individual tubulin subunits. **B:** A cross section through a cilium reveals a core of microtubules called an axoneme. The axoneme consists of 2 central microtubules surrounded by 9 microtubule doublets. In the doublets, microtubule A is complete and consists of 13 subunits, whereas microtubule B shares 2 or 3 heterodimers with A. When activated by ATP, the dynein arms link adjacent tubules and provide for the sliding of doublets against each other. **C:** Centrioles consist of 9 microtubule triplets linked together in a pinwheel-like arrangement. In the triplets, microtubule A is complete and consists of 13 subunits, whereas tubules B and C share tubulin subunits. Under normal circumstances, these organelles are found in pairs with the centrioles disposed at right angles to one another.

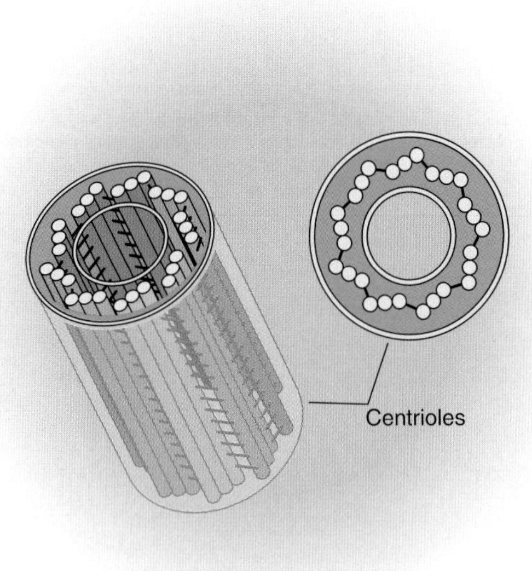

Figure 2–35. Drawing of a centrosome with its granular protein material surrounding a pair of centrioles, one shown at a right angle to the other. Each centriole is made of 9 bundles of microtubles, with 3 microtubules per bundle.

genes and have different chemical and immunologic properties. This diversity of keratin is related to the various roles these proteins play in the epidermis, nails, hooves, horns, feathers, scales, and the like that provide animals with defense against abrasion and loss of water and heat.

Vimentin filaments are characteristic of cells of mesenchymal origin. (Mesenchyme is an embryonic tissue.) Vimentin is a single protein (56–58 kDa) and may copolymerize with desmin or glial fibrillary acidic protein.

Desmin (**skeletin**) is found in smooth muscle and in the Z disks of skeletal and cardiac muscle (53–55 kDa).

Glial filaments (**glial fibrillary acidic protein**) are characteristic of astrocytes but are not found in neurons, muscle, mesenchymal cells, or epithelia (51 kDa).

Neurofilaments consist of at least 3 high-molecular-weight polypeptides (68, 140, and 210 kDa). Intermediate filament proteins have different chemical structures and different roles in cellular function.

MEDICAL APPLICATION

The presence of a specific type of intermediate filament in tumors can reveal which cell originated the tumor, information important for diagnosis and treatment (see Table 1–1). Identification of intermediate filament proteins by means of immunocytochemical methods is a routine procedure.

Cytoplasmic Deposits

Cytoplasmic deposits are usually transitory components of the cytoplasm, composed mainly of accumulated metabolites or other substances. The accumulated molecules occur in several forms, one of them being lipid droplets in adipose tissue, adrenal cortex cells, and liver cells (Figure 2–38). Carbohydrate accumulations are also visible in several cells in the form of glycogen. After impregnation with lead salts, glycogen appears as collections of electron-dense particles (Figure 2–39). Proteins are stored in glandular cells such as **secretory granules** or **secretory vesicles** (Figure 2–29); under stimulation, these proteins are periodically released into the extracellular medium.

Deposits of colored substances—**pigments**—are often found in cells (Figure 2–40). They may be synthesized by the cell (eg, in the skin melanocytes) or come from outside the body (eg, carotene). One of the most common pigments is **lipofuscin,** a yellowish-brown substance present mainly in permanent cells (eg, neurons, cardiac muscle) that increases in quantity with age. Its chemical constitution is complex. It is believed that granules of lipofuscin derive from secondary lysosomes and represent deposits of indigestible substances. A widely distributed pigment, **melanin,** is abundant in the epidermis and in the pigment layer of the retina in the form of dense intracellular membrane-limited granules.

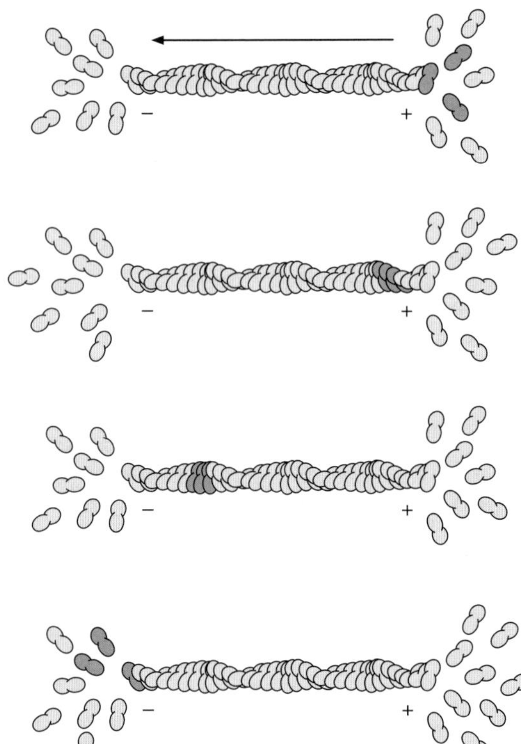

Figure 2–36. **The cytosolic actin filament. Actin dimers are added to the plus (+) end and removed at the minus (−) end, dynamically lengthening or shortening the filament, as required by the cell.** (Redrawn and reproduced, with permission, from Junqueira LC, Carneiro J: *Biologia Celular e Molecular,* 6th ed. Editora Guanabara, 1997.)

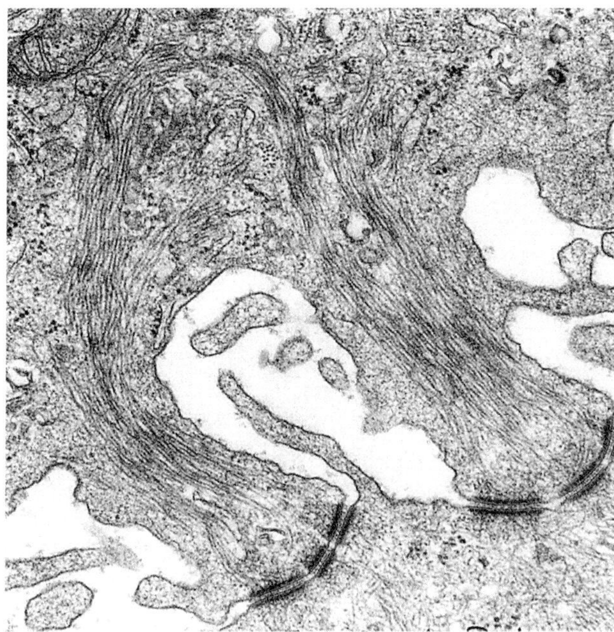

Figure 2–37. Electron micrograph of a skin epithelial cell showing intermediate filaments of keratin associated with desmosomes.

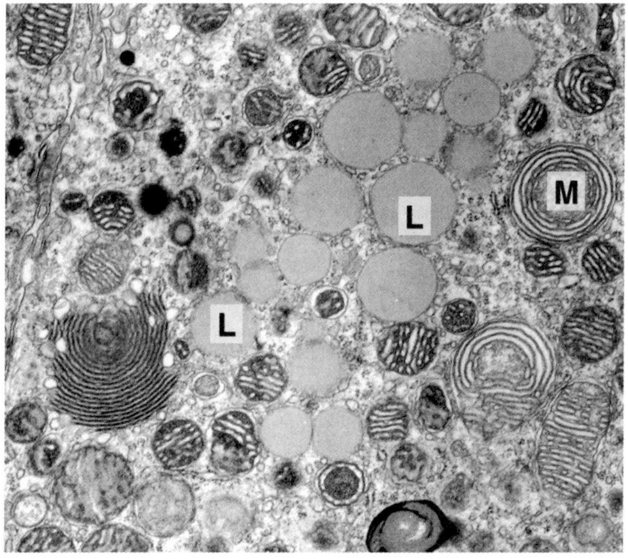

Figure 2–38. Section of adrenal gland showing lipid droplets (L) and abundant anomalous mitochondria (M). ×19,000.

Cytosol

At one time, it was believed that the cytoplasm intervening between the discrete organelles and deposits was unstructured. This belief was reinforced by the use of homogenization and centrifugation of the homogenates to yield fractions consisting of recognizable membrane-bound organelles. The final supernatant produced by this process, after the separation of organelles, is called the **cytosol.** The cytosol constitutes about half the total volume of the cell. Homogenization of cells disrupts a delicate **microtrabecular lattice** that incorporates microfilaments of actin, microtubules, intermediate filaments, enzymes,

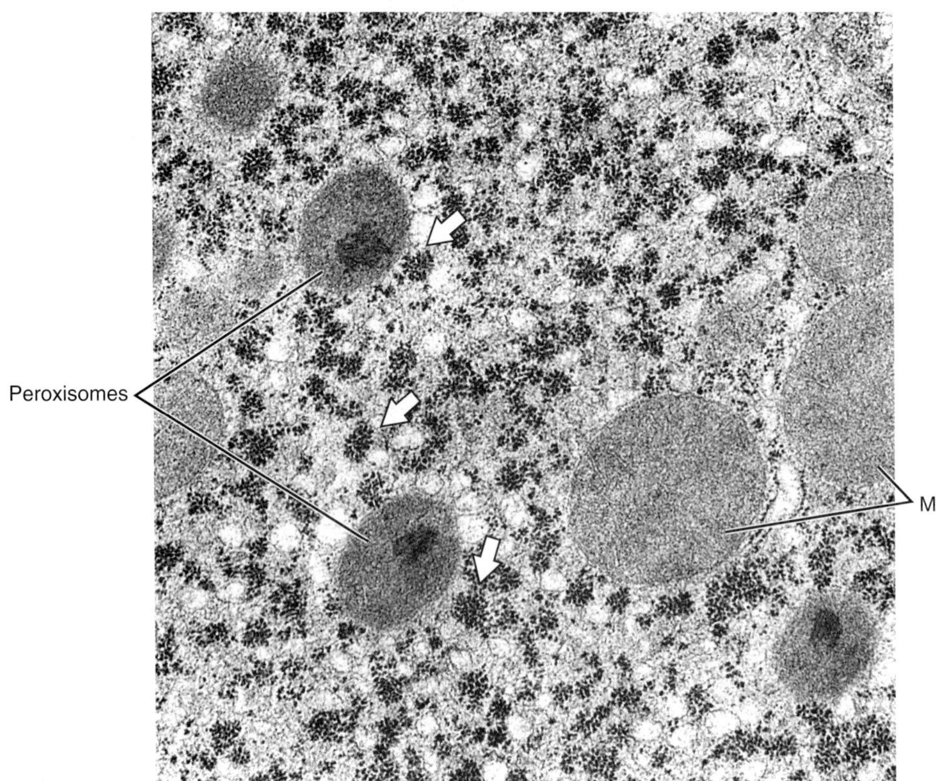

Figure 2–39. Electron micrograph of a section of a liver cell showing glycogen deposits as accumulations of electron-dense particles (arrows). The dark structures with a dense core are peroxisomes. Mitochondria (M) are also shown. ×30,000.

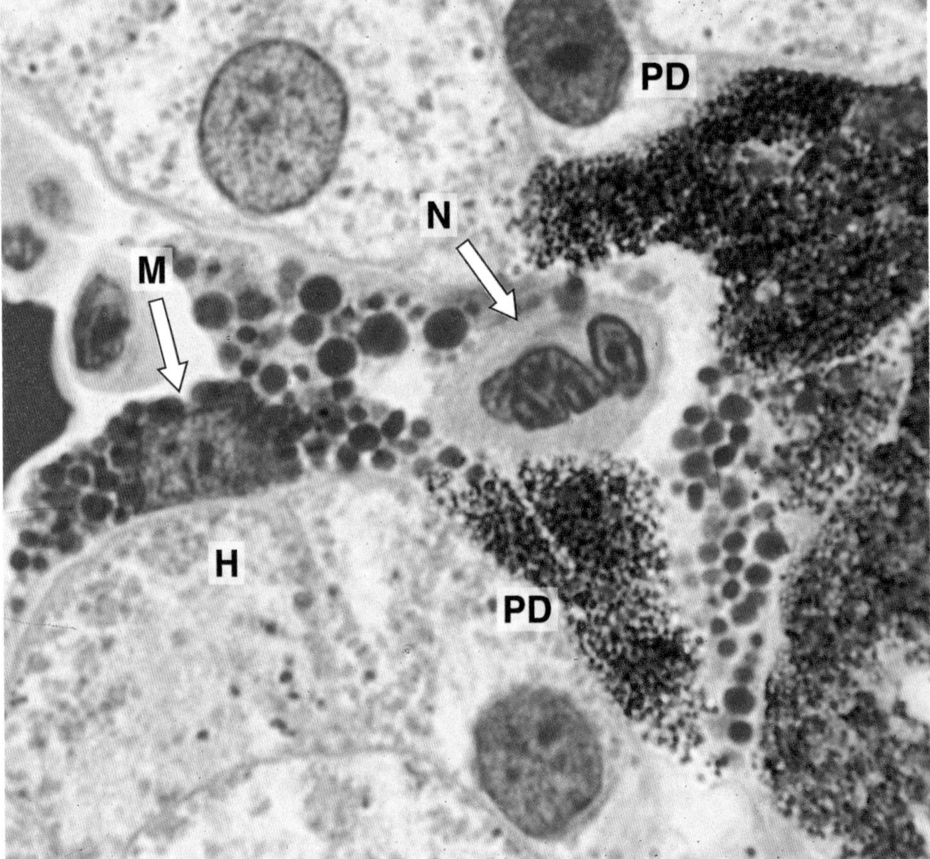

Figure 2–40. Section of amphibian liver shows cells with pigment deposit (PD) in the cytoplasm, a macrophage (M), hepatocytes (H), and a neutrophil leukocyte (N). In this resin-embedded material it is possible to see mitochondria (pale red) and lysosomes (blue) in the cytoplasm of the hepatocytes. Only with resin embedding is it possible to obtain such information. Giemsa stain. Medium magnification.

Table 2–5. Some human and animal diseases related to altered cellular components.

Cell Component Involved	Disease	Molecular Defect	Morphologic Change	Clinical Consequence
Mitochondrion	Mitochondrial cytopathy	Defect of oxidative phosphorylation	Increase in size and number of muscle mitochondria	High basal metabolism without hyperthyroidism
Microtubule	Immotile cilia syndrome	Lack of dynein in cilia and flagella	Lack of arms of the doublet microtubules	Immotile cilia and flagella with male sterility and chronic respiratory infection
	Mouse (*Acomys*) diabetes	Reduction of tubulin in pancreatic β cells	Reduction of microtubules in β cells	High blood sugar content (diabetes)
Lysosome	Metachromatic leukodystrophy	Lack of lysosomal sulfatase	Accumulation of lipid (cerebroside) in tissues	Motor and mental impairment
	Hurler disease	Lack of lysosomal α-L-iduronidase	Accumulation of dermatan sulfate in tissues	Growth and mental retardation
Golgi complex	I-cell disease	Phosphotransferase deficiency	Inclusion-particle storage in several cells	Psychomotor retardation, bone abnormalities

and other soluble constituents into a structured cytosol. The cytosol coordinates the intracellular movements of organelles and provides an explanation for the viscosity of the cytoplasm. Soluble (not membrane-bound) enzymes, such as those of the glycolytic pathway, for example, function more efficiently when organized in a sequence instead of having to rely on random collisions with their substrates. The cytosol provides a framework for this organization. It contains thousands of enzymes that produce building blocks for larger molecules and break down small molecules to liberate energy. All machinery to synthesize proteins (rRNA, mRNA, transfer RNA, enzymes, and other factors) is contained in the cytosol.

Cell Components & Diseases

MEDICAL APPLICATION

Many diseases are related to molecular alterations in specific cell components. In several of these diseases, structural changes can be detected by light or electron microscopy or by cytochemical techniques. Table 2–5 lists some of these diseases and emphasizes the importance of understanding the many cell components in pathobiology.

REFERENCES

Afzelius BA, Eliasson R: Flagellar mutants in man: on the heterogeneity of the immotile-cilia syndrome. J Ultrastruct Res 1979;69:43.

Aridor M, Balch WE: Integration of endoplasmic reticulum signaling in health and disease. Nat Med 1999;5:745.

Barrit GJ: *Communication Within Animal Cells.* Oxford Univ Press, 1992.

Becker WM et al: *The World of the Cell,* 4th ed. Benjamin/Cummings, 2000.

Bretscher MS: The molecules of the cell membrane. Sci Am 1985;253:100.

Brinkley BR: Microtubule organizing centers. Annu Rev Cell Biol 1985;1:145.

Brown MS et al: Recycling receptors: the round-trip itinerary of migrant membrane proteins. Cell 1983;32:663.

Cooper, GM: *The Cell: A Molecular Approach.* ASM Press/Sinauer Associates, Inc., 1997.

DeDuve C: *A Guided Tour of the Living Cell.* Freeman, 1984.

DeDuve C: Microbodies in the living cell. Sci Am 1983;248:74.

Dustin P: *Microtubules,* 2nd ed. Springer-Verlag, 1984.

Farquhar MG: Progress in unraveling pathways of Golgi traffic. Annu Rev Cell Biol 1985;1:447.

Fawcett D: *The Cell,* 2nd ed. Saunders, 1981.

Krstíc RV: *Ultrastructure of the Mammalian Cell.* Springer-Verlag, 1979.

Mitchison TJ, Cramer LP: Actin-based cell motility and cell locomotion. Cell 1996;84:371.

Osborn M, Weber K: Intermediate filaments: cell-type-specific markers in differentiation and pathology. Cell 1982;31:303.

Pfeffer SR, Rothman JE: Biosynthetic protein transport and sorting in the endoplasmic reticulum. Annu Rev Biochem 1987;56:829.

Rothman J: The compartmental organization of the Golgi apparatus. Sci Am 1985;253:74.

Simons K, Ikonen E: How cells handle cholesterol. Science 2000;290:1721.

Tzagoloff A: *Mitochondria.* Plenum, 1982.

Weber K, Osborn M: The molecules of the cell matrix. Sci Am 1985;253:110.

The Cell Nucleus

3

The nucleus contains a blueprint for all cell structures and activities, encoded in the DNA of the chromosomes. It also contains the molecular machinery to replicate its DNA and to synthesize and process the 3 types of RNA—ribosomal (rRNA), messenger, and transfer. Mitochondria have a small DNA genome and produce RNAs to be used in this organelle, but the genome is so small that it is not sufficient even for the mitochondrion itself. On the other hand, the nucleus does not produce proteins; the numerous protein molecules needed for the activities of the nucleus are imported from the cytoplasm.

The nucleus frequently appears as a rounded or elongated structure, usually in the center of the cell (Figure 3–1). Its main components are the **nuclear envelope, chromatin** (Figures 3–2 and 3–3), **nucleolus,** and **nuclear matrix.** The size and morphologic features of nuclei in a specific normal tissue tend to be uniform. In contrast, the nuclei in cancer cells have an irregular shape, variable size, and atypical chromatin patterns.

Nuclear Envelope

Electron microscopy shows that the nucleus is surrounded by 2 parallel unit membranes separated by a narrow space (40–70 nm) called the **perinuclear cisterna** (Figures 3–2 and 3–4). Together, the paired membranes and the intervening space make up the nuclear envelope. Closely associated with the internal membrane of the nuclear envelope is a protein structure called the **fibrous lamina** (Figure 3–4), which helps to stabilize the nuclear envelope. The fibrous lamina is composed of 3 main proteins called **lamins A, B,** and **C.** In nondividing cells, chromosomes are associated with the fibrous lamina (Figure 3–5). The pattern of association is regular from cell to cell within a tissue, supporting the conclusion that chromosomes have a definite localization within the nucleus. Polyribosomes are attached to the outer membrane, showing that the nuclear envelope is a part of the endoplasmic reticulum. Proteins synthesized in the polyribosomes attached to the nuclear envelope are temporarily segregated in the perinuclear cisterna. At sites where the inner and outer membranes of the nuclear envelope fuse, there are gaps, the **nuclear pores** (Figures 3–6 and 3–7), that provide controlled pathways between the nucleus and the cytoplasm. The pores are not open but show an octagonal **pore complex** made of more than 100 proteins (Figure 3–8). Because the nuclear envelope is impermeable to ions and molecules of all sizes, the exchange of substances between the nucleus and the cytoplasm is

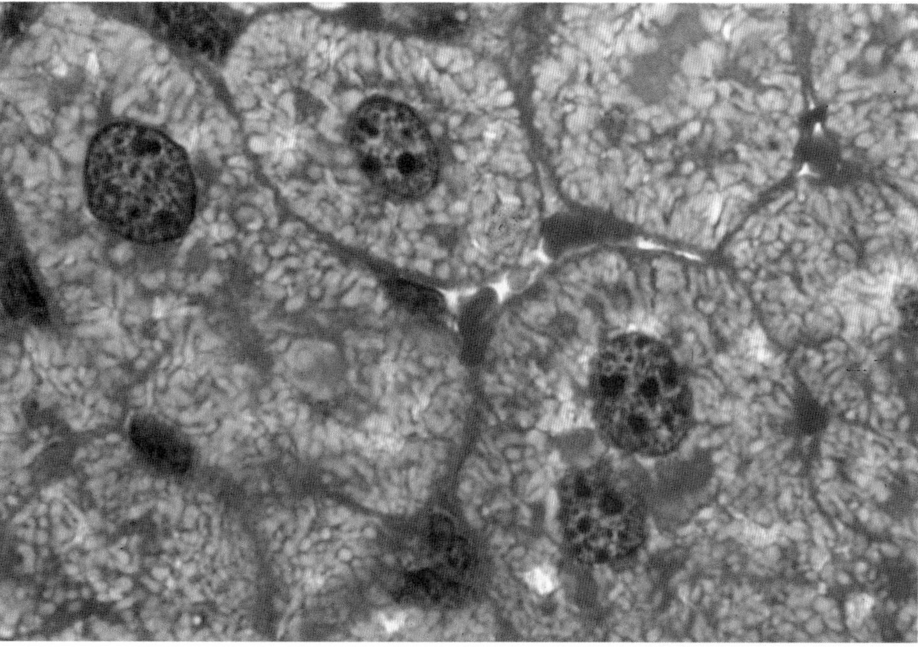

Figure 3–1. Liver cells (hepatocytes). Several dark-stained nuclei are shown. Note the apparent nuclear membrane consisting mainly of a superficial condensation of chromatin. Several nucleoli are seen inside the nuclei, suggesting intense protein synthesis. One hepatocyte contains 2 nuclei. Pararosaniline–toluidine blue (PT) stain. Medium magnification.

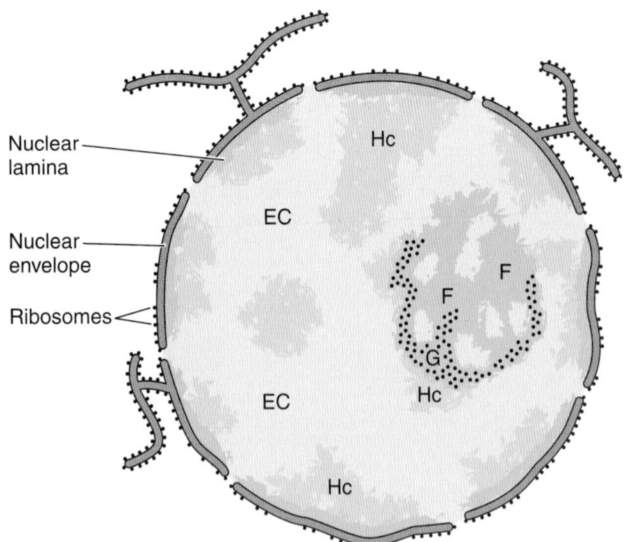

Figure 3–2. Schematic representation of a cell nucleus. The nuclear envelope is made of 2 membranes of the endoplasmic reticulum, enclosing a perinuclear cisterna. Where the two membranes fuse, they form nuclear pores. Ribosomes are attached to the outer nuclear membrane. Heterochromatin clumps are associated with the nuclear lamina, whereas the euchromatin (EC) appears dispersed in the interior of the nucleus. In the nucleolus, note the associated chromatin (arrows), heterochromatin (Hc), the pars granulosa (G), and the pars fibrosa (F).

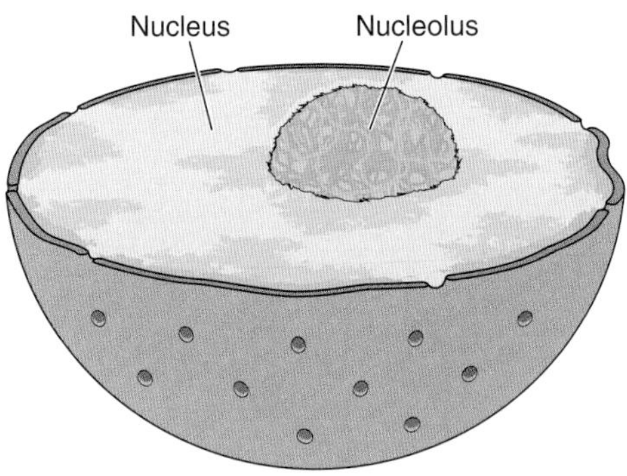

Figure 3–3. Three-dimensional representation of a cell nucleus to show the distribution of the nuclear pores, the heterochromatin (dark regions), the euchromatin (light regions), and a nucleolus. Note that there is no chromatin closing the pores. The number of nuclear pores varies greatly from cell to cell.

made only through the nuclear pores. Ions and molecules with a diameter up to 9 nm pass freely through the nuclear pore without consuming energy. But molecules and molecular complexes larger than 9 nm are transported by an active process, mediated by receptors, which uses energy from adenosine triphosphate (ATP) and takes place in 2 stages. First, proteins with one or several nuclear signal locations become attached to specific cytosolic proteins, originating a complex, which is temporarily attached to the nuclear pore complex without using energy. In the

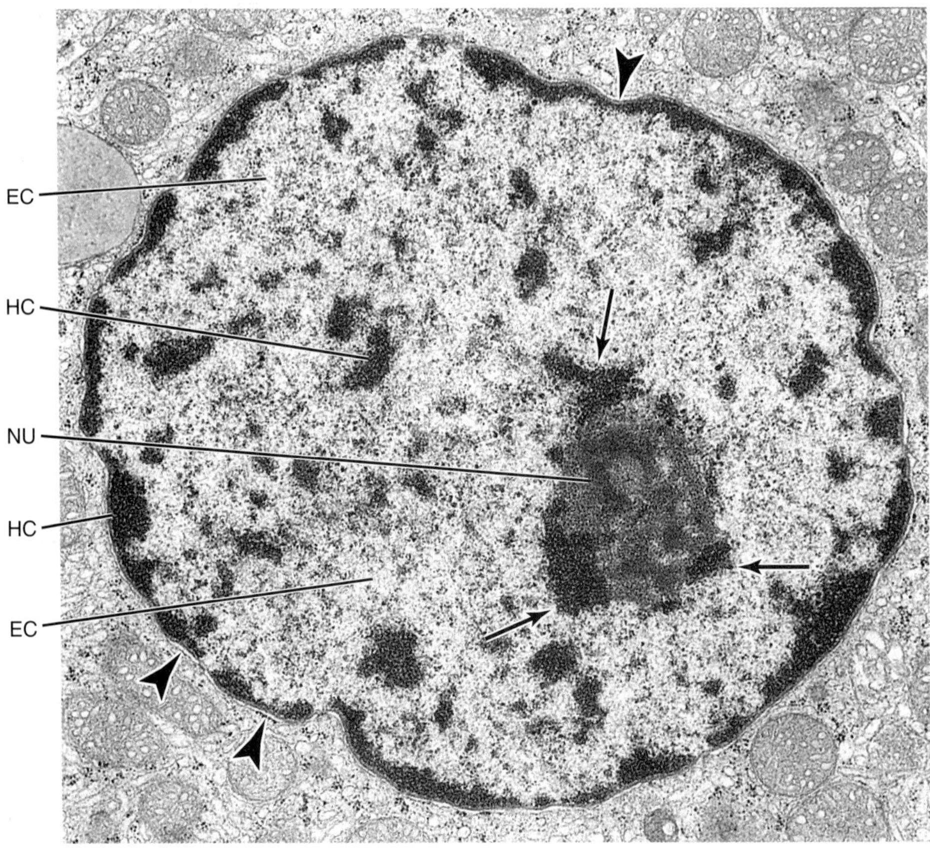

Figure 3–4. Electron micrograph of a nucleus, showing the heterochromatin (HC) and euchromatin (EC). Unlabeled arrows indicate the nucleolus-associated chromatin around the nucleolus (NU). Arrowheads indicate the perinuclear cisterna. Underneath the cisterna is a layer of heterochromatin, the main component of the so-called **nuclear membrane** seen under the light microscope. ×26,000.

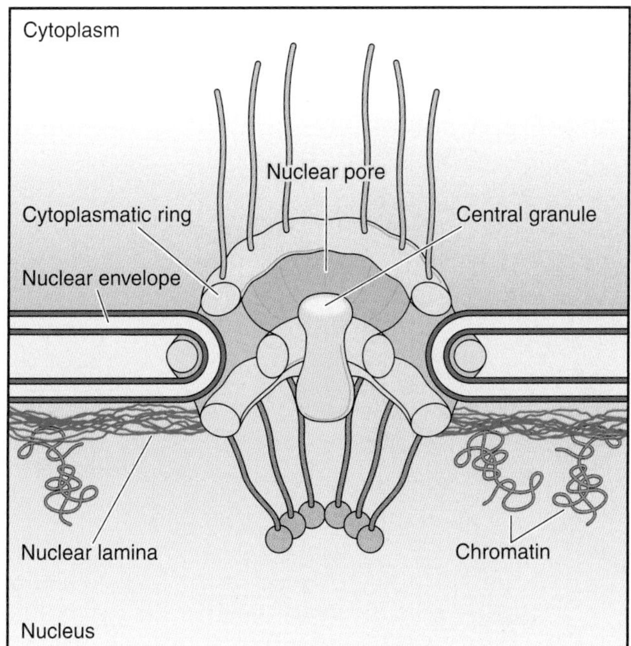

Figure 3–5. Illustration to show the structure, the localization, and the relationship of the nuclear lamina with chromosomes. The drawing also shows that the nuclear pore complex is made of 2 protein rings in an octagonal organization. From the cytoplasmic ring, long filaments penetrate the cytosol, and from the intranuclear ring arise filaments that constitute a basketlike structure. The presence of the central cylindrical granule in the nuclear pore is not universally accepted.

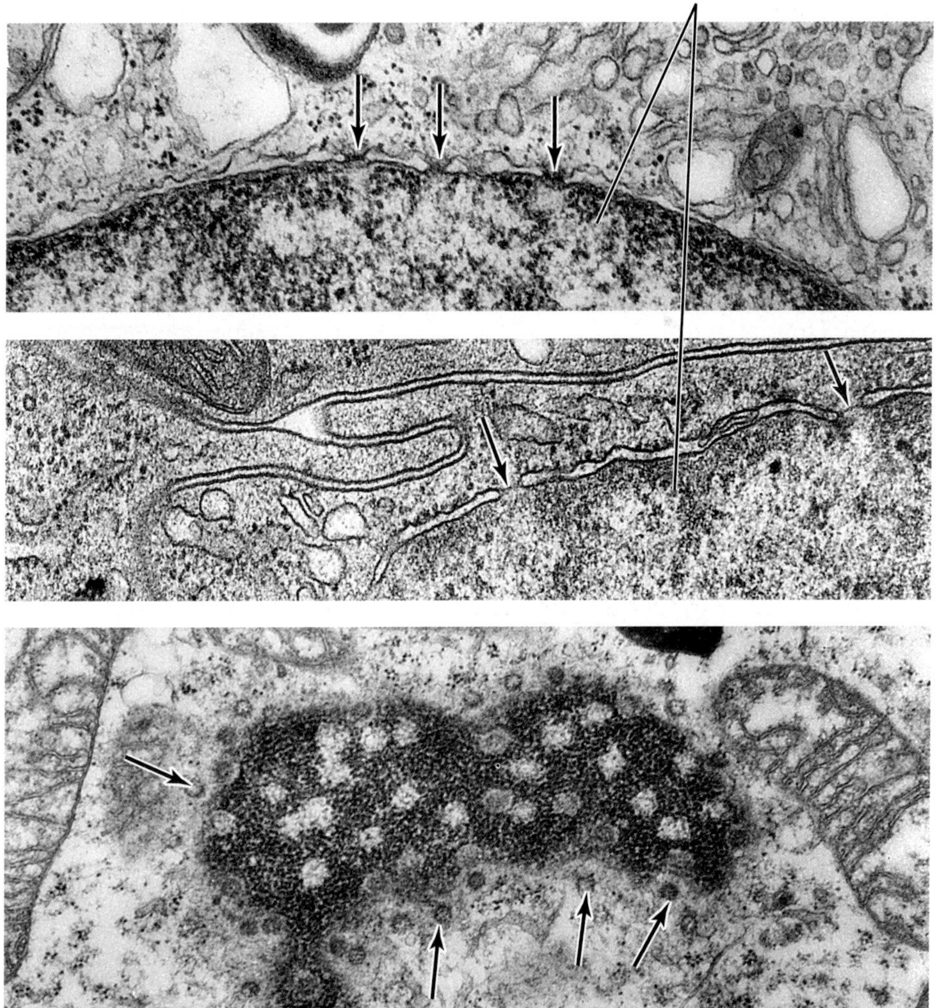

Figure 3–6. Electron micrographs of nuclei showing their envelopes composed of 2 membranes and the nuclear pores (arrows). The two upper pictures are of transverse sections; the bottom is of a tangential section. Chromatin, frequently condensed below the nuclear envelope, is not usually seen in the pore regions. ×80,000.

Figure 3–7. Electron micrograph obtained by cryofracture of a rat intestine cell, showing the two components of the nuclear envelope and the nuclear pores. (Courtesy of P Pinto da Silva.)

second stage, proteins with nuclear signal locations are transferred to the nucleus, using energy from ATP, and the cytosolic protein remains in the cytoplasm. Probably, at least part of the ATP energy is utilized to open the nuclear pore complex to make possible the passage of large molecules. Less is known about the transfer of molecules and molecular complexes, some as large as ribosome subunits, from the nucleus to the cytoplasm.

Chromatin

Chromatin, in nondividing nuclei, is in fact the chromosomes in a different degree of uncoiling. According to the degree of chromosome condensation, 2 types of chromatin can be distinguished with both the light and electron microscopes (Figures 3–2 and 3–4). **Heterochromatin** (Gr. *heteros,* other, + *chroma,* color), which is electron-dense, appears as coarse granules in the electron microscope and as basophilic clumps in the light microscope. **Euchromatin** is the less coiled portion of the chromosomes, visible as a finely dispersed granular material in the electron microscope and as lightly stained basophilic areas in the light microscope. The proportion of heterochromatin to euchromatin accounts for the light-to-dark appearance of nuclei in tissue sections as seen in light and electron microscopes. The intensity of nuclear staining of the chromatin is frequently used to distinguish and identify different tissues and cell types in the light microscope.

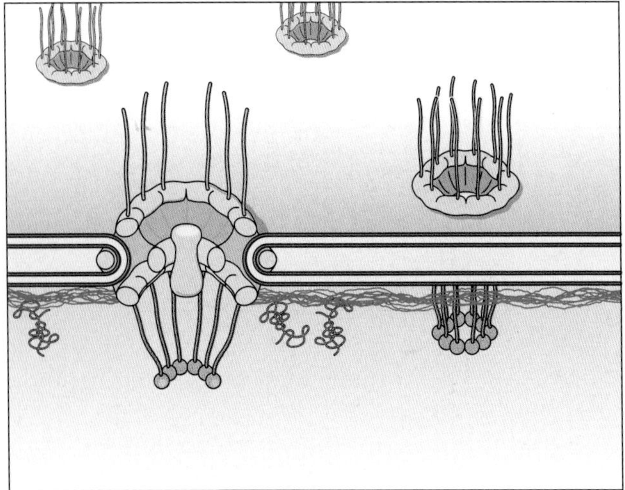

Figure 3–8. Simplified representation of 2 nuclear pore complexes. In this model, the final nuclear portion is seen to be a more continuous structure, in the shape of a ring.

Chromatin is composed mainly of coiled strands of DNA bound to basic proteins (histones); its structure is schematically presented in Figure 3–5. The basic structural unit of chromatin is the nucleosome (Figure 3–9), which consists of a core of 4 types of histones: 2 copies each of histones H2A, H2B, H3, and H4, around which are wrapped 166 DNA base pairs. An additional 48–base pair segment forms a link between adjacent nucleosomes, and another type of histone (H1 or H5) is bound to this DNA. This organization of chromatin has been referred to as "beads-on-a-string." Nonhistone proteins are also associated with chromatin, but their arrangement is less well understood.

The next higher order of organization of chromatin is the 30-nm fiber (Figure 3–10). In this structure, nucleosomes become coiled around an axis, with 6 nucleosomes per turn, to form the 30-nm chromatin fiber. There are higher orders of coiling, especially in the condensation of chromatin during mitosis and meiosis.

The chromatin pattern of a nucleus has been considered a guide to the cell's activity. In general, cells with light nuclei are more active than those with condensed, dark nuclei. In light-stained nuclei (with few heterochromatin clumps), more DNA

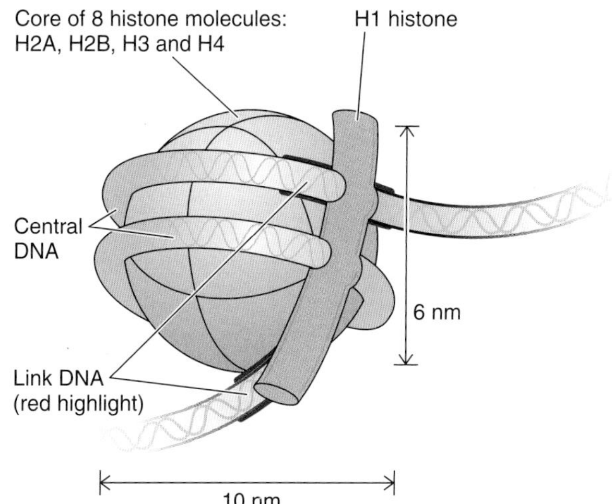

Figure 3–9. Schematic representation of a nucleosome. This structure consists of a core of 4 types of histones (2 copies of each)—H2A, H2B, H3, and H4—and one molecule of H1 or H5 located outside the DNA filament.

surface is available for the transcription of genetic information. In dark-stained nuclei (rich in heterochromatin), the coiling of DNA makes less surface available.

Careful study of the chromatin of mammalian cell nuclei reveals a heterochromatin mass that is frequently observed in female cells but not in male cells. This chromatin clump is the **sex chromatin** and is one of the two X chromosomes present in female cells. The X chromosome that constitutes the sex chromatin remains tightly coiled and visible, whereas the other X chromosome is uncoiled and not visible. Evidence suggests that the sex chromatin is genetically inactive. The male has one X chromosome and one Y chromosome as sex determinants; the X chromosome is uncoiled, and therefore no sex chromatin is visible. In human epithelial cells, sex chromatin appears as a small granule attached to the nuclear envelope. The cells lining the internal surface of the cheek are frequently used to study sex chromatin. Blood smears are also often used, in which case the sex chromatin appears as a drumsticklike appendage to the nuclei of the neutrophilic leukocytes (Figure 3–11).

MEDICAL APPLICATION

The study of sex chromatin discloses the genetic sex in patients whose external sex organs do not permit assignment of gender, as in hermaphroditism and pseudohermaphroditism. Sex chromatin helps the study of other anomalies involving the sex chromosomes—eg, Klinefelter syndrome, in which testicular abnormalities, azoospermia (absence of spermatozoa), and other symptoms are associated with the presence of XXY chromosomes.

The study of chromosomes made considerable progress after the development of methods that induce cells to divide, arrest mitotic cells during metaphase, and cause cell rupture. Mitosis can be induced by phytohemagglutinin (in cell cultures) and can be arrested in metaphase by colchicine. Cells are immersed in a hypotonic solution, which causes swelling, after which cells are flattened and broken between a glass slide and a coverslip.

The pattern of chromosomes obtained in a human cell after staining is illustrated in Figure 3–12. In addition to the X and Y sex chromosomes, the remaining chromosomes are customarily grouped according to their size and morphologic characteristics, in 22 successively numbered pairs.

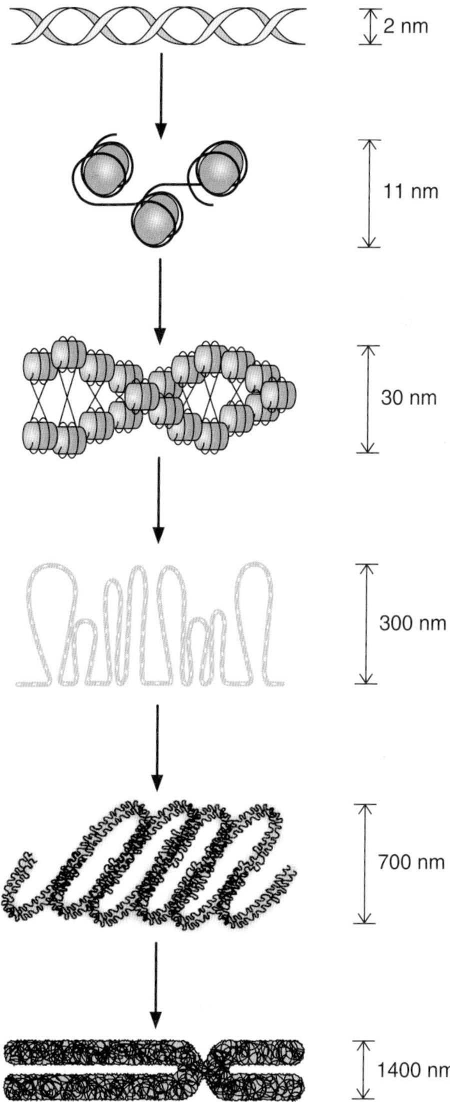

Figure 3–10. The orders of chromatin packing believed to exist in the metaphase chromosome. Starting at the top, the 2-nm DNA double helix is shown; next is the association of DNA with histones to form filaments of nucleosomes of 11 nm and 30 nm. Through further condensation, filaments with diameters of 300 nm and 700 nm are formed. Finally, the bottom drawing shows a metaphase chromosome, which exhibits the maximum packing of DNA.

MEDICAL APPLICATION

The number and characteristics of chromosomes encountered in an individual are known as the ***karyotype*** *(Figure 3–12). The study of karyotypes has revealed chromosomal alterations associated with tumors, leukemias, and several types of genetic diseases.*

The development of techniques that reveal segmentation of chromosomes in transverse, differentially stained bands permitted a more precise identification of individual chromosomes and the study of gene deletions and translocations. These techniques are based mainly on the study of chromosomes previously treated with saline or enzyme solution and stained with fluorescent dyes or Giemsa's blood-staining technique. In situ hybridization is also a valuable technique for localizing DNA sequences (genes) in chromosomes.

Nucleolus

The nucleolus is a spherical structure (Figure 3–13), up to 1 mm in diameter, that is rich in rRNA and protein. It is usually basophilic when stained with hematoxylin and eosin. As seen with the electron microscope, the nucleolus consists of 3 distinct components: (1) From one to several pale-staining regions are composed of **nucleolar organizer DNA**—sequences of bases that code for rRNA (Figure 3–14). In the human genome, 5 pairs of chromosomes contain nucleolar organizers. (2) Closely associated with the nucleolar organizers are densely packed 5- to 10-nm ribonucleoprotein fibers composing the **pars fibrosa,** which consists of primary transcripts of rRNA genes. (3) The **pars granulosa** consists of 15- to 20-nm granules (maturing ribosomes; see Figure 3–14). Proteins, synthesized in the cytoplasm, become associated with rRNAs in the nucleolus; ribosome subunits then migrate into the cytoplasm. Heterochromatin is often attached to the nucleolus (**nucleolus-associated chromatin**), but the functional significance of the association is not known. The rRNAs are synthesized and modified inside the nucleus. In the nucleolus they receive proteins and are organized into small and large ribosomal subunits, which migrate to the cytoplasm through the nuclear pores.

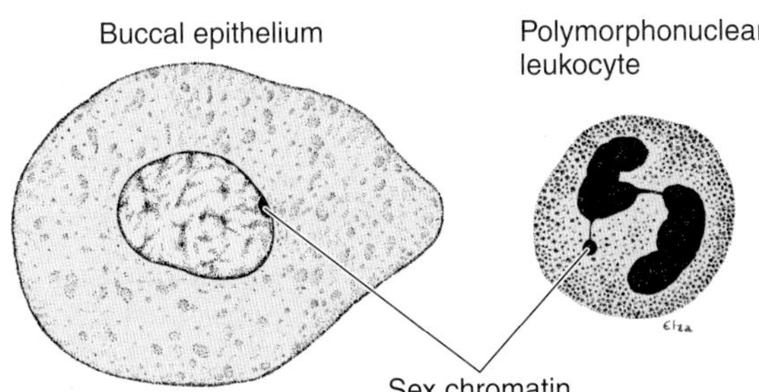

Figure 3–11. Morphologic features of sex chromatin in human female oral (buccal) epithelium and in a polymorphonuclear leukocyte. In the epithelium, sex chromatin appears as a small, dense granule adhering to the nuclear envelope. In the leukocyte, it has a drumstick shape.

MEDICAL APPLICATION

Large nucleoli are encountered in embryonic cells during their proliferation, in cells that are actively synthesizing proteins, and in rapidly growing malignant tumors. The nucleolus disperses during the prophase of cell division but reappears in the telophase stage of mitosis.

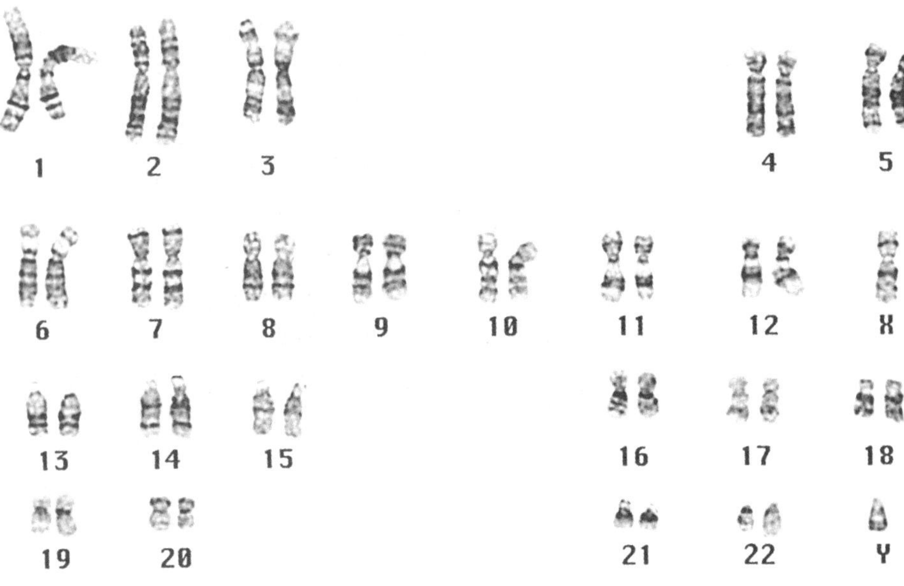

Figure 3–12. Human karyotype preparation made by means of a banding technique. Each chromosome has a particular pattern of banding that facilitates its identification and also the relationship of the banding pattern to genetic anomalies. The chromosomes are grouped in numbered pairs according to their morphologic characteristics.

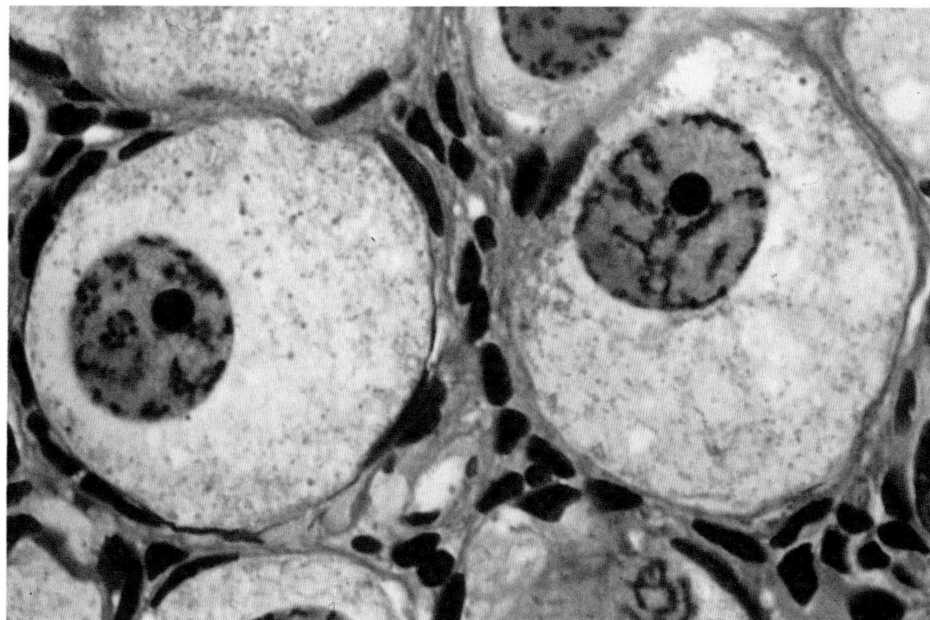

Figure 3–13. Photomicrograph of 2 primary oocytes, each one with its pale cytoplasm and round, dark-stained nucleus. In each nucleus the nucleolus, very darkly stained, is clearly seen. The sectioned chromosomes are also seen, because they are condensed. These cells stopped at the first meiotic division. Meiosis will proceed just before ovulation (extrusion of the oocyte from the ovary; see Chapter 23).

Nuclear Matrix

The nuclear matrix is the component that fills the space between the chromatin and the nucleoli in the nucleus. It is composed mainly of proteins (some of which have enzymatic activity), metabolites, and ions. When its nucleic acids and other soluble components are removed, a continuous fibrillar structure remains, forming the **nucleoskeleton.** The fibrous lamina of the nuclear envelope is part of the nuclear matrix. The nucleoskeleton probably contributes to the formation of a protein base to which DNA loops are bound.

CELL DIVISION

Cell division, or mitosis (Gr. *mitos,* a thread), can be observed with the light microscope. During this process, the parent cell divides, and each of the daughter cells receives a chromosomal set identical to that of the parent cell. Essentially, a longitudinal duplication of the chromosomes takes place, and these chromosomes are distributed to the daughter cells. The phase between 2 mitoses is called **interphase,** during which the nucleus appears as it is normally observed in microscope preparations. The process of mitosis is subdivided into phases to facilitate its study (Figures 3–15, 3–16, and 3–17).

The **prophase** of mitosis is characterized by the gradual coiling of nuclear chromatin (uncoiled chromosomes), giving rise to several individual rod- or hairpin-shaped bodies (coiled chromosomes) that stain intensely. At the end of prophase, the nuclear envelope is broken by phosphorylation (addition of PO_4^{3-}) of the nuclear lamina proteins, originating vesicles that remain in the cytoplasm. The centrosomes with their centrioles separate, and a centrosome migrates to each pole of the cell. The duplication of the centrosomes and centrioles starts in the interphase, before mitosis. Simultaneously with the centrosome migration, the microtubules of the mitotic spindle appear between the two centrosomes, and the nucleolus disintegrates.

During **metaphase,** chromosomes migrate by the activity of microtubules to the equatorial plane of the cell, where each divides longitudinally to form 2 chromosomes called sister chromatids. The chromatids attach to the microtubules of the mitotic spindle (Figures 3–18 and 3–19) at an electron-dense,

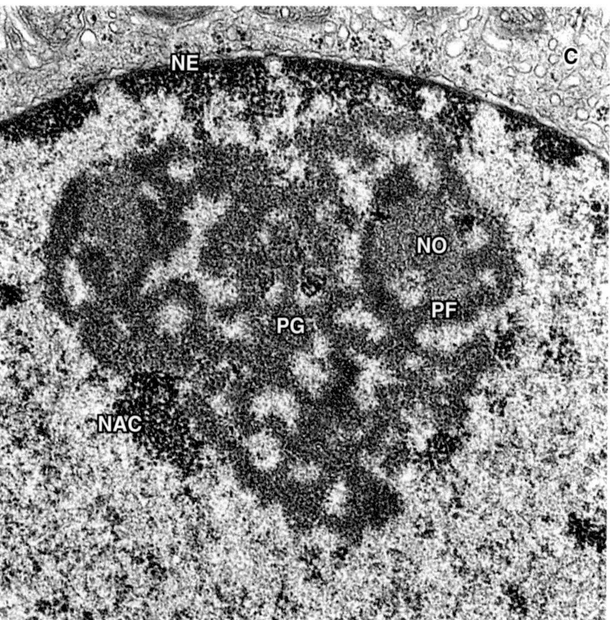

Figure 3–14. Electron micrograph of a nucleolus. The nucleolar organizer DNA (NO), pars fibrosa (PF), pars granulosa (PG), nucleolus-associated chromatin (NAC), nuclear envelope (NE), and cytoplasm (C) are shown.

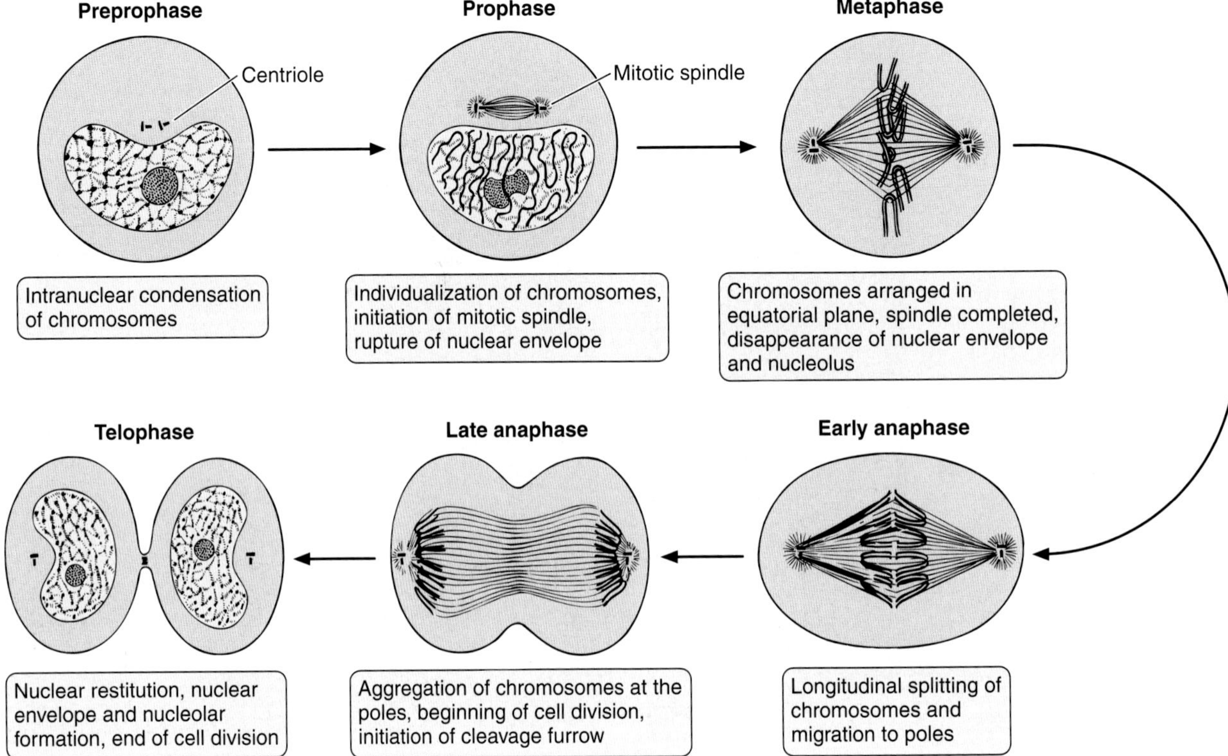

Figure 3–15. Phases of mitosis.

DNA-protein plaque, the **kinetochore** (Gr. *kinetos,* moving, + *chora,* central region), located close to the **centromere** (Gr. *kentron,* center, + *meros,* part) of each chromatid.

In **anaphase,** the sister chromatids separate from each other and migrate toward the opposite poles of the cell, pulled by microtubules. Throughout this process, the centromeres move away from the center, pulling along the remainder of the chromosome. The centromere is the constricted region of a mitotic chromosome that holds the two sister chromatids together until the beginning of anaphase.

Telophase is characterized by the reappearance of nuclei in the daughter cells. The chromosomes revert to their semidispersed state, and the nucleoli, chromatin, and nuclear envelope reappear. While these nuclear alterations are taking place, a constriction develops at the equatorial plane of the parent cell and progresses until the cytoplasm and its organelles are divided in two. This constriction is produced by microfilaments of actin associated with myosin that accumulate in a beltlike shape beneath the cell membrane.

Most tissues undergo constant cell turnover because of continuous cell division and the ongoing death of cells. Nerve tissue and cardiac muscle cells are exceptions, since they do not multiply postnatally and therefore cannot regenerate. The turnover rate of cells varies greatly from one tissue to another—rapid in the epithelium of the digestive tract and the epidermis, slow in the pancreas and the thyroid gland.

THE CELL CYCLE

Mitosis is the visible manifestation of cell division, but other processes, not so easily observed with the light microscope, play a fundamental role in cell multiplication. Principal among these is the phase in which DNA replicates. This process can be analyzed by introducing labeled radioactive DNA precursors (eg, ^{3}H-thymidine) into the cell and tracing them by means of biochemical and autoradiographic methods. DNA replication has been shown to occur during **interphase,** when no visible phenomena of cell division can be seen with the microscope. This alternation between mitosis and interphase, known as the **cell cycle,** occurs in all tissues with cell turnover. A careful study of the cell cycle reveals that it can be divided into 2 stages: mitosis, consisting of the 4 phases already described (prophase, metaphase, anaphase, and telophase), and interphase (Figures 3–20 and 3–21).

Interphase is itself divided into 3 phases: G_1 (presynthesis), S (DNA synthesis), and G_2 (post–DNA duplication). The sequence of these phases and the approximate times involved are illustrated in Figures 3–20 and 3–21. The S phase is characterized by the synthesis of DNA and the beginning of the duplications of the centrosomes with their centrioles. During the G_1 phase, there is an intense synthesis of RNA and proteins, including proteins that control the cell cycle, and the cell volume, previously reduced to one-half by mitosis, is restored to its nor-

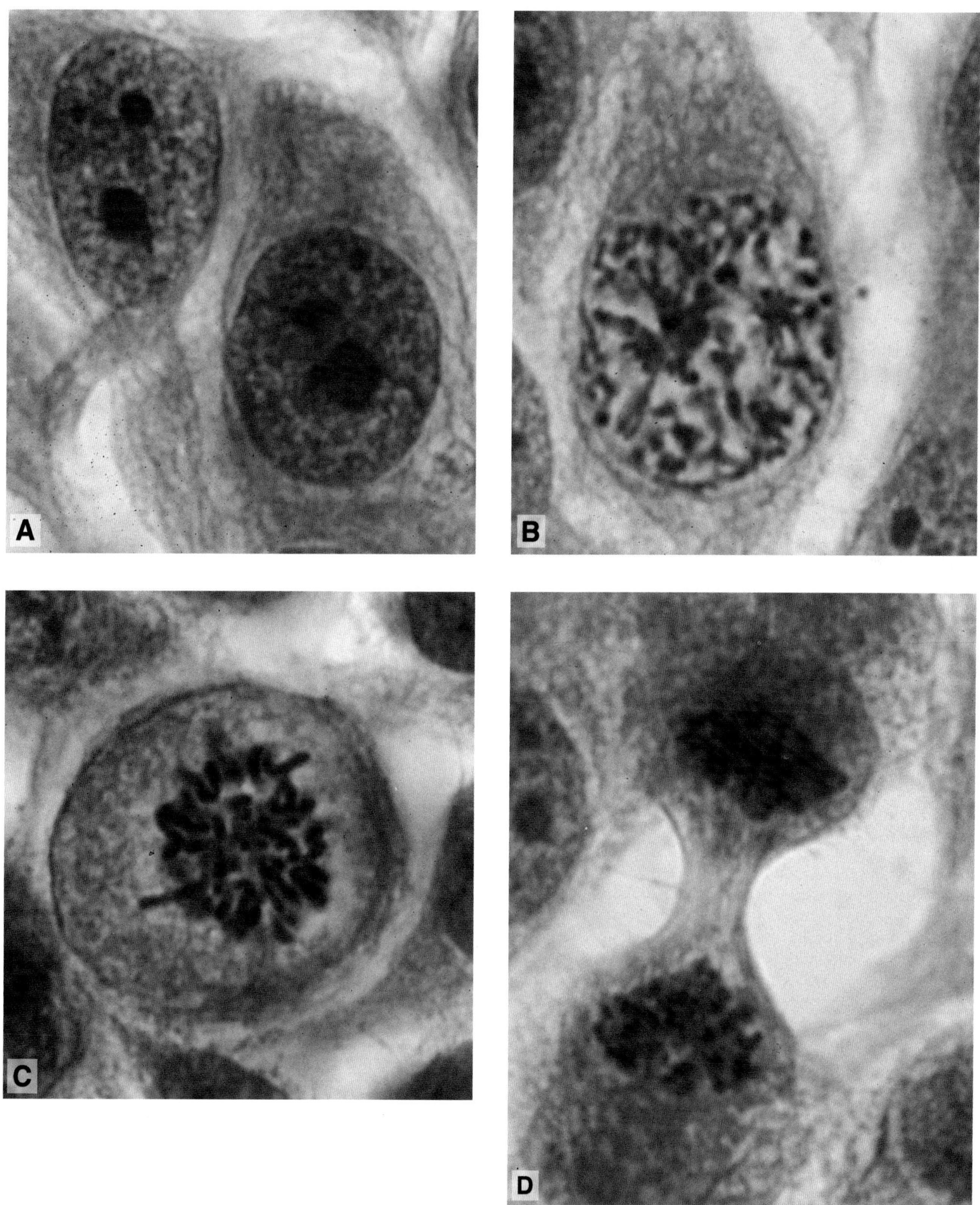

Figure 3–16. Photomicrograph of cultured cells to show cell division. Picrosirius-hematoxylin stain. Medium magnification. **A:** Interphase nuclei. Note the chromatin and nucleoli inside each nucleus. **B:** Prophase. No distinct nuclear envelope, no nucleoli. Condensed chromosomes. **C:** Metaphase. The chromosomes are located in a plate at the cell equator. **D:** Late anaphase. The chromosomes are located in both cell poles, to distribute the DNA equally between the daughter cells.

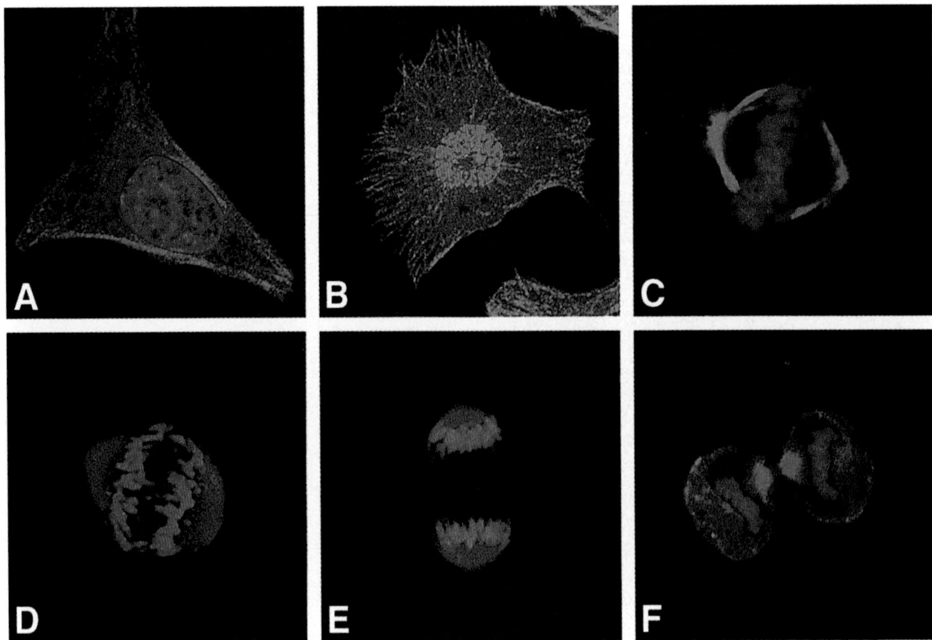

Figure 3–17. Images obtained with a confocal laser scanning microscope from cultured cells. An interphase nucleus and several nuclei are in several phases of mitosis. DNA appears red, and microtubules in the cytoplasm are blue. Medium magnification. **A:** Interphase. A nondividing cell. **B:** Prophase. The blue structure over the nucleus is the centrosome. Note that the chromosomes are becoming visible because of their condensation. The cytoplasm is acquiring a round shape typical of cells in mitosis. **C:** Metaphase. The chromosomes are organized in an equatorial plane. **D:** Anaphase. The chromosomes are pulled to the cell poles through the activity of microtubules. **E:** Early telophase. The two sets of chromosomes have arrived at the cell poles to originate the two daughter cells, which will contain sets of chromosomes similar to those in the mother cell. **F:** Telophase. The cytoplasm is being divided by a constriction in the cell equator. Note that the daughter cells are round and smaller than the mother cell. Soon they will increase in size and become elongated. (Courtesy of R Manelli-Oliveira, R Cabado, and G Machado-Santelli.)

mal size. In cells that are not continuously dividing, the cell cycle activities may be temporarily or permanently suspended. Cells in such a state (eg, muscle, nerve) are referred to as being in G_0 phase.

Regulation of the mammalian cell cycle is complex. It is known that cultured cells deprived of serum stop proliferating and arrest in G_0. The essential components provided by serum are highly specific proteins called **growth factors,** which are required only in very low concentrations.

MEDICAL APPLICATION

Some growth factors are being used in medicine. One example is erythropoietin, which stimulates proliferation, differentiation, and survival of red blood cell precursors in the bone marrow.

The cell cycle is also regulated by a variety of signals that inhibit progression through the cycle. DNA damage arrests the cell cycle not only in G_2 but also at a checkpoint in G_1 (Figure 3–21). G_1 arrest may permit repair of the damage to take place before the cell enters S phase, where the damaged DNA would be replicated. In mammalian cells, arrest at the G_1 checkpoint is mediated by the action of a protein known as p53. The gene encoding p53 is often mutated in human cancers, thus reducing the cell's ability to repair damaged DNA. Inheritance of damaged DNA by daughter cells results in an increased frequency of mutations and general instability of the genome, which may contribute to the development of cancer.

Processes that occur during the G_2 phase are the accumulation of energy to be used during mitosis, the synthesis of tubulin to be assembled in mitotic microtubules, and the synthesis of chromosomal nonhistone proteins. In G_2 there is also a checkpoint where the cell remains until all DNA synthesized with defects is corrected. In G_2 there is an accumulation of the protein complex **MPF** (maturation promoting factor) that induces the beginning of mitosis, the condensation of the chromosomes, the rupture of the nuclear envelope, and other events related to mitosis.

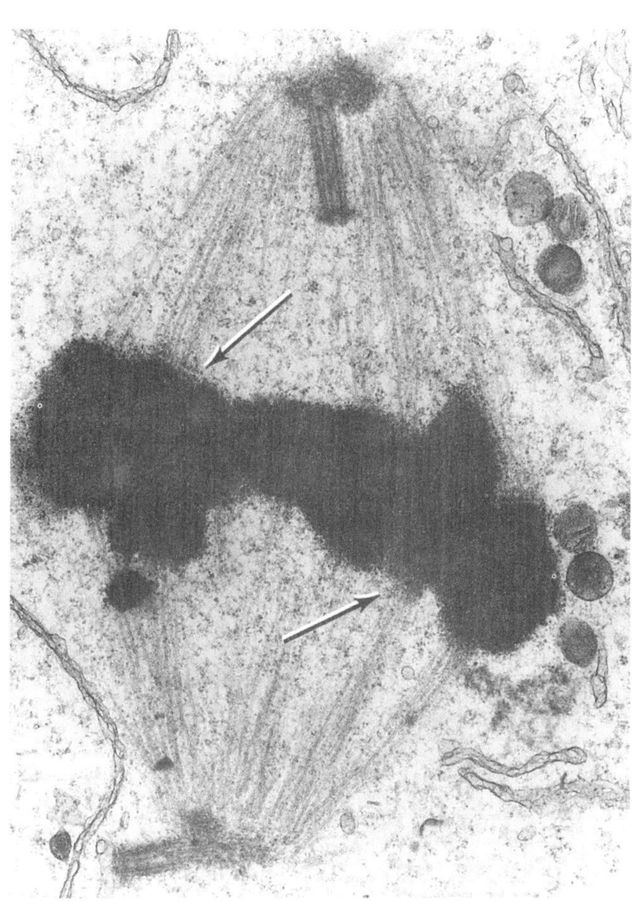

Figure 3–18. Electron micrograph of a section of a rooster spermatocyte in metaphase. The figure shows the two centrioles in each pole, the mitotic spindle formed by microtubules, and the chromosomes in the equatorial plane. The arrows show the insertion of microtubules in the centromeres. Reduced from ×19,000. (Courtesy of R McIntosh.)

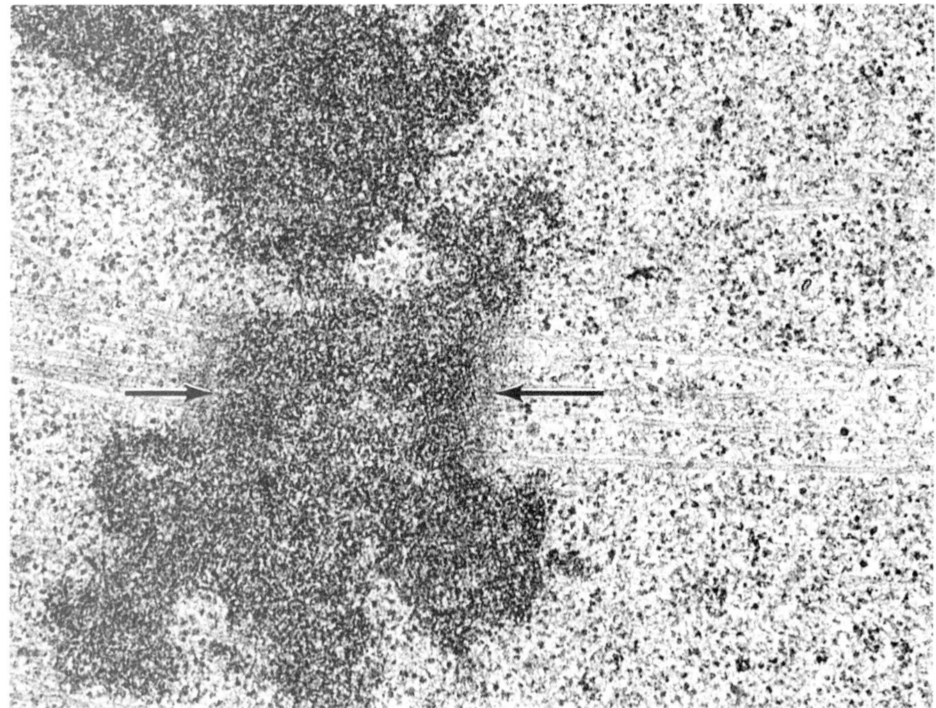

Figure 3–19. Electron micrograph of the metaphase of a human lung cell in tissue culture. Note the insertion of microtubules in the centromeres (arrows) of the densely stained chromosomes. Reduced from ×50,000. (Courtesy of R McIntosh.)

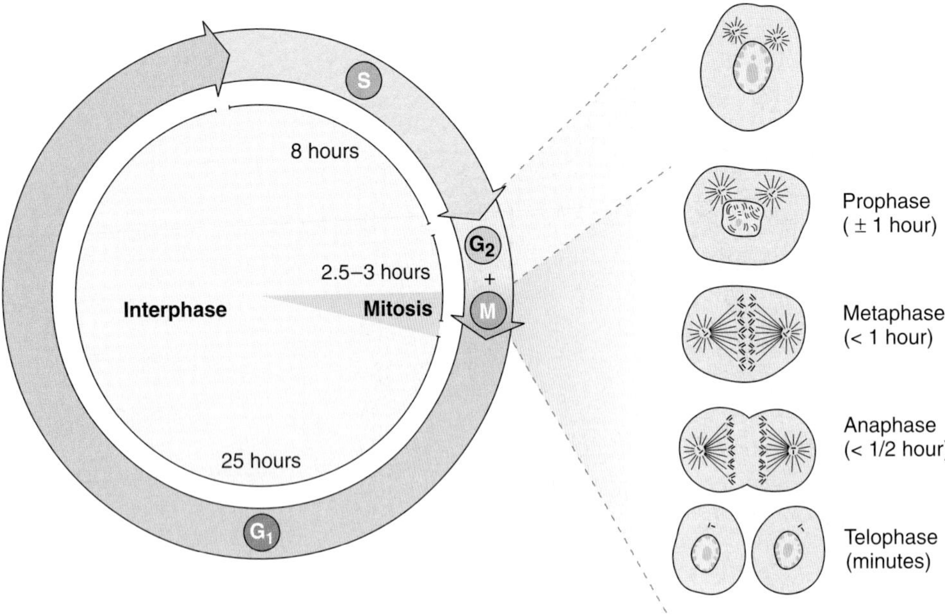

Figure 3–20. Phases of the cell cycle in bone tissue. The G_1 phase (presynthesis) varies in duration, which depends on many factors, including the rate of cell division in the tissue. In bone tissue, G_1 lasts 25 h. The S phase (DNA synthesis) lasts about 8 h. The G_2-plus-mitosis phase lasts 2.5–3 h. (The times indicated are courtesy of RW Young.)

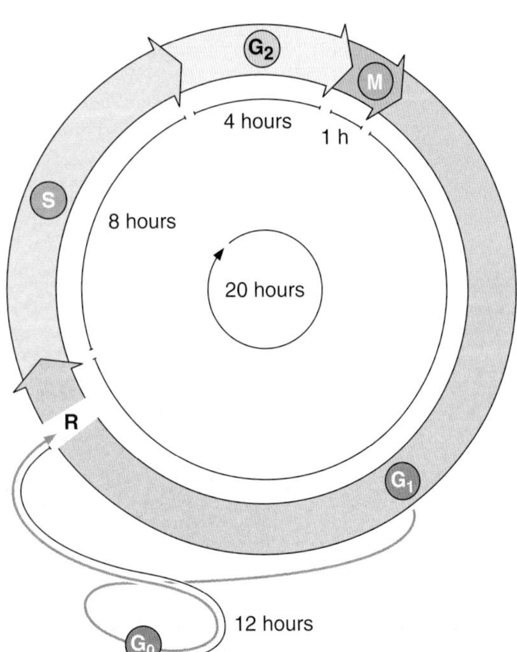

Figure 3–21. The 4 phases of the cell cycle. In G_1 the cell either continues the cycle or enters a quiescent phase called G_0. From this phase, most cells can return to the cycle, but some stay in G_0 for a long time or even for their entire lifetime. The checking or restriction point (R) in G_1 stops the cycle under conditions unfavorable to the cell. When the cell passes this restriction point, it continues the cycle through the synthetic phase (S) and the G_2 phase, originating 2 daughter cells in mitosis (M) except when interrupted by another restriction point (not shown) in G_2.

MEDICAL APPLICATION

Rapidly growing tissues (eg, intestinal epithelium) frequently contain cells in mitosis, whereas slowly growing tissues do not. The increased number of mitotic figures and abnormal mitoses in tumors is an important characteristic that distinguishes malignant from benign tumors. The organism has elaborate regulatory systems that control cell reproduction by either stimulating or inhibiting mitosis. Normal cell proliferation and differentiation are controlled by a group of genes called ***proto-oncogenes;*** *altering the structure or expression of these genes promotes the production of tumors. Altered proto-oncogenes are present in tumor-producing viruses and are probably derived from cells. Altered oncogene activity can be induced by a change in the DNA sequence (mutation), an increase in the number of genes (gene amplification), or gene rearrangement, in which genes are relocated near an active promoter site. Altered oncogenes have been associated with several tumors and hematologic neoplasia. Proteins that stimulate mitotic activity in various cell types include nerve growth factor, epithelial growth factor, fibroblast growth factor, and precursors of erythrocyte growth factor (erythropoietin); there is an extensive and rapidly growing list of these proteins (see Chapter 13).*

Cell proliferation is usually regulated by precise mechanisms that can, when necessary, stimulate or retard mitosis according to the needs of the organism. Several factors (eg, chemical substances, certain

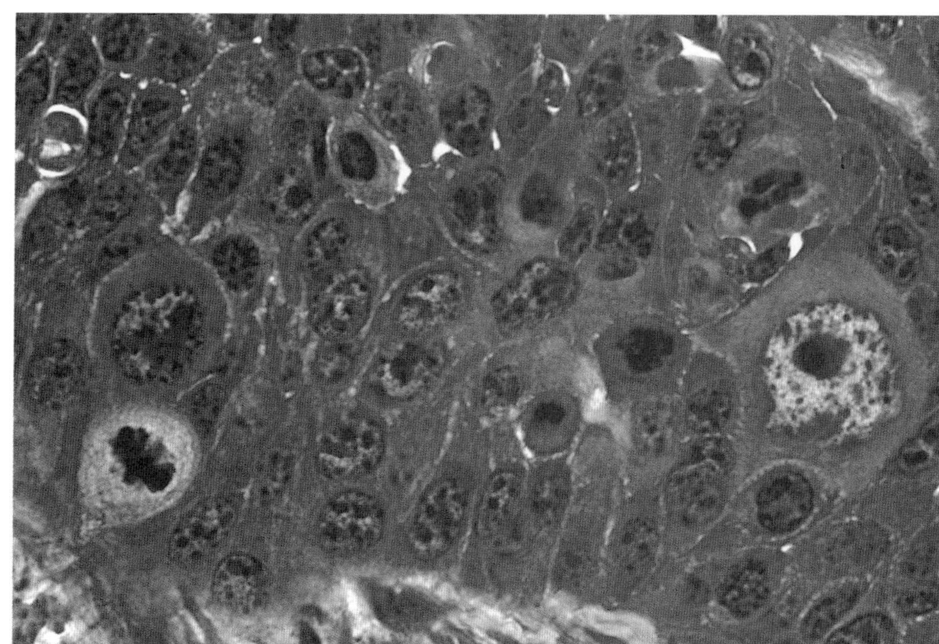

Figure 3–22. Section of a malignant epithelial skin tumor (squamous cell carcinoma). An increase in the number of cells in mitosis and diversity of nuclear morphology are signs of malignancy. PT stain. Medium magnification.

types of radiation, viral infections) can induce DNA damage, mutation, and abnormal cell proliferation that bypasses normal regulation mechanisms for controlled growth and results in the formation of tumors.

*The term **tumor,** initially used to denote any localized swelling in the body caused by inflammation or abnormal cell proliferation, is now usually used as a synonym for **neoplasm** (Gr.* neos, *new,* + plasma, *thing formed). Neoplasm can be defined as an abnormal mass of tissue formed by uncoordinated cell proliferation. Neoplasms are either benign or malignant according to their characteristics of slow growth and no invasiveness (benign) or rapid growth and great capacity to invade other tissues and organs (malignant). **Cancer** is the common term for all malignant tumors (Figures 3–22 and 3–23).*

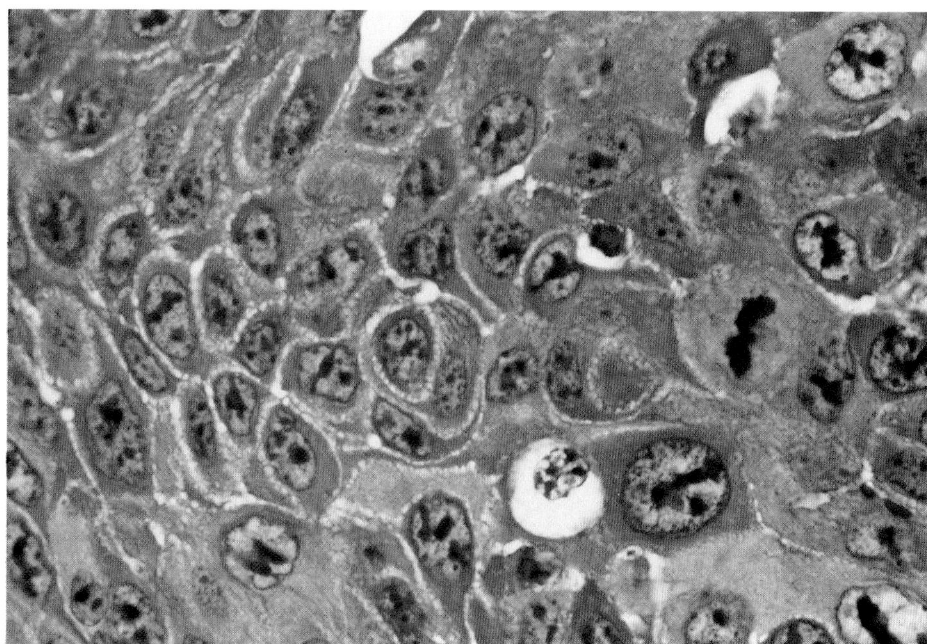

Figure 3–23. Section of a fast-growing malignant epithelial skin tumor showing an increased number of cells in mitosis and great diversity of nuclear morphology. PT stain. Medium magnification.

APOPTOSIS

Cell proliferation for renewal and growth is a process of self-evident physiologic significance. Less evident, but no less important for body functions and health, is the process of programmed cell death called **apoptosis.** A few examples of apoptosis will illustrate its significance.

Most T lymphocytes originating in the thymus have the ability to attack and destroy body components and would cause serious damage if they entered the blood circulation. Inside the thymus, T lymphocytes receive signals that activate the apoptotic program encoded in their chromosomes. These T lymphocytes are destroyed by apoptosis before leaving the thymus (see Chapter 14).

Apoptosis was first discovered in developing embryos, where programmed cell death is an essential process for shaping the embryo (morphogenesis). Later investigators observed that apoptosis is also a common event in the tissues of normal adults.

In apoptosis, the cell and its nucleus become compact, decreasing in size. At this stage the apoptotic cell shows a dark-stained nucleus (pyknotic nucleus), easily identified with the light microscope (Figure 3–24). Next, the chromatin is cut into pieces by DNA endonucleases. During apoptosis the cell shows cytoplasmic large vesicles (blebs) that detach from the cell surface (Figure 3–25). These detached fragments are contained within the plasma membrane, which is changed in such a way that all cell remnants are readily engulfed, or phagocytosed, mainly by macrophages. However, the apoptotic fragments do not elicit in macrophages the synthesis of the molecules that trigger the inflammatory process (see below).

MEDICAL APPLICATION

Most cells of the body can activate their apoptotic program when major changes occur in their DNA—for example, just before a tumor appears, when a number of mutations have already accumulated in the DNA. In this way, apoptosis prevents the proliferation of malignant cells that develop as a result of accumulated mutations in the DNA. To form a clone and develop into a tumor, the malignant cell needs to deactivate the genes that control the apoptotic process.

MEDICAL APPLICATION

The accidental death of cells, a pathologic process, is called ***necrosis.*** *Necrosis can be caused by microorganisms, viruses, chemicals, and other harmful agents. Necrotic cells swell; their organelles increase in volume; and finally they burst, releasing their contents into the extracellular space. Macrophages engulf the debris of necrotic cells by phagocytosis and then secrete molecules that activate other immunodefensive cells to promote inflammation.*

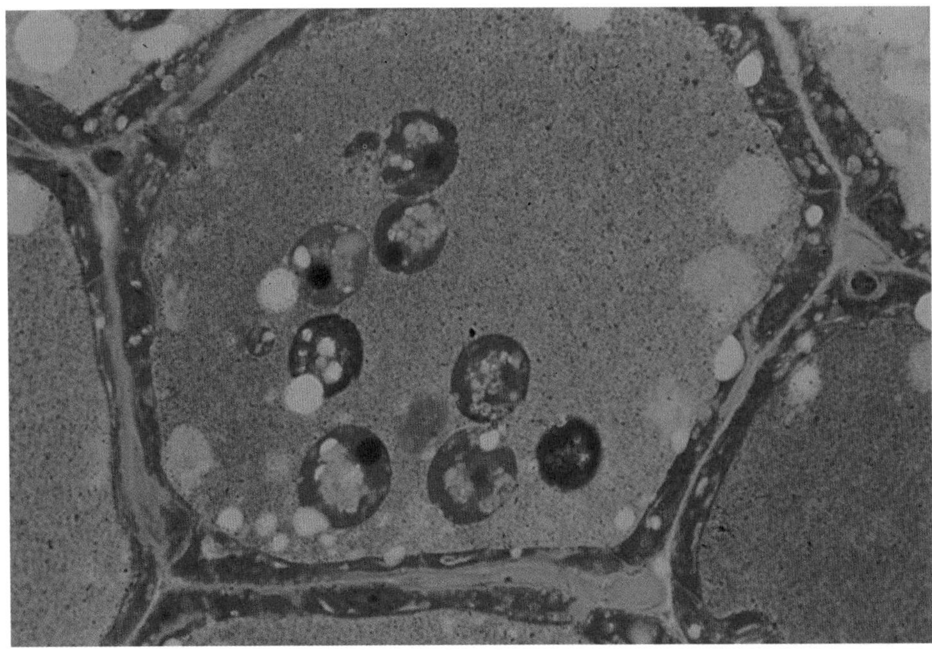

Figure 3–24. Section of a mammary gland from an animal whose lactation was interrupted for 5 days. Note atrophy of the epithelial cells and dilation of the alveolar lumen, which contains several detached cells in the process of apoptosis, as seen from the nuclear alterations. PT stain. Medium magnification.

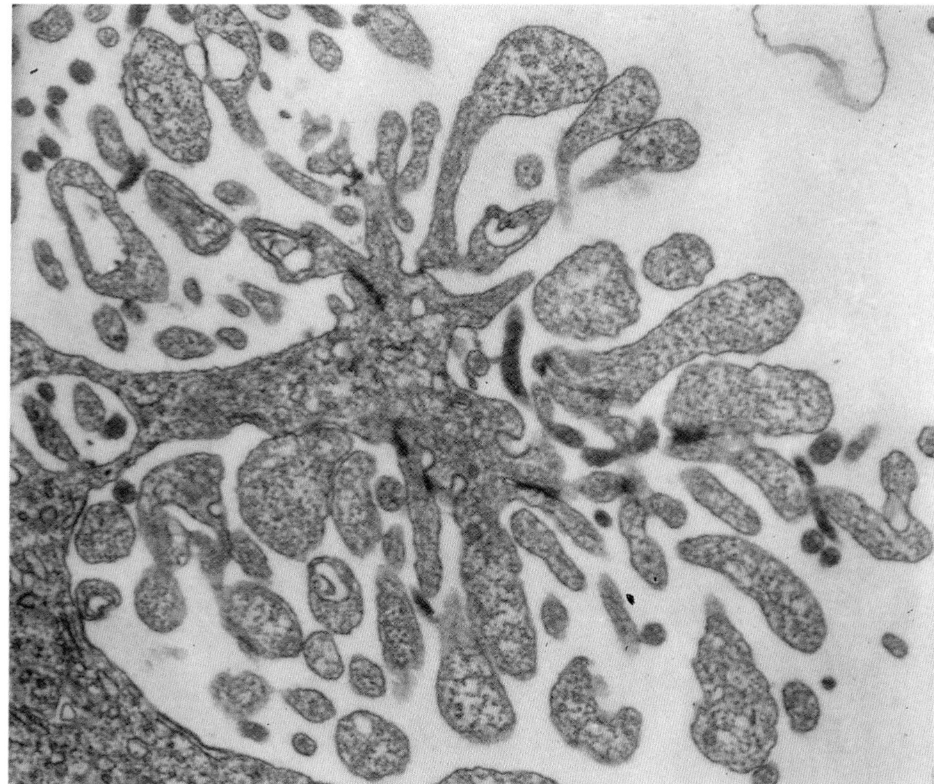

Figure 3–25. Electron micrograph of a cell in apoptosis showing that its cytoplasm is undergoing a process of fragmentation in blebs that preserve their plasma membranes. These blebs are phagocytized by macrophages without eliciting an inflammatory reaction. No cytoplasmic substances are released into the extracellular space.

REFERENCES

Cooper GM: *The Cell: A Molecular Approach.* ASM Press/Sinauer Associates, Inc., 1997.

Doye V, Hurt E: From nucleoporins to nuclear pore complexes. Curr Opin Cell Biol 1997;9:401.

Duke RC et al: Cell suicide in health and disease. Sci Am 1996;275(6):48.

Fawcett D: *The Cell,* 2nd ed. Saunders, 1981.

Goodman SR: *Medical Cell Biology.* Lippincott, 1994.

Jordan EG, Cullis CA (editors): *The Nucleolus.* Cambridge Univ Press, 1982.

Kornberg RD, Klug A: The nucleosome. Sci Am 1981;244:52.

Krstíc RV: *Ultrastructure of the Mammalian Cell.* Springer-Verlag, 1979.

Lloyd D et al: *The Cell Division Cycle.* Academic Press, 1982.

Mélèse T, Xue Z: The nucleolus: an organelle formed by the act of building a ribosome. Curr Opin Cell Biol 1995;7:319.

Trent RJ: *Molecular Medicine. An Introductory Text for Students.* Churchill Livingstone, 1993.

Watson JD et al: *Recombinant DNA,* 2nd ed. Scientific American Books, 1992.

Epithelial Tissue 4

Despite its complexity, the human body is composed of only **4 basic types of tissue:** epithelial, connective, muscular, and nervous. These tissues, which are formed by cells and molecules of the extracellular **matrix,** exist not as isolated units but rather in association with one another and in variable proportions, forming different organs and systems of the body. The main characteristics of these basic types of tissue are shown in Table 4–1. Also of great functional importance are the free cells found in body fluids such as blood and lymph.

Connective tissue is characterized by the abundance of extracellular material produced by its cells; muscle tissue is composed of elongated cells that have the specialized function of contraction; and nerve tissue is composed of cells with elongated processes extending from the cell body that have the specialized functions of receiving, generating, and transmitting nerve impulses. Organs can be divided into **parenchyma,** which is composed of the cells responsible for the main functions typical of the organ, and **stroma,** which is the supporting tissue. Except in the brain and spinal cord, the stroma is made of connective tissue.

Epithelial tissues are composed of closely aggregated polyhedral cells with very little extracellular substance. These cells have strong adhesion and form cellular sheets that cover the surface of the body and line its cavities.

Table 4–1. Main characteristics of the four basic types of tissues.

Tissue	Cells	Extracellular Matrix	Main Functions
Nervous	Intertwining elongated processes	None	Transmission of nervous impulses
Epithelial	Aggregated polyhedral cells	Only small amount	Lining of surface or body cavities, glandular secretion
Muscle	Elongated contractile cells	Moderate amount	Movement
Connective	Several types of fixed and wandering cells	Abundant amount	Support and protection

The principal functions of epithelial (Gr. *epi,* upon, + *thele,* nipple) tissues are the covering and lining of surfaces (eg, skin), absorption (eg, the intestines), secretion (eg, the epithelial cells of glands), sensation (eg, neuroepithelium), and contractility (eg, myoepithelial cells). Because epithelial cells line all external and internal surfaces of the body, everything that enters or leaves the body must cross an epithelial sheet.

THE FORMS & CHARACTERISTICS OF EPITHELIAL CELLS

The forms and dimensions of epithelial cells range from high **columnar** to **cuboidal** to low **squamous** cells. Their common polyhedral form results from their juxtaposition in cellular layers or masses. A similar phenomenon might be observed if a large number of inflated rubber balloons were compressed into a limited space. Epithelial cell nuclei have a distinctive shape, varying from spherical to elongated or elliptic. The nuclear form often corresponds roughly to the cell shape; thus, cuboidal cells have spherical nuclei, and squamous cells have flattened nuclei. The long axis of the nucleus is always parallel to the main axis of the cell.

Since the boundaries between cells are frequently indistinguishable with the light microscope, the form of the cell nucleus is a clue to the shape and number of cells. Nuclear form is also of value in determining whether the cells are arranged in layers, a primary morphologic criterion for classifying epithelia.

Basal Laminae & Basement Membranes

All epithelial cells in contact with subjacent connective tissue have, at their basal surfaces, a sheetlike extracellular structure called the **basal lamina.** This structure is visible only with the electron microscope, where it appears as a dense layer, 20–100 nm thick, consisting of a delicate network of fine fibrils (**lamina densa**). In addition, basal laminae may have electron-lucent layers on one or both sides of the lamina densa, called **laminae rarae** or **laminae lucidae.** The main components of basal laminae are **type IV collagen,** the glycoproteins **laminin** and **entactin,** and **proteoglycan** (eg, the heparan sulfate proteoglycan called perlecan). The precise molecular composition of these components varies between and within tissues. Basal laminae are attached to the underlying connective tissues by anchoring fibrils formed by collagen type VII (Figures 4–1 and 4–2).

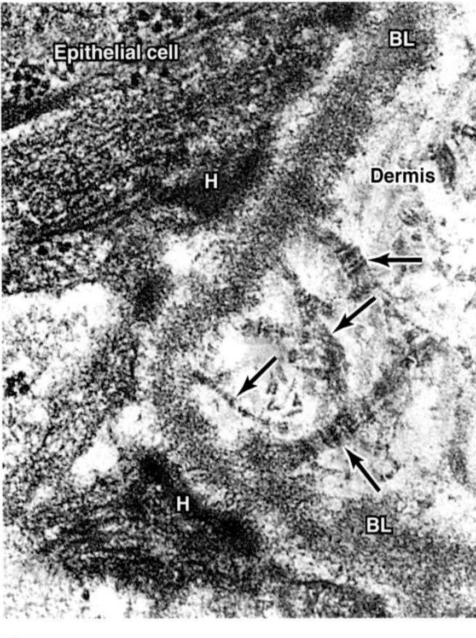

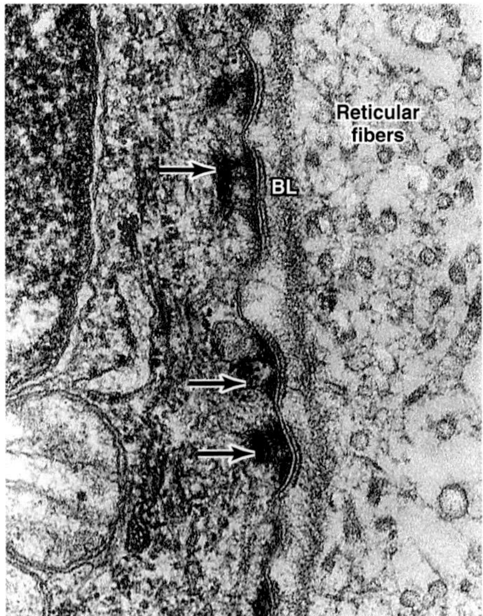

Figure 4–1. **A:** Section of human skin showing hemidesmosomes (H) at the epithelial–connective tissue junction. Note the anchoring fibrils (arrows) that apparently insert into the basal lamina (BL). The characteristically irregular spacing of these fibrils distinguishes them from collagen fibrils. ×54,000. (Courtesy of FM Guerra Rodrigo.) **B:** Section of skin showing the basal lamina (BL) and hemidesmosomes (arrows). This is a typical example of a basement membrane formed by a basal lamina and a reticular lamina (to the right of the basal lamina in this micrograph). ×80,000.

Basal laminae are found not only in epithelial tissues but also where other cell types come into contact with connective tissue. Around muscle, adipose, and Schwann cells, basal laminae provide a barrier that limits or regulates exchanges of macromolecules between connective tissue and other tissues. Basal laminae are also found between adjacent epithelial layers, such as in lung alveoli and in the renal glomerulus (Figure 4–2A). In these cases, the basal lamina is thicker as a result of fusion of the basal laminae of each epithelial cell layer.

The components of basal laminae are secreted by epithelial, muscle, adipose, and Schwann cells. In some instances, reticular fibers are closely associated with the basal lamina, forming the **reticular lamina** (Figures 4–1B and 4–2B). Connective tissue cells produce the reticular fibers.

Basal laminae have many functions. In addition to simple structural and filtering functions, they are also able to influence cell polarity; regulate cell proliferation and differentiation by binding with growth factors; influence cell metabolism; organize the proteins in adjacent plasma membrane (affecting signal transduction); and serve as pathways for cell migration. The basal lamina seems to contain the information necessary for certain cell-to-cell interactions, such as the reinnervation of denervated muscle cells. The presence of the basal lamina around a muscle cell is necessary for the establishment of new neuromuscular junctions.

The term **basement membrane** is used to specify a periodic acid–Schiff (PAS)-positive layer, visible with the light microscope, beneath epithelia and in the kidney glomerulus (Figure 4–3) and lung alveoli. The basement membrane is usually formed by the fusion of either 2 basal laminae (Figure 4–2A) or a basal lamina and a reticular lamina (Figure 4–2B) and is therefore thicker. Not all investigators agree upon the use of the terms

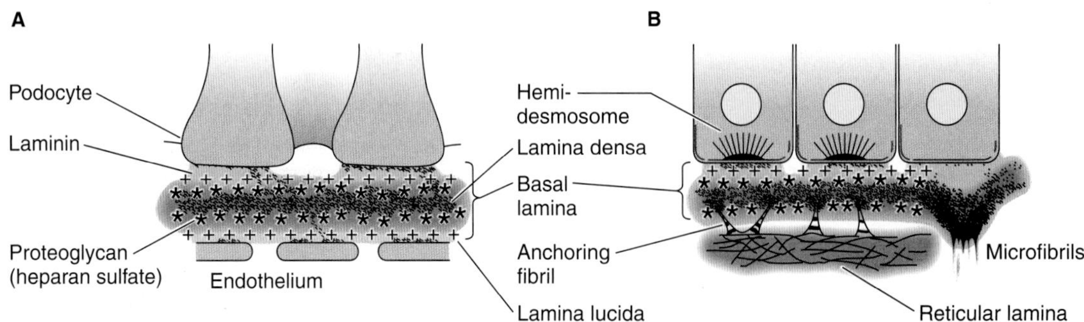

Figure 4–2. Two types of basement membranes. **A:** The thickness of this type of membrane results from fusion of 2 basal laminae produced by an epithelial and an endothelial cell layer, as found in the kidney glomerulus (shown here) and in the alveoli of the lung. It consists of a thick central **lamina densa** with a **lamina lucida (lamina rara)** on either side. **B:** The more common type of basement membrane that separates and binds epithelia to connective tissue is formed by association of the **basal** and **reticular laminae.** Note the presence of the anchoring fibrils formed by type VII collagen, which binds the basal lamina to the subjacent collagen.

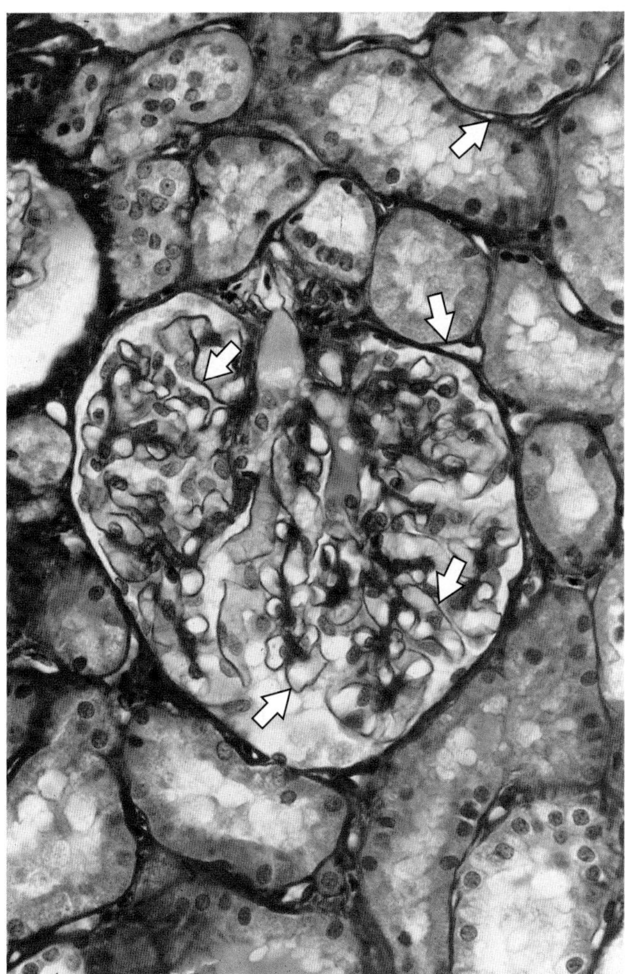

Figure 4–3. Kidney section showing the collagen type IV of the glomerular and tubular basement membranes (arrows). In the glomeruli the basement membrane, besides having a supporting function, has an important role as a filter. Picrosirius-hematoxylin (PSH) stain. Medium magnification.

basement membrane and basal lamina; the two terms are often used indiscriminately, causing confusion. In this book, "basal lamina" is used to denote the lamina densa and the variable presence of laminae rarae, which are structures seen with the electron microscope. "Basement membrane" is used to denote the thicker structures seen with the light microscope.

Intercellular Junctions

Several membrane-associated structures contribute to cohesion and communication between cells. They are present in most tissues but are prominent in epithelia. For this reason they are described in this chapter. Epithelial cells are extremely cohesive, and relatively strong mechanical forces are necessary to separate them. Intercellular adhesion is especially marked in epithelial tissues that are subjected to traction and pressure (eg, the skin). Adhesion is due in part to the binding action of a family of transmembrane glycoproteins called cadherins. Cadherins lose their adhesiveness in the absence of Ca^{2+}.

In addition to the cohesive effects of those intercellular macromolecules the lateral membranes of many epithelial cells exhibit several specializations that form **intercellular junctions.** These junctions serve not only as sites of **adhesion** but also as **seals** to prevent the flow of materials through the intercellular space and to provide a mechanism for communication between adjacent cells. In some epithelia the various junctions are present in a definite order from the apex toward the base of the cell.

Tight junctions, or **zonulae occludens** (singular, **zonula occludens**), are the most apical of the junctions. The Latin terminology gives important information about the geometry of the junction. "Zonula" refers to the fact that the junction forms a band completely encircling the cell, and "occludens" refers to the membrane fusions that close off the intercellular space. In properly stained thin sections viewed in the electron microscope, the outer leaflets of adjacent membranes are seen to fuse, giving rise to a local pentalaminar appearance. One to several of these fusion sites may be observed, depending on the epithelium (Figures 4–4 and 4–5). After cryofracture (Figure 4–6), the replicas show anastomosing ridges and grooves that form a netlike structure corresponding to the fusion sites observed in conventional thin sections. The number of ridges and grooves,

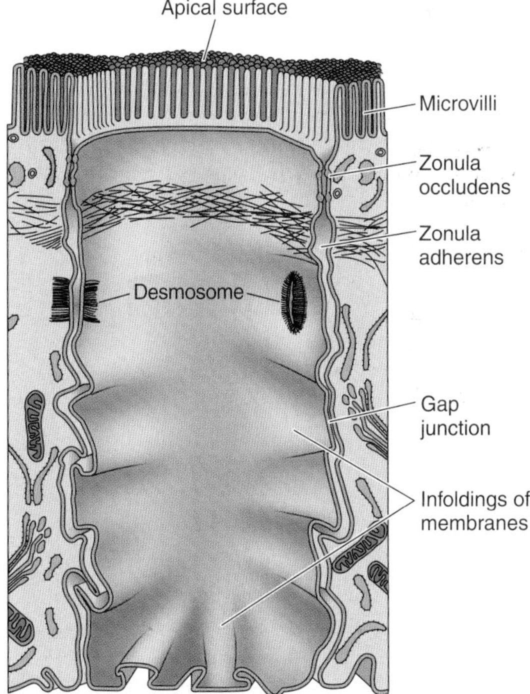

Figure 4–4. The main structures that participate in cohesion among epithelial cells. The drawing shows 3 cells from the intestinal epithelium. The cell in the middle was emptied of its contents to show the inner surface of its membrane. The zonula occludens and zonula adherens form a continuous ribbon around the cell apex, whereas the desmosomes and gap junctions make spotlike plaques. Multiple ridges form the zonula occludens, where the outer laminae of apposed membranes fuse. (Redrawn and reproduced, with permission, from Krstíc RV: *Ultrastructure of the Mammalian Cell.* Springer-Verlag, 1979.)

or fusion sites, has a high correlation with the leakiness of the epithelium. Epithelia with one or very few fusion sites (eg, proximal renal tubule) are more permeable to water and solutes than are epithelia with numerous fusion sites (eg, urinary bladder).

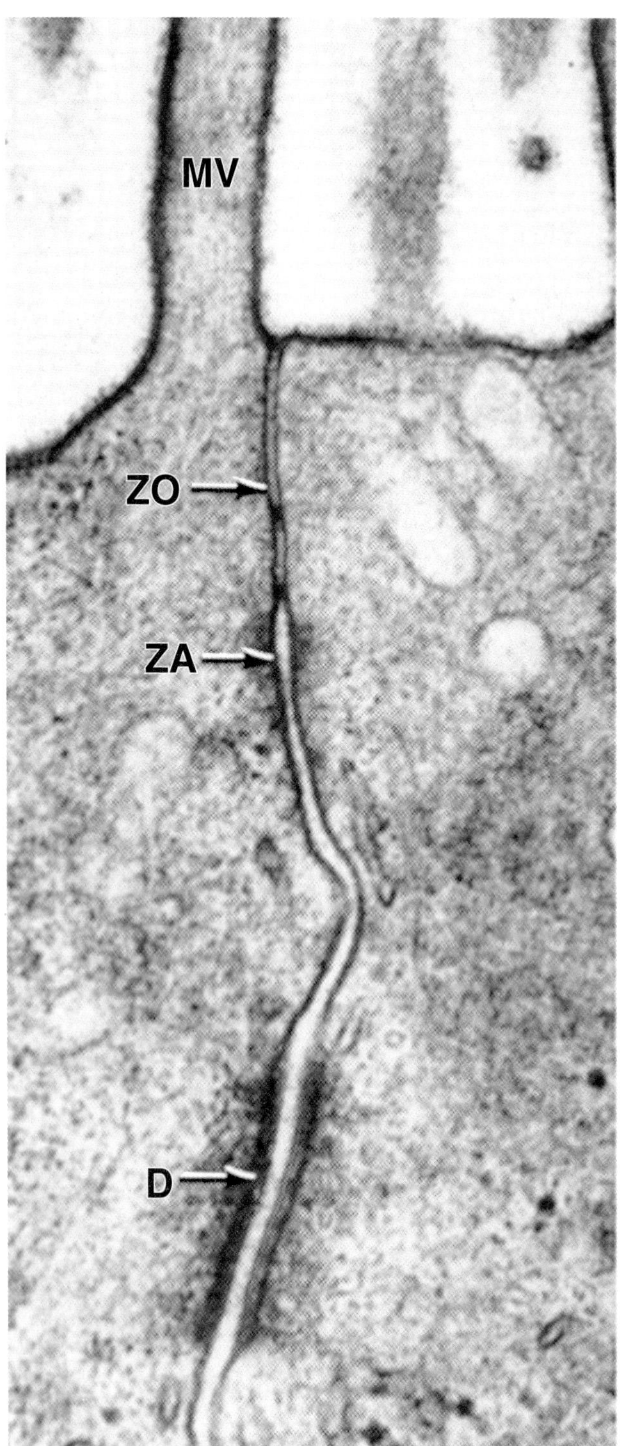

Figure 4–5. Electron micrograph of a section of epithelial cells in the large intestine showing a junctional complex with its zonula occludens (ZO), zonula adherens (ZA), and desmosome (D). Also shown is a microvillus (MV). ×80,000.

Thus, the principal function of the tight junction is to form a seal that prevents the flow of materials between epithelial cells (paracellular pathway) in either direction (from apex to base or from base to apex; see Figure 4–25). In this way, zonula occludens participates in the formation of functional compartments delimited by sheets of epithelial cells.

In many epithelia, the next type of junction encountered is the **zonula adherens** (Figures 4–4 and 4–5). This junction encircles the cell and provides for the adhesion of one cell to its neighbor. A noteworthy feature of this junction is the insertion of numerous actin filaments into electron-dense plaques of material on the cytoplasmic surfaces of the junctional membranes. The filaments arise from the **terminal web,** a web of actin filaments, intermediate filaments, and spectrin.

A **gap junction** can occur almost anywhere along the lateral membranes of most epithelial cells. Gap junctions are found in nearly all mammalian tissues; skeletal muscle is a major exception. They are characterized, in conventional electron micrographs, by the close (2-nm) apposition of adjacent cell membranes (Figure 4–7A and C). After cryofracture, aggregates of intramembrane particles are found in circular patches in the plasma membrane (Figure 4–7B).

Gap junction protein units, called **connexins,** form hexamers with a hydrophilic pore about 1.5 nm in diameter in the center. This individual unit of the gap junction is called a **connexon,** and connexins in adjacent cell membranes are aligned to form a hydrophilic channel between the two cells (Figure 4–7A). Molecular cloning studies have demonstrated that connexins are a family of related proteins that are distributed differently and form channels with differing physiologic properties. Gap junctions permit the exchange between cells of molecules with molecular mass <1500 Da. In addition, signaling molecules such as some hormones, cyclic AMP and GMP, and ions can move through gap junctions, causing the cells in many tissues to act in a coordinated manner rather than as independent units. A typical example is heart muscle cells, where gap junctions are greatly responsible for the heart's coordinated beat.

Gap junctions between previously isolated cells can be formed rapidly. Metabolic inhibitors—especially those that block oxidative phosphorylation—can inhibit the formation of junctions or can undo junctions already present between cells. New junctions can be formed in the absence of protein synthesis, however, in which case, connexins may form from subunits diffusely scattered in the plasma membrane.

The final type of junction is the **desmosome** (Gr. *desmos,* band, + *soma,* body), or **macula adherens** (Figures 4–4 and 4–5). The desmosome is a complex disk-shaped structure at the surface of one cell that is matched with an identical structure at the surface of the adjacent cell. The cell membranes in this region are very straight and are usually somewhat farther apart (> 30 nm) than the usual 20 nm. On the cytosolic side of the membrane of each cell and separated from it by a short distance is a circular plaque of material called an **attachment plaque,** made up of at least 12 proteins. In epithelial cells, groups of intermediate keratin filaments are inserted into the attachment plaque or make hairpin turns and return to the cytoplasm. Because intermediate filaments of the cytoskeleton are very strong, desmosomes provide a firm adhesion among the cells. In nonepithelial cells, the intermediate filaments attached to desmosomes are made not of keratin but of other proteins, such as desmin or vimentin. Proteins of the cadherin family participate in the adhesion provided by desmosomes. In vitro this adhesiveness is abolished by the removal of Ca^{2+}.

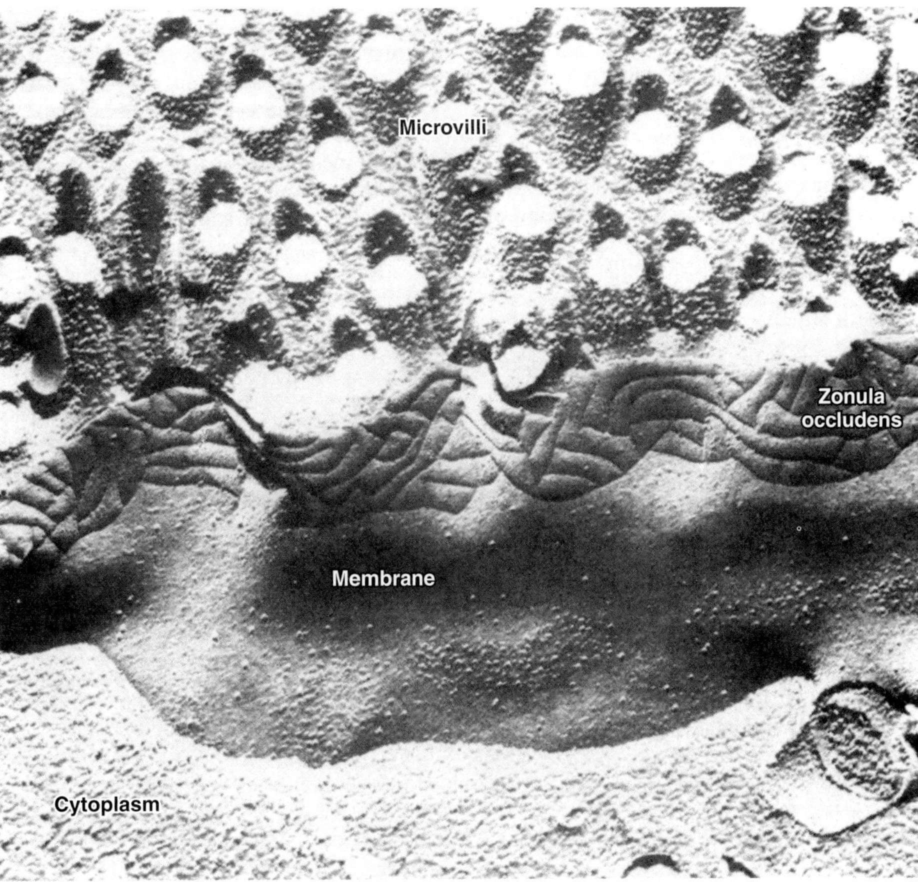

Figure 4–6. Electron micrograph of a small-intestine epithelial cell after cryofracture. In the upper portion, the microvilli are fractured transversely; in the lower portion, the fracture crosses through the cytoplasm of the intestinal epithelial cell. The grooves, which actually lie in the lipid (middle) layer of each plasmalemma, reveal that the membranes of adjoining cells were fused in the zonula occludens. ×100,000. (Courtesy of P Pinto da Silva.)

In the contact zone between certain epithelial cells and the basal lamina, **hemidesmosomes** (Gr. *hemi,* half, + *desmos* + *soma*) can often be observed. These structures take the form of half a desmosome and bind the epithelial cell to the subjacent basal lamina (Figure 4–1B). However, in desmosomes the attachment plaques contain mainly cadherins, whereas in hemidesmosomes the plaques are made of **integrins,** a family of transmembrane proteins that are receptor sites for the extracellular macromolecules laminin and collagen type IV.

From the functional point of view, junctions between cells can be classified as **adhering junctions** (zonulae adherentes, hemidesmosomes, and desmosomes), **impermeable junctions** (zonulae occludentes), and **communicating junctions** (gap junctions).

SPECIALIZATIONS OF THE CELL SURFACE

The free surface of some epithelial cells shows specializations to increase cell surface area or to move foreign particles.

Microvilli

When viewed in the electron microscope, most cells from the several tissues are seen to have cytoplasmic projections. These projections may be short or long fingerlike extensions or folds that pursue a sinuous course, and they range in number from a few to many. In absorptive cells, such as the lining epithelium of the small intestine and the cells of the proximal renal tubule, orderly arrays of many hundreds of microvilli (Gr. *mikros,* small, + L. *villus,* tuft of hair) are encountered (Figures 4–8 and 4–9). Each microvillus is about 1 μm high and 0.08 μm wide. In these absorptive cells the glycocalyx is thicker than that in most cells. The complex of microvilli and glycocalyx is easily seen in the light microscope and is called the **brush,** or **striated, border.**

Within the microvilli are clusters of actin filaments (Figure 4–9) that are cross-linked to each other and to the surrounding plasma membrane by several other proteins.

Stereocilia

Stereocilia are long, nonmotile processes of cells of the epididymis and ductus deferens that are actually longer branched microvilli and should not be confused with true cilia. Stereocilia increase the cell surface area, facilitating the movement of molecules into and out of the cell.

Cilia & Flagella

Cilia are elongated, motile structures on the surface of some epithelial cells, 5–10 μm long and 0.2 μm in diameter. They are surrounded by the cell membrane and contain a central pair of isolated microtubules surrounded by 9 more pairs of microtubules. The two microtubules of the peripheral pairs are joined to each other (Figure 4–10).

Cilia are inserted into **basal bodies,** which are electron-dense structures at the apical pole just below the cell membrane (Figure 4–10). Basal bodies have a structure analogous to that of the centrioles (see Chapter 2).

A

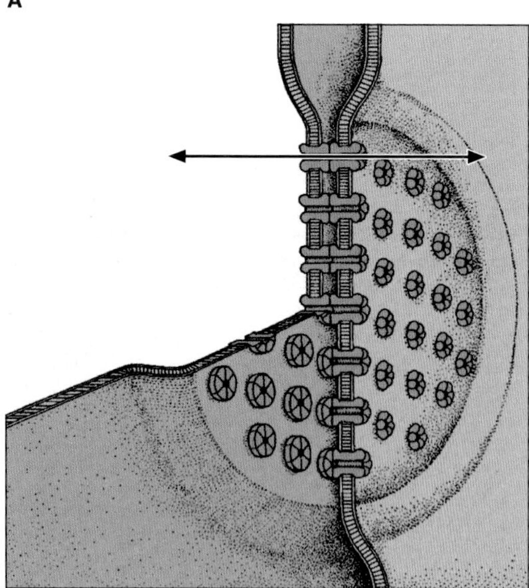

B

C

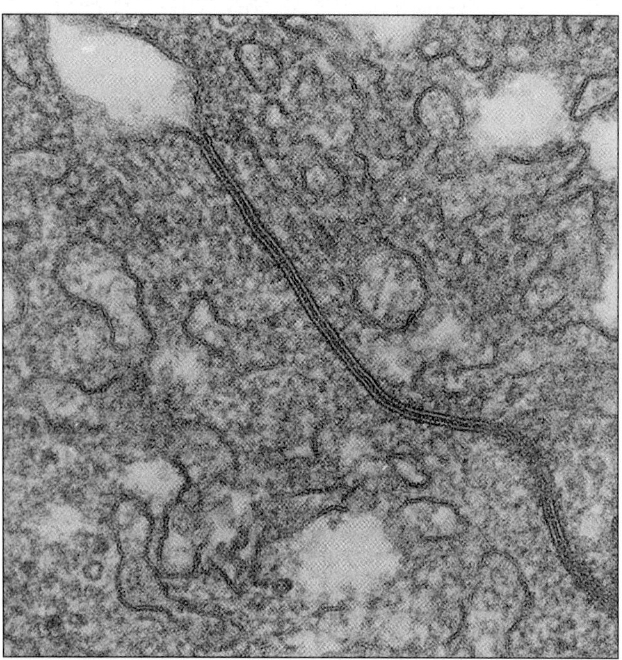

Figure 4–7. **A:** Model of a gap junction (oblique view) depicting the structural elements that allow the exchange of nutrients and signal molecules between cells without loss of material into the intercellular space. The communicating pipes are formed by pairs of abutting particles, which are in turn composed of 6 dumbbell-shaped protein subunits that span the lipid bilayer of each cell membrane. The channel passing through the cylindrical bridges (arrow in **A**) is about 1.5 nm in diameter, limiting the size of the molecules that can pass through it. Fluids and tracers in the intercellular space can permeate the gap junction by flowing around the protein bridges. (Reproduced, with permission, from Staehelin LA, Hull BE: Junctions between living cells. Sci Am 1978;238:41. Copyright © 1978 by Scientific American, Inc. All rights reserved.) **B:** Gap junction between living cells as seen on a cryofracture preparation. The junction appears as a plaquelike agglomeration of intramembrane protein particles. ×45,000. (Courtesy of P Pinto da Silva). **C:** Gap junction between 2 rat liver cells. At the junction, 2 apposed membranes are separated by a 2-nm-wide electron-dense space, or gap. ×193,000. (Courtesy of MC Williams.)

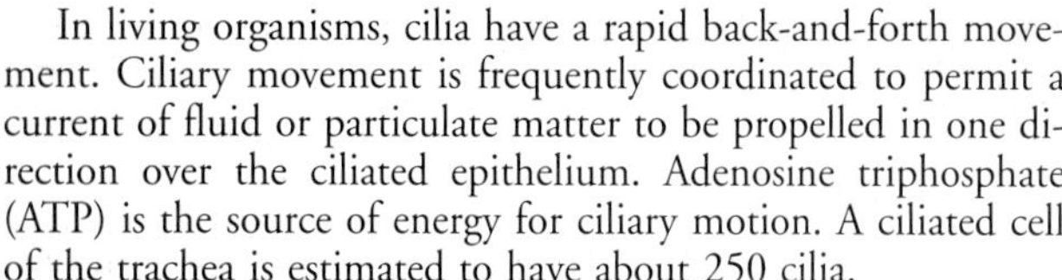

In living organisms, cilia have a rapid back-and-forth movement. Ciliary movement is frequently coordinated to permit a current of fluid or particulate matter to be propelled in one direction over the ciliated epithelium. Adenosine triphosphate (ATP) is the source of energy for ciliary motion. A ciliated cell of the trachea is estimated to have about 250 cilia.

Flagella, present in the human body only in spermatozoa, are similar in structure to cilia but are much longer and are limited to one flagellum per cell.

TYPES OF EPITHELIA

Epithelia are divided into 2 main groups according to their structure and function: **covering epithelia** and **glandular epithelia.** This is an arbitrary division, for there are covering epithelia in which all cells secrete (eg, the surface epithelium of the stomach) or in which glandular cells are sparse among covering cells (eg, mucous cells in the small intestine or trachea).

Covering Epithelia

Covering epithelia are tissues in which the cells are organized in layers that cover the external surface or line the cavities of the body. They can be classified according to the number of cell layers and the morphologic features of the cells in the surface layer (Table 4–2). **Simple epithelium** (Figure 4–11) contains only one layer of cells, and **stratified epithelium** contains more than one layer (Figure 4–12).

Simple epithelium can, according to cell shape, be **squamous, cuboidal,** or **columnar** (Figures 4–13 through 4–16). The

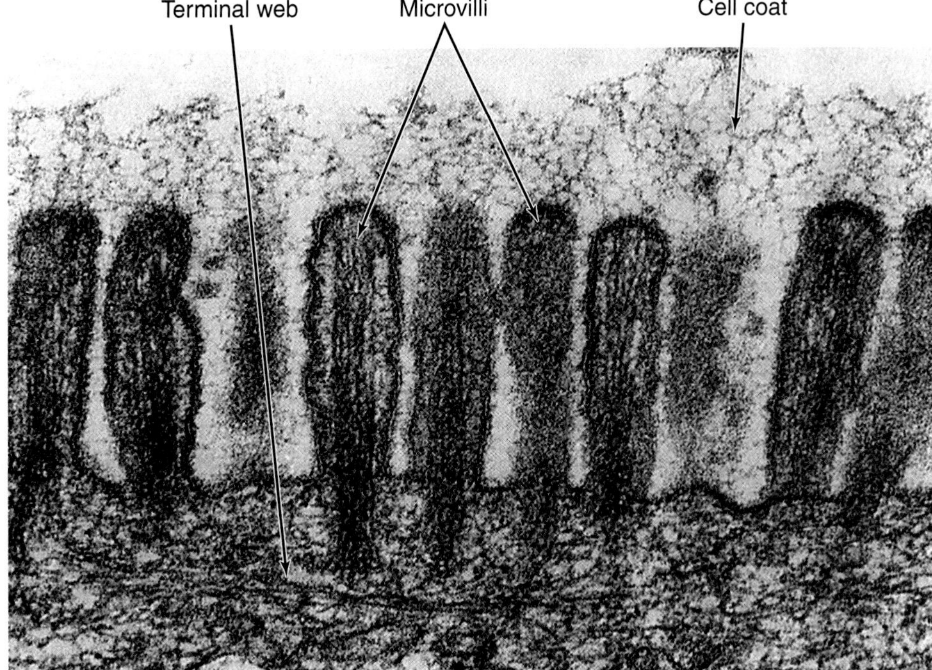

Figure 4–8. Electron micrograph of the apical region of an intestinal epithelial cell. Note the terminal web composed of a horizontal network that contains mainly actin microfilaments. The vertical microfilaments that constitute the core of the microvilli are clearly seen. An extracellular cell coat (glycocalyx) is bound to the plasmalemma of the microvilli. ×45,000.

Figure 4–9. Electron micrograph of a section from the apical region of a cell from the intestinal lining showing cross-sectioned microvilli. In their interiors, note the microfilaments in a cross section. The surrounding unit membrane can be clearly discerned and is covered by a layer of glycocalyx, or cell coat. ×100,000.

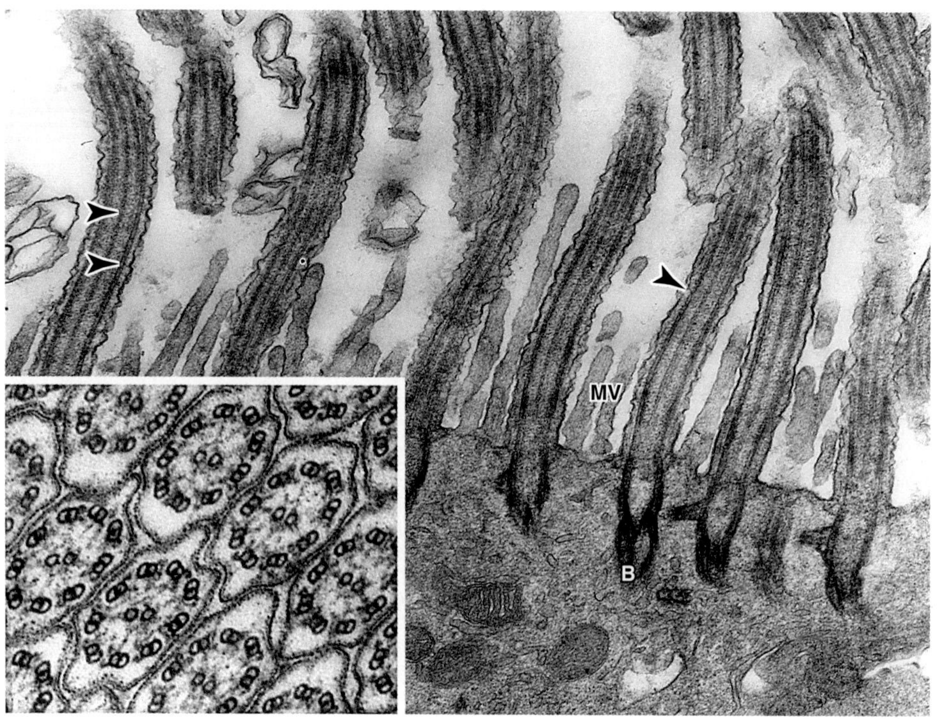

Figure 4–10. Electron micrograph of the apical portion of a ciliated epithelial cell. Cilia are seen in longitudinal section. At the left, arrowheads point to the central and peripheral microtubules of the axoneme. The arrowhead at right indicates the plasma membrane surrounding the cilium. Each cilium has a basal body (B) from which it grows. Microvilli (MV) are shown. ×59,000. **Inset:** Cilia in cross section. The 9 + 2 array of microtubules in each cilium is evident. ×80,000. (Reproduced, with permission, from Junqueira LCU, Salles LMM: Ultra-Estrutura e Função Celular. Edgard Blücher, 1975.)

Table 4–2. Common types of covering epithelia in the human body.

Number of Cell Layers	Cell Form	Examples of Distribution	Main Function
Simple (one layer)	Squamous	Lining of vessels (endothelium). Serous lining of cavities; pericardium, pleura, peritoneum (mesothelium).	Facilitates the movement of the viscera (mesothelium), active transport by pinocytosis (mesothelium and endothelium), secretion of biologically active molecules (mesothelium)
	Cuboidal	Covering the ovary, thyroid.	Covering, secretion.
	Columnar	Lining of intestine, gallbladder.	Protection, lubrication, absorption, secretion.
Pseudostratified (layers of cells with nuclei at different levels; not all cells reach surface but all adhere to basal lamina)		Lining of trachea, bronchi, nasal cavity.	Protection, secretion; cilia-mediated transport of particles trapped in mucus out of the air passages.
Stratified (two or more layers)	Squamous keratinized (dry)	Epidermis.	Protection; prevents water loss.
	Squamous nonkeratinized (moist)	Mouth, esophagus, larynx, vagina, anal canal.	Protection, secretion; prevents water loss.
	Cuboidal	Sweat glands, developing ovarian follicles.	Protection, secretion.
	Transitional	Bladder, ureters, renal calyces.	Protection, distensibility.
	Columnar	Conjunctiva.	Protection.

A Simple squamous epithelium

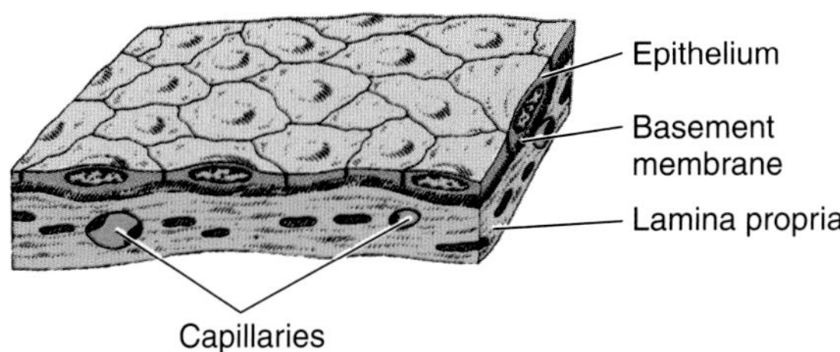

B Simple cuboidal epithelium

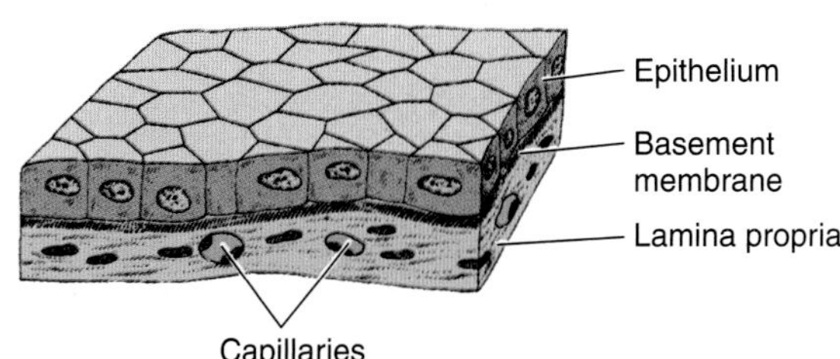

C Simple ciliated columnar epithelium

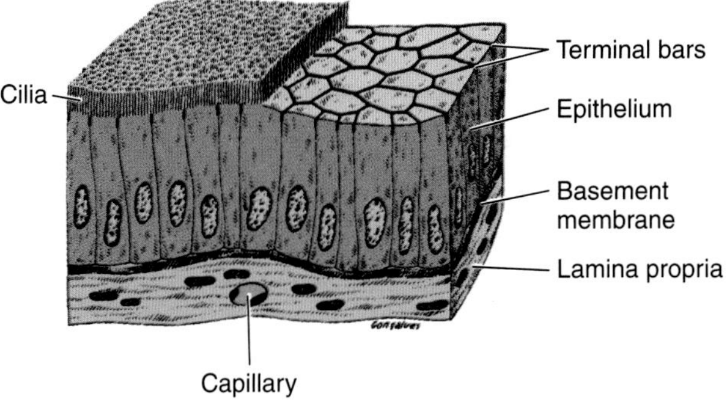

Figure 4–11. Diagrams of simple epithelial tissue. **A:** Simple squamous epithelium. **B:** Simple cuboidal epithelium. **C:** Simple ciliated columnar epithelium. All are separated from the subjacent connective tissue by a basement membrane. In **C,** note the terminal bars that correspond in light microscopy to the zonula occludens and the zonula adherens of the junctional complex.

A Stratified squamous epithelium

Epithelium
Basement membrane
Lamina propria

B Transitional epithelium

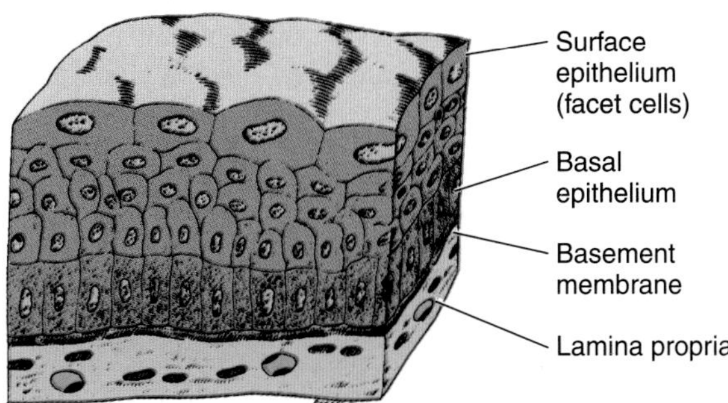

C Ciliated pseudostratified epithelium

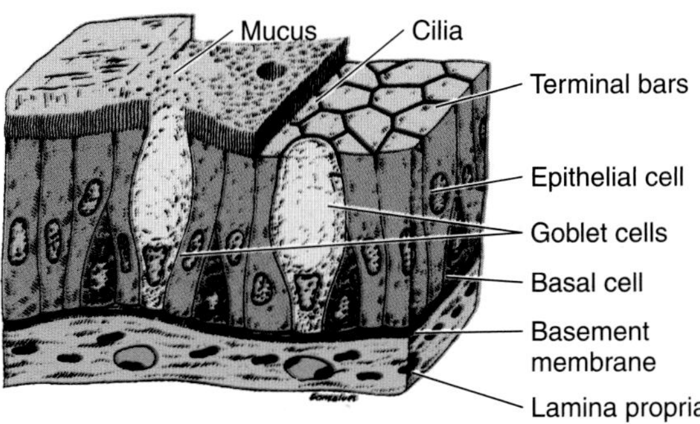

Figure 4–12. Diagrams of stratified and pseudostratified epithelial tissue. **A:** Stratified squamous epithelium. **B:** Transitional epithelium. **C:** Ciliated pseudostratified epithelium. The goblet cells secrete mucus, which forms a continuous mucous layer over the ciliary layer.

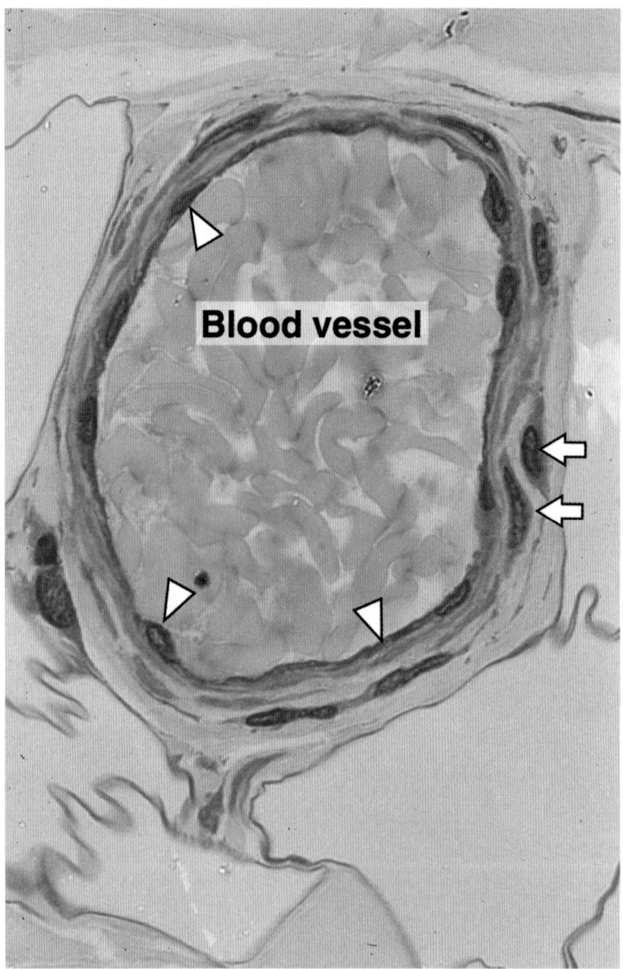

Figure 4–13. Section of a vein. All blood vessels are lined with a simple squamous epithelium called endothelium (arrowheads). Smooth muscle cells in the vein wall are indicated by arrows. Pararosaniline–toluidine blue (PT) stain. Medium magnification.

endothelium that lines blood and lymph vessels and the mesothelium that lines certain body cavities, such as the pleural and peritoneal cavities, and covers the viscera are examples of simple squamous epithelium (Figures 4–13 and 4–14).

An example of cuboidal epithelium is the surface epithelium of the ovary, and an example of columnar epithelium is the lining of the small intestine.

Stratified epithelium (Figures 4–17, 4–18, and 4–19) is classified according to the cell shape of its superficial layer: **squamous, cuboidal, columnar,** and **transitional.** Pseudostratified epithelium forms a separate group, discussed below.

Stratified squamous keratinized epithelium is found mainly in the skin. Its cells form many layers, and the cells closer to the underlying connective tissue are usually cuboidal or columnar. The cells become irregular in shape and flatten as they get progressively closer to the surface, where they are thin and squamous (see Chapter 18 for more detailed information).

Stratified squamous nonkeratinized epithelium (Figure 4–17) lines wet cavities (eg, mouth, esophagus, vagina), in contrast to the skin, whose surface is dry. A surface layer of flattened living cells that retain their nuclei characterizes stratified squamous nonkeratinized epithelium. This is not the case with the keratinized variety of this epithelium, in which the surface cells are dead and their nuclei are not discernible.

Stratified columnar epithelium is rare; it is present in the human body only in small areas, such as the ocular conjunctiva and the large ducts of salivary glands.

Transitional epithelium, which lines the urinary bladder, the ureter, and the upper part of the urethra, is characterized by a surface layer of domelike cells that are neither squamous nor columnar (Figures 4–18 and 4–19). The form of these cells changes according to the degree of distention of the bladder. This type of epithelium is discussed in detail in Chapter 19.

Pseudostratified epithelium is so called because the nuclei appear to lie in various layers. Although all cells are attached to the basal lamina, some do not reach the surface. The best-known example of this tissue is the ciliated pseudostratified columnar epithelium in the respiratory passages (Figure 4–12).

Two other types of epithelium warrant brief mention. **Neuroepithelial cells** are cells of epithelial origin with specialized sensory functions (eg, cells of taste buds). **Myoepithelial cells** are branched cells that contain myosin and a large number

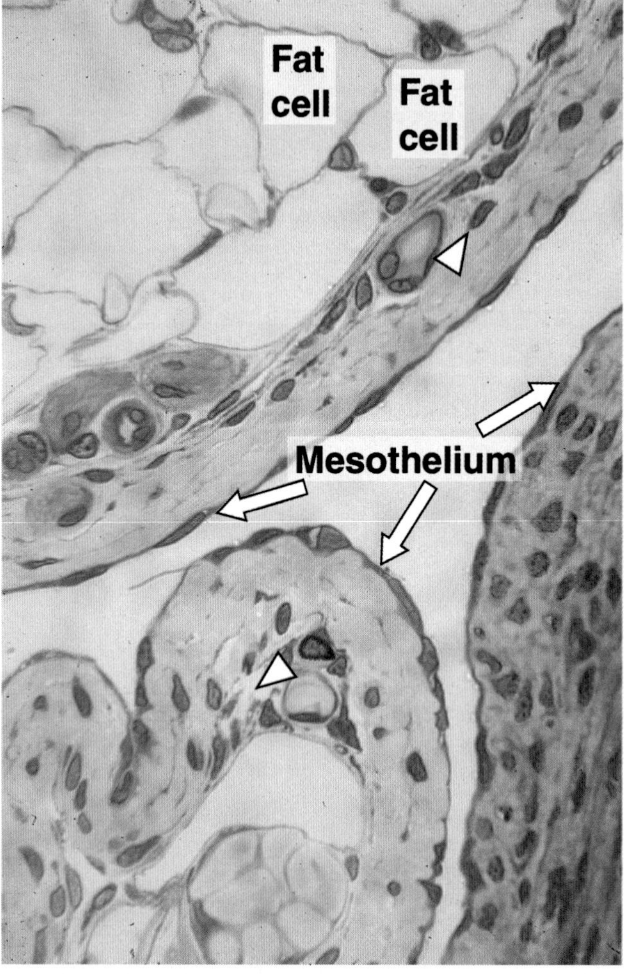

Figure 4–14. Simple squamous epithelium covering the peritoneum (mesothelium). Some blood capillaries are indicated by arrows. PT stain. Medium magnification.

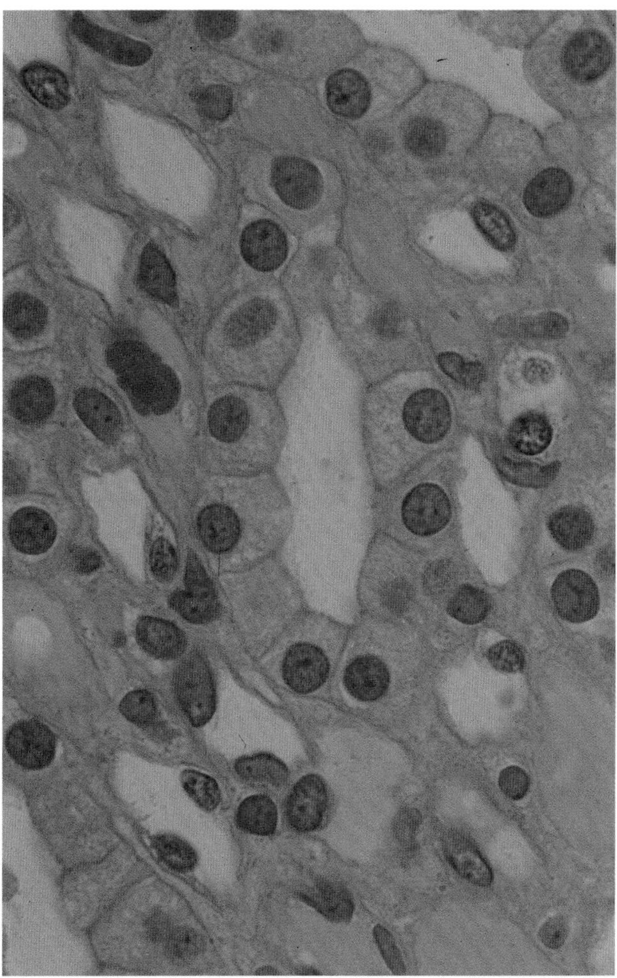

Figure 4–15. Simple cuboidal epithelium from kidney collecting tubules. Cells of these tubules are responsive to the antidiuretic hormone and control the resorption of water from the glomerular filtrate, thus affecting urine density and helping retain the water content of the body. PT stain. Low magnification.

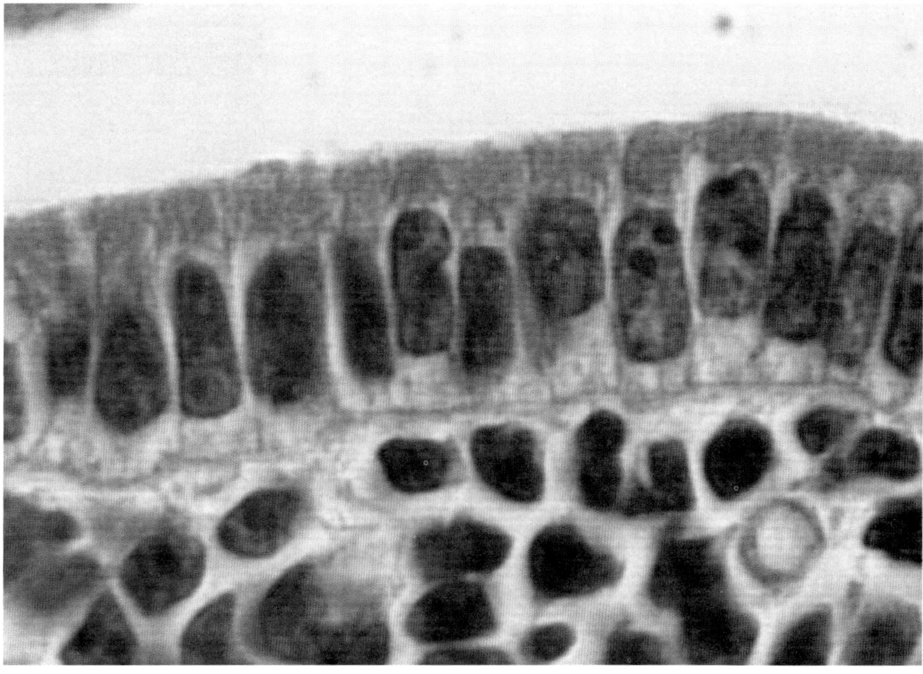

Figure 4–16. Simple columnar epithelium that covers the inner cavity of the uterus. Note that the epithelium rests on the loose connective tissue of the lamina propria. The epithelium and the lamina propria constitute the mucosa. H&E stain. Medium magnification.

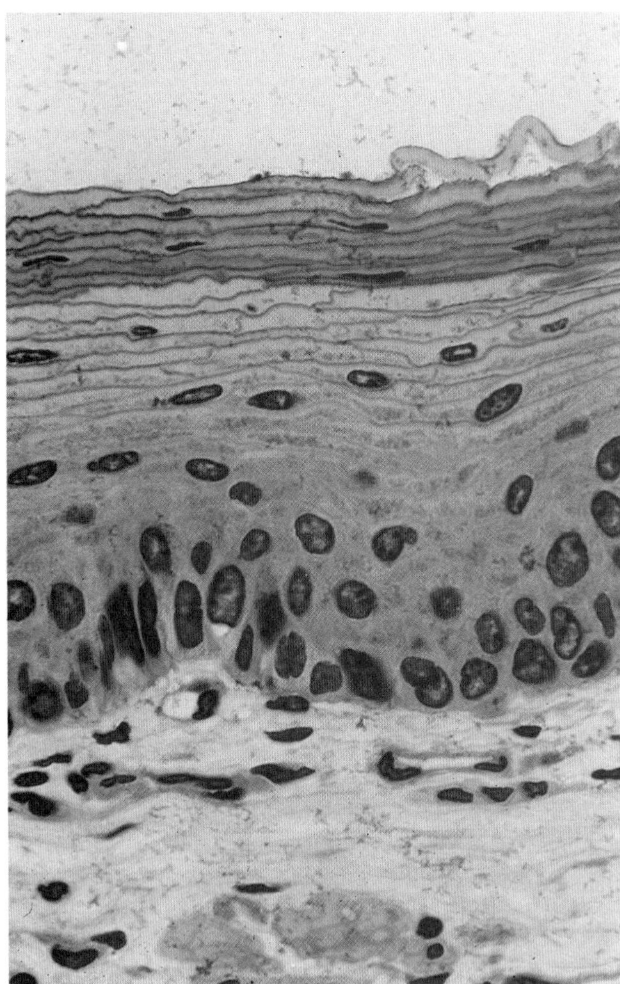

Figure 4–17. Stratified squamous nonkeratinized (moist) epithelium of the esophagus. PT stain. Medium magnification.

of actin filaments. They are specialized for contraction, mainly of the acini of the mammary, sweat, and salivary glands.

Glandular Epithelia

Glandular epithelia are tissues formed by cells specialized to produce secretion. The molecules to be secreted are generally stored in the cells in small membrane-bound vesicles called **secretory granules.**

Glandular epithelial cells may synthesize, store, and secrete proteins (eg, pancreas), lipids (eg, adrenal, sebaceous glands), or complexes of carbohydrates and proteins (eg, salivary glands). The mammary glands secrete all 3 substances. Less common are the cells of glands that have low synthesizing activity (eg, sweat glands) and that secrete mostly substances transferred from the blood to the lumen of the gland.

Types of Glandular Epithelia

The epithelia that form the glands of the body can be classified according to various criteria. Unicellular glands consist of isolated glandular cells, and multicellular glands are composed of clusters of cells. An example of a unicellular gland is the **goblet cell** of the lining of the small intestine (Figure 4–20) or of the respiratory tract. The term "gland," however, is usually used to designate large, complex aggregates of glandular epithelial cells, such as in the salivary glands and the pancreas.

Glands always arise from covering epithelia by means of cell proliferation and invasion of subjacent connective tissue, followed by further differentiation (Figure 4–21). **Exocrine** (Gr. *exo,* outside, + *krinein,* to separate) glands retain their connection with the surface epithelium from which they originated. This connection takes the form of tubular ducts lined with epithelial cells through which the glandular secretions pass to reach the surface. **Endocrine** (Gr. *endon,* within, + *krinein*) glands are those whose connection with the surface from which they originated was obliterated during development. These glands are therefore ductless, and their secretions are picked up and transported to their site of action by the bloodstream rather than by a duct system.

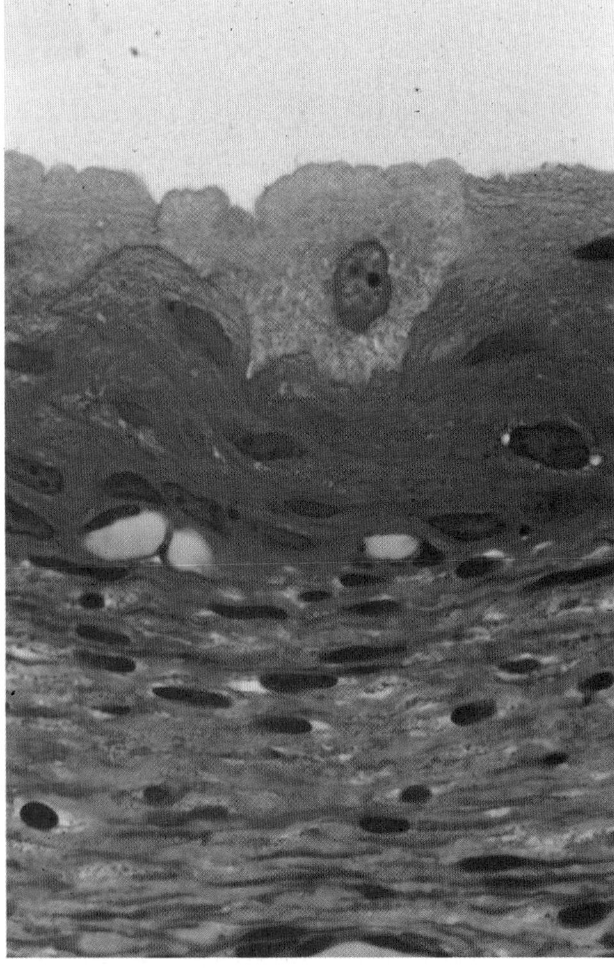

Figure 4–18. Stratified transitional epithelium of the urinary bladder (above) and the connective tissue of the lamina propria (below) with several fibroblasts. PT stain. Medium magnification.

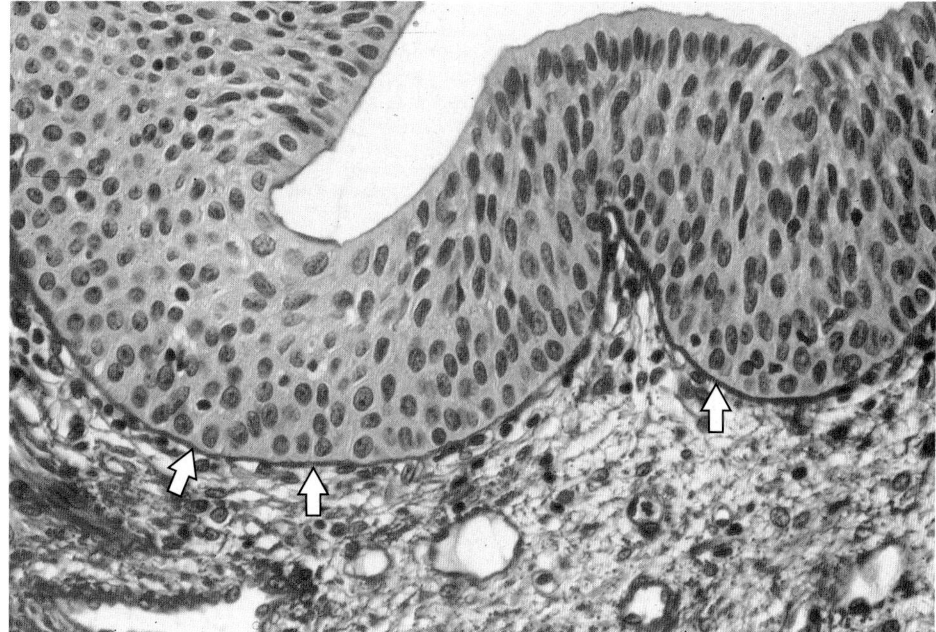

Figure 4–19. Stratified transitional epithelium of the urethra. The basement membrane between the epithelium and the underlying loose connective tissue is indicated by arrows. PSH stain. Medium magnification.

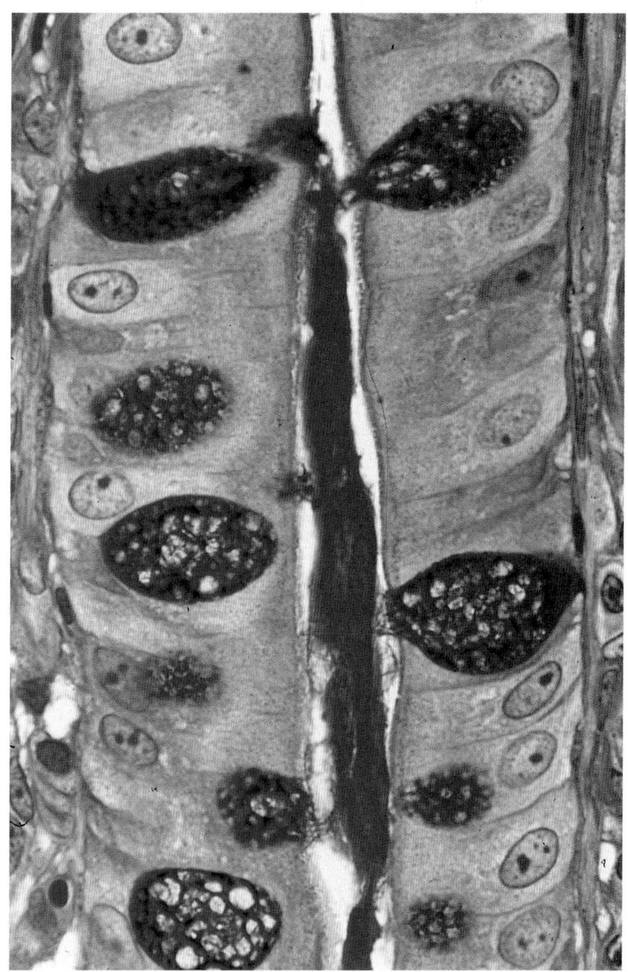

Figure 4–20. Section of large intestine showing goblet cells secreting mucus to the extracellular space. The mucus precursor stored in the cytoplasm of the goblet cells is also stained in a dark color. PAS-PT stain. Medium magnification.

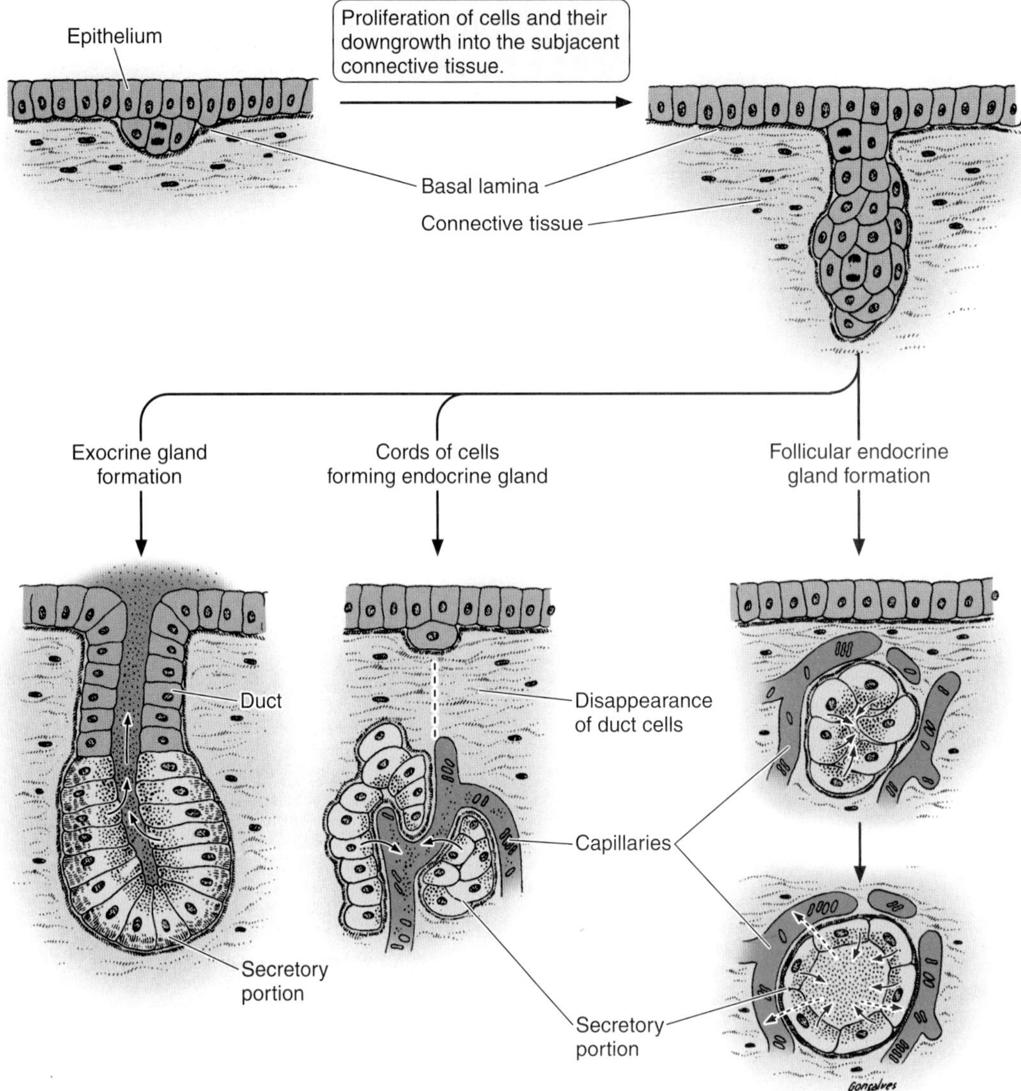

Figure 4–21. Formation of glands from covering epithelia. Epithelial cells proliferate and penetrate connective tissue. They may—or may not—maintain contact with the surface. When contact is maintained, exocrine glands are formed; without contact, endocrine glands are formed. The cells of endocrine glands can be arranged in cords or in follicles. The lumens of the follicles accumulate large quantities of secretions; cells of the cords store only small quantities of secretions in their cytoplasm. (Redrawn and reproduced, with permission, from Ham AW: *Histology,* 6th ed. Lippincott, 1969.)

Two types of endocrine glands can be differentiated according to cell grouping. In the first type, the agglomerated cells form anastomosing cords interspersed between dilated blood capillaries (eg, adrenal gland, parathyroid, anterior lobe of the pituitary; see Figure 4–21). In the second type, the cells line a vesicle or follicle filled with noncellular material (eg, the thyroid gland; Figure 4–21).

Exocrine glands have a **secretory portion,** which contains the cells responsible for the secretory process, and **ducts,** which transport the secretion to the exterior of the gland (Figure 4–22). **Simple glands** have only one unbranched duct, whereas **compound glands** have ducts that branch repeatedly. The cellular organization within the secretory portion differentiates the glands further. Simple glands can be tubular, coiled tubular, branched tubular, or acinar. Compound glands can be tubular, acinar, or tubuloacinar (Figure 4–22). Some organs have both endocrine and exocrine functions, and one cell type may function both ways—eg, in the liver, where cells that secrete bile into the duct system also secrete some of their products into the bloodstream. In other organs, some cells are specialized in exocrine secretion and others are specialized in endocrine secretion; in the pancreas, for example, the acinar cells secrete digestive enzymes into the intestinal lumen, whereas the islet cells secrete insulin and glucagon into the blood.

According to the way in which the secretory products leave the cell, glands can be classified as **merocrine** (Gr. *meros,* part, +

krinein) or **holocrine** (Gr. *holos,* whole, + *krinein*). In merocrine glands (eg, the pancreas), the secretory granules leave the cell by exocytosis with no loss of other cellular material. In holocrine glands (eg, sebaceous glands), the product of secretion is shed with the whole cell—a process that involves destruction of the secretion-filled cells. In an intermediate type—the **apocrine** (Gr. *apo,* away from, + *krinein*) gland—the secretory product is discharged together with parts of the apical cytoplasm (Figure 4–23).

Multicellular glands usually have a surrounding capsule of connective tissue and septa that divides the gland into lobules. These lobules then subdivide, and in this way the connective tissue separates and binds the glandular components together. Blood vessels and nerves also penetrate and subdivide in the gland.

GENERAL BIOLOGY OF EPITHELIAL TISSUES

Underlying the covering epithelial tissues that line the body cavities is a layer of connective tissue, the **lamina propria,** which is bound to the epithelium by the basal lamina. The lamina propria not only serves to support the epithelium but also binds it to neighboring structures. The area of contact between epithelium and lamina propria is increased by irregularities in the connective tissue surface in the form of evaginations called **papillae** (L. diminutive of *papula,* nipple; singular, **papilla**). Papillae occur most frequently in epithelial tissues subject to stress, such as the skin and the tongue.

Polarity

An important feature of epithelia is their polarity; ie, they have a free, or apical, surface and a basal surface that rests on a basal lamina. Since blood vessels do not penetrate an epithelium, all nutrients must pass out of the capillaries in the underlying lamina propria. These nutrients and precursors of products of the epithelial cells then diffuse across the basal lamina and are taken up through the basolateral surface of the epithelial cell, usually by an energy-dependent process. Receptors for chemical messengers (eg, hormones, neurotransmitters) that influence the activity of epithelial cells are localized in the basolateral membranes. In absorptive epithelial cells, the apical cell membrane

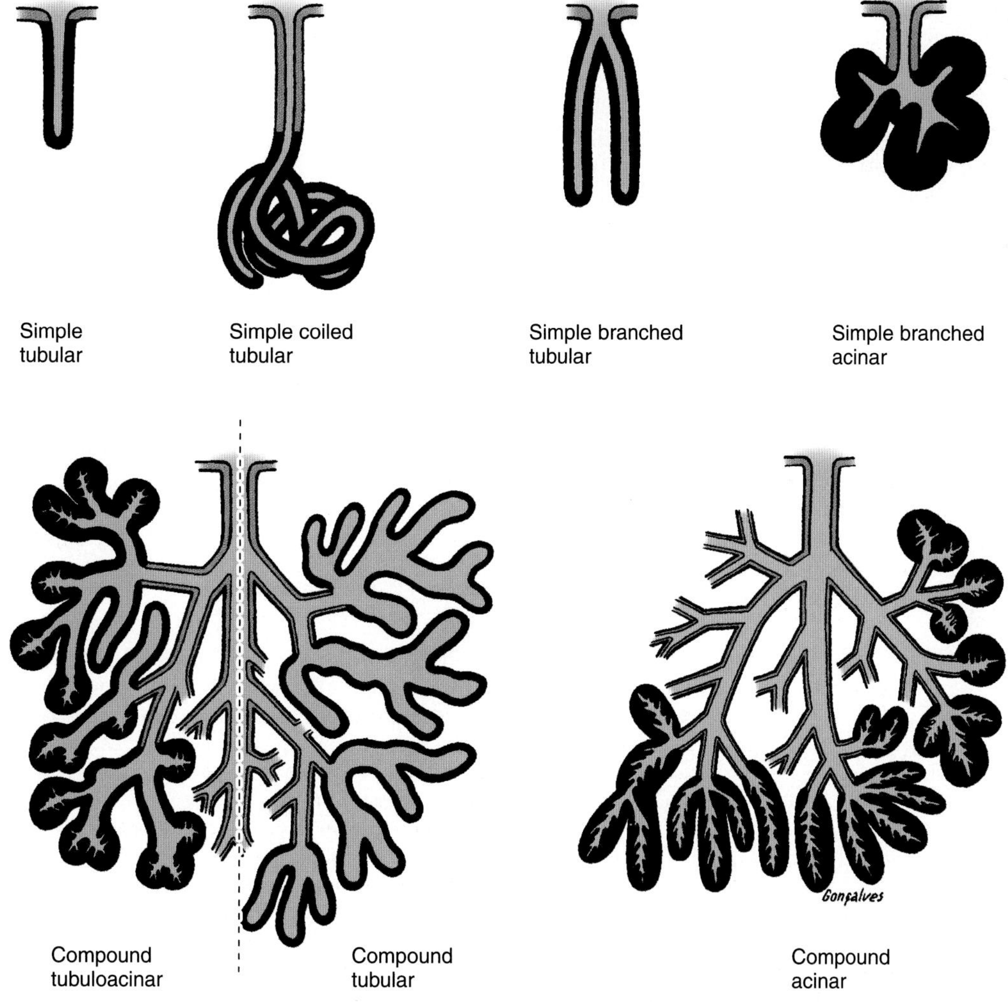

Figure 4–22. Principal types of exocrine glands. The part of the gland formed by secretory cells is shown in black; the remainder shows the ducts. The compound glands have branching ducts.

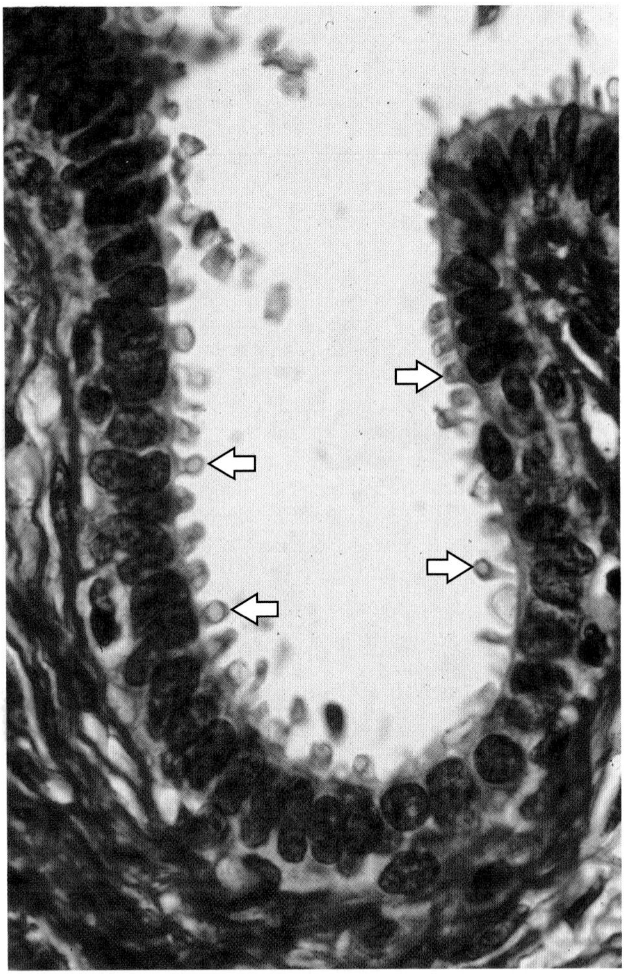

Figure 4–23. Section of the secreting portion of a mammary gland; apocrine secretion is characterized by the discharge of the secretion product with part of the cytoplasm (arrows). PSH stain. Medium magnification.

contains, as integral membrane proteins, enzymes such as disaccharidases and peptidases, which complete the digestion of molecules to be absorbed. Tight junctions help to prevent the integral membrane proteins of the various cell membrane regions from intermingling.

Innervation

Most epithelial tissues receive a rich supply of sensory nerve endings from nerve plexuses in the lamina propria. Everyone is aware of the exquisite sensitivity of the cornea, the epithelium covering the anterior surface of the eye. This sensitivity is due to the great number of sensory nerve fibers that ramify between corneal epithelial cells.

Renewal of Epithelial Cells

Epithelial tissues are labile structures whose cells are renewed continuously by means of mitotic activity. The renewal rate is variable; it can be fast in such tissues as the intestinal epithelium, which is replaced every week, or slow, as in the liver and the pancreas. In stratified and pseudostratified epithelial tissues, mitosis takes place within the germinal layer, which contains the stem cells closest to the basal lamina.

Metaplasia

MEDICAL APPLICATION

Under certain abnormal conditions, one type of epithelial tissue may undergo transformation into another type. This process is called ***metaplasia*** *(Gr.* metaplasis, *transformation). The following examples illustrate this process.*

In heavy cigarette smokers, the ciliated pseudostratified epithelium lining the bronchi can be transformed into stratified squamous epithelium.

In individuals with chronic vitamin A deficiency, epithelial tissues of the type found in the bronchi and urinary bladder are gradually replaced by stratified squamous epithelium.

Metaplasia is not restricted to epithelial tissue; it may also occur in connective tissue. Metaplasia is reversible.

Control of Glandular Activity

Usually, glands are sensitive to both neural and endocrine control. However, one form of control frequently dominates the other. For example, exocrine secretion in the pancreas depends mainly on stimulation by the hormones secretin and cholecystokinin (Gr. *chole,* bile, + *kystis,* bladder, + *kinein,* to move). In contrast, the salivary glands are principally under neural control.

The neural and endocrine control of glands occurs through the action of chemical substances called **chemical messengers.**

Cells That Transport Ions

All cells have the ability to transport certain ions against a concentration and electrical-potential gradient, using ATP as an energy source. This is called **active transport,** to distinguish it from passive diffusion down a concentration gradient. In mammals, the sodium ion (Na^+) concentration in the extracellular fluid is 140 mmol/L, whereas the intracellular concentration is 5–15 mmol/L. In addition, the interior of the cells is electrically negative with respect to the extracellular environment. Under these conditions, Na^+ would tend to diffuse down both an electrical gradient and a concentration gradient. The cell uses the energy stored in ATP to actively extrude Na^+ by means of Mg^{2+}-activated Na^+/K^+-ATPase (**sodium pump**), thereby maintaining the required low intracellular sodium concentration.

Some epithelial cells (eg, proximal and distal renal tubules, striated ducts of salivary glands) use the sodium pump to transfer sodium across the epithelium, from its apex to its base; this

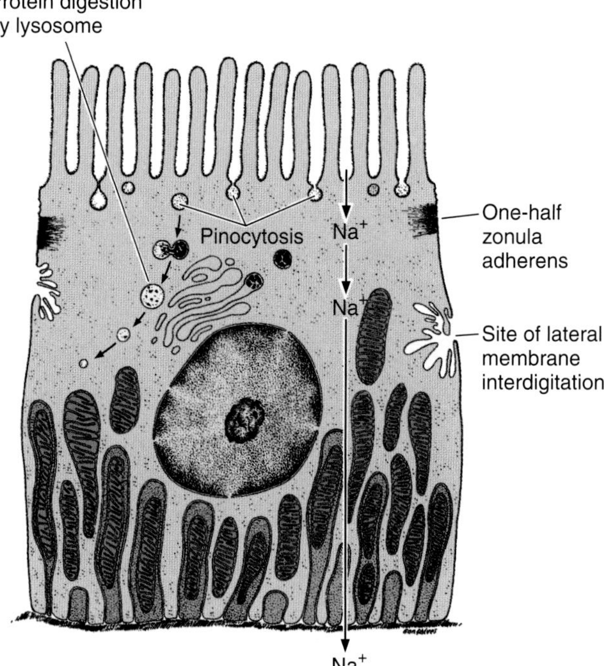

Figure 4–24. Ultrastructure of a proximal convoluted tubule cell of the kidney. Invaginations of the basal cell membrane outline regions filled with elongated mitochondria. This typical disposition is present in ion-transporting cells. Interdigitations from neighboring cells interlock with those of this cell. Protein being absorbed by pinocytosis and digested by lysosomes is shown in the upper left portion of the diagram. Sodium ions diffuse passively through the apical membranes of renal epithelial cells. These ions are then actively transported out of the cells by Na^+/K^+-ATPase located in the basolateral membranes of the cells. Energy for this sodium pump is supplied by nearby mitochondria.

is known as **transcellular transport.** The apical surface of the proximal renal tubule cell is freely permeable to Na^+. To maintain electrical and osmotic balance, equimolar amounts of chloride and water follow the Na^+ ion into the cell. The basal surfaces of these cells are elaborately folded (Figure 4–24); many long invaginations of the basal plasma membrane are seen in electron micrographs. In addition, there is elaborate interdigitation of basal processes between adjacent cells. It has been shown that Mg^{2+}-activated Na^+/K^+-ATPase is localized in these invaginations of the basal plasma membrane but is also present in the lateral membranes. Located between the invaginations are vertically oriented mitochondria that supply the energy (ATP) for the active extrusion of Na^+ from the base of the cell. Chloride and water again follow passively. In this way, sodium is returned to the circulation and not lost in massive amounts in the urine.

Tight junctions play an important role in the transport process. Because of their relative impermeability to ions, water, and larger molecules, they prevent back-diffusion of materials already transported across the epithelium. Otherwise, a great deal of energy would be wasted.

Ion transport and the consequent flow of fluid may occur in opposite directions (ie, apical → basal, basal → apical) in different epithelial tissues. In the intestine, proximal convoluted tubules of the kidney, striated ducts of the salivary glands, gallbladder, etc, the flow is from the apex of the cell to its basal region. Flow is in the opposite direction in other epithelial sheets, such as the choroid plexus and ciliary body. In both cases, the tight junctions seal the apical portions of the cells and provide for inner and outer tissue compartments (Figure 4–25).

Cells That Transport by Pinocytosis

In most cells of the body tissues, extracellular molecules are internalized in the cytoplasm by pinocytotic vesicles that form abundantly on the plasmalemma. This activity is clearly observed in the simple squamous epithelia that line the blood and lymphatic capillaries (endothelia) or the body cavities (mesothelia). These cells have few organelles other than the abundant pinocytotic vesicles found on the cell surfaces and in the cytoplasm. These observations, in conjunction with results obtained by injection of electron-dense colloidal particles (eg, ferritin, colloidal gold, thorium) followed by observation with the electron microscope, indicate that the vesicles transporting the injected materials flow in both directions through the cells.

Serous Cells

The acinar cells of the pancreas and parotid salivary glands are examples of the serous cell type. They are polyhedral or pyramidal, with central, rounded nuclei and well-defined polarity. In the basal region, serous cells exhibit an intense basophilia, which results from local accumulation of rough endoplasmic reticulum in the form of parallel arrays of cisternae studded with abundant polyribosomes (Figure 4–26). In the apical region lies a well-developed Golgi complex and many rounded, protein-rich, membrane-bound vesicles called **secretory granules.** In cells that produce digestive enzymes (eg, pancreatic acinar cells), these vesicles are called **zymogen granules** (Figures 4–26, 4–27, and 4–28). From the Golgi cisternae pinch off large membrane-bound **immature secretory granules** (Figure 4–28). They become denser as water is removed, forming the **mature secretory granules,** which accumulate until the cell is stimulated to secrete. When the cells release their secretory products, the membranes of secretory granules fuse to the cell membrane, and the granule contents spill out of the cell in a process called **exocytosis.** Because their surfaces have the same electrical charge, membrane lipid bilayers repel each other. Therefore, fusion of cell membranes is a rather complex process assisted and controlled by proteins. The movements of secretory granules, as well as all other cytoplasmic structures, are under the influence of cytoskeletal and motor proteins of the cytosol. Under the light microscope, serous secreting cells show marked basophilia in the basal portion of the cytoplasm (Figure 4–29) owing to the presence of many ribosomes, most attached to the rough endoplasmic reticulum. The apical cytoplasm appears filled with light-staining secretory vesicles.

Mucus-Secreting Cells

The most thoroughly studied mucus-secreting cell is the **goblet cell** of the intestines. This cell is characterized by the presence

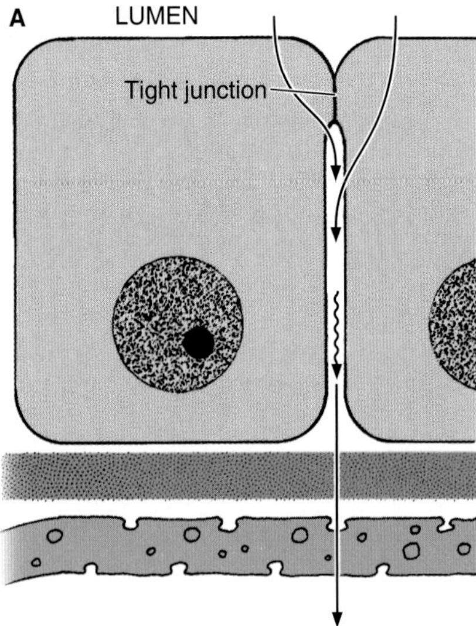

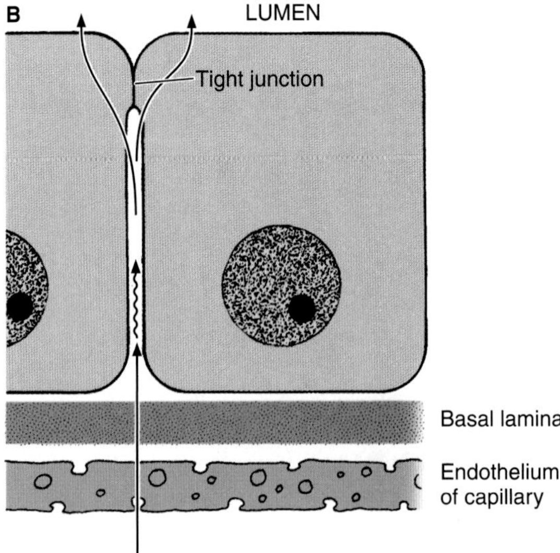

Figure 4–25. Ion and fluid transport can occur in different directions, depending on which tissue is involved. **A:** The direction of transport is from the lumen to the blood vessel, as in the gallbladder and intestine. This process is called **absorption.** **B:** Transport is in the opposite direction, as in the choroid plexus, ciliary body, and sweat gland. This process is called **secretion.** Note that the presence of occluding junctions is necessary to maintain compartmentalization and consequent control over ion distribution.

of numerous large, lightly staining granules containing strongly hydrophilic glycoproteins called **mucins.** Secretory granules fill the extensive apical pole of the cell, and the nucleus is usually located in the cell base. This region is rich in rough endoplasmic reticulum (Figures 4–30 and 4–31). The Golgi complex, located just above the nucleus, is exceptionally well developed, indicative of its important function in this cell. Data obtained by autoradiography suggest that, in this cell, proteins are synthesized in the cell base where most rough endoplasmic reticulum is located. Monosaccharides are added to the core protein by enzymes—**glycosyltransferases**—located in the endoplasmic reticulum and in the Golgi apparatus. When mucins are released from the cell, they become highly hydrated and form mucus, a viscous, elastic, protective lubricating gel.

The goblet cell of the intestines (Figure 4–32) is only one of several types of cells that synthesize mucin glycoproteins. Other types are found in the stomach, salivary glands, respiratory tract, and genital tract. These mucous cells show great variability in their morphologic features and in the chemical nature of their secretions. For example, in salivary glands, mucous secretory cells have a diverse structure (Figure 4–33) and frequently are present with serous secretory cells in the same acinus (Figure 4–34).

The Diffuse Neuroendocrine System (DNES)

Studies initially performed in the digestive system revealed the presence of endocrine cells interspersed among nonendocrine cells. The cytoplasm of the endocrine cells contains either polypeptide hormones or the biogenic amines epinephrine, norepinephrine, or 5-hydroxytryptamine (serotonin). In some cases, more than one of these compounds is present in the same cell. Many, but not all, of these cells are able to take up amine precursors and exhibit amino acid decarboxylase activity. These characteristics explain the acronym APUD (amine precursor uptake and decarboxylation) by which they are known. Because some of these cells stain with silver salts, they are also called **argentaffin** and **argyrophil** cells.

Because some of these endocrine cells do not concentrate amine precursors, the APUD designation has largely been replaced by DNES (diffuse neuroendocrine system). DNES cells are derived from the embryonic nervous system and can be identified and localized by immunocytochemical methods or other cytochemical techniques for specific amines. These cells are widespread throughout the organism and include about 35 types of cells in the respiratory, urinary, and gastrointestinal systems; thyroid; and hypophysis. Some DNES cells are known as **paracrine cells** because they produce chemical signals that diffuse into the surrounding extracellular fluid to regulate the function of neighboring cells without passing through the vascular system. Many of the polypeptide hormones and amines produced by DNES cells also act as chemical mediators in the nervous system. DNES polypeptide-secreting cells generally have distinctive, dense granules about 100–400 nm in diameter (Figure 4–35).

MEDICAL APPLICATION

***Apudomas** are tumors derived from polypeptide-secreting cells of the DNES. Clinical symptoms result from hypersecretion of the specific chemical messenger involved. The diagnosis is usually confirmed by using immunocytochemical methods on sections of the tumor biopsies.*

Myoepithelial Cells

Several exocrine glands (eg, sweat, lachrymal, salivary, mammary) contain stellate or spindle-shaped myoepithelial cells (Figure 4–36). These cells embrace gland acini as an octopus might embrace a rounded boulder. They are more longitudinally arranged along ducts. Myoepithelial cells are located between the basal lamina and the basal pole of secretory or ductal cells. They are connected to each other and to the epithelial cells by gap junctions and desmosomes. The cytoplasm contains numerous actin filaments, as well as myosin. Myoepithelial cells also contain intermediate filaments that belong to the keratin family, which confirms their epithelial origin. The function of myoepithelial cells is to contract around the secretory or conducting

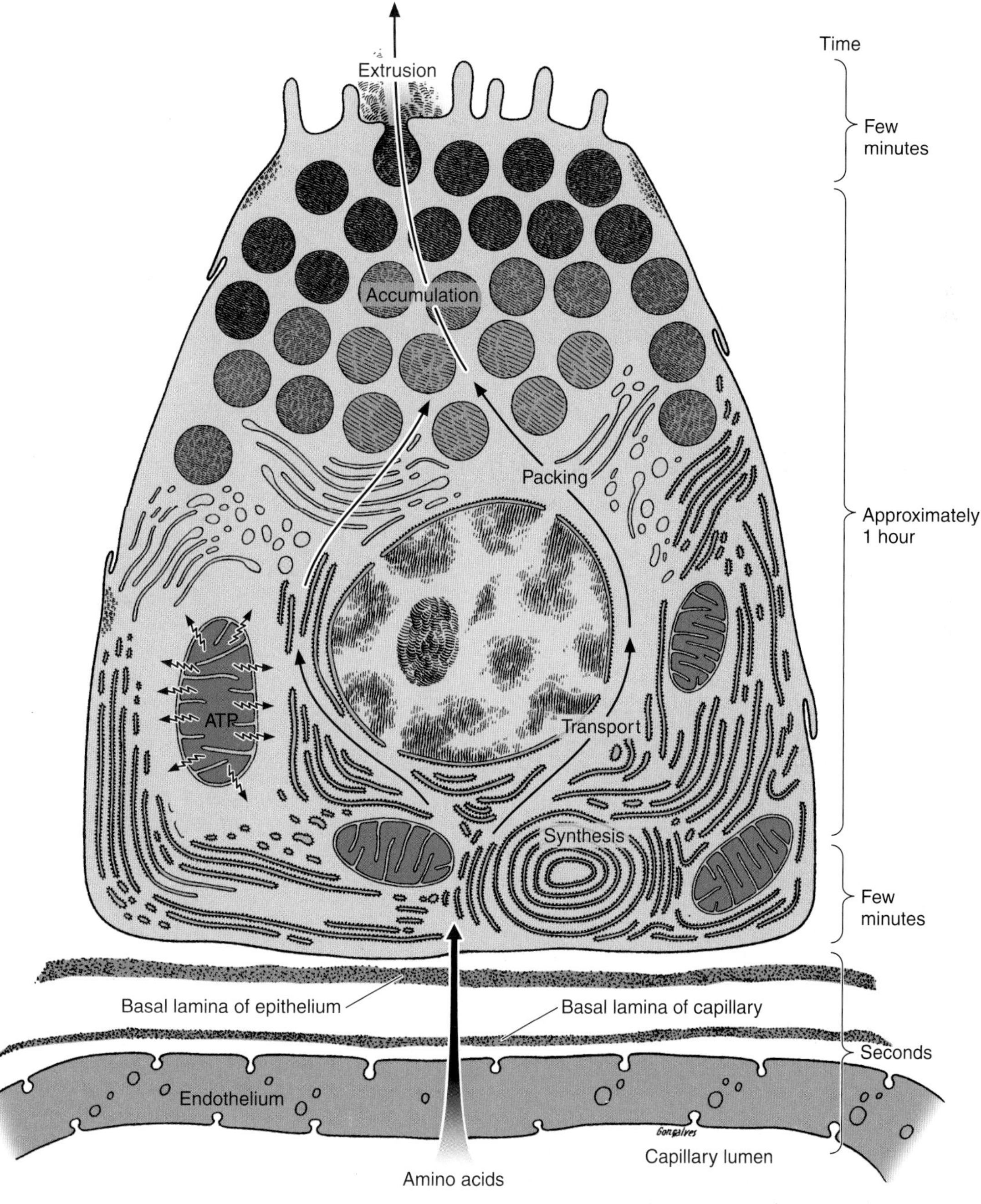

Figure 4–26. Diagram of a serous (pancreatic acinar) cell. Note its evident polarity, with abundant basal rough endoplasmic reticulum. The Golgi complex and zymogen granules are in the apical region. To the right is a scale indicating the approximate time necessary for each step.

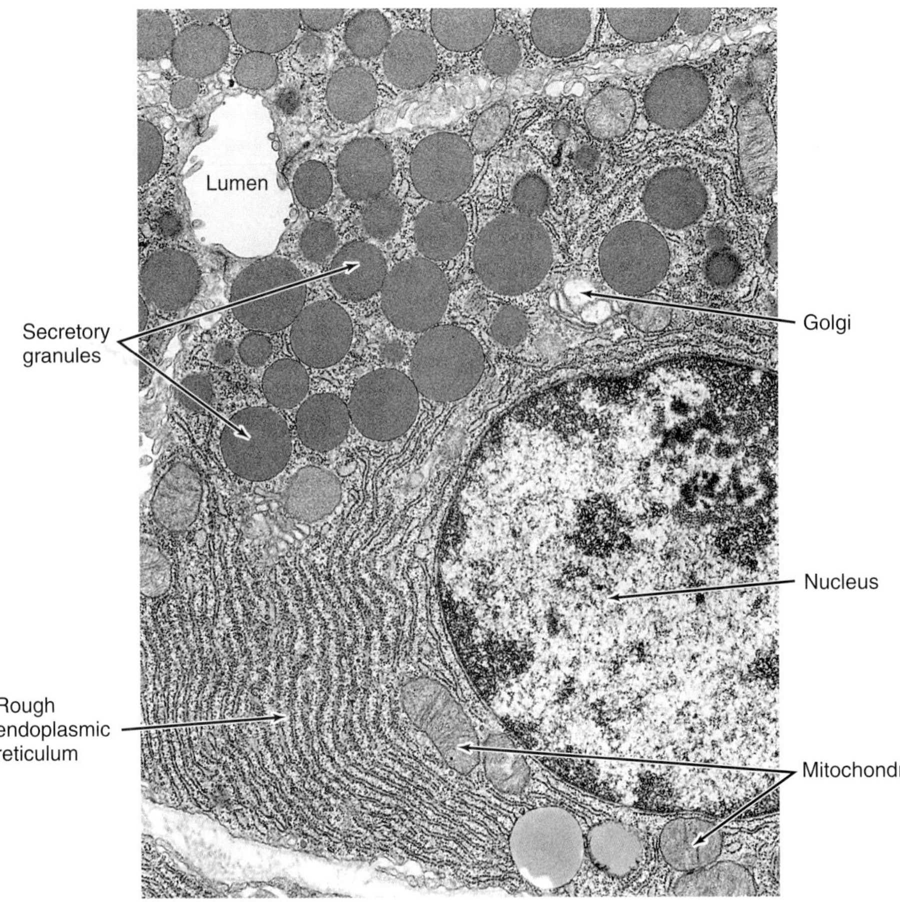

Figure 4–27. Electron micrograph of a pancreatic cell. Note the nucleus, mitochondria, Golgi complex, secretory (zymogen) granules in various stages of condensation, and rough endoplasmic reticulum. ×13,000. (Courtesy of KR Porter.)

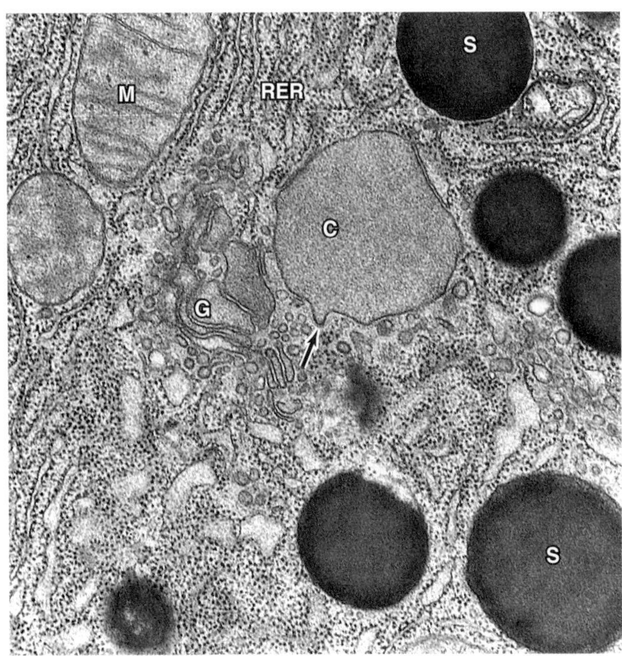

Figure 4–28. Electron micrograph of part of a pancreatic acinar cell showing a condensing vacuole (C), which is presumed to be receiving a small quantity of secretory product (arrow) from the Golgi complex (G). M, mitochondrion; RER, rough endoplasmic reticulum; S, mature condensed secretory (zymogen) granule. ×40,000.

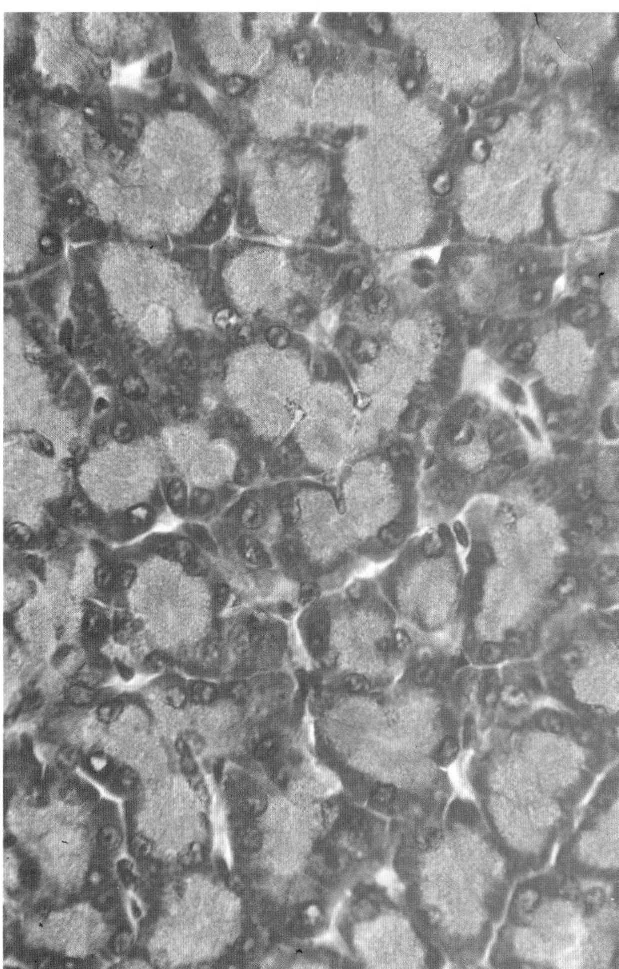

Figure 4–29. Serous secretory cells of the pancreas disposed in acini. Note the basophilic cell basal region rich in RNA and the apex with the light-stained secretory vesicles. PT stain. Medium magnification.

portion of the gland and thus to help propel secretory products toward the exterior.

Steroid-Secreting Cells

Cells that secrete steroids are found in various organs of the body (eg, testes, ovaries, adrenals). They are endocrine cells specialized for synthesizing and secreting steroids with hormonal activity and have the following characteristics (Figure 4–37):

1. They are polyhedral or rounded acidophilic cells with a central nucleus and a cytoplasm that is usually—but not invariably—rich in lipid droplets.
2. The cytoplasm of steroid-secreting cells contains an exceptionally rich smooth endoplasmic reticulum, which takes the form of anastomosing tubules. Smooth endoplasmic reticulum contains the enzymes necessary to synthesize cholesterol from acetate and other substrates and to transform the pregnenolone produced in the mitochondria into androgens, estrogens, and progestogens.
3. The spherical or elongated mitochondria that are present usually contain tubular cristae rather than the shelflike cristae that are common in mitochondria of other epithelial cells. In addition to being the main site of energy production for cell function, these organelles have the necessary enzymatic equipment not only to cleave the cholesterol side chain and produce pregnenolone but also to participate in subsequent reactions that result in steroid hormones. The process of steroid synthesis results, therefore, from close collaboration between smooth endoplasmic reticulum and mitochondria, a striking example of cooperation between cell organelles (Figure 21–4). This process also explains the close proximity observed between these two organelles in steroid-secreting cells.

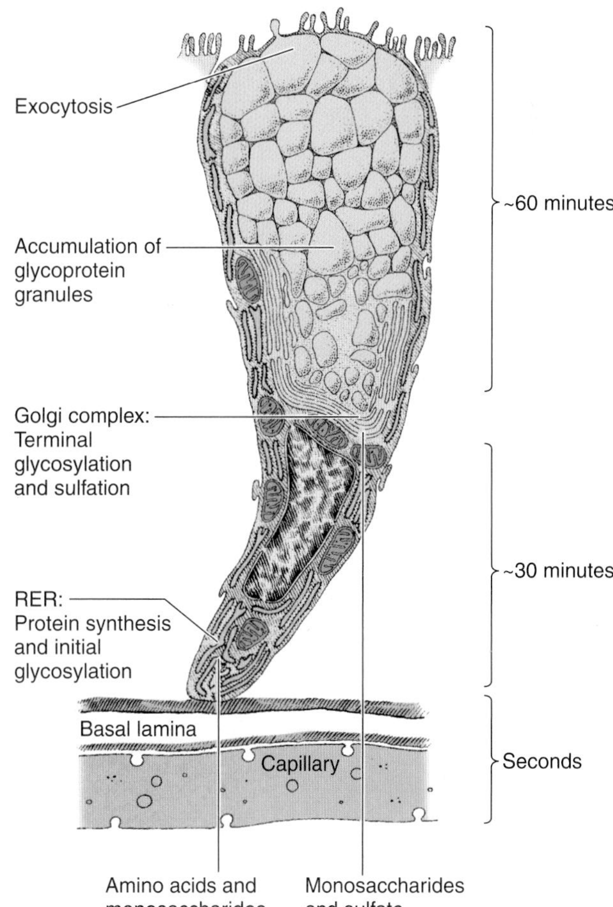

Figure 4–30. Diagram of a mucus-secreting intestinal goblet cell showing a typically constricted base, where the mitochondria and rough endoplasmic reticulum (RER) are located. The protein part of the glycoprotein complex is synthesized in the endoplasmic reticulum. A well-developed Golgi complex is present in the supranuclear region. (Redrawn after Gordon and reproduced, with permission, from Ham AW: *Histology,* 6th ed. Lippincott, 1969.)

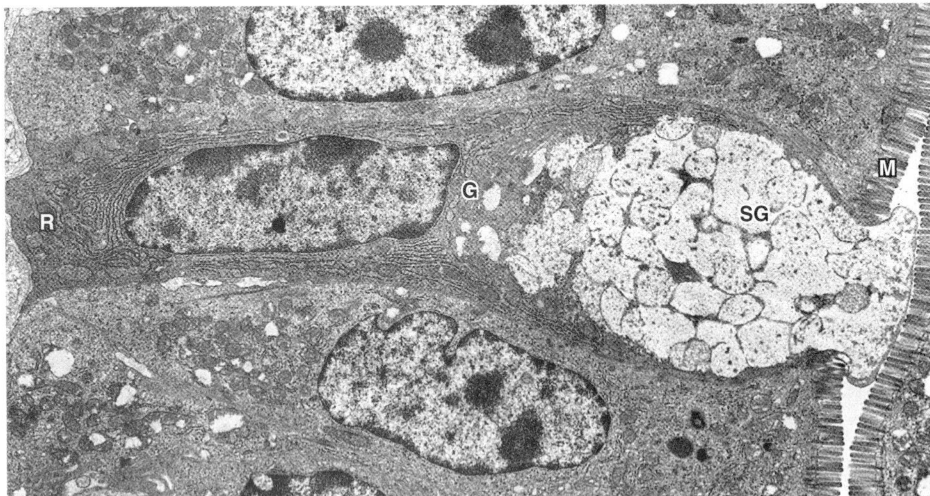

Figure 4–31. Electron micrograph of a goblet cell from the small intestine. The rough endoplasmic reticulum is present mainly in the basal portion of the cell (R), while the cell apex is filled with light secretory vesicles or granules (SG) some of which are being discharged. The Golgi complex (G) lies just above the nucleus. Typical columnar absorptive cells with microvillar borders (M) lie adjacent to the goblet cell. ×7000. (Reproduced, with permission, from Junqueira LCU, Salles LMM. Ultra-Estrutura e Função Celular, Edgard Blücher, 1975.)

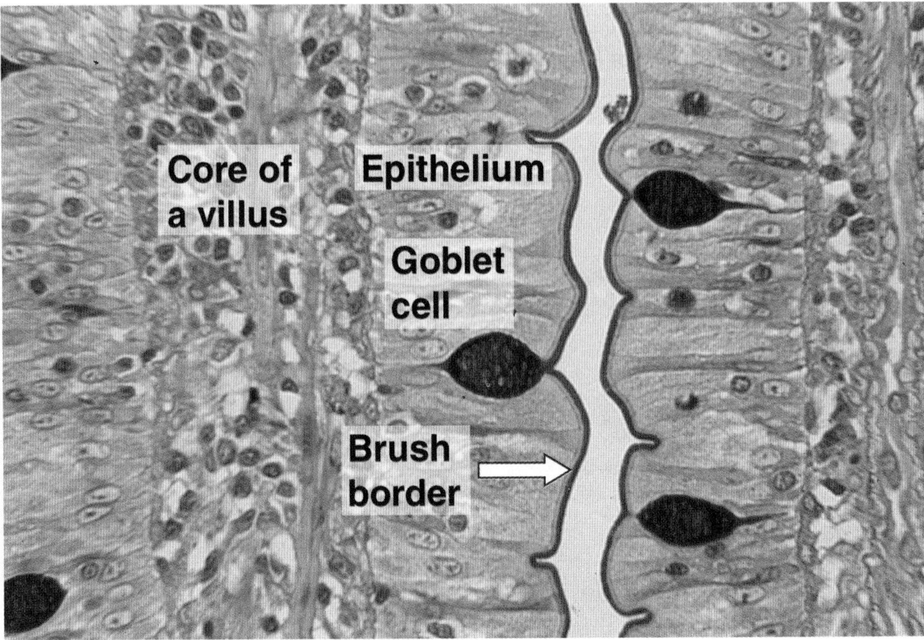

Figure 4–32. Intestinal villi stained by the PAS technique, a procedure that detects some polysaccharides. Note the positive reaction in the goblet cells and brush border, which consists of microvilli. Counterstained with hematoxylin.

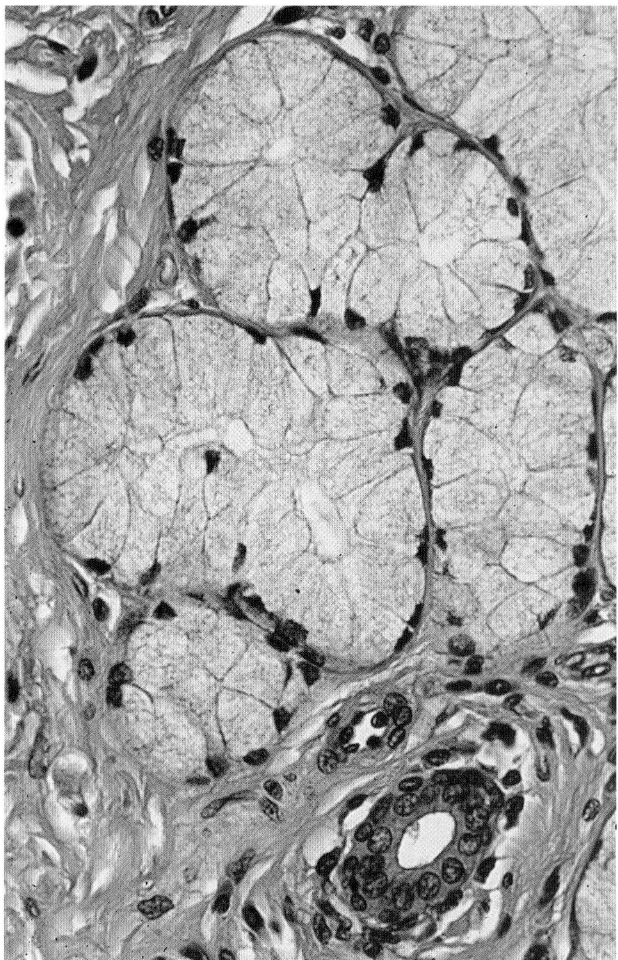

Figure 4–33. Esophageal mucous secretory gland with characteristic irregular, clear cytoplasm and basal nuclei. Loose connective tissue surrounds a secretory duct.

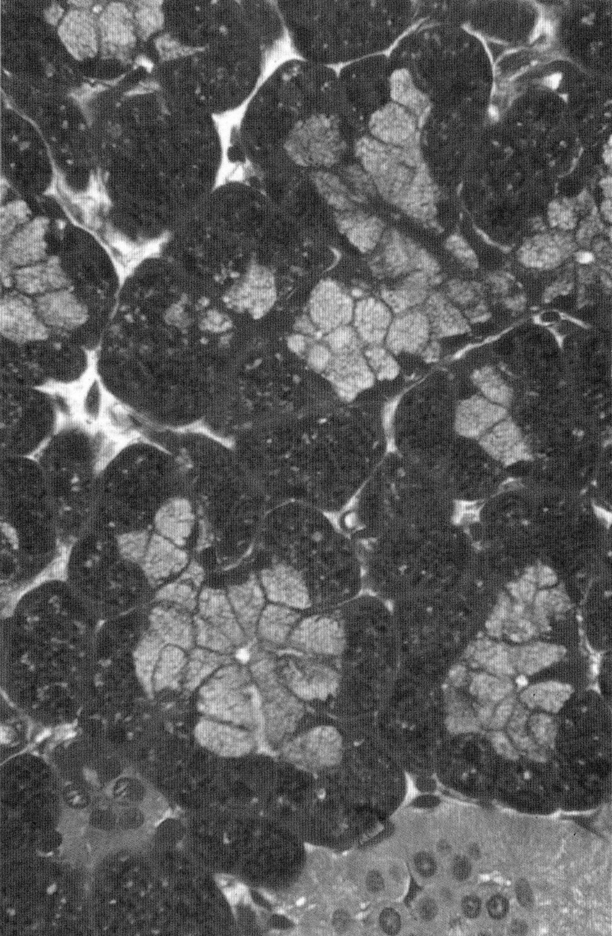

Figure 4–34. Submandibular salivary gland showing 2 types of secretory epithelial cells in a compound tubuloacinar gland. The light cells are mucous and the dark cells are serous. PT stain. Medium magnification.

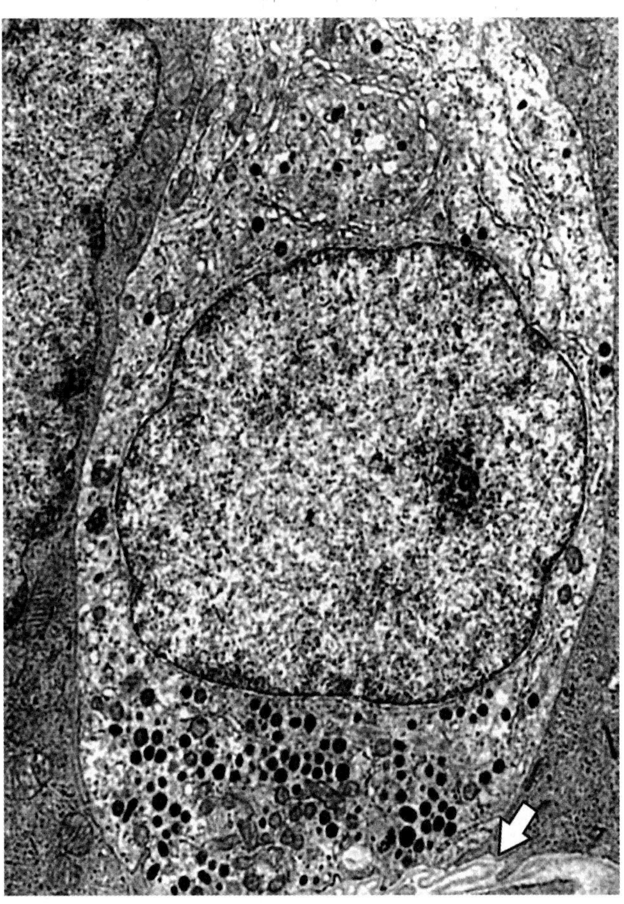

Figure 4–35. Electron micrograph of a cell of the diffuse neuroendocrine system. Note the accumulation of secretory granules in the basal region of the cell. The Golgi complex seen in the upper part of the micrograph shows some secretory granules; it is here that these granules first appear. The arrow indicates the basal lamina.

Epithelial Cell–Derived Tumors

MEDICAL APPLICATION

Both benign and malignant tumors can arise from most types of epithelial cells. A ***carcinoma*** *(Gr.* karkinos, *cancer, +* oma, *tumor) is a malignant tumor of epithelial cell origin. Malignant tumors derived from glandular epithelial tissue are usually called* ***adenocarcinomas*** *(Gr.* adenos, *gland, +* karkinos*); these are by far the most common tumors in adults. In children up to age 10, most tumors develop (in decreasing order) from hematopoietic organs, nerve tissues, connective tissues, and epithelial tissues. This proportion gradually changes, and after age 45, more than 90% of all tumors are of epithelial origin.*

Carcinomas composed of differentiated cells reflect cell-specific morphologic features and behaviors (eg, the production of keratins, mucins, and hormones). Undifferentiated carcinomas are often difficult to diagnose by morphologic analysis alone. Since these carcinomas usually contain keratins, the detection of keratins by immunocytochemistry often helps to determine the diagnosis and treatment of these tumors.

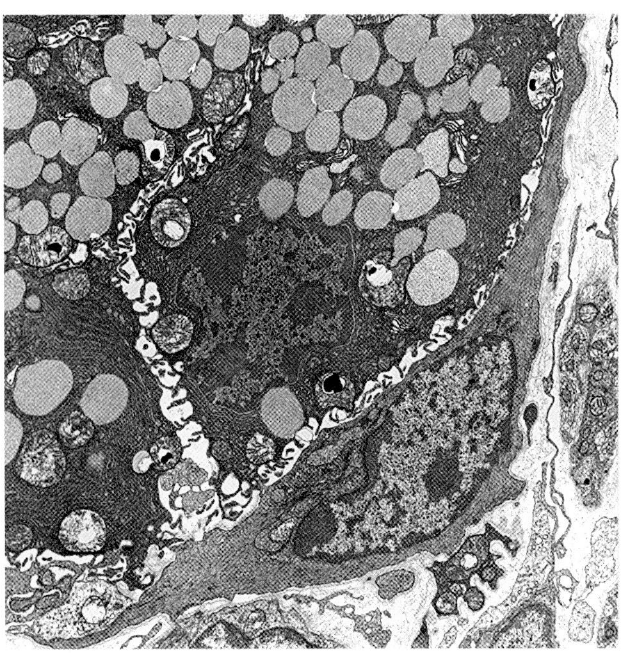

Figure 4–36. Electron micrograph of salivary gland showing secretory cells in the upper left; in the lower right is a myoepithelial cell that embraces the secretory acinus. Contraction of the myoepithelial cell compresses the acinus and aids in the expulsion of secretory products.

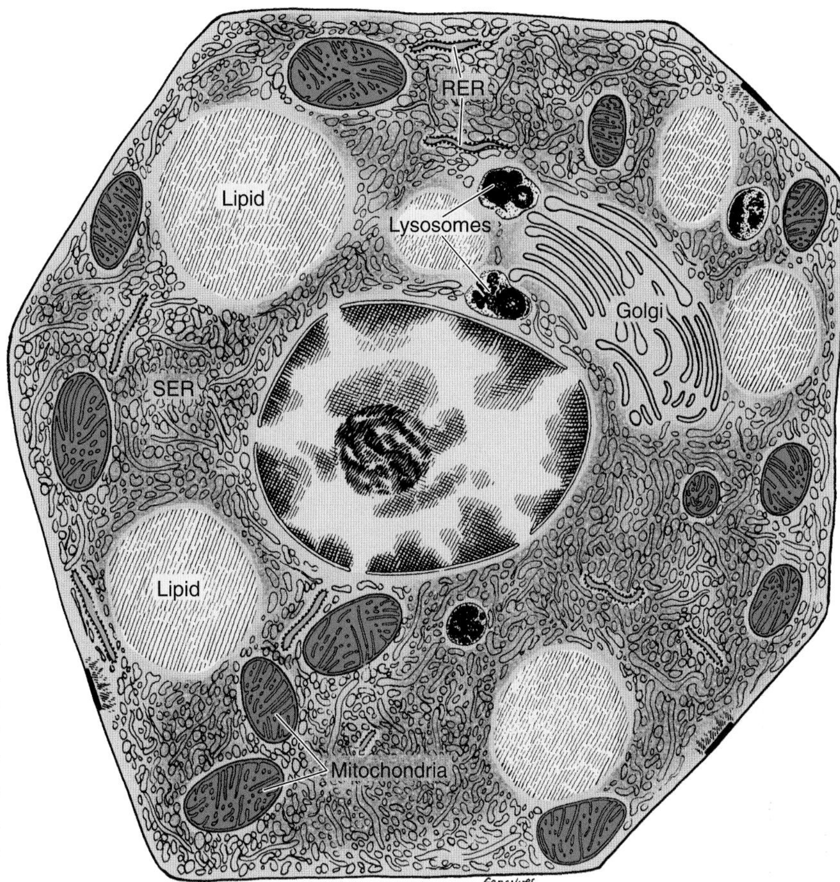

Figure 4–37. Diagram of the ultrastructure of a hypothetical steroid-secreting cell. Note the abundance of the smooth endoplasmic reticulum (SER), lipid droplets, Golgi complex, and lysosomes. The numerous mitochondria have mainly tubular cristae. They not only produce the energy necessary for the activity of the cell but are also involved in steroid hormone synthesis. Rough endoplasmic reticulum (RER) is also shown.

REFERENCES

Balda MS, Matter K: Transmembrane proteins of tight junctions. Cell Dev Biol 2000;11:281.

Bertram JS: Cellular communications via gap junctions. Sci & Med 2000;7(2):18.

Darnell J et al: *Molecular Cell Biology,* 2nd ed. Scientific American Books, 1990.

Farquhar MG, Palade GE: Junctional complexes in various epithelia. J Cell Biol 1963;17:375.

Fawcett D: *The Cell,* 2nd ed. Saunders, 1981.

Hall PF: Cellular organization for steroidogenesis. Int Rev Cytol 1984;86:53.

Hertzberg EL et al: Gap junctional communication. Annu Rev Physiol 1981;43:479.

Hull BE, Staehelin LA: The terminal web: a reevaluation of its structure and function. J Cell Biol 1979;81:67.

Jamieson JD, Palade, GE: Intracellular transport of secretory protein in the pancreatic exocrine cell. 4. Metabolic requirements. J Cell Biol 1968;39:589.

Kefalides NA: *Biology and Chemistry of Basement Membranes.* Academic Press, 1978.

Krstíc RV: *Illustrated Encyclopedia of Human Histology.* Springer-Verlag, 1984.

Mooseker MS: Organization, chemistry, and assembly of the cytoskeletal apparatus of the intestinal brush border. Annu Rev Cell Biol 1985; 1:209.

Simons K, Fuller SD: Cell surface polarity in epithelia. Annu Rev Cell Biol 1985;1:243.

Staehelin LA, Hull BE: Junctions between living cells. Sci Am 1978;238:41.

Connective Tissue

5

The several types of connective tissues are responsible for providing and maintaining form in the body. Functioning in a mechanical role, they provide a matrix that connects and binds the cells and organs and ultimately gives support to the body.

Structurally, connective tissue is formed by 3 classes of components: cells, fibers, and ground substance. Unlike the other tissues (epithelium, muscle, and nerve), which are formed mainly by cells, the major constituent of connective tissue is the **extracellular matrix.** Extracellular matrices consist of different combinations of **protein fibers** (collagen, reticular, and elastic) and **ground substance.** Ground substance is a highly hydrophilic, viscous complex of anionic macromolecules (glycosaminoglycans and proteoglycans) and multiadhesive glycoproteins (laminin, fibronectin, and others) that imparts strength and rigidity to the matrix by binding to receptor proteins (**integrins**) on the surface of cells and to the other matrix components. Besides its conspicuous structural function, the molecules of connective tissue serve other important biological functions, such as being a reservoir of hormones controlling cell growth and differentiation.

The connective tissue matrix also serves as the medium through which nutrients and metabolic wastes are exchanged between cells and their blood supply.

The wide variety of connective tissue types in the body reflects variations in the composition and amount of the 3 components (cells, fibers, and ground substance) which are responsible for the remarkable structural, functional, and pathologic diversity of connective tissue. Fibers, predominantly composed of collagen, constitute tendons, aponeuroses, capsules of organs, and membranes that envelop the central nervous system (**meninges**). They also make up the trabeculae and walls inside several organs, forming the most resistant component of the **stroma,** or supporting tissue of organs.

The connective tissues originate from the **mesenchyme,** an embryonic tissue formed by elongated cells, the **mesenchymal cells.** These cells are characterized by an oval nucleus with prominent nucleoli and fine chromatin. They possess many thin cytoplasmic processes and are immersed in an abundant and viscous extracellular substance containing few fibers. The mesenchyme develops mainly from the middle layer of the embryo, the **mesoderm.** Mesodermal cells migrate from their site of origin, surrounding and penetrating developing organs. In addition to being the point of origin of all types of connective tissue cells, mesenchyme develops into other types of structures, such as blood cells, endothelial cells, and smooth muscle cells.

CELLS OF THE CONNECTIVE TISSUE

Some cells of connective tissue are produced locally and remain in the connective tissue; others, such as leukocytes, come from other areas and can be transient inhabitants of connective tissue (Figure 5–1). The following cell types are found in connective tissue: fibroblasts, macrophages, mast cells, plasma cells, adipose cells, and leukocytes. The various functions of these cells are summarized in Table 5–1.

Fibroblasts

Fibroblasts synthesize collagen, elastin, glycosaminoglycans, proteoglycans and multiadhesive glycoproteins. Fibroblasts are the most common cells in connective tissue (Figure 5–2) and are responsible for the synthesis of extracellular matrix components. Two stages of activity—active and quiescent—are observed in these cells. Cells with intense synthetic activity are morphologically distinct from the quiescent fibroblasts that are scattered within the matrix they have already synthesized. Some histologists reserve the term **fibroblast** to denote the active cell and **fibrocyte** to denote the quiescent cell.

The active fibroblast has an abundant and irregularly branched cytoplasm. Its nucleus is ovoid, large, and pale-staining, with fine chromatin and a prominent nucleolus. The cytoplasm is rich in rough endoplasmic reticulum, and the Golgi complex is well developed (Figures 5–3, 5–4, and 5–5).

The quiescent fibroblast, or fibrocyte (Figure 5–3), is smaller than the active fibroblast and tends to be spindle-shaped. It has fewer processes; a smaller, darker, elongated nucleus; an acidophilic cytoplasm; and a small amount of rough endoplasmic reticulum.

Fibroblasts synthesize proteins, such as collagen and elastin, that form collagen, reticular, and elastic fibers, and the glycosaminoglycans, proteoglycans, and glycoproteins of the extracellular matrix. Fibroblasts also are involved in the production of **growth factors** that influence cell growth and differentiation. In adults, fibroblasts in connective tissue rarely undergo division; however, mitoses are observed when the organism requires additional fibroblasts.

MEDICAL APPLICATION

The regenerative capacity of the connective tissue is clearly observed when tissues are destroyed by inflammation or traumatic injury. In these cases, the spaces left after injury to tissues whose cells do not divide (eg, cardiac muscle) are filled by connective

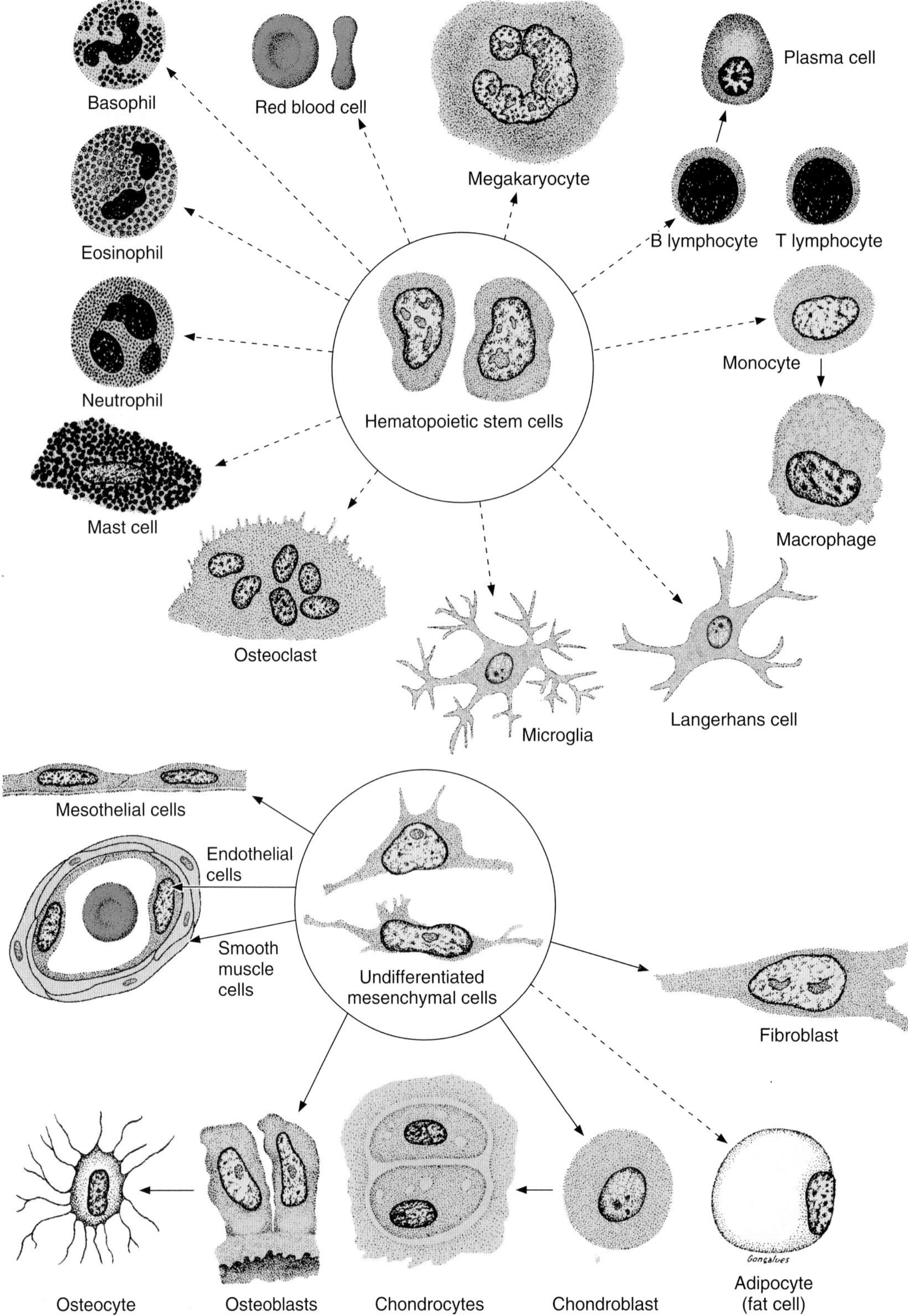

Figure 5–1. Simplified representation of the connective tissue cell lineage derived from the multipotential embryonic mesenchyme cell. Dotted arrows indicate that intermediate cell types exist between the examples illustrated. Note that the cells are not drawn in proportion to actual sizes, eg, adipocyte, megakaryocyte, and osteoclast cells are significantly larger than the other cells illustrated.

Table 5–1. Functions of connective tissue cells.

Cell Type	Representative Product or Activity	Representative Function
Fibroblast, chondroblast, osteoblast, odontoblast	Production of fibers and ground substance	Structural
Plasma cell	Production of antibodies	Immunologic (defense)
Lymphocyte (several types)	Production of immunocompetent cells	Immunologic (defense)
Eosinophilic leukocyte	Participation in allergic and vasoactive reactions, modulation of mast cell activities and the inflammatory process	Immunologic (defense)
Neutrophilic leukocyte	Phagocytosis of foreign substances, bacteria	Defense
Macrophage	Secretion of cytokines and other molecules, phagocytosis of foreign substances and bacteria, antigen processing and presentation to other cells	Defense
Mast cell and basophilic leukocyte	Liberation of pharmacologically active molecules (eg, histamine)	Defense (participate in allergic reactions)
Adipose (fat) cell	Storage of neutral fats	Energy reservoir, heat production

tissue, which forms a scar. The healing of surgical incisions depends on the reparative capacity of connective tissue. The main cell type involved in repair is the fibroblast.

*When it is adequately stimulated, such as during wound healing, the fibrocyte reverts to the fibroblast state, and its synthetic activities are reactivated. In such instances the cell reassumes the form and appearance of a fibroblast. The **myofibroblast,** a cell with features of both fibroblasts and smooth muscle, is also observed during wound healing. These cells have most of the morphologic characteristics of fibroblasts but contain increased amounts of actin microfilaments and myosin and behave like smooth muscle cells. Their activity is responsible for wound closure after tissue injury, a process called **wound contraction.***

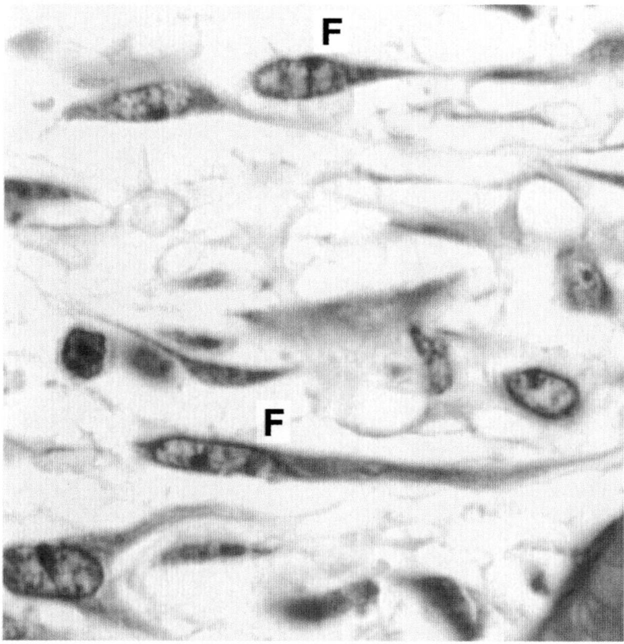

Figure 5–2. Section of rat skin. A connective tissue layer (dermis) shows several fibroblasts (F), which are the elongated cells. H&E stain. Medium magnification.

Macrophages and the Mononuclear Phagocyte System

Macrophages were discovered and initially characterized by their phagocytic ability. Macrophages have a wide spectrum of morphologic features that correspond to their state of functional activity and to the tissue they inhabit.

When a vital dye such as trypan blue or India ink is injected into an animal, macrophages engulf and accumulate the dye in their cytoplasm in the form of granules or vacuoles visible in the light microscope (Figure 5–6).

In the electron microscope, they are characterized by an irregular surface with pleats, protrusions, and indentations, a morphologic expression of their active pinocytotic and phagocytic activities. They generally have a well-developed Golgi complex, many lysosomes, and a prominent rough endoplasmic reticulum (Figures 5–7 and 5–8).

Macrophages derive from bone marrow precursor cells that divide, producing **monocytes** that circulate in the blood. In a second step, these cells cross the wall of venules and capillaries to penetrate the connective tissue, where they mature and acquire morphologic features of **macrophages.** Therefore, monocytes and macrophages are the same cell in different stages of maturation. Tissue macrophages can proliferate locally, producing more such cells.

Macrophages, which are distributed throughout the body, are present in most organs and constitute the **mononuclear**

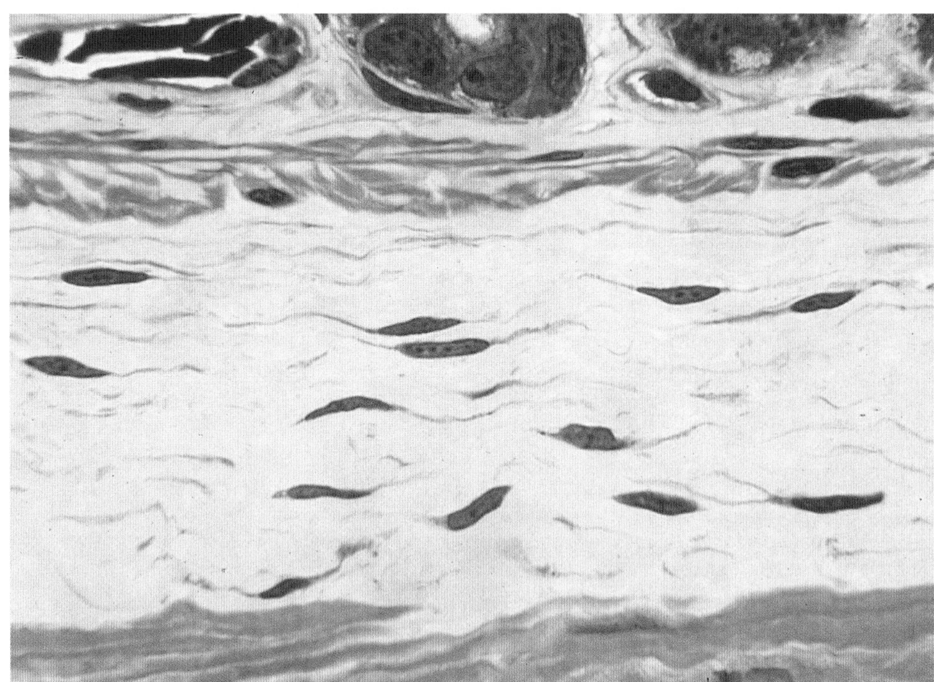

Figure 5–3. Quiescent fibroblasts are elongated cells with thin cytoplasmic extensions and condensed chromatin. Pararosaniline-toluidine blue (PT) stain. Medium magnification.

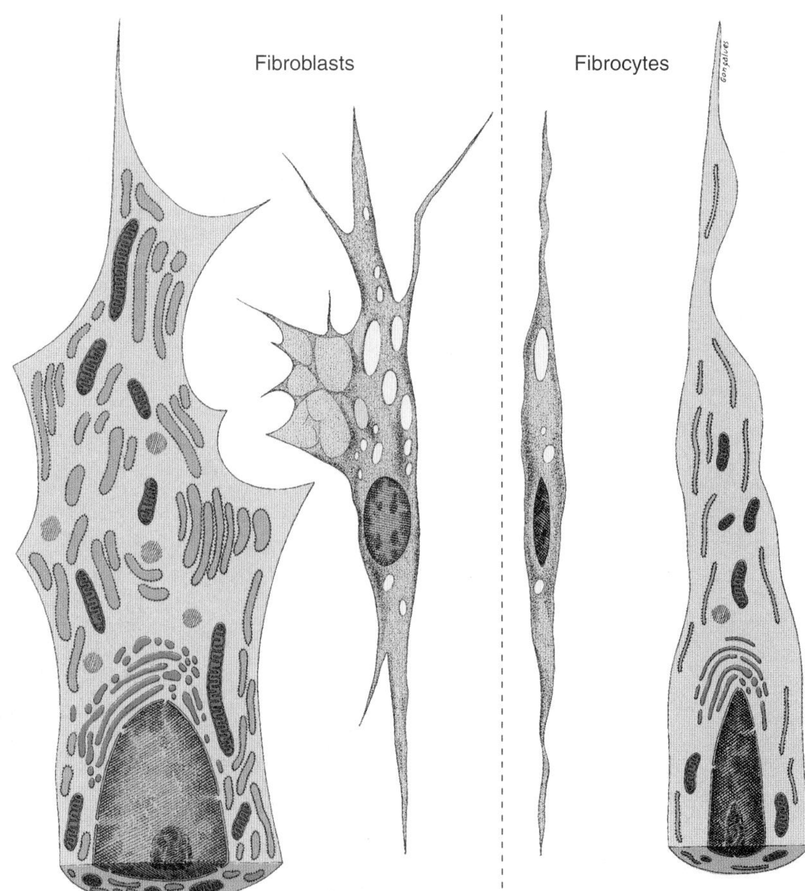

Figure 5–4. Active (**left**) and quiescent (**right**) fibroblasts. External morphologic characteristics and ultrastructure of each cell are shown. Fibroblasts that are actively engaged in synthesis are richer in mitochondria, lipid droplets, Golgi complex, and rough endoplasmic reticulum than are quiescent fibroblasts (fibrocytes).

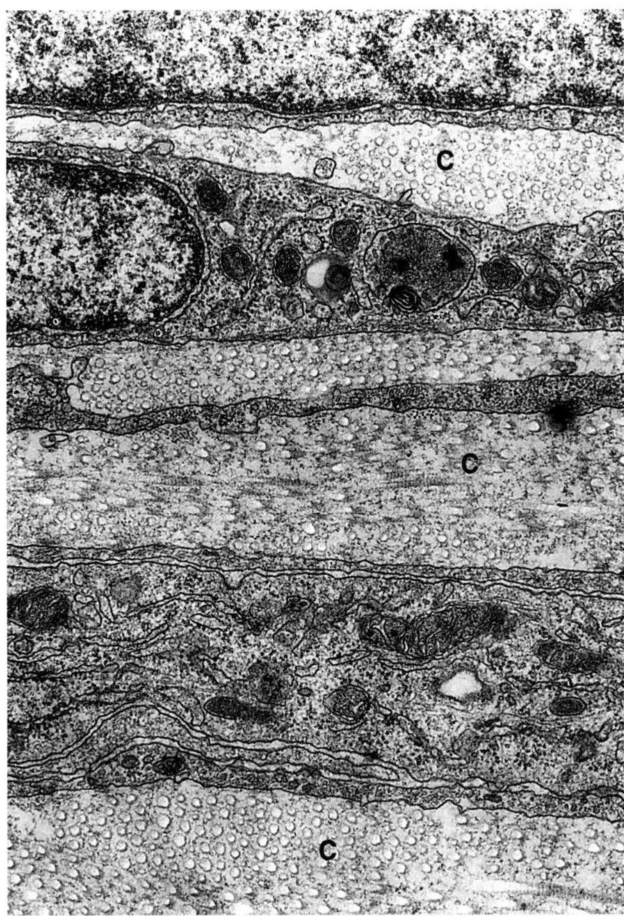

Figure 5–5. Electron micrograph revealing portions of several flattened fibroblasts in dense connective tissue. Abundant mitochondria, rough endoplasmic reticulum, and vesicles distinguish these cells from the less active fibrocytes. Multiple strata of collagen fibrils (C) lie among the fibroblasts. ×30,000.

phagocyte system (Table 5–2). They are long-living cells and may survive for months in the tissues. In certain regions, macrophages have special names, eg, Kupffer cells in the liver, microglial cells in the central nervous system, Langerhans cells of the skin, and osteoclasts in bone tissue. The process of monocyte-to-macrophage transformation results in an increase in protein synthesis and cell size. Increases in the Golgi complex and in the number of lysosomes, microtubules, and microfilaments are also apparent. Macrophages measure between 10 and 30 μm and usually have an oval or kidney-shaped nucleus located eccentrically.

MEDICAL APPLICATION

When adequately stimulated, macrophages may increase in size and are arranged in clusters forming ***epithelioid*** *cells (named for their vague resemblance to epithelial cells), or several may fuse to form* ***multinuclear giant cells.*** *Both cell types are usually found only in pathologic conditions (Figure 5–9).*

Macrophages act as defense elements. They phagocytize cell debris, abnormal extracellular matrix elements, neoplastic cells, bacteria, and inert elements that penetrate the organism.

Macrophages are also antigen-presenting cells that participate in the processes of partial digestion and presentation of antigen to other cells (see Chapter 14). A typical example of an antigen-processing cell is the macrophage present in the skin epidermis, called the Langerhans cell (see Chapter 18). Although macrophages are the main antigen-presenting cells, under certain circumstances many other cell types, such as fibroblasts, endothelial cells, astrocytes, and thyroid epithelial cells, are also able to perform this function. Macrophages also participate in cell-mediated resistance to infection by bacteria, viruses, protozoans, fungi, and metazoans (eg, parasitic worms); in cell-mediated resistance to tumors; and in extrahepatic bile production, iron and fat metabolism, and the destruction of aged erythrocytes.

When macrophages are stimulated (by injection of foreign substances or by infection), they change their morphologic characteristics and metabolism. They are then called ***activated macrophages*** *and acquire characteristics not present in their nonactivated state. These activated macrophages, in addi-*

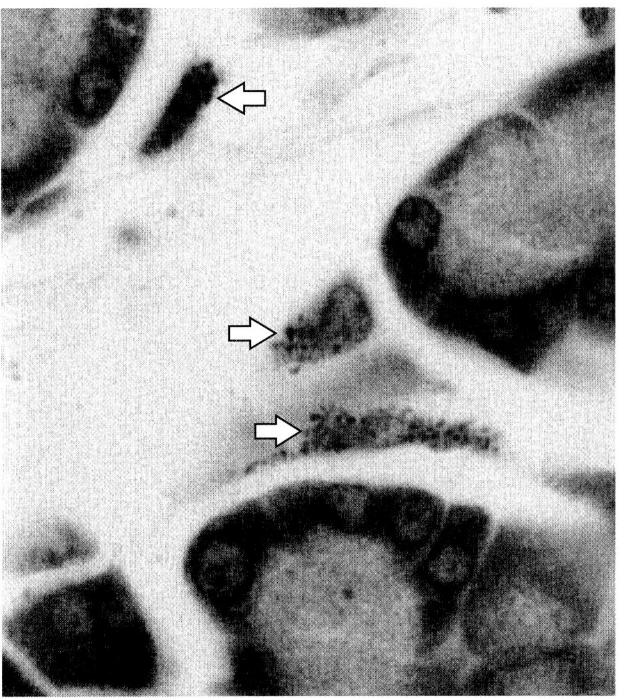

Figure 5–6. Section of pancreas from a rat injected with the vital dye trypan blue. Note that 3 macrophages (arrows) have engulfed and accumulated the dye in the form of granules. H&E stain. Low magnification.

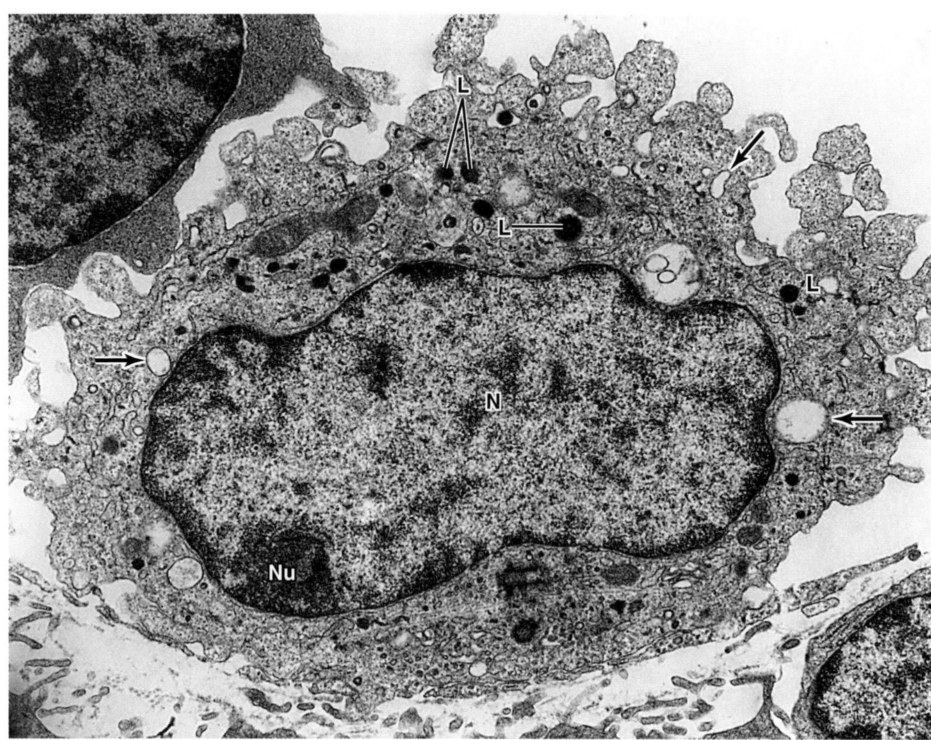

Figure 5–7. Electron micrograph of a macrophage. Note the secondary lysosomes (L), the nucleus (N), and the nucleolus (Nu). The arrows indicate phagocytic vacuoles.

tion to showing an increase in their capacity for phagocytosis and intracellular digestion, exhibit enhanced metabolic and lysosomal enzyme activity.

Macrophages also have an important role in removing cell debris and damaged extracellular components formed during the physiologic involution process. For example, during pregnancy the uterus increases in size. Immediately after parturition, the uterus suffers an involution during which some of its tissues are destroyed by the action of macrophages. Macrophages are also secretory cells that produce an impressive array of substances, including enzymes (eg, collagenase) and cytokines that participate in defensive and reparative functions, and they exhibit increased tumor cell–killing capacity (Figure 5–8).

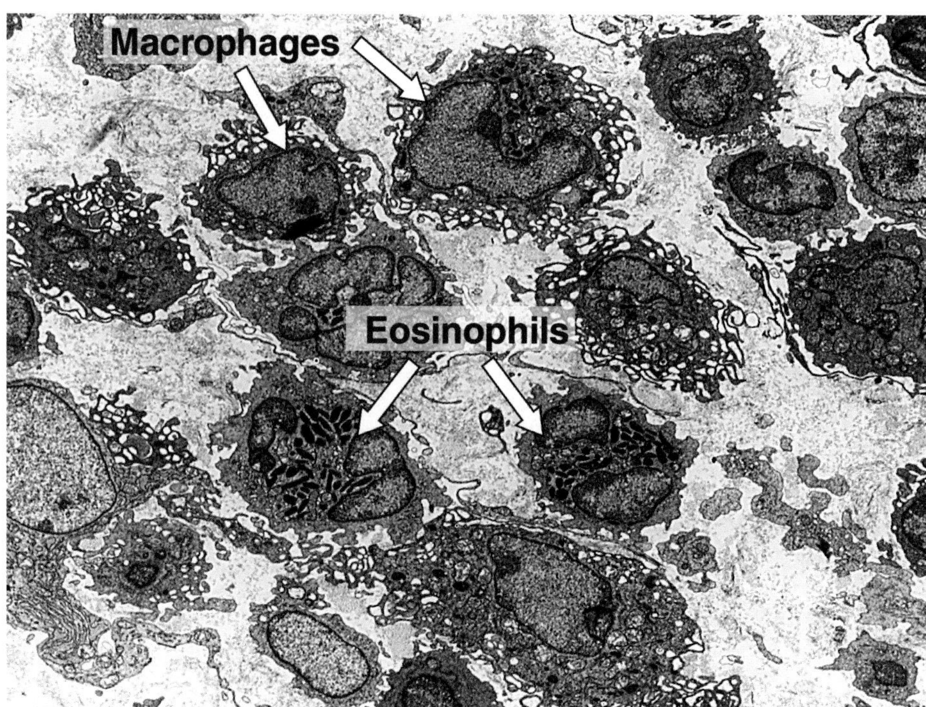

Figure 5–8. Electron micrograph of several macrophages and 2 eosinophils in a region adjacent to a tumor. This figure illustrates the participation of macrophages in tissue reaction to tumor invasion.

Table 5–2. Distribution and main functions of the cells of the mononuclear phagocyte system.

Cell Type	Location	Main Function
Monocyte	Blood	Precursor of macrophages
Macrophage	Connective tissue, lymphoid organs, lungs, bone marrow	Production of cytokines, chemotactic factors, and several other molecules that participate in inflammation (defense), antigen processing and presentation
Kupffer cell	Liver	Same as macrophages
Microglia cell	Nerve tissue of the central nervous system	Same as macrophages
Langerhans cell	Skin	Antigen processing and presentation
Dendritic cell	Lymph nodes	Antigen processing and presentation
Osteoclast	Bone (fusion of several macrophages)	Digestion of bone
Multinuclear giant cell	Connective tissue (fusion of several macrophages)	Segregation and digestion of foreign bodies

Mast Cells

Mast cells are oval to round connective tissue cells, 20–30 μm in diameter, whose cytoplasm is filled with basophilic secretory granules. The rather small, spherical nucleus is centrally situated; it is frequently obscured by the cytoplasmic granules (Figure 5–10).

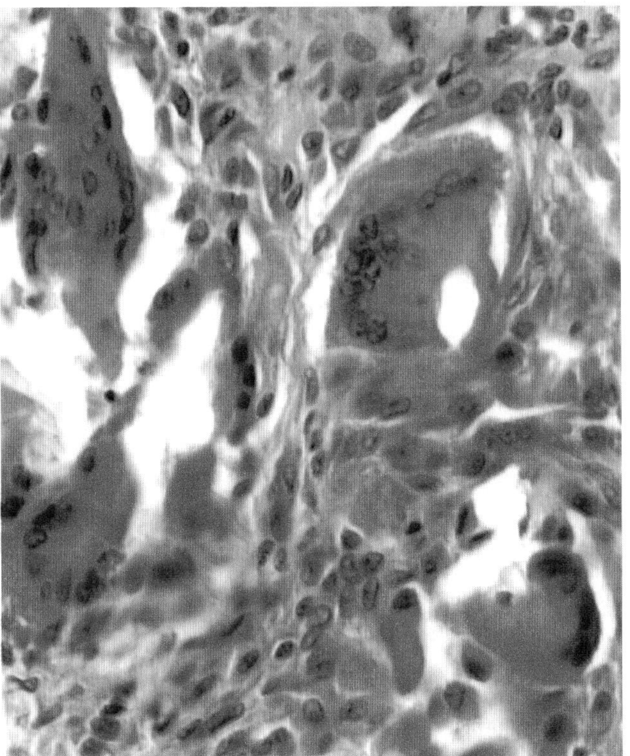

Figure 5–9. Section of rat skin showing multinuclear giant cells surrounded by macrophages. H&E stain.

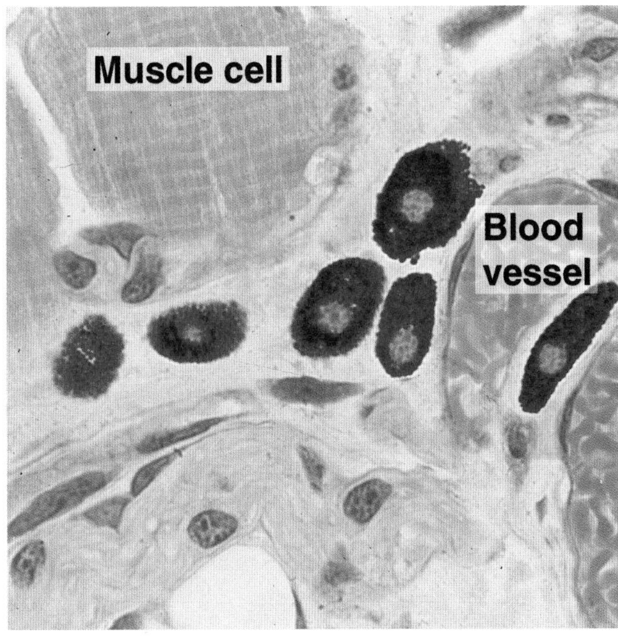

Figure 5–10. Section of rat tongue. Several mast cells in the connective tissue surround muscle cells and blood vessels. PT stain. Medium magnification.

The secretory granules are 0.3–2.0 μm in diameter. Their interior is heterogeneous in appearance, with a prominent scroll-like substructure (Figure 5–11) that contains pre-formed mediators such as histamine and proteoglycans. The principal function of mast cells is the storage of chemical mediators of the inflammatory response.

Mast cell granules are metachromatic because of the high content of acidic radicals in the glycosaminoglycans. **Metachromasia** is a property of certain molecules that changes the color of some basic aniline dyes (eg, toluidine blue). The structure containing the metachromatic molecules takes on a different color (purple-red) from that of the applied dye (blue). Other constituents of mast cell granules are histamine, which promotes an

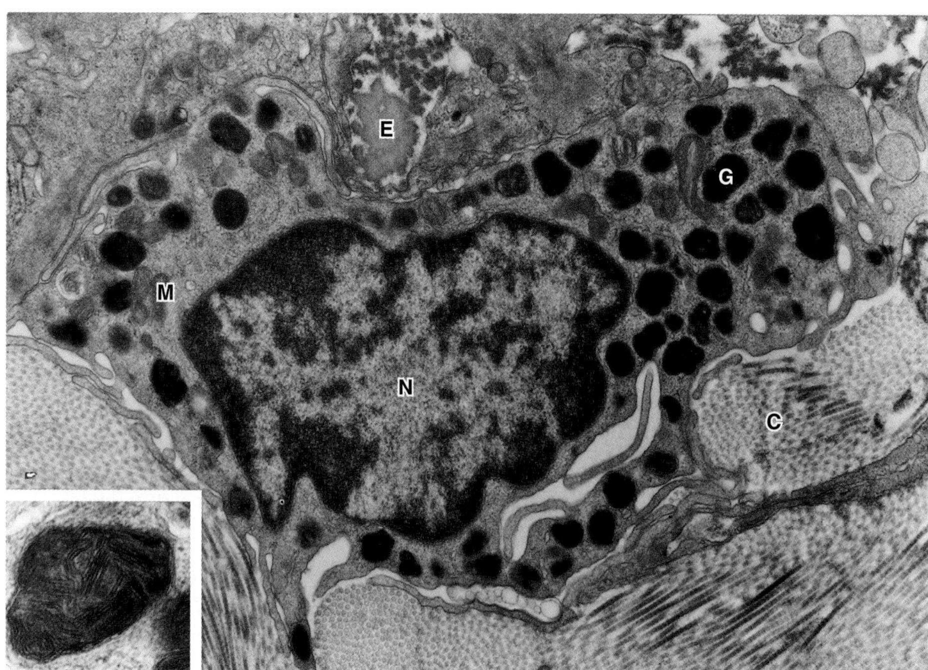

Figure 5–11. Electron micrograph of a human mast cell. The granules (G) contain heparin and histamine. Note the characteristic scroll-like structures within the granules. M, mitochondrion; C, collagen fibrils; E, elastic fibril; N, nucleus. ×14,700. **Inset:** Higher magnification view of a mast cell granule. ×44,600. (Courtesy of MC Williams.)

increase in vascular permeability that is important in inflammation, neutral proteases, and eosinophil chemotactic factor of anaphylaxis (ECF-A). Mast cells also release leukotrienes (C4, D4, E4) or slow-reacting substance of anaphylaxis, also called SRS-A, but these substances are not stored in the cell. Rather, they are synthesized from membrane phospholipids and immediately released to the extracellular microenvironment upon appropriate stimulation, such as interaction with fibroblasts. The molecules produced by mast cells act locally in paracrine secretion.

Although they have similar morphology, there are at least 2 populations of mast cells in connective tissues. One type is called the **connective tissue mast cell,** found in the skin and peritoneal cavity. The second type, the **mucosal mast cell,** is present in the intestinal mucosa and in the lungs. These two cell populations also differ in their granular content.

Mast cells originate from progenitor cells in the bone marrow. These progenitor cells circulate in the blood, cross the wall of venules and capillaries, and penetrate the tissues, where they proliferate and differentiate. Although they are, in many respects, similar to basophilic leukocytes, they have a separate stem cell.

The surface of mast cells contains specific receptors for IgE, a type of immunoglobulin produced by plasma cells. Most IgE molecules are bound to the surface of mast cells and blood basophils; very few remain in the plasma.

MEDICAL APPLICATION

*Release of the chemical mediators stored in mast cells promotes the allergic reactions known as **immediate hypersensitivity reactions,** because they occur within a few minutes after penetration by an antigen of an individual previously sensitized to the same or a very similar antigen. There are many examples of immediate hypersensitivity reaction; a dramatic one is **anaphylactic shock,** a potentially fatal condition. The process of anaphylaxis consists of the following sequential events: The first exposure to an antigen (allergen), such as bee venom, results in production of the IgE class of immunoglobulins (antibodies) by plasma cells. IgE is avidly bound to the surface of mast cells. A second exposure to the antigen results in binding of the antigen to IgE on the mast cells. This event triggers release of the mast cell granules, liberating histamine, leukotrienes, ECF-A, and heparin (Figure 5–12). Degranulation of mast cells also occurs as a result of the action of the complement molecules that participate in the immunologic reaction cited in Chapter 14.*

Histamine causes contraction of smooth muscle (mainly of the bronchioles) and dilates and increases permeability (mainly in postcapillary venules). Any liberated histamine is inactivated immediately after release. Leukotrienes produce slow contractions in smooth muscle, and ECF-A attracts blood eosinophils. Heparin is a blood anticoagulant, but blood clotting remains normal in humans during anaphylactic shock. Mast cells are widespread in the human body but are particularly abundant in the dermis and in the digestive and respiratory tracts.

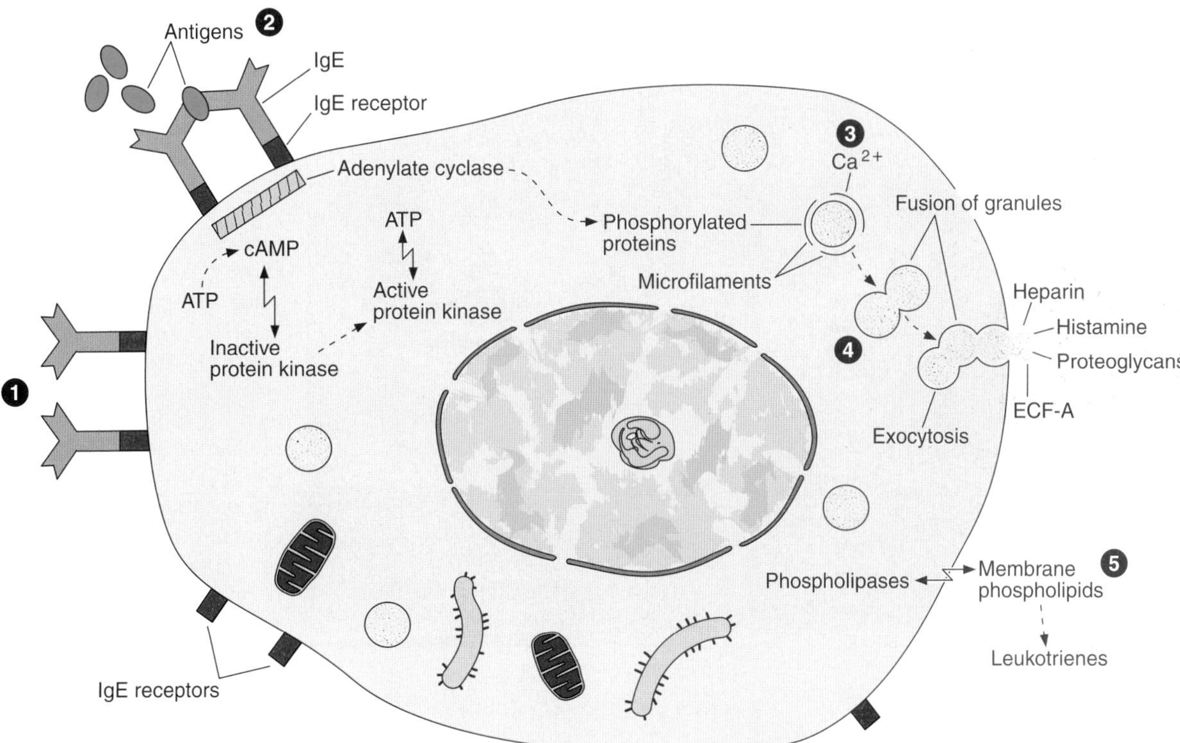

Figure 5–12. Mast-cell secretion. **1:** IgE molecules are bound to the surface receptors. **2:** After a second exposure to an antigen (eg, bee venom), IgE molecules bound to surface receptors are cross-linked by the antigen. This activates adenylate cyclase and results in the phosphorylation of certain proteins. **3:** At the same time, Ca^{2+} enters the cell. **4:** These events lead to intracellular fusion of specific granules and exocytosis of their contents. **5:** In addition, phospholipases act on membrane phospholipids to produce leukotrienes. The process of extrusion does not damage the cell, which remains viable and synthesizes new granules. ECF-A, eosinophil chemotactic factor of anaphylaxis.

Plasma Cells

Plasma cells are large, ovoid cells that have a basophilic cytoplasm due to their richness in rough endoplasmic reticulum (Figures 5–13, 5–14, and 5–15). The juxtanuclear Golgi complex and the centrioles occupy a region that appears pale in regular histologic preparations.

The nucleus of the plasma cell is spherical and eccentrically placed, containing compact, coarse heterochromatin alternating with lighter areas of approximately equal size. This configuration resembles the face of a clock, with the heterochromatin clumps corresponding to the numerals. Thus, the nucleus of a plasma cell is commonly described as having a clock-face appearance. There are few plasma cells in most connective tissues. Their average life is short, 10–20 days.

MEDICAL APPLICATION

Plasma cells are derived from B lymphocytes and are responsible for the synthesis of the antibodies. Antibodies are immunoglobulins produced in response to penetration by antigens. Each antibody is specific for the one antigen that gave rise to its production and reacts specifically with molecules possessing similar epitopes (see Chapter 14). The results of the antibody-antigen reaction are variable. The capacity of the reaction to neutralize harmful effects caused by antigens is important. An antigen that is a toxin (eg, tetanus, diphtheria) may lose its capacity to do harm when it combines with its respective antibody.

Adipose Cells

Adipose cells (adipocytes; L. *adeps,* fat, + Gr. *kytos*) are connective tissue cells that have become specialized for storage of neutral fats or for the production of heat. Often called **fat cells,** they are discussed in detail in Chapter 6.

Leukocytes

The normal connective tissue contains leukocytes that migrate from the blood vessels by diapedesis. Leukocytes (Gr. *leukos,* white, + *kytos*), or white blood corpuscles, are the wandering cells of the connective tissue. They migrate through the walls of capillaries and post-

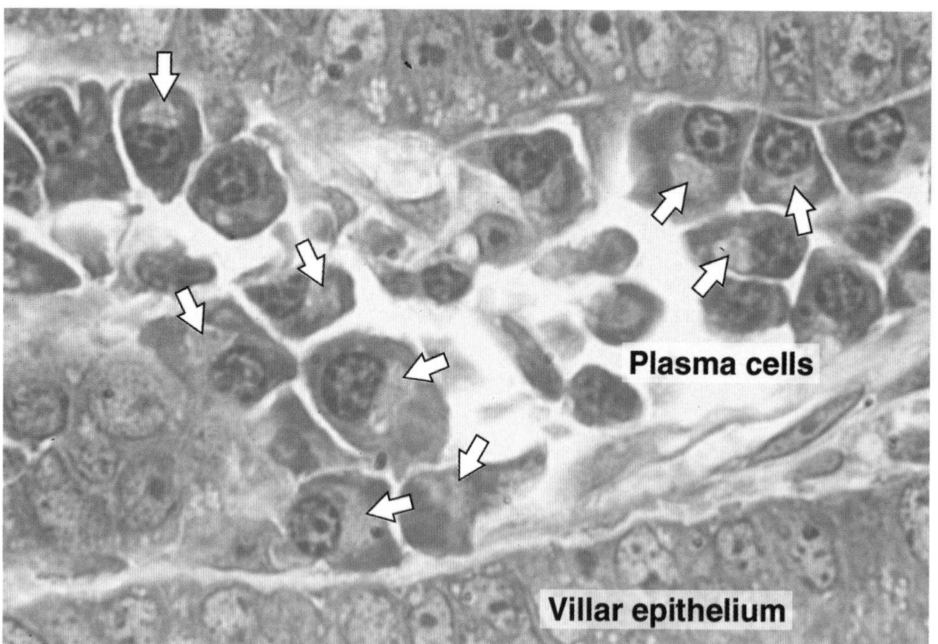

Figure 5–13. Portion of a chronically inflamed intestinal villus. The plasma cells are characterized by their size and abundant basophilic cytoplasm (rough endoplasmic reticulum) and are involved in the synthesis of antibodies. A large Golgi complex (arrows) is where the terminal glycosylation of the antibodies (glycoproteins) occurs. Plasma cells produce antibodies of importance in immune reactions. PT stain. Medium magnification.

capillary venules from the blood to connective tissues, by a process called **diapedesis.** This process increases greatly during inflammation (Figure 5–16). Inflammation is a vascular and cellular defensive reaction against foreign substances, in most cases pathogenic bacteria or irritating chemical substances. The classic signs of inflammation were first described by Celsus (first century A.D.) as redness and swelling with heat and pain (*rubor et tumor cum calore et dolore*). Much later, disturbed function (*functio laesa*) was added as the fifth cardinal sign.

Inflammation begins with the local release of **chemical mediators of inflammation,** substances of various origin (mainly from cells and blood plasma proteins) that induce some of the events characteristic of inflammation, eg, **increase of blood flow** and **vascular permeability, chemotaxis,** and **phagocytosis.**

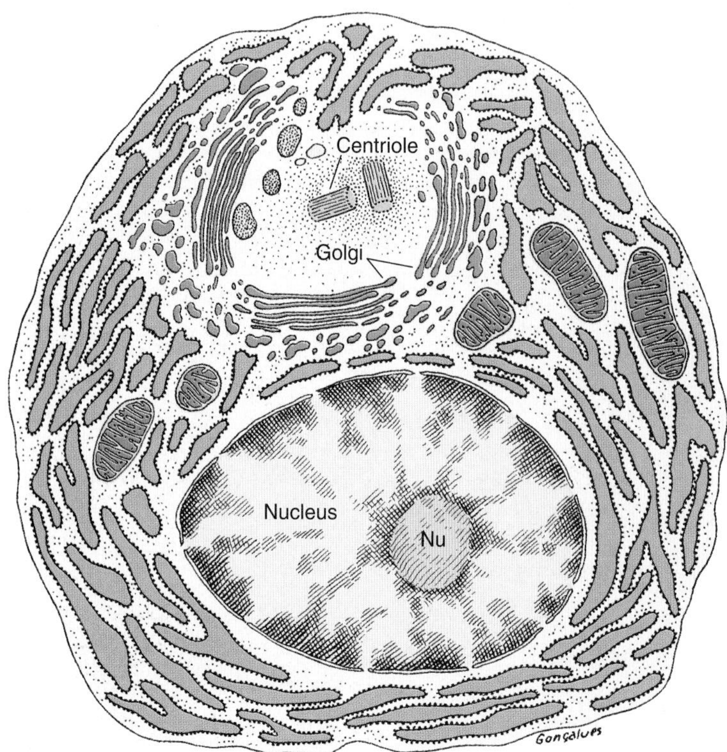

Figure 5–14. Ultrastructure of a plasma cell. The cell contains a well-developed rough endoplasmic reticulum, with dilated cisternae containing immunoglobulins (antibodies). In plasma cells, the secreted proteins do not aggregate into secretory granules. Nu, nucleolus. (Redrawn and reproduced, with permission, from Ham AW: *Histology,* 6th ed. Lippincott, 1969.)

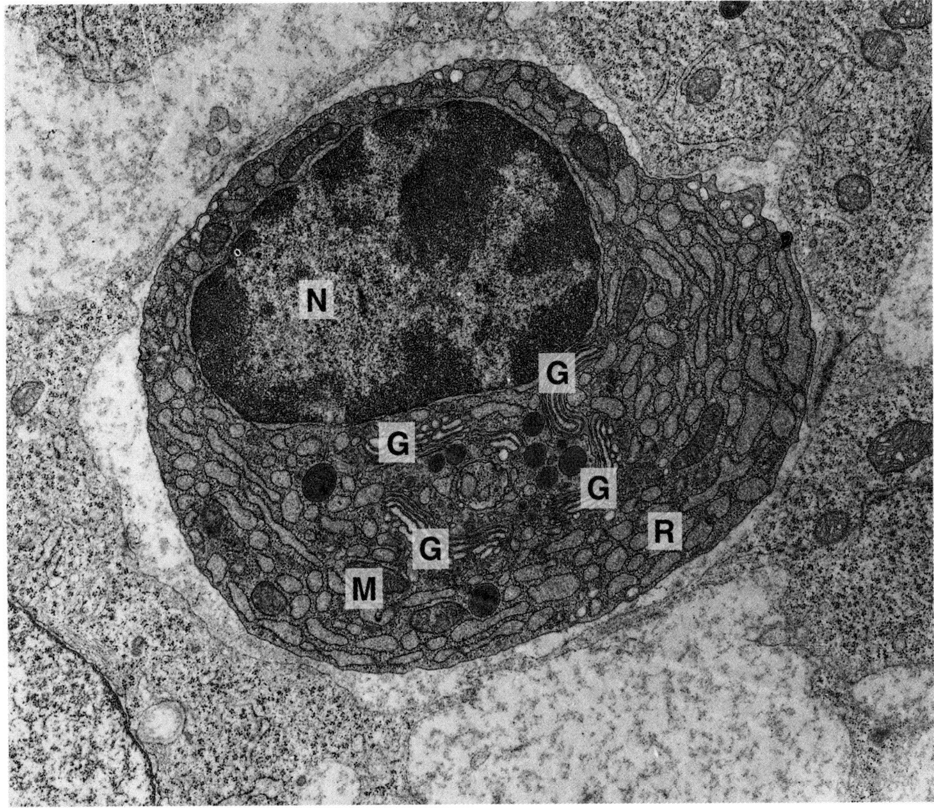

Figure 5–15. Electron micrograph of a plasma cell showing an abundance of rough endoplasmic reticulum (R). Note that many cisternae are dilated. Four profiles of the Golgi complex (G) are observed near the nucleus (N). M, mitochondria. (Courtesy of P. Abrahamsohn.)

MEDICAL APPLICATION

Increased vascular permeability is caused by the action of vasoactive substances; an example is histamine, which is liberated from mast cells and basophilic leukocytes. Increases in blood flow and vascular permeability are responsible for local swelling (edema), redness, and heat. Pain is due mainly to the action of chemical mediators on nerve endings. ***Chemotaxis*** *(Gr.* chemeia, *alchemy, +* taxis, *orderly arrangement), the phenomenon by which specific cell types are attracted by some molecules, is respon-*

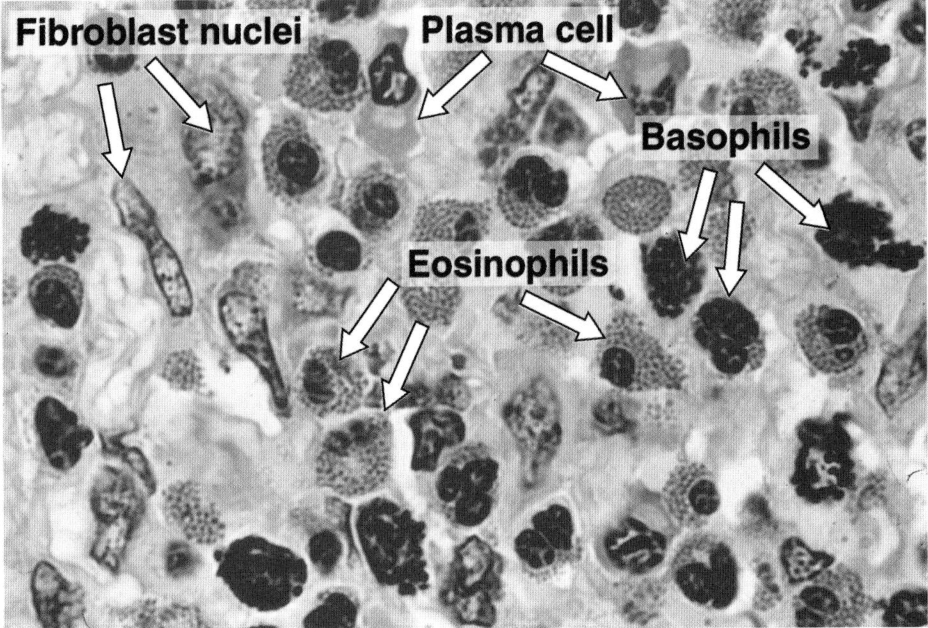

Figure 5–16. Section of an inflamed intestinal lamina propria. Inflammation was caused by nematode parasitism. Aggregated eosinophils and plasma cells function mainly in the connective tissue by modulating the inflammatory process. Giemsa stain. Low magnification.

sible for the migration of large quantities of specific cell types to regions of inflammation. As a consequence of chemotaxis, leukocytes cross the walls of venules and capillaries by diapedesis, invading the inflamed areas.

Leukocytes do not return to the blood after having resided in connective tissue, except for the lymphocytes that circulate continuously in various compartments of the body (blood, lymph, connective tissues, lymphatic organs). A detailed analysis of the structure and functions of leukocytes is presented in Chapter 12.

FIBERS

The connective tissue fibers are formed by proteins that polymerize into elongated structures. The 3 main types of connective tissue fibers are **collagen, reticular,** and **elastic.** Collagen and reticular fibers are formed by the protein **collagen,** and elastic fibers are composed mainly of the protein **elastin.** These fibers are distributed unequally among the types of connective tissue. Actually, there are 2 systems of fibers: the collagen system, consisting of collagen and reticular fibers, and the elastic system, consisting of the elastic, elaunin, and oxytalan fibers. In many cases, the predominant fiber type is responsible for conferring specific properties on the tissue.

The collagens constitute a family of proteins selected during evolution for the execution of several (mainly structural) functions. During the process of evolution of multicellular organisms, a family of structural proteins that were modified by environmental influences and the functional requirements of the animal organism developed to acquire varying degrees of rigidity, elasticity, and strength. These proteins are known collectively as **collagen,** and the chief examples among its various types are present in the skin, bone, cartilage, smooth muscle, and basal lamina.

Collagen is the most abundant protein in the human body, representing 30% of its dry weight. The collagens of vertebrates are produced by several cell types and are distinguishable by their molecular compositions, morphologic characteristics, distribution, functions, and pathologies (Table 5–3). According to their structure and functions, they can be classified in the following groups.

COLLAGENS THAT FORM LONG FIBRILS

The molecules of long fibril–forming collagens aggregate to form fibrils clearly visible in the electron microscope (Figure 5–17). These are collagens type I, II, III, V, and XI. Collagen type I is the most abundant and has a widespread distribution. It occurs in tissues as structures that are classically designated as **collagen fibers** and that form such structures as bones, dentin, tendons, organ capsules, and dermis.

FIBRIL-ASSOCIATED COLLAGENS

Fibril-associated collagens are short structures that bind collagen fibrils to one another and to other components of the extracellular matrix. They are collagens type IX, XII, and XIV.

COLLAGENS THAT FORM NETWORKS

Network-forming collagen is type IV collagen, whose molecules assemble in a meshwork that constitutes the structural component of the basal lamina.

COLLAGENS THAT FORM ANCHORING FIBRILS

Anchoring collagen is type VII collagen, present in the anchoring fibrils that bind collagen fibers to the basal lamina.

Collagen synthesis, an activity originally believed to be restricted to fibroblasts, chondroblasts, osteoblasts, and odontoblasts, has now been shown to be widespread, with many cell types producing this protein. The principal amino acids that make up collagen are glycine (33.5%), proline (12%), and hydroxyproline (10%). Collagen contains 2 amino acids that are characteristic of this protein: **hydroxyproline** and **hydroxylysine.**

The protein unit that polymerizes to form collagen fibrils is the elongated molecule called **tropocollagen,** which measures 280 nm in length and 1.5 nm in width. Tropocollagen consists of 3 subunit polypeptide chains intertwined in a triple helix (Figure 5–18). Differences in the chemical structure of these polypeptide chains are responsible for the various types of collagen.

In collagen types I, II, and III, tropocollagen molecules aggregate into microfibrillar subunits that are packed together to form **fibrils.** Hydrogen bonds and hydrophobic interactions are important in the aggregation and packing of these units. In a subsequent step, this structure is reinforced by the formation of covalent cross-links, a process catalyzed by the activity of the enzyme lysyl oxidase.

Collagen fibrils are thin, elongated structures that have a variable diameter (ranging from 20 to 90 nm) and can be several micrometers in length; they have transverse striation with a characteristic periodicity of 64 nm (Figure 5–18). The transverse striations of the collagen fibrils are determined by the overlapping arrangement of the tropocollagen molecules (Figure 5–19). The dark bands retain more of the lead-based stain used in electron-microscope studies, because their more numerous free chemical groups react more intensely with the lead solution than do the light bands. In collagen types I and III, these fibrils associate to form fibers. In collagen type I, the fibers can associate to form bundles (Figure 5–19). Collagen type II (present in cartilage) occurs as fibrils but does not form fibers or bundles (Figure 5–20). Collagen type IV, present in all basement membranes, does not form either fibrils or fibers. Because of its molecular configuration, collagen type IV has a "chicken-wire" organization.

Biosynthesis of Collagen Type I

Because collagen type I is widely distributed in the body, its synthesis has been thoroughly studied. Collagen synthesis involves several steps, which are summarized in Figure 5–21:

1. Polypeptide α chains are assembled on polyribosomes bound to rough endoplasmic reticulum membranes and injected into the cisternae as **preprocollagen** molecules. The signal peptide is clipped off, forming **procollagen.**
2. Hydroxylation of proline and lysine occurs after these amino acids are incorporated into polypeptide chains. Hydroxylation begins after the peptide chain has reached a certain minimum length and is still bound to the ribosomes. The two enzymes involved are **peptidyl proline hydroxylase** and **peptidyl lysine hydroxylase.**
3. Glycosylation of hydroxylysine occurs after its hydroxylation. Different collagen types have different amounts of carbohydrate in the form of galactose or glycosylgalactose linked to hydroxylysine.
4. Each α chain is synthesized with an extra length of peptides called **registration peptides** on both amino- and carboxyl-

Table 5–3. Collagen types.

Type	Molecule Composition	Structure	Optical Microscopy	Representative Tissues	Main Function
Collagen that forms fibrils					
I	$[\alpha 1\ (I)]_2\ [\alpha 2\ (I)]$	300-nm molecule, 67-nm banded fibrils	Thick, highly picrosirius birefringent, nonargyrophilic fibers	Skin, tendon, bone, dentin	Resistance to tension
II	$[\alpha 1\ (II)]_3$	300-nm molecule, 67-nm banded fibrils	Loose aggregates of fibrils, birefringent	Cartilage, vitreous body	Resistance to pressure
III	$[\alpha 1\ (III)]_3$	67-nm banded fibrils	Thin, weakly birefringent, argyrophilic fibers	Skin, muscle, blood vessels, frequently together with type I	Structural maintenance in expansible organs
V	$[\alpha 1\ (V)]_3$	390-nm molecule, N-terminal globular domain	Frequently forms fiber together with type I	Fetal tissues, skin, bone, placenta, most interstitial tissues	Participates in type I collagen function
XI	$[\alpha 1\ (XI)]\ [\alpha 2\ (XI)]\ [\alpha 3\ (XI)]$	300-nm molecule	Small fibers	Cartilage	Participates in type II collagen function
Fibril-associated collagen					
IX	$[\alpha 1\ (IX)]\ [\alpha 2\ (IX)]\ [\alpha 3\ (IX)]$	200-nm molecule	Not visible, detected by immunocytochemistry	Cartilage, vitreous body	Bound glycosaminoglycans; associated with type II collagen
XII	$[\alpha 1\ (XII)]_3$	Large N-terminal domain; interacts with type I collagen	Not visible, detected by immunocytochemistry	Embryonic tendon and skin	Interacts with type I collagen
XIV	$[\alpha 1\ (XIV)]_3$	Large N-terminal domain; cross-shaped molecule	Not visible; detected by immunocytochemistry	Fetal skin and tendon	
Collagen that forms anchoring fibrils					
VII	$[\alpha 1\ (VII)]_3$	450 nm, globular domain at each end	Not visible, detected by immunocytochemistry	Epithelia	Anchors skin epidermal basal lamina to underlying stroma
Collagen that forms networks					
IV	$[\alpha 1\ (VII)]_2\ [\alpha 1\ (IV)]$	Two-dimensional cross-linked network	Not visible, detected by immunocytochemistry	All basement membranes	Support of delicate structures, filtration

terminal ends. Registration peptides probably ensure that the appropriate α chains ($\alpha 1$, $\alpha 2$) assemble in the correct position as a triple helix. In addition, the extra peptides make the resulting **procollagen molecule** soluble and prevent its premature intracellular assembly and precipitation as collagen fibrils. Procollagen is transported as such out of the cell to the extracellular environment.

5. Outside the cell, specific proteases called **procollagen peptidases** remove the registration peptides. The altered protein, known as **tropocollagen,** is capable of assembling into polymeric collagen fibrils. The hydroxyproline residues contribute to the stability of the tropocollagen triple helix, forming hydrogen bonds between its polypeptide chains.
6. Collagen fibrils aggregate spontaneously to form fibers. Proteoglycans and structural glycoproteins play an important role in the aggregation of tropocollagen to form fibrils and in the formation of fibers from fibrils.
7. Fibrillar structure is reinforced by the formation of covalent cross-links between tropocollagen molecules. This process is catalyzed by the action of the enzyme **lysyl oxidase,** which also acts in the extracellular space.

The other fibrillar collagens are probably formed according to the same pattern described for collagen type I, with only minor differences.

The synthesis of collagen involves a cascade of unique posttranslational biochemical modifications of the original procolla-

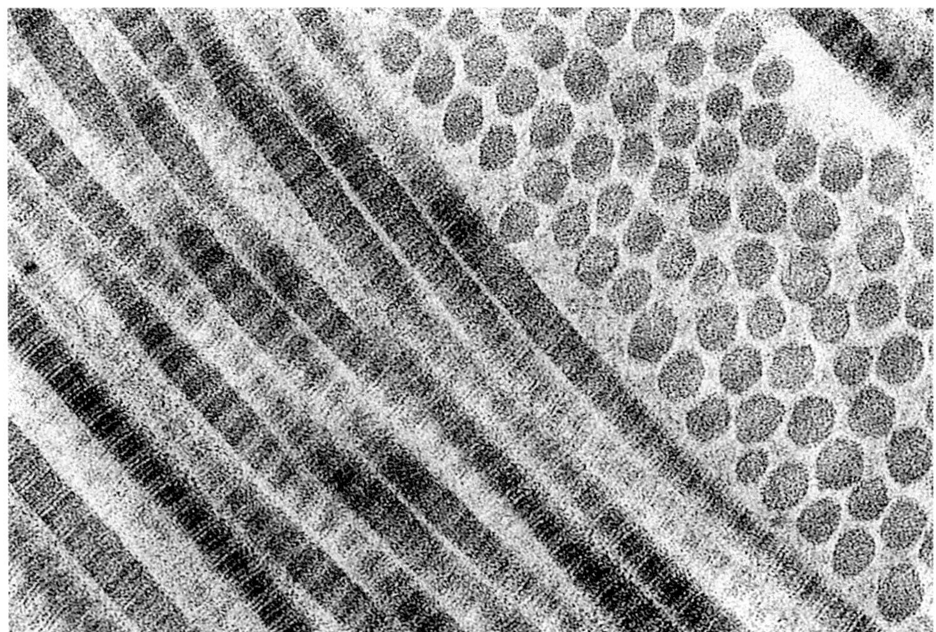

Figure 5–17. Electron micrograph of human collagen fibrils in cross and longitudinal sections. Each fibril consists of regular alternating dark and light bands that are further divided by cross-striations. Ground substance completely surrounds the fibrils. ×100,000.

MEDICAL APPLICATION

Collagen synthesis depends on the expression of several genes and several post-translational events. It should not be surprising, therefore, that a large number of pathologic conditions are directly attributable to insufficient or abnormal collagen synthesis.

Certain mutations in the α1 (I) or α2 (I) genes lead to osteogenesis imperfecta. Many cases of osteogenesis imperfecta are due to deletions of all or part of the α1 (I) gene. However, a single amino acid change is sufficient to cause certain forms of this disease, particularly mutations involving glycine. Glycine must be at every third position for the collagen triple helix to form.

In addition to these disorders, several diseases result from an overaccumulation of collagen. In ***progressive systemic sclerosis,*** *almost all organs may present an excessive accumulation of collagen (**fibrosis**). This occurs mainly in the skin, digestive tract, muscles, and kidneys, causing hardening and functional impairment of the implicated organs.*

Keloid *is a local swelling caused by abnormal amounts of collagen that form in scars of the skin. Keloids, which occur most often in individuals of black African descent, can be a troublesome clinical problem to manage; not only can they be disfiguring, but excision is almost always followed by recurrence.*

Vitamin C (ascorbic acid) deficiency leads to scurvy, a disease characterized by the degeneration of connective tissue. Without this vitamin, fibroblasts synthesize defective collagen, and the defective fibers are not replaced. This process leads to a general degeneration of connective tissue that becomes more pronounced in areas where collagen renewal takes place at a faster rate. The periodontal ligament that holds teeth in their sockets has a relatively high collagen turnover; consequently, this ligament is markedly affected by scurvy, which leads to a loss of teeth. Ascorbic acid is a cofactor for proline hydroxylase, which is essential for the normal synthesis of collagen. Table 5–4 lists a few examples of the many disorders caused by collagen biosynthesis failure.

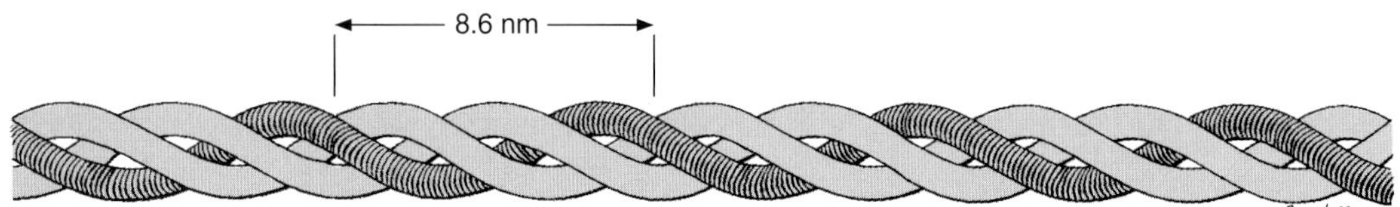

Figure 5–18. In the most abundant form of collagen, type I, each molecule (tropocollagen) is composed of two α1 and one α2 peptide chains, each with a molecular mass of approximately 100 kDa, intertwined in a right-handed helix and held together by hydrogen bonds and hydrophobic interactions. Each complete turn of the helix spans a distance of 8.6 nm. The length of each tropocollagen molecule is 280 nm, and its width is 1.5 nm.

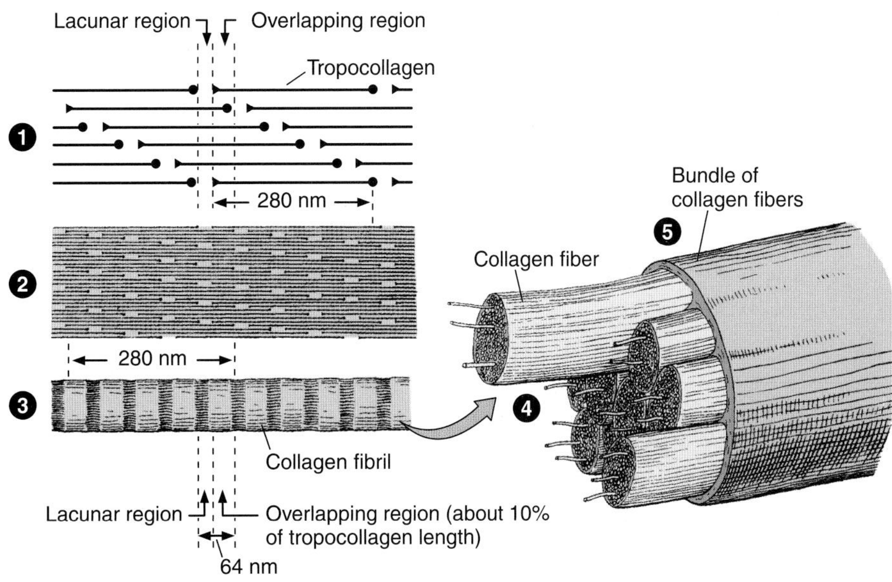

Figure 5–19. Schematic drawing of an aggregate of collagen molecules (tropocollagen), fibrils, fibers, and bundles. There is a stepwise overlapping arrangement of rodlike tropocollagen subunits, each measuring 280 nm (**1**). This arrangement results in the production of alternating lacunar and overlapping regions (**2**) that cause the cross-striations characteristic of collagen fibrils and confer a 64-nm periodicity of dark and light bands when the fibril is observed in the electron microscope (**3**). Fibrils aggregate to form fibers (**4**), which aggregate to form bundles (**5**) routinely called collagen fibers. Collagen type III usually does not form bundles.

gen polypeptide. All these modifications are critical to the structure and function of normal mature collagen. Because there are so many steps in collagen biosynthesis, there are many points at which the process can be interrupted or changed by faulty enzymes or by disease processes.

Collagen renewal is in general a very slow process. In some organs, such as tendons and ligaments, the collagen is very stable, whereas in others, as in the periodontal ligament, the turnover of collagen is very high. To be renewed, the collagen must first be degraded. Degradation is initiated by specific enzymes called **collagenases.** These enzymes cut the collagen molecule into 2 parts that are susceptible to further degradation by nonspecific proteases (enzymes that degrade proteins).

Fibers of Collagen Type I

Collagen fibers made of collagen type I are the most numerous fibers in connective tissue. Although fresh collagen fibers are colorless strands, when they are present in great numbers the tissues in which they occur (eg, tendons, aponeuroses) are white.

The orientation of the elongated tropocollagen molecules in collagen fibers makes them birefringent. When fibers containing collagen are stained with an acidic dye composed of elongated

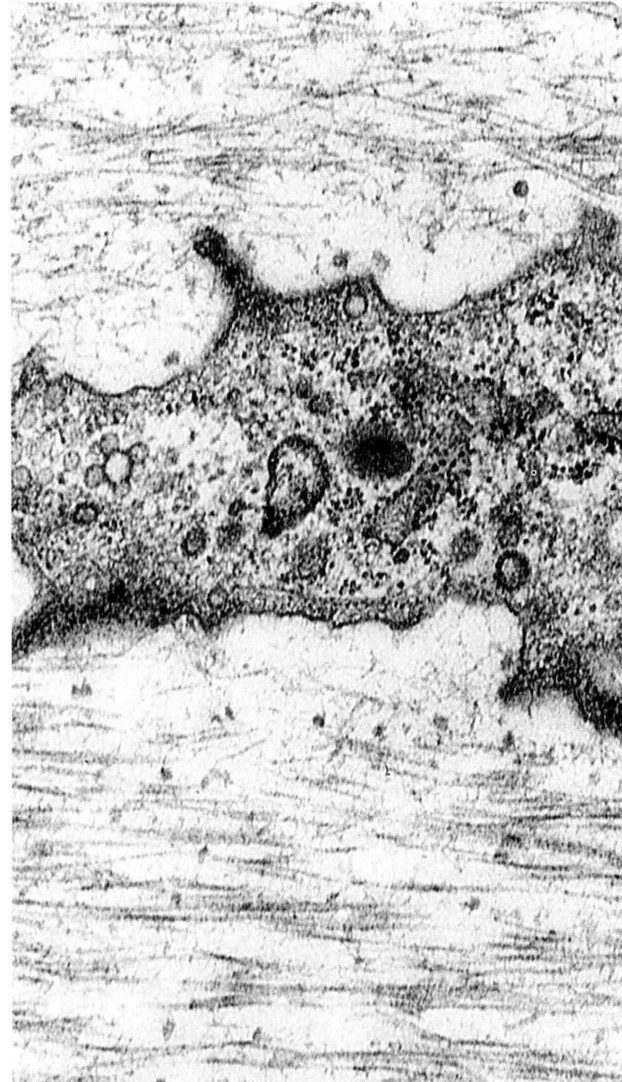

Figure 5–20. Electron micrograph of hyaline cartilage matrix showing the fine collagen fibrils of collagen type II interspersed with abundant ground substance. Transverse striations of the fibrils are barely visible because of the interaction of collagen with chondroitin sulfate. In the center is a portion of a chondrocyte. Compare the appearance of these fibrils with those of fibrocartilage (see Figure 7–8 in Chapter 7).

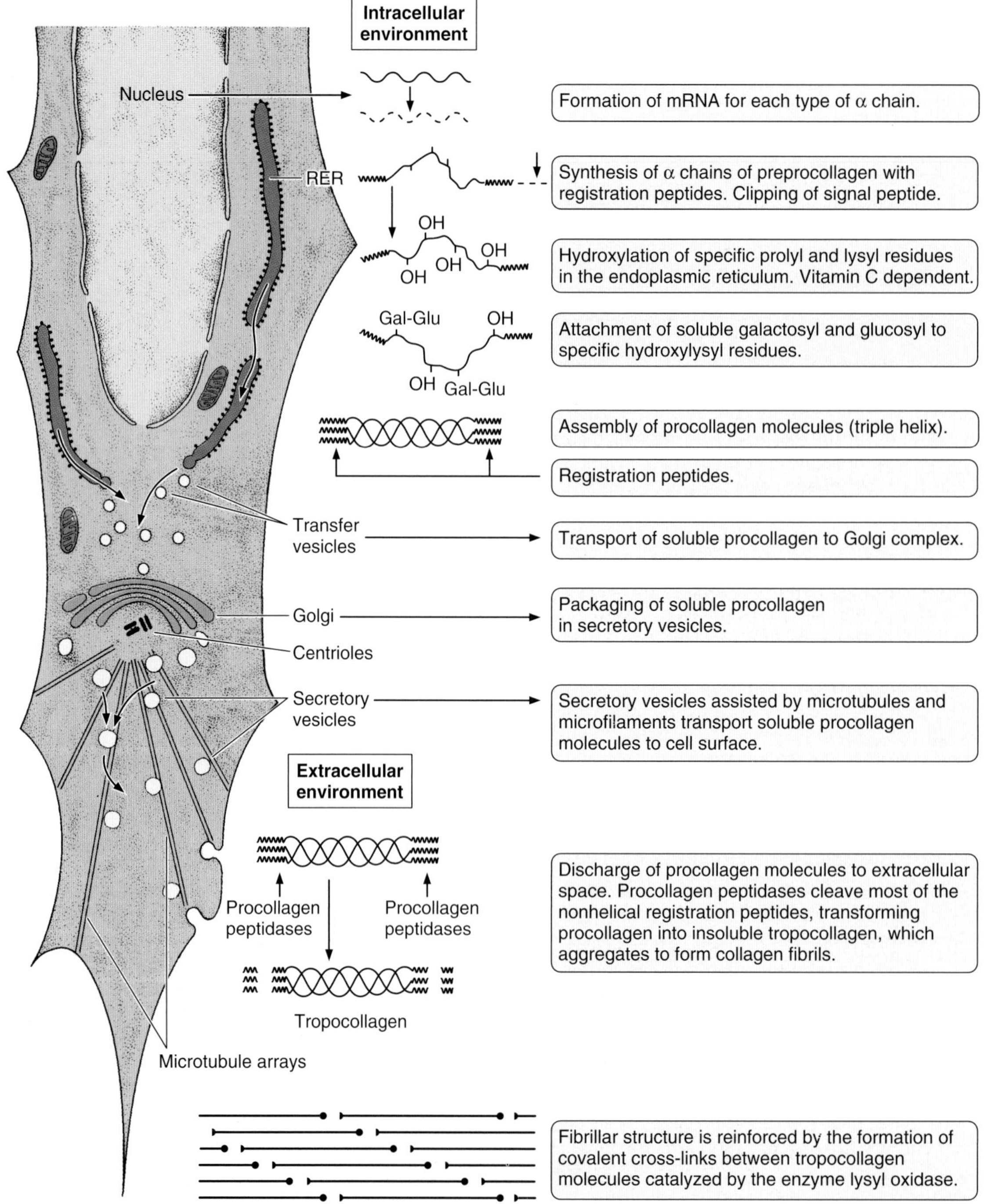

Figure 5–21. Collagen synthesis. The assembly of the triple helix and the hydroxylation and glycosylation of procollagen molecules are simultaneous processes that begin as soon as the 3 chains cross the membrane of the rough endoplasmic reticulum (RER). Because collagen synthesis depends on the expression of several genes and on several post-translation events, many collagen diseases have been described.

molecules (eg, Sirius red) that bind to collagen in an array parallel to its molecules, the collagen's normal birefringence increases considerably, producing a strong yellow color (Figure 5–22). Because this increase in birefringence occurs only in oriented molecular structures such as collagen, it is used as a specific method for collagen detection.

In many parts of the body, collagen fibers are organized in parallel to each other, forming **collagen bundles** (Figure 5–23). Because of the long and tortuous course of collagen bundles, their morphologic characteristics are better studied in spread preparations than in histologic sections (Figure 5–24). Mesentery is frequently used for this purpose; when spread on a slide,

Table 5–4. Examples of clinical disorders resulting from defects in collagen synthesis.

Disorder	Defect	Symptoms
Ehlers-Danlos type IV	Faulty transcription or translation of type III	Aortic and/or intestinal rupture
Ehlers-Danlos type VI	Faulty lysine hydroxylation	Augmented skin elasticity, rupture of eyeball
Ehlers-Danlos type VII	Decrease in procollagen peptidase activity	Increased articular mobility, frequent luxation
Scurvy	Lack of vitamin C (cofactor for proline hydroxylase)	Ulceration of gums, hemorrhages
Osteogenesis imperfecta	Change of one nucleotide in genes for collagen type I	Spontaneous fractures, cardiac insufficiency

this structure is sufficiently thin to let the light pass through; it can be stained and examined directly under the microscope. Mesentery consists of a central portion of connective tissue lined on both surfaces by a simple squamous epithelium, the mesothelium. The collagen fibers in a spread preparation appear as elongated and tortuous cylindrical structures of indefinite length, with a diameter that varies from 1 to 20 μm.

In the light microscope, collagen fibers are acidophilic; they stain pink with eosin, blue with Mallory's trichrome stain, green with Masson's trichrome stain, and red with Sirius red.

Reticular Fibers

Reticular fibers consist mainly of collagen type III. Reticular fibers are extremely thin, with a diameter between 0.5 and 2 μm, and they form an extensive network in certain organs. They are not visible in hematoxylin-and-eosin (H&E) preparations but can be easily stained black by impregnation with silver salts. Because of their affinity for silver salts, these fibers are called **argyrophilic** (Gr. *argyros,* silver, + *philein,* to love) (Figure 5–25).

Reticular fibers are also periodic acid–Schiff (PAS)-positive. Both PAS positivity and argyrophilia are considered to be due to the high content of sugar chains associated with these fibers. Reticular fibers contain 6–12% hexoses as opposed to 1% in collagen fibers. Immunocytochemical and histochemical evidence reveals that reticular fibers (in contrast to collagen fibers, which consist of collagen type I) are composed mainly of collagen type III in association with other types of collagen, glycoproteins, and proteoglycans. They are formed by loosely packed, thin (average 35-nm) fibrils (Figure 5–26) bound together by abundant small interfibrillar bridges probably composed of proteoglycans and glycoproteins. Because of their small diameter, reticular fibers show a green color when stained with Sirius red and observed by means of polarizing microscopy.

Reticular fibers are particularly abundant in smooth muscle, endoneurium, and the framework of hematopoietic (or hemopoietic) organs (eg, spleen, lymph nodes, red bone marrow) and constitute a network around the cells of parenchymal organs (eg, liver, endocrine glands). The small diameter and the loose disposition of reticular fibers create a flexible network in organs that are subjected to changes in form or volume, such as the arteries, spleen, liver, uterus, and intestinal muscle layers.

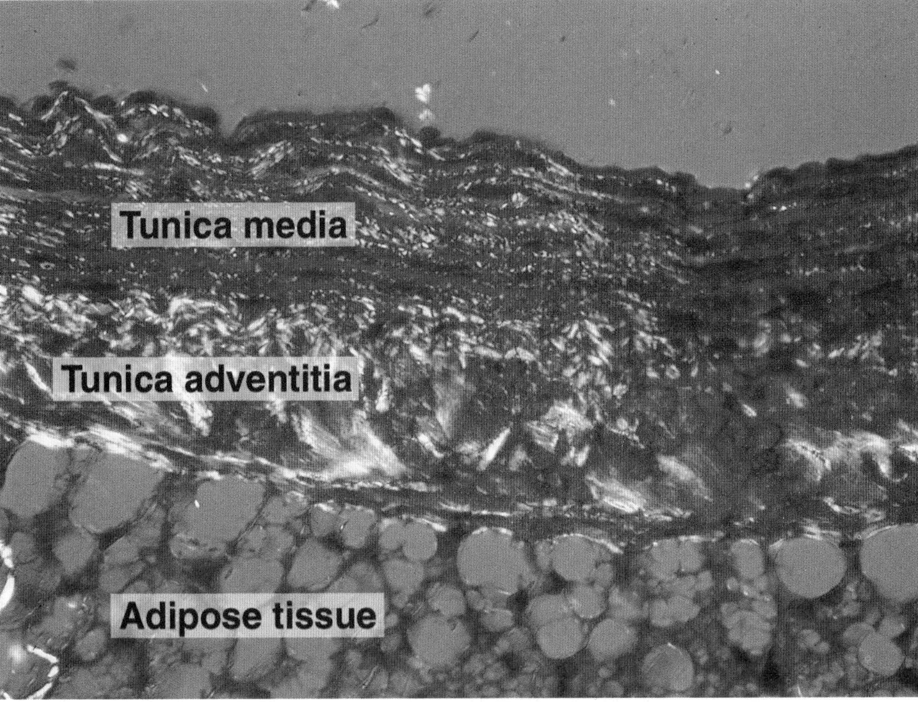

Figure 5–22. Section of a muscular artery stained with picrosirius and observed with polarization optics. The upper tunica media (muscular layer) contains reticular fibers consisting mainly of collagen type III. The lower layer (tunica adventitia) contains thick fibers and bundles of collagen type I. Deficiencies of collagen type III may result in rupture of the arterial wall. Medium magnification.

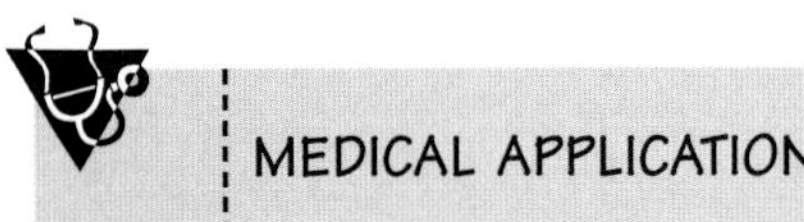

MEDICAL APPLICATION

Ehlers-Danlos type IV disease, a deficiency of collagen type III, is characterized by ruptures in arteries and the intestine (Table 5–4), both structures rich in reticular fibers.

The Elastic Fiber System

The elastic fiber system is composed of 3 types of fibers—oxytalan, elaunin, and elastic. The structures of the elastic fiber system develop through 3 successive stages (Figures 5–27 and 5–28). In the first stage, oxytalan, the fiber consists of a bundle of 10-nm microfibrils composed of various glycoproteins, including one with a large molecule called **fibrillin.** Fibrillin is a family of proteins related to the scaffolding necessary to the deposition of elastin. Defective fibrillin results in the formation of fragmented elastic fibrils. The **oxytalan** (Gr. *oxys,* thin) fibers can be found in the zonule fibers of the eye) and where the dermis connects the elastic system to the basal lamina. In the second stage of development, an irregular deposition of the protein **elastin** appears between the oxytalan microfibrils, forming the **elaunin** (Gr. *elaunem,* to drive) fibers. These structures are found around sweat glands and in the dermis. During the third stage, elastin gradually accumulates until it occupies the center of the fiber bundles, which are further surrounded by a thin sheath of microfibrils. These are the **elastic fibers,** the most numerous component of the elastic fiber system.

Oxytalan fibers are not elastic and are highly resistant to pulling forces, whereas the elastic fibers, which are rich in the protein elastin, stretch easily in response to tension. The elastic fiber system, by using different proportions of microfibrils and elastin, constitutes a family of fibers whose variable functional characteristics are adapted to local tissue requirements.

Proelastin is a globular molecule (molecular mass 70 kDa) produced by fibroblasts in connective tissue and by smooth muscle cells in blood vessels. Proelastin polymerizes, producing elastin, the amorphous rubberlike glycoprotein that predominates in mature fibers. Elastin is resistant to boiling, acid and alkali extraction, and digestion by the usual proteases. It is easily hydrolyzed by pancreatic **elastase.**

The amino acid composition of elastin resembles that of collagen, because both are rich in glycine and proline. Elastin contains 2 unusual amino acids, **desmosine** and **isodesmosine,** formed by covalent reactions among 4 lysine residues. These reactions effectively cross-link elastin and are thought to account for the rubberlike qualities of this protein, which forms fibers at least 5 times more extensible than rubber. Figure 5–29 presents a model that illustrates the elasticity of elastin.

Elastin also occurs in a nonfibrillar form as **fenestrated membranes** (elastic laminae) present in the walls of some blood vessels.

MEDICAL APPLICATION

Mutations in the fibrillin gene result in Marfan syndrome, a disease characterized by a lack of resistance in the tissues rich in elastic fibers. Because the large arteries are rich in components of the elastic system and because the blood pressure is high in the aorta, patients with this disease often experience aortic rupture, a life-threatening condition.

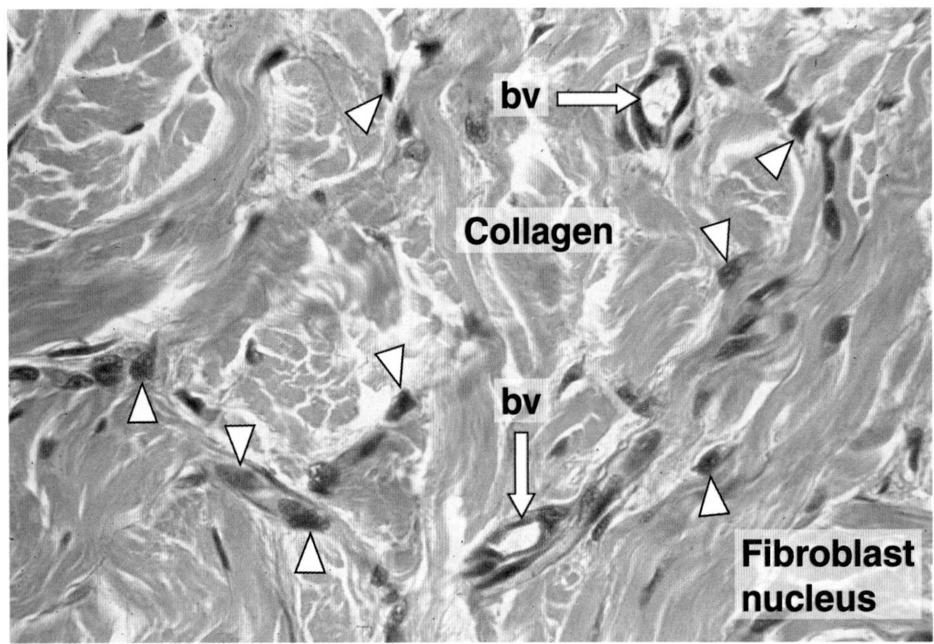

Figure 5–23. Dense irregular connective tissue from human dermis contains thick bundles of collagen fibers, fibroblast nuclei (arrowheads), and a few small blood vessels (bv). H&E stain. Medium magnification.

GROUND SUBSTANCE

The intercellular ground substance is a highly hydrated, complex mixture of glycosaminoglycans, proteoglycans, and multiadhesive glycoproteins. The complex molecular mixture of the ground substance is colorless and transparent. It fills the space between cells and fibers of the connective tissue and, because it is viscous, acts as both a lubricant and a barrier to the penetration of invaders. When adequately fixed for histologic analysis, its components aggregate and precipitate in the tissues as granular material that is observed in electron-microscope preparations as electron-dense filaments or

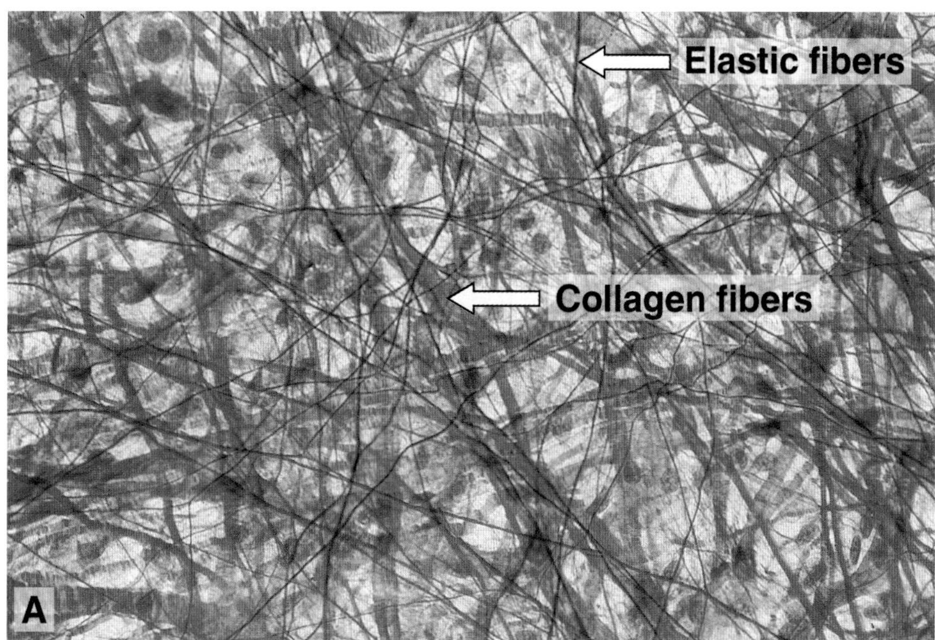

Figure 5–24. **A:** Total preparation of young rat mesentery showing red picrosirius-stained nonanastomosing bundles of collagen fibers, while the elastic fibers appear as thin, dark anastomosing fibers stained by orcein. Collagen and elastic fibers provide structure and elasticity, respectively, to the mesentery. Medium magnification. **B:** The same preparation observed with polarizing microscopy. Collagen bundles of various thicknesses are observed. In the superimposed regions, the bundles of collagen are a dark color. Medium magnification.

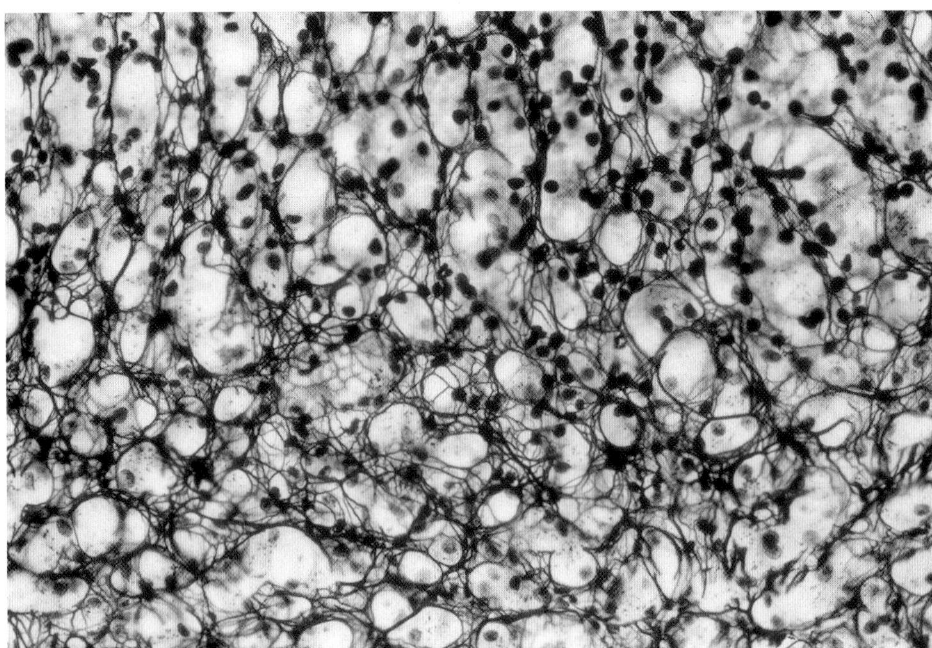

Figure 5–25. Section of an adrenal cortex, silver stained to show reticular fibers. This is a thick section made to emphasize the networks formed by these fibers, which consist of collagen type III. Nuclei are black, and cytoplasm is unstained. Medium magnification.

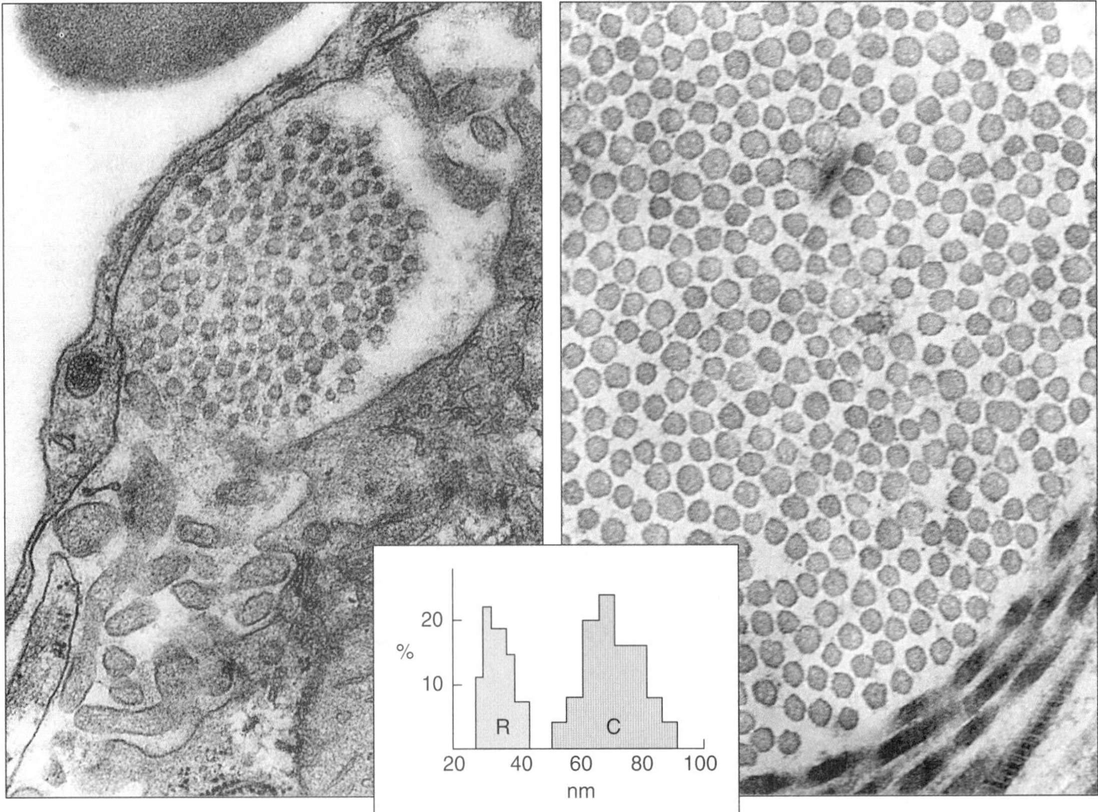

Figure 5–26. Electron micrograph of cross sections of reticular (**left**) and collagen (**right**) fibers. Note that each fiber type is composed of numerous smaller collagen fibrils. Reticular fibrils (R) are significantly narrower in diameter than collagen fibrils of collagen fibers (C; see histogram **inset**); in addition, the constituent fibrils of the reticular fibers reveal an abundant surface-associated granularity not present on regular collagen fibrils (**right**). ×70,000.

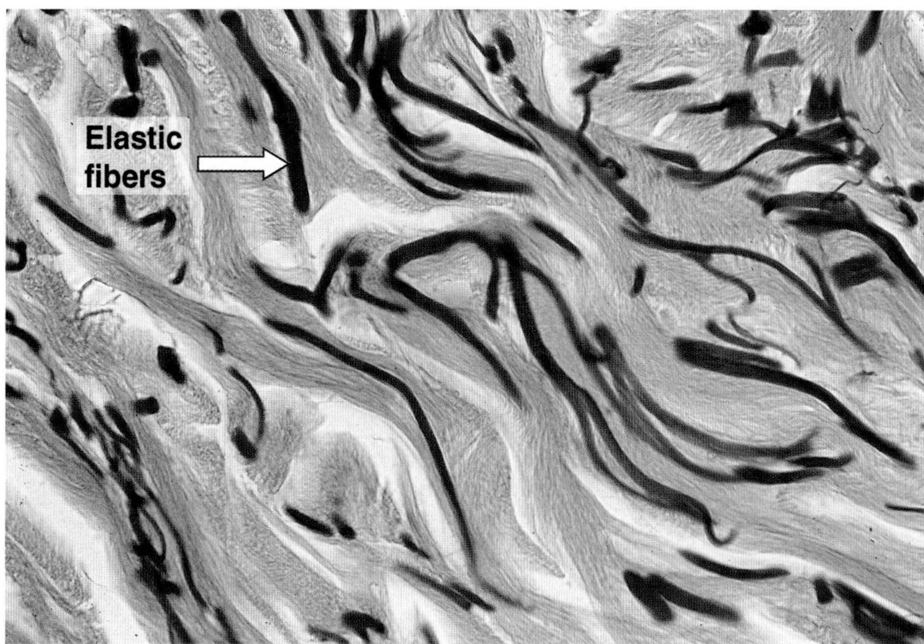

Figure 5–27. Skin dermis, selectively stained for elastic fibers. Dark elastic fibers are interspersed with pale red collagen fibers. The elastic fibers are responsible for skin's elasticity. Medium magnification.

granules (Figures 5–30 and 5–31). The ground substance is formed mainly of 3 classes of components: **glycosaminoglycans, proteoglycans,** and **multiadhesive glycoproteins.**

Glycosaminoglycans (originally called **acid mucopolysaccharides**) are linear polysaccharides formed by repeating disaccharide units usually composed of a uronic acid and a hexosamine. The hexosamine can be **glucosamine** or **galactosamine,** and the uronic acid can be **glucuronic** or **iduronic acid.** With the exception of hyaluronic acid, these linear chains are bound covalently to a protein core (Figure 5–32), forming a **proteoglycan molecule.** Because of the abundance of hydroxyl, carboxyl, and sulfate groups in the carbohydrate moiety of most glycosaminoglycans, the glycosaminoglycans are intensely hydrophilic and act as polyanions. With the exception of hyaluronic acid, all other glycosaminoglycans are sulfated to some degree in the adult state. The carbohydrate portion of proteoglycans constitutes 80–90% of the weight of this macromolecule. Because of these characteristics, proteoglycans can bind to a great number of cations (usually sodium) by electrostatic (ionic) bonds. Proteoglycans are intensely hydrated structures with a thick layer of solvation water surrounding the molecule. When fully hydrated, proteoglycans fill a much larger volume (domain) than they do in their anhydrous state and are highly viscous.

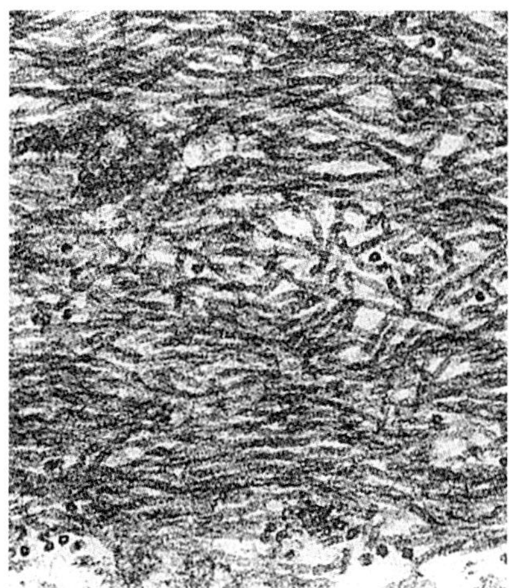

A. Oxytalan

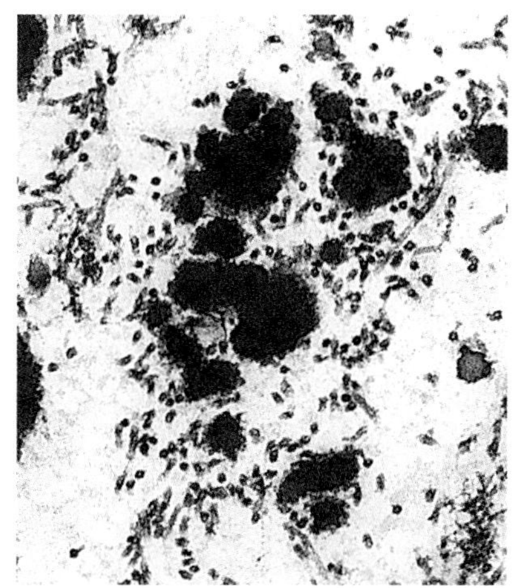

B. Elaunin

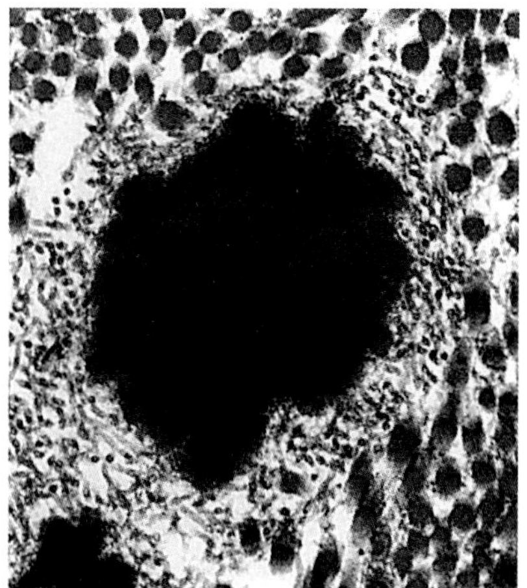

C. Elastic fibers

Figure 5–28. Electron micrographs of developing elastic fibers. **A:** In early stages of formation, developing fibers consist of numerous small glycoprotein microfibrils. **B:** With further development, amorphous aggregates of elastin are found among the microfibrils. **C:** The amorphous elastin accumulates, ultimately occupying the center of an elastic fiber delineated by microfibrils. Note the collagen fibrils, seen in cross section. (Courtesy of GS Montes.)

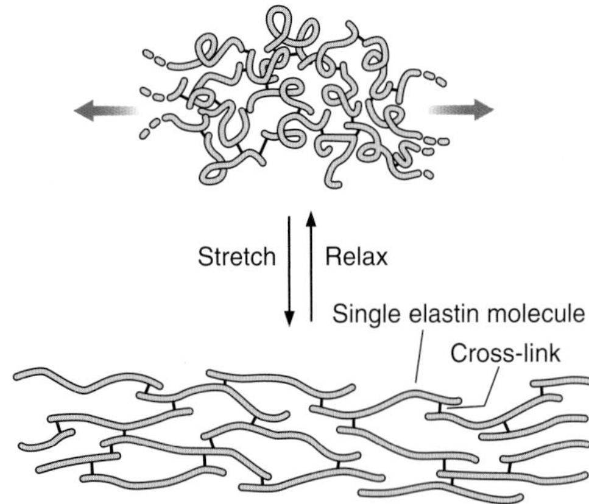

Figure 5–29. **Elastin molecules are joined by covalent bonds to generate an extensive cross-linked network. Because each elastin molecule in the network can expand and contract like a random coil, the entire network can stretch and recoil like a rubber band.** (Reproduced, with permission, from Alberts B et al: *Molecular Biology of the Cell.* Garland, 1983.)

The proteoglycans are composed of a core protein associated with the 4 main glycosaminoglycans: **dermatan sulfate, chondroitin sulfates, keratan sulfate,** and **heparan sulfate.** Table 5–5 shows the chemical composition and tissue distribution of the glycosaminoglycans and proteoglycans. Proteoglycan is a 3-dimensional structure that can be pictured as a test tube brush, with the wire stem representing the protein core and the bristles representing the glycosaminoglycans (Figure 5–32). In cartilage, the proteoglycan molecules have been shown to be bound to a hyaluronic acid chain, forming larger molecules—proteoglycan aggregates. The acidic groups of proteoglycans cause these molecules to bind to the basic amino acid residues of collagen. Proteoglycans are distinguished for their diversity and include a family of cell-surface and extracellular matrix macromolecules. A given matrix may contain several different types of core proteins, and each may contain different numbers of glycosaminoglycans with different lengths and composition. One of the most important extracellular matrix proteoglycans is **aggrecan,** the dominant proteoglycan in cartilage. In the aggrecan, several molecules of proteoglycans (containing chondroitin sulfate chains) are noncovalently associated by its core protein to a molecule of hyaluronic acid. Cell-surface proteoglycans are attached to the surface of many types of cells, particularly epithelial cells. Two examples are **syndecan** and **fibroglycan.** The core protein of cell-surface proteoglycans spans the plasma membrane and contains a short cytosolic extension. A small number of heparan sulfate or chondroitin sulfate chains of glycosaminoglycans are attached to the extracellular extension of the core protein (Figure 5–33).

Besides acting as structural components of the extracellular matrix and anchoring cells to the matrix, both extracellular and surface proteoglycans also bind many protein growth factors (eg, TGF-β, transforming growth factor of fibroblasts).

The synthesis of proteoglycans begins in the rough endoplasmic reticulum with the synthesis of the protein moiety of the molecule. Glycosylation is initiated in the rough endoplasmic reticulum and completed in the Golgi complex, where sulfation also occurs (see Chapter 2).

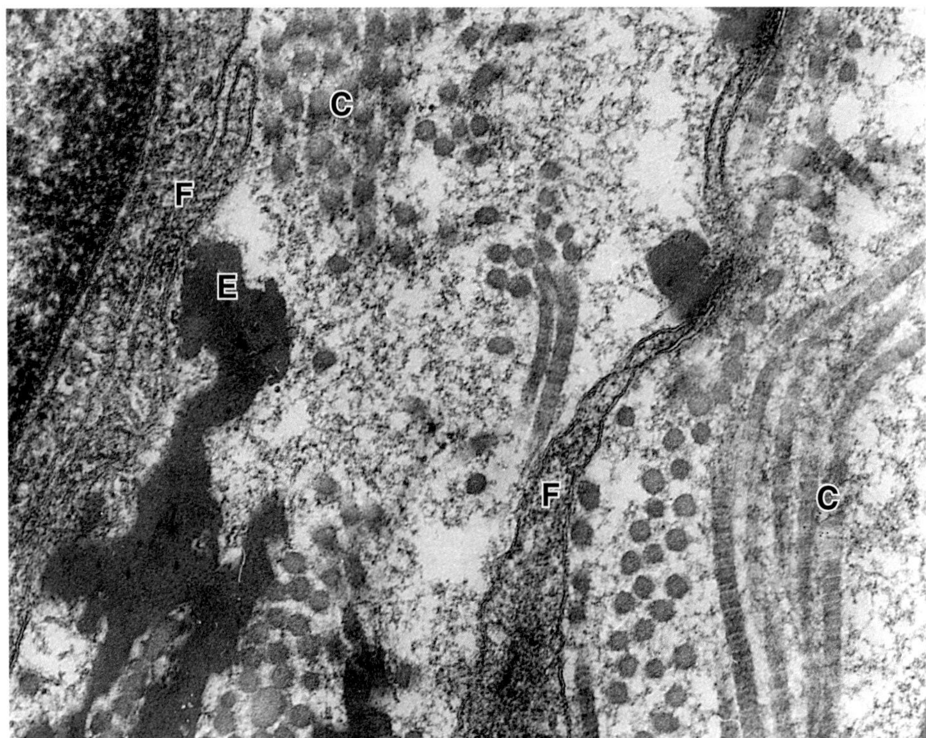

Figure 5–30. Electron micrograph showing the structural organization of the connective tissue matrix. The ground substance is a fine granular material that fills the spaces between the collagen (C) and elastic (E) fibers and surrounds fibroblast cells and processes (F). The granularity of ground substance is an artifact of the glutaraldehyde–tannic acid fixation procedure. ×100,000.

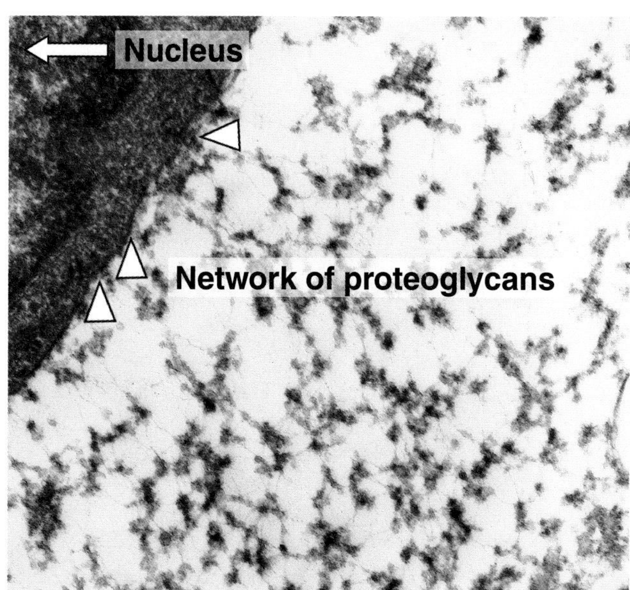

Figure 5–31. Extracellular matrix of mouse endometrium after fixation in the presence of Safranin O. A network of proteoglycans fills the intercellular spaces. Some proteoglycan filaments are in close contact with the cell surface (arrows). Medium magnification. (Courtesy of C. Greca and T. Zorn.)

MEDICAL APPLICATION

The degradation of proteoglycans is carried out by several cell types and depends on the presence of several lysosomal enzymes. Several disorders have been described in which a deficiency in lysosomal enzymes causes glycosaminoglycan degradation to be blocked, with the consequent accumulation of these compounds in tissues. The lack of specific hydrolases in the lysosomes has been found to be the cause of several disorders in humans, including Hurler syndrome, Hunter syndrome, Sanfilippo syndrome, and Morquio syndrome.

*Because of their high viscosity, intercellular substances act as a barrier to the penetration of bacteria and other microorganisms. Bacteria that produce **hyaluronidase,** an enzyme that hydrolyzes hyaluronic acid and other glycosaminoglycans, have great invasive power because they reduce the viscosity of the connective tissue ground substance.*

Multiadhesive glycoproteins are compounds that contain a protein moiety to which carbohydrates are attached. In contrast to proteoglycans, the protein moiety usually predominates, and these molecules do not contain the linear polysaccharides formed by repeating disaccharides containing hexosamines. Instead, the carbohydrate moiety of glycoproteins is frequently a branched structure.

Several glycoproteins have been isolated from connective tissue, and they play an important role not only in the interaction between neighboring adult and embryonic cells but also in the adhesion of cells to their substrate. **Fibronectin** (L. *fibra,* fiber, + *nexus,* interconnection) is a glycoprotein synthesized by fibroblasts and some epithelial cells. This molecule, with a molecular mass of 222–240 kDa, has binding sites for cells, collagen, and glycosaminoglycans. Interactions at these sites help to mediate normal cell adhesion and migration (Figure 5–34). Fibronectin is distributed as a network in the intercellular spaces of many tissues (Figures 5–34 and 5–35). **Laminin** is a large glycoprotein that participates in the adhesion of epithelial cells to the basal lamina, a structure rich in laminin (Figures 5–34 and 5–36).

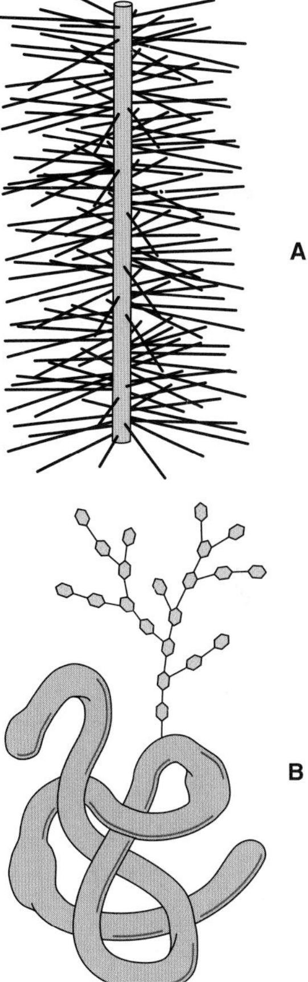

Figure 5–32. The molecular structure of proteoglycans and glycoproteins. **A:** Proteoglycans contain a core of protein (vertical rod in drawing) to which molecules of glycosaminoglycans (GAGs) are covalently bound. A GAG is an unbranched polysaccharide made up of repeating disaccharides; one component is an amino sugar, and the other is uronic acid. Proteoglycans contain a greater amount of carbohydrate than do glycoproteins. **B:** Glycoproteins are globular protein molecules to which branched chains of monosaccharides are covalently attached. (Reproduced, with permission, from Junqueira LCU, Carneiro J: *Biologia Celular e Molecular,* 7[a] ed. Editora Guanabara Koogan. Rio de Janeiro, 2000.)

Table 5–5. Composition and distribution of glycosaminoglycans in connective tissue and their interactions with collagen fibers.

	Repeating Disaccharides			
Glycosaminoglycan	**Hexuronic Acid**	**Hexosamine**	**Distribution**	**Electrostatic Interaction with Collagen**
Hyaluronic acid	D-glucuronic acid	D-glucosamine	Umbilical cord, synovial fluid, vitreous humor, cartilage	
Chondroitin 4-sulfate	D-glucuronic acid	D-galactosamine	Cartilage, bone, cornea, skin, notochord, aorta	High levels of interaction, mainly with collagen type II
Chondroitin 6-sulfate	D-glucuronic acid	D-galactosamine	Cartilage, umbilical cord, skin, aorta (media)	High levels of interaction, mainly with collagen type II
Dermatan sulfate	L-iduronic acid or D-glucuronic acid	D-galactosamine	Skin, tendon, aorta (adventitia)	Low levels of interaction, mainly with collagen type I
Heparan sulfate	D-glucuronic acid or L-iduronic acid	D-galactosamine	Aorta, lung, liver, basal laminae	Intermediate levels of interaction, mainly with collagen types III and IV
Keratan sulfate (cornea)	D-galactose	D-galactosamine	Cornea	None
Keratan sulfate (skeleton)	D-galactose	D-glucosamine	Cartilage, nucleus pulposus, annulus fibrosus	None

Cells interact with extracellular matrix components by using cell-surface molecules (**matrix receptors**) that bind to collagen, fibronectin, and laminin. These receptors are the **integrins,** a family of transmembrane linker proteins (Figures 5–37 and 5–38). Integrins bind their ligands in the extracellular matrix with relatively low affinity, allowing cells to explore their environment without losing attachment to it or becoming glued to it. Clearly, integrins must interact with the cytoskeleton, usually the actin microfilaments. The interactions between integrins, extracellular matrix, and cytoskeleton elements are mediated by several intracellular proteins, such as **paxilin, vinculin,** and **talin.** The interactions that integrins mediate between the extracellular matrix and the cytoskeleton operate in both directions and play an important role in orienting both the cells and the extracellular matrix in tissues (Figure 5–37).

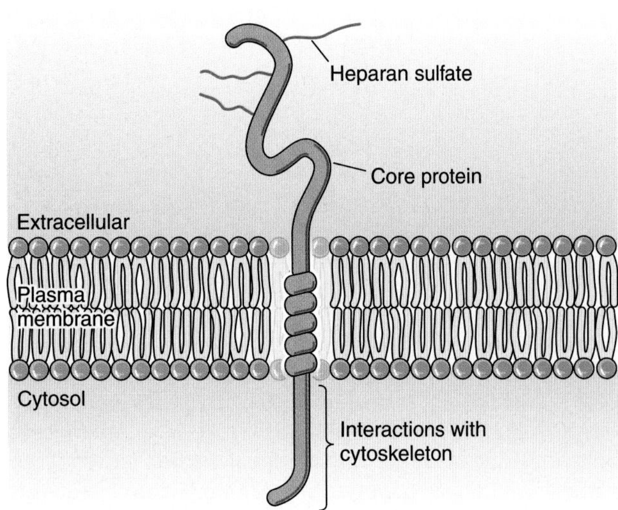

Figure 5–33. Schematic diagram of cell-surface syndecan proteoglycan. The core protein spans the plasma membrane through the cytoplasmic domain. The syndecan proteoglycans possess 3 heparan sulfate chains and sometimes chondroitin sulfate.

MEDICAL APPLICATION

The participation of fibronectin and laminin in both embryonic development and the increased ability of cancer cells to invade other tissues has been postulated. The importance of fibronectin is shown by the fact that mice whose fibronectin has been inactivated die during early embryogenesis.

In connective tissue, in addition to the ground substance, there is a very small quantity of fluid—called **tissue fluid**—that is similar to blood plasma in its content of ions and diffusible substances. Tissue fluid contains a small percentage of plasma proteins of low molecular weight that pass through the capillary walls as a result of the hydrostatic pressure of the blood. Although only a small proportion of connective tissue consists of plasma proteins, it is estimated that because of its wide distribution, as much as one third of the plasma proteins of the body are stored in the intercellular connective tissue matrix.

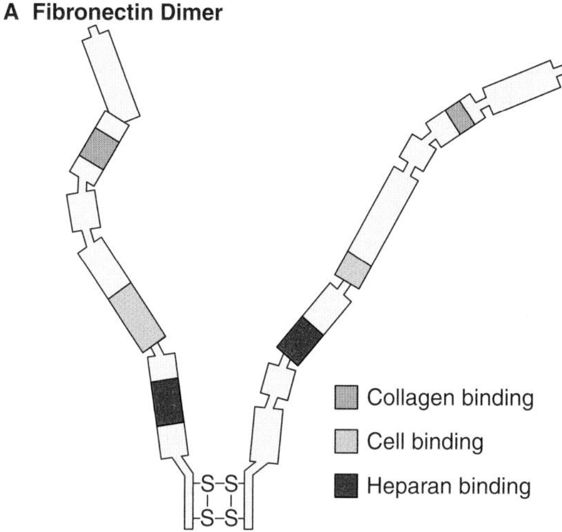

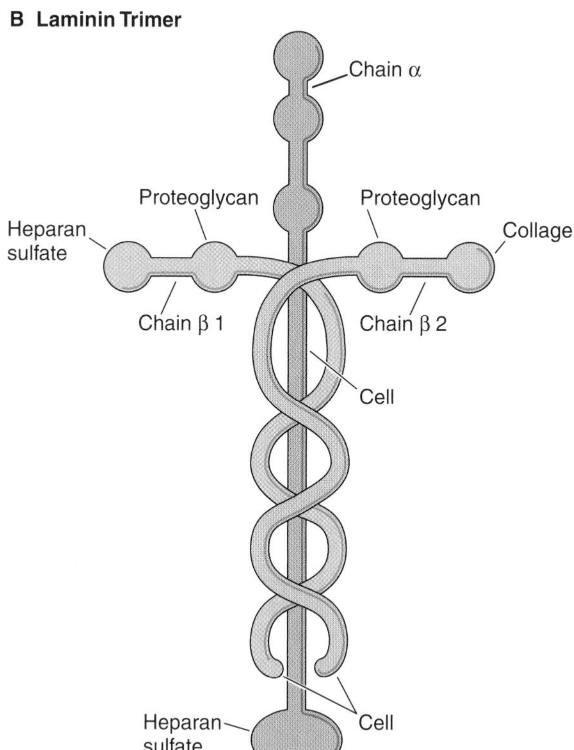

Figure 5–34. **A:** The structure of fibronectin. Fibronectin is a dimer bound by S—S groups, formed by serially disposed coiled sites, that bind to type I collagen, heparan sulfate, other proteoglycans, and cell membrane receptors. **B:** The structure of laminin, which is formed by 3 intertwined polypeptides in the shape of a cross. The figure shows sites on the molecule with a high affinity for cell membrane receptors and type IV collagen and heparan sulfate, which are components of basal laminae. Laminin thus promotes adhesion of cells to basal laminae. (Reproduced, with permission, from Junqueira LCU, Carneiro J: *Biologia Celular e Molecular,* 7ª ed. Editora Guanabara Koogan. Rio de Janeiro, 2000.)

MEDICAL APPLICATION

Edema is promoted by the accumulation of water in the extracellular spaces. Water in the intercellular substance of connective tissue comes from the blood, passing through the capillary walls into the intercellular regions of the tissue. The capillary wall is only slightly permeable to macromolecules but permits the passage of water and small molecules, including low-molecular-weight proteins.

Blood brings to connective tissue the various nutrients required by its cells and carries metabolic waste products away to the detoxifying and excretory organs, such as liver and kidneys.

Two forces act on the water contained in the capillaries: the hydrostatic pressure of the blood, a consequence of the pumping action of the heart, which forces water to pass through the capillary walls; and the colloid osmotic pressure of the blood plasma, which draws water back into the capillaries (Figure 5–39). Osmotic pressure is due mainly to plasma proteins.

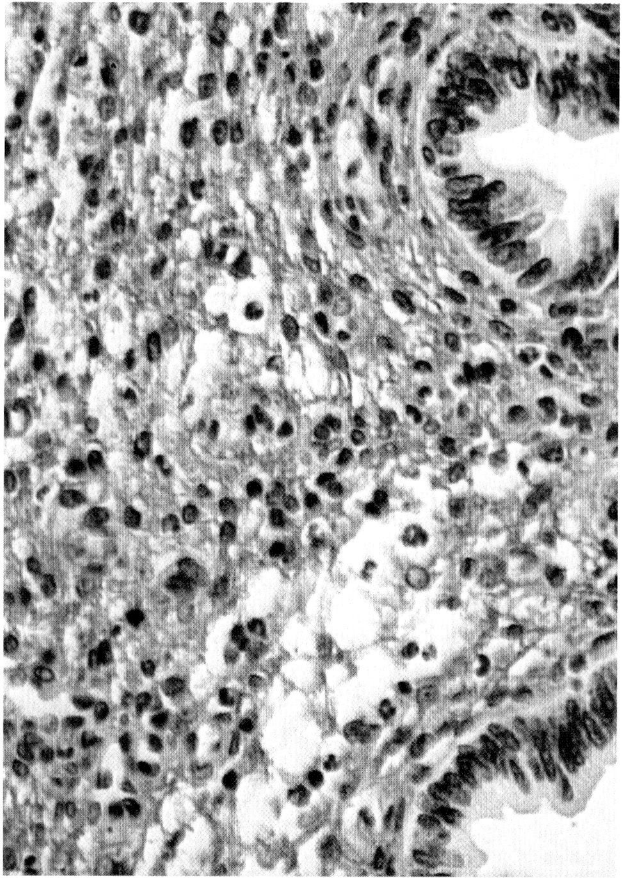

Figure 5–35. Transverse section of mouse endometrium. Immunocytochemical staining shows the distribution of fibronectin in the endometrial stroma. Medium magnification. (Courtesy of D. Tenório and T. Zorn.)

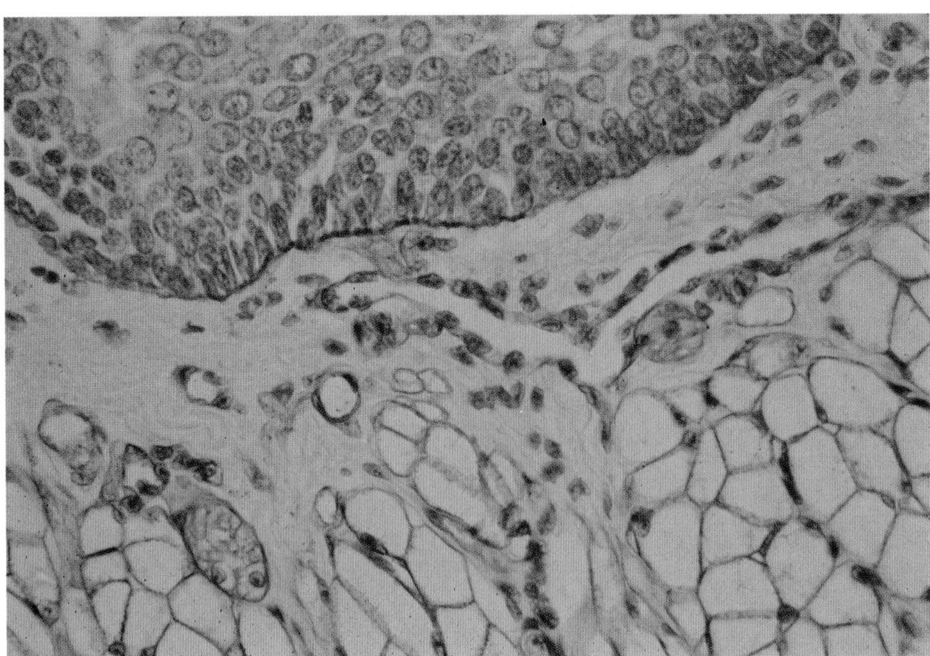

Figure 5–36. Transverse section of tongue. Immunocytochemical staining shows the distribution of laminin basement membranes in epithelial layer, capillary blood vessels, nerve fibers, and striated muscle. Medium magnification.

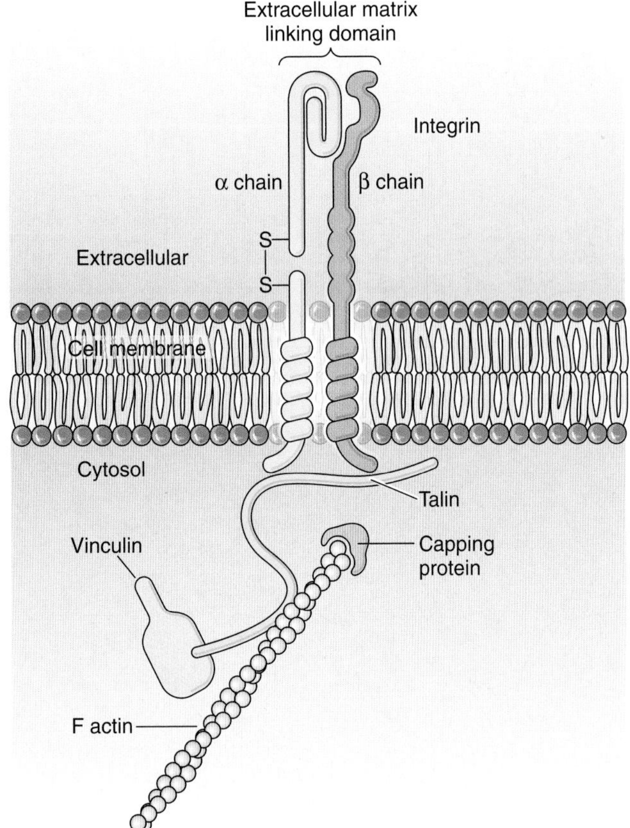

Figure 5–37. Integrin cell-surface matrix receptor. By binding to a matrix protein and to the actin cytoskeleton (via α-actinin) inside the cell, the integrin serves as a transmembrane link. The molecule is a heterodimer, with α and β chains. The head portion may protrude some 20 nm from the surface of the cell membrane into the extracellular matrix.

Because the ions and low-molecular-weight compounds that pass easily through the capillary walls have approximately the same concentration inside and outside these blood vessels, the osmotic pressures they exert are approximately equal on either side of the capillaries and cancel each other. The colloid osmotic pressure exerted by the blood protein macromolecules—which are unable to pass through the capillary walls—is not counterbalanced by outside pressure and tends to bring water back into the blood vessel.

Normally, water passes through capillary walls to the surrounding tissues at the arterial end of a capillary, because the hydrostatic pressure there is greater than the colloid osmotic pressure; the hydrostatic pressure, however, decreases along the length of the capillary toward the venous end. As the hydrostatic pressure falls, osmotic pressure rises because of the progressive increase in the concentration of proteins, which is caused by the passage of water from the capillaries. As a result of this increase in protein concentration and decrease in hydrostatic pressure, osmotic pressure becomes greater than hydrostatic pressure at the venous end of the capillary, and water is drawn back into the capillary (Figure 5–39). In this way, metabolites circulate in the connective tissue, feeding its cells.

The quantity of water drawn back is less than that which passes out through the capillaries. The water that remains in the connective tissue returns to the blood through the lymphatic vessels. The smallest lymphatic vessels are the lymphatic capillaries, which originate in connective tissue with closed ends. Lymphatic vessels drain into veins at the base of the neck (see Chapter 11).

Because of the equilibrium that exists between the water entering and the water leaving the intercellular substance of connective tissue, there is little free water in the tissue.

In several pathologic conditions, the quantity of tissue fluid may increase considerably, causing **edema.** In tissue sections, this condition is characterized by enlarged spaces, caused by the increase in liquid, between the components of the connective tissue. Macroscopically, edema is characterized by an increase in volume that yields easily to localized pressure, causing a depression that slowly disappears (pitting edema).

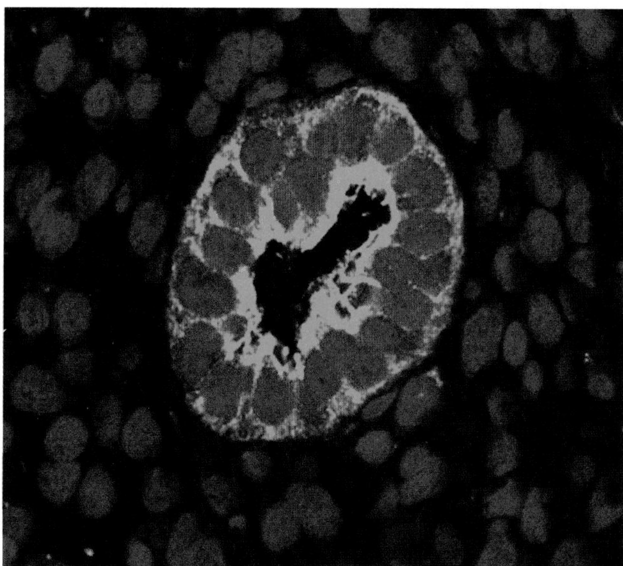

Figure 5–38. Fluorescent micrograph of integrin $\alpha 2$ in the mouse endometrium. Integrin $\alpha 2$ (green) is observed in the cytoplasm of uterine gland cells. The nuclei (red) were stained with fluorescent propidium iodide. Medium magnification. (Courtesy of F. Costa and P. Abrahamsohn.)

Edema may result from venous or lymphatic obstruction or from a decrease in venous blood flow (eg, congestive heart failure). It may also be caused by the obstruction of lymphatic vessels due to parasitic plugs or tumor cells and chronic starvation; protein deficiency results in a lack of plasma proteins and a decrease in colloid osmotic pressure. Water therefore accumulates in the connective tissue and is not drawn back into the capillaries.

Another possible cause of edema is increased permeability of the blood capillary endothelium resulting from chemical or mechanical injury or the release of certain substances produced in the body (eg, histamine).

TYPES OF CONNECTIVE TISSUE

There are several types of connective tissue that consist of the basic components already described: fibers, cells, and ground substance. The names given to the various types denote either the component that predominates in the tissue or a structural characteristic of the tissue. Figure 5–40 illustrates the main types of connective tissue.

Connective Tissue Proper

There are 2 classes of connective tissue proper: loose and dense (Figure 5–41).

Loose connective tissue supports many structures that are normally under pressure and low friction. It is a very common type of connective tissue; it fills spaces between groups of muscle cells, supports epithelial tis-

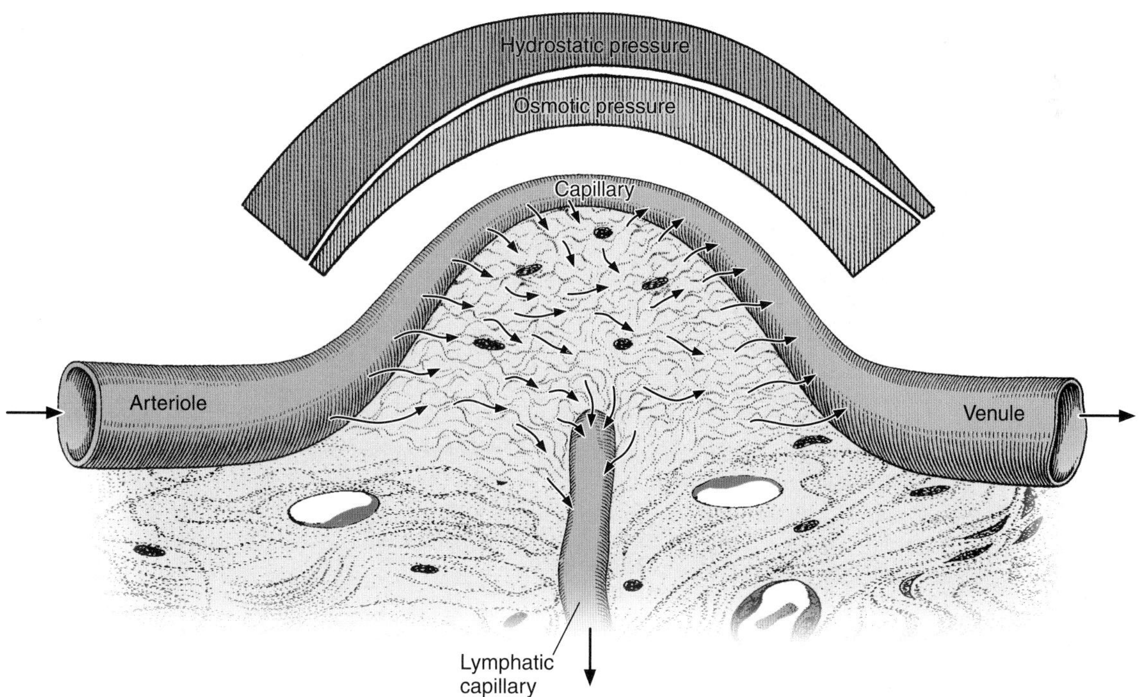

Figure 5–39. Movement of fluid through connective tissue. There is a decrease in hydrostatic pressure and an increase in osmotic pressure from the arterial to the venous ends of blood capillaries (upper part of drawing). Fluid leaves the capillary through its arterial end and repenetrates the blood at the venous end. Some fluid is drained by the lymphatic capillaries.

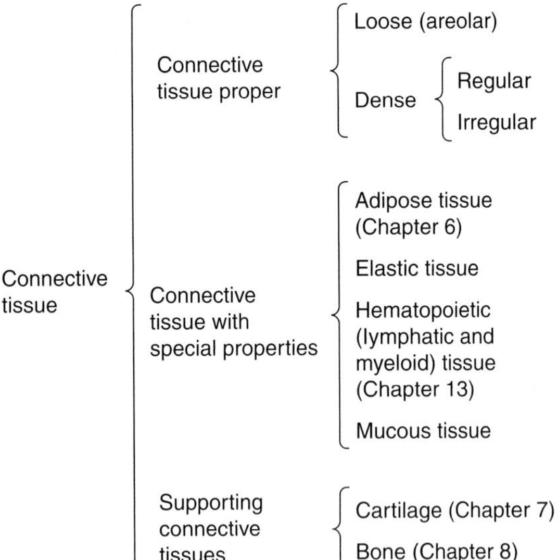

Figure 5–40. Simplified scheme classifying the principal types of connective tissue, which are discussed in the chapters indicated.

sue, and forms a layer that sheathes the lymphatic and blood vessels. Loose connective tissue is also found in the papillary layer of the dermis, in the hypodermis, in the serosal linings of peritoneal and pleural cavities, and in glands and the mucous membranes (wet membranes that line the hollow organs) supporting the epithelial cells.

Loose connective tissue (Figure 5–42) comprises all the main components of connective tissue proper. There is no predominant element in this tissue. The most numerous cells are fibroblasts and macrophages, but all the other types of connective tissue cells are also present. A moderate amount of collagen, elastic, and reticular fibers appear in this tissue. Loose connective tissue has a delicate consistency; it is flexible, well vascularized, and not very resistant to stress.

Dense connective tissue is adapted to offer resistance and protection. It consists of the same components found in loose connective tissue, but there are fewer cells and a clear predominance of collagen fibers (Figures 5–43, 5–44, and 5–45). Dense connective tissue is less flexible and far more resistant to stress than is loose connective tissue. It is known as **dense irregular** connective tissue when the collagen fibers are arranged in bundles without a definite orientation (Figure 5–44). The collagen fibers form a 3-dimensional network in dense irregular tissue and provide resistance to stress from all directions. This type of tissue is encountered in such areas as the dermis.

The collagen bundles of **dense regular** connective tissue are arranged according to a definite pattern. The collagen fibers of this tissue are aligned with the linear orientation of fibroblasts in response to prolonged stresses exerted in the same direction; they consequently offer great resistance to traction forces.

Tendons are the most common example of dense regular connective tissue. These elongated cylindrical structures attach striated muscle to bone; by virtue of their richness in collagen fibers, they are white and inextensible. They have parallel, closely packed bundles of collagen separated by a small quantity of intercellular ground substance. Their fibrocytes contain elongated nuclei parallel to the fibers and sparse cytoplasmic folds that envelop portions of the collagen bundles. The cytoplasm of these fibrocytes is rarely revealed in H&E stains, not only because it is sparse but also because it stains the same color as the fibers (Figures 5–45, 5–46, and 5–47).

The collagen bundles of the tendons (**primary bundles**) aggregate into larger bundles (**secondary bundles**) that are enveloped by loose connective tissue containing blood vessels and nerves. Externally, the tendon is surrounded by a sheath of dense connective tissue. In some tendons, this sheath is made up of 2 layers, both lined by squamous cells of mesenchymal origin. One layer is attached to the tendon, and the other lines the neighboring structures. A cavity containing a viscous fluid (similar to the fluid of synovial joints) is formed between the two layers. This fluid, which contains water, proteins, glycosaminoglycans, glycoproteins, and ions, is a lubricant that permits an easy sliding movement of the tendon within its sheath.

Figure 5–41. Section of rat skin in the process of repair of a lesion. The subepithelial connective tissue (dermis) is loose connective tissue formed soon after the lesion occurs. In this area, the cells, most of which are fibroblasts, are abundant. The deepest part of the dermis consists of dense irregular connective tissue, which contains many randomly oriented thick collagen fibers, scarce ground substance, and few cells. H&E stain. Medium magnification.

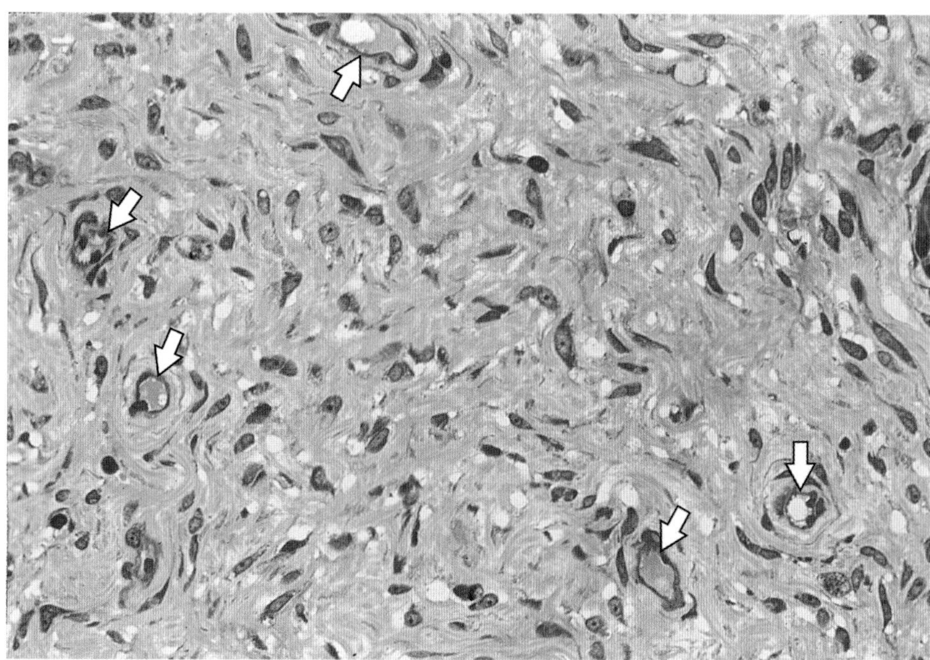

Figure 5–42. Section of loose connective tissue. Many fibroblast nuclei are interspersed with irregularly distributed collagen fibers. Small blood vessels are indicated by arrows. H&E stain. Medium magnification.

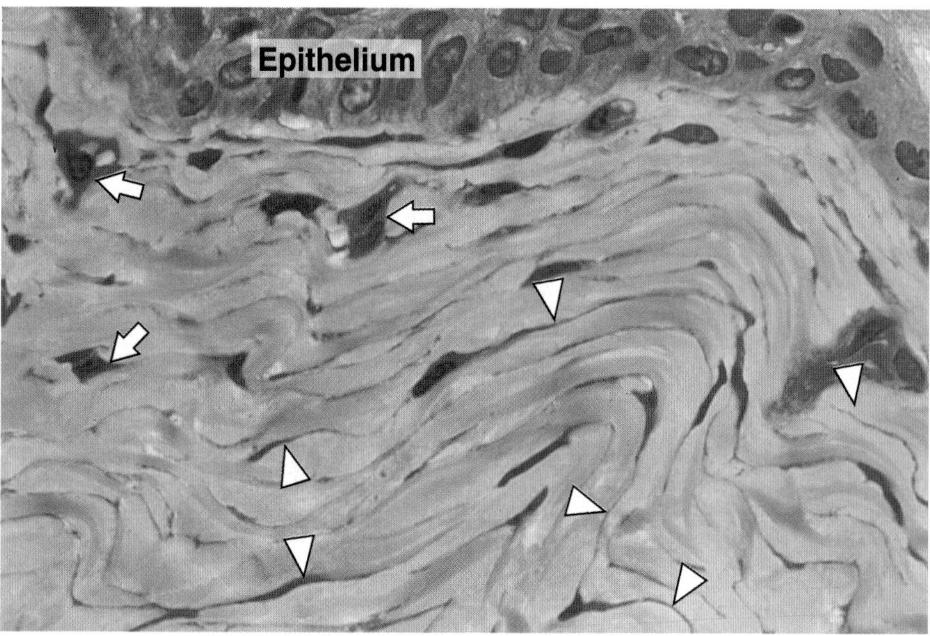

Figure 5–43. Section of immature dense irregular collagen tissue. This figure shows numerous fibroblasts (arrow) with many thin cytoplasmic extensions (arrowheads). As these cells are pressed by collagen fibers, the appearance of their cytoplasm depends on the section orientation; when the section is parallel to the cell surface, parts of the cytoplasm are visible. PT stain. Medium magnification.

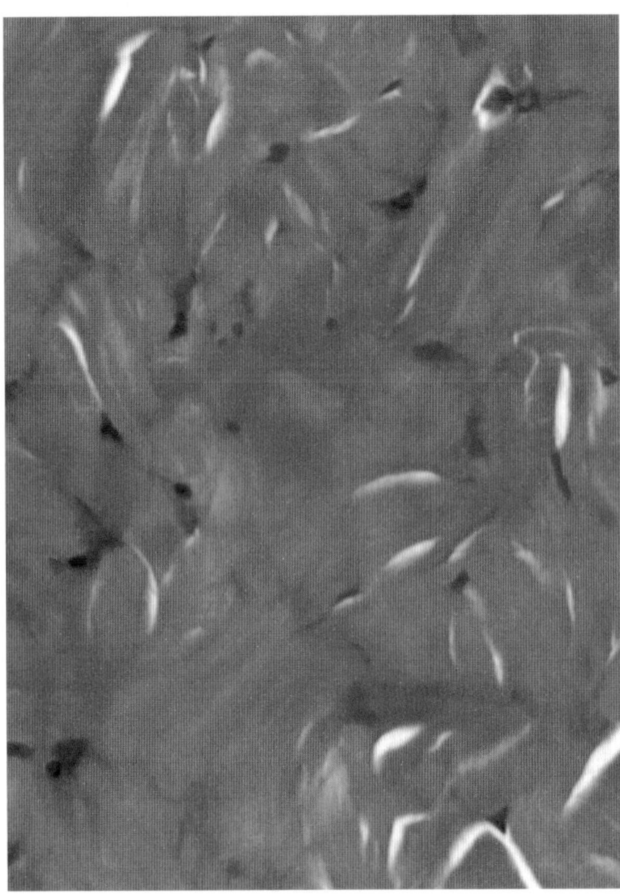

Figure 5–44. Dense irregular connective tissue contains many randomly oriented bundles of collagen fibers. H&E stain. Medium magnification.

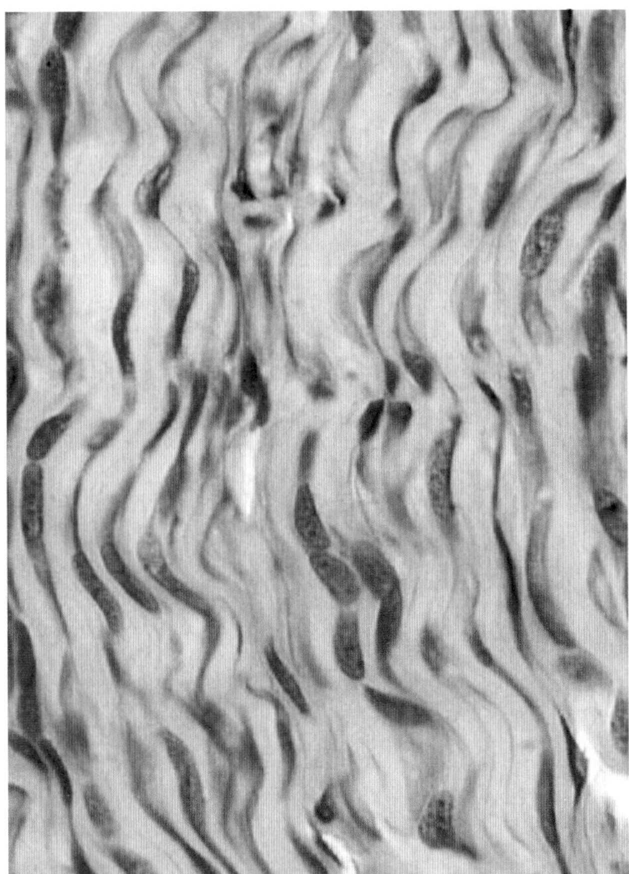

Figure 5–45. Longitudinal section of dense regular connective tissue (tendon). Bundles of collagen fibers fill the spaces between the elongated fibroblasts. H&E stain. Medium magnification.

Elastic Tissue

Elastic tissue is composed of bundles of thick, parallel elastic fibers. The space between these fibers is occupied by thin collagen fibers and flattened fibroblasts. The abundance of elastic fibers in this tissue confers on it a typical yellow color and great elasticity. Elastic tissue, which occurs infrequently, is present in the yellow ligaments of the vertebral column and in the suspensory ligament of the penis.

The very delicate reticular tissue forms 3-dimensional networks that support cells. Reticular tissue is a specialized loose connective tissue consisting of reticular fibers intimally associated with specialized fibroblasts called reticular cells (Figure 5–48). Reticular tissue provides the

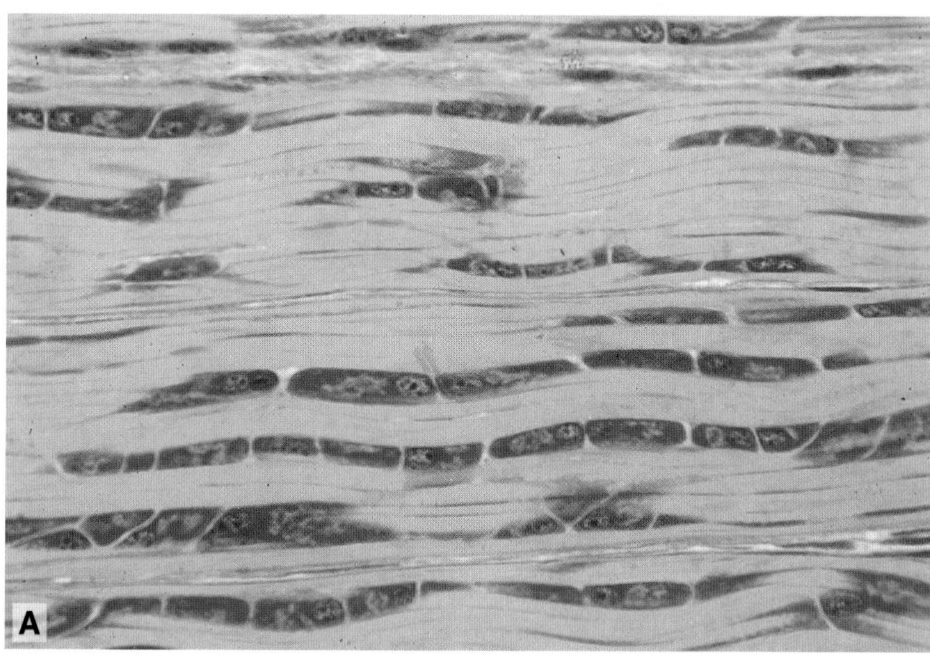

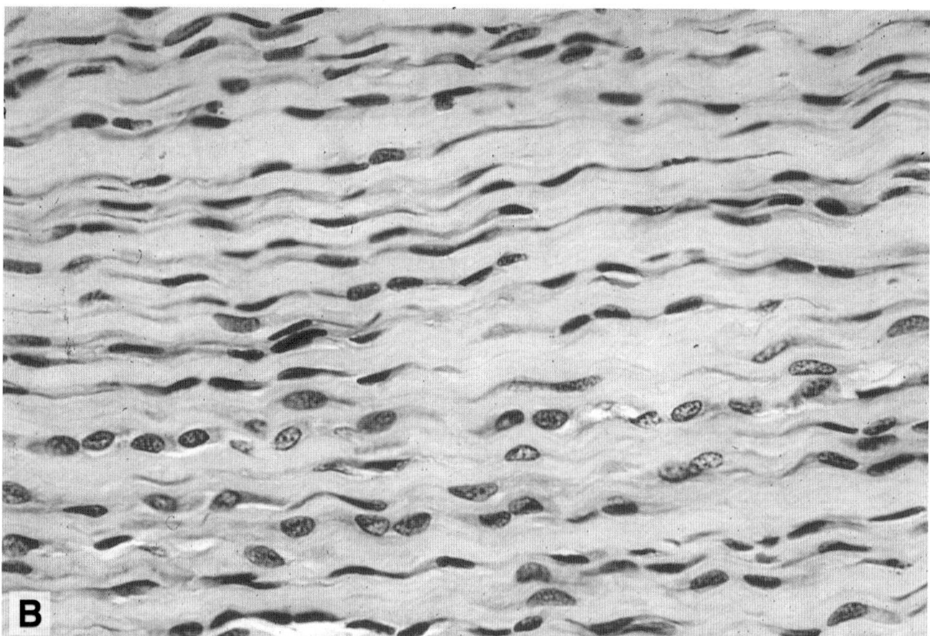

Figure 5–46. Longitudinal section of dense regular connective tissue from a tendon. **A:** Thick bundles of parallel collagen fibers fill the intercellular spaces between fibroblasts. Low magnification. **B:** Higher magnification view of a tendon of a young animal. Note active fibroblasts with prominent Golgi regions and dark cytoplasm rich in RNA. PT stain.

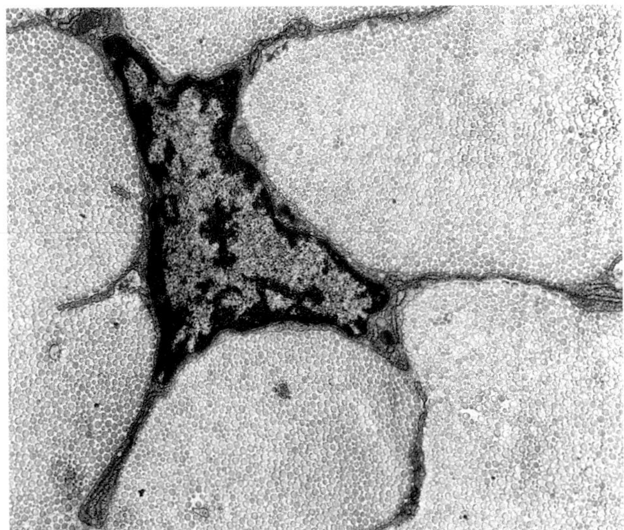

Figure 5–47. Electron micrograph of a fibrocyte in dense regular connective tissue. The sparse cytoplasm of the fibrocytes is divided into numerous thin cytoplasmic processes that interdigitate among the collagen fibers. ×25,000.

architectural framework that creates a special microenvironment for hematopoietic organs and lymphoid organs (bone marrow, lymph nodules and nodes, and spleen). The reticular cells are dispersed along this framework and partially cover the reticular fibers and ground substance with cytoplasmic processes. The resulting cell-lined trabecular system creates a spongelike structure (Figure 5–48) within which cells and fluids are freely mobile.

In addition to the reticular cells, cells of the mononuclear phagocyte system are strategically dispersed along the trabeculae. These cells monitor the slow flow of materials through the sinuslike spaces and remove invaders by phagocytosis.

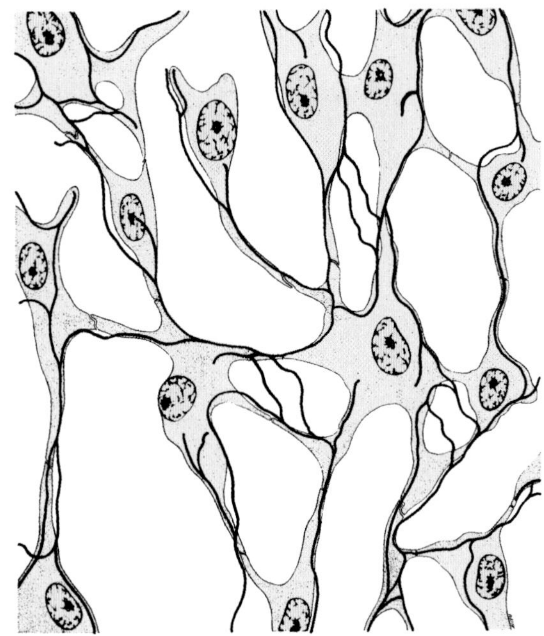

Figure 5–48. Reticular connective tissue showing only the attached cells and the fibers (free cells are not represented). Reticular fibers are enveloped by the cytoplasm of reticular cells; the fibers, however, are extracellular, being separated from the cytoplasm by the cell membrane. Within the sinuslike spaces, cells and tissue fluids of the organ are freely mobile.

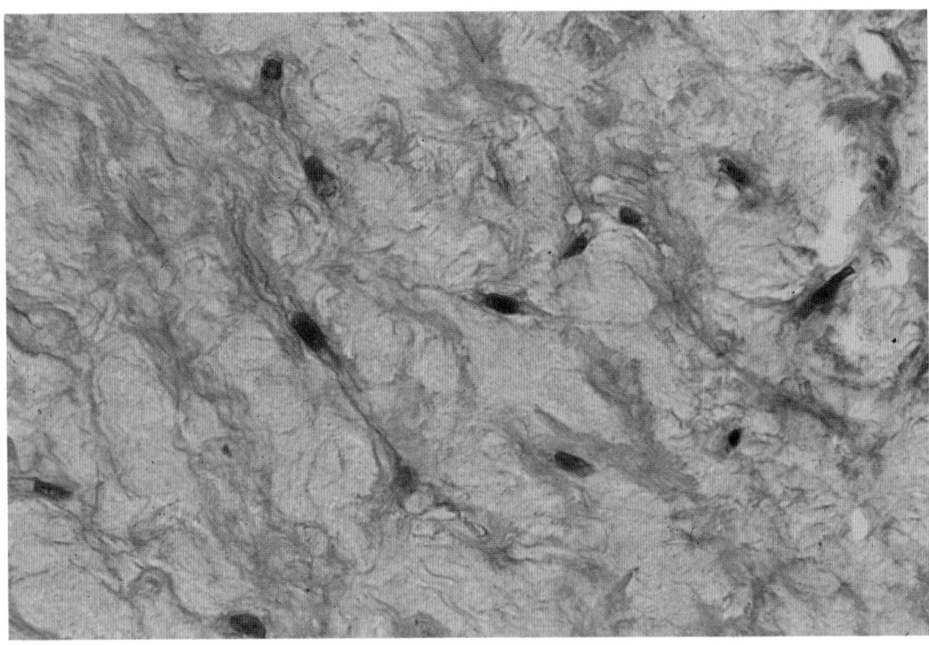

Figure 5–49. Mucous tissue of an embryo showing fibroblasts immersed in a very loose extracellular matrix composed mainly of molecules of the ground substances. H&E stain. Medium magnification.

The mucous tissue is found mainly in the umbilical cord. Mucous tissue has an abundance of ground substance composed chiefly of hyaluronic acid (Figure 5–49). It is a jellylike tissue containing very few fibers. The cells in this tissue are mainly fibroblasts. Mucous tissue is the principal component of the umbilical cord, where it is referred to as **Wharton's jelly.** It is also found in the pulp of young teeth.

REFERENCES

Deyl Z, Adam M: *Connective Tissue Research: Chemistry, Biology and Physiology.* Liss, 1981.

Gay S, Miller EJ: *Collagen in the Physiology and Pathology of Connective Tissue.* Gustav Fischer, 1978.

Greca CP et al: Ultrastructural cytochemical characterization of collagen-associated proteoglycans in the endometrium of mice. Anat Rec 2000;259:413.

Hay ED (editor): *Cell Biology of Extracellular Matrix,* 2nd ed. Plenum, 1991.

Hogaboam C, Kunkel SL et al: Novel role of transmembrane SCF for mast cell activation and eotaxin production in mast cell-fibroblast interaction. J Immunol 1998;160:6166.

Jamur MC, Grodzki ACG et al: Immunomagnetic isolation of rat bone marrow derived and peritoneal mast cells. J Histochem Cytochem 1997;45:1715.

Junqueira LCU et al: Picrosirius staining plus polarization microscopy, a specific method for collagen detection in tissue sections. Histochem J 1979;11:447.

Junqueira LCU, Montes GS: Biology of collagen proteoglycan interaction. Arch Histol Jpn 1983;6:589.

Kefalides NA et al: Biochemistry and metabolism of basement membranes. Int Rev Cytol 1979;1:167.

Krstíc RV: *Illustrated Encyclopedia of Human Histology.* Springer-Verlag, 1984.

Mathews MB: *Connective Tissue, Macromolecular Structure and Evolution.* Springer-Verlag, 1975.

Mercalafe DD et al: Mast cells. Physiol Rev 1997;77:1033.

Montes GS et al: Collagen distribution in tissues. In: *Ultrastructure of the Connective Tissue Matrix.* Ruggieri A, Motta PM (editors). Martinus Nijhoff, 1984.

Montes GS, Junqueira LCU: The use of the picrosirius-polarization method for the study of biopathology of collagen. Mem Inst Oswaldo Cruz. 1991;86 (suppl):1.

Prockop DJ et al: The biosynthesis of collagen and its disorders. N Engl J Med 1979;301:77.

Sandberg LB et al: Elastin structure, biosynthesis, and relation to disease state. N Engl J Med 1981;304:566.

Van Furth R (editor): *Mononuclear Phagocytes: Functional Aspects.* 2 vols. Martinus Nijhoff, 1980.

Yamada KM, Miyamoto S: Integrin transmembrane signaling and cytoskeletal control. Curr Opin Cell Biol 1995;143:2323.

Adipose Tissue 6

Adipose tissue is a special type of connective tissue in which adipose (L. *adeps,* fat) cells (**adipocytes**) predominate. These cells can be found isolated or in small groups within the connective tissue itself; most are found in large aggregates, making up the adipose tissues that are spread throughout the body. Adipose tissue is, in a sense, one of the largest organs in the body. In men of normal weight, adipose tissue represents 15–20% of the body weight; in women of normal weight, 20–25% of body weight.

Adipose tissue is the largest repository of energy (in the form of triglycerides) in the body. The other organs that store energy (in the form of glycogen) are the liver and skeletal muscle. Since eating is a periodic activity and the supply of glycogen is limited, there must be a large store of calories that can be mobilized between meals. Because triglycerides are of lower density than glycogen and have a higher caloric value (9.3 kcal/g for triglycerides versus 4.1 kcal/g for carbohydrates), adipose tissue is a very efficient storage tissue. It is in a state of continuous turnover and is sensitive to both nervous and hormonal stimuli. Subcutaneous layers of adipose tissue help to shape the surface of the body, whereas deposits in the form of pads act as shock absorbers, chiefly in the soles and palms. Since fat is a poor heat conductor, it contributes to the thermal insulation of the body. Adipose tissue also fills up spaces between other tissues and helps to keep some organs in place. Recently, it was observed that adipose tissue secretes various types of molecules that may be carried by the blood to influence distant organs. There are two known types of adipose tissue that have different locations, structures, colors, and pathologic characteristics. **Unilocular (common,** or **yellow) adipose tissue** is composed of cells that, when completely developed, contain one large central droplet of yellow fat in their cytoplasm. **Multilocular** (or **brown**) adipose tissue is composed of cells that contain numerous lipid droplets and abundant brown mitochondria. Both types of adipose tissue have a rich blood supply.

UNILOCULAR ADIPOSE TISSUE

The color of unilocular adipose tissue varies from white to dark yellow, depending on the diet; it is due mainly to the presence of carotenoids dissolved in fat droplets of the cells. Almost all adipose tissue in adults is of this type. It is found throughout the human body except for the eyelids, the penis, the scrotum, and the entire auricle of the external ear but the lobule. Age and sex determine the distribution and density of adipose deposits.

In the newborn, unilocular adipose tissue has a uniform thickness throughout the body. As the baby matures, the tissue tends to disappear from some parts of the body and increase in others. Its distribution is partly regulated by sex hormones and adrenocortical hormones, which control the accumulation of fat and are largely responsible for male or female body contour.

Unilocular adipose cells are spherical when isolated but are polyhedral in adipose tissue, where they are closely packed. Each cell is between 50 and 150 μm in diameter. Since lipid droplets are removed by the alcohol and xylol used in routine histologic techniques, each cell appears in standard microscope preparations as a thin ring of cytoplasm surrounding the vacuole left by the dissolved lipid droplet—the **signet ring cell.** Consequently, these cells have eccentric and flattened nuclei (Figure 6–1). The rim of cytoplasm that remains after removal of the stored triglycerides (neutral fats) may rupture and collapse, distorting the tissue structure.

The thickest portion of the cytoplasm surrounds the nucleus of these cells and contains a Golgi complex, mitochondria, poorly developed cisternae of the rough endoplasmic reticulum, and free polyribosomes. The rim of cytoplasm surrounding the lipid droplet contains cisternae of smooth endoplasmic reticulum and numerous pinocytotic vesicles. Electron-microscope studies reveal that each adipose cell usually possesses minute lipid droplets in addition to the single large droplet seen with the light microscope; the droplets are not enveloped by a membrane but show many vimentin intermediate filaments in their periphery. Each adipose cell is surrounded by a basal lamina.

Unilocular adipose tissue is subdivided into incomplete lobules by a partition of connective tissue containing a rich vascular bed and network of nerves. Reticular fibers form a fine interwoven network that supports individual fat cells and binds them together.

Although blood vessels are not always apparent in tissue sections, adipose tissue is richly vascularized. If the amount of cytoplasm in fat cells is taken into consideration, the ratio of blood volume to cytoplasm volume is greater in adipose tissue than in striated muscle.

Storage and Mobilization of Lipids

The unilocular adipose tissue is a large depot of energy for the organism. The lipids stored in adipose cells are chiefly triglycerides, ie, esters of fatty acids and glycerol. Fatty acids stored by these cells have their origin in dietary fats that are brought to

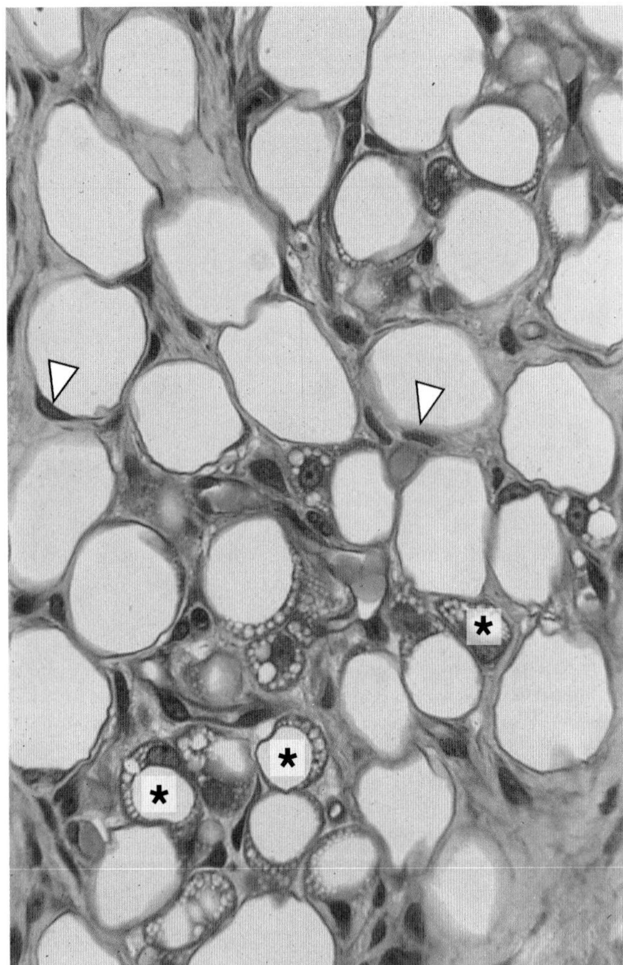

Figure 6–1. Photomicrograph of unilocular adipose tissue of a young mammal. Arrowheads show nuclei of adipocytes (fat cells) compressed against the cell membrane. Note that, although most cells are unilocular, there are several cells (asterisks) with small lipid droplets in their cytoplasm, an indication that their differentiation is not yet complete. Pararosaniline–toluidine blue (PT) stain. Medium magnification.

adipose tissue cells in the form of chylomicron triglycerides, in triglycerides synthesized in the liver and transported to adipose tissue in the form of **very low-density lipoproteins (VLDL)**, and by the synthesis of free fatty acids and glycerol from glucose to form triglycerides in adipose cells.

Chylomicrons (Gr. *chylos,* juice, + *micros,* small) are particles up to 3 μm in diameter, formed in intestinal epithelial cells and transported in blood plasma and mesenteric lymph. They consist of a central core, composed mainly of triglycerides and a small quantity of cholesterol esters, surrounded by a stabilizing monolayer consisting of apolipoproteins, cholesterol, and phospholipids. VLDL have proportionately more lipid in their surface layer because they are smaller (providing a greater surface-to-volume ratio), have different apolipoproteins at the surface, and contain a higher proportion of cholesterol esters to triglycerides than do chylomicrons. Chylomicrons and VLDL are hydrolyzed at the luminal surfaces of blood capillaries of adipose tissue by lipoprotein lipase, an enzyme synthesized by the adipocyte and transferred to the capillary cell membrane. Free fatty acids enter the adipocyte by mechanisms that are not completely understood. Both an active transport system and free diffusion seem to be involved. The numerous pinocytotic vesicles seen at the surfaces of adipocytes are probably not involved. The fatty acids cross the following layers (in order) in passing from the endothelium into the adipose cell: (1) capillary endothelium, (2) capillary basal lamina, (3) connective tissue ground substance, (4) adipocyte basal lamina, and (5) adipocyte plasma membrane. The movement of fatty acids across the cytoplasm into the lipid droplet is incompletely understood but may utilize specific carrier proteins (Figure 6–2). Within the adipocyte, the fatty acids combine with glycerol phosphate, an intermediate product of glucose metabolism, to form triglyceride molecules. These are then deposited in the triglyceride droplets. Mitochondria and smooth endoplasmic reticulum are organelles that participate actively in the process of lipid uptake and storage.

Adipose cells can synthesize fatty acids from glucose, a process accelerated by insulin. Insulin also stimulates the uptake of glucose into the adipose cells and increases the synthesis of lipoprotein lipase.

Stored lipids are mobilized by humoral and neurogenic mechanisms, resulting in the liberation of fatty acids and glycerol into the blood. Triglyceride lipase, an enzyme known as **hormone-sensitive lipase,** is activated by adenylate cyclase when the tissue is stimulated by norepinephrine. Norepinephrine is liberated at the endings of the postganglionic sympathetic nerves present in adipose tissue. The activated enzyme breaks down triglyceride molecules, which are located mainly at the surface of the lipid droplets. The relatively insoluble fatty acids are transported in association with serum albumin to other tissues of the body, whereas the more soluble glycerol remains free and is taken up by the liver.

Growth hormone, glucocorticoids, prolactin, corticotropin, insulin, and thyroid hormone also have roles at various stages in the metabolism of adipose tissue.

Adipose tissue also functions as a secretory organ. It synthesizes several molecules that are carried by the blood or remain attached to the endothelium of capillaries around the adipose cells (eg, the lipoprotein lipase). The most studied substance produced by adipose cells is **leptin,** a protein made of 164 amino acids. Several cells in the brain and other tissues have receptors for leptin. This molecule participates in the regulation of the amount of adipose tissue in the body and in food ingestion. It acts mainly in the hypothalamus to decrease food intake and increase energy consumption.

The sympathetic division of the autonomic nervous system richly innervates both unilocular and multilocular adipose tissues. In unilocular adipose tissue, nerve endings are found only in the walls of blood vessels; the adipocytes are not directly innervated. Release of the neurotransmitter norepinephrine activates the hormone-sensitive lipase described above. This innervation plays an important role in the mobilization of fats.

In response to body needs, lipids are not mobilized uniformly in all parts of the body. Subcutaneous, mesenteric, and retroperitoneal deposits are the first to be mobilized, whereas adipose tissue in the hands, feet, and retro-orbital fat pads resists long periods of starvation. After such periods of starvation, unilocular adipose tissue loses nearly all its fat and contains polyhedral or spindle-shaped cells with very few lipid droplets.

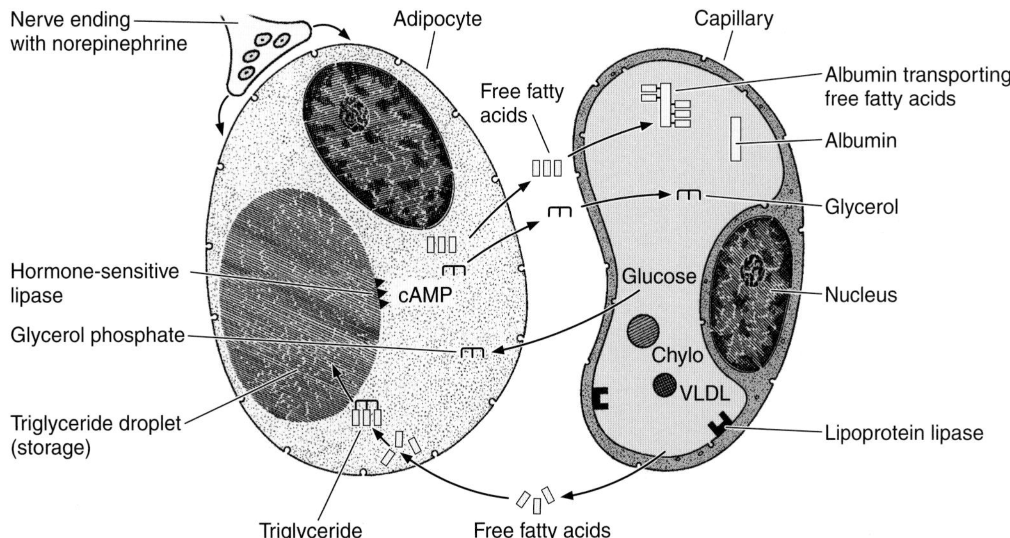

Figure 6–2. The process of lipid storage and release by the adipocyte. Triglycerides are transported in blood from the intestine and liver by lipoproteins known as chylomicrons (Chylo) and very low-density lipoproteins (VLDL). In adipose tissue capillaries, these lipoproteins are partly broken down by lipoprotein lipase, releasing free fatty acids and glycerol. The free fatty acids diffuse from the capillary into the adipocyte, where they are re-esterified to glycerol phosphate, forming triglycerides. These resulting triglycerides are stored in droplets until needed. Norepinephrine from nerve endings stimulates the cyclic AMP (cAMP) system, which activates hormone-sensitive lipase. Hormone-sensitive lipase hydrolyzes stored triglycerides to free fatty acids and glycerol. These substances diffuse into the capillary, where free fatty acids are bound to the hydrophobic moiety of albumin for transport to distant sites for use as an energy source.

MEDICAL APPLICATION

*Obesity in adults may result from an excessive accumulation of fat in unilocular tissue cells that become larger than usual (**hypertrophic obesity**). An increase in the number of adipocytes causes **hyperplastic obesity**.*

Histogenesis of Unilocular Adipose Tissue

Adipose cells develop from mesenchymally derived lipoblasts. These cells have the appearance of fibroblasts but are able to accumulate fat in their cytoplasm. Lipid accumulations are isolated from one another at first but soon fuse to form the single larger droplet that is characteristic of unilocular tissue cells (Figure 6–3).

The human being is one of the few mammals born with fat stores, which begin to accumulate at the 30th week of gestation. After birth, the development of new adipose cells is common around small blood vessels, where undifferentiated mesenchymal cells are usually found.

It is believed that during a finite postnatal period, nutritional and other influences can result in an increase in the number of adipocytes, but the cells do not increase in number after that period. They accumulate more lipids only under conditions of excess caloric intake (overfeeding). This early increase in the number of adipocytes may predispose an individual to hyperplastic obesity in later life.

MULTILOCULAR ADIPOSE TISSUE

Multilocular adipose tissue is also called **brown fat** because of its color, which is due to both the large number of blood capillaries in this tissue and the numerous mitochondria (containing colored cytochromes) in the cells. Compared with unilocular tissue, which is present throughout the body, brown adipose tissue has a more limited distribution. (Because it is more abundant in hibernating animals, it was at one time called the **hibernating gland.**)

In rats and several other mammals, multilocular adipose tissue is found mainly around the shoulder girdle. In the human embryo and newborn, this tissue is encountered in several areas and remains restricted to these locations after birth (Figure 6–4). In humans, this tissue appears to be important mainly in the first months of postnatal life, when it produces heat and thus protects the newborn against cold. It is greatly reduced in adulthood.

Multilocular tissue cells are polygonal and smaller than cells of unilocular adipose tissue. Their cytoplasm contains a great number of lipid droplets of various sizes (Figures 6–5 and 6–6), a spherical and central nucleus, and numerous mitochondria with abundant long cristae.

Multilocular adipose tissue resembles an endocrine gland in that its cells assume an almost epithelial arrangement of closely

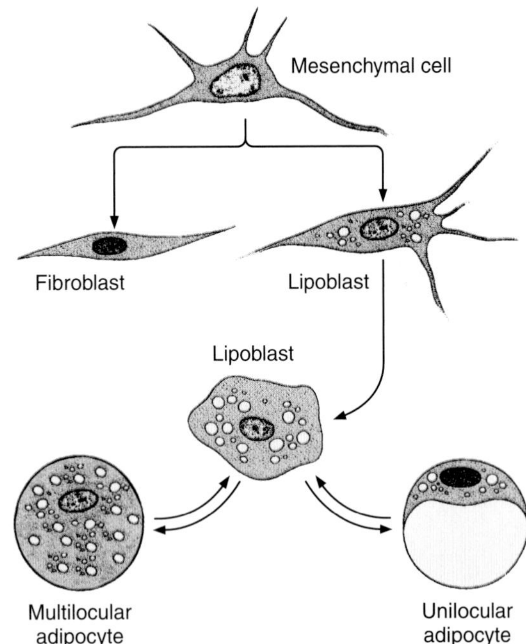

Figure 6–3. Development of fat cells. Undifferentiated mesenchymal cells are transformed into lipoblasts that accumulate fat and thus give rise to mature fat cells. When a large amount of lipid is mobilized by the body, mature unilocular fat cells return to the lipoblast stage. Undifferentiated mesenchymal cells also give rise to a variety of other cell types, including fibroblasts. The mature fat cell is larger than that shown here in relation to the other cell types.

packed masses associated with blood capillaries. This tissue is subdivided by partitions of connective tissue into lobules that are better delineated than are unilocular adipose tissue lobules. Cells of this tissue receive direct sympathetic innervation.

Function of the Multilocular Adipose Cells

The main function of the multilocular adipose cells is to produce heat. The physiology of multilocular adipose tissue is best understood in the study of hibernating species.

In animals ending their hibernation period, or in newborn mammals (including humans) that are exposed to a cold environment, nerve impulses liberate norepinephrine into the tissue. This neurotransmitter activates the hormone-sensitive lipase present in adipose cells, promoting hydrolysis of triglycerides to fatty acids and glycerol. Liberated fatty acids are metabolized, with a consequent increase in oxygen consumption and heat production, elevating the temperature of the tissue and warming the blood passing through it. Heat production is increased, because the mitochondria in cells of this tissue have a transmembrane protein called **thermogenin** in their inner membrane. Thermogenin permits the backflow of protons previously transported to the intermembranous space without passing through the ATP-synthetase system in the mitochondrial globular units. Consequently, the energy generated by proton flow is not used to synthesize ATP but is dissipated as heat. Warmed blood circulates throughout the body, heating the body and carrying fatty acids not metabolized in the adipose tissue. Other organs use these fatty acids.

Histogenesis of Multilocular Adipose Tissue

Multilocular adipose tissue develops differently from unilocular tissue. The mesenchymal cells that constitute this tissue resem-

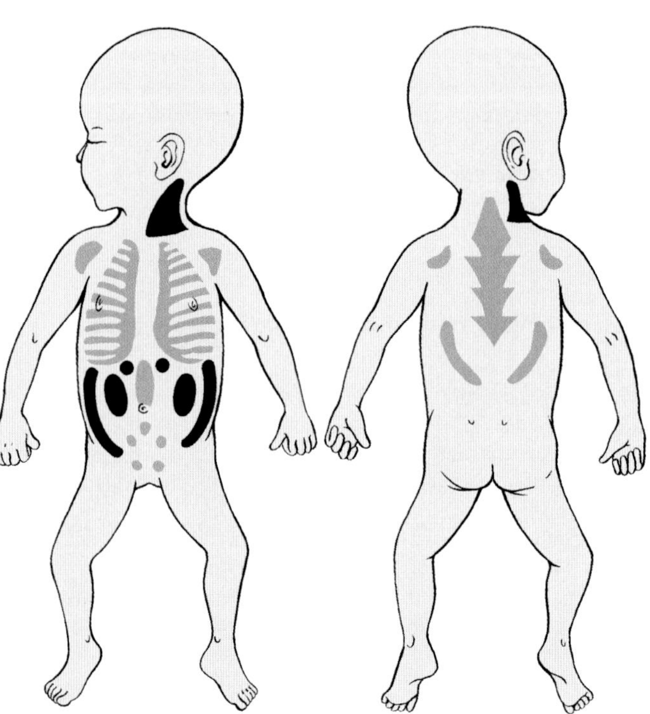

Figure 6–4. Distribution of adipose tissue. In a human newborn, multilocular adipose tissue constitutes 2–5% of the body weight and is distributed as shown. The black areas indicate multilocular adipose tissue; shaded areas are a mixture of multilocular and unilocular adipose tissue. (Modified, redrawn, and reproduced, with permission, from Merklin RJ: Growth and distribution of human fetal brown fat. Anat Rec 1974;178:637.)

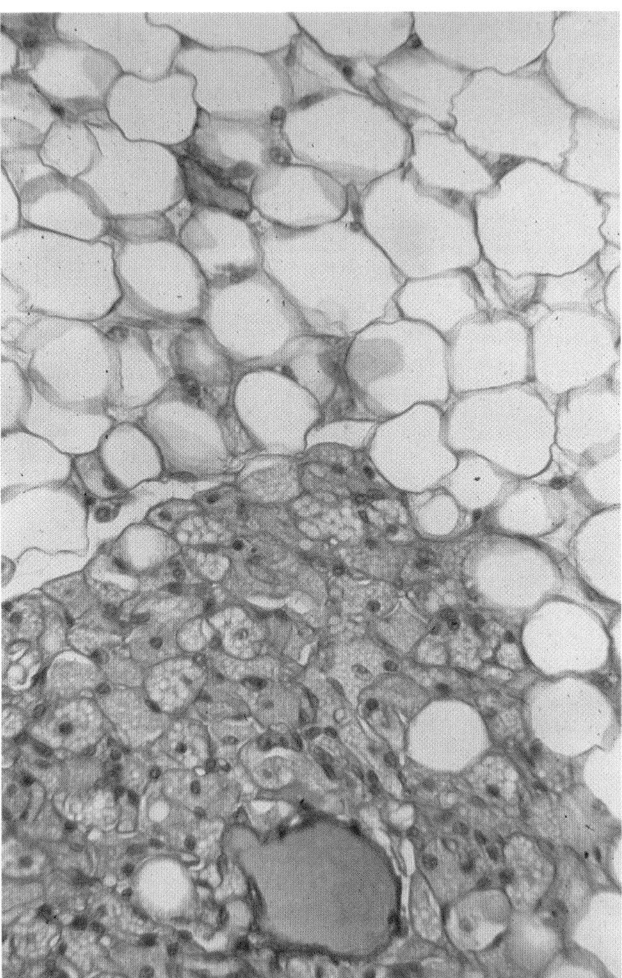

Figure 6–5. Photomicrograph of multilocular adipose tissue (lower portion) with its characteristic cells containing central spherical nuclei and multiple lipid droplets. For comparison, the upper part of the photomicrograph shows unilocular tissue. PT stain. Medium magnification.

ble epithelium (thus suggesting an endocrine gland) before they accumulate fat. Apparently, there is no formation of multilocular adipose tissue after birth, and one type of adipose tissue is not transformed into another.

Tumors of Adipose Tissues

MEDICAL APPLICATION

Unilocular adipocytes can generate very common benign tumors called ***lipomas.*** *Malignant adipocyte-derived tumors* ***(liposarcomas)*** *are not frequent in humans.*

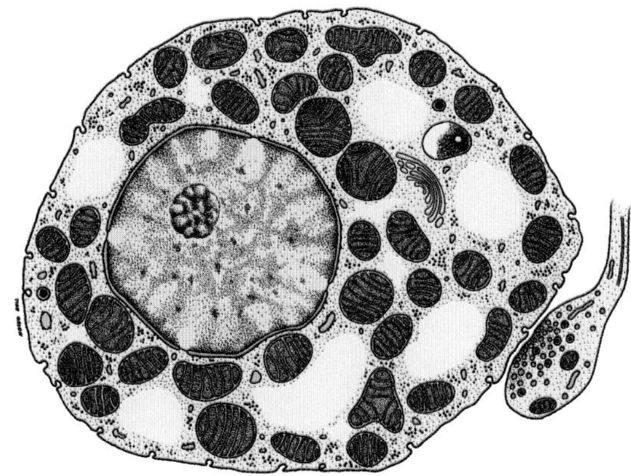

Figure 6–6. Multilocular adipose tissue. Note the central nucleus, multiple fat droplets, and abundant mitochondria. A sympathetic nerve ending is shown at the lower right.

REFERENCES

Angel A et al (editors): *The Adipocyte and Obesity: Cellular and Molecular Mechanisms.* Raven Press, 1983.

Forbes GB: The companionship of lean and fat. Basic Life Sci 1993;60:1.

Matarese G: Leptin and the immune system: how nutritional status influences the immune response. Eur Cytokine Netw 2000;11:7.

Matson CA et al: Leptin and regulation of body adiposity. Diabetes Rev 1999;4:488.

Napolitano L: The differentiation of white adipose cells: an electron microscope study. J Cell Biol 1963;8:663.

Nedergaard J, Lindberg O: The brown fat cell. Int Rev Cytol 1982;4:310.

Schubring C et al: Leptin, the ob gene product, in female health and disease. Eur J Obstet Gynecol Reprod Biol 2000;88:121.

Cartilage

7

Cartilage is characterized by an extracellular matrix enriched with glycosaminoglycans and proteoglycans, macromolecules that interact with collagen and elastic fibers. Variations in the composition of these matrix components produce 3 types of cartilage adapted to local biomechanical needs.

Cartilage is a specialized form of connective tissue in which the firm consistency of the extracellular matrix allows the tissue to bear mechanical stresses without permanent distortion. Another function of cartilage is to support soft tissues. Because it is smooth-surfaced and resilient, cartilage is a shock-absorbing and sliding area for joints and facilitates bone movements. Cartilage is also essential for the development and growth of long bones both before and after birth (see Chapter 8).

Cartilage consists of cells called **chondrocytes** (Gr. *chondros,* cartilage, + *kytos,* cell) and an extensive **extracellular matrix** composed of fibers and ground substance. Chondrocytes synthesize and secrete the extracellular matrix, and the cells themselves are located in matrix cavities called **lacunae.** Collagen, hyaluronic acid, proteoglycans, and small amounts of several glycoproteins are the principal macromolecules present in all types of cartilage matrix. Elastic cartilage, characterized by its great pliability, contains significant amounts of the protein elastin in the matrix.

Since collagen and elastin are flexible, the firm gel-like consistency of cartilage depends on electrostatic bonds between collagen fibers and the glycosaminoglycan side chains of matrix proteoglycans. It also depends on the binding of water (solvation water) to the negatively charged glycosaminoglycan chains that extend from the proteoglycan core proteins.

As a consequence of various functional requirements, 3 forms of cartilage have evolved, each exhibiting variation in matrix composition. In the matrix of **hyaline cartilage,** the most common form, type II collagen is the principal collagen type (Figure 7–1). The more pliable and distensible **elastic cartilage** possesses, in addition to collagen type II, an abundance of elastic fibers within its matrix. **Fibrocartilage,** present in regions of the body subjected to pulling forces, is characterized by a matrix containing a dense network of coarse type I collagen fibers.

In all 3 forms, cartilage is avascular and is nourished by the diffusion of nutrients from capillaries in adjacent connective tissue (perichondrium) or by synovial fluid from joint cavities. In some instances, blood vessels traverse cartilage to nourish other tissues, but these vessels do not supply nutrients to the cartilage. As might be expected of cells in an avascular tissue, chondrocytes exhibit low metabolic activity. Cartilage has no lymphatic vessels or nerves.

The **perichondrium** (Figures 7–2 and 7–4) is a sheath of dense connective tissue that surrounds cartilage in most places, forming an interface between the cartilage and the tissue supported by the cartilage. The perichondrium harbors the vascular supply for the avascular cartilage and also contains nerves and lymphatic vessels. Articular cartilage, which covers the surfaces of the bones of movable joints, is devoid of perichondrium and

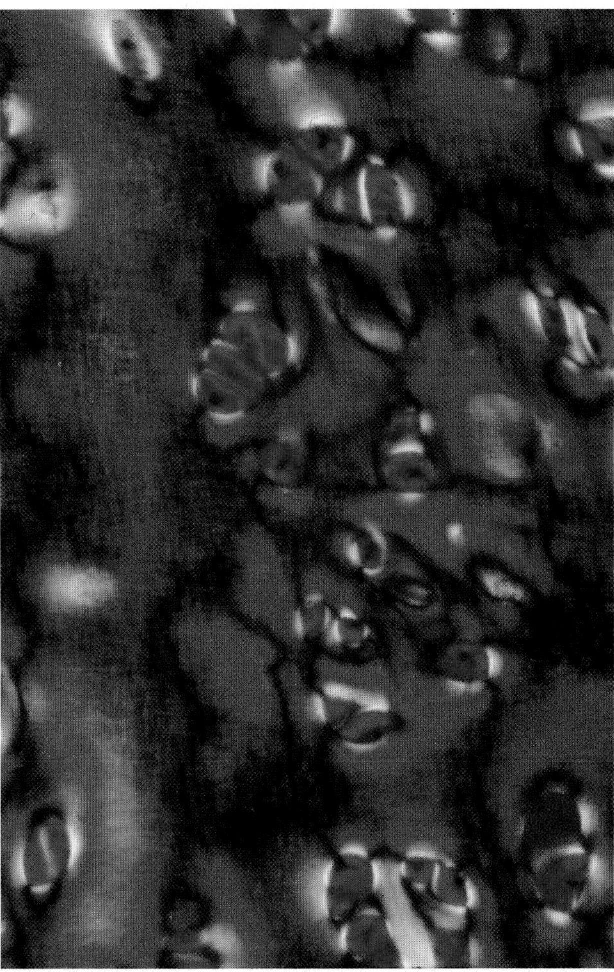

Figure 7–1. Photomicrograph of hyaline cartilage. The extracellular matrix was previously digested with papain to enhance the oriented aggregates of variously disposed collagen type II fibrils. These aggregates appear as black areas. Picrosirius–polarized light stain. Medium magnification.

is sustained by the diffusion of oxygen and nutrients from the synovial fluid.

HYALINE CARTILAGE

Hyaline cartilage (Figure 7–2) is the most common and best studied of the 3 forms. Fresh hyaline cartilage is bluish-white and translucent. In the embryo, it serves as a temporary skeleton until it is gradually replaced by bone.

In adult mammals, hyaline cartilage is located in the articular surfaces of the movable joints, in the walls of larger respiratory passages (nose, larynx, trachea, bronchi), in the ventral ends of ribs, where they articulate with the sternum, and in the **epiphyseal plate,** where it is responsible for the longitudinal growth of bone (see Chapter 8).

Matrix

Forty percent of the dry weight of hyaline cartilage consists of collagen embedded in a firm, hydrated gel of proteoglycans and structural glycoproteins. In routine histology preparations, the collagen is indiscernible for 2 reasons: the collagen is in the form of fibrils, which have submicroscopic dimensions; and the refractive index of the fibrils is almost the same as that of the ground substance in which they are embedded. Hyaline cartilage contains primarily type II collagen (Figure 7–1). However, small amounts of collagen types IX, X, XI, and others are frequently present.

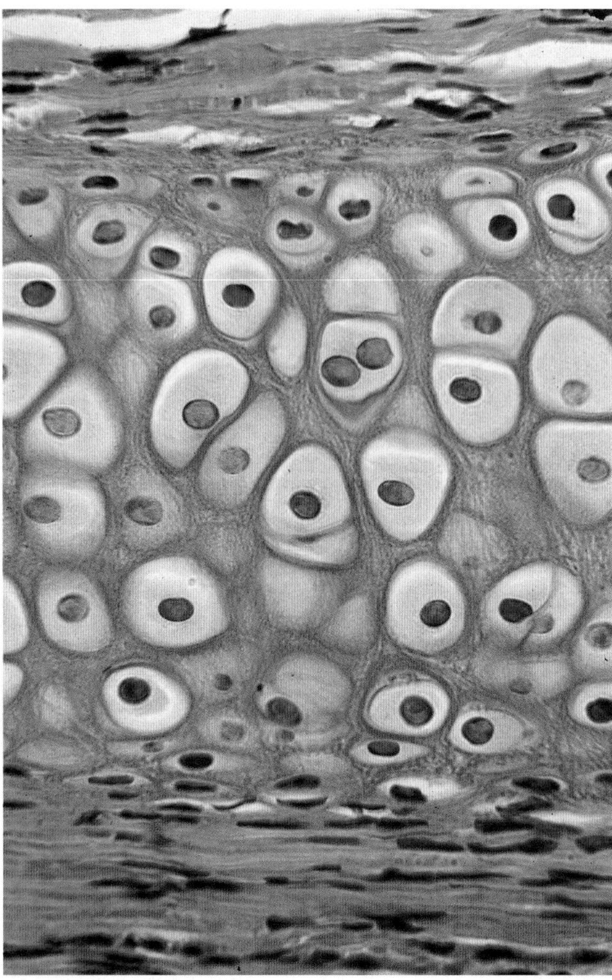

Figure 7–2. Photomicrograph of hyaline cartilage. Chondrocytes are located in matrix lacunae, and most belong to isogenous groups. The upper and lower parts of the figure show the perichondrium stained pink. Note the gradual differentiation of cells from the perichondrium into chondrocytes. H&E stain. Low magnification.

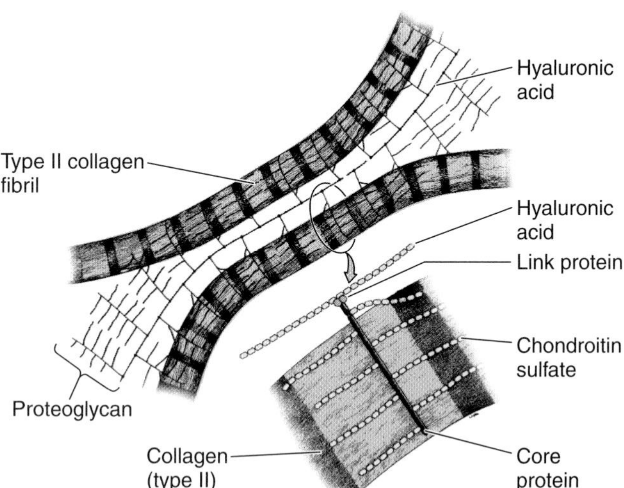

Figure 7–3. Schematic representation of molecular organization in cartilage matrix. Link proteins noncovalently bind the protein core of proteoglycans to the linear hyaluronic acid molecules. The chondroitin sulfate side chains of the proteoglycan electrostatically bind to the collagen fibrils, forming a cross-linked matrix. The oval outlines the area shown larger in the lower part of the figure.

Cartilage proteoglycans contain chondroitin 4-sulfate, chondroitin 6-sulfate, and keratan sulfate, covalently linked to core proteins. Up to 200 of these proteoglycans are noncovalently associated with long molecules of hyaluronic acid, forming **proteoglycan aggregates** that interact with collagen (Figure 7–3). The aggregates can be up to 4 μm in length. Structurally, proteoglycans resemble bottlebrushes, the protein core being the stem and the radiating glycosaminoglycan chains the bristles.

The high content of solvation water bound to the negative charges of glycosaminoglycans acts as a shock absorber or biomechanical spring; this is of great functional importance, especially in articular cartilages (see Chapter 8).

In addition to type II collagen and proteoglycan, an important component of cartilage matrix is the structural glycoprotein **chondronectin,** a macromolecule that binds specifically to glycosaminoglycans and collagen type II, mediating the adherence of chondrocytes to the extracellular matrix. The cartilage matrix surrounding each chondrocyte is rich in glycosaminoglycan and poor in collagen. This peripheral zone, called the **territorial,** or **capsular,** matrix, stains differently from the rest of the matrix (Figures 7–2 and 7–4).

Perichondrium

Except in the articular cartilage of joints, all hyaline cartilage is covered by a layer of dense connective tissue, the perichondrium, which is essential for the growth and maintenance of cartilage (Figures 7–2 and 7–4). It is rich in collagen type I fibers and

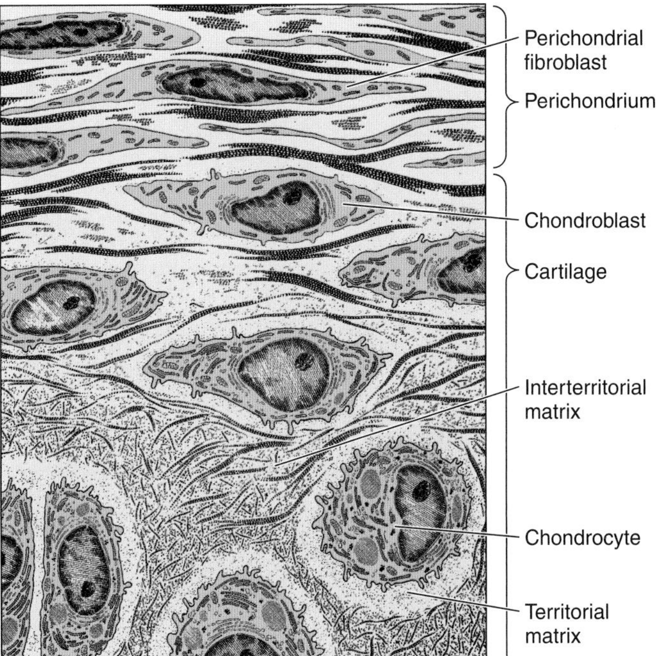

Figure 7–4. Diagram of the area of transition between the perichondrium and the hyaline cartilage. As perichondrial cells differentiate into chondrocytes, they become round, with an irregular surface. Cartilage (interterritorial) matrix contains numerous fine collagen fibrils except around the periphery of the chondrocytes, where the matrix consists primarily of glycosaminoglycans; this peripheral region is called the territorial, or capsular, matrix.

contains numerous fibroblasts. Although cells in the inner layer of the perichondrium resemble fibroblasts, they are chondroblasts and easily differentiate into chondrocytes.

Chondrocytes

At the periphery of hyaline cartilage, young chondrocytes have an elliptic shape, with the long axis parallel to the surface. Farther in, they are round and may appear in groups of up to 8 cells originating from mitotic divisions of a single chondrocyte. These groups are called **isogenous** (Gr. *isos,* equal, + *genos,* family).

Cartilage cells and the matrix shrink during routine histologic preparation, resulting in both the irregular shape of the chondrocytes and their retraction from the capsule. In living tissue, and in properly prepared sections, the chondrocytes fill the lacunae completely (Figure 7–5).

Chondrocytes synthesize collagens and the other matrix molecules.

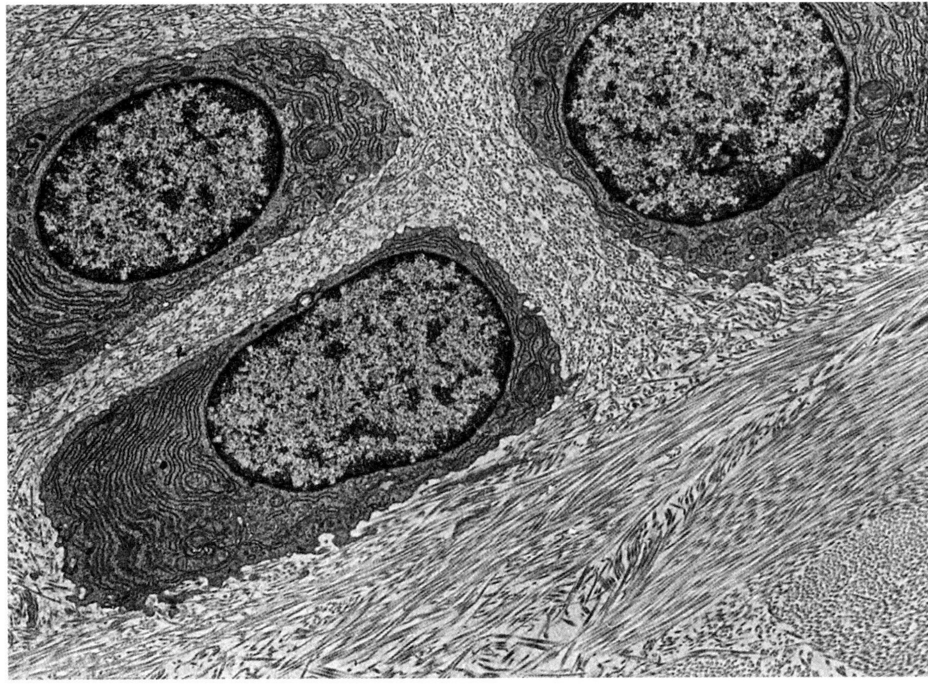

Figure 7–5. Electron micrograph of fibrocartilage from a young animal, showing 3 chondrocytes in their lacunae. Note the abundance of rough endoplasmic reticulum. Chondrocytes synthesize the cartilage matrix. Fine collagen fibers, sectioned in several places, are prominent around the chondrocytes. ×3750.

Because cartilage is devoid of blood capillaries, chondrocytes respire under low oxygen tension. Hyaline cartilage cells metabolize glucose mainly by anaerobic glycolysis to produce lactic acid as the end product. Nutrients from the blood cross the perichondrium to reach more deeply placed cartilage cells. Mechanisms include diffusion and transport of water and solute promoted by the pumping action of intermittent cartilage compression and decompression. Because of this, the maximum width of the cartilage is limited.

Chondrocyte function depends on a proper hormonal balance. The synthesis of sulfated glycosaminoglycans is accelerated by growth hormone, thyroxin, and testosterone and is slowed by cortisone, hydrocortisone, and estradiol. Cartilage growth depends mainly on the hypophyseal growth hormone **somatotropin.** This hormone does not act directly on cartilage cells but promotes the synthesis of **somatomedin C** in the liver. Somatomedin C acts directly on cartilage cells, promoting their growth.

MEDICAL APPLICATION

*Cartilage cells can give rise to benign (**chondroma**) or malignant (**chondrosarcoma**) tumors.*

Histogenesis

Cartilage derives from the mesenchyme (Figure 7–6). The first modification observed is the rounding up of the mesenchymal cells, which retract their extensions, multiply rapidly, and form mesenchymal condensations of chondroblasts. The cells formed by this direct differentiation of mesenchymal cells, now called **chondroblasts,** have a ribosome-rich basophilic cytoplasm. Synthesis and deposition of the matrix then begin to separate the chondroblasts from one another. During development, the differentiation of cartilage takes place from the center outward; therefore, the more central cells have the characteristics of chondrocytes, whereas the peripheral cells are typical chondroblasts. The superficial mesenchyme develops into the perichondrium.

Growth

The growth of cartilage is attributable to 2 processes: **interstitial growth,** resulting from the mitotic division of preexisting chondrocytes; and **appositional growth,** resulting from the differentiation of perichondrial cells. In both cases, the synthesis of matrix contributes to the growth of the cartilage. Interstitial growth is the less important of the two processes. It occurs only during the early phases of cartilage formation, when it increases tissue mass by expanding the cartilage matrix from within. Interstitial growth also occurs in the epiphyseal plates of long bones and within articular cartilage. In the epiphyseal plates, interstitial growth is important in increasing the length of long bones and in providing a cartilage model for endochondral bone formation (see Chapter 8). In articular cartilage, as the cells and matrix near the articulating surface are gradually worn away, the cartilage must be replaced from within, since there is no perichondrium there to add cells by apposition. In cartilage found elsewhere in the body, interstitial growth becomes less pronounced, as the matrix becomes increasingly rigid from the cross-linking of matrix molecules. Cartilage then grows in girth only by apposition. Chondroblasts of the perichondrium proliferate

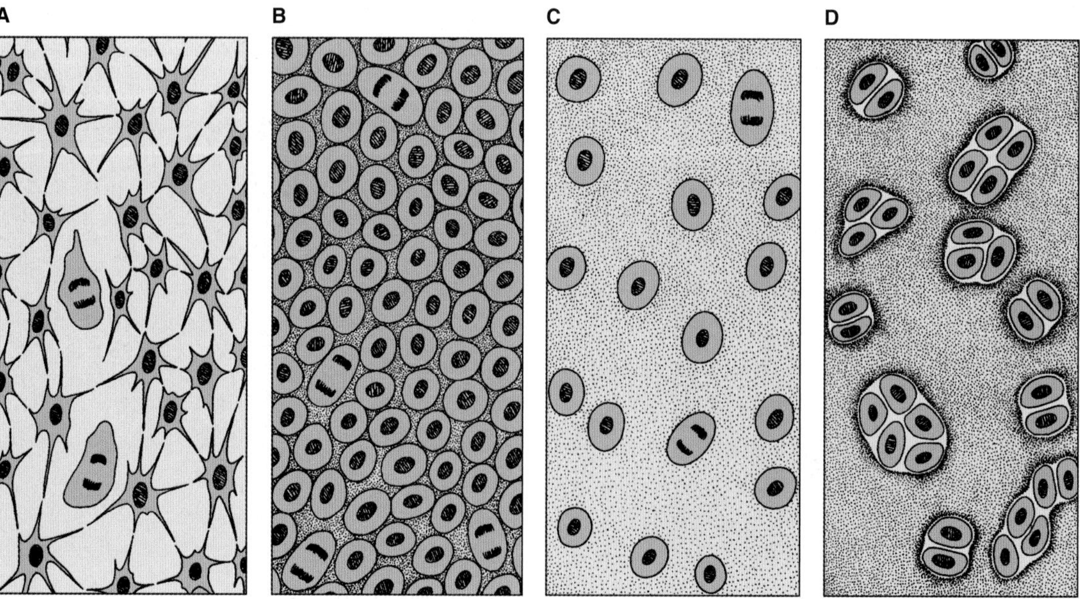

Figure 7–6. Histogenesis of hyaline cartilage. **A:** The mesenchyme is the precursor tissue of all types of cartilage. **B:** Mitotic proliferation of mesenchymal cells gives rise to a highly cellular tissue. **C:** Chondroblasts are separated from one another by the formation of a great amount of matrix. **D:** Multiplication of cartilage cells gives rise to isogenous groups, each surrounded by a condensation of territorial (capsular) matrix.

and become chondrocytes once they have surrounded themselves with cartilaginous matrix and are incorporated into the existing cartilage (Figures 7–2 and 7–4).

Degenerative Changes

MEDICAL APPLICATION

In contrast to other tissues, hyaline cartilage is more susceptible to degenerative aging processes. Calcification of the matrix, preceded by an increase in the size and volume of the chondrocytes and followed by their death, is a common process in some cartilage. Asbestiform degeneration, frequent in aged cartilage, is due to the formation of localized aggregates of thick, abnormal collagen fibrils.

Poor Regeneration of Cartilage Tissue

Except in young children, damaged cartilage regenerates with difficulty and often incompletely, by activity of the perichondrium, which invades the injured area and generates new cartilage. In extensively damaged areas—and occasionally in small areas—the perichondrium produces a scar of dense connective tissue instead of forming new cartilage.

ELASTIC CARTILAGE

Elastic cartilage is found in the auricle of the ear, the walls of the external auditory canals, the auditory (eustachian) tubes, the epiglottis, and the cuneiform cartilage in the larynx.

Elastic cartilage is essentially identical to hyaline cartilage except that it contains an abundant network of fine elastic fibers in addition to collagen type II fibrils. Fresh elastic cartilage has a yellowish color owing to the presence of elastin in the elastic fibers (Figure 7–7).

Elastic cartilage is frequently found to be gradually continuous with hyaline cartilage. Like hyaline cartilage, elastic cartilage possesses a perichondrium.

FIBROCARTILAGE

Fibrocartilage is a tissue intermediate between dense connective tissue and hyaline cartilage. It is found in intervertebral disks, in attachments of certain ligaments to the cartilaginous surface of bones, and in the symphysis pubis. Fibrocartilage is always associated with dense connective tissue, and the border areas between these two tissues are not clear-cut, showing a gradual transition.

Fibrocartilage contains chondrocytes, either singly or in isogenous groups, usually arranged in long rows separated by coarse collagen type I fibers (Figure 7–8). Because it is rich in collagen type I, the fibrocartilage matrix is acidophilic.

In fibrocartilage, the numerous collagen fibers either form irregular bundles between the groups of chondrocytes or are aligned in a parallel arrangement along the columns of chondrocytes (Figure 7–8). This orientation depends on the stresses acting on fibrocartilage, since the collagen bundles take up a direction parallel to those stresses. There is no identifiable perichondrium in fibrocartilage.

INTERVERTEBRAL DISKS

Each intervertebral disk is situated between 2 vertebrae and held to them by means of ligaments. The disks have 2 components: the fibrous annulus fibrosus and the nucleus pulposus. The intervertebral disk acts as a lubricated cushion that prevents adjacent vertebrae from being eroded by abrasive forces during movement of the spinal column. The nucleus pulposus serves as a shock absorber to cushion the impact between vertebrae.

The **annulus fibrosus** has an external layer of dense connective tissue, but it is mainly composed of overlapping laminae of fibrocartilage in which collagen bundles are orthogonally arranged in adjacent layers. The multiple lamellae, with the

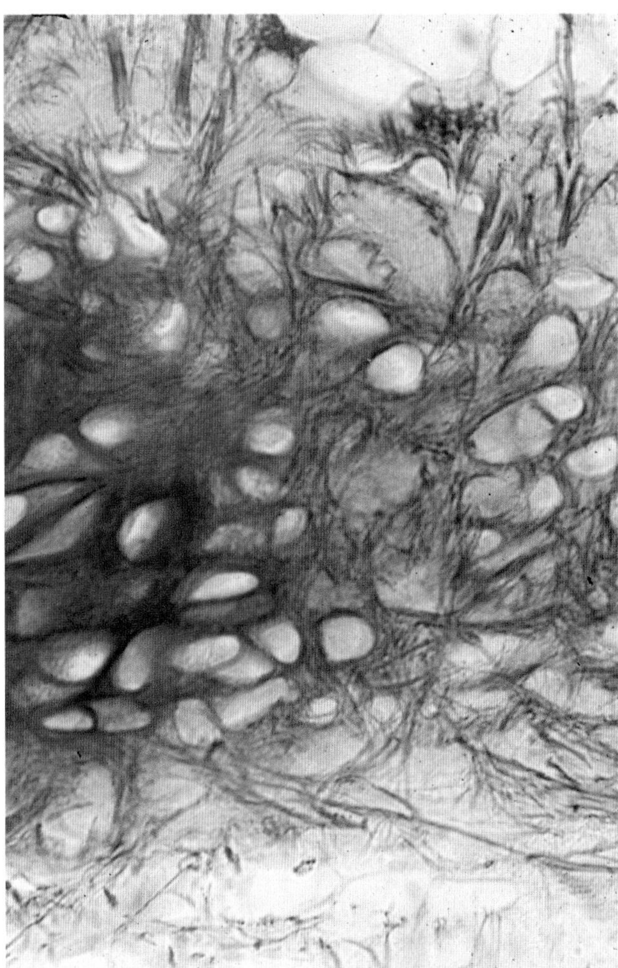

Figure 7–7. Photomicrograph of elastic cartilage, stained for elastic fibers. Cells are not stained. This flexible cartilage is present, for example, in the auricle of the ear and in the epiglottis. Resorcin stain. Medium magnification.

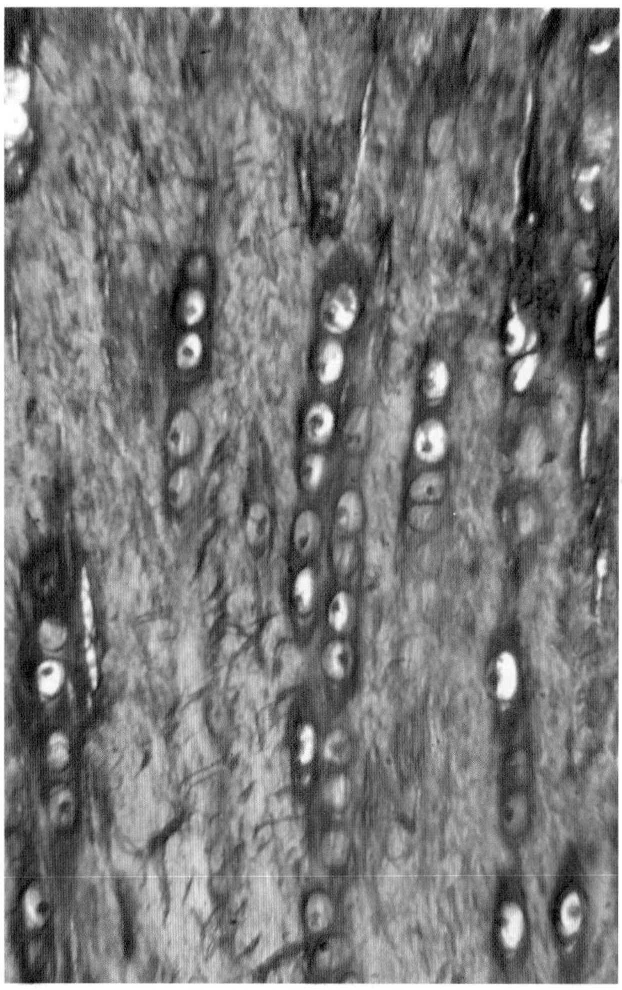

Figure 7–8. Photomicrograph of fibrocartilage. Note the rows of chondrocytes separated by collagen fibers. Fibrocartilage is frequently found in the insertion of tendons on the epiphyseal hyaline cartilage. Picrosirius-hematoxylin stain. Medium magnification.

90-degree registration of type I collagen fibers in adjacent layers, provide the disk with unusual resilience that enables it to withstand the pressures generated by impinging vertebrae.

The **nucleus pulposus** is situated in the center of the annulus fibrosus. It is derived from the embryonic notochord and consists of a few rounded cells embedded in a viscous matrix rich in hyaluronic acid and type II collagen fibrils. In children, the nucleus pulposus is large, but it gradually becomes smaller with age and is partially replaced by fibrocartilage.

Herniation of the Intervertebral Disk

MEDICAL APPLICATION

Rupture of the annulus fibrosus, which most frequently occurs in the posterior region where there are fewer collagen bundles, results in expulsion of the nucleus pulposus and a concomitant flattening of the disk. As a consequence, the disk frequently dislocates or slips from its position between the vertebrae. If it moves toward the spinal cord, it can compress the nerves and result in severe pain and neurologic disturbances. The pain accompanying a slipped disk may be perceived in areas innervated by the compressed nerve fibers—usually the lower lumbar region.

REFERENCES

Chakrabarti B, Park JW: Glycosaminoglycans: structure and interaction. CRC Crit Rev Biochem 1980;8:225.

Eyre DR, Muir H: The distribution of different molecular species of collagen in fibrous, elastic and hyaline cartilages of the pig. Biochem J 1975;51:595.

Hall BK (editor): *Cartilage,* Vol 1: *Structure, Function, and Biochemistry.* Academic Press, 1983.

Jasin, HE: Structure and function of the articular cartilage surface. Scand J Rheumatol 1995;101:51.

Junqueira LCU et al: Quantitation of collagen-proteoglycan interaction in tissue sections. Connect Tissue Res 1980;7:91.

Perka C et al: Matrix-mixed culture: new methodology for chondrocyte culture and preparation of cartilage transplants. J Biomed Mater Res 2000;49:305.

Reddy AH (editor): *Extracellular Matrix Structure and Functions.* Liss, 1985.

Stockwell RA: *Biology of Cartilage Cells.* Cambridge Univ Press, 1979.

Zambrano NZ et al: Collagen arrangement in cartilages. Acta Anat 1982;113:26.

Bone

8

As the main constituent of the adult skeleton, bone tissue supports fleshy structures, protects such vital organs as those in the cranial and thoracic cavities, and harbors the bone marrow, where blood cells are formed. Bone also serves as a reservoir of calcium, phosphate, and other ions that can be released or stored in a controlled fashion to maintain constant concentrations of these important ions in body fluids.

In addition, bones form a system of levers that multiply the forces generated during skeletal muscle contraction and transform them into bodily movements. This mineralized tissue confers mechanical and metabolic functions to the skeleton.

Bone is a specialized connective tissue composed of intercellular calcified material, the **bone matrix,** and 3 cell types: **osteocytes** (Gr. *osteon,* bone, + *kytos,* cell), which are found in cavities (**lacunae**) within the matrix (Figure 8–1); **osteoblasts** (*osteon* + Gr. *blastos,* germ), which synthesize the organic components of the matrix; and **osteoclasts** (*osteon* + Gr. *klastos,* broken), which are multinucleated giant cells involved in the resorption and remodeling of bone tissue.

Since metabolites are unable to diffuse through the calcified matrix of bone, the exchanges between osteocytes and blood capillaries depend on communication through the **canaliculi** (L. *canalis,* canal), which are thin, cylindrical spaces that perforate the matrix (Figure 8–2).

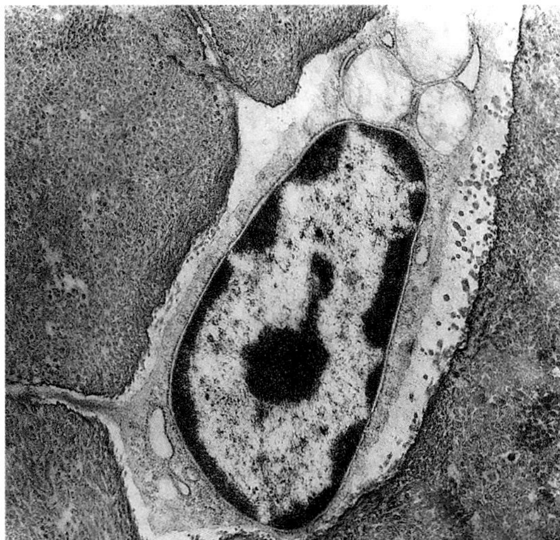

Figure 8–1. Section of bone tissue showing an osteocyte with its cytoplasmic processes surrounded by matrix. The ultrastructure of the cell nucleus and cytoplasm is compatible with a low level of protein synthesis.

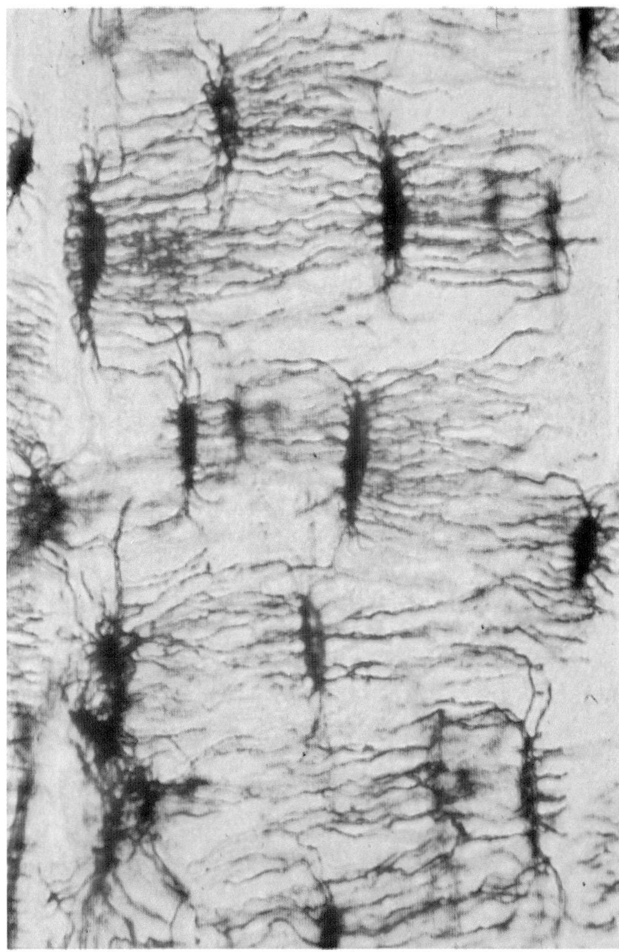

Figure 8–2. Photomicrograph of dried bone ground very thin. The lacunae and canaliculi filled with air deflect the light and appear dark, showing the communication between these structures through which nutrients derived from blood vessels flow. Medium magnification.

All bones are lined on both internal and external surfaces by layers of tissue containing osteogenic cells—**endosteum** on the internal surface and **periosteum** on the external surface.

Because of its hardness, bone is difficult to section with the microtome, and special techniques must be used for its study. A common technique that permits the observation of the cells and organic matrix is based on the decalcification of bone preserved by standard fixatives. The mineral is removed by immersion in a solution containing a calcium-chelating substance (eg, ethylenediaminetetraacetic acid [EDTA]). The decalcified tissue is then embedded, sectioned, and stained.

BONE CELLS

Osteoblasts

Osteoblasts are responsible for the synthesis of the organic components of bone matrix (type I collagen, proteoglycans, and glycoproteins). Deposition of the inorganic components of bone also depends on the presence of viable osteoblasts. Osteoblasts are exclusively located at the surfaces of bone tissue, side by side, in a way that resembles simple epithelium (Figure 8–3). When they are actively engaged in matrix synthesis, osteoblasts have a cuboidal to columnar shape and basophilic cytoplasm. When their synthesizing activity declines, they flatten, and cytoplasmic basophilia declines.

Some osteoblasts are gradually surrounded by newly formed matrix and become **osteocytes.** During this process a space called a **lacuna** is formed. Lacunae are occupied by osteocytes and their extensions, along with a small amount of extracellular noncalcified matrix.

During matrix synthesis, osteoblasts have the ultrastructure of cells actively synthesizing proteins for export. Osteoblasts are polarized cells. Matrix components are secreted at the cell surface, which is in contact with older bone matrix, producing a layer of new (but not yet calcified) matrix, called **osteoid,** between the osteoblast layer and the previously formed bone (Figure 8–3). This process, **bone apposition,** is completed by subsequent deposition of calcium salts into the newly formed matrix.

Osteocytes

Osteocytes, which derive from osteoblasts, lie in the lacunae (Figure 8–3) situated between lamellae (L. diminutive of *lamina,* leaf) of matrix. Only one osteocyte is found in each lacuna. The thin, cylindrical matrix canaliculi house cytoplasmic processes of osteocytes. Processes of adjacent cells make contact via gap junctions, and molecules are passed via these structures from cell to cell. Some molecular exchange between osteocytes and blood vessels also takes place through the small amount of extracellular substance located between osteocytes (and their processes) and the bone matrix. This exchange can provide nourishment for a chain of about 15 cells.

When compared with osteoblasts, the flat, almond-shaped osteocytes exhibit a significantly reduced rough endoplasmic reticulum (Figure 8–1) and Golgi complex and more condensed nuclear chromatin. These cells are actively involved in the maintenance of the bony matrix, and their death is followed by resorption of this matrix.

MEDICAL APPLICATION

The fluorescent antibiotic tetracycline interacts with great affinity with recently deposited mineralized bone matrix. Based on this interaction, a method was developed to measure the rate of bone apposition—an important parameter in the study of bone growth and the diagnosis of bone growth diseases. Tetracycline is administered twice to patients, with an interval of 5 days between injections. A bone biopsy is then performed, and the sections are studied by means of fluorescence microscopy. The distance between the two fluorescent layers is proportional to the rate of bone apposition. This procedure is of diagnostic importance in such diseases as ***osteomalacia,*** *in which mineralization is impaired, and* ***osteitis fibrosa cystica,*** *in which increased osteoclast activity results in removal of bone matrix and fibrous degeneration.*

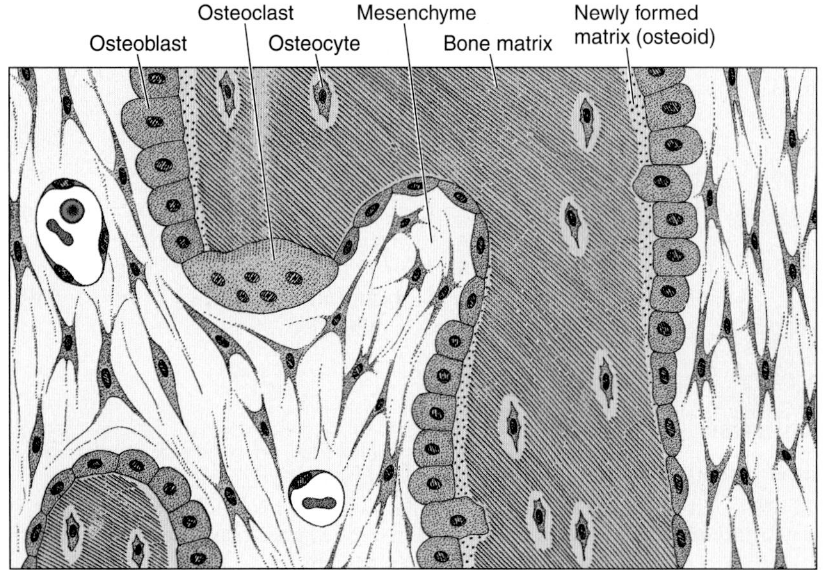

Figure 8–3. Events that occur during intramembranous ossification. Osteoblasts are synthesizing collagen, which forms a strand of matrix that traps cells. As this occurs, the osteoblasts gradually differentiate to become osteocytes. The lower part of the drawing shows an osteoblast being trapped in newly formed bone matrix.

Osteoclasts

Osteoclasts are very large, branched motile cells. Dilated portions of the cell body (Figure 8–4) contain from 5 to 50 (or more) nuclei. In areas of bone undergoing resorption, osteoclasts lie within enzymatically etched depressions in the matrix known as **Howship's lacunae.** Osteoclasts are derived from the fusion of bone marrow–derived cells.

In active osteoclasts, the surface-facing bone matrix is folded into irregular, often subdivided projections, forming a **ruffled border.** Surrounding the ruffled border is a cytoplasmic zone—the **clear zone**—that is devoid of organelles, yet rich in actin filaments. This zone is a site of adhesion of the osteoclast to the bone matrix and creates a microenvironment in which bone resorption occurs (Figure 8–5).

The osteoclast secretes collagenase and other enzymes and pumps protons into a subcellular pocket (the microenvironment referred to above), promoting the localized digestion of collagen and dissolving calcium salt crystals. Osteoclast activity is controlled by cytokines (small signaling proteins that act as local mediators) and hormones. Osteoclasts have receptors for calcitonin, a thyroid hormone, but not for parathyroid hormone.

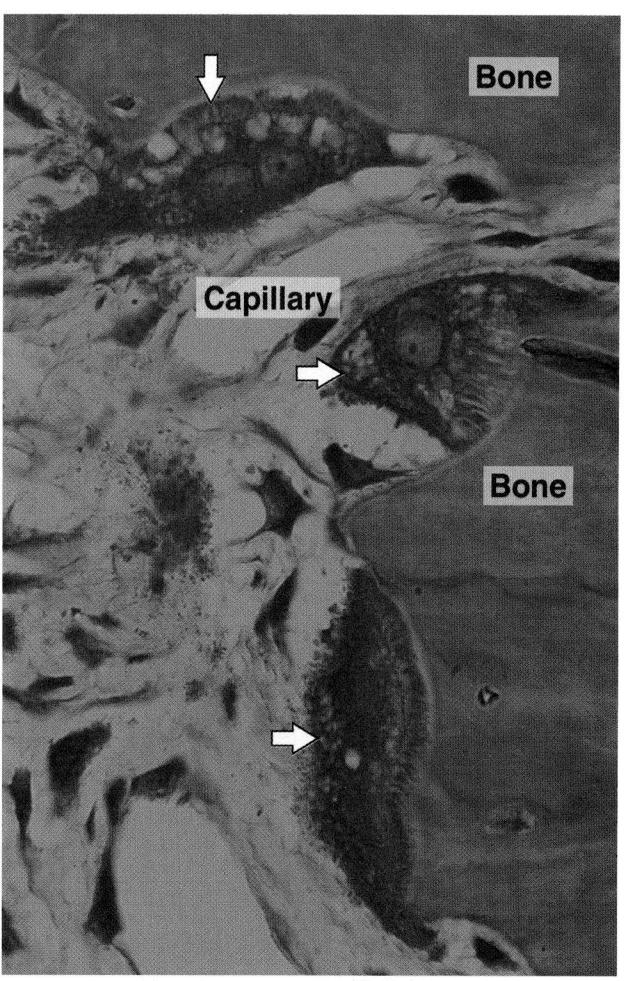

Figure 8–4. Section showing 3 osteoclasts (arrows) digesting bone tissue. The osteoclast is a large cell with several nuclei and a ruffled border close to the bone matrix. Note the clear compartment where the process of bone erosion occurs. This compartment is acidified by a proton pump localized in the osteoclast membrane. It is the place of decalcification and matrix digestion and can be compared to a giant extracellular lysosome. Chondroclasts found in eroded regions of epiphyseal calcified cartilage are similar in shape to osteoclasts.

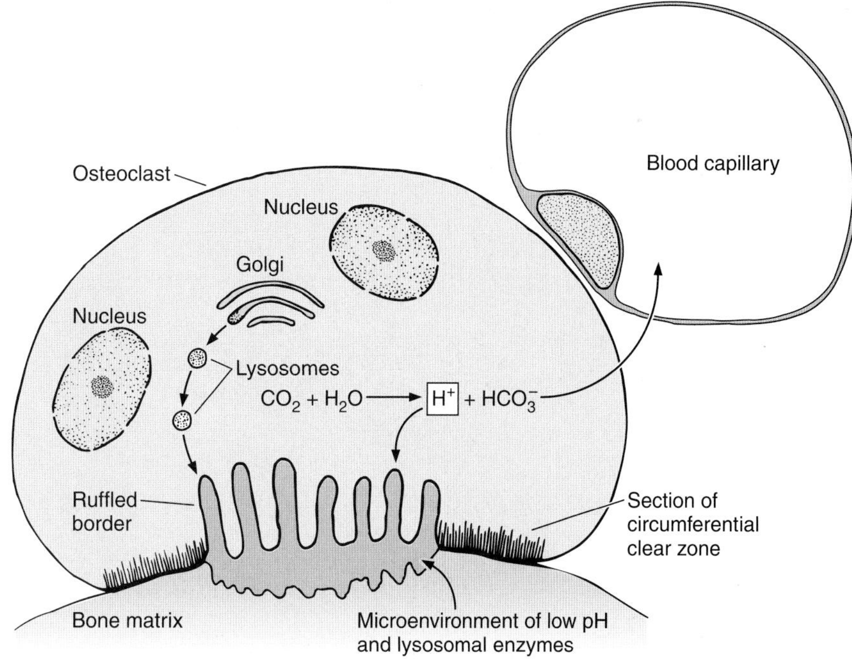

Figure 8–5. Bone resorption. Lysosomal enzymes packaged in the Golgi complex and hydrogen ions produced are released into the confined microenvironment created by the attachment between bone matrix and the osteoclast's peripheral clear zone. The acidification of this confined space facilitates the dissolution of calcium phosphate from bone and is the optimal pH for the activity of lysosomal hydrolases. Bone matrix is thus removed and the products of bone resorption are taken up by the osteoclast's cytoplasm, probably digested further, and transferred to blood capillaries.

However, osteoblasts have receptors for parathyroid hormone and, when activated by this hormone, produce a cytokine called osteoclast stimulating factor.

Ruffled borders are related to the activity of osteoclasts.

MEDICAL APPLICATION

In the genetic disease ***osteopetrosis,*** *which is characterized by dense, heavy bones ("marble bones"), the osteoclasts lack ruffled borders, and bone resorption is defective.*

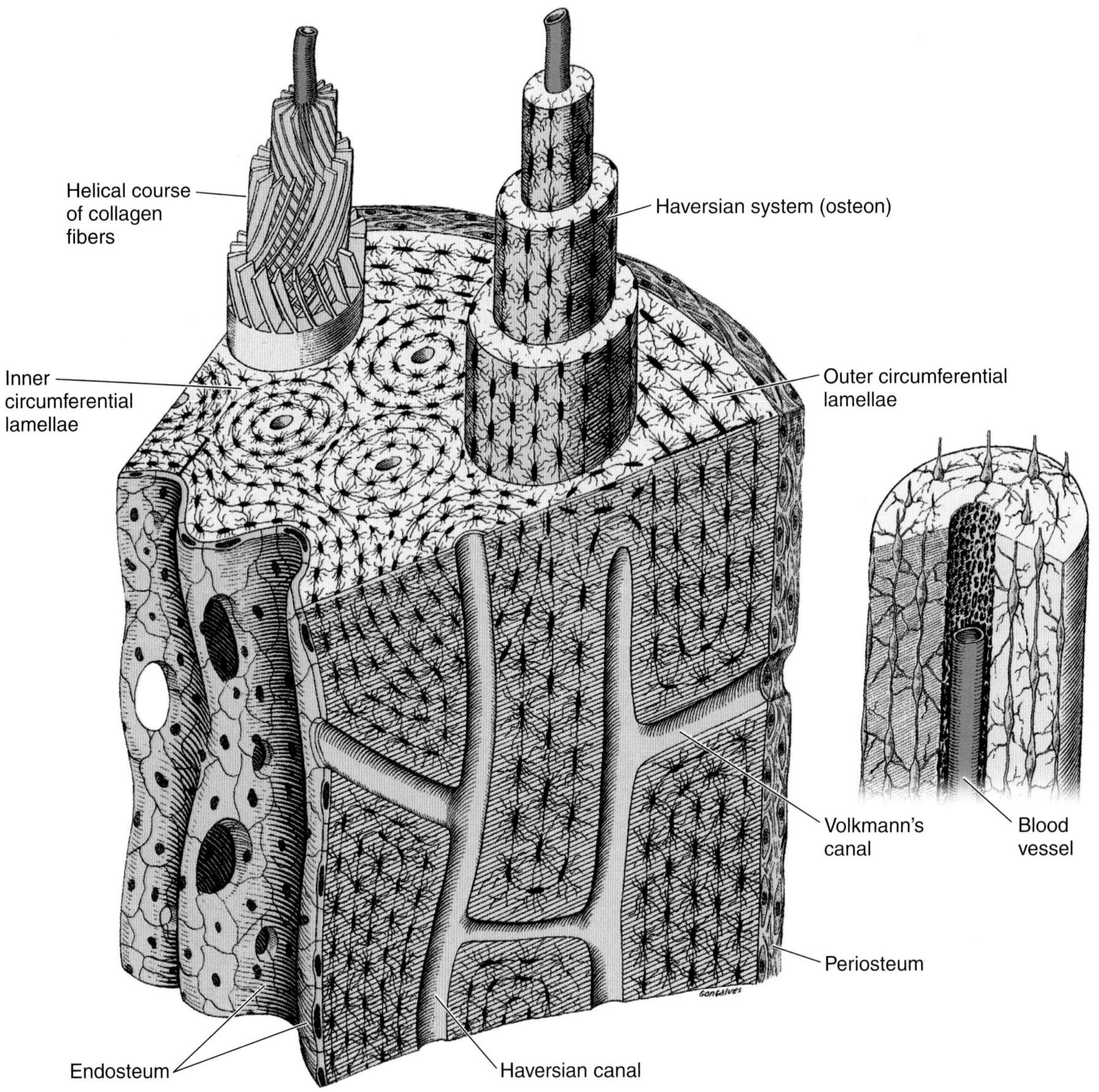

BONE MATRIX

Inorganic matter represents about 50% of the dry weight of bone matrix. Calcium and phosphorus are especially abundant, but bicarbonate, citrate, magnesium, potassium, and sodium are also found. X-ray diffraction studies have shown that calcium and phosphorus form hydroxyapatite crystals with the composition $Ca_{10}(PO_4)_6(OH)_2$. However, these crystals show imperfections and are not identical to the hydroxyapatite found in the rock minerals. Significant quantities of amorphous (noncrystalline) calcium phosphate are also present. In electron micrographs, hydroxyapatite crystals of bone appear as plates that lie alongside

Helical course of collagen fibers
Haversian system (osteon)
Inner circumferential lamellae
Outer circumferential lamellae
Volkmann's canal
Blood vessel
Periosteum
Endosteum
Haversian canal

Figure 8–6. Schematic drawing of the wall of a long-bone diaphysis showing 3 types of lamellar bone: haversian system and outer and inner circumferential lamellae. (For interstitial lamellae, see Figure 8–10.) The protruding haversian system on the left shows the orientation of collagen fibers in each lamella. At the right is a haversian system showing lamellae, a central blood capillary (there are also small nerves, not shown), and many osteocytes with their processes.

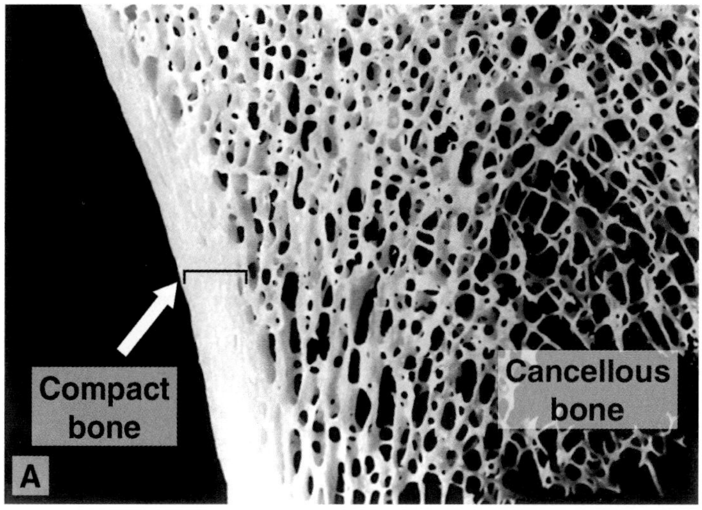

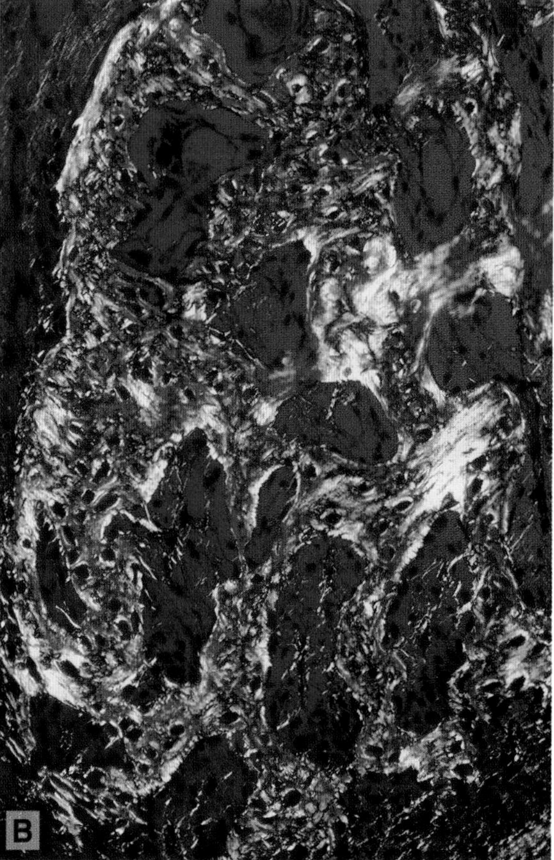

Figure 8–7. **A:** Thick section of bone illustrating the cortical compact bone and the lattice of trabeculae of cancellous bone. (Courtesy of DW Fawcett.) **B:** Section of cancellous (spongy) bone with its characteristic random disposition of collagen fibers. Picrosirius–polarized light (PSP) stain. Low magnification.

the collagen fibrils but are surrounded by ground substance. The surface ions of hydroxyapatite are hydrated, and a layer of water and ions forms around the crystal. This layer, the **hydration shell,** facilitates the exchange of ions between the crystal and the body fluids.

The organic matter in bone matrix is type I collagen and ground substance, which contains proteoglycan aggregates and several specific structural glycoproteins. Bone glycoproteins may be responsible for promoting calcification of bone matrix. Other tissues containing type I collagen are not normally calcified and do not contain these glycoproteins. Because of its high collagen content, decalcified bone matrix intensely binds stains for collagen fibers.

The association of minerals with collagen fibers is responsible for the hardness and resistance of bone tissue. After a bone is decalcified, its shape is preserved, but it becomes as flexible as a tendon. Removal of the organic part of the matrix—which is mainly collagenous—also leaves the bone with its original shape; however, it becomes fragile, breaking and crumbling easily when handled.

PERIOSTEUM & ENDOSTEUM

External and internal surfaces of bone are covered by layers of bone-forming cells and connective tissue called periosteum and endosteum.

The **periosteum** consists of an outer layer of collagen fibers and fibroblasts (Figure 8–6). Bundles of periosteal collagen fibers, called **Sharpey's fibers,** penetrate the bone matrix, binding the periosteum to bone. The inner, more cellular layer of the periosteum is composed of fibroblastlike cells called **osteoprogenitor cells,** with the potential to divide by mitosis and differentiate into osteoblasts. Autoradiographic studies demonstrate that these cells take up ^{3}H-thymidine, which is subsequently encountered in osteoblasts. Osteoprogenitor cells play a prominent role in bone growth and repair.

The **endosteum** (Figure 8–6) lines all internal cavities within the bone and is composed of a single layer of flattened osteoprogenitor cells and a very small amount of connective tissue. The endosteum is therefore considerably thinner than the periosteum.

The principal functions of periosteum and endosteum are nutrition of osseous tissue and provision of a continuous supply of new osteoblasts for repair or growth of bone.

TYPES OF BONE

Gross observation of bone in cross section shows dense areas without cavities—corresponding to **compact bone**—and areas with numerous interconnecting cavities—corresponding to **cancellous (spongy) bone** (Figure 8–7). Under the microscope, however, both compact bone and the trabeculae separating

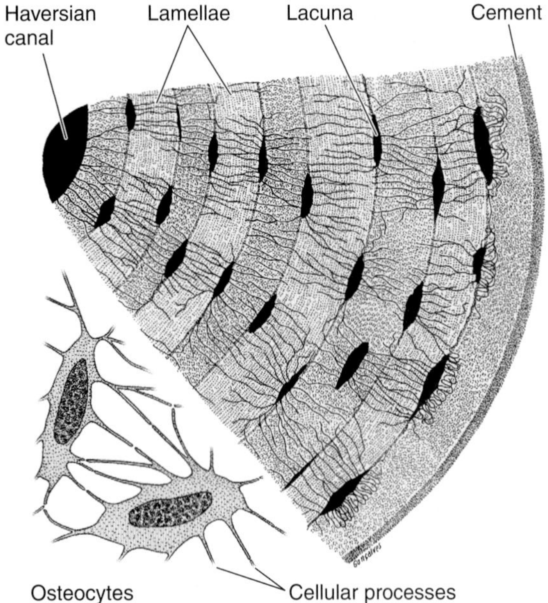

Figure 8–8. Schematic drawing of 2 osteocytes and part of a haversian system. Collagen fibers of contiguous lamellae are sectioned at different angles. Note the numerous canaliculi that permit communication between lacunae and with the haversian canals. Although it is not apparent in this simplified diagram, each lamella consists of multiple parallel arrays of collagen fibers. In adjacent lamellae, the collagen fibers are oriented in different directions. The presence of large numbers of lamellae with differing fiber orientations provides the bone with great strength, despite its light weight. (Redrawn and reproduced, with permission, from Leeson TS, Leeson CR: *Histology,* 2nd ed. Saunders, 1970.)

the cavities of cancellous bone have the same basic histologic structure.

In long bones, the bulbous ends—called **epiphyses** (Gr. *epiphysis,* an excrescence)—are composed of spongy bone covered by a thin layer of compact bone. The cylindrical part—**diaphysis** (Gr. *diaphysis,* a growing between)—is almost totally composed of compact bone, with a small component of spongy bone on its inner surface around the bone marrow cavity. Short bones usually have a core of spongy bone completely surrounded by compact bone. The flat bones that form the calvaria have 2 layers of compact bone called **plates** (tables), separated by a layer of spongy bone called the **diploë.**

Microscopic examination of bone shows 2 varieties: **primary, immature,** or **woven bone** and **secondary, mature,** or **lamellar bone.** Primary bone is the first bone tissue to appear in embryonic development and in fracture repair and other repair processes. It is characterized by random disposition of fine collagen fibers, in contrast to the organized lamellar disposition of collagen in secondary bone.

Primary Bone Tissue

Primary bone tissue is usually temporary and is replaced in adults by secondary bone tissue except in a very few places in the body, eg, near the sutures of the flat bones of the skull, in tooth sockets, and in the insertions of some tendons.

In addition to the irregular array of collagen fibers, other characteristics of primary bone tissue are a lower mineral content (it is more easily penetrated by x-rays) and a higher proportion of osteocytes than that in secondary bone tissue.

Secondary Bone Tissue

Secondary bone tissue is the variety usually found in adults. It characteristically shows collagen fibers arranged in lamellae

Figure 8–9. Lamellar (secondary) bone in which the collagen fibers can be parallel to each other (at left) or organized concentrically around neurovascular channels, to constitute the haversian systems, or osteons (in most of the figure). Among the numerous haversian systems are some interstitial lamellae. PSP stain. Low magnification.

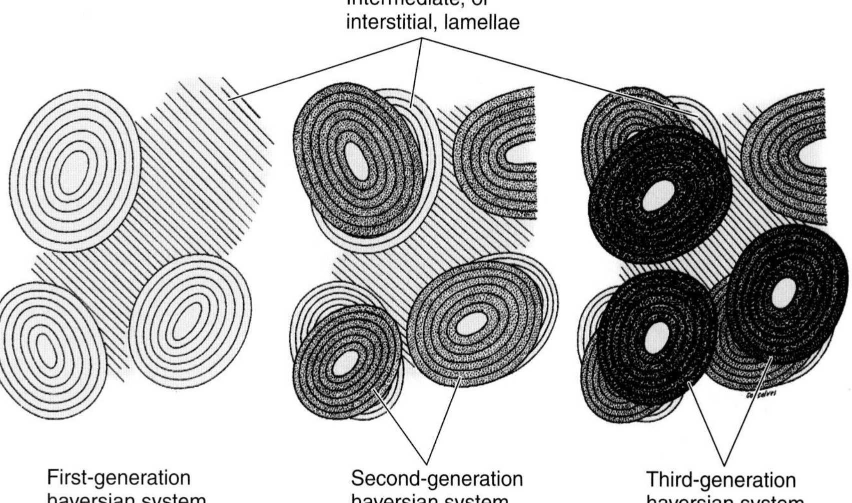

Figure 8–10. Schematic drawing of diaphyseal bone remodeling showing 3 generations of haversian systems and their successive contributions to the formation of intermediate, or interstitial, lamellae. Remodeling is a continuous process responsible for bone adaptations, especially during growth.

(3–7 μm thick) that are parallel to each other or concentrically organized around a vascular canal. The whole complex of concentric lamellae of bone surrounding a canal containing blood vessels, nerves, and loose connective tissue is called a **haversian system,** or **osteon** (Figures 8–6 and 8–8). Lacunae containing osteocytes are found between and occasionally within the lamellae. In each lamella, collagen fibers are parallel to each other. Surrounding each haversian system is a deposit of amorphous material called the **cementing substance** that consists of mineralized matrix with few collagen fibers.

In compact bone (eg, the diaphysis of long bones), the lamellae exhibit a typical organization consisting of **haversian systems, outer circumferential lamellae, inner circumferential lamellae,** and **interstitial lamellae** (Figures 8–6 and 8–9).

Inner circumferential lamellae are located around the marrow cavity, and outer circumferential lamellae are located immediately beneath the periosteum. There are more outer than inner lamellae.

Between the two circumferential systems are numerous haversian systems, including triangular or irregularly shaped groups of parallel lamellae called **interstitial** (or **intermediate**) **lamellae.** These structures are lamellae left by haversian systems destroyed during growth and remodeling of bone (Figure 8–10).

Each haversian system is a long, often bifurcated cylinder parallel to the long axis of the diaphysis. It consists of a central canal surrounded by 4–20 concentric lamellae (Figure 8–11). Each endosteum-lined canal contains blood vessels, nerves, and loose connective tissue. The haversian canals communicate with the marrow cavity, the periosteum, and one another through transverse or oblique Volkmann's canals (Figure 8–6). Volkmann's canals do not have concentric lamellae; instead, they perforate the lamellae. All vascular canals found in bone tissue come into existence when matrix is laid down around preexisting blood vessels.

Examination of haversian systems with polarized light shows bright anisotropic layers alternating with dark isotropic layers (Figure 8–11). When observed under polarized light at right angles to their length, collagen fibers are birefringent (anisotropic). The alternating bright and dark layers are due to the changing orientation of collagen fibers in the lamellae. In each lamella, fibers are parallel to each other and follow a helical course. The

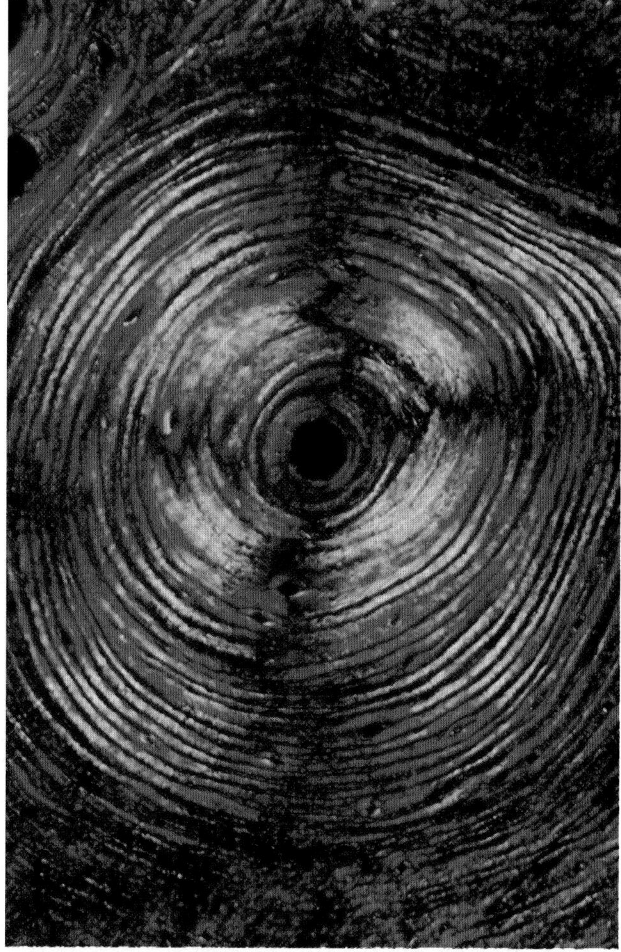

Figure 8–11. Section of a haversian system, or osteon. Note the alternation of clear and dark circles resulting from the alternation in the direction of the collagen fibers. The collagen fibers appear bright when cut longitudinally and dark when cross-sectioned. In the center of the osteon is a channel. PSP stain. Medium magnification.

pitch of the helix is, however, different for different lamellae, so that at any given point, fibers from adjacent lamellae intersect at approximately right angles (Figure 8–6).

Because bone tissue is under constant remodeling, there is great variability in the diameter of haversian canals. Each system is formed by successive deposits of lamellae, starting inward from the periphery, so that younger systems have larger canals. In mature haversian systems, the most recently formed lamella is the one closest to the central canal.

HISTOGENESIS

Bone can be formed in 2 ways: by direct mineralization of matrix secreted by osteoblasts (**intramembranous ossification**) or by deposition of bone matrix on a preexisting cartilage matrix (**endochondral ossification**).

In both processes, the bone tissue that appears first is primary, or woven. Primary bone is a temporary tissue and is soon replaced by the definitive lamellar, or secondary, bone. During bone growth, areas of primary bone, areas of resorption, and areas of secondary bone appear side by side. This combination of bone synthesis and removal (**remodeling**) occurs not only in growing bones but also throughout adult life, although its rate of change in adults is considerably slower.

Intramembranous Ossification

Intramembranous ossification, the source of most of the flat bones, is so called because it takes place within condensations of mesenchymal tissue. The frontal and parietal bones of the skull—as well as parts of the occipital and temporal bones and the mandible and maxilla—are formed by intramembranous ossification. This process also contributes to the growth of short bones and the thickening of long bones.

In the mesenchymal condensation layer, the starting point for ossification is called a **primary ossification center.** The process begins when groups of cells differentiate into osteoblasts. Osteoblasts produce bone matrix and calcification follows, resulting in the encapsulation of some osteoblasts, which then become osteocytes (Figure 8–12). These islands of developing bone form walls that delineate elongated cavities containing capillaries, bone marrow cells, and undifferentiated cells. Several such groups arise almost simultaneously at the ossification center, so that the fusion of the walls gives the bone a spongy structure. The connective tissue that remains among the bone walls is penetrated by growing blood vessels and additional undifferentiated mesenchymal cells, giving rise to the bone marrow cells.

The ossification centers of a bone grow radially and finally fuse together, replacing the original connective tissue. The fontanelles of newborn infants, for example, are soft areas in the skull that correspond to parts of the connective tissue that are not yet ossified.

In cranial flat bones there is a marked predominance of bone formation over bone resorption at both the internal and external surfaces. Thus, 2 layers of compact bone (internal and external plates) arise, whereas the central portion (diploë) maintains its spongy nature.

The portion of the connective tissue layer that does not undergo ossification gives rise to the endosteum and the periosteum of intramembranous bone.

Endochondral Ossification

Endochondral (Gr. *endon,* within, + *chondros,* cartilage) ossification takes place within a piece of hyaline cartilage whose shape resembles a small version, or model, of the bone to be formed. This type of ossification (Figures 8–13 and 8–14) is principally responsible for the formation of short and long bones.

Endochondral ossification of a long bone consists of the following sequence of events. Initially, the first bone tissue appears as a hollow bone cylinder that surrounds the mid portion of the cartilage model. This structure, the **bone collar,** is produced by intramembranous ossification within the local perichondrium. In the next step, the local cartilage undergoes a degenerative process of programmed cell death with cell enlargement (hypertrophy) and matrix calcification, resulting in a 3-dimensional structure formed by the remnants of the calcified cartilage matrix (Figure 8–15). This process begins at the central portion of the cartilage model (diaphysis), where blood vessels penetrate

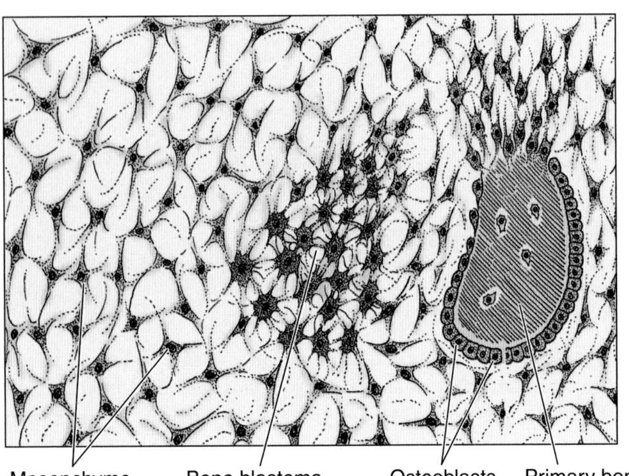

Figure 8–12. The beginning of intramembranous ossification. Mesenchymal cells round up and form a blastema, from which osteoblasts differentiate, producing primary bone tissue.

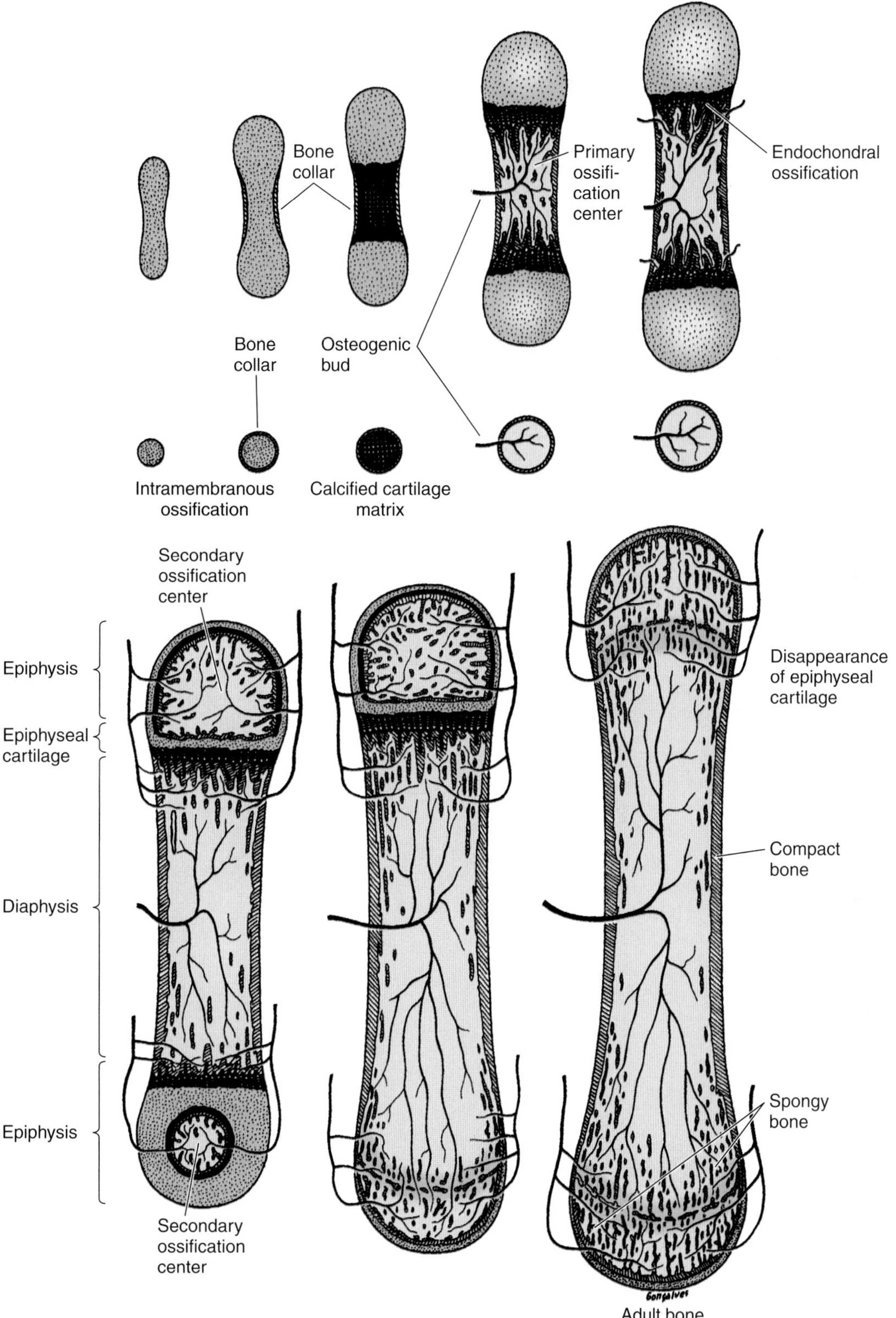

Figure 8–13. Formation of a long bone on a model made of cartilage. Hyaline cartilage is stippled; calcified cartilage is black, and bone tissue is indicated by oblique lines. The 5 small drawings in the middle row represent cross sections through the middle regions of the figures shown in the upper row. Note the formation of the bone collar and primary and secondary ossification centers. Epiphyseal fusion with diaphysis, with disappearance of the epiphyseal cartilage, occurs at different times in the same bone. (Redrawn and reproduced, with permission, from Bloom W, Fawcett DW: *A Textbook of Histology,* 9th ed. Saunders, 1968.)

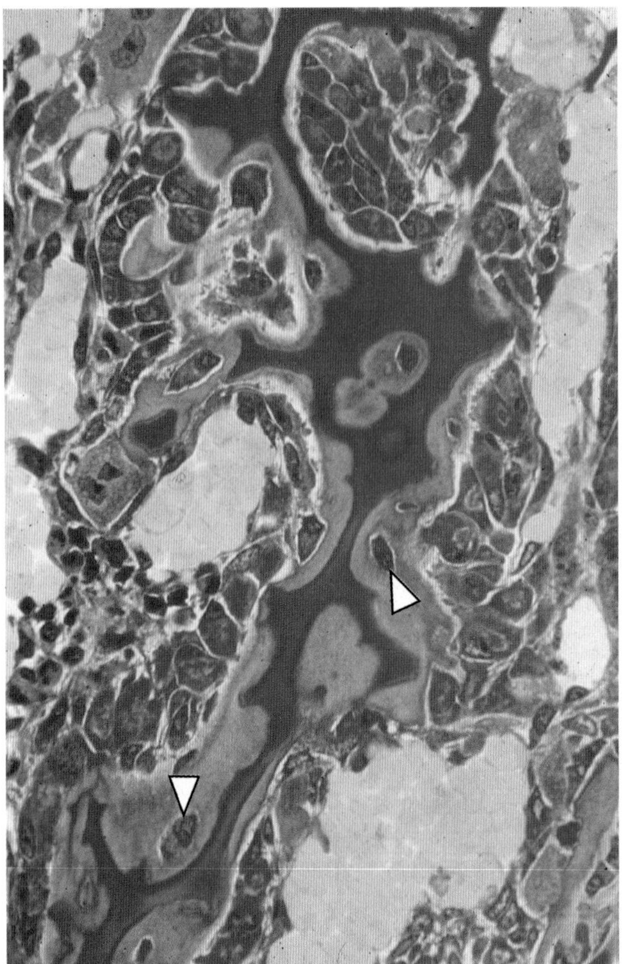

Figure 8–14. A small portion of an epiphyseal plate showing endochondral ossification. Remnants of calcified cartilage matrix (dark purple) appear covered by light-stained bone tissue. The newly formed bone is surrounded by osteoblasts. Some osteoblasts that were captured by the osseous matrix become osteocytes (arrowheads). Pararosaniline–toluidine blue (PT) stain. Medium magnification.

through the bone collar previously perforated by osteoclasts, bringing osteoprogenitor cells to this region. Next, osteoblasts adhere to the calcified cartilage matrix and produce continuous layers of primary bone that surround the cartilaginous matrix remnants. At this stage, the calcified cartilage appears basophilic, and the primary bone is eosinophilic. In this way the **primary ossification center** is produced (Figure 8–13). Then, **secondary ossification centers** appear at the swellings in the extremities of the cartilage model (epiphyses). During their expansion and remodeling, the primary and secondary ossification centers produce cavities that are gradually filled with bone marrow.

In the secondary ossification centers, cartilage remains in 2 regions: the **articular cartilage,** which persists throughout adult life and does not contribute to bone growth in length, and the **epiphyseal cartilage,** also called **epiphyseal plate,** which connects the two epiphyses to the diaphysis (Figures 8–15 and 8–16). The epiphyseal cartilage is responsible for the growth in length of the bone, and it disappears in adults, which is why bone growth ceases in adulthood.

The closure of the epiphyses follows a chronologic order according to each bone and is complete at about 20 years of age. Through x-ray examination of the growing skeleton, it is possible to determine the "bone age" of a young person, noting which epiphyses are open and which are closed. Once the epiphyses have closed, growth in length of bones becomes impossible, although widening may still occur.

Epiphyseal cartilage is divided into 5 zones (Figure 8–16), starting from the epiphyseal side of cartilage: (1) The **resting zone** consists of hyaline cartilage without morphologic changes in the cells. (2) In the **proliferative zone,** chondrocytes divide rapidly and form columns of stacked cells parallel to the long axis of the bone. (3) The **hypertrophic cartilage zone** contains large chondrocytes whose cytoplasm has accumulated glycogen. The resorbed matrix is reduced to thin septa between the chon-

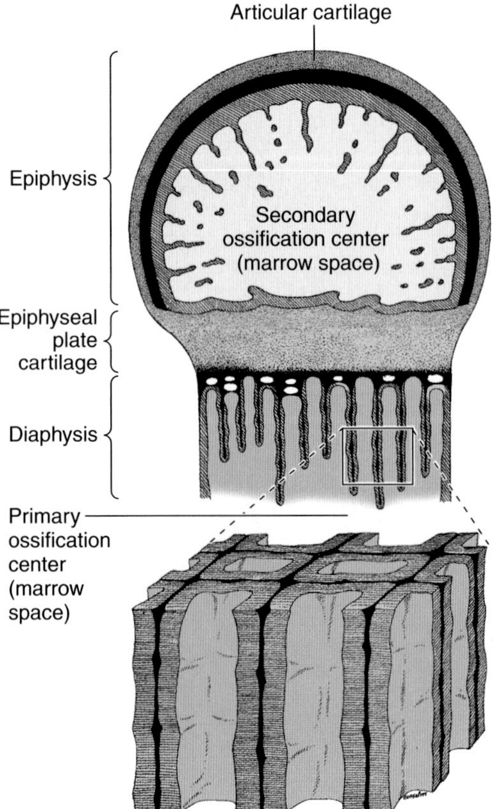

Figure 8–15. Schematic drawings showing the 3-dimensional shape of bone in the epiphyseal plate area. Hyaline cartilage is stippled; calcified cartilage is black, and bone tissue is shown as yellow hatched areas. The upper drawing shows the region represented 3-dimensionally in the lower drawing. (Redrawn and reproduced, with permission, from Ham AW: *Histology,* 6th ed. Lippincott, 1969.)

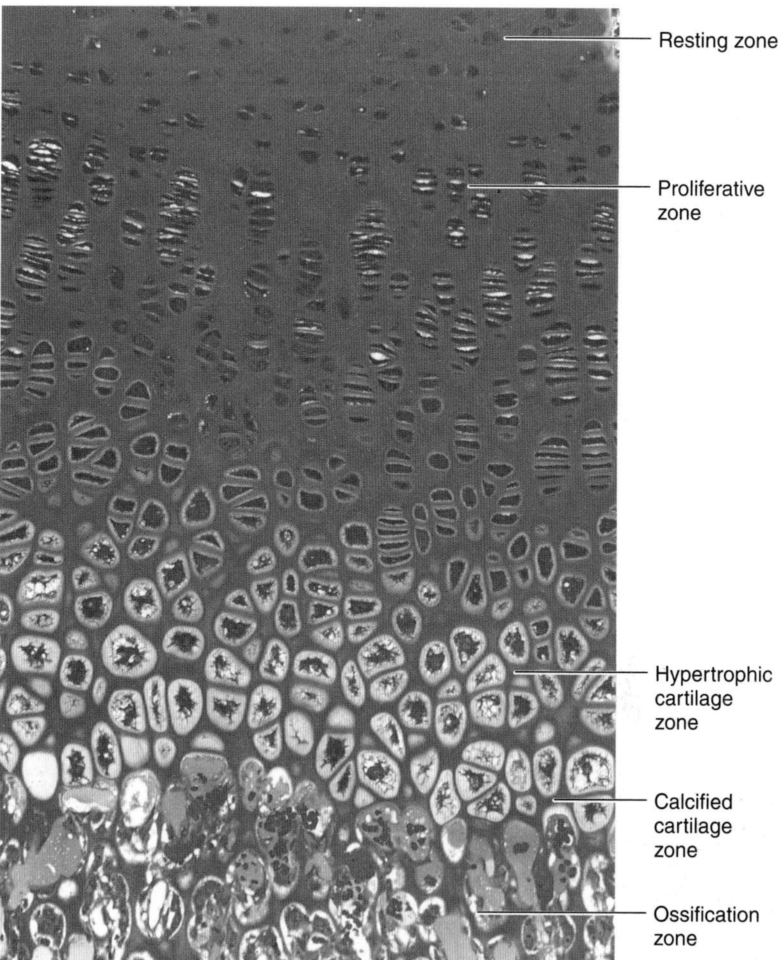

Figure 8–16. Photomicrograph of the epiphyseal plate, showing its 5 zones, the changes that take place in the cartilage, and the formation of bone. PT stain. Low magnification.

drocytes. (4) Simultaneous with the death of chondrocytes in the **calcified cartilage zone,** the thin septa of cartilage matrix become calcified by the deposit of hydroxyapatite (Figures 8–15 and 8–16). (5) In the **ossification zone,** endochondral bone tissue appears. Blood capillaries and osteoprogenitor cells formed by mitosis of cells originating from the periosteum invade the cavities left by the chondrocytes. The osteoprogenitor cells form osteoblasts, which are distributed in a discontinuous layer over the septa of calcified cartilage matrix. Ultimately, the osteoblasts deposit bone matrix over the 3-dimensional calcified cartilage matrix (Figures 8–17 through 8–20).

In summary, growth in length of a long bone occurs by proliferation of chondrocytes in the epiphyseal plate adjacent to the epiphysis. At the same time, chondrocytes of the diaphyseal side of the plate hypertrophy; their matrix becomes calcified, and the cells die. Osteoblasts lay down a layer of primary bone on the calcified cartilage matrix. Because the rates of these two opposing events (proliferation and destruction) are approximately equal, the epiphyseal plate does not change thickness. Instead, it is displaced away from the middle of the diaphysis, resulting in growth in length of the bone.

Mechanisms of Calcification

No hypothesis to explain the events occurring during calcium phosphate deposition on bone matrix is yet generally accepted.

It is known that calcification begins by the deposition of calcium salts on collagen fibrils, a process induced by proteoglycans and high-affinity calcium-binding glycoproteins. The deposition of calcium salts is probably accelerated by the ability of osteoblasts to concentrate them in intracytoplasmic vesicles and to release these vesicles, when necessary, to the extracellular medium (matrix vesicles).

Calcification is aided, in some unknown way, by alkaline phosphatase, which is produced by osteoblasts and is present at ossification sites.

BONE GROWTH & REMODELING

Bone growth is generally associated with partial resorption of preformed tissue and the simultaneous laying down of new bone (exceeding the rate of bone loss). This process permits the shape of the bone to be maintained while it grows. The rate of bone remodeling (**bone turnover**) is very active in young children, where it can be 200 times faster than that in adults. Bone remodeling in adults is a dynamic physiologic process that occurs simultaneously in multiple locations of the skeleton, not related to bone growth.

Cranial bones grow mainly because of the formation of bone tissue by the periosteum between the sutures and on the external bone surface. At the same time, resorption takes place on the internal surface. Since bone is an extremely plastic tissue, it responds to the growth of the brain and forms a skull of adequate size. The skull will be small if the brain does not develop completely and will be larger than normal in a person suffering from hydrocephalus, a disorder characterized by abnormal accumulation of spinal fluid and dilatation of the cerebral ventricles.

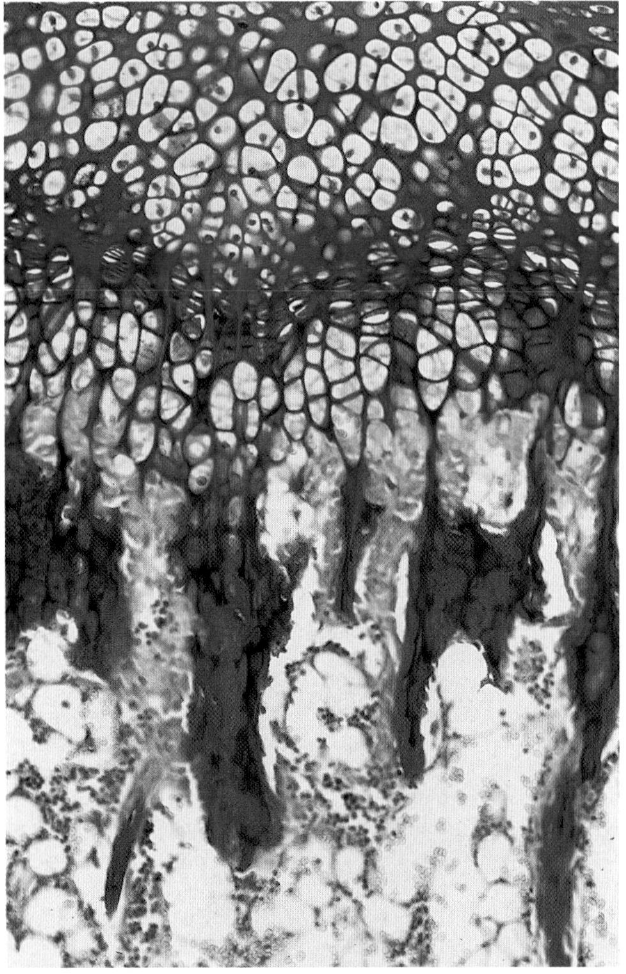

Figure 8–17. Higher magnification of the epiphyseal plate showing details of the endochondral ossification. Cartilage matrix (purple) is covered by recently formed bone tissue (red). Bone marrow and fat cells fill up the space left by the new bone. Picrosirius-hematoxylin (PSH) stain. Medium magnification.

Fracture Repair

MEDICAL APPLICATION

When a bone is fractured, bone matrix is destroyed and bone cells adjoining the fracture die. The damaged blood vessels produce a localized hemorrhage and form a blood clot.

During repair, the blood clot, cells, and damaged bone matrix are removed by macrophages. The periosteum and the endosteum around the fracture respond with intense proliferation producing a tissue that surrounds the fracture and penetrates between the extremities of the fractured bone (Figure 8–21).

Primary bone is then formed by endochondral and intramembranous ossification, both processes contributing simultaneously to the healing of fractures. Repair progresses in such a way that irregularly formed trabeculae of primary bone temporarily unite the extremities of the fractured bone, forming a ***bone callus*** *(Figure 8–21).*

Stresses imposed on the bone during repair and during the patient's gradual return to activity serve to remodel the bone callus. If these stresses are identical to those that occurred during the growth of the bone—and therefore influence its structure—the primary bone tissue of the callus is gradually resorbed and replaced by secondary tissue, remodeling the bone and restoring its original structure (Figure 8–21). Unlike other connective tissues, bone tissue heals without forming a scar.

INTERNAL STRUCTURE OF BONES

Despite its hardness, bone is capable of changes in its internal structure according to the various stresses to which it is subjected. For example, the positions of the teeth in the jawbone can be modified by lateral pressures produced by orthodontic appliances. Bone is formed on the side where traction is applied and is resorbed where pressure is exerted (on the opposite side). In this way, teeth move within the jawbone while the alveolar bone is being remodeled.

METABOLIC ROLE OF BONE TISSUE

The skeleton contains 99% of the total calcium of the body and acts as a calcium reservoir. The concentration of calcium in the blood and tissues is quite stable because of a continuous interchange between blood calcium and bone calcium.

Bone calcium is mobilized by 2 mechanisms, one rapid and the other slow. The first is the simple transfer of ions from hydroxyapatite crystals to interstitial fluid—from which, in turn, calcium passes into the blood. This purely physical mechanism takes place mainly in spongy bone. The younger, slightly calcified lamellae that exist even in adult bone (because of continu-

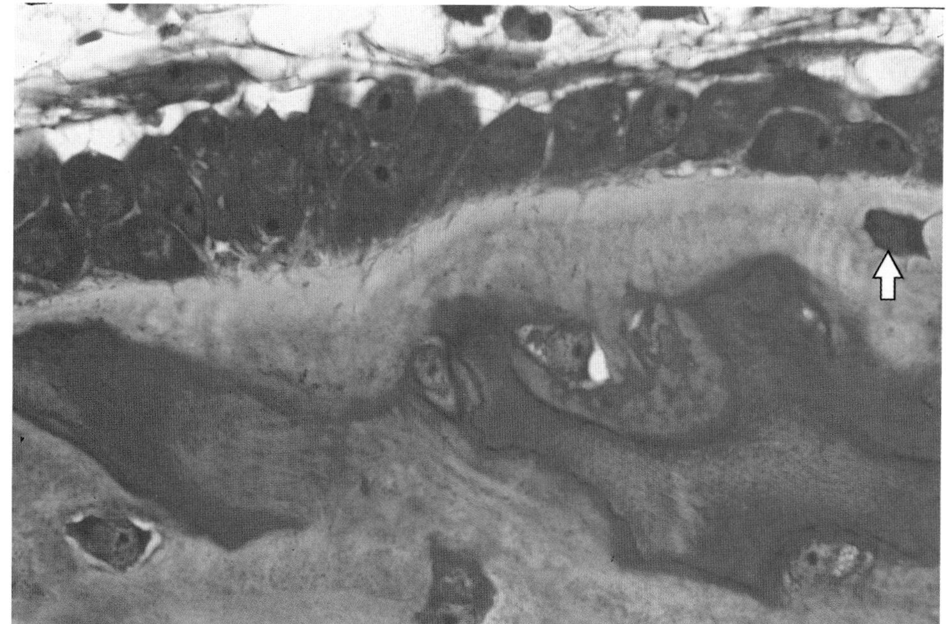

Figure 8–18. Photomicrograph of endochondral ossification. In the upper region is a row of osteoblasts with intense cytoplasmic basophilia, a feature to be expected in cells synthesizing a glycoprotein (collagen). Note an osteoblast being captured in the bone matrix (arrow). Between the layer of osteoblasts and the calcified bone matrix is a pale region made of noncalcified bone matrix called osteoid. PT stain. Medium magnification.

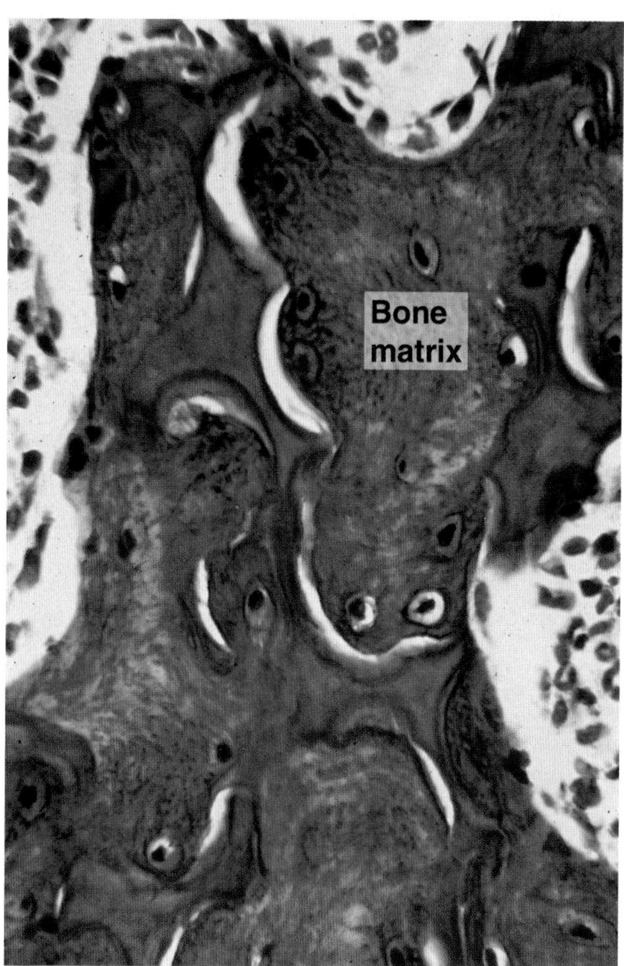

Figure 8–19. Section of endochondral ossification. The osseous matrix, rich in collagen type I, is specifically stained with picrosirius-hematoxylin. The cartilaginous matrix, containing collagen type II, stains blue with hematoxylin because of its high content of chondroitin sulfate. Medium magnification.

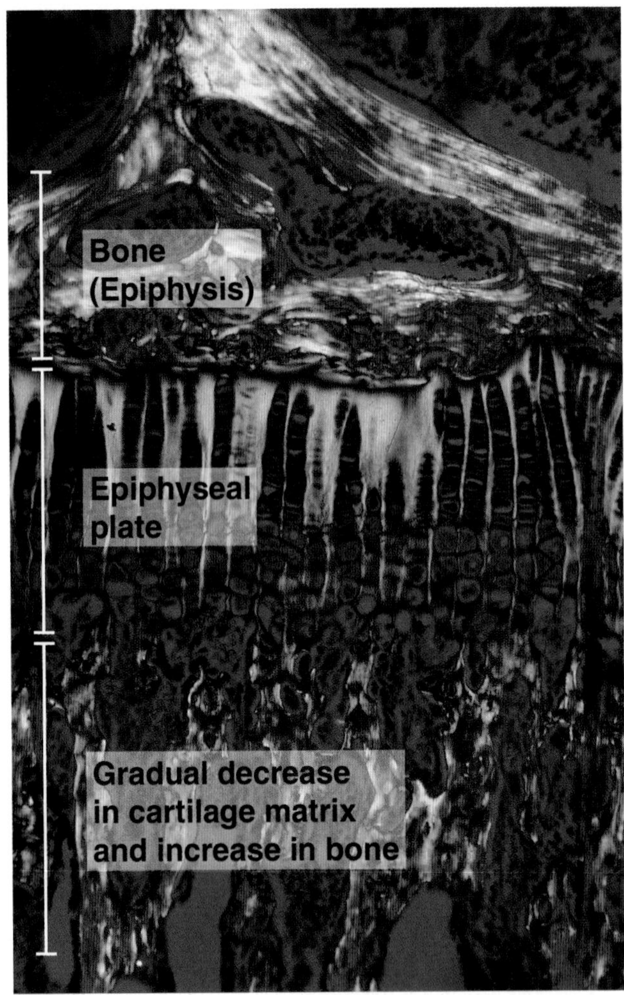

Figure 8–20. Section of the extremity of a long bone showing the epiphysis, epiphyseal plate and newly formed bone tissue. PSP stain. Low magnification.

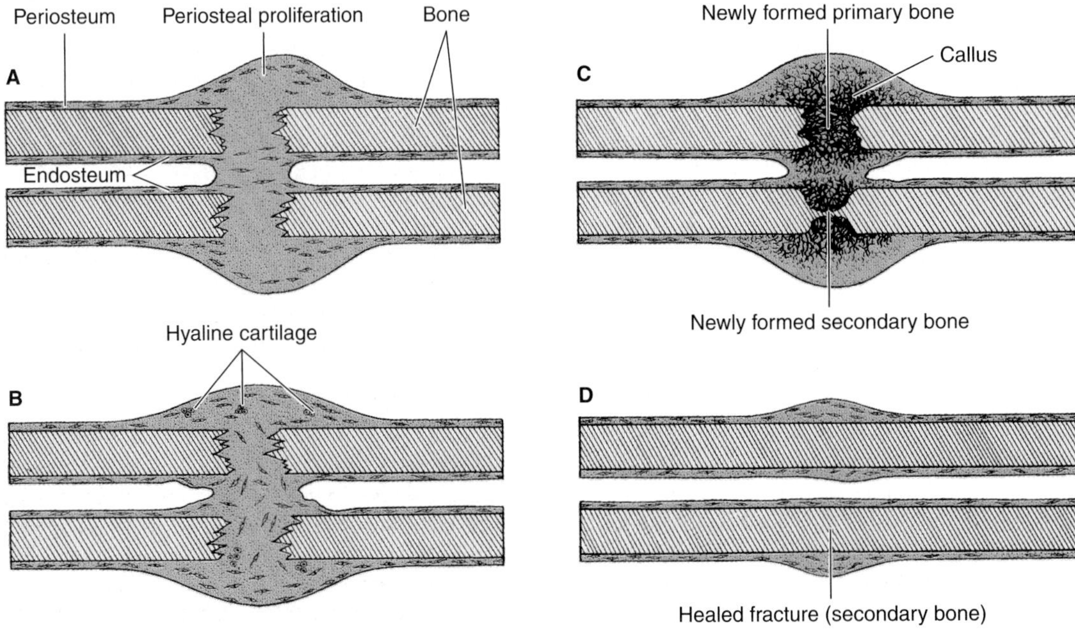

Figure 8–21. Repair of a fractured bone by formation of new bone tissue through periosteal and endosteal cell proliferation.

ous remodeling) receive and lose calcium more readily. These lamellae are more important for the maintenance of calcium concentration in the blood than are the older, greatly calcified lamellae, whose role is mainly that of support and protection.

The second mechanism for controlling blood calcium level depends on the action of hormones on bone. **Parathyroid hormone** promotes osteoclastic resorption of the bone matrix with the consequent liberation of calcium. This hormone acts primarily on osteoblast receptors. The activated osteoblasts stop producing bone and start the secretion of an **osteoclast-stimulating factor.**

Another hormone, **calcitonin,** which is synthesized mainly by the parafollicular cells of the thyroid gland, inhibits matrix resorption. Calcitonin has an inhibitory effect on osteoclast activity.

MEDICAL APPLICATION

Since the concentration of calcium in tissues and blood must be kept constant, nutritional deficiency of calcium results in decalcification of bones; decalcified bones are more likely to fracture and are more transparent to x-rays.

Decalcification of bone may also be caused by excessive production of parathyroid hormone (hyperparathyroidism), which results in increased osteoclastic activity, intense resorption of bone, elevation of blood Ca^{2+} and PO_4^{3-} levels, and abnormal deposits of calcium in several organs, mainly the kidneys and arterial walls.

The opposite occurs in ***osteopetrosis*** *(L.* petra, *stone), a disease caused by a defect in osteoclast function that results in overgrowth, thickening, and hardening of bones. This process produces obliteration of the bone marrow cavities, depressing blood cell formation with consequent anemia and frequent infections that may be fatal.*

Effects of Nutritional Deficiencies on Bone Tissue

Especially during growth, bone is sensitive to nutritional factors. Deficiency of calcium leads to incomplete calcification of the organic bone matrix, owing either to the lack of calcium in the diet or to the lack of the steroid prohormone vitamin D, which is important for the absorption of Ca^{2+} and PO_4^{3-} by the small intestine.

Calcium deficiency in children causes ***rickets,*** *a disease in which the bone matrix does not calcify normally and the epiphyseal plate becomes distorted by the normal strains of body weight and muscular activity. Ossification processes at this level are consequently hindered, and the bones not only grow more slowly but also become deformed.*

Calcium deficiency in adults gives rise to ***osteomalacia*** (osteon + *Gr.* malakia, *softness), which is characterized by deficient calcification of recently formed bone and partial decalcification of already calcified matrix. Osteomalacia should not be confused with* ***osteoporosis.*** *In osteomalacia, there is a decrease in the amount of calcium per unit of bone matrix. Osteoporosis, frequently found in immobilized patients and in postmenopausal women, is an imbalance in skeletal turnover so that bone resorption exceeds bone formation.*

Hormones Acting on Bone Tissue

In addition to parathyroid hormone and calcitonin, several other hormones act on bone. The anterior lobe of the pituitary synthesizes growth hormone, which stimulates the liver to produce somatomedins. This, in turn, has an overall growth effect, especially on the epiphyseal cartilage. Consequently, lack of growth hormone during the growing years causes ***pituitary dwarfism;*** *an excess of growth hormone causes excessive growth of the long bones, resulting in* ***gigantism.*** *Adult bones cannot increase in length when stimulated by an excess of somatomedins because of the lack of epiphyseal cartilage, but they do increase in width by periosteal growth. In adults, an increase in growth hormone causes* ***acromegaly,*** *a disease in which the bones—mainly the long ones—become very thick.*

The sex hormones, both male (androgens) and female (estrogens), have a complex effect on bones and are, in a general way, stimulators of bone formation. They influence the time of appearance and development of ossification centers and accelerate the closure of epiphyses.

Precocious sexual maturity caused by sex hormone–producing tumors retards bodily growth, since the epiphyseal cartilage is quickly replaced by bone (closure of epiphysis). In hormone deficiencies caused by abnormal development of the gonads, epiphyseal cartilage remains functional for a longer period of time, resulting in tall stature. Thyroid hormone deficiency in children, as in ***cretinism,*** *is associated with* ***dwarfism.*** *Recently, evidence was found for a participation of the central nervous system in the regulation of bone formation during bone remodeling in adult mice. This regulatory mechanism involves the hormone leptin produced by adipose tissue and may thus explain the observation that bones of obese people have an increased mass, containing a higher concentration of calcium.*

Bone Tumors

Although bone tumors are uncommon (0.5% of all cancer deaths), bone cells may escape the normal controls of proliferation to become benign (eg, ***osteoblastoma, osteoclastoma****) or malignant (eg,* ***osteosarcoma****) tumors. Osteosarcomas show pleomorphic (Gr.* pleion, *more, +* morphe, *form) and mitotically active osteoblasts associated with osteoid. Most cases of this aggressive malignant tumor occur in adolescents and young adults. The lower end of the femur, the upper tibia, and the upper humerus*

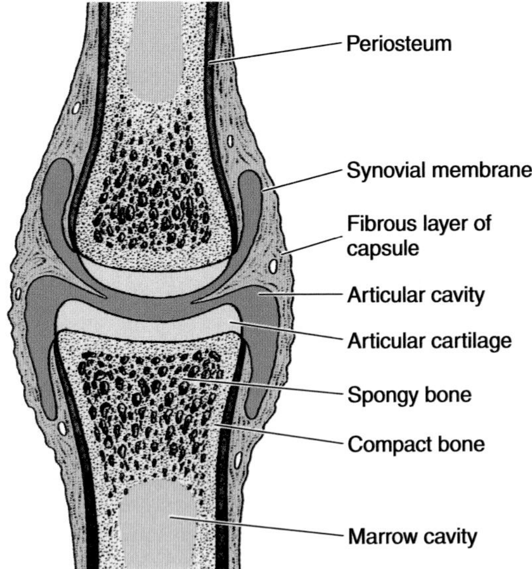

Figure 8–22. Schematic drawing of a diarthrosis. The capsule is formed by 2 parts: the external fibrous layer and the synovial layer (synovial membrane) that lines the articular cavity except for the cartilaginous areas (blue).

are the most common locations. In addition to the tumors originating from bone cells, the skeleton is often the site of metastases from malignant tumors originating in other organs. The most frequent bone metastases are from breast, lung, prostate, kidney, and thyroid tumors.

JOINTS

Joints are regions where bones are capped and surrounded by connective tissues that hold the bones together and determine the type and degree of movement between them. Joints may be classified as **diarthroses,** which permit free bone movement, and **synarthroses** (Gr. *syn,* together, + *arthrosis,* articulation), in which very limited or no movement occurs. There are 3 types of synarthroses, based on the type of tissue uniting the bone surfaces: **synostosis, synchondrosis,** and **syndesmosis.**

In synostosis (*syn* + *osteon* + Gr. *osis,* condition), bones are united by bone tissue and no movement takes place. In older adults, this type of synarthrosis unites the skull bones, which, in children and young adults, are united by dense connective tissue.

Synchondroses (*syn* + *chondros*) are articulations in which the bones are joined by hyaline cartilage. The epiphyseal plates of growing bones are one example, and in the adult human, synchondrosis unites the first rib to the sternum.

As with synchondrosis, a syndesmosis permits a certain amount of movement. The bones are joined by an interosseous ligament of dense connective tissue (eg, the pubic symphysis).

Diarthroses (Figures 8–22 and 8–23) are joints that generally unite long bones and have great mobility, such as the elbow and knee joints. In a diarthrosis, ligaments and a capsule of connective tissue maintain the contact at the ends of the bone. The capsule encloses a sealed **articular cavity** that contains **synovial fluid,** a colorless, transparent, viscous fluid. Synovial fluid is a blood plasma dialysate with a high concentration of hyaluronic acid produced by cells of the synovial layer. The sliding of articular surfaces covered by hyaline cartilage (Figure 8–22) and having no perichondrium is facilitated by the lubricating synovial fluid, which also supplies nutrients and oxygen to the avascular articular cartilage.

The collagen fibers of the articular surface cartilage are disposed as gothic arches, a convenient arrangement to distribute the forces generated by pressure in this tissue (Figure 8–24).

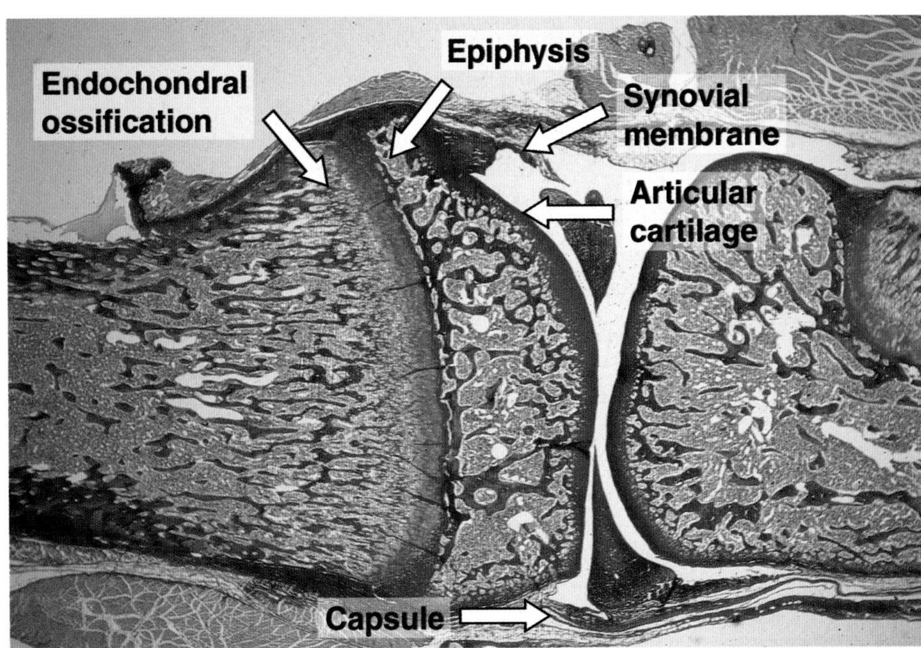

Figure 8–23. Photomicrograph of a diarthrosis. Section of a guinea pig knee. PSH stain. Low magnification.

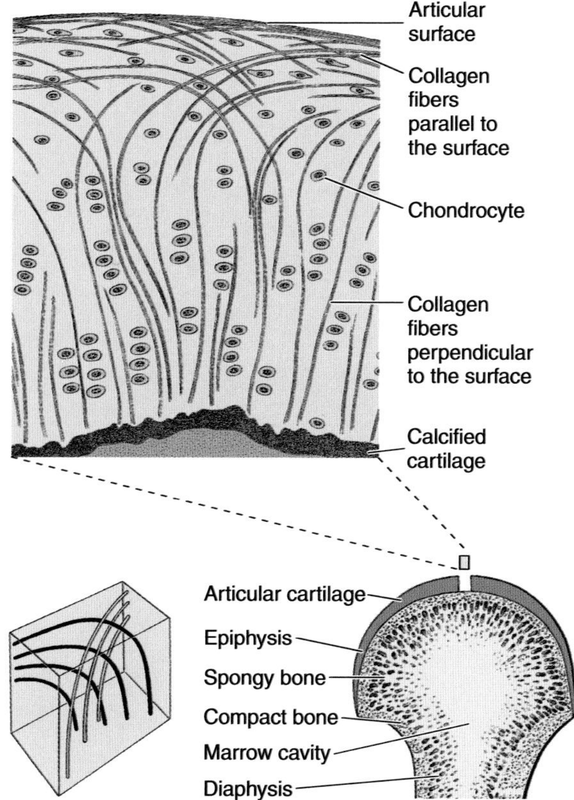

Figure 8–24. Articular surfaces of a diarthrosis are covered by hyaline cartilage that is devoid of perichondrium. The upper drawing shows that in this cartilage, collagen fibers are first perpendicular and then bend gradually, becoming parallel to the cartilage surface. Deeply located chondrocytes are globular and are arranged in vertical rows. Superficially placed chondrocytes are flattened; they are not organized in groups. The lower left drawing shows the organization of collagen fibers in articular cartilage in 3 dimensions.

The resilient articular cartilage is also an efficient absorber of the intermittent mechanical pressures to which many joints are subjected. A similar mechanism is seen in intervertebral disks (Figure 8–25). Proteoglycan molecules, found isolated or aggregated in a network, contain a large amount of water. These matrix components, rich in highly branched hydrophilic glycosaminoglycans, function as a biomechanical spring. When pressure is applied, water is forced out of the cartilage matrix into the synovial fluid. When water is expelled, another mechanism that contributes to cartilage resilience enters into play. This is the reciprocal electrostatic repulsion of the negatively charged carboxyl and sulfate groups in the glycosaminoglycan molecules. These charges are also responsible for separating the glycosaminoglycan branches and thus creating spaces to be occupied by water. When the pressure is released, water is attracted back into the interstices of the glycosaminoglycan branches. These water movements are brought about by the use of the joint. They are essential for nutrition of the cartilage and for facilitating the interchange of O_2, CO_2, and other molecules between the synovial fluid and the articular cartilage.

The capsules of diarthroses (Figure 8–22) vary in structure according to the joint. Generally, however, this capsule is composed of 2 layers, the external **fibrous layer** and the internal **synovial layer** (Figure 8–26).

The synovial layer is formed by 2 types of cells. One resembles fibroblasts and the other has the aspect and behavior of macrophages (Figure 8–27). The fibrous layer is made of dense connective tissue.

MEDICAL APPLICATION

Obesity imposes significant strain on the articular cartilage, accelerating its degeneration. Joint problems are far more frequent in obese individuals.

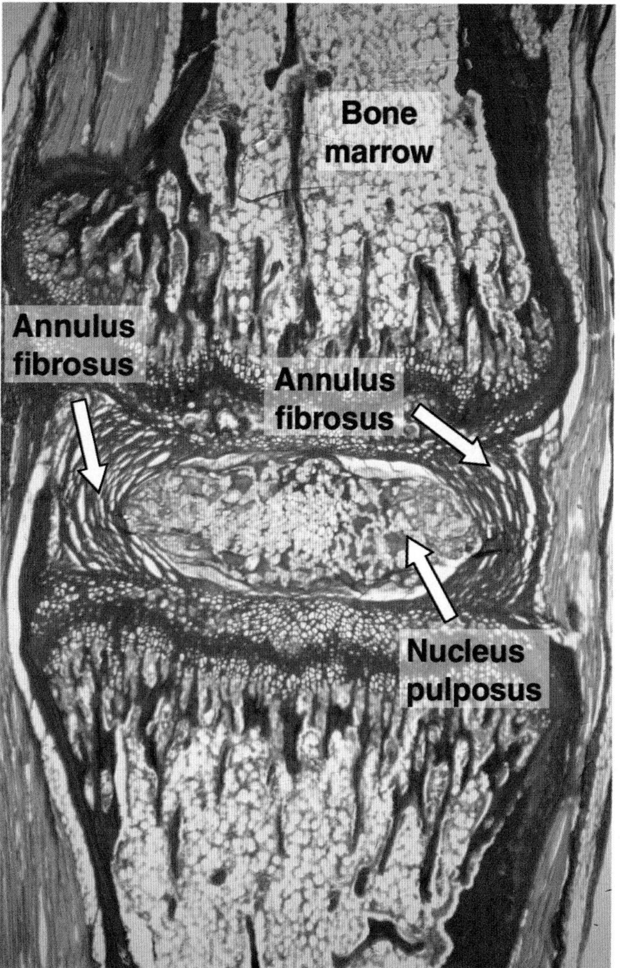

Figure 8–25. Example of a special type of joint. Section of a rat tail showing in the center the intervertebral disk consisting of concentric layers of fibrocartilage (annulus fibrosus) surrounding the nucleus pulposus (see Chapter 7). The nucleus pulposus is formed by residual cells of the notochord immersed in abundant viscous intercellular matrix. PSH stain. Low magnification.

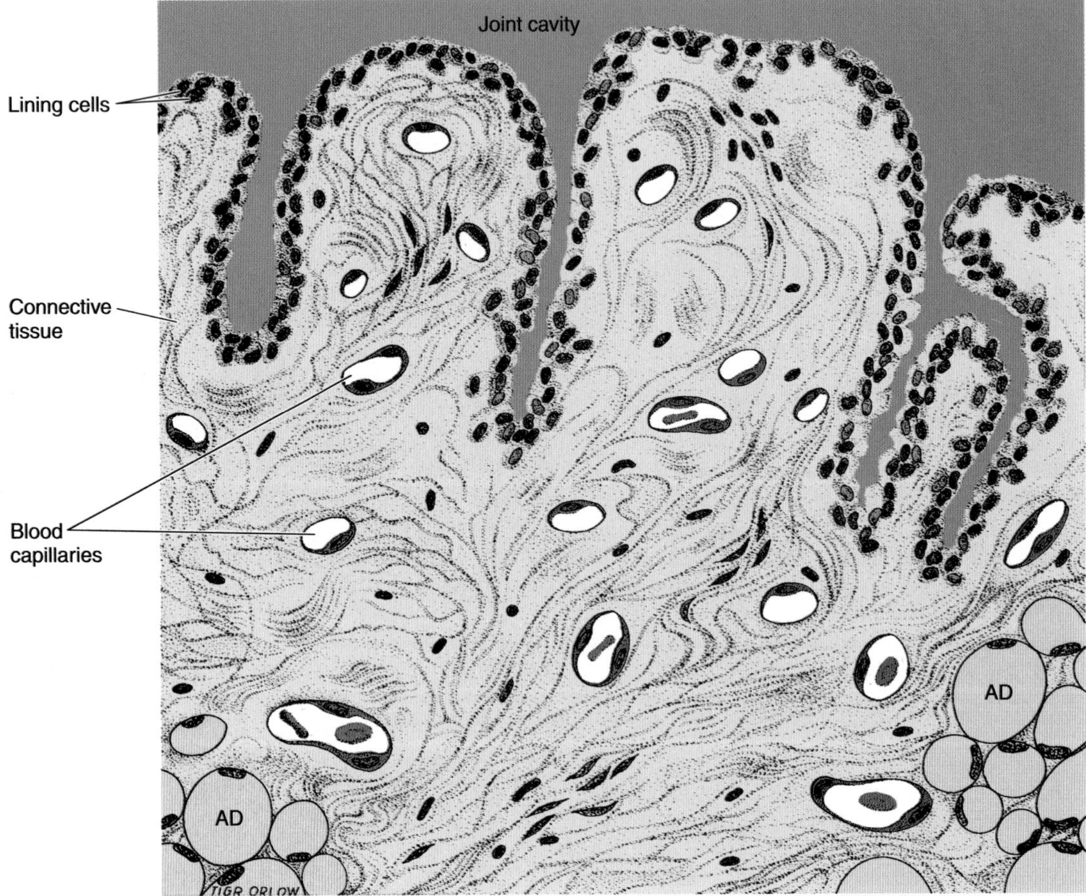

Figure 8–26. **Histologic structure of the synovial membrane, with its lining cells in epithelioid arrangement. There is no basal lamina between the lining cells and the underlying connective tissue. This tissue is rich in blood capillaries and contains a variable number of adipose cells (AD).** (Reproduced, with permission, from Cossermelli W: *Reumatologia Basica,* Sarvier, 1971.)

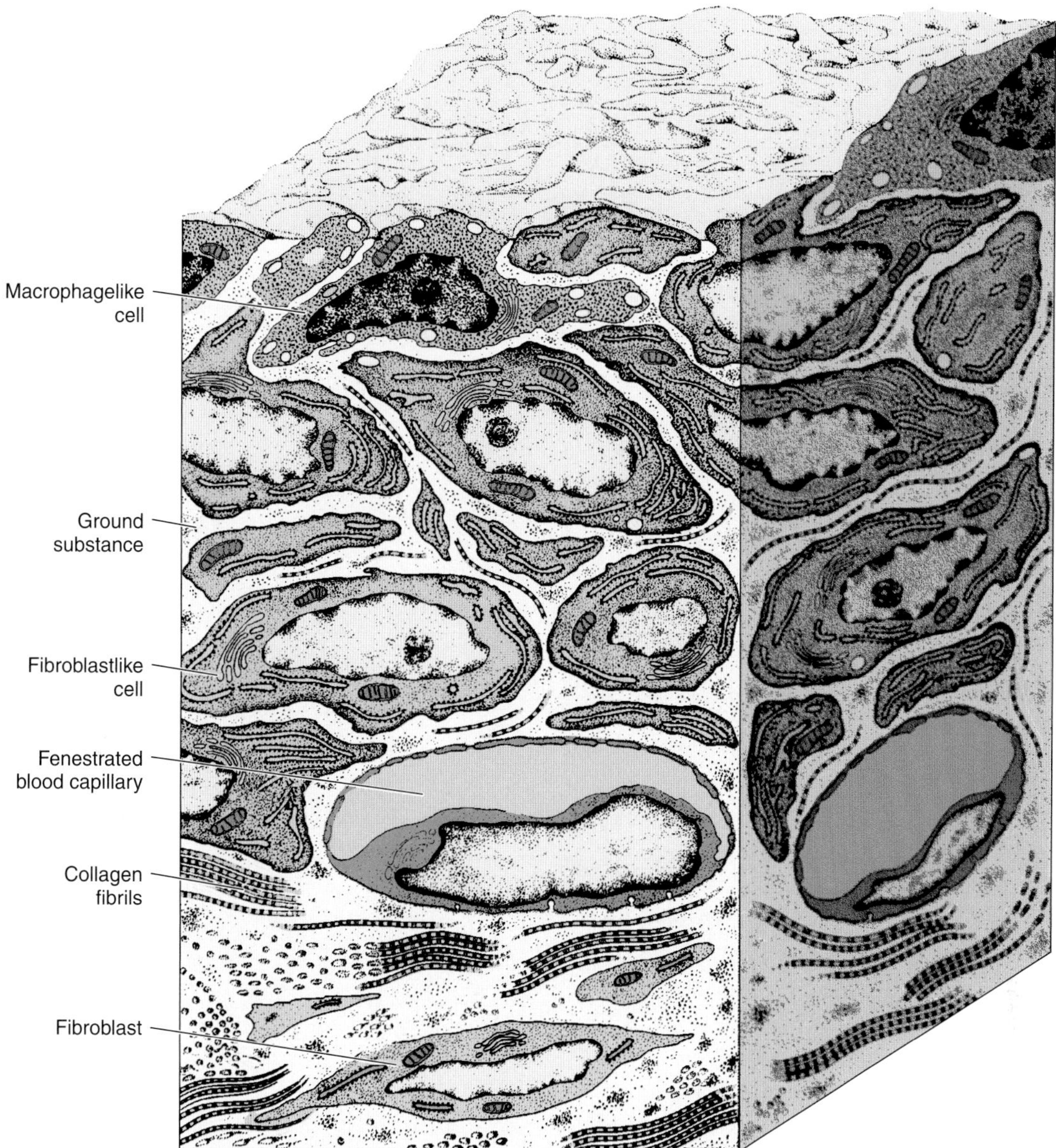

Figure 8–27. Schematic representation of the ultrastructure of synovial membrane. The two covering cell types are separated by a small amount of connective tissue ground substance. No basal lamina is seen separating the lining cells from the connective tissue. Blood capillaries are of the fenestrated type, which facilitates exchange of substances between blood and synovial fluid.

REFERENCES

Demers C, Handy RC: Bone morphogenetic proteins. Sci & Med 1999;6(6):8.

Ducy P et al: The osteoblast: a sophisticated fibroblast under central surveillance. Science 2000;289:1421.

Ducy P et al: Leptin inhibits bone formation through a hypothalamic relay: a central control of bone mass. Cell 2000;100:197.

Ghadially FN: *Fine Structure of Synovial Joints.* Butterworth, 1983.

Gunness M, Hock JM: Anabolic effect of parathyroid hormone on cancellous and cortical bone histology. Bone 1993;14:277.

Levick JR: Synovial fluid hydraulics. Sci & Med 1996;3(5):52.

Mundy GR et al: The effects of cytokines and growth factors on osteoblastic cells. Bone 1995;17:71S.

Roach HI, Clark NM: Physiological cell death of chondrocytes in vivo is not confined to apoptosis. New observations on the mammalian growth plate. J Bone Joint Surg Br 2000;82:601.

Ross PD et al: Bone mass and beyond: risk factors for fractures. Calcified Tissue International 1993;53:S134.

Teltelbaum SL: Bone resorption by osteoclasts. Science 2000;289:1504.

Urist MR: *Fundamental and Clinical Bone Physiology.* Lippincott, 1980.

Nerve Tissue & the Nervous System 9

The human nervous system is by far the most complex system in the human body and is formed by a network of more than 100 million nerve cells (**neurons**), assisted by many more glial cells. Each neuron has, on average, at least a thousand interconnections with other neurons, forming a very complex system for communication.

Neurons are grouped as **circuits.** Like electronic circuits, neural circuits are highly specific combinations of elements that make up systems of various sizes and complexities. Although a neural circuit may be single, in most cases it is a combination of two or more circuits that interact to generate a function. A neural function is a set of coordinated processes intended to produce a definite result. A number of elementary circuits may be combined to form higher-order systems.

Nerve tissue is distributed throughout the body as an integrated communications network. Anatomically, the nervous system is divided into the **central nervous system,** consisting of the brain and the spinal cord; and the **peripheral nervous system,** composed of nerve fibers and small aggregates of nerve cells called **nerve ganglia** (Figure 9–1).

Structurally, nerve tissue consists of two cell types: **nerve cells,** or **neurons,** which usually show numerous long processes; and several types of **glial cells** (Gr. *glia,* glue), which have short processes, support and protect neurons, and participate in neu-

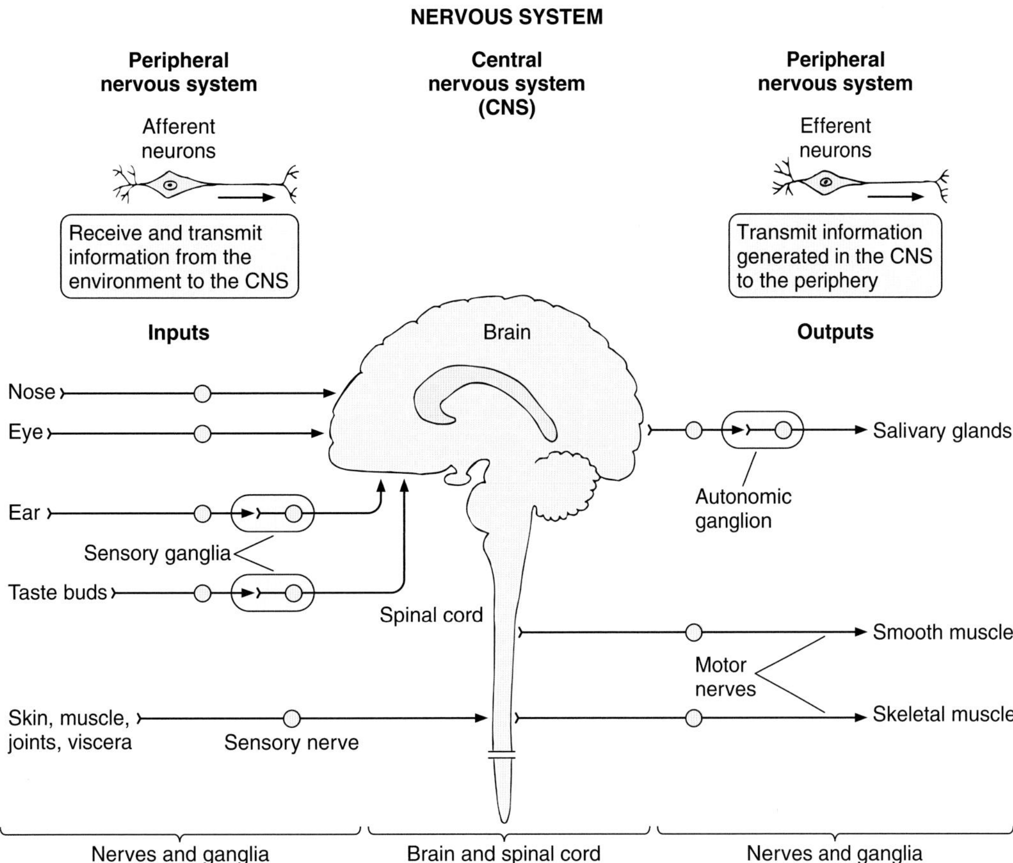

Figure 9–1. The general functional organization of the central and peripheral nervous systems.

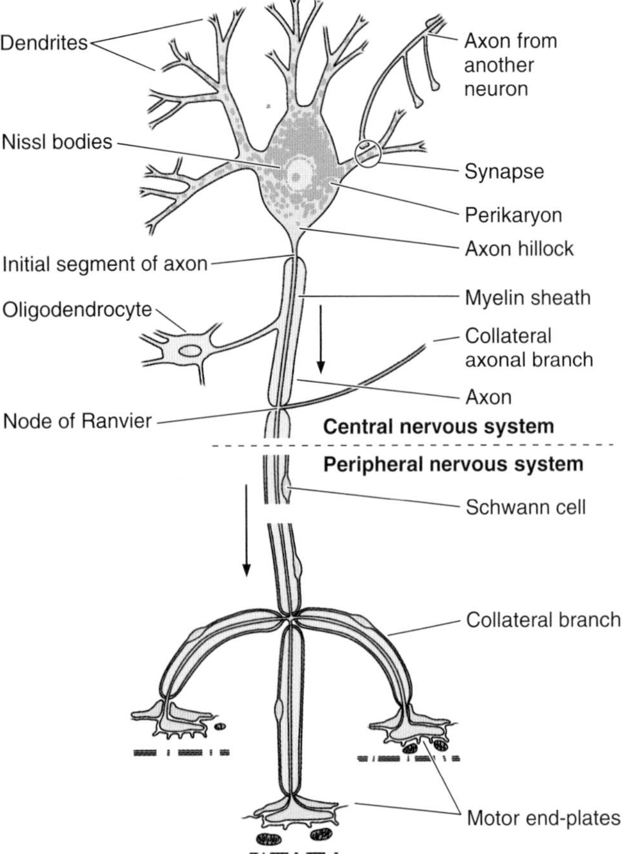

Figure 9–2. Motor neuron. The myelin sheath is produced by oligodendrocytes in the central nervous system and by Schwann cells in the peripheral nervous system. The neuronal cell body has an unusually large, euchromatic nucleus with a well-developed nucleolus. The perikaryon contains Nissl bodies, which are also found in large dendrites. An axon from another neuron is shown at upper right. It has 3 end bulbs, one of which forms a synapse with the neuron. Note also the 3 motor end-plates, which transmit the nerve impulse to striated skeletal muscle fibers. Arrows show the direction of the nerve impulse.

ral activity, neural nutrition, and the defense processes of the central nervous system.

The study of nerve tissue has recently progressed rapidly owing to the use of markers that identify neurons and glia cells and the use of molecules that flow in a retrograde direction, permitting a more precise study of neuronal circuits.

Neurons respond to environmental changes (**stimuli**) by altering electrical potentials that exist between the inner and outer surfaces of their membranes. Cells with this property (eg, neurons, muscle cells, some gland cells) are called **excitable,** or **irritable.** Neurons react promptly to stimuli with a modification of electrical potential that may be restricted to the place that received the stimulus or may be spread (propagated) throughout the neuron by the plasma membrane. This propagation, called the **action potential,** or **nerve impulse,** is capable of traveling long distances; it transmits information to other neurons, muscles, and glands.

By creating, analyzing, identifying, and integrating information, the nervous system generates two great classes of functions: stabilization of the intrinsic conditions (eg, blood pressure, O_2 and CO_2 content, pH, blood glucose levels, and hormone levels) of the organism within normal ranges; and behavioral patterns (eg, feeding, reproduction, defense, interaction with other living creatures).

DEVELOPMENT OF NERVE TISSUE

Nerve tissues develop from embryonic ectoderm that is induced to differentiate by the underlying notochord. First, a neural plate forms; then the edges of the plate thicken, forming the neural groove. The edges of the groove grow toward each other and ul-

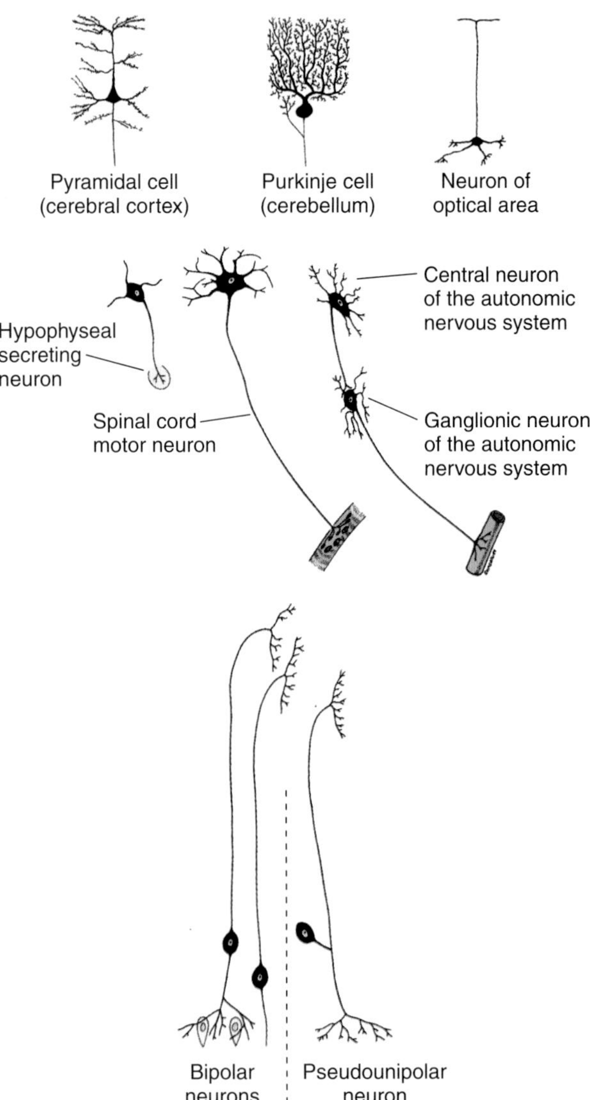

Figure 9–3. Diagrams of several types of neurons. The morphologic characteristics of neurons are very complex. All neurons shown here, except for the bipolar and pseudounipolar neurons, which are not very numerous in nerve tissue, are of the common multipolar variety.

timately fuse, forming the neural tube. This structure gives rise to the entire central nervous system, including neurons, glial cells, ependymal cells, and the epithelial cells of the choroid plexus.

Cells lateral to the neural groove form the **neural crest.** These cells undergo extensive migrations and contribute to the formation of the peripheral nervous system, as well as a number of other structures. Neural crest derivatives include (1) chromaffin cells of the adrenal medulla (see Chapter 21); (2) melanocytes of skin and subcutaneous tissues (see Chapter 18); (3) odontoblasts (see Chapter 15); (4) cells of the pia mater and the arachnoid; (5) sensory neurons of cranial and spinal sensory ganglia; (6) postganglionic neurons of sympathetic and parasympathetic ganglia; (7) Schwann cells of peripheral axons; and (8) satellite cells of peripheral ganglia.

NEURONS

Nerve cells, or neurons, are responsible for the reception, transmission, and processing of stimuli; the triggering of certain cell activities; and the release of neurotransmitters and other informational molecules.

Most neurons consist of 3 parts (Figure 9–2): the **dendrites,** which are multiple elongated processes specialized in receiving stimuli from the environment, sensory epithelial cells, or other neurons; the **cell body,** or **perikaryon** (Gr. *peri,* around, + *karyon,* nucleus), which is the trophic center for the whole nerve cell and is also receptive to stimuli; and the **axon** (from Greek, meaning axis), which is a single process specialized in generating or conducting nerve impulses to other cells (nerve, muscle, and gland cells). Axons may also receive information from other neurons; this information mainly modifies the transmission of action potentials to other neurons. The distal portion of the axon is usually branched and constitutes the **terminal arborization.** Each branch of this arborization terminates on the next cell in dilatations called **end bulbs (boutons)**, which interact with other neurons or non-nerve cells, forming structures called **synapses.** Synapses transmit information to the next cell in the circuit.

Neurons and their processes are extremely variable in size and shape (Figure 9–3). Cell bodies can be spherical, ovoid, or angular; some are very large, measuring up to 150 μm in diameter—large enough to be visible to the naked eye. Other nerve cells are among the smallest cells in the body; for example, the cell bodies of granule cells of the cerebellum are only 4–5 μm in diameter.

According to the size and shape of their processes, most neurons can be placed in one of the following categories (Figures 9–3 and 9–4): **multipolar neurons,** which have more than two cell processes, one process being the axon and the others dendrites; **bipolar neurons,** with one dendrite and one axon; and **pseudounipolar neurons,** which have a single process that is close to the perikaryon and divides into two branches. The process then forms a T shape, with one branch extending to a peripheral ending and the other toward the central nervous system (Figure 9–4). In pseudounipolar neurons, stimuli that are picked up by the dendrites travel directly to the axon terminal without passing through the perikaryon.

During the maturation process of pseudounipolar neurons, the central (axon) and the peripheral (dendrite) fibers fuse, becoming one single fiber. In these neurons, the cell body does not seem to be involved in the conduction of impulses, although it does synthesize many molecules, including neurotransmitters that migrate to the peripheral fibers.

Most neurons of the body are multipolar. Bipolar neurons are found in the cochlear and vestibular ganglia as well as in the retina and the olfactory mucosa. Pseudounipolar neurons are found in the spinal ganglia (the sensory ganglia located in the dorsal roots of the spinal nerves). They are also found in most cranial ganglia.

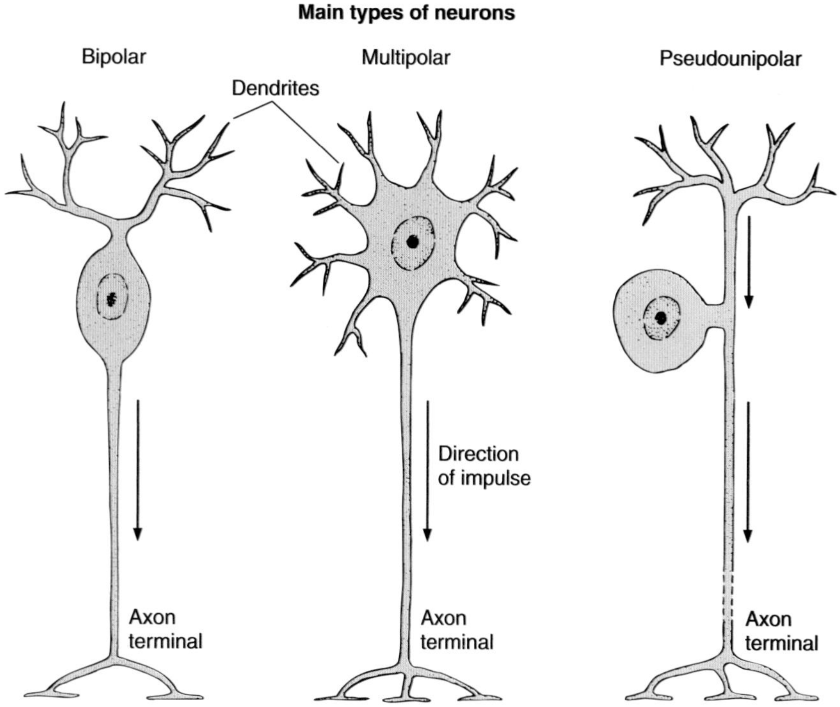

Main types of neurons
Bipolar
Multipolar
Pseudounipolar
Dendrites
Direction of impulse
Axon terminal
Axon terminal
Axon terminal

Figure 9–4. Simplified view of the 3 main types of neurons, according to their morphologic characteristics.

Neurons can also be classified according to their functional roles. **Motor (efferent) neurons** control effector organs such as muscle fibers and exocrine and endocrine glands. **Sensory (afferent) neurons** are involved in the reception of sensory stimuli from the environment and from within the body. **Interneurons** establish relationships among other neurons, forming complex functional networks or circuits (as in the retina).

During mammalian evolution there has been a great increase in the number and complexity of interneurons. Highly developed functions of the nervous system cannot be ascribed to simple neuron circuits; rather, they depend on complex interactions established by the integrated functions of many neurons.

In the central nervous system, nerve cell bodies are present only in the gray matter. White matter contains neuronal processes but no nerve cell bodies. In the peripheral nervous system, cell bodies are found in ganglia and in some sensory regions (eg, olfactory mucosa).

CELL BODY

The cell body, also called **perikaryon,** is the part of the neuron that contains the nucleus and surrounding cytoplasm, exclusive of the cell processes (Figure 9–2). It is primarily a trophic center, although it also has receptive capabilities. The perikaryon of

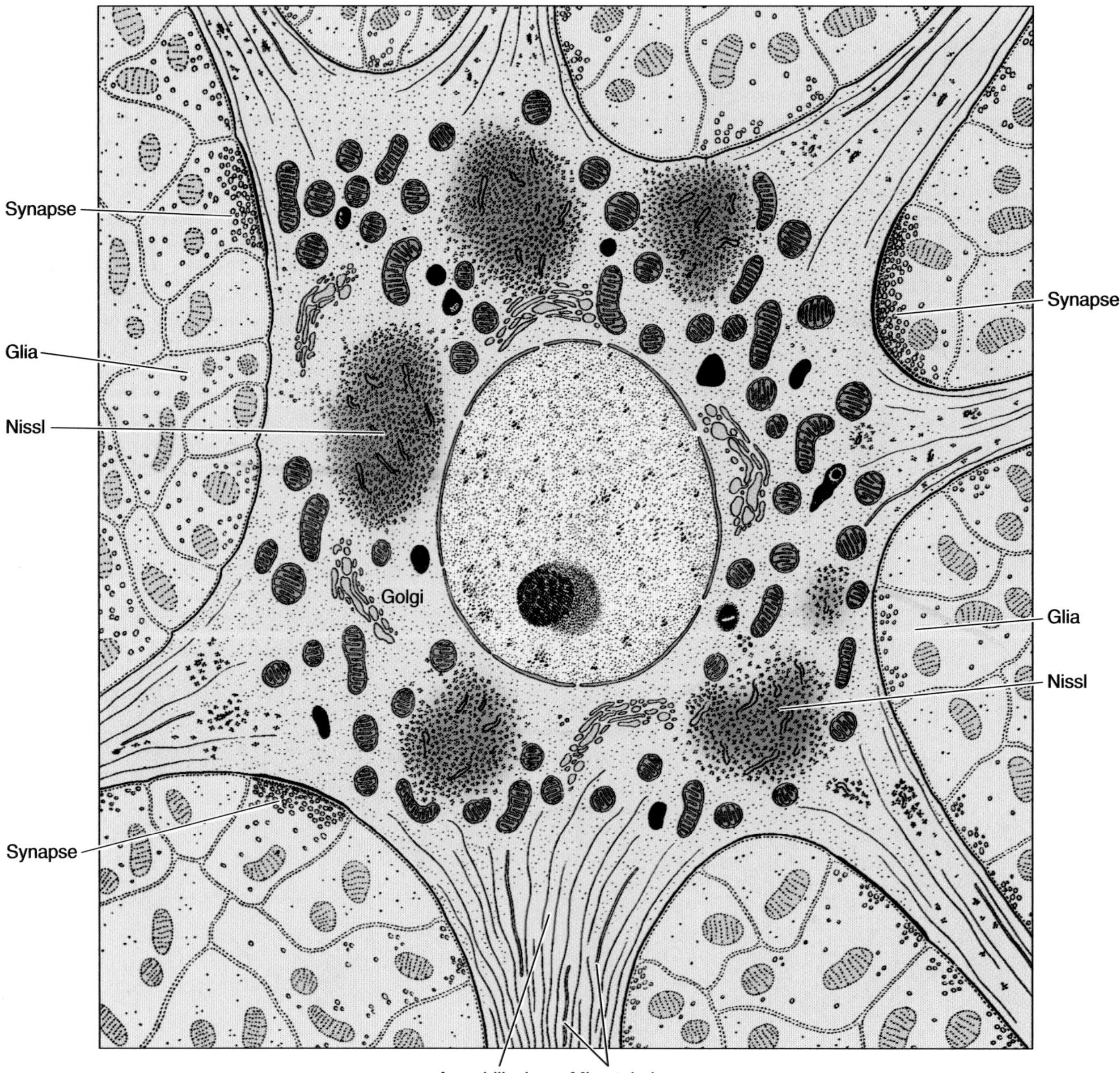

Figure 9–5. Ultrastructure of a neuron. The neuronal surface is completely covered either by synaptic endings of other neurons or by processes of glial cells. At synapses, the neuronal membrane is thicker and is called the postsynaptic membrane. The neuronal process devoid of ribosomes (lower part of figure) is the axon hillock. The other processes of this cell are dendrites.

most neurons receives a great number of nerve endings that convey excitatory or inhibitory stimuli generated in other nerve cells.

Most nerve cells have a spherical, unusually large, euchromatic (pale-staining) nucleus with a prominent nucleolus. Binuclear nerve cells are seen in sympathetic and sensory ganglia. The chromatin is finely dispersed, reflecting the intense synthetic activity of these cells.

The cell body (Figure 9–5) contains a highly developed rough endoplasmic reticulum organized into aggregates of parallel cisternae. In the cytoplasm between the cisternae are numerous polyribosomes, suggesting that these cells synthesize both structural proteins and proteins for transport. When appropriate stains are used, rough endoplasmic reticulum and free ribosomes appear under the light microscope as basophilic granular areas called **Nissl bodies** (Figures 9–2 and 9–6). The number of Nissl bodies varies according to neuronal type and functional state. They are particularly abundant in large nerve cells such as motor neurons (Figure 9–6). The **Golgi complex** is located only in the cell body and consists of multiple parallel arrays of smooth cisternae arranged around the periphery of the nucleus (Figure 9–5). Mitochondria are especially abundant in the axon terminals. They are scattered throughout the cytoplasm of the cell body.

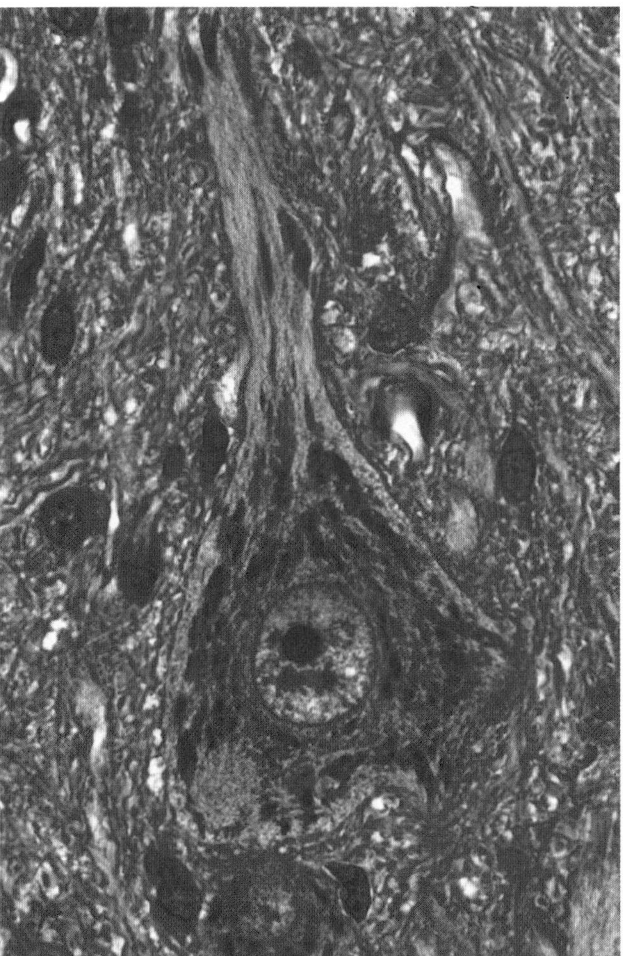

Figure 9–6. Photomicrograph of a motor neuron, a very large cell, from the spinal cord. The cytoplasm contains a great number of Nissl bodies. The large cell process is a dendrite. Note the large, round, stained nucleus, with a central dark-stained nucleolus. Pararosaniline–toluidine blue (PT) stain. Medium magnification.

Neurofilaments (intermediate filaments with a diameter of 10 nm) are abundant in perikaryons and cell processes. Neurofilaments bundle together as a result of the action of certain fixatives. When impregnated with silver, they form **neurofibrils** that are visible with the light microscope. The neurons also contain microtubules that are identical to those found in many other cells. Nerve cells occasionally contain inclusions of pigments, such as **lipofuscin,** which is a residue of undigested material by lysosomes.

DENDRITES

Dendrites (Gr. *dendron,* tree) are usually short and divide like the branches of a tree (Figure 9–4). They receive many synapses and are the principal signal reception and processing sites on neurons. Most nerve cells have numerous dendrites, which considerably increase the receptive area of the cell. The arborization of dendrites makes it possible for one neuron to receive and integrate a great number of axon terminals from other nerve cells. It has been estimated that up to 200,000 axonal terminations establish functional contact with the dendrites of a Purkinje cell of the cerebellum (Figure 9–3). That number may be even higher in other nerve cells. Bipolar neurons, with only one dendrite, are uncommon and are found only in special sites. Unlike axons, which maintain a constant diameter from one end to the other, dendrites become thinner as they subdivide into branches. The cytoplasmic composition of the dendrite base, close to the neuron body, is similar to that of the perikaryon but is devoid of Golgi complexes. Most synapses impinging on neurons are located in **dendrite spines,** which are usually mushroom-shaped structures (an expanded head connected to the dendrite shaft by a narrower neck) measuring 1 to 3 μm long and less than 1 μm in diameter. These spines play relevant functions and occur in vast numbers, estimated to be on the order of 10^{14} for the human cerebral cortex. Dendrite spines are the first processing locale for synaptic signals arriving on a neuron. The processing apparatus is contained in a complex of proteins attached to the cytosolic surface of the postsynaptic membrane, which is visible under the electron microscope and received the name postsynaptic membrane long before its function was disclosed. Dendritic spines participate in the plastic changes that underlie adaptation, learning, and memory. They are dynamic structures with a morphologic plasticity based on the cytoskeletal protein actin, which is related to the development of the synapses and their functional adaptation in adults.

AXONS

Most neurons have only one axon; a very few have no axon at all. An axon is a cylindrical process that varies in length and diameter according to the type of neuron. Although some neurons have short axons, axons are usually very long processes. For example, axons of the motor cells of the spinal cord that innervate the foot muscles may have a length of up to 100 cm (about 40 inches). All axons originate from a short pyramid-shaped region, the **axon hillock,** that usually arises from the perikaryon (Figure 9–5). The plasma membrane of the axon is called the **axolemma** (*axon* + Gr. *eilema,* sheath); its contents are known as **axoplasm.**

In neurons that give rise to a myelinated axon, the portion of the axon between the axon hillock and the point at which myelination begins is called the **initial segment.** This is the site where various excitatory and inhibitory stimuli impinging on the neuron are algebraically summed, resulting in the decision to propagate—or not to propagate—an action potential, or nerve impulse. It is known that several types of ion channels are localized in the initial segment and that these channels are important in generating the change in electrical potential that constitutes the action potential. In contrast to dendrites, axons have a constant diameter and do not branch profusely. Occasionally, the axon, shortly after its departure from the cell body, gives rise to a branch that returns to the area of the nerve cell body. All axon branches are known as **collateral branches** (Figure 9–2). Axonal cytoplasm (axoplasm) possesses mitochondria, microtubules, neurofilaments, and some cisternae of smooth endoplasmic reticulum. The absence of polyribosomes and rough endoplasmic reticulum emphasizes the dependence of the axon on the perikaryon for its maintenance. If an axon is severed, its peripheral parts degenerate and die.

There is a lively bidirectional transport of small and large molecules along the axon.

Macromolecules and organelles that are synthesized in the cell body are transported continuously by an **anterograde flow** along the axon to its terminals.

Anterograde flow occurs at 3 distinct speeds. A slow stream (a few millimeters per day) transports proteins and microfilaments. A flow of intermediate speed transports mitochondria, and a fast stream (100 times more rapid) transports the substances contained in vesicles that are needed at the axon terminal during neurotransmission.

Simultaneously with anterograde flow, a **retrograde flow** in the opposite direction transports several molecules, including material taken up by endocytosis (including viruses and toxins), to the cell body. This process is used to study the pathways of neurons; peroxidase or another marker is injected in regions with axon terminals, and its distribution is followed after a certain period of time.

Motor proteins related to axon flow include **dynein,** a protein with ATPase activity present in microtubules (related to retrograde flow); and **kinesin,** a microtubule-activated ATPase that, when attached to vesicles, promotes anterograde flow in the axon.

MEMBRANE POTENTIALS

The nerve cells have molecules in their membranes with the role of pumps and channels that transport ions into and out of the cytoplasm. The axolemma or limiting membrane of the axon pumps Na^+ out of the axoplasm, maintaining a concentration of Na^+ that is only a tenth of that in the extracellular fluid. In contrast, the concentration of K^+ is maintained at a level many times greater than that prevailing in the extracellular environment. Therefore, there is a potential difference across the axolemma of −65 mV with the inside negative to the outside. This is the **resting membrane potential.** When a neuron is stimulated, ion channels open and there is a sudden influx of extracellular Na^+ (an ion whose concentration is much higher in the extracellular fluid than in the cytoplasm) that changes the rest-

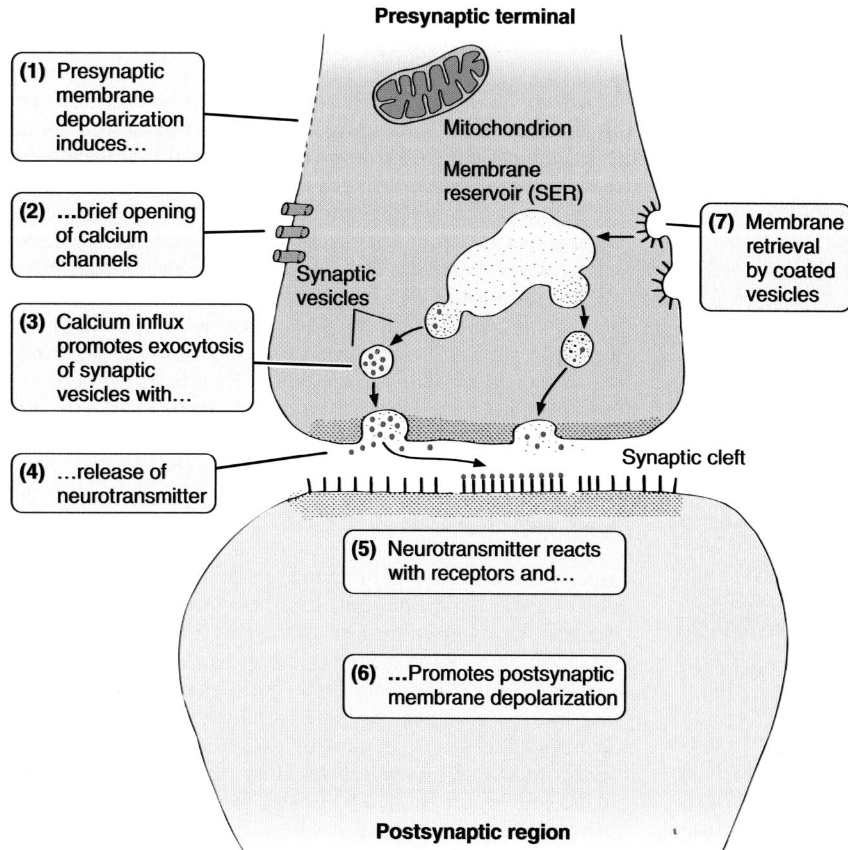

Figure 9–7. The main functional aspects of the two parts of the synapse: the presynaptic axon terminal and the postsynaptic region of the next neuron in the circuit. Numbers indicate the sequence of events during its activity. SER, smooth endoplasmic reticulum.

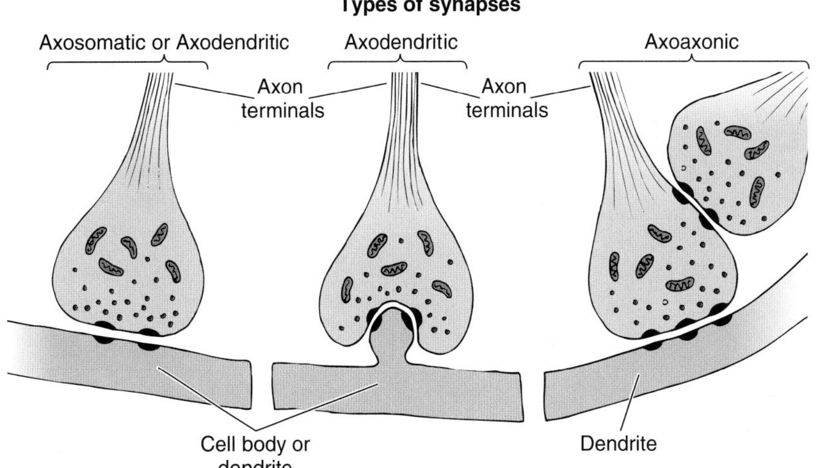

Figure 9–8. **Types of synapses. The axon terminals usually transmit the nerve impulse to a dendrite or to a nerve cell body; less frequently, they make a synapse with another axon.** (Redrawn, with permission, from Cormack DH: *Essential Histology.* Lippincott, 1993.)

ing potential from −65 mV to +30 mV. The cell interior becomes positive in relation to the extracellular environment that determines the beginning of the **action potential** or **nerve impulse.** However, the +30 mV potential closes the sodium channels, and the axonal membrane again becomes impermeable to this ion. In axons, in a few milliseconds, the opening of potassium channels modifies this ionic situation. As a result of the elevated intracellular concentration of potassium, this ion leaves the axon by diffusion, and the membrane potential returns to −65 mV, ending the action potential. The duration of these events is very short (about 5 ms) and takes place in a very small membrane area. However, the action potential propagates along the membrane; that is, the electrical disturbance opens neighboring sodium channels and, in sequence, potassium channels. In this way the action potential propagates at a high speed along the axon. When the action potential arrives at the nerve ending, it promotes discharge of stored neurotransmitter that stimulates or inhibits another neuron or a non-neural cell, such as a muscle or gland cell.

MEDICAL APPLICATION

Local anesthetics are hydrophobic molecules that bind to sodium channels, inhibiting sodium transport and, consequently, also the action potential responsible for the nerve impulse.

SYNAPTIC COMMUNICATION

The synapse (Gr. *synapsis,* union) is responsible for the unidirectional transmission of nerve impulses. Synapses are sites of functional contact between neurons or between neurons and other effector cells (eg, muscle and gland cells). The function of the synapse is to convert an electrical signal (impulse) from the presynaptic cell into a chemical signal that acts on the **postsynaptic** cell. Most synapses transmit information by releasing **neurotransmitters** during the signaling process. Neurotransmitters are chemicals that, when combined with a receptor protein, either open or close ion channels or initiate second-messenger cascades. **Neuromodulators** are chemical messengers that do not act directly on synapses but modify neuron sensitivity to synaptic stimulation or inhibition. Some neuromodulators are neuropeptides or steroids produced in the nerve tissue, others are circulating steroids. The synapse itself is formed by an axon terminal (**presynaptic terminal**) that delivers the signal; a region on the surface of another cell where a new signal is generated (**postsynaptic terminal**); and a thin intercellular space called the **synaptic cleft** (Figure 9–7). If an axon forms a synapse with a cell body, it is called an **axosomatic synapse;** with a dendrite, **axodendritic;** or with an axon, **axoaxonic** (Figure 9–8).

Although most synapses are **chemical synapses** and use chemical messengers, a few synapses transmit ionic signals through gap junctions that cross the pre- and postsynaptic membranes, thereby conducting neuronal signals directly. These synapses are called **electrical synapses.**

The presynaptic terminal always contains **synaptic vesicles** with neurotransmitters and numerous **mitochondria** (Figures 9–7 and 9–9).

Neurotransmitters are generally synthesized in the cell body; they are then stored in vesicles in the presynaptic region of a synapse. During transmission of a nerve impulse, they are released into the synaptic cleft by **exocytosis.** The extra membrane that collects at the presynaptic region as a result of exocytosis of the synaptic vesicles is recycled by **endocytosis.** Retrieved membrane fuses with the smooth endoplasmic reticulum of the presynaptic compartment to be reused in the formation of new synaptic vesicles (Figure 9–7). Some neurotransmitters are synthesized in the presynaptic compartment, using enzymes and precursors brought by axonal transport.

The first neurotransmitters to be described were acetylcholine and norepinephrine. A norepinephrine-releasing axon terminal is shown in Figure 9–10. Most neurotransmitters are amines, amino acids, or small peptides (neuropeptides). Inorganic substances such as nitric oxide have also been shown to act as neurotransmitters. Several peptides that act as neurotransmitters are used elsewhere in the body, eg, as hormones in the digestive tract. Neuropeptides are important in regulating feelings and drives, such as pain, pleasure, hunger, thirst, and sex (Figure 9–11).

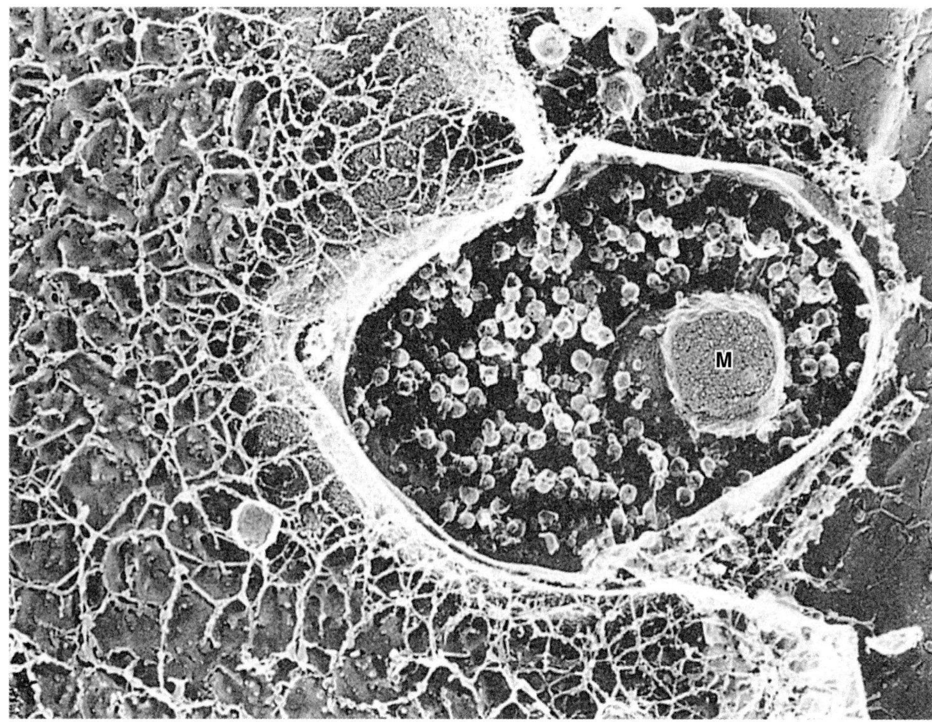

Figure 9–9. Electron micrograph of a rotary-replicated freeze-etched synapse. Synaptic vesicles surround a mitochondrion (M) in the axon terminal. ×25,000. (Reproduced, with permission, from Heuser JE, Salpeter SR: Organization of acetylcholine receptors in quick-frozen, deep-etched and rotary-replicated Torpedo postsynaptic membrane. J Cell Biol 1979;82:150.)

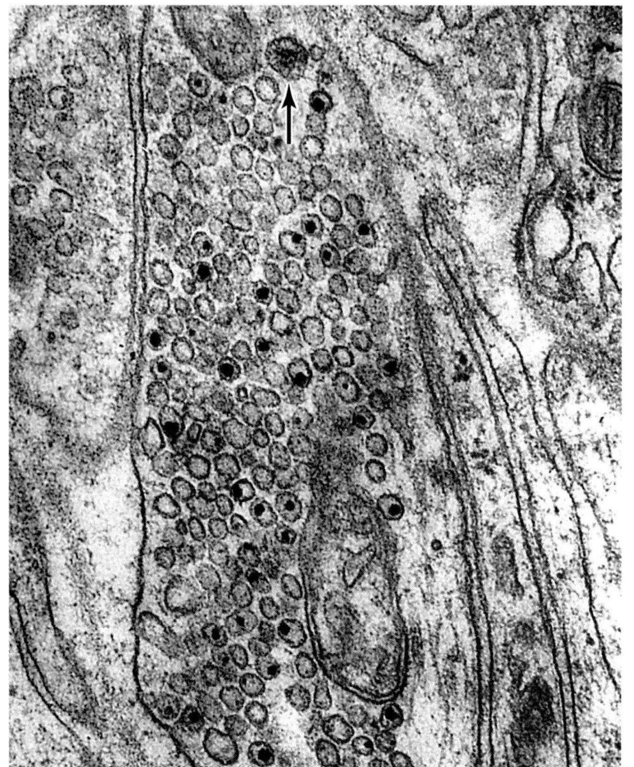

Figure 9–10. Adrenergic nerve ending. There are many 50-nm-diameter vesicles (arrow) with dark, electron-dense cores containing norepinephrine. ×40,000. (Courtesy of A Machado.)

Sequence of Events During Chemical Synapse Transmission

The events that take place during chemical synapse transmission are illustrated in Figure 9–7. Nerve impulses that sweep rapidly (in milliseconds) along the cell membrane promote an explosive electrical activity (depolarization) that is propagated along the cell membrane. This impulse briefly opens calcium channels in the presynaptic region, promoting a calcium influx that triggers the exocytosis of synaptic vesicles. The neurotransmitters released at the sites of exocytosis react with receptors present at the postsynaptic region, promoting a transient electrical activity (depolarization) at the postsynaptic membrane. These synapses are called **excitatory,** because their activity promotes impulses in the postsynaptic cell membrane. In some synapses the neurotransmitter-receptor interaction has an opposite effect, promoting **hyperpolarization** with no transmission of the nerve impulse. These are called **inhibitory** synapses. Thus, synapses can excite or inhibit impulse transmission and thereby regulate nerve activity (Figure 9–12).

Once used, neurotransmitters are removed quickly by enzymatic breakdown, diffusion, or endocytosis mediated by specific receptors on the presynaptic membrane. This removal of neurotransmitters is functionally important because it prevents an undesirable sustained stimulation of the postsynaptic neuron.

GLIAL CELLS & NEURONAL ACTIVITY

Glial cells are 10 times more abundant in the mammalian brain than neurons; they surround both cell bodies and their axonal and dendrite processes that occupy the interneuronal spaces.

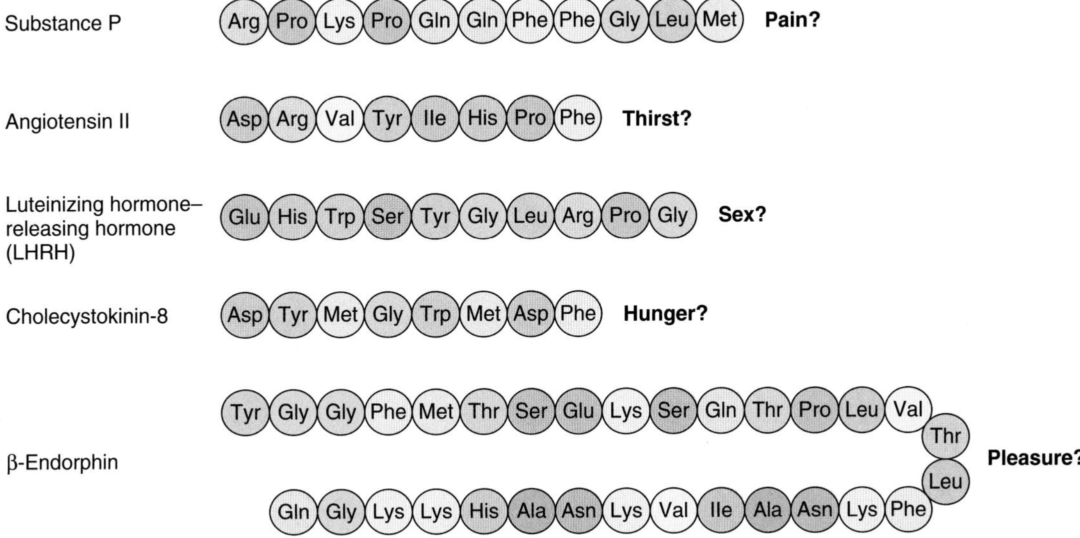

Figure 9–11. Amino acid sequence of some neuropeptides and the sensations and drives in which they probably participate. (Reproduced, with permission, from Alberts B et al: *Molecular Biology of the Cell,* 2nd ed. Garland Press, 1993.)

Nerve tissue has only a very small amount of extracellular matrix, and glial cells (Table 9–1) furnish a microenvironment suitable for neuronal activity.

Oligodendrocytes

Oligodendrocytes (Gr. *oligos,* small, + *dendron* + *kytos,* cell) produce the myelin sheath that provides the electrical insulation of neurons in the central nervous system (Figures 9–13 and 9–14). These cells have processes that wrap around axons, producing a myelin sheath as shown in Figure 9–15.

Schwann Cells

Schwann cells have the same function as oligodendrocytes but are located around axons in the peripheral nervous system. One Schwann cell forms myelin around a segment of one axon, in contrast to the ability of oligodendrocytes to branch and serve more than one neuron and its processes. Figure 9–27 shows how the Schwann cell membrane wraps around the axon.

Astrocytes

Astrocytes (Gr. *astron,* star, + *kytos*) are star-shaped cells with multiple radiating processes. These cells have bundles of intermediate filaments made of **glial fibrillary acid protein** that reinforce their structure. Astrocytes bind neurons to capillaries and to the pia mater (a thin connective tissue that covers the central nervous system). Astrocytes with few long processes are called **fibrous astrocytes** and are located in the white matter; **protoplasmic astrocytes,** with many short-branched processes, are found in the gray matter (Figures 9–13, 9–14, and 9–16). Astrocytes are by far the most numerous glial cells and exhibit an exceptional morphologic and functional diversity.

In addition to their supporting function, astrocytes participate in controlling the ionic and chemical environment of neurons. Some astrocytes develop processes with expanded **end-feet** that are linked to endothelial cells. It is believed that through the end-feet, astrocytes transfer molecules and ions from the blood to the neurons. Expanded processes are also present at the

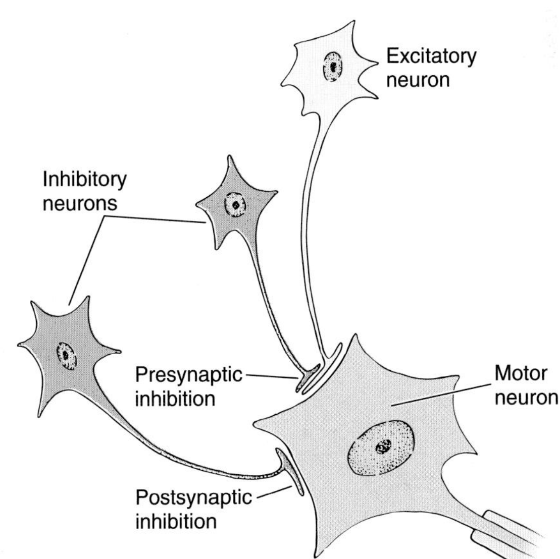

Figure 9–12. Examples of excitatory and inhibitory synapses in a motor neuron. (Redrawn, with permission, from Ganong WF: *Review of Medical Physiology,* 15th ed. Originally published by Appleton & Lange, 1991. Copyright © 2001 by the McGraw-Hill Companies, Inc.)

Table 9–1. Origin and principal functions of neuroglial cells.

Glial Cell Type	Origin	Location	Main Functions
Oligodendrocyte	Neural tube	Central nervous system	Myelin production, electric insulation
Schwann cell	Neural tube	Peripheral nerves	Myelin production, electric insulation
Astrocyte	Neural tube	Central nervous system	Structural support, repair processes Blood-brain barrier, metabolic exchanges
Ependymal cell	Neural tube	Central nervous system	Lining cavities of central nervous system
Microglia	Bone marrow	Central nervous system	Macrophagic activity

external surface of the central nervous system, where they make a continuous layer. Furthermore, when the central nervous system is damaged, astrocytes proliferate to form cellular scar tissue.

Astrocytes also play a role in regulating the numerous functions of the central nervous system. Astrocytes in vitro exhibit adrenergic receptors, amino acid receptors (eg, γ-aminobutyric acid [GABA]), and peptide receptors (including natriuretic peptide, angiotensin II, endothelins, vasoactive intestinal peptide, and thyrotropin-releasing hormone). The presence of these and other receptors on astrocytes provides them with the ability to respond to several stimuli.

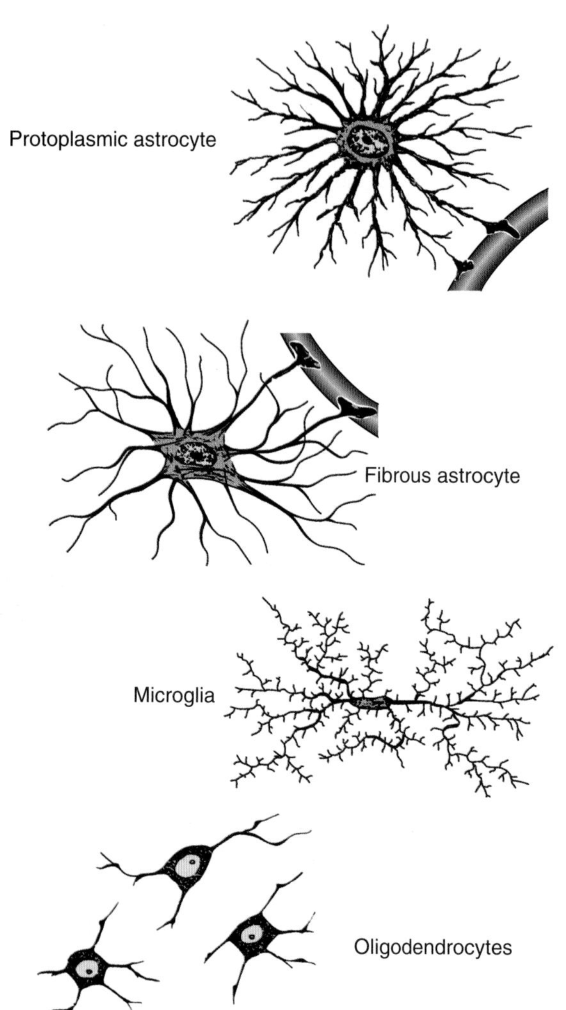

Figure 9–13. Drawings of neuroglial cells as seen in slides stained by metallic impregnation. Note that only astrocytes exhibit vascular end-feet, which cover the walls of blood capillaries.

Astrocytes can influence neuronal survival and activity through their ability to regulate constituents of the extracellular environment, absorb local excess of neurotransmitters, and release metabolic and neuroactive molecules. The latter molecules include peptides of the angiotensinogen family, vasoactive endothelins, opioid precursors called **enkephalins,** and the potentially neurotrophic somatostatin. On the other hand, there is some evidence that astrocytes transport energy-rich compounds from the blood to the neurons and also metabolize glucose to lactate, which is then supplied to the neurons.

Finally, astrocytes are in direct communication with one another via gap junctions, forming a network through which information can flow from one point to another, reaching distant sites. For example, by means of gap junctions and the release of various cytokines, astrocytes can interact with oligodendrocytes to influence myelin turnover in both normal and abnormal conditions.

Ependymal Cells

Ependymal cells are low columnar epithelial cells lining the ventricles of the brain and central canal of the spinal cord. In some locations, ependymal cells are ciliated, which facilitates the movement of cerebrospinal fluid.

Microglia

Microglia (Gr. *micros,* small, + *glia*) are small elongated cells with short irregular processes (Figures 9–13 and 9–14). They can be recognized in routine hematoxylin-and-eosin (H&E) preparations by their dense elongated nuclei, which contrast with the spherical nuclei of other glial cells. Microglia, phagocytic cells that represent the mononuclear phagocytic system in nerve tissue, are derived from precursor cells in the bone marrow. They are involved with inflammation and repair in the adult central nervous system, and they produce and release neutral proteases and oxidative radicals. When activated, microglia retract their processes and assume the morphologic characteristics of macrophages, becoming phagocytic and acting as antigen-presenting cells (see Chapter 14). Microglia secrete a number of immunoregulatory cytokines and dispose of unwanted cellular debris caused by central nervous system lesions.

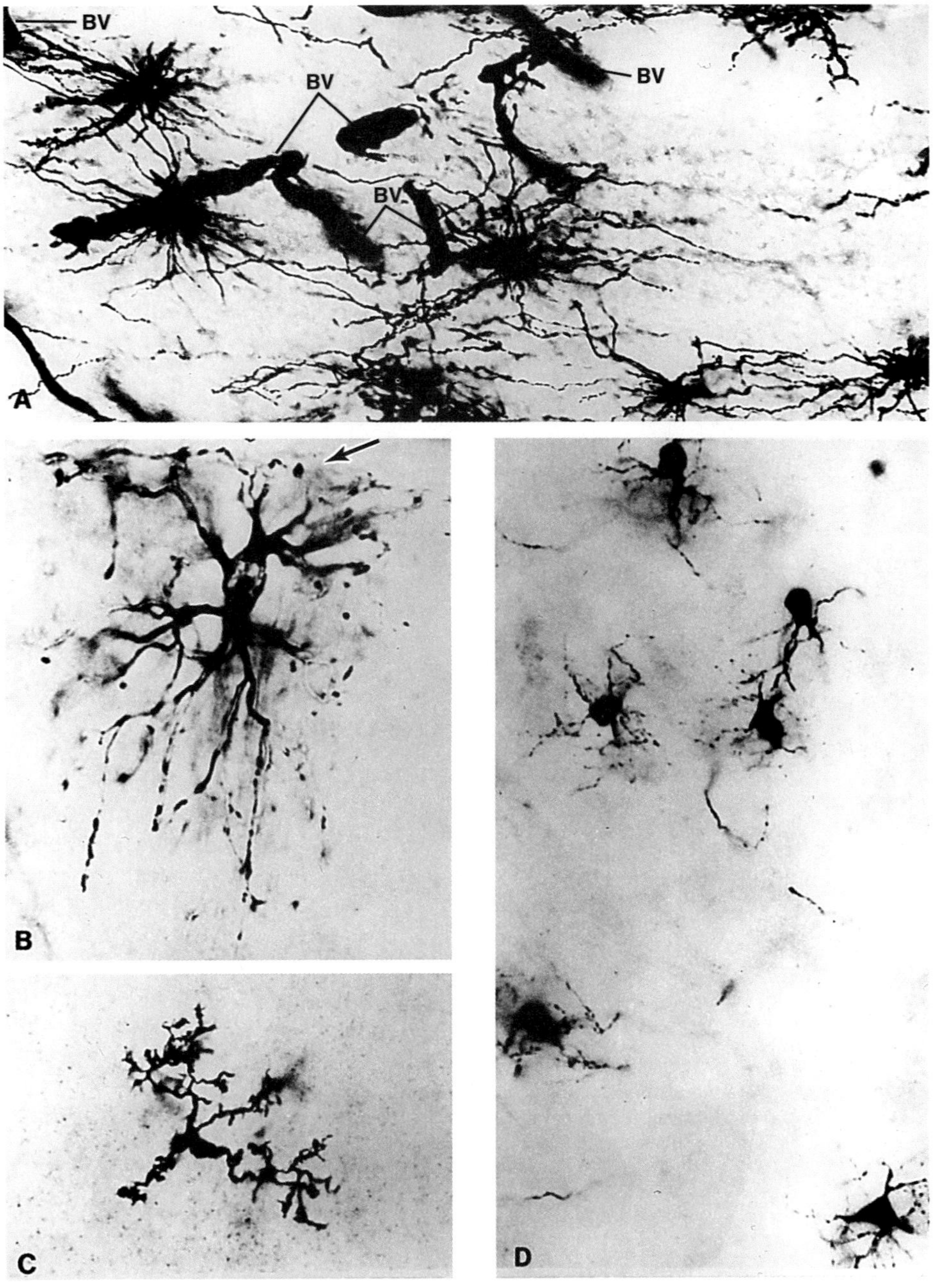

Figure 9–14. Photomicrographs (prepared with Golgi stain) of glial cells from the cerebral cortex. **A:** Fibrous astrocytes, showing blood vessels (BV). ×1000. **B:** Protoplasmic astrocyte showing brain surface (arrow). ×1900. **C:** Microglial cell. ×1700. **D:** Oligodendrocytes. ×1900. (Reproduced, with permission, from Jones E, Cowan WM: The nervous tissue. In: *Histology: Cell and Tissue Biology,* 5th ed. Weiss L [editor]. Elsevier, 1983.)

MEDICAL APPLICATION

In multiple sclerosis, the myelin sheath is destroyed by an unknown mechanism with severe neurologic consequences. In this disease, microglia phagocytose and degrade myelin debris by receptor-mediated phagocytosis and lysosomal activity. In addition, AIDS dementia complex is caused by HIV-1 infection of the central nervous system. Overwhelming experimental evidence indicates that microglia are infected by HIV-1. A number of cytokines, such as interleukin-1 and tumor necrosis factor-α, activate and enhance HIV replication in microglia.

THE CENTRAL NERVOUS SYSTEM

The central nervous system consists of the **cerebrum, cerebellum,** and **spinal cord.** It has virtually no connective tissue and is therefore a relatively soft, gel-like organ.

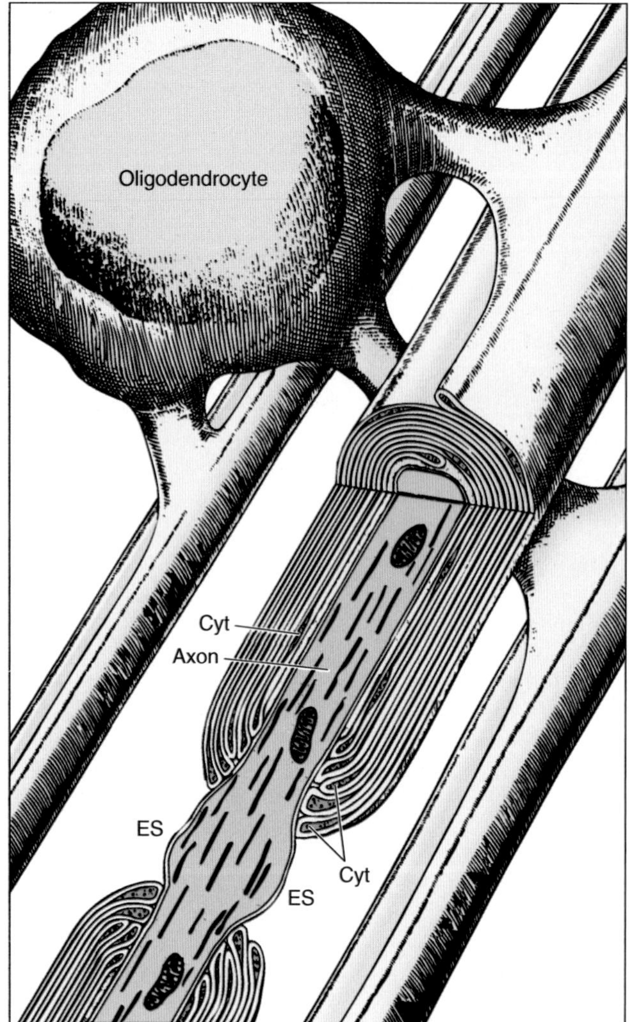

Figure 9–15. Myelin sheath of the central nervous system. The same oligodendrocyte forms myelin sheaths for several (3–50) nerve fibers. In the central nervous system, processes of other cells sometimes cover the nodes of Ranvier, or there is considerable extracellular space (ES) at that point. The axolemma shows a thickening where the cell membrane of the oligodendrocyte comes into contact with it. This limits the diffusion of materials into the periaxonal space between the axon and the myelin sheath. At upper left is a surface view of the cell body of an oligodendrocyte. Cyt, cytoplasm of the oligodendrocyte. (Redrawn and reproduced, with permission, from Bunge et al: J Biophys Biochem Cytol 1961;10:67.)

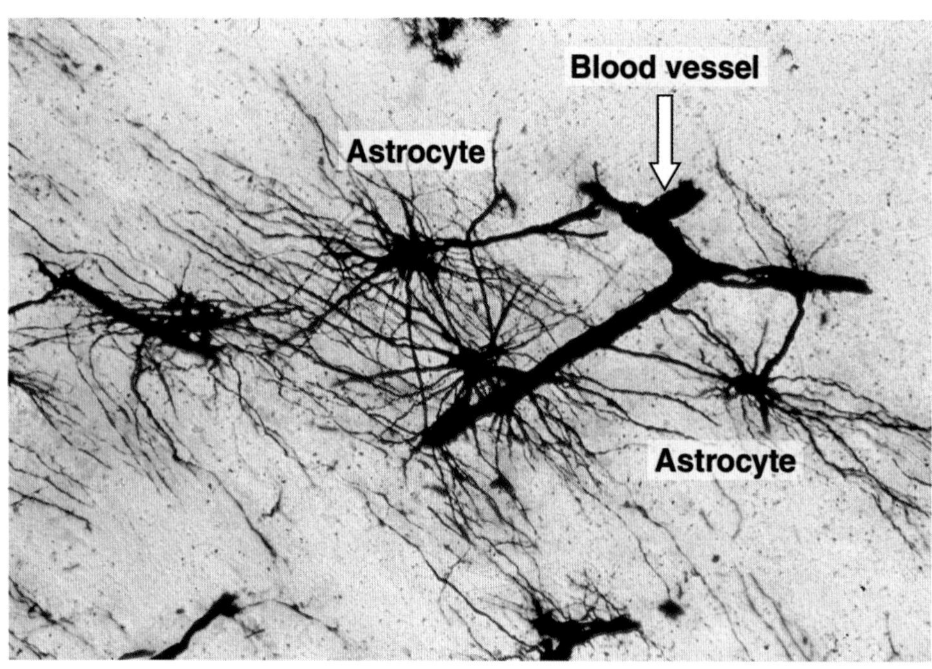

Figure 9–16. Brain section prepared with Rio Hortega silver stain showing fibrous astrocytes with their processes ending on the external surface of blood vessels. Medium magnification.

When sectioned, the cerebrum, cerebellum, and spinal cord show regions of white (**white matter**) and gray (**gray matter**). The differential distribution of myelin in the central nervous system is responsible for these differences: The main component of white matter is myelinated axons (Figure 9–17) and the myelin-producing oligodendrocytes. White matter does not contain neuronal cell bodies.

Gray matter contains neuronal cell bodies, dendrites, and the initial unmyelinated portions of axons and glial cells. This is the region where synapses occur. Gray matter is prevalent at the surface of the cerebrum and cerebellum, forming the **cerebral and cerebellar cortex** (Figures 9–18, 9–19, and 9–20), whereas white matter is present in more central regions. Aggregates of neuronal cell bodies forming islands of gray matter embedded in the white matter are called **nuclei.** In the **cerebral cortex,** the gray matter has 6 layers of cells with different forms and sizes. Neurons of some regions of the cerebral cortex register **afferent** (**sensory**) impulses; in other regions, **efferent** (**motor**) neurons generate motor impulses that control voluntary movements. Cells of the cerebral cortex are related to the integration of sensory information and the initiation of voluntary motor responses.

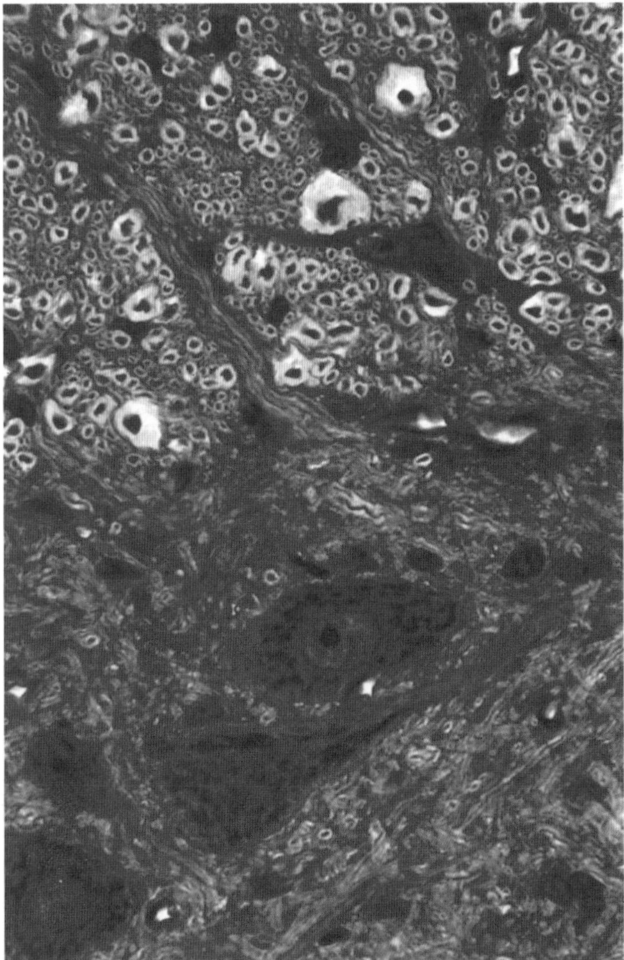

Figure 9–17. Cross section of the spinal cord in the transition between gray matter (below) and white matter (above). Note the neuronal bodies and abundant cell processes in the gray matter, whereas the white matter consists mainly of nerve fibers whose myelin sheath was dissolved by the histologic procedure. PT stain. Medium magnification.

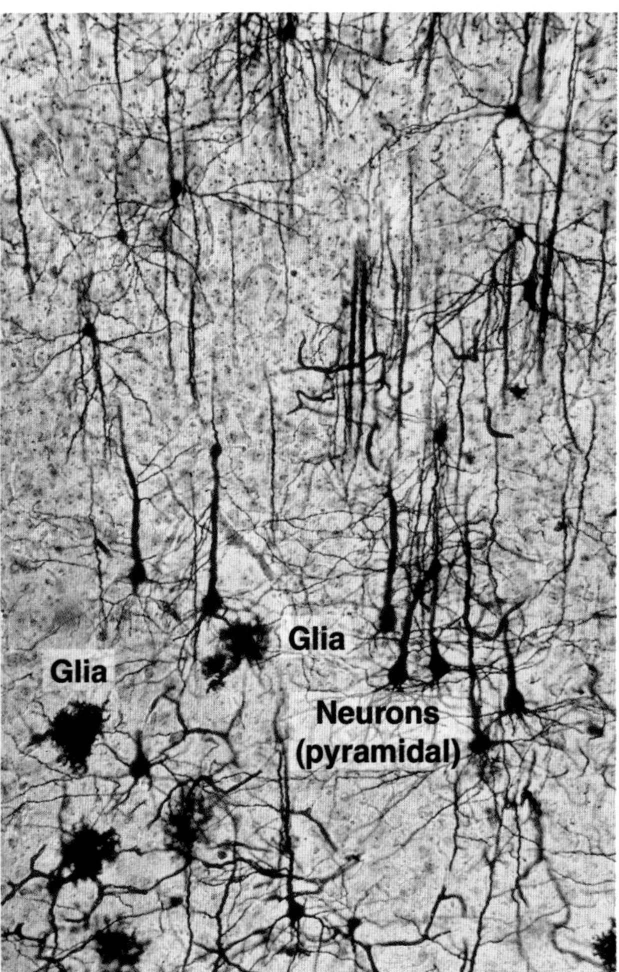

Figure 9–18. Silver-stained section of cerebral cortex showing many pyramid-shaped neurons with their processes and a few glial cells. Medium magnification.

The **cerebellar cortex** has 3 layers (Figures 9–19 and 9–20): an outer molecular layer, a central layer of large Purkinje cells, and an inner granule layer. The Purkinje cells have a conspicuous cell body and their dendrites are highly developed, assuming the aspect of a fan (Figure 9–3). These dendrites occupy most of the molecular layer and are the reason for the sparseness of nuclei. The granule layer is formed by very small neurons (the smallest in the body), which are compactly disposed, in contrast to the less cell-dense molecular layer (Figure 9–19).

In cross sections of the **spinal cord,** white matter is peripheral and gray matter is central, assuming the shape of an H (Figure 9–21). In the horizontal bar of this H is an opening, the **central canal,** which is a remnant of the lumen of the embryonic neural tube. Ependymal cells line it. The gray matter of the legs of the H forms the **anterior horns.** These contain motor neurons whose axons make up the ventral roots of the spinal nerves. Gray matter also forms the posterior horns (the arms of the H), which receive sensory fibers from neurons in the spinal ganglia (dorsal roots).

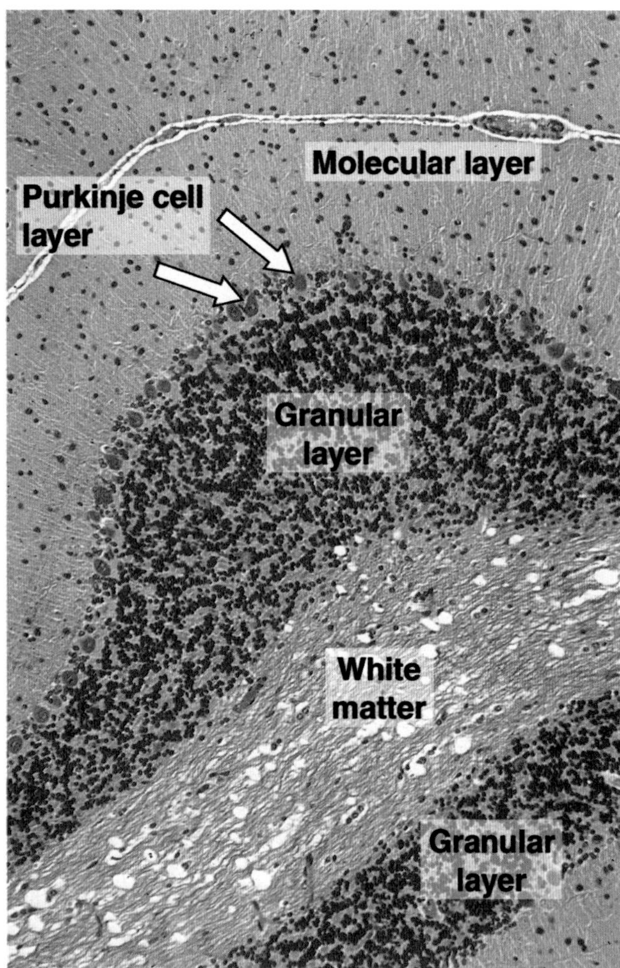

Figure 9–19. Photomicrograph of the cerebellum. The staining procedure used (H&E) does not reveal the unusually large dendritic arborization of the Purkinje cell, which is illustrated in Figure 9–3. Low magnification.

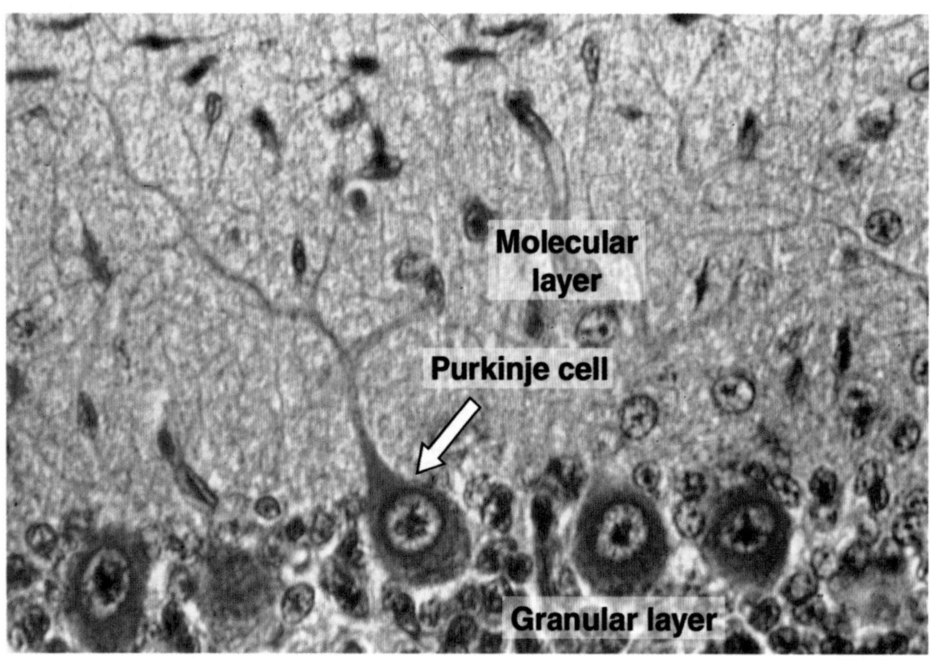

Figure 9–20. Section of the cerebellum with distinct Purkinje cells. One Purkinje cell shows part of its rich dendritic arborization. H&E stain. Medium magnification.

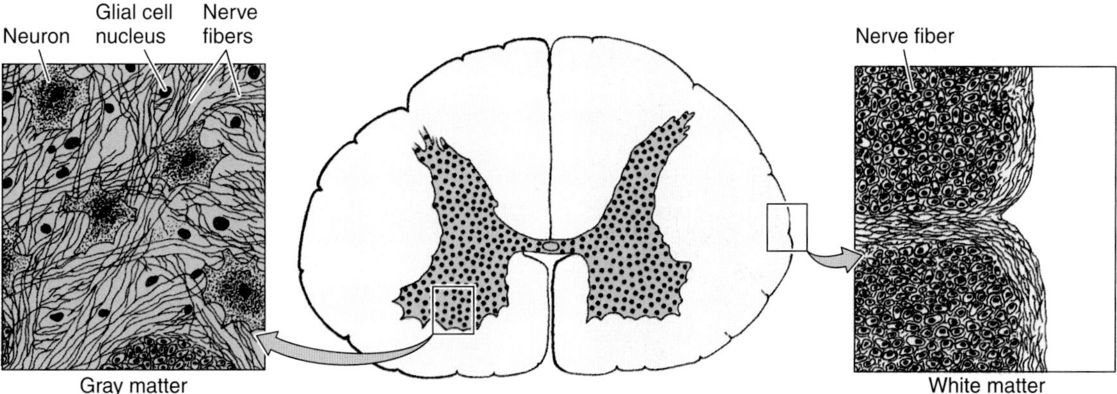

Figure 9–21. Cross section through the spinal cord.

Spinal cord neurons are large and multipolar, especially in the anterior horns, where large motor neurons are found (Figures 9–22 and 9–23).

MENINGES

The skull and the vertebral column protect the central nervous system. It is also encased in membranes of connective tissue called the **meninges** (Figure 9–24). Starting with the outermost layer, the meninges are the **dura mater, arachnoid,** and **pia mater.** The arachnoid and the pia mater are linked together and are often considered a single membrane called the **pia-arachnoid.**

Dura Mater

The dura mater is the external layer and is composed of dense connective tissue continuous with the periosteum of the skull. The dura mater that envelops the spinal cord is separated from the periosteum of the vertebrae by the epidural space, which contains thin-walled veins, loose connective tissue, and adipose tissue.

The dura mater is always separated from the arachnoid by the thin subdural space. The internal surface of all dura mater, as well as its external surface in the spinal cord, is covered by simple squamous epithelium of mesenchymal origin.

Arachnoid

The arachnoid (Gr. *arachnoeides,* cobweblike) has two components: a layer in contact with the dura mater and a system of trabeculae connecting the layer with the pia mater. The cavities between the trabeculae form the **subarachnoid space,** which is filled with cerebrospinal fluid and is completely separated from the **subdural space.** This space forms a hydraulic cushion that protects the central nervous system from trauma. The subarachnoid space communicates with the ventricles of the brain.

The arachnoid is composed of connective tissue devoid of blood vessels. The same type of simple squamous epithelium that covers the dura mater covers its surfaces. Since the arachnoid has fewer trabeculae in the spinal cord, it can be more clearly distinguished from the pia mater in that area.

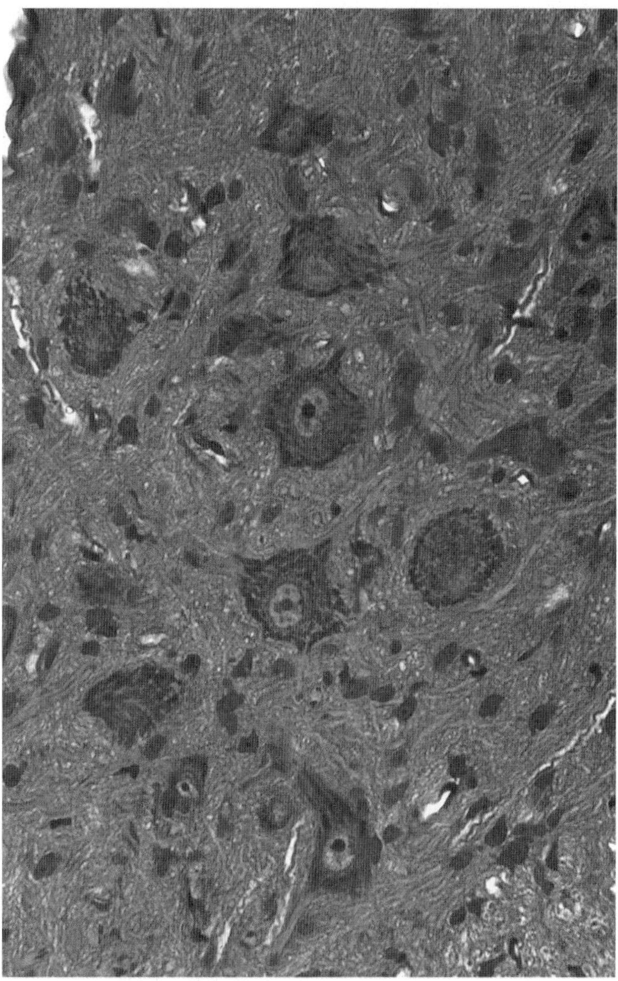

Figure 9–22. Section of the gray matter of the spinal cord showing several motor neurons with their basophilic bodies (Nissl bodies). Nucleoli are seen in some nuclei. The neurons are surrounded by a mesh of neuronal and glial processes. PT stain. Medium magnification.

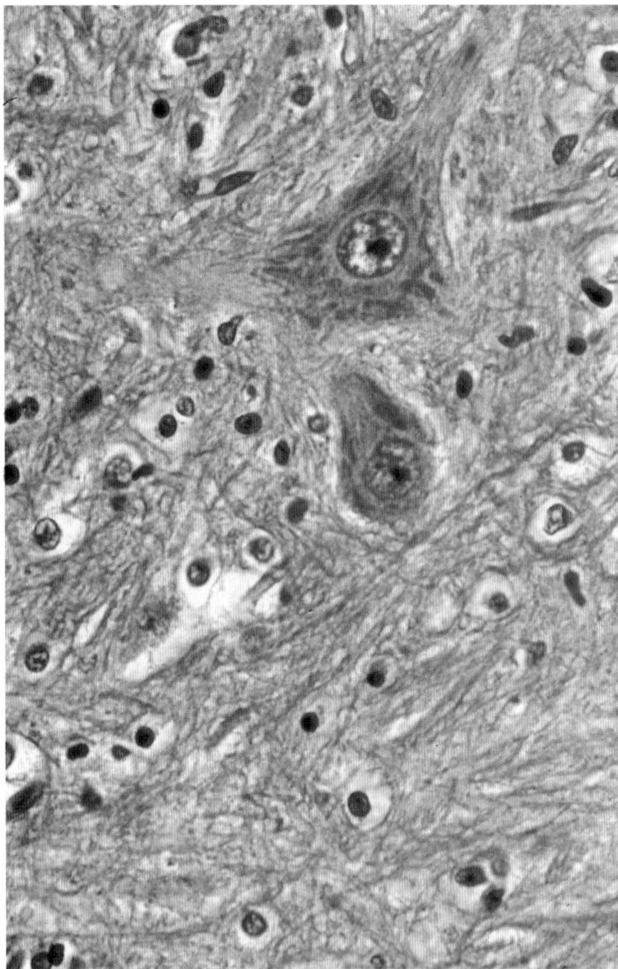

Figure 9–23. Section of spinal cord gray matter. The meshwork of cell neuron and glial processes appears distinctly. The small nuclei are from glia cells. Note that these cells are more numerous than neurons. H&E stain. Medium magnification.

In some areas, the arachnoid perforates the dura mater, forming protrusions that terminate in venous sinuses in the dura mater. These protrusions, which are covered by endothelial cells of the veins, are called **arachnoid villi.** Their function is to reabsorb cerebrospinal fluid into the blood of the venous sinuses.

Pia Mater

The pia mater is a loose connective tissue containing many blood vessels. Although it is located quite close to the nerve tissue, it is not in contact with nerve cells or fibers. Between the pia mater and the neural elements is a thin layer of neuroglial processes, adhering firmly to the pia mater and forming a physical barrier at the periphery of the central nervous system. This barrier separates the central nervous system from the cerebrospinal fluid (Figure 9–24).

The pia mater follows all the irregularities of the surface of the central nervous system and penetrates it to some extent along with the blood vessels. Squamous cells of mesenchymal origin cover pia mater.

Blood vessels penetrate the central nervous system through tunnels covered by pia mater—the **perivascular spaces.** The pia mater disappears before the blood vessels are transformed into capillaries. In the central nervous system, the blood capillaries are completely covered by expansions of the neuroglial cell processes (Figure 9–24).

Blood-Brain Barrier

The blood-brain barrier is a functional barrier that prevents the passage of some substances, such as antibiotics and chemical and bacterial toxic matter, from the blood to nerve tissue.

The blood-brain barrier results from the reduced permeability that is a property of blood capillaries of nerve tissue. Occluding junctions, which provide continuity between the endothelial cells of these capillaries, represent the main structural component of the barrier. The cytoplasm of these endothelial cells does not have the fenestrations found in many other locations, and very few pinocytotic vesicles are observed. The expansions of neuroglial cell processes that envelop the capillaries are partly responsible for their low permeability.

CHOROID PLEXUS & CEREBROSPINAL FLUID

The choroid plexus consists of invaginated folds of pia mater, rich in dilated fenestrated capillaries, that penetrate the interior of the brain ventricles. It is found in the roofs of the third and fourth ventricles and in part in the walls of the lateral ventricles.

The choroid plexus is composed of loose connective tissue of the pia mater, covered by a simple cuboidal or low columnar epithelium (Figure 9–25) made of ion-transporting cells (see Chapter 4).

The main function of the choroid plexus is to elaborate cerebrospinal fluid, which contains only a small amount of solids and completely fills the ventricles, central canal of the spinal cord, subarachnoid space, and perivascular space. Cerebrospinal fluid is important for the metabolism of the central nervous system and acts as a protective device against mechanical shocks.

Cerebrospinal fluid is clear, has a low density (1.004–1.008 g/mL), and is very low in protein content. A few desquamated cells and 2 to 5 lymphocytes per milliliter are also present. Cerebrospinal fluid is continuously produced and circulates through the ventricles, from which it passes into the subarachnoid space. There, arachnoid villi provide the main pathway for absorption of cerebrospinal fluid into the venous circulation. (There are no lymphatic vessels in brain nerve tissue.)

MEDICAL APPLICATION

A decrease in the absorption of cerebrospinal fluid or a blockage of outflow from the ventricles results in the condition known as ***hydrocephalus*** *(Gr.* hydro, *water,* + kephale, *head), which promotes a progressive enlargement of the head followed by mental impairment and muscular weakness.*

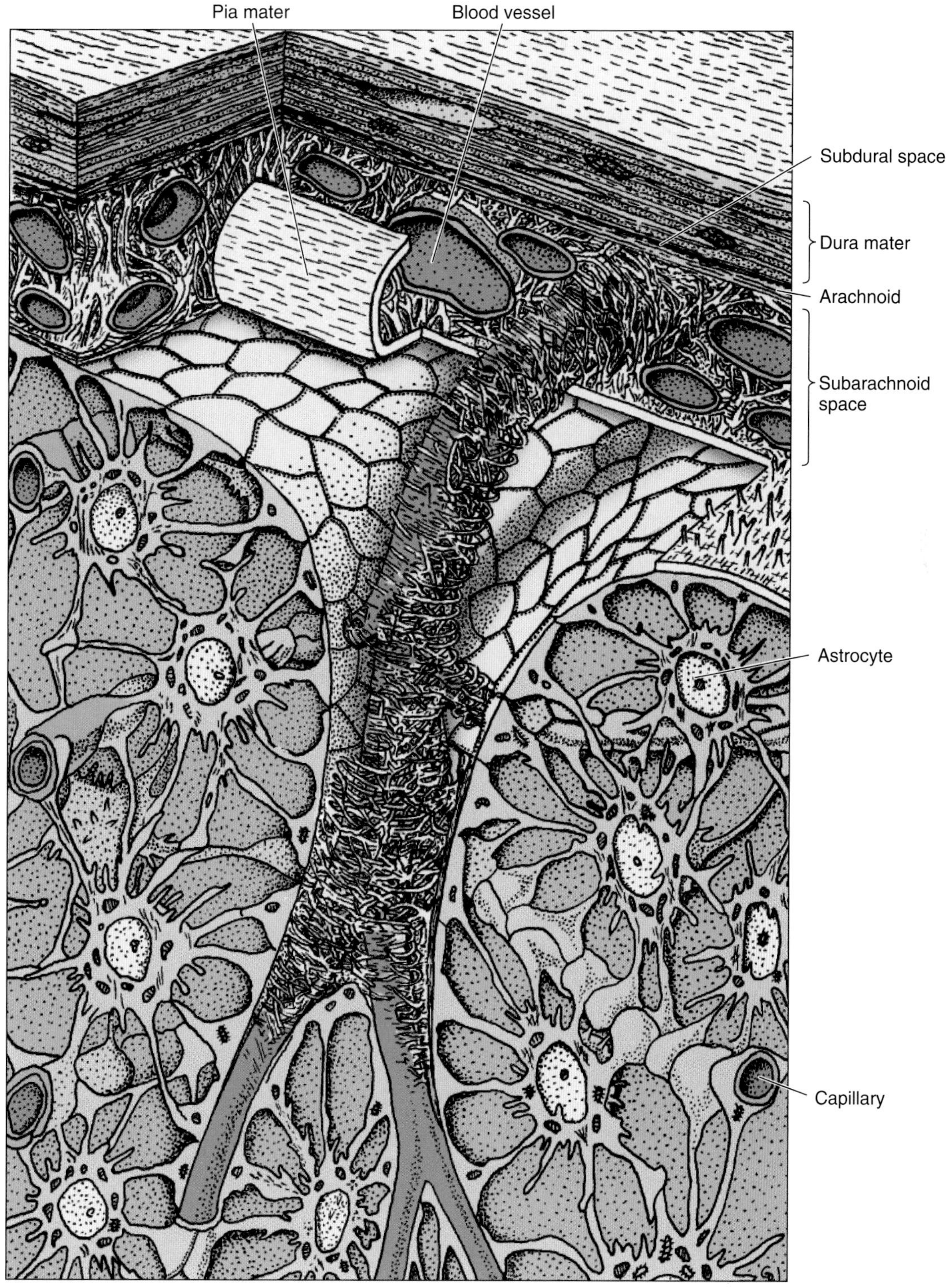

Figure 9–24. The structure of the meninges, with the superposition of pia mater, arachnoid, and dura mater. Astrocytes form a 3-dimensional net around the neurons (not shown). Note that the footlike processes of the astrocytes form a continuous layer that involves the blood vessels that contribute to the blood-brain barrier. (Reproduced, with permission, from Krstíc RV: *Microscopic Human Anatomy.* Springer-Verlag, 1991.)

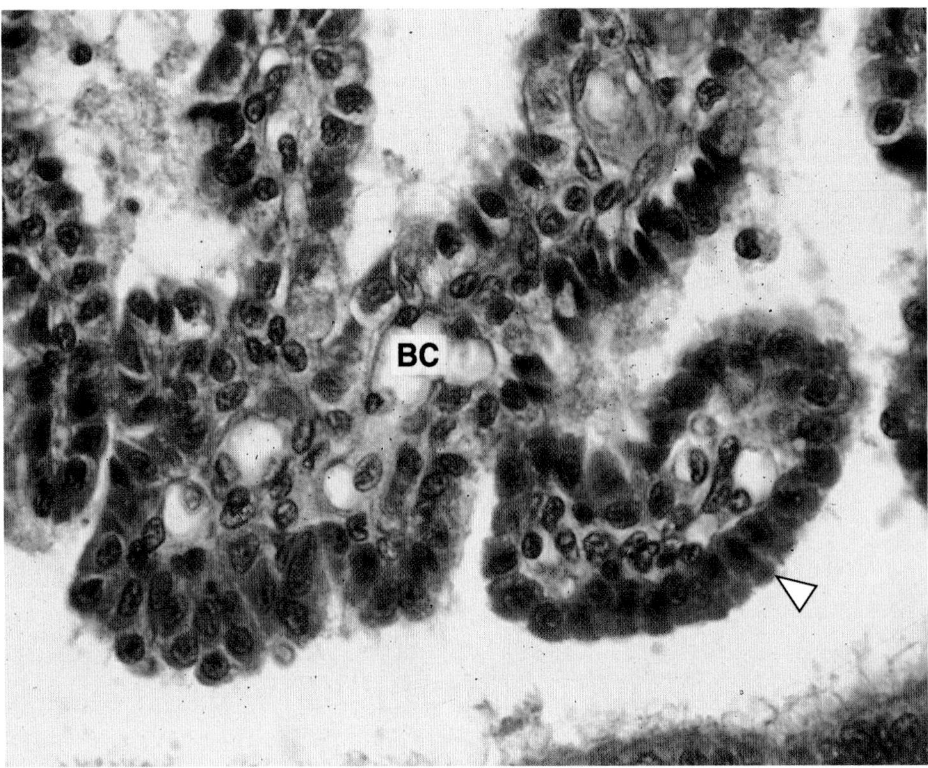

Figure 9–25. Photomicrograph of choroid plexus section. The choroid plexus presents a core of loose connective tissue rich in blood capillaries (BC) covered by a simple cubic epithelium (arrowhead). H&E stain. Medium magnification.

PERIPHERAL NERVOUS SYSTEM

The main components of the peripheral nervous system are the **nerves, ganglia,** and **nerve endings.** Nerves are bundles of nerve fibers surrounded by connective tissue sheaths.

NERVE FIBERS

Nerve fibers consist of axons enveloped by a special sheath derived from cells of ectodermal origin. Groups of nerve fibers constitute the tracts of the brain, spinal cord, and peripheral nerves. Nerve fibers exhibit differences in their enveloping sheaths, related to whether the fibers are part of the central or the peripheral nervous system.

Single or multiple folds of a sheath cell cover most axons in adult nerve tissue. In peripheral nerve fibers, the sheath cell is the **Schwann cell,** and in central nerve fibers it is the **oligodendrocyte.** Axons of small diameter are usually **unmyelinated nerve fibers** (Figures 9–26, 9–28, and 9–29). Progressively thicker axons are generally sheathed by increasingly numerous concentric wrappings of the enveloping cell, forming the **myelin sheaths.** These fibers are known as **myelinated nerve fibers** (Figures 9–27, 9–28, and 9–29).

Myelinated Fibers

In myelinated fibers of the peripheral nervous system, the plasmalemma of the covering Schwann cell winds and wraps around

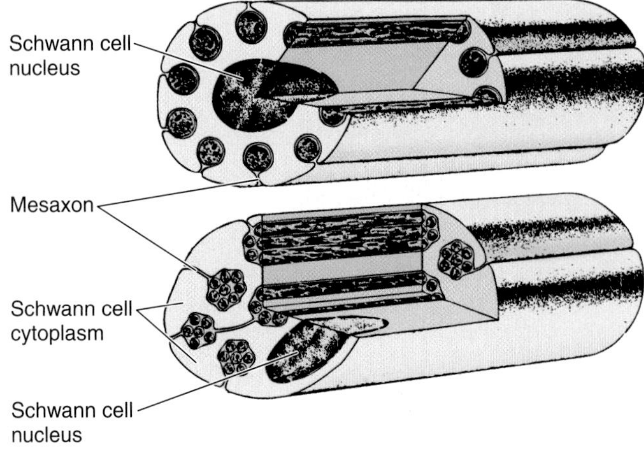

Figure 9–26. **Upper:** The most frequent type of unmyelinated nerve fiber, in which isolated axons are surrounded by a Schwann cell and each axon has its own mesaxon. **Lower:** Many very thin axons are sometimes found together, surrounded by the Schwann cell. In such cases, there is one mesaxon for several axons.

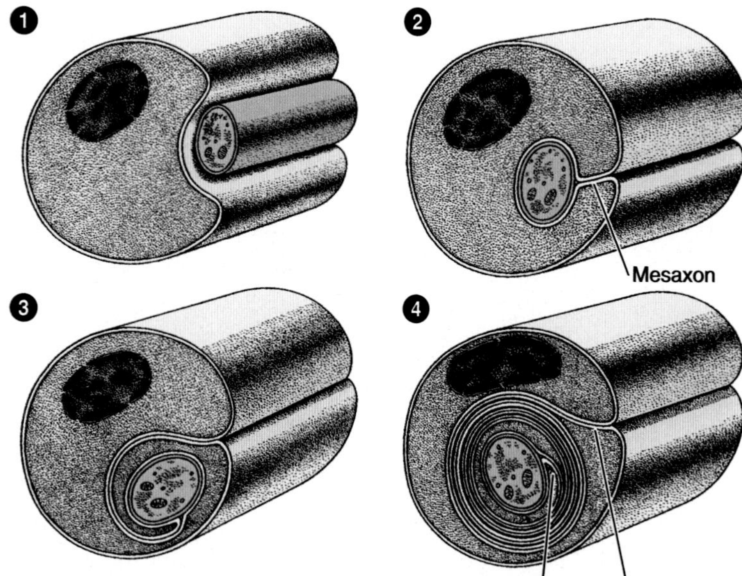

Figure 9–27. Four consecutive phases of myelin formation in peripheral nerve fibers.

the axon (Figures 9–27, 9–28, and 9–30). The layers of membranes of the sheath cell unite and form **myelin,** a whitish lipoprotein complex whose lipid component can be partly removed by standard histologic procedures.

Myelin consists of many layers of modified cell membranes. These membranes have a higher proportion of lipids than do other cell membranes. The myelin sheath shows gaps along its path called the **nodes of Ranvier** (Figures 9–28 and 9–31); these represent the spaces between adjacent Schwann cells along the length of the axon. Interdigitating processes of Schwann cells partially cover the node. The distance between two nodes is called an **internode** and consists of

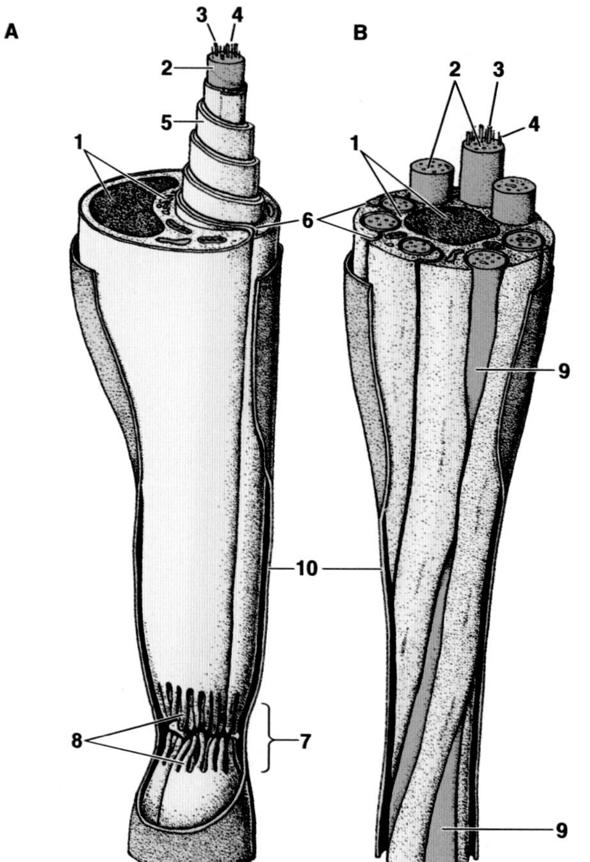

Figure 9–28. Ultrastructural features of myelinated (**A**) and unmyelinated (**B**) nerve fibers. (1) Nucleus and cytoplasm of a Schwann cell; (2) axon; (3) microtubule; (4) neurofilament; (5) myelin sheath; (6) mesaxon; (7) node of Ranvier; (8) interdigitating processes of Schwann cells at the node of Ranvier; (9) side view of an unmyelinated axon; (10) basal lamina. (Slightly modified and reproduced, with permission, from Krstíc RV: *Ultrastructure of the Mammalian Cell.* Springer-Verlag, 1979.)

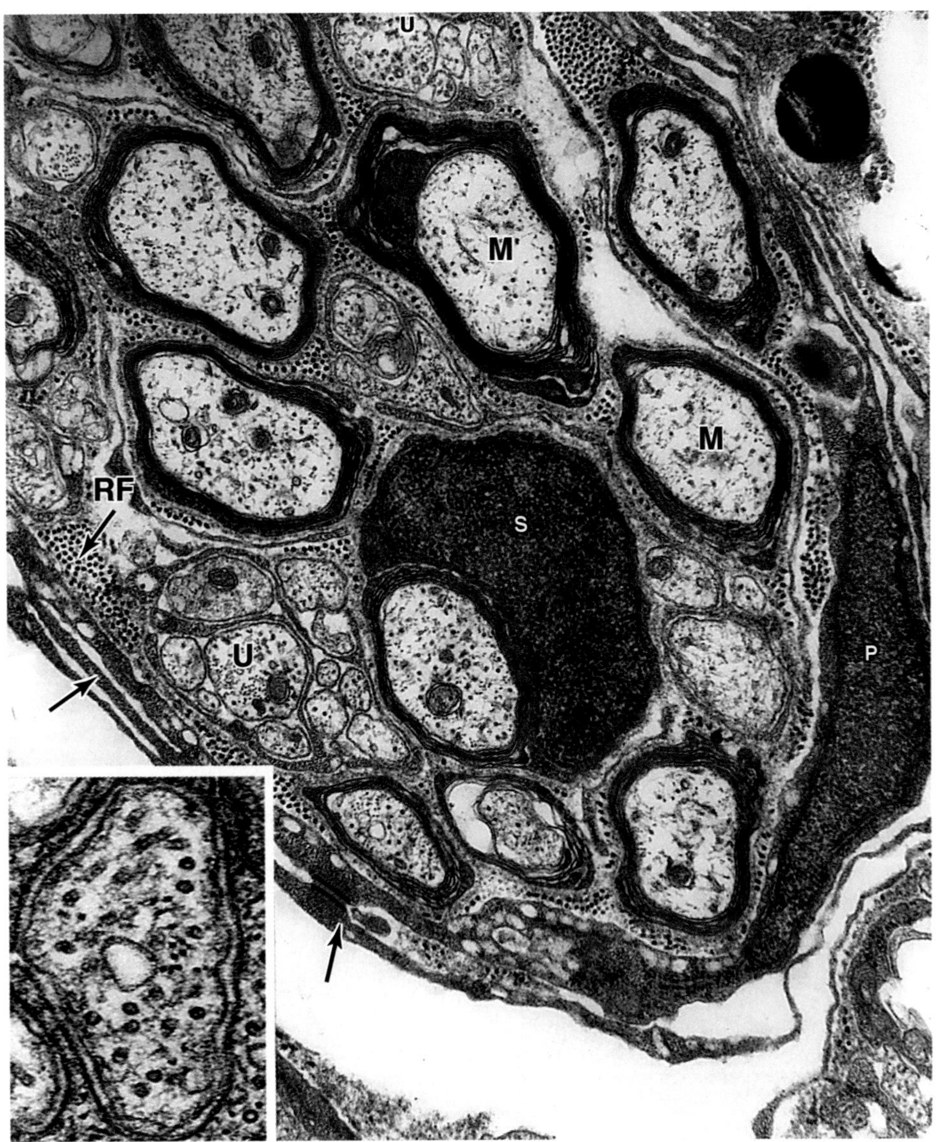

Figure 9–29. Electron micrograph of a peripheral nerve containing both myelinated (M) and unmyelinated (U) nerve fibers. The reticular fibers (RF) seen in cross section belong to the endoneurium. Near the center of the figure is a Schwann cell nucleus (S). The perineurial cells (P [over a nucleus], arrows) form a barrier that controls access of materials to nerve tissue. ×30,000. **Inset:** Part of an axon, where numerous neurofilaments and microtubules are seen in cross section. ×60,000.

one Schwann cell. The length of the internode varies between 1 and 2 mm.

There are no Schwann cells in the central nervous system; there, the processes of the oligodendrocytes form the myelin sheath. Oligodendrocytes differ from Schwann cells in that different branches of one cell can envelop segments of several axons (Figure 9–15).

Unmyelinated Fibers

In both the central and peripheral nervous systems, not all axons are sheathed in myelin. In the peripheral system, all unmyelinated axons are enveloped within simple clefts of the Schwann cells (Figure 9–26). Unlike their association with individual myelinated axons, each Schwann cell can sheathe many unmyelinated axons. Unmyelinated nerve fibers do not have nodes of Ranvier, because abutting Schwann cells are united to form a continuous sheath.

The central nervous system is rich in unmyelinated axons; unlike those in the peripheral system, these axons are not sheathed. In the brain and spinal cord, unmyelinated axonal processes run free among the other neuronal and glial processes.

NERVES

In the peripheral nervous system, the nerve fibers are grouped in bundles to form the nerves. Except for a few very thin nerves made up of unmyelinated fibers, nerves have a whitish, homogeneous, glistening appearance because of their myelin and collagen content.

Nerves (Figures 9–32 through 9–36) have an external fibrous coat of dense connective tissue called **epineurium,** which also fills the space between the bundles of nerve fibers. Each bundle is surrounded by the **perineurium,** a sleeve formed by layers of flattened epitheliumlike cells. The cells of each layer of the perineurial sleeve are joined at their edges by tight junctions, an arrangement that makes the perineurium a barrier to the passage of most macromolecules and has the important function of protecting the nerve fibers from aggression. Within the perineurial sheath run the Schwann cell–sheathed axons and their enveloping connective tissue, the **endoneurium** (Figure 9–33). The endoneurium consists of a thin layer of reticular fibers, produced by Schwann cells.

The nerves establish communication between brain and spinal cord centers and the sense organs and effectors (muscles, glands, etc). They possess afferent and efferent fibers to and from the central nervous system. **Afferent** fibers carry the information obtained from the interior of the body and the environment to the central nervous system. **Efferent** fibers carry impulses from the central nervous system to the effector organs commanded by these centers. Nerves possessing only sensory fibers are called **sensory nerves;** those composed only of fibers carrying impulses to the effectors are called **motor nerves.** Most nerves have both sensory and motor fibers and are called **mixed nerves;** these nerves have both myelinated and unmyelinated axons (Figure 9–29).

GANGLIA

Ganglia are ovoid structures containing neuronal cell bodies and glial cells supported by connective tissue. Because they serve as relay stations to transmit nerve impulses, one nerve enters and

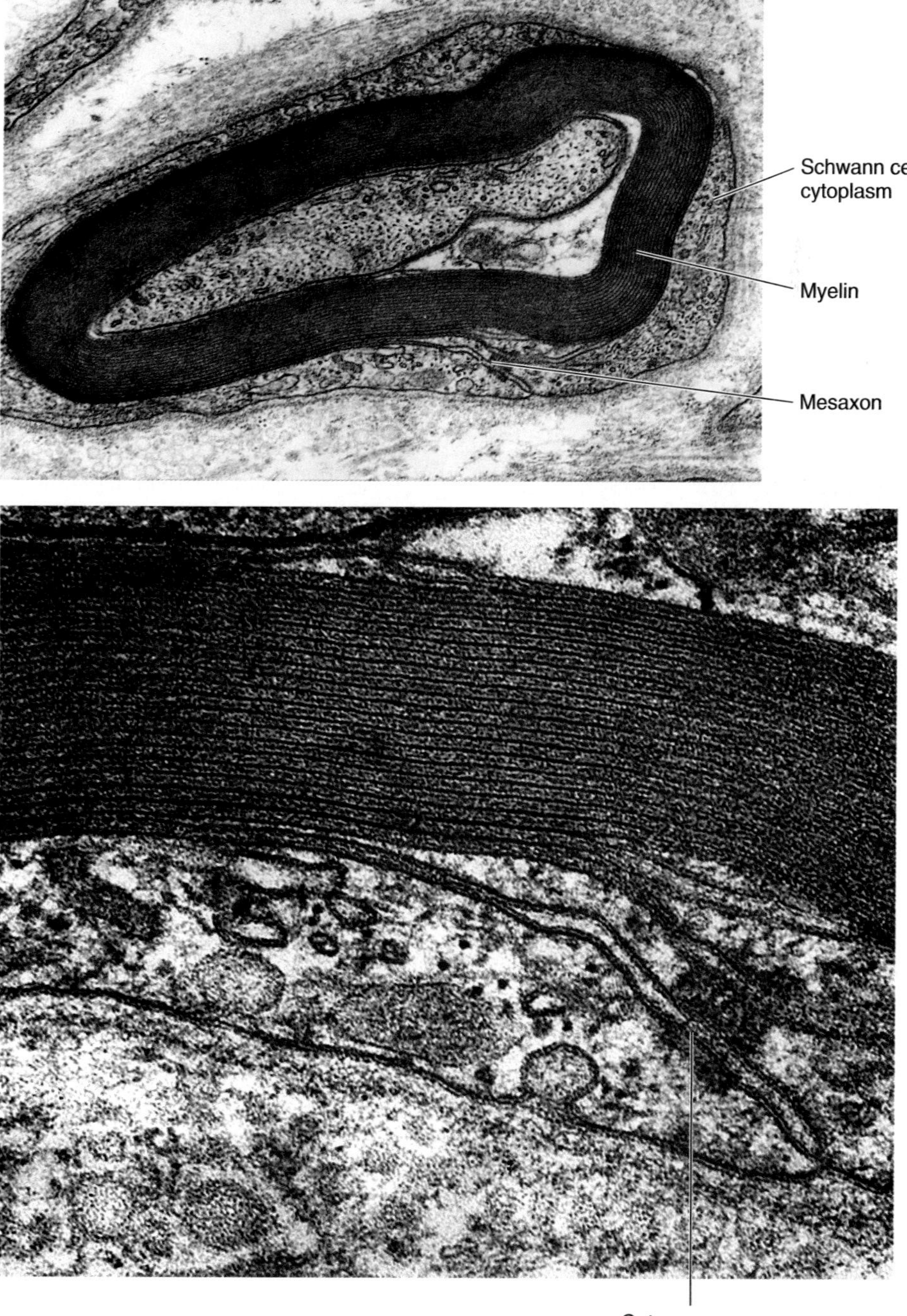

Figure 9–30. Electron micrographs of a myelinated nerve fiber. **Top:** ×20,000. **Bottom:** ×80,000.

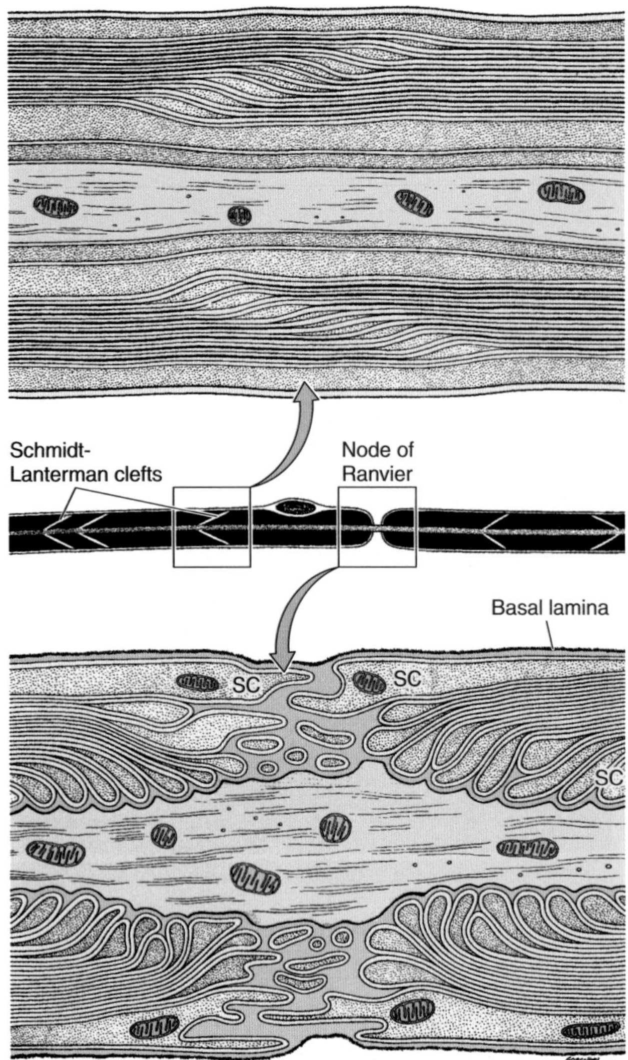

Figure 9–31. **The center drawing shows a myelinated peripheral nerve fiber as seen under the light microscope. The process is the axon enveloped by the myelin sheath and by the cytoplasm of Schwann cells. A Schwann cell nucleus, the Schmidt-Lanterman clefts, and a node of Ranvier are shown. The upper drawing shows the ultrastructure of the Schmidt-Lanterman cleft. The cleft is formed by Schwann cell cytoplasm that is not displaced to the periphery during myelin formation. The lower drawing shows the ultrastructure of a node of Ranvier. Note the appearance of loose interdigitating processes of the outer leaf of the Schwann cells' cytoplasm (SC) and the close contact of the axolemma. This contact acts as a sort of barrier to the movement of materials in and out of the periaxonal space between the axolemma and the membrane of the Schwann cell. The basal lamina around the Schwann cell is continuous. Covering the nerve fiber is a connective tissue layer—mainly reticular fibers—that belong to the endoneurial sheath of the peripheral nerve fibers.**

another exits from each ganglion. The direction of the nerve impulse determines whether the ganglion will be a **sensory** or an **autonomic** ganglion.

Sensory Ganglia

Sensory ganglia receive afferent impulses that go to the central nervous system. Two types of sensory ganglia exist. Some are associated with cranial nerves (**cranial ganglia**); others are associated with the dorsal root of the spinal nerves and are called **spinal ganglia.** The latter comprise large neuronal cell bodies (Figure 9–37) with prominent fine Nissl bodies surrounded by abundant small glial cells called **satellite cells.**

A connective tissue framework and capsule support the ganglion cells. The neurons of these ganglia are pseudounipolar and relay information from the ganglion's nerve endings to the gray matter of the spinal cord via synapses with local neurons.

Autonomic Ganglia

Autonomic ganglia appear as bulbous dilatations in autonomic nerves. Some are located within certain organs, especially in the walls of the digestive tract, where they constitute the **intramural ganglia.** These ganglia are devoid of connective tissue capsules, and their cells are supported by the stroma of the organ in which they are found.

Autonomic ganglia usually have multipolar neurons. As with craniospinal ganglia, autonomic ganglia have neuronal perikaryons with fine Nissl bodies.

A layer of satellite cells frequently envelops the neurons of autonomic ganglia. In intramural ganglia, only a few satellite cells are seen around each neuron.

AUTONOMIC NERVOUS SYSTEM

The autonomic (Gr. *autos,* self, + *nomos,* law) nervous system is related to the control of smooth muscle, the secretion of some glands, and the modulation of cardiac rhythm. Its function is to make adjustments in certain activities of the body to maintain a constant internal environment (**homeostasis**). Although the autonomic nervous system is by definition a motor system, fibers that receive sensation originating in the interior of the organism accompany the motor fibers of the autonomic system.

The term "autonomic" is not correct—although it is widely used—inasmuch as most of the functions of the autonomic ner-

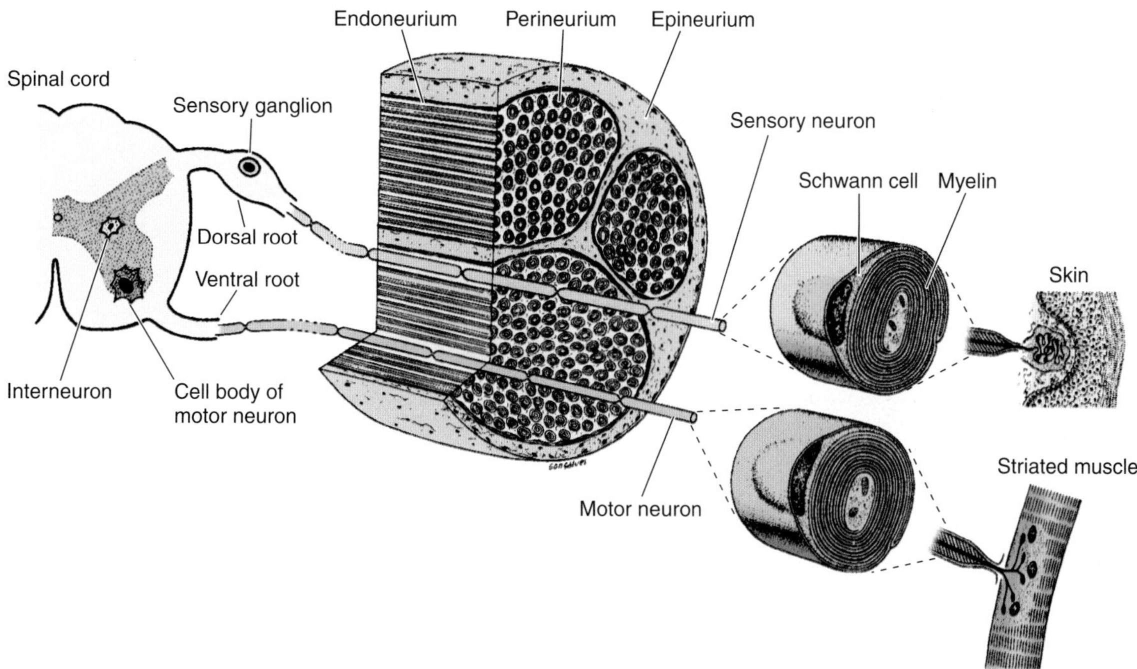

Figure 9–32. Schematic representation of a nerve and a reflex arc. In this example, the sensory stimulus starts in the skin and passes to the spinal cord via the dorsal root ganglion. The sensory stimulus is transmitted to an interneuron that activates a motor neuron that innervates skeletal muscle. Examples of the operation of this reflex are withdrawal of the finger from a hot surface and the knee-jerk reflex. (Slightly modified, redrawn, and reproduced, with permission, from Ham AW: *Histology,* 6th ed. Lippincott, 1969.)

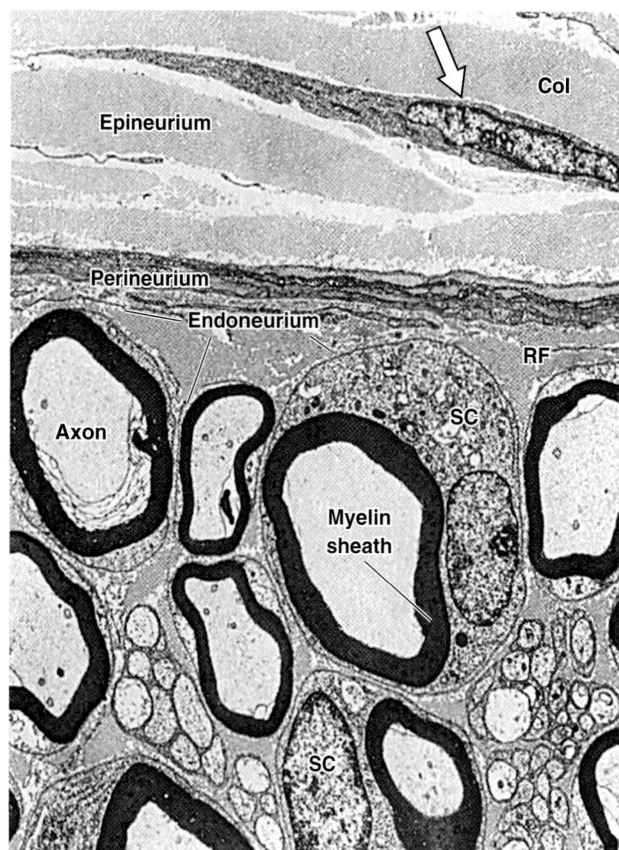

Figure 9–33. Electron micrograph of a cross section through a nerve, showing the epineurium, the perineurium, and the endoneurium. The epineurium is a dense connective tissue rich in collagen fibers (Col) and fibroblasts (arrow). The perineurium is made up of several layers of flat cells tightly joined together to form a barrier to the penetration of the nerve by macromolecules. The endoneurium is composed mainly of reticular fibers (RF) synthesized by Schwann cells (SC). ×1200.

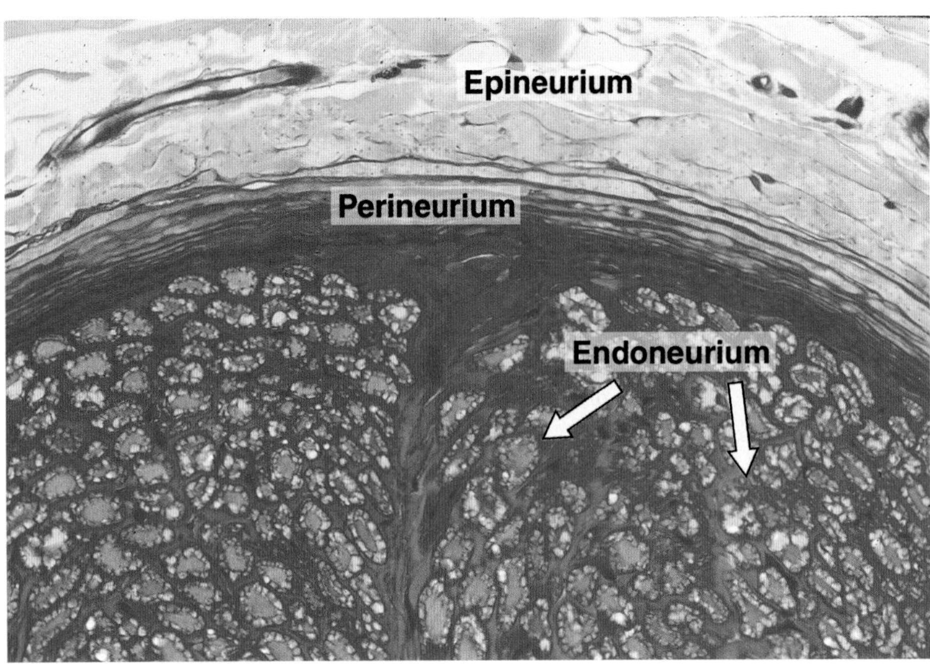

Figure 9–34. Cross section of a thick nerve showing the epineurium, perineurium, and endoneurium. The myelin sheath that envelops each axon was partially removed by the histologic technique. PT stain. Medium magnification.

vous system are not autonomous at all; they are organized and regulated in the central nervous system. The concept of the autonomic nervous system is mainly functional. Anatomically, it is composed of collections of nerve cells located in the central nervous system, fibers that leave the central nervous system through cranial or spinal nerves, and nerve ganglia situated in the paths of these fibers. The term "autonomic" covers all the neural elements concerned with visceral function. In fact, the so-called autonomic functions are as dependent on the central nervous system as are the motor neurons that trigger muscle contractions.

The autonomic nervous system is a two-neuron network. The first neuron of the autonomic chain is located in the central nervous system. Its axon forms a synapse with the second multipolar neuron in the chain, located in a ganglion of the peripheral nervous system. The nerve fibers (axons) of the first neuron are called **preganglionic fibers;** the axons of the second neuron to the effectors—muscle or gland—are called **postganglionic fibers.** The chemical mediator present in the synaptic vesicles of all preganglionic endings and at anatomically parasympathetic postganglionic endings is **acetylcholine,** which is released from the terminals by nerve impulses.

The adrenal medulla is the only organ that receives preganglionic fibers, because the majority of the cells, after migra-

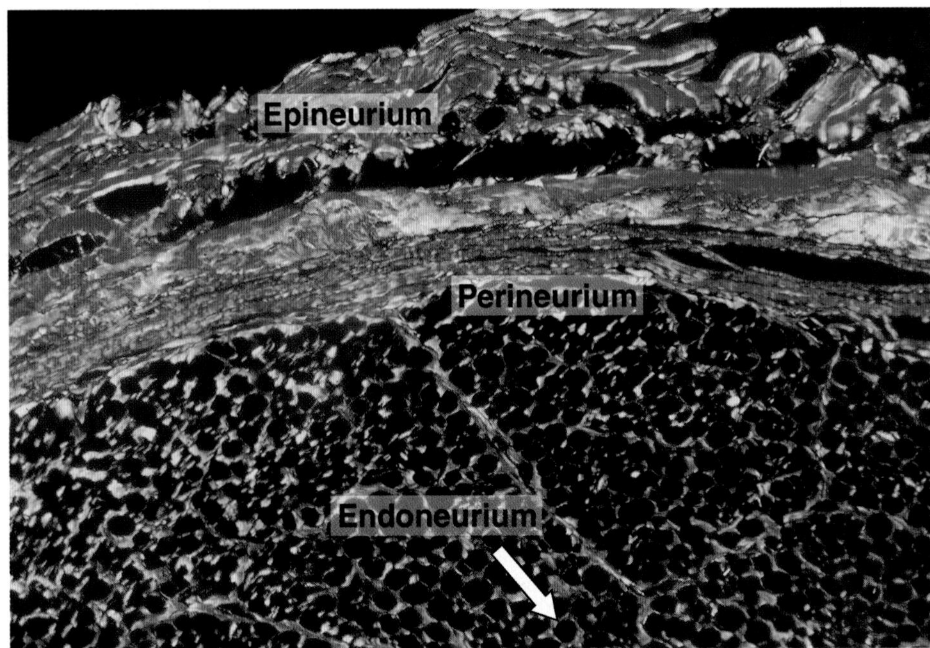

Figure 9–35. Cross section of a thick nerve stained to show its collagenous components. Picrosirius–polarized light stain. Medium magnification.

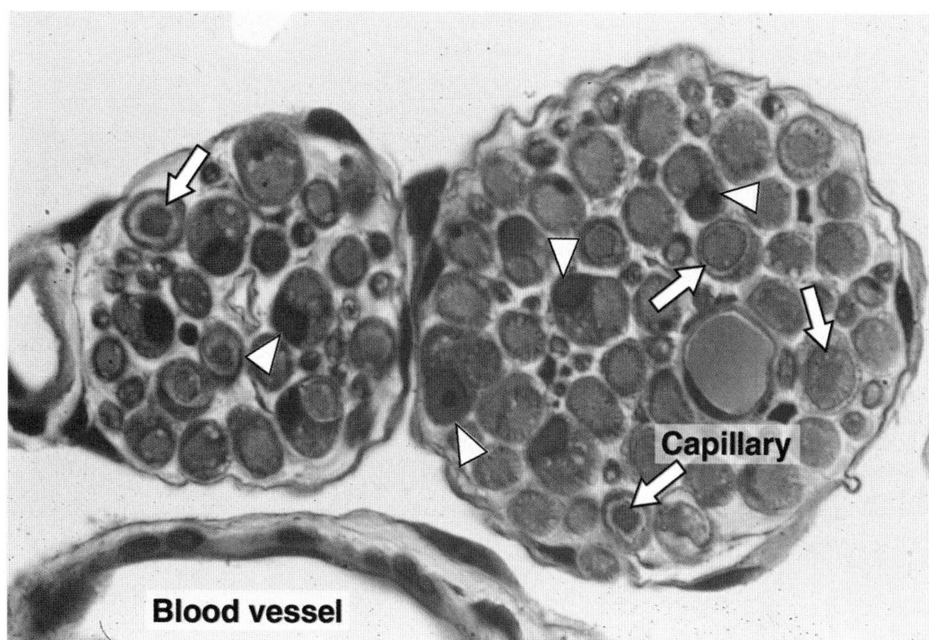

Figure 9–36. Cross sections of two small nerves with a thin covering layer. Note the Schwann cell nuclei (arrowheads) and the axons (arrows). PT stain. Medium magnification.

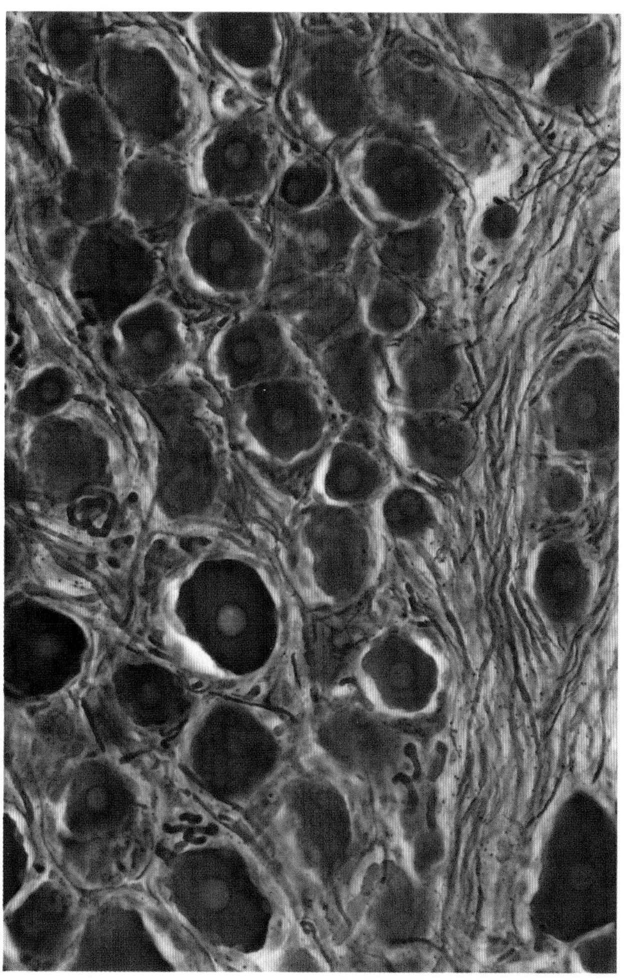

Figure 9–37. Silver-impregnated sensory ganglion consisting of pseudounipolar neurons. Medium magnification.

tion into the gland, differentiate into secretory cells rather than ganglion cells.

The autonomic nervous system is composed of two parts that differ both anatomically and functionally: the sympathetic system and the parasympathetic system (Figure 9–38). Nerve fibers that release acetylcholine are called **cholinergic.** Cholinergic fibers include all the preganglionic autonomic fibers (sympathetic as well as parasympathetic) and postganglionic parasympathetic fibers to smooth muscles, heart, and exocrine glands (Figure 9–38).

Sympathetic System

The nuclei (formed by a collection of nerve cell bodies) of the sympathetic system are located in the thoracic and lumbar segments of the spinal cord. Therefore, the sympathetic system is also called the **thoracolumbar division** of the autonomic nervous system. The axons of these neurons—preganglionic fibers—leave the central nervous system by way of the ventral roots and white communicating rami of the thoracic and lumbar nerves. The chemical mediator of the postganglionic fibers

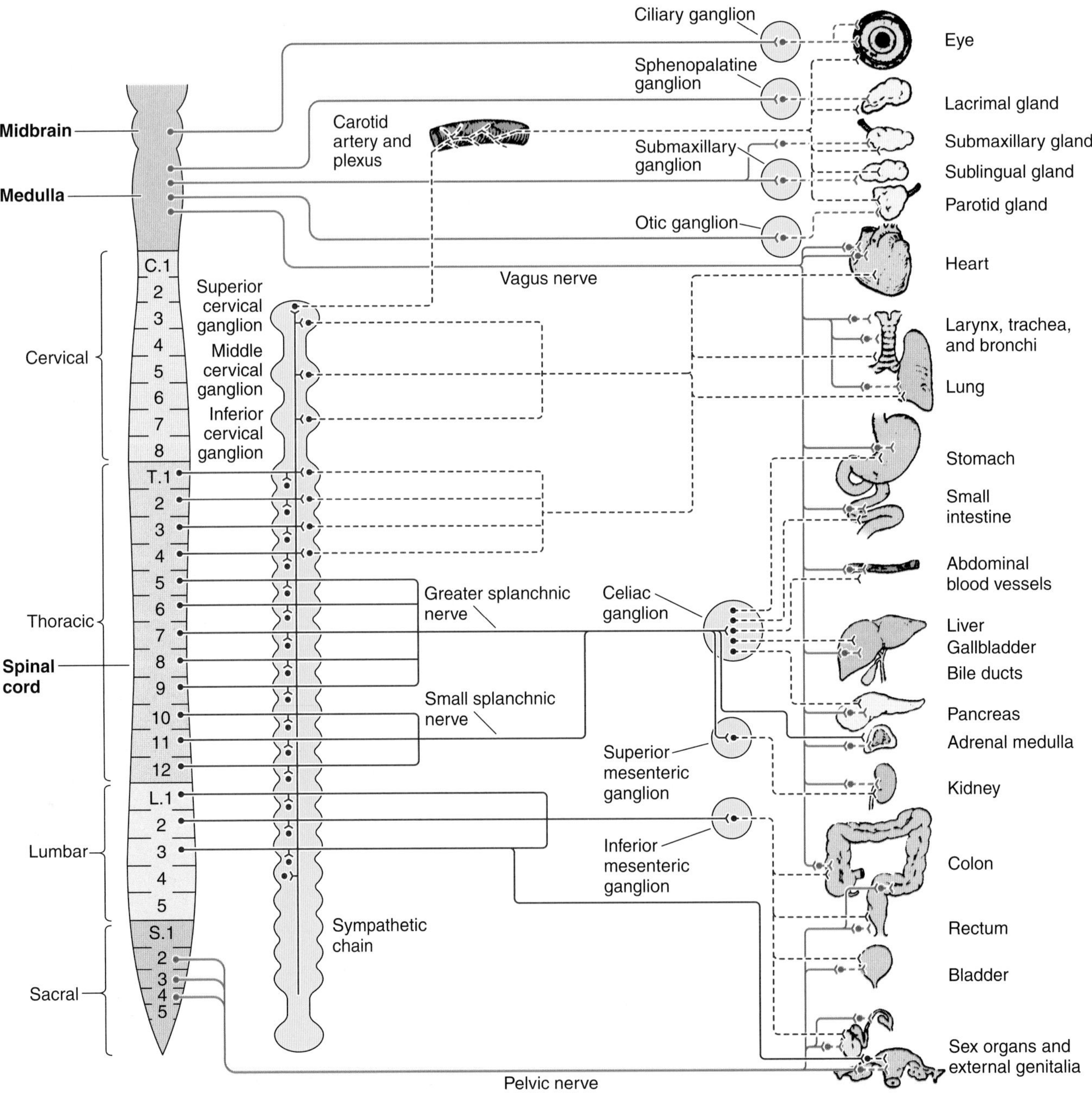

Figure 9–38. Diagram of the efferent autonomic pathways. Preganglionic neurons are shown as solid lines, postganglionic neurons as dotted lines. The blue lines are parasympathetic fibers; the red lines are sympathetic fibers. (Slightly modified and reproduced, with permission, from Youmans W: *Fundamentals of Human Physiology,* 2nd ed. Year Book, 1962.)

of the sympathetic system is **norepinephrine,** which is also produced by the adrenal medulla. Nerve fibers that release norepinephrine are called **adrenergic** (a word derived from noradrenalin, another term for norepinephrine). Adrenergic fibers innervate sweat glands and blood vessels of skeletal muscle. Cells of the adrenal medulla release epinephrine and norepinephrine in response to preganglionic sympathetic stimulation.

Parasympathetic System

The parasympathetic system has its nuclei in the medulla and midbrain and in the sacral portion of the spinal cord. The preganglionic fibers of these neurons leave through 4 of the cranial nerves (III, VII, IX, and X) and also through the second, third, and fourth sacral spinal nerves. The parasympathetic system is therefore also called the craniosacral division of the autonomic system.

The second neuron of the parasympathetic series is found in ganglia smaller than those of the sympathetic system; it is always located near or within the effector organs. These neurons are usually located in the walls of organs (eg, stomach, intestines), in which case the preganglionic fibers enter the organs and form a synapse there with the second neuron in the chain.

The chemical mediator released by the pre- and postganglionic nerve endings of the parasympathetic system, **acetylcholine,** is readily inactivated by acetylcholinesterase—one of the reasons parasympathetic stimulation has both a more discrete and a more localized action than does sympathetic stimulation.

Distribution

Most of the organs innervated by the autonomic nervous system receive both sympathetic and parasympathetic fibers (Figure 9–38). Generally, in organs where one system is the stimulator, the other has an inhibitory action.

DEGENERATION & REGENERATION OF NERVE TISSUE

MEDICAL APPLICATION

Although it has been shown that neurons can divide in the brain of adult birds, mammalian neurons usually do not divide, and their degeneration represents a permanent loss. Neuronal processes in the central nervous system are, within very narrow limits, replaceable by growth through the synthetic activity of their perikaryons. Peripheral nerve fibers can also regenerate if their perikaryons are not destroyed.

Death of a nerve cell is limited to its perikaryon and processes. The neurons functionally connected to the dead neuron do not die, except for those with only one link. In this latter instance, the isolated neuron undergoes ***transneuronal degeneration.***

In contrast to nerve cells, neuroglia of the central nervous system—and Schwann cells and ganglionic satellite cells of the peripheral nervous system—are able to divide by mitosis. Spaces in the central nervous system left by nerve cells lost by disease or injury are invaded by neuroglia.

Since nerves are widely distributed throughout the body, they are often injured. When a nerve axon is transected, degenerative changes take place, followed by a reparative phase.

In a wounded nerve fiber, it is important to distinguish the changes occurring in the proximal segment from those in the distal segment. The proximal segment maintains its continuity with the trophic center (perikaryon) and frequently regenerates. The distal segment, separated from the nerve cell body, degenerates (Figure 9–39).

Axonal injury causes several changes in the perikaryon: ***chromatolysis,*** *ie, dissolution of Nissl substances with a consequent decrease in cytoplasmic basophilia; an increase in the volume of the perikaryon; and migration of the nucleus to a peripheral position in the perikaryon. The proximal segment of the axon degenerates close to the wound for a short distance, but growth starts as soon as debris is removed by macrophages. Macrophages produce interleukin-1, which stimulates Schwann cells to secrete substances that promote nerve growth.*

In the nerve stub distal to the injury, both the axon (now separated from its trophic center) and the myelin sheath degenerate completely, and their remnants, excluding their connective tissue and perineurial sheaths, are removed by macrophages. While these regressive changes take place, Schwann cells proliferate within the remaining connective tissue sleeve, giving rise to solid cellular columns. These rows of Schwann cells serve as guides to the sprouting axons formed during the reparative phase.

After the regressive changes, the proximal segment of the axon grows and branches, forming several filaments that progress in the direction of the columns of Schwann cells. Only fibers that penetrate these Schwann cell columns will continue to grow and reach an effector organ (Figure 9–39).

When there is an extensive gap between the distal and proximal segments, or when the distal segment disappears altogether (as in the case of amputation of a limb), the newly grown nerve fibers may form a swelling, or ***neuroma,*** *that can be the source of spontaneous pain (Figure 9–39).*

Regeneration is functionally efficient only when the fibers and the columns of Schwann cells are directed to the correct place. The possibility is good, however, since each regenerating fiber gives origin to several processes, and each column of Schwann cells receives processes from several regenerating fibers. In an injured mixed nerve, however, if regenerating sensory fibers grow into columns connected to motor end-plates that were occupied by motor fibers, the function of the muscle will not be reestablished.

Figure 9–39. Main changes that take place in an injured nerve fiber. **A:** Normal nerve fiber, with its perikaryon and effector cell (striated skeletal muscle). Note the position of the neuron nucleus and the quantity and distribution of Nissl bodies. **B:** When the fiber is injured, the neuronal nucleus moves to the cell periphery, and Nissl bodies become greatly reduced in number. The nerve fiber distal to the injury degenerates along with its myelin sheath. Debris is phagocytosed by macrophages. **C:** The muscle fiber shows a pronounced denervation atrophy. Schwann cells proliferate, forming a compact cord penetrated by the growing axon. The axon grows at the rate of 0.5–3 mm/day. **D:** Here, the nerve fiber regeneration was successful. Note that the muscle fiber was also regenerated after receiving nerve stimuli. **E:** When the axon does not penetrate the cord of Schwann cells, its growth is not organized. (Redrawn and reproduced, with permission, from Willis RA, Willis AF: *The Principles of Pathology and Bacteriology,* 3rd ed. Butterworth, 1972.)

Neuronal Plasticity

Despite its general stability, the nervous system exhibits some plasticity in adults. Plasticity is very high during embryonic development, when an excess of nerve cells is formed, and the cells that do not establish correct synapses with other neurons are eliminated. Several studies made in adult mammals have shown that, after an injury, the neuronal circuits may be reorganized by the growth of neuronal processes, forming new synapses to replace the ones lost by injury. Thus, new communications are established with some degree of functional recovery. This property of nerve tissue is known as **neuronal plasticity.** The regenerative processes in the nervous system are controlled by several growth factors produced by neurons, glial cells, Schwann cells, and target cells. These growth factors form a family of molecules called **neurotrophins.**

Neural Stem Cells

In several tissues of adult organs, there is a stem cell population that may generate new cells continuously or in response to injury. This population remains constant in the tissues: After the cell divisions, only some daughter cells differentiate whereas others remain as stem cells, thereby maintaining a stable pool of stem cells. Because neurons do not divide to replace the ones lost by accident or disease, the subject of neural stem cells is now under intense investigation. The pool of neural stem cells may constitute a reserve of cells that under correct stimulation could replace lost neurons. Some regions of the brain and spinal cord of adult mammals retain stem cells that can generate astrocytes, neurons, and oligodendrocytes. Recently, it was shown that neural stem cells can even generate cells not related to the nerve tissue. This observation demonstrates that neural stem cells have a great potential for differentiation.

Tumors of the Nervous System

MEDICAL APPLICATION

Virtually all cells of the nerve tissue generate tumors. Glial cells produce gliomas, immature nerve cells produce ***medulloblastomas,*** *and Schwann cells produce* ***schwannomas.*** *Because adult neurons do not divide, they do not produce tumors.*

REFERENCES

Bothwell M: Functional interactions of neurotrophins and neurotrophin receptors. Annu Rev Neurosci 1995;18:223.

Clarke DL et al: Generalized potential of adult neural stem cells. Science 2000;288:1660.

Compagnone NA, Mellon SH: Neurosteroids: biosynthesis and function of these novel neuromodulators. Front Neuroendocrinol 2000;21:1.

Gage FH: Mammalian neural stem cell. Science 2000;287:1433.

Giulian D, Carpuz M: Neuroglial secretion products and their impact on the nervous system. Adv Neurology 1993;59:315.

Halpain S: Actin and the agile spine: how and why do dendritic spines dance? Trends Neurosci 2000;23(4):141.

Häusser M et al: Diversity and dynamics of dendritic signaling. Science 2000;290:739.

Heuser JE, Reese TS: Structural changes after transmitter release at the frog neuromuscular junction. J Cell Biol 1981;88:564.

Kahn MA, de Vellis J: Growth factors in the CNS and their effects on oligodendroglia. Prog Brain Res 1995;105:145.

Kempermann G, Gage FH: New nerve cells for the adult brain. Sci Am 1999;280:38.

Kennedy MB: Signal-processing machines at the postsynaptic density. Science 2000;290:750.

Lancaster IC Jr: Nitric oxide in cells. Am Sci 1992;80:248.

Matus A: Actin-based plasticity in dendritic spines. Science 2000;290:754.

McKay R: Stem cells and the cellular organization of the brain. J Neurosci Res 2000;59:298.

Momma S et al: Get to know your stem cells. Curr Opin Neurobiol 2000;10:45.

Morell P, Norton WT: Myelin. Sci Am 1980;242:88.

Murphy S (editor): *Astrocytes: Pharmacology and Function.* Academic Press, 1993.

Patterson PH: Cytokines in Alzheimers disease and multiple sclerosis. Curr Opin Neurobiol 1995;5:642.

Reichardt LF, Kelly RB: A molecular description of nerve terminal function. Annu Rev Biochem 1983;52:871.

Rodgers RJ et al: Animal models of anxiety: an ethological perspective. Braz J Med Biol Res 1997;30:289.

Saffell JL et al: Axonal growth mediated by cell adhesion molecules requires activation of fibroblast growth factor receptors. Biochem Soc Trans 1995;23:469.

Sears TA (editor): *Neuronal-Glial Cell Interrelationships.* Springer-Verlag, 1982.

Thoenen H: Neurotrophins and neuronal plasticity. Science 1995;270:593.

Tsutsui K et al: Novel brain function: biosynthesis and actions of neurosteroids in neurons. Neurosci Res 2000;36:261.

Waxman SG: The neuron as a dynamic electrogenic machine: modulation of sodium-channel expression as a basis for functional plasticity in neurons. Philos Trans R Soc London B Biol Sci 2000;355:199.

Zigova T, Sanberg PR: Neural stem cell for brain repair. Sci & Med 1999;6:18.

Muscle Tissue

10

Muscle tissue is composed of differentiated cells containing contractile proteins. The structural biology of these proteins generates the forces necessary for cellular contraction, which drives movement within certain organs and the body as a whole. Most muscle cells are of mesodermal origin, and they are differentiated mainly by a gradual process of lengthening, with simultaneous synthesis of myofibrillar proteins.

Three types of muscle tissue in mammals can be distinguished on the basis of morphologic and functional characteristics (Figure 10–1), and each type of muscle tissue has a structure adapted to its physiologic role. **Skeletal muscle** is composed of bundles of very long, cylindrical, multinucleated cells that show cross-striations. Their contraction is quick, forceful, and usually under voluntary control. It is caused by the interaction of thin actin filaments and thick myosin filaments whose molecular configuration allows them to slide upon one another. The forces necessary for sliding are generated by weak interactions in the bridges that bind actin to myosin. **Cardiac muscle** also has cross-striations and is composed of elongated, branched individual cells that lie parallel to each other. At sites of end-to-end contact are the **intercalated disks,** structures found only in cardiac muscle. Contraction of cardiac muscle is involuntary, vigorous, and rhythmic. **Smooth muscle** consists of collections of fusiform cells that do not show striations in the light microscope. Their contraction process is slow and not subject to voluntary control.

Some muscle cell organelles have names that differ from their counterparts in other cells. The cytoplasm of muscle cells (excluding the myofibrils) is called **sarcoplasm** (Gr. *sarkos,* flesh, + *plasma,* thing formed), and the smooth endoplasmic reticulum is called **sarcoplasmic reticulum.** The **sarcolemma** (*sarkos* + Gr. *lemma,* husk) is the cell membrane, or plasmalemma.

SKELETAL MUSCLE

Skeletal muscle consists of **muscle fibers,** bundles of very long (up to 30 cm) cylindrical multinucleated cells with a diameter of 10–100 μm. Multinucleation results from the fusion of embryonic mononucleated myoblasts (muscle cell precursors). The oval nuclei are usually found at the periphery of the cell under the cell membrane. This characteristic nuclear location is helpful in distinguishing skeletal muscle from cardiac and smooth muscle, both of which have centrally located nuclei.

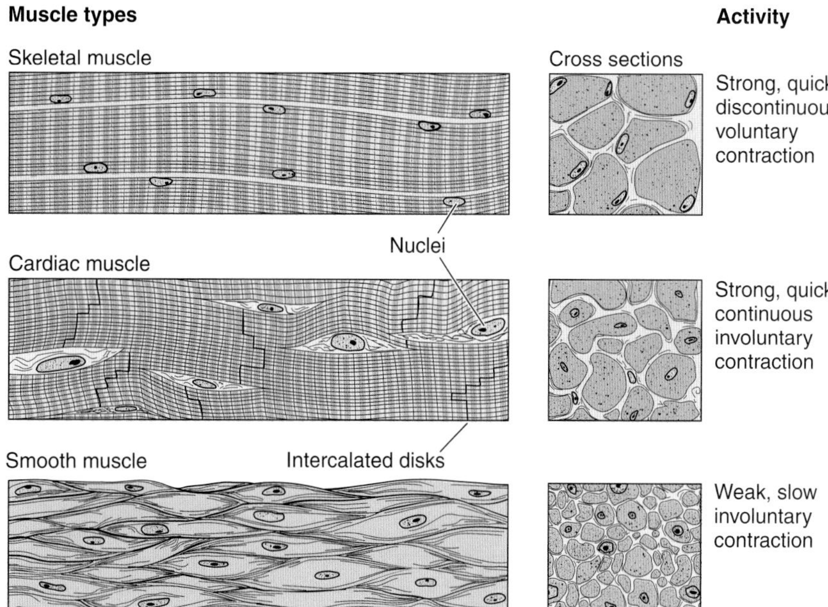

Figure 10–1. Structure of the 3 muscle types. The drawings at right show these muscles in cross section. Skeletal muscle is composed of large, elongated, multinucleated fibers. Cardiac muscle is composed of irregular branched cells bound together longitudinally by intercalated disks. Smooth muscle is an agglomerate of fusiform cells. The density of the packing between the cells depends on the amount of extracellular connective tissue present.

MEDICAL APPLICATION

The variation in diameter of skeletal muscle fibers depends on such factors as the specific muscle and the age and sex, state of nutrition, and physical training of the individual. It is a common observation that exercise enlarges the musculature and decreases fat depots. The increase in muscle thus obtained is caused by formation of new myofibrils and a pronounced growth in the diameter of individual muscle fibers. This process, characterized by augmentation of cell volume, is called ***hypertrophy*** *(Gr.* hyper, *above, +* trophe, *nourishment); tissue growth by an increase in the number of cells is termed* ***hyperplasia*** *(*hyper + *Gr.* plasis, *molding). Hyperplasia does not occur in either skeletal or cardiac muscle but does take place in smooth muscle, whose cells have not lost the capacity to divide by mitosis. Hyperplasia is rather frequent in organs such as the uterus, where both hyperplasia and hypertrophy occur during pregnancy.*

Organization of Skeletal Muscle

The masses of fibers that make up the various types of muscle are not grouped in random fashion but are arranged in regular bundles surrounded by the **epimysium** (Gr. *epi,* upper, + *mys,* muscle), an external sheath of dense connective tissue surrounding the entire muscle (Figures 10–2, 10–3, and 10–4). From the epimysium, thin septa of connective tissue extend inward, surrounding the bundles of fibers within a muscle. The connective tissue around each bundle of muscle fibers is called the **perimysium** (Gr. *peri,* around, + *mys*). Each muscle fiber is itself surrounded by a delicate layer of connective tissue, the **endomysium** (Gr. *endon,* within, + *mys*), composed mainly of a basal lamina and reticular fibers.

One of the most important roles of connective tissue is to mechanically transmit the forces generated by contracting muscle cells, because in most instances, individual muscle cells do not extend from one end of a muscle to the other.

Blood vessels penetrate the muscle within the connective tissue septa and form a rich capillary network (Figure 10–5) that runs between and parallel to the muscle fibers. The capillaries are of the continuous type, and lymphatic vessels are found in the connective tissue.

Some muscles taper off at their extremities, where a myotendinous junction is formed. The electron microscope shows

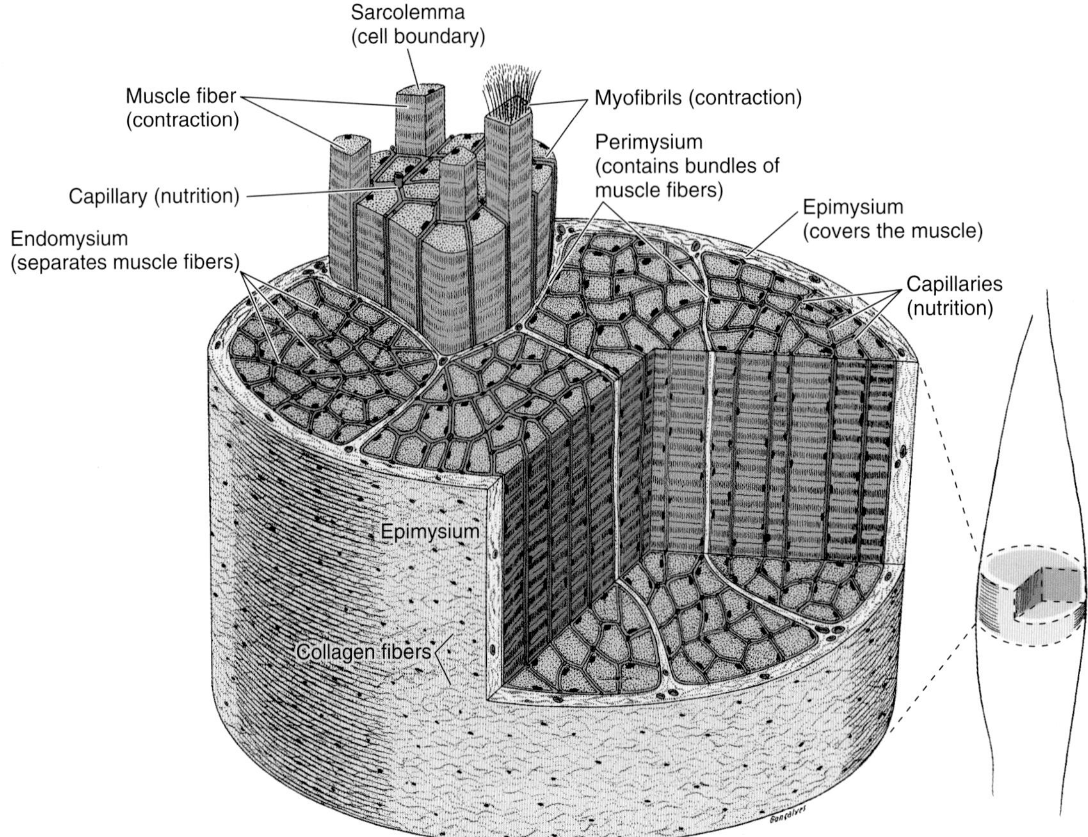

Figure 10–2. Structure and function of skeletal muscle. The drawing at right shows the area of muscle detailed in the enlarged segment. Color highlights endomysium, perimysium, and epimysium.

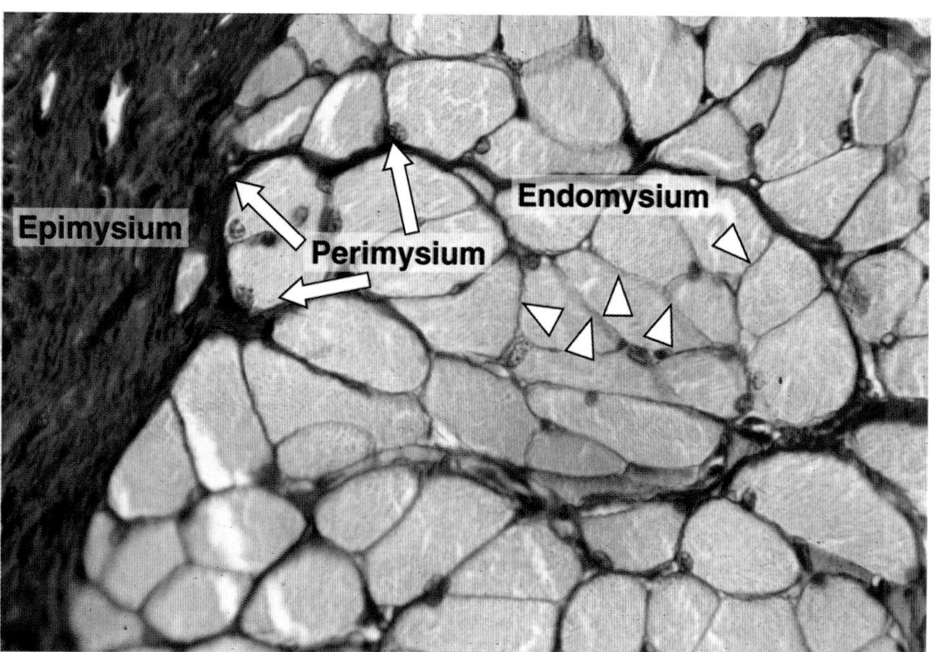

Figure 10–3. Cross section of striated muscle stained to show collagens type I and III and cell nuclei. The endomysium is indicated by arrowheads and the perimysium by arrows. At left is a piece of epimysium. Picrosirius-hematoxylin stain. High magnification.

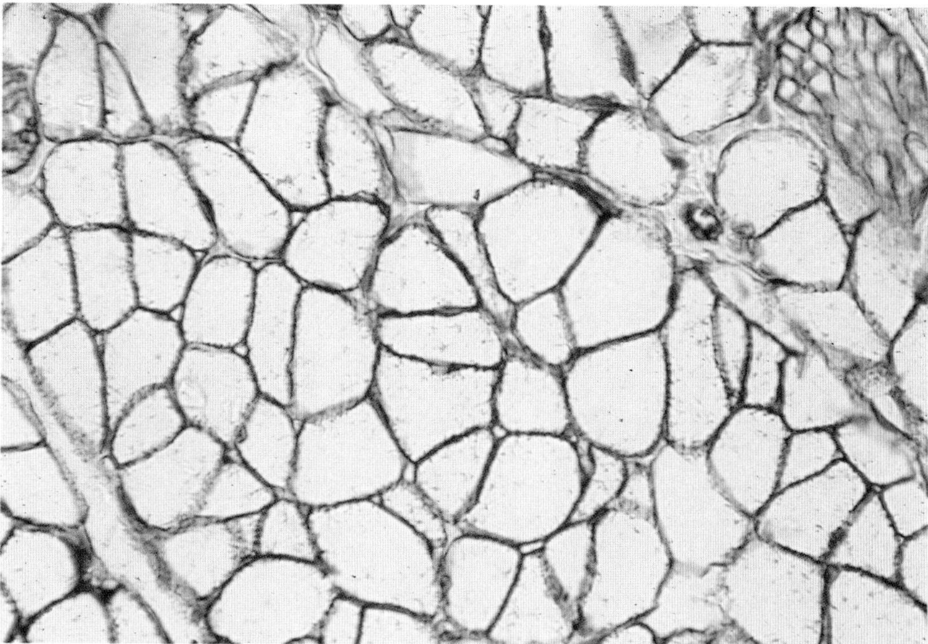

Figure 10–4. Cross section of striated muscle immunohistochemically stained for laminin, a protein component of the endomysium, which appears in various shades of brown. In the upper right corner is a slightly oblique section of a small nerve. Laminin is also present around nerve fibers.

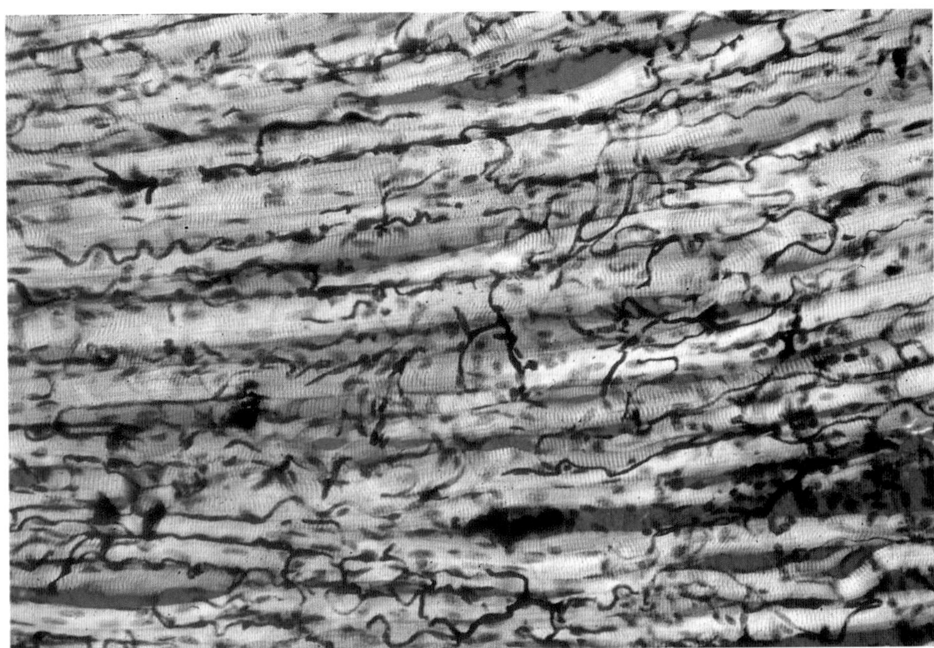

Figure 10–5. Longitudinal section of striated muscle fibers. The blood vessels were injected with a plastic material before the animal was killed. Note the extremely rich network of blood capillaries around the muscle fibers. Giemsa stain. Photomicrograph of low magnification made under polarized light.

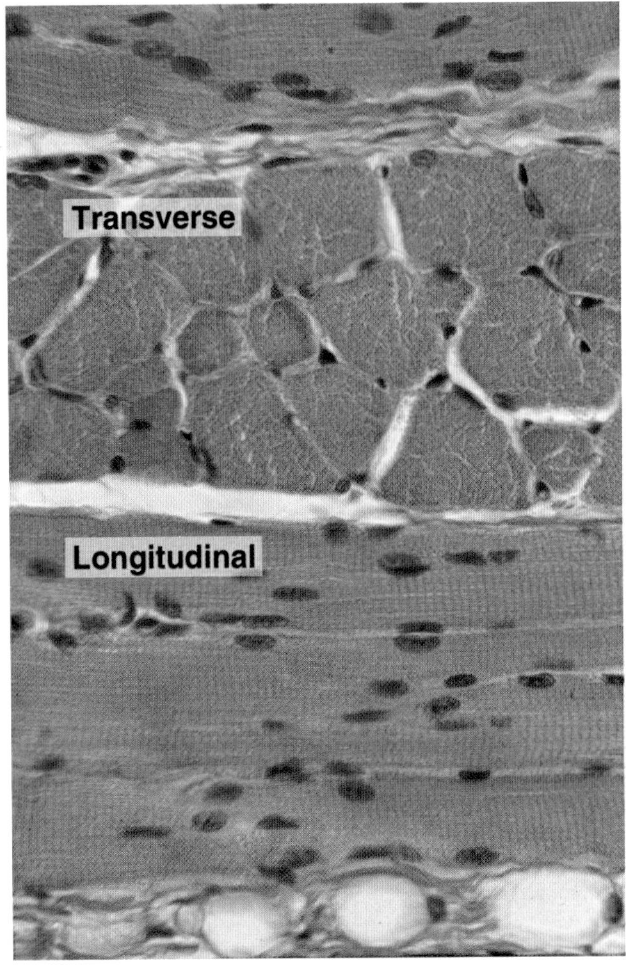

Figure 10–6. Striated skeletal muscle in longitudinal section (lower) and in cross section (upper). The nuclei can be seen in the periphery of the cell, just under the cell membrane, particularly in the cross sections of these striated fibers. H&E stain. Medium magnification.

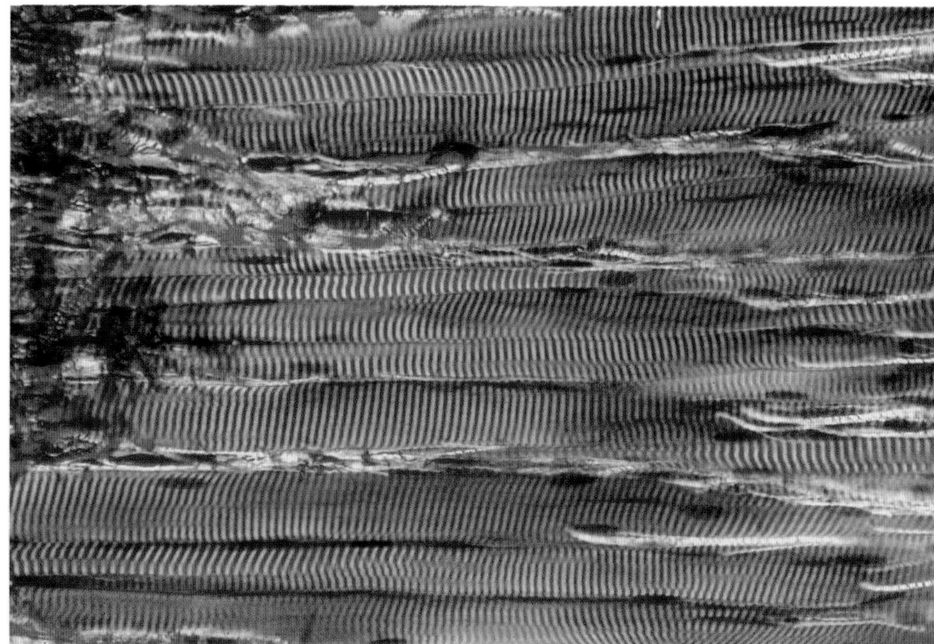

Figure 10–7. Striated skeletal muscle in longitudinal section. In the left side of the photomicrograph the insertion of collagen fibers with the muscle is clearly seen. Picrosirius–polarized light (PSP) stain. Medium magnification.

that in this transitional region, collagen fibers of the tendon insert themselves into complex infoldings of the plasmalemma of the muscle fibers (Figure 10–7).

Organization of Skeletal Muscle Fibers

As observed with the light microscope, longitudinally sectioned muscle fibers show cross-striations of alternating light and dark bands (Figures 10–6 through 10–9). The darker bands are called **A bands** (**anisotropic,** ie, are birefringent in polarized light); the lighter bands are called **I bands** (**isotropic,** ie, do not alter polarized light). In the electron microscope, each I band is bisected by a dark transverse line, the **Z line.** The smallest repetitive subunit of the contractile apparatus, the **sarcomere** (*sarkos* + Gr. *mere,* part), extends from Z line to Z line (Figures 10–10 and 10–11) and is about 2.5 μm long in resting muscle.

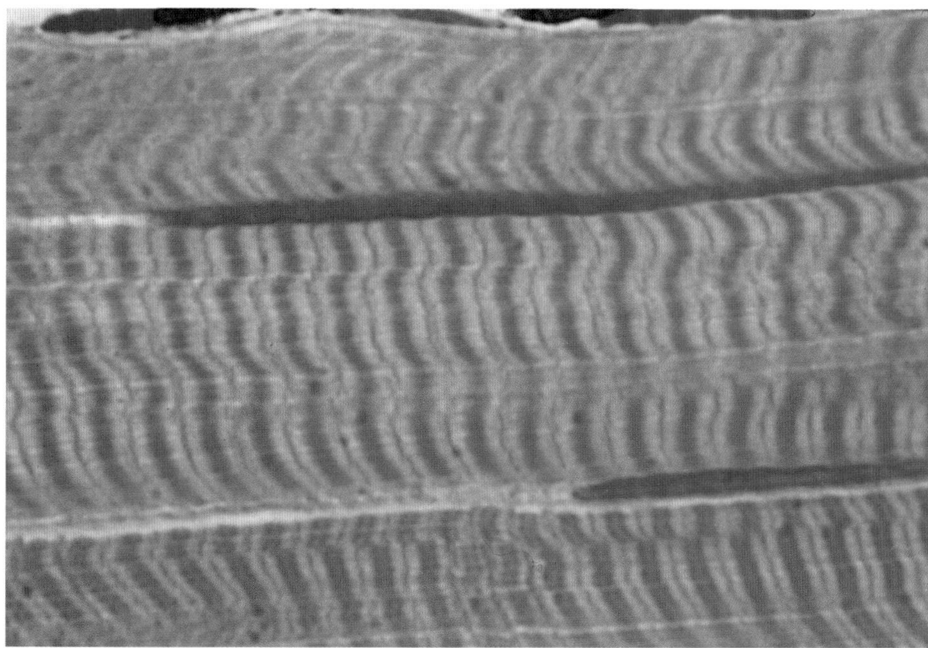

Figure 10–8. Longitudinal section of skeletal muscle fibers. Note the dark-stained A bands and the light-stained I bands, which are crossed by Z lines. Giemsa stain. High magnification.

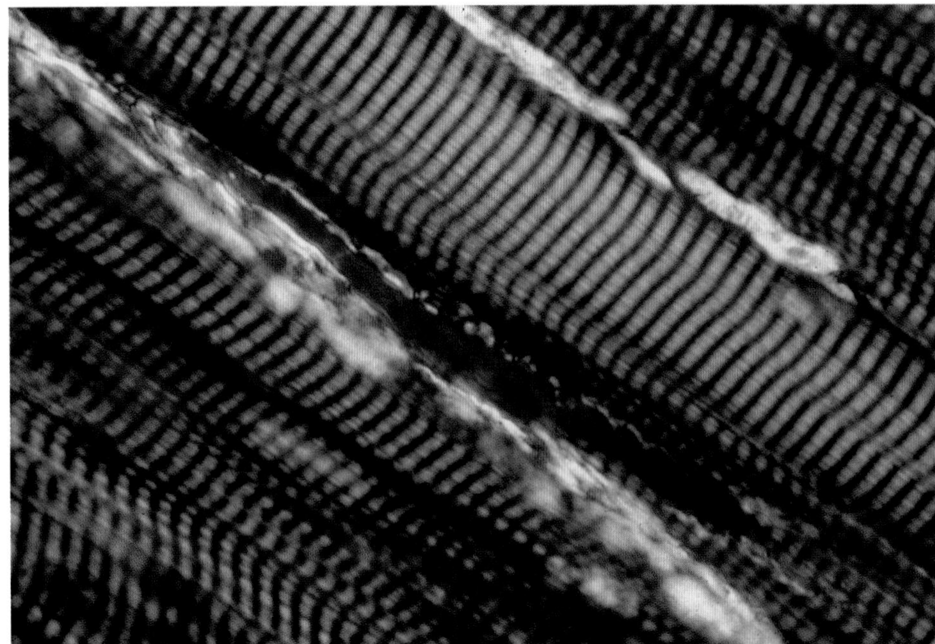

Figure 10–9. Skeletal muscle in longitudinal section. Note the striation in the muscle cells and the moderate amount of collagen (yellow). PSP stain. High magnification.

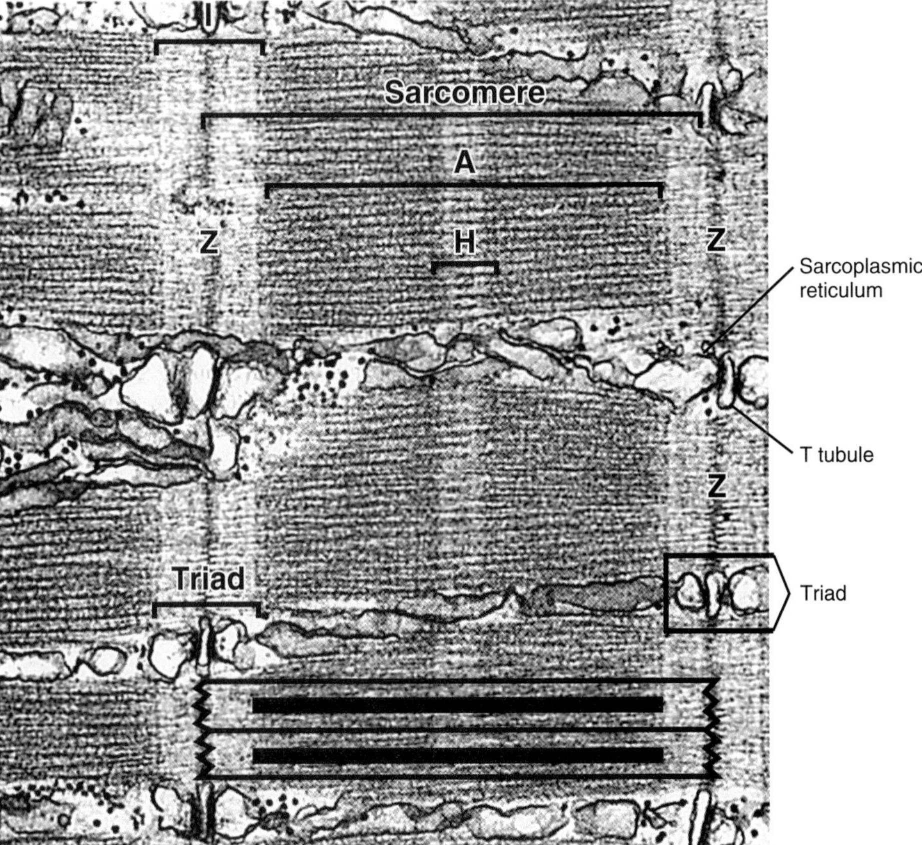

Figure 10–10. Electron micrograph of skeletal muscle of a tadpole. Note the sarcomere with its A, I, and H bands and Z line. The position of the thick and thin filaments in the sarcomere is shown schematically in the lower part of the figure. As illustrated here, triads in amphibian muscle are aligned with the Z line in each sarcomere. In mammalian muscle, however, each sarcomere exhibits 2 triads, one at each A–I band interface (see Figure 10–16). ×35,000. (Courtesy of KR Porter.)

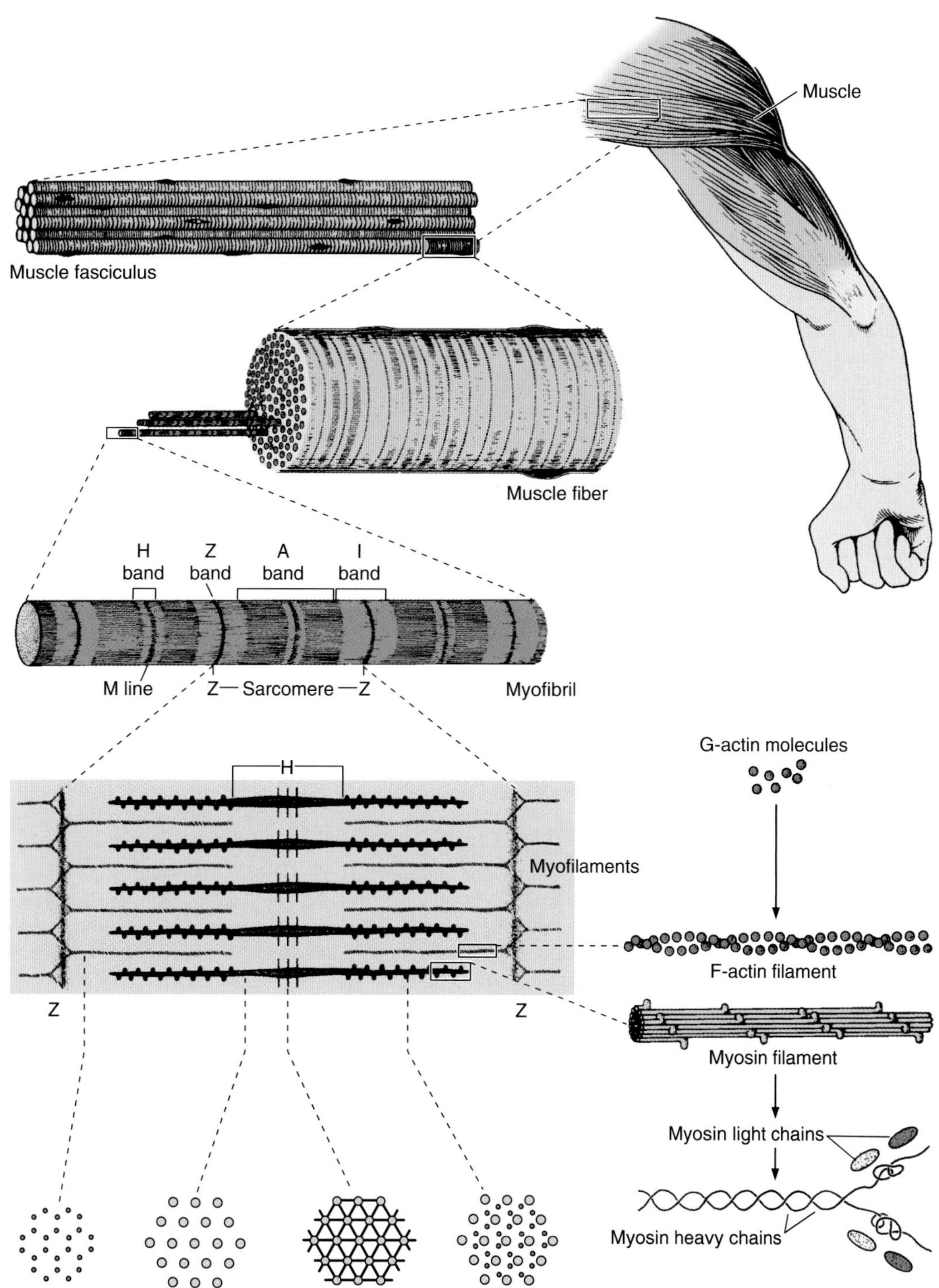

Figure 10–11. **Structure and position of the thick and thin filaments in the sarcomere. The molecular structure of these components is shown at right.** (Drawing by Sylvia Colard Keene. Reproduced, with permission, from Bloom W, Fawcett DW: *A Textbook of Histology,* 9th ed, Saunders, 1968.)

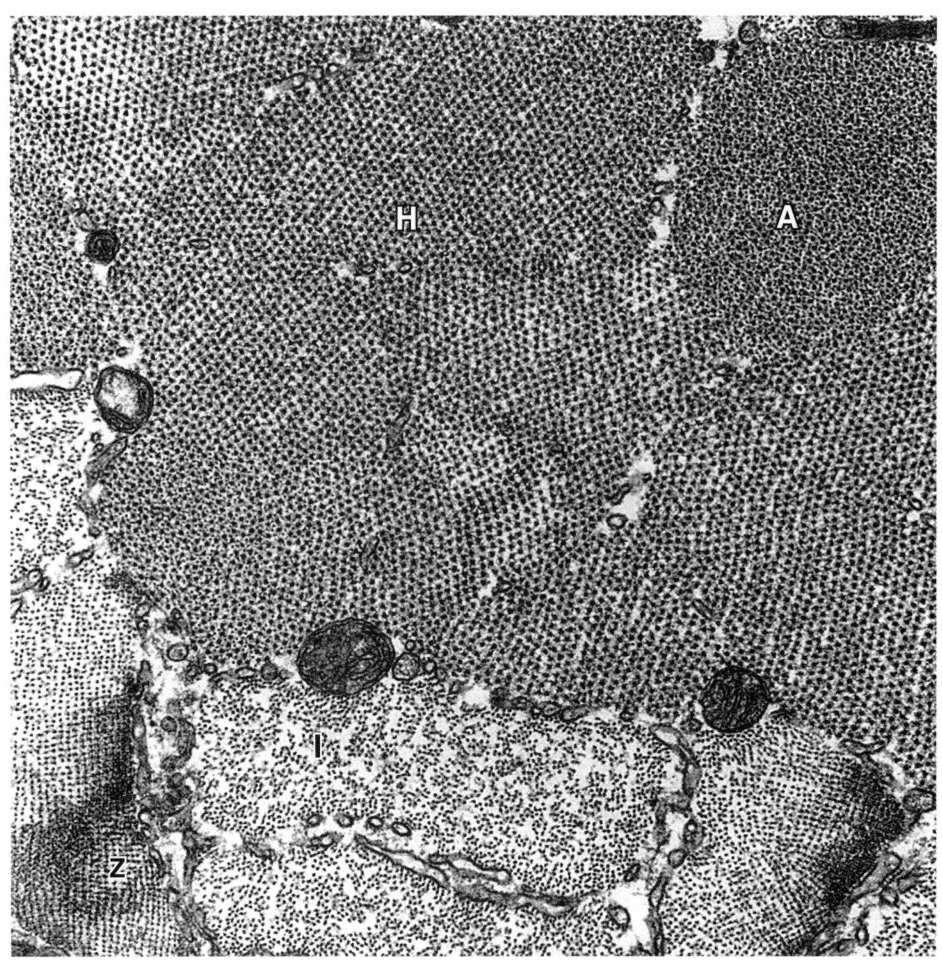

Figure 10–12. Transverse section of skeletal muscle myofibrils illustrating some of the features diagrammed in Figure 10–11. I, I band; A, A band; H, H band; Z, Z line. ×36,000.

The sarcoplasm is filled with long cylindrical filamentous bundles called **myofibrils.** The myofibrils, which have a diameter of 1–2 μm and run parallel to the long axis of the muscle fiber, consist of an end-to-end chainlike arrangement of sarcomeres (Figures 10–10 and 10–11). The lateral registration of sarcomeres in adjacent myofibrils causes the entire muscle fiber to exhibit a characteristic pattern of transverse striations.

Studies with the electron microscope reveal that this sarcomere pattern is due mainly to the presence of 2 types of filaments—thick and thin—that lie parallel to the long axis of the myofibrils in a symmetric pattern.

The thick filaments are 1.6 μm long and 15 nm wide; they occupy the A band, the central portion of the sarcomere. The thin filaments run between and parallel to the thick filaments and have one end attached to the Z line (Figures 10–10 and 10–11). Thin filaments are 1.0 μm long and 8 nm wide. As a result of this arrangement, the I bands consist of the portions of the thin filaments that do not overlap the thick filaments. The A bands are composed mainly of thick filaments in addition to portions of overlapping thin filaments. Close observation of the A band shows the presence of a lighter zone in its center, the **H band,** that corresponds to a region consisting only of the rodlike portions of the myosin molecule (Figures 10–10 and 10–11). Bisecting the H band is the **M line,** a region where lateral connections are made between adjacent thick filaments (Figure 10–11). The major protein of the M line is creatine kinase. Creatine kinase catalyzes the transfer of a phosphate group from phosphocreatine (a storage form of high-energy phosphate groups) to ADP, thus supplying ATP for muscle contraction.

Thin and thick filaments overlap for some distance within the A band. As a consequence, a cross section in the region of filament overlap shows each thick filament surrounded by 6 thin filaments in the form of a hexagon (Figures 10–11 and 10–12).

Striated muscle filaments contain several proteins; the 4 main proteins are actin, tropomyosin, troponin, and myosin. Thin filaments are composed of the first 3 proteins, whereas thick filaments consist primarily of myosin. Myosin and actin together represent 55% of the total protein of striated muscle.

Actin is present as long filamentous (F-actin) polymers consisting of 2 strands of globular (G-actin) monomers, 5.6 nm in diameter, twisted around each other in a double helical formation (Figure 10–11). A notable characteristic of all G-actin molecules is their structural asymmetry. When G-actin molecules polymerize to form F-actin, they bind back to front, producing a filament with distinguishable polarity (Figure 10–13). Each G-actin monomer contains a binding site for myosin (Figure 10–14). Actin filaments, which anchor perpendicularly on the Z line, exhibit opposite polarity on each side of the line

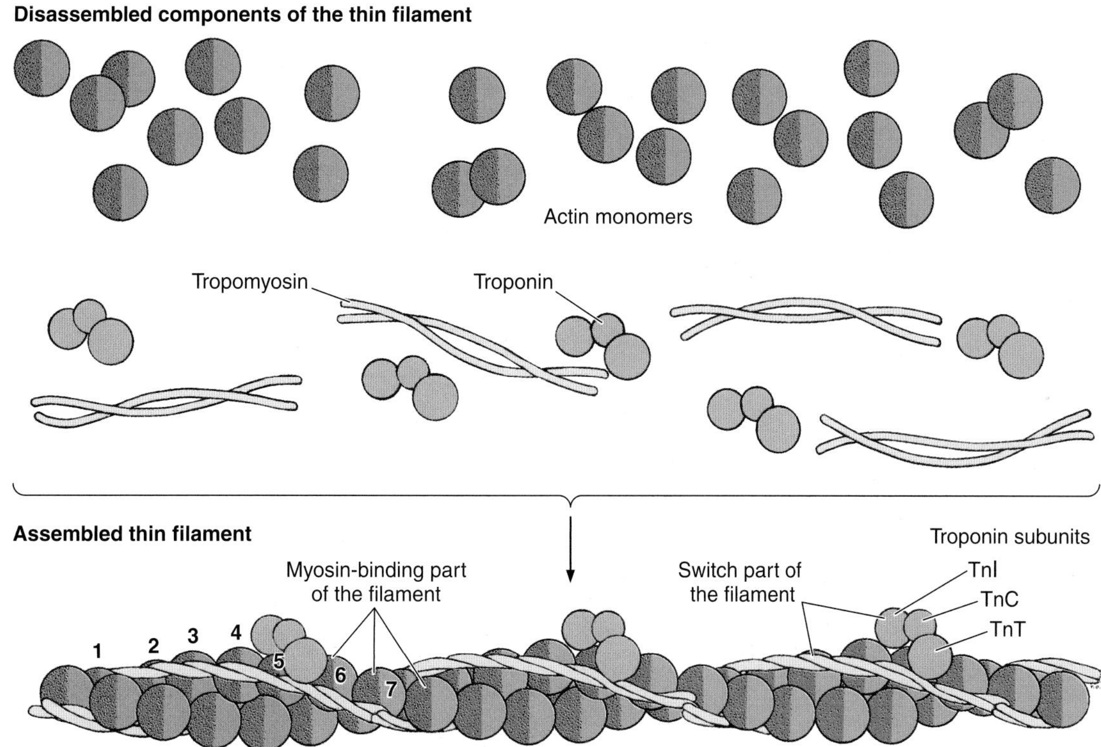

Figure 10–13. Schematic representation of the thin filament, showing the spatial configuration of 3 major protein components—actin, tropomyosin, and troponin. The individual components in the upper part of the drawing are shown in polymerized form in the lower part. The globular actin molecules are polarized and polymerize in one direction. Note that each tropomyosin molecule extends over 7 actin molecules. TnI, TnC, and TnT are troponin subunits.

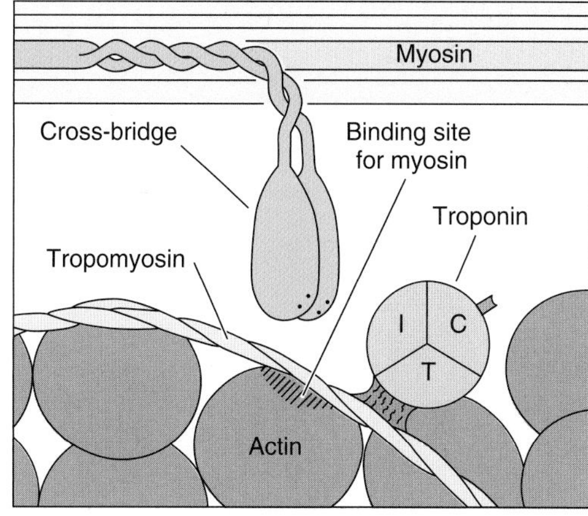

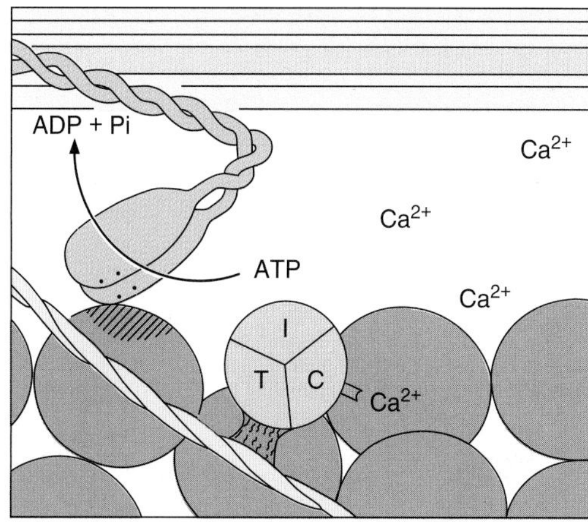

Figure 10–14. Muscle contraction, initiated by the binding of Ca^{2+} to the TnC unit of troponin, which exposes the myosin binding site on actin (cross-hatched area). In a second step, the myosin head binds to actin and the ATP breaks down into ADP, yielding energy, which produces a movement of the myosin head. As a consequence of this change in myosin, the bound thin filaments slide over the thick filaments. This process, which repeats itself many times during a single contraction, leads to a complete overlapping of the actin and myosin and a resultant shortening of the whole muscle fiber. I, T, C are troponin subunits. (Reproduced, with permission, from Ganong WF: *Review of Medical Physiology,* 20th ed. McGraw-Hill, 2001.)

(Figure 10–11). The protein α-actinin, a major component of the Z line, is thought to anchor the actin filaments to this region. α-Actinin and desmin (an intermediate filament protein) are believed to tie adjacent sarcomeres together, thus keeping the myofibrils in register.

Tropomyosin, a long, thin molecule about 40 nm in length, contains 2 polypeptide chains. These molecules are bound head to tail, forming filaments that run over the actin subunits alongside the outer edges of the groove between the two twisted actin strands (Figure 10–13).

Troponin is a complex of 3 subunits: **TnT,** which strongly attaches to tropomyosin; **TnC,** which binds calcium ions; and **TnI,** which inhibits the actin-myosin interaction. A troponin complex is attached at one specific site on each tropomyosin molecule (Figure 10–13).

In thin filaments, each tropomyosin molecule spans 7 G-actin molecules and has one troponin complex bound to its surface (Figure 10–13).

Myosin is a much larger complex (molecular mass ~500 kDa). Myosin can be dissociated into 2 identical heavy chains and 2 pairs of light chains. Myosin heavy chains are thin, rodlike molecules (150 nm long and 2–3 nm thick) made up of 2 heavy chains twisted together. Small globular projections at one end of each heavy chain form the heads, which have ATP binding sites as well as the enzymatic capacity to hydrolyze ATP (ATPase activity) and the ability to bind to actin. The 4 light chains are associated with the head (Figure 10–11). Several hundred myosin molecules are arranged within each thick filament with their rodlike portions overlapping and their globular heads directed toward either end (Figure 10–11).

Analysis of thin sections of striated muscle shows the presence of cross-bridges between thin and thick filaments. These bridges are known to be formed by the head of the myosin molecule plus a short part of its rodlike portion. These bridges are involved in the conversion of chemical energy into mechanical energy (Figure 10–14).

Sarcoplasmic Reticulum & Transverse Tubule System

The depolarization of the sarcoplasmic reticulum membrane, which results in the release of Ca^{2+} ions, is initiated at a specialized myoneural junction on the surface of the muscle cell. Surface-initiated depolarization signals would have to diffuse throughout the cell to effect the release of Ca^{2+} from internal sarcoplasmic reticulum cisternae. In larger muscle cells, the diffusion of the depolarization signal would lead to a wave of contraction, with peripheral myofibrils contracting before more centrally positioned myofibrils do. To provide for a uniform contraction, skeletal muscle possesses a system of **transverse (T) tubules** (Figure 10–15). These fingerlike invaginations of the sarcolemma form a complex anastomosing network of tubules that encircle the boundaries of the A–I bands of each sarcomere in every myofibril (Figures 10–16 and 10–17).

Adjacent to opposite sides of each T tubule are expanded **terminal cisternae** of the sarcoplasmic reticulum. This specialized complex, consisting of a T tubule with 2 lateral portions of sarcoplasmic reticulum, is known as the **triad** (Figures 10–10, 10–16, and 10–17). At the triad, depolarization of the sarcolemma-derived T tubules is transmitted to the sarcoplasmic reticulum membrane.

As described above, muscle contraction depends on the availability of Ca^{2+} ions, and muscle relaxation is related to an absence of Ca^{2+}. The sarcoplasmic reticulum specifically regulates

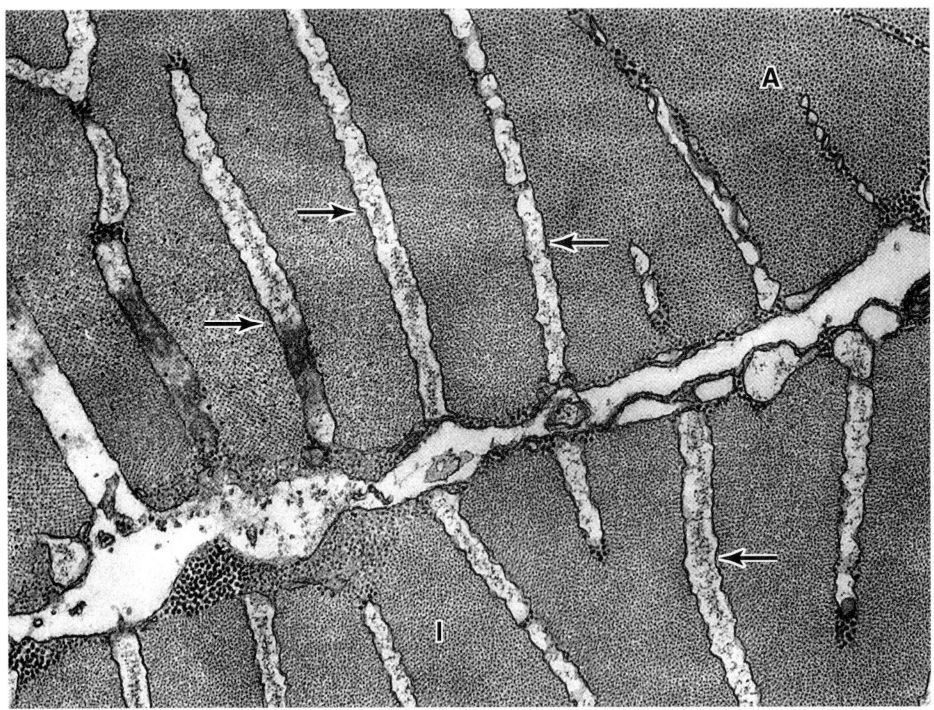

Figure 10–15. Electron micrograph of a transverse section of fish muscle, showing the surface of 2 cells limiting an intercellular space. Note the invaginations of the sarcolemma, forming the tubules of the T system (arrows). The dark, coarse granules in the cytoplasm (lower left) are glycogen particles. The section passes through the A band (upper right), showing thick and thin filaments. The I band is sectioned (lower left), showing only thin filaments. ×60,000. (Courtesy of KR Porter.)

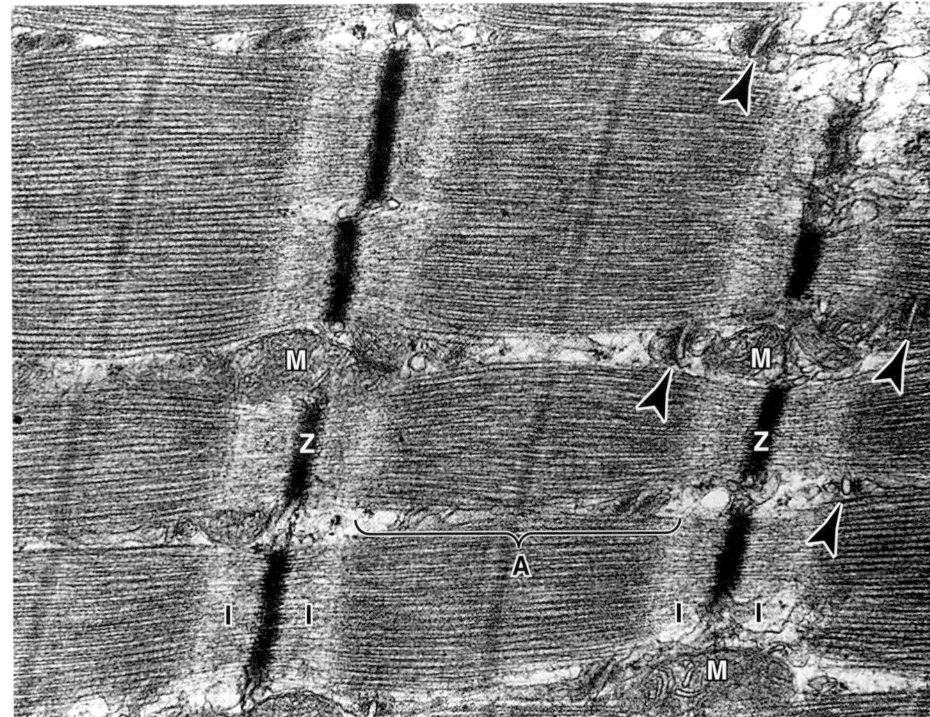

Figure 10–16. Electron micrograph of a longitudinal section of the skeletal muscle of a monkey. Note the mitochondria (M) between adjacent myofibrils. The arrowheads indicate triads—2 for each sarcomere in this muscle—located at the A–I band junction. A, A band; I, I band; Z, Z line. ×40,000. (Reproduced, with permission, from Junqueira LCU, Salles LMM: *Ultra-Estrutura e Função Celular.* Edgard Blücher, 1975.)

calcium flow, which is necessary for rapid contraction and relaxation cycles. The sarcoplasmic reticulum system consists of a branching network of smooth endoplasmic reticulum cisternae surrounding each myofibril (Figure 10–17). After a neurally mediated depolarization of the sarcoplasmic reticulum membrane, Ca^{2+} ions concentrated within the sarcoplasmic reticulum cisternae are passively released into the vicinity of the overlapping thick and thin filaments, whereupon they bind to troponin and allow bridging between actin and myosin. When the membrane depolarization ends, the sarcoplasmic reticulum acts as a calcium sink and actively transports the Ca^{2+} back into the cisternae, resulting in the cessation of contractile activity.

Mechanism of Contraction

Resting sarcomeres consist of partially overlapping thick and thin filaments. During contraction, both the thick and thin filaments retain their original length. Since contraction is not caused by a shortening of individual filaments, it must be the result of an increase in the amount of overlap between the filaments. The **sliding filament** hypothesis of muscle contraction has received the most widespread acceptance.

The following is a brief description of how actin and myosin interact during a contraction cycle. At rest, ATP binds to the ATPase site on the myosin heads, but the rate of hydrolysis is very slow. Myosin requires actin as a cofactor to break down ATP rapidly and release energy. In a resting muscle, myosin cannot associate with actin, because the binding sites for myosin heads on actin molecules are covered by the troponin-tropomyosin complex on the F-actin filament (Figure 10–14, top). When sufficiently high concentrations of calcium ions are available, however, they bind to the TnC subunit of troponin. The spatial configuration of the 3 troponin subunits changes and drives the tropomyosin molecule deeper into the groove of the actin helix (Figure 10–14). This exposes the myosin-binding site on the globular actin components, so that actin is free to interact with the head of the myosin molecule.

The binding of calcium ions to the TnC unit corresponds to the stage at which myosin-ATP is converted into the active complex. As a result of bridging between the myosin head and the G-actin subunit of the thin filament, the ATP is split into ADP and Pi, and energy is released. This activity leads to a deformation, or bending, of the head and a part of the rodlike portion (hinge region) of the myosin (Figure 10–14). Since the actin is bound to the myosin, movement of the myosin head pulls the actin past the myosin filament. The result is that the thin filament is drawn farther into the A band.

Although a large number of myosin heads extend from the thick filament, at any one time during the contraction only a small number of heads align with available actin-binding sites. As the bound myosin heads move the actin, however, they provide for alignment of new actin-myosin bridges. The old actin-myosin bridges detach only after the myosin binds a new ATP molecule; this action also resets the myosin head and prepares it for another contraction cycle. If no ATP is available, the actin-myosin complex becomes stable; this accounts for the extreme muscular rigidity (**rigor mortis**) that occurs after death. A single muscle contraction is the result of hundreds of bridge-forming and bridge-breaking cycles. The contraction activity that leads to a complete overlap between thin and thick filaments continues until Ca^{2+} ions are removed and the troponin-tropomyosin complex again covers the myosin binding site.

During contraction, the I band decreases in size as thin filaments penetrate the A band. The H band—the part of the A

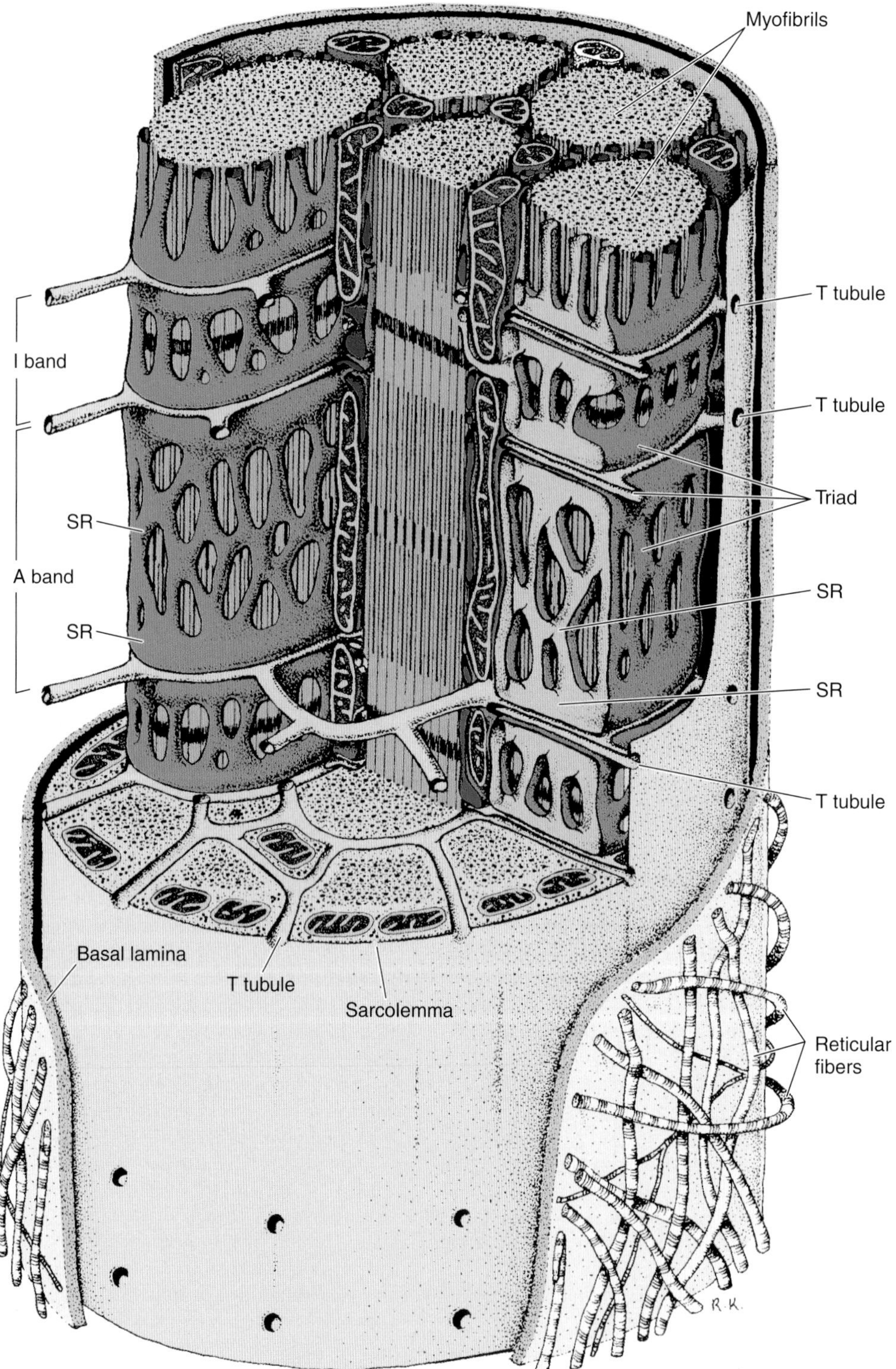

Figure 10–17. Segment of mammalian skeletal muscle. The sarcolemma and muscle fibrils are partially cut, showing the following components: The invaginations of the T system occur at the level of transition between the A and I bands twice in every sarcomere. They associate with terminal cisternae of the sarcoplasmic reticulum (SR), forming triads. Abundant mitochondria lie between the myofibrils. The cut surface of the myofibrils shows the thin and thick filaments. Surrounding the sarcolemma are a basal lamina and reticular fibers. (Reproduced, with permission, from Krstíc RV: *Ultrastructure of the Mammalian Cell.* Springer-Verlag, 1979.)

band with only thick filaments—diminishes in width as the thin filaments completely overlap the thick filaments. A net result is that each sarcomere, and consequently the whole cell (fiber), is greatly shortened (Figure 10–18).

Innervation

Myelinated motor nerves branch out within the perimysial connective tissue, where each nerve gives rise to several terminal twigs. At the site of innervation, the nerve loses its myelin sheath and forms a dilated termination that sits within a trough on the muscle cell surface. This structure is called the **motor end-plate,** or **myoneural junction** (Figure 10–18). At this site, the axon is covered by a thin cytoplasmic layer of Schwann cells. Within the axon terminal are numerous mitochondria and synaptic vesicles, the latter containing the neurotransmitter **acetylcholine.** Between the axon and the muscle is a space, the **synaptic cleft,** in which lies an amorphous basal lamina matrix. At the junction, the sarcolemma is thrown into numerous deep **junctional folds.** In the sarcoplasm below the folds lie several nuclei and numerous mitochondria, ribosomes, and glycogen granules.

When an action potential invades the motor end-plate, acetylcholine is liberated from the axon terminal, diffuses through the cleft, and binds to acetylcholine receptors in the sarcolemma of the junctional folds. Binding of the transmitter makes the sarcolemma more permeable to sodium, which results in **membrane depolarization.** Excess acetylcholine is hydrolyzed by the enzyme cholinesterase bound to the synaptic cleft basal lamina. Acetylcholine breakdown is necessary to avoid prolonged contact of the transmitter with receptors present in the sarcolemma.

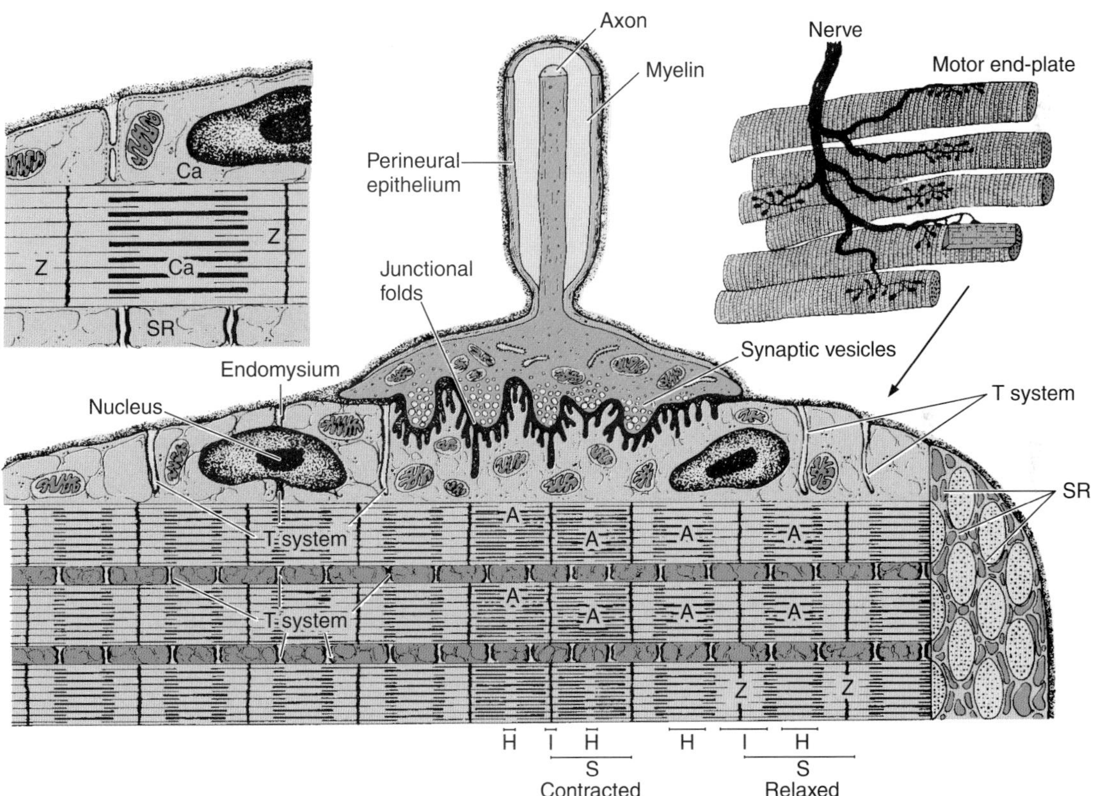

Figure 10–18. Ultrastructure of the motor end-plate and the mechanism of muscle contraction. The drawing at the upper right shows branching of a small nerve with a motor end-plate for each muscle fiber. The structure of one of the bulbs of an end-plate is highly enlarged in the center drawing. Note that the axon terminal bud contains synaptic vesicles. The region of the muscle cell membrane covered by the terminal bud has clefts and ridges called **junctional folds.** The axon loses its myelin sheath and dilates, establishing close, irregular contact with the muscle fiber. Muscle contraction begins with the release of acetylcholine from the synaptic vesicles of the end-plate. This neurotransmitter causes a local increase in the permeability of the sarcolemma. The process is propagated to the rest of the sarcolemma, including its invaginations (all of which constitute the T system), and is transferred to the sarcoplasmic reticulum (SR). The increase of permeability in this organelle liberates calcium ions (drawing at upper left) that trigger the sliding filament mechanism of muscle contraction. Thin filaments slide between the thick filaments and reduce the distance between the Z lines, thereby reducing the size of all bands except the A band. H, H band; S, sarcomere.

The depolarization initiated at the motor end-plate is propagated along the surface of the muscle cell and deep into the fibers via the transverse tubule system. At each triad, the depolarization signal is passed to the sarcoplasmic reticulum and results in the release of Ca^{2+}, which initiates the contraction cycle. When depolarization ceases, the Ca^{2+} is actively transported back into the sarcoplasmic reticulum cisternae, and the muscle relaxes.

MEDICAL APPLICATION

Myasthenia gravis *is an autoimmune disorder characterized by progressive muscular weakness caused by a reduction in the number of functionally active acetylcholine receptors in the sarcolemma of the myoneural junction. This reduction is caused by circulating antibodies that bind to the acetylcholine receptors in the junctional folds and inhibit normal nerve-muscle communication. As the body attempts to correct the condition, membrane segments with affected receptors are internalized, digested by lysosomes, and replaced by newly formed receptors. These receptors, however, are again made unresponsive to acetylcholine by the same antibodies, and the disease follows its progressive course.*

A single nerve fiber (**axon**) can innervate one muscle fiber, or it may branch and be responsible for innervating 160 or more muscle fibers. In the case of multiple innervation, a single nerve fiber and all the muscles it innervates are called a **motor unit.** Individual striated muscle fibers do not show graded contraction—they contract either all the way or not at all. To vary the force of contraction, the fibers within a muscle bundle should not all contract at the same time. Since muscles are broken up into motor units, the firing of a single nerve motor axon will generate tension proportional to the number of muscle fibers innervated by that axon. Thus, the number of motor units and the variable size of each unit can control the intensity of a muscle contraction. The ability of a muscle to perform delicate movements depends on the size of its motor units. For example, because of the fine control required by eye muscles, each of their fibers is innervated by a different nerve fiber. In larger muscles exhibiting coarser movements, such as those of the limb, a single, profusely branched axon innervates a motor unit that consists of more than 100 individual muscle fibers.

Muscle Spindles & Golgi Tendon Organs

All human striated muscles contain encapsulated proprioceptors (L. *proprius,* one's own, + *capio,* to take) known as **muscle spindles** (Figure 10–19). These structures consist of a connective tissue capsule surrounding a fluid-filled space that contains a few long, thick muscle fibers and some short, thinner fibers (collectively called **intrafusal fibers**). Several sensory nerve fibers penetrate the muscle spindles, where they detect changes in the length (distension) of extrafusal muscle fibers and relay this information to the spinal cord. Here, reflexes of varying complexity are activated to maintain posture and to regulate the activity of opposing muscle groups involved in motor activities such as walking.

In tendons, near the insertion sites of muscle fibers, a connective tissue sheath encapsulates several large bundles of collagen fibers that are continuous with the collagen fibers that make up the myotendinous junction. Sensory nerves penetrate the connective tissue capsule. These structures, known as **Golgi tendon organs** (Figure 10–20), contribute to proprioception by detecting tensional differences in tendons.

Because these structures are sensitive to increases in tension, they help to regulate the amount of effort required to perform movements that call for variable amounts of muscular force.

System of Energy Production

Skeletal muscle cells are highly adapted for discontinuous production of intense mechanical work through the release of chem-

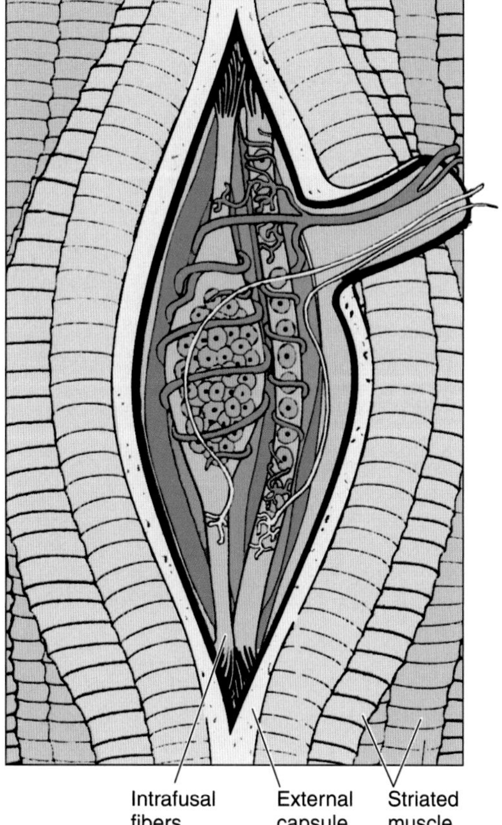

Figure 10–19. Muscle spindle showing afferent and efferent nerve fibers that make synapses with the intrafusal fibers (modified muscle fibers). Note the complex nerve terminal on the intrafusal fibers. The two types of intrafusal fibers, one with a small diameter and the other with a dilation filled with nuclei, are shown. Muscle spindles participate in the nervous control of body posture and the coordinate action of opposing muscles.

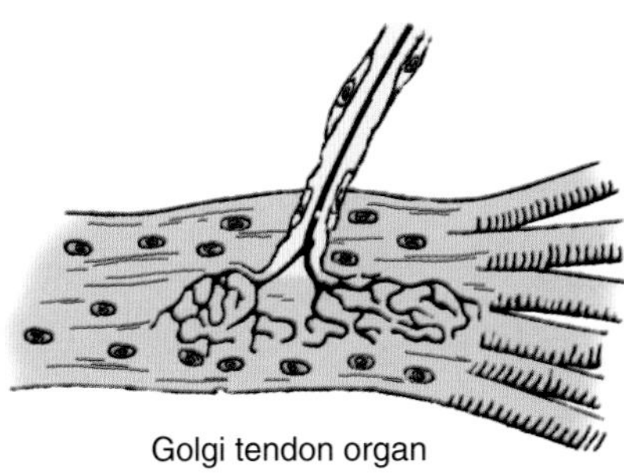

Figure 10–20. Drawing of a Golgi tendon organ. This structure collects information about differences in tension among tendons and relays data to the central nervous system, where they are processed and help to coordinate fine muscular contractions.

ical energy and must have depots of energy to cope with bursts of activity. The most readily available energy is stored in the form of ATP and phosphocreatine, both of which are energy-rich phosphate compounds. Chemical energy is also available in glycogen depots, which constitute about 0.5–1% of muscle weight. Muscle tissue obtains energy to be stored in phosphocreatine and ATP from the breakdown of fatty acids and glucose. Fatty acids are broken down to acetate by the enzymes of β-oxidation, located in the mitochondrial matrix. Acetate is then further oxidized by the citric acid cycle, with the resulting energy being conserved in the form of ATP. When skeletal muscles are subjected to a short-term (sprint) exercise, they rapidly metabolize glucose (coming mainly from muscle glycogen stores) to lactate, causing an oxygen debt that is repaid during the recovery period. The lactate formed during this type of exercise causes cramping and pain in skeletal muscles.

Based on their morphologic, histochemical, and biochemical characteristics, muscle fibers can be classified as type I (slow) and type II (quick). **Type I fibers** are rich in sarcoplasm, which contains myoglobin (accounting for the dark red color; see below). They are related to continuous contraction, and their energy is derived from oxidative phosphorylation of fatty acids. **Type II fibers** are related to rapid discontinuous contraction. They contain less myoglobin (producing a light red color). Type II fibers can be further divided into types IIA, IIB, and IIC, according to their activity and chemical characteristics (mainly, the stability of the actomyosin-ATPase they contain). Type IIB fibers have the fastest action and depend more than the others on glycolysis as a source of energy. The classification of muscle fibers has clinical significance for the diagnosis of muscle diseases, or myopathies (*mys* + Gr. *pathos,* suffering). In humans, skeletal muscles are frequently composed of mixtures of these various types of fibers.

The differentiation of muscle into red, white, and intermediate fiber types is controlled by its innervation. In experiments where the nerves to red and white fibers are cut, crossed, and allowed to regenerate, the myofibers change their morphologic and physiologic characteristics to conform to the innervating nerve. Simple denervation of muscle will lead to fiber atrophy and paralysis.

Other Components of the Sarcoplasm

Glycogen is found in abundance in the sarcoplasm in the form of coarse granules (Figure 10–15). It serves as a depot of energy that is mobilized during muscle contraction.

Another component of the sarcoplasm is **myoglobin** (Figure 10–21); this oxygen-binding protein, which is similar to hemoglobin, is principally responsible for the dark red color of some muscles. Myoglobin acts as an oxygen-storing pigment, which is necessary for the high oxidative phosphorylation level in this type of fiber. For obvious reasons, it is present in great amounts in the muscle of deep-diving ocean mammals (eg, seals, whales). Muscles that must maintain activity for prolonged periods usually are red and have a high myoglobin content.

Mature muscle cells have negligible amounts of rough endoplasmic reticulum and ribosomes, an observation that is consistent with the low level of protein synthesis in this tissue.

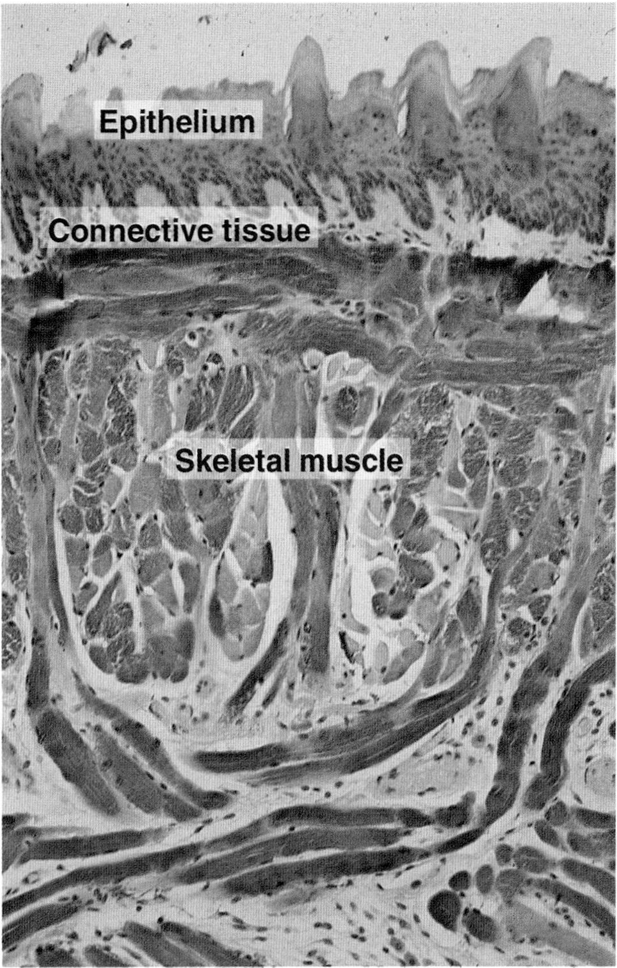

Figure 10–21. Section of tongue, an organ rich in striated skeletal muscle fibers. These fibers appear brown because the section was immunohistologically stained to show myoglobin. The light-colored areas among and above the muscle fibers contain connective tissue. In the upper region of the section, stratified and cornified epithelium can be seen. Nuclei are stained by hematoxylin. Low magnification.

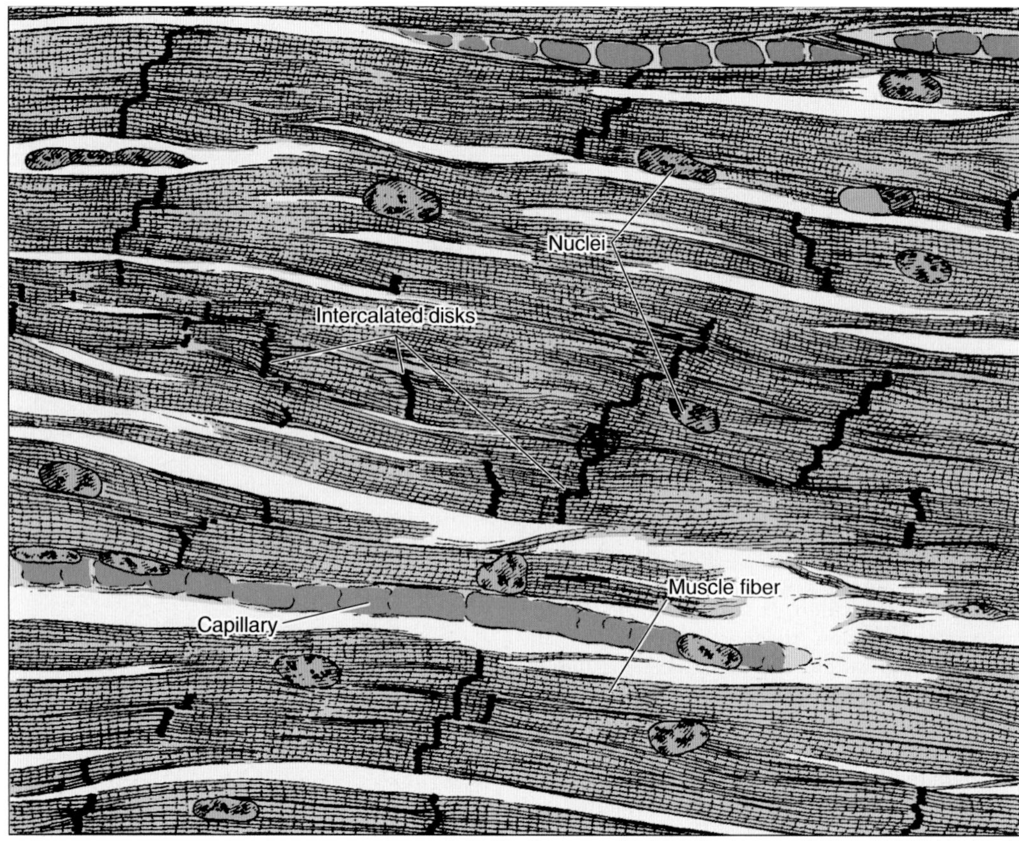

Figure 10–22. Drawing of a section of heart muscle, showing central nuclei, cross-striation, and intercalated disks.

CARDIAC MUSCLE

During embryonic development, the splanchnic mesoderm cells of the primitive heart tube align into chainlike arrays. Rather than fusing into syncytial (Gr. *syn,* together, + *kytos,* cell) cells, as in the development of skeletal muscle, cardiac cells form complex junctions between their extended processes. Cells within a chain often bifurcate, or branch, and bind to cells in adjacent chains. Consequently, the heart consists of tightly knit bundles of cells, interwoven in a fashion that provides for a characteristic wave of contraction that leads to a wringing out of the heart ventricles.

Mature cardiac muscle cells are approximately 15 μm in diameter and from 85 to 100 μm in length. They exhibit a cross-striated banding pattern identical to that of skeletal muscle. Unlike multinucleated skeletal muscle, however, each cardiac muscle cell possesses only one or two centrally located pale-staining nuclei. Surrounding the muscle cells is a delicate sheath of endomysial connective tissue containing a rich capillary network.

A unique and distinguishing characteristic of cardiac muscle is the presence of dark-staining transverse lines that cross the chains of cardiac cells at irregular intervals (Figures 10–22 and 10–23). These **intercalated disks** represent junctional complexes found at the interface between adjacent cardiac muscle cells (Figures 10–24, 10–25, and 10–26). The junctions may appear as straight lines or may exhibit a steplike pattern. Two regions can be distinguished in the steplike junctions—a **transverse portion,** which runs across the fibers at right angles, and a **lateral portion,** which runs parallel to the myofilaments. There are 3 main junctional specializations within the disk. **Fasciae adherentes,** the most prominent membrane specialization in transverse portions of the disk, serve as anchoring sites for actin filaments of the terminal sarcomeres. Essentially, they represent **hemi-Z bands. Maculae adherentes** (desmosomes) are also present in the transverse portion and bind the cardiac cells together to prevent their pulling apart under constant contractile activity. On the lateral portions of the disk, **gap junctions** provide ionic continuity between adjacent cells (Figure 10–26). The significance of ionic coupling is that chains of individual cells act as a syncytium, allowing the signal to contract to pass in a wave from cell to cell.

The structure and function of the contractile proteins in cardiac cells are virtually the same as in skeletal muscle. The T-tubule system and sarcoplasmic reticulum, however, are not as regularly arranged in the cardiac myocytes. The T tubules are more numerous and larger in ventricular muscle than in skeletal muscle. Cardiac T tubules are found at the level of

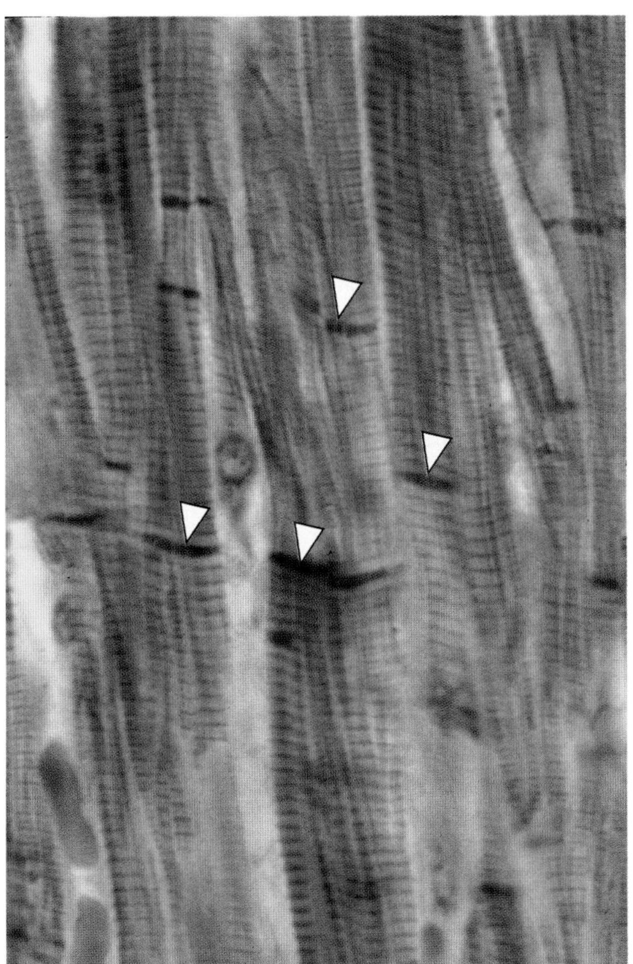

Figure 10–23. Photomicrograph of cardiac muscle. Note the cross-striation and the intercalated disks (arrowheads). Pararosaniline–toluidine blue (PT) stain. High magnification.

the Z band rather than at the A–I junction (as in mammalian skeletal muscle). The sarcoplasmic reticulum is not as well developed and wanders irregularly through the myofilaments. As a consequence, discrete myofibrillar bundles are not present.

Triads are not common in cardiac cells, because the T tubules are generally associated with only one lateral expansion of sarcoplasmic reticulum cisternae. Thus, heart muscle characteristically possesses **diads** composed of one T tubule and one sarcoplasmic reticulum cisterna.

Cardiac muscle cells contain numerous mitochondria, which occupy 40% or more of the cytoplasmic volume (Figure 10–27), reflecting the need for continuous aerobic metabolism in heart muscle. By comparison, only about 2% of skeletal muscle fiber is occupied by mitochondria. Fatty acids, transported to cardiac muscle cells by lipoproteins, are the major fuel of the heart. Fatty acids are stored as triglycerides in the numerous lipid droplets seen in cardiac muscle cells. A small amount of glycogen is present and can be broken down to glucose and used for energy production during periods of stress. Lipofuscin pigment granules (aging pigment), often seen in long-lived cells, are found near the nuclear poles of cardiac muscle cells.

A few differences in structure exist between atrial and ventricular muscle. The arrangement of myofilaments is the same in the two types of cardiac muscle, but atrial muscle has markedly fewer T tubules, and the cells are somewhat smaller. Membrane-limited granules, each about 0.2–0.3 μm in diameter, are found at both poles of cardiac muscle nuclei and in association with Golgi complexes in this region. These granules (Figure 10–28) are most abundant in muscle cells of the right atrium (approximately 600 per cell), but they are also found in the left atrium, the ventricles, and several other places in the body. These atrial granules contain the high-molecular-weight precursor of a polypeptide hormone known as **atrial natriuretic factor.** Atrial natriuretic factor acts on the kidneys to cause sodium and water loss (natriuresis and diuresis). This hormone thus opposes the actions of aldosterone and antidiuretic hormone, whose effects on kidneys result in sodium and water conservation.

The rich autonomic nerve supply to the heart and the rhythmic impulse-generating and conducting structures are discussed in Chapter 11.

SMOOTH MUSCLE

Smooth muscle is composed of elongated, nonstriated cells (Figure 10–29), each of which is enclosed by a basal lamina and a network of reticular fibers (Figures 10–30, 10–31, and 10–32). The last two components serve to combine the force generated by each smooth muscle fiber into a concerted action, eg, peristalsis in the intestine.

Smooth muscle cells are fusiform; ie, they are largest at their midpoints and taper toward their ends. They may range in size from 20 μm in small blood vessels to 500 μm in the pregnant uterus. During pregnancy, uterine smooth muscle cells undergo a marked increase in size and number. Each cell has a single nucleus located in the center of the broadest part of the cell. To achieve the tightest packing, the narrow part of one cell lies adjacent to the broad parts of neighboring cells. Such an arrangement viewed in cross section shows a range of diameters, with only the largest profiles containing a nucleus (Figures 10–29). The borders of the cell become scalloped when smooth muscle contracts, and the nucleus becomes folded or has the appearance of a corkscrew (Figure 10–33).

Concentrated at the poles of the nucleus are mitochondria, polyribosomes, cisternae of rough endoplasmic reticulum, and the Golgi complex. Pinocytotic vesicles are frequent near the cell surface (Figure 10–32).

A rudimentary sarcoplasmic reticulum is present; it consists of a closed system of membranes, similar to the sarcoplasmic reticulum of striated muscle. T tubules are not present in smooth muscle cells.

The characteristic contractile activity of smooth muscle is related to the structure and organization of its actin and myosin filaments, which do not exhibit the paracrystalline organization present in striated muscles. In smooth muscle cells, bundles of myofilaments crisscross obliquely through the cell, forming a latticelike network. These bundles consist of thin filaments (5–7 nm) containing actin and tropomyosin and thick filaments (12–16 nm) consisting of myosin. Both structural and biochemical studies reveal that smooth muscle actin and myosin

Figure 10–24. Longitudinal section of portions of 2 cardiac muscle cells. The transversely oriented parts of the intercalated disk consist of a fascia adherens and numerous desmosomes. The longitudinal parts (arrows) contain gap junctions. Mitochondria (M) are numerous. Fibrils of reticular fibers are seen between the two cells. ×18,000. (Reproduced, with permission, from Junqueira LCU, Salles LMM: *Ultra-Estrutura e Função Celular.* Edgard Blücher, 1975.)

contract by a sliding filament mechanism similar to that which occurs in striated muscles.

An influx of Ca^{2+} is involved in the initiation of contraction in smooth muscle cells. The myosin of smooth muscle, however, interacts with actin only when its light chain is phosphorylated. For this reason, and because the tropomyosin complex of skeletal muscle is absent, the contraction mechanism in smooth muscle differs somewhat from skeletal and cardiac muscle. Ca^{2+} in a smooth muscle complexes with **calmodulin,** a calcium-binding protein that is also involved in the contraction of nonmuscle cells. The Ca^{2+}-calmodulin complex activates myosin light-chain kinase, the enzyme responsible for the phosphorylation of myosin.

Factors other than calcium affect the activity of myosin light-chain kinase and thus influence the degree of contraction of smooth muscle cells. Contraction or relaxation may be regulated by hormones that act via cyclic AMP (cAMP). When levels of cAMP increase, myosin light-chain kinase is activated, myosin is phosphorylated, and the cell contracts. A decrease in cAMP has the opposite effect, decreasing contractility. The action of sex hormones on uterine smooth muscle is another example of nonneural control. Estrogens increase cAMP and promote the phosphorylation of myosin and the contractile activity of uterine smooth muscle. Progesterone has the opposite effect: It decreases cAMP, promotes dephosphorylation of myosin, and relaxes uterine musculature.

Smooth muscle cells have an elaborate array of 10-nm intermediate filaments coursing through their cytoplasm. **Desmin** (**skeletin**) has been identified as the major protein of intermediate filaments in all smooth muscles, and **vimentin**

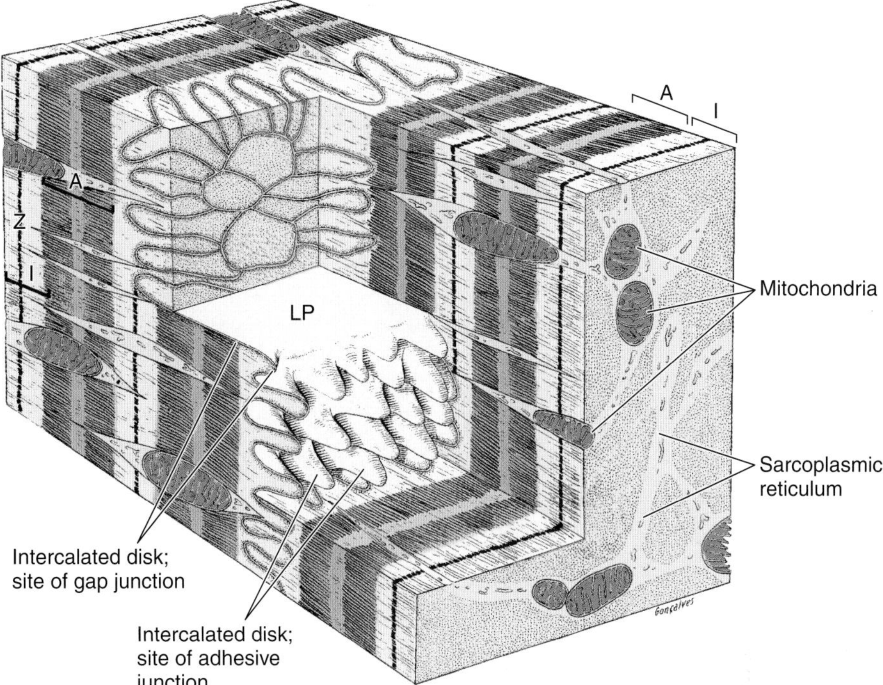

Figure 10–25. Ultrastructure of heart muscle in the region of an intercalated disk. Contact between cells is accomplished by interdigitation in the transverse region; contact is broad and flat in the longitudinal plane (LP). A, A band; I, I band; Z, Z line. (Redrawn and reproduced, with permission, from Marshall JM: The heart. In: *Medical Physiology,* 13th ed, Vol 2, Mountcastle VB [editor]. Mosby, 1974. Based on the results of Fawcett DW, McNutt NS: J Cell Biol 1969;42:1, modified from Poche R, Lindner E: Zellforsch Mikrosk Anat 1955; 43:104.)

is an additional component in vascular smooth muscle. Two types of **dense bodies** (Figure 10–33) appear in smooth muscle cells. One is membrane-associated; the other is cytoplasmic. Both contain α-actinin and are thus similar to the Z lines of striated muscles. Both thin and intermediate filaments insert into dense bodies that transmit contractile force to adjacent smooth muscle cells and their surrounding network of reticular fibers.

The degree of innervation in a particular bundle of smooth muscle depends on the function and the size of that muscle. Smooth muscle is innervated by both sympathetic and parasympathetic nerves of the autonomic system. Elaborate neuromuscular junctions like those in skeletal muscle are not present in smooth muscle. Frequently, autonomic nerve axons terminate in a series of dilatations in the endomysial connective tissue.

In general, smooth muscle occurs in large sheets such as those found in the walls of hollow viscera, eg, the intestines, uterus, and ureters. Their cells possess abundant gap junctions and a relatively poor nerve supply. These muscles function in syncytial fashion and are called **visceral smooth muscles.** In contrast, the **multiunit smooth muscles** have a rich innervation and can produce such precise and graded contractions as those occurring in the iris of the eye.

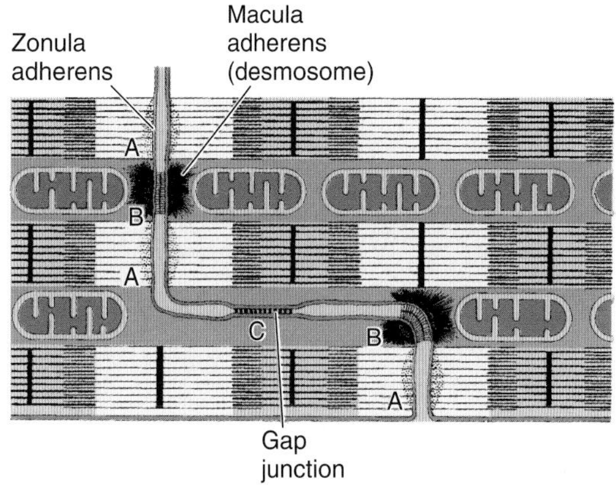

Figure 10–26. Junctional specializations making up the intercalated disk. Fasciae (or zonulae) adherentes (A) in the transverse portions of the disk anchor actin filaments of the terminal sarcomeres to the plasmalemma. Maculae adherentes, or desmosomes (B), found primarily in the transverse portions of the disk, bind cells together, preventing their separation during contraction cycles. Gap junctions (C), restricted to longitudinal portions of the disk—the area subjected to the least stress—ionically couple cells and provide for the spread of contractile depolarization.

Figure 10–27. Electron micrograph of a longitudinal section of heart muscle. Note the striation pattern and the alternation of myofibrils and mitochondria rich in cristae. Note the sarcoplasmic reticulum (SR), which is the specialized calcium-storing smooth endoplasmic reticulum. ×30,000.

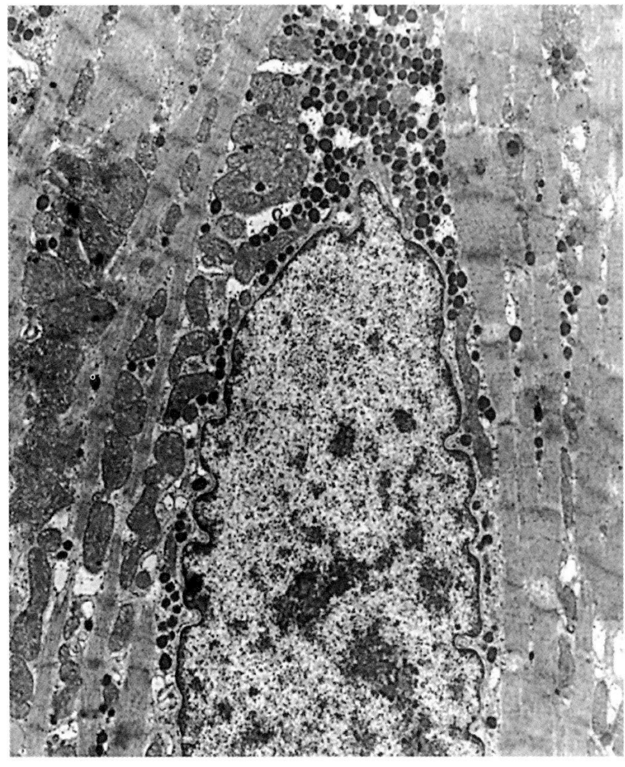

Figure 10–28. Electron micrograph of an atrial muscle cell showing the presence of natriuretic granules aggregated at the nuclear pole. (Courtesy of JC Nogueira.)

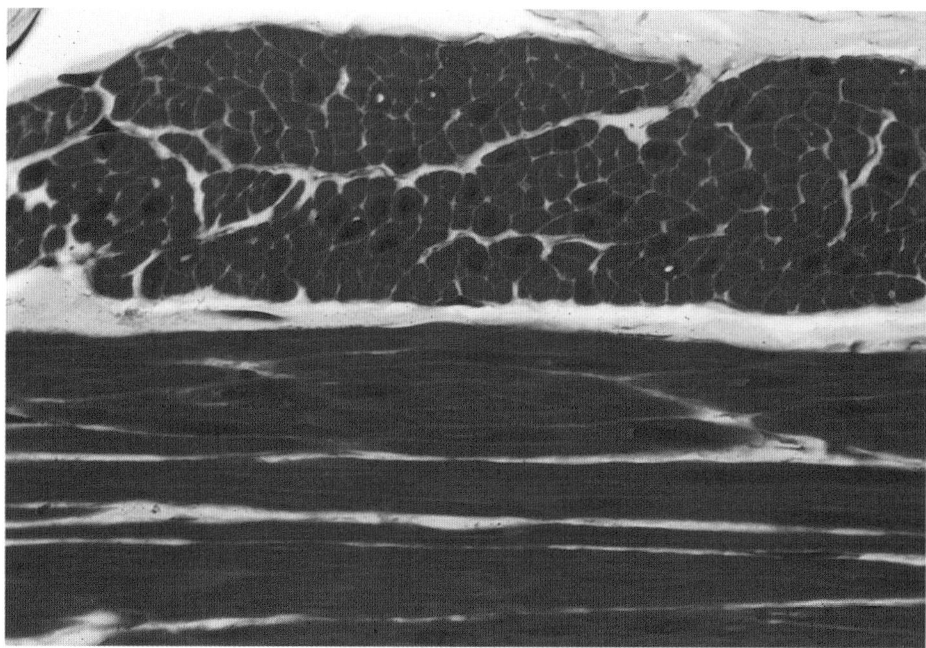

Figure 10–29. Photomicrographs of smooth muscle cells in cross section (upper) and in longitudinal section (lower). Note the centrally located nuclei. In many cells the nuclei were not included in the section. PT stain. Medium magnification.

Smooth muscle usually has spontaneous activity in the absence of nervous stimuli. Its nerve supply, therefore, has the function of modifying activity rather than, as in skeletal muscle, initiating it. Smooth muscle receives both adrenergic and cholinergic nerve endings that act antagonistically, stimulating or depressing its activity. In some organs, the cholinergic endings activate and the adrenergic nerves depress; in others, the reverse occurs.

In addition to contractile activity, smooth muscle cells also synthesize collagen, elastin, and proteoglycans, which are extracellular products normally associated with the function of fibroblasts.

REGENERATION OF MUSCLE TISSUE

The 3 types of adult muscle have different potentials for regeneration after injury.

Cardiac muscle has virtually no regenerative capacity beyond early childhood. Defects or damage (eg, infarcts) in heart muscle are generally replaced by the proliferation of connective tissue, forming myocardial scars.

In skeletal muscle, although the nuclei are incapable of undergoing mitosis, the tissue can undergo limited regeneration. The source of regenerating cells is believed to be the **satellite cells.** The latter are a sparse population of mononucleated spindle-shaped cells that lie within the basal lamina surrounding each mature muscle fiber. Because of their intimate apposition with the surface of the muscle fiber, they can be identified only with the electron microscope. They are considered to be inactive myoblasts that persist after muscle differentiation. After injury or certain other stimuli, the normally quiescent satellite cells become activated, proliferating and fusing to form new skeletal muscle fibers. A similar activity of satellite cells has been implicated in muscle hypertrophy, where

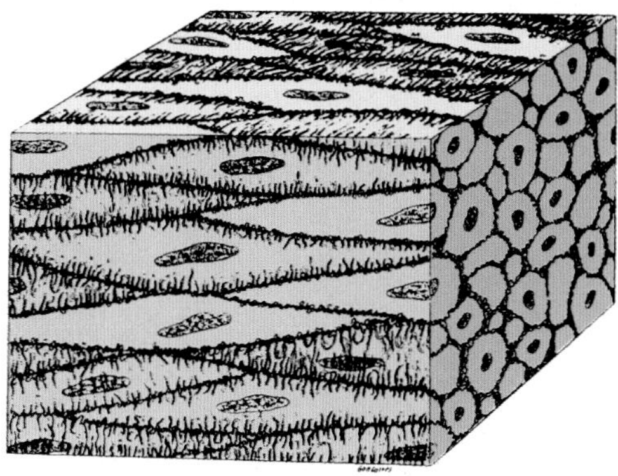

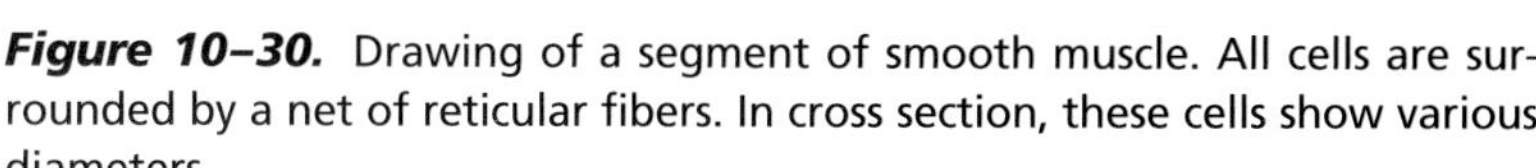

Figure 10–30. Drawing of a segment of smooth muscle. All cells are surrounded by a net of reticular fibers. In cross section, these cells show various diameters.

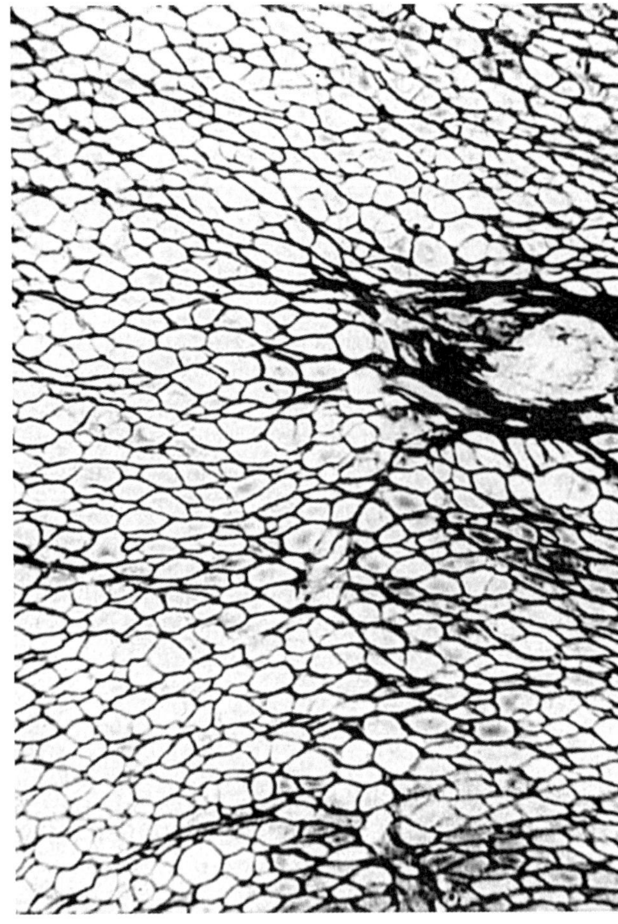

Figure 10–31. Transverse section of smooth muscle impregnated with silver to stain the reticular fibers. These fibers form a network that surrounds the muscle cells that are not stained by this method. At the right is an arteriole surrounded by thicker collagen fibers. ×300.

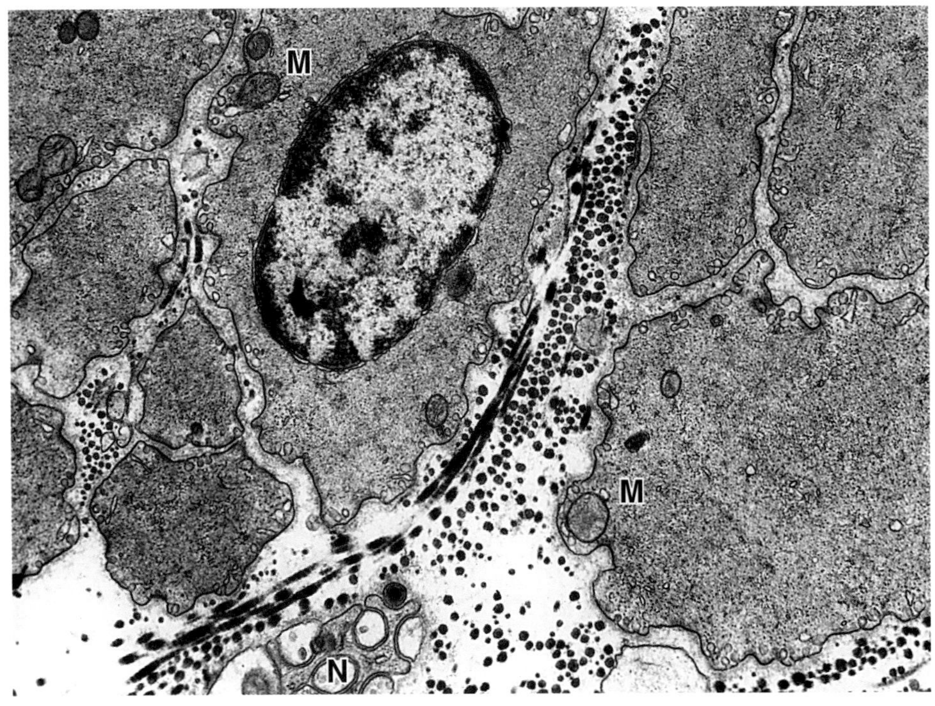

Figure 10–32. Electron micrograph of a transverse section of smooth muscle. The cells are sectioned at various diameters and have many subsurface vesicles in their cytoplasm. Thick and thin filaments are not organized into myofibrils, and there are few mitochondria (M). Note the collagen fibrils of the reticular fibers and a small unmyelinated nerve (N) between the cells. ×6650.

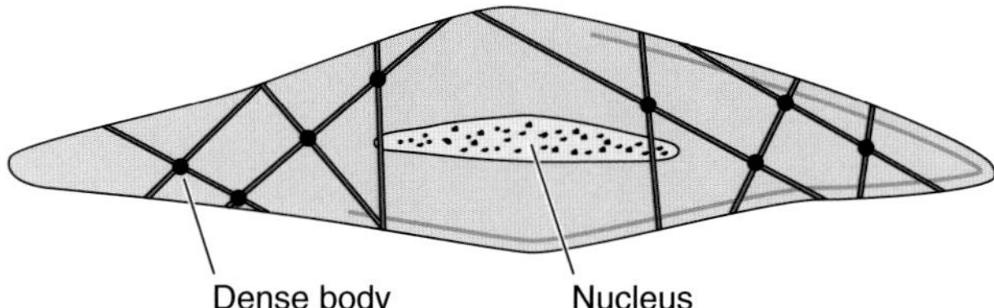

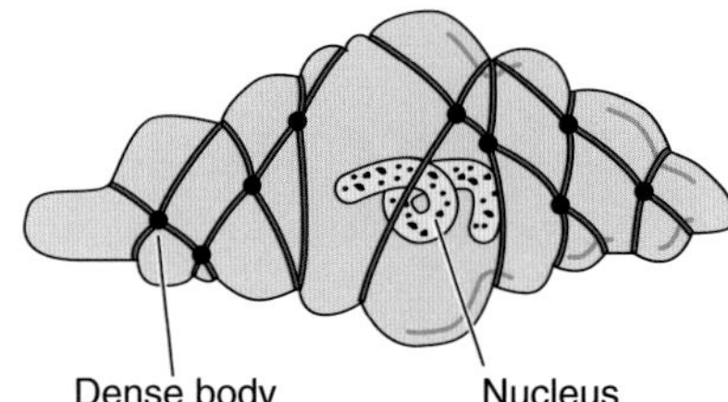

Figure 10–33. Smooth muscle cells relaxed and contracted. Cytoplasmic filaments insert on dense bodies located in the cell membrane and deep in the cytoplasm. Contraction of these filaments decreases the size of the cell and promotes the contraction of the whole muscle. During the contraction the cell nucleus is deformed.

they fuse with their parent fibers to increase muscle mass after extensive exercise. The regenerative capacity of skeletal muscle is limited, however, after major muscle trauma or degeneration.

Smooth muscle is capable of an active regenerative response. After injury, viable mononucleated smooth muscle cells and pericytes from blood vessels (see Chapter 11) undergo mitosis and provide for the replacement of the damaged tissue.

REFERENCES

Campion DR: The muscle satellite cell: a review. Int Rev Cytol 1984;87:225.

Cantin M, Genest J: The heart as an endocrine gland. Sci Am 1986;254:76.

Cohen C: The protein switch of muscle contraction. Sci Am 1975;233:36.

Grounds MD: Age-associated changes in the response of skeletal muscle cells to exercise and regeneration. Ann N Y Acad Sci 1998;854:78.

Huxley HE: Molecular basis of contraction in cross-striated muscles and relevance to motile mechanisms in other cells. In Stracher A (editor): *Muscle and Nonmuscle Motility,* Vol 1. Academic Press, 1983.

Vierck J et al: Satellite cell regulation following myotrauma caused by resistance exercise. Cell Biol Intl 2000;24:263.

The Circulatory System 11

The circulatory system comprises both the blood and lymphatic vascular systems. The blood vascular system is composed of the following structures:

The **heart,** an organ whose function is to pump the blood.

The **arteries,** a series of efferent vessels that become smaller as they branch, and whose function is to carry the blood, with nutrients and oxygen, to the tissues.

The **capillaries,** the smallest blood vessels, constituting a complex network of thin tubules that anastomose profusely and through whose walls the interchange between blood and tissues takes place.

The **veins,** which result from the convergence of the capillaries into a system of channels. These channels become larger as they approach the heart, toward which they convey the blood to be pumped again.

The **lymphatic vascular system** begins in the **lymphatic capillaries,** closed-ended tubules that anastomose to form vessels of steadily increasing size; these vessels terminate in the **blood vascular system** emptying into the large veins near the heart. One of the functions of the lymphatic system is to return the fluid of the tissue spaces to the blood. The internal surface of all components of the blood and lymphatic systems is lined by a single layer of a squamous epithelium, called endothelium.

It is customary to divide the circulatory system into the **macrovasculature,** vessels with more than 0.1 mm in diameter (large arterioles, muscular and elastic arteries, and muscular veins), and the **microvasculature** (arterioles, capillaries, and postcapillary venules) visible only with a microscope (Figure 11–1). The microvasculature is particularly important as the site of interchanges between the blood and surrounding tissues under normal conditions and in the event of inflammatory processes.

Capillaries have structural variations to permit different levels of metabolic exchange between blood and surrounding tissues. Capillaries are composed of a single layer of **endothelial cells** rolled up in the form of a tube. The average diameter of capillaries varies from 7 to 9 μm, and their length is usually not more than 50 μm. The total length of capillaries in the human body has been estimated

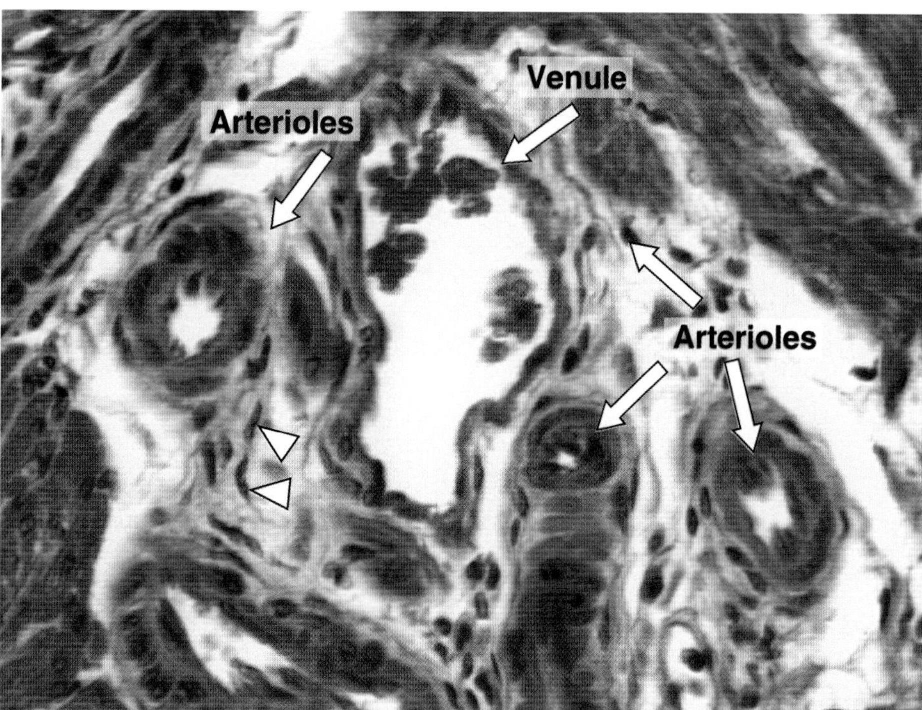

Figure 11–1. Small blood vessels from the microvasculature (arterioles and venules) surrounded by components of connective tissue. The arrowheads point to fibroblasts. H&E stain. Low magnification.

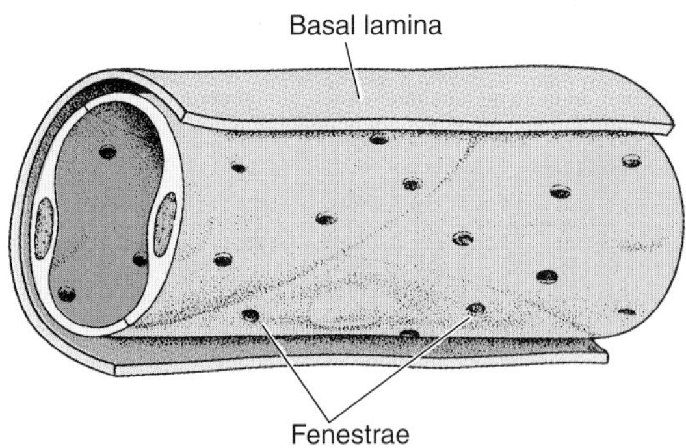

Figure 11–2. Three-dimensional representation of the structure of a capillary with fenestrae in its wall. The transverse section shows that, in this example, the capillary wall is formed by 2 endothelial cells. Note the basal lamina surrounding endothelial cells.

MEDICAL APPLICATION

Junctions between endothelial cells of venules are the loosest. At these locations there is a characteristic loss of fluid from the circulatory system during the inflammatory response, leading to edema.

at 96,000 km (60,000 miles). When cut transversely, their walls are observed to consist of portions of 1–3 cells (Figure 11–2). The external surfaces of these cells usually rest on a basal lamina, a product of endothelial origin.

In general, endothelial cells are polygonal and elongated in the direction of blood flow. The nucleus causes the cell to bulge into the capillary lumen. Its cytoplasm contains few organelles, including a small Golgi complex, mitochondria, free ribosomes, and a few cisternae of rough endoplasmic reticulum (Figure 11–3). Junctions of the zonula occludentes type are present between most endothelial cells and are of physiologic importance. Such junctions offer variable permeability to the macromolecules that play a significant role in both normal and pathologic conditions.

At various locations along capillaries and postcapillary venules are cells of mesenchymal origin with long cytoplasmic processes that partly surround the endothelial cells. These cells are called **pericytes** (Gr. *peri,* around, + *kytos,* cell). They are enclosed in their own basal lamina, which may fuse with that of the endothelial cells. The presence of myosin, actin, and tropomyosin in pericytes strongly suggests that these cells also have a contractile function. After tissue injuries, pericytes proliferate and differentiate to form new blood vessels and connective tissue cells, thus participating in the repair process.

Blood capillaries can be grouped into 4 types, depending on the continuity of both the endothelial sheet and the basal lamina.

1. The **continuous,** or **somatic, capillary** (Figure 11–4) is characterized by the absence of fenestrae in its wall. This type of capillary is found in all kinds of muscle tissue, connective tissue, exocrine glands, and nervous tissue. In some places, but not in the nervous system, numerous pinocytotic vesicles are present on both surfaces of endothelial cells. Pinocytotic vesicles also appear as isolated vesicles in the cytoplasm of these cells and are responsible for the transport of macromolecules in both directions across the endothelial cytoplasm.

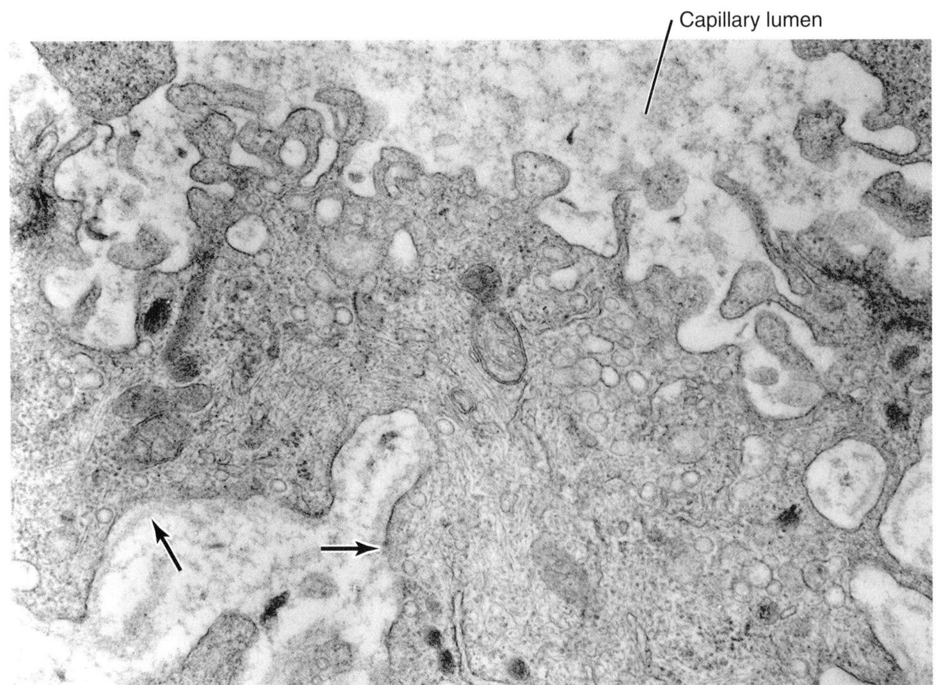

Figure 11–3. Electron micrograph of a section of a continuous capillary. Note the ruffled appearance of its interior surface, the large and small pinocytic vesicles, and numerous microfilaments in the cytoplasm. Arrows show the basal lamina. Medium magnification.

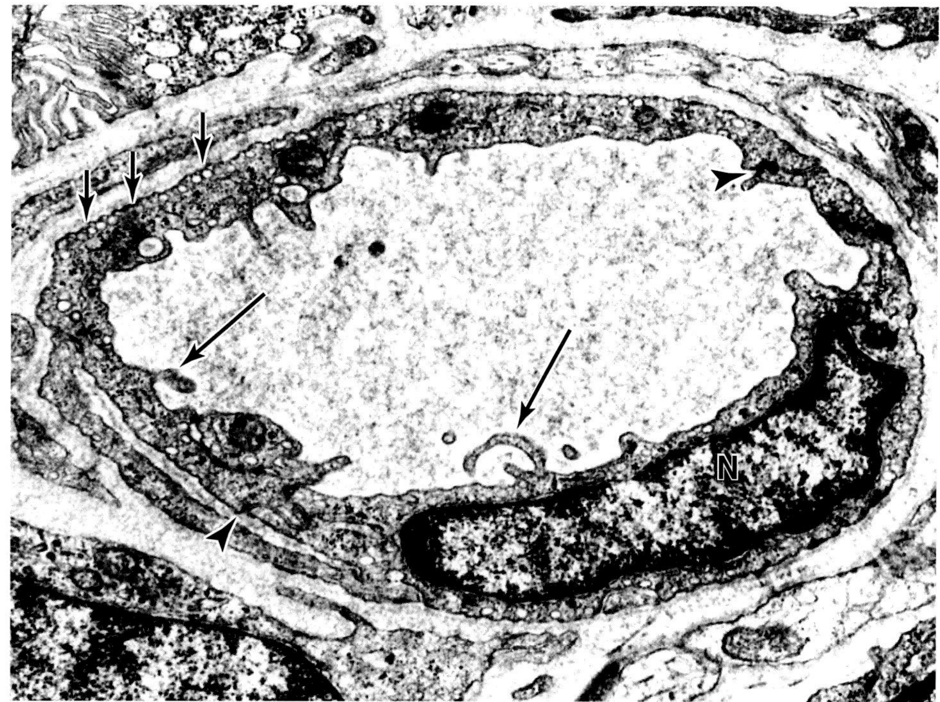

Figure 11–4. Electron micrograph of a transverse section of a continuous capillary. Note the nucleus (N) and the junctions between neighboring cells (arrowheads). Numerous pinocytotic vesicles are evident (small arrows). The large arrows show large vesicles being formed by infoldings of broad sheets of the endothelial cell cytoplasm. ×10,000.

2. The **fenestrated,** or **visceral, capillaries** are characterized by the presence of large fenestrae in the walls of endothelial cells that are obliterated by a diaphragm that is thinner than a cell membrane (Figures 11–2 and 11–5). The diaphragm does not have the trilaminar structure of a unit membrane. The basal lamina of the fenestrated capillaries is continuous.

Fenestrated capillaries are found in tissues where rapid interchange of substances occurs between the tissues and the blood, as in the kidney, the intestine, and the endocrine glands. Macromolecules experimentally injected into the bloodstream can cross the capillary wall through the fenestrae to enter the tissue spaces.

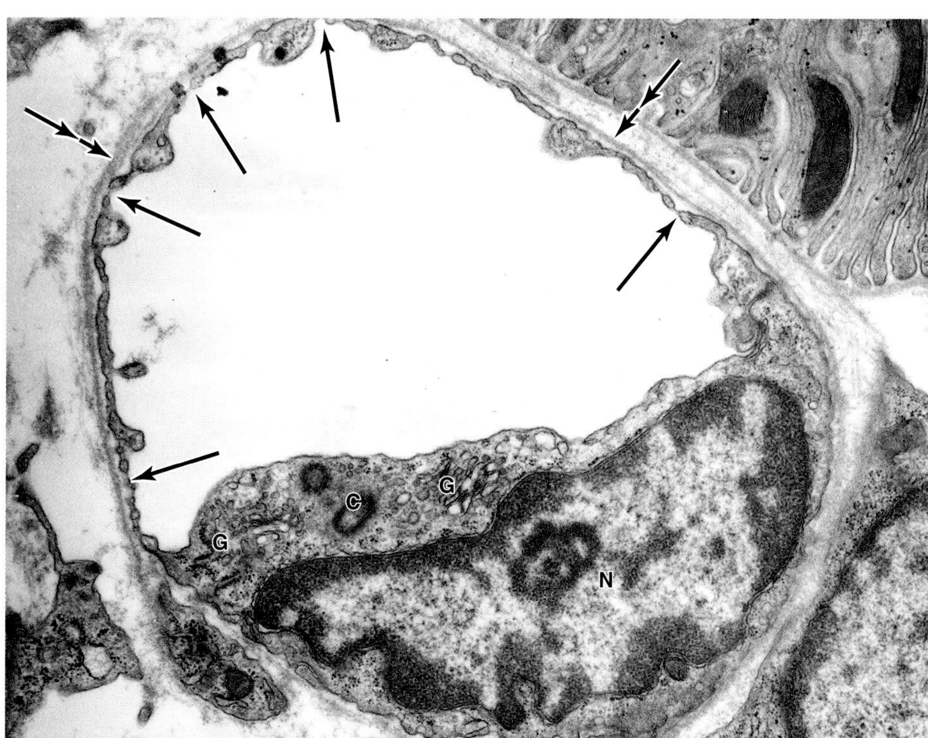

Figure 11–5. A fenestrated capillary in the kidney. Arrows indicate fenestrae closed by diaphragms. In this cell the Golgi complex (G), nucleus (N), and centrioles (C) can be seen. Note the continuous basal lamina on the outer surface of the endothelial cell (double arrows). Medium magnification. (Courtesy of J Rhodin.)

3. The third type of capillary is characteristic of the renal glomerulus. This is a fenestrated capillary devoid of diaphragm. In this type of capillary, the blood is separated from the tissues only by a very thick and continuous basal lamina that underlies the fenestrae (see Chapter 19).
4. The fourth type of capillary is the **discontinuous sinusoidal capillary,** which has the following characteristics:
   a. The capillaries have a tortuous path and greatly enlarged diameter (30–40 μm), which slows the circulation of blood.
   b. The endothelial cells form a discontinuous layer and are separated from one another by wide spaces.
   c. The cytoplasm of the endothelial cells shows multiple fenestrations without diaphragms.
   d. Macrophages are located either among or outside the cells of the endothelium.
   e. The basal lamina is discontinuous.

Sinusoidal capillaries are found mainly in the liver and in hematopoietic organs such as the bone marrow and spleen. The interchange between blood and tissues is greatly facilitated by the structure of the capillary wall.

Capillaries anastomose freely, forming a rich network that interconnects the small arteries and veins (Figure 11–6). The arterioles branch into small vessels surrounded by a discontinuous layer of smooth muscle, the **metarterioles** (Figure 11–6), which branch into capillaries. Constriction of metarterioles helps to regulate the circulation in capillaries when it is not necessary for the tissue to have blood flow throughout the entire capillary network. In some tissues, there are arteriovenous anastomoses (Figure 11–6) that enable the arterioles to empty directly into venules. This is an additional mechanism that contributes to regulation of the capillary circulation. These interconnections are abundant in skeletal muscle and in the skin of the hands and feet. When vessels of the arteriovenous anastomosis contract, all the blood must pass through the capillary network. When they relax, some blood flows directly to a vein instead of circulating in the capillaries. Capillary circulation is controlled by neural and hormonal stimulation. The richness of the capillary network is related to the metabolic activity of the tissues. Tissues with high metabolic rates, such as the kidney, liver, and cardiac and skeletal muscle, have an abundant capillary network; the opposite is true of tissues with low metabolic rates, such as smooth muscle and dense connective tissue.

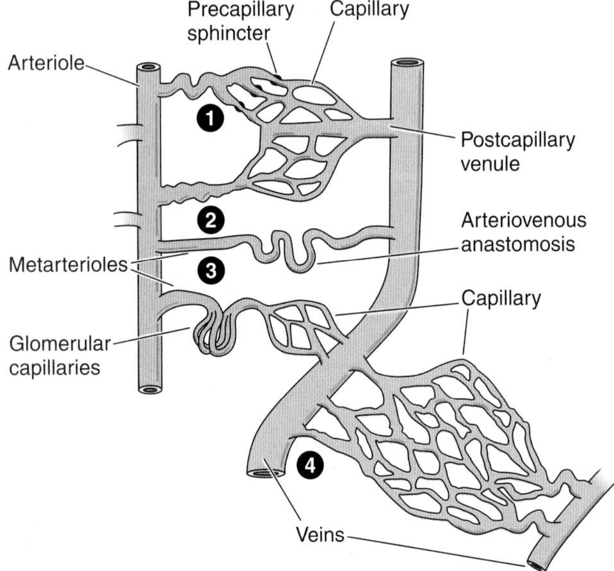

Figure 11–6. Types of microcirculation formed by small blood vessels. (**1**) The usual sequence of arteriole → metarteriole → capillary → venule and vein. (**2**) An arteriovenous anastomosis. (**3**) An arterial portal system, as is present in the kidney glomerulus. (**4**) A venous portal system, as is present in the liver. (Reproduced, with permission, from Krstíc RV: *Illustrated Encyclopedia of Human Histology.* Springer-Verlag, 1984.)

The total diameter of the capillaries is approximately 800 times larger than that of the aorta. The velocity of blood in the aorta averages 320 mm/s; in the capillaries, about 0.3 mm/s. Because of their thin walls and slow blood flow, capillaries are a favorable place for the exchange of water, solutes, and macromolecules between blood and tissues.

Endothelial cells are functionally diverse according to the vessel they line. The capillaries are often referred to as **exchange vessels,** since it is at these sites that oxygen, carbon dioxide, substrates, and metabolites are transferred from blood to the tissues and from the tissues to blood. The mechanisms responsible for the interchange of materials between blood and tissue are not completely known. They depend on the kind of molecule and also on the structural characteristics and arrangement of endothelial cells in each type of capillary.

Small molecules, both hydrophobic and hydrophilic (eg, oxygen, carbon dioxide, and glucose), can diffuse or be actively transported across the plasmalemma of capillary endothelial cells. These substances are then transported by diffusion through the endothelial cytoplasm to the opposite cell surface, where they are discharged into the extracellular space. Water and some other hydrophilic molecules, less than 1.5 nm in diameter and below 10 kDa in molecular mass, can cross the capillary wall by diffusing through the intercellular junctions (paracellular pathway). The pores of fenestrated capillaries, the spaces among endothelial cells of sinusoid capillaries, and the pinocytotic vesicles are other pathways for the passage of large molecules.

Besides their role in the exchanges between the blood and the tissues, endothelial cells perform several other functions:

Conversion of angiotensin I (Gr. *angeion,* vessel, + *tendere,* to stretch) to angiotensin II (see Chapter 19).

Conversion of bradykinin, serotonin, prostaglandins, norepinephrine, thrombin, etc, to biologically inert compounds.

Lipolysis of lipoproteins by enzymes located on the surface of endothelial cells, to yield triglycerides and cholesterol (substrates for steroid-hormone synthesis and membrane structure).

Production of vasoactive factors that affect the vascular tone, such as endothelins, vasoconstrictive agents, and nitric oxide, a relaxing factor.

Growth factor such as VEGFs (vascular endothelial growth factors) play pivotal roles in formation of the vascular system during embryonic development, in regulation of capillary growth under normal and pathologic conditions in adults, and in maintenance of the normal vasculature.

MEDICAL APPLICATION

The endothelium also has an antithrombogenic action, preventing blood coagulation. When endothelial cells are damaged by atherosclerotic lesions, for example, the uncovered subendothelial connective tissue induces the aggregation of blood platelets. This aggregation initiates a cascade of events that produce fibrin from blood fibrinogen. An intravascular coagulum, or ***thrombus*** *(plural, thrombi), is formed that may grow until there is complete obstruction of the local vascular flow. From this thrombus, solid masses called* ***emboli*** *(singular, embolus) may detach and be carried by the blood to obstruct distant blood vessels. In both cases, the vascular flow may stop, a potentially life-threatening condition. Thus, the integrity of the endothelial layer preventing the contact between platelets and the subendothelial connective tissue, is an important antithrombogenic mechanism (see Chapter 12).*

Note that, although morphologically similar, the endothelial cells of different blood vessels have different functional properties.

Blood Vessels with Diameters Above a Certain Size

All blood vessels above a certain diameter have a number of structural features in common and present a general plan of construction. However, the same type of vessel can show remarkable structural variations. On the other hand, the distinction between different types often is not clear-cut because the transition from one type of vessel to another is gradual.

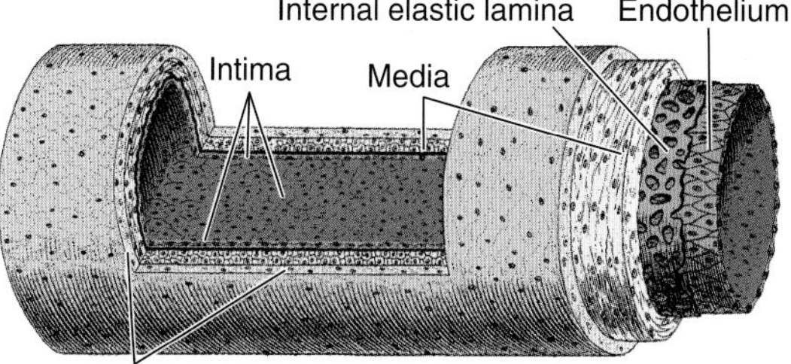

Figure 11–7. Drawing of a medium-sized muscular artery, showing its layers. Although the usual histologic preparations cause the layers to appear thicker than those shown here, the drawing is actually similar to the in vivo architecture of the vessel. At the moment of death, the artery experiences an intense contraction; consequently, the lumen is reduced, the internal elastic membrane undulates, and the muscular tunica thickens.

Blood vessels are usually composed of the following layers, or tunics (L. *tunica,* coat), as shown in Figures 11–7 and 11–8.

Tunica Intima

The intima shows one layer of endothelial cells supported by a subendothelial layer of loose connective tissue containing occasional smooth muscle cells. In arteries, the intima is separated from the media by an **internal elastic lamina,** the most external component of the intima. This lamina, composed of elastin, has gaps (fenestrae) that allow the diffusion of substances to nourish cells deep in the vessel wall. As a result of the absence of blood pressure and the contraction of the vessel at death, the tunica intima of the arteries generally has an undulated appearance in tissue sections (Figures 11–8 and 11–12).

Tunica Media

The media consists chiefly of concentric layers of helically arranged smooth muscle cells (Figure 11–8). Interposed among

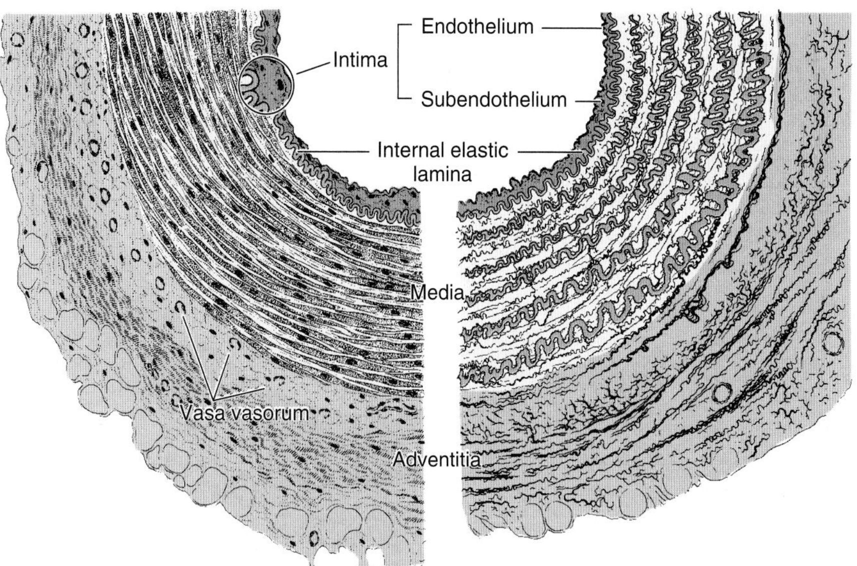

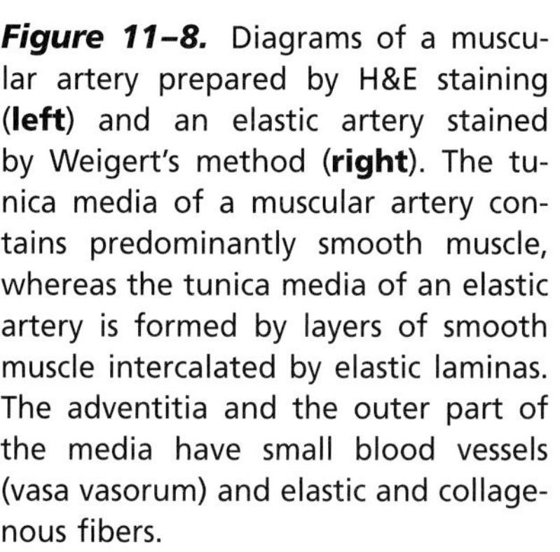

Figure 11–8. Diagrams of a muscular artery prepared by H&E staining (**left**) and an elastic artery stained by Weigert's method (**right**). The tunica media of a muscular artery contains predominantly smooth muscle, whereas the tunica media of an elastic artery is formed by layers of smooth muscle intercalated by elastic laminas. The adventitia and the outer part of the media have small blood vessels (vasa vasorum) and elastic and collagenous fibers.

the smooth muscle cells are variable amounts of elastic fibers and lamellae, reticular fibers (collagen type III), proteoglycans, and glycoproteins. Smooth muscle cells are the cellular source of this extracellular matrix. In arteries, the media has a thinner **external elastica lamina,** which separates it from the tunica adventitia.

Tunica Adventitia

The adventitia consists principally of collagen and elastic fibers (Figures 11–7 and 11–8). Collagen in the adventitia is type I. The adventitial layer gradually becomes continuous with the connective tissue of the organ through which the vessel runs.

Vasa Vasorum

Large vessels usually have vasa vasorum ("vessels of the vessel"), which are arterioles, capillaries, and venules that branch profusely in the adventitia and the outer part of the media. The vasa vasorum provide metabolites to the adventitia and the media, since in larger vessels the layers are too thick to be nourished solely by diffusion from the blood in the lumen. Vasa vasorum are more frequent in veins than in arteries (Figures 11–8 and 11–13). In arteries of intermediate and large diameter, the intima and the most internal region of the media are devoid of vasa vasorum. These layers receive oxygen and nutrition by diffusion from the blood that circulates into the lumen of the vessel.

Innervation

Most blood vessels that contain smooth muscle in their walls are supplied with a profuse network of unmyelinated sympathetic nerve fibers (**vasomotor nerves**) whose neurotransmitter is norepinephrine. Discharge of norepinephrine from these nerves results in vasoconstriction. Because these efferent nerves generally do not enter the media of arteries, the neurotransmitter must diffuse for several micrometers to affect smooth muscle cells of the media. Gap junctions between smooth muscle cells of the media propagate the response to the neurotransmitter to the inner layers of muscle cells. In veins, nerve endings are found in both the adventitia and the media, but the overall density of innervation is less than that encountered in arteries. Arteries in skeletal muscle also receive a cholinergic vasodilator nerve supply. Acetylcholine released by these vasodilator nerves acts on the endothelium to produce nitric oxide, which diffuses into the smooth muscle cells, activating a cyclic GMP system of intracellular messengers. The muscle cells then relax, and the vessel lumen is dilated.

For didactic purposes, the arterial blood vessels are classified according to their diameter into arterioles, arteries of medium diameter (muscular arteries), and larger (elastic) arteries.

Arterioles

The arterioles are generally less than 0.5 mm in diameter and have relatively narrow lumens (Figures 11–9 and 11–17). The subendothelial layer is very thin. In the very small arterioles, the internal elastic lamina is absent, and the media is generally composed of one or two circularly arranged layers of smooth muscle cells; it shows no external elastic lamina (Figures 11–9 and 11–17). Above the arterioles are small arteries in which the tunica media is more developed, and the lumens are larger than those of the arterioles (Figures 11–10, 11–11, and 11–12). In both arterioles and small arteries, the tunica adventitia is very thin.

Medium (Muscular) Arteries

The muscular arteries may control the affluence of blood to the organs by contracting or relaxing the smooth muscle cells of the tunica media. The intima have a subendothelial layer that is somewhat thicker than that of the arterioles (Figures 11–7 and 11–13). The internal elastic lamina, the most external component of the intima, is prominent (Figure 11–13), and the tunica media may contain up to 40 layers of smooth muscle cells. These cells are

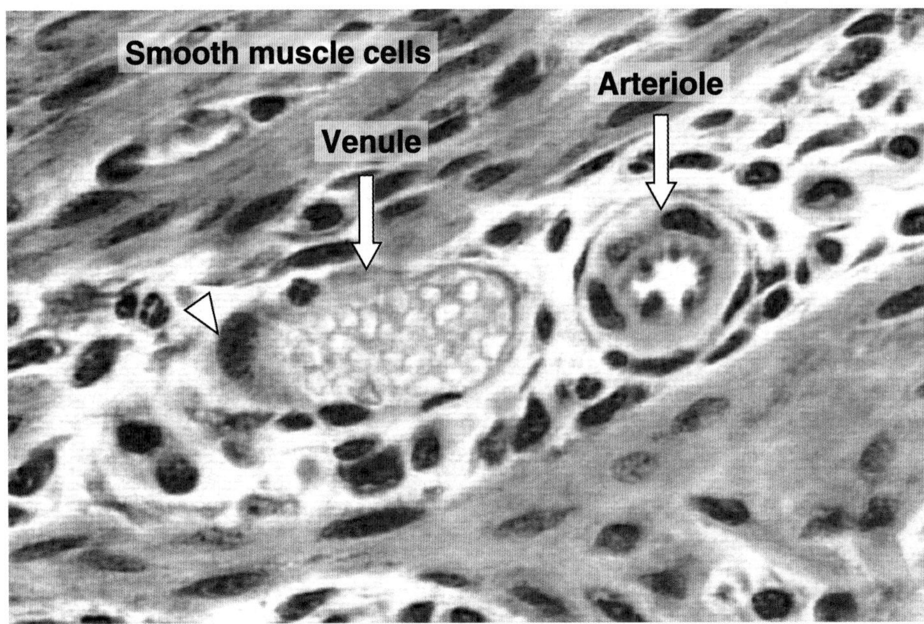

Figure 11–9. Cross section through an arteriole and its accompanying venule from the myometrium of mouse uterus. Note the elongated, large nucleus (arrow) of a pericyte surrounding the venule wall. Toluidine blue stain. High magnification.

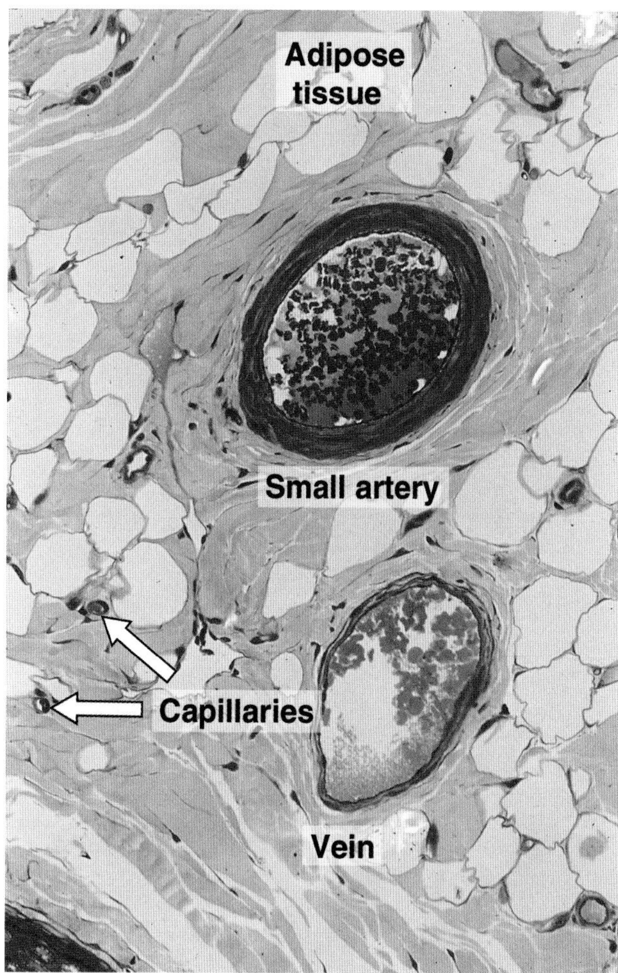

Figure 11–10. Cross section through a small artery and its accompanying muscular vein. Because of vasodilatation, the arteriole is unusually filled with blood. At this stage the internal elastic lamina is not distinguished. Many other small arterial branches and capillaries can be seen in the surrounding connective tissue. Pararosaniline–toluidine blue (PT) stain. Medium magnification.

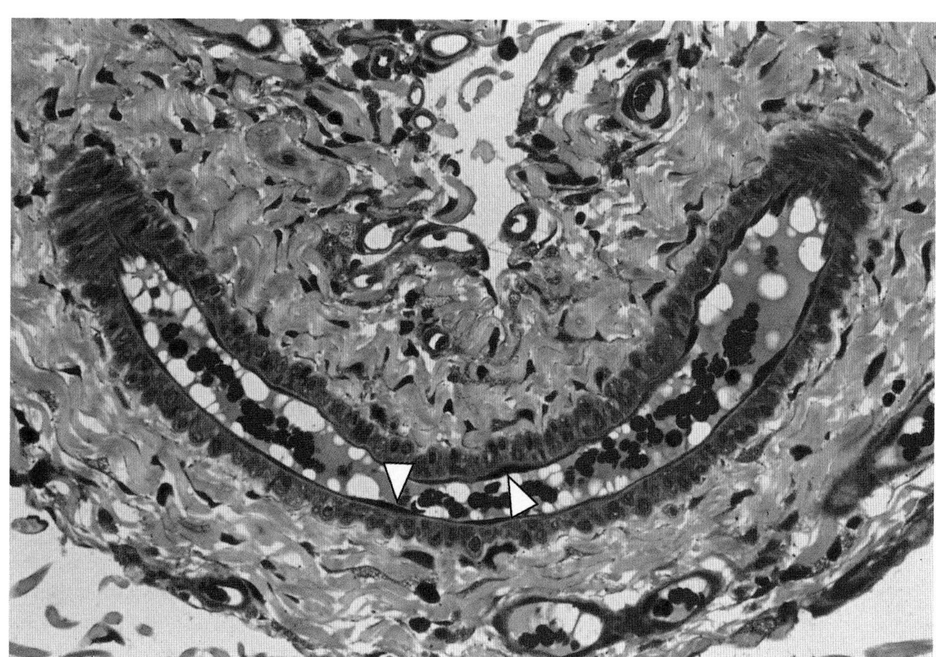

Figure 11–11. Oblique section of a small artery from the mesentery. Note the transverse section of the smooth muscle cells of the media and the endothelial layer covering the lumen of the vessel (arrowheads). PT stain. Medium magnification.

intermingled with various numbers of elastic lamellae (depending on the size of the vessel) as well as reticular fibers and proteoglycans, all synthesized by the smooth muscle fibers. An external elastic lamina, the last component of the media, is present only in the larger muscular arteries. The adventitia consists of connective tissue. Lymphatic capillaries, vasa vasorum, and nerves are also found in the adventitia, and these structures may penetrate to the outer part of the media.

Large Elastic Arteries

Large elastic arteries help to stabilize the blood flow. The elastic arteries include the aorta and its large branches. They have a yellowish color from the accumulation of elastin in the media (Figures 11–8 and 11–14). The intima is thicker than the corresponding tunic of a muscular artery. An internal elastic lamina, although present,

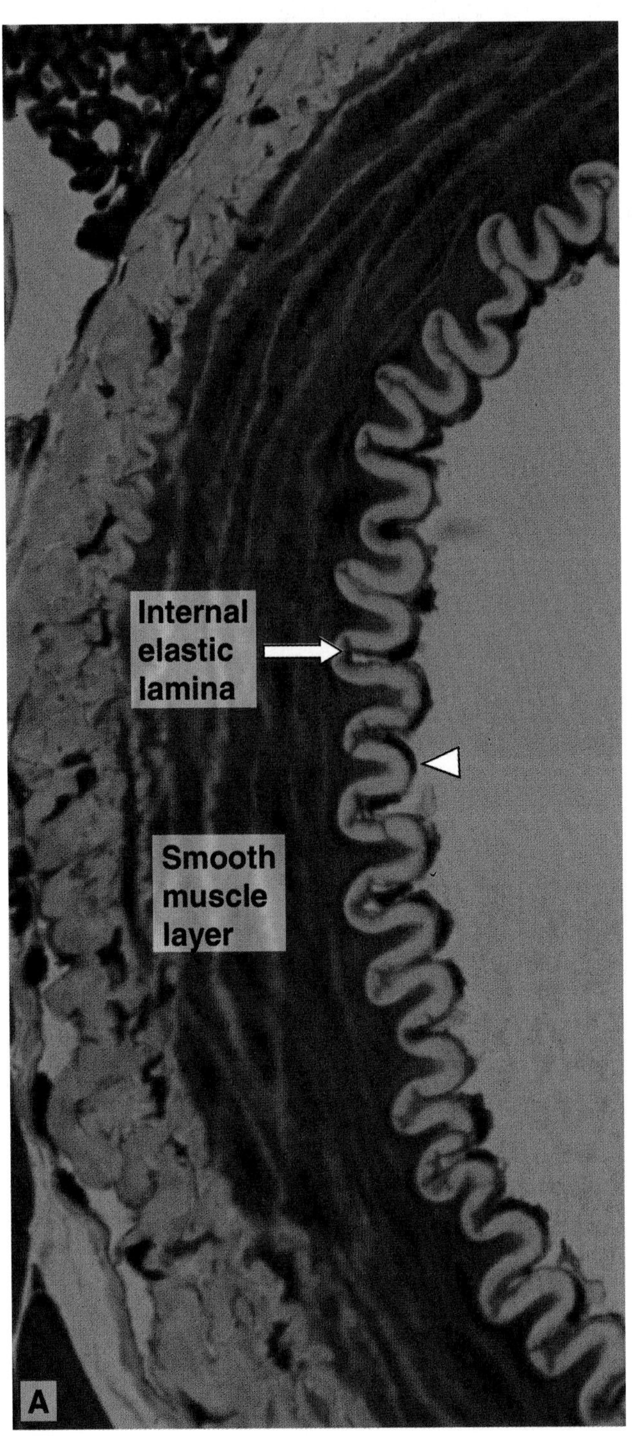

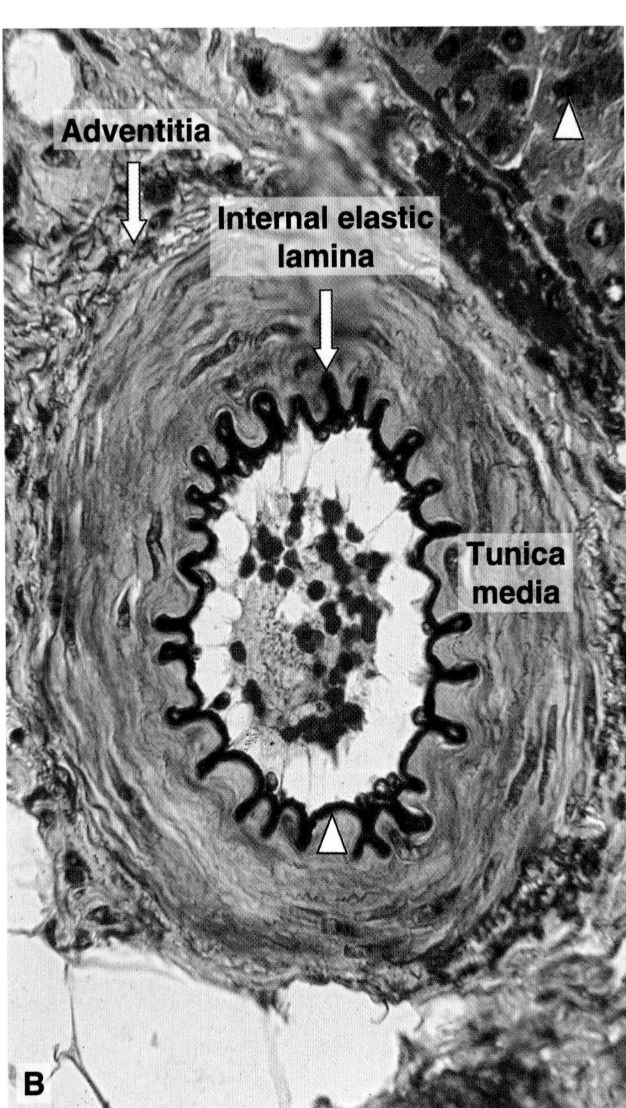

Figure 11–12. Cross sections of small arteries. **A:** The elastic lamina is not stained and is seen as a pallid lamina of scalloped appearance just below the endothelium (arrowhead). Medium magnification. **B:** A small artery with a distinctly stained internal elastic lamina (arrowhead). From a preparation of the late George Gomori. Low magnification.

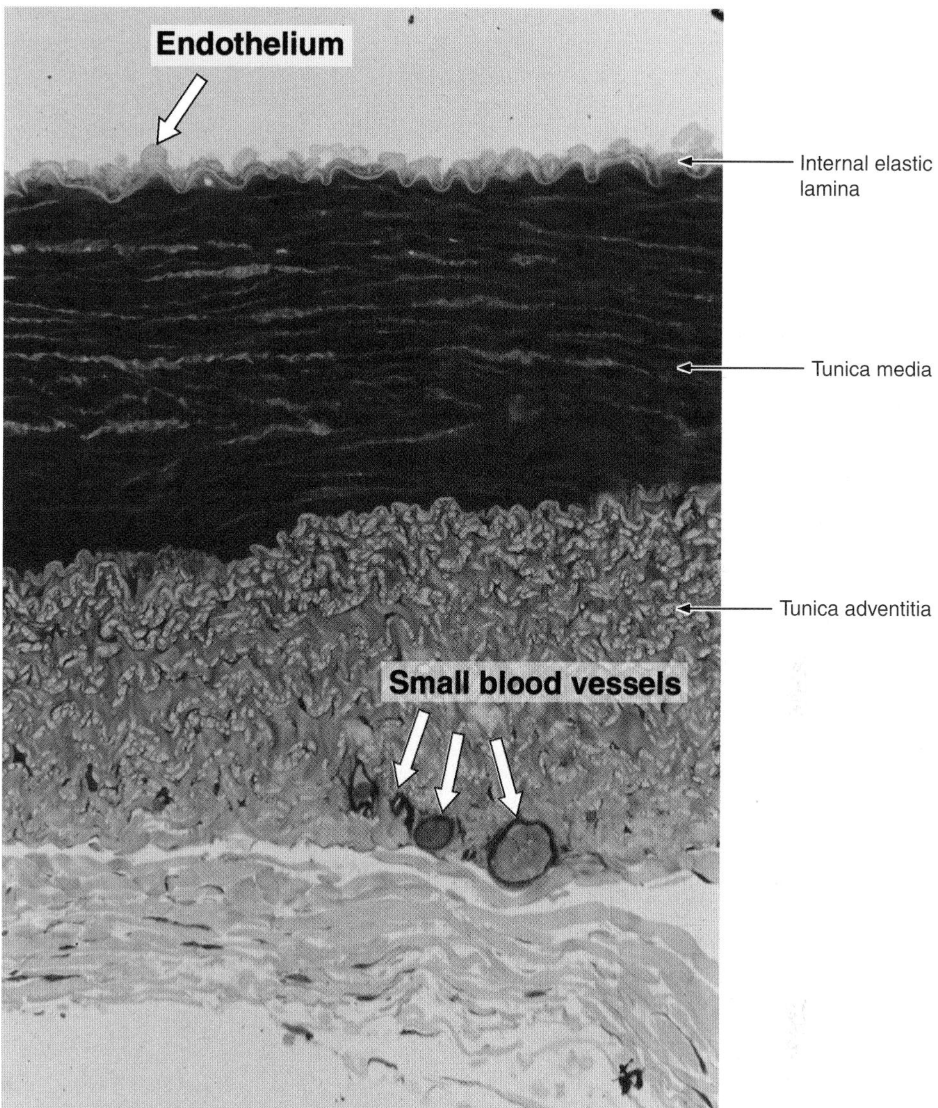

Figure 11–13. Transverse section showing part of a muscular (medium caliber) artery. Small blood vessels (vasa vasorum) are observed in the tunica adventitia.

may not be easily discerned, since it is similar to the elastic laminae of the next layer. The media consists of elastic fibers and a series of concentrically arranged, perforated elastic laminae whose number increases with age (there are 40 in the newborn, 70 in the adult). Between the elastic laminae are smooth muscle cells, reticular fibers, proteoglycans, and glycoproteins. The tunica adventitia is relatively underdeveloped.

The several elastic laminae contribute to the important function of making the blood flux more uniform. During ventricular contraction (**systole**), the elastic laminae of large arteries are stretched and reduce the pressure change. During ventricular relaxation (**diastole**), ventricular pressure drops to a low level, but the elastic rebound of large arteries helps to maintain arterial pressure. As a consequence, arterial pressure and blood velocity decrease and become less variable as the distance from the heart increases (Figure 11–15).

Arterial Degenerative Alterations

MEDICAL APPLICATION

Arteries undergo progressive and gradual changes from birth to death, and it is difficult to say where the normal growth processes end and the processes of involution begin. Each artery exhibits its own aging pattern.

Atherosclerotic lesions are characterized by focal thickening of the intima, proliferation of smooth muscle cells and extracellular connective tissue elements, and the deposit of cholesterol in smooth muscle cells and macrophages. When heavily loaded *with lipid, these cells are referred to as **foam cells** and form the macroscopically visible fatty streaks and plaques that characterize **atherosclerosis.** These changes may extend to the inner part of the tunica media, and the thickening may become so great as to occlude the vessel. Coronary arteries are among those most predisposed to atherosclerosis. Uniform thickening of the intima is believed to be a normal phenomenon of aging.*

*Certain arteries irrigate only definite areas of specific organs, and obstruction of the blood supply results in **necrosis** (death of tissues from a lack of metabolites). These **infarcts** occur commonly in the heart, kidneys, cerebrum, and certain other organs. In other regions (such as the skin), arteries anastomose frequently, and the obstruction of one artery does not lead to tissue necrosis, because the blood flow is maintained.*

*When the media of an artery is weakened by an embryonic defect, disease, or lesion, the wall of the artery may dilate extensively. As this process of dilatation progresses, it becomes an **aneurysm.** Rupture of the aneurysm brings severe consequences and may cause death.*

Carotid Bodies

The carotid bodies, encountered near the bifurcation of the common carotid artery, are chemoreceptors sensitive to carbon dioxide and oxygen concentrations in the blood. These structures are richly irrigated by fenestrated capillaries that surround type I and type II cells. The type II cells are supporting cells, whereas type I cells contain numerous dense-core vesicles that store dopamine, serotonin, and

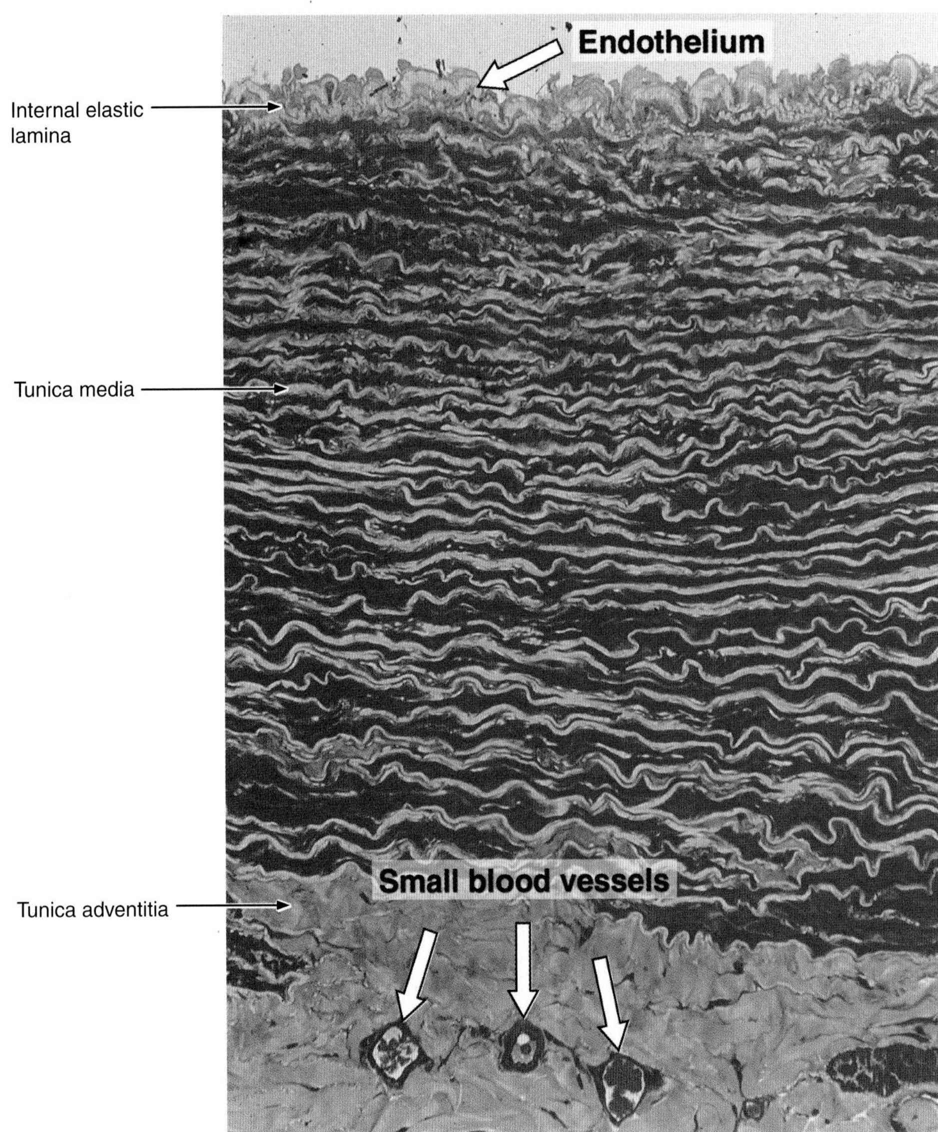

Figure 11–14. Transverse section showing part of a large elastic artery showing a well developed tunica media containing several elastic laminas. PT stain. Medium magnification.

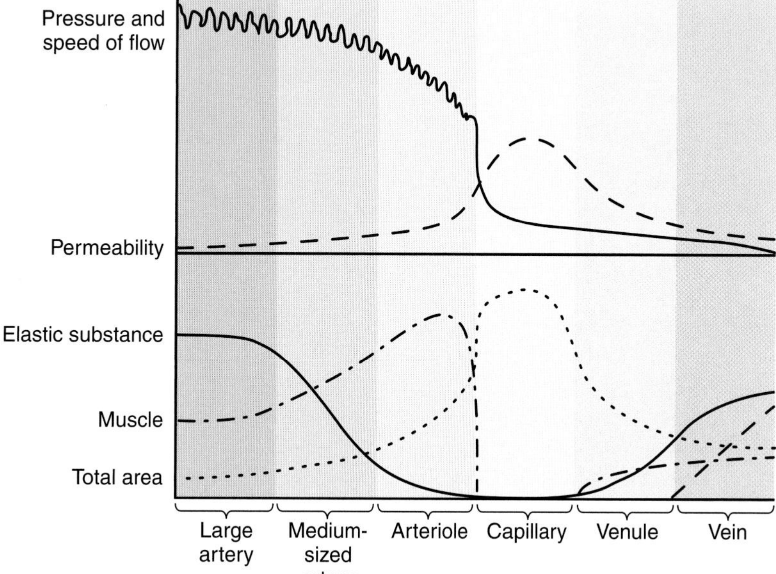

Figure 11–15. Graph showing the relationship between the characteristics of blood circulation (**left**) and the structure of the blood vessels (**bottom**). The arterial blood pressure and speed of flow decrease and become more constant as the distance from the heart increases. This decrease coincides with a reduction in the number of elastic fibers and an increase in the number of smooth muscle cells in the arteries. The graph illustrates the gradual changes in the structure of vessels and their biophysical properties. (Reproduced, with permission, from Cowdry EV: *Textbook of Histology.* Lea & Febiger, 1944.)

adrenaline (Figure 11–16). Most of the nerves of the carotid body are afferent fibers (carry impulses to the central nervous system). The carotid bodies are sensitive to low oxygen tension, high carbon dioxide concentration, and low arterial blood pH. Whether the afferent nerve endings or type I cells are the principal chemoreceptor elements is controversial. Aortic bodies located on the arch of the aorta are similar in structure to carotid bodies and are believed to have a similar function.

Carotid Sinuses

Carotid sinuses are slight dilatations of the internal carotid arteries. These sinuses contain baroreceptors that detect changes in blood pressure and relay the information to the central nervous system. The arterial media layer of the sinus is thinner to allow it to respond to changes in blood pressure. The intima and the adventitia are very rich in nerve endings. The afferent nerve impulses are processed in the brain to control vasoconstriction and maintain normal blood pressure.

Arteriovenous Anastomoses

Arteriovenous anastomoses participate in the regulation of blood flow in certain regions of the body by allowing direct communication between arterioles and venules. The luminal diameters of anastomotic vessels vary with the physiologic condition of the organ. Changes in diameter of these vessels regulate blood pressure, flow, and temperature and the conservation of heat in particular areas. In addition to these direct connections, there are more complex structures, the **glomera** (singular, **glomus**), mainly in fingerpads, fingernail beds, and ears. When the arteriole penetrates the connective tissue capsule of the glomus, it loses an internal elastic membrane and develops a thick muscular wall and small lumen. Arteriovenous anastomoses are believed to participate in physiologic phenomena such as regulation of local blood flow and blood pressure. All arteriovenous anastomoses are richly innervated by the sympathetic and parasympathetic nervous systems.

Postcapillary Venules and Capillaries

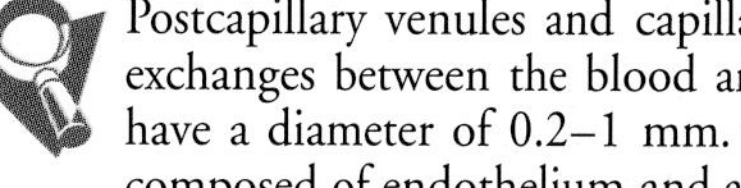

Postcapillary venules and capillaries participate in the exchanges between the blood and the tissues. Venules have a diameter of 0.2–1 mm. Their tunica intima is composed of endothelium and a very thin subendothelial layer. The media in small venules may contain only contractile pericytes. These vessels are called **postcapillary** or **pericytic venules.** Their luminal diameter is up to 50 μm. However, most venules are muscular, with at least a few smooth muscle cells in their walls (Figures 11–1 and 11–9). Postcapillary venules have several features in common with capillaries, eg, participation in inflammatory processes and exchange of cells and molecules between blood and tissues. Venules may also influence blood flow in the arterioles by producing and secreting diffusible vasoactive substances.

Veins

The majority of veins are **small** or **medium-sized** (Figures 11–10 and 11–17), with a diameter of 1–9 mm. The intima usually has a thin subendothelial layer, which may be absent at times. The media consists of small bundles of smooth muscle cells intermixed with reticular fibers and a delicate network of elastic fibers. The collagenous adventitial layer is well developed.

The big venous trunks, close to the heart, are large veins. Large veins have a well-developed tunica intima, but the media is much thinner, with few layers of smooth muscle cells and abundant connective tissue. The adventitial layer is the thickest and best-developed tunic in veins; it frequently contains longitudinal bundles of smooth muscle (Figure 11–18). These veins, particularly the largest ones, have valves in their interior (Figure 11–19). The valves consist of 2 semilunar folds of the tunica intima that project into the lumen. They are composed of connective tissue rich in elastic fibers and are lined on both sides by endothelium. The valves, which are especially numerous in veins of the limbs, direct the venous blood toward the heart. The propulsive force of the heart is reinforced by contraction of skeletal muscles that surround these veins.

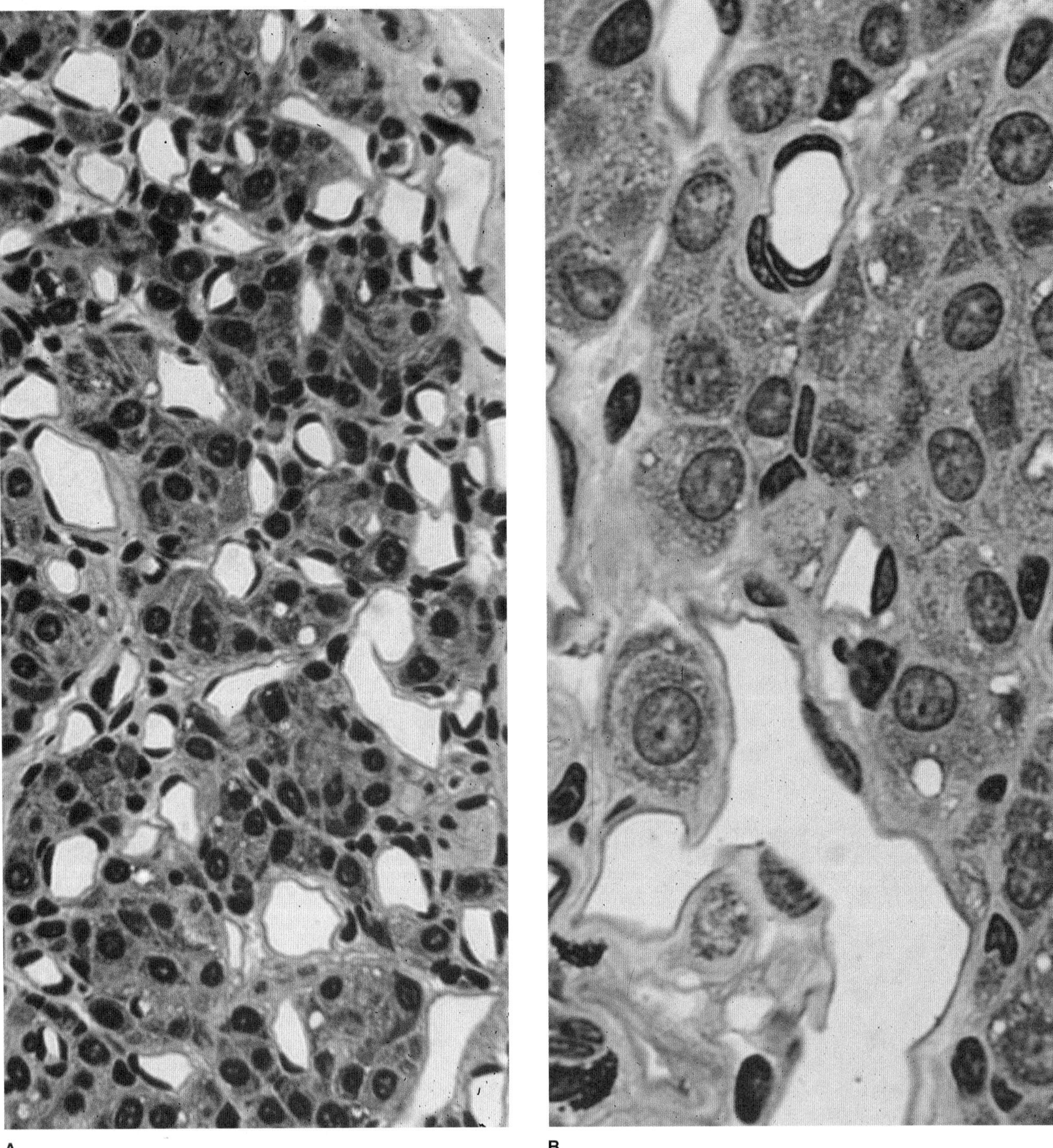

Figure 11–16. Sections of a carotid body, which is a highly vascularized structure sensitive to hypoxia. Its main cells have dense-core granules containing catecholamines that are surrounded by glia-like sustentacular cells. PT stain. **A:** Low magnification. **B:** Medium magnification.

Heart

The heart is a muscular organ that contracts rhythmically, pumping the blood through the circulatory system. It is also responsible for producing a hormone called **atrial natriuretic factor.** Its walls consist of 3 tunics: the internal, or endocardium; the middle, or myocardium; and the external, or pericardium (*peri* + Gr. *kardia,* heart). The fibrous central region of the heart, called, rather inappropriately, the **fibrous skeleton,** serves as the base of the valves as well as the site of origin and insertion of the cardiac muscle cells.

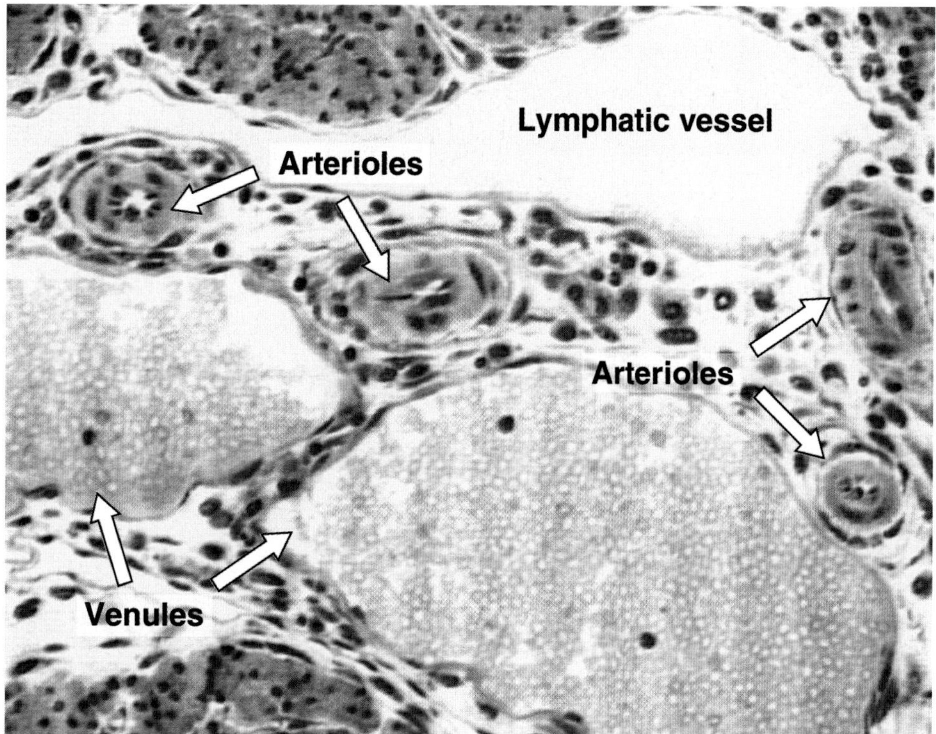

Figure 11–17. Cross section of 2 venules and 4 small arterioles. The walls of the arteries are thicker than the walls of the veins. A lymphatic vessel can be seen at the top. Note the cross sections of smooth muscle cells and the field of loose connective tissue that surrounds the vessels. Toluidine blue stain. Medium magnification.

The **endocardium** is homologous with the intima of blood vessels. It consists of a single layer of squamous endothelial cells resting on a thin subendothelial layer of loose connective tissue that contains elastic and collagen fibers as well as some smooth muscle cells. Connecting the myocardium to the subendothelial layer is a layer of connective tissue (often called the **subendocardial layer**) that contains veins, nerves, and branches of the impulse-conducting system of the heart (Purkinje cells).

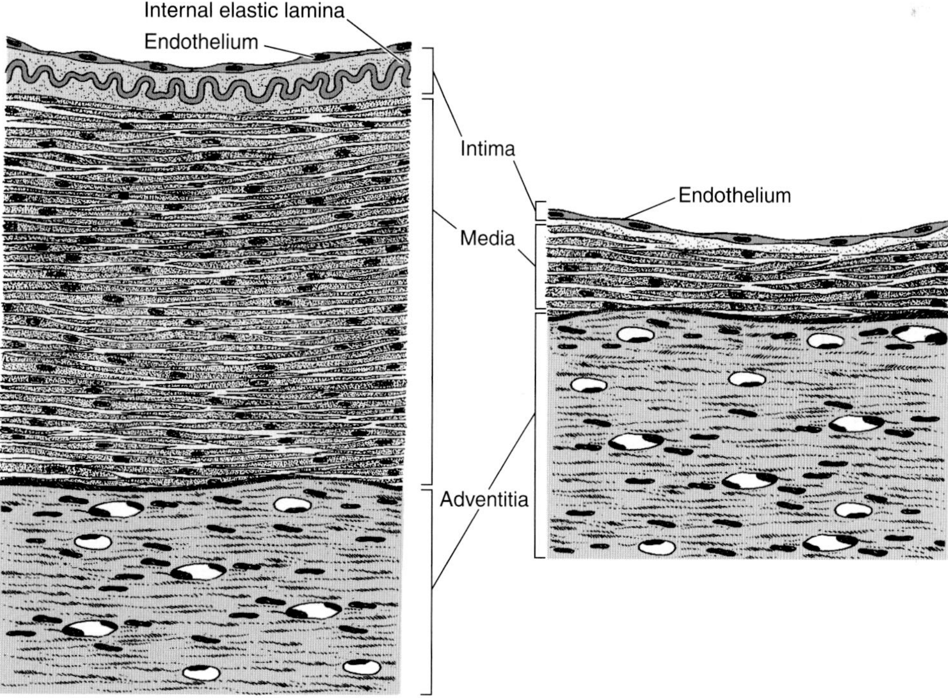

Figure 11–18. Diagram comparing the structure of a muscular artery (**left**) and accompanying vein (**right**). Note that the tunica intima and the tunica media are highly developed in the artery but not in the vein.

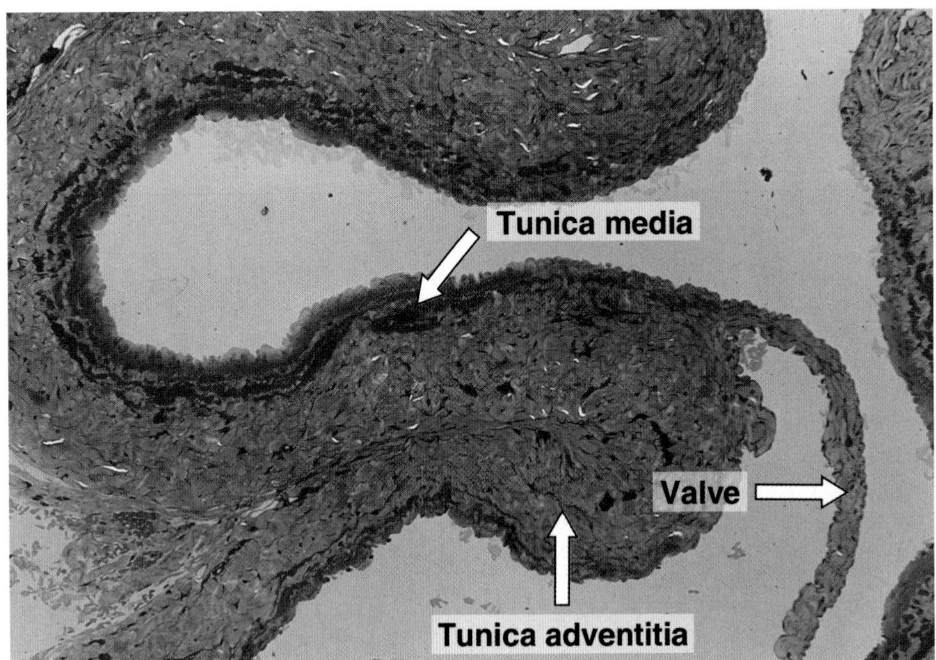

Figure 11–19. Section showing part of a large vein. The vein has a very thin muscular tunica media that contrasts with the thick adventitia composed of dense connective tissue. Note the presence a valve. PT stain. Medium magnification.

The **myocardium** is the thickest of the tunics of the heart and consists of cardiac muscle cells (see Chapter 10) arranged in layers that surround the heart chambers in a complex spiral. A large number of these layers insert themselves into the fibrous cardiac skeleton. The arrangement of these muscle cells is extremely varied, so that in histologic preparations of a small area, cells are seen to be oriented in many directions. The heart is covered externally by simple squamous epithelium (mesothelium) supported by a thin layer of connective tissue that constitutes the **epicardium.** A subepicardial layer of loose connective tissue contains veins, nerves, and nerve ganglia. The adipose tissue that generally surrounds the heart accumulates in this layer. The epicardium corresponds to the visceral layer of the **pericardium,** the serous membrane in which the heart lies. Between the visceral layer (epicardium) and the parietal layer is a small amount of fluid that facilitates the heart's movements.

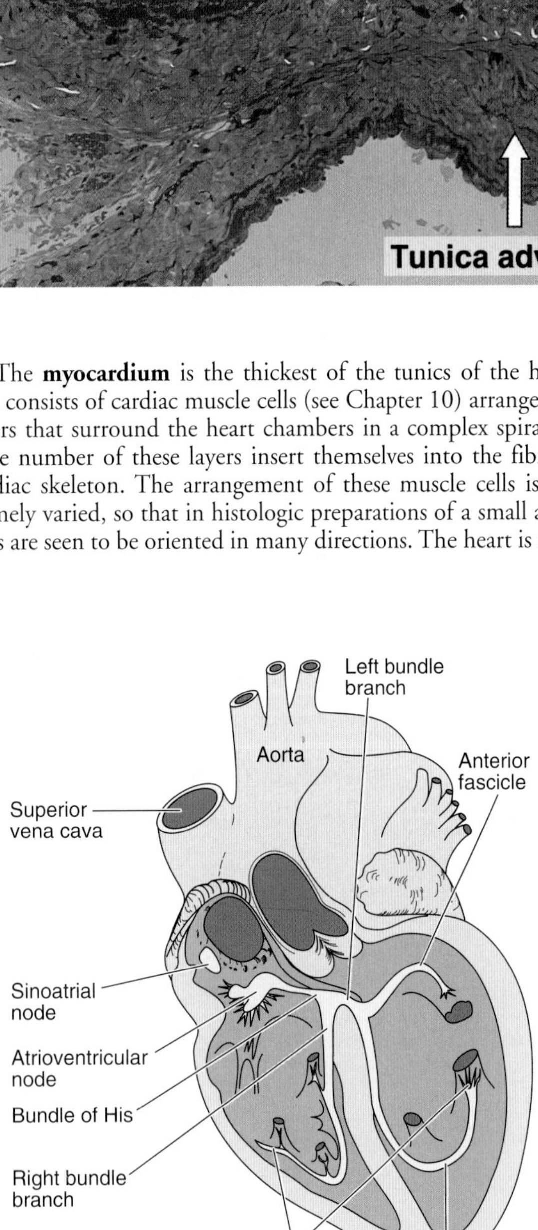

Figure 11–20. Diagram of the heart, showing the impulse-generating and -conducting system.

The cardiac fibrous skeleton is composed of dense connective tissue. Its principal components are the **septum membranaceum,** the **trigona fibrosa,** and the **annuli fibrosi.** These structures consist of dense connective tissue, with thick collagen fibers oriented in various directions. Certain regions contain nodules of fibrous cartilage.

The cardiac valves consist of a central core of dense fibrous connective tissue (containing both collagen and elastic fibers), lined on both sides by endothelial layers. The bases of the valves are attached to the annuli fibrosi of the fibrous skeleton.

The heart has a specialized system to generate a rhythmic stimulus that is spread to the entire myocardium. This system (Figures 11–20 and 11–21) consists of 2 nodes located in the atrium, the **sinoatrial node** and **atrioventricular node,** and by the **atrioventricular bundle.** The atrioventricular bundle originates from the node of the same name and branches to both ventricles. The cells of the impulse-conducting system are functionally integrated by gap junctions. The sinoatrial node is a mass of modified cardiac muscle cells that are fusiform, are smaller than atrial muscle cells, and have fewer myofibrils. The cells of the atrioventricular node are similar to those of the sinoatrial node, but their cytoplasmic projections branch in various directions, forming a network.

The **atrioventricular** bundle is formed by cells similar to those of the atrioventricular node. Distally, however, these cells become larger than ordinary cardiac muscle cells and acquire a distinctive appearance. These so-called **Purkinje cells** have one or two central nuclei, and their cytoplasm is rich in mitochondria

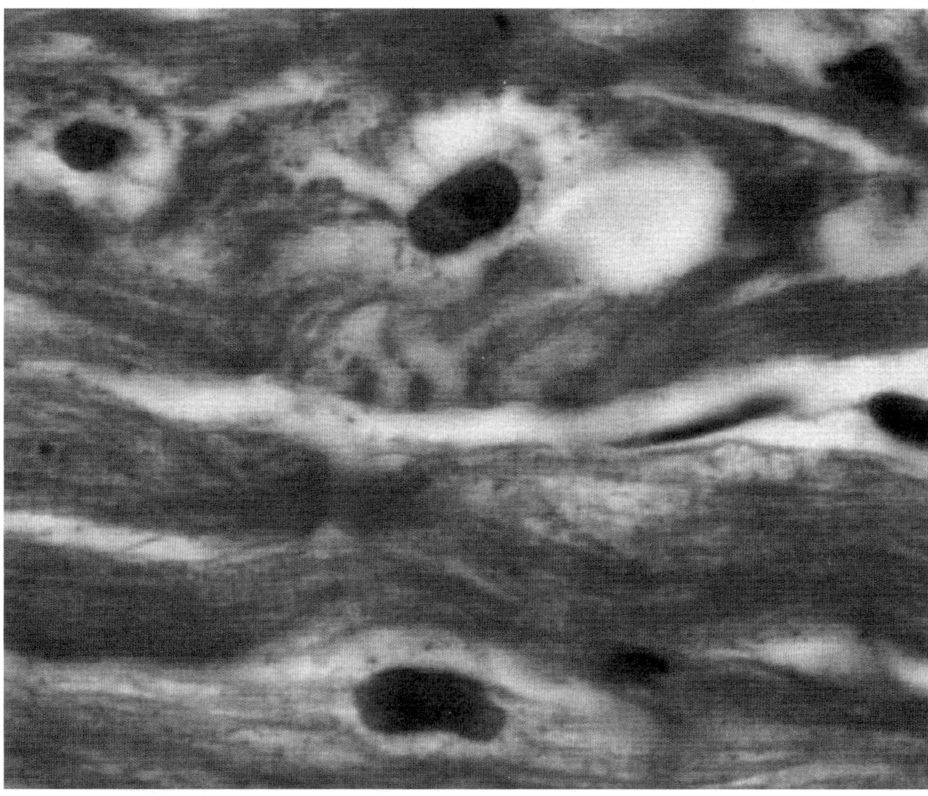

Figure 11–21. Purkinje cells of the impulse-conducting system. They are characterized by a reduced number of myofibrils that are present mainly in the periphery of the muscle cell. The light area around the nuclei of the conducting cells is caused by a local accumulation of glycogen. High magnification. H&E stain.

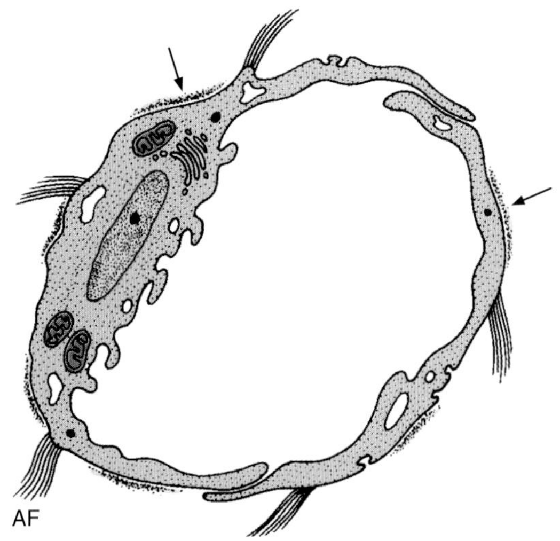

Figure 11–22. Structure of a lymphatic capillary at the electron microscope level. Note the overlapping free borders of endothelial cells, the discontinuous basal lamina (arrows), and the attachment of anchoring fibrils (AF). (Courtesy of J James.)

and glycogen. The myofibrils are sparse and restricted to the periphery of the cytoplasm (Figure 11–21). After traveling in the subendocardic layer, they penetrate the ventricle and became intramyocardic. This arrangement is important because it allows the stimulus to get into the innermost layers of the ventricular musculature.

Both the parasympathetic and sympathetic divisions of the autonomic system contribute to innervation of the heart and form widespread plexuses at the base of the heart. Ganglionic nerve cells and nerve fibers are present in the regions close to the sinoatrial and atrioventricular nodes. Although these nerves do not affect generation of the heartbeat, a process attributed to the sinoatrial (pacemaker) node, they do affect heart rhythm, such as during physical exercise and emotional stress. Stimulation of the parasympathetic division (vagus nerve) slows the heartbeat, whereas stimulation of the sympathetic nerve accelerates the rhythm of the pacemaker.

Between the muscular fibers of the myocardium are numerous afferent free nerve endings that are related to sensibility and pain. Partial obstruction of the coronary arteries reduces the supply of oxygen to the myocardium and causes pain (angina pectoris). The same sensorial enervation occurs during a heart attack, which is very painful because many muscular fibers die as a result of the low levels of oxygen.

Lymphatic Vascular System

The lymphatic vascular system returns the extracellular liquid to the bloodstream. In addition to blood vessels, the human body has a system of endothelium-lined thin-walled channels that collect fluid from the tissue spaces and return it to the blood. This fluid is called lymph; unlike the blood, it circulates in only one direction, toward the heart. The **lymphatic capillaries** originate in the various tissues as thin, closed-ended vessels that consist of a single layer of endothelium and an incomplete basal lamina. Lymphatic capillaries are held open by numerous microfibrils of the elastic fiber system, which also bind them firmly to the surrounding connective tissue (Figures 11–17, 11–22, and 11–23).

The thin lymphatic vessels gradually converge and ultimately end up as 2 large trunks—the **thoracic duct** and the **right lymphatic duct**—that empty into the junction of the left internal jugular vein with the left subclavian vein and into the confluence of the right subclavian vein and the right internal jugular vein. Interposed in the path of the lymphatic vessels are lymph nodes, whose morphologic characteristics and functions are discussed in Chapter 14. With rare exceptions, such as the

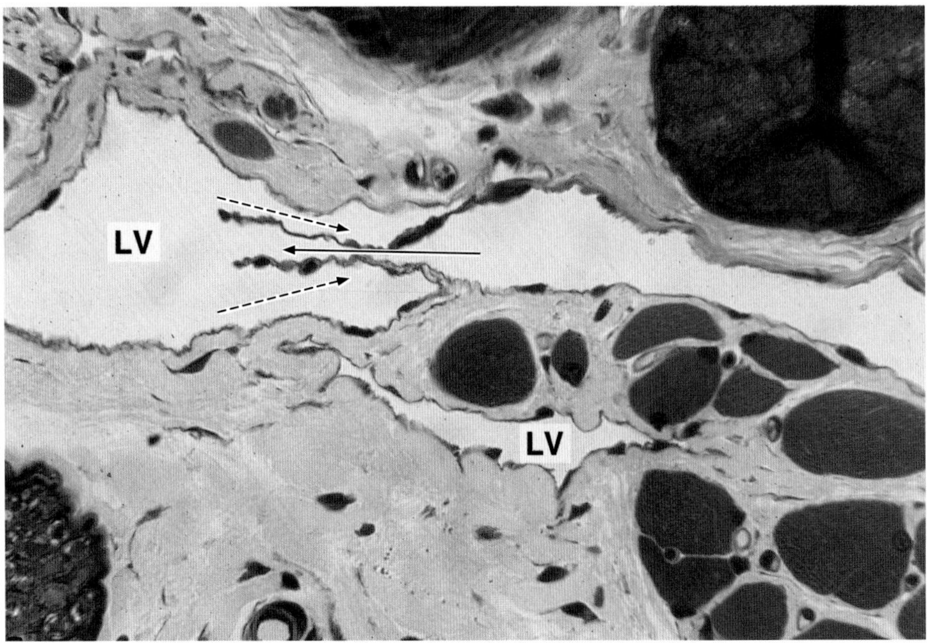

Figure 11–23. Two lymphatic vessels (LV). The vessel on top was sectioned longitudinally and shows a valve, the structure responsible for the unidirectional flow of lymph. The solid arrow shows the direction of the lymph flow, and the dotted arrows show how the valves prevent lymph backflow. The lower small vessel presents a very thin wall. PT stain. Medium magnification.

central nervous system and the bone marrow, a lymphatic system is found in almost all organs.

The lymphatic vessels have a structure similar to that of veins except that they have thinner walls and lack a clear-cut separation between layers (intima, media, adventitia). They also have more numerous internal valves (Figure 11–23). The lymphatic vessels are dilated and assume a nodular, or beaded, appearance between the valves.

As in veins, lymphatic circulation is aided by the action of external forces (eg, contraction of surrounding skeletal muscle) on their walls. These forces act discontinuously, and unidirectional lymph flow is mainly a result of the presence of many valves in these vessels. Contraction of smooth muscle in the walls of larger lymphatic vessels also helps to propel lymph toward the heart.

The structure of the large **lymphatic ducts** (thoracic duct and right lymphatic duct) is similar to that of veins, with reinforced smooth muscle in the middle layer. In this layer, the muscle bundles are longitudinally and circularly arranged, with longitudinal fibers predominating. The adventitia is relatively underdeveloped. Like arteries and veins, large lymphatic ducts contain vasa vasorum and a rich neural network.

The function of the lymphatic system is to return the fluid of the tissue spaces to the blood. Upon entering the lymphatic capillaries, this fluid contributes to the formation of the liquid part of the lymph; by passing through the lymphoid organs, it contributes to the circulation of lymphocytes and other immunologic factors.

REFERENCES

Boegehold MA: Shear-dependent release of venular nitric oxide: effect on arteriolar tone in rat striated muscle. Am J Physiol 1996;271:H387.

Cantin M et al: Immunocytochemical localization of atrial natriuretic factor in the heart and salivary glands. Histochemistry 1984;80:113.

Challice CE, Viragh S (editors): *Ultrastructure of the Mammalian Heart.* Academic Press, 1973.

Cliff WJ: *Blood Vessels.* Cambridge Univ Press, 1976.

Johnson PC: *Peripheral Circulation.* Wiley, 1978.

Joyce NE et al: Contractile proteins in pericytes. J Cell Biol 1985;100:1387.

Leak LV: Normal anatomy of the lymphatic vascular system. In: *Handbuch der Allgemeine Pathologie.* Meessen H (editor). Springer-Verlag, 1972.

Li X, Erickson U: Novel VEGF family members: VEGF-B, VEGF-C and VEGF-D. Int J Biochem Cell Biol 2001;33:421.

Masuda H, Kalka C, Asahara T: Endothelial progenitor cells for regeneration. Hum Cell 2000;13:153.

Rhodin JAG: Architecture of the vessel wall. In: *Handbook of Physiology.* Section 2: *Cardiovascular System.* Vol 2. American Physiological Society, 1980.

Richardson JB, Beaulines A: The cellular site of action of angiotensin. J Cell Biol 1971;51:419.

Simionescu N: Cellular aspects of transcapillary exchange. Physiol Rev 1983;63:1536.

Thorgeirsson G, Robertson AL Jr: The vascular endothelium: pathobiologic significance. Am J Pathol 1978;93:802.

Wagner D, Marder J: Biosynthesis of von Willebrand protein by human endothelial cells: processing steps and their intracellular localization. J Cell Biol 1984;99:2123.

Blood Cells

12

Blood (about 5.5 L in a man) consists of the cells and fluid that flow in a regular unidirectional movement within the closed circulatory system. Blood is propelled mainly by the rhythmic contractions of the heart and is made up of 2 parts: **formed elements,** or blood cells, and **plasma** (Gr. *plasma,* thing formed), the liquid in which the formed elements are suspended. The formed elements are **erythrocytes** (red blood cells); **platelets;** and **leukocytes** (white blood cells).

If blood is removed from the circulatory system, it will clot. This clot contains formed elements and a clear yellow liquid called **serum,** which separates from the coagulum.

Blood that is collected and kept from coagulating by the addition of anticoagulants (eg, heparin, citrate) separates, when centrifuged, into layers that reflect its heterogeneity (Figure 12–1). The **hematocrit** is an estimate of the volume of packed erythrocytes per unit volume of blood. The normal value is 40–50% in men and 35–45% in women.

The translucent, yellowish, somewhat viscous supernatant obtained when whole blood is centrifuged is the plasma. The formed elements of the blood separate into 2 easily distinguishable layers. The lower layer represents 42–47% of the entire volume of blood in the hematocrit tube. It is red and is made up of erythrocytes. The layer immediately above (1% of the blood volume), which is white or grayish in color, is called the **buffy coat** and consists of leukocytes. These elements separate because the leukocytes are less dense than the erythrocytes. Covering the leukocytes is a fine layer of platelets not distinguishable by the naked eye.

Leukocytes, which have diversified functions (Table 12–1), are one of the body's chief defenses against infection. They circulate throughout the body via the blood vascular system, but while suspended in the blood they are round and inactive. Crossing the wall of venules and capillaries, these cells penetrate the tissues, where they display their defensive capabilities. The blood is a distributing vehicle, transporting oxygen (Figure 12–2), carbon dioxide (CO_2), metabolites, and hormones, among other substances. O_2 is bound mainly to the hemoglobin of the erythrocytes, whereas CO_2, in addition to being bound to the proteins of the erythrocytes (mainly hemoglobin), is carried in solution in the plasma as CO_2 or HCO_3^-.

The plasma transports nutrients from their site of absorption or synthesis, distributing them to various areas of the organism. It also transports metabolic residues, which are removed from the blood by the excretory organs. Blood, as the distributing vehicle for the hormones, permits the exchange of chemical messages between distant organs for normal cellular function. It further participates in the regulation of body temperature and in acid-base and osmotic balance.

Composition of Plasma

Plasma is an aqueous solution containing substances of low or high molecular weight that make up 10% of its volume. The plasma proteins account for 7% of the volume and the inorganic salts for 0.9%; the remainder of the 10% consists of several organic compounds—eg, amino acids, vitamins, hormones, lipoproteins—of various origins.

Through the capillary walls, the low-molecular-weight components of plasma are in equilibrium with the interstitial fluid of the tissues. The composition of plasma is usually an indicator of the mean composition of the extracellular fluids in general.

The main plasma proteins are **albumin; alpha, beta,** and **gamma globulins; lipoproteins,** and proteins that participate in blood coagulation, such as **prothrombin** and **fibrinogen.**

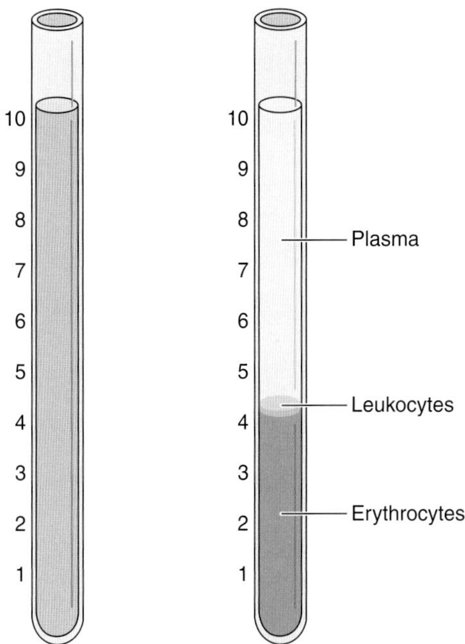

Figure 12–1. Hematocrit tubes with blood. **Left:** Before centrifugation. **Right:** After centrifugation. The erythrocytes represent 43% of the blood volume in the centrifuged tube. Between the sedimented erythrocytes and the supernatant light-colored plasma is a thin layer of leukocytes called the buffy coat.

Table 12–1. Products and functions of the blood cells.

Cell Type	Main Products	Main Functions
Erythrocyte	Hemoglobin	CO_2 and O_2 transport
Leukocytes Neutrophil (terminal cell)	Specific granules and modified lysosomes (azurophilic granules)	Phagocytosis of bacteria
Eosinophil (terminal cell)	Specific granules, pharmacologically active substances	Defense against parasitic helminths; modulation of inflammatory processes
Basophil (terminal cell)	Specific granules containing histamine and heparin	Release of histamine and other inflammation mediators
Monocyte (not terminal cell)	Granules with lysosomal enzymes	Generation of mononuclear-phagocyte system cells in tissues; phagocytosis and digestion of protozoa and virus and senescent cells
B lymphocyte	Immunoglobulins	Generation of antibody-producing terminal cells (plasma cells)
T lymphocyte	Substances that kill cells. Substances that control the activity of other leukocytes (interleukins)	Killing of virus-infected cells
Natural killer cell (lacks T- and B-cell markers)	Attacks virus-infected cells and cancer cells without previous stimulation	Killing of some tumor and virus-infected cells
Platelet	Blood-clotting factors	Clotting of blood

Albumin, the most abundant component, has a fundamental role in maintaining the osmotic pressure of the blood.

Staining of Blood Cells

Blood cells are generally studied in smears or films prepared by spreading a drop of blood in a thin layer on a microscope slide. The blood should be evenly distributed over the slide and allowed to dry rapidly in air. In such films the cells are clearly visible and distinct from one another. Their cytoplasm is spread out, facilitating observation of their nuclei and cytoplasmic organization.

Blood smears are routinely stained with special mixtures of red (acidic) and blue (basic) dyes. These mixtures also contain **azures,** dyes that are useful in staining some structures of blood cells known as **azurophilics** (azure + Gr. *philein,* to love). Some of these special mixtures (eg, Giemsa, Wright's, Leishman's) are named for the investigators who introduced their own modifications into the original mixture.

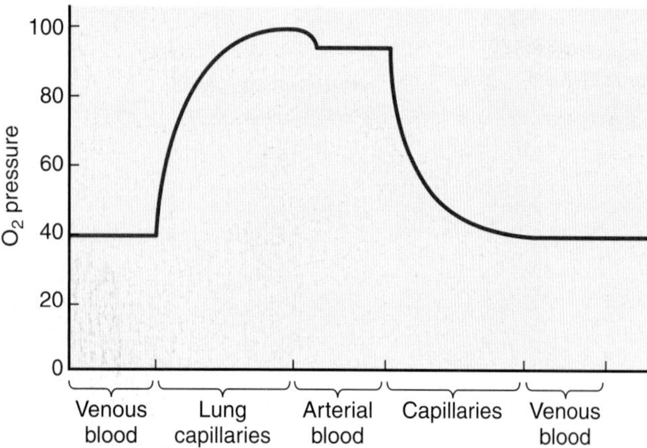

Figure 12–2. Blood O_2 content in each type of blood vessel. The amount of O_2 (O_2 pressure) is highest in arteries and lung capillaries; it decreases in tissue capillaries, where exchange takes place between blood and tissues.

Erythrocytes

Erythrocytes (red blood cells), which are anucleate, are packed with the O_2-carrying protein hemoglobin. Under normal conditions, these corpuscles never leave the circulatory system.

Most mammalian erythrocytes are biconcave disks without nuclei (Figure 12–3). When suspended in an isotonic medium, human erythrocytes are 7.5 μm in diameter, 2.6 μm thick at the rim, and 0.8 μm thick in the center. The biconcave shape provides erythrocytes with a large surface-to-volume ratio, thus facilitating gas exchange.

The normal concentration of erythrocytes in blood is approximately 3.9–5.5 million per microliter in women and 4.1–6 million per microliter in men.

MEDICAL APPLICATION

A decreased number of erythrocytes in the blood is usually associated with ***anemia.*** *An increased number of erythrocytes (****erythrocytosis,*** *or* ***polycythemia****) may be a physiologic adaptation. It is found, for example, in people who live at high altitudes, where O_2 tension is low. Polycythemia (Gr.* polys, *many,* + kytos, *cell,* + haima, *blood), which is*

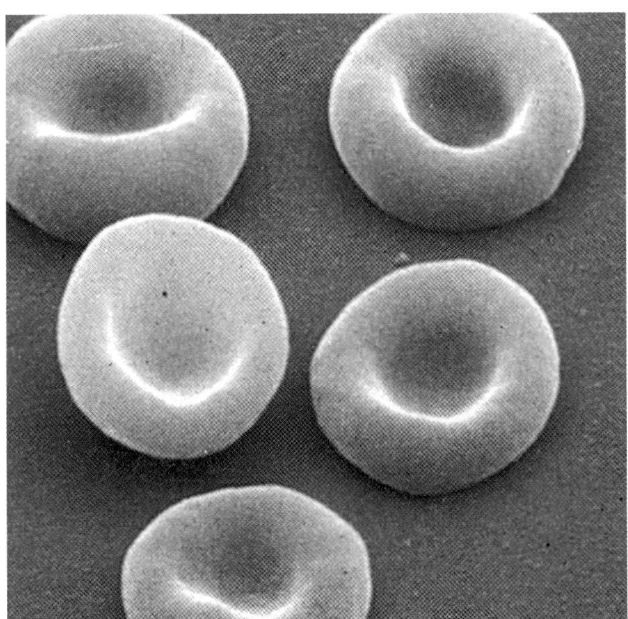

Figure 12–3. Scanning electron micrograph of normal human erythrocytes. Note their biconcave shape. ×3300.

often associated with diseases of varying degrees of severity, increases blood viscosity; when severe, it can impair circulation of blood through the capillaries. Polycythemia might be better characterized as an increased hematocrit, ie, an increased volume occupied by erythrocytes.

Erythrocytes with diameters greater than 9 μm are called ***macrocytes,*** *and those with diameters less than 6 μm are called* ***microcytes.*** *The presence of a high percentage of erythrocytes with great variations in size is called* ***anisocytosis*** *(Gr.* aniso, *uneven, +* kytos*).*

The erythrocyte is quite flexible, a property that permits it to adapt to the irregular shapes and small diameters of capillaries. Observations in vivo show that when traversing the angles of capillary bifurcations, erythrocytes containing normal adult hemoglobin (HbA) are easily deformed and frequently assume a cuplike shape.

Erythrocytes are surrounded by a plasmalemma; because of its ready availability, this is the best-known membrane of any cell. It consists of about 40% lipid (eg, phospholipids, cholesterol, glycolipids), 50% protein, and 10% carbohydrate. About half the proteins span the lipid bilayer and are known as **integral membrane proteins** (see Chapter 2). Several peripheral proteins are associated with the inner surface of the erythrocyte membrane. The peripheral proteins seem to serve as a membrane skeleton that determines the shape of the erythrocyte. One protein associated with the surface of the erythrocyte membrane is the cytoskeletal spectrin, which links several membrane components with other cytoskeletal elements, forming a meshwork that reinforces the erythrocyte membrane. This meshwork also permits the flexibility of the membrane necessary for the large changes in shape that occur when the erythrocyte passes through capillaries. Because erythrocytes are not rigid, the viscosity of blood normally remains low.

In their interiors, erythrocytes contain a 33% solution of hemoglobin, the O_2-carrying protein that accounts for their acidophilia. In addition, there are enzymes of the glycolytic and hexose-monophosphate-shunt pathways of glucose metabolism.

MEDICAL APPLICATION

Inherited alterations in hemoglobin molecules are responsible for several pathologic conditions, of which ***sickle cell disease*** *is an example. This inherited disorder is caused by a mutation of one nucleotide (****point mutation****) in the DNA of the gene for the β chain of hemoglobin. The triplet GAA (for glutamic acid) is changed to GUA, which specifies valine. As a result, the translated hemoglobin differs from the normal one by the presence of valine in the place of glutamic acid. The consequences of this substitution of a single amino acid are profound, however. When the altered hemoglobin (called HbS) is deoxygenated, as occurs in venous capillaries, it polymerizes and forms rigid aggregates that give the erythrocyte a characteristic sickle shape (Figure 12–4). The sickled erythrocyte is inflexible and fragile and has a shortened life span that leads to anemia. It increases the blood viscosity and can damage the wall of blood vessels, promoting blood coagulation. Blood flow through the capillaries is retarded or even stopped, leading to severe* O_2 *shortage (****anoxia****) in tissues.*

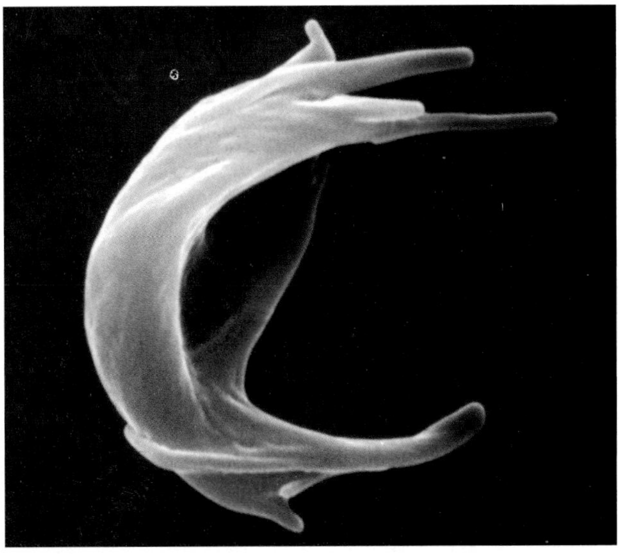

Figure 12–4. Scanning electron micrograph of a distorted erythrocyte from a person who is homozygous for the HbS gene (sickle cell disease). ×6500.

*Another disease of the erythrocytes is **hereditary spherocytosis,** characterized by spheroidal erythrocytes that are more vulnerable to sequestration and destruction by macrophages, causing anemia and other symptoms. In some cases, spherocytosis is related to deficiency or defects in the spectrin molecule. Surgical removal of the spleen usually ameliorates the symptoms of hereditary spherocytosis because of the removal of a large portion of the body's macrophages.*

Combined with O_2 or CO_2, hemoglobin forms **oxyhemoglobin** or **carbaminohemoglobin,** respectively. The reversibility of these combinations is the basis for the gas-transporting capability of hemoglobin. The combination of hemoglobin with carbon monoxide (**carboxyhemoglobin**) is irreversible, however, and causes a reduced capacity to transport O_2.

MEDICAL APPLICATION

***Anemia** is a pathologic condition characterized by blood concentrations of hemoglobin below normal values. Although anemias are usually associated with a decreased number of erythrocytes, it is also possible for the number of cells to be normal but for each cell to contain a reduced amount of hemoglobin (**hypochromic anemia**). Anemia may be caused by loss of blood (hemorrhage); insufficient production of erythrocytes by the bone marrow; production of erythrocytes with insufficient hemoglobin, usually related to iron deficiency in the diet; or accelerated destruction of blood cells.*

Erythrocytes recently released by the bone marrow into the bloodstream often contain residual ribosomal RNA, which, in the presence of some supravital dyes (eg, brilliant cresyl blue), can be precipitated and stained. Under these conditions, the younger erythrocytes, called **reticulocytes,** may have a few granules or a netlike structure in their cytoplasm.

Reticulocytes normally constitute about 1% of the total number of circulating erythrocytes; this is the rate at which erythrocytes are replaced daily by the bone marrow. Increased numbers of reticulocytes indicate a demand for increased O_2^- carrying capacity, which may be caused by such factors as hemorrhage or a recent ascent to high altitude.

Erythrocytes lose their mitochondria, ribosomes, and many cytoplasmic enzymes during their maturation. The source of energy for erythrocytes is glucose, which is anaerobically degraded to lactate. Because erythrocytes do not have a nucleus or other organelles necessary for protein synthesis, they do not synthesize hemoglobin.

Human erythrocytes survive in the circulation for about 120 days. Worn-out erythrocytes are removed from the circulation mainly by macrophages of the spleen and bone marrow. The signal for removal seems to be the appearance of defective complex oligosaccharides attached to integral membrane proteins of the plasmalemma.

Sometimes—mainly in disease states—nuclear fragments (containing DNA) remain in the erythrocyte after extrusion of its nucleus, which occurs late in its development (Chapter 13).

Leukocytes

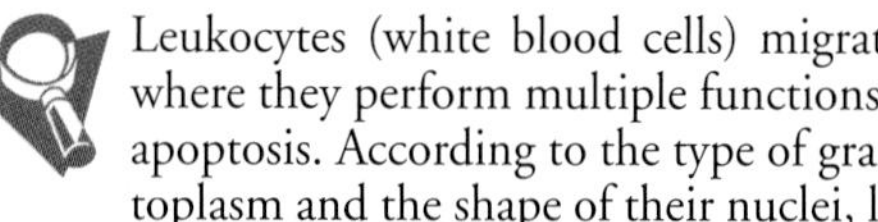

Leukocytes (white blood cells) migrate to the tissues, where they perform multiple functions and most die by apoptosis. According to the type of granules in their cytoplasm and the shape of their nuclei, leukocytes are divided into 2 groups: **granulocytes** (polymorphonuclear leukocytes) and **agranulocytes** (mononuclear leukocytes). Both granulocytes and agranulocytes are spherical (Figure 12–5) while suspended in blood plasma, but some become ameboid after leaving the blood vessels and invading the tissues. Their estimated sizes mentioned below refer to blood smears, in which the cells are spread and appear larger than they actually are in the blood.

Granulocytes (L. *granulum,* granule, + Gr. *kytos*) possess 2 types of granules: the **specific** granules that bind neutral, basic, or acidic components of the dye mixture and have specific functions and the **azurophilic granules.** Azurophilic granules stain purple and are lysosomes. Specific and azurophilic granules contain the enzymes listed in Table 12–2. Granulocytes have nuclei with 2 or more lobes and include the **neutrophils, eosinophils,** and **basophils** (Figure 12–5). All granulocytes are nondividing terminal cells with a life span of a few days, dying by apoptosis (programmed cell death) in the connective tissue. It is estimated that billions of neutrophils die by apoptosis each day in the adult human. The resulting cellular debris is removed by macrophages and does not elicit an inflammatory response. Being nondividing terminal cells, granulocytes do not synthesize much protein. Their Golgi complex and rough endoplasmic reticulum are poorly developed. They have few mitochondria (low energy metabolism) and depend more on glycolysis; this is why they contain glycogen and can function in regions scarce in oxygen, such as inflamed areas.

Agranulocytes do not have specific granules, but they do contain azurophilic granules (lysosomes) that bind the azure dyes of the stain. The nucleus is round or indented. This group includes **lymphocytes** and **monocytes** (Figure 12–5).

The differential count of blood leukocytes is presented in Table 12–3.

Leukocytes are involved in the cellular and humoral defense of the organism against foreign material. In suspension in the circulating blood, they are spherical, nonmotile cells, but they are capable of becoming flattened and motile on encountering a solid substrate. Leukocytes leave the venules and capillaries by passing between endothelial cells and penetrating the connective tissue by **diapedesis** (Gr. *dia,* through, + *pedesis,* to leap), a process that accounts for the unidirectional flow of granulocytes and monocytes from the blood to the tissues. (Lymphocytes recirculate.) Diapedesis is increased in individuals infected by microorganisms. Inflamed areas release chemicals originating mainly from cells and microorganisms, which increase diapedesis. The attraction of specific cells by chemical mediators is called **chemotaxis,** a significant event in inflammation through which leukocytes rapidly concentrate in places where their defensive properties are needed.

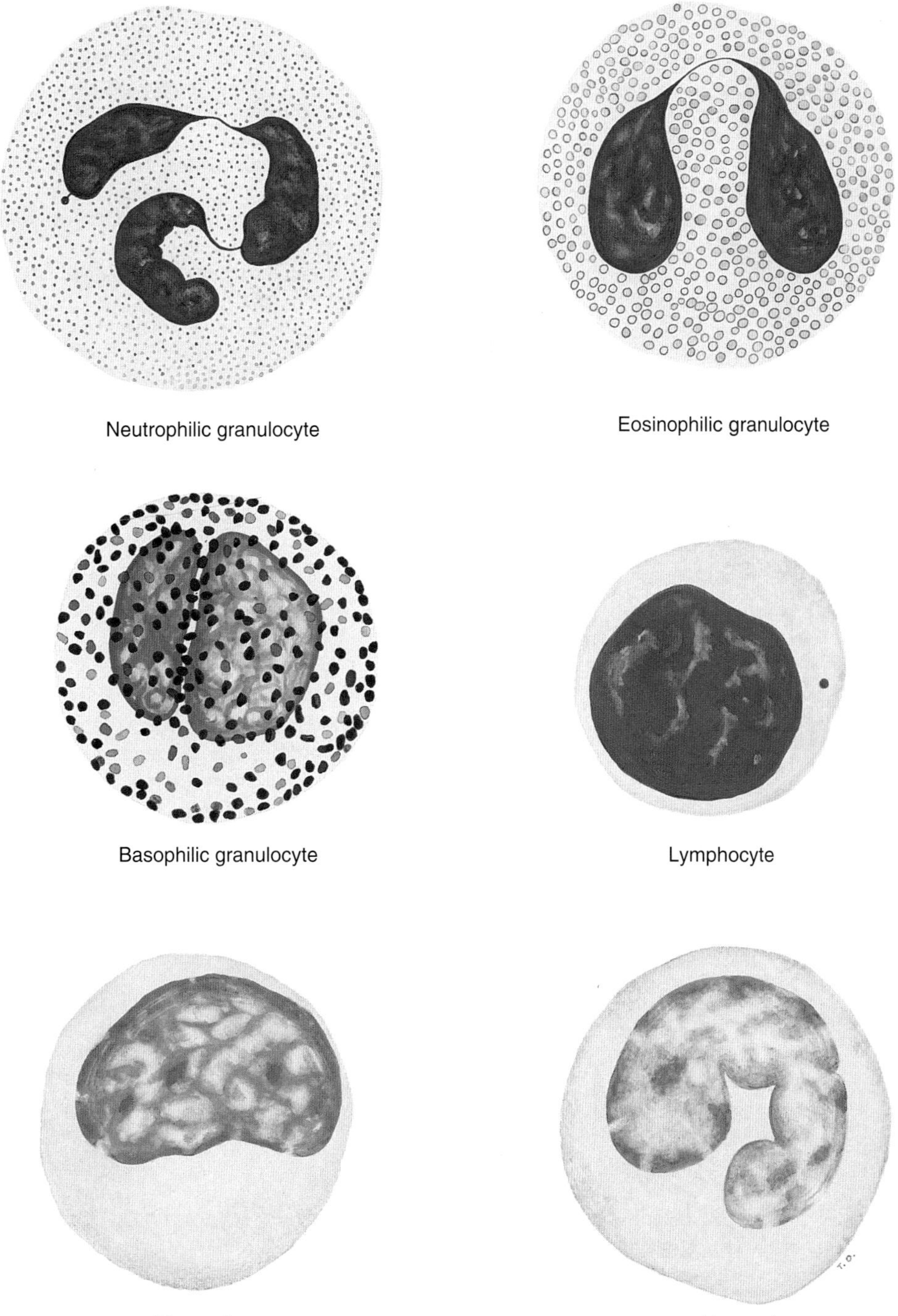

Figure 12–5. The 5 types of human leukocytes. Neutrophils, eosinophils, and basophils have granules that stain specifically with certain dyes and are called granulocytes. Lymphocytes and monocytes are agranulocytes; they may show azurophilic granules, which are also present in other leukocytes.

Table 12–2. Granule composition in human granulocytes.

Cell Type	Specific Granules	Azurophilic Granules
Neutrophil	Alkaline phosphatase Collagenase Lactoferrin Lysozyme Several nonenzymatic antibacterial basic proteins	Acid phosphatase α-Mannosidase Arylsulfatase β-Galactosidase β-Glucuronidase Cathepsin 5′-Nucleotidase Elastase Collagenase Myeloperoxidase Lysozyme Cationic antibacterial proteins
Eosinophil	Acid phosphatase Arylsulfatase β-Glucuronidase Cathepsin Phospholipase RNAase Eosinophilic peroxidase Major basic protein	
Basophil	Eosinophilic chemotactic factor Heparin Histamine Peroxidase	

The number of leukocytes in the blood varies according to age, sex, and physiologic conditions. In normal adults, there are roughly 6000–10,000 leukocytes per microliter of blood (Table 12–3).

Neutrophils (Polymorphonuclear Leukocytes)

Neutrophils constitute 60–70% of circulating leukocytes. They are 12–15 μm in diameter (in blood smears), with a nucleus consisting of 2–5 (usually 3) lobes linked by fine threads of chromatin (Figures 12–5, 12–6, and 12–7).

MEDICAL APPLICATION

Immature neutrophils that have recently entered the blood circulation have a nonsegmented nucleus in the shape of a horseshoe (band forms). An increased number of band neutrophils in the blood indicates a higher production of neutrophils, probably in response to a bacterial infection.

*Neutrophils with more than 5 lobes are called **hypersegmented** and are typically old cells. Although the maturation of the neutrophil parallels the increase in the number of nuclear lobes under normal conditions, in some pathologic conditions, young cells appear with 5 or more lobes.*

In females, the inactive X chromosome appears as a drumsticklike appendage on one of the lobes of the nucleus (Figure 12–5). However, this characteristic is not obvious in all neutrophils in a blood smear. The cytoplasm of the neutrophil contains 2 main types of granules. The more abundant granules are the **specific granules,** which are small, near the limit of resolution of the light microscope (Figure 12–6).

The second granule population in neutrophils consists of **azurophilic granules,** which are lysosomes 0.5 μm in diameter. Neutrophils also contain glycogen in their cytoplasm.

Glycogen is broken down into glucose to yield energy via the glycolytic pathway of glucose oxidation. The citric acid cycle is less important, as might be expected in view of the paucity of mitochondria in these cells. The ability of neutrophils to survive in an anaerobic environment is highly advantageous, since they can kill bacteria and help clean up debris in poorly oxygenated regions, eg, inflamed or necrotic tissue.

Neutrophils are short-lived cells with a half-life of 6–7 hours in blood and a life span of 1–4 days in connective tissues, where they die by apoptosis. They are active phagocytes of bacteria and other small particles. Neutrophils are inactive and spherical while circulating but show an active ameboid movement upon adhering to a solid substrate, such as collagen in the extracellular matrix.

MEDICAL APPLICATION

Neutrophils look for bacteria to engulf by pseudopodia and internalize them in vacuoles called phagosomes, whose membrane is derived from the neutrophil plasmalemma. Immediately thereafter, specific granules fuse with and discharge their contents into the phagosomes. By means of proton pumps in the phagosome membrane, the pH of the vacuole is lowered to about 5.0, a favorable pH for

Table 12–3. Number and percentage of blood corpuscles (blood count).

Corpuscle Type	Approximate Number per μL[a]	Approximate Percentage
Erythrocyte	Female: 3.9–5.5 × 10^6/μL Male: 4.1–6 × 10^6/μL	
Reticulocyte		1% of the erythrocyte count
Leukocyte	6000–10,000	
Neutrophil	5000	60–70%
Eosinophil	150	2–4%
Basophil	30	0.5%
Lymphocyte	2400	28%
Monocyte	350	5%
Platelet	300,000	

[a]Some publications give these values per cubic milliliter (mm^3). Microliters and cubic milliters are identical units.

maximal activity of lysosomal enzymes. Azurophilic granules then discharge their enzymes into the acid environment, killing and digesting the microorganisms.

During phagocytosis, a burst of O_2 consumption leads to the formation of superoxide (O_2^-) anions and hydrogen peroxide (H_2O_2). O_2^- is a short-lived free radical formed by the gain of one electron by O_2. It is a highly reactive radical that kills microorganisms ingested by neutrophils. Together with myeloperoxidase and halide ions, it forms a powerful killing system. Other strong oxidizing agents (eg, hypochlorite) can inactivate proteins. Lysozyme has the function of specifically cleaving a bond in the peptidoglycan that forms the cell wall of some gram-positive bacteria, thus causing their death. Lactoferrin avidly binds iron; because iron is a crucial element in bacterial nutrition, lack of its availability leads to bacterial death. The acid environment of phagocytic vacuoles can itself cause the death of certain microorganisms. A combination of these mechanisms will kill most microorganisms, which are then digested by lysosomal enzymes. Dead neutrophils, bacteria, semidigested material, and tissue fluid form a viscous, usually yellow collection of fluid called ***pus.***

Several neutrophil hereditary dysfunctions have been described. In one of them, actin does not polymerize normally, and the neutrophils are sluggish. In another, there is a failure to produce O_2^-, H_2O_2, and hypochlorite, and microbial killing power is reduced. This dysfunction results from a deficiency of NADPH oxidase, leading to a deficient respiratory burst. Children with these dysfunctions are subject to persistent bacterial infections. More severe infections result when neutrophil dysfunction and macrophage dysfunction occur simultaneously.

Eosinophils

Eosinophils are far less numerous than neutrophils, constituting only 2–4% of leukocytes in normal blood. In blood smears, this cell is about the same size as a neutrophil and contains a characteristic bilobed nucleus (Figures 12–8 and 12–9). The main identifying characteristic is the presence of many large and elongated refractile specific granules (about 200 per cell) that are stained by eosin.

Eosinophilic specific granules have a crystalline core (**internum**) that lies parallel to the long axis of the granule (Figure 12–10). It contains a protein—called the **major basic protein**—with a large number of arginine residues. This protein constitutes 50% of the total granule protein and accounts for the eosinophilia of these granules. The major basic protein also seems to function in the killing of parasitic worms such as schistosomes. The less dense material surrounding the internum is known as the **externum,** or **matrix.**

MEDICAL APPLICATION

*An increase in the number of eosinophils in blood (**eosinophilia**) is associated with allergic reactions*

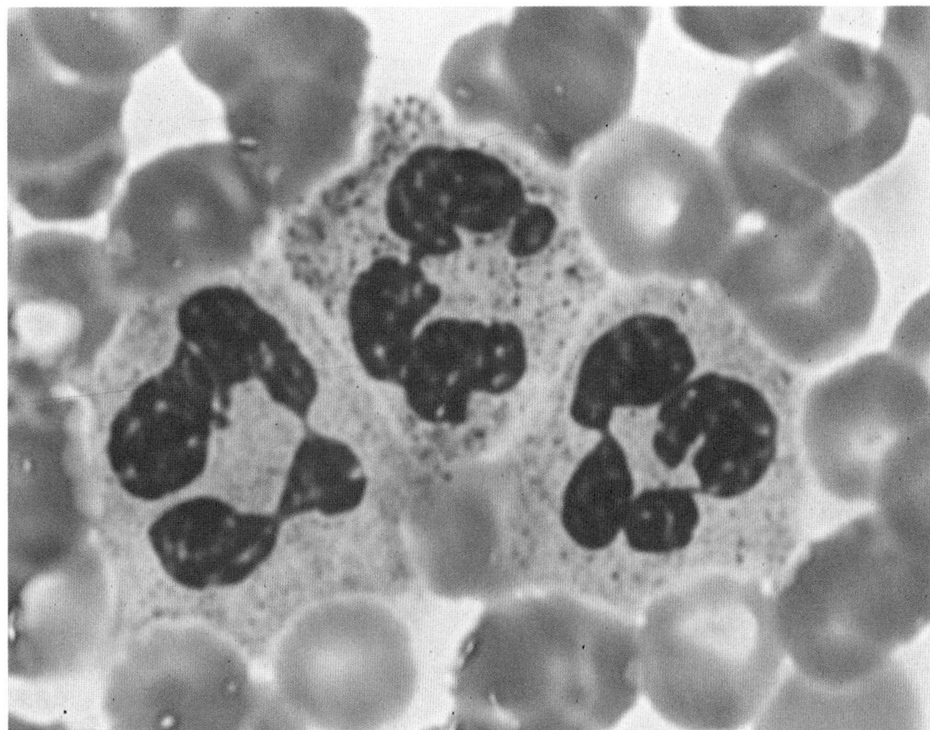

Figure 12–6. Photomicrograph of a blood smear showing 3 neutrophils and several erythrocytes. Each neutrophil has only one nucleus, with a variable number of lobes. Giemsa stain. High magnification.

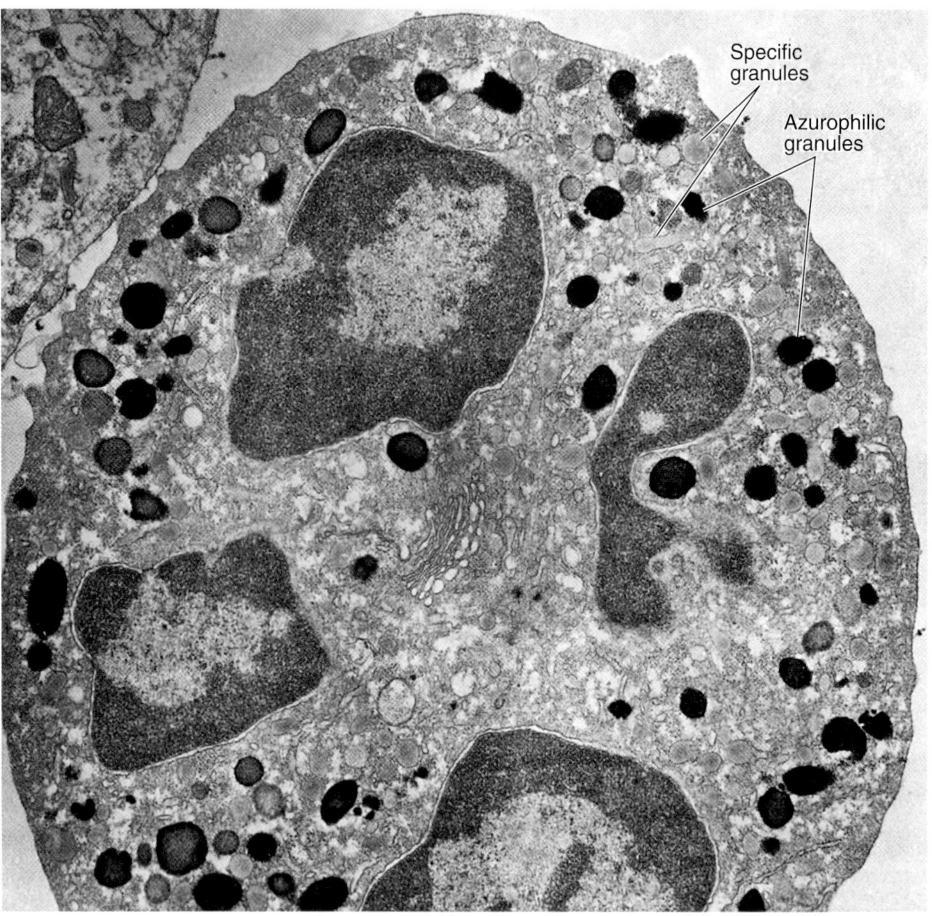

Figure 12–7. Electron micrograph of a human neutrophil stained for peroxidase. The cytoplasm contains 2 types of granules: the small, pale, peroxidase-negative specific granules and the larger, dense, peroxidase-positive azurophilic granules. The nucleus is lobulated, and the Golgi complex is small. Rough endoplasmic reticulum and mitochondria are not abundant, because this cell is in the terminal stage of its differentiation. ×27,000. (Reproduced, with permission, from Bainton DF: Selective abnormalities of azurophil and specific granules of human neutrophilic leukocytes. Fed Proc 1981;40:1443.)

and helminthic (parasitic) infections. In tissues, eosinophils are found in the connective tissues underlying epithelia of the bronchi, gastrointestinal tract, uterus, and vagina, and surrounding the parasitic worms. In addition, these cells produce substances that modulate inflammation by inactivating the leukotrienes and histamine produced by other cells. They also phagocytose antigen-antibody complexes.

Corticosteroids (hormones from the adrenal cortex) produce a rapid decrease in the number of blood eosinophils, probably by interfering with their release from the bone marrow into the bloodstream.

Basophils

Basophils make up less than 1% of blood leukocytes and are therefore difficult to find in smears of normal blood. They are about 12–15 μm in diameter. The nucleus is divided into irregular lobes, but the overlying specific granules usually obscure the division.

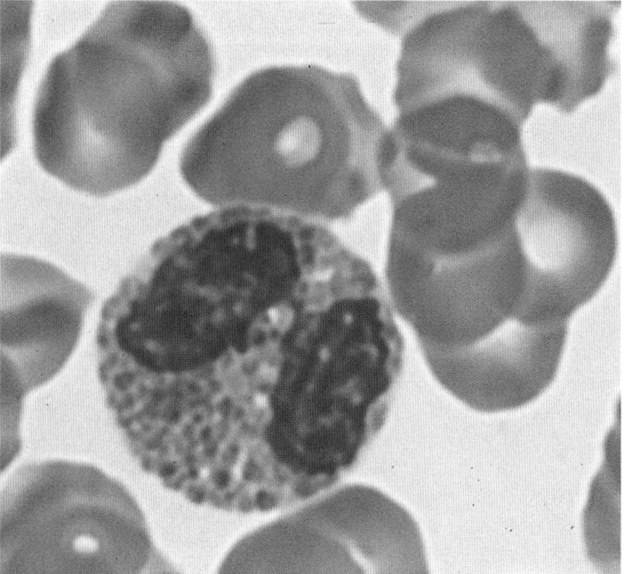

Figure 12–8. Photomicrograph of an eosinophil. Note its typical bilobed nucleus and coarse cytoplasmic granules. Giemsa stain. High magnification.

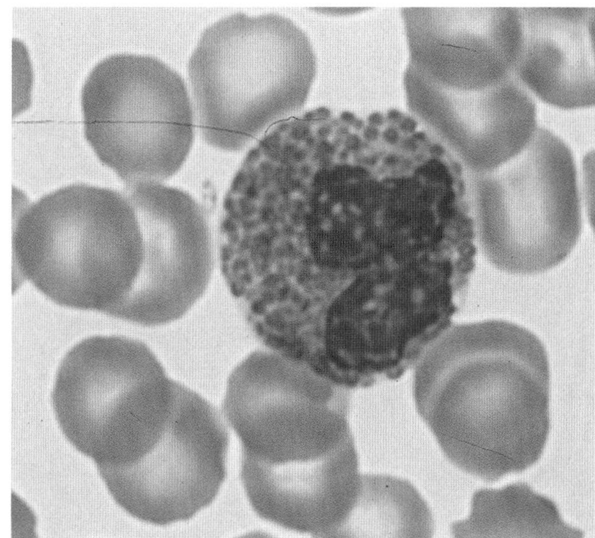

Figure 12–9. Eosinophil with its bilobed nucleus and coarse cytoplasmic granules. Giemsa stain. High magnification.

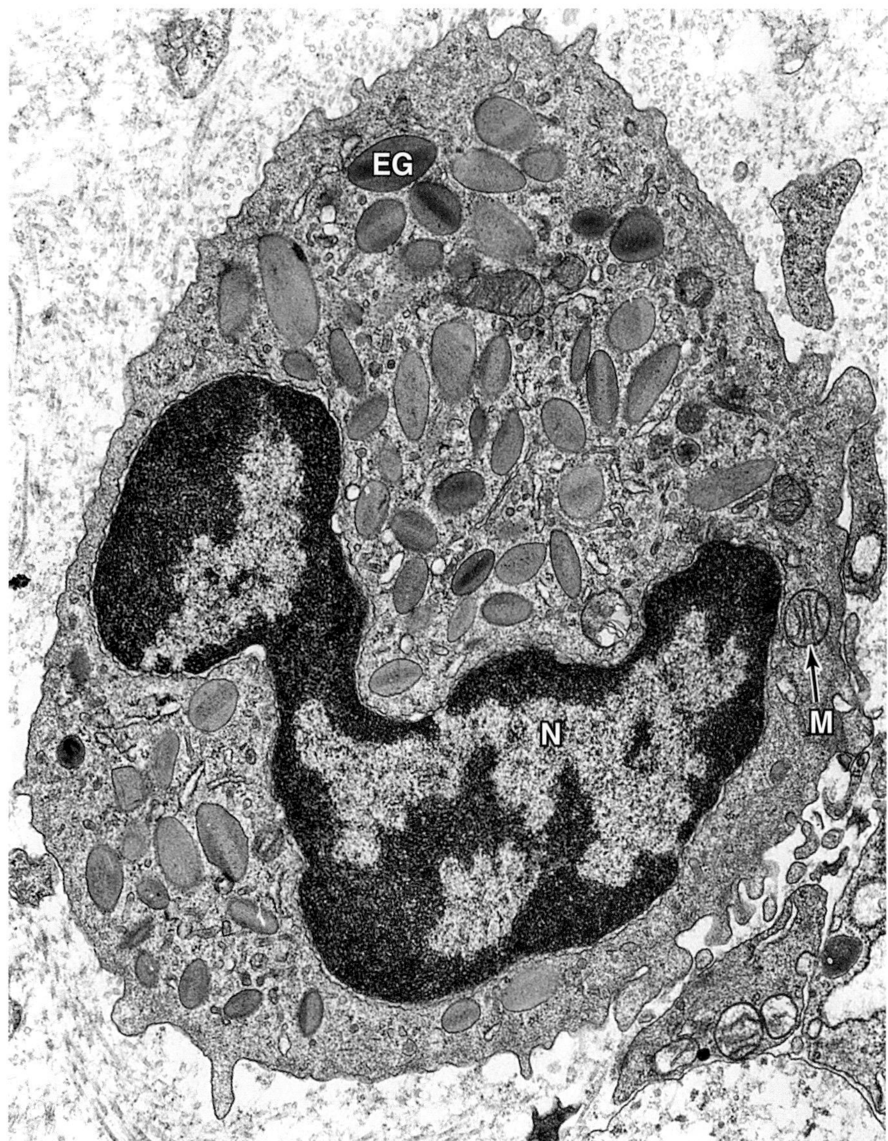

Figure 12–10. Electron micrograph of an eosinophil. Typical eosinophilic granules are clearly seen. Each granule has a disk-shaped electron-dense crystalline core that appears surrounded by a matrix enveloped by a unit membrane. EG, eosinophil granule; N, nucleus; M, mitochondria. ×20,000.

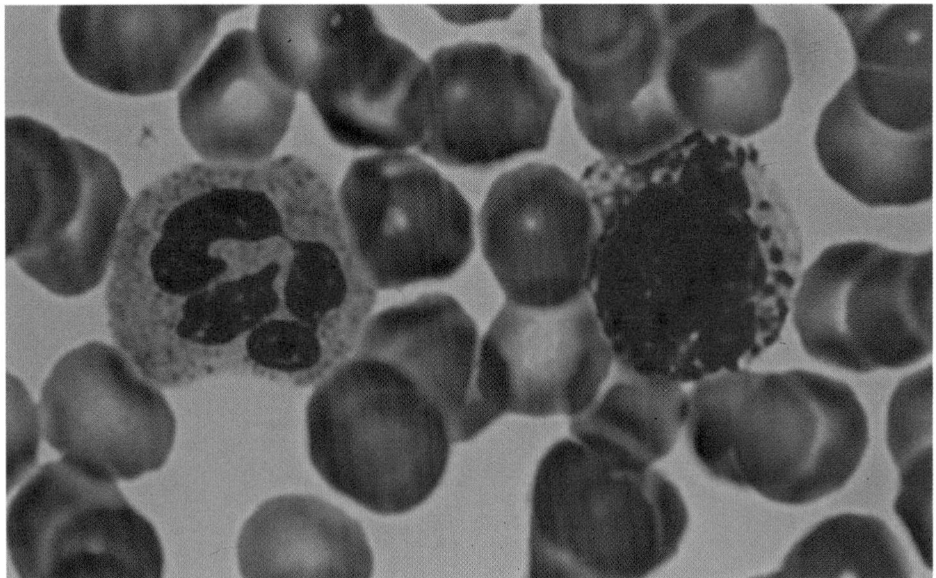

Figure 12–11. Two leukocytes and several erythrocytes. The cell on the right is a basophil. The cell on the left is a neutrophil. In the basophil there are many cytoplasmic granules over the nucleus. Giemsa stain. High magnification.

The specific granules (0.5 μm in diameter) stain metachromatically (change the color of the stain used) with the basic dye of the usual blood stains (Figures 12–11 and 12–12). This metachromasia is due to the presence of heparin. Specific granules in basophils are fewer and more irregular in size and shape than the granules of the other granulocytes (Figure 12–13). Basophilic specific granules contain heparin and histamine. Basophils may supplement the functions of mast cells in immediate hypersensitivity reactions by migrating (under special circumstances) into connective tissues.

There is some similarity between granules of basophils and those of mast cells. Both are metachromatic and contain heparin and histamine. Basophils can liberate their granule content in response to certain antigens, as can mast cells (see Chapter 5). Despite the similarities they present, mast cells and basophils are not the same, for even in the same species they have different structures, and they originate from different stem cells in the bone marrow.

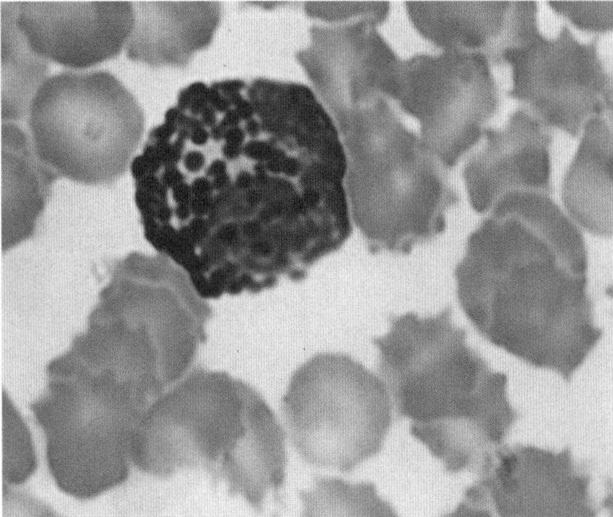

Figure 12–12. A basophil with many granules covering the cell nucleus. This makes it difficult to see the nucleus clearly. Some erythrocytes were deformed during the smear preparation. Giemsa stain. High magnification.

MEDICAL APPLICATION

*In the dermatologic disease called **cutaneous basophil hypersensitivity,** basophils are the major cell type at the site of inflammation.*

Lymphocytes

Lymphocytes constitute a family of spherical cells with similar morphologic characteristics. They can be classified into several groups according to distinctive surface molecules (markers), which can be distinguished by immunocytochemical methods. They also have diverse functional roles, all related to immune reactions in defending against invading microorganisms, foreign macromolecules, and cancer cells (see Chapter 14).

Lymphocytes with diameters of 6–8 μm are known as **small lymphocytes.** A small number of **medium-sized lymphocytes** and **large lymphocytes** with diameters up to 18 μm are present in the circulating blood (Figure 12–14). This difference has functional significance in that some larger lymphocytes are believed to be cells activated by specific antigens. The small lymphocyte (Figure 12–15), which is predominant in the blood, has a spherical nucleus, sometimes with an indentation. Its chromatin is condensed and appears as coarse clumps, so that the nucleus is intensely stained in the usual preparations, a characteristic that facilitates identification of the lymphocyte. In blood smears, the nucleolus of the lymphocyte is not visible, but it can be demon-

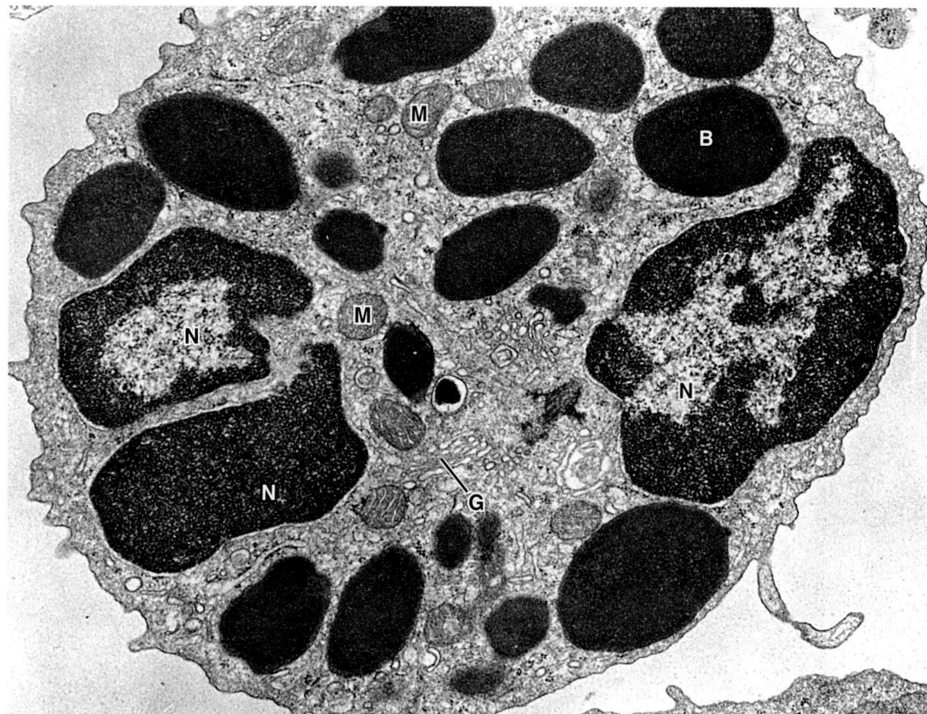

Figure 12–13. Electron micrograph of a rabbit basophil. The lobulated nucleus (N) appears as 3 separated portions. Note the basophilic granule (B), mitochondria (M), and Golgi complex (G). ×16,000. (Reproduced, with permission, from Terry RW et al: Lab Invest 1969;21:65.)

strated by special staining techniques and with the electron microscope.

The cytoplasm of the small lymphocyte is scanty, and in blood smears it appears as a thin rim around the nucleus. It is slightly basophilic, assuming a light blue color in stained smears. It may contain a few azurophilic granules. The cytoplasm of the small lymphocyte has a few mitochondria and a small Golgi complex; it contains free polyribosomes (Figure 12–16).

Lymphocytes vary in life span; some live only a few days, and others survive in the circulating blood for many years. Lymphocytes are the only type of leukocytes that return from the tissues back to the blood, after diapedesis. Further information regarding different types of lymphocytes and their individual participation in immunologic responses is presented in Chapter 14.

Monocytes

Monocytes are bone marrow–derived agranulocytes with diameters varying from 12 to 20 μm. The nucleus is oval, horseshoe-, or kidney-shaped and is generally eccentrically placed (Figure 12–17). The chromatin is less condensed than that in lymphocytes. Because of their delicate chromatin distribution, the nuclei of monocytes stain lighter than do those of large lymphocytes.

The cytoplasm of the monocyte is basophilic and frequently contains very fine azurophilic granules (lysosomes), some of which are at the limit of the light microscope's resolution. These granules are distributed through the cytoplasm, giving it a bluish-gray color in stained smears. In the electron microscope, one or two nucleoli are seen in the nucleus, and a small quantity of rough endoplasmic reticulum, polyribosomes, and many small mitochondria are observed. A Golgi complex involved in the formation of the lysosomal granules is present in the cytoplasm. Many microvilli and pinocytotic vesicles are found at the cell surface (Figure 12–18).

Blood monocytes are not terminal cells; rather, they are precursor cells of the mononuclear phagocyte system (see Chapter 5). After crossing capillary walls and entering connective tissues, monocytes differentiate into macrophages.

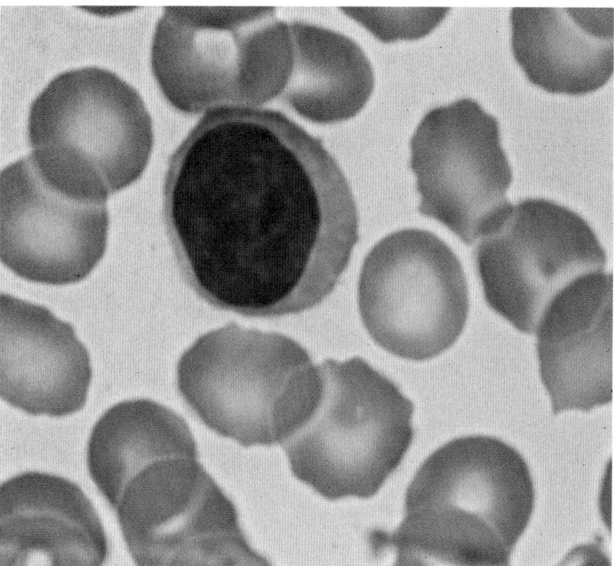

Figure 12–14. Photomicrograph of a large lymphocyte and several erythrocytes. The nucleus of this cell is round, and the cytoplasm is devoid of specific granules. Giemsa stain. High magnification.

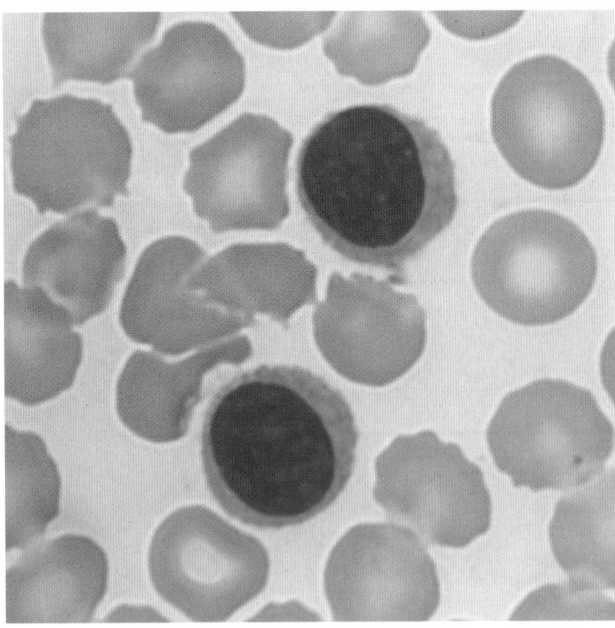

Figure 12–15. Two small lymphocytes with their round, dark-stained nuclei. Giemsa stain. High magnification.

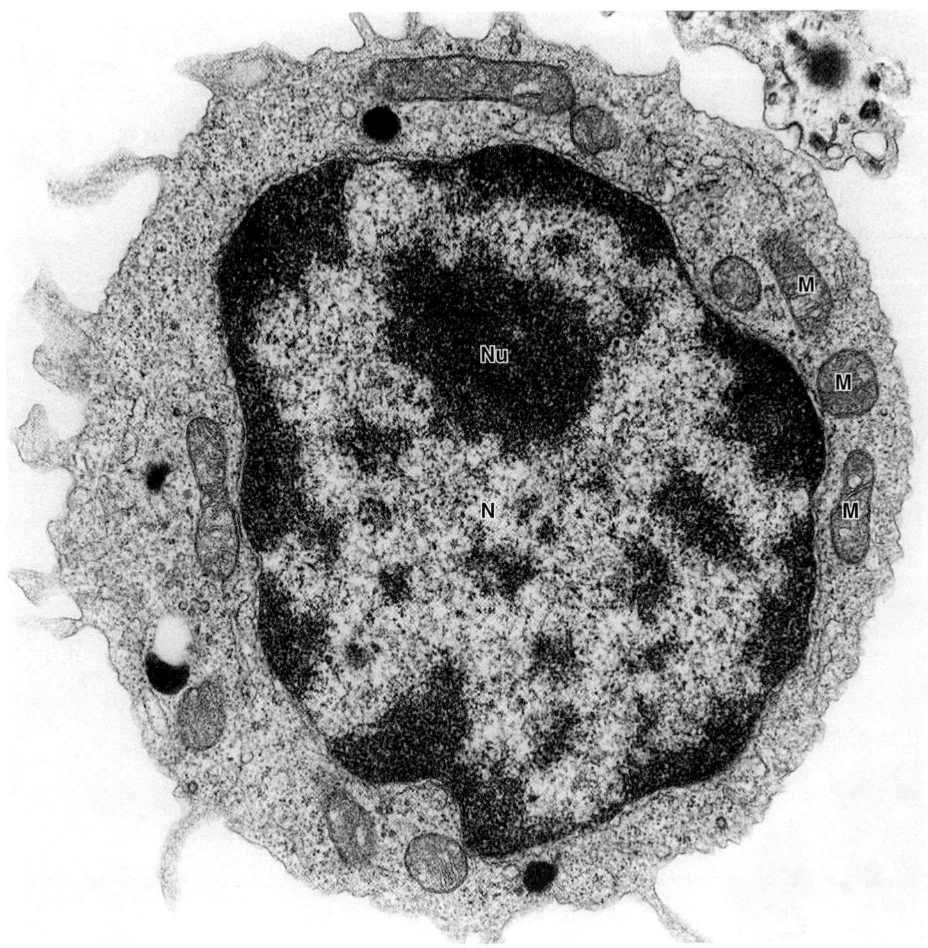

Figure 12–16. Electron micrograph of a human blood lymphocyte. This cell has little rough endoplasmic reticulum but a moderate quantity of free polyribosomes. Note the nucleus (N), the nucleolus (Nu), and the mitochondria (M). Reduced from ×22,000.

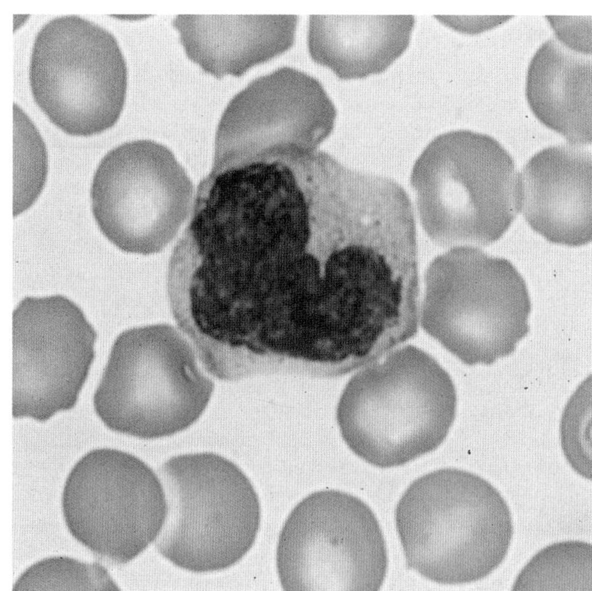

Figure 12–17. Photomicrograph of a monocyte. This cell type has a kidney-shaped nucleus with delicately stained chromatin. The cytoplasm is slightly basophilic. Giemsa stain. High magnification.

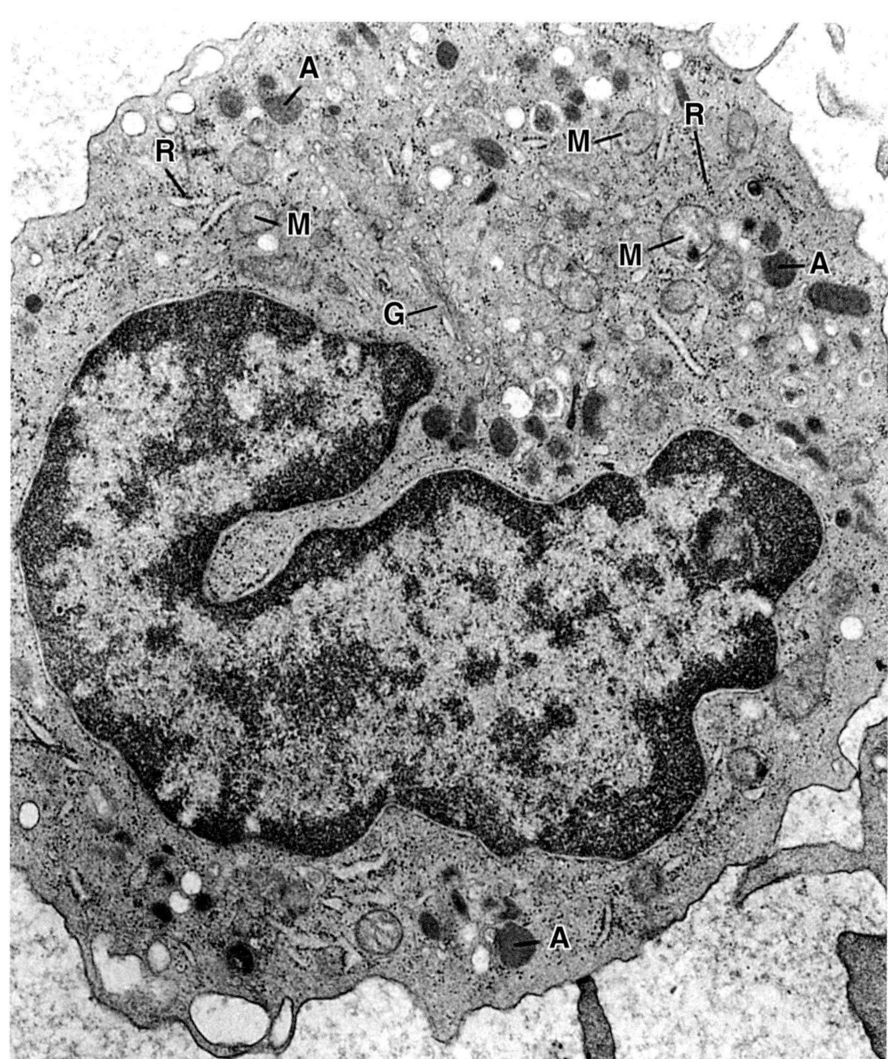

Figure 12–18. Electron micrograph of a human monocyte. Note the Golgi complex (G), the mitochondria (M), and the azurophilic granule (A). Rough endoplasmic reticulum is poorly developed. There are some free ribosomes (R). ×22,000. (Courtesy of DF Bainton and MG Farquhar.)

Platelets

Blood platelets (**thrombocytes**) are nonnucleated, disklike cell fragments 2–4 μm in diameter. Platelets originate from the fragmentation of giant polyploid **megakaryocytes** that reside in the bone marrow. Platelets promote blood clotting and help repair gaps in the walls of blood vessels, preventing loss of blood. Normal platelet counts range from 200,000 to 400,000 per microliter of blood. Platelets have a life span of about 10 days.

In stained blood smears, platelets often appear in clumps. Each platelet has a peripheral light blue–stained transparent zone, the **hyalomere,** and a central zone containing purple granules, called the **granulomere.**

Platelets contain a system of channels, the **open canalicular system,** that connect to invaginations of the platelet plasma membrane (Figure 12–19). This arrangement is probably of functional significance in facilitating the liberation of active molecules stored in platelets. Around the periphery of the platelet lies a **marginal bundle** of microtubules; this bundle helps to maintain the platelet's ovoid shape. In the hyalomere, there are also a number of electron-dense irregular tubes known as the **dense tubular system.** Actin and myosin molecules in the hyalomere can assemble to form a contractile system that functions in platelet movement and aggregation. A cell coat rich in glycosaminoglycans and glycoproteins, 15–20 nm thick, lies outside the plasmalemma and is involved in platelet adhesion.

The central granulomere possesses a variety of membrane-bound granules and a sparse population of mitochondria and glycogen particles (Figure 12–19). **Dense bodies** (**delta granules**), 250–300 nm in diameter, contain calcium ions, pyrophosphate, ADP, and ATP. These granules also take up and store serotonin (5-hydroxytryptamine) from the plasma. **Alpha granules** are a little larger (300–500 nm in diameter) and contain fibrinogen, platelet-derived growth factor, and several other platelet-specific proteins. Small vesicles, 175–250 nm in diameter, have been shown to contain only lysosomal enzymes and have been termed **lambda granules.** Most of the azurophilic granules seen with the light microscope in the granulomere of platelets are alpha granules.

Platelet Functions

The role of platelets in controlling hemorrhage can be summarized as follows.

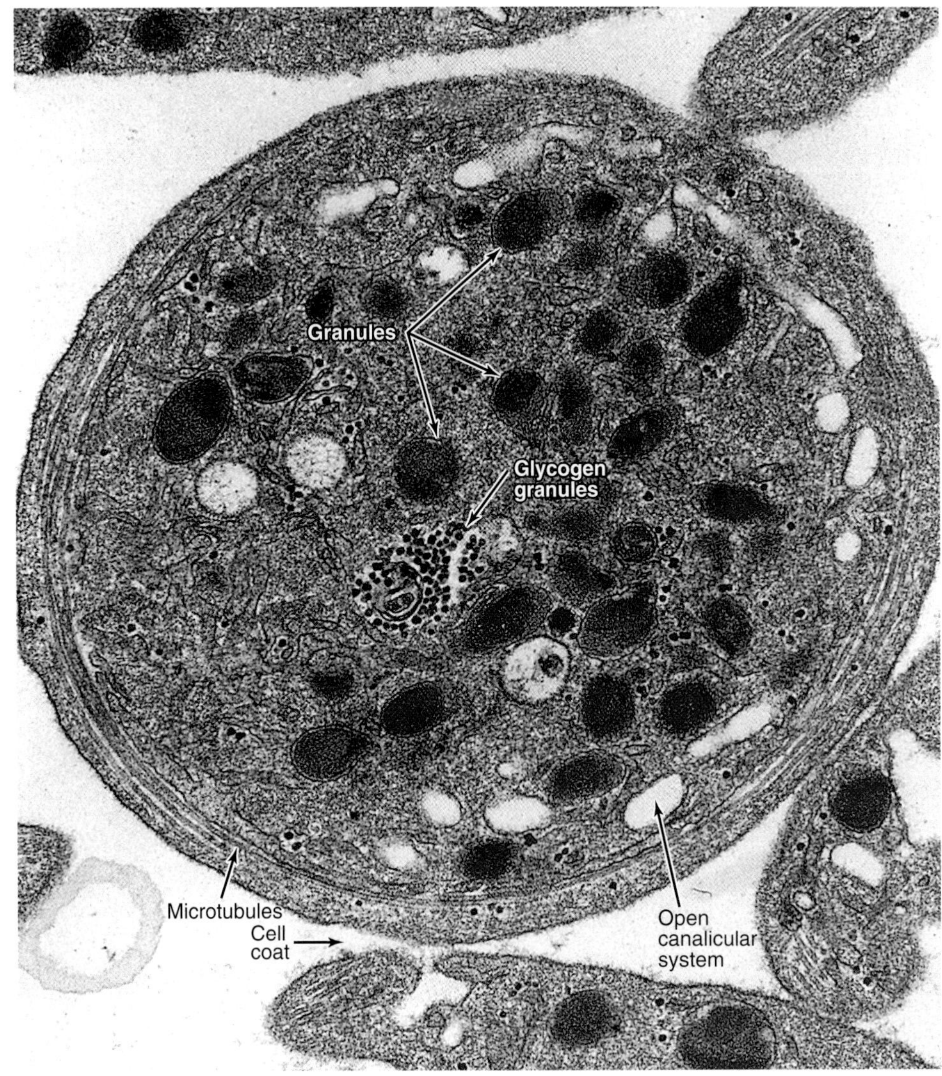

Figure 12–19. Electron micrograph of human platelets. ×40,740. (Courtesy of M Harrison.)

Primary aggregation—Discontinuities in the endothelium, produced by injuries, are followed by platelet aggregation to the exposed collagen, via collagen-binding protein in platelet membrane. Thus, a **platelet plug** is formed as a first step to stop bleeding.

Secondary aggregation—Platelets in the plug release an adhesive glycoprotein and ADP. Both are potent inducers of platelet aggregation, increasing the size of the platelet plug.

Blood coagulation—During platelet aggregation, factors from the blood plasma, damaged blood vessels, and platelets promote the sequential interaction (**cascade**) of approximately 13 plasma proteins, giving rise to a polymer, **fibrin,** that forms a 3-dimensional network of fibers trapping red cells, leukocytes, and platelets to form a **blood clot,** or **thrombus.**

Clot retraction—The clot that initially bulges into the blood vessel lumen contracts because of the interaction of platelet actin, myosin, and ATP.

Clot removal—Protected by the clot, the vessel wall is restored by new tissue formation. The clot is then removed, mainly by the proteolytic enzyme **plasmin,** formed, through the activation of the plasma proenzyme **plasminogen,** by endothelium-produced **plasminogen activators.** Enzymes released from platelet lambda granules also contribute to clot removal.

MEDICAL APPLICATION

Hemophilia A and B are clinically identical, differing only in the deficient factor. Both are due to sex-linked recessive inherited disorders. Blood from hemophiliac patients does not coagulate normally: the blood clotting time is prolonged. Persons with this disease bleed severely even after mild injuries, such as a skin cut, and may bleed to death after more severe injuries. The blood plasma of patients with hemophilia A is deficient in clotting factor VIII or contains a defective factor VIII, one of the plasma proteins involved in fibrin generation; in hemophilia B, the defect is in factor IX. In severe cases the blood is incoagulable. There are spontaneous hemorrhages in body cavities, such as major joints and the urinary tract. Generally, only males are affected by hemophilia A, because the recessive gene to factor VIII is in the X chromosome. Females may have one defective X chromosome, but the other one is usually normal. Females develop hemophilia only when they have the abnormal gene in both X chromosomes, a rare event. However, women with a defective X chromosome may transmit the disease to their male children.

REFERENCES

Bainton DF: Sequential degranulation of the 2 types of polymorphonuclear leukocyte granules during phagocytosis of microorganisms. J Cell Biol 1973;58:249.

Collins T: Adhesion molecules in leukocyte emigration. Sci & Med 1995;2:28.

Comenzo, RL, Berkman EM: Hematopoietic stem and progenitor cells from blood. Transfusion 1995;35:335.

Gompertz S, Stockley RA: Inflammation—role of the neutrophil and the eosinophil. Semin Respir Infect 2000;15:14.

Sampson AP: The role of eosinophils and neutrophils in inflammation. Clin Exp Allergy 2000;30(suppl 1):22.

Williams WJ et al: *Hematology,* 5th ed. McGraw-Hill, 1995.

Zucker-Franklin D et al: *Atlas of Blood Cells: Function and Pathology.* Vols 1 and 2. Lea & Febiger, 1981.

Hematopoiesis 13

Mature blood cells have a relatively short life span, and consequently the population must be continuously replaced with the progeny of stem cells produced in the **hematopoietic** (Gr. *haima,* blood, + *poiesis,* a making) organs. In the earliest stages of embryogenesis, blood cells arise from the yolk sac mesoderm. Sometime later, the liver and spleen serve as temporary hematopoietic tissues, but by the second month the clavicle has begun to ossify and begins to develop bone marrow in its core. As the prenatal ossification of the rest of the skeleton accelerates, the bone marrow becomes an increasingly important hematopoietic tissue.

After birth and on into childhood, erythrocytes, granular leukocytes, monocytes, and platelets are derived from stem cells located in bone marrow. The origin and maturation of these cells are termed, respectively, **erythropoiesis** (Gr. *erythros,* red, + *poiesis*), **granulopoiesis, monocytopoiesis,** and **megakaryocytopoiesis.** The bone marrow also produces cells that migrate to the lymphoid organs, producing the various types of lymphocytes discussed in Chapter 14.

Before attaining maturity and being released into the circulation, blood cells go through specific stages of differentiation and maturation. Because these processes are continuous, cells with characteristics that lie between the various stages are frequently encountered in smears of blood or bone marrow.

STEM CELLS, GROWTH FACTORS, & DIFFERENTIATION

Stem cells are **pluripotential** cells capable of self-renewal. Some of their daughter cells form specific, irreversibly differentiated cell types, and other daughter cells remain stem cells. A constant number of pluripotential stem cells is maintained in a pool, and cells recruited for differentiation are replaced with daughter cells from the pool.

Hematopoietic stem cells are isolated by using fluorescence-labeled antibodies to mark specific cell-surface antigens and a fluorescence-activated, cell-sorting instrument. Stem cells are also studied by means of experimental techniques that permit analysis of hematopoiesis in vivo and in vitro.

In vivo techniques include injecting the bone marrow of normal donor mice into lethally irradiated mice whose hematopoietic cells have been destroyed. In these animals, the transplanted bone marrow cells develop colonies of hematopoietic cells in the spleen.

In vitro techniques involve the use of a semisolid tissue culture medium made with a layer of cells derived from bone marrow stroma. This medium creates favorable microenvironmental conditions for hematopoiesis. Data from an extensive series of experiments show that under these favorable microenvironmental conditions, stimulation by growth factors influences the development of the various types of blood cells.

Pluripotential Hematopoietic Stem Cells

It is believed that all blood cells arise from a single type of stem cell in the bone marrow. Because this cell can produce all blood cell types, it is called a **pluripotential stem cell** (Figure 13–1). These cells proliferate and form one cell lineage that will become lymphocytes (**lymphoid cells**) and another lineage that will form the **myeloid cells** that develop in bone marrow (granulocytes, monocytes, erythrocytes, and megakaryocytes). Early in their development, lymphoid cells migrate from the bone marrow to the thymus, lymph nodes, spleen, and other lymphoid structures, where they proliferate (see Chapter 14).

Progenitor and Precursor Cells

The proliferating stem cells form daughter cells with reduced potentiality. These **unipotential** or **bipotential progenitor cells** generate **precursor cells** (**blasts**) in which the morphologic characteristics differentiate for the first time, suggesting the mature cell types they will become (see Figures 13–1 and 13–5). In contrast, stem and progenitor cells cannot be morphologically distinguished and resemble large lymphocytes. Stem cells divide at a rate sufficient to maintain their relatively small population. The rate of cell division is accelerated in progenitor and precursor cells, and large numbers of differentiated, mature cells are produced (3×10^9 erythrocytes and 0.85×10^9 granulocytes/kg/day in human bone marrow). Whereas progenitor cells can divide and produce both progenitor and precursor cells, precursor cells produce only mature blood cells.

Hematopoiesis is therefore the result of simultaneous, continuous proliferation and differentiation of cells derived from stem cells whose potentiality is reduced as differentiation progresses. This process can be observed in both in vivo and in vitro studies, in which colonies of cells derived from stem cells with various potentialities appear. Colonies derived from a myeloid stem cell can produce erythrocytes, granulocytes, monocytes, and megakaryocytes, all in the same colony.

In these experiments, however, some colonies produce only red blood cells (erythrocytes). Other colonies produce granulocytes and monocytes. Cells forming colonies are called **colony-forming cells** (**CFC**), or **colony-forming units** (**CFU**). The convention in naming these various cell colonies is to use the initial letter of the cell each colony produces. Thus, MCFC denotes a monocyte-forming colony, ECFC forms erythrocytes, MGCFC forms monocytes and granulocytes, and so on.

Phase	Stem Cells	Progenitor Cells	Precursor Cells (Blasts)	Mature Cells
Early morphologic	Not morphologically distinguishable; have the general aspect of lymphocytes		Beginning of morphologic differentiation	Clear morphologic differentiation
Mitotic activity	Low mitotic activity; self-renewing; scarce in bone marrow	High mitotic activity; self-renewing; common in marrow and lymphoid organs; mono- or bipotential	High mitotic activity; not self-renewing; common in marrow and lymphoid organs; monopotential	No mitotic activity; abundant in blood and hematopoietic organs

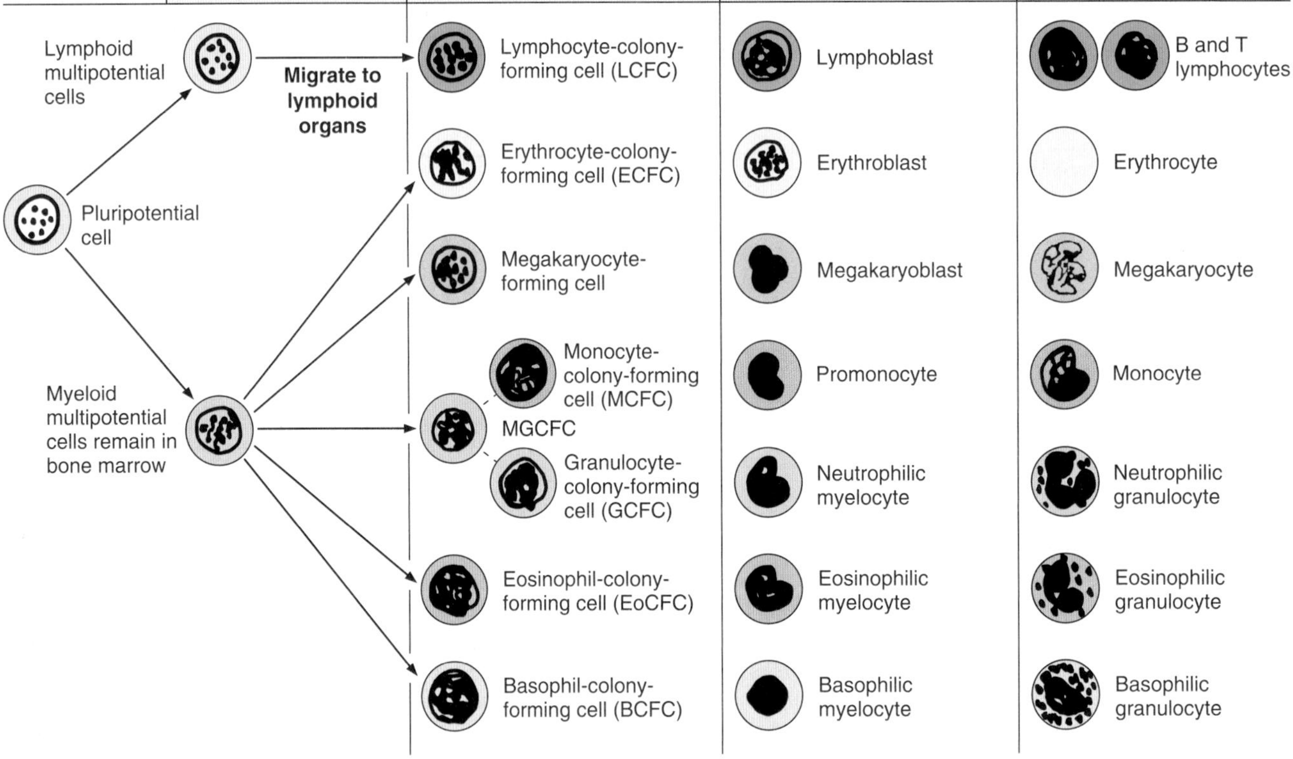

Pluripotential cell

Lymphoid multipotential cells — Migrate to lymphoid organs — Lymphocyte-colony-forming cell (LCFC) — Lymphoblast — B and T lymphocytes

Myeloid multipotential cells remain in bone marrow

Erythrocyte-colony-forming cell (ECFC) — Erythroblast — Erythrocyte

Megakaryocyte-forming cell — Megakaryoblast — Megakaryocyte

MGCFC — Monocyte-colony-forming cell (MCFC) — Promonocyte — Monocyte

MGCFC — Granulocyte-colony-forming cell (GCFC) — Neutrophilic myelocyte — Neutrophilic granulocyte

Eosinophil-colony-forming cell (EoCFC) — Eosinophilic myelocyte — Eosinophilic granulocyte

Basophil-colony-forming cell (BCFC) — Basophilic myelocyte — Basophilic granulocyte

Figure 13–1. Differentiation of pluripotential stem cells during hematopoiesis. See also Figure 13–5.

Hematopoiesis depends on favorable microenvironmental conditions and the presence of growth factors. The microenvironmental conditions are furnished by cells of the stroma of hematopoietic organs, which produce an adequate extracellular matrix. A general view of hematopoiesis shows that, as this process takes place, both the potential for differentiation and the self-renewing capacity of the initial cells gradually decrease. In contrast, the mitotic response to growth factors gradually increases, attaining its maximum in the middle of the process. From that point on, mitotic activity decreases, morphologic characteristics and functional activity develop, and mature cells are formed (Table 13–1). Once the necessary environmental con-

Table 13–1. Changes in properties of hematopoietic cells during differentiation.

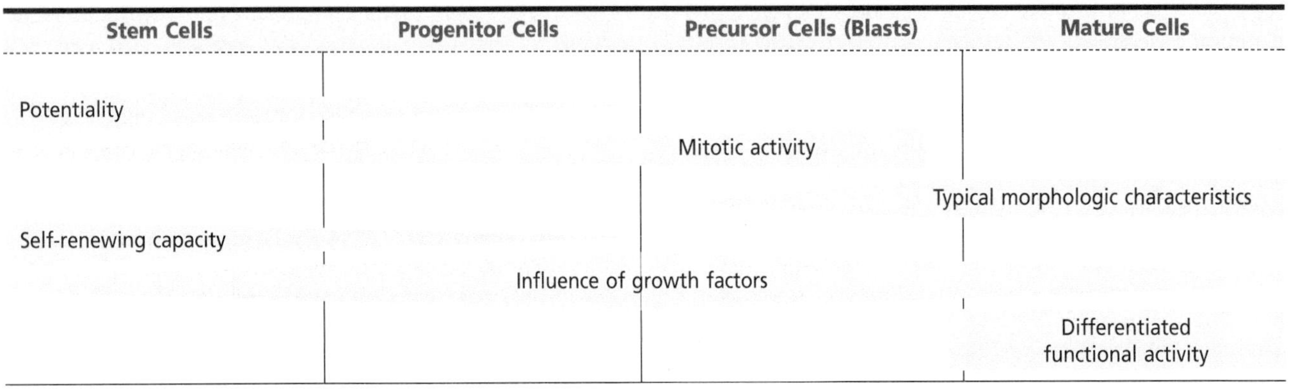

Stem Cells	Progenitor Cells	Precursor Cells (Blasts)	Mature Cells
Potentiality		Mitotic activity	
			Typical morphologic characteristics
Self-renewing capacity			
	Influence of growth factors		
			Differentiated functional activity

ditions are present, the development of blood cells depends on factors that affect cell proliferation and differentiation. These substances are called **growth factors, colony-stimulating factors** (**CSF**), or **hematopoietins** (**poietins**). Growth factors, which have differing chemical compositions and complex, overlapping functions, act mainly by stimulating proliferation (mitogenic activity) of immature (mostly progenitor and precursor) cells, supporting the differentiation of maturing cells and enhancing the functions of mature cells.

The three functions just described may be present in the same growth factor, but they may be expressed with different levels of intensity in different growth factors. The isolation and cloning of genes for several growth factors permits both the mass production of growth factors and the study of their effects in vivo and in vitro. The main characteristics of the five best-characterized growth factors are presented in Table 13–2.

MEDICAL APPLICATION

Growth factors have been used clinically to increase marrow cellularity and blood cell counts. The use of growth factors to stimulate the proliferation of leukocytes is opening broad new applications for clinical therapy. Potential therapeutic uses of growth factors include increasing the number of blood cells in diseases or induced conditions (eg, chemotherapy, irradiation) that result in low blood counts; increasing the efficiency of marrow transplants by enhancing cell proliferation; enhancing host defenses in patients with malignancies and infectious and immunodeficient diseases; and enhancing the treatment of parasitic diseases.

Hematopoietic diseases are usually caused by suppression or enhancement of some undifferentiated cell production, with a consequent reduction or overproduction of hematopoietic cells. In some diseases, however, suppression and enhancement of proliferation of more than one type of stem cell can occur, sequentially or simultaneously. In such cases, there are reduced numbers of some cell types (eg, aplastic anemia, a disorder characterized by decreased production of hematopoietic cells) coinciding with increased numbers of others (eg, leukemia, the abnormal proliferation of leukocytes).

The initial experiments with normal bone marrow transplanted to lethally irradiated mice established the basis for bone marrow transplantation, now frequently used to treat some disorders of hematopoietic cell growth.

BONE MARROW

Under normal conditions, the production of blood cells by the bone marrow is adjusted to the body's needs, increasing its activity several-fold in a very short time. Bone marrow is found in the medullary canals of long bones and in the cavities of cancellous bones (Figure 13–2). Two types of bone marrow have been described according to their appearance on gross examination: **red,** or **hematogenous, bone marrow,** whose color is produced by the presence of blood and blood-forming cells; and **yellow bone marrow,** whose color is produced by the presence of a great number of adipose cells. In newborns, all bone marrow is red and is therefore active in the production of blood cells. As the child grows, most of the bone marrow changes gradually into the yellow variety. Under certain conditions, such as severe bleeding or hypoxia, yellow bone marrow is replaced by red bone marrow.

Table 13–2. Main characteristics of the five best-known hematopoietic growth factors (colony-forming substances).

Name	Human Gene Location and Producing Cells	Main Biologic Activity
Granulocyte (G-CSF)	Chromosome 17 Macrophages Endothelium Fibroblasts	Stimulates formation (in vitro and in vivo) of granulocytes. Enhances metabolism of granulocytes. Stimulates malignant (leukemic) cells.
Granulocyte + macrophage (GM-CSF)	Chromosome 5 T lymphocytes Endothelium Fibroblasts	Stimulates in vitro and in vivo production of granulocytes and macrophages.
Macrophage (M-CSF)	Chromosome 5 Macrophages Endothelium Fibroblasts	Stimulates formation of macrophages in vitro. Increases antitumor activity of macrophages.
Interleukin 3 (IL-3)	Chromosome 5 T lymphocytes	Stimulates in vivo and in vitro production of all myeloid cells.
Erythropoietin (EPO)	Chromosome 7 Renal interstitial cells (outer cortex)	Stimulates red blood cell formation in vivo and in vitro.

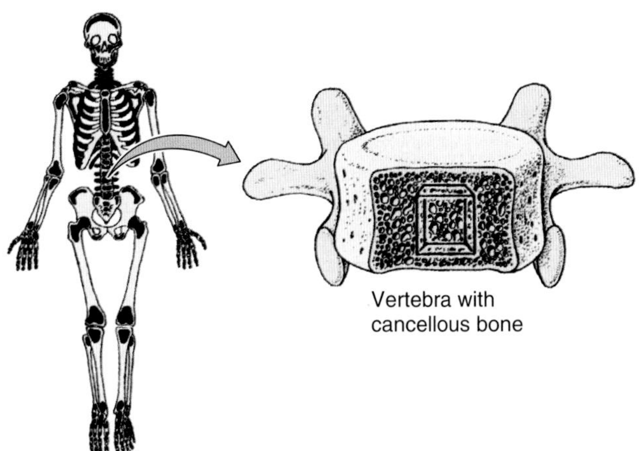

Figure 13–2. Distribution of red bone marrow (hematopoietic active) in the adult. This type of bone marrow tends to be located in cancellous bone tissue. (Reproduced, with permission, from Krstíc RV: *Human Microscopic Anatomy.* Springer-Verlag, 1991.)

Red Bone Marrow

Red bone marrow (Figure 13–3) is composed of a **stroma** (from Greek, meaning bed), **hematopoietic cords,** and **sinusoidal capillaries.** The stroma is a 3-dimensional meshwork of reticular cells and a delicate web of reticular fibers containing hematopoietic cells and macrophages. The matrix of bone marrow contains collagen types I and III, fibronectin, laminin, and proteoglycans. Laminin, fibronectin, and another cell-binding substance, **hemonectin,** interact with cell receptors to bind cells to the matrix. The sinusoids are formed by a layer of endothelial cells.

An external discontinuous layer of reticular cells and a loose net of reticular fibers reinforce the sinusoidal capillaries. The release of mature bone cells from the marrow is controlled by **releasing factors** produced in response to the needs of the organism. Several substances with releasing activity have been described, including the C3 component of **complement** (a series of immunologically active blood proteins), hormones (glucocorticoids and androgens), and some bacterial toxins. The release of cells from the marrow is illustrated in Figure 13–4.

The main functions of red bone marrow are the production of blood cells, destruction of worn-out red blood cells, and storage (in macrophages) of iron derived from the breakdown of hemoglobin.

BONE MARROW AS A SOURCE OF STEM CELLS FOR OTHER TISSUES

MEDICAL APPLICATION

Contrary to previous observations, red bone marrow is rich in stem cells that can produce several tissues, not just blood cells. With their great potential for differentiation, these cells make it possible to generate specialized cells that are not rejected by the body because they are produced from stem cells from the marrow of the same person. The procedure is to collect bone marrow stem cells, cultivate them in appropriate medium to direct their differentiation to the cell type needed for transplantation, and then use the cells originating in tissue culture to replace the cells needed by the patient. In this case the donor and the recipient are the same person and the histocompatibility is complete, excluding the possibility of rejection. Even though these studies are just beginning, the results so far are promising.

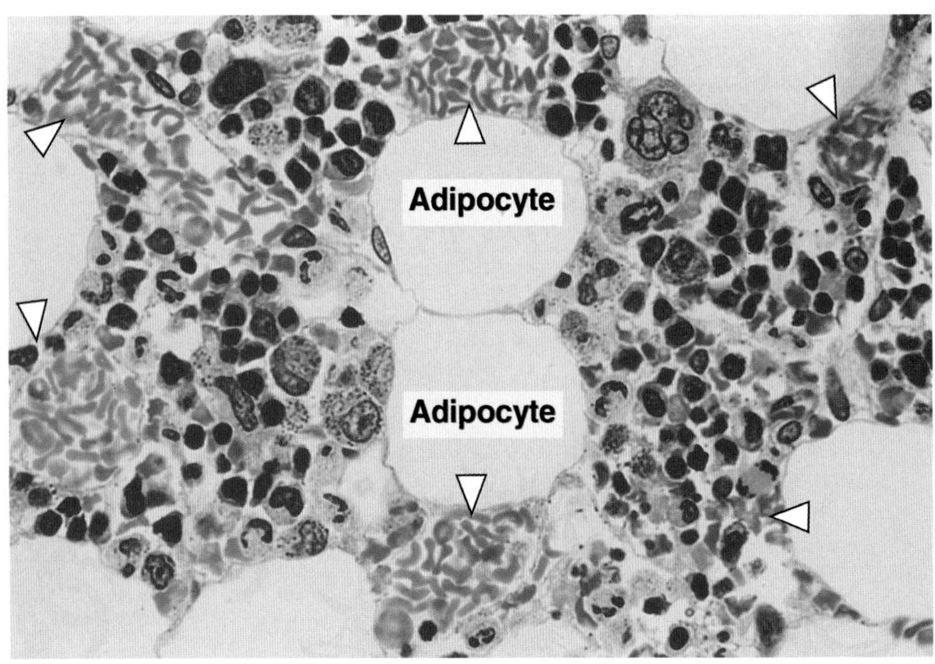

Figure 13–3. Section of active bone marrow (red bone marrow) showing some of its components. Five blood sinusoid capillaries containing many erythrocytes are indicated by arrowheads. Note the thinness of the blood capillary wall. Giemsa stain. Medium magnification.

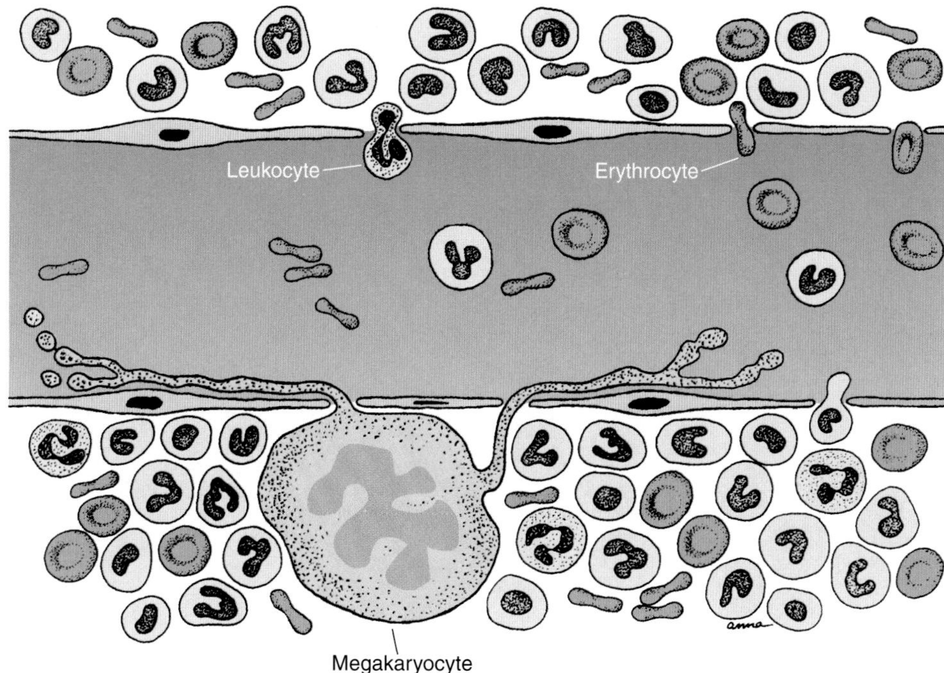

Figure 13–4. Drawing showing the passage of erythrocytes, leukocytes, and platelets across a sinusoid capillary in red bone marrow. Because erythrocytes (unlike leukocytes) do not have sufficient motility to cross the wall of the sinusoid, they are believed to enter the sinusoid by a pressure gradient that exists across its wall. Leukocytes, after the action of releasing substances, cross the wall of the sinusoid by their own activity. Megakaryocytes form thin processes that cross the wall of the sinusoid and fragment at their tips, liberating the platelets.

MATURATION OF ERYTHROCYTES

A **mature** cell is one that has differentiated to the stage at which it has the capability of carrying out all its specific functions. The basic process in maturation is the synthesis of hemoglobin and the formation of an enucleated, biconcave, small corpuscle, the erythrocyte. During maturation of the erythrocyte, several major changes take place (Figures 13–5 through 13–9). Cell volume decreases, and the nucleoli diminish in size until they become invisible in the light microscope. The nuclear diameter decreases, and the chromatin becomes increasingly more dense until the nucleus presents a pyknotic appearance (Figure 13–10) and is finally extruded from the cell (see Figure 13–18). There is a gradual decrease in the number of polyribosomes (basophilia decreases), with a simultaneous increase in the amount of hemoglobin (an acidophilic protein) within the cytoplasm. Mitochondria and other organelles gradually disappear (Figure 13–6).

There are 3 to 5 intervening cell divisions between the proerythroblast and the mature erythrocyte. The development of an erythrocyte from the first recognizable cell of the series to the release of reticulocytes into the blood takes approximately 7 days. The hormone erythropoietin and substances such as iron, folic acid, and cyanocobalamin (vitamin B_{12}) are essential for the production of erythrocytes. **Erythropoietin** is a glycoprotein produced in the kidneys that stimulates the production of mRNA for **globin,** the protein component of the hemoglobin molecule.

Differentiation

The differentiation and maturation of erythrocytes involve the formation (in order) of proerythroblasts, basophilic erythroblasts, polychromatophilic erythroblasts, orthochromatophilic erythroblasts (normoblasts), reticulocytes, and erythrocytes (Figure 13–5).

The first recognizable cell in the erythroid series is the **proerythroblast.** It is a large cell with loose, lacy chromatin and clearly visible nucleoli; its cytoplasm is basophilic. The next stage is represented by the **basophilic erythroblast** (*erythros* + Gr. *blastos,* germ), with a strongly basophilic cytoplasm and a condensed nucleus that has no visible nucleolus. The basophilia of these two cell types is caused by the large number of polyribosomes involved in the synthesis of hemoglobin. During the next stage, polyribosomes decrease, and areas of the cytoplasm begin to be filled with hemoglobin. At this stage, staining causes several colors to appear in the cell—the **polychromatophilic** (Gr. *polys,* many, + *chroma,* color, + *philein,* to love) **erythroblast.** In the next stage, the nucleus continues to condense and no cytoplasmic basophilia is evident, resulting in a uniformly acidophilic cytoplasm—the **orthochromatophilic** (Gr. *orthos,* correct, + *chroma* + *philein*) **erythroblast.** At a given moment, this cell puts forth a series of cytoplasmic protrusions and expels its nucleus, encased in a thin layer of cytoplasm. The remaining cell still has a small number of polyribosomes that, when treated with the dye brilliant cresyl blue, aggregate to form a stained network. This cell is the **reticulocyte,** which soon loses its polyribosomes and becomes a mature erythrocyte.

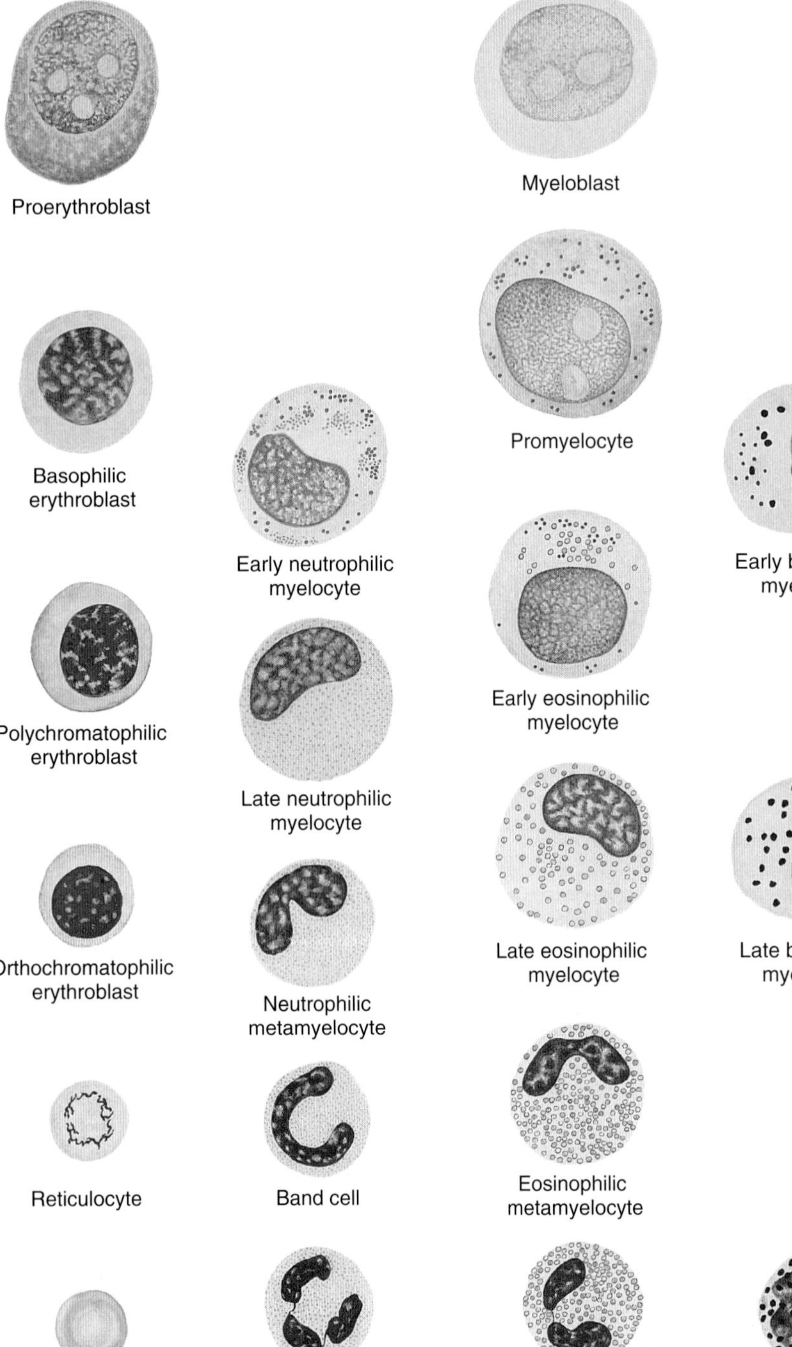

Figure 13–5. Stages in the development of erythrocytes and granulocytes.

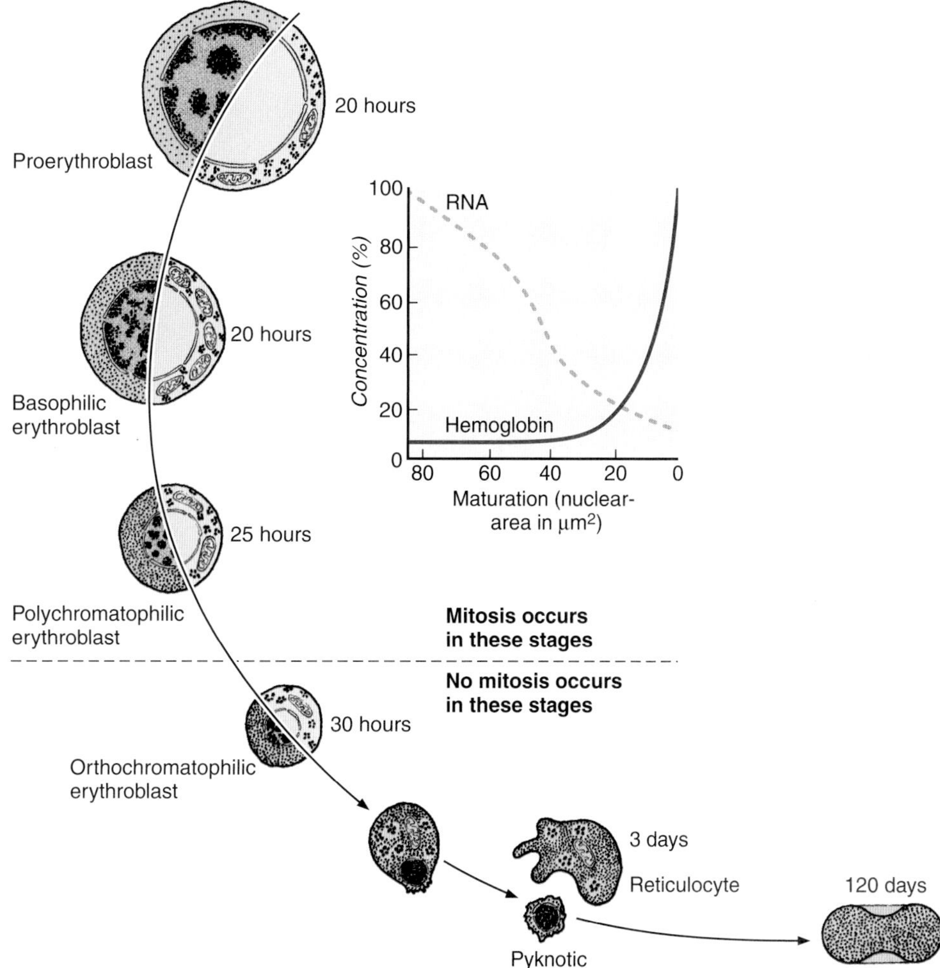

Figure 13–6. Summary of erythrocyte maturation. The stippled part of the cytoplasm (on the left) shows the continuous increase in hemoglobin concentration from proerythroblast to erythrocyte. There is also a gradual decrease in nuclear volume and an increase in chromatin condensation, followed by extrusion of a pyknotic nucleus. The times are the average life span of each cell type. In the graph, 100% represents the highest recorded concentrations of hemoglobin and RNA.

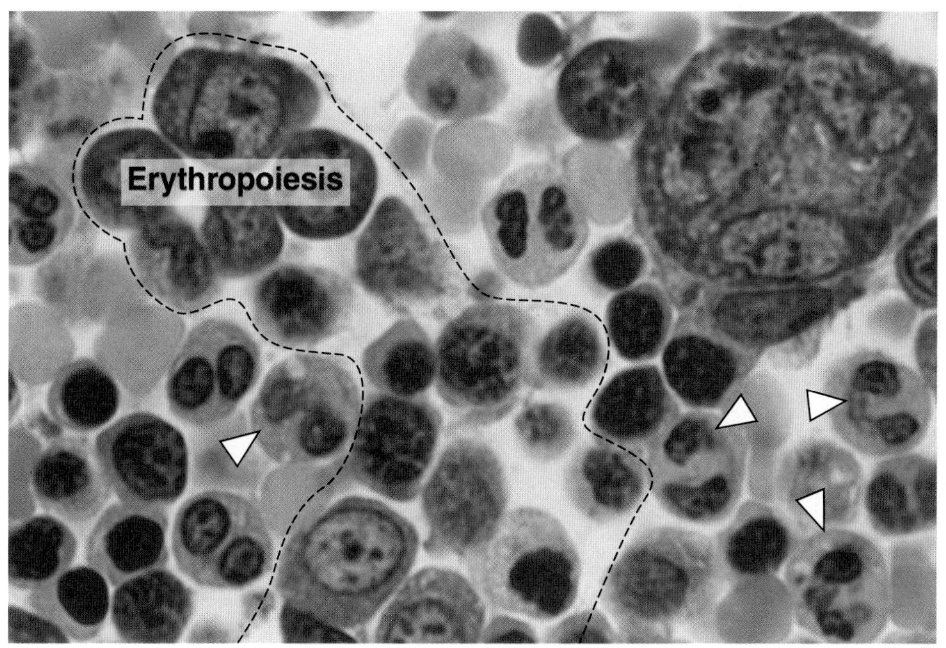

Figure 13–7. Section of red bone marrow showing an immature megakaryocyte in the upper right corner. There is also a large group of erythropoietic cells (delimited by a broken line) and sparse immature neutrophils (arrowheads). Pararosaniline–toluidine blue (PT) stain. High magnification.

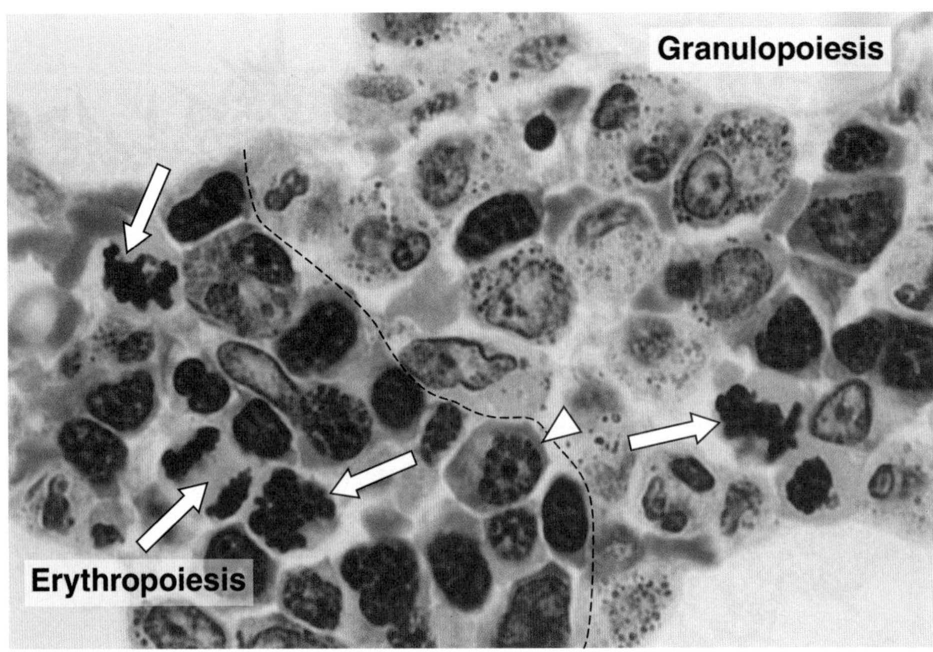

Figure 13–8. Section of stimulated red bone marrow. Note 4 mitotic figures (arrows) and a plasma cell (arrowhead). Regions of erythropoiesis and of granulopoiesis are also evident. Most immature granulocytes are in the myelocyte stage: their cytoplasm contains large, dark-stained azurophilic granules and small, less darkly stained specific granules. Giemsa stain. High magnification.

GRANULOPOIESIS

The maturation process of granulocytes takes place with cytoplasmic changes characterized by the synthesis of a number of proteins that are packed in 2 organelles: the **azurophilic** and **specific granules.** These proteins are produced in the rough endoplasmic reticulum and the Golgi complex in 2 successive stages (Figure 13–11). The first stage results in the production of the **azurophilic granules,** which stain with basic dyes in the Wright or Giemsa methods and contain enzymes of the lysosomal system. In the second stage, a change in synthetic activity takes place with the production of several proteins that are packed in the **specific granules.** These granules contain different proteins in each of the three types of granulocytes and are utilized for the various activities of each type of granulocyte. Evidently, a shift in gene expression occurs in this process, permitting neutrophils to specialize in bacterial destruction and eosinophils and basophils to become involved in the regulation of inflammation. The different stages of maturation and the morphologic changes that occur during this process are shown in Figures 13–5, 13–8, and 13–9.

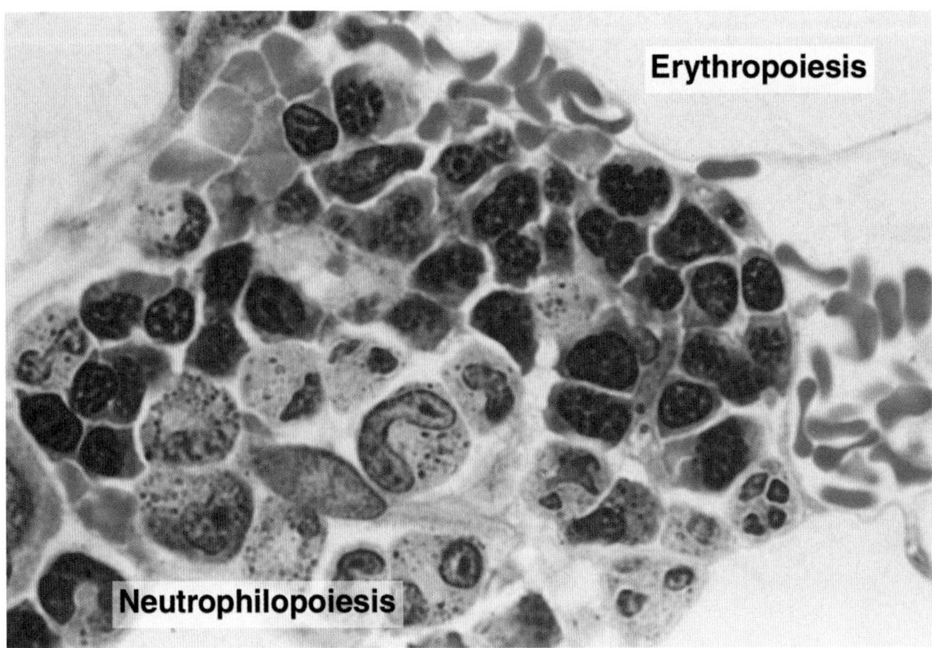

Figure 13–9. Section of red bone marrow with a group of erythropoietic cells (upper right) and a group of neutrophilopoietic cells (lower left). The immature granulocytes shown have mostly azurophilic granules in their cytoplasm and therefore are myelocytes. PT stain. High magnification.

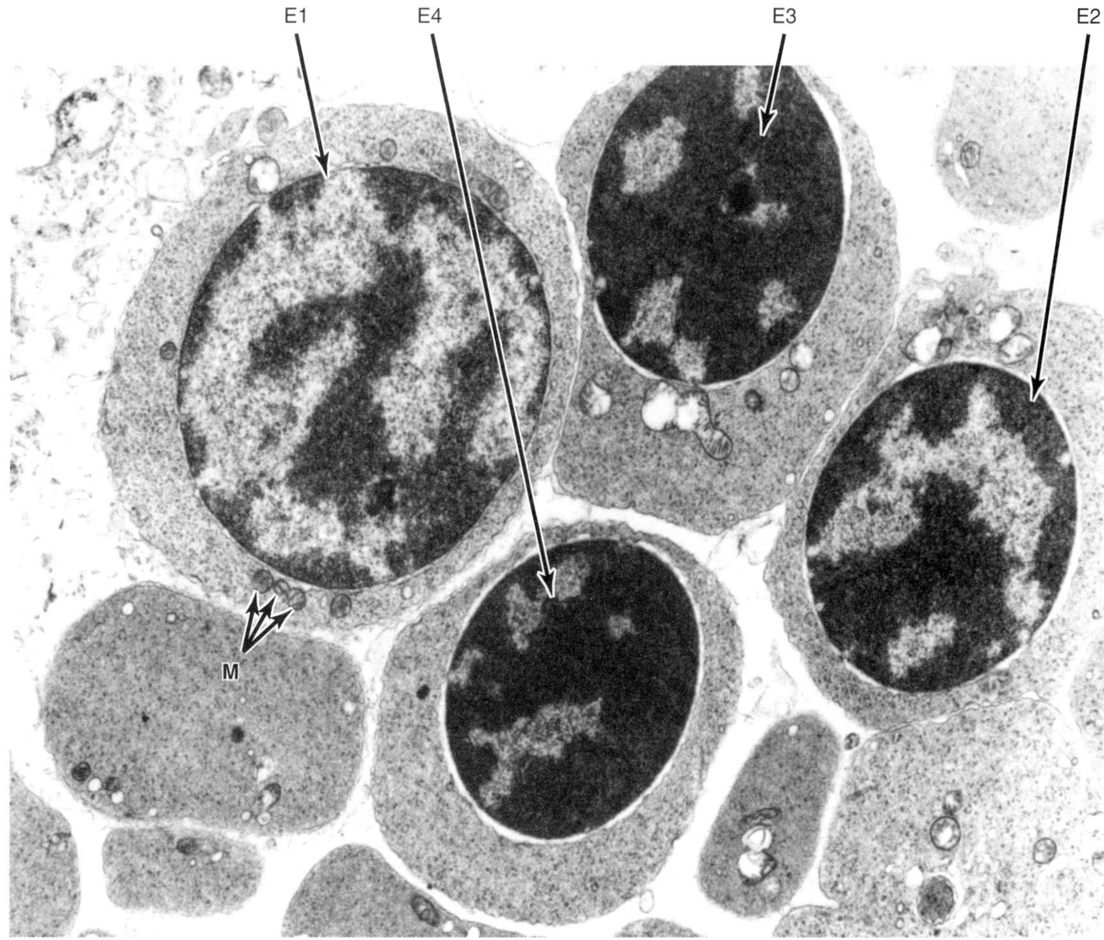

Figure 13–10. Electron micrograph of red bone marrow. Four erythroblasts in successive stages of maturation are seen (E1, E2, E3, and E4). As the cell matures, its chromatin becomes gradually condensed, the accumulation of hemoglobin increases the electron density of the cytoplasm, and the mitochondria (M) decrease in number. ×11,000.

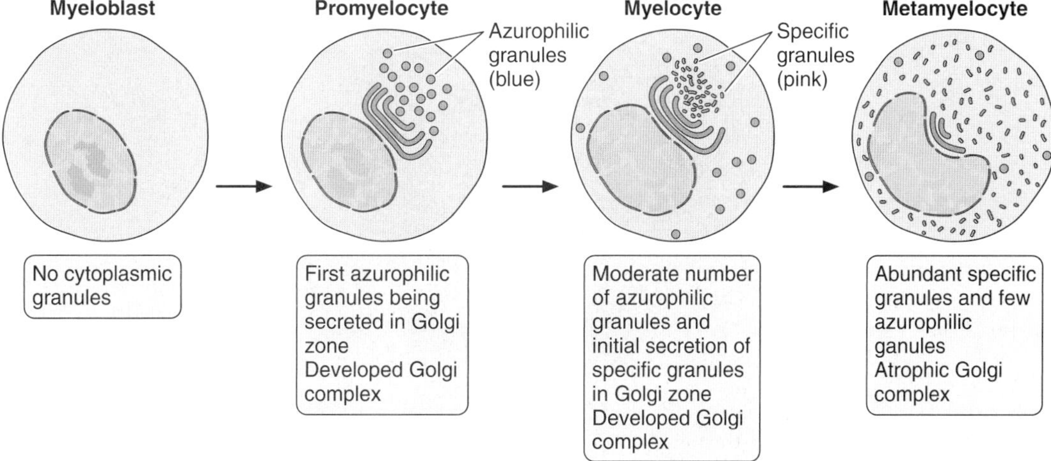

Figure 13–11. Drawing illustrating the sequence of gene expression in the maturation of granulocytes. Azurophilic granules are blue; specific granules are pink.

MATURATION OF GRANULOCYTES

The **myeloblast** is the most immature recognizable cell in the myeloid series (Figure 13–5). It has a finely dispersed chromatin, and nucleoli can be seen. In the next stage, the **promyelocyte** (L. *pro,* before, + Gr. *myelos,* marrow, + *kytos,* cell) is characterized by its basophilic cytoplasm and azurophilic granules. These granules contain lysosomal enzymes and myeloperoxidase. The promyelocyte gives rise to the three known types of granulocyte. The first sign of differentiation appears in the myelocytes, in which specific granules gradually increase in quantity and eventually occupy most of the cytoplasm. These **neutrophilic** (Figures 13–11 and 13–12), **basophilic,** and **eosinophilic** (Figure 13–13) **myelocytes** mature with further condensation of the nucleus and a considerable increase in their specific granule content. Before its complete maturation, the neutrophilic granulocyte passes through an intermediate stage in which its nucleus has the form of a curved rod (band cell). This cell appears in quantity in the blood after strong stimulation of hematopoiesis.

MEDICAL APPLICATION

The appearance of large numbers of immature neutrophils (band cells) in the blood is called a ***shift to the left*** *and is clinically significant, usually indicating bacterial infection.*

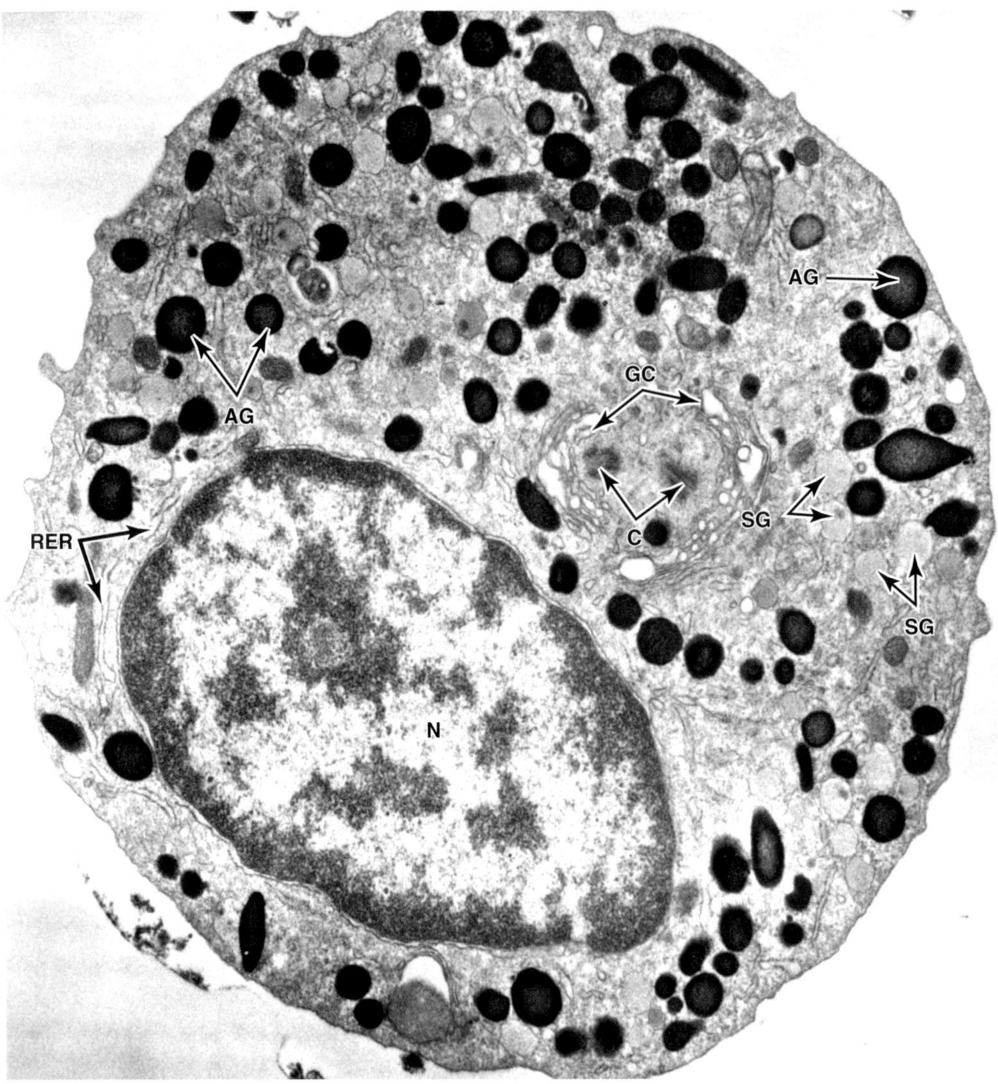

Figure 13–12. Neutrophilic myelocyte from normal human bone marrow treated with peroxidase. At this stage, the cell is smaller than the promyelocyte, and the cytoplasm contains 2 types of granules: large, peroxidase-positive azurophilic granules (AG) and smaller specific granules (SG), which do not stain for peroxidase. Note that the peroxidase reaction product is present only in azurophilic granules and is not seen in the rough endoplasmic reticulum (RER) or Golgi cisternae (GC), which are located around the centriole (C). N, nucleus. ×15,000. (Courtesy of DF Bainton.)

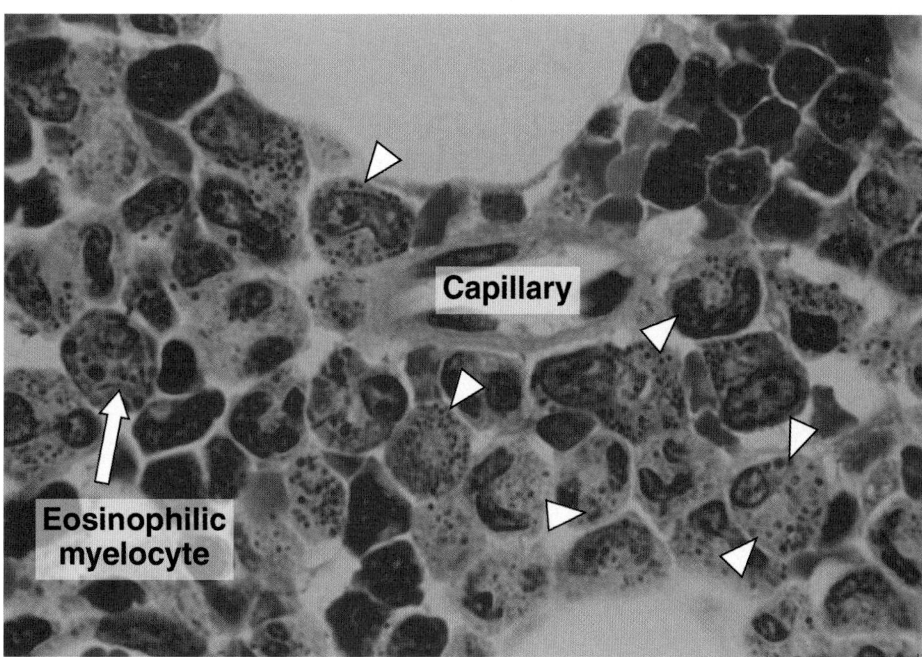

Figure 13–13. Bone marrow with neutrophilic (arrowheads) and eosinophilic myelocytes. Giemsa stain. High magnification.

KINETICS OF NEUTROPHIL PRODUCTION

The total time taken for a myeloblast to emerge as a mature neutrophil in the circulation is about 11 days. Under normal circumstances, 5 mitotic divisions occur in the myeloblast, promyelocyte, and neutrophilic myelocyte stages of development.

Neutrophils pass through several functionally and anatomically defined compartments (Figure 13–14).

The **medullary formation compartment** can be subdivided into a mitotic compartment (~3 days) and a maturation compartment (~4 days).

A **medullary storage compartment** acts as a buffer system, capable of releasing large numbers of mature neutrophils on demand. Neutrophils remain in this compartment for about 4 days.

The **circulating compartment** consists of neutrophils suspended in plasma and circulating in blood vessels.

The **marginating compartment** is composed of neutrophils that are present in blood but do not circulate. These neutrophils are in capillaries and are temporarily excluded from the circulation by vasoconstriction, or—especially in the lungs—they may be at the periphery of vessels, adhering to the endothelium, and not in the main bloodstream.

The marginating and circulating compartments are of about equal size, and there is a constant interchange of cells between them. The half-life of a neutrophil in these two compartments is 6–7 hours. The medullary formation and storage compartments together are about 10 times as large as the circulating and marginating compartments.

Neutrophils and other granulocytes enter the connective tissues by passing through intercellular junctions found between endothelial cells of capillaries and postcapillary venules (**diapedesis**). The connective tissues form a fifth compartment for neutrophils, but its size is not known. Neutrophils reside here for 1–4 days and then die by apoptosis, whether or not they have performed their major function of phagocytosis.

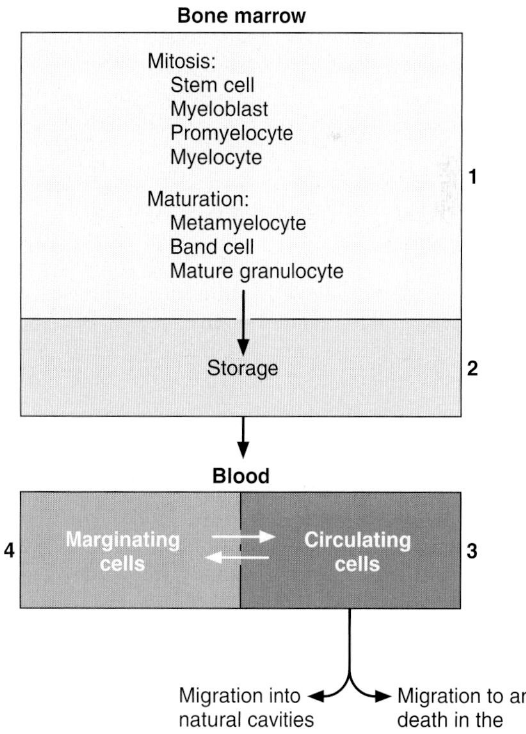

Figure 13–14. Functional compartments of neutrophils. **1:** Medullary formation compartment. **2:** Medullary storage (reserve) compartment. **3:** Circulating compartment. **4:** Marginating compartment. The size of each compartment is roughly proportional to the number of cells.

MEDICAL APPLICATION

Changes in the number of neutrophils in the blood must be evaluated by taking all these compartments into consideration. Thus, ***neutrophilia,*** *an increase in the number of neutrophils in the circulation, does not necessarily imply an increase in neutrophil production. Intense muscular activity or the administration of epinephrine causes neutrophils in the marginating compartment to move into the circulating compartment, causing an apparent neutrophilia even though neutrophil production has not increased. However, glucocorticoids (adrenal gland hormones) increase the mitotic activity of neutrophil precursors in the marrow and increase the blood count of neutrophils.*

Neutrophilia may also result from liberation of greater numbers of neutrophils from the medullary storage compartment. This type of neutrophilia is transitory and is followed by a recovery period during which no neutrophils are released.

The neutrophilia that occurs during the course of bacterial infections is due to an increase in production of neutrophils and a shorter duration of these cells in the medullary storage compartment. In such cases, immature forms such as band cells, neutrophilic metamyelocytes, and even myelocytes may appear in the bloodstream. The neutrophilia that occurs during infection is of longer duration than that which occurs as a result of intense muscular activity.

MATURATION OF LYMPHOCYTES & MONOCYTES

Study of the precursor cells of lymphocytes and monocytes is difficult, because these cells do not contain specific cytoplasmic granules or nuclear lobulation, both of which facilitate the distinction between young and mature forms of granulocytes. Lymphocytes and monocytes are distinguished mainly on the basis of size, chromatin structure, and the presence of nucleoli in smear preparations. As lymphocyte cells mature, their chromatin becomes more compact, nucleoli become less visible, and the cells decrease in size. In addition, subsets of the lymphocyte series acquire distinctive cell-surface receptors during differentiation that can be detected by immunocytochemical techniques.

Lymphocytes

Circulating lymphocytes originate mainly in the thymus and the peripheral lymphoid organs (eg, spleen, lymph nodes, tonsils). However, all lymphocyte progenitor cells originate in the bone marrow. Some of these lymphocytes migrate to the thymus, where they acquire the full attributes of T lymphocytes. Subsequently, T lymphocytes populate specific regions of peripheral lymphoid organs. Other bone marrow lymphocytes differentiate into B lymphocytes in the bone marrow and then migrate to peripheral lymphoid organs, where they inhabit and multiply in their own special compartments.

The first identifiable progenitor of lymphoid cells is the **lymphoblast,** a large cell capable of incorporating ^{3}H-thymidine and dividing 2 or 3 times to form **prolymphocytes.** Prolymphocytes are smaller and have relatively more condensed chromatin but none of the cell-surface antigens that mark prolymphocytes as T or B lymphocytes. In the bone marrow and in the thymus, these cells synthesize cell-surface receptors characteristic of their lineage, but they are not recognizable as distinct B or T lymphocytes in routine histologic procedures. Using immunocytochemical techniques makes the distinction.

Monocytes

The **monoblast** is a committed progenitor cell that is virtually identical to the **myeloblast** in its morphologic characteristics. Further differentiation leads to the **promonocyte,** a large cell (up to 18 μm in diameter) with a basophilic cytoplasm and a large, slightly indented nucleus. The chromatin is lacy, and nucleoli are evident. Promonocytes divide twice in the course of their development into **monocytes.** A large amount of rough endoplasmic reticulum is present, as is an extensive Golgi complex in which granule condensation can be seen to be taking place. These granules are **primary lysosomes,** which are observed as fine **azurophilic granules** in blood monocytes. Mature monocytes enter the bloodstream, circulate for about 8 hours, and then enter the connective tissues, where they mature into **macrophages** and function for several months.

MEDICAL APPLICATION

Abnormal bone marrow can produce diseases based on cells derived from that tissue. ***Leukemias*** *are malignant clones of leukocyte precursors. They occur in lymphoid tissue (****lymphocytic leukemias****) and in bone marrow (****myelogenous*** *and* ***monocytic leukemias****). In these diseases, there is usually a release of large numbers of immature cells into the blood. The symptoms of leukemias are a consequence of this shift in cell proliferation, with a lack of some cell types and excessive production of others (which are often abnormal in function). The patient is usually anemic and prone to infection.*

A clinical technique that is helpful in the study of leukemias and other bone marrow disturbances is ***bone marrow aspiration.*** *A needle is introduced through compact bone (usually the sternum), and a sample of marrow is withdrawn. The sample is spread on a microscope slide and stained. The use of labeled monoclonal antibodies specific to proteins in the membranes of precursor blood cells aids in identifying cell types derived from these stem cells and contributes to a more precise diagnosis of the various types of leukemia.*

ORIGIN OF PLATELETS

In adults, platelets originate in the red bone marrow by fragmentation of the cytoplasm of mature **megakaryocytes** (Gr. *megas,* big, + *karyon,* nucleus, + *kytos*), which, in turn, arise by differentiation of **megakaryoblasts.**

Megakaryoblasts

The megakaryoblast is 15–50 μm in diameter and has a large ovoid or kidney-shaped nucleus (Figure 13–15) with numerous nucleoli. The nucleus becomes highly polyploid (ie, it contains up to 30 times as much DNA as a normal cell) before platelets begin to form. The cytoplasm of this cell is homogeneous and intensely basophilic.

Megakaryocytes

The megakaryocyte (Figures 13–15 through 13–19) is a giant cell (35–150 μm in diameter) with an irregularly lobulated nucleus, coarse chromatin, and no visible nucleoli. The cytoplasm contains numerous mitochondria, a well-developed rough endoplasmic reticulum, and an extensive Golgi complex. Platelets have conspicuous granules, originating from the Golgi complex, that contain biologically active substances, such as platelet-derived growth factor, fibroblast growth factor, von Willebrand's factor (which promotes adhesion of platelets to endothelial cells), and platelet factor IV (which stimulates blood coagulation). With maturation of the megakaryocyte, numerous invaginations of the plasma membrane ramify throughout the cytoplasm, forming the **demarcation membranes** (Figure 13–20). This system defines areas of a megakaryocyte's cytoplasm that shed platelets, extruding them into the circulation.

MEDICAL APPLICATION

In certain forms of ***thrombocytopenic purpura,*** *a disease in which the number of blood platelets is reduced, the platelets appear to be bound to the cytoplasm of the megakaryocytes, indicating a defect in the liberation mechanism of these corpuscles. The life span of platelets is approximately 10 days.*

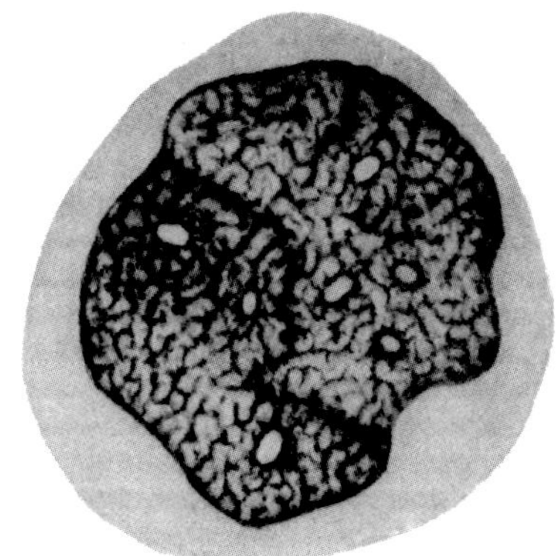

Megakaryoblast

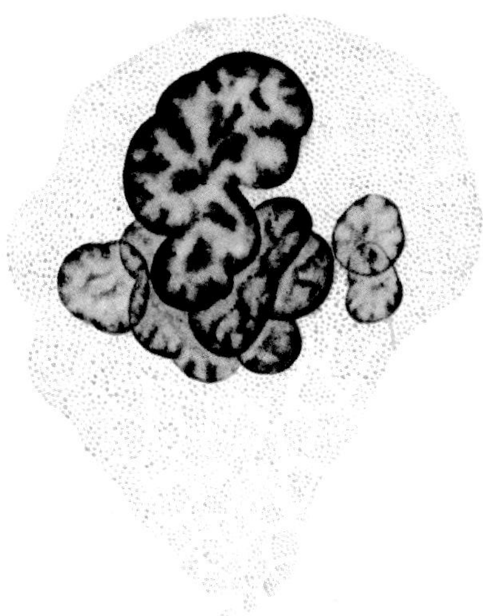

Megakaryocyte

Platelets

Figure 13–15. Cells of the megakaryocyte series shown in a bone marrow smear. Note the formation of platelets at the lower end of the megakaryocyte.

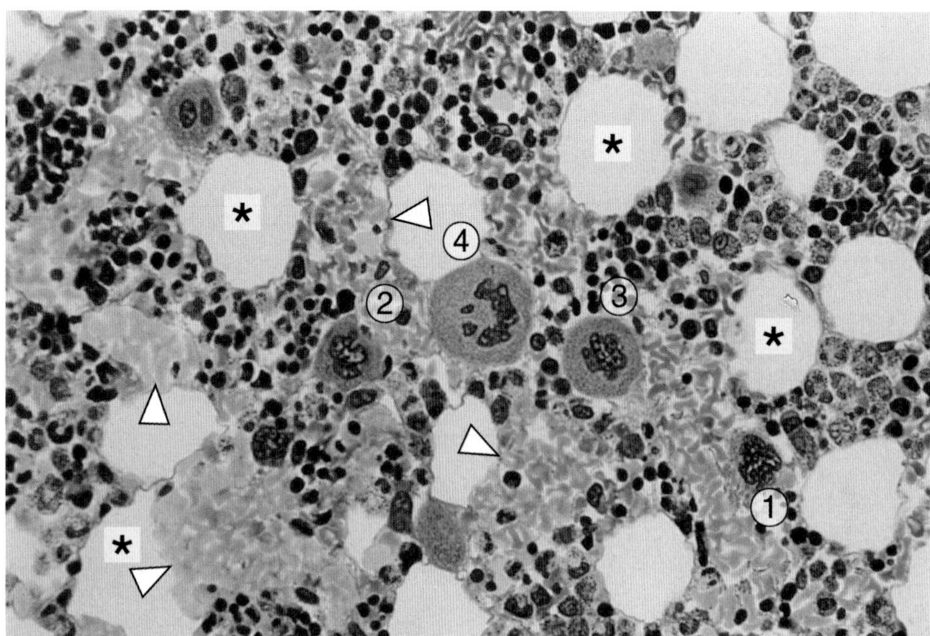

Figure 13–16. Section of bone marrow showing various stages of megakaryocyte development (**1–4**), several adipocytes (*), and blood sinusoids (arrowheads). PT stain. Medium magnification.

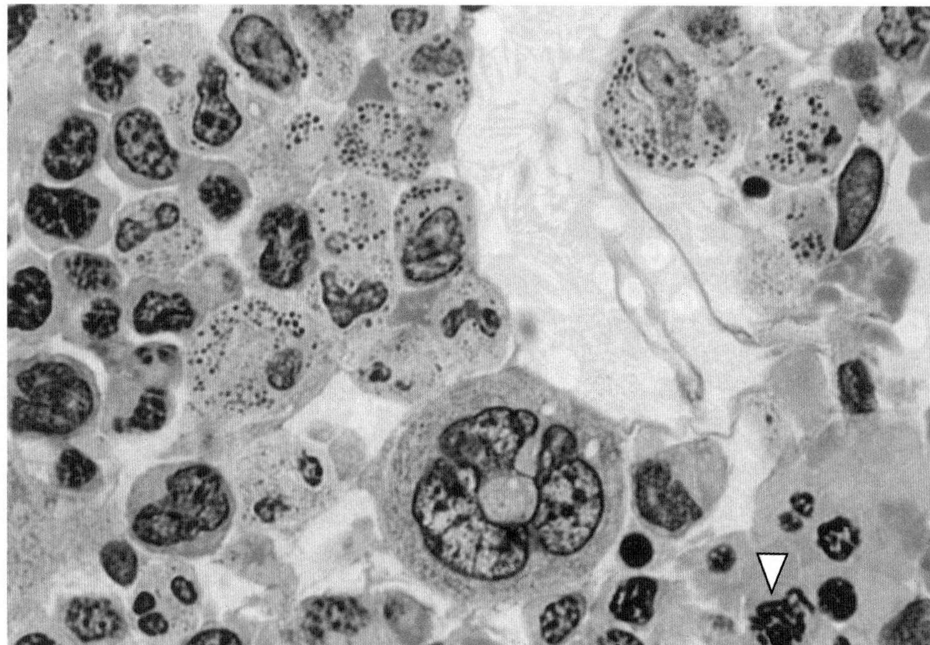

Figure 13–17. Section of bone marrow with an adult megakaryocyte and several granulocytes, mainly neutrophils in the myelocyte stage with many azurophilic granules and few, less darkly stained specific granules. A mitotic figure is indicated by an arrowhead. Giemsa stain. High magnification.

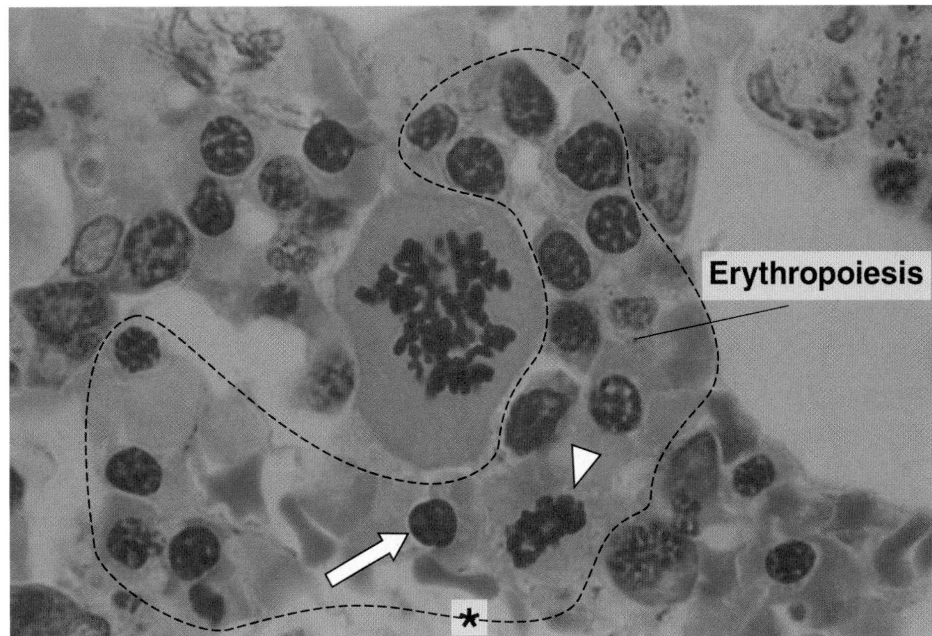

Figure 13–18. A megakaryocyte in mitosis (center) surrounded by erythropoietic cells with a mitotic figure (arrowhead). The arrow indicates an erythroblast extruding its nucleus. Giemsa stain. High magnification.

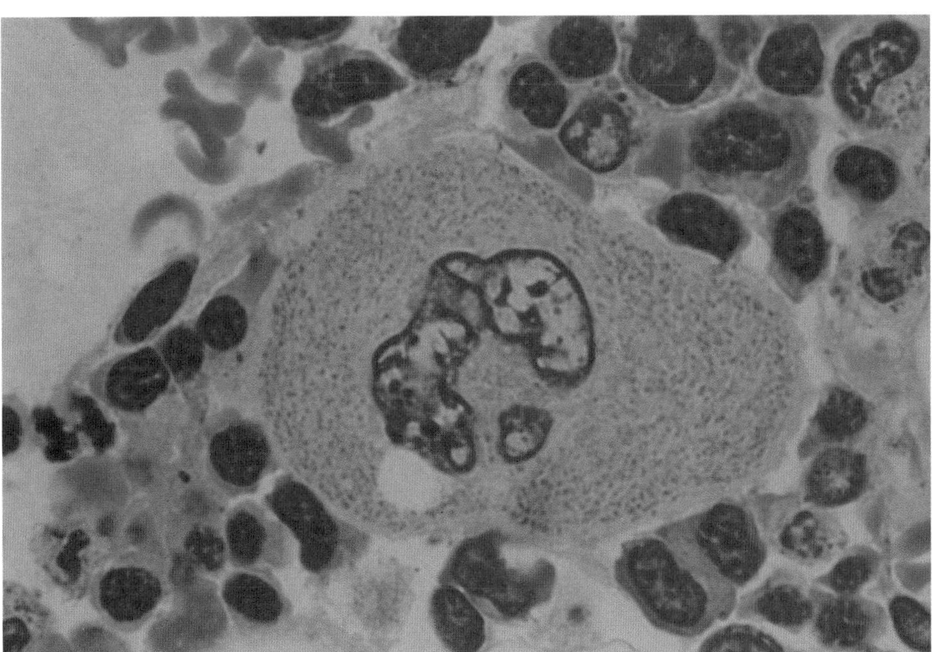

Figure 13–19. A megakaryocyte in a section of red bone marrow. This cell has only one nucleus. A small part of the nucleus appears separated because the irregularly shaped nucleus was cut into 2 pieces. Note the characteristic size and granular cytoplasm of this cell type. Giemsa stain. High magnification.

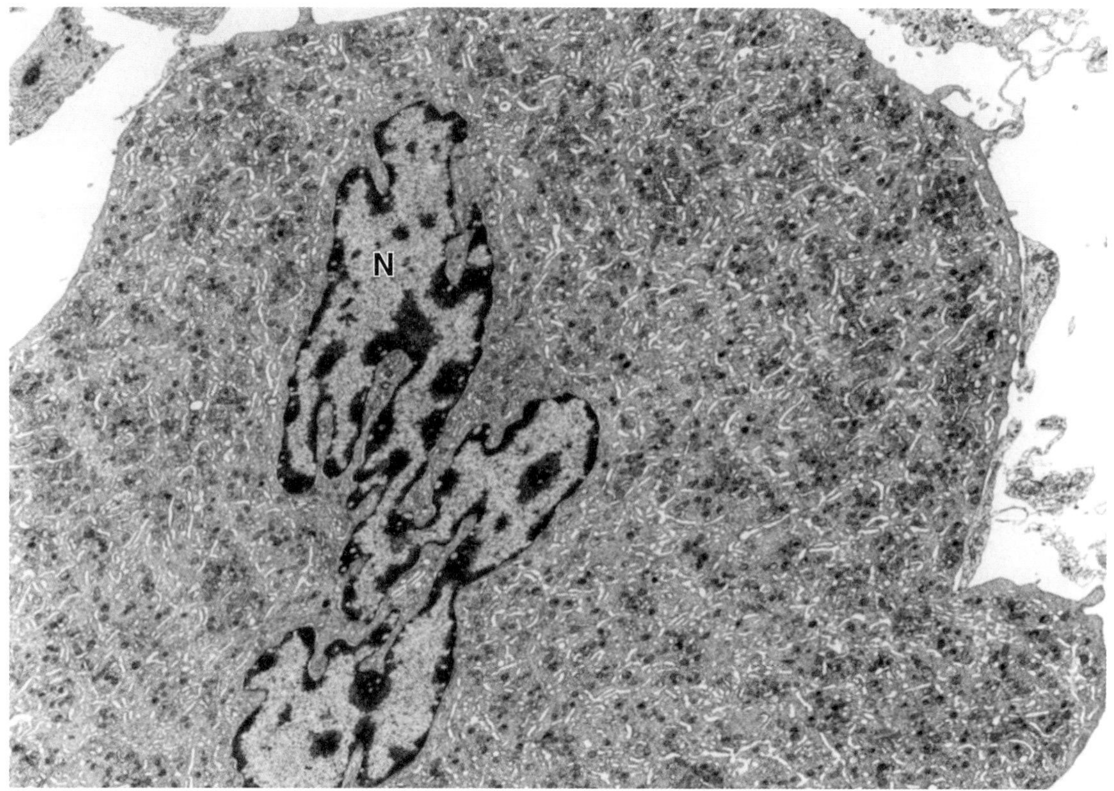

Figure 13–20. Electron micrograph of a megakaryocyte showing a lobulated nucleus (N) and numerous cytoplasmic granules. The demarcation membranes are visible as tubular profiles. ×4900. (Reproduced, with permission, from Junqueira LCU, Salles LMM: *Ultra-Estrutura e Função Celular.* Edgard Blücher, 1975.)

REFERENCES

Becker RP, DeBruyn PP: The transmural passage of blood cells into myeloid sinusoids and the entry of platelets into the sinusoidal circulation. Am J Anat 1976;145:183.

Berman I: The ultrastructure of erythroblastic islands and reticular cells in mouse bone marrow. J Ultrastruct Res 1967;17:291.

Brazelton TR et al: From marrow to brain: expression of neuronal phenotypes in adult mice. Science 2000;290:1775.

Chandrasoma P, Taylor CR: *Concise Pathology.* 3rd ed. Originally published by Appleton & Lange. Copyright © 1998 by the McGraw-Hill Companies, Inc.

Evatt BL et al: *Megakaryocyte Biology and Precursors: In Vitro Cloning and Cellular Properties.* Elsevier/North-Holland, 1981.

Fleischmann RA et al: Totipotent hematopoietic stem cells: normal self-renewal and differentiation after transplantation between mouse fetuses. Cell 1982;30:351.

Foucar K: *Bone Marrow Pathology.* American Society of Clinical Pathologists (ASCP) Press, 1995.

Krstíc RV: *Human Microscopic Anatomy.* Springer-Verlag, 1991.

Pennington DG: The cellular biology of megakaryocytes. Blood Cells 1979;5:5.

Simmons PJ et al: The mobilization of primitive hemopoietic progenitors into the peripheral blood. Stem Cells 1994;12:187.

Tavassoli M, Yoffey JM: *Bone Marrow Structure and Function.* Liss, 1983.

Williams WJ et al (editors): *Hematology,* 5th ed. McGraw-Hill, 1995.

The Immune System & Lymphoid Organs

14

The immune system comprises structures and cells that are distributed throughout the body; their principal function is to protect the body from invasion and damage by microorganisms and foreign substances. Cells of the immune system have the ability to distinguish "self" (the organism's own macromolecules) from "nonself" (foreign substances) and to coordinate the destruction or inactivation of foreign substances (eg, individual molecules), parts of microorganisms, or even cancer cells that originate in the body. On occasion, the immune system reacts against normal body tissues, causing **autoimmune diseases.** The immune system includes both individual structures (eg, lymph nodes, spleen) and free cells, such as lymphocytes, granulocytes, and cells of the mononuclear phagocyte system that are present in the blood, lymph, and connective tissues. Another important component of the immune system is the **antigen-presenting cells,** found not only in the lymphoid tissues but also in other organs, such as the skin (which is heavily exposed to foreign antigens). The cells of the immune system communicate with each other and with cells of other systems mainly through signaling proteins known as **cytokines.**

Lymphoid Organs

The main anatomical structures that participate in the immune response are the lymphoid organs: the thymus, spleen, and lymph nodes. Lymphoid nodules, smaller collections of lymphoid tissue that are formed mainly of nodular aggregates, are present in the mucosa of the digestive system (tonsils, Peyer's patches, and appendix), respiratory system, reproductive system, and urinary system, forming the mucosa-associated lymphoid tissue (**MALT**). The wide distribution of lymphoid structures and the constant circulation of lymphoid cells in the blood, lymph, and connective tissues provide the body with an elaborate, efficient system of surveillance and defense by immunocompetent cells (Figure 14–1).

All lymphocytes originate in the bone marrow; however, T lymphocytes mature further in the thymus, whereas B lymphocytes leave the bone marrow as mature cells. For this reason, the bone marrow and the thymus are called the **primary** or **central lymphoid organs.** Lymphocytes migrate from these organs to the blood and **peripheral lymphoid organs** (spleen, lymph nodes, solitary nodules, tonsils, appendix, and Peyer's patches of the ileum) where they proliferate and complete their differentiation.

Basic Types of Immune Reactions

In **cellular immunity** (Figure 14–2) cells react against and kill microorganisms, foreign cells (from tumors and transplants), and virus-infected cells. This category of immunity is mediated mainly by T lymphocytes, or T cells. **Humoral immunity** (Figure 14–2) is related to the presence of circulating glycoproteins called **antibodies** that inactivate or destroy foreign substances. The antibodies are produced by plasma cells derived from B lymphocytes, or B cells.

The immune reaction can also be **innate** or **adaptive** (also called **acquired immunity**). The innate reaction is the simplest; its response is fast, nonspecific, and does not depend on previous contact with the pathogen. The main cells responsible for the innate reaction (Figure 14–3) are phagocytes (macrophages and neutrophils) and natural killer cells (discussed later). Even though they are not a part of the immune system, the barriers formed by epithelial cells of the internal linings and the skin protect the organism against the penetration of foreign molecules and microorganisms. The adaptive reaction is a refinement of the more primitive innate reaction. The adaptive reaction is more efficient, highly specific, and has memory. A second attack by the same pathogen (Gr. *pathos,* disease, + *genin,* to produce) is dealt with faster and more efficiently than the first attack. Because it involves not only the same cells as the innate reaction but also the formation of antibodies, the adaptive reaction starts more slowly.

Immunogens & Antigens

The foreign (nonself) substance encountered by the immune system acts as an **immunogen**—ie, a substance that elicits a response from the host. The response may be cellular, humoral, or (most commonly) both. Immunogens may be present in whole cells, such as bacteria or tumor cells, or in macromolecules, such as proteins, polysaccharides, or nucleoproteins. More specifically, an **antigen** (Gr. *anti,* against, + *genin*) is an immunogen that can react with an antibody, even if it is not capable of eliciting an immune reaction. Because most immunogens are also antigens, in this book the term "antigen" will be used. The specificity of the humoral immune response (B cells) is determined by small molecular domains—**antigenic determinants** or **epitopes**—of the antigen, whereas the specificity of the cellular immune response (T cells) is determined by small peptides associated with major histocompatibility complex (MHC) molecules on the membrane of antigen-presenting cells. An antigen that has many epitopes (eg, a bacterial cell) will elicit a wide spectrum of humoral and cellular responses.

Antibodies

Antibodies, also called **immunoglobulins,** are circulating plasma glycoproteins that interact specifically with the antigenic

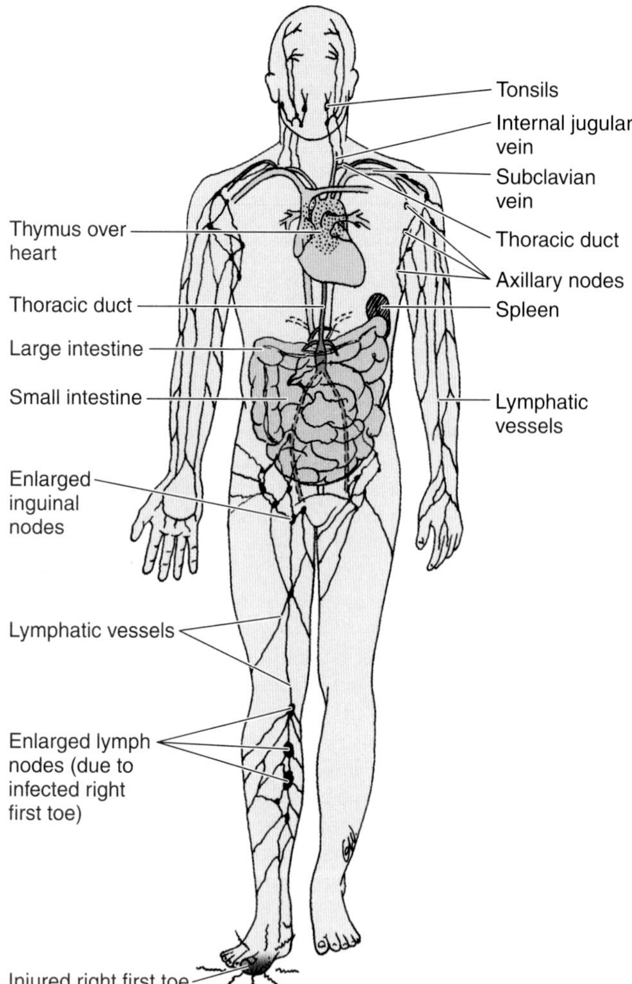

Figure 14–1. Distribution of lymphoid organs and lymphatic vessels in the body. As an example of the functions of lymphatic system, an infection of the first toe is shown with enlargement of the lymph nodes that collect lymph from the infected region. This enlargement is mainly due to the proliferation of B lymphocytes and their differentiation into antibody-secreting plasma cells. The infected toe becomes red, warm, painful, and swollen.

determinant that elicited their formation. Antibodies are secreted by plasma cells that arise by proliferation and differentiation of B lymphocytes. One important function of an antibody molecule is to combine specifically with the epitope it recognizes and then to signal other components of the immune system that this is a foreign invader to be eliminated. Some antibodies are able to agglutinate cells and to precipitate soluble antigens. Agglutination of microorganisms and harmful molecules localizes an invader and facilitates phagocytosis. Antigens bound to immunoglobulin G or M (IgG or IgM) activate the complement system, a group of plasma proteins, leading to lysis of microorganisms. The activated complement also stimulates phagocytosis of bacteria and other invaders. Neutrophils and macrophages have receptors for the Fc region of antigen-bound IgG; this antibody can attach the complex antigen-invader to these phagocytic cells.

Five classes of immunoglobulins (Table 14–1) are recognized in humans:

IgG, the most abundant class, constitutes 75% of serum immunoglobulins. Because it also serves as a model for the other classes, it will be described in detail. IgG consists of 2 identical light chains and 2 identical heavy chains (Figure 14–4), bound by disulfide bonds and noncovalent forces. When isolated, the two carboxyl-terminal portions of the heavy chains crystallize easily and are called **Fc** (fragment crystallizable) fragments. The Fc regions of several immunoglobulins react with receptors of many different cells. The 4 amino-terminal segments (2 formed of light chains and 2 of heavy chains) constitute the **Fab** (fragment antigen-binding) fragments of the immunoglobulins. The Fab segments are variable in amino acid sequence and are thus responsible for the exquisite specificity of the immune response. IgG is the only immunoglobulin that crosses the placental barrier and is incorporated into the circulatory system of the fetus, protecting the newborn against infection.

IgA is found in small amounts in blood. It is the main immunoglobulin found in tears, colostrum, and saliva; in nasal, bronchial, intestinal, and prostatic secretions; and in the vaginal fluid. It is found in secretions as a dimer called **secretory IgA,** which is composed of 2 molecules of monomeric IgA united by a polypeptide chain called **protein J** and combined with another protein, the **secretory,** or **transport, component.** Because it is resistant to several enzymes, secretory IgA provides protection against the proliferation of microorganisms in body secretions. IgA monomers and protein J are secreted by plasma cells in the mucous membranes that line the digestive, respiratory, and urinary passages; the secretory component is synthesized by the mucosa epithelial cells.

IgM constitutes 10% of blood immunoglobulins and usually exists as a pentamer with a molecular mass of 900 kDa. Together with IgD, it is the major immunoglobulin found on the surfaces of B lymphocytes. These two classes of immunoglobulins have both membrane-bound and circulating forms. Lymphocyte membrane-bound IgM and IgD serve as receptors for specific antigens. The result of this interaction is the proliferation and further differentiation of B lymphocytes, producing antibody-secreting plasma cells. Circulating IgM is also effective in activating the **complement system,** a group of plasma proteins synthesized in the liver.

IgE usually exists as a monomer. This immunoglobulin has a great affinity for receptors located in the plasma membranes of mast cells and basophils. Immediately after its secretion by plasma cells, IgE attaches to these cells and virtually disappears from the blood plasma. When the antigen that elicited the production of a specific IgE is again encountered, the antigen-antibody complex formed on the surface of a mast cell or basophil triggers the production and liberation of several biologically active substances, such as histamine, heparin, leukotrienes, and eosinophil-chemotactic factor of anaphylaxis. An **allergic reaction** is thus mediated by the activity of IgE and the antigens (**allergens**) that stimulate its production (see Mast Cells in Chapter 5).

The properties and activities of **IgD** are not completely understood. It has a molecular mass of 180 kDa, and its concentration in blood plasma constitutes only 0.2% of the immunoglobulins. IgD is found on the plasma membranes of B lymphocytes (together with IgM) and is involved in the differentiation of these cells.

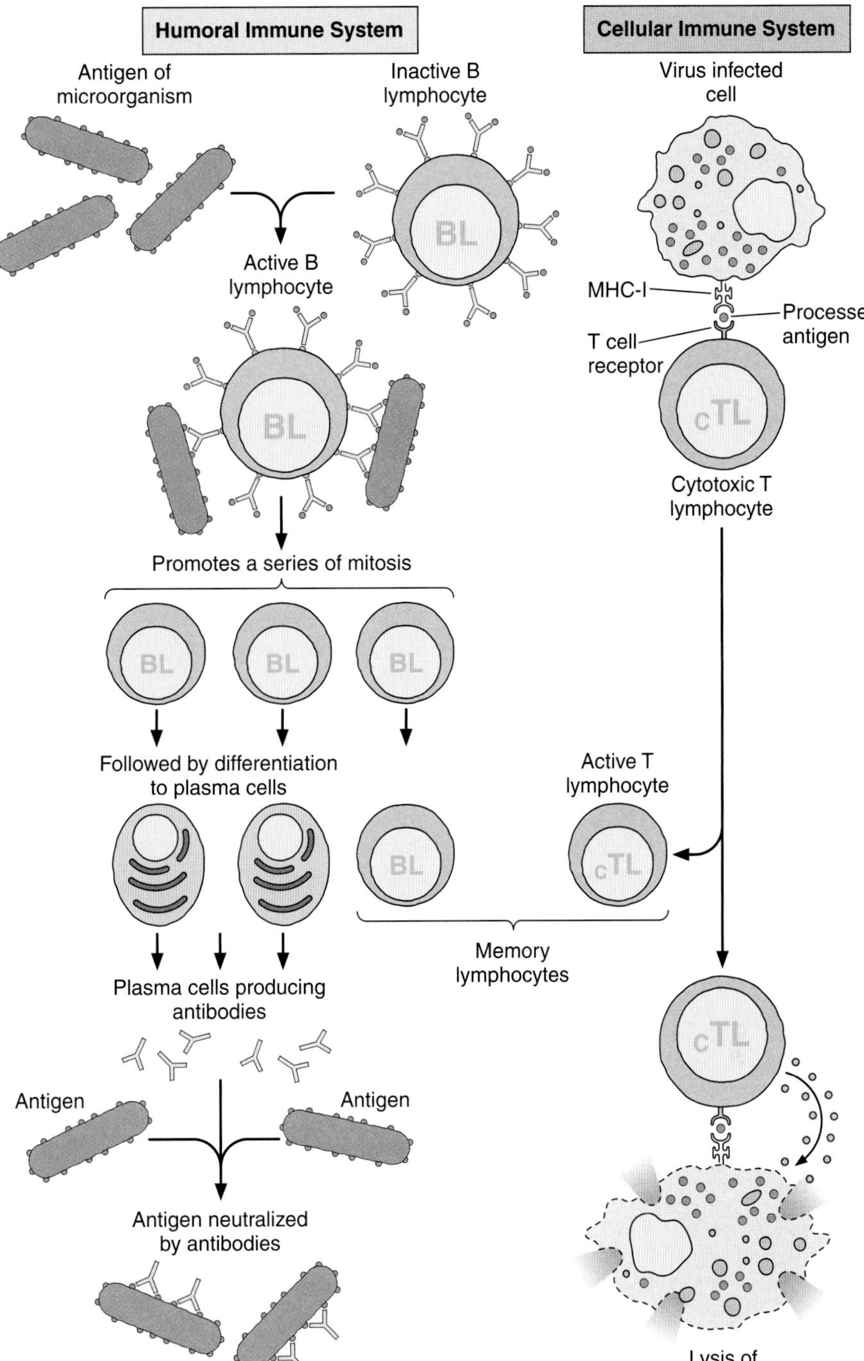

Figure 14–2. Comparison between humoral and cellular adaptive immunity reactions. **Left:** The antigen of the microorganism reacts with antibodies present on the surface of B lymphocytes, activating them. These activated lymphocytes proliferate and differentiate into memory B lymphocytes and plasma cells, which secrete antibodies to neutralize the invading microorganism. **Right:** Cytotoxic (cytolytic) T lymphocytes are activated by contact with an infected cell that presents complexes of viral antigens with MHC-I molecules on its surface. This activation results in the production of cytotoxic memory T lymphocytes and T cytotoxic lymphocytes, which produce perforin that lyses the infected cell.

B & T Lymphocytes

A fundamental division of lymphocytes into 2 classes and several subclasses (Table 14–2 and Figures 14–5 and 14–6) can be made on the basis of their site of differentiation and the presence of distinctive receptors in their membrane. In B lymphocytes these receptors are immunoglobulins, and in T lymphocytes they are special molecules called T-cell receptors (**TCR**). B lymphocytes originate and mature in the bone marrow and are carried by the blood to secondary lymphoid structures, where they nest, proliferate when activated, and differentiate into antibody-secreting **plasma cells.** B cells constitute 5–10% of the circulating blood lymphocytes; each is covered by 150,000 molecules of IgM that are receptors for specific antigens. Some activated B cells do not become plasma cells; instead, they generate **B memory cells,** which react rapidly to a second exposure to the same antigen.

T cells constitute 65–75% of blood lymphocytes. They originate in the bone marrow and migrate to the thymus, where

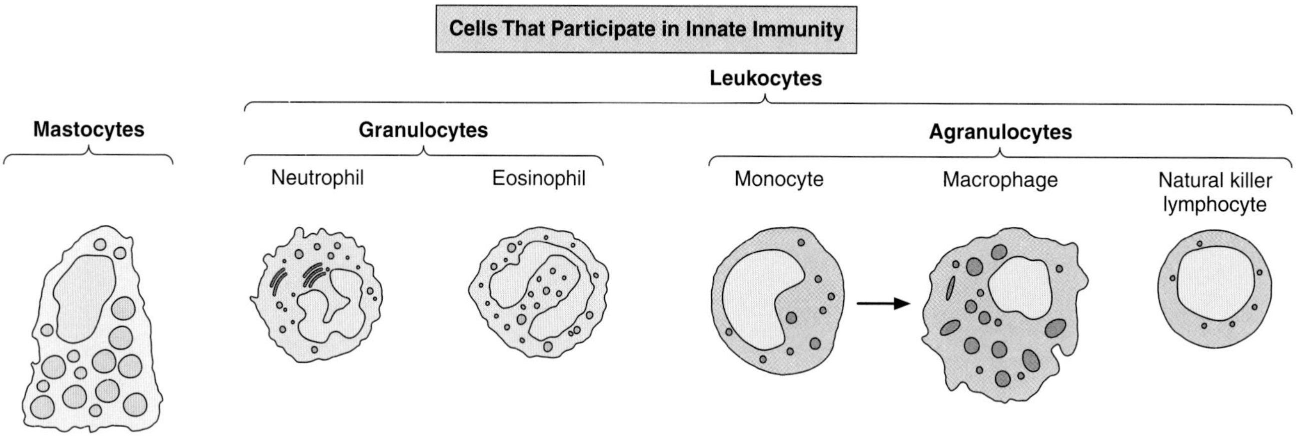

Cells That Participate in Innate Immunity

Leukocytes

Mastocytes | Granulocytes: Neutrophil, Eosinophil | Agranulocytes: Monocyte, Macrophage, Natural killer lymphocyte

Figure 14–3. The main cells that participate in innate immunity. Mast cell degranulation liberates compounds that modulate inflammation (see Chapter 5). Neutrophils act mainly by phagocytizing and destroying bacteria. Eosinophils participate in allergic reactions and parasitic (worm) invasion.

they proliferate and are carried by the blood to other lymphoid tissues. Three main subpopulations of T cells have been recognized: **helper, cytotoxic,** and **memory cells.** Helper cells stimulate the differentiation of B cells into plasma cells. Cytotoxic cells can act against foreign cells or virus-infected cells by means of 2 mechanisms. In one, they produce proteins called **perforins** that create holes in the cell membrane, with consequent cell lysis. In the other, they kill the cell by activating certain genes that induce programmed cell death, or **apoptosis** (see Chapter 3). Memory T cells react rapidly to the reintroduction of antigens (pathogens) and stimulate production of cytotoxic T cells.

MEDICAL APPLICATION

Helper T cells are killed by the retrovirus that causes the immunodeficiency syndrome known as AIDS, crippling the immunity of infected patients and rendering them susceptible to opportunistic infections—microorganisms that usually do not infect healthy individuals.

Table 14–1. Summary of classes of antibodies.

	IgG	IgM	IgA	IgD	IgE
	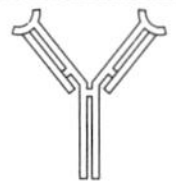	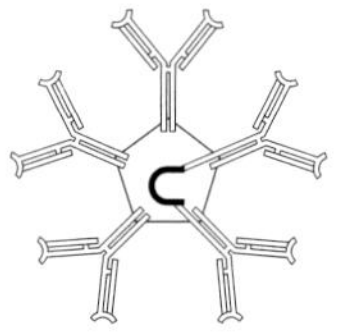	 Secretory component	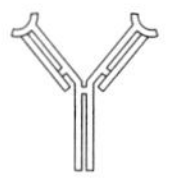	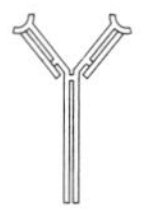
Structure	Monomere	Pentamere	Dimer with secretory component	Monomere	Monomere
Antibody percentage in the serum	80%	5–10%	10–15%	0.2%	0.002%
Site	Blood, lymph intestinal lumen	B lymphocytes surface (as a monomere)	Produced in B lymphocytes of the lamina propria and presents as dimers in secretions (saliva, milk, tears, etc)	Presents only on surface of B lymphocytes	Bound to the surface of mastocytes and basophils
Known functions	Activates phagocytosis, neutralizes antigens, protects newborn	First antibodies to be produced in an initial immune response	Protects the surface of mucosas for it resists proteolysis	Functions as a receptor to antigens triggering B cell activation	Participates in allergy and lyses parasitic worms

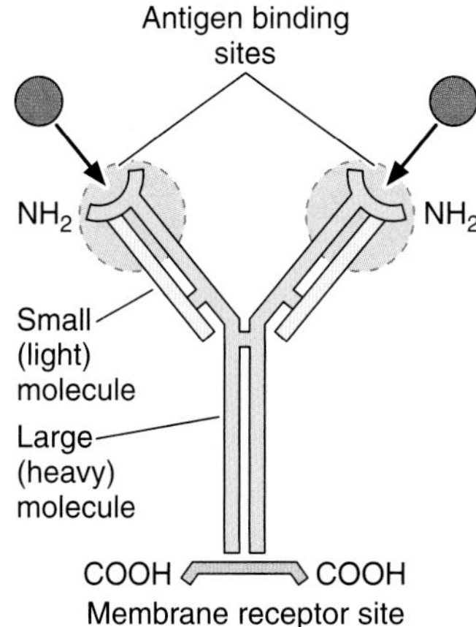

Figure 14–4. Schematic representation of an antibody molecule. Its variable portions near the NH_2 end, formed by the small and large molecules, bind the antigen. The carboxyl extremity may react with cell surface receptors.

B and T cells are not uniformly distributed in the lymphoid system (Table 14–3); they occupy special regions in nonthymic lymphoid structures. Although B and T cells are morphologically indistinguishable in either light or electron microscopes, they can be distinguished by immunocytochemical methods. They exhibit different surface proteins (markers) that permit the identification and detection of differentiated subpopulations of these two types of cells. When stimulated by antigens, B and T cells proliferate through several mitotic cycles, a process called clone selection and expansion.

In addition to B and T cells, there are **natural killer,** or **NK,** lymphocytes that lack the marker molecules characteristic of B and T cells. In the circulating blood, 10–15% of the lymphocytes are NK cells. They are called natural killer cells because they attack virus-infected cells and cancer cells, without previous stimulation.

Antigen-Presenting Cells

Antigen-presenting cells (**APCs**) are found in most tissues. They originate in the bone marrow and constitute a heterogeneous cell population that includes dendritic cells, macrophages, Langerhans cells of the skin, and B lymphocytes. In a process known as **antigen processing,** proteins are reduced to small peptides and attached to MHC molecules. The expression of class II MHC molecules characterizes APCs. Antigen processing, not restricted to APCs, is a necessary preliminary step for activation of T cells, because these cells are blind to native proteins and other antigens. Note that T cells can recognize only peptides associated with MHC molecules, whereas B cells can directly recognize, and react to, proteins, peptides, lipids, polysaccharides, and many small molecules.

Endocytosed exogenous proteins (Figure 14–7) are digested in the endosome-lysosome system, and the resulting small peptides (10–30 amino acids) form a complex with class II MHC (MHC-II) molecules. Proteins derived from pathogens (viruses, some bacteria, and some protozoans) that live inside infected host cells are digested by proteasomes (multicatalytic proteases, described in Chapter 2) to small peptides (8–11 amino acids) that are transported to the endoplasmic reticulum cisternae, where they form a complex with class I MHC (MHC-I) molecules (Figure 14–7). Both class I and class II complexes are then transported to the cell membrane, where they are inspected by T lymphocytes. CD4+ T cells interact with complexes of peptides with MHC-II molecules, whereas CD8+ T cells interact with complexes of peptides with MHC-I molecules. CD4 and CD8 are surface protein molecules present in some T cells, utilized as markers of these particular T-cell types.

Dendritic Cells

Dendritic cells have different names according to their location. They are present in the interstitium of many organs, are abundant in T-cell areas of lymphoid organs, and are present in the epidermis as Langerhans cells. Dendritic cell precursors are seeded through the blood into lymphoid and nonlymphoid organs, where they lodge as immature dendritic cells. These cells are characterized by their high ability to capture and process antigens. Inflammatory conditions induce the migration of dendritic cells from the bone marrow through the blood or lymph to the peripheral lymphoid organs, where they proceed to the T-cell areas. In this way, dendritic cells are able to present T cells with antigen that has been captured in peripheral tissues. The ability of dendritic cells to be attracted to sites of antigen challenge and travel to peripheral organs is a crucial function of these APCs. For example, antigens entering via the skin are picked up by Langerhans cells and transported via lymph vessels to the satellite lymph node, where the immune reaction is initiated. Dendritic cells in other organs may also pick up antigens and take them to the spleen through the blood circulation.

Table 14–2. Lymphocyte types and main functions.*

Type	Main Function
B lymphocyte	Carries membrane receptors (IgM). When activated by specific antigens, proliferates by mitosis, differentiating into plasma cells that secrete large amounts of antibodies.
B memory lymphocyte	Activated B cell that is primed to respond more rapidly and to a greater extent upon subsequent exposure to the same antigen.
T cytotoxic lymphocyte	Carries TCRs. Specialized to recognize antigens associated with MHC-I on the surface of other cells. Produces perforin and other proteins that kill foreign cells, virus-infected cells, and some tumor cells.
T helper lymphocyte	Carries TCRs. Modulates other T and B cells, stimulating their activities.
T memory lymphocyte	Carries TCRs. Is primed to respond more rapidly and to a greater extent upon subsequent exposure to the same antigen.
NK lymphocyte	Lacks T- and B-cell receptors. Attacks virus-infected cells and cancer cells without previous stimulation.

*TCR, T-cell receptor; MHC-I, class I major histocompatibility complex; NK, natural killer.

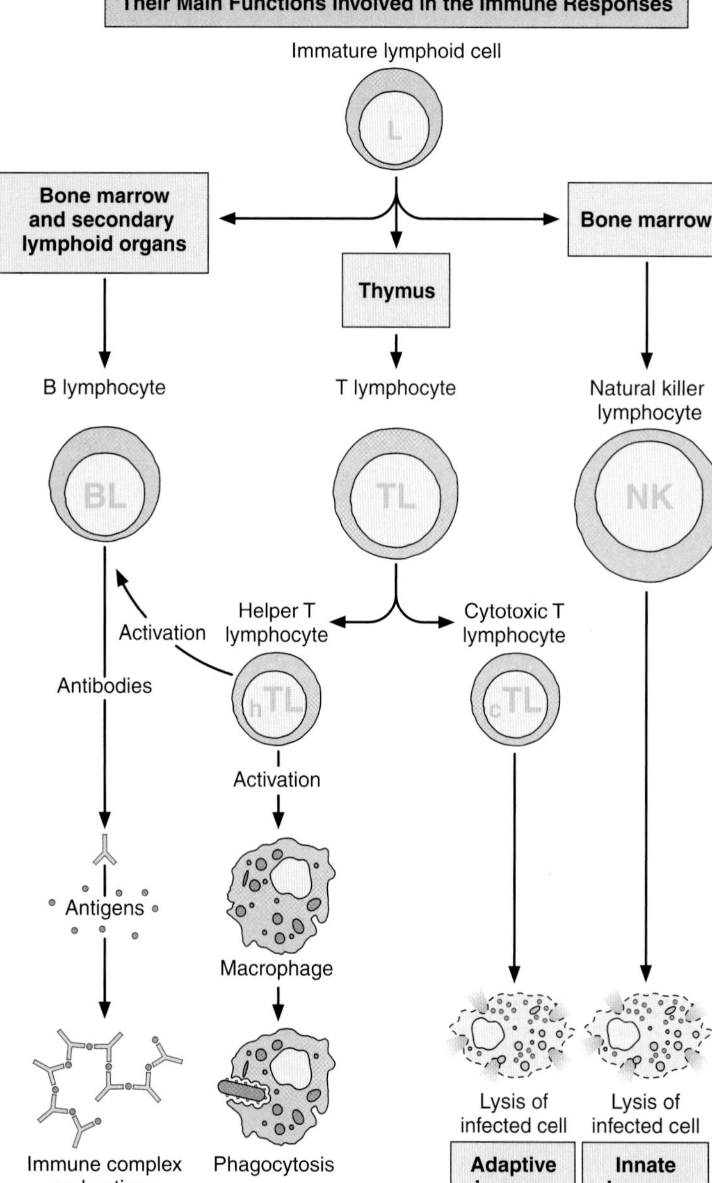

Figure 14–5. Origin, development, and activities of the main types of lymphocytes. Note that NK lymphocytes circulating in the blood derive directly from the bone marrow and act in innate immunity by killing infected cells. Immature T lymphocytes are transported by the blood circulation from the bone marrow to the thymus, where they complete their maturation. B lymphocytes leave the bone marrow already mature, to seed the secondary lymphoid organs, such as lymph nodes and spleen.

Note that in the lymph nodes, spleen, and other lymphoid tissues, there are cells with similar morphologic characteristics, called **follicular dendritic cells,** that are functionally different. Follicular dendritic cells are not derived from bone marrow and are unable to endocytose and process antigens. However, they are very efficient in trapping antigens complexed to antibodies. They retain these antigens for long periods of time on their surface membrane, where the antigens are recognized by B cells, thus helping to maintain the immunologic memory.

Major Histocompatibility Complex

The immune system distinguishes self from nonself mainly by the presence on cell surfaces of MHC molecules. In humans they are also called HLA, from *h*uman *l*eukocyte *a*ntigen, because they were discovered in leukocytes. These molecules fall into 2 classes (Figure 14–8): MHC-I is present in all cells, whereas MHC-II is more restricted in distribution, being found only in APCs. The MHC molecules constitute an intracellular system that places the MHC-processed antigen complex on the cell membrane for inspection by T lymphocytes. B lymphocytes are activated directly by free antigens. These lymphocytes do not require the help of MHC complexes to fulfill their function. MHC molecules are the expression of several genes that have a structure that is unique to each individual. This is the main reason tissue grafts and organ transplants are often rejected if they are not made between identical twins who possess identical MHC and other molecules.

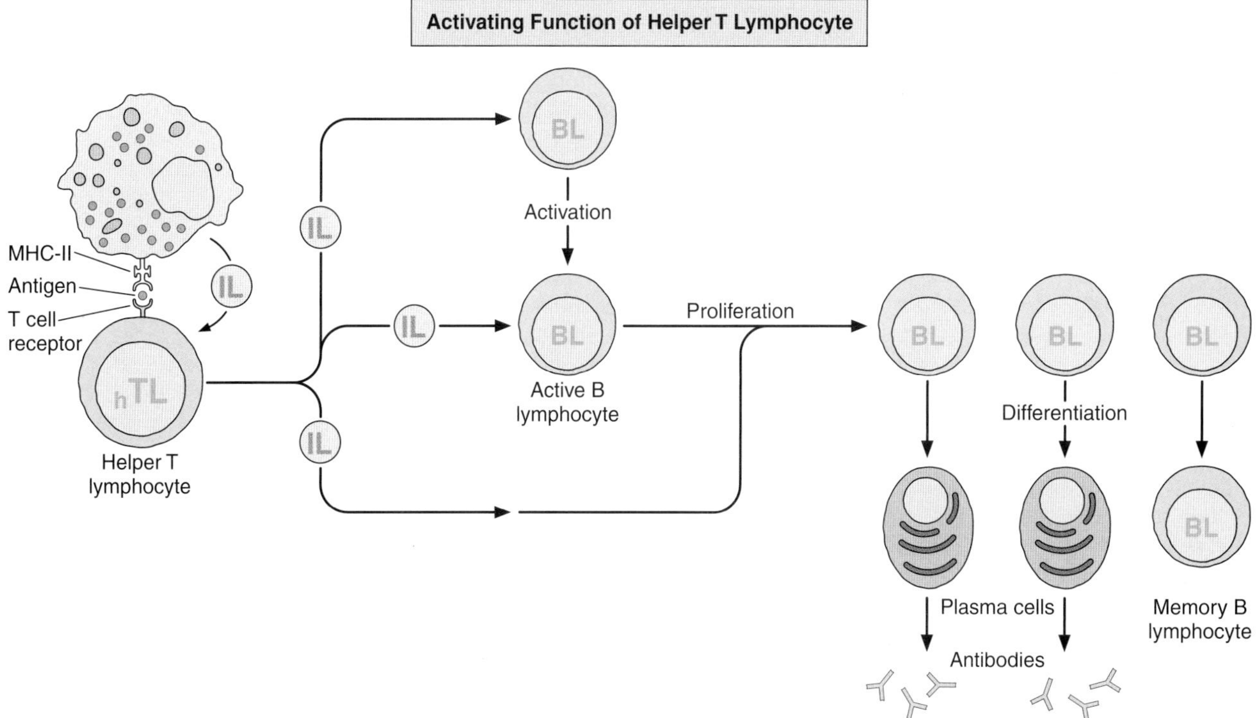

Figure 14–6. A helper T lymphocyte is activated by its interaction with a processed antigen located on the surface of a macrophage. The activated T lymphocyte produces interleukins that induce the activation, proliferation, and differentiation of B lymphocytes. This results in the production of memory B lymphocytes and antibody-producing plasma cells.

In common with other integral membrane proteins, MHC-I and MHC-II proteins are synthesized by polyribosomes and introduced in the rough endoplasmic reticulum. When the chains are completed, they project into the cisterna of the rough endoplasmic reticulum with their carboxyl ends remaining embedded in the membrane of the rough endoplasmic reticulum (Figure 14–7). Then MHC proteins of the two classes follow different pathways to reach the surface cell membrane. The main difference is that MHC-I proteins do not pass through the endosome-lysosome system of vesicles, whereas MHC-II proteins enter this pathway, where they may form complexes with processed antigens (Figure 14–7). The vesicles containing MHC-I proteins may receive polypeptides processed in proteasomes; these vesicles are integrated in the cell membrane, exposing its processed antigen.

Table 14–3. Approximate percentage of lymphocytes in lymphoid organs.

Lymphoid Organ	T Lymphocytes, %	B Lymphocytes, %
Thymus	100	0
Bone marrow	10	90
Spleen	45	55
Lymph nodes	60	40
Blood	80	20

ORGAN TRANSPLANTATION

Tissue grafts and organ transplants are classified as **autografts** when the transplanted tissues or organs are taken from the individual receiving them; **isografts** when taken from an identical twin; **homografts** when taken from an individual (related or unrelated) of the same species; and **heterografts** when taken from an animal of a different species.

The body readily accepts autografts and isografts as long as an efficient blood supply is established. There is no rejection in such cases, because the transplanted cells are genetically identical to those of the host and present the same MHC on their surfaces. The organism recognizes the grafted cells as self (same MHC) and produces no cellular or humoral reactions.

Homografts and heterografts, on the other hand, contain cells whose membranes have MHC-I molecules that are foreign to the host; they are therefore recognized and treated as such. Transplant rejection is due mainly to the activity of NK lymphocytes and cytotoxic T lymphocytes that penetrate the transplant and destroy the transplanted cells.

Cytokines

The high complexity of the immune system is finely regulated by a large number of molecules, mainly **cytokines,** which are a group of peptides or glycoproteins with low molecular masses (between 8 and 80 kDa). They influence both the cellular (Figure 14–9) and humoral mechanisms. Cytokines act on many cells that have receptors for them—not only cells of the immune system, but also cells of other systems, such as the nervous system and endocrine system. They are primarily produced by cells of

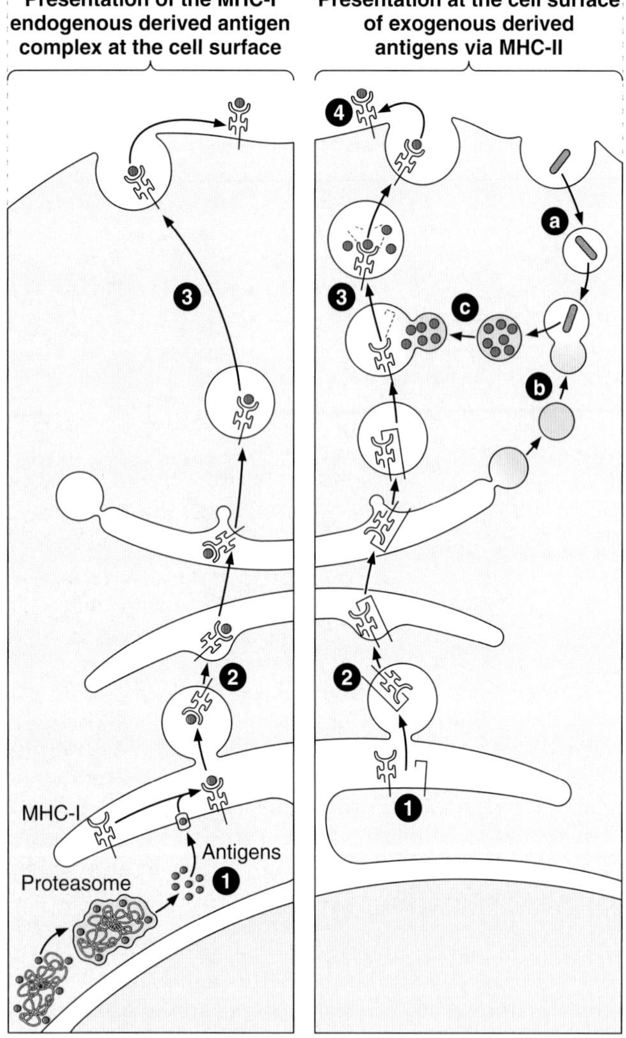

Figure 14–7. **Left:** The sequence of events by which antigens from microorganisms that infect host cells (viruses, some bacteria, some protozoans) are processed, bound to MHC-I molecules in post-Golgi vesicles, and presented as complexes at the cell surface. **(1)** Proteins derived from parasites already inside the cell are digested by proteasomes and transferred to the rough endoplasmic reticulum (RER) where they associate with the locally synthesized MHC-I. **(2)** The MHC-I complex (MHC-I plus antigen) is transferred to the Golgi region. **(3)** Golgi vesicles transport the MHC-I ( ) plus antigen (∘) to the outer surface of the cell membrane presenting the antigen produced by the cytoplasmic microorganism to the lymphocytes. **Right:** The processing of antigens introduced in the cell by phagocytosis and processed in the endosome-lysosome system. Through this pathway the antigens form complexes with MHC-II molecules. **(1)** Synthesis of MHC-II ( ) and Li ( ) protein and coupling of both ( ) take place. **(2)** Transfer of MHC-II plus Li to the Golgi region leads to the formation of a Golgi vesicle. **(3)** Fusion of the Golgi vesicle to a secondary lysosome with simultaneous lysis of the Li component occurs. This removal permits the fusion of the previously processed antigen to the MHC-II receptor portion. **(4)** The MHC-II–antigen complex is exposed at the cell surface where it can react with a T lymphocyte, thereby activating it. **(a)** Endocytosis of microorganisms produces a phagosome. **(b)** Transfer of lysosomal enzymes derived from the Golgi region to the phagosome produces a digesting vesicle (secondary lysosome), where the antigen is processed. **(c)** The processed antigen is coupled to the Golgi vesicle.

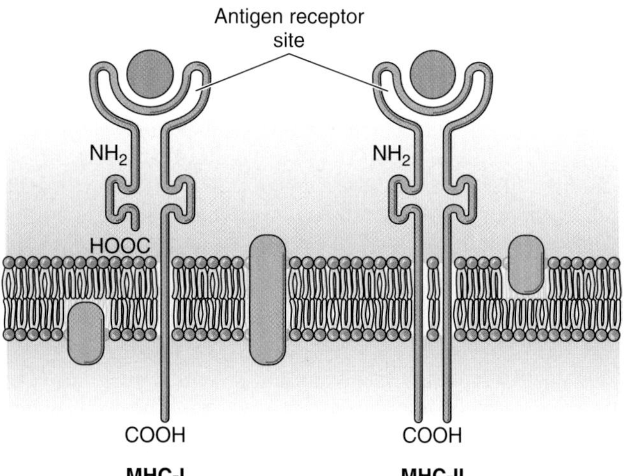

Figure 14–8. Schematic representation of the 2 types of MHC molecules. Note that the MHC-I molecule has only one transmembrane segment.

the immune system, mainly macrophages and leukocytes, but are also synthesized by other cell types, such as endothelial cells and fibroblasts. Cytokines that interact with leukocytes are called **interleukins** (Table 14–4). They activate leukocytes and stimulate their multiplication and differentiation. The term **lymphokine** is used to designate cytokines produced by lymphocytes. The cytokines synthesized by monocytes or macrophages are known as **monokines.** A great variety of cytokines with multiple effects have been described. These intercellular signaling molecules regulate not only local and systemic immune responses, but also inflammatory response, wound healing, hematopoiesis, and other biologic processes.

Cytokines can act on the cells that produced them (**autocrine** mechanism), on cells located at a short distance (**paracrine** mechanism), or on distant cells (**endocrine** mechanism)(Figure 14–9).

Chemotaxins, or **chemokines,** are cytokines that induce the attraction of leukocytes to sites of inflammation.

Interferons are glycoprotein cytokines produced by any cell infected by viruses. Interferons react with receptors of neighboring macrophages, fibroblasts, and lymphocytes, inducing them to produce substances that inhibit the production of viruses.

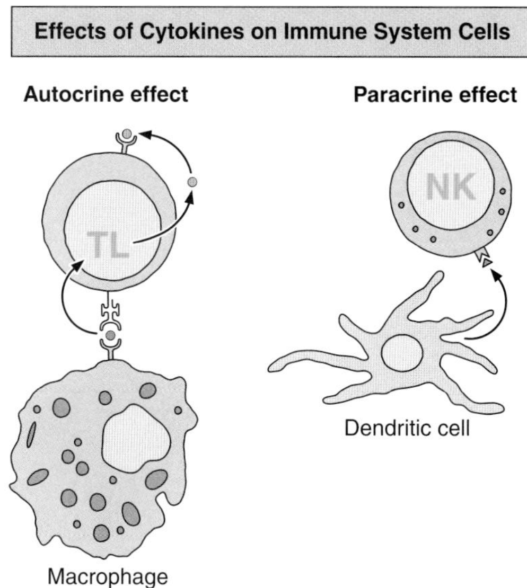

Figure 14–9. Two examples of cytokine actions (autocrine and paracrine). Cytokines also act on distant tissues (endocrine action).

The cytokine called **tumor necrosis factor** (Figure 14–10) has multiple local and general actions. Thus, tumor necrosis factor stimulates the expression of adhesion molecules and the secretion of chemiokines by macrophages, promotes apoptosis of target cells, and has systemic effects, such as fever.

Complement System

The complement system is formed by a pool of approximately 20 proteins produced mainly in the liver, each one designated by the letter "C" followed by a number. The name derives from the fact that this system "complements" the activity of some immunity processes.

The complement system can be activated by 2 mechanisms that change the structure of one initial component, triggering a process that is gradually transmitted to the next components, forming a cascade of events (Figure 14–12).

One significant property of the activated complement system is that molecules to which macrophages have receptors are placed on the surface of bacteria. This facilitates the phagocytosis and destruction of the invading bacteria. The process of preparing the bacteria to attach more easily to macrophages is called **opsonization** of the bacteria (Figure 14–11). Another role of the complement system is to produce a complex that causes damage to the bacterial membrane (Figure 14–12).

Diseases of the Immune System

The immune system comprises very extensive and complex mechanisms, with the participation of many cell types and signaling molecules. This explains the high diversity of immune diseases described so far. These diseases can be grouped into 3 general types of reaction:

1. The disease can be the result of an abnormal and intense reaction in an attempt to neutralize the effects of some antigens. This exaggerated intolerance produces the numerous processes called **allergic reactions.**
2. When there is a depression of the reaction against antigens, the pathologic process is broadly named **immunodeficiency,** which can be caused by deficiency in components of the complement system, by defects in the phagocytic activity of macrophages and neutrophils, or by abnormalities in B and T lymphocytes.
3. The presence of T lymphocytes that attack self antigens cause **autoimmune diseases.** In this case, tissues are affected or even destroyed by T cells produced in the organism against itself.

THYMUS

The thymus is a lymphoepithelial organ located in the mediastinum; it attains its peak development during youth. Whereas

Table 14–4. Interleukins*

	Origins	Some Functions
IL-1	Macrophages, keratinocytes, etc	Proinflammatory endogenous pyrogen; activates fibroblasts, granulocytes, osteoclasts; makes T cells responsive to signals
IL-2	T cells	Proliferation of T, B, and natural killer cells
IL-3	T cells	Proliferation of early hematopoietic cells (multi-CSF)
IL-4	T cells, mast cells	Governs B-cell isotype switching to IgG and IgE
IL-5	T cells, mast cells, possibly B cells	Eosinophil differentiation and proliferation; IgA production
IL-6	Macrophages, T cells, fibroblasts	Proinflammatory; B-cell differentiation; thymocyte growth
IL-7	Bone marrow stroma	B-cell differentiation and maturation
IL-8	Keratinocytes, fibroblasts, monocytes	Neutrophil chemotaxis and activation
IL-9	T cells	Proliferation of T cells, thymocytes, mast cells
IL-10	T cells, mast cells, possibly B cells	Inhibition of cytokine synthesis in various cells; proliferation of mast cells

*IL, interleukin; CSF, colony-stimulating factor.

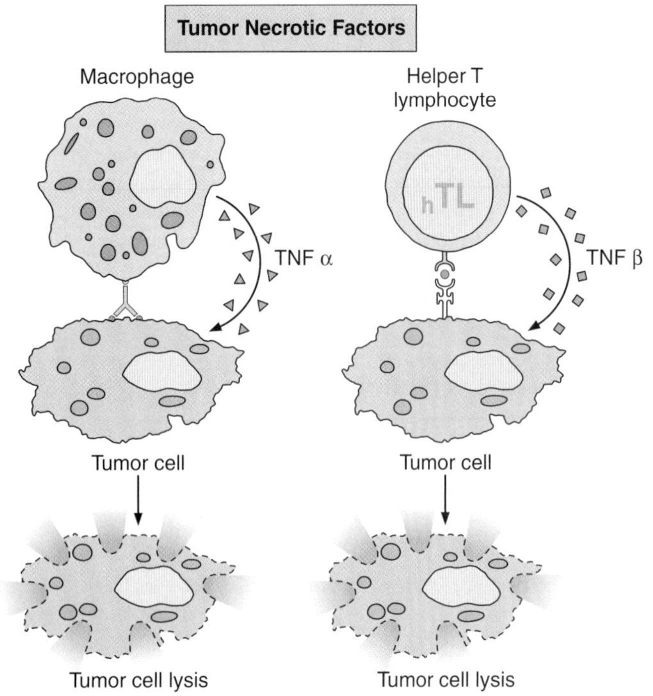

Tumor Necrotic Factors

Macrophage

Helper T lymphocyte

hTL

TNF α

TNF β

Tumor cell

Tumor cell

Tumor cell lysis

Tumor cell lysis

Figure 14–10. Activities of 2 tumor necrosis factors (TNF), produced by macrophages and helper T lymphocytes, that kill tumor cells.

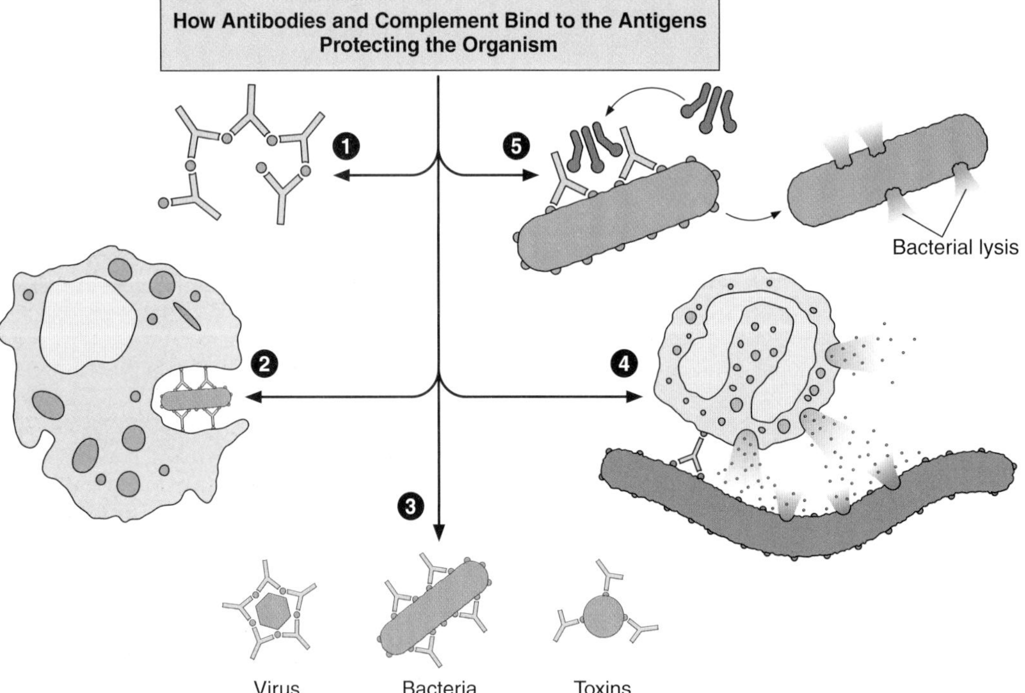

Figure 14–11. Drawings showing several mechanisms of antigen inactivation that protect the organism. These include **(1)** agglutination, in which antibodies bind to antigens, forming aggregates and reducing the amount of free antigens; **(2)** opsonization, in which the binding of antibodies to the microorganism stimulates phagocytosis; **(3)** neutralization, whereby the binding of antibody to microorganisms blocks their adhesion to cells and inactivates toxins; **(4)** cytotoxicity mediated by cells, which involves antibodies adhering to the surface of worms activating cells of the immune system (macrophages and eosinophils) and inducing them to liberate chemical agents that attack the surface of the animal; **(5)** complement activation, in which the binding of antibodies to the initial protein of the complement system triggers the complement cascade and causes cell lysis.

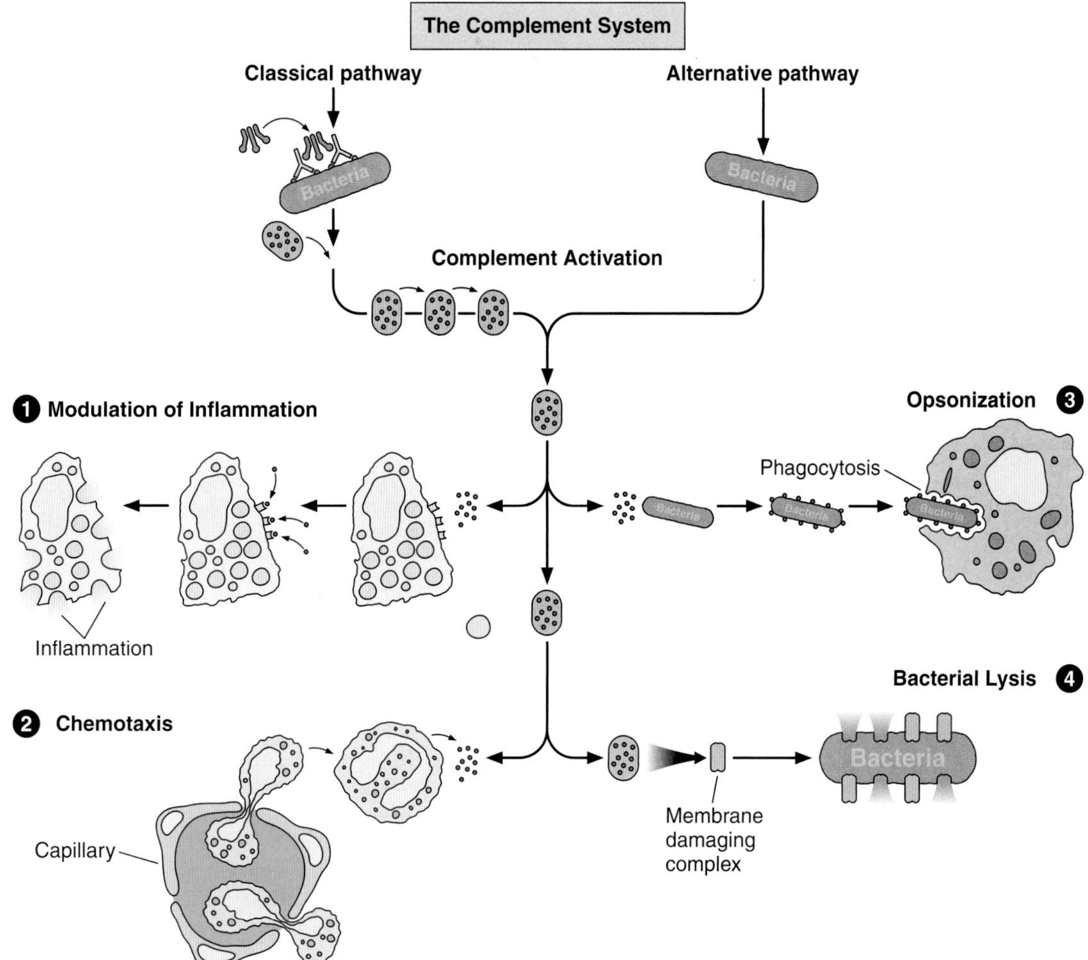

Figure 14–12. The 2 pathways of complement activation. In the alternative pathway, the microorganism acts directly on the component C3, triggering its activation, which is followed by the cascade. C3 activation promotes the degranulation of mast cells, liberating several compounds that modulate inflammation. By the activity of C3, molecules are bound to the surface of bacteria, making them more susceptible to phagocytosis, a process called **opsonization.** Activation of C5 promotes the attraction of neutrophils (bacteria-killing leukocytes) to the site of inflammation (**chemotaxis**). Activation of C6 to C9 complex promotes bacterial lysis. The alternative pathway is innate, because it is triggered by the simple presence of a microorganism. The classical pathway is adaptive, because it depends on binding of the antigen (in the microorganism) to a previously synthesized antibody. From C3 on, the effects of both classical and alternative pathways are the same. Activation of the complement system can result in several processes that kill bacteria and regulate inflammation. The main processes are: **(1)** modulation of inflammation due to the degranulation of M cells, **(2)** promoting the attraction of neurophils to a site of inflammation (chemotaxis), **(3)** making bacteria more susceptible to phagocytosis (opsonization), **(4)** damage of bacterial plasma membranes (lysis).

nonthymic lymphoid organs originate exclusively from mesenchyme (mesoderm), the thymus has a dual embryonic origin. Its lymphocytes arise from mesenchymal cells that invade an epithelial primordium that has developed from the endoderm of the third and fourth pharyngeal pouches.

The thymus has a connective tissue capsule that penetrates the parenchyma and divides it into lobules (Figure 14–13). Each lobule has a peripheral dark zone known as the **cortex** and a central light zone called the **medulla** (L. *medius,* middle).

The **cortex** (Figure 14–14) is composed of an extensive population of T lymphocytes, dispersed epithelial reticular cells, and few macrophages. Because the cortex is richer in small lymphocytes than the medulla, it stains more darkly. The epithelial reticular cells are stellate cells with light-staining oval nuclei. Usually, they are joined to similar adjacent cells by desmosomes (Figure 14–15). Bundles of intermediate keratin filaments (tonofibrils) in their cytoplasm are evidence of the epithelial origin of these cells.

Some investigators have described several types of thymus epithelial reticular cells according to their location and structure. The functions of each type of epithelial reticular cell are still not completely understood.

Because of the proliferation of lymphocytes in the cortex, immature T lymphocytes are produced in quantity and accumulate in this region. Although most of these lymphocytes die in the cortex by apoptosis and are removed by macrophages, a small number migrate to the medulla and enter the bloodstream through the walls of venules. These cells migrate to nonthymic lymphoid structures and accumulate in specific sites as T lymphocytes.

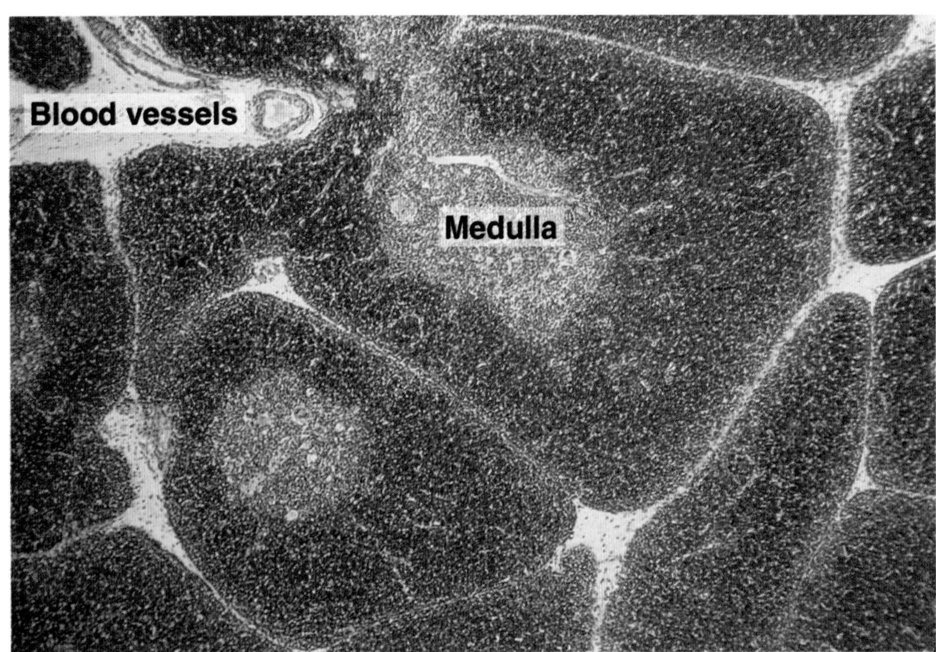

Figure 14–13. Photomicrograph of a section of thymus showing the lobules. Two lobules show the dark cortical and the light medullary zones. At the upper left are blood vessels and the connective tissue capsule. Pararosaniline-toluidine blue (PT) stain. Low magnification.

The **medulla** (Figure 14–16) contains **Hassall's corpuscles,** which are characteristic of this region (Figure 14–17). These structures are concentrically arranged, flattened epithelial reticular cells that become filled with keratin filaments, degenerate, and sometimes calcify. Their function is unknown. The medulla has the same cell population as the cortex, with a larger number of epithelial reticular cells.

Vascularization

Arteries enter the thymus through the capsule (Figure 14–18); they branch and penetrate the organ more deeply, following the septa of connective tissue. Arterioles leave the septa to penetrate the parenchyma along the border between the cortical and medullary zones. These arterioles give off capillaries that penetrate the cortex in an arched course; they finally reach the medulla, where they drain into venules. The medulla is supplied with capillary branches of the arterioles in the medullary-cortical border. The capillaries of the medulla drain into venules, which also receive capillaries returning from the cortical zone.

Thymus capillaries have a nonfenestrated endothelium and a very thick basal lamina. These capillaries are particularly impermeable to proteins, preventing most circulating antigens from reaching the thymus cortex where T lymphocytes are being formed.

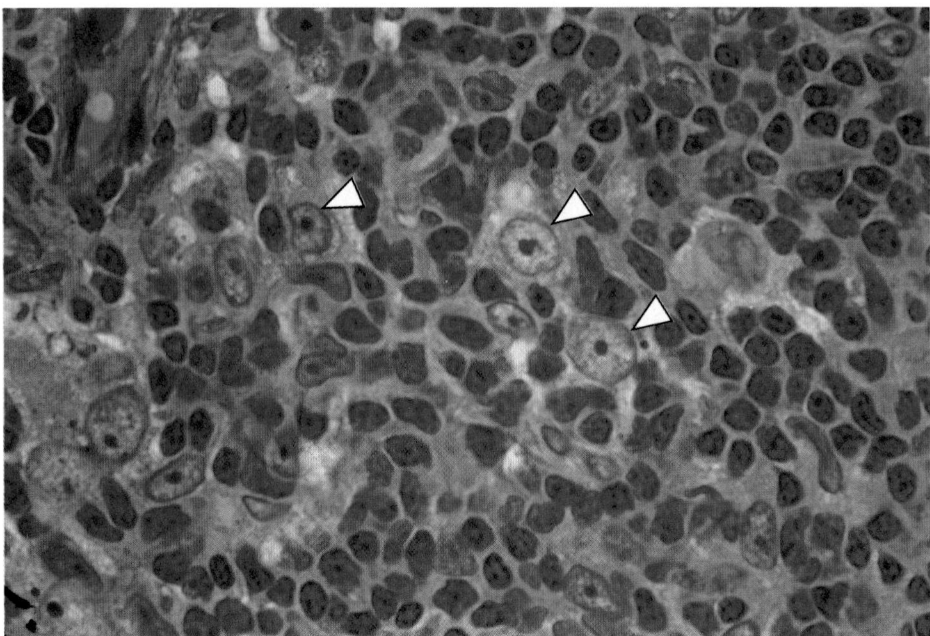

Figure 14–14. Thymus cortical zone showing epithelial reticular cells with visible nucleoli (arrowheads) surrounded by dark-stained T lymphocytes undergoing differentiation. PT stain. Medium magnification.

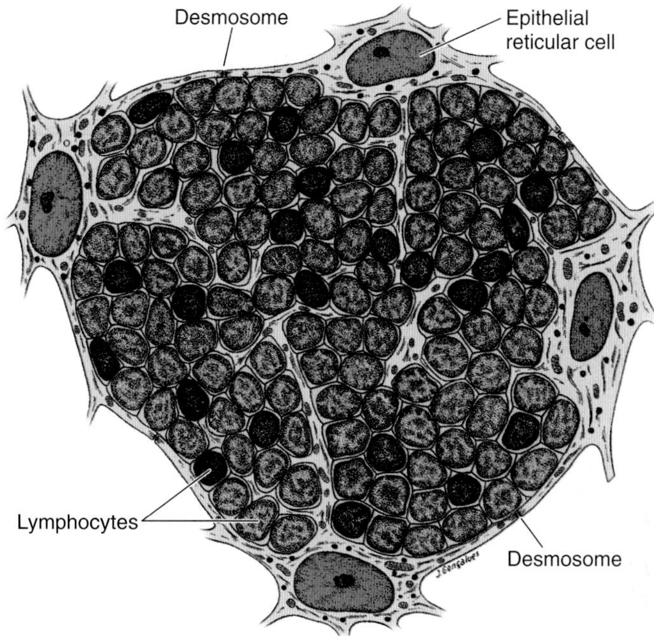

Figure 14–15. The relationship between epithelial reticular cells and thymus lymphocytes. Note the desmosomes and the long processes of epithelial reticular cells extending among the lymphocytes.

Medullary veins penetrate the connective tissue septa and leave the thymus through its capsule. There is no blood-thymus barrier in the medulla.

The thymus has no afferent lymphatic vessels and does not constitute a filter for the lymph, as do lymph nodes. The few lymphatic vessels encountered in the thymus are all efferent; they are located in the walls of blood vessels and in the connective tissue of the septa and the capsule.

Histophysiology

The thymus shows its maximum development in relation to body weight immediately after birth; it undergoes involution after puberty (Figure 14–19). Cells derived from the bone marrow continually repopulate it. Stem cells committed to originate T cells arise in the bone marrow and migrate to the thymus during both fetal and adult life. After penetrating the thymus, the developing T cells or thymocytes initially populate the cortex.

The thymus is the site of terminal differentiation and selection of T lymphocytes. During this process the thymus lymphocytes undergo numerous mitoses. However, more than 95% of them are eliminated by apoptosis. The lymphocytes eliminated are those that do not react to antigens and are therefore useless, and those that react with self antigens. When lymphocytes that react with self antigens are not eliminated, they will cause **autoimmune diseases.**

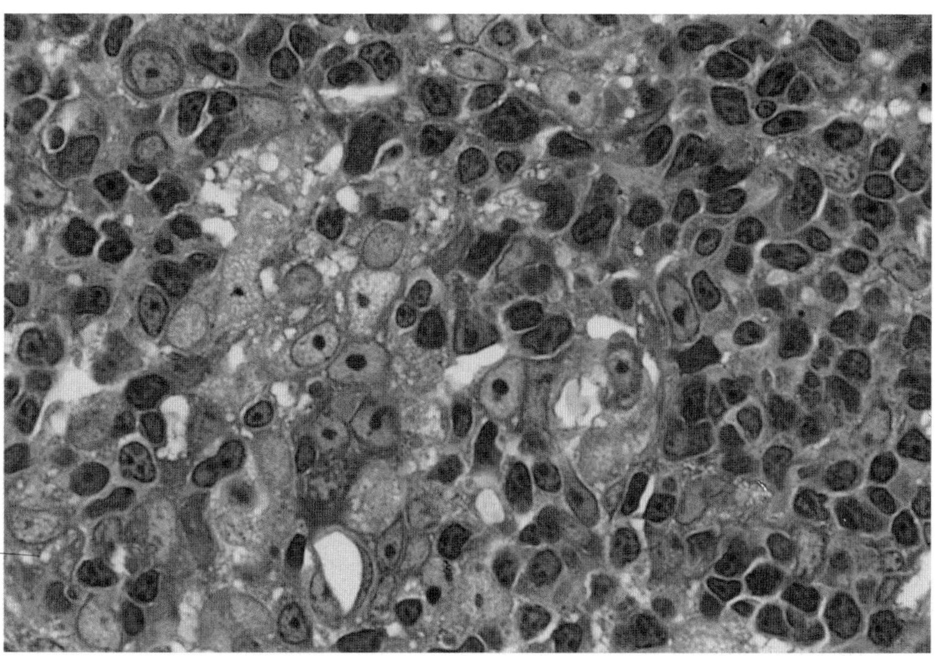

Figure 14–16. Photomicrograph of the medullary zone of the thymus. The large numbers of epithelial reticular cells with their large and light-stained nuclei are responsible for the light color of the thymus medulla. This zone also contains mature T lymphocytes. PT stain. Medium magnification.

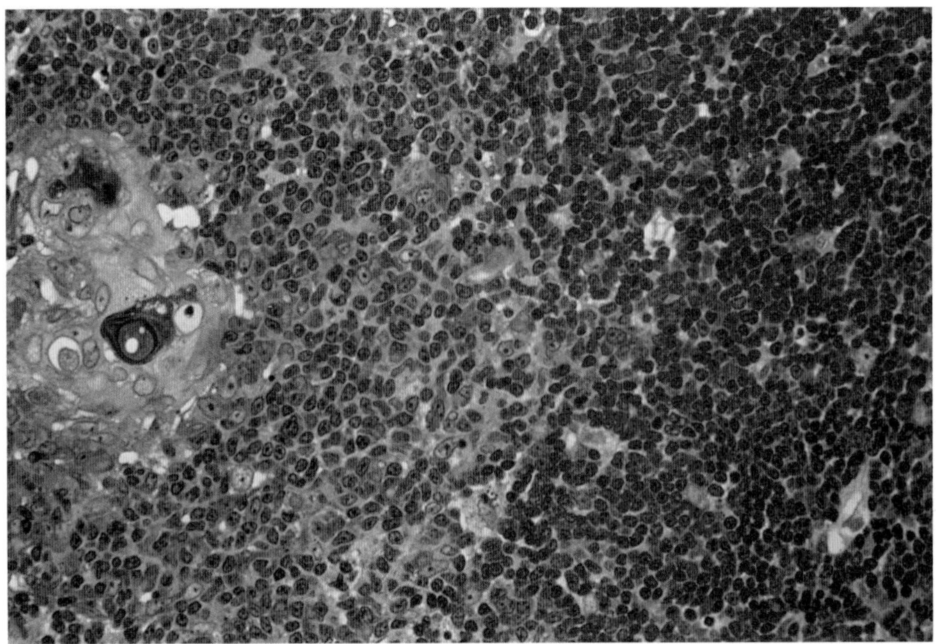

Figure 14–17. Photomicrograph of a portion of the cortical zone, identified by its dark staining (right), and a portion of medulla, identified by its lighter staining and the presence of a Hassall corpuscle (left). These corpuscles exist only in the medulla. PT stain. Medium magnification.

The resulting lymphocytes produced in the thymus are the T lymphocytes that react with nonself antigens and are essential for normal adaptive immune reactions.

In mammals, the main thymus-dependent areas (rich in T cells) are the paracortical zones of lymph nodes, some parts of Peyer's patches, and the periarterial sheaths in the white pulp of the spleen.

The thymus produces several protein growth factors that stimulate proliferation and differentiation of T lymphocytes. They seem to be paracrine secretions, acting in the thymus. Four factors have been identified: thymosin-α, thymopoietin, thymolin, and thymus humoral factor. The thymus is also subject to the effects of several hormones. Injections of some adrenocorticosteroids cause a reduction in lymphocyte number and mitotic rate and atrophy of the cortical layer of the thymus. **Adrenocorticotropic hormone,** produced by the anterior pituitary, achieves the same effect by stimulating the activity of the adrenal cortex. Male and female sex hormones also accelerate thymus involution, whereas castration has the opposite effect.

LYMPH NODES

Lymph nodes are encapsulated spherical or kidney-shaped organs composed of lymphoid tissue that are distributed through-

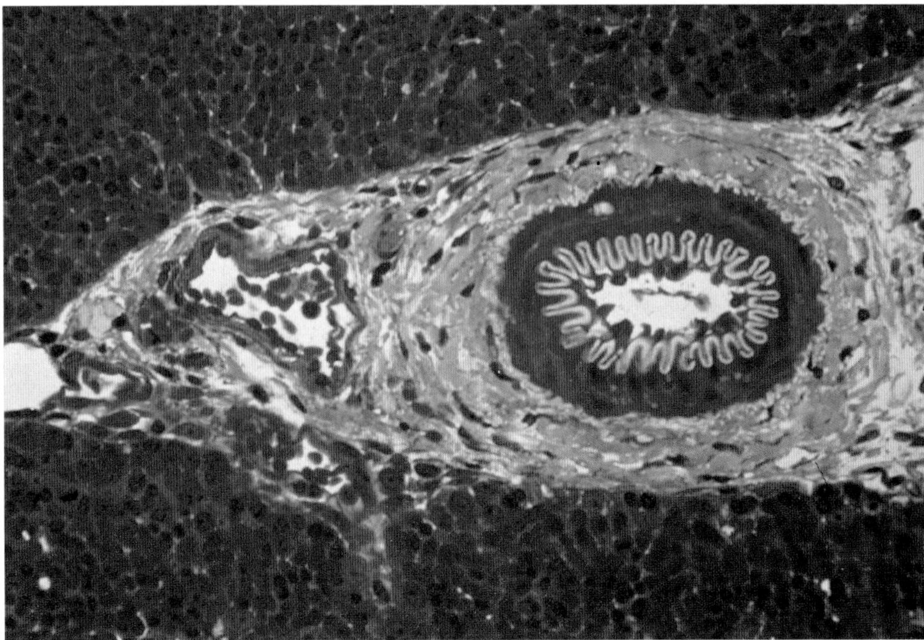

Figure 14–18. Thymus cortex showing the connective tissue capsule that penetrates into the organ. Note an artery and a vein in the capsule. PT stain. Low magnification.

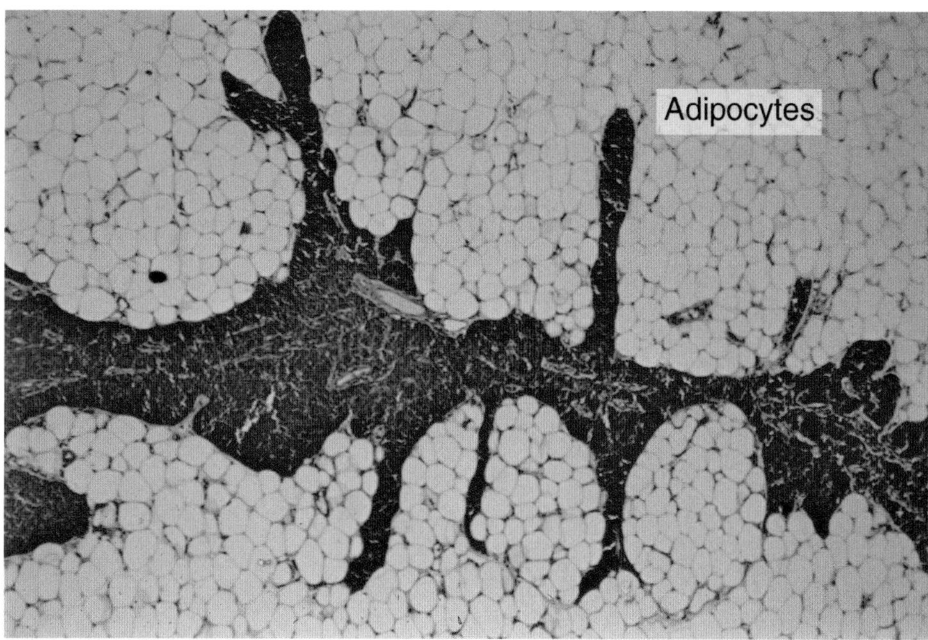

Figure 14–19. Section of the thymus of an elderly adult. Severe atrophy of the parenchyma, which was partially replaced by adipose tissue, can be seen. PT stain. Low magnification.

out the body along the course of the lymphatic vessels. The nodes are found in the axilla and the groin, along the great vessels of the neck, and in large numbers in the thorax and abdomen, especially in mesenteries. Lymph nodes constitute a series of in-line filters that are important in the body's defense against microorganisms and the spread of tumor cells. All tissue fluid–derived lymph is filtered by at least one node before returning to the circulation. Lymph nodes have a convex side and a concave depression, the **hilum,** through which arteries and nerves enter and veins and lymphatic vessels leave the organ (Figure 14–20). A connective tissue **capsule** surrounds the lymph node, sending trabeculae into its interior. Each node contains an **outer cortex,** an **inner cortex,** and a **medulla** (Figures 14–20 and 14–21).

OUTER CORTEX

At the surface of the outer cortex is the **subcapsular sinus,** which is limited on its outer boundary by the capsule and on

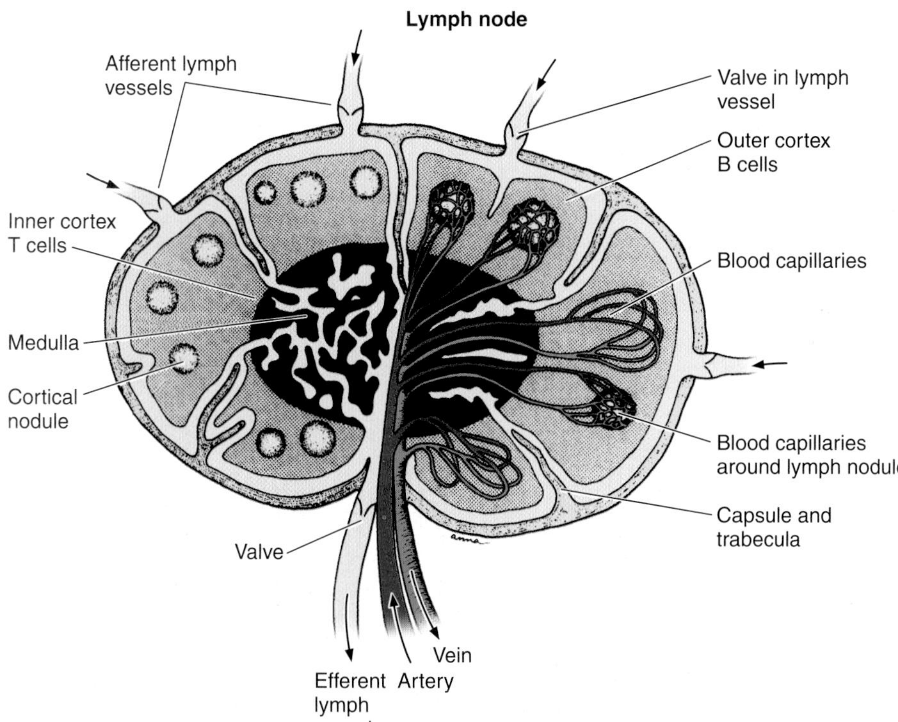

Figure 14–20. Schematic representation of the structure of a lymph node. Note the outer and inner cortex, the medulla, and the blood and lymph circulation. Also note that the lymph enters through the convex side of the node and leaves through the hilum. The lymph percolates through the node, exposing its contents to the action of defensive cells (macrophages, lymphocytes, APCs).

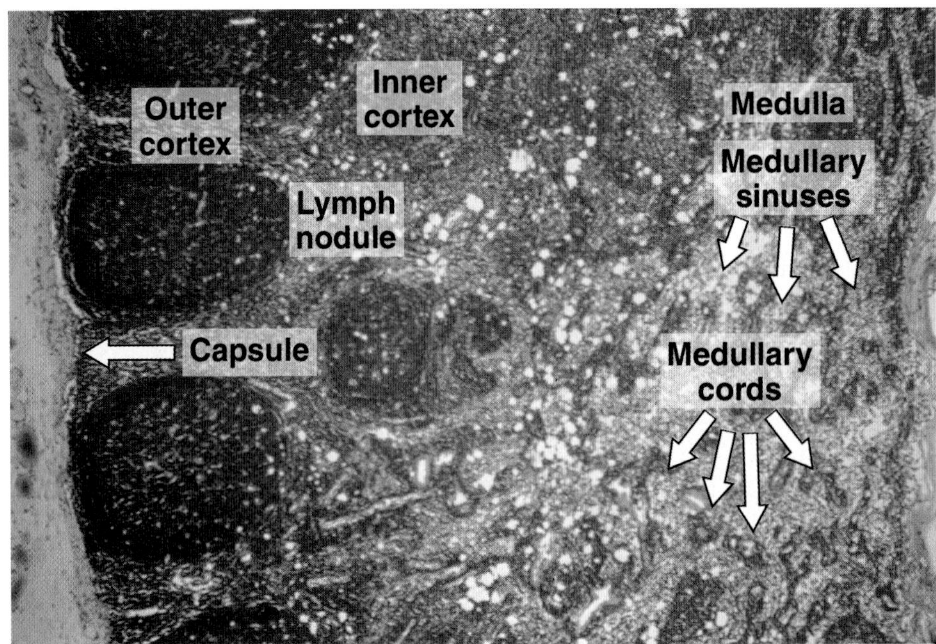

Figure 14–21. Section of a lymph node showing the structure of the cortex and the medulla. The medullary cords and sinuses are clearly visible. PT stain. Low magnification.

its inner boundary by the outer cortex (Figure 14–22). It is formed of a loose network of macrophages and reticular cells and fibers. The subcapsular sinus communicates with the medullary sinuses through **intermediate sinuses** that run parallel to the capsular trabeculae. The **outer cortex** is formed of a network of reticular cells and fibers whose meshwork is populated by B cells. Within the cortical lymphoid tissue are spherical structures called **lymphoid nodules** (Figure 14–23). These nodules are rich in B lymphocytes that react with antigens, increase in size, and proliferate by mitosis, resulting in large basophilic cells with prominent nucleoli called **immunocytes** (Figure 14–24). Some nodules present lighter-stained central zones called **germinal centers.** The germinal centers usually show several cells in mitosis and are rich in immunocytes. These cells produce antibody-synthesizing plasma cells.

Inner Cortex

The inner cortex is a continuation of the outer cortex and contains few, if any, nodules but many T lymphocytes.

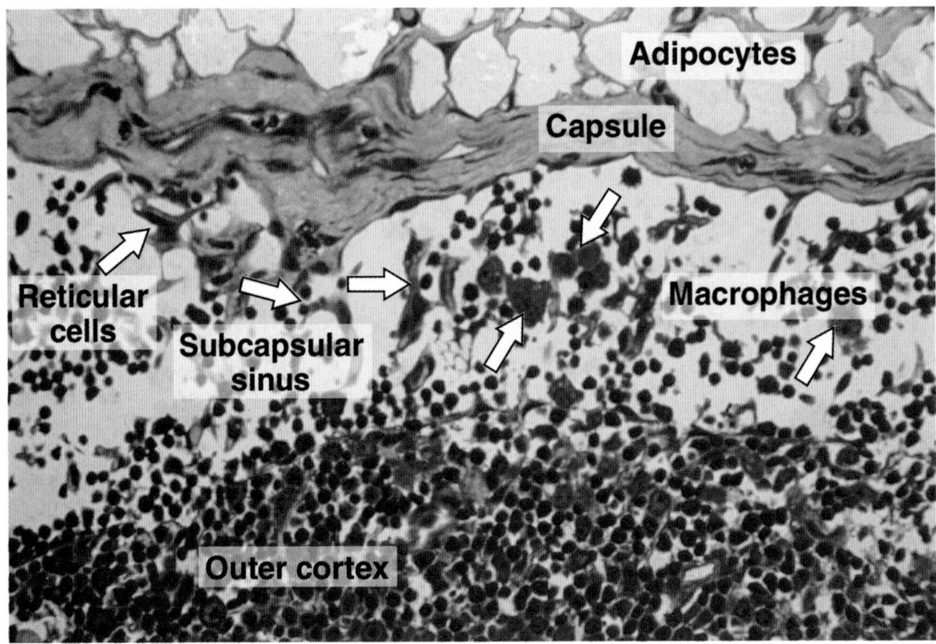

Figure 14–22. Section of a lymph node showing the capsule, subcapsular sinuses, and a small portion of the outer cortex. Some adipose tissue is seen outside the capsule. PT stain. Medium magnification.

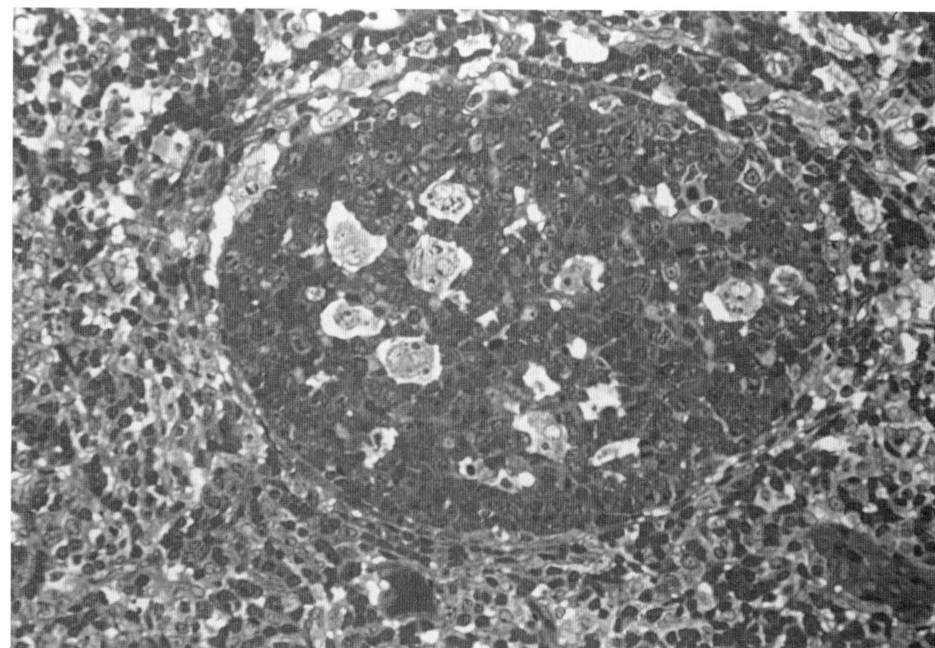

Figure 14–23. Section of the cortex of a lymph node showing a lymphoid nodule that was strongly activated by the injection of an antigen. Note the large number of macrophages (light structures) surrounded by B lymphocytes. PT stain. Medium magnification.

Medulla

The medulla is composed of the **medullary cords** (Figure 14–25), which are branched extensions of the inner cortex that contain B lymphocytes and some plasma cells. The medullary cords are separated by dilated, capillary-like structures called **medullary lymphoid sinuses** (Figure 14–25). These sinuses are irregular spaces containing lymph; like the subcapsular and intermediate sinuses, they are partially lined by reticular cells and macrophages. Reticular cells and fibers (Figures 14–26 and 14–27) frequently bridge the sinus in a loose network.

MEDICAL APPLICATION

The functions of B and T lymphocytes are most notable in immunodeficiency diseases, which are caused by defects in the B cells, T cells, or both. Figure 14–28 illustrates the correlation between pathologic conditions and changes in the lymph nodes.

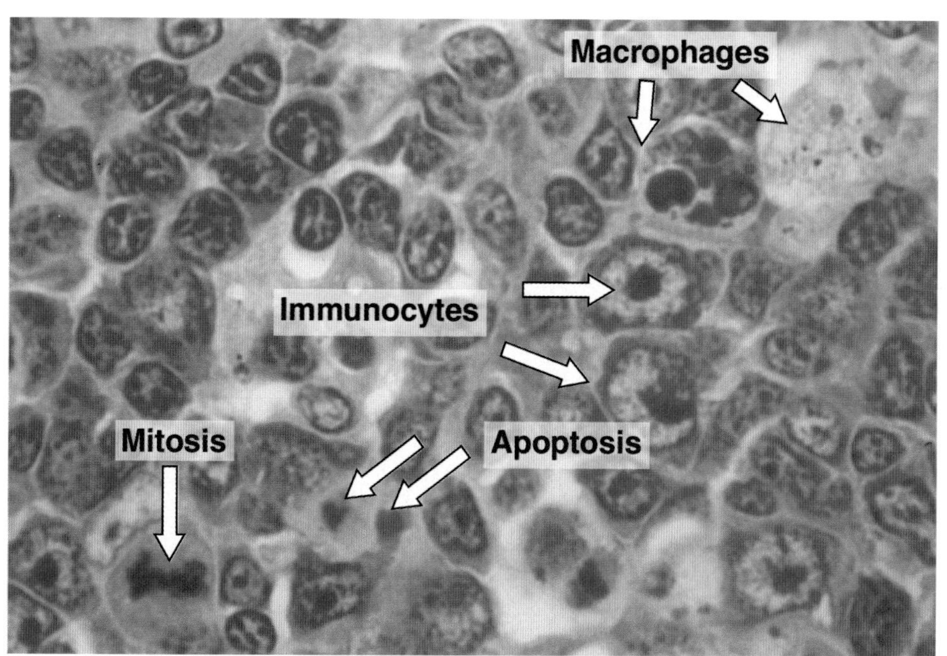

Figure 14–24. Photomicrograph of the cortex of a lymph node activated by injection of an antigen. The main cells are macrophages, which capture and process antigens, and activated B lymphocytes, also called immunocytes. Immunocytes are large cells with basophilic cytoplasm, and large nucleoli that are in the process of multiplication to produce antibodysecreting plasma cells. Some cells are dying by apoptosis (arrows). PT stain. High magnification.

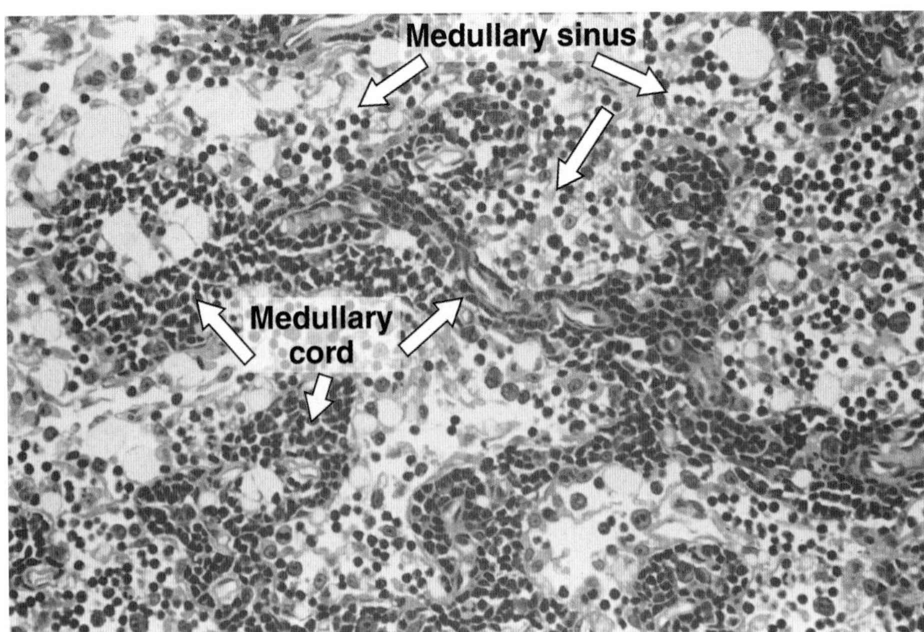

Figure 14–25. Photomicrograph of the medulla of a lymph node; the medullary sinuses are separated by medullary cords. PT stain. Medium magnification.

Lymph & Blood Circulation

Afferent lymphatic vessels cross the capsule of each node and pour lymph into the subcapsular sinus. From there, lymph passes through intermediate sinuses that run parallel to the trabeculae of the capsule and into the interior of the node, where they reach the medullary sinuses. The complex architecture of both the subcapsular and the medullary sinuses slows the flow of lymph through the node, facilitating the uptake and digestion of foreign materials by macrophages and dendritic cells. These antigen-presenting cells are characterized by abundant cytoplasmic processes. They are formed in the bone marrow and are carried by the circulation to lymph nodes. Lymph that infiltrates the node flows slowly from cortex to medulla and is collected by **efferent lymphatic vessels** at the hilum. Valves in both the afferent and efferent vessels aid the unidirectional flow of lymph (Figure 14–20).

Penetration of blood vessels into lymph nodes is limited to small arteries that enter at the hilum and form capillaries in the lymphoid nodules. In the nodules, small veins originate and exit at the hilum (Figure 14–20).

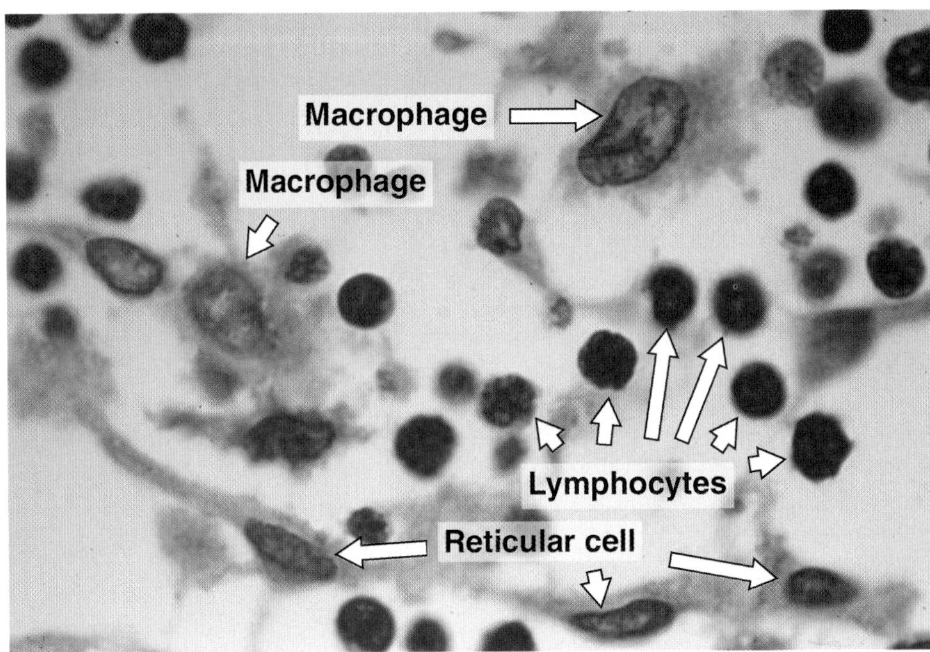

Figure 14–26. Photomicrograph of a medullary sinus of a lymph node. The main cells are macrophages, reticular cells, and lymphocytes. PT stain. High magnification.

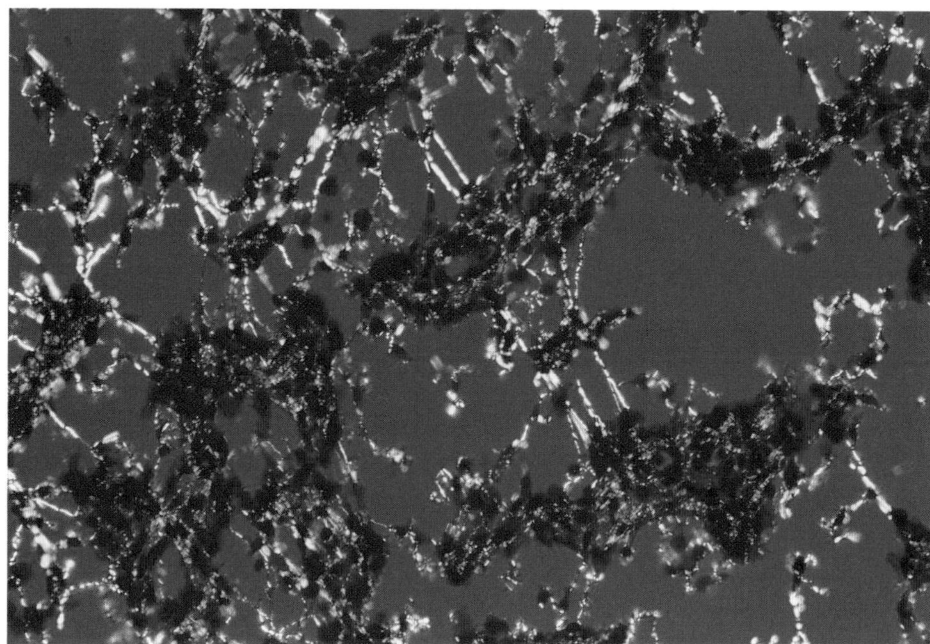

Figure 14–27. Section of the medullary region of a lymph node stained with picrosirius and photographed under polarized light. Note the abundance of reticular fibers made of collagen type III. Medium magnification.

Histophysiology

Lymph flows through the lymph nodes to be cleared of foreign particles before its return to the blood circulation. Because the nodes are distributed throughout the body, lymph formed in tissues must cross at least one node before entering the bloodstream.

MEDICAL APPLICATION

Each node receives lymph from, and is said to be a ***satellite node*** *of, a limited region of the body. Malignant tumors often metastasize via these nodes.*

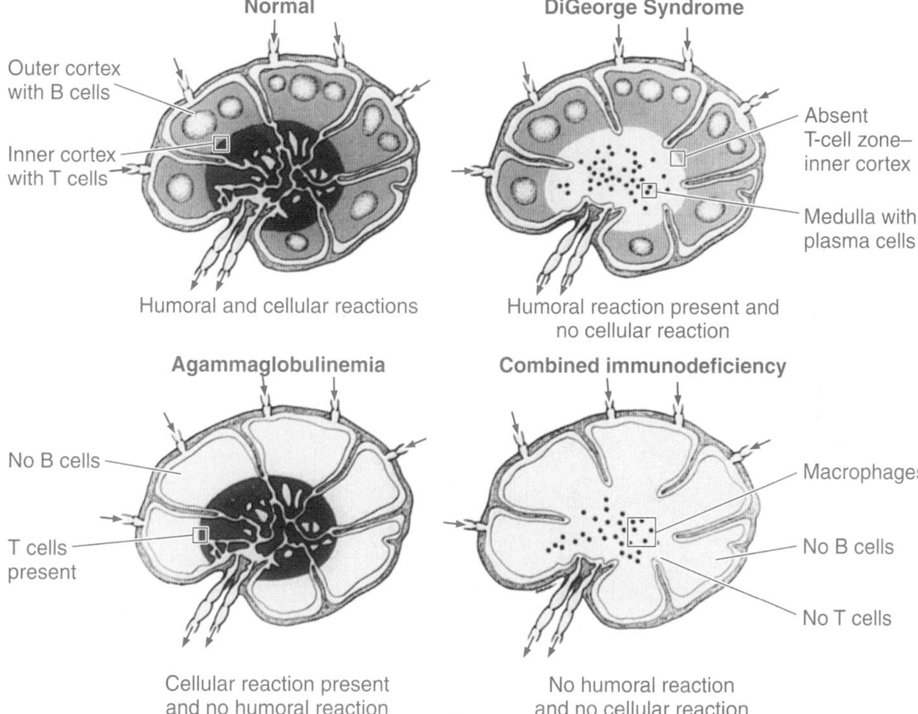

Figure 14–28. Pathologic alterations in lymph nodes related to deficiency of B cells, T cells, or both. (Redrawn, with permission, from Chandrasoma P, Taylor CR: *Concise Pathology.* Appleton & Lange, 1991.)

As lymph flows through the sinuses, 99% or more of the antigens and other debris are removed by the phagocytotic activity of macrophages. Infection and antigenic stimulation cause the affected lymph nodes to enlarge and form multiple germinal centers with active cell proliferation. Although plasma cells constitute only 1–3% of the cell population in resting nodes, their numbers increase greatly in lymph nodes that are stimulated (eg, by inflammation), and they partially account for the enlargement of those structures.

Recirculation of Lymphocytes: A Communication System

Lymphocytes leave the lymph nodes by efferent lymphatic vessels and eventually reach the bloodstream; all lymph formed in the body drains back into the blood. Lymphocytes return to the lymph nodes by leaving the blood through specific blood vessels, the **postcapillary,** or **high endothelial, venules** (Figure 14–29). These venules have an unusual endothelial lining of tall cuboidal cells, and lymphocytes are capable of traveling between them. Some lymphocytes are long-living cells and can recirculate many times in this way. High endothelial venules are also present in other lymphoid organs, such as the appendix, tonsils, and Peyer's patches, but not in the spleen. Although recirculation of lymphocytes also occurs in several places, it is most prominent in the lymph nodes.

The homing behavior of lymphocytes is due to complementary molecules on their surface and on the tall endothelial cells of postcapillary venules. Through recirculation, locally stimulated lymphocytes (eg, in an infected finger) from satellite lymph nodes will inform other lymphoid organs and prepare the organism for a generalized immune response against the infection. The continuous recirculation of lymphocytes results in a constant monitoring of all parts of the body by cells that inform the immune system of the presence of foreign antigens. During passage through the lymphoid organs, lymphocytes meet antigens on the membrane of APCs that have migrated there from infected sites.

MEDICAL APPLICATION

Figure 14–28 shows how deficiencies of B and T lymphocytes affect the structure of lymph nodes and the corresponding effects on immunologic reactions.

SPLEEN

The spleen is the largest accumulation of lymphoid tissues in the body. Because of its abundance of phagocytotic cells and the close contact between these cells and circulating blood, the spleen is an important defense against microorganisms that penetrate the circulation. It is also the site of destruction of aged erythrocytes. As is true of all other lymphoid organs, the spleen is a production site for activated lymphocytes, which pass into the blood. The spleen reacts promptly to antigens carried in the blood and is an important immunologic blood filter and antibody-forming organ.

General Structure

A capsule of dense connective tissue that sends out trabeculae that divide the parenchyma, or splenic pulp, into incomplete compartments (Figure 14–30) surrounds the spleen. At the hilum on the medial surface of the spleen, the capsule gives rise to a number of trabeculae that carry nerves and arteries into the splenic pulp. Veins derived from the parenchyma and lymphatic vessels that originate in the trabeculae leave through the hilum. The splenic pulp has no lymphatic vessels.

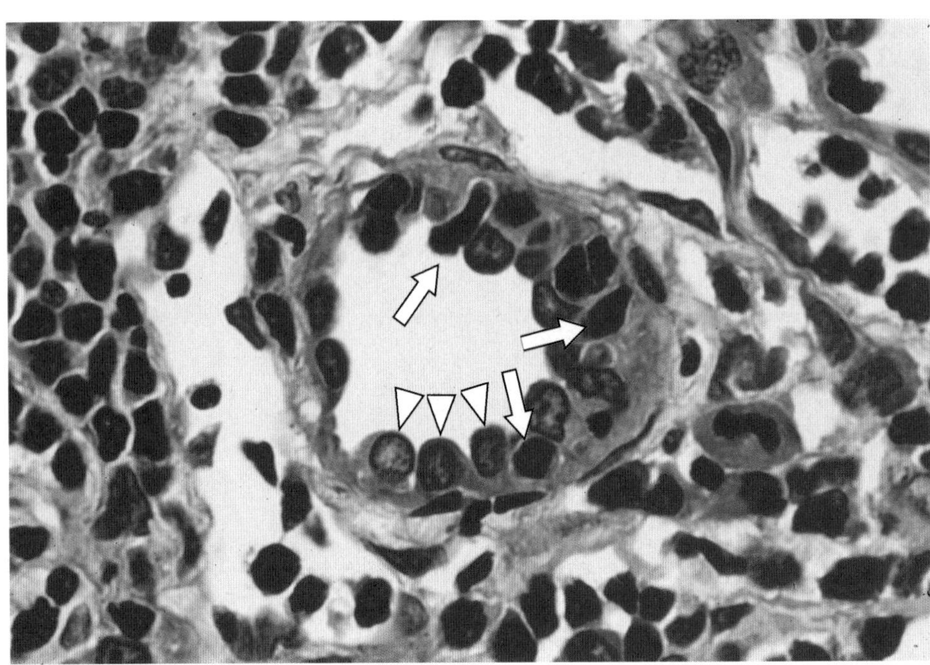

Figure 14–29. Photomicrograph of a high endothelial venule in a lymph node. Arrowheads indicate high endothelial cells. The venule is crossed by lymphocytes (arrows). PT stain. High magnification.

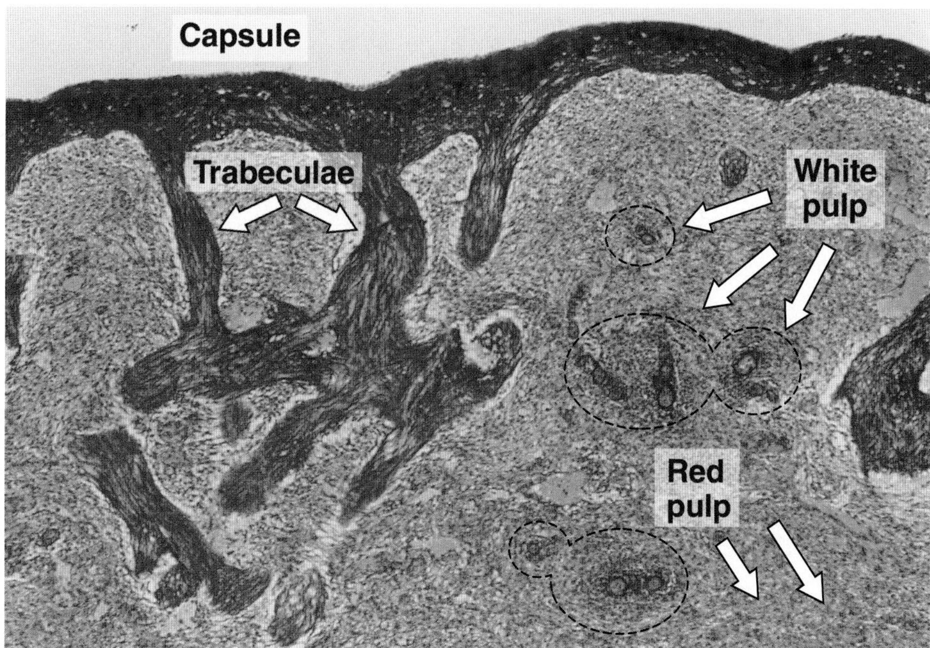

Figure 14–30. Section of spleen showing its capsule sending trabeculae to the interior of the organ. Note the white pulp with its arterioles. The red pulp occupies most of the microscopic field. Picrosirius stain. Low magnification.

In humans, the connective tissue of the capsule and trabeculae contains only a few smooth muscle cells. The spleen, like other lymphoid structures, is formed of a network of reticular tissue that contains lymphoid cells, macrophages, and APCs.

Splenic Pulp

On the surface of a cut through an unfixed spleen, one can observe white spots in the parenchyma. These are lymphoid nodules and are part of the **white pulp.** The nodules appear within the **red pulp,** a dark red tissue that is rich in blood (Figures 14–30 and 14–31). Examination under a low-power microscope reveals that the red pulp is composed of elongated structures, the splenic cords (**Billroth's cords**), that lie between the sinusoids (Figure 14–34).

Blood Circulation

The splenic artery divides as it penetrates the hilum, branching into **trabecular arteries,** vessels of various sizes that follow the course of the connective tissue trabeculae. When they leave the trabeculae to enter the parenchyma, the arteries are immediately enveloped by a sheath of T lymphocytes called the **periarterial lymphatic sheath** (**PALS**). These vessels are known as **central arteries** or **white pulp arteries** (Figure 14–32). Although the lymphocytic sheath (white pulp) thickens along its course to form a number of lymphoid nodules in which the vessel occupies an eccentric position (Figures 14–31 and 14–32), the vessel is still called the central artery. During its course through the white pulp, the artery also divides into numerous

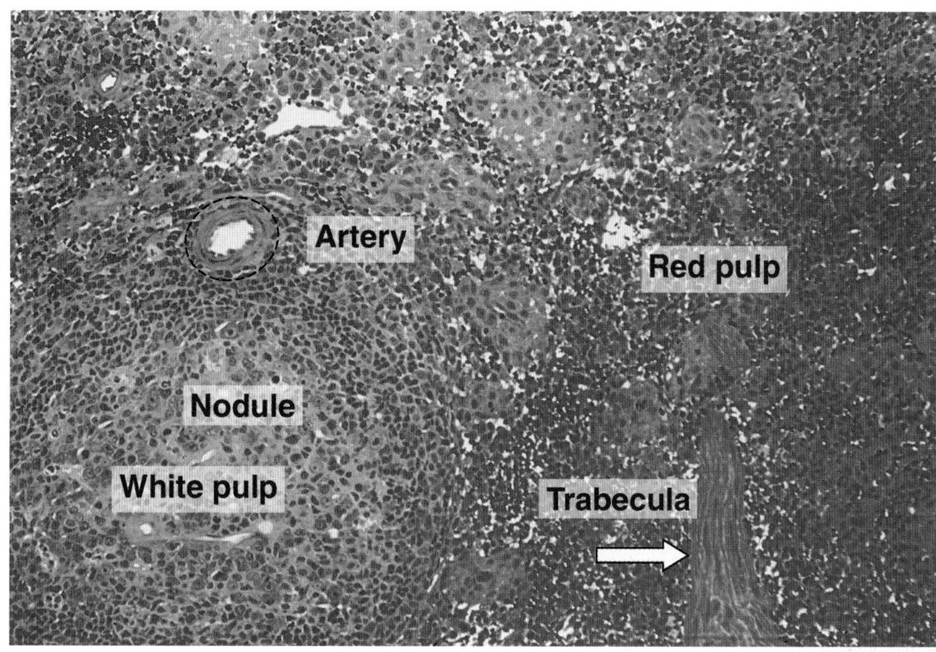

Figure 14–31. Section of spleen. At left is a region of white pulp with its nodule and arteriole. At right are the red pulp and a trabecula of connective tissue. PT stain. Low magnification.

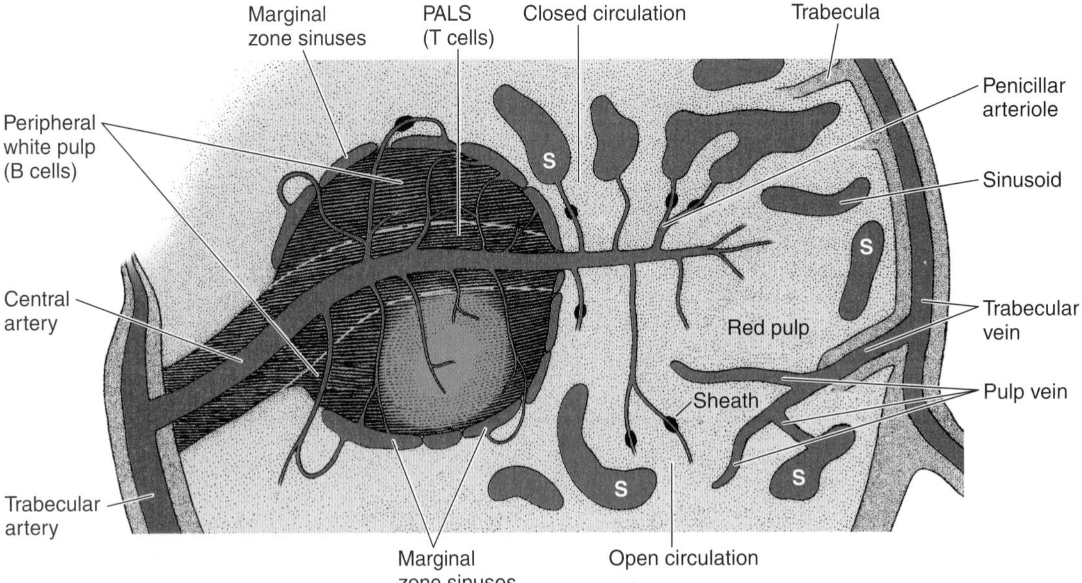

Figure 14–32. Schematic view of the blood circulation of the spleen. Theories of open and closed circulation are represented. Splenic sinuses (S) are indicated. PALS, periarterial lymphatic sheath. (Redrawn and reproduced, with permission, from Greep RO, Weiss L: *Histology,* 3rd ed. McGraw-Hill, 1973.)

radial branches that supply the surrounding lymphoid tissue (Figure 14–32).

After leaving the white pulp, the central artery subdivides to form straight **penicillar arterioles** with an outside diameter of approximately 24 μm. Near their termination, some of the penicillar arterioles are surrounded by a sheath of reticular cells, lymphoid cells, and macrophages.

Beyond the sheath, the vessels continue as simple arterial capillaries that carry blood to the sinusoids (red pulp sinuses). These sinusoids occupy the spaces between the red pulp cords. The manner in which blood flows from the arterial capillaries of the red pulp to the interior of the sinusoids has not yet been completely explained. Some investigators suggest that the capillaries open directly into the sinusoids, and others maintain that the blood passes through the spaces between the red pulp cord cells and then moves on to be collected by the sinusoids (Figure 14–33). The first theory suggests a **closed circulation;** ie, the blood always remains inside the vessels. According to the second theory, the circulation would open into the parenchyma of the red pulp (Billroth's cords), and the blood would pass through the space between the cells to reach the sinusoids (**open circulation**). Current evidence suggests that blood circulation in the human spleen is of the open type.

From the sinusoids, blood proceeds to the red pulp veins that join together and enter the trabeculae, forming the **trabecular veins** (Figure 14–32). The splenic vein originates from these vessels and emerges from the hilum of the spleen. The trabecular veins do not have individual muscle walls; their walls are composed of trabecular tissue. They can be considered channels hollowed out in the trabecular connective tissue and lined by endothelium.

White Pulp

White pulp consists of lymphoid tissue that sheathes both the central arteries and the lymphoid nodules appended to the sheaths. The lymphoid cells surrounding the central arteries are mainly T lymphocytes and form the PALS (Figure 14–32). Lymphoid nodules consist mainly of B lymphocytes.

Between the white pulp and the red pulp lies a **marginal zone** consisting of many sinuses and loose lymphoid tissue. Few lymphocytes but many active macrophages can be found there. The marginal zone contains an abundance of blood antigens and thus plays a major role in the immunologic activities of the spleen.

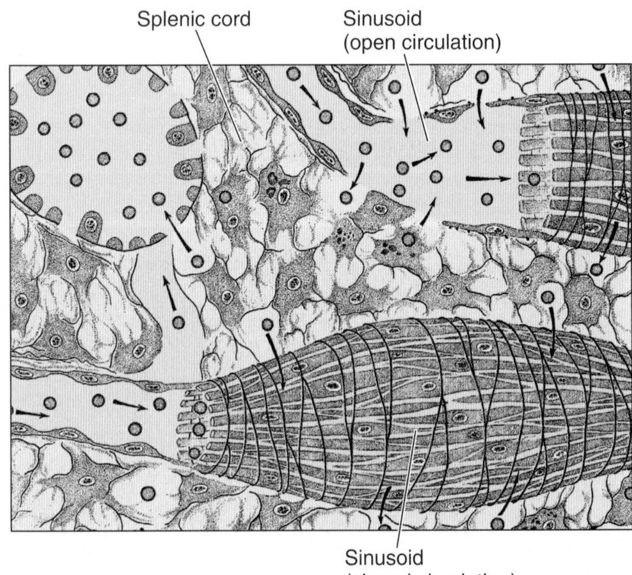

Figure 14–33. Structure of the red pulp of the spleen, showing splenic sinusoids and splenic cords with reticular cells and macrophages (some macrophages contain ingested material). The disposition of the reticular fibers in the red pulp is illustrated. In the splenic cords they form a 3-dimensional network; in the sinusoids they are mainly perpendicular to the long axis of the sinusoid. Both the open and closed theories of circulation are illustrated. Arrows indicate blood flow and options for movement of blood cells.

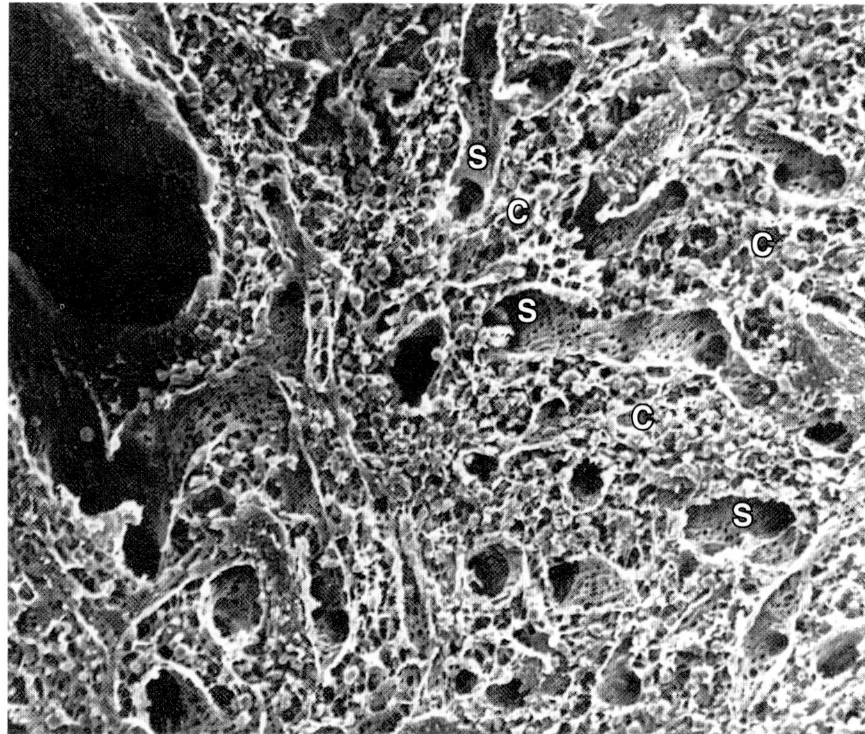

Figure 14–34. **General view of splenic red pulp with a scanning electron microscope. Note the sinusoids (S) and the splenic cords (C). ×360.** (Reproduced, with permission, from Miyoshi M, Fujita T: Stereo-fine structure of the splenic red pulp. A combined scanning and transmission electron microscope study on dog and rat spleen. Arch Histol Jpn 1971; 33:225.)

The lymphocytes of the central portion of the PALS are **thymus-dependent,** whereas the marginal zones and the nodules—the **peripheral white pulp—**are populated by B lymphocytes.

Red Pulp

The red pulp contains splenic cords and sinusoids (Figures 14–34 and 14–35). The splenic cords are composed of a loose network of reticular cells supported by reticular fibers (collagen type III). In addition, the splenic cords contain macrophages, T and B lymphocytes, plasma cells, and many blood cells (erythrocytes, platelets, and granulocytes).

Elongated endothelial cells line the sinusoids of the spleen with long axes parallel to the long axes of the sinusoids. These cells are enveloped in reticular fibers set mainly in a transverse direction, much like the hoops on a barrel (Figure 14–33).

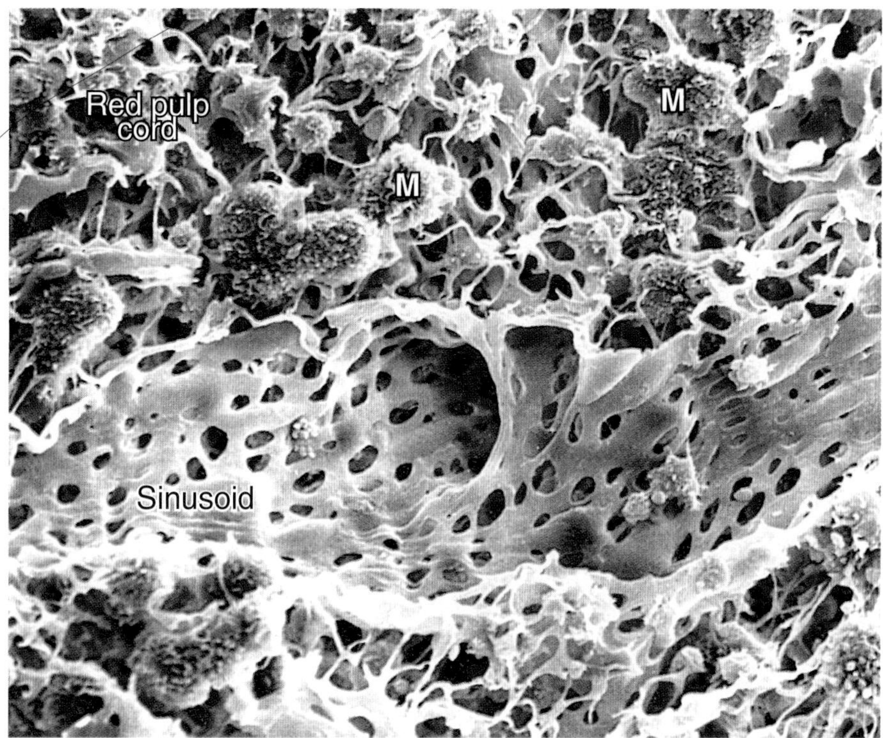

Figure 14–35. **Scanning electron micrograph of the red pulp of the spleen showing sinusoids, red pulp cords, and macrophages (M). Note the multiple fenestrations in the endothelial cells of the sinusoids. ×1600.** (Reproduced, with permission, from Miyoshi M, Fujita T: Stereo-fine structure of the splenic red pulp. A combined scanning and transmission electron microscope study on dog and rat spleen. Arch Histol Jpn 1971; 33:225.)

The transverse and longitudinal fibers join to form a network enveloping the sinusoid cells and macrophages that occupy the spaces between neighboring endothelial cells. Around the sinusoid is an incomplete basal lamina.

Because the spaces between endothelial cells of the splenic sinusoids are 2–3 μm in diameter or smaller (Figure 14–35), only flexible cells are able to pass easily from the red pulp cords to the lumen of the sinusoids.

As already stated, the secondary lymphoid structures have regions that are populated by cells of the T or B family.

Histophysiology

The best-known functions of the spleen are the production of lymphocytes, the destruction of erythrocytes (Figure 14–36), the defense of the organism against invaders that enter the bloodstream, and the storage of blood.

Production of Lymphocytes

The white pulp of the spleen produces lymphocytes that migrate to the red pulp and reach the lumens of the sinusoids, where they are incorporated into the blood.

MEDICAL APPLICATION

In certain pathologic conditions (eg, leukemia), the spleen may reinitiate the production of granulocytes and erythrocytes, a function present during fetal life, and undergo a process known as ***myeloid metaplasia*** *(the occurrence of myeloid tissues in extramedullary sites).*

Destruction of Erythrocytes

Erythrocytes have an average life span of 120 days, after which they are destroyed, mainly in the spleen. A reduction in their flexibility and changes in their membrane seem to be the signals for destruction of erythrocytes. Degenerating erythrocytes are also removed in the bone marrow.

Macrophages in the splenic cords engulf and digest the erythrocytes that frequently fragment in the extracellular space. The hemoglobin they contain is broken down into several parts. The protein, globin, is hydrolyzed to amino acids that are reused in protein synthesis. Iron is released from heme and, together with transferrin, is transported in blood to the bone marrow, where it is reused in erythropoiesis. Iron-free heme is metabolized to **bilirubin,** which is excreted in the bile by liver cells. After surgical removal of the spleen (splenectomy), there is an increase in abnormal erythrocytes, seen to have deformed shapes in blood smears. There is also an increase in the number of blood platelets, suggesting that the spleen normally removes aged platelets.

Defense Against Invaders

Because it contains both B and T lymphocytes as well as APCs and phagocytic cells, the spleen is important in the immune defense of the body. In the same way that lymph nodes are a filter for the lymph, the spleen is a filter for the blood. Of all the phagocytotic cells of the organism, those of the spleen are most active in the phagocytosis of living organisms (bacteria and viruses) and inert particles that find their way into the bloodstream.

MUCOSA-ASSOCIATED LYMPHOID TISSUE

The digestive, respiratory, and genitourinary tracts are common sites of microbial invasion because their lumens are open to the external environment. To protect the organism, lymphoid aggregates (lymphoid nodules) and diffuse collections of lymphoid tissue (Figures 14–37 through 14–40) occur in the mucosa and submucosa of these tracts, in some places forming conspicuous structures such as the tonsils and the Peyer's patches of the small intestine (see Chapter 15).

The skin also contains many cells of the immune system (lymphocytes, macrophages, Langerhans cells). The lymphoid tissue of the skin and mucosa constitute an efficient system located in a key position to protect the organism against environmental pathogens.

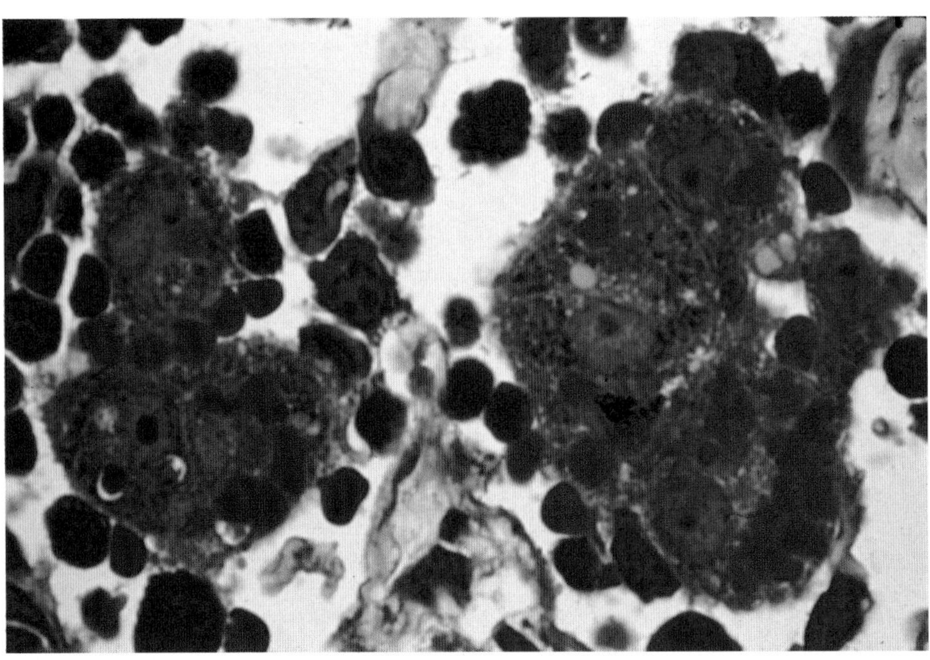

Figure 14–36. Photomicrograph of 5 spleen macrophages in active phagocytosis of erythrocytes. Note that the erythrocytes in the cytoplasm of the macrophages are in different stages of degradation. PT stain. High magnification.

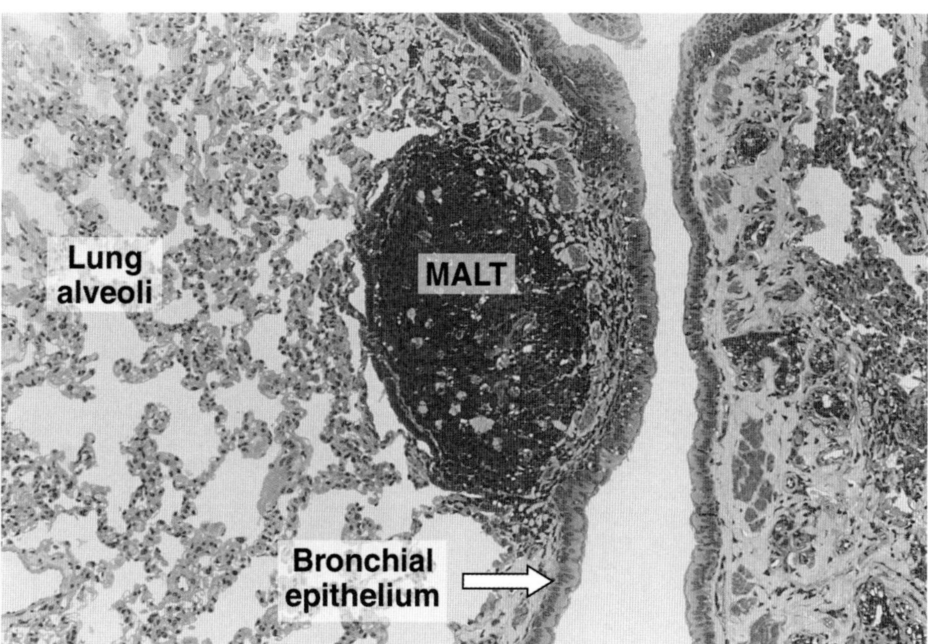

Figure 14–37. Section of lung showing a collection of lymphocytes in the connective tissue of the bronchiolar mucosa, an example of mucosa-associated lymphoid tissue (MALT). PT stain. Low magnification.

TONSILS

Tonsils are organs composed of aggregates of incompletely encapsulated lymphoid tissues that lie beneath, and in contact with, the epithelium of the initial portion of the digestive tract. Depending on their location, tonsils in the mouth and pharynx are called **palatine, pharyngeal,** or **lingual tonsils.** They produce lymphocytes, many of which infiltrate the epithelium.

Palatine Tonsils

The two palatine tonsils are located in the lateral walls of the oral part of the pharynx. Under the squamous stratified epithelium, the dense lymphoid tissue in these tonsils forms a band that contains lymphoid nodules, generally with germinal centers. Each tonsil has 10–20 epithelial invaginations that penetrate the parenchyma deeply, forming **crypts,** whose lumens contain desquamated epithelial cells, live and dead lymphocytes, and bacteria. Crypts may appear as purulent spots in tonsillitis. Separating the lymphoid tissue from subjacent structures is a band of dense connective tissue, the **capsule** of the tonsil. This capsule usually acts as a barrier against spreading tonsillar infections.

Pharyngeal Tonsil

The pharyngeal tonsil is a single tonsil situated in the superior-posterior portion of the pharynx. It is covered by ciliated

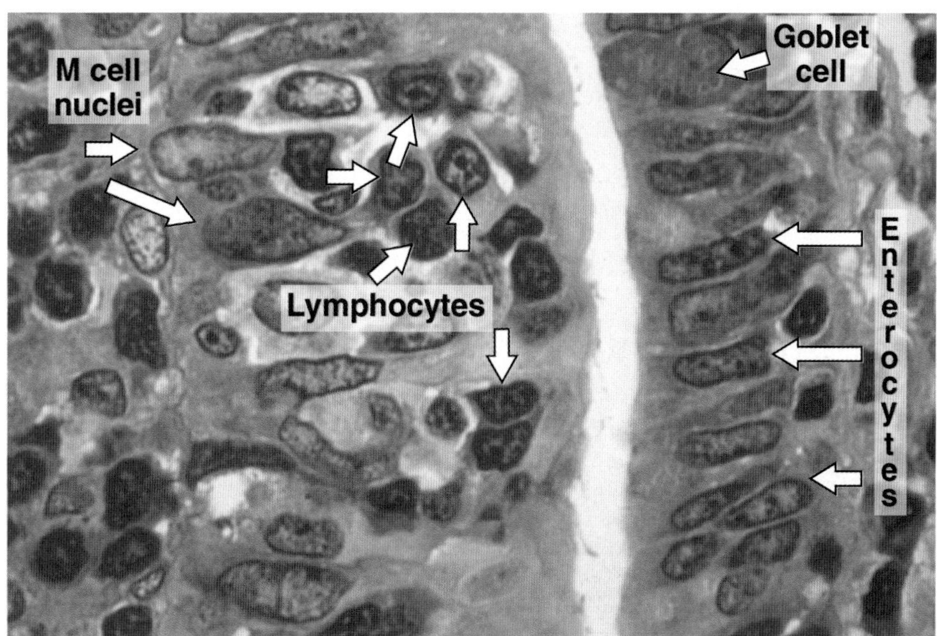

Figure 14–38. Section of Peyer's patch of the small intestine showing the epithelial covering of enterocytes and goblet cells (right), the intestinal lumen (center), and the covering of the patch with a row of M cells and groups of lymphocytes (left). The small dark nuclei belong to B and T lymphocytes, and the large pale-stained nuclei belong to M cells. PT stain. Medium magnification.

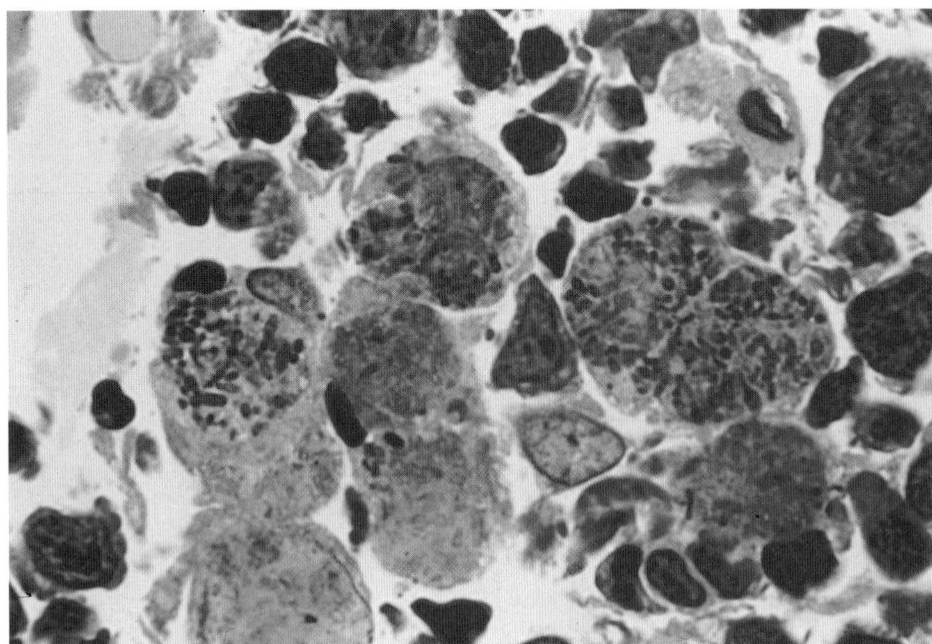

Figure 14–39. Lymph nodule of a Peyer's patch. Note a group of macrophages containing bacteria in various stages of digestion. Lymphocytes are seen around the macrophages. PT stain. High magnification.

pseudostratified columnar epithelium typical of the respiratory tract, and areas of stratified epithelium can also be observed.

The pharyngeal tonsil is composed of pleats of mucosa and contains diffuse lymphoid tissue and nodules. It has no crypts, and its capsule is thinner than those of the palatine tonsils.

Hypertrophy of the pharyngeal tonsil resulting from chronic inflammation is called **adenoids.**

Lingual Tonsils

The lingual tonsils are smaller and more numerous than the palatine and pharyngeal tonsils. They are situated at the base of the tongue (see Chapter 15) and are covered by stratified squamous epithelium. Each lingual tonsil has a single crypt.

REFERENCES

Abbas AK et al: *Cellular and Molecular Immunology,* 2nd ed. Saunders, 2000.

Alberts B et al: The immune system. In: *Molecular Biology of the Cell,* 3rd ed. Garland, 1994.

Austyn JM, Wood KJ: *Principles of Cellular and Molecular Immunology.* Oxford Univ Press, 1993.

Cella M et al: Origin, maturation and antigen presenting function of dendritic cells. Curr Opin Immunol 1997;9:10.

Parslow TG et al: *Medical Immunology,* 10th ed. McGraw-Hill, 2001.

Rajewsky K: B-cell differentiation: clonal selection and learning in the antibody system. Nature 1996;381:751.

Sainte-Marie G, Peng FS: High endothelial venules of the rat lymph node, a review and a question: Is their activity antigen specific? Anat Rec 1996;245:593.

Tough DF, Sprent J: Lifespan of lymphocytes. Immunol Res 1995;14:252.

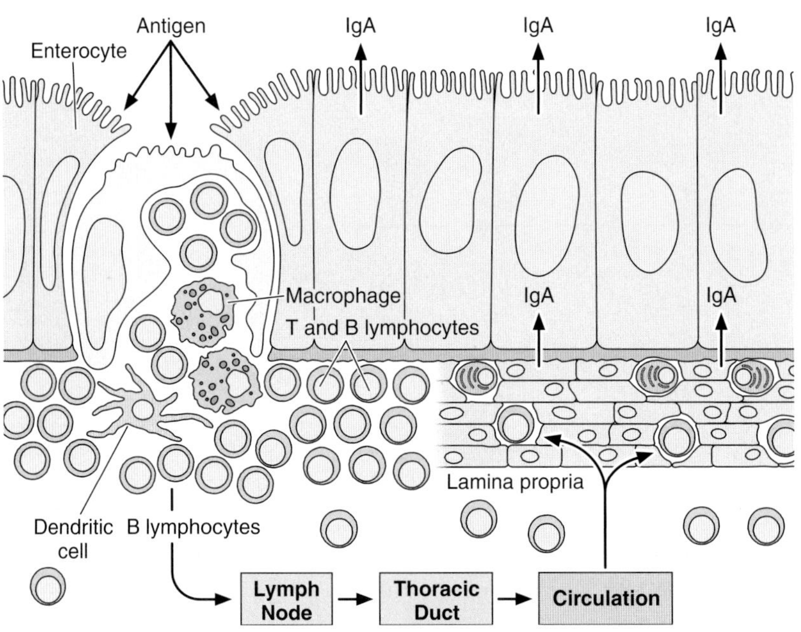

Figure 14–40. General view of the mucosa immunity in the intestine. Luminal antigens are captured by dome-shaped M cells present in the covering of Peyer's patches and transported to subjacent lymphocytes, macrophages, and dendritic cells. Macrophages and dendritic cells migrate to neighboring lymph nodes, where they stimulate B and T lymphocytes, which then enter the lymphatic circulation and later the blood circulation (lymph flows to the blood). The stimulated lymphocytes home in other tissues, including the mucosa lamina propria, where plasma cells produce considerable amounts of IgA. The lymphoid cells of the lamina propria of intestinal mucosa are a major antibody producer, because of their extension and close contact with antigens introduced into the digestive tract.

Digestive Tract 15

The digestive system consists of the digestive tract—oral cavity, esophagus, stomach, small and large intestines, rectum, and anus—and its associated glands—salivary glands, liver, and pancreas. Its function is to obtain from ingested food the molecules necessary for the maintenance, growth, and energy needs of the body. Large molecules such as proteins, fats, complex carbohydrates, and nucleic acids are broken down into small molecules that are easily absorbed through the lining of the digestive tract, mostly in the small intestine. Water, vitamins, and minerals are also absorbed from ingested food. In addition, the inner layer of the digestive tract is a protective barrier between the content of the tract's lumen and the internal milieu of the body.

The first step in the complex process known as digestion occurs in the mouth, where food is moistened by saliva and ground by the teeth into smaller pieces; saliva also initiates the digestion of carbohydrates. Digestion continues in the stomach and small intestine, where the food—transformed into its basic components (eg, amino acids, monosaccharides, free fatty acids, monoglycerides)—is absorbed. Water absorption occurs in the large intestine, causing the undigested contents to become semisolid.

GENERAL STRUCTURE OF THE DIGESTIVE TRACT

The entire gastrointestinal tract presents certain common structural characteristics. It is a hollow tube composed of a lumen whose diameter varies, surrounded by a wall made up of 4 principal layers: the **mucosa, submucosa, muscularis,** and **serosa.** The structure of these layers is summarized below and illustrated in Figure 15–1.

The **mucosa** comprises an **epithelial lining;** a **lamina propria** of loose connective tissue rich in blood and lymph vessels and smooth muscle cells, sometimes also containing glands and lymphoid tissue; and the **muscularis mucosae,** usually consisting of a thin inner circular layer and an outer longitudinal layer of smooth muscle cells separating the mucosa from the submucosa. The mucosa is frequently called a **mucous membrane.**

The **submucosa** is composed of dense connective tissue with many blood and lymph vessels and a **submucosal** (also called **Meissner's) nerve plexus.** It may also contain glands and lymphoid tissue.

The **muscularis** contains smooth muscle cells that are spirally oriented and divided into 2 sublayers according to the main direction the muscle cells follow. In the internal sublayer (close to the lumen), the orientation is generally circular; in the external sublayer, it is mostly longitudinal. The muscularis also contains the **myenteric** (or **Auerbach's**) **nerve plexus,** which lies between the two muscle sublayers, and blood and lymph vessels in the connective tissue between the muscle sublayers.

The **serosa** is a thin layer of loose connective tissue, rich in blood and lymph vessels and adipose tissue, and a simple squamous covering epithelium (**mesothelium**). In the abdominal cavity, the serosa is continuous with the mesenteries (thin membranes covered by mesothelium on both sides), which support the intestines, and with the peritoneum, a serous membrane that lines the cavity wall. In places where the digestive organ is bound to other organs or structures, however, the serosa is replaced by a thick adventitia, consisting of connective tissue containing vessels and nerves, without the mesothelium.

The main functions of the epithelial lining of the digestive tract are to provide a selectively permeable barrier between the contents of the tract and the tissues of the body, to facilitate the transport and digestion of food, to promote the absorption of the products of this digestion, and to produce hormones that affect the activity of the digestive system. Cells in this layer produce mucus for lubrication and protection.

The abundant lymphoid nodules in the lamina propria and the submucosal layer protect the organism (in association with the epithelium) from bacterial invasion. The necessity for this immunologic support is obvious, because the entire digestive tract—with the exception of the oral cavity, esophagus, and anal canal—is lined by a simple thin, vulnerable epithelium. The lamina propria, located just below the epithelium, is a zone rich in macrophages and lymphoid cells, some of which actively produce antibodies. These antibodies are mainly immunoglobulin A (IgA) and are bound to a secretory protein produced by the epithelial cells of the intestinal lining and secreted into the intestinal lumen. This complex protects against viral and bacterial invasion.

The IgA present in the respiratory, digestive, and urinary tracts is resistant to proteolytic enzymes and can therefore coexist with the proteases present in the lumen.

The muscularis mucosae promotes the movement of the mucosa independent of other movements of the digestive tract, increasing its contact with the food. The contractions of the muscularis, generated and coordinated by nerve plexuses, propel and mix the food in the digestive tract. These plexuses are composed mainly of nerve cell aggregates (multipolar visceral neurons) that form small parasympathetic ganglia. A rich network of pre- and postganglionic fibers of the autonomic nervous system and some visceral sensory fibers in these ganglia permit communication between them. The number of these ganglia along the digestive tract is variable; they are more numerous in regions of greatest motility.

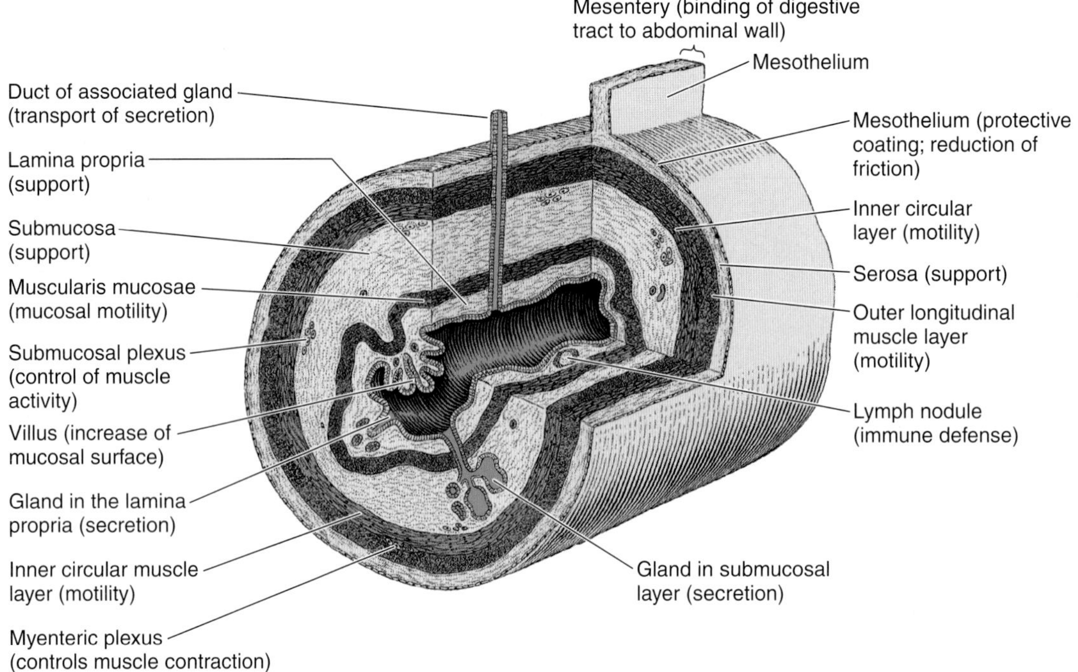

Figure 15–1. Schematic structure of a portion of the digestive tract with various components and their functions. (Redrawn and reproduced, with permission, from Bevelander G: *Outline of Histology,* 7th ed. Mosby, 1971.)

MEDICAL APPLICATION

*In certain diseases, such as **Hirschsprung disease** (congenital megacolon) or **Chagas' disease** (**Trypanosoma cruzi** infection), the plexuses in the digestive tract are severely injured and most of their neurons are destroyed. This results in disturbances of digestive tract motility, with frequent dilatations in some areas. The abundant innervation from the autonomic nervous system that the digestive tract receives provides an anatomic explanation of the widely observed action of emotional stress on the digestive tract—a phenomenon of importance in psychosomatic medicine.*

THE ORAL CAVITY

The oral cavity is lined with stratified squamous epithelium, keratinized or nonkeratinized, depending on the region. The keratin layer protects the oral mucosa from damage during masticatory function and is present mostly in the gingiva (gum) and hard palate. The lamina propria in these regions has several papillae and rests directly on bony tissue Nonkeratinized squamous epithelium covers the soft palate, lips, cheeks, and the floor of the mouth. The lamina propria has papillae, similar to those in the dermis of the skin, and is continuous with a submucosa containing diffuse small salivary glands. In the lips, a transition from the oral nonkeratinized epithelium to the keratinized epithelium of the skin can be observed.

The soft palate has a core of skeletal muscle, numerous mucous glands, and lymphoid nodules in its submucosa.

Tongue

The tongue is a mass of striated muscle covered by a mucous membrane whose structure varies according to the region. The muscle fibers cross one another in 3 planes; they are grouped in bundles, usually separated by connective tissue. Because the connective tissue of the lamina propria penetrates the spaces between the muscular bundles, the mucous membrane is strongly adherent to the muscle. The mucous membrane is smooth on the lower surface of the tongue. The tongue's dorsal surface is irregular, covered anteriorly by a great number of small eminences called **papillae.** The posterior one third of the dorsal surface of the tongue is separated from the anterior two thirds by a V-shaped boundary. Behind this boundary, the surface of the tongue shows small bulges composed mainly of 2 types of small lymphoid aggregations: small collections of lymphoid nodules and the lingual tonsils, where lymphoid nodules aggregate around invaginations (crypts) of the mucous membrane (Figure 15–2).

Papillae

Papillae are elevations of the oral epithelium and lamina propria that assume various forms and functions. There are 4 types (see Figure 15–2):

Filiform Papillae

Filiform papillae have an elongated conical shape; they are quite numerous and are present over the entire surface of the tongue.

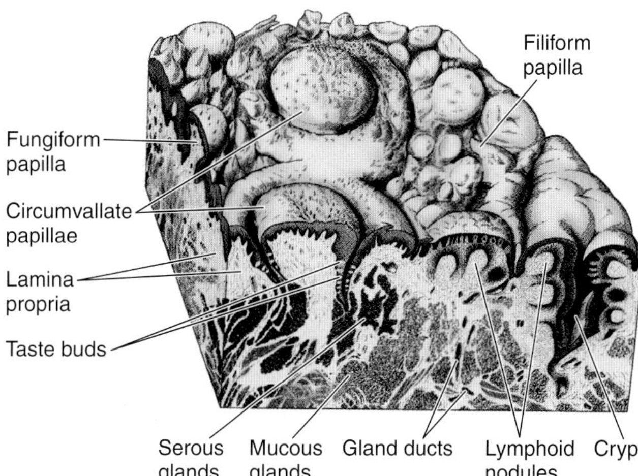

Figure 15–2. Surface of the tongue on the region close to its V-shaped boundary, between the anterior and posterior portions. Note the lymphoid nodules (lingual tonsil), glands, and papillae.

Their epithelium, which does not contain taste buds, is keratinized.

Fungiform Papillae

Fungiform papillae resemble mushrooms in that they have a narrow stalk and a smooth-surfaced, dilated upper part. These papillae, which contain scattered taste buds on their upper surfaces, are irregularly interspersed among the filiform papillae.

Foliate Papillae

Foliate papillae are poorly developed in humans. They consist of 2 or more parallel ridges and furrows on the dorsolateral surface of the tongue and contain many taste buds.

Circumvallate Papillae

Circumvallate papillae are 7–12 extremely large circular papillae whose flattened surfaces extend above the other papillae. They are distributed in the V region in the posterior portion of the tongue. Numerous serous (von Ebner's) glands drain their contents into the deep groove that encircles the periphery of each papilla. This moatlike arrangement provides a continuous flow of fluid over the great number of taste buds present along the sides of these papillae. The glands also secrete a lipase that probably prevents the formation of a hydrophobic layer over the taste buds that would hinder their function. This flow of secretions is important in removing food particles from the vicinity of the taste buds so that they can receive and process new gustatory stimuli. Along with this local role, lingual lipase is active in the stomach and can digest up to 30% of dietary triglycerides. Other small mucous salivary glands dispersed throughout the lining of the oral cavity act in the same way as the serous glands associated with this type of papilla to prepare the taste buds in other parts of the oral cavity, such as the soft palate and anterior portion of the tongue, to respond to taste stimuli.

There are at least 4 qualities in human taste perception: saltiness, sourness, sweetness, and bitterness. All qualities can be elicited from all the regions of the tongue that contain taste buds, specialized structures that contain the taste cells, the detectors of tastants (substances capable of eliciting taste). Taste buds are onion-shaped structures, each one containing 50–100 cells. The bud rests in the basal lamina, and in its apical portion the taste cells project microvilli that poke through an opening called the taste pore. Most of the cells are actually the taste cells, while others have a supportive function, secreting an amorphous material that surrounds the microvilli in the taste pore. Undifferentiated basal cells are responsible for the replacement of all the cell types. Tastants dissolved in saliva contact the taste cells through the pore, interacting with taste receptors (sweet and bitter tastes) or ion channels (salty and sour tastes) on the surface of the cells. The result is a depolarization of the taste cells, leading to the release of neurotransmitters which will, in turn, stimulate afferent nerve fibers connected to the taste cells (Figure 15–3). This information will be processed by central gustatory neurons. It is believed that each tasting stimulus generates a unique pattern of activity across a large set of neurons, which explains taste discrimination. The receptors for bitter tastants, recently identified,

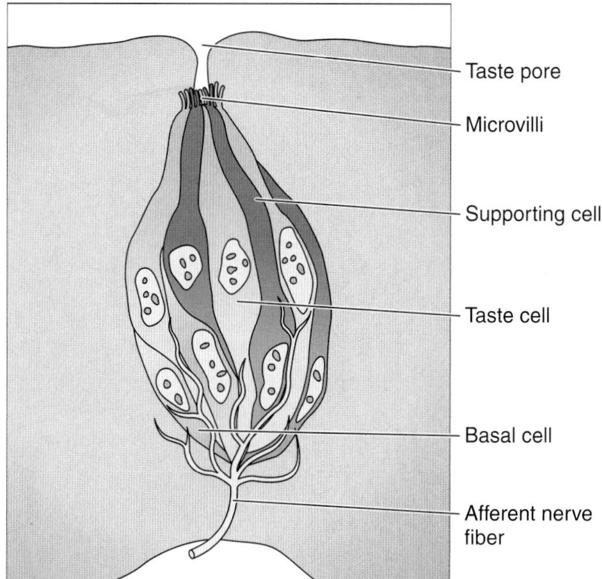

Figure 15–3. Drawing of a taste bud, showing the taste cells and the taste pore. The drawing also illustrates several cell types (basal, taste, and supporting) and afferent nerve fibers that, upon stimulation, will transmit the sensory information to the central gustatory neurons.

belong to a family estimated to have 40–80 members. In the near future other families of taste receptors will certainly be identified as well.

Pharynx

The pharynx, a transitional space between the oral cavity and the respiratory and digestive systems, forms an area of communication between the nasal region and the larynx. The pharynx is lined by stratified nonkeratinized squamous epithelium in the region continuous with the esophagus and by ciliated pseudostratified columnar epithelium containing goblet cells in the regions close to the nasal cavity.

The pharynx contains the tonsils (described in Chapter 14). The mucosa of the pharynx also has many small mucous salivary glands in its lamina propria, composed of dense connective tissue. The constrictor and longitudinal muscles of the pharynx are located outside this layer.

Teeth & Associated Structures

In adult humans, there are normally 32 **permanent teeth.** These teeth are disposed in 2 bilaterally symmetric arches in the maxillary and mandibular bones, with 8 teeth in each quadrant: 2 incisors, 1 canine, 2 premolars, and 3 permanent molars. Twenty of the permanent teeth are preceded by **deciduous** (**baby**) **teeth;** the remainder (the permanent molars) have no deciduous precursors.

Each tooth is composed of a portion that projects above the **gingiva**—the **crown**—and one or more **roots** below the gingiva that hold the teeth in bony sockets called **alveoli,** one for each tooth (Figure 15–4). The crown is covered by the extremely hard **enamel** and the roots by another mineralized tissue, the **cementum.** These two coverings meet at the **cervix** of the tooth. The bulk of a tooth is composed of another calcified material, **dentin,** which surrounds a soft connective tissue-filled space known as the **pulp cavity** (Figure 15–4). The pulp cavity has a coronary portion (the pulp chamber) and a root portion (the root canal), extending to the apex of the root, where an orifice (**apical foramen**) permits the entrance and exit of blood vessels, lymphatics, and nerves of the pulp cavity. The **periodontal ligament** is a fibrous connective tissue with bundles of collagen fibers inserted into the cementum and alveolar bone, fixing the tooth firmly in its bony socket (alveolus).

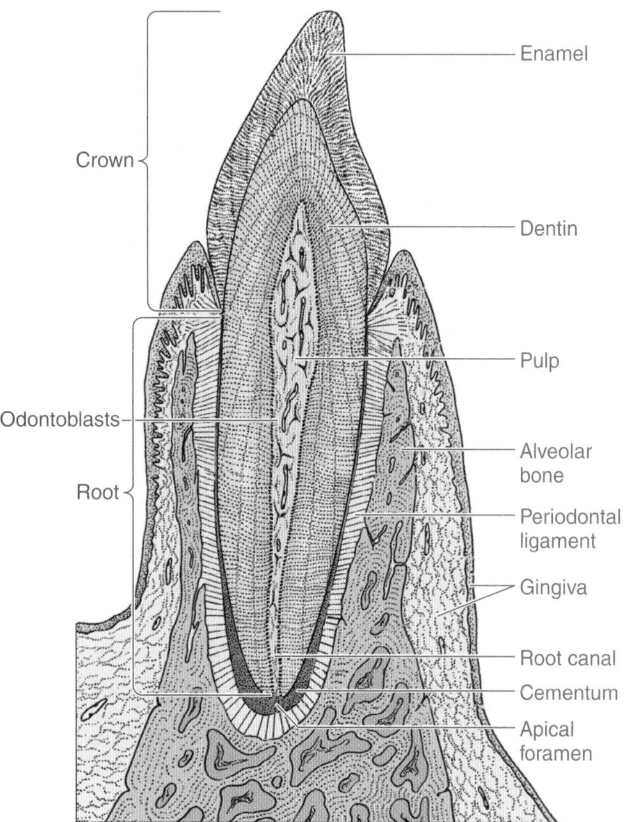

Figure 15–4. Diagram of a sagittal section from an incisor tooth in position in the mandibular bone. (Redrawn and reproduced, with permission, from Leeson TS, Leeson CR: *Histology,* 2nd ed. Saunders, 1970.)

Dentin

Dentin is a calcified tissue that is harder than bone because of its higher content of calcium salts (70% of dry weight). It is composed mainly of type I collagen fibrils, glycosaminoglycans, phosphoproteins, phospholipids, and calcium salts in the form of **hydroxyapatite** crystals. The organic matrix of dentin is secreted by **odontoblasts,** pulp cells that line the internal surface of the tooth (Figures 15–5 and 15–7). The odontoblast is a slender polarized cell that produces organic matrix only at the dentinal surface. These cells have the structure of polarized protein-secreting cells with secretion granules in the apical cytoplasm and a basal nucleus. Odontoblasts have slender, branched apical extensions that penetrate perpendicularly through the width of the dentin—the **odontoblast processes** (Tomes' fibers). These processes gradually become longer as the dentin becomes thicker, running in small canals called **dentinal tubules** that are extensively branched near the junction between dentin and enamel (Figure 15–6). Odontoblast processes have a diameter of 3–4 μm near the cell body but gradually become thinner at their distal ends, close to the enamel or cementum.

The matrix produced by odontoblasts is initially unmineralized and is called **predentin** (Figure 15–7). The mineralization of developing dentin begins when membrane-limited vesicles—**matrix vesicles**—appear, produced by odontoblasts. Because of a high content of calcium and phosphate ions, they facilitate the

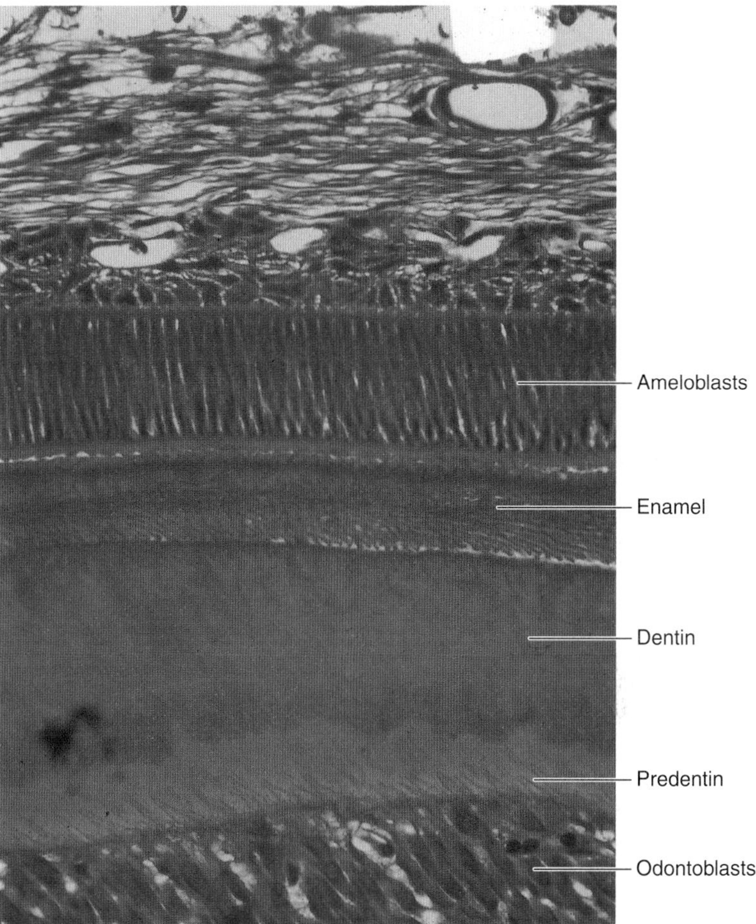

Figure 15–5. Photomicrograph of a section of an immature tooth, showing dentin and enamel. The ameloblasts (enamel-secreting cells) and odontoblasts (predentin-secreting cells) are both disposed as palisades. Pararosaniline–toluidine blue (PT) stain. Medium magnification.

appearance of fine crystals of hydroxyapatite that grow and serve as nucleation sites for further mineral deposition on the surrounding collagen fibrils.

Dentin is sensitive to several stimuli, such as heat, cold, trauma, and acidic pH, and all of these stimuli are perceived as pain. Although the pulp is highly innervated, dentin has a few unmyelinated nerve fibers penetrating its internal (pulpar) portion. According to the hydrodynamic theory, the different stimuli can cause the movement of fluids inside dentinal tubules, which stimulate the nerve fibers located near odontoblast processes.

MEDICAL APPLICATION

Unlike bone, dentin persists as a mineralized tissue long after destruction of the odontoblasts. It is therefore possible to maintain teeth whose pulp and odontoblasts have been destroyed by infection (canal treatment). In adult teeth, destruction of the covering enamel by erosion from use or dental caries (tooth decay) usually triggers a reaction in odontoblasts that causes them to resume the synthesis of dentin components.

Enamel

Enamel is the hardest component of the human body. It consists of about 96% mineral, up to 1% organic material, and 3% water as the remainder. As in other mineralized tissues, the inorganic component of enamel is mostly hydroxyapatite crystals. Other ions, such as strontium, magnesium, lead, and fluoride, if present during enamel synthesis, can be incorporated or adsorbed by the crystals.

MEDICAL APPLICATION

The susceptibility of enamel crystals to dissolution in acidic pH is the basis for dental caries, and some of these crystals (fluorapatite, for example) are less susceptible than hydroxyapatite.

Enamel is produced by cells of ectodermal origin, whereas most of the other structures of teeth derive from mesodermal or neural crest cells. The organic enamel matrix is composed not of collagen fibrils but of at least 2 heterogeneous classes of proteins called **amelogenins** and **enamelins.** The roles of these proteins in the organization of the mineral component of enamel are under intensive investigation.

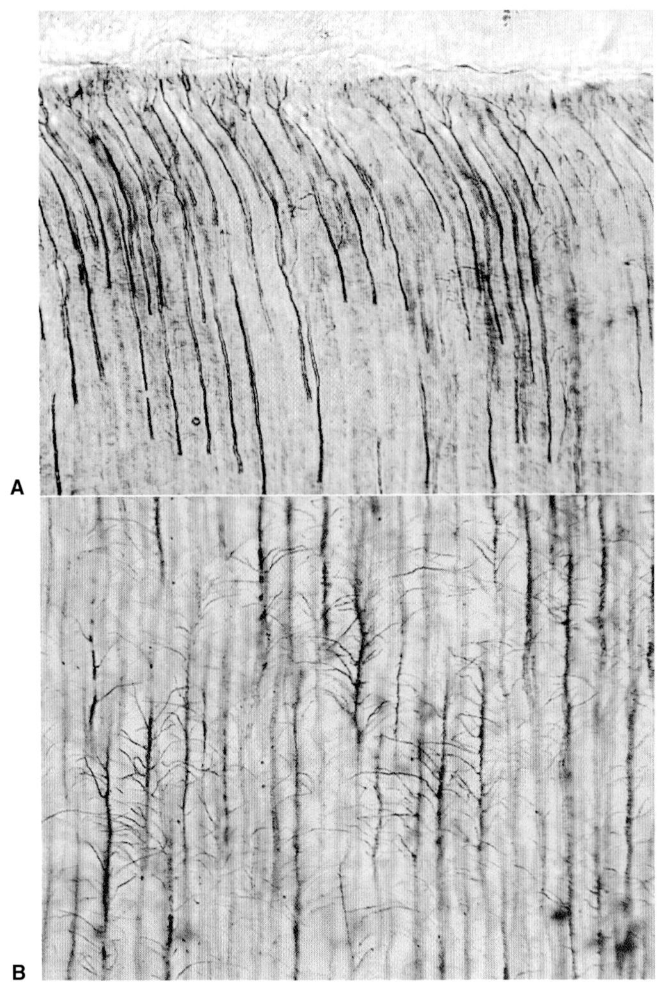

Figure 15–6. Photomicrograph of a section of a tooth, showing the dentin tubules occupied by odontoblast processes. **A:** Initial portion, close to the enamel. **B:** Middle portion. The processes branch into delicate extensions. High magnification.

Enamel consists of elongated rods or columns—**enamel rods (prisms)**—that are bound together by **interrod enamel.** Both interrod enamel and enamel rods are formed of hydroxyapatite crystals; they differ only in the orientation of the crystals. Each rod extends through the entire thickness of the enamel layer and has a sinuous track; the arrangement of rods in groups is very important for enamel's mechanical properties.

Enamel matrix is secreted by cells called **ameloblasts** (Figure 15–5). These tall columnar cells possess numerous mitochondria in the region below the nucleus. Rough endoplasmic reticulum and a well-developed Golgi complex are found above the nucleus. Each ameloblast has an apical extension, known as a **Tomes' process,** containing numerous secretory granules that contain the proteins that make up the enamel matrix. After finishing the synthesis of enamel, ameloblasts form a protective epithelium that covers the crown until the eruption of the tooth. This protective function is very important to prevent several enamel defects.

Pulp

Tooth pulp consists of loose connective tissue. Its main components are odontoblasts, fibroblasts, thin collagen fibrils, and a ground substance that contains glycosaminoglycans (Figure 15–7).

Pulp is a highly innervated and vascularized tissue. Blood vessels and myelinated nerve fibers enter the apical foramen and divide into numerous branches. Some nerve fibers lose their myelin sheaths and extend for a short distance into the dentinal tubules. Pulp fibers are sensitive to pain, the only sensory modality recognized in teeth.

Periodontium

The periodontium comprises the structures responsible for maintaining the teeth in the maxillary and mandibular bones. It consists of the **cementum, periodontal ligament, alveolar bone,** and **gingiva.**

Cementum

Cementum covers the dentin of the root and is similar in composition to bone, although haversian systems and blood vessels are absent. It is thicker in the apical region of the root, where there are **cementocytes,** cells with the appearance of osteocytes. Like osteocytes, they are encased in lacunae; unlike those cells, however, cementocytes do not communicate through canaliculi, and their nourishment comes from the periodontal ligament. Like bone tissue, cementum is labile and reacts to the stresses to which it is subjected by resorbing old tissue or producing new tissue. Continuous production of cementum in the apex compensates for the physiologic wear of the teeth and maintains close contact between the roots of the teeth and their sockets.

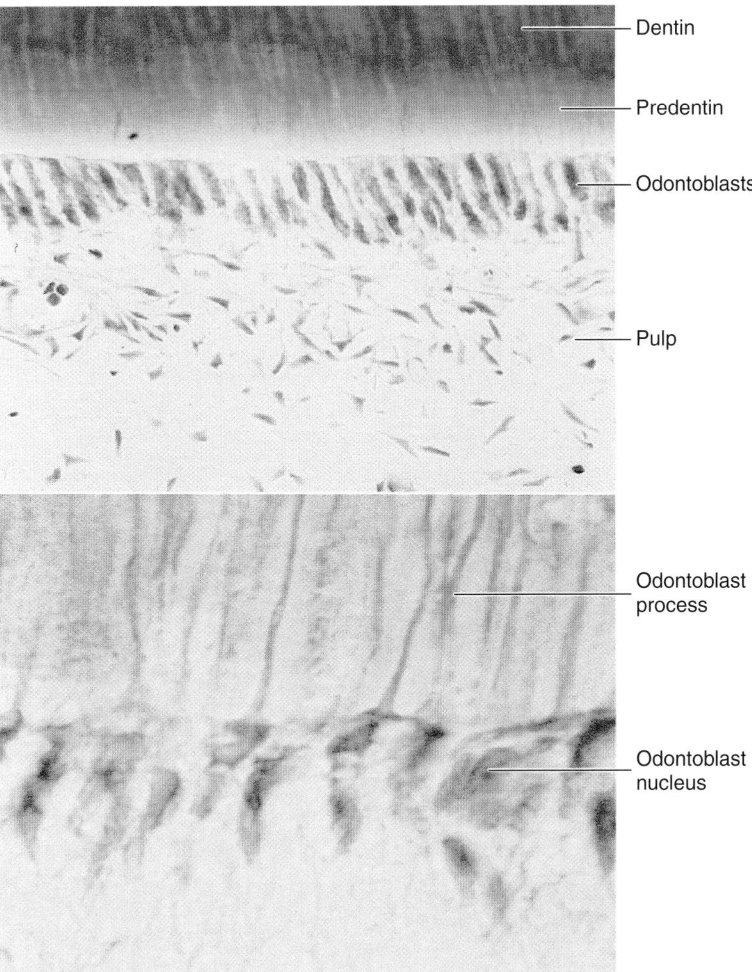

Figure 15–7. Photomicrograph of the dental pulp, in which fibroblasts are abundant. In the upper region are the odontoblasts, from which the odontoblast processes derive. The predentin layer is blue, and the dentin is red. Mallory's stain. **Upper:** Medium magnification; **lower:** high magnification.

MEDICAL APPLICATION

Compared with bone, the cementum has lower metabolic activity because it is not irrigated by blood vessels. This feature allows the movement of teeth by orthodontic appliances without significant root resorption.

Periodontal Ligament

The periodontal ligament is composed of a special type of connective tissue whose fibers penetrate the cementum of the tooth and bind it to the bony walls of its socket while permitting limited movement of the tooth (Figure 15–8A and B). Its fibers are organized to support the pressures exerted during mastication. This avoids transmission of pressure directly to the bone, a process that would cause the bone's localized resorption.

Collagen of the periodontal ligament has characteristics that resemble those of immature tissue. It has a high turnover rate (as demonstrated by autoradiography) and a large soluble collagen content. The space between its fibers is filled with glycosaminoglycans.

MEDICAL APPLICATION

*The high rate of collagen renewal in the periodontal ligament allows processes affecting protein or collagen synthesis, eg, protein or vitamin C deficiency (**scurvy**), to cause atrophy of this ligament. As a consequence, teeth become loose in their sockets; in extreme cases they fall out. This relative plasticity of the periodontal ligament is important because it allows orthodontic intervention, which can produce extensive changes in the disposition of teeth in the mouth.*

Alveolar Bone

The alveolar bone is in immediate contact with the periodontal ligament. It is an immature type of bone (primary bone) in which the collagen fibers are not arranged in the typical lamellar pattern of adult bone. Many of the collagen fibers of the periodontal ligament are arranged in bundles that penetrate this bone and the cementum, forming a connecting bridge between the two structures (**Sharpey's fibers**) (Figure 15–8B). The bone closest to the roots of the teeth forms the socket. Vessels run through

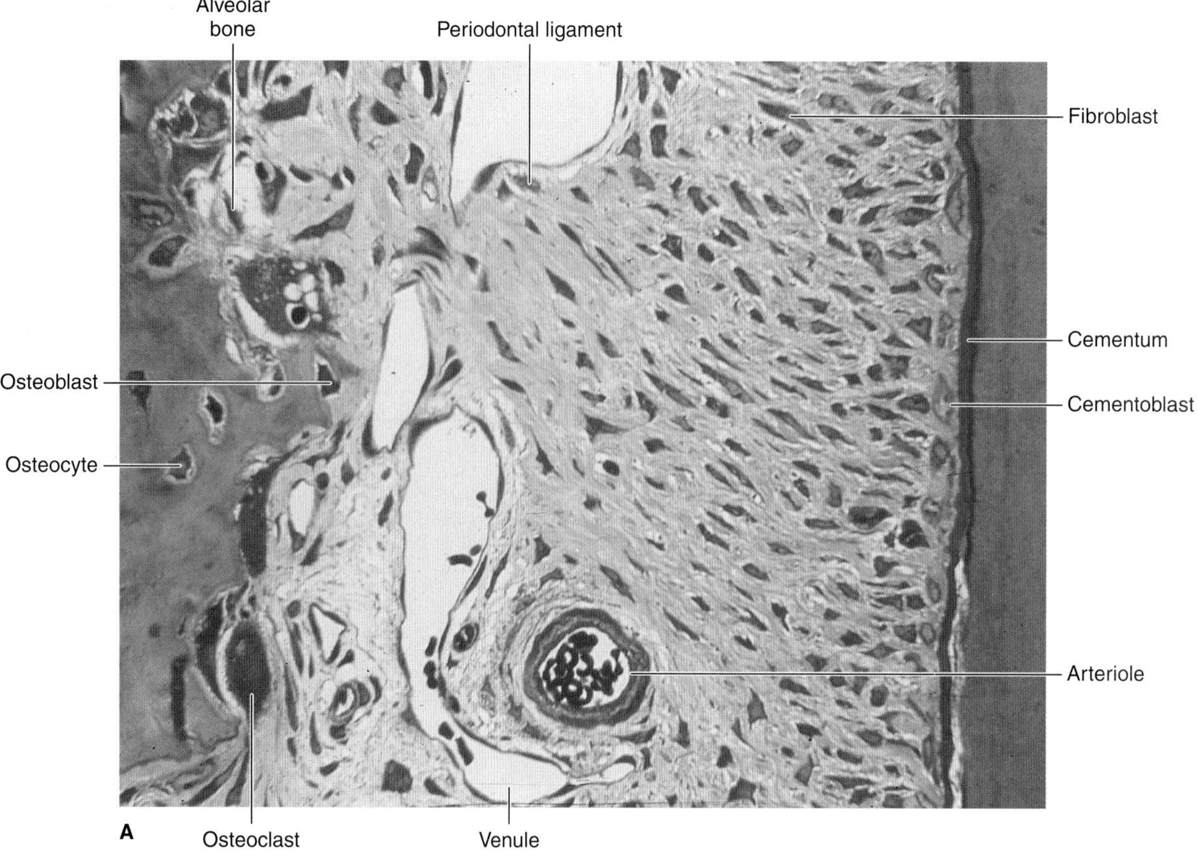

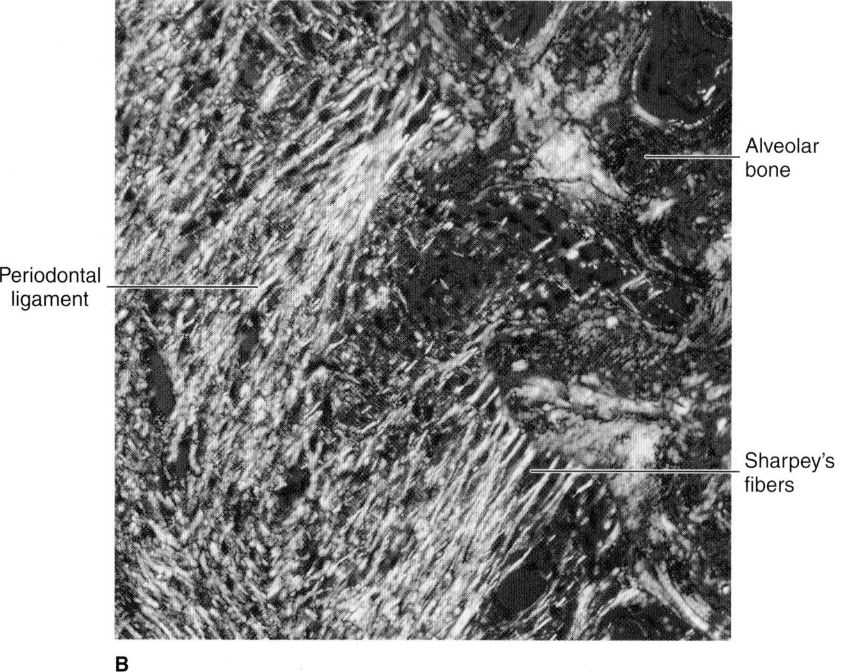

Figure 15–8. **A:** Section showing the insertion of a tooth to the alveolar bone via periodontal ligament. Because this material was obtained from a young animal, the bone is being continuously reformed to adapt to the tooth's eruption; this explains the presence of osteoclasts. The ligament is formed by oriented fibroblasts. PT stain. Medium magnification. **B:** Picrosirius polarization of the periodontal ligament showing the oriented collagen bundles (yellow) inserted in the alveolar bone. Medium magnification.

the alveolar bone and penetrate the periodontal ligament along the root, forming the **perforating vessels.** Some vessels and nerves run to the apical foramen of the root to enter the pulp.

GINGIVA

The gingiva is a mucous membrane firmly bound to the periosteum of the maxillary and mandibular bones. It is composed of stratified squamous epithelium and lamina propria containing numerous connective tissue papillae. A very specialized part of this epithelium, named **junctional epithelium,** is bound to the tooth enamel by means of a cuticle that resembles a thick basal lamina and forms the **epithelial attachment of Gottlieb.** The epithelial cells are attached to this cuticle by hemidesmosomes. Between the enamel and the epithelium is the **gingival sulcus,** a small deepening up to 3 mm surrounding the crown.

MEDICAL APPLICATION

The depth of the gingival sulcus, measured during clinical examination, is very important and may indicate periodontal disease.

ESOPHAGUS

The part of the gastrointestinal tract called the **esophagus** is a muscular tube whose function is to transport foodstuffs from the mouth to the stomach. It is covered by nonkeratinized stratified squamous epithelium (Figure 15–9). In general, it has the same layers as the rest of the digestive tract. In the submucosa are groups of small mucus-secreting glands, the **esophageal glands,** whose secretion facilitates the transport of foodstuffs and protects the mucosa. In the lamina propria of the region near the stomach are groups of glands, the **esophageal cardiac glands,** that also secrete mucus. At the distal end of the esophagus, the muscular layer consists of only smooth muscle cells; in the mid portion, a mixture of striated and smooth muscle cells; and at the proximal end, only striated muscle cells. Only that portion of the esophagus that is in the peritoneal cavity is covered by serosa. The rest is covered by a layer of loose connective tissue, the adventitia, which blends into the surrounding tissue.

STOMACH

The stomach, like the small intestine, is a mixed exocrine-endocrine organ that digests food and secretes hormones. It is a dilated segment of the digestive tract whose main functions are to continue the digestion of carbohydrates initiated in the mouth, add an acidic fluid to the ingested food, transform it by

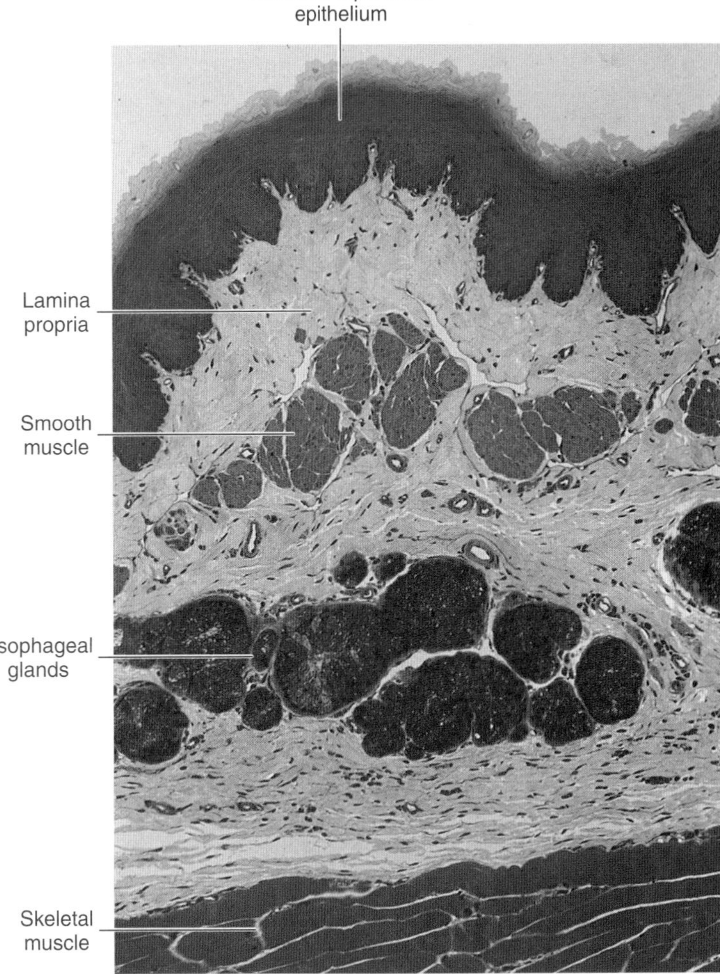

Figure 15–9. Photomicrograph of a section of the upper region of the esophagus. Mucous esophageal glands are in the submucosa; striated skeletal muscle is in the muscularis. PAS and PT stain. Low magnification.

muscular activity into a viscous mass (**chyme**), and promote the initial digestion of proteins with the enzyme **pepsin.** It also produces a gastric lipase that digests triglycerides with the help of lingual lipase. Gross inspection reveals 4 regions: **cardia, fundus, body,** and **pylorus** (Figure 15–10). Because the fundus and body are identical in microscopic structure, only 3 histologic regions are recognized. The mucosa and submucosa of the undistended stomach lie in longitudinally directed folds known as **rugae.** When the stomach is filled with food, these folds flatten out.

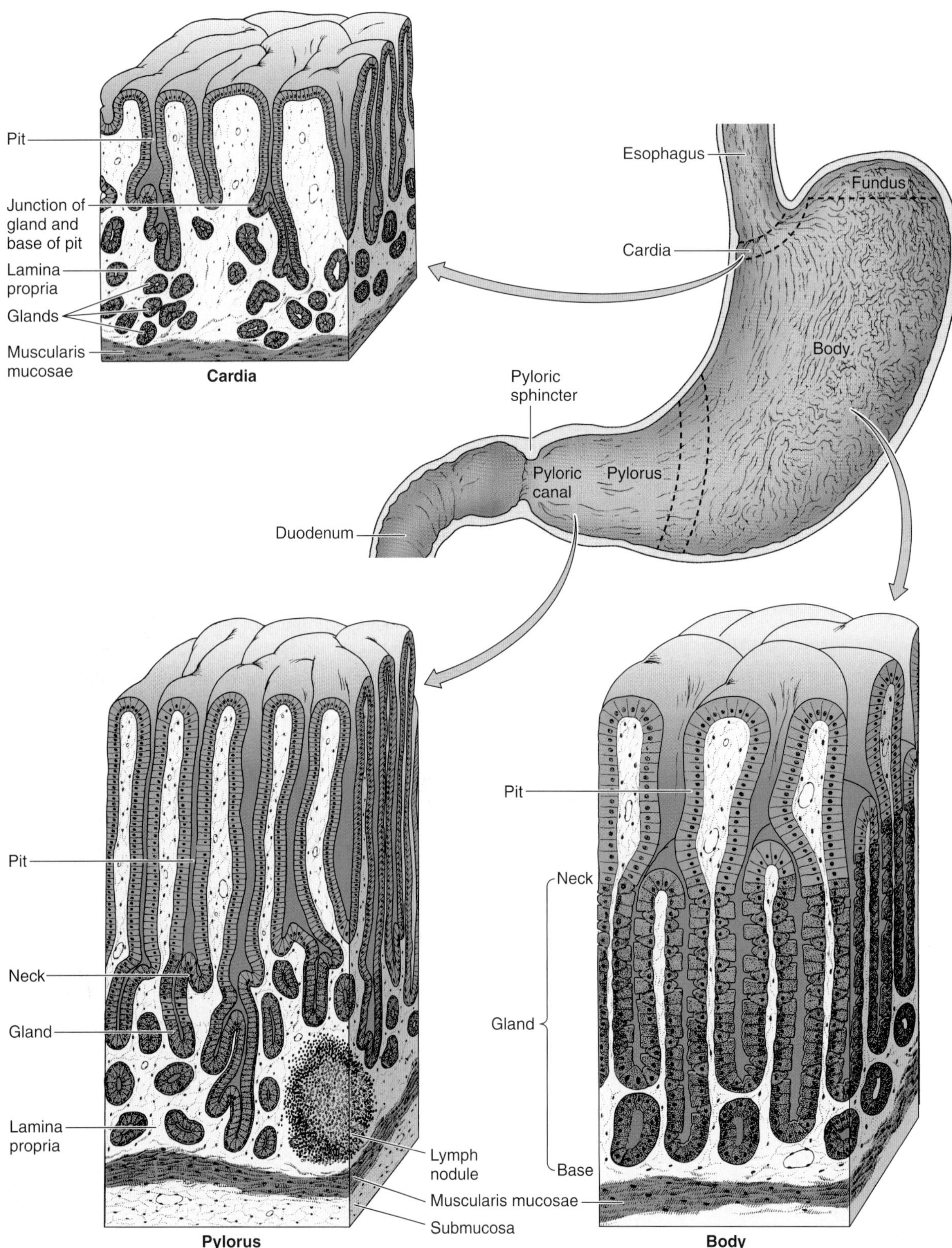

Figure 15–10. Regions of the stomach and their histologic structure.

Mucosa

The gastric mucosa consists of a **surface epithelium** that invaginates to various extents into the lamina propria, forming **gastric pits.** Emptying into the gastric pits are branched, tubular glands (cardiac, gastric, and pyloric) characteristic of each region of the stomach. The **lamina propria** of the stomach is composed of loose connective tissue interspersed with smooth muscle and lymphoid cells. Separating the mucosa from the underlying submucosa is a layer of smooth muscle, the **muscularis mucosae.**

When the luminal surface of the stomach is viewed under low magnification, numerous small circular or ovoid invaginations of the epithelial lining are observed. These are the openings of the gastric pits (Figures 15–10 and 15–11). The epithelium covering the surface and lining the pits is a simple columnar epithelium, and all the cells secrete an alkaline mucus (Figures 15–11 and 15–12). When released from these cells, the mucus forms a thick gel layer that protects them from the effects of the strong acid secreted by the stomach. The mucus firmly adherent to the epithelial surface is very effective in protection, while the superficial luminal mucous layer is more soluble, partially digested by pepsin and mixed with the luminal contents.

Tight junctions around surface and pit cells also form part of the barrier to acid. Like the hydrochloric acid, pepsin, lipases (lingual and gastric), and bile must also be considered as endogenous aggressors to the epithelial lining.

MEDICAL APPLICATION

Stress and other psychosomatic factors; ingested substances such as aspirin, nonsteroidal anti-inflammatory drugs or ethanol; the hyperosmolality of meals; and some microorganisms (eg, Helicobacter pylori*) can disrupt this epithelial layer and lead to ulceration. The initial ulceration may heal, or it may be further aggravated by the local aggressive agents, leading to additional gastric and duodenal ulcers. Processes that enable the gastric mucosa to rapidly repair superficial damage incurred by several factors play a very important role in the defense mechanism, as does an adequate blood flow that supports gastric physiologic activity. Any imbalance between aggression and protection may lead to pathologic alterations. As an example, aspirin and ethanol irritate the mucosa partly by reducing mucosal blood flow. Several anti-inflammatory drugs inhibit the production of prostaglandins of the E type, which are very important substances for the alkalinization of the mucous layer and, consequently, important for protection.*

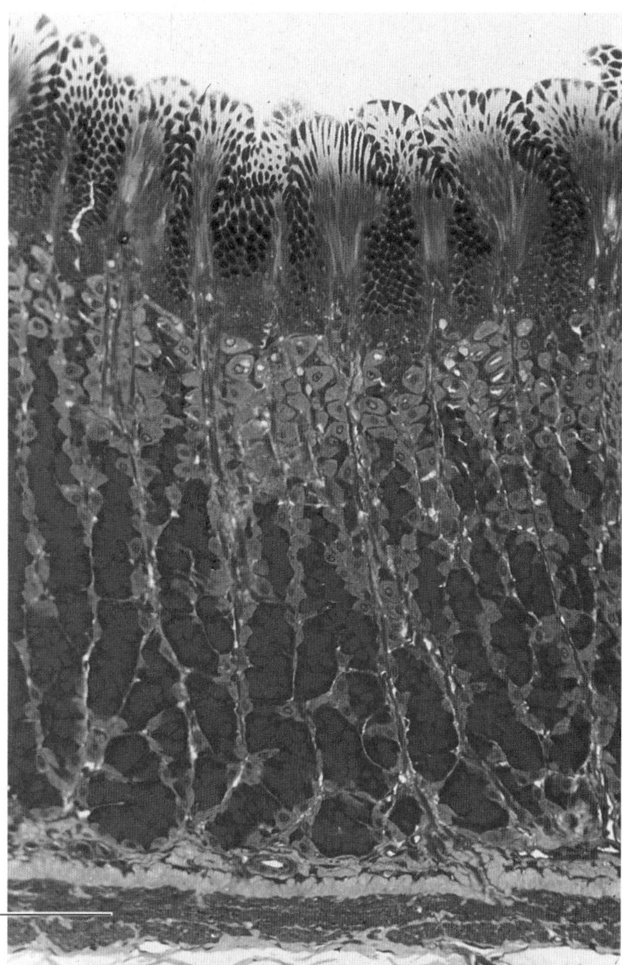

Figure 15–11. Photomicrograph of a section of the gastric glands in the fundus of the stomach. Note the superficial mucus-secreting epithelium. Parietal cells (light-stained) predominate in the mid and upper regions of the glands; chief (zymogenic) cells (dark-stained) predominate in the lower region of the gland. MM, muscularis mucosae. PT stain. Low magnification.

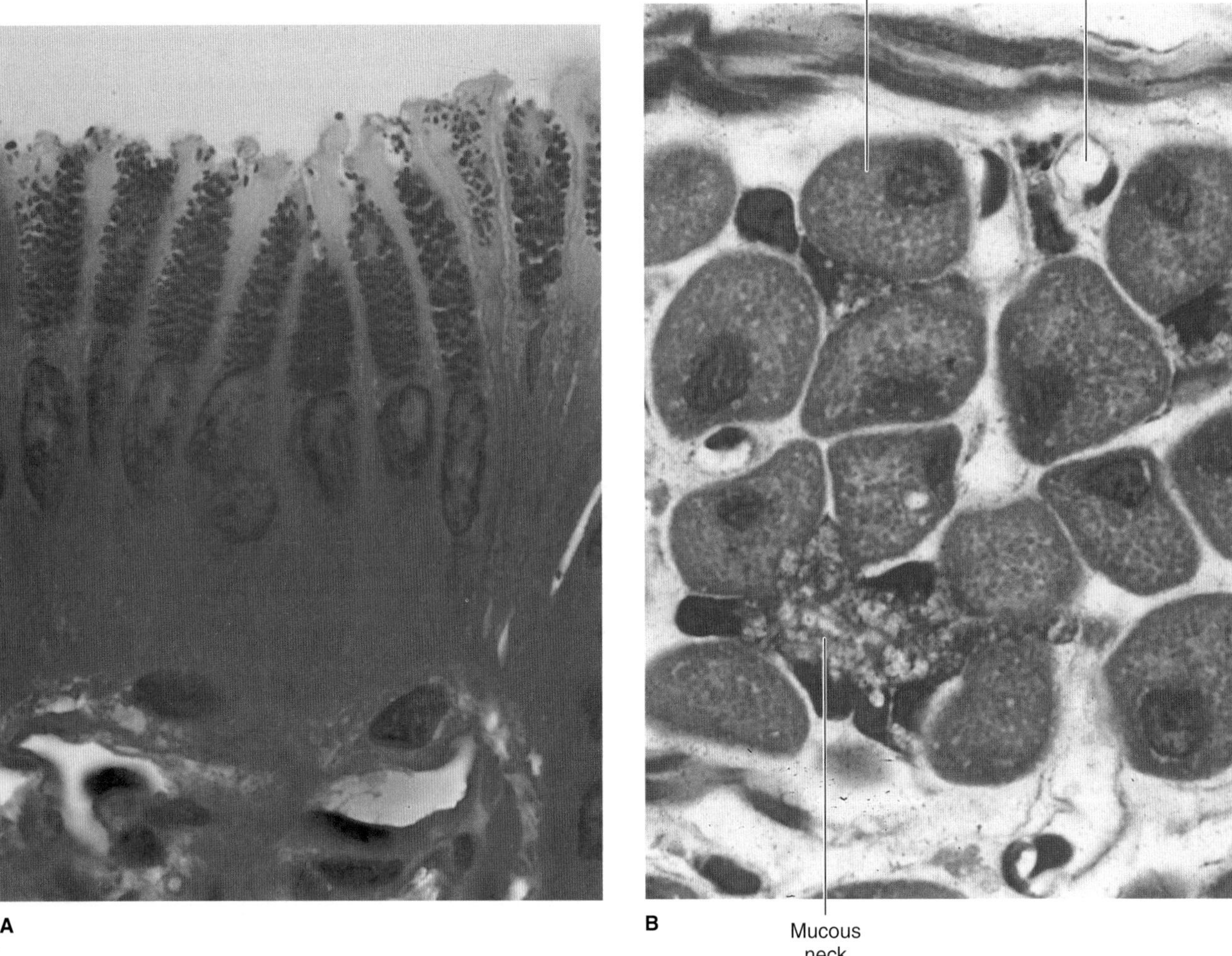

Figure 15–12. Photomicrograph of a mucus-secreting surface epithelium (**A**) and mucous neck cells intercalated between oxyntic (parietal) cells located in the mid portion of the gastric gland (**B**). Abundant capillaries can be seen. PT stain. Medium magnification.

Cardia

The cardia is a narrow circular band, 1.5–3 cm in width, at the transition between the esophagus and the stomach (Figure 15–10). Its mucosa contains simple or branched tubular cardiac glands. The terminal portions of these glands are frequently coiled, often with large lumens. Most of the secretory cells produce mucus and lysozyme (an enzyme that attacks bacterial walls), but a few hydrochloride-producing parietal cells can be found. These glands are similar in structure to the cardiac glands of the terminal portion of the esophagus.

Fundus & Body

The lamina propria of the fundus and body is filled with branched, tubular **gastric** (**fundic**) **glands,** 3 to 7 of which open into the bottom of each gastric pit. The distribution of epithelial cells in gastric glands is not uniform (Figures 15–10 and 15–11). The **neck** of the glands consists of stem, mucous neck, and oxyntic (parietal) cells (Figure 15–12B); the **base** of the glands contains parietal, chief (zymogenic), and enteroendocrine cells.

Stem Cells

Found in the neck region but few in number, stem cells are low columnar cells with oval nuclei near the bases of the cells. These cells have a high rate of mitosis; some of them move upward to replace the pit and surface mucous cells, which have a turnover time of 4–7 days. Other daughter cells migrate more deeply into the glands and differentiate into mucous neck cells and parietal, chief, and enteroendocrine cells. These cells are replaced much more slowly than are surface mucous cells.

Mucous Neck Cells

Mucous neck cells are present in clusters or as single cells between parietal cells in the necks of gastric glands. Their mucus secretion is quite different from that of the surface epithelial mucous cells. They are irregular in shape, with the nucleus at the base of the cell and the secretory granules near the apical surface.

Oxyntic (Parietal) Cells

Parietal cells are present mainly in the upper half of gastric glands; they are scarce in the base. They are rounded or pyramidal cells, with one centrally placed spherical nucleus and intensely eosinophilic cytoplasm (Figures 15–11, 15–12B, and 15–13). The most striking features of the active secreting cell seen in the electron microscope are an abundance of mitochondria (eosinophilic) and a deep, circular invagination of the apical plasma membrane, forming the **intracellular canaliculus** (Figures 15–13, 15–14, and 15–15). In the resting cell, a number of tubulovesicular structures can be seen in the apical region just below the plasmalemma (Figure 15–15, left). At this stage, the cell has few microvilli. When stimulated to produce hydrochloric acid, tubulovesicles fuse with the cell membrane to form the canaliculus and more microvilli, thus providing a generous increase in the surface of the cell membrane (Figure 15–15, right).

Parietal cells secrete hydrochloric acid (actually, H^+ and Cl^-), 0.16 mol/L; potassium chloride, 0.07 mol/L; traces of other electrolytes; and gastric intrinsic factor (see below). The ion H^+ originates from the dissociation of the H_2CO_3 produced by the action of **carbonic anhydrase,** an enzyme abundant in oxyntic cells. Once produced, H_2CO_3 dissociates in the cytoplasm into H^+ and HCO_3^- (Figure 15–16). The active cell also secretes KCl in the canaliculus, which dissociates into K^+ and Cl^-; the K^+ is exchanged for H^+ by the action of the H^+/K^+ pump, while the Cl^- forms HCl. The presence of abundant mitochondria in the parietal cells indicates that their metabolic processes, particularly the pumping of H^+/K^+, are highly energy consuming.

MEDICAL APPLICATION

In cases of atrophic gastritis, both parietal and chief cells are much less numerous, and the gastric juice has little or no acid or pepsin activity. In humans, oxyntic cells are the site of production of ***intrinsic factor,*** *a glycoprotein that binds avidly to vitamin* B_{12}*. In other species, however, the intrinsic factor may be produced by other cells.*

The complex of vitamin B_{12} *with intrinsic factor is absorbed by pinocytosis into the cells in the ileum; this explains why a lack of intrinsic factor can lead to vitamin* B_{12} *deficiency. This condition results in a disorder of the erythrocyte-forming mechanism known as* ***pernicious anemia,*** *usually caused by* ***atrophic gastritis.*** *In a certain percentage of cases, pernicious anemia seems to be an autoimmune disease, because antibodies against parietal cell proteins are often detected in the blood of patients with the disease.*

The secretory activity of parietal cells is initiated by various mechanisms. One mechanism is through the cholinergic nerve endings (parasympathetic stimulation). Histamine and a polypeptide called **gastrin,** both secreted in the gastric mucosa, act strongly to

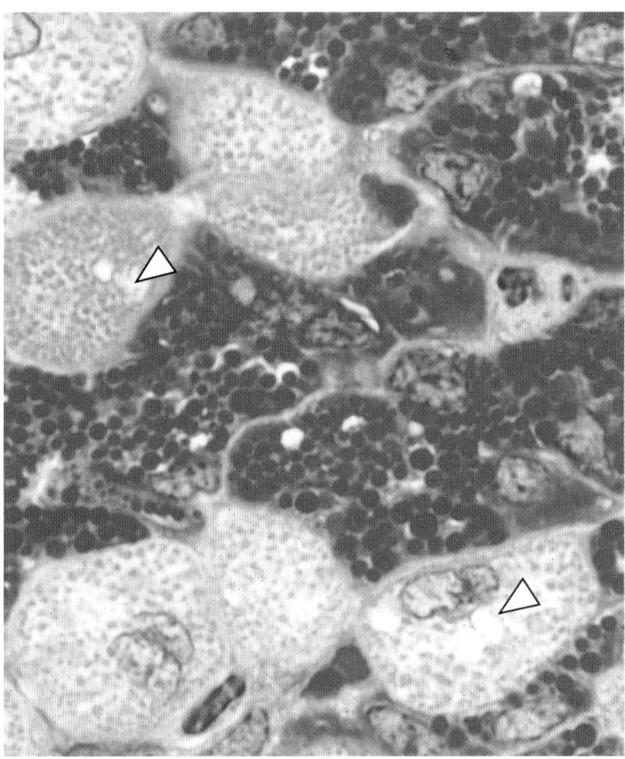

Figure 15–13. Photomicrograph of the basal portion of the gastric gland in the fundus. This section shows parietal cells rich in mitochondria and their characteristic intracellular canaliculi (arrowheads). Chief cells show red secretory granules in their cytoplasm. Pararosaniline toluidine staining.

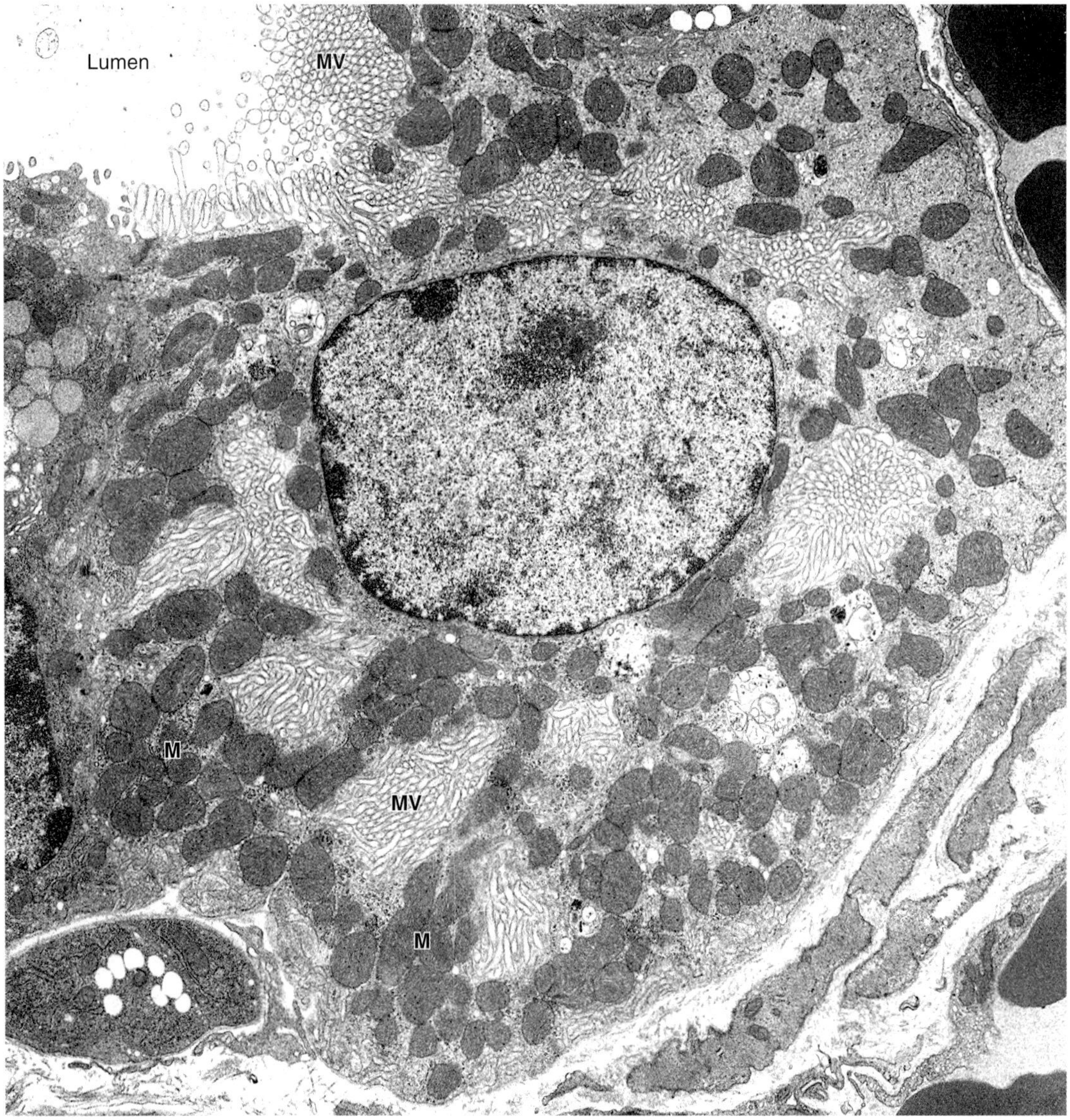

Figure 15–14. Electron micrograph of an active parietal cell. Note the microvilli (MV) protruding into the intracellular canaliculi and the abundant mitochondria (M). ×10,200. (Courtesy of S Ito.)

stimulate the production of hydrochloric acid. Gastrin also has a trophic effect on the gastric mucosa, stimulating growth.

CHIEF (ZYMOGENIC) CELLS

Chief cells predominate in the lower region of the tubular glands (Figure 15–13) and have all the characteristics of protein-synthesizing and -exporting cells. Their basophilia is due to the abundant rough endoplasmic reticulum. The granules in their cytoplasm contain the inactive enzyme **pepsinogen.** The precursor pepsinogen is rapidly converted into the highly active proteolytic enzyme **pepsin** after being released into the acid environment of the stomach. There are 7 different pepsins in the human gastric juice, which are aspartate endoproteinases of relatively broad specificity active at pH <5. In humans, chief cells also produce the enzyme **lipase.**

ENTEROENDOCRINE CELLS

Enteroendocrine cells, discussed more extensively below, are found near the bases of gastric glands (Figures 15–17 and 15–18). In the fundus of the stomach, **5-hydroxytryptamine** (serotonin) is one of the principal secretory products. Other products of enteroendocrine cells in the gastrointestinal tract are shown in Table 15–1.

MEDICAL APPLICATION

*Tumors called **carcinoids,** which arise from these cells, are responsible for the clinical symptoms caused by overproduction of serotonin. Serotonin increases gut motility, but high levels of this hormone/neurotransmitter have been related to mucosal vasoconstriction and damage.*

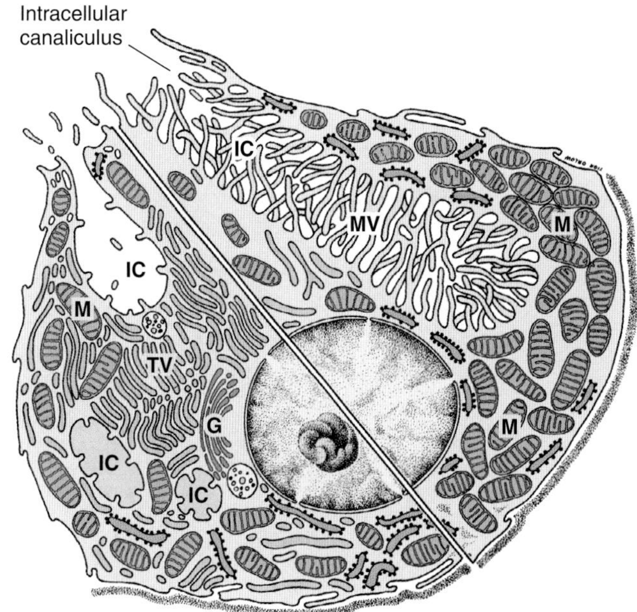

Figure 15–15. Composite diagram of a parietal cell, showing the ultrastructural differences between a resting cell (**left**) and an active cell (**right**). Note that the tubulovesicles (TV) in the cytoplasm of the resting cell fuse to form microvilli (MV) that fill up the intracellular canaliculi (IC). G, Golgi complex; M, mitochondria. (Based on the work of Ito S, Schofield GC. J Cell Biol 1974;63:364.)

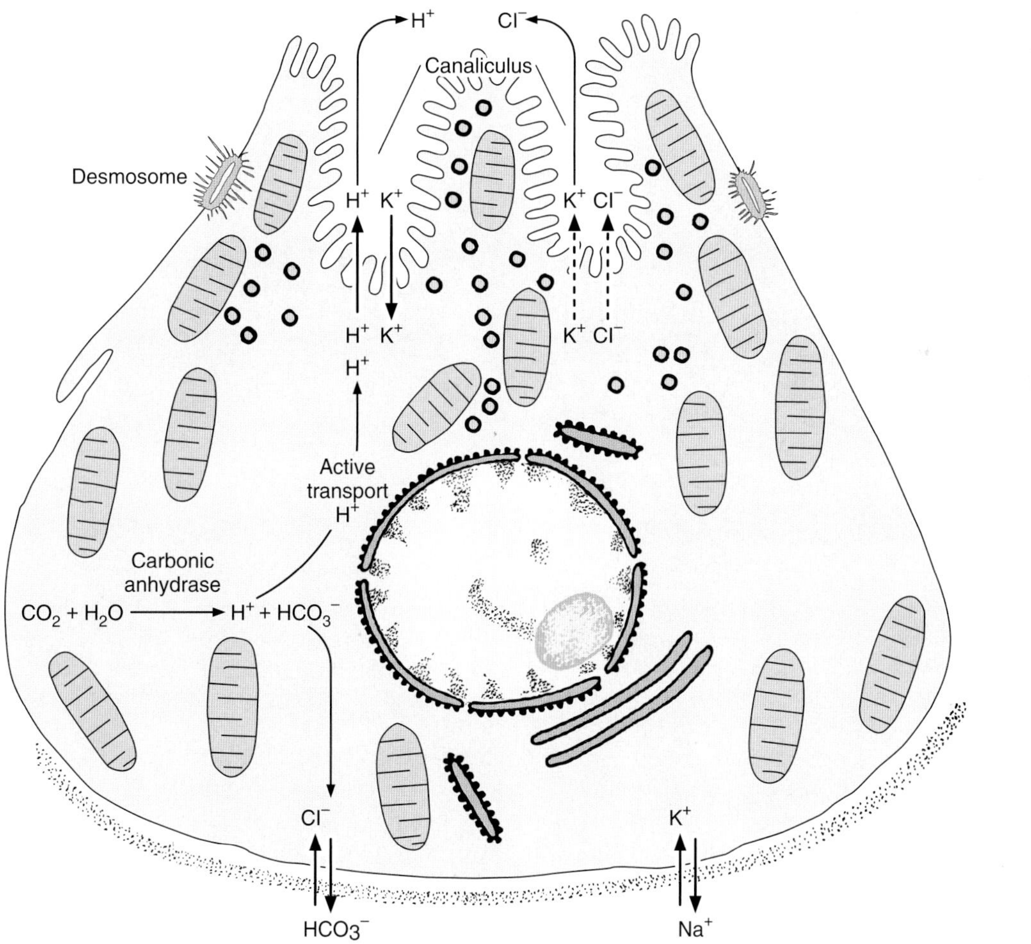

Figure 15–16. Diagram of a parietal cell, showing the main steps in the synthesis of hydrochloric acid. Active transport by ATPase is indicated by arrows and diffusion is indicated by dotted arrows. Under the action of carbonic anhydrase, blood CO_2 produces carbonic acid. Carbonic acid dissociates into a bicarbonate ion and a proton H^+, which is pumped into the stomach lumen in exchange for K^+. A high concentration of intracellular K^+ is maintained by the Na^+,K^+ ATPase, while HCO_3^- is exchanged for Cl^- by an antiport. The tubulovesicles of the cell apex are seen to be related to hydrochloric acid secretion, because their number decreases after parietal cell stimulation. The bicarbonate ion returns to the blood and is responsible for a measurable increase in blood pH during digestion.

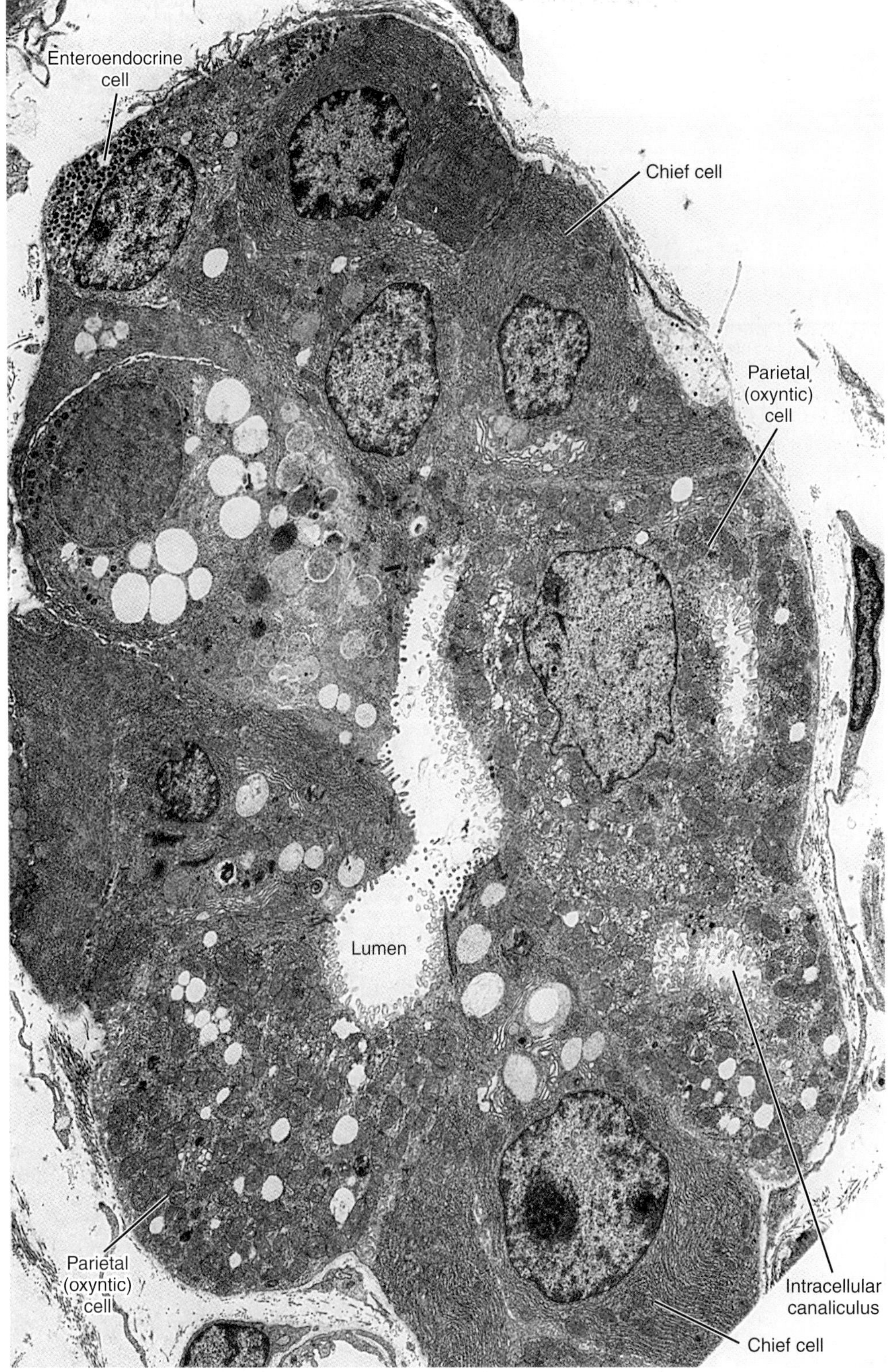

Figure 15–17. Electron micrograph of a section of gastric gland in the fundus of the stomach. Note the lumen and the parietal cells, containing abundant mitochondria; chief cells, with extensive rough endoplasmic reticulum; and enteroendocrine cells (closed type), with basal secretory granules. ×5300.

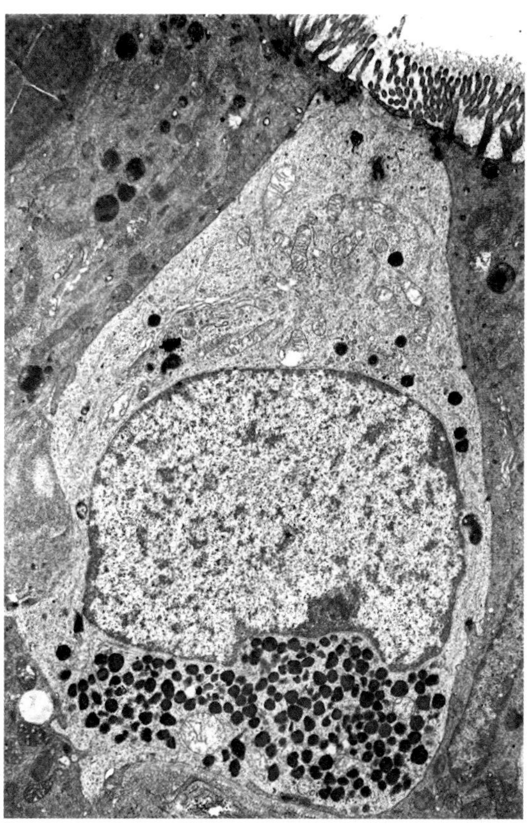

Figure 15–18. Electron micrograph of an enteroendocrine cell (open type) of the human duodenum. Note the microvilli in its apex. ×6900. (Courtesy of AGE Pearse.)

Table 15–1. Principal enteroendocrine cells in the gastrointestinal tract.

Cell Type and Location	Hormone Produced	Major Action
A—stomach	Glucagon	Hepatic glycogenolysis
G—pylorus	Gastrin	Stimulation of gastric acid secretion
S—small intestine	Secretin	Pancreatic and biliary bicarbonate and water secretion
K—small intestine	Gastric inhibitory polypeptide	Inhibition of gastric acid secretion
L—small intestine	Glucagon-like substance (glicentin)	Hepatic glycogenolysis
I—small intestine	Cholecystokinin	Pancreatic enzyme secretion, gallbladder contraction
D—pylorus, duodenum	Somatostatin	Local inhibition of other endocrine cells
Mo—small intestine	Motilin	Increased gut motility
EC—digestive tract	Serotonin, substance P	Increased gut motility
D_1—digestive tract	Vasoactive intestinal polypeptide	Ion and water secretion, increased gut motility

Pylorus

The pylorus (from Latin, meaning gatekeeper) has deep gastric pits into which the branched, tubular **pyloric glands** open. Compared with the glands in the cardiac region, pyloric glands have longer pits and shorter coiled secretory portions (Figure 15–19). These glands secrete mucus as well as appreciable amounts of the enzyme lysozyme. **Gastrin (G) cells** (which release **gastrin**) are intercalated among the mucous cells of pyloric glands. Gastrin stimulates the secretion of acid by the parietal cells of gastric glands and has a trophic effect on gastric mucosa. Other enteroendocrine cells (**D cells**) secrete **somatostatin,** which inhibits the release of some other hormones, including gastrin.

Other Layers of the Stomach

The **submucosa** is composed of dense connective tissue containing blood and lymph vessels; it is infiltrated by lymphoid cells, macrophages, and mast cells. The **muscularis** is composed of smooth muscle fibers oriented in 3 main directions. The external layer is longitudinal, the middle layer is circular, and the internal layer is oblique. At the pylorus, the middle layer is greatly thickened to form the **pyloric sphincter.** The stomach is covered by a thin **serosa.**

SMALL INTESTINE

The small intestine is the site of terminal food digestion, nutrient absorption, and endocrine secretion. The processes of digestion are completed in the small intestine, where the nutrients (products of digestion) are absorbed by cells of the epithelial lining. The small intestine is relatively long—approximately 5 m—and consists of 3 segments: **duodenum, jejunum,** and **ileum.** These segments have many characteristics in common and will be discussed together.

Mucous Membrane

Viewed with the naked eye, the lining of the small intestine shows a series of permanent folds, **plicae circulares (Kerckring's valves)**, consisting of mucosa and submucosa and having a semilunar, circular, or spiral form. The plicae are most developed in, and consequently a characteristic of, the jejunum. They do not constitute a significant feature of the duodenum and ileum, although they are frequently present. **Intestinal villi** are 0.5- to 1.5-mm-long outgrowths of the mucosa (epithelium plus lamina propria) projecting into the lumen of the small intestine. In the duodenum they are leaf-shaped, gradually assuming fingerlike shapes as they reach the ileum (Figures 15–20 and 15–27).

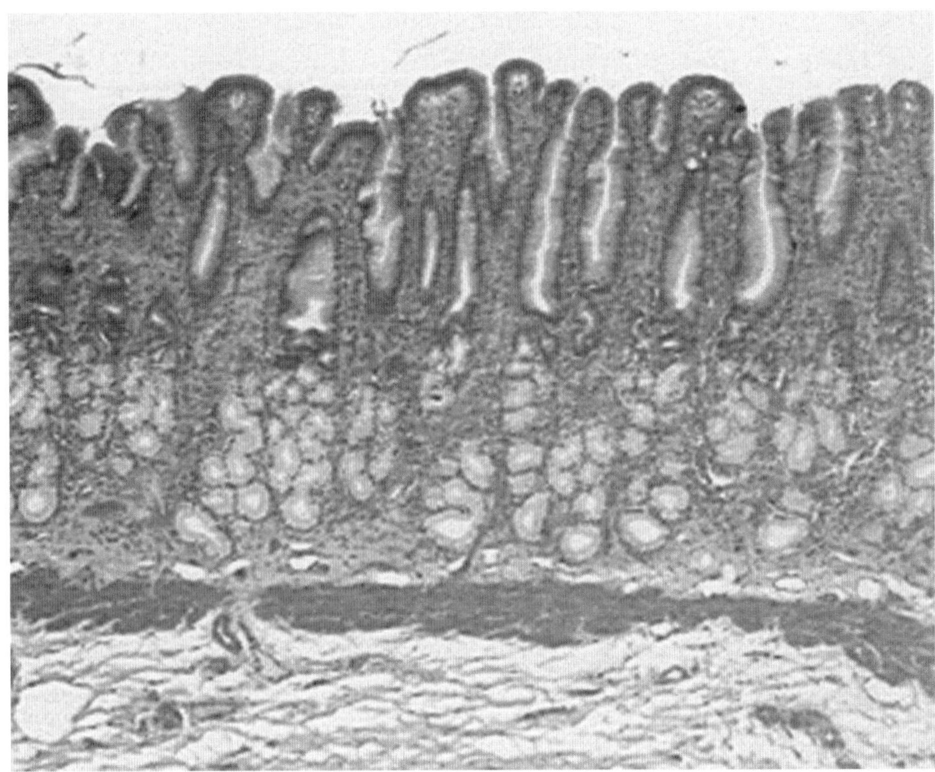

Figure 15–19. Photomicrograph of a section of the pyloric region of the stomach. Note the deep gastric pits with short pyloric glands in the lamina propria. H&E stain. Low magnification.

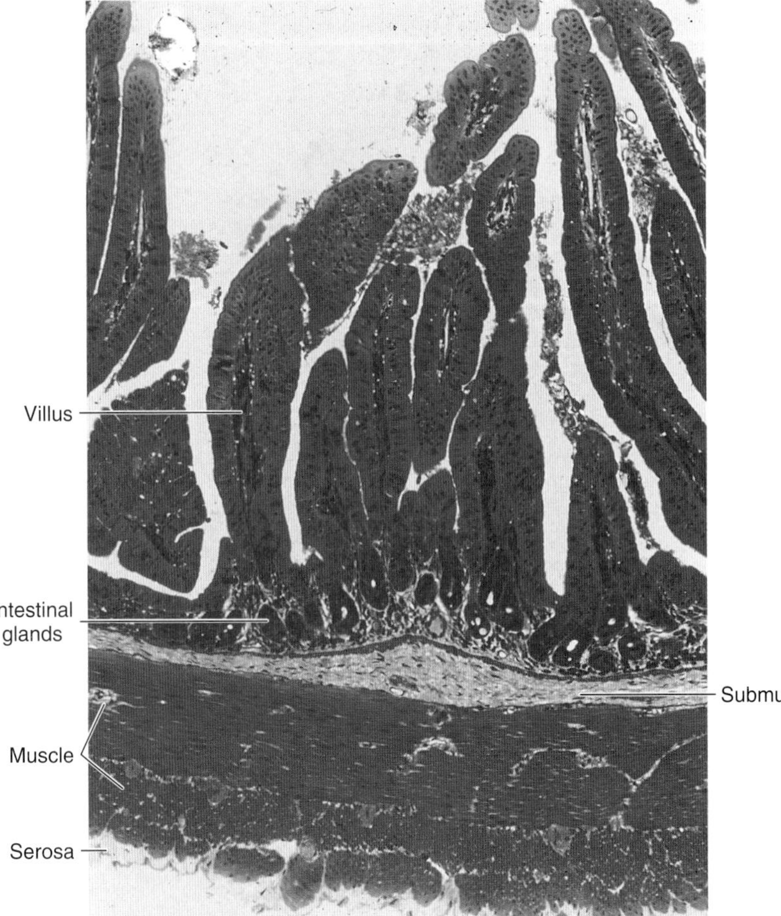

Figure 15–20. Photomicrograph of the small intestine. Note the villi, intestinal glands, submucosa, muscle layers, and serosa. PT stain. Low magnification.

Between the villi are small openings of simple tubular glands called **intestinal glands** (also inappropriately called **crypts**), or **glands of Lieberkühn** (Figures 15–20 and 15–27).

The epithelium of the villi is continuous with that of the glands. The intestinal glands contain stem cells, some absorptive cells, goblet cells, Paneth's cells, and enteroendocrine cells.

Absorptive cells are tall columnar cells, each with an oval nucleus in the basal half of the cell. At the apex of each cell is a homogeneous layer called the **striated (brush) border** (Figure 15–21). When viewed with the electron microscope, the striated border is seen to be a layer of densely packed **microvilli** (Figures 15–22 and 15–23). Each microvillus is a cylindrical protrusion of the apical cytoplasm that is approximately 1 μm tall by 0.1 μm in diameter and consists of the cell membrane enclosing a core of actin microfilaments associated with other cytoskeletal proteins (Figures 15–22 and 15–24). Each absorptive cell is estimated to have an average of 3000 microvilli, and 1 mm^2 of mucosa contains about 200 million of these structures. Microvilli have the important physiologic function of increasing the area of contact between the intestinal surface and the nutrients. The presence of plicae, villi, and microvilli greatly increases the surface of the intestinal lining—an important characteristic in an organ in which absorption occurs so intensely. It has been calculated that plicae increase the intestinal surface 3-fold, the villi increase it 10-fold, and the microvilli increase it 20-fold. Together, these processes are responsible for a 600-fold increase in the intestinal surface, resulting in a total area of 200 m^2.

The more important function of the columnar intestinal cells is to absorb the nutrient molecules produced by the digestive process. Disaccharidases and peptidases secreted by absorptive cells and bound to microvilli in the brush border hydrolyze the disaccharides and dipeptides into monosaccharides and amino acids that are easily absorbed through secondary active transport. Lipid digestion occurs mainly as a result of the action of pancreatic lipase and bile. In humans, most of the lipid absorption takes place in the duodenum and upper jejunum. Figures 15–25 and 15–26 illustrate current concepts of this process of absorption.

MEDICAL APPLICATION

Deficiencies of disaccharidases have been described in human diseases characterized by digestive disturbances. Some of the enzymatic deficiencies seem to be of genetic origin.

The absorption of nutrients is also greatly hindered in disorders marked by atrophy of the intestinal mucosa caused by infections or nutritional deficiencies, producing the ***malabsorption syndrome.***

Goblet cells are interspersed between the absorptive cells (Figures 15–21 and 15–27). They are less abundant in the duodenum and increase in number as they approach the ileum. These cells produce acid glycoproteins of the mucin type that are hydrated and cross-linked to form mucus, whose main function is to protect and lubricate the lining of the intestine.

Paneth's cells in the basal portion of the intestinal glands are exocrine cells with secretory granules in their apical cytoplasm. Researchers using immunocytochemical methods have detected lysozyme—an enzyme that digests the cell walls of some bacteria—in the large eosinophilic secretory granules of these cells (Figures 15–27 through 15–30). Lysozyme has antibacterial activity and may play a role in controlling the intestinal flora.

M (microfold) cells are specialized epithelial cells overlying the lymphoid follicles of Peyer's patches. These cells are characterized by the presence of numerous basal membrane invaginations that form pits containing many intraepithelial lymphocytes and antigen-presenting cells (macrophages). M cells can endocytose antigens and transport them to the underlying macrophages and lymphoid cells, which then migrate to other compartments of the lymphoid system (nodes), where immune responses to foreign antigens are initiated. M cells represent an important link in the intestinal immunologic system (Figures 15–31, 15–32, and 15–33). The basement membrane under M cells is discontinuous, facilitating transit between the lamina propria and M cells (Figure 15–33).

The very large mucosal surface of the gastrointestinal tract is exposed to many potentially invasive microorganisms. Secretory immunoglobulins of the IgA class (discussed earlier) are the first line of defense. Another protective device is the intercellular tight junctions that make the epithelial cells a barrier to the penetration of microorganisms. In addition—and probably serving as the main protective barrier—the gastrointestinal tract contains antibody-secreting plasma cells, macrophages, and a very large number of lymphocytes (Figures 15–31 and 15–33), located in both the mucosa and the submucosa. Together, these cells are called the gut-associated lymphatic tissue (GALT).

Endocrine Cells of the Intestine

In addition to the cells discussed above, the intestine contains some widely distributed cells with characteristics of the **diffuse neuroendocrine system.** The main results obtained so far are summarized in Table 15–1.

Upon stimulation these cells release their secretory granules by exocytosis, and the hormones may then exert paracrine (local) or endocrine (blood-borne) effects. Polypeptide-secreting cells of the digestive tract fall into 2 classes: the **open type,** in which the apex of the cell presents microvilli and contacts the lumen of the organ (Figure 15–18); and the **closed type,** in which the cellular apex is covered by other epithelial cells (Figure 15–17). In the small intestine, endocrine cells of the open type are more slender than the neighboring absorptive cells, possessing irregular microvilli in the apical surface and small secretory granules in the cytoplasm. It has been suggested that in the open type, the chemical contents of the digestive tract might act on its microvilli and thereby influence secretion of these cells. Although the picture of gastrointestinal endocrinology is still incomplete, the activity of the digestive system is clearly controlled by the nervous system and modulated by a complex system of locally produced peptide hormones.

Lamina Propria Through Serosa

The lamina propria of the small intestine is composed of loose connective tissue with blood and lymph vessels, nerve fibers, and smooth muscle cells.

The lamina propria penetrates the core of the intestinal villi, taking along blood and lymph vessels, nerves, connective tissue,

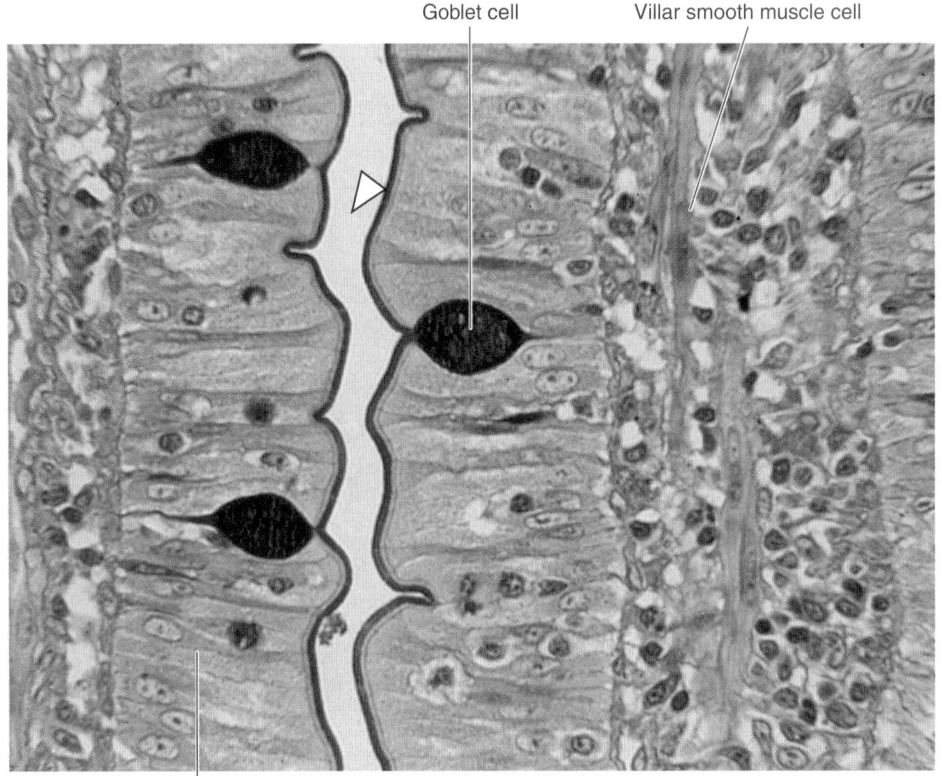

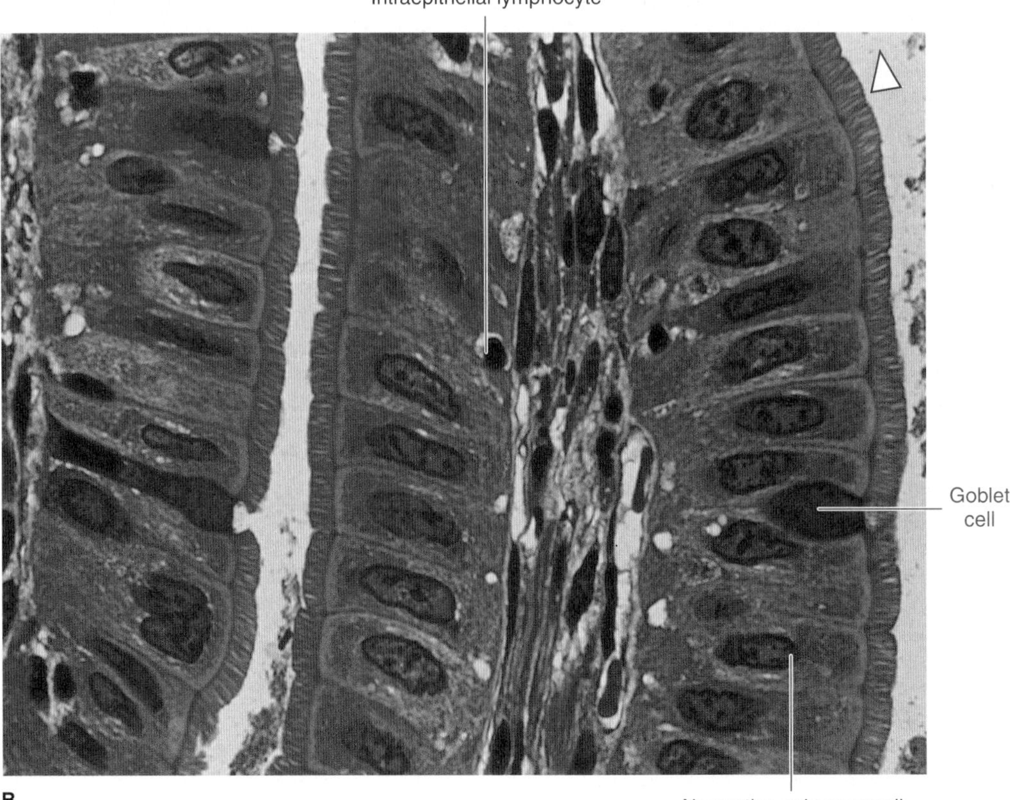

Figure 15–21. Photomicrograph of the epithelium covering the small intestine. **A:** Columnar epithelial cells with the brush border (arrowhead) interspersed with mucus-secreting goblet cells. The PAS-hematoxylin staining gives a positive reaction for the glycoproteins present in mucus and the brush border. Medium magnification. **B:** Numerous absorptive cells with their brush borders and the clearly visible intercellular limits. PT stain. High magnification.

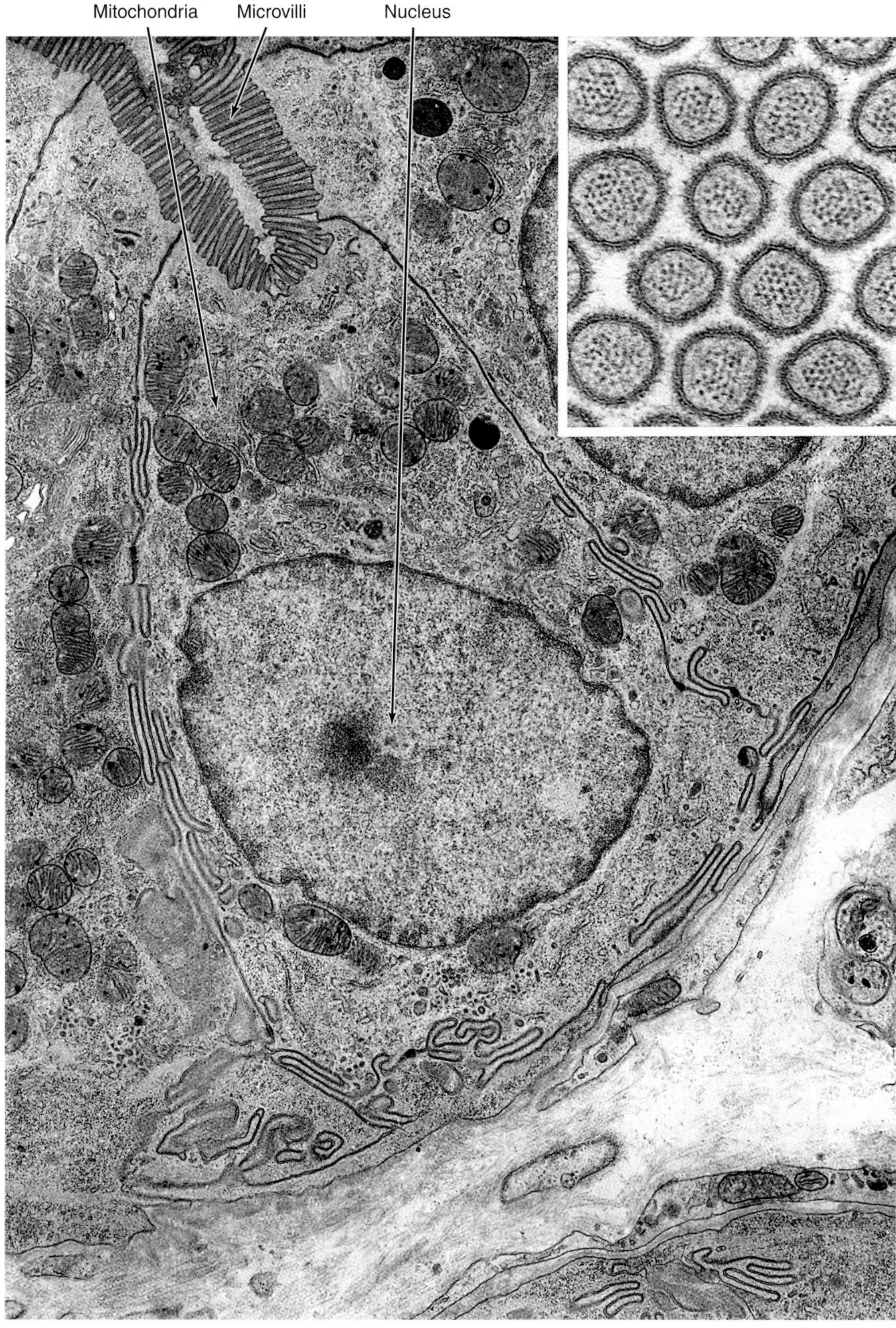

Figure 15–22. Electron micrograph of an absorptive epithelial cell of the small intestine. Note the accumulation of mitochondria in its apex. The luminal surface is covered with microvilli (shown in transverse section in the **inset**). Actin filaments, sectioned transversely, constitute the principal structural feature in the core of the microvilli. ×6300. (Courtesy of KR Porter.)

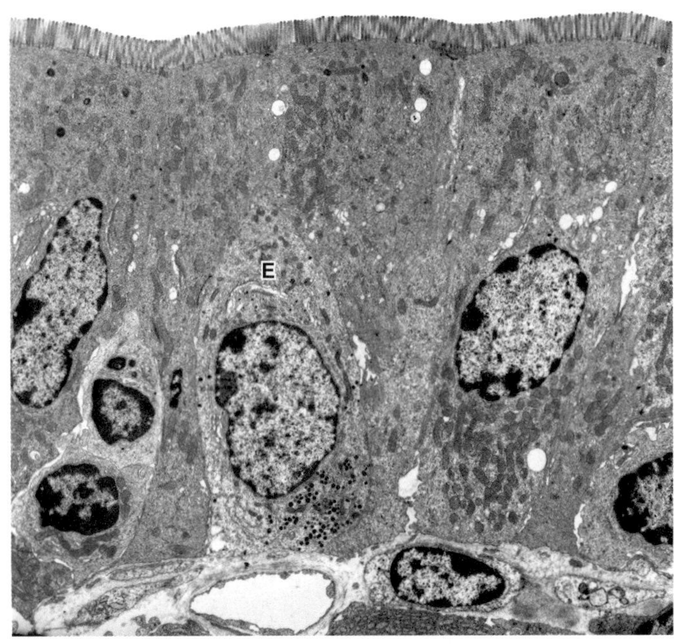

Figure 15–23. Electron micrograph of epithelium of the small intestine. Abundant microvilli at the cell apex can be seen to form the brush border. At the left are 2 lymphocytes migrating in the epithelium. In the center is an enteroendocrine cell (E) with its basal secretory granules. ×1850.

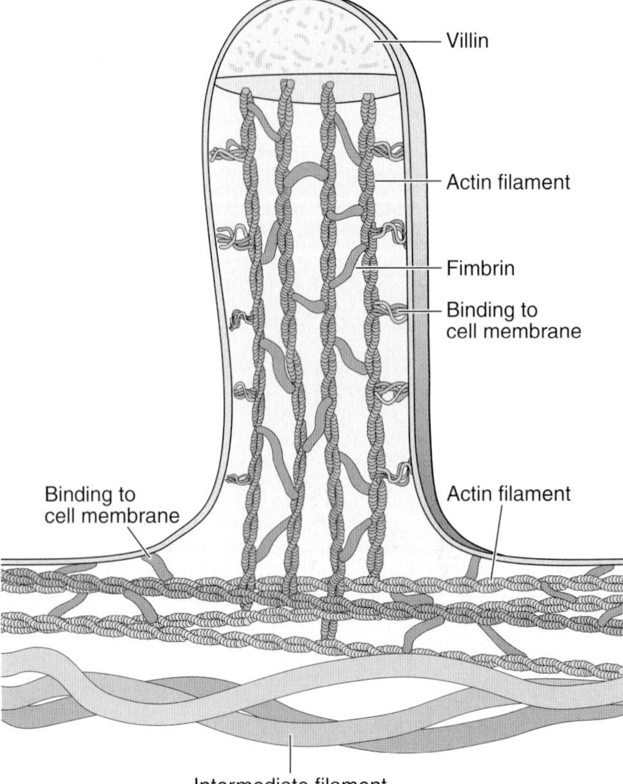

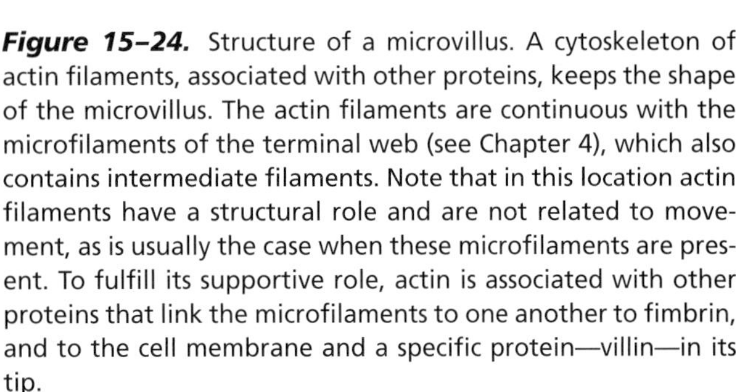

Figure 15–24. Structure of a microvillus. A cytoskeleton of actin filaments, associated with other proteins, keeps the shape of the microvillus. The actin filaments are continuous with the microfilaments of the terminal web (see Chapter 4), which also contains intermediate filaments. Note that in this location actin filaments have a structural role and are not related to movement, as is usually the case when these microfilaments are present. To fulfill its supportive role, actin is associated with other proteins that link the microfilaments to one another to fimbrin, and to the cell membrane and a specific protein—villin—in its tip.

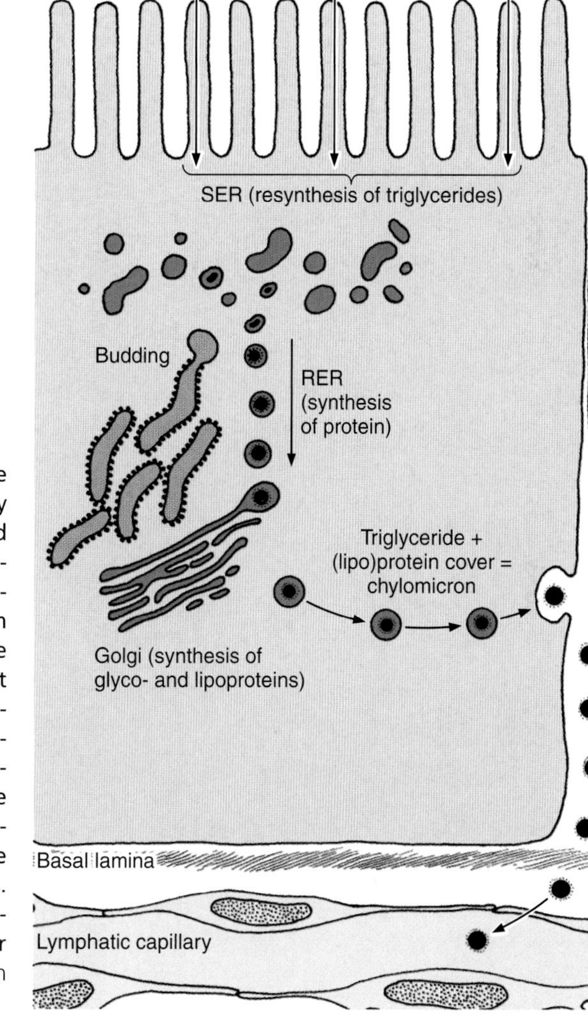

Figure 15–25. Lipid absorption in the small intestine. Lipase promotes the hydrolysis of lipids to monoglycerides and fatty acids in the intestinal lumen. These compounds are stabilized in an emulsion by the action of bile acids. The products of hydrolysis cross the microvilli membranes passively and are collected in the cisternae of the smooth endoplasmic reticulum (SER), where they are resynthesized to triglycerides. These triglycerides are surrounded by a thin layer of proteins that form particles called chylomicrons (0.2–1 μm in diameter). Chylomicrons are transferred to the Golgi complex and then migrate to the lateral membrane, cross it by a process of membrane fusion (exocytosis), and flow into the extracellular space in the direction of the blood and lymphatic vessels. Most chylomicrons go to the lymph; a few go to the blood vessels. The long-chain lipids (>C12) go mainly to the lymphatic vessels. Fatty acids of fewer than 10–12 carbon atoms are not re-esterified to triglycerides but leave the cell directly and enter the blood vessels. RER, rough endoplasmic reticulum. (Based on results of Friedman HI, Cardell RR Jr: Anat Rec 1977;188:77.)

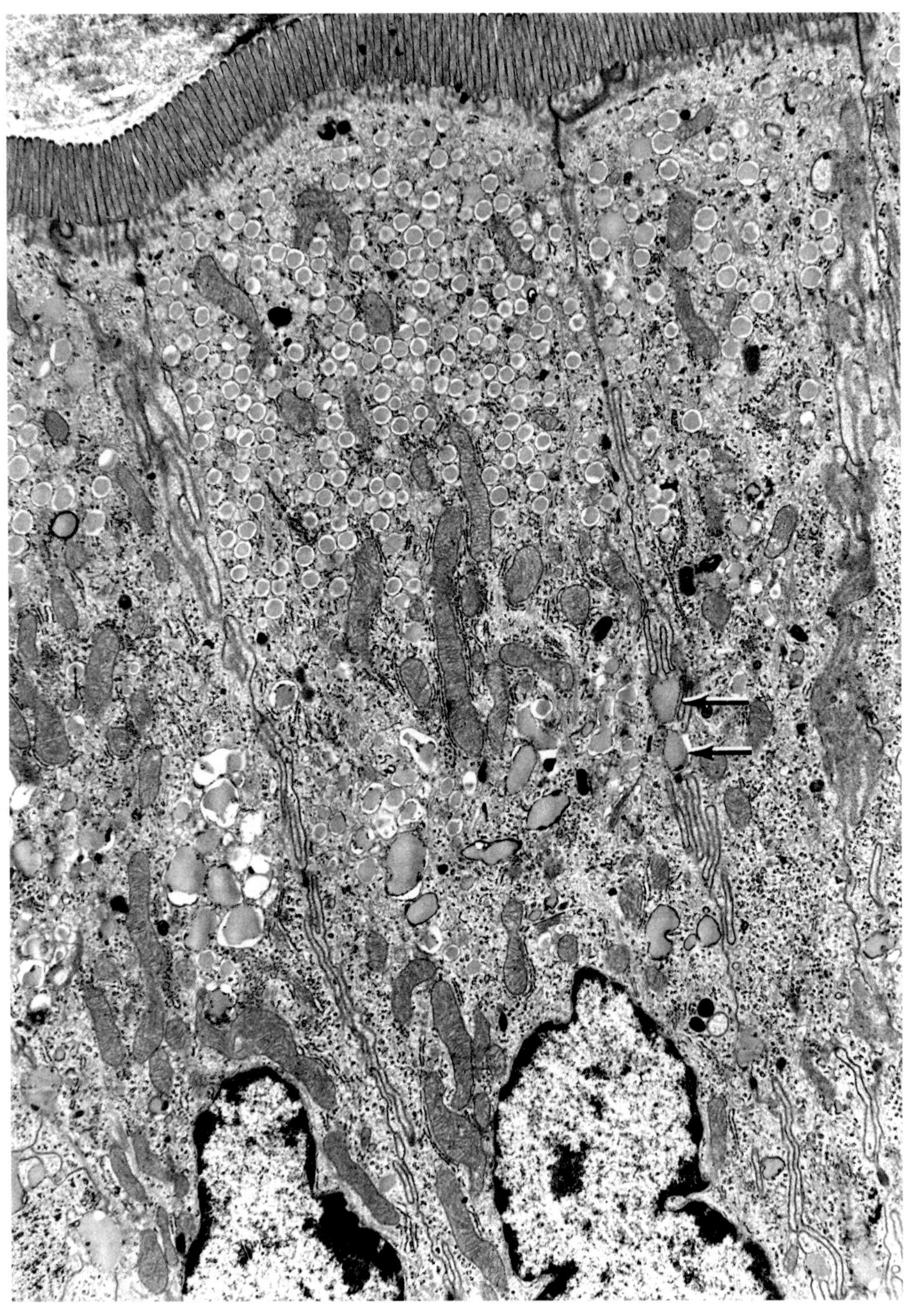

Figure 15–26. Electron micrograph of intestinal epithelium in the lipid-absorption phase. Note the accumulation of lipid droplets in vesicles of the smooth endoplasmic reticulum. These vesicles fuse near the nucleus, forming larger lipid droplets that migrate laterally and cross the cell membrane to the extracellular space (arrows). ×5000. (Courtesy of HI Friedman.)

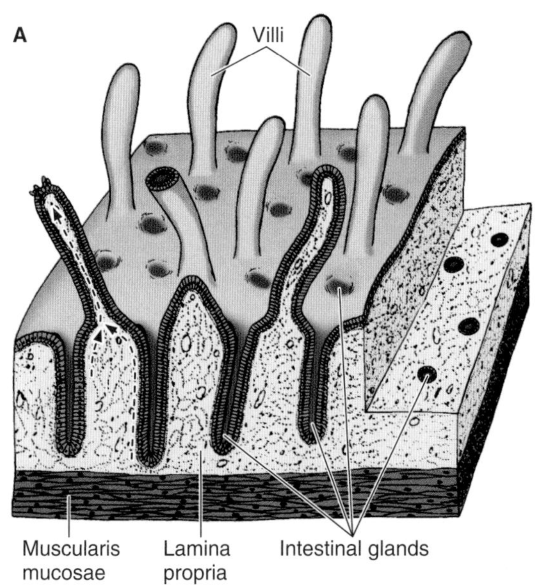

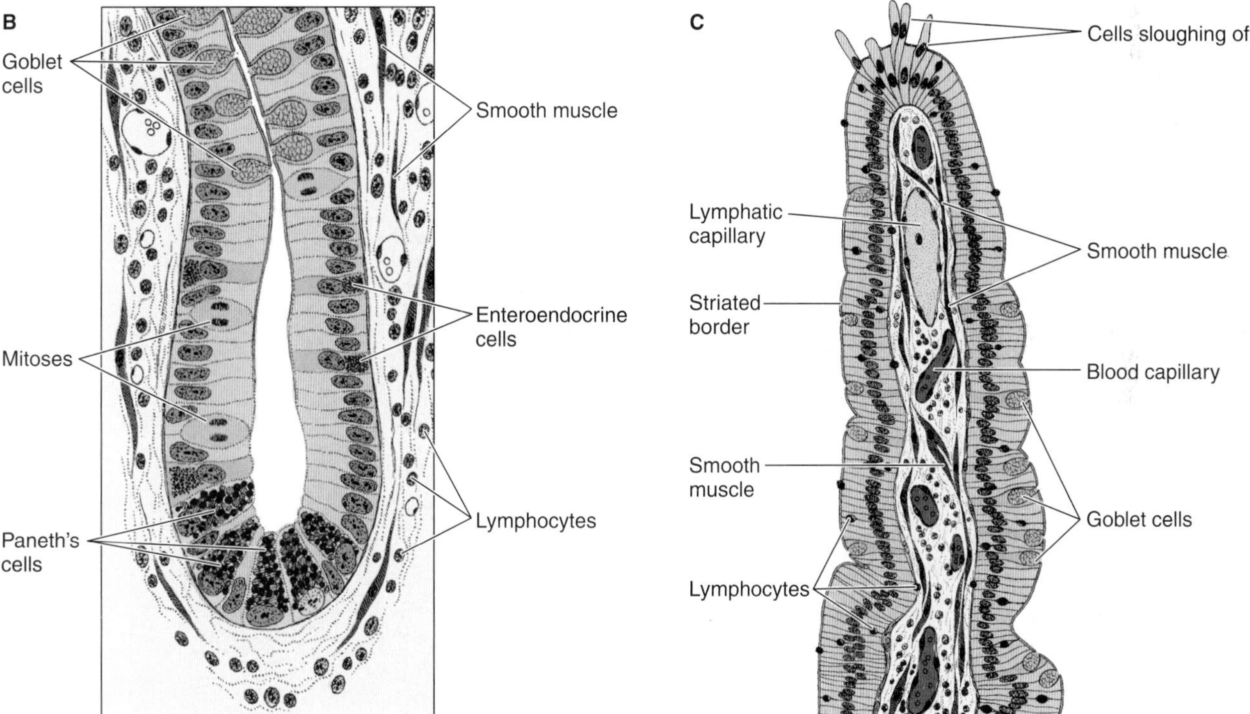

Figure 15–27. Schematic diagrams illustrating the structure of the small intestine. **A:** The small intestine under low magnification. In the villus to the left, note the desquamation of epithelial cells. Because of constant mitotic activity of the cells from the closed end of the glands and the upward migration of these cells (dashed arrows), the intestinal epithelium is continuously renewed. Note the intestinal glands of Lieberkühn. **B:** The intestinal glands have a lining of intestinal epithelium and goblet cells (upper portion). At the lower level, the immature epithelial cells are frequently seen in mitosis; note also the presence of Paneth's and enteroendocrine cells. As the immature cells progress upward, they differentiate and develop microvilli, seen as a brush border in the electron microscope. (See Figure 15–22.) Cell proliferation and cell differentiation occur simultaneously in the closed end of these glands. **C:** A villus tip showing the columnar covering epithelium with its striated border and a moderate number of goblet cells. Blood capillaries, a lymphatic capillary, smooth muscle cells, and lymphocytes can be seen in the connective tissue core of the villus. Great numbers of lymphocytes are in the epithelial layer. Cells are sloughing off at the apex of the villus. (Redrawn and reproduced, with permission, from Ham AW: *Histology,* 6th ed. Lippincott, 1969.)

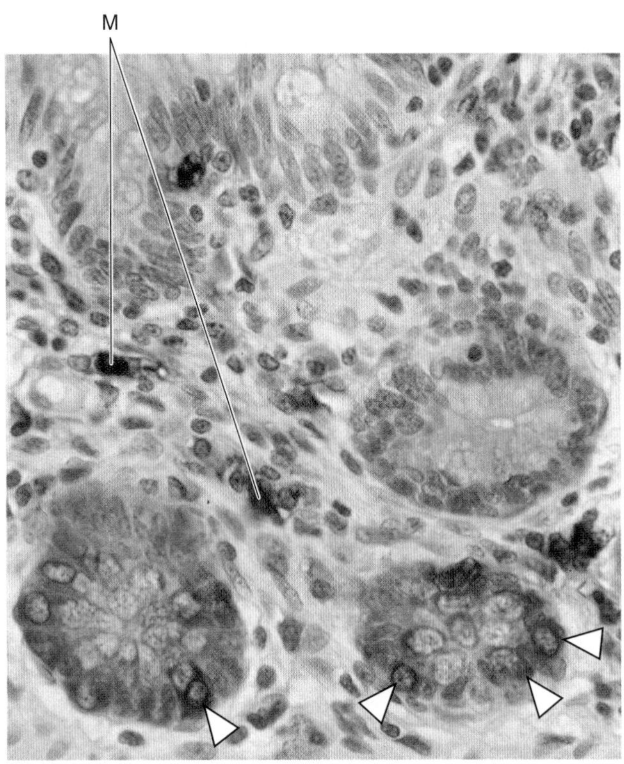

Figure 15–28. Section treated immunohistochemically to demonstrate the presence of lysozyme in the Paneth's cells of the small intestine (arrowheads) and the macrophages (M) of the connective tissue. Medium magnification.

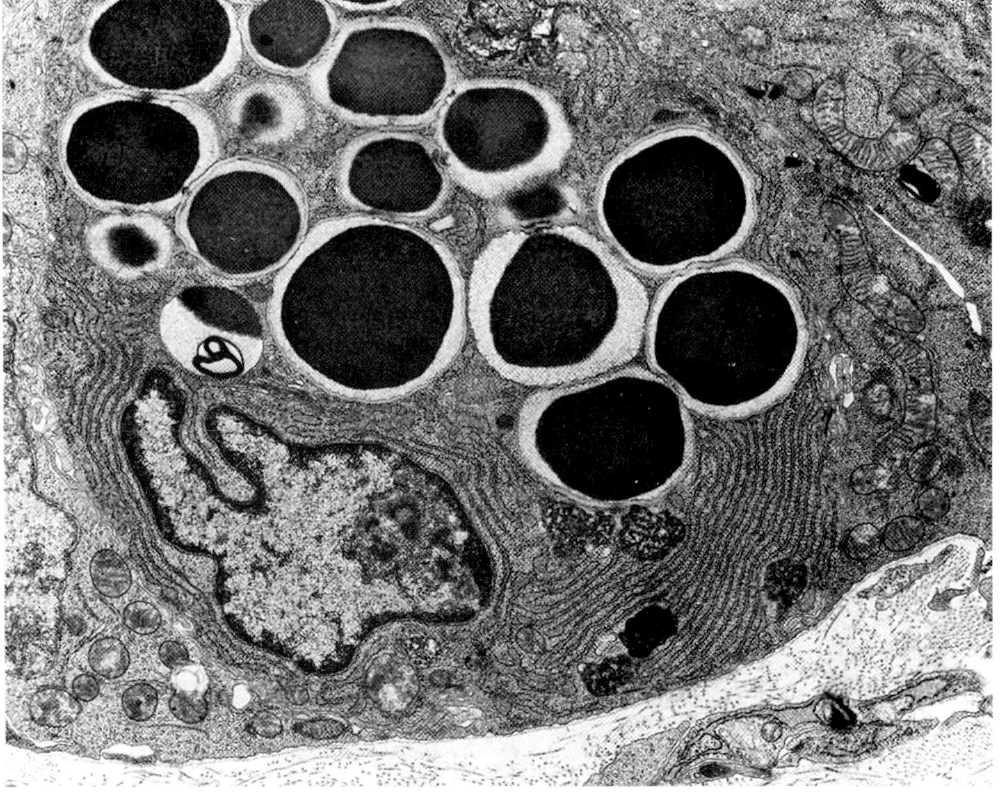

Figure 15–29. Electron micrograph of a Paneth's cell. Note the basal nucleus with prominent nucleolus, abundant rough endoplasmic reticulum, and large secretory granules with a protein core surrounded by a halo of polysaccharide-rich material. These granules contain lysozyme, a lytic enzyme involved in the regulation of intestinal bacteria. ×3000.

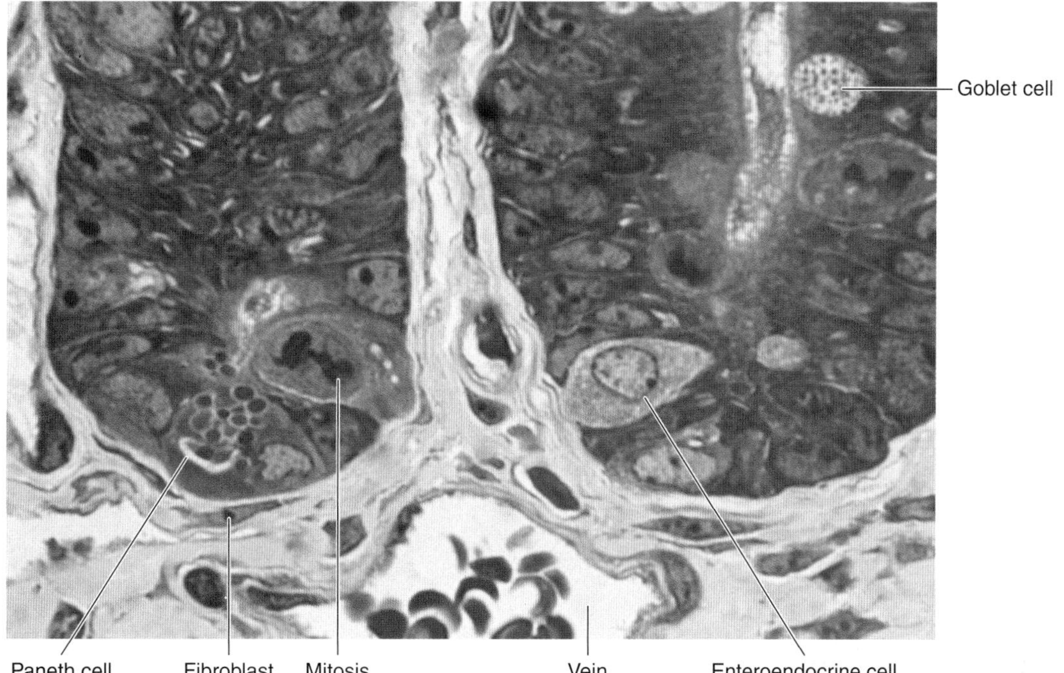

Figure 15–30. Photomicrograph of the basal portion of 2 glands (crypts) of the small intestine. Note the enteroendocrine cell, Paneth cell, goblet cell, and a cell in mitosis.

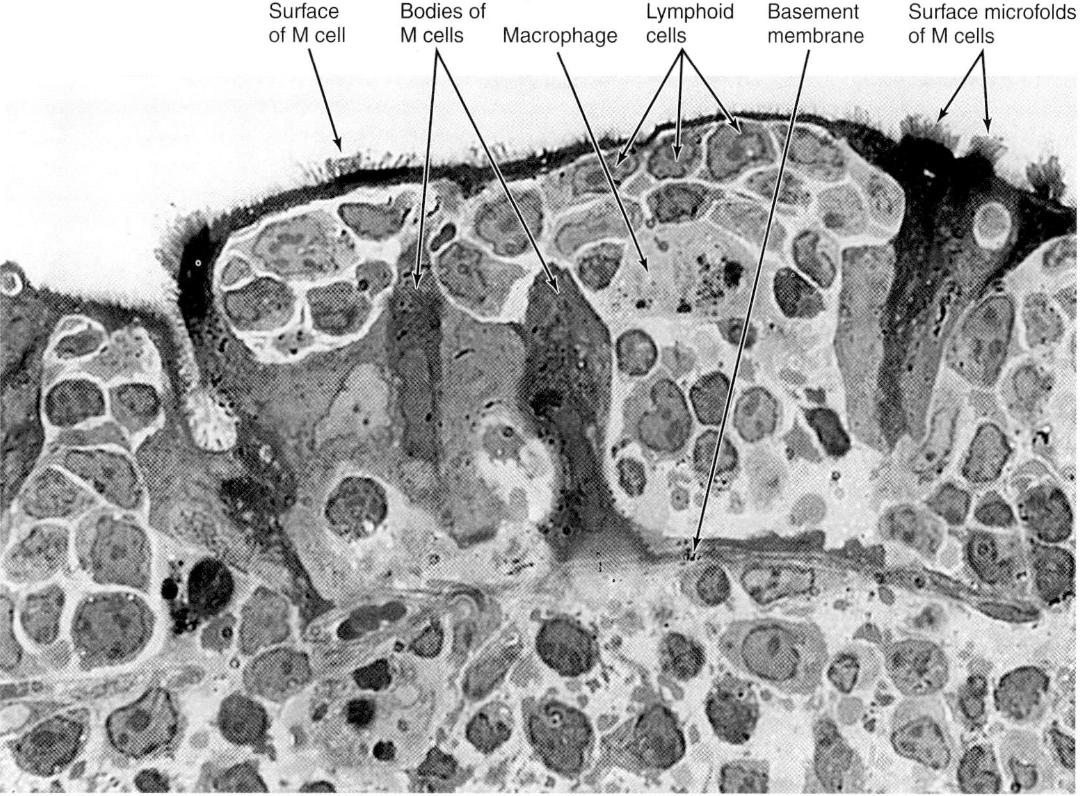

Figure 15–31. Electron micrograph from a region of intestine where a lymphoid nodule is covered by the intestinal mucosa. Note the presence of M cells that form a special compartment containing lymphoid cells. A macrophage (an antigen-presenting cell) is also in the compartment. (Courtesy of M Neutra.)

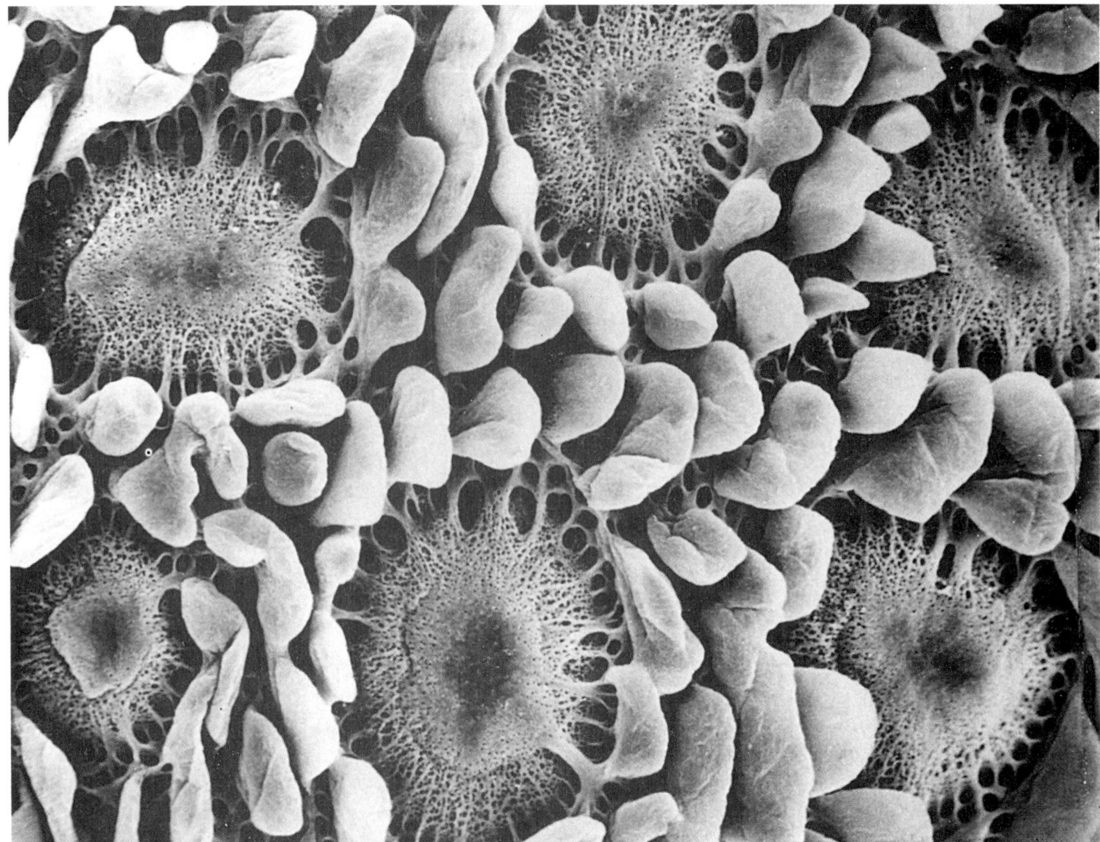

Figure 15–32. Scanning electron micrograph of the intestinal surface after removal of the mucosal epithelium, showing the basement membrane. Note that this layer is continuous when covering the remnants of the intestinal villi but assumes the structure of a sieve when covering the lymphoid follicles above Peyer's patches. This configuration permits an easier means for immunogenic materials to reach underlying lymphoid tissues. (Courtesy of S McClugage.)

and smooth muscle cells. The smooth muscle cells are responsible for the rhythmic movements of the villi, which are important for absorption (see Figures 15–27 and 15–34).

The muscularis mucosae does not present any peculiarities in this organ. The **submucosa** contains, in the initial portion of the duodenum, clusters of ramified, coiled tubular glands that open into the intestinal glands. These are the **duodenal** (or **Brunner's) glands** (Figure 15–35). Their cells are of the mucous type. The product of secretion of the glands is distinctly alkaline (pH 8.1–9.3). It acts to protect the duodenal mucous membrane against the effects of the acid gastric juice and to bring the intestinal contents to the optimum pH for pancreatic enzyme action.

The lamina propria and the submucosa of the small intestine contain aggregates of lymphoid nodules known as **Peyer's patches,** an important component of the GALT. Each patch consists of 10–200 nodules and is visible to the naked eye as an oval area on the antimesenteric side of the intestine. There are about 30 patches in the human, most of them in the ileum. When viewed from the luminal surface, each Peyer's patch appears as a dome-shaped area devoid of villi (Figure 15–32). Instead of absorptive cells, its covering epithelium consists of **M cells** (Figure 15–31). The muscularis is well developed in the intestines, composed of an internal circular layer and an external longitudinal layer (Figure 15–36). The appearance of the smooth muscle cells in these layers in histological sections will depend on the plane of the section (transverse or longitudinal).

Vessels & Nerves

The blood vessels that nourish the intestine and remove absorbed products of digestion penetrate the muscularis and form a large plexus in the submucosa (Figure 15–34). From the submucosa, branches extend through the muscularis mucosae and lamina propria and into the villi. Each villus receives, according to its size, one or more branches that form a capillary network just below its epithelium. At the tips of the villi, one or more venules arise from these capillaries and run in the opposite direction, reaching the veins of the submucosal plexus. The lymph vessels of the intestine begin as closed tubes in the core of the villi. These capillaries (**lacteals**), despite being larger than the blood capillaries, are difficult to observe because their walls are so close together that they appear to be collapsed. Lacteals run to the region of lamina propria above the muscularis mucosae, where they form a plexus. From there they are directed to the submucosa, where they surround lymphoid nodules. Lacteals anastomose repeatedly and leave the intestine along with the blood vessels. They are especially important for the absorption of lipids, because blood circulation does not easily accept the lipoproteins produced by the tall columnar cells during this process.

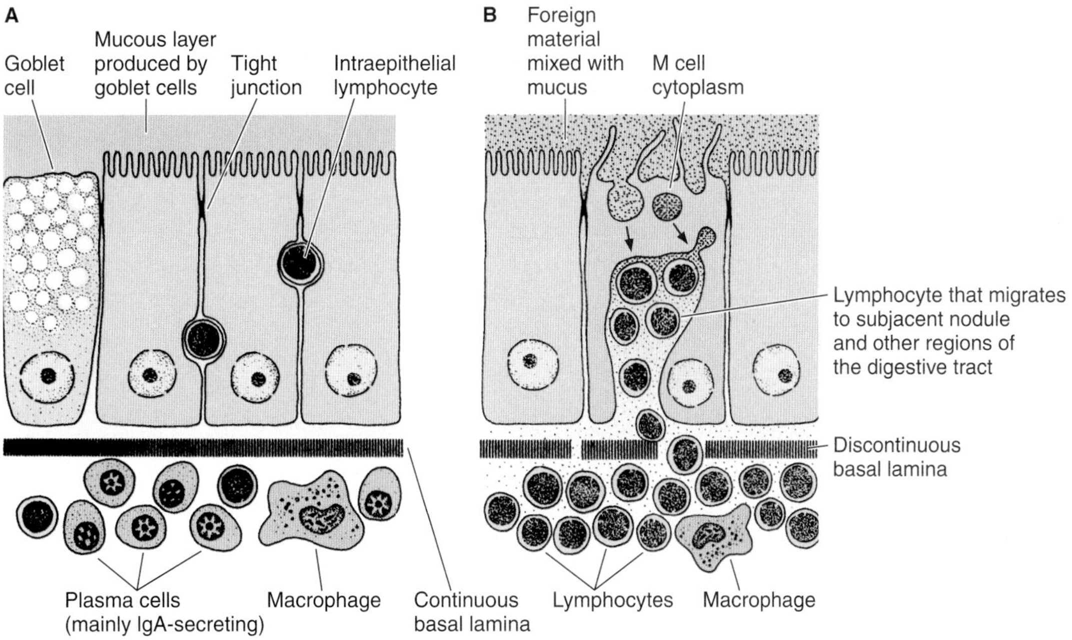

A
Goblet cell
Mucous layer produced by goblet cells
Tight junction
Intraepithelial lymphocyte
Plasma cells (mainly IgA-secreting)
Macrophage
Continuous basal lamina

B
Foreign material mixed with mucus
M cell cytoplasm
Lymphocyte that migrates to subjacent nodule and other regions of the digestive tract
Discontinuous basal lamina
Lymphocytes
Macrophage

Figure 15–33. Some aspects of immunologic protection of the intestine. **A:** A condition that is more frequent in the upper tract, such as in the jejunum. There are many IgA-secreting plasma cells, scattered lymphocytes, and some macrophages. Note that the lymphocytes in the epithelial lining are located outside the epithelial cells, and below the tight junctions. **B:** A condition that is more frequent in the ileum, where aggregates of lymphocytes are located under M cells. The M cells transfer foreign material (microorganisms and macromolecules) to lymphocytes located deep in the cavities of the M cells. Lymphocytes spread the information received from this foreign material to other regions of the digestive tract, and probably to other organs, through blood and lymph.

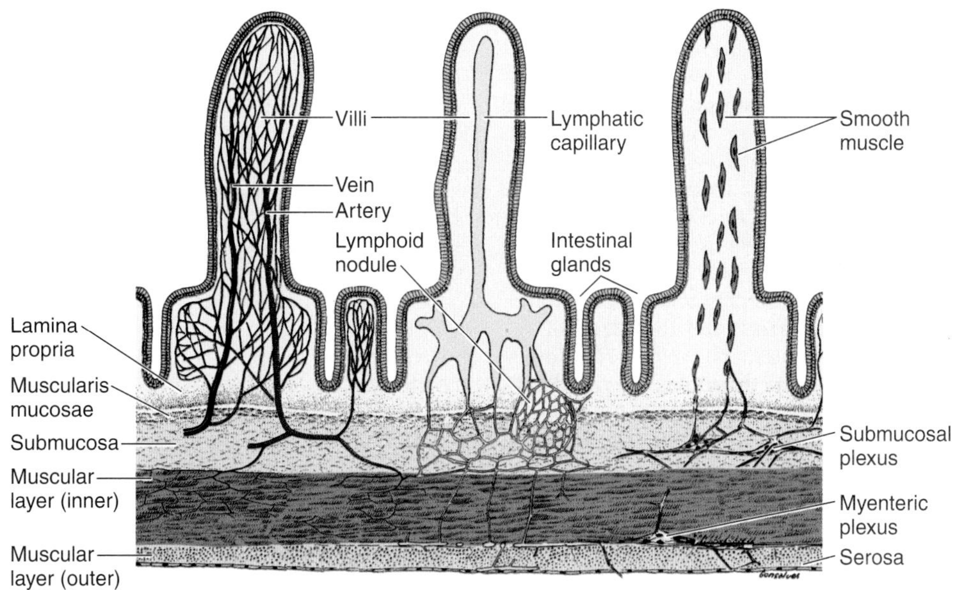

Figure 15–34. Blood circulation (left), lymphatic circulation (center), and innervation (right) of the small intestine. The smooth muscle system for contracting the villi is illustrated in the villus on the right.

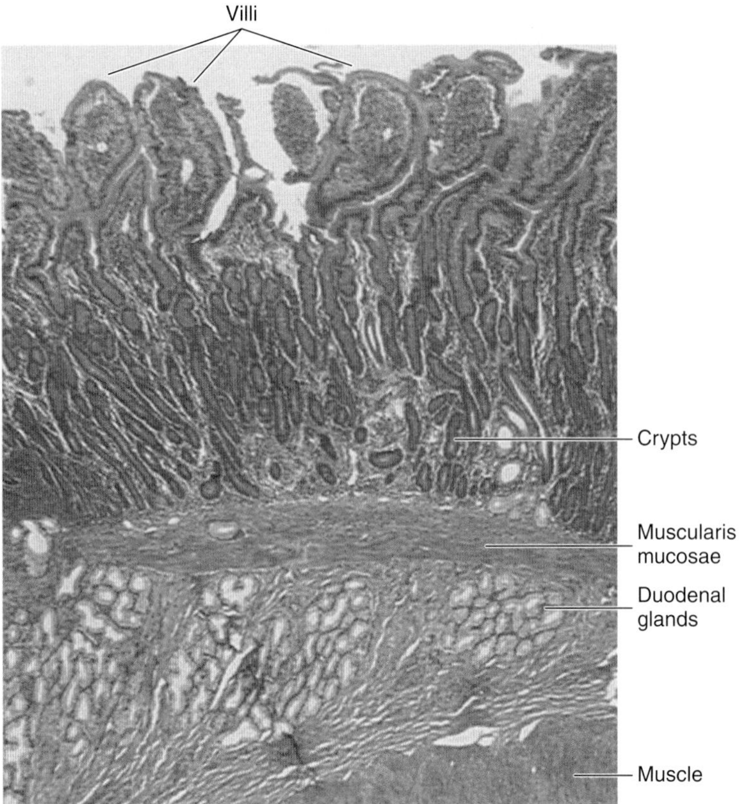

Figure 15–35. Photomicrograph of the duodenum, showing villi and duodenal glands in the submucosa. H&E stain. Low magnification.

Another process important for intestinal function is the rhythmic movement of the villi. This movement is the result of the contraction of smooth muscle cells running vertically between the muscularis mucosae and the tip of the villi (see villus at right in Figure 15–34). These contractions occur at the rate of several strokes per minute and have a pumping action on the villi that propel the lymph to the mesenteric lymphatics.

The innervation of the intestines is formed by both an **intrinsic component** and an **extrinsic component.** The intrinsic component comprises groups of neurons that form the myenteric (Auerbach's) nerve plexus (Figures 15–34 and 15–37) between the outer longitudinal and inner circular layers of the muscularis and the **submucosal (Meissner's) plexus** in the submucosa. The plexuses contain some sensory neurons that receive information from nerve endings near the epithelial layer and in the smooth muscle layer regarding the composition of the intestinal content (chemoreceptors) and the degree of expansion of the intestinal wall (mechanoreceptors). The other nerve cells are effectors and innervate the muscle layers and hormone-secreting cells. The intrinsic innervation formed by these plexuses is responsible for the intestinal contractions that occur in the total absence of the extrinsic innervation. The extrinsic innervation is formed by parasympathetic cholinergic nerve fibers that stimulate the activity of the intestinal smooth muscle and by sympathetic adrenergic nerve fibers that depress intestinal smooth muscle activity.

LARGE INTESTINE

The large intestine consists of a mucosal membrane with no folds except in its distal (rectal) portion. No villi are present in this portion of the intestine (Figure 15–38). The intestinal glands are long and characterized by a great abundance of goblet and absorptive cells and a small number of enteroendocrine cells (Figures 15–38 and 15–39). The absorptive cells are columnar and have short, irregular microvilli (Figure 15–40). The large intestine is well suited to its main functions: absorption of water, formation of the fecal mass, and production of mucus. Mucus is a highly hydrated gel that not only lubricates the intestinal surface but also covers bacteria and particulate matter. The absorption of water is passive, following the active transport of sodium out of the basal surfaces of the epithelial cells (Figure 15–40).

The lamina propria is rich in lymphoid cells and in nodules that frequently extend into the submucosa. This richness in lymphoid tissue (GALT) is related to the abundant bacterial population of the large intestine. The muscularis comprises longitudinal and circular strands. This layer differs from that of the small intestine, because fibers of the outer longitudinal layer congregate in 3 thick longitudinal bands called **teniae coli.** In the intraperitoneal portions of the colon, the serous layer is characterized by small, pendulous protuberances composed of adipose tissue—the **appendices epiploicae.**

In the anal region, the mucous membrane forms a series of longitudinal folds, the **rectal columns of Morgagni.** About 2 cm above the anal opening, the intestinal mucosa is replaced by stratified squamous epithelium. In this region, the lamina propria contains a plexus of large veins that, when excessively dilated and varicose, produce hemorrhoids.

Cell Renewal in the Gastrointestinal Tract

The epithelial cells of the entire gastrointestinal tract are constantly being cast off and replaced with new ones formed through mitosis of stem cells. These stem cells are located in the basal

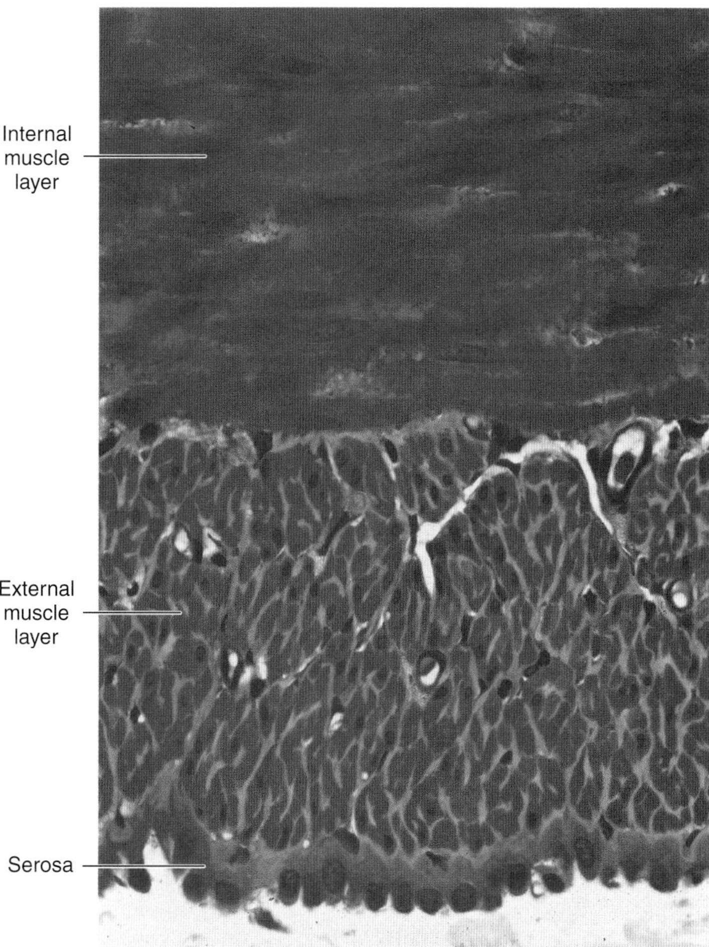

Figure 15–36. Transverse section of the small intestinal wall showing the two smooth muscle layers and serosa. The inner layer is predominantly circular while the other layer is longitudinal. The serosa is composed of a thin connective tissue covered by the mesothelium. PT stain. Medium magnification.

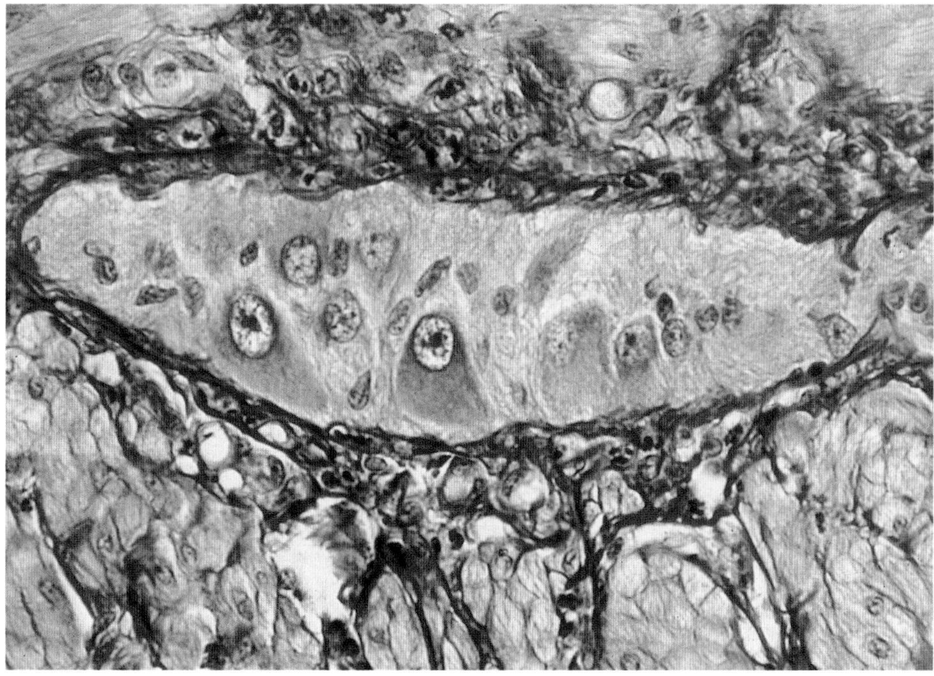

Figure 15–37. Photomicrograph of a group of neurons (with large nuclei) and satellite cells (with small nuclei) constituting a component of the myenteric plexus between 2 smooth muscle layers. Note the red stained collagen fibers. Picrosirius-hematoxylin. Medium magnification.

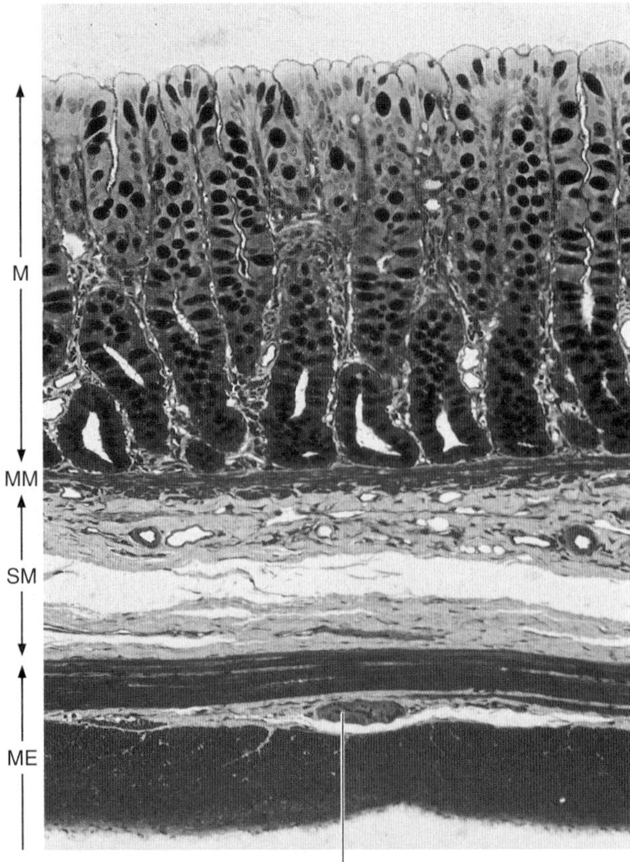

Figure 15–38. Photomicrograph of a section of large intestine with its various layers. Note the absence of villi. M, mucosa; MM, muscularis mucosae; SM, submucosa; ME, muscularis externa. PT stain. Low magnification.

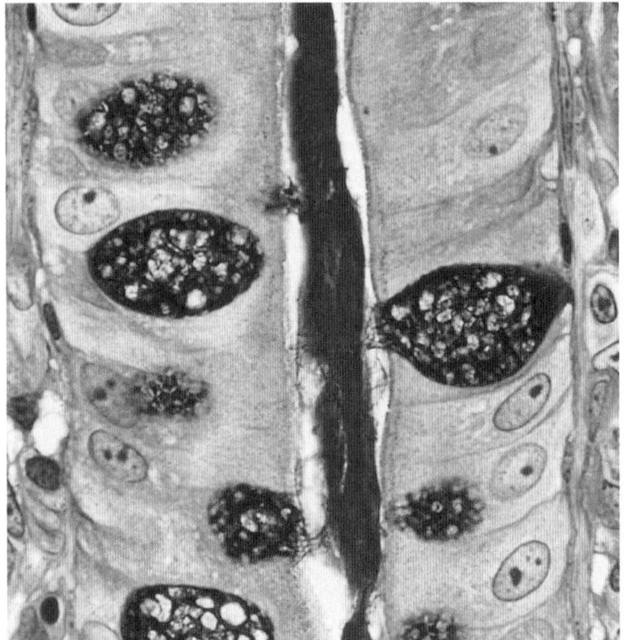

Figure 15–39. Section of a large intestinal gland showing its absorptive and mucous goblet cells. Observe that the goblet cells are secreting and beginning to fill the lumen of the gland with its secretions. The microvilli in the absorptive cells participate in the process of water absorption. PT stain. High magnification.

layer of the esophageal epithelium, the neck of gastric glands, the lower half of the intestinal glands (Figure 15–41), and the bottom third of the crypts of the large intestine. From this proliferative zone in each region, cells move to the maturation area, where they undergo structural and enzymatic maturation, providing the functional cell population of each region. In the small intestine the cells die by apoptosis in the tip of the villi.

MEDICAL APPLICATION

The high rate of cell renewal explains why the intestine is affected promptly by the administration of antimitotic drugs, as in cancer chemotherapy. The epithelial cells continue to be lost at the tips of the villi, but the drugs inhibit cell proliferation. This inhibition promotes atrophy of the epithelium, which results in defective absorption of nutrients, excessive fluid loss, and diarrhea. Paneth's cells of the intestinal glands have a much slower turnover rate, living about 30 days before being replaced.

APPENDIX

The appendix is an evagination of the cecum; it is characterized by a relatively small, narrow, and irregular lumen that is caused

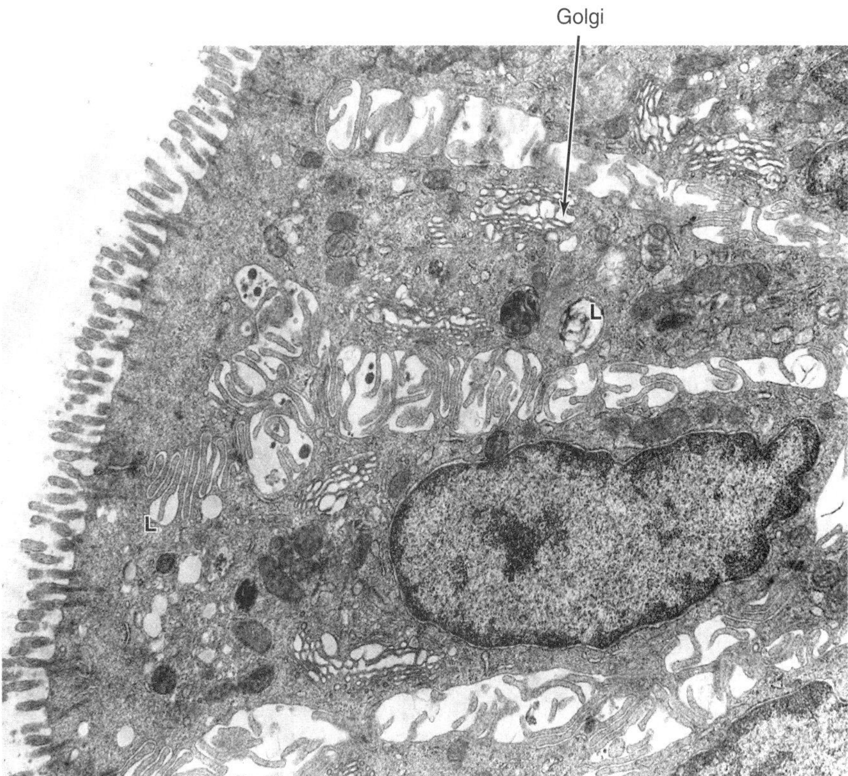

Figure 15–40. Electron micrograph of epithelial cells of the large intestine. Note the microvilli at the luminal surface, the well-developed Golgi complex, and dilated intercellular spaces filled by interdigitating membrane leaflets, a sign of active water transport. ×3900.

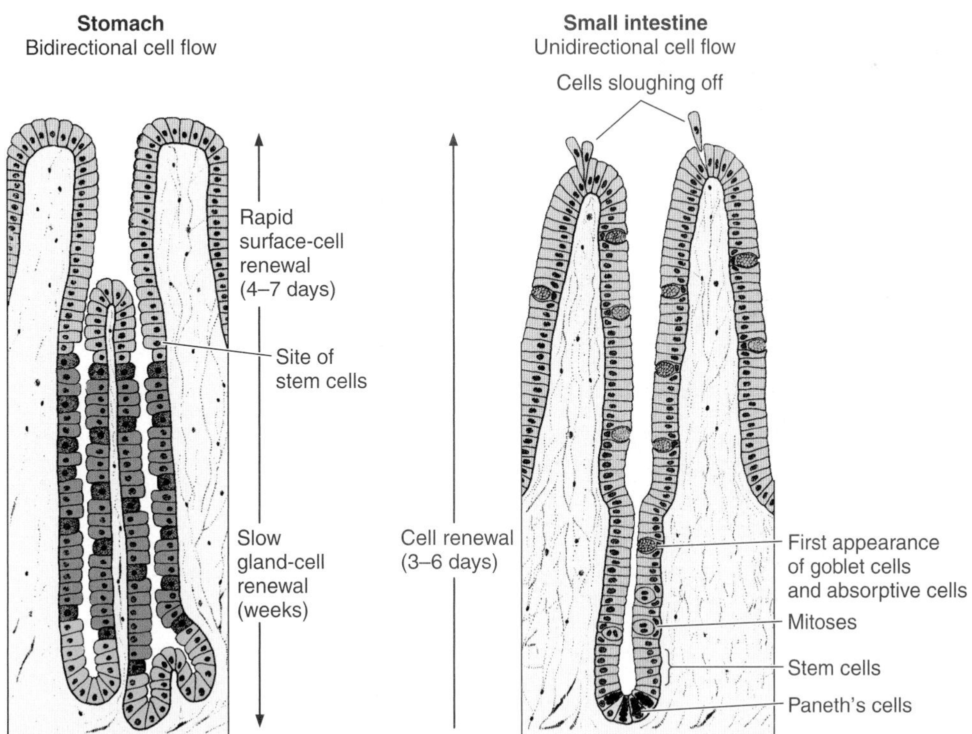

Figure 15–41. Regeneration of the epithelial lining of the stomach and small intestine. Note differences in the location of stem cells.

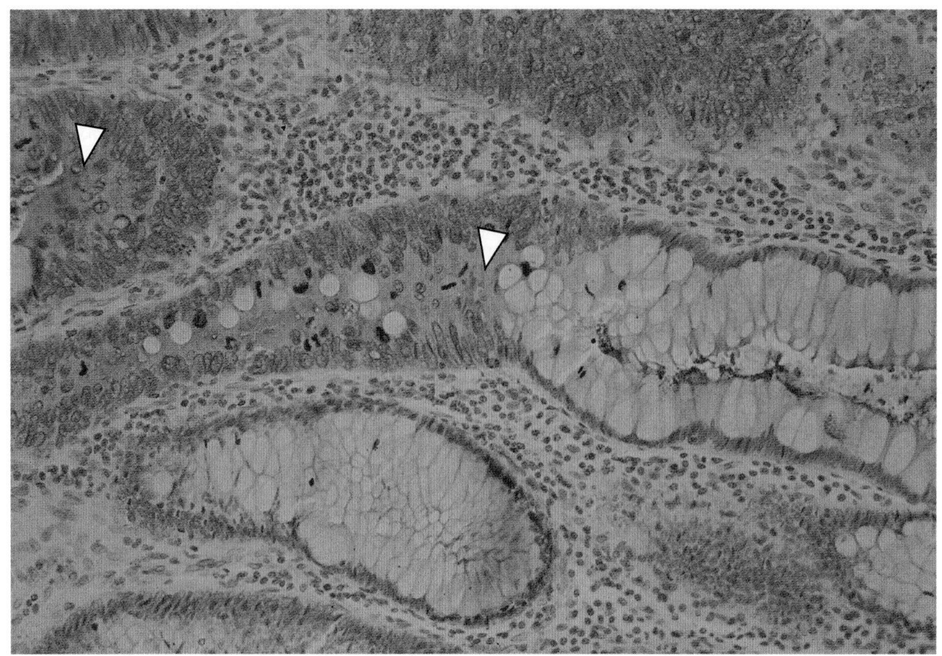

Figure 15–42. Section of large intestine treated with an immunocytochemical procedure to detect carcinoembryonic antigen (CEA), a protein present in malign tumors mainly of the digestive tract and breast. Note the transition from normal unstained mucous gland cells to the stained tumor cells (arrowheads). Counterstained with hematoxylin. Medium magnification.

by the presence of abundant lymphoid follicles in its wall. Although its general structure is similar to that of the large intestine, it contains fewer and shorter intestinal glands and has no teniae coli.

MEDICAL APPLICATION

*Because the appendix is closed-ended, its contents are not renewed rapidly, and it frequently becomes a site of inflammation (**appendicitis**). The inflammation can progress to the point of destruction of this structure, with consequent infection of the peritoneal cavity.*

Cancer of the Digestive Tract

MEDICAL APPLICATION

*Approximately 90–95% of malignant tumors of the digestive system are derived from intestinal or gastric epithelial cells. Malignant tumors of the large bowel are derived almost exclusively from its glandular epithelium (**adenocarcinomas**) and are the second most common cause of cancer deaths in the United States. Some proteins such as the carcinoembryonic antigen produced exclusively by malign cells are very important for the diagnosis of cancer (Figure 15–42).*

REFERENCES

Allen A, Flemström G, Garner A, Kivilaakso E: Gastroduodenal mucosal protection. Physiol Rev 1993;73:823.

Cheng H, Leblond CP: Origin, differentiation and renewal of the four main epithelial cell types in the mouse small intestine. 5. Unitarian theory of the origin of the four epithelial cell types. Am J Anat 1974;141:537.

Forte JG et al: Mechanism of gastric H^+ and Cl^- transport. Annu Rev Physiol 1980;42:111.

Friedman HL, Cardell RR Jr: Alterations in the endoplasmic reticulum and Golgi complex of intestinal epithelial cells during fat absorption and after termination of this process: a morphological and morphometric study. Anat Rec 1977;188:77.

Gabella G: Innervation of the gastrointestinal tract. Int Rev Cytol 1979;59:130.

Grube D: The endocrine cells of the digestive system: amines, peptides and modes of action. Anat Embryol (Berl) 1986;175:151.

Hoedemseker PJ et al: Further investigations about the site of production of Castles gastric intrinsic factor. Lab Invest 1966;15:1163.

Jankowski A et al: Maintenance of normal intestinal mucosae: function, structure and adaptation. Gut 1994;35:S1.

Klockars M, Reitamo S: Tissue distribution of lysozyme in man. J Histochem Cytochem 1975;23:932.

Madara JL, Trier JS: The functional morphology of the mucosa of the small intestine. In Johnson LR (editor): *Physiology of the Gastrointestinal Tract,* Vol 2. Raven Press, 1994.

McClugage SG et al: Porosity of the basement membrane overlying Peyer's patches in rats and monkeys. Gastroenterology 1986;91:1128.

Moog F: The lining of the small intestine. Sci Am 1981;245:154.

Mooseker MS, Tilney LG: Organization of an actin filament-membrane complex: Filament polarity and membrane attachment in the microvilli of intestinal epithelial cells. J Cell Biol 1975;67:725.

Owen D: Normal histology of the stomach. Am J Surg Pathol 1986;10:48.

Pabst R: The anatomical basis for the immune function of the gut. Anat Embryol (Berl) 1986;176:135.

Pfeiffer CJ et al: *Gastrointestinal Ultrastructure.* Academic Press, 1974.

Sachs G: The gastric H^+,K^+ ATPase: regulation and structure/function of the acid pump of the stomach. In Johnson LR (editor): *Physiology of the Gastrointestinal Tract,* Vol 2. Raven Press, 1994.

Smith DV, Margolskee RF: Making sense of taste. Sci Am 2001;284(3):33.

Organs Associated with the Digestive Tract

16

The organs associated with the digestive tract include the salivary glands, the pancreas, the liver, and the gallbladder. The main functions of the salivary glands are to wet and lubricate the oral mucosa and the ingested food, to initiate the digestion of carbohydrates and lipids (by means of amylase and lingual lipase activities, respectively), and to secrete germicidal protective substances such as the immunoglobulin IgA, lysozyme, and lactoferrin. The saliva also has a very important buffering function and forms a protective pellicle on the teeth by means of calcium-binding proline-rich salivary proteins. In some species (but not in humans), saliva is very important for evaporative cooling.

MEDICAL APPLICATION

The hypofunction of the major salivary glands due to diseases or radiotherapy is associated with caries, atrophy of the oral mucosa and speech difficulties.

The main functions of the pancreas are to produce digestive enzymes that act in the small intestine and to secrete hormones such as insulin and glucagon into the bloodstream. Both are very important for the metabolism of the absorbed nutrients. The liver produces bile, an important fluid in the digestion of fats. The liver plays a major role in lipid, carbohydrate, and protein metabolism and inactivates and metabolizes many toxic substances and drugs. It also participates in iron metabolism and the synthesis of blood proteins and the factors necessary for blood coagulation. The gallbladder absorbs water from the bile and stores the bile in a concentrated form.

SALIVARY GLANDS

Exocrine glands in the mouth produce saliva, which has digestive, lubricating, and protective functions. In addition to the small glands scattered throughout the oral cavity, there are 3 pairs of large salivary glands: the **parotid, submandibular (submaxillary)**, and **sublingual glands.** In humans, the minor salivary glands secrete 10% of the total volume of saliva, but they account for approximately 70% of the mucus secreted.

A capsule of connective tissue, rich in collagen fibers, surrounds the large salivary glands. The parenchyma of the glands consists of secretory endpieces and a branching duct system arranged in lobules, separated by septae of connective tissue originating from the capsule. The secretory endpieces present 2 types of secretory cells—serous and mucous (Figure 16–1)—as well as the nonsecretory myoepithelial cells (described in Chapter 4). This secretory portion is followed by a duct system whose components modify and conduct the saliva to the oral cavity.

Serous cells are usually pyramidal in shape, with a broad base resting on the basal lamina and a narrow apical surface with short, irregular microvilli facing the lumen (Figure 16–2). They exhibit characteristics of polarized protein-secreting cells. Adjacent secretory cells are joined together by junctional complexes and usually form a spherical mass of cells called **acinus,** with a lumen in the center (Figure 16–1). This structure can be likened to a grape attached to its stem; the stem corresponds to the duct system.

Mucous cells are usually cuboidal to columnar in shape; their nuclei are oval and pressed toward the bases of the cells. They exhibit the characteristics of mucus-secreting cells (Figures 16–1, 16–3, and 16–4), containing glycoproteins important for the moistening and lubricating functions of the saliva. Most of these glycoproteins are called mucins and contain 70–80% carbohydrate moieties in their structure. Mucous cells are most often organized as **tubules,** consisting of cylindrical arrays of secretory cells surrounding a lumen.

In the human **submandibular gland,** serous and mucous cells are arranged in a characteristic pattern. The mucous cells form tubules, but their ends are capped by serous cells, which constitute the **serous demilunes** (Figures 16–1 and 16–4).

Myoepithelial cells, described in Chapter 4, are found within the basal lamina of the secretory endpieces and intercalated ducts (to a lesser extent), which form the initial portion of the duct system (Figure 16–1). Myoepithelial cells surrounding the secretory portion are well developed and branched (and are sometimes called **basket cells**), whereas those associated with intercalated ducts are spindle-shaped and lie parallel to the length of the duct. Although the contraction of these cells accelerates the secretion of saliva, their main function seems to be the prevention of endpiece distension during secretion due to the increase in intraluminal pressure.

In the **duct system,** secretory endpieces empty into the **intercalated ducts,** lined by cuboidal epithelial cells. Several of these short ducts join to form **striated ducts** (Figure 16–1), characterized by radial striations that extend from the bases of the cells to the level of the nuclei. When viewed in the electron microscope, the striations are seen to consist of infoldings of the basal plasma membrane with numerous elongated mitochondria

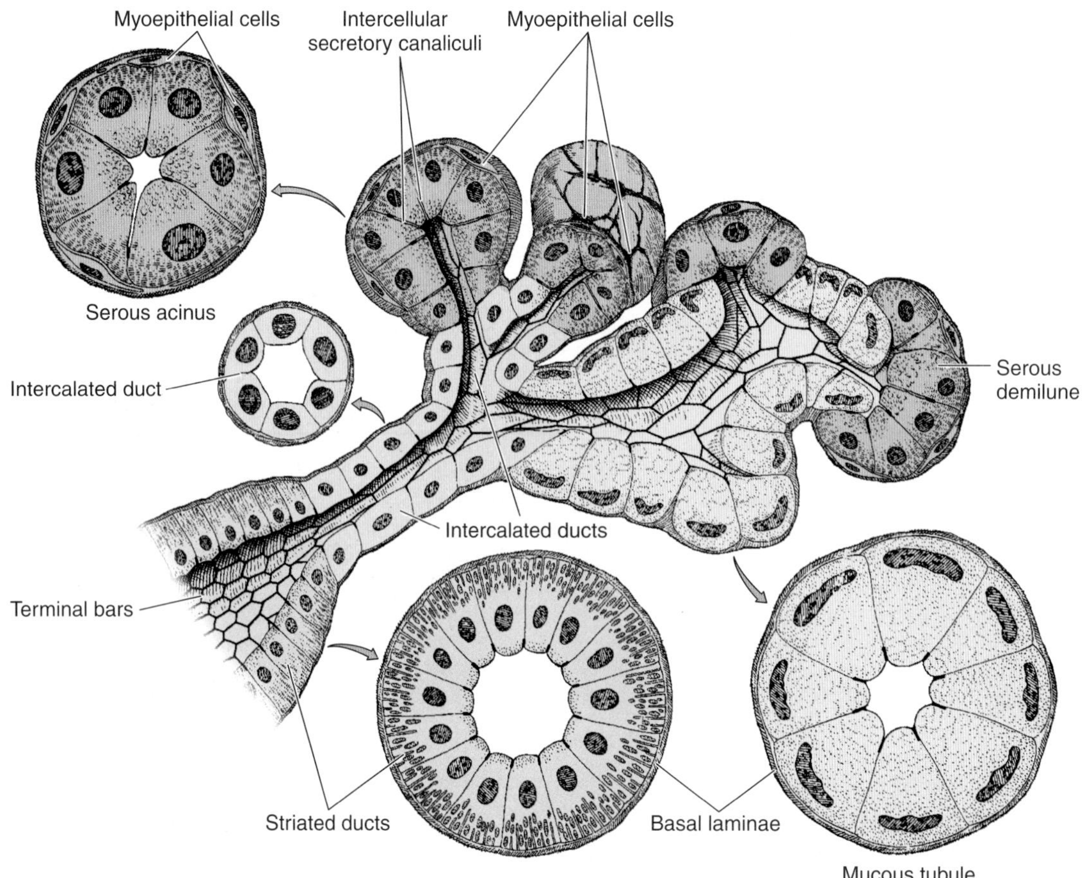

Figure 16–1. The structure of the submandibular (submaxillary) gland. The secretory portions are composed of pyramidal serous (light purple) and mucous (light yellow) cells. Serous cells are typical protein-secreting cells, with rounded nuclei, accumulation of rough endoplasmic reticulum in the basal third, and an apex filled with protein-rich secretory granules. The nuclei of mucous cells, flattened with condensed chromatin, are located near the bases of the cells. The short intercalated ducts are lined with cuboidal epithelium. The striated ducts are composed of columnar cells with characteristics of ion-transporting cells, such as basal membrane invaginations and mitochondrial accumulation. Myoepithelial cells are shown in the serous secretory endpieces.

that are aligned parallel to the infolded membranes; this structure is characteristic of ion-transporting cells. Intercalated and striated ducts are also called intralobular ducts because of their location within the lobule.

The striated ducts of each lobule converge and drain into ducts located in the connective tissue septae separating the lobules, where they become **interlobular,** or **excretory, ducts.** They are initially lined with stratified cuboidal epithelium, but more distal parts of the excretory ducts are lined with stratified columnar epithelium. The main duct of each major salivary gland ultimately empties into the oral cavity and is lined with nonkeratinized-stratified squamous epithelium.

Vessels and nerves enter the large salivary glands at the hilum and gradually branch into the lobules. A rich vascular and nerve plexus surrounds the secretory and ductal components of each lobule. The capillaries surrounding the secretory endpieces are very important for the secretion of saliva, stimulated by the autonomic nervous system. Parasympathetic stimulation, usually through the smell or taste of food, provokes a copious watery secretion with relatively little organic content. Sympathetic stimulation produces small amounts of viscous saliva, rich in organic material.

MEDICAL APPLICATION

This latter secretion is often associated with a "dry mouth" sensation.

Parotid Gland

The parotid gland is a branched acinar gland; its secretory portion is composed exclusively of serous cells (Figure 16–2) containing secretory granules that are rich in proteins and have a

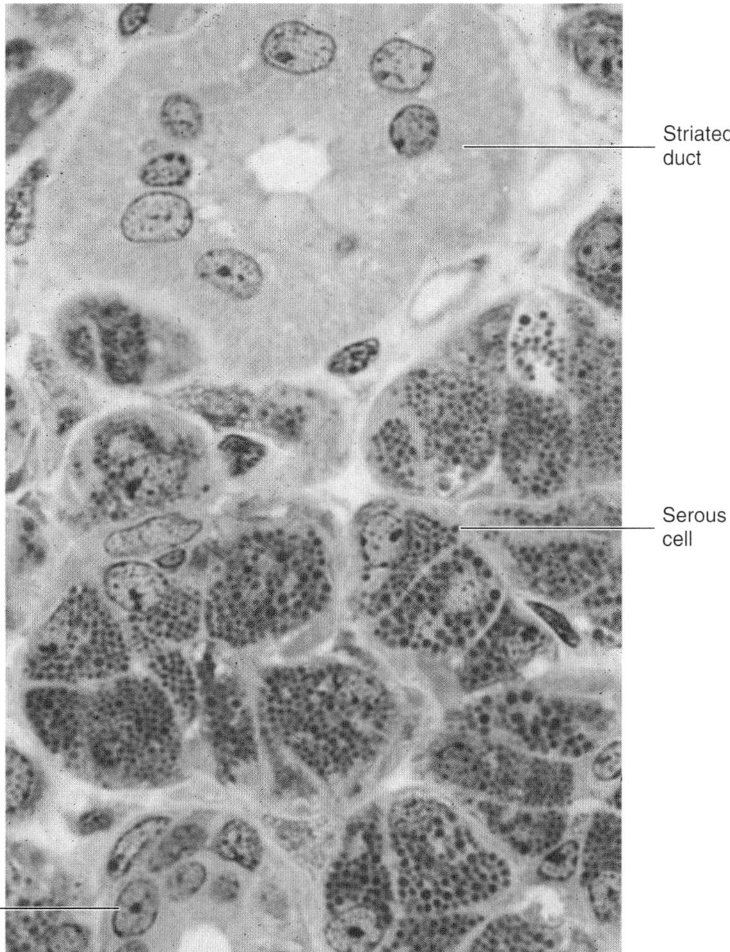

Figure 16–2. Photomicrograph of a parotid gland. Its secretory portion consists of serous amylase–producing cells that store this enzyme in secretory granules. Intralobular (intercalated and striated) ducts are also present. Pararosaniline–toluidine blue (PT) stain. Medium magnification.

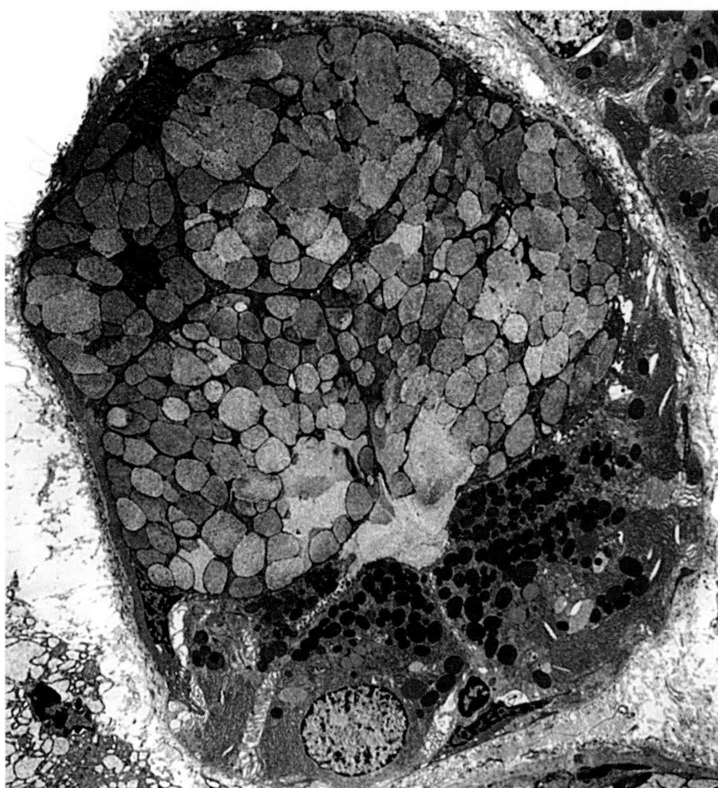

Figure 16–3. Electron micrograph of a mixed acinus from a human submandibular gland. Note the difference between the serous (lower part) and mucous (upper part) secretory granules. ×2500. (Courtesy of JD Harrison.)

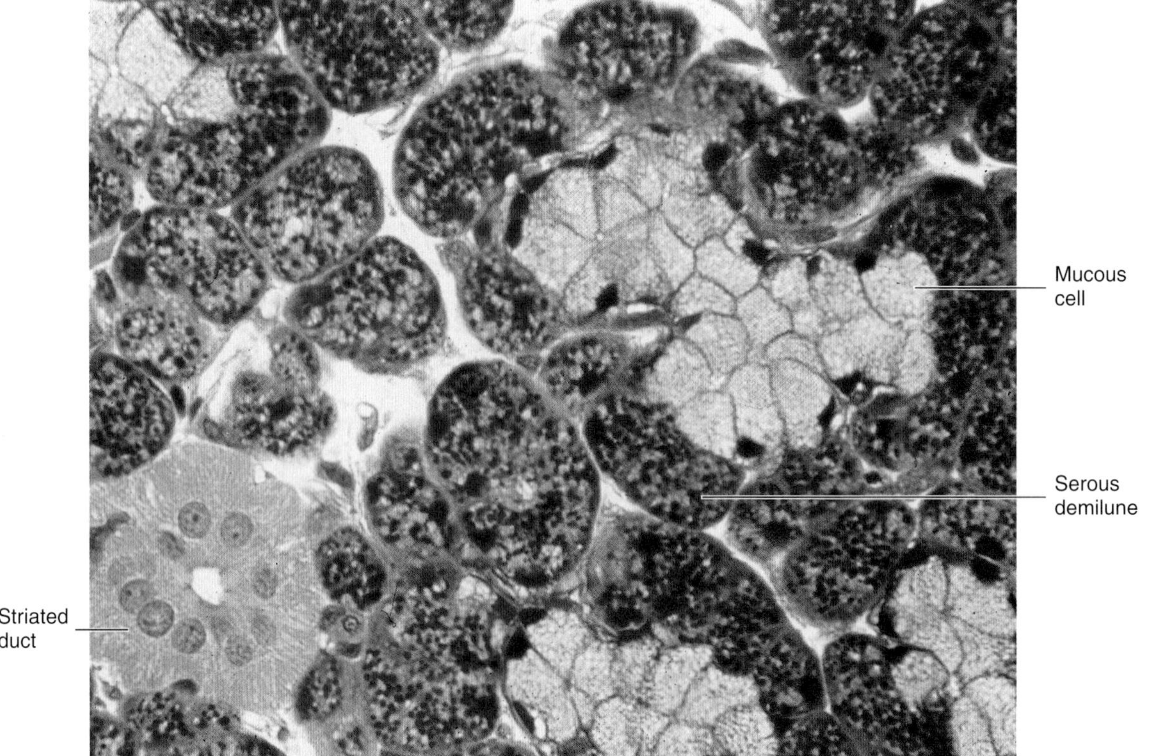

Figure 16–4. Photomicrograph of a submandibular gland. Note the presence of dense serous cells forming demilunes and pale-staining mucous cells grouped along the tubular portion of this tubuloacinar gland. Medium magnification.

high amylase activity. This activity is responsible for most of the hydrolysis of ingested carbohydrates. The digestion begins in the mouth and continues for a short time in the stomach, before the gastric juice acidifies the food and thus decreases amylase activity considerably.

As in other large salivary glands, the connective tissue contains many plasma cells and lymphocytes. The plasma cells secrete IgA, which forms a complex with a **secretory component** synthesized by the serous acinar, intercalated duct, and striated duct cells. The IgA-rich secretory complex released into the saliva is resistant to enzymatic digestion and constitutes an immunologic defense mechanism against pathogens in the oral cavity.

Submandibular (Submaxillary) Gland

The submandibular gland is a branched tubuloacinar gland (Figures 16–3 and 16–4); its secretory portion contains both mucous and serous cells. The serous cells are the main component of this gland and are easily distinguished from mucous cells by their rounded nuclei and basophilic cytoplasm. In humans, 90% of the endpieces of the submandibular gland are serous acinar, whereas 10% consist of mucous tubules with serous demilunes. The presence of extensive lateral and basal membrane infoldings toward the vascular bed increases the ion-transporting surface area 60 times, facilitating electrolyte and water transport. Because of these folds, the cell boundaries are indistinct. Serous cells are responsible for the weak amylolytic activity present in this gland and its saliva. The cells that form the demilunes in the submandibular gland secrete the enzyme **lysozyme,** whose main activity is to hydrolyze the walls of certain bacteria. Some acinar and intercalated duct cells in large salivary glands also secrete lactoferrin, which binds iron, a nutrient necessary for bacterial growth.

Sublingual Gland

The sublingual gland, like the submandibular gland, is a branched tubuloacinar gland formed of serous and mucous cells. Mucous cells predominate in this gland; serous cells are present exclusively on demilunes of mucous tubules (Figure 16–5). As in the submandibular gland, cells that form the demilunes in this gland secrete lysozyme.

PANCREAS

The pancreas is a mixed exocrine-endocrine gland that produces digestive enzymes and hormones. The enzymes are stored and released by cells of the exocrine portion, arranged in acini. The hormones are synthesized in clusters of endocrine epithelial cells known as islets of Langerhans (Figure 16–6; see also Chapter 21). The exocrine portion of the pancreas is a compound acinar gland (Figure 16–6), similar in structure to the parotid gland. In histologic sections, a distinction between the two glands can be made based on the absence of striated ducts and the presence of the islets of Langerhans in the pancreas. Another characteristic detail is that in the pancreas the initial portions of intercalated ducts penetrate the lumens of the acini. Nuclei, surrounded by a pale cytoplasm, belong to **centroacinar cells** that constitute the

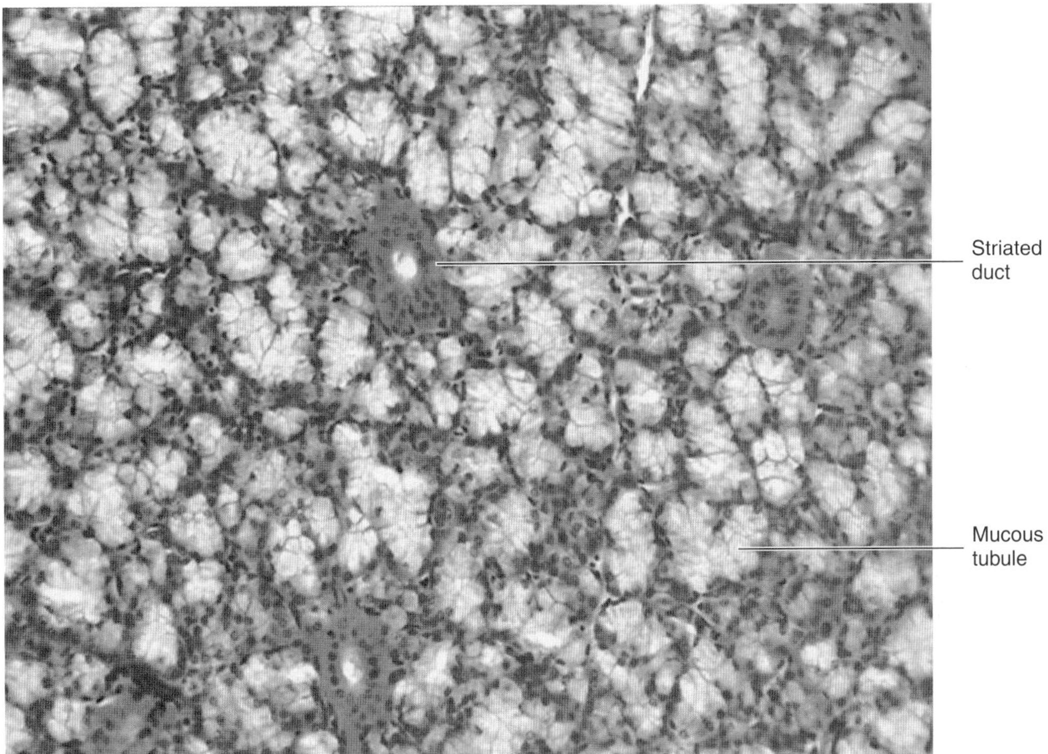

Figure 16–5. Photomicrograph of a sublingual gland showing the predominance of mucous cells. H&E stain. Low magnification.

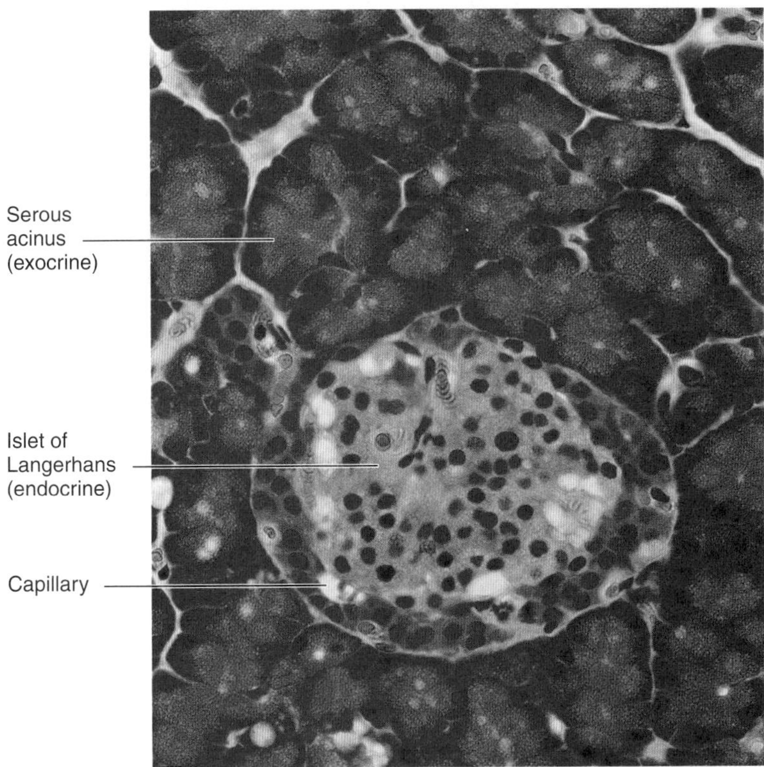

Figure 16–6. Photomicrograph of a pancreas showing the exocrine portion (acini) and the endocrine portion (islet of Langerhans). The acini contain secretory cells with basophilic cytoplasm. Different types of endocrine cells are seen in the islet. PT stain. Medium magnification.

intra-acinar portion of the intercalated duct (Figures 16–7 and 16–8). These cells are found only in pancreatic acini. Intercalated ducts are tributaries of larger interlobular ducts lined by columnar epithelium. There are no striated ducts in the pancreatic duct system.

The exocrine pancreatic acinus is composed of several serous cells surrounding a lumen (Figures 16–7 and 16–8). These cells are highly polarized, with a spherical nucleus, and are typical protein-secreting cells. The number of zymogen granules present in each cell varies according to the digestive phase and attains its maximum in animals that have fasted (Figures 16–9 and 16–10).

A thin capsule of connective tissue covers the pancreas and sends septa into it, separating the pancreatic lobules. The acini are surrounded by a basal lamina that is supported by a delicate sheath of reticular fibers. The pancreas also has a rich capillary network, essential for the secretory process.

In addition to water and ions, the human exocrine pancreas secretes several proteases (**trypsinogens 1, 2,** and **3, chymotrypsinogen, proelastases 1** and **2, protease E, kallikreinogen, procarboxipeptidases A1, A2, B1,** and **B2**), **amylase, lipases** (**triglyceride lipase, colipase,** and **carboxyl ester hydrolase**), **phospholipase A_2,** and **nucleases** (**deoxyribonuclease ribonuclease**). The majority of the enzymes are stored as proenzymes in the secretory granules of acinar cells, being activated in the lumen of the small intestine after secretion. This is very important for the protection of the pancreas.

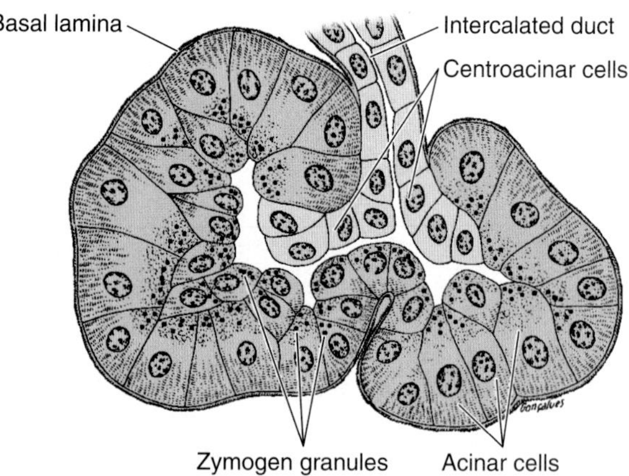

Figure 16–7. Schematic drawing of the structure of pancreatic acini. Acinar cells are pyramidal, with granules at their apex and rough endoplasmic reticulum at their base. The intercalated duct partly penetrates the acini. These duct cells are known as centroacinar cells. Note the absence of myoepithelial cells.

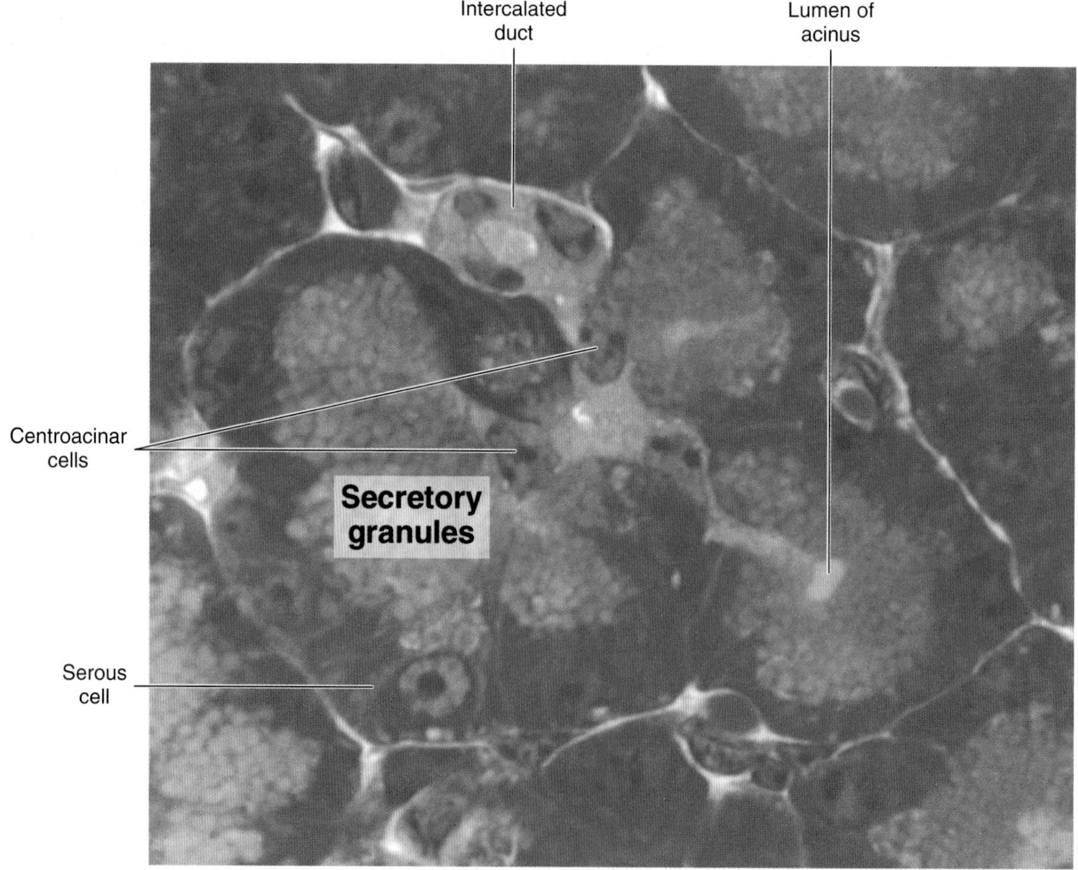

Figure 16–8. Section of the exocrine pancreas showing its main components. PT stain. Medium magnification.

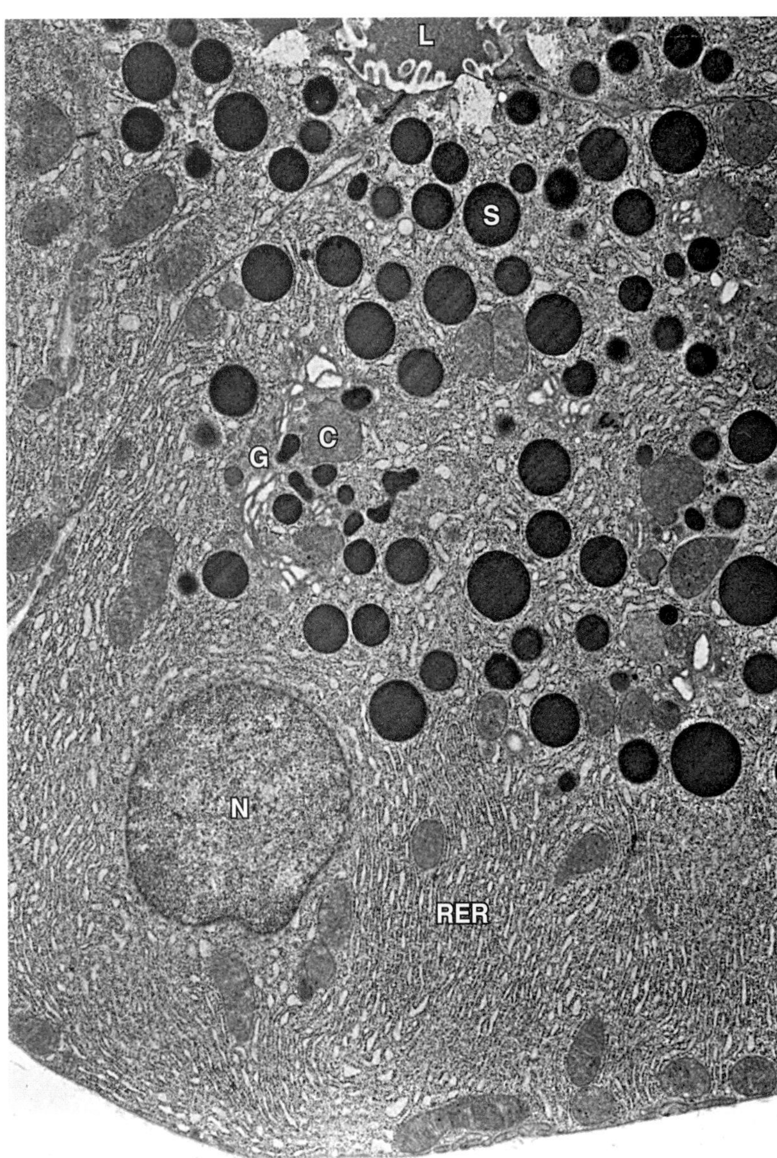

Figure 16–9. Electron micrograph of an acinar cell from a rat pancreas. Note the nucleus (N) surrounded by numerous cisternae of rough endoplasmic reticulum (RER) near the base of the cell. The Golgi complex (G) is situated at the apical pole of the nucleus and is associated with several condensing vacuoles (C) and numerous mature secretory (zymogen) granules (S). The lumen (L) contains proteins recently released from the cell by exocytosis. ×8000.

MEDICAL APPLICATION

In acute pancreatitis, the proenzymes may be activated and digest the whole pancreas, leading to very serious complications.

Pancreatic secretion is controlled mainly through 2 hormones—**secretin** and **cholecystokinin** (previously called **pancreozymin**)—that are produced by enteroendocrine cells of the duodenal mucosa. Stimulation of the vagus nerve (parasympathetic stimulation) will also produce pancreatic secretion.

Secretin promotes secretion of an abundant fluid, poor in enzyme activity and rich in bicarbonate. It is secreted mainly by the small interlobular duct cells and serves to neutralize the acidic **chyme** (partially digested food) so that pancreatic enzymes can function at their optimal neutral pH range. Cholecystokinin promotes secretion of a less abundant but enzyme-rich fluid. This hormone acts mainly in the extrusion of zymogen granules. The integrated action of both these hormones provides for a heavy secretion of enzyme-rich pancreatic juice.

MEDICAL APPLICATION

In conditions of extreme malnutrition such as ***kwashiorkor,*** *pancreatic acinar cells and other active protein-secreting cells atrophy and lose much of their rough endoplasmic reticulum. The production of digestive enzymes is also hindered.*

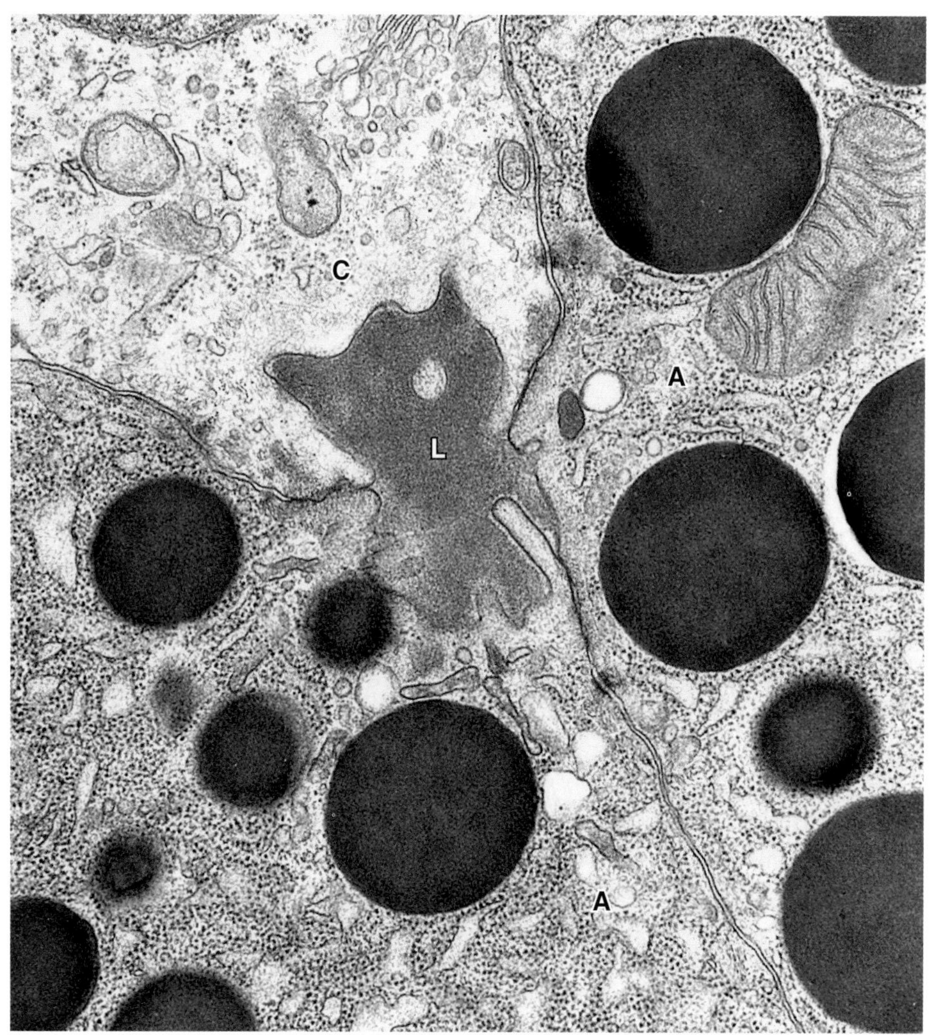

Figure 16–10. Electron micrograph of the apex of 2 pancreatic acinar cells (A) and a centroacinar cell (C) from a rat pancreas. Note the lack of secretory granules and the very scant rough endoplasmic reticulum in the centroacinar cell as compared with the acinar cell. L, acinar lumen. ×30,000.

LIVER

The liver is the second-largest organ of the body (the largest is the skin) and the largest gland, weighing about 1.5 kg. It is situated in the abdominal cavity beneath the diaphragm. The liver is the organ in which nutrients absorbed in the digestive tract are processed and stored for use by other parts of the body. It is thus an interface between the digestive system and the blood. Most of its blood (70–80%) comes from the portal vein; the smaller percentage is supplied by the hepatic artery. All the materials absorbed via the intestines reach the liver through the portal vein, except the complex lipids (**chylomicrons**), which are transported mainly by lymph vessels. The position of the liver in the circulatory system is optimal for gathering, transforming, and accumulating metabolites and for neutralizing and eliminating toxic substances. Elimination occurs in the bile, an exocrine secretion of the liver that is important for lipid digestion. The liver also has the very important function of producing plasma proteins, such as albumin, and other carrier proteins.

Stroma

The liver is covered by a thin connective tissue capsule (**Glisson's capsule**) that becomes thicker at the **hilum,** where the portal vein and the hepatic artery enter the organ and where the right and left hepatic ducts and lymphatics exit. These vessels and ducts are surrounded by connective tissue all the way to their termination (or origin) in the portal spaces between the liver lobules. At this point, a delicate reticular fiber network that supports the hepatocytes and sinusoidal endothelial cells of the liver lobules is formed.

The Liver Lobule

The basic structural component of the liver is the liver cell, or **hepatocyte** (Gr. *hepar,* liver, + *kytos,* cell). These epithelial cells are grouped in interconnected plates. In light-microscope sections, structural units called **liver lobules** can be seen (Figure 16–11). The liver lobule is formed of a polygonal mass of tissue about 0.7 × 2 mm in size (Figures 16–11 and 16–12). In certain animals (eg, pigs), the lobules are separated from each other by a layer of connective tissue. This is not the case in humans, where the lobules are in close contact along most of their length, making it difficult to establish the exact limits between different lobules. In some peripheral regions the lobules are demarcated by connective tissue containing bile ducts, lymphatics, nerves, and blood vessels. These regions, the **portal spaces,** are

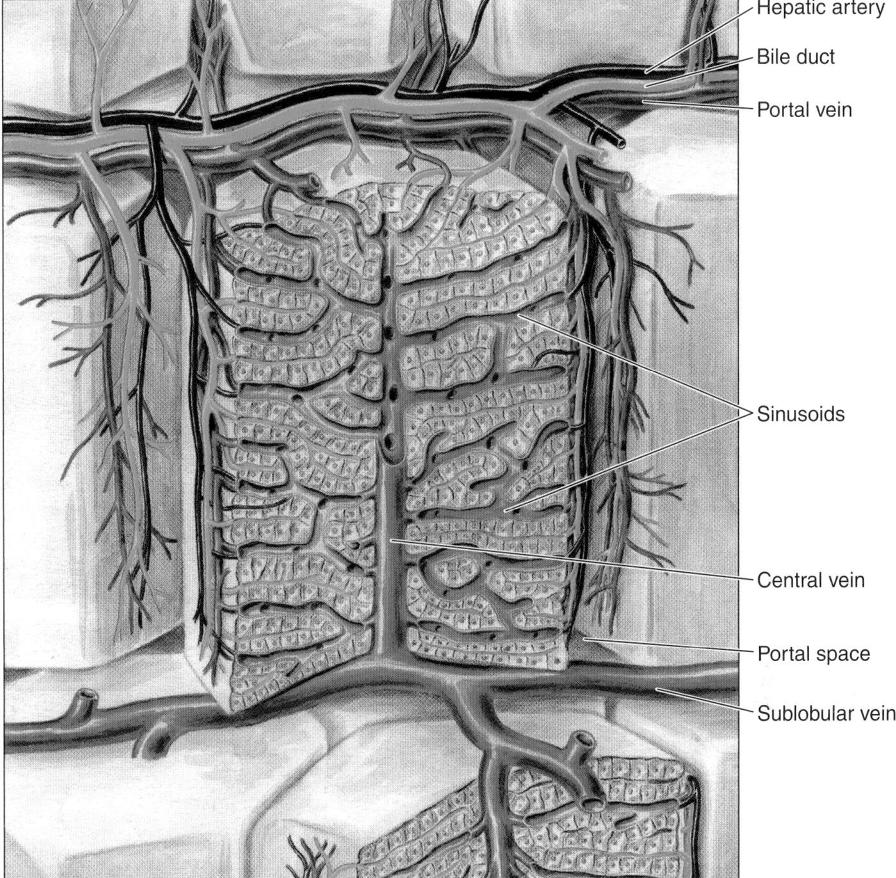

Figure 16–11. Schematic drawing of the structure of the liver. The liver lobule in the center is surrounded by the portal space (dilated here for clarity). Arteries, veins, and bile ducts occupy the portal spaces. Nerves, connective tissue, and lymphatic vessels are also present but are (again, for clarity) not shown in this illustration. In the lobule, note the radial disposition of the plates formed by hepatocytes; the sinusoidal capillaries separate the plates. The bile canaliculi can be seen between the hepatocytes. The sublobular (intercalated) veins drain blood from the lobules. (Redrawn and reproduced, with permission, from Bourne G: *An Introduction to Functional Histology*. Churchill, 1953.)

present at the corners of the lobules. The human liver contains 3–6 portal spaces per lobule, each with a venule (a branch of the portal vein), an arteriole (a branch of the hepatic artery), a duct (part of the bile duct system), and lymphatic vessels. The venule contains blood from the superior and inferior mesenteric and splenic veins. The arteriole contains blood from the celiac trunk of the abdominal aorta. The duct, lined by cuboidal epithelium, carries bile synthesized by the parenchymal cells

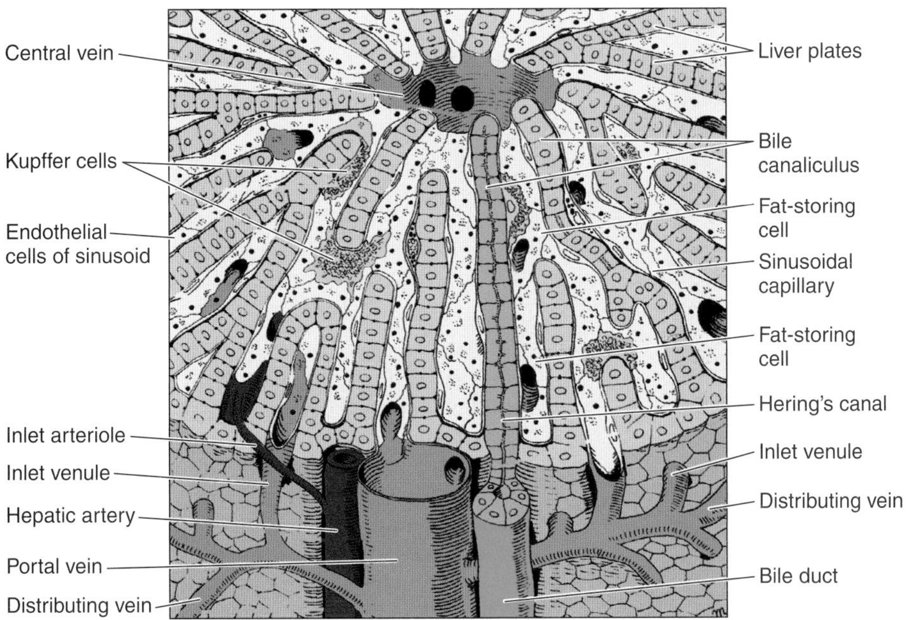

Figure 16–12. Three-dimensional aspect of the normal liver. In the upper center is the central vein; in the lower center, the portal vein. Note the bile canaliculus, liver plates, Hering's canal, Kupffer cells, sinusoid, fat-storing cell, and sinusoid endothelial cells. (Courtesy of M Muto.)

(hepatocytes) and eventually empties into the hepatic duct. One or more lymphatics carry lymph, which eventually enters the blood circulation. All these structures are embedded in a sheath of connective tissue (Figure 16–13B).

The hepatocytes in the liver lobule are radially disposed and are arranged like the bricks of a wall. These cellular plates are directed from the periphery of the lobule to its center and anastomose freely, forming a labyrinthine and spongelike structure (Figure 16–12). The space between these plates contains capillaries, the **liver sinusoids** (Figures 16–11, 16–12, and 16–13A). As discussed in Chapter 11, sinusoidal capillaries are irregularly dilated vessels composed solely of a discontinuous layer of fenestrated endothelial cells. The fenestrae are about 100 nm in diameter and are grouped in clusters (Figure 16–14).

The endothelial cells are separated from the underlying hepatocytes by a discontinuous basal lamina (depending on the species) and a subendothelial space known as the **space of Disse,** which contains microvilli of the hepatocytes (see Figures 16–14, 16–18, and 16–21). Blood fluids readily percolate through the endothelial wall and make intimate contact with the surface of the hepatocytes, permitting an easy exchange of macromolecules from the sinusoidal lumen to the hepatocytes and vice versa. This exchange is physiologically important not only because of the large number of macromolecules (eg, lipoproteins, albumin, fibrinogen) secreted into the blood by hepatocytes but also because the liver takes up and catabolizes many of these large molecules. The sinusoid is surrounded and supported by a delicate sheath of reticular fibers (Figure 16–13C). In addition to the endothelial cells, the sinusoids contain macrophages known as **Kupffer cells** (Figure 16–15). These cells are found on the luminal surface of the endothelial cells. Their main functions are to metabolize aged erythrocytes, digest hemoglobin, secrete proteins related to immunologic processes, and destroy bacteria that eventually enter the portal blood through the large intestine. Kupffer cells account for 15% of the liver cell population. Most of them are located in the periportal region of the liver lobule, where they are very active in phagocytosis. In the space of Disse (perisinusoidal space), fat-storing cells, also called Ito's cells, contain vitamin A–rich lipid inclusions. In the healthy liver, these cells have several functions, such as uptake, storage, and release of retinoids, synthesis and secretion of several extracellular matrix proteins and proteoglycans, secretion of growth factors and cytokines, and the regulation of the sinusoidal lumen diameter in response to different regulators (eg, prostaglandins, thromboxane A_2).

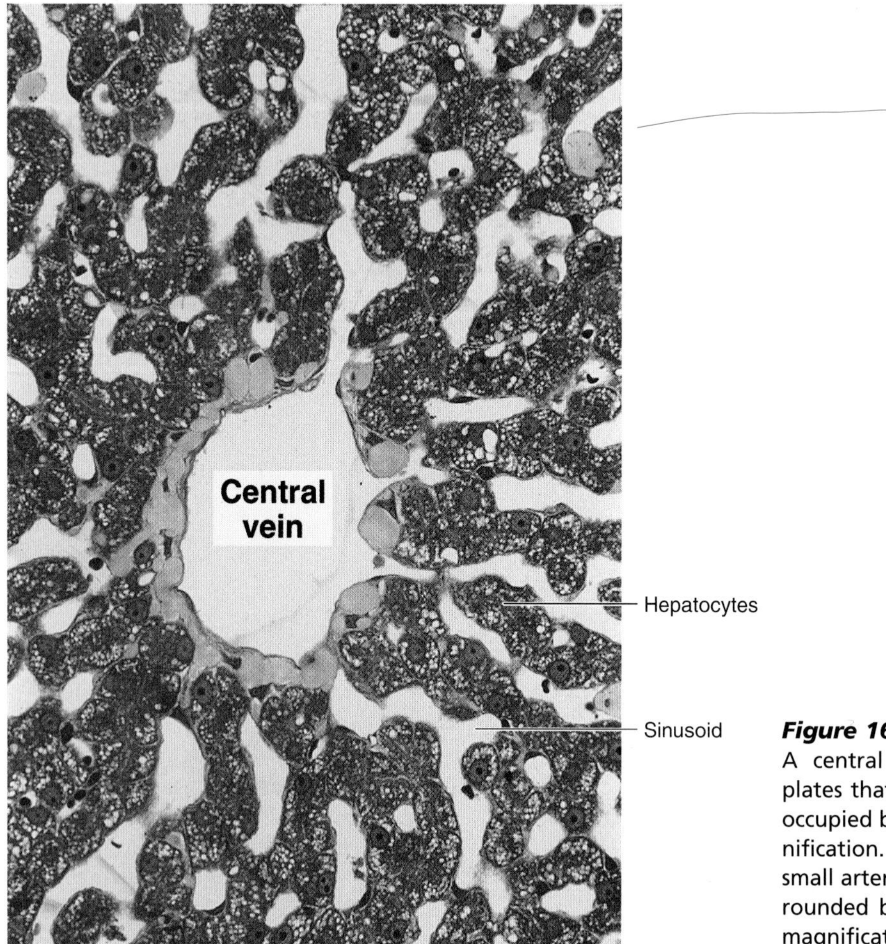

Figure 16–13. Photomicrograph of the liver. **A:** A central (centrolobular) vein. Note the liver plates that anastomose freely, limiting the space occupied by the sinusoids. PT stain. Medium magnification. **B:** A portal space with its characteristic small artery, vein, lymph vessel, and bile duct surrounded by connective tissue. PT stain. Medium magnification. **C:** Collagen III reticular fibers in the lobule, forming a scaffold for the hepatic tissue. Silver impregnation. Medium magnification.

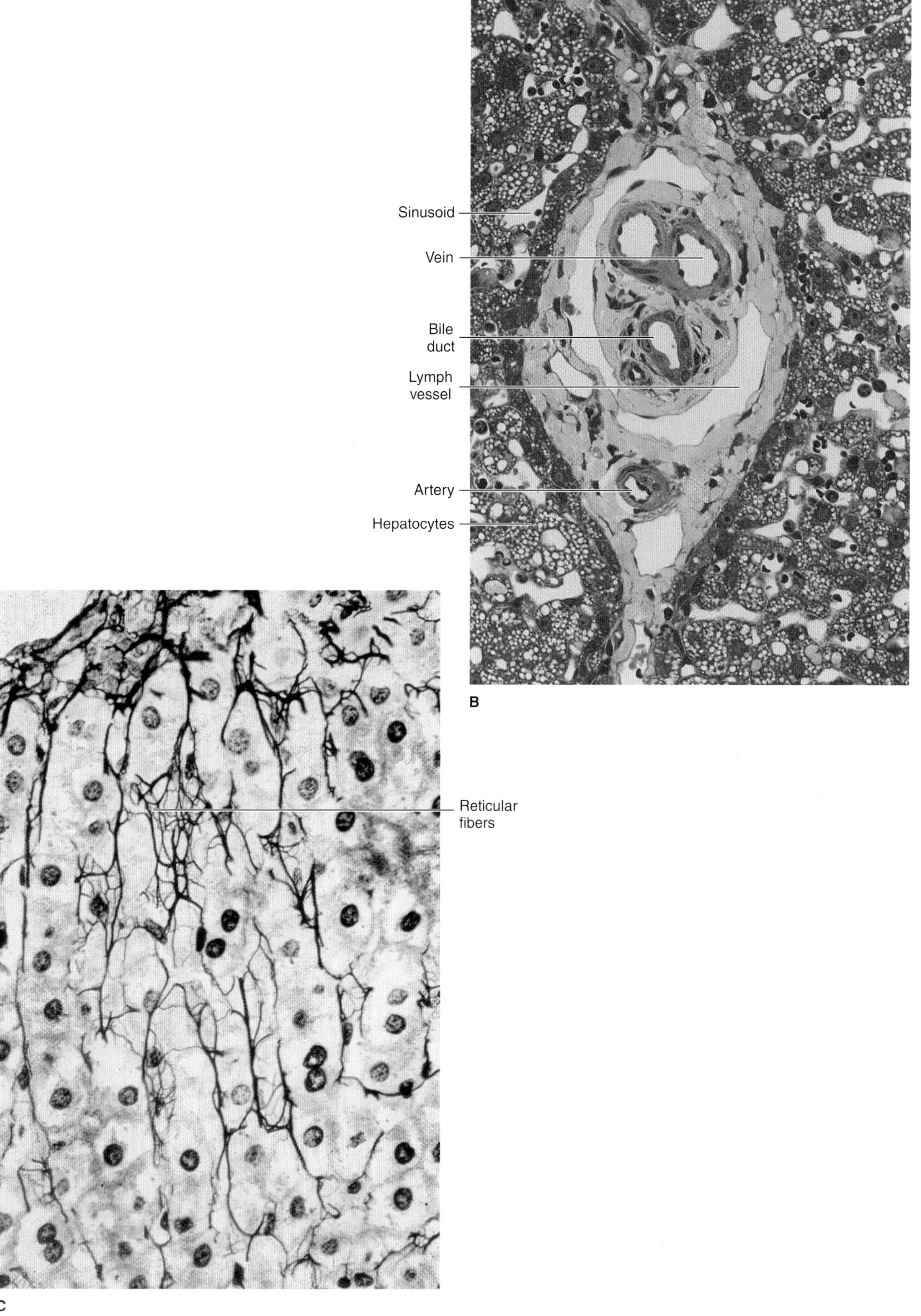
Sinusoid
Vein
Bile
duct
Lymph
vessel
Artery
Hepatocytes
B
Reticular
fibers
C

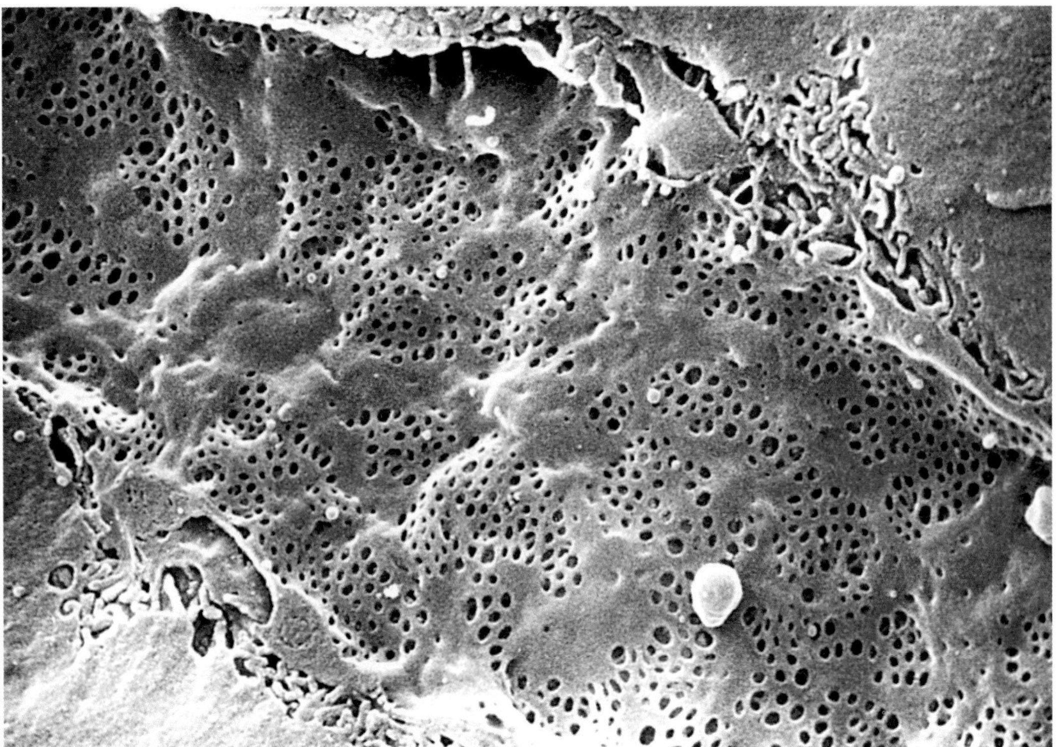

Figure 16–14. Scanning electron micrograph of the endothelial lining of a sinusoidal capillary in rat liver, showing the grouped fenestrations in its wall. At the borders, edges of cut hepatocytes are present, with their villi protruding into spaces of Disse. ×6500. (Courtesy of E Wisse.)

MEDICAL APPLICATION

In chronically diseased liver, Ito's cells proliferate and acquire the features of myofibroblasts, with or without the lipid droplets. Under these conditions, these cells are found close to the damaged hepatocytes and play a major role in the development of fibrosis, including the fibrosis secondary to alcoholic liver disease.

Blood Supply

The liver is unusual in that it receives blood from 2 sources: 80% of the blood derives from the **portal vein,** which carries oxygen-poor, nutrient-rich blood from the abdominal viscera, and 20% derives from the **hepatic artery,** which supplies oxygen-rich blood (Figures 16–11 and 16–12).

Portal Vein System

The portal vein branches repeatedly and sends small **portal venules** to the portal spaces. The portal venules branch into the **distributing veins** that run around the periphery of the lobule. From the distributing veins, small **inlet venules** empty into the **sinusoids.** The sinusoids run radially, converging in the center of the lobule to form the **central,** or **centrolobular, vein** (Figures 16–11 and 16–12). This vessel has thin walls consisting only of endothelial cells supported by a sparse population of collagen fibers (Figure 16–13A). As the central vein progresses along the lobule, it receives more and more sinusoids and gradually increases in diameter. At its end, it leaves the lobule at its base by merging with the larger **sublobular vein** (Figure 16–11). The sublobular veins gradually converge and fuse, forming the two or more large **hepatic veins** that empty into the inferior vena cava.

The portal system conveys blood from the pancreas and spleen and blood containing nutrients absorbed in the intestines. Nutrients are accumulated and transformed in the liver. Toxic substances are also neutralized and eliminated in the liver.

Arterial System

The hepatic artery branches repeatedly and forms the **interlobular arteries.** Some of these arteries irrigate the structures of the portal, and others form arterioles (inlet arterioles; see Figure 16–12) that end directly in the sinusoids at various distances from the portal spaces, thus providing a mixture of arterial and portal venous blood in the sinusoids. The main function of the arterial system is to supply an adequate amount of oxygen to hepatocytes.

Blood flows from the periphery to the center of the **liver lobule.** Consequently, oxygen and metabolites, as well as all other toxic or nontoxic substances absorbed in the intestines, reach the peripheral cells first and then reach the central cells of the lobule. This direction of blood flow partly explains why the behavior of the perilobular cells differs from that of the centrolobular cells (Figure 16–16). This duality of behavior of the hepatocyte is particularly evident in pathologic specimens, where

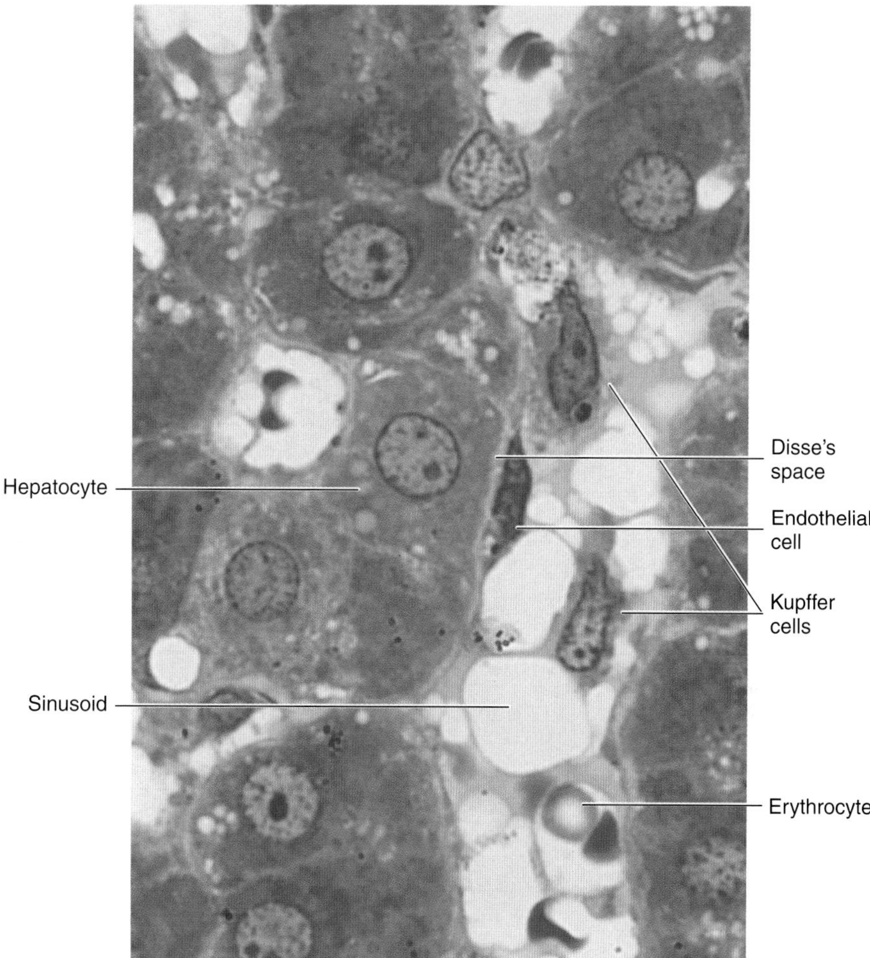

Figure 16–15. Liver section showing sinusoid capillaries with their endothelial cells close to the hepatocytes. The thin slit between the hepatocytes and the endothelium is the space of Disse. Kupffer cells can be seen inside the sinusoid. PT stain. High magnification.

changes are seen in either the central cells or the peripheral cells of the lobule.

The Hepatocyte

Hepatocytes are polyhedral, with 6 or more surfaces, and have a diameter of 20–30 μm. In sections stained with hematoxylin and eosin (H&E), the cytoplasm of the hepatocyte is eosinophilic, mainly because of the large number of mitochondria and some smooth endoplasmic reticulum. Hepatocytes located at different distances from the portal triads show differences in structural, histochemical, and biochemical characteristics. The surface of each hepatocyte is in contact with the wall of the sinusoids, through the space of Disse, and with the surfaces of other hepatocytes. Wherever 2 hepatocytes abut, they delimit a tubular space between them known as the **bile canaliculus** (Figures 16–12, 16–17, 16–18, and 16–19).

The canaliculi, the first portions of the bile duct system, are tubular spaces 1–2 μm in diameter. They are limited only by the plasma membranes of 2 hepatocytes and have a small number of microvilli in their interiors (Figures 16–18 and 16–19). The cell membranes near these canaliculi are firmly joined by tight junctions (described in Chapter 4). Gap junctions are frequent between hepatocytes and are sites of intercellular communication, an important process in the coordination of these cells' physiologic activities. The bile canaliculi form a complex anastomosing network progressing along the plates of the liver lobule and terminating in the region of the portal spaces (Figures 16–11 and 16–12). The bile flow therefore progresses in a direction opposite to that of the blood, ie, from the center of the lobule to its periphery. At the periphery, bile enters the **bile ductules,** or **Hering's canals** (Figures 16–12 and 16–20), composed of cuboidal cells. After a short distance, the ductules cross the limiting hepatocytes of the lobule and end in the **bile ducts** in the portal spaces (Figures 16–11, 16–12, and 16–20). Bile ducts are lined by cuboidal or columnar epithelium and have a distinct connective tissue sheath. They gradually enlarge and fuse, forming right and left **hepatic ducts,** which subsequently leave the liver.

The surface of the hepatocyte that faces the space of Disse bears many microvilli that protrude into that space, but there is always a space between them and the cells of the sinusoidal wall (Figures 16–18 and 16–21). The hepatocyte has one or two rounded nuclei with one or two nucleoli. Some of the nuclei are polyploid; ie, they contain some even multiples of the haploid number of chromosomes. Polyploid nuclei are characterized by their greater size, which is proportional to their ploidy. The hepatocyte has an abundant endoplasmic reticulum—both smooth and rough (Figures 16–18 and 16–22). In the hepatocyte, the rough endoplasmic reticulum forms aggregates dispersed in the cytoplasm; these are often called **basophilic bodies.** Several proteins (eg, blood albumin, fibrinogen) are

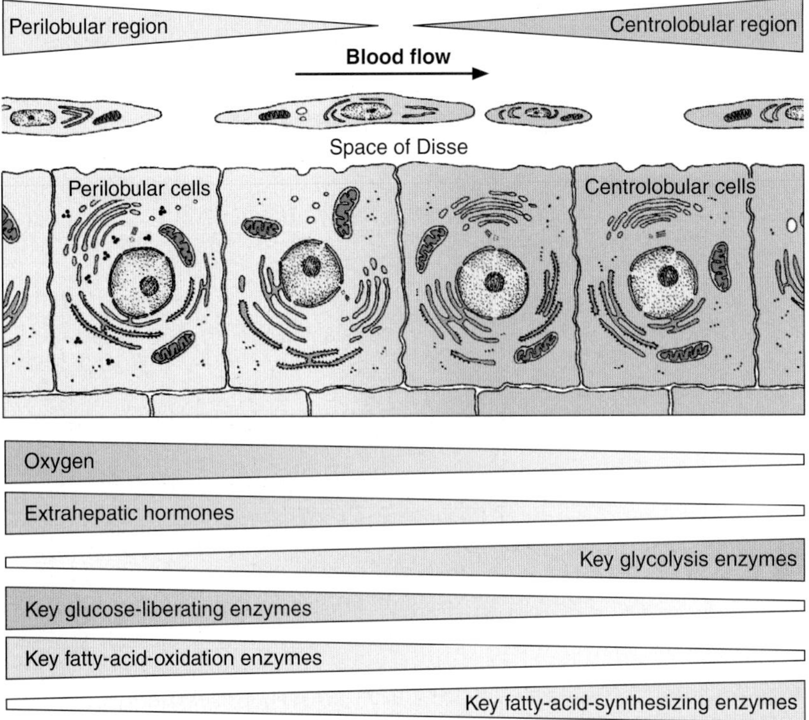

Figure 16–16. The heterogeneity of hepatocytes from the perilobular to the centrolobular regions. Cells in the perilobular region are the first to alter the incoming blood and the first to be affected by it. Cells in the middle are the next to respond to the blood, and those in the centrolobular region see portal vein blood that has already been altered by cells in previous regions. For example, after feeding, peripheral lobular cells are the first to receive incoming glucose and to store it as glycogen (shown on the figure as dots grouped in threes). Any glucose passing these cells would probably be picked up by cells in the next region. In the fasting state, perilobular (peripheral) cells would be the first to respond to glucose-poor blood by breaking down glycogen and releasing it as glucose. In this event, the cells in intermediate and centrolobular regions would not respond to the fasting condition until the glycogen in peripheral cells was depleted. This zonal arrangement would account for some of the differences in the selective damage of hepatocytes caused by various noxious agents or disease conditions. (Courtesy of A Brecht.)

synthesized on polyribosomes in these structures. Various important processes take place in the smooth endoplasmic reticulum, which is distributed diffusely throughout the cytoplasm. This organelle is responsible for the processes of oxidation, methylation, and conjugation required for inactivation or detoxification of various substances before their excretion from the body. The smooth endoplasmic reticulum is a labile system that reacts promptly to the molecules received by the hepatocyte.

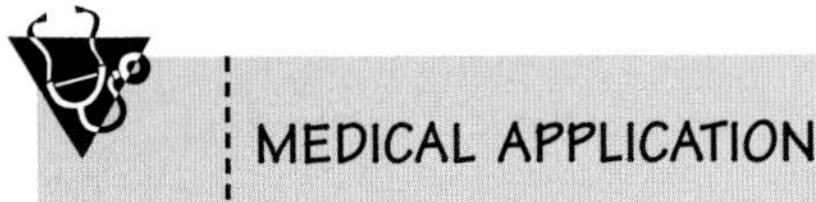

MEDICAL APPLICATION

One of the main processes occurring in the smooth endoplasmic reticulum is the conjugation of hydrophobic (water-insoluble) toxic bilirubin by glucuronyltransferase to form a water-soluble nontoxic bilirubin glucuronide. This conjugate is excreted by hepatocytes into the bile. When bilirubin or bilirubin glucuronide is not excreted, various diseases characterized by jaundice can result (see Figure 16–25).

*One of the frequent causes of jaundice in newborns is the often underdeveloped state of the smooth endoplasmic reticulum in their hepatocytes (**neonatal hyperbilirubinemia**). The current treatment for these cases is exposure to blue light from ordinary fluorescent tubes, which transforms unconjugated bilirubin into a water-soluble photoisomer that can be excreted by the kidneys.*

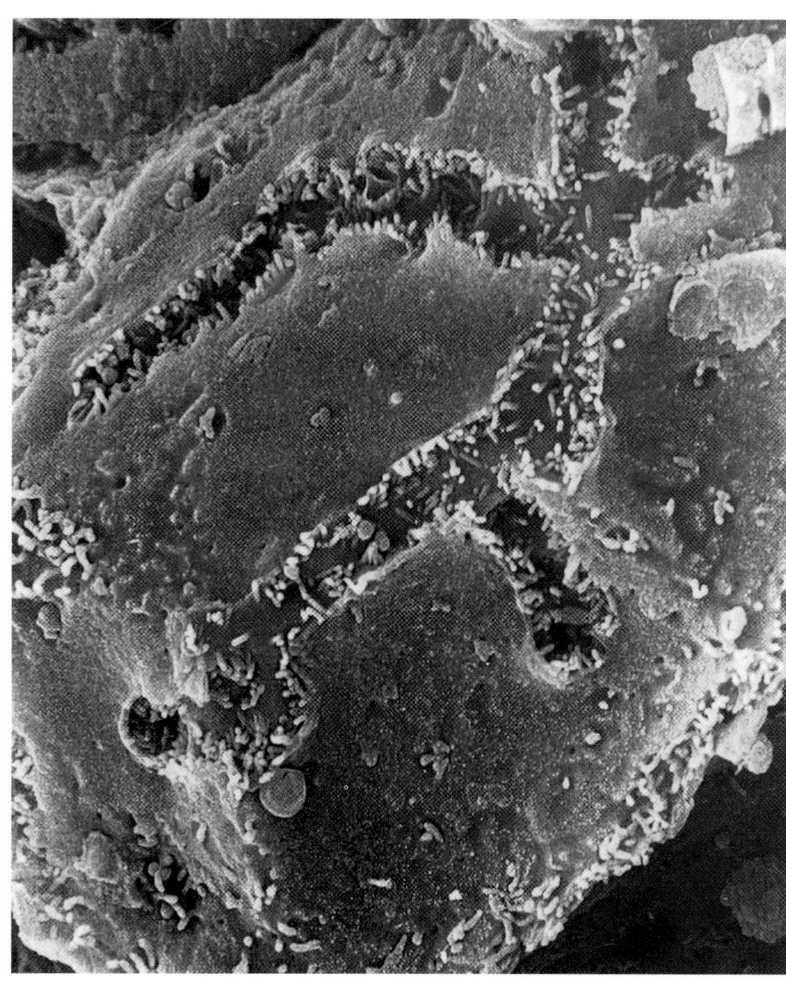

Figure 16–17. Scanning electron micrograph of branching bile canaliculi in the liver. Note the microvilli lining the internal surface. (Reproduced, with permission, from Motta P et al: *The Liver: An Atlas of Scanning Electron Microscopy.* Igaku-Shonin, 1978.)

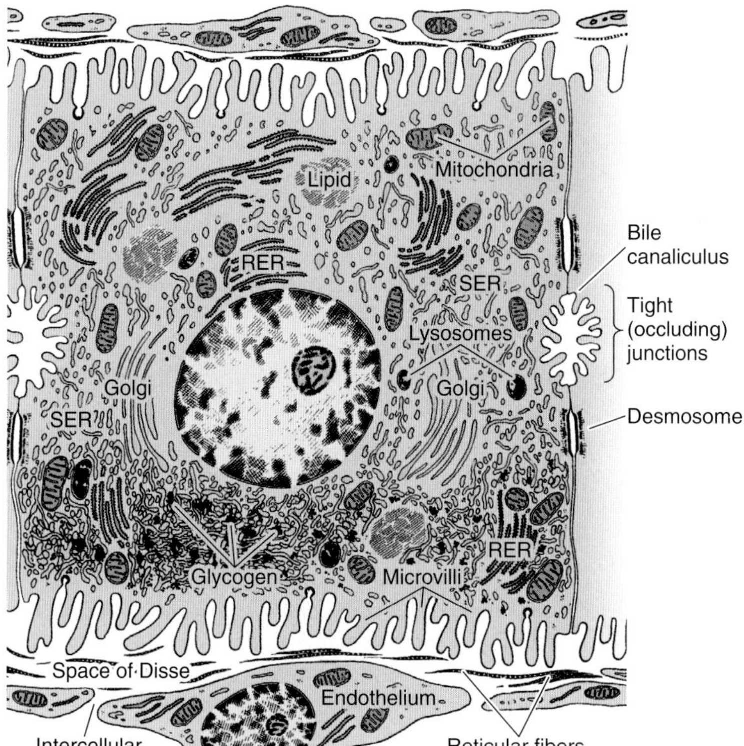

Figure 16–18. Ultrastructure of a hepatocyte. RER, rough endoplasmic reticulum; SER, smooth endoplasmic reticulum. ×10,000.

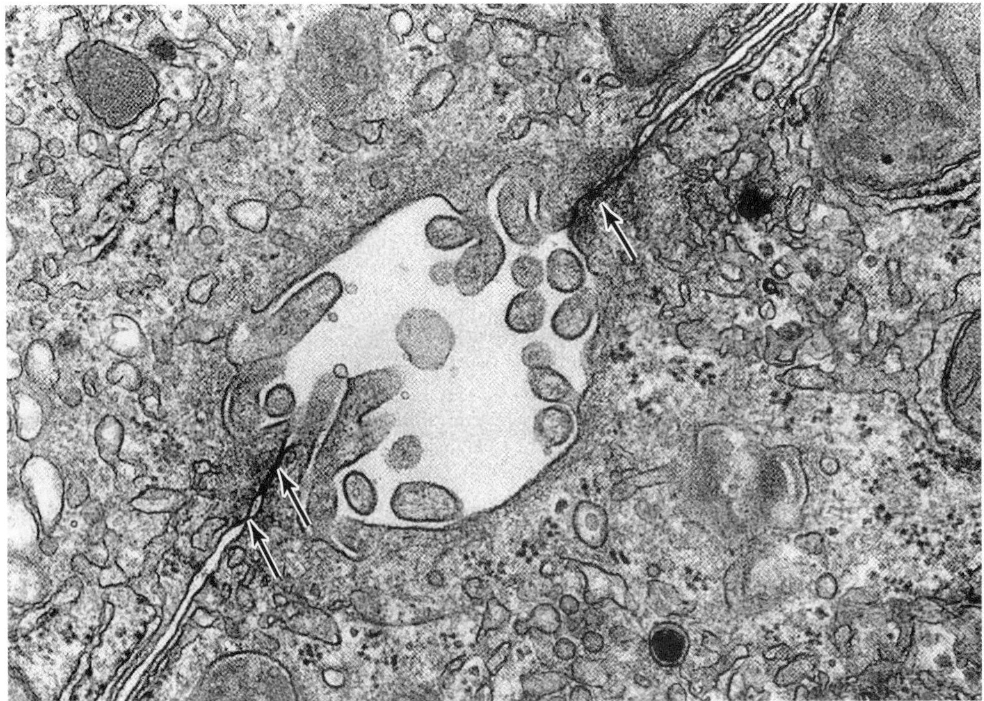

Figure 16–19. Electron micrograph of a bile canaliculus in rat liver. Note the microvilli in its lumen and the junctional complexes (arrows) that seal off this space from the remaining extracellular space. ×54,000. (Courtesy of SL Wissig.)

The hepatocyte frequently contains glycogen. This polysaccharide appears in the electron microscope as coarse, electron-dense granules that frequently collect in the cytosol close to the smooth endoplasmic reticulum (Figures 16–18 and 16–22). The amount of glycogen present in the liver conforms to a diurnal rhythm; it also depends on the nutritional state of the individual. Liver glycogen is a depot for glucose and is mobilized if the blood glucose level falls below normal. In this way, hepatocytes maintain a steady level of blood glucose, one of the main sources of energy for use by the body.

Each hepatocyte has approximately 2000 mitochondria. Another common cellular component is the lipid droplet, whose numbers vary greatly. Hepatocyte lysosomes are important in the turnover and degradation of intracellular organelles. Like lysosomes, peroxisomes are enzyme-containing organelles abundant in hepatocytes. Some of their functions are the oxidation of excess fatty acids, breakdown of the hydrogen peroxide generated by this oxidation (by means of catalase activity), breakdown of excess purines (AMP, GMP) to uric acid, and participation in the synthesis of cholesterol, bile acids, and some lipids used to make myelin. Golgi complexes in the liver are also numerous—up to 50 per cell. The functions of this organelle include the formation of lysosomes and the secretion of plasma proteins (eg, albumin, proteins of the complement system), glycoproteins (eg, transferrin), and lipoproteins (eg, very low-density lipoproteins).

Bile canaliculi
Bile duct
Hepatocytes
Bile ductule

Figure 16–20. The confluence of bile canaliculi and bile ductules, which are lined by cuboidal epithelium. The ductules merge with bile ducts in the portal spaces.

MEDICAL APPLICATION

A variety of rare inherited disorders of peroxisome function occur in humans, most involving mutations of the enzymes found within peroxisomes. For example, X-linked adrenoleukodystrophy (X-ALD) results from a failure to metabolize fatty acids properly, resulting in the deterioration of the myelin sheaths of neurons. An attempt to find an effective treatment was the subject of the 1992 film Lorenzo's Oil.

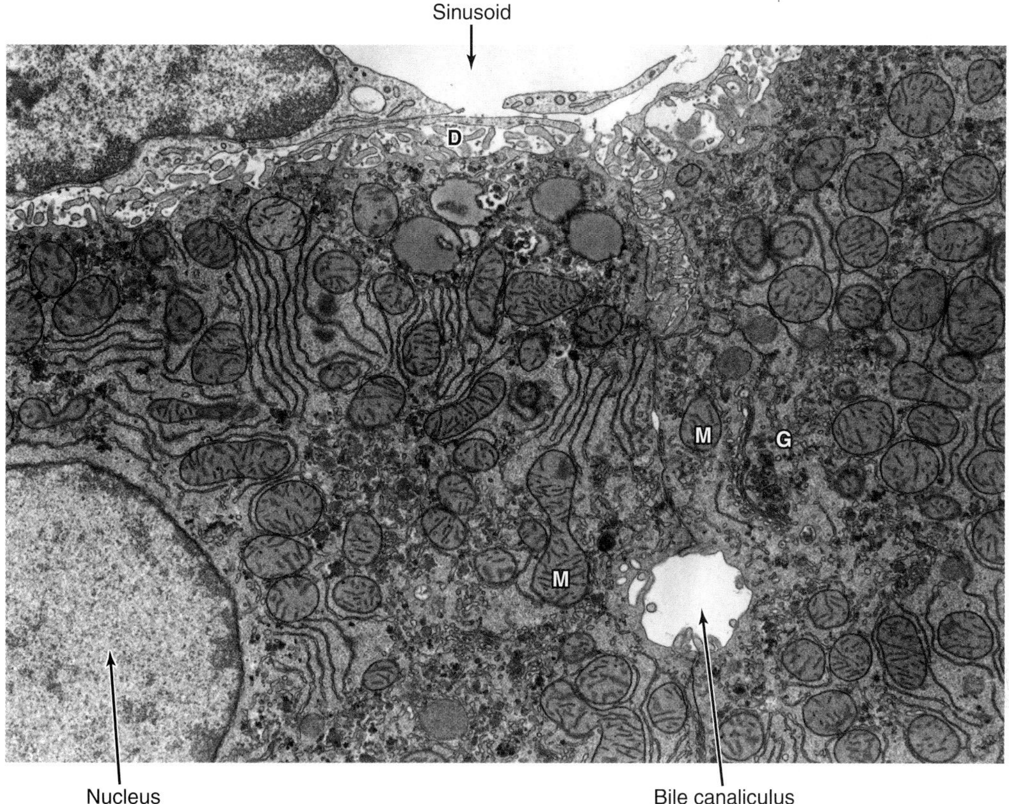

Figure 16–21. Electron micrograph of the liver. Note the two adjacent hepatocytes with a bile canaliculus between them. The hepatocytes contain numerous mitochondria (M) and smooth and rough endoplasmic reticulum. A prominent Golgi complex (G) is near the bile canaliculus. The sinusoid is lined by endothelial cells with large open fenestrae. The space of Disse (D) is occupied by numerous microvilli projecting from the hepatocytes. ×9200. (Courtesy of D Schmucker.)

The hepatocyte is probably the most versatile cell in the body. It is a cell with both endocrine and exocrine functions; it also synthesizes and accumulates certain substances, detoxifies others, and transports still others.

In addition to synthesizing proteins for its own maintenance, the hepatocyte produces various plasma proteins for export—among them albumin, prothrombin, fibrinogen, and lipoproteins. These proteins are synthesized on polyribosomes attached to the rough endoplasmic reticulum. Usually, the hepatocyte does not store proteins in its cytoplasm as secretory granules but continuously releases them into the bloodstream (Figure 16–23). About 5% of the protein exported by the liver is produced by the cells of the macrophage system (Kupffer cells); the remainder is synthesized in the hepatocytes.

Bile secretion is an exocrine function in the sense that hepatocytes promote the uptake, transformation, and excretion of blood components into the bile canaliculi. Bile has several other essential components in addition to water and electrolytes: bile acids, phospholipids, cholesterol, and bilirubin. The secretion of bile acids is illustrated in Figure 16–24. About 90% of these substances are derived by absorption from the distal intestinal epithelium and are transported by the hepatocyte from the blood to bile canaliculi (enterohepatic recirculation). About 10% of bile acids are synthesized in the smooth endoplasmic reticulum of the hepatocyte by conjugation of cholic acid (synthesized by the liver from cholesterol) with the amino acid glycine or taurine, producing glycocholic and taurocholic acids. Bile acids have an important function in emulsifying the lipids in the digestive tract, promoting easier digestion by lipases and subsequent absorption.

MEDICAL APPLICATION

Abnormal proportions of bile acids may lead to the formation of gallstones (cholelithiasis). Gallstones can block bile flow and cause jaundice—the presence of bile pigments in blood—from the rupture of tight junctions around the bile canaliculi.

Bilirubin, most of which results from the breakdown of hemoglobin, is formed in the mononuclear phagocyte system (which includes the Kupffer cells of the liver sinusoids) and is transported to the hepatocytes. In the smooth endoplasmic reticulum of the hepatocyte, hydrophobic bilirubin is conjugated to glucuronic acid, forming water-soluble bilirubin glucuronide

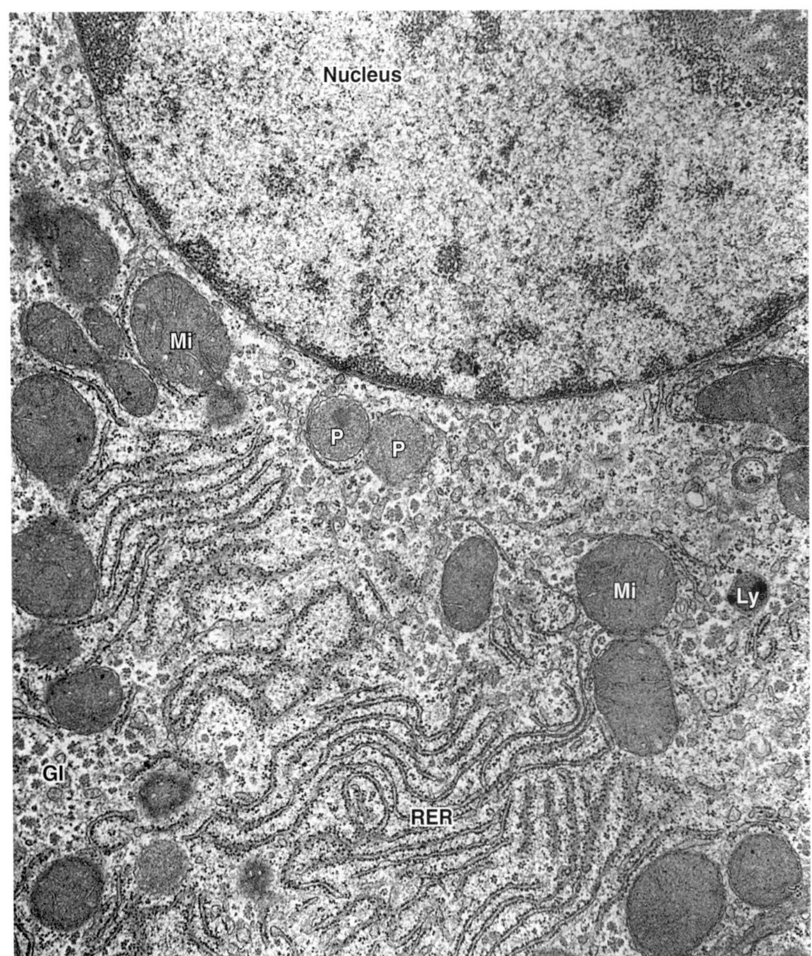

Figure 16–22. Electron micrograph of a hepatocyte. In the cytoplasm, below the nucleus, are mitochondria (Mi), rough endoplasmic reticulum (RER), glycogen (Gl), lysosomes (Ly), and peroxisomes (P). ×6600.

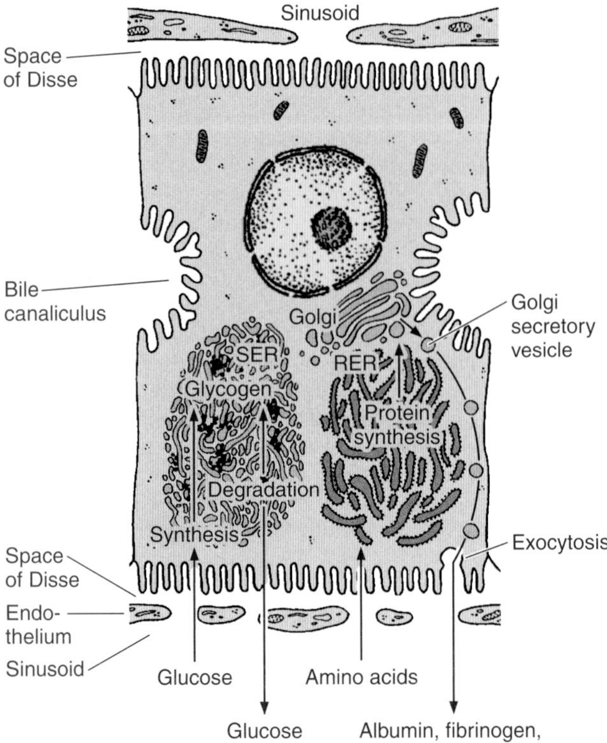

Figure 16–23. Protein synthesis and carbohydrate storage in the liver. Carbohydrate is stored as glycogen, usually associated with the smooth endoplasmic reticulum (SER). When glucose is needed, glycogen is degraded. In several diseases, glycogen degradation is depressed, resulting in abnormal intracellular accumulations of glycogen. Proteins produced by hepatocytes are synthesized in the rough endoplasmic reticulum (RER), which explains why hepatocyte lesions or starvation lead to a decrease in the amounts of albumin, fibrinogen, and prothrombin in a patient's blood. The impairment of protein synthesis leads to several complications, since most of these proteins are carriers, important for the blood's osmotic pressure and for coagulation.

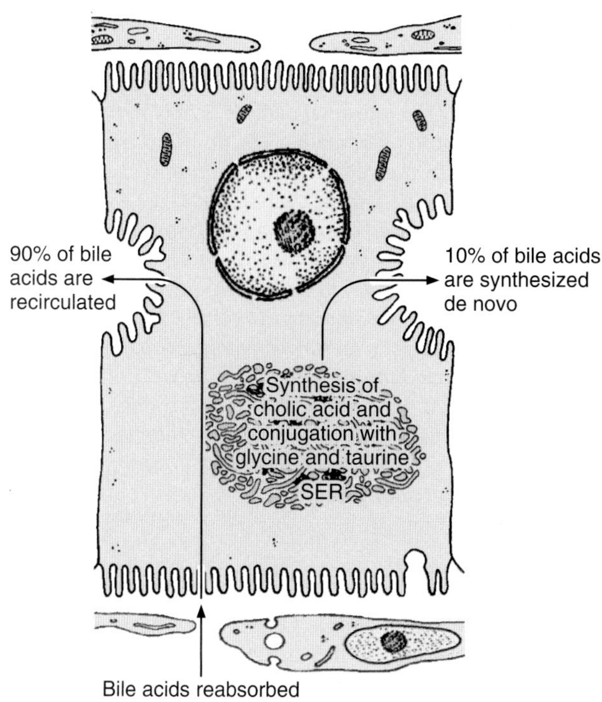

Figure 16–24. Mechanism of secretion of bile acids. About 90% of bile acids are derived from the intestinal epithelium and transported to the liver. The remaining 10% are synthesized in the liver by the conjugation of cholic acid with the amino acids glycine and taurine. This process occurs in the smooth endoplasmic reticulum (SER).

(Figure 16–25). In a further step, **bilirubin glucuronide** is secreted into the bile canaliculi.

Lipids and carbohydrates are stored in the liver in the form of triglycerides and glycogen (Figure 16–23). This capacity to store metabolites is important, because it supplies the body with energy between meals. Figure 16–23 shows how carbohydrates are stored. The liver also serves as the major storage compartment for vitamins, especially vitamin A. Vitamin A originates in the diet, reaching the liver along with other dietary lipids in the form of chylomicrons. In the liver, vitamin A is stored in Ito's cells.

The hepatocyte is also responsible for converting lipids and amino acids into glucose by means of a complex enzymatic process called **gluconeogenesis** (Gr. *glykys,* sweet, + *neos,* new, + *genesis,* production). It is also the main site of amino acid deamination, resulting in the production of urea. Urea is transported through the blood to the kidney and is excreted by that organ.

Various drugs and substances can be inactivated by oxidation, methylation, or conjugation. The enzymes participating in these processes are located mainly in the smooth endoplasmic reticulum. Glucuronyltransferase, an enzyme that conjugates glucuronic acid to bilirubin, also causes conjugation of several other compounds such as steroids, barbiturates, antihistamines, and anticonvulsants. Under certain conditions, drugs that are inactivated in the liver can induce an increase in the hepatocyte's smooth endoplasmic reticulum, thus improving the detoxification capacity of the organ.

MEDICAL APPLICATION

The administration of barbiturates to laboratory animals results in rapid development of smooth endoplasmic reticulum in hepatocytes. Barbiturates can also increase synthesis of glucuronyltransferase. This finding has led to the use of barbiturates in the treatment of glucuronyltransferase deficiencies.

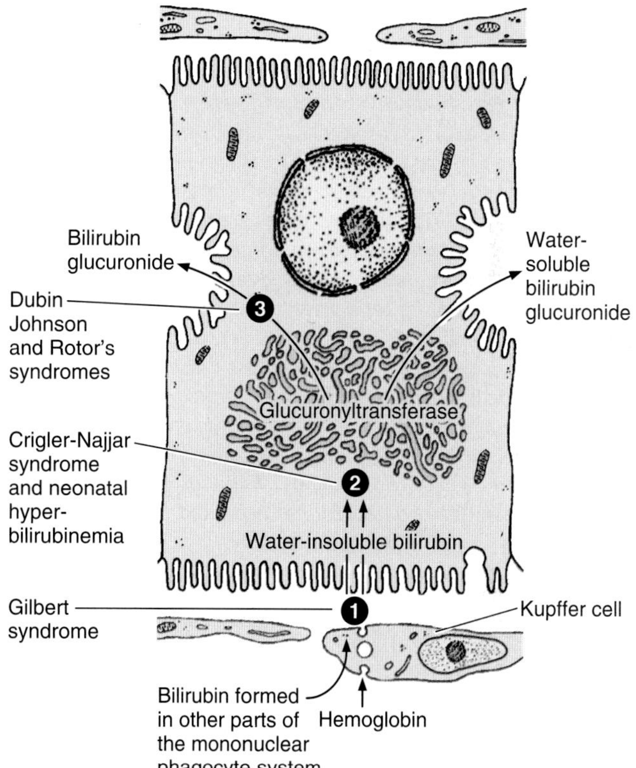

Figure 16–25. The secretion of bilirubin. The water-insoluble form of bilirubin is derived from the metabolism of hemoglobin in macrophages. Glucuronyltransferase activity in the hepatocytes causes bilirubin to be conjugated with glucuronide in the smooth endoplasmic reticulum, forming a water-soluble compound. When bile secretion is blocked, the yellow bilirubin or bilirubin glucuronide is not excreted; it accumulates in the blood, and jaundice results. Several defective processes in the hepatocytes can cause diseases that produce jaundice: a defect in the capacity of the cell to trap and absorb bilirubin **(1)**, the inability of the cell to conjugate bilirubin because of a deficiency in glucuronyltransferase **(2)**, or problems in the transfer and excretion of bilirubin glucuronide into the bile canaliculi **(3)**. One of the most frequent causes of jaundice, however—unrelated to hepatocyte activity—is the obstruction of bile flow as a result of gallstones or tumors of the pancreas.

Liver Regeneration

Despite its slow rate of cell renewal, the liver has an extraordinary capacity for regeneration. The loss of hepatic tissue by surgical removal or from the action of toxic substances triggers a mechanism by which hepatocytes begin to divide, continuing until the original mass of tissue is restored. In humans, this capacity is considerably restricted but is still important, because parts of a liver can be used in surgical liver transplantation.

MEDICAL APPLICATION

The regenerated liver tissue is usually well organized, exhibiting the typical lobular arrangement and replacing the functions of the destroyed tissue. However, when there is continuous or repeated damage to hepatocytes over a long period of time, the multiplication of liver cells is followed by a pronounced increase in the amount of connective tissue (Figure 16–26). Instead of normal liver tissue there is the formation of nodules of several sizes, most of them visible to the naked eye, composed of a central mass of disorganized hepatocytes surrounded by a great amount of connective tissue very rich in collagenous fibers. This disorder, called ***cirrhosis,*** *is a progressive and irreversible process, causes liver failure, and is usually fatal. This type of fibrosis is diffuse, affecting the entire liver. Cirrhosis is the end result of several conditions that affect the liver architecture.*

Cirrhosis is a consequence of any sustained progressive injury to hepatocytes produced by several agents, such as ethanol, drugs or other chemicals, hepatitis virus (mainly types B, C, or D), and autoimmune liver disease. In some regions of the world, infection by the intestinal parasite Schistosoma *is a frequent cause of cirrhosis. Eggs from the parasite are carried by the venous blood and trapped in the liver sinusoids, damaging the hepatocytes.*

Alcohol-induced liver damage is responsible for most of the cases of cirrhosis, because ethanol is metabolized primarily in the liver. Some of the putative pathogenic mechanisms in alcohol-induced liver damage are oxygen radical formation (probably due to lipid peroxidation), the generation of acetaldehyde, proinflammatory, and profibrogenic cytokines. Ethanol also alters hepatic regeneration through an unknown mechanism, favoring the development of cirrhosis.

BILIARY TRACT

The bile produced by the hepatocyte flows through the **bile canaliculi, bile ductules,** and **bile ducts.** These structures gradually merge, forming a network that converges to form the **hepatic duct.** The hepatic duct, after receiving the **cystic duct** from the gallbladder, continues to the duodenum as the **common bile duct** (**ductus choledochus**).

The hepatic, cystic, and common bile ducts are lined with a mucous membrane of simple columnar epithelium. The lamina propria is thin and surrounded by an inconspicuous layer of smooth muscle. This muscle layer becomes thicker near the duodenum and finally, in the intramural portion, forms a sphincter that regulates bile flow (sphincter of Oddi).

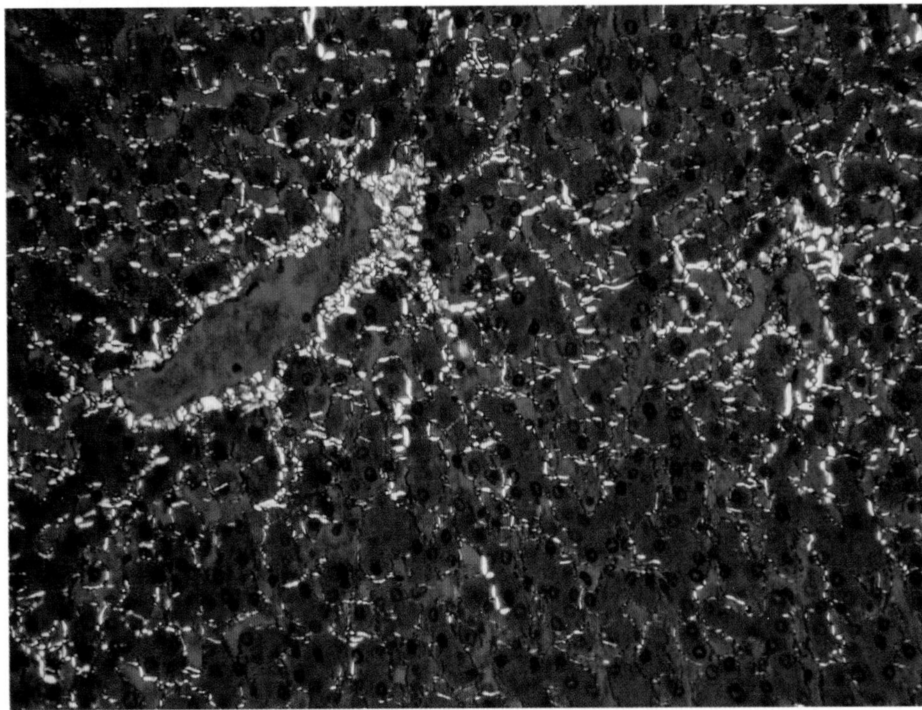

Figure 16–26. Section of human liver with cirrhosis induced by the local inflammatory action of the eggs of a nematode (*Schistosoma*). Collagen content was increased severalfold, resulting in circulatory disturbance. Picrosirius stain and polarizing microscopy. Medium magnification.

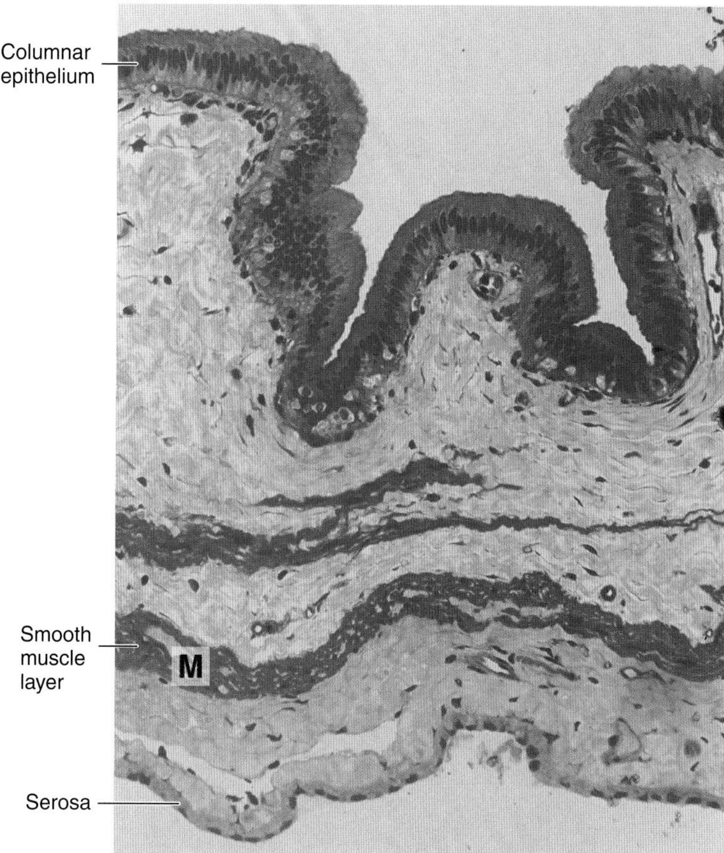

Figure 16–27. Photomicrograph of a section of gallbladder. Note the lining of columnar epithelium and the smooth muscle layer (M). PT stain. Low magnification.

GALLBLADDER

The gallbladder is a hollow, pear-shaped organ attached to the lower surface of the liver. It can store 30–50 mL of bile. The wall of the gallbladder consists of a mucosa composed of simple columnar epithelium and lamina propria, a layer of smooth muscle, a perimuscular connective tissue layer, and a serous membrane (Figure 16–27).

The mucosa has abundant folds that are particularly evident when the gallbladder is empty. The epithelial cells are rich in mitochondria (Figure 16–28). All these cells are capable of secreting small amounts of mucus. Tubuloacinar mucous glands near the cystic duct are responsible for the production of most of the mucus present in bile.

The main function of the gallbladder is to store bile, concentrate it by absorbing its water, and release it when necessary into the digestive tract. This process depends on an active sodium-transporting mechanism in the gallbladder's epithelium. Water absorption is an osmotic consequence of the sodium pump. Contraction of the smooth muscle of the gallbladder is induced by **cholecystokinin,** a hormone produced by enteroendocrine cells (I cells) located in the epithelial lining of the small intestine. Release of cholecystokinin is, in turn, stimulated by the presence of dietary fats in the small intestine.

Tumors of the Digestive Glands

MEDICAL APPLICATION

Most malignant tumors of the liver derive from hepatic parenchyma or epithelial cells of the bile duct. The pathogenesis of hepatocellular carcinoma is not completely understood, but it is believed to be associated with a variety of acquired disorders, such as chronic viral hepatitis (B or C) and cirrhosis. In the exocrine pancreas, most tumors arise from ductal epithelial cells; the mortality rate from pancreatic tumors is high.

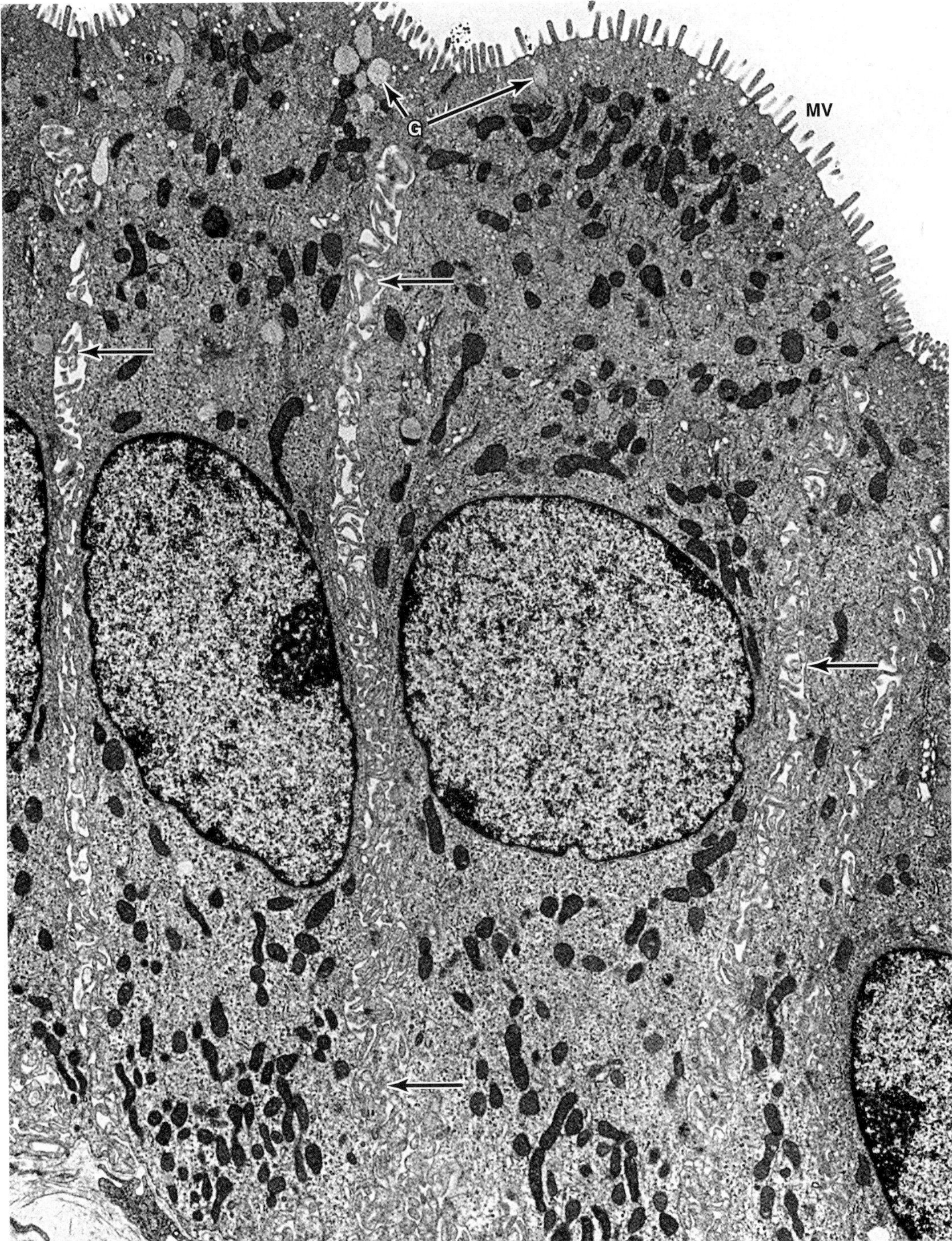

Figure 16–28. Electron micrograph of the gallbladder of a guinea pig. Note the microvilli (MV) on the surface of the cell and the secretory granules (G) containing mucus. Arrows indicate the intercellular spaces. These epithelial cells transport sodium chloride from the lumen to the subjacent connective tissue. Water follows passively, causing the bile to become concentrated. ×5600.

REFERENCES

Pancreas & Salivary Glands

Cook DI et al: Secretion by the major salivary glands. In: Johnson LR (editor): *Physiology of the Gastrointestinal Tract,* Vol 2. Raven Press, 1994.

Lowe, ME: The structure and function of pancreatic enzymes. In: Johnson LR (editor): *Physiology of the Gastrointestinal Tract,* Vol 2. Raven Press, 1994.

Mason DK, Chisholm DM: *Salivary Glands in Health and Disease.* Saunders, 1975.

McDaniel ML et al: Cytokines and nitric oxide in islet inflammation and diabetes. Proc Soc Exp Biol Med 1996;211:24.

Liver & Biliary Tract

Geerts A et al: Fat-storing (Ito) cell biology. In: Arias IM et al (editors): *The Liver: Biology and Pathobiology.* Raven Press, 1994.

Gerber MA, Swan NT: Histology of the liver. Am J Surg Pathol 1987;11:709.

Ito T, Shibasaki S: Electron microscopic study on the hepatic sinusoidal wall and the fat-storing cells in the human normal liver. Arch Histol Jpn 1968;29:137.

Jones AL, Fawcett CW: Hypertrophy of the agranular endoplasmic reticulum in hamster liver induced by phenobarbital. J Histochem Cytochem 1966;14:215.

Maddrey WC. Alcohol-induced liver disease. Clin Liver Dis 2000;4:115.

Minato Y et al: The role of fat-storing cells in Disse space fibrogenesis in alcoholic liver disease. Hepatology 1983;3:559.

Rouiller C (editor): *The Liver: Morphology, Biochemistry, Physiology.* 2 vols. Academic Press, 1963, 1964.

Trutman M, Sasse D: The lymphatics of the liver. Anat Embryol 1994;190:201.

The Respiratory System

17

The respiratory system includes the **lungs** and a system of tubes that link the sites of gas exchange with the external environment. A **ventilation mechanism,** consisting of the thoracic cage, intercostal muscles, diaphragm, and elastic and collagen components of the lungs, is important in the movement of air through the lungs. The respiratory system is customarily divided into 2 principal regions (Figure 17–1): a **conducting portion,** consisting of the nasal cavity, nasopharynx, larynx, trachea, bronchi (Gr. *bronchos,* windpipe), bronchioles, and terminal bronchioles; and a **respiratory portion** (where gas exchange takes place), consisting of respiratory bronchioles, alveolar ducts, and alveoli. **Alveoli** are specialized saclike structures that make up the greater part of the lungs. They are the main sites for the principal function of the lungs—the exchange of O_2 and CO_2 between inspired air and blood.

The conducting portion serves 2 main functions: to provide a conduit through which air can travel to and from the lungs and to condition the inspired air. To ensure an uninterrupted supply of air, a combination of cartilage, elastic and collagen fibers, and smooth muscle provides the conducting portion with rigid structural support and the necessary flexibility and extensibility.

Respiratory Epithelium

Most of the conducting portion is lined with ciliated pseudostratified columnar epithelium that contains a rich population of goblet cells and is known as **respiratory epithelium** (Figure 17–2). Typical respiratory epithelium consists of 5 cell types (as seen in the electron microscope). **Ciliated columnar cells** constitute the most abundant type. Each cell has about 300 cilia on its apical surface (Figures 17–2, 17–3, and 17–4); beneath the cilia, in addition to basal bodies, are numerous small mitochondria which supply ATP for ciliary beating.

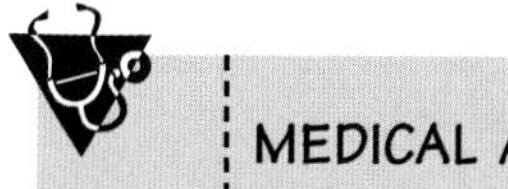

MEDICAL APPLICATION

Immotile cilia syndrome, *a disorder that causes infertility in men and chronic respiratory-tract infections in both sexes, is caused by immobility of cilia and flagella induced, in some cases, by deficiency of* ***dynein,*** *a protein normally present in the cilia. Dynein participates in the ciliary movement (see Chapter 2).*

The next most abundant cells in the respiratory epithelium are the **mucous goblet cells** (Figure 17–2). The apical portion of these cells (described in Chapter 4) contains the mucous droplets composed of glycoproteins. The remaining columnar cells are known as **brush cells** (Figure 17–4) because of the numerous microvilli on their apical surface. Brush cells have

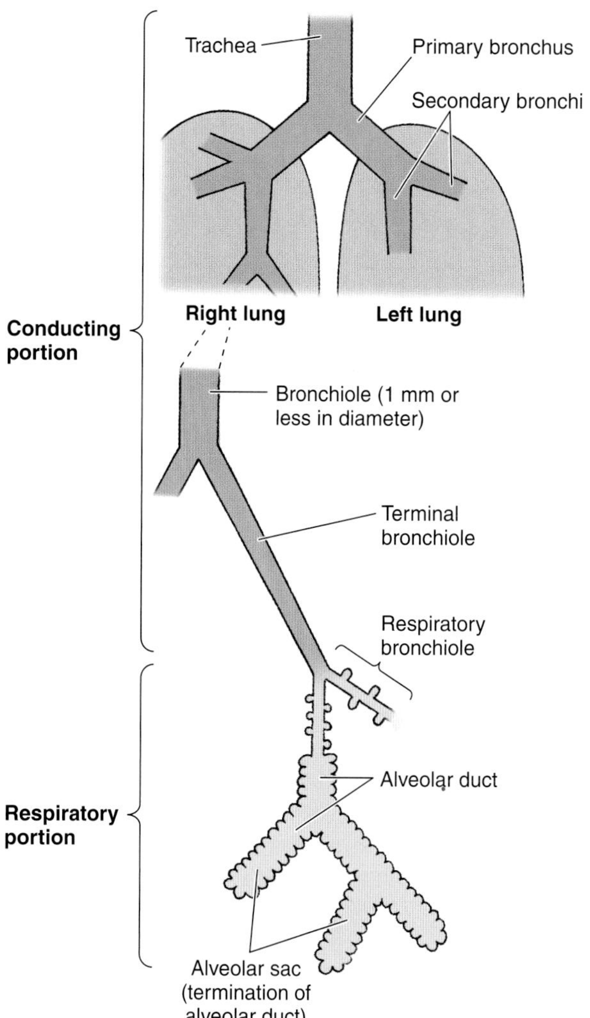

Figure 17–1. The main divisions of the respiratory tract. The natural proportions of these structures have been altered for clarity; the respiratory bronchiole, for example, is in reality a short transitional structure.

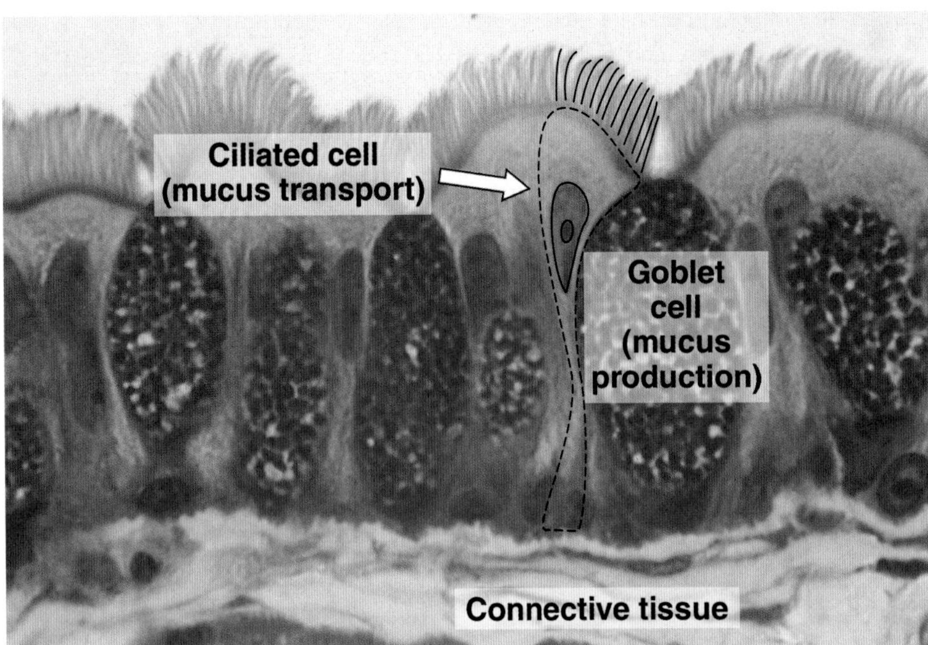

Figure 17–2. Photomicrograph illustrating the main components of the respiratory epithelium. Pararosaniline–toluidine blue (PT) stain. High magnification.

afferent nerve endings on their basal surfaces and are considered to be sensory receptors. **Basal** (**short**) **cells** are small rounded cells that lie on the basal lamina but do not extend to the luminal surface of the epithelium. These cells are believed to be generative stem cells that undergo mitosis and subsequently differentiate into the other cell types. The last cell type is the **small granule cell,** which resembles a basal cell except that it possesses numerous granules 100–300 nm in diameter with dense cores. Histochemical studies reveal that these cells constitute a population of cells of the diffuse neuroendocrine system (see Chapter 4). All cells of the ciliated pseudostratified columnar epithelium touch the basement membrane (Figure 17–2).

MEDICAL APPLICATION

From the nasal cavity through the larynx, portions of the epithelium are stratified squamous. This type of epithelium is evident in regions exposed to direct airflow or physical abrasion (eg, oropharynx, epiglottis, vocal folds); it provides more protection from attrition than does typical respiratory epithelium. If airflow currents are altered or new abrasive sites develop, the affected areas can convert from typical ciliated pseudostratified columnar epithelium to stratified squamous epithelium. Similarly, in smokers, the proportion of ciliated cells to goblet cells is altered to aid in clearing the increased particulate and gaseous pollutants (eg, CO, SO_2). Although the greater numbers of goblet cells in a smoker's epithelium provide for a more rapid clearance of pollutants, the reduction in ciliated cells caused by excessive intake of CO results in decreased movement of the mucous layer and frequently leads to congestion of the smaller airways.

NASAL CAVITY

The nasal cavity consists of 2 structures: the external **vestibule** and the internal **nasal fossae.**

Vestibule

The vestibule is the most anterior and dilated portion of the nasal cavity. The outer integument of the nose enters the **nares** (nostrils) and continues partway up the vestibule. Around the inner surface of the nares are numerous sebaceous and sweat glands, in addition to the thick short hairs, or **vibrissae,** that filter out large particles from the inspired air. Within the vestibule, the epithelium loses its keratinized nature and undergoes a transition into typical respiratory epithelium before entering the nasal fossae.

Nasal Fossae

Within the skull lie 2 cavernous chambers separated by the osseous **nasal septum.** Extending from each lateral wall are 3 bony shelflike projections known as **conchae.** Of the superior, middle, and inferior conchae, only the middle and inferior projections are covered with respiratory epithelium. The superior conchae are covered with a specialized **olfactory epithelium.** The narrow, ribbonlike passages created by the conchae improve the conditioning of the inspired air by increasing the surface area of respiratory epithelium and by creating turbulence in the airflow. The result is increased contact between air streams and the mucous layer. Within the lamina propria of the conchae are large venous plexuses known as **swell bodies.** Every 20–30 minutes, the swell bodies on one side of the nasal fossae become engorged with blood, resulting in distention of the conchal mucosa and a concomitant decrease in the flow of air. During this time, most of the air is directed through the other nasal fossa. These periodic intervals of occlusion reduce airflow, allowing the respiratory epithelium to recover from desiccation.

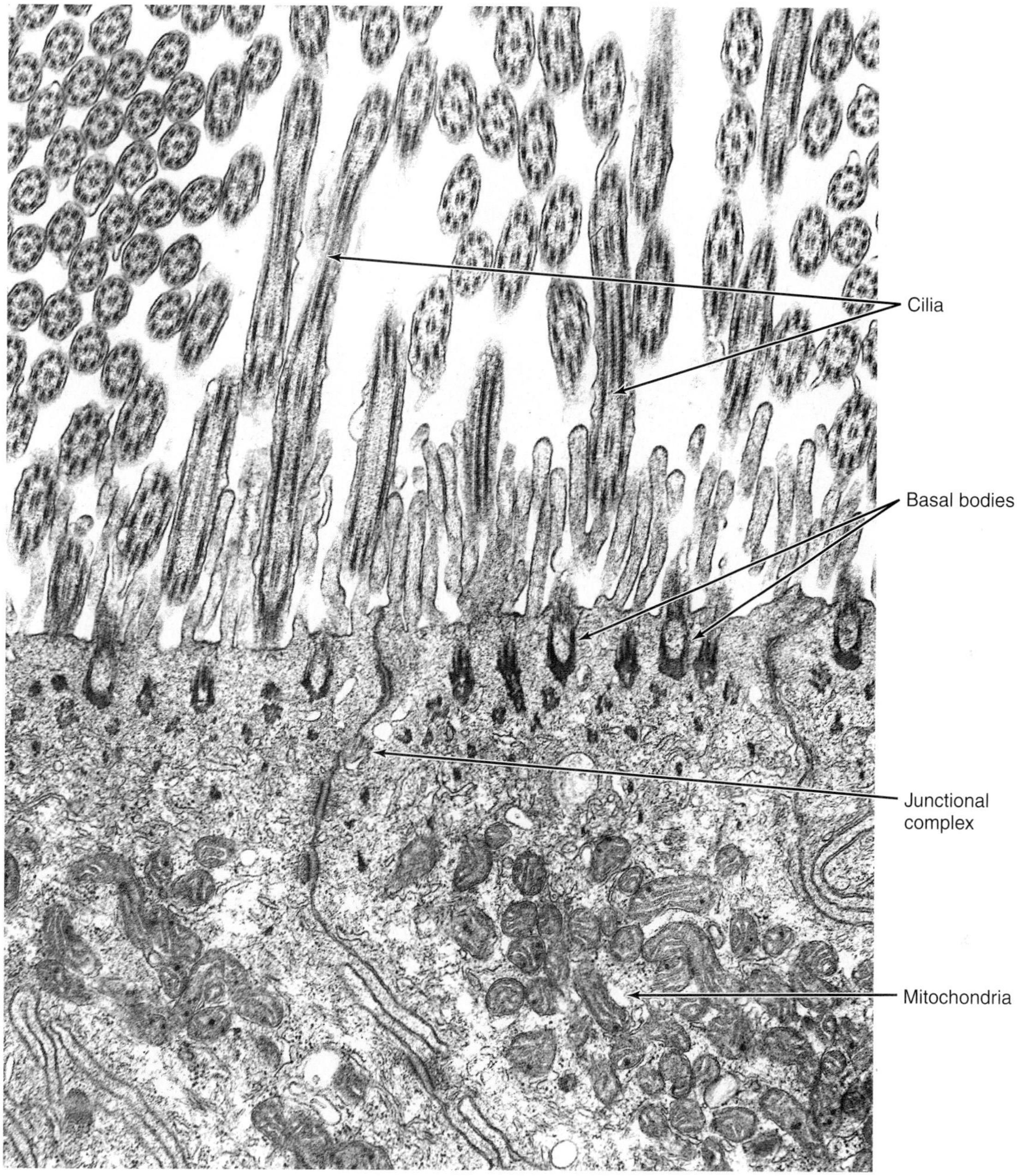

Figure 17–3. Electron micrograph of ciliated columnar cells of the respiratory epithelium, showing the ciliary microtubules in transverse and oblique section. In the cell apex are the U-shaped basal bodies that serve as the source of, and anchoring sites for, the ciliary axonemes. The local accumulation of mitochondria is related to energy production for ciliary movement. Note the junctional complex. ×9200.

MEDICAL APPLICATION

Allergic reactions and inflammation can cause abnormal engorgement of swell bodies in both fossae, severely restricting the air flow.

In addition to swell bodies, the nasal cavity has a rich vascular system with a complex organization. Large vessels form a close-meshed latticework next to the periosteum, from which arcading branches lead toward the surface. Blood in arcading vessels flows forward from the rear region in a direction counter to the flow of inspired air. As a result, the incoming air is efficiently warmed by a countercurrent system.

Smell (Olfaction)

The olfactory chemoreceptors are located in the **olfactory epithelium,** a specialized area of the mucous membrane in the

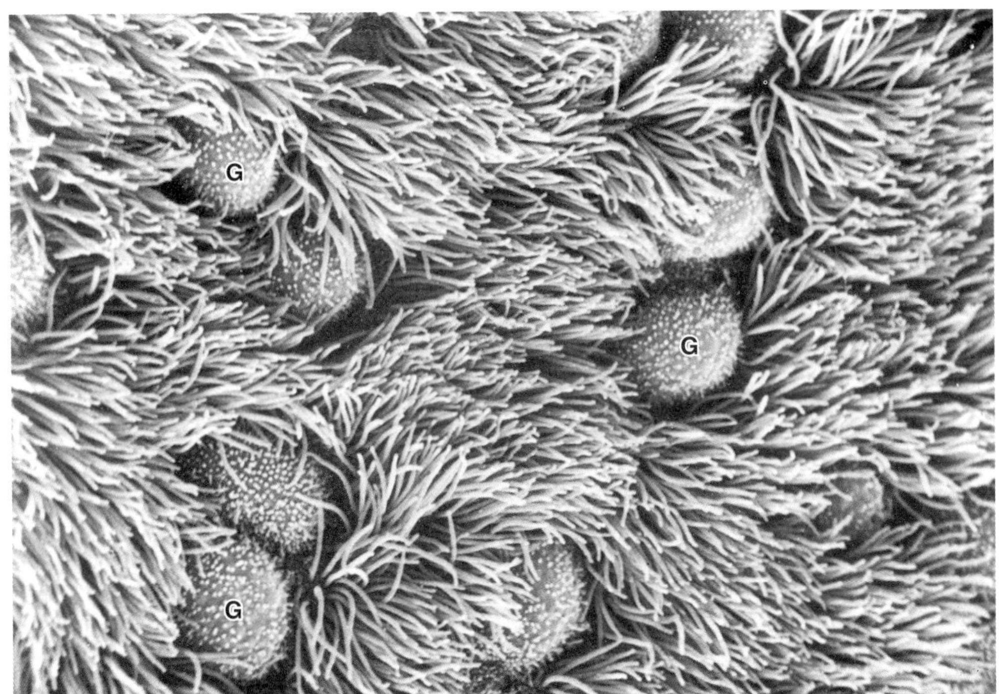

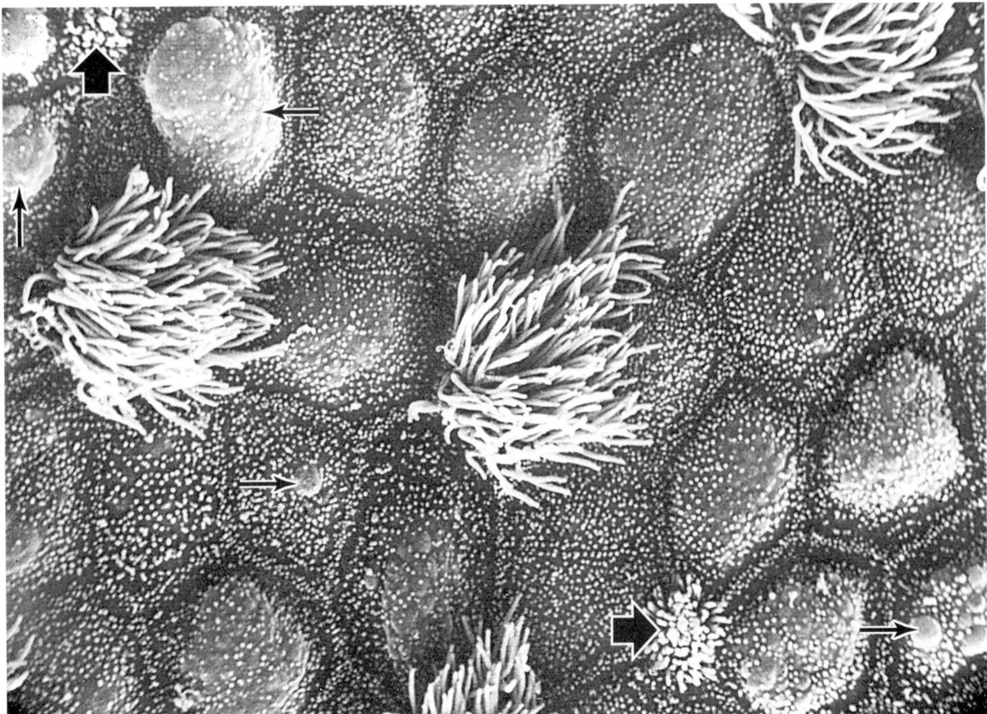

Figure 17–4. Scanning electron micrographs of the surface of respiratory mucosa. **Top:** Most of the surface is covered with cilia. G, goblet cells. ×2500. **Bottom:** Subsurface accumulations of mucus are evident in the goblet cells (thin arrows). Thick arrowheads indicate brush cells. ×3000. (Reproduced, with permission, from Andrews P: A scanning electron microscopic study of the extrapulmonary respiratory tract. Am J Anat 1974;139:421.)

superior conchae, located in the roof of the nasal cavity. In humans, it is about 10 cm^2 in area and up to 100 μm in thickness. It is a pseudostratified columnar epithelium composed of 3 types of cells (Figure 17–5).

The **supporting cells** have broad, cylindrical apexes and narrower bases. On their free surface are microvilli submerged in a fluid layer. Well-developed junctional complexes bind the supporting cells to the adjacent olfactory cells. The cells contain a light yellow pigment that is responsible for the color of the olfactory mucosa.

The **basal cells** are small; they are spherical or cone-shaped and form a single layer at the base of the epithelium.

Between the basal cells and the supporting cells are the **olfactory cells**—bipolar neurons distinguished from the supporting cells by the position of their nuclei, which lie below the nuclei of the supporting cells. Their apexes (dendrites) possess elevated and dilated areas from which arise 6–8 cilia. These cilia are very long, nonmotile (Figure 17–5), and respond to odoriferous substances by generating a receptor potential. They increase the receptor surface considerably. The afferent axons of these bipolar neurons unite in small bundles directed toward the central nervous system, where they synapse with neurons of the brain **olfactory lobe.**

The lamina propria of the olfactory epithelium possesses the glands of Bowman. Their secretion produces a fluid environment around the olfactory cilia that may clear the cilia, facilitating the access of new odoriferous substances.

Conditioning of Air

A major function of the conducting portion is to condition the inspired air. Before it enters the lungs, inspired air is cleansed, moistened, and warmed. To carry out these functions, the mucosa of the conducting portion is lined with a specialized **respiratory epithelium,** and there are numerous mucous and serous glands as well as a rich superficial vascular network in the lamina propria.

As the air enters the nose, large **vibrissae** (specialized hairs) remove coarse particles of dust. Once the air reaches the **nasal fossae,** particulate and gaseous impurities are trapped in a layer of mucus. This mucus, in conjunction with serous secretions, also serves to moisten the incoming air, protecting the delicate alveolar lining from desiccation. A rich superficial vascular network also warms the incoming air.

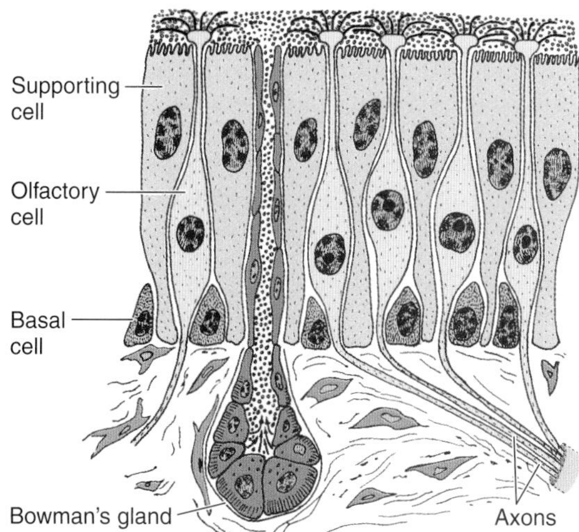

Figure 17–5. Olfactory mucosa showing the 3 cell types (supporting, olfactory, and basal) and a Bowman's gland.

PARANASAL SINUSES

The paranasal sinuses are closed cavities in the frontal, maxillary, ethmoid, and sphenoid bones. They are lined with a thinner respiratory epithelium that contains few goblet cells. The lamina propria contains only a few small glands and is continuous with the underlying periosteum. The paranasal sinuses communicate with the nasal cavity through small openings. The mucus produced in these cavities drains into the nasal passages as a result of the activity of its ciliated epithelial cells.

MEDICAL APPLICATION

Sinusitis *is an inflammatory process of the sinuses that may persist for long periods of time, mainly because of obstruction of drainage orifices. Chronic sinusitis and bronchitis are components of immotile cilia syndrome, which is characterized by defective ciliary action.*

NASOPHARYNX

The nasopharynx is the first part of the pharynx, continuing caudally with the oropharynx, the oral portion of this organ. It is lined with respiratory epithelium in the portion that is in contact with the soft palate.

LARYNX

The larynx is an irregular tube that connects the pharynx to the trachea. Within the lamina propria lie a number of laryngeal cartilages. The larger cartilages (thyroid, cricoid, and most of the arytenoids) are hyaline. The smaller cartilages (epiglottis, cuneiform, corniculate, and the tips of the arytenoids) are elastic cartilages. In addition to their supporting role (maintenance of an open airway), these cartilages serve as a valve to prevent swallowed food or fluid from entering the trachea. They also participate in producing sounds for phonation.

The **epiglottis,** which projects from the rim of the larynx, extends into the pharynx and has both a lingual and a laryngeal surface. The entire lingual surface and the apical portion of the laryngeal surface are covered with stratified squamous epithelium. Toward the base of the epiglottis on the laryngeal surface, the epithelium undergoes a transition into ciliated pseudostratified columnar epithelium. Mixed mucous and serous glands are found beneath the epithelium.

Below the epiglottis, the mucosa forms 2 pairs of folds that extend into the lumen of the larynx. The upper pair constitutes the **false vocal cords** (vestibular folds), covered with typical respiratory epithelium beneath which lie numerous serous glands within the lamina propria. The lower pair of folds constitutes the **true vocal cords.** Large bundles of parallel elastic fibers that compose the **vocal ligament** lie within the vocal folds, which are covered with a stratified squamous epithelium. Parallel to the ligaments are bundles of skeletal muscle, the **vocalis muscles,** which regulate the tension of the fold and its ligaments. As air is forced between the folds, these muscles provide the means for sounds of different frequencies to be produced.

TRACHEA

The trachea (Figure 17–6) is lined with a typical respiratory mucosa (Figures 17–2 and 17–4). In the lamina propria are 16–20 C-shaped rings of hyaline cartilage that keep the tracheal lumen open and numerous seromucous glands that produce a more fluid mucus. The open ends of these cartilage rings are located on the posterior surface of the trachea. A fibroelastic ligament and bundle of smooth muscle bind to the perichondrium and bridge the open ends of these C-shaped cartilages. The ligament prevents overdistention of the lumen, and the muscle allows regulation of the lumen.

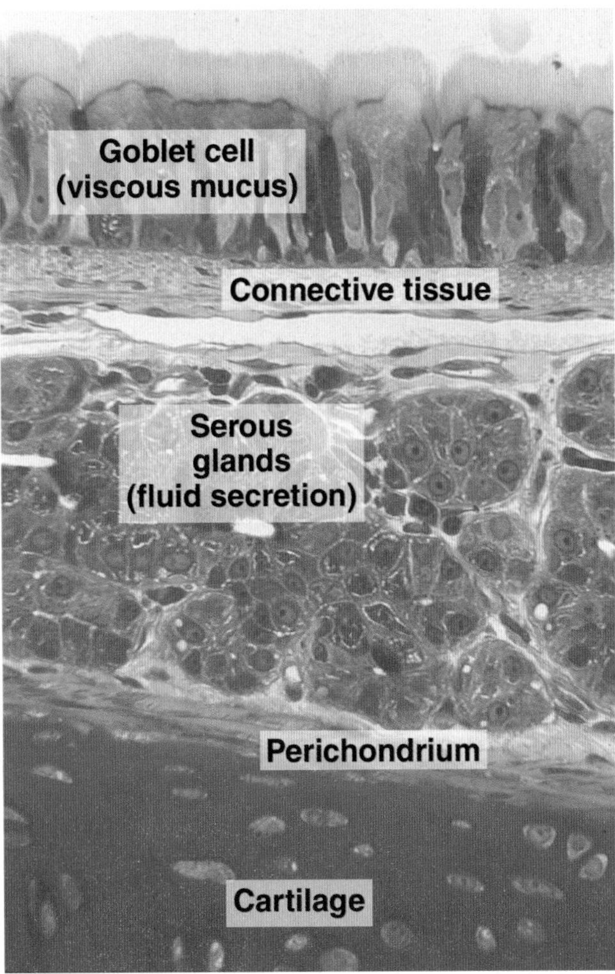

Figure 17–6. Section of trachea showing the respiratory epithelium with goblet cells and columnar ciliated cells. Also shown are serous glands in the lamina propria and hyaline cartilage. The mucous fluid produced by the goblet cells and by the glands forms a layer that permits the ciliary movement to propel foreign particles out of the respiratory system. PT stain. Medium magnification.

Contraction of the muscle and the resultant narrowing of the tracheal lumen are used in the cough reflex. The smaller bore of the trachea after contraction provides for increased velocity of expired air, which aids in clearing the air passage.

BRONCHIAL TREE

The trachea divides into 2 **primary bronchi** (Figure 17–1) that enter the lungs at the hilum. At each hilum, arteries enter, and veins and lymphatic vessels leave. These structures are surrounded by dense connective tissue and form a unit called the **pulmonary root.**

After entering the lungs, the primary bronchi course downward and outward, giving rise to 3 bronchi in the right lung and 2 in the left lung (Figure 17–1), each of which supplies a pulmonary lobe. These **lobar bronchi** divide repeatedly, giving rise to smaller bronchi, whose terminal branches are called **bronchioles.** Each bronchiole enters a pulmonary lobule, where it branches to form 5–7 **terminal bronchioles.**

The pulmonary lobules are pyramid-shaped, with the apex directed toward the pulmonary hilum. Each lobule is delineated by a thin connective tissue septum, best seen in the fetus. In adults, these septa are frequently incomplete, resulting in a poor delineation of the lobules.

The primary bronchi generally have the same histologic appearance as the trachea. Proceeding toward the respiratory portion, the histologic organization of both the epithelium and the underlying lamina propria becomes simplified. It must be stressed that this simplification is gradual; no abrupt transition can be observed between the bronchi and bronchioles. For this reason, the division of the bronchial tree into bronchi, bronchioles, etc, is to some extent artificial, although this division has both pedagogic and practical value.

Bronchi

Each primary bronchus branches dichotomously 9–12 times, with each branch becoming progressively smaller until it reaches a diameter of about 5 mm. Except for the organization of cartilage and smooth muscle, the mucosa of the bronchi is structurally similar to the mucosa of the trachea (Figure 17–6). The bronchial cartilages are more irregular in shape than those found in the trachea; in the larger portions of the bronchi, the cartilage rings completely encircle the lumen. As bronchial diameter decreases, the cartilage rings are replaced with isolated plates, or islands, of hyaline cartilage. Beneath the epithelium, in the bronchial lamina propria, is a smooth muscle layer consisting of crisscrossing bundles of spirally arranged smooth muscle (Figures 17–7 through 17–9). Bundles of smooth muscle become more prominent near the respiratory zone. Contraction of this muscle layer after death is responsible for the folded appearance of the bronchial mucosa observed in histologic section. The lamina propria is rich in elastic fibers and contains an abundance of mucous and serous glands (Figure 17–9) whose ducts open into the bronchial lumen. Numerous lymphocytes (Figures 17–10 and 17–11) are found both within the lamina propria and among the epithelial cells. Lymphatic nodules are present and are particularly numerous at the branching points of the bronchial tree.

Bronchioles

Bronchioles, intralobular airways with diameters of 5 mm or less, have neither cartilage nor glands in their mucosa; there are only scattered goblet cells within the epithelium of the initial seg-

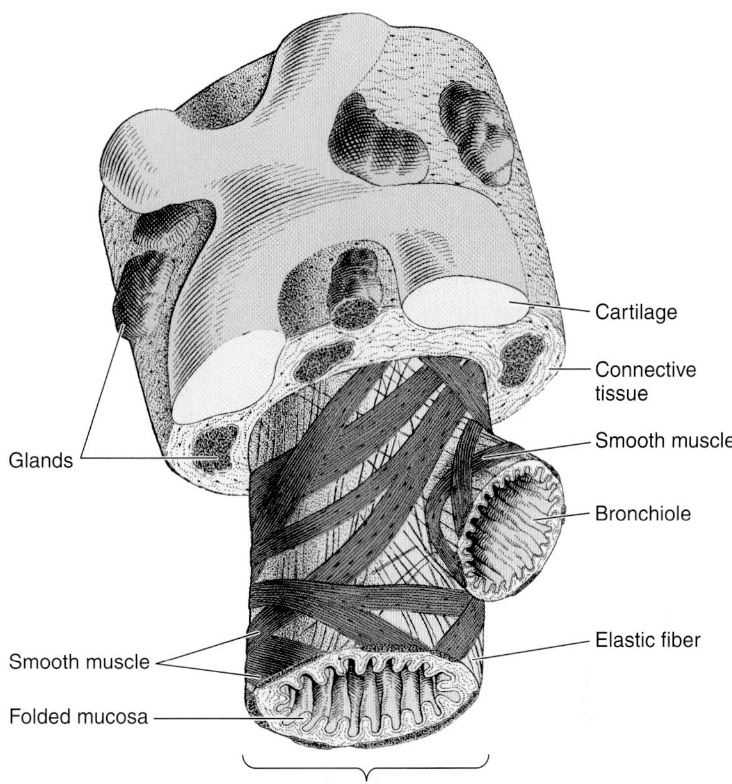

Figure 17–7. Structure of a bronchus. Smooth muscle is present in the entire bronchiolar tree, including the respiratory bronchiole. Contraction of this muscle induces folding of the mucosa. The elastic fibers in the bronchus continue into the bronchiole. The lower portion of the drawing represents a region with its connective tissue removed to show the presence of elastic fibers and smooth muscle. The adventitia is not shown.

ments. In the larger bronchioles, the epithelium is ciliated pseudostratified columnar, which decreases in height and complexity to become ciliated simple columnar or cuboidal epithelium in the smaller terminal bronchioles. The epithelium of terminal bronchioles also contains **Clara cells** (Figures 17–12 and 17–18). These cells, which are devoid of cilia, have secretory granules in their apex and are known to secrete proteins that protect the bronchiolar lining against oxidative pollutants and inflammation.

Bronchioles also exhibit specialized regions called **neuroepithelial bodies.** These are formed by groups of 80–100 cells that contain secretory granules and receive cholinergic nerve endings. Their function is poorly understood, but they are probably chemoreceptors that react to changes in gas composition within

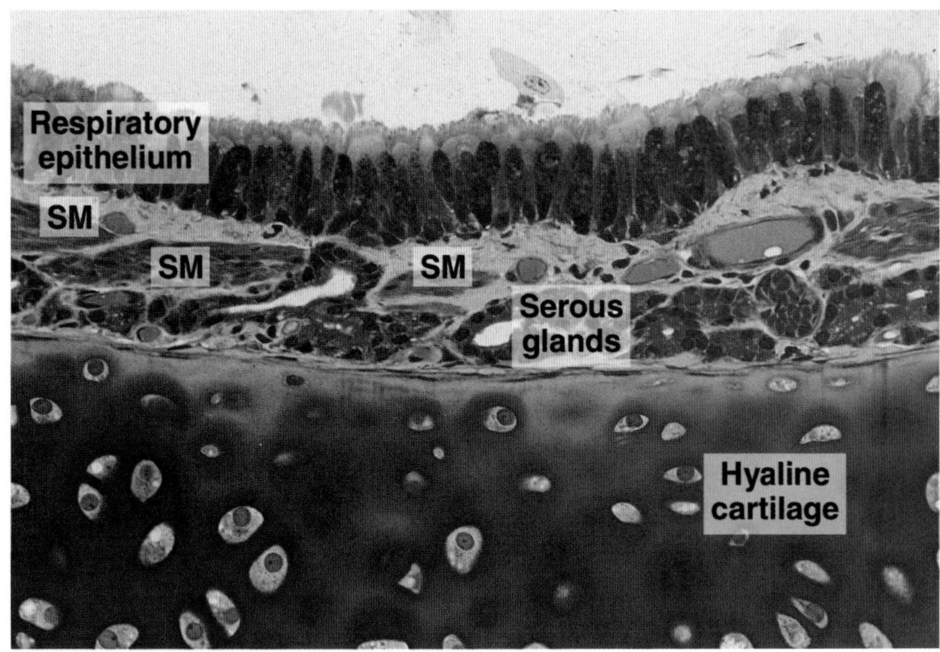

Figure 17–8. Section of a bronchus wall showing the respiratory epithelium with goblet cells and columnar ciliated cells. The connective tissue of the lamina propria contains serous glands and smooth muscle (SM). In the lower half of the photomicrograph is a large piece of hyaline cartilage. PT stain. Medium magnification.

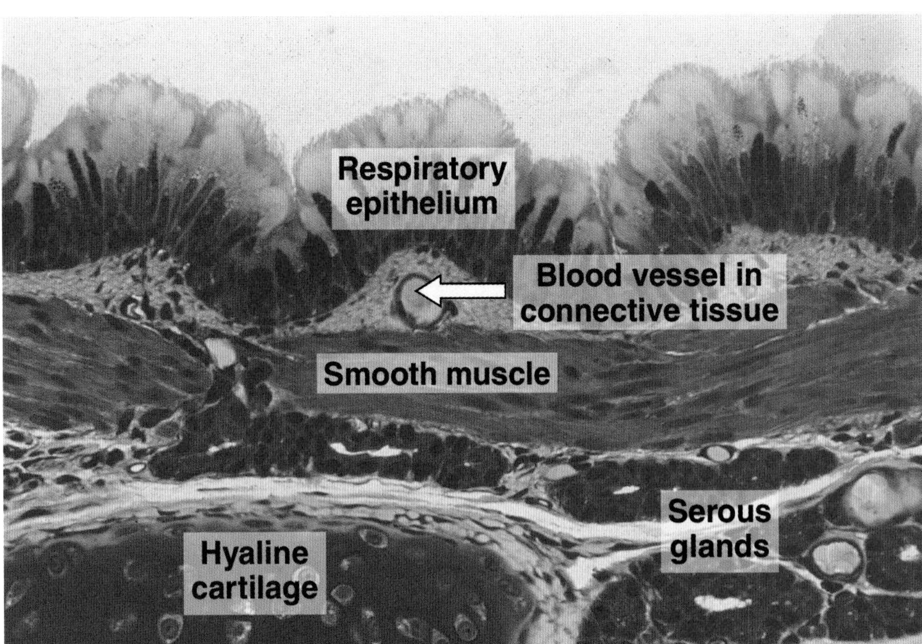

Figure 17–9. Large bronchus. Note the distinct layer of smooth muscle that influences the flux of air in the respiratory system. PT stain. Medium magnification.

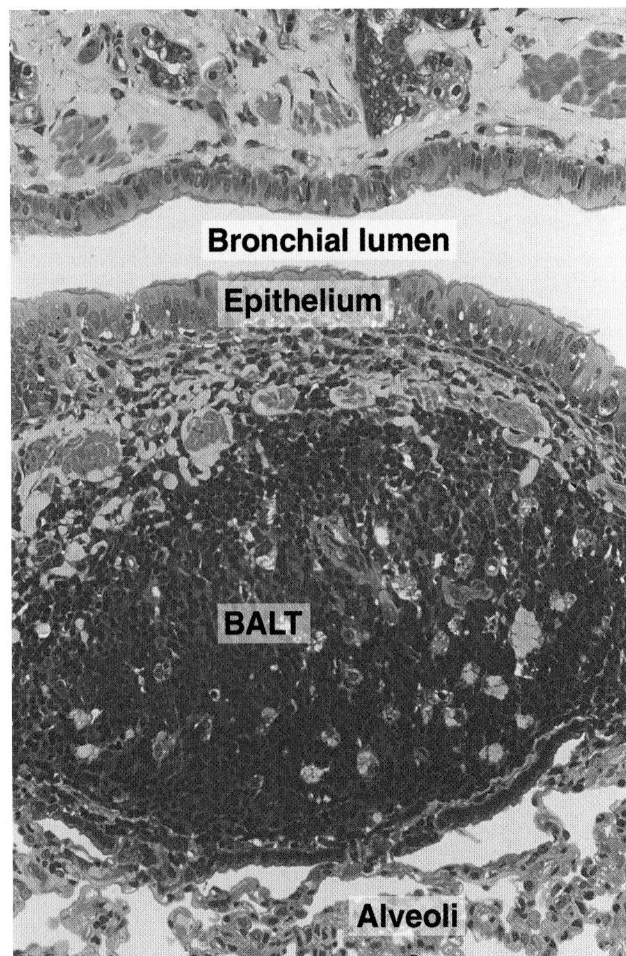

Figure 17–10. Section of a bronchus wall with bronchus-associated lymphoid tissue (BALT), a component of the diffuse mucosa-associated lymphoid tissue (MALT), whose distribution and functions are described in Chapter 14. PT stain. Medium magnification.

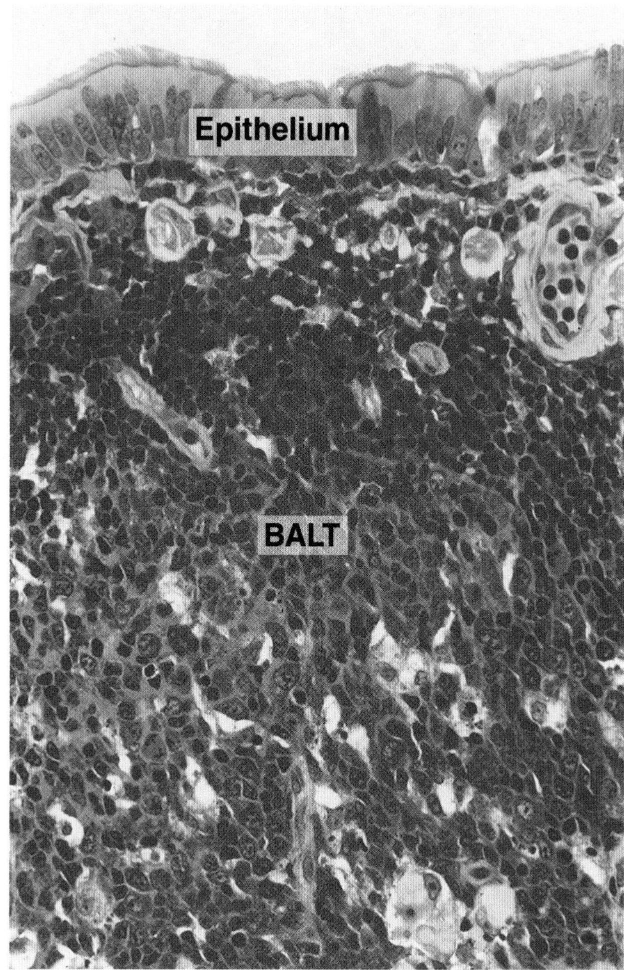

Figure 17–11. The same section as in Figure 17–10 but at higher magnification. The respiratory epithelium and its cilia are more evident. The lower part of the figure shows an activated region (germinal center) with light-stained areas, most with antigen-presenting macrophages. PT stain. High magnification.

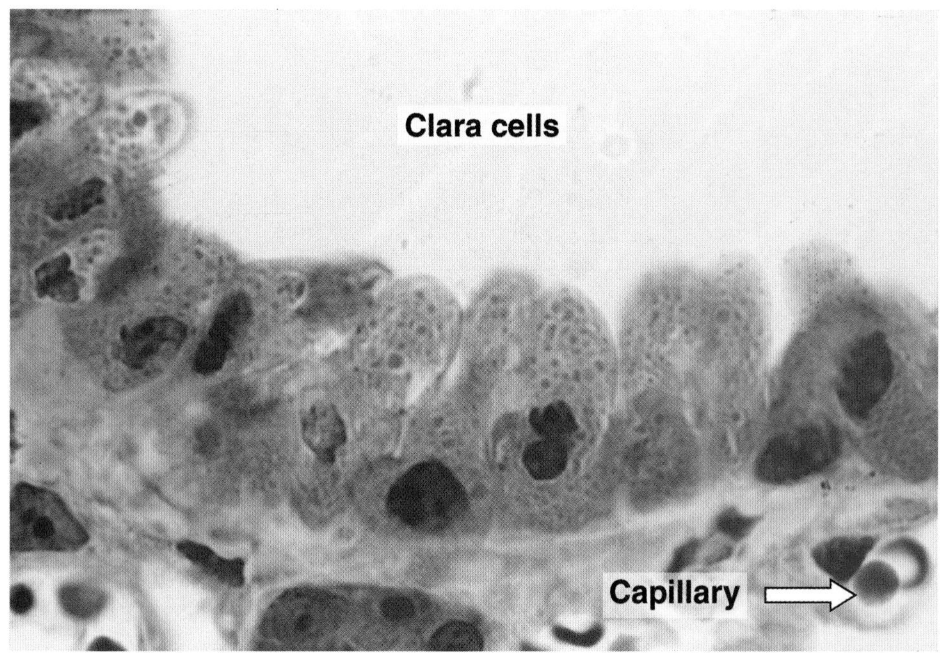

Figure 17–12. Clara cells in the epithelium of a terminal bronchiole. These cells show secretory granules and a bulging apical cytoplasm. PT stain. High magnification.

the airway. They also seem involved in the reparative process of airway epithelial cell renewal after injury.

Bronchiolar lamina propria is composed largely of smooth muscle and elastic fibers. The musculature of both the bronchi and the bronchioles is under the control of the vagus nerve and the sympathetic nervous system. Stimulation of the vagus nerve decreases the diameter of these structures; sympathetic stimulation produces the opposite effect.

MEDICAL APPLICATION

The increase in bronchiole diameter in response to stimulation of the sympathetic nervous system explains why epinephrine and other sympathomimetic drugs are frequently used to relax smooth muscle during asthma attacks. When the thickness of the bronchial walls is compared with that of the bronchiolar walls, it can be seen that the bronchiolar muscle layer is more developed. Increased airway resistance in asthma is believed to be due mainly to contraction of bronchiolar smooth muscle.

Respiratory Bronchioles

Each terminal bronchiole (Figure 17–13) subdivides into 2 or more respiratory bronchioles that serve as regions of transition between the conducting and respiratory portions of the respiratory system (Figures 17–14 and 17–15). The respiratory bronchiolar mucosa is structurally identical to that of the terminal bronchioles, except that their walls are interrupted by numerous saclike alveoli where gas exchange occurs. Portions of the respiratory bronchioles are lined with ciliated cuboidal epithelial cells and Clara cells, but at the rim of the alveolar openings the bronchiolar epithelium becomes continuous with the squamous alveolar lining cells (type I alveolar cells; see below). Proceeding distally along these bronchioles, the alveoli increase greatly in number, and the distance between them is markedly reduced. Between alveoli, the bronchiolar epithelium consists of ciliated cuboidal epithelium; however, the cilia may be absent in more distal portions. Smooth muscle and elastic connective tissue lie beneath the epithelium of respiratory bronchioles.

Alveolar Ducts

Proceeding distally along the respiratory bronchioles, the number of alveolar openings into the bronchiolar wall becomes ever greater until the wall consists of nothing else, and the tube is now called an **alveolar duct** (Figure 17–16). Both the alveolar ducts and the alveoli (Figures 17–17 and 17–18) are lined with extremely attenuated squamous alveolar cells. In the lamina propria surrounding the rim of the alveoli is a network of smooth muscle cells. These sphincterlike smooth muscle bundles appear as knobs between adjacent alveoli. Smooth muscle disappears at the distal ends of alveolar ducts. A rich matrix of elastic and reticular fibers provides the only support of the duct and its alveoli.

Alveolar ducts open into **atria** that communicate with **alveolar sacs,** 2 or more of which arise from each atrium. Elastic and reticular fibers form a complex network encircling the openings of atria, alveolar sacs, and alveoli. The elastic fibers enable the alveoli to expand with inspiration and to contract passively with expiration. The reticular fibers serve as a support that prevents overdistention and damage to the delicate capillaries and thin alveolar septa.

Alveoli

Alveoli are saclike evaginations (about 200 μm in diameter) of the respiratory bronchioles, alveolar ducts, and alveolar sacs. Alveoli are responsible for the spongy structure of the lungs (Figure 17–15). Structurally, alveoli resemble small pockets that are

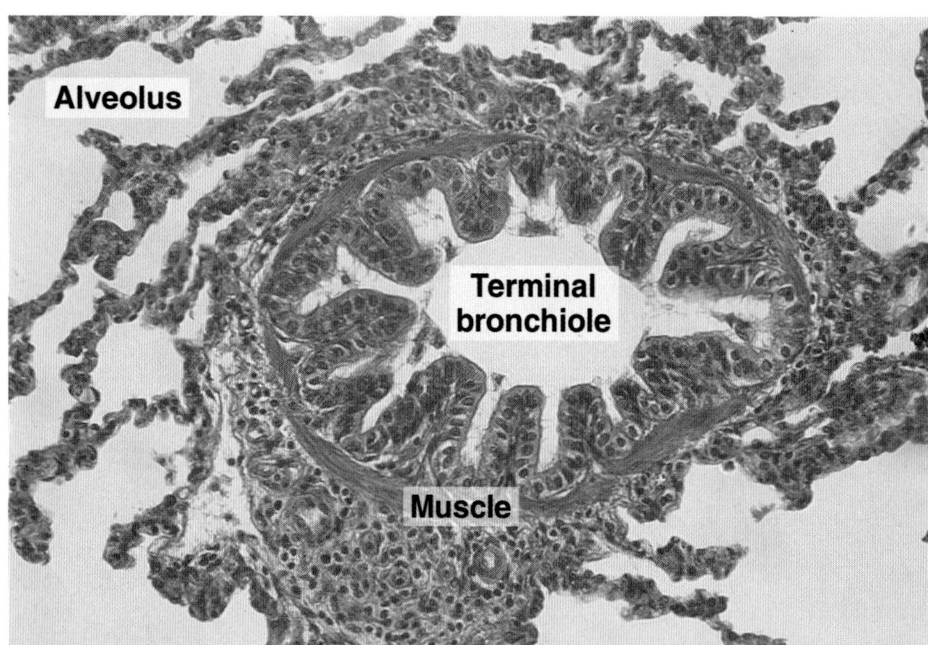

Figure 17–13. Photomicrograph of a section from the wall of a terminal bronchiole. Note that no cartilage is present, but there is an incomplete ring of smooth muscle. PT stain. Low magnification.

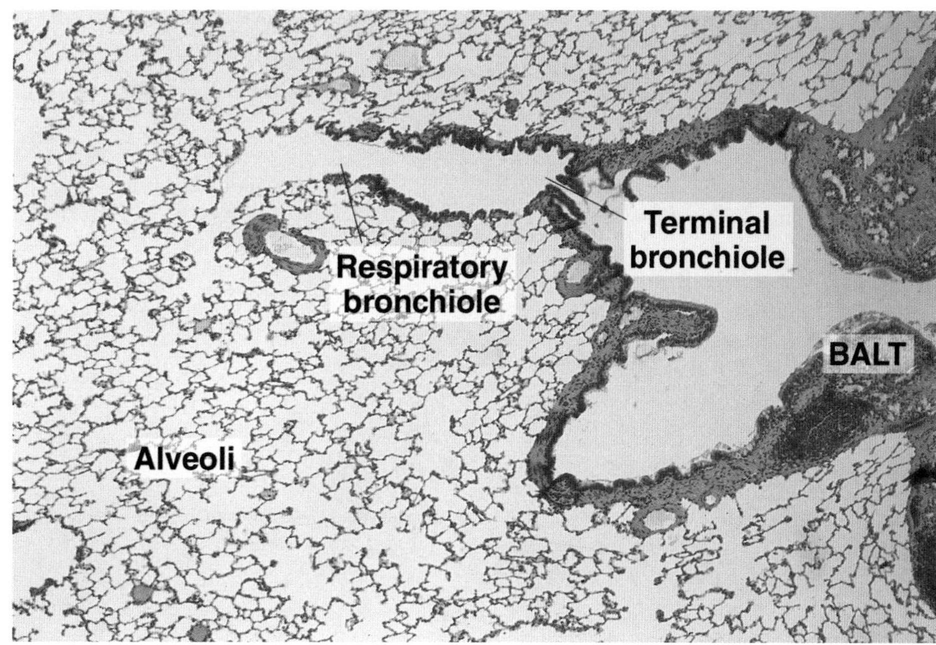

Figure 17–14. Section of a terminal bronchiole with a small portion of a respiratory bronchiole continuous with an alveolar duct and many alveoli. PT stain. Low magnification.

open on one side, similar to the honeycombs of a beehive. Within these cuplike structures, O_2 and CO_2 are exchanged between the air and the blood. The structure of the alveolar walls is specialized for enhancing diffusion between the external and internal environments. Generally, each wall lies between 2 neighboring alveoli and is therefore called an **interalveolar septum,** or **wall.** An interalveolar septum (Figures 17–19 through 17–23) consists of 2 thin squamous epithelial layers between which lie capillaries, elastic and reticular fibers, and connective tissue matrix and cells. The capillaries and connective tissue constitute the **interstitium.** Within the interstitium of the interalveolar septum is found the richest capillary network in the body.

Air in the alveoli is separated from capillary blood by 3 components referred to collectively as the **blood-air barrier:** the surface lining and cytoplasm of the alveolar cells; the fused basal laminae of the closely apposed alveolar and endothelial cells; and the cytoplasm of the endothelial cells (Figure 17–20). The total thickness of these layers varies from 0.1 to 1.5 μm. Within the interalveolar septum, anastomosing pulmonary capillaries are supported by a meshwork of reticular and elastic fibers. These fibers, which are arranged to permit expansion and contraction of the interalveolar septum, are the primary means of structural support of the alveoli. The basement membrane, leukocytes, macrophages, and fibroblasts can also be found within the

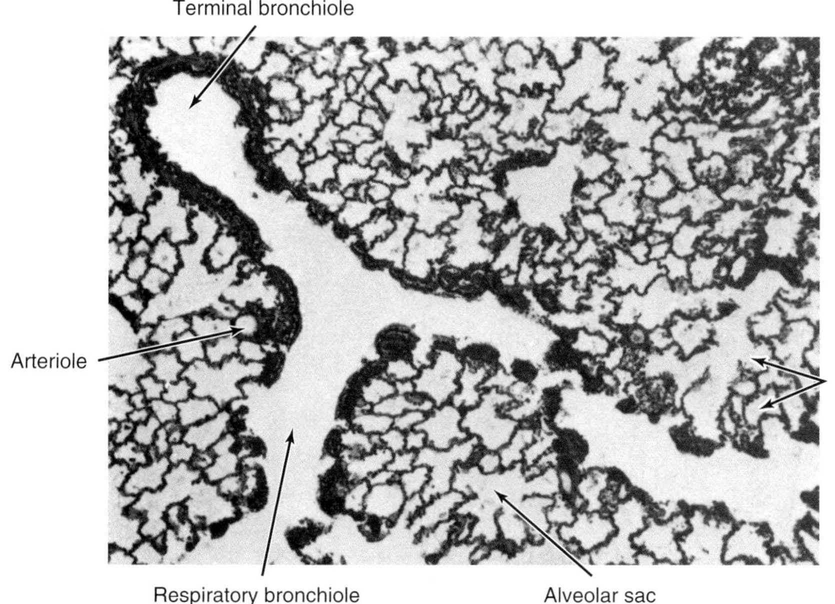

Figure 17–15. Photomicrograph of a thick section of lung showing a terminal bronchiole dividing into 2 respiratory bronchioles, in which alveoli appear. The spongelike appearance of the lung is due to the abundance of alveoli and alveolar sacs. H&E stain. Low magnification.

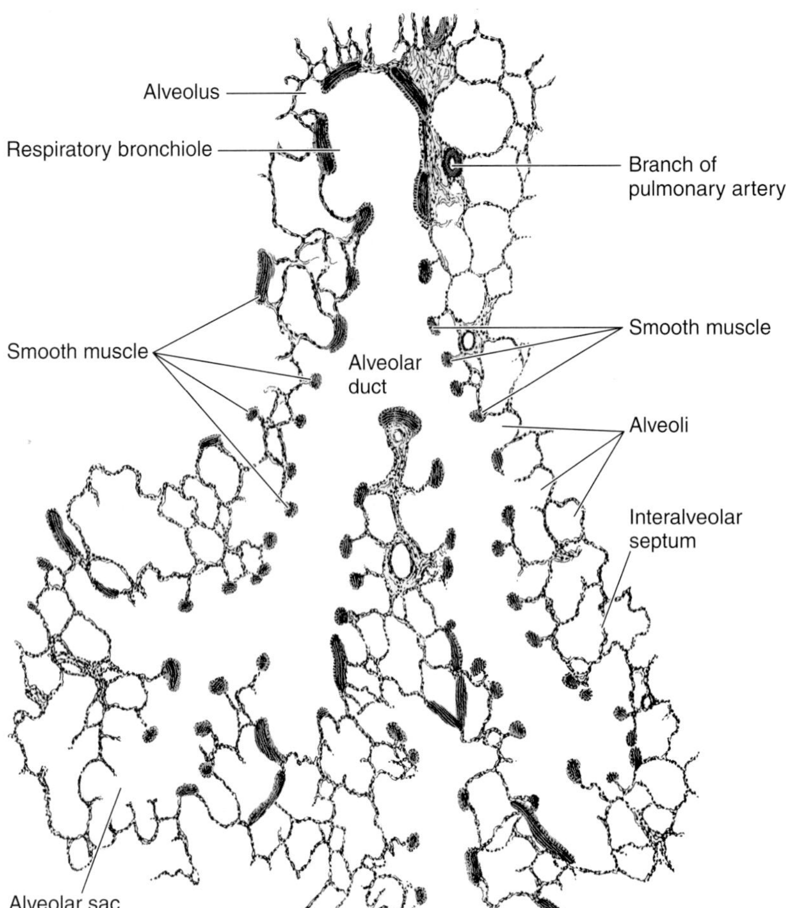

Figure 17–16. Diagram of a portion of the bronchial tree. Note that the smooth muscle in the alveolar duct disappears in the alveoli. (Redrawn from Baltisberger.)

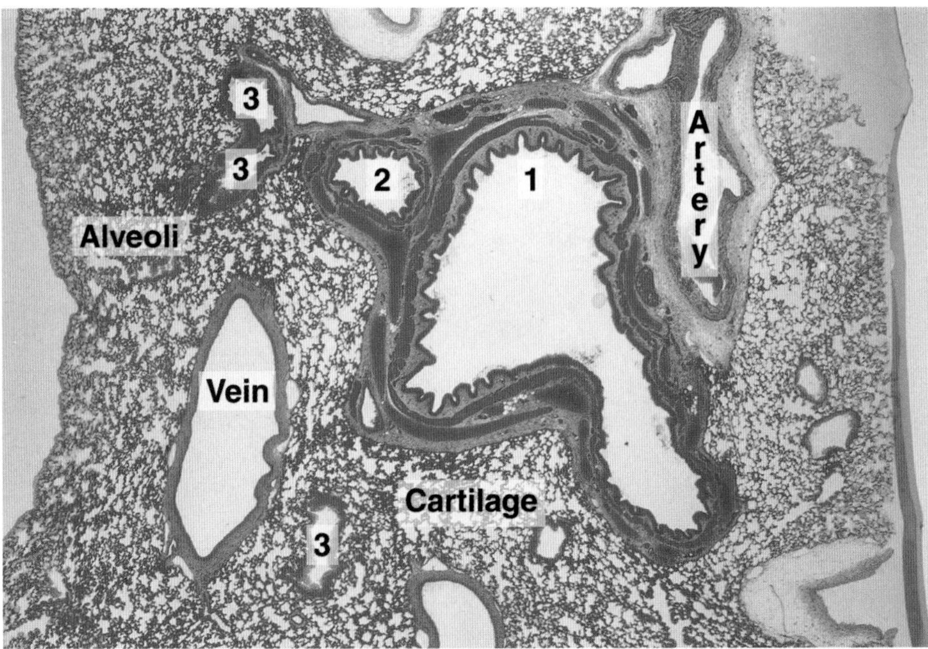

Figure 17–17. Bird's-eye view of a lung section showing branching of bronchioles with different sizes of bronchioles (1, 2, 3), large blood vessels, and alveoli. PT stain. Low magnification.

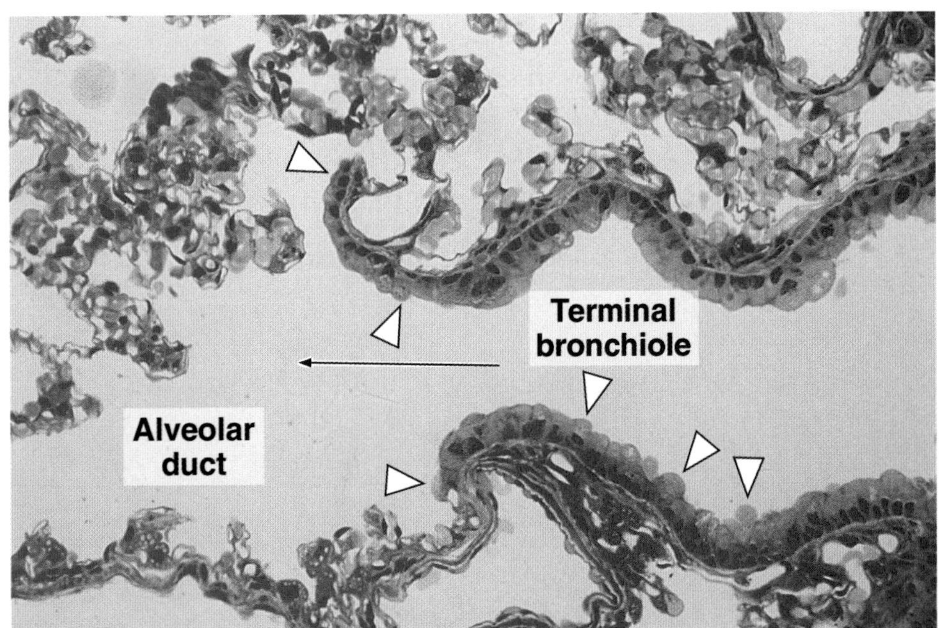

Figure 17–18. Transition of a terminal bronchiole into an alveolar duct (arrow). Note the Clara cells (arrowheads). PT stain. Medium magnification.

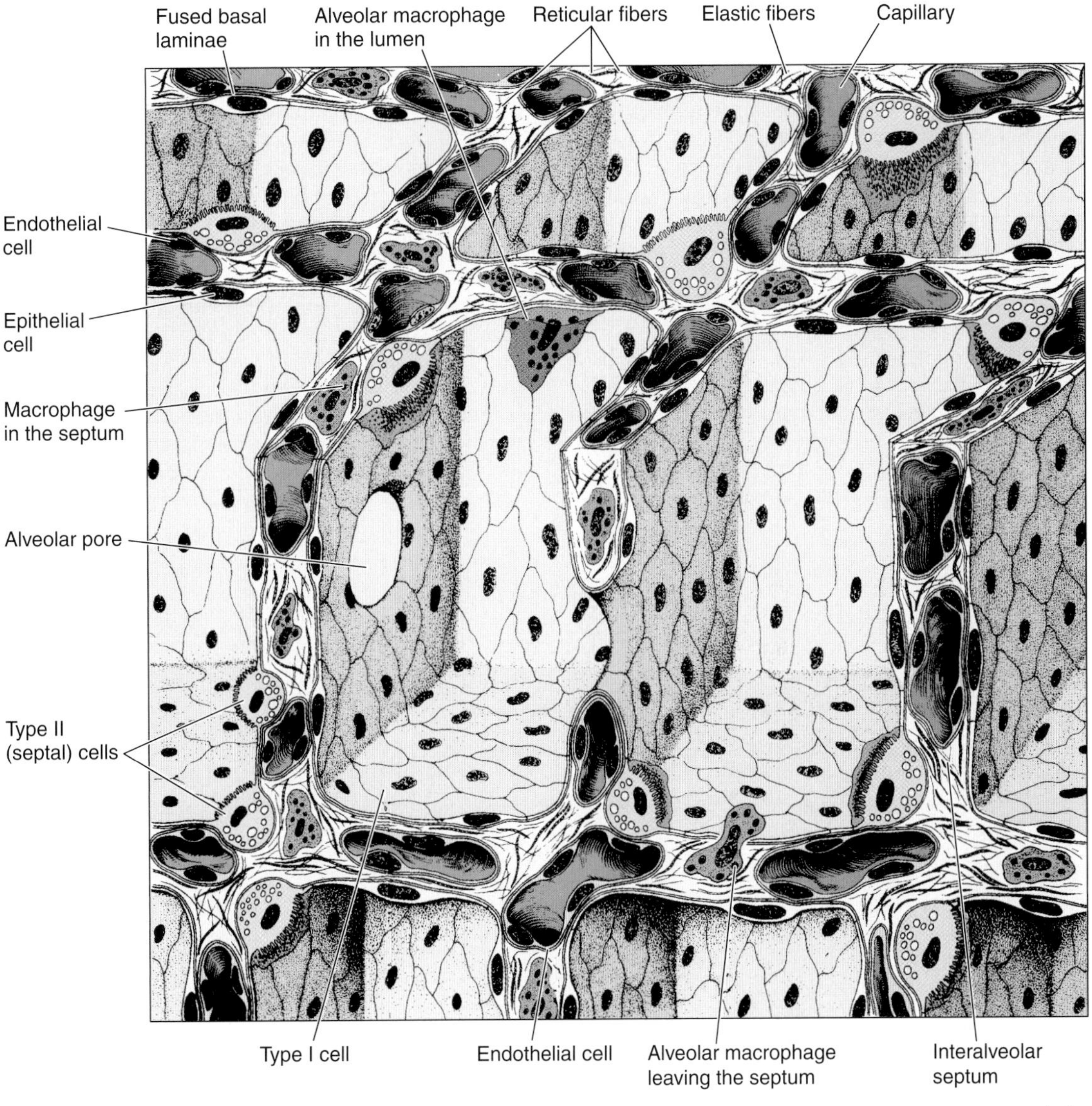

Fused basal laminae
Alveolar macrophage in the lumen
Reticular fibers
Elastic fibers
Capillary
Endothelial cell
Epithelial cell
Macrophage in the septum
Alveolar pore
Type II (septal) cells
Type I cell
Endothelial cell
Alveolar macrophage leaving the septum
Interalveolar septum

Figure 17–19. Three-dimensional schematic diagram of pulmonary alveoli showing the structure of the interalveolar septum. Note the capillaries, connective tissue, and macrophages. These cells can also be seen in—or passing into—the alveolar lumen. Alveolar pores are numerous. Type II cells are identified by their abundant apical microvilli. The alveoli are lined with a continuous epithelial layer of type I cells.

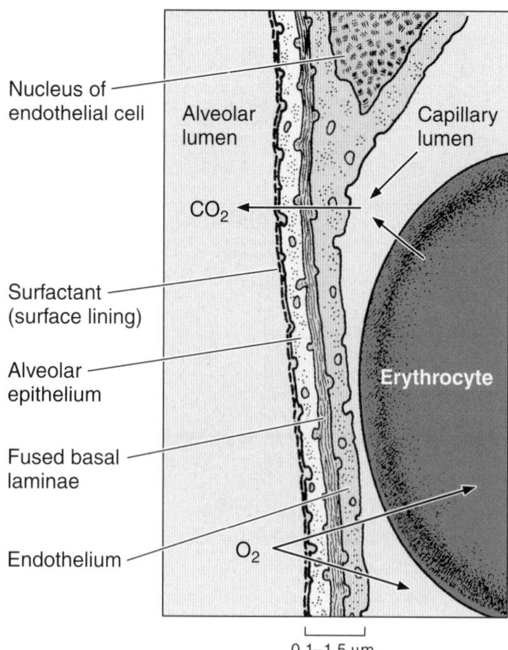

Figure 17–20. Portion of the interalveolar septum showing the blood-air barrier. To reach the erythrocyte, O_2 traverses the surface lining, the alveolar epithelium cytoplasm, and the plasma. In some locations, there is loose interstitial tissue between the epithelium and the endothelium. (Modified and reproduced, with permission, from Ganong WF: *Review of Medical Physiology,* 8th ed. Lange, 1977.)

interstitium of the septum (Figure 17–19). The fusion of 2 basal laminae produced by the endothelial cells and the epithelial (alveolar) cells of the interalveolar septum form the basement membrane (Figures 17–20, 17–21 and 17–23).

O_2 from the alveolar air passes into the capillary blood through the blood-air barrier; CO_2 diffuses in the opposite direction. Liberation of CO_2 from H_2CO_3 is catalyzed by the enzyme **carbonic anhydrase** present in erythrocytes. The approximately 300 million alveoli in the lungs considerably increase their internal exchange surface, which has been calculated to be approximately 140 m^2 (Figures 17–17 and 17–20).

Capillary endothelial cells are extremely thin and can be easily confused with type I alveolar epithelial cells. The endothelial lining of the capillaries is continuous and not fenestrated (Figure 17–21). Clustering of the nuclei and other organelles allows the remaining areas of the cell to become extremely thin, increasing the efficiency of gas exchange. The most prominent feature of the cytoplasm in the flattened portions of the cell is numerous pinocytotic vesicles.

Type I cells, or **squamous alveolar cells,** are extremely attenuated cells that line the alveolar surfaces. Type I cells make up 97% of the alveolar surfaces (type II cells make up the remaining 3%). These cells are so thin (sometimes only 25 nm) that the electron microscope was needed to prove that all alveoli are covered with an epithelial lining (Figures 17–19 and 17–21). Organelles such as the Golgi complex, endoplasmic reticulum, and mitochondria are grouped around the nucleus, reducing the thickness of the blood-air barrier and leaving large areas of cytoplasm virtually free of organelles. The cytoplasm in the thin portion contains abundant pinocytotic vesicles, which may play a role in the turnover of surfactant (described below) and the removal of small particulate contaminants from the outer surface. In addition to desmosomes, all type I epithelial cells have occluding junctions that prevent the leakage of tissue fluid into the alveolar air space (Figure 17–24). The main role of these cells is to provide a barrier of minimal thickness that is readily permeable to gases.

Type II cells are interspersed among the type I alveolar cells with which they have occluding and desmosomal junctions (Figures 17–25 and 17–26). Type II cells are rounded cells that are usually found in groups of 2 or 3 along the alveolar surface at points where the alveolar walls unite and form angles. These cells, which rest on the basement membrane, are part of the epithelium, with the same origin as the type I cells that line the alveolar walls. They divide by mitosis to replace their own population

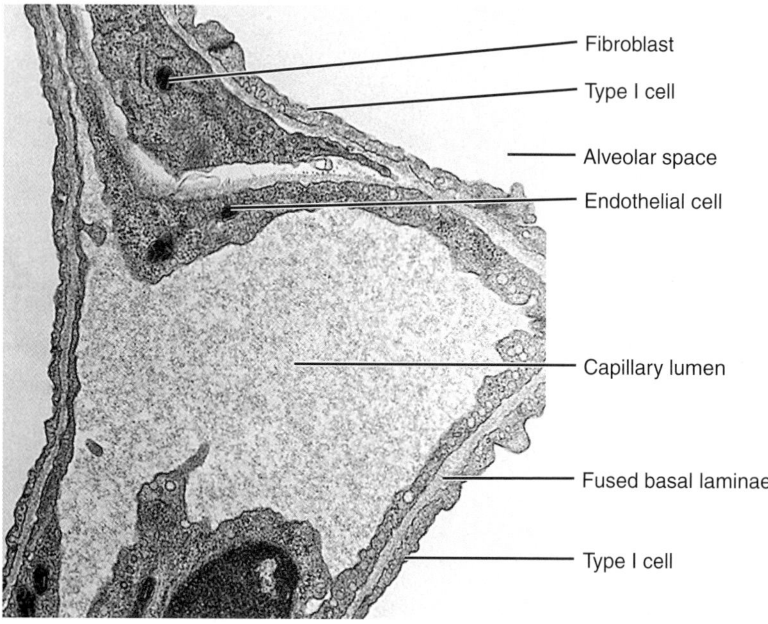

Figure 17–21. Electron micrograph of the interalveolar septum. Note the capillary lumen, alveolar spaces, alveolar type I epithelial cells, fused basal laminae, and a fibroblast. ×30,000. (Courtesy of MC Williams.)

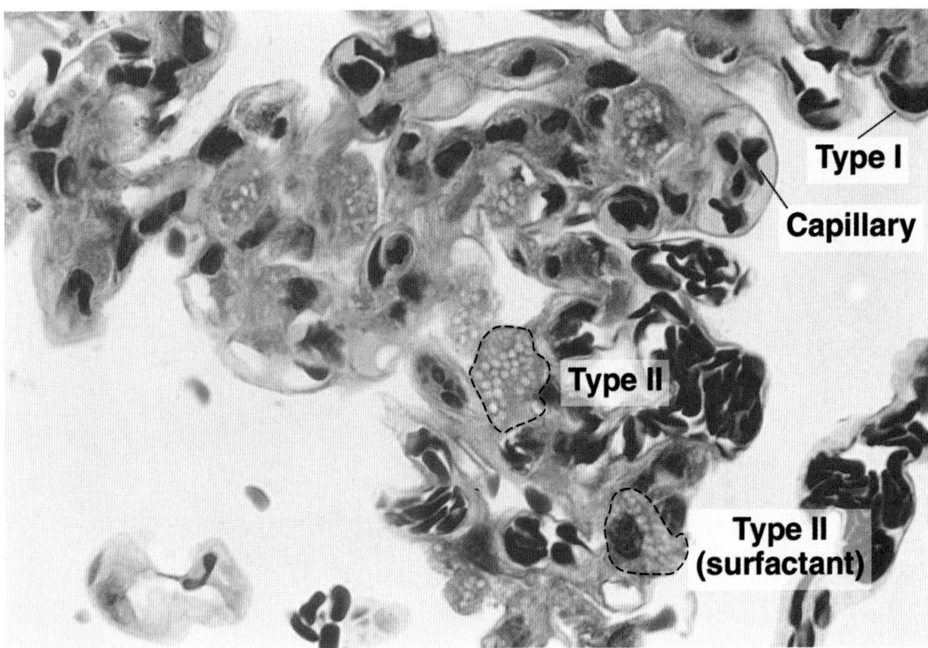

Figure 17–22. Alveoli and interalveolar septum showing blood capillaries and epithelial cells type I and type II. PT stain. Medium magnification.

and also the type I population. In histologic sections, they exhibit a characteristic vesicular or foamy cytoplasm. These vesicles are caused by the presence of **lamellar bodies** (Figures 17–25 and 17–26) that are preserved and evident in tissue prepared for electron microscopy. Lamellar bodies, which average 1–2 μm in diameter, contain concentric or parallel lamellae limited by a unit membrane. Histochemical studies show that these bodies, which contain phospholipids, glycosaminoglycans, and proteins, are continuously synthesized and released at the apical surface of the cells. The lamellar bodies give rise to a material that spreads over the alveolar surfaces, providing an extracellular alveolar coating, **pulmonary surfactant,** that lowers alveolar surface tension.

The surfactant layer consists of an aqueous, proteinaceous hypophase covered with a monomolecular phospholipid film that is primarily composed of **dipalmitoyl phosphatidylcholine** and **phosphatidylglycerol.** Surfactant also contains several types of proteins. Pulmonary surfactant serves several major functions in the economy of the lung, but it primarily aids in reducing the surface tension of the alveolar cells. The reduction of surface tension means that less inspiratory force is needed to inflate the alveoli, and thus the work of breathing is reduced. In addition, without surfactant, alveoli would tend to collapse during expiration. In fetal development, surfactant appears in the last weeks of gestation and coincides with the appearance of lamellar bodies in the type II cells.

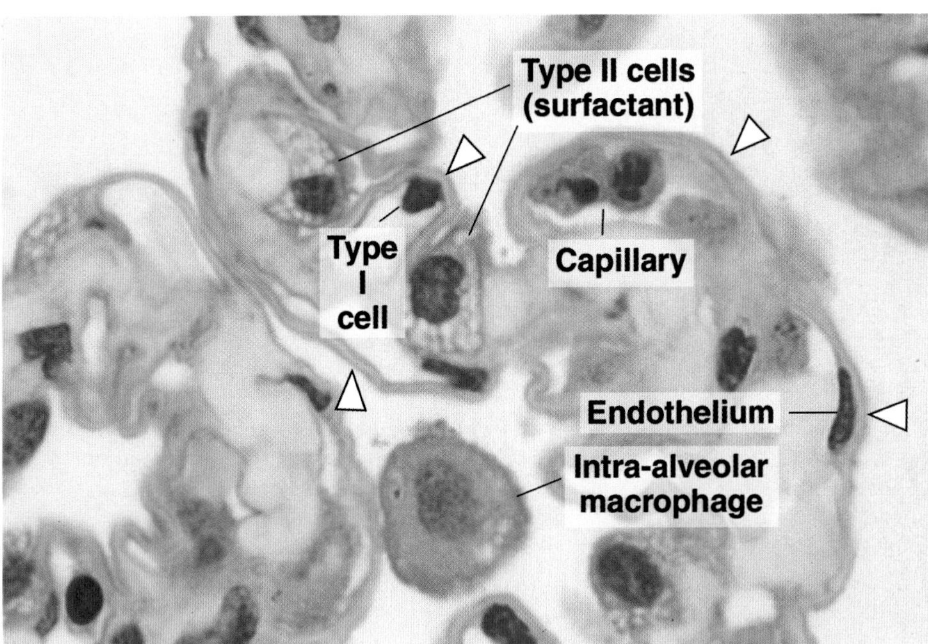

Figure 17–23. Section of a lung fixed by intra-alveolar injection of fixative. Observe in the interalveolar septum 3-laminar structures (arrowheads) constituted by a central basement membrane and 2 very thin cytoplasmic layers. These layers are formed by the cytoplasm of epithelial cell type I and the cytoplasm of capillary endothelial cells. PT stain. High magnification.

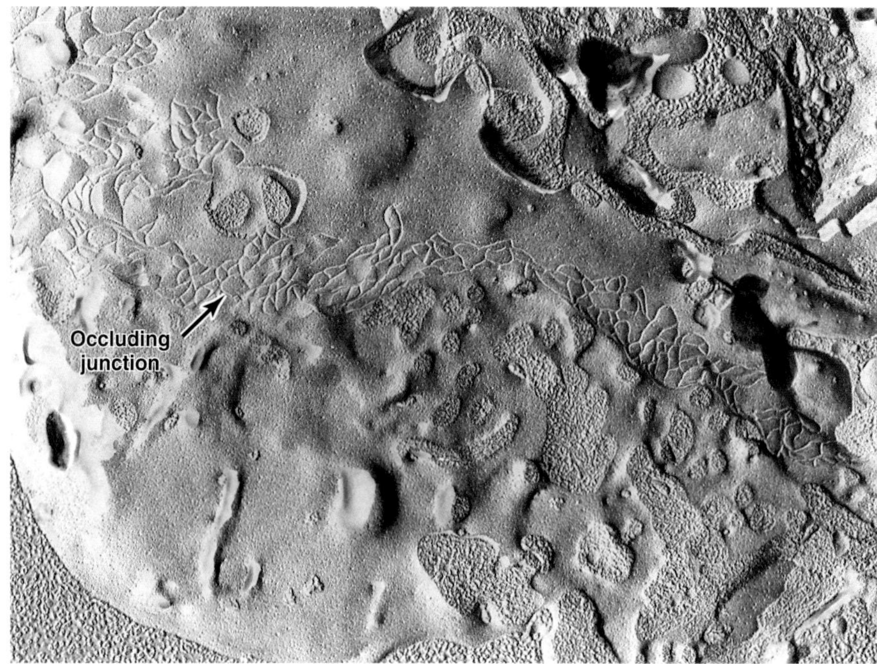

Figure 17–24. Cryofracture preparation showing an occluding junction between 2 type I epithelial cells of the alveolar lining. ×25,000. (Reproduced, with permission, from Schneeberger EE: *Lung Liquids.* Ciba Foundation Symposium no. 38. Elsevier/North-Holland, 1976.)

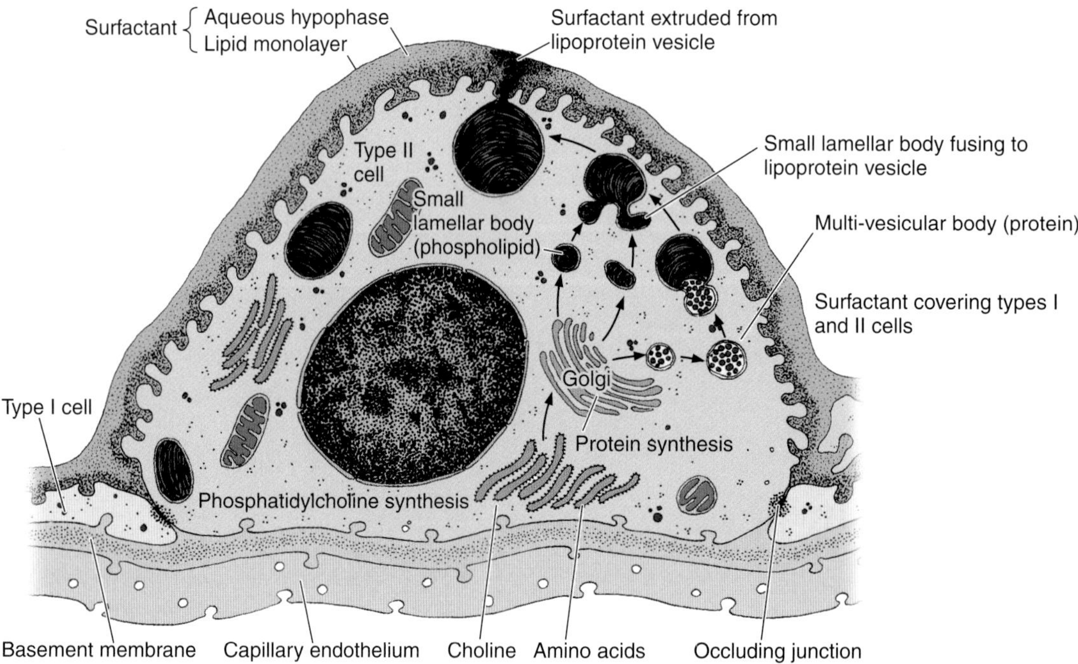

Figure 17–25. Secretion of surfactant by a type II cell. Surfactant is a protein-lipid complex synthesized in the rough endoplasmic reticulum and Golgi complex and stored in the lamellar bodies. It is continuously secreted by means of exocytosis (arrows) and forms an overlying monomolecular film of lipid covering an underlying aqueous hypophase. Occluding junctions around the margins of the epithelial cells prevent leakage of tissue fluid into the alveolar lumen.

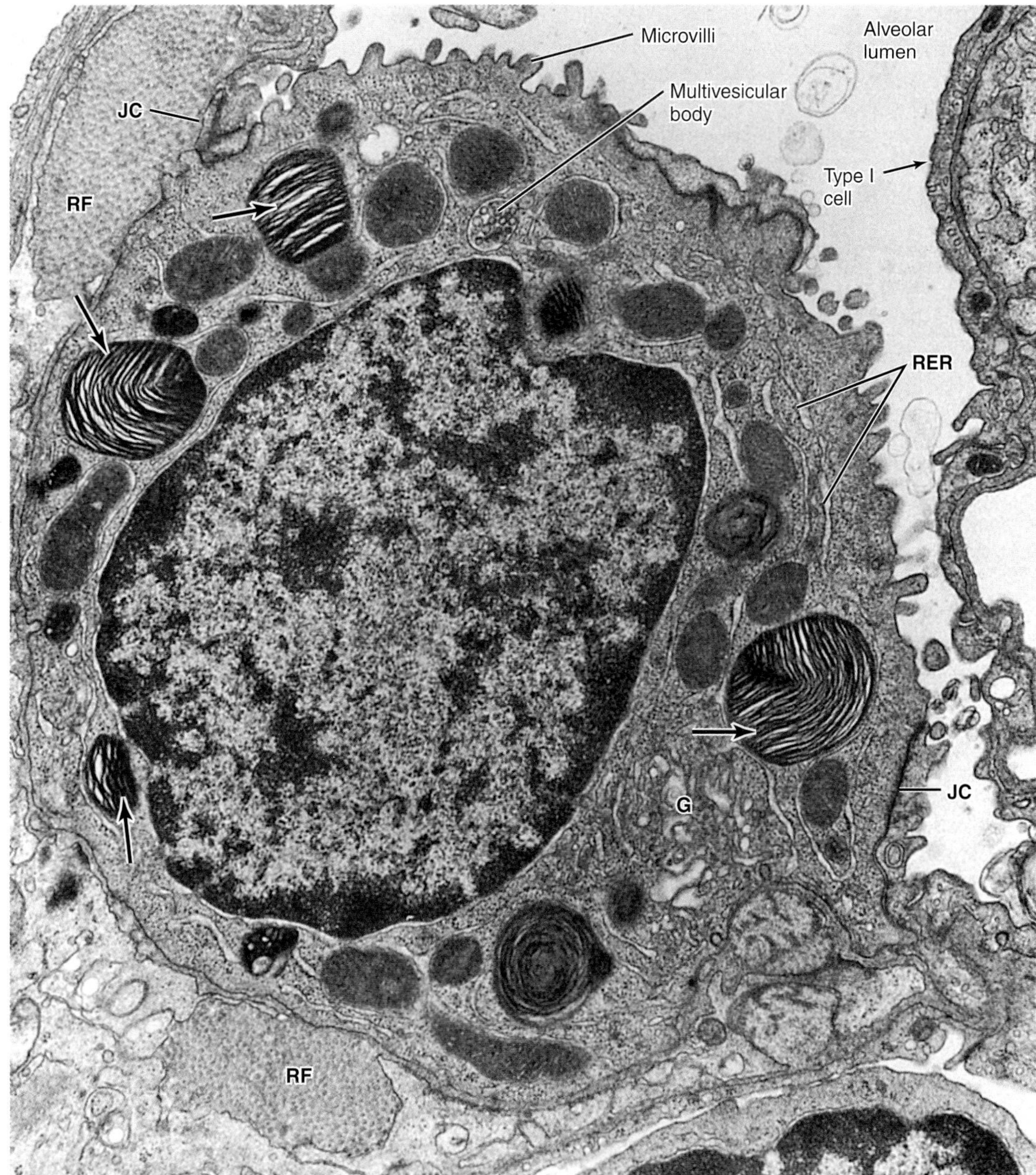

Figure 17–26. Electron micrograph of a type II cell protruding into the alveolar lumen. Arrows indicate lamellar bodies containing newly synthesized pulmonary surfactant. RER, rough endoplasmic reticulum; G, Golgi complex; RF, reticular fibers. Note the microvilli of the type II cell and the junctional complexes (JC) with the type I epithelial cell. ×17,000. (Courtesy of MC Williams.)

MEDICAL APPLICATION

*The **respiratory distress syndrome** of the newborn is a life-threatening disorder of the lungs caused by a deficiency of surfactant. It is principally associated with prematurity and is the leading cause of mortality among premature infants. The incidence of respiratory distress syndrome varies inversely with gestation age. The immature lung is deficient in both the amount and composition of surfactant. In the normal newborn, the onset of breathing is associated with a massive release of stored surfactant, which reduces the surface tension of the alveolar cells. This means that less inspiratory force is needed to inflate the alveoli, and thus the work of breathing is reduced. Microscopically, the alveoli are collapsed, and the respiratory bronchioles and alveolar ducts are dilated and contain edema fluid. A fibrin-rich eosinophilic material called hyaline membrane lines the alveolar ducts. This explains why respiratory distress syndrome was initially named hyaline membrane disease. Fortunately, synthesis of surfactant can be induced by administration of glucocorticoids, a medication used in cases of respiratory distress syndrome. Recently, surfactant has also been suggested to have a bactericidal effect, aiding in the removal of potentially dangerous bacteria that reach the alveoli.*

The surfactant layer is not static but is constantly being turned over. The lipoproteins are gradually removed from the surface by the pinocytotic vesicles of the squamous epithelial cells, by macrophages, and by type II alveolar cells.

Alveolar lining fluids are also removed via the conducting passages as a result of ciliary activity. As the secretions pass up through the airways, they combine with bronchial mucus, forming a **bronchoalveolar fluid,** which aids in the removal of particulate and noxious components from the inspired air. The bronchoalveolar fluid contains several lytic enzymes (eg, lysozyme, collagenase, β-glucuronidase) that are probably derived from the alveolar macrophages.

Lung Macrophages

Alveolar macrophages, also called **dust cells,** are found in the interior of the interalveolar septum and are often seen on the surface of the alveolus (Figure 17–23). Numerous carbon- and dust-laden macrophages in the connective tissue around major blood vessels or in the pleura probably represent cells that have never passed through the epithelial lining. The phagocytized debris within these cells was most likely passed from the alveolar lumen into the interstitium by the pinocytotic activity of type I alveolar cells. The alveolar macrophages that scavenge the outer surface of the epithelium within the surfactant layer are carried to the pharynx, where they are swallowed.

MEDICAL APPLICATION

*In congestive heart failure, the lungs become congested with blood, and erythrocytes pass into the alveoli, where they are phagocytized by alveolar macrophages. In such cases, these macrophages are called **heart failure cells** when present in the lung and sputum; they are identified by a positive histochemical reaction for iron pigment (hemosiderin).*

Increased production of collagen is common, and many diseases that lead to respiratory distress are known to be associated with lung fibrosis. In these pathologic conditions the collagen present is type I.

Alveolar Pores

The interalveolar septum contains pores, 10–15 μm in diameter, that connect neighboring alveoli (Figure 17–19). These pores equalize air pressure in the alveoli and promote the collateral circulation of air when a bronchiole is obstructed.

Alveolar-Lining Regeneration

Inhalation of NO_2 destroys most of the cells lining the alveoli (type I and type II cells). The action of this compound or other toxic substances with the same effect is followed by an increase in the mitotic activity of the remaining type II cells. The normal turnover rate of type II cells is estimated to be 1% per day and results in a continuous renewal of both its own population and that of type I cells.

MEDICAL APPLICATION

***Emphysema** is a chronic lung disease characterized by enlargement of the air space distal to the bronchioles, with destruction of the interalveolar wall. Emphysema usually develops gradually and results in respiratory insufficiency. The major cause of emphysema is cigarette smoking. Even moderate emphysema is rare in nonsmokers. Probably irritation produced by cigarette smoking stimulates the destruction, or impairs the synthesis, of elastic fibers and other components of the interalveolar septum.*

PULMONARY BLOOD VESSELS

Circulation in the lungs includes both nutrient (systemic) and functional (pulmonary) vessels. Pulmonary arteries and veins represent the functional circulation. Pulmonary arteries are thin-walled as a result of the low pressures (25 mm Hg systolic, 5 mm Hg diastolic) encountered in the pulmonary circuit. Within the lung the pulmonary artery branches, accompanying the bronchial tree (Figure 17–27). Its branches are surrounded by adventitia of the bronchi and bronchioles. At the level of the

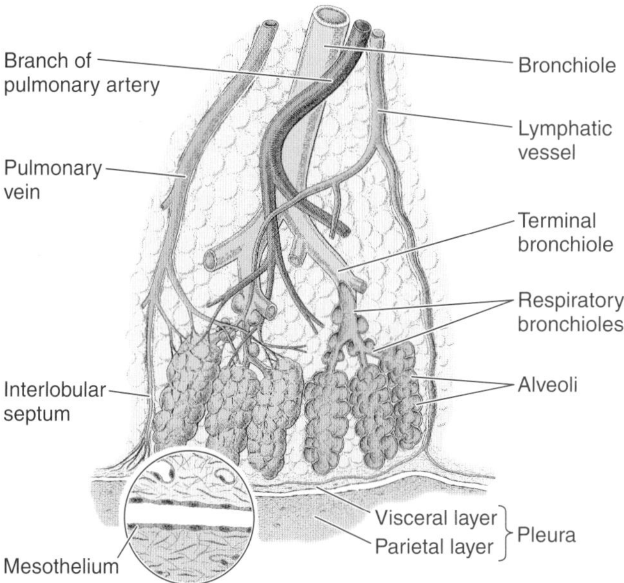

Figure 17–27. Blood and lymph circulation in a pulmonary lobule. Both vessels and bronchi are enlarged out of proportion in this drawing. In the interlobular septum, only one vein (on the left) and one lymphatic vessel (on the right) are shown, although both actually coexist in both regions. At the lower left, an enlargement of the pleura shows its mesothelial lining. (Modified and reproduced, with permission, from Ham AW: *Histology,* 6th ed. Lippincott, 1969.)

alveolar duct, the branches of this artery form a capillary network in the interalveolar septum and in close contact with the alveolar epithelium. The lung has the best-developed capillary network in the body, with capillaries between all alveoli, including those in the respiratory bronchioles.

Venules that originate in the capillary network are found singly in the parenchyma, somewhat removed from the airways; they are supported by a thin covering of connective tissue and enter the interlobular septum (Figure 17–27). After veins leave a lobule, they follow the bronchial tree toward the hilum.

Nutrient vessels follow the bronchial tree and distribute blood to most of the lung up to the respiratory bronchioles, at which point they anastomose with small branches of the pulmonary artery.

PULMONARY LYMPHATIC VESSELS

The lymphatic vessels (Figure 17–27) follow the bronchi and the pulmonary vessels; they are also found in the interlobular septum, and they all drain into lymph nodes in the region of the hilum. This lymphatic network is called the **deep network** to distinguish it from the **superficial network,** which includes the lymphatic vessels in the visceral pleura. The lymphatic vessels of the superficial network drain toward the hilum. They either follow the entire length of the pleura or penetrate the lung tissue via the interlobular septum.

Lymphatic vessels are not found in the terminal portions of the bronchial tree or beyond the alveolar ducts.

NERVES

Both parasympathetic and sympathetic efferent fibers innervate the lungs; general visceral afferent fibers, carrying poorly localized pain sensations, are also present. Most of the nerves are found in the connective tissues surrounding the larger airways.

PLEURA

The pleura (Figure 17–27) is the serous membrane covering the lung. It consists of 2 layers, parietal and visceral, that are continuous in the region of the hilum. Both membranes are composed of mesothelial cells resting on a fine connective tissue layer that contains collagen and elastic fibers. The elastic fibers of the visceral pleura are continuous with those of the pulmonary parenchyma.

The parietal and visceral layers define a cavity entirely lined with squamous mesothelial cells. Under normal conditions, this pleural cavity contains only a film of liquid that acts as a lubricant, facilitating the smooth sliding of one surface over the other during respiratory movements.

In certain pathologic states, the pleural cavity can become a real cavity, containing liquid or air. The walls of the pleural cavity, like all serosal cavities (peritoneal and pericardial), are quite permeable to water and other substances—hence the high frequency of fluid accumulation (pleural effusion) in this cavity in pathologic conditions. This fluid is derived from the blood plasma by exudation. Conversely, under certain conditions, liquids or gases in the pleural cavity can be rapidly absorbed.

RESPIRATORY MOVEMENTS

During inhalation, contraction of the intercostal muscles elevates the ribs, and contraction of the diaphragm lowers the bottom of the thoracic cavity, increasing its diameter and resulting in pulmonary expansion. The bronchi and bronchioles increase in diameter and length during inhalation. The respiratory portion also enlarges, mainly as a result of expansion of the alveolar ducts; the alveoli enlarge only slightly. The elastic fibers of the pulmonary parenchyma are stretched by this expansion. Retraction of the lungs is passive during exhalation. Retraction is the result of muscle relaxation and the action of elastic fibers, which had been under tension.

DEFENSE MECHANISMS

MEDICAL APPLICATION

The respiratory system has an exceptionally large area that is exposed to both blood-borne microorganisms and the external environment. Because it is consequently very susceptible to the invasion of airborne infective and irritating noninfective agents, it is not surprising that the respiratory system presents an elaborate array of defense mechanisms. Particles larger than 10 μm are retained in the nasal passages, and particles of 2 to 10 μm are trapped by the mucus-coated ciliated epithelium. The cough reflex can eliminate these particles by expectoration or swal-

*lowing. Smaller particles are removed by alveolar macrophages. In addition to these nonspecific mechanisms, elaborate immunologic processes occur in abundant lymphoid tissues of the bronchus, mainly in nodules containing T and B lymphocytes that interact with lung macrophages. This important component (Figure 17–10) of the immune system is called **BALT** (bronchus-associated lymphatic tissue).*

Tumors of the Lung

MEDICAL APPLICATION

The incidence of lung tumors is higher in men but is increasing in women, probably because of cigarette smoking. There is conclusive evidence that squamous cell carcinoma, the principal lung tumor type, is related to the effects of cigarette smoking on the bronchial and bronchiolar epithelial lining. Chronic smoking induces the transformation of the respiratory epithelium into a stratified squamous epithelium, an initial step in its eventual differentiation into a tumor.

REFERENCES

Arsalane K et al: Clara cell specific protein (CC16) expression after acute lung inflammation induced by intratracheal lipopolysaccharide administration. Am J Respir Crit Care Med 2000;161:1624.

Breeze RG, Wheeldon EG: The cells of the pulmonary airways. Am Rev Respir Dis 1977;116:705.

Camner P et al: Evidence for congenital nonfunctional cilia in the tracheobronchial tract in two subjects. Am Rev Respir Dis 1975;112:807.

Cummings G (editor): *Cellular Biology of the Lung.* Ettore Majorana International Science Service, 1982.

Elia J et al: Response of bronchiolar Clara cells induced by a domestic insecticide. Analysis of CC10 kDa protein content. Histochem Cell Biol 2000;113:125.

Gehr P et al: The normal human lung: ultrastructure and morphometric estimation of diffusion capacity. Respir Physiol 1978;32:121.

Kikkawa Y, Smith F: Cellular and biochemical aspects of pulmonary surfactant in health and disease. Lab Invest 1983;49:122.

Reynolds SD et al: Neuroepithelial bodies of pulmonary airways serve as a reservoir of progenitor cells capable of epithelial regeneration. Am J Pathol 2000;156:269.

Takashima T: *Airway Secretion: Physiological Bases for the Control of Mucous Hypersecretion.* Marcel Dekker, 1994.

Thurlbeck WM, Abell RM (editors): *The Lung: Structure, Function, and Disease.* Williams & Wilkins, 1978.

Skin

18

The skin is the heaviest single organ of the body, accounting for about 16% of total body weight and, in adults, presenting 1.2–2.3 m^2 of surface to the external environment. It is composed of the **epidermis,** an epithelial layer of ectodermal origin, and the **dermis,** a layer of connective tissue of mesodermal origin. Based on the comparative thickness of the epidermis, **thick** and **thin** skin can be distinguished (Figures 18–1 and 18–2). The junction of dermis and epidermis is irregular, and projections of the dermis called **papillae** interdigitate with evaginations of the epidermis known as **epidermal ridges.** In 3 dimensions, these interdigitations may be of the peg-and-socket variety (thin skin) or formed of ridges and grooves (thick skin). Epidermal derivatives include hairs, nails, and sebaceous and sweat glands. Beneath the dermis lies the **hypodermis** (Gr. *hypo,* under, + *derma,* skin), or **subcutaneous tissue,** a loose connective tissue that may contain a pad of adipose cells, the **panniculus adiposus.** The hypodermis, which is not considered part of the skin, binds skin loosely to the subjacent tissues and corresponds to the superficial fascia of gross anatomy.

The external layer of the skin is relatively impermeable to water; this prevents water loss by evaporation and allows for terrestrial life. The skin functions as a receptor organ in continuous communication with the environment and protects the organism from impact and friction injuries. **Melanin,** a pigment produced and stored in the cells of the epidermis, provides further protective action against the sun's UV rays. Glands of the skin, blood vessels, and adipose tissue participate in thermoregulation, body metabolism, and the excretion of various substances. Under the action of solar radiation, vitamin D_3 is formed from precursors synthesized by the organism. Because skin is elastic, it can expand to cover large areas in conditions associated with swelling, such as edema and pregnancy.

Upon close observation, certain portions of human skin show ridges and grooves arranged in distinctive patterns. These ridges first appear during intrauterine life: at 13 weeks in the tips of the fingers and later in the volar surfaces of the hands and feet (palm and sole). The patterns assumed by ridges and intervening sulci are known as **dermatoglyphics.** They are unique for each individual, appearing as loops, arches, whorls, or combinations of these forms. These configurations, which are used for personal identification (fingerprints), are probably determined by multiple genes; the field of dermatoglyphics has come to be of considerable medical and anthropologic as well as legal interest.

EPIDERMIS

The epidermis consists mainly of a stratified squamous keratinized epithelium, but it also contains 3 less abundant cell types: **melanocytes, Langerhans cells,** and **Merkel's cells.** The keratinizing epidermal cells are called **keratinocytes.** It is customary to distinguish between the **thick skin** (**glabrous,** or smooth and nonhairy) found on the palms and soles and the **thin skin** (hairy) found elsewhere on the body. The designations "thick" and "thin" refer to the thickness of the epidermal layer, which varies

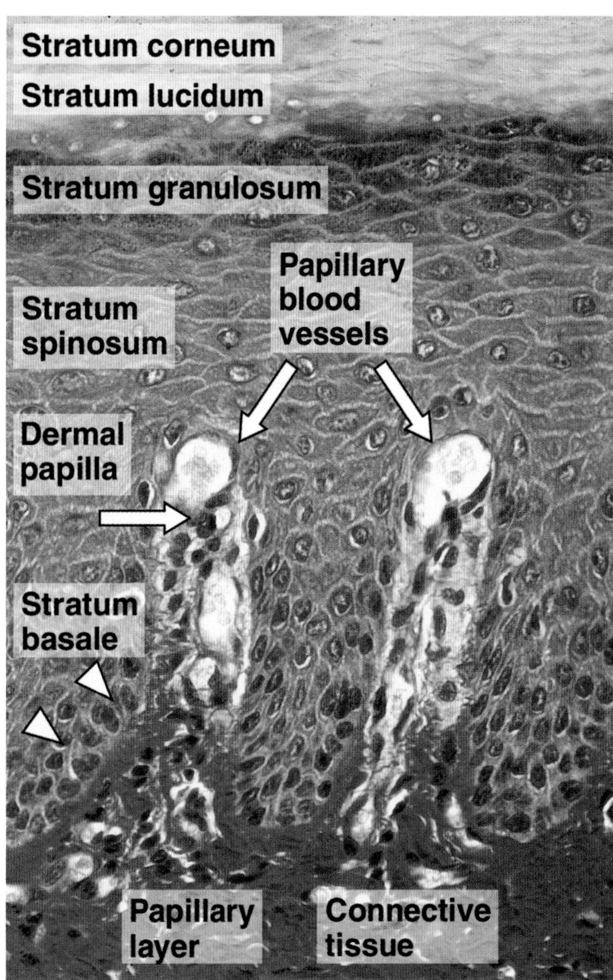

Figure 18–1. Photomicrograph of a section of thick skin. Note the blood vessels in the dermal papillae improving the nutrition of the thick epithelium. Picrosirius-hematoxylin stain. Medium magnification.

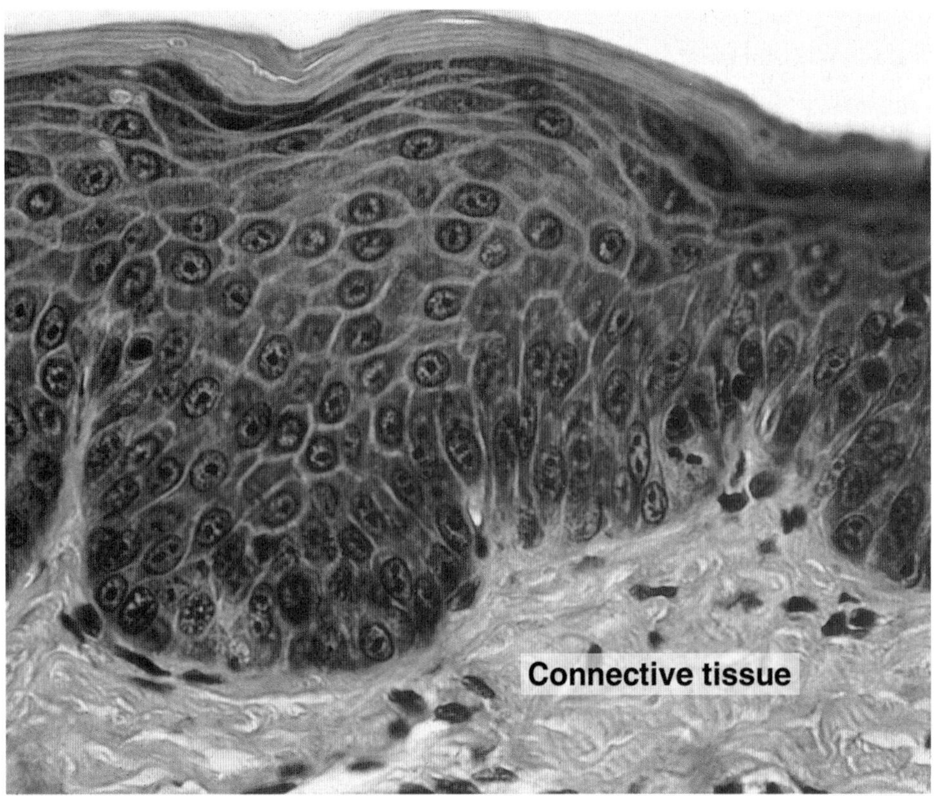

Figure 18–2. Photomicrograph of a section of thin skin. The stratum corneum is thinner than that in thick skin, and the keratinized plates are organized in a more compact way.

between 75 and 150 μm for thin skin and 400 and 600 μm for thick skin. Total skin thickness (epidermis plus dermis) also varies according to site. For example, skin on the back is about 4 mm thick, whereas that of the scalp is about 1.5 mm thick.

From the dermis outward, the epidermis consists of 5 layers of keratin-producing cells (keratinocytes).

Stratum Basale (Stratum Germinativum)

The stratum basale consists of a single layer of basophilic columnar or cuboidal cells resting on the basement membrane at the dermal-epidermal junction (Figure 18–1). Desmosomes bind the cells of this layer together in their lateral and upper surfaces. Hemidesmosomes, found in the basal plasmalemma, help bind these cells to the basal lamina. The stratum basale, containing stem cells, is characterized by intense mitotic activity and is responsible, in conjunction with the initial portion of the next layer, for constant renewal of epidermal cells. The human epidermis is renewed about every 15–30 days, depending on age, the region of the body, and other factors. All cells in the stratum basale contain intermediate keratin filaments about 10 nm in diameter. As the cells progress upward, the number of filaments increases until they represent half the total protein in the stratum corneum.

Stratum Spinosum

The stratum spinosum (Figures 18–1, 18–3, and 18–4) consists of cuboidal, or slightly flattened, cells with a central nucleus and a cytoplasm whose processes are filled with bundles of keratin filaments. These bundles converge into many small cellular extensions, terminating with desmosomes located at the tips of these spiny projections (Figure 18–5). The cells of this layer are firmly bound together by the filament-filled cytoplasmic spines and desmosomes that punctuate the cell surface, giving a spine-studded appearance. These keratin bundles, visible under the light microscope, are called **tonofilaments;** they end at and insert into the cytoplasmic densities of the desmosomes. The filaments play an important role in maintaining cohesion among cells and resisting the effects of abrasion. The epidermis of areas subjected to continuous friction and pressure (such as the soles of the feet) has a thicker stratum spinosum with more abundant tonofilaments and desmosomes.

All mitoses are confined to what is termed the **malpighian layer,** which consists of both the stratum basale and the stratum spinosum. Only the malpighian layer contains epidermal stem cells.

Stratum Granulosum

The stratum granulosum consists of 3–5 layers of flattened polygonal cells whose cytoplasm is filled with coarse basophilic granules (Figure 18–1) called **keratohyalin granules.** The proteins of these granules contain a phosphorylated histidine-rich protein as well as proteins containing cystine. The numerous phosphate groups account for the intense basophilia of keratohyalin granules, which are not surrounded by a membrane.

Another characteristic structure in the cells of the granular layer of epidermis that can be seen with the electron microscope is the membrane-coated **lamellar granule,** a small (0.1–0.3 μm) ovoid or rodlike structure containing lamellar disks that are formed by lipid bilayers. These granules fuse with the cell membrane and discharge their contents into the intercellular spaces of the stratum granulosum, where they are deposited in the form of sheets containing lipid. The function of this extruded material is similar to that of intercellular cement in that it acts as a

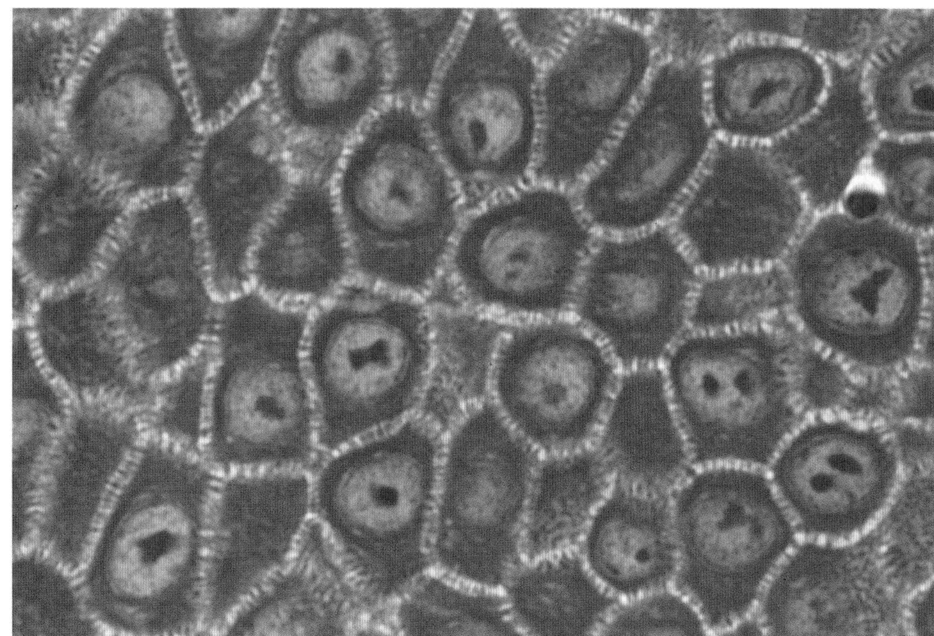

Figure 18–3. Stratum spinosum of the skin from the sole of the foot (thick skin) showing the spiny projections that strongly bind the cells of this layer together to resist abrasion. Pararosaniline–toluidine blue stain. Medium magnification.

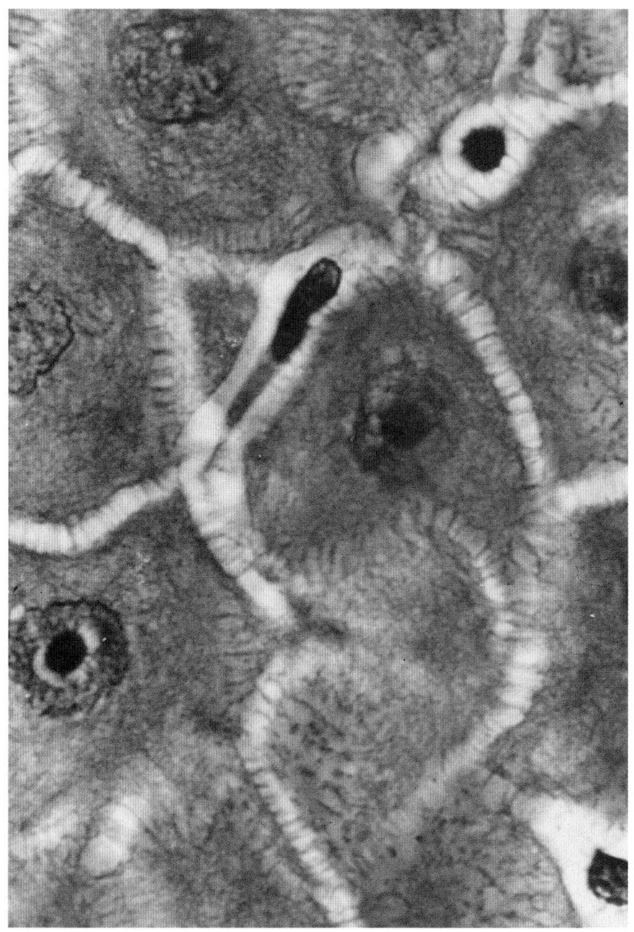

Figure 18–4. High magnification of cells from the stratum spinosum. This section was processed to identify keratin by immunocytochemistry and shows the bundles of keratin filaments (tonofilaments) in the cells and their spines (intercellular bridges).

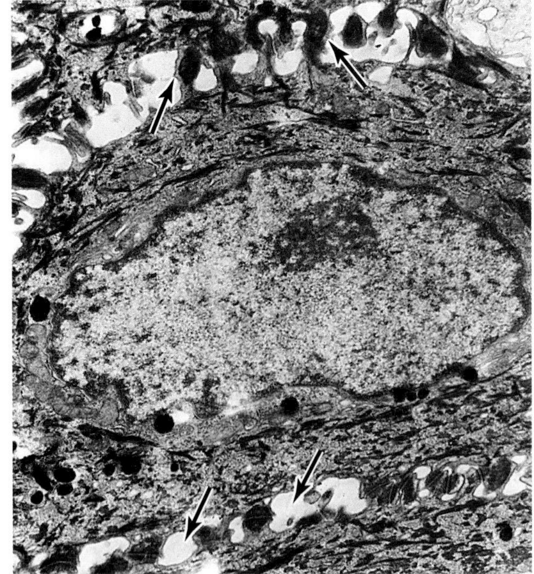

A

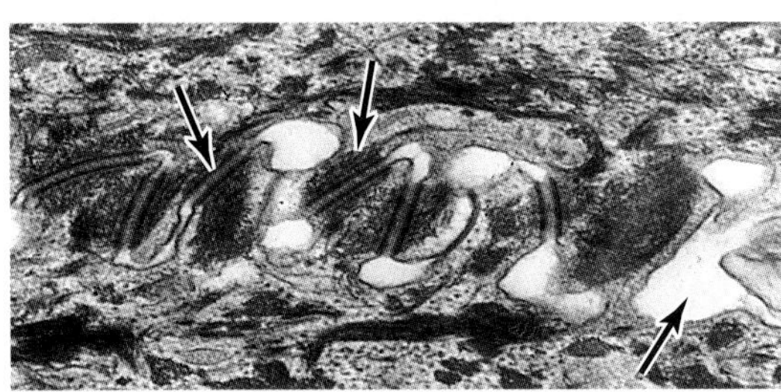

B

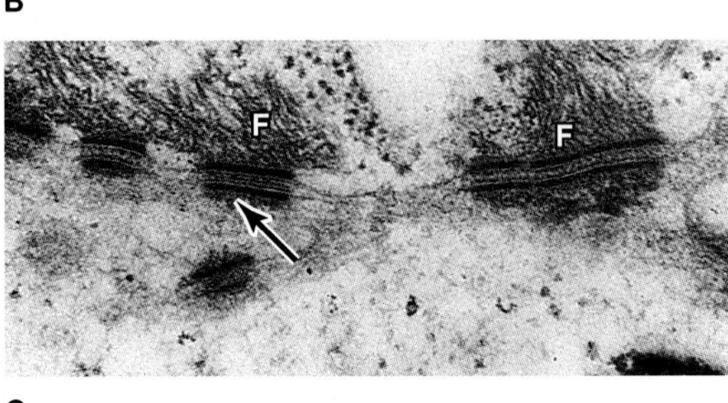

C

Figure 18–5. Electron micrographs of the stratum spinosum. **A:** A cell of the stratum spinosum with melanin granules and the cytoplasm full of tonofilaments. The arrows show the spines with their desmosomes. ×8400. **B** and **C:** The desmosomes from **A,** in greater detail. Note that a dense substance appears between the cell membranes and that bundles of cytoplasmic tonofilaments (F) insert themselves on the desmosomes. **B,** ×36,000; **C,** ×45,000. (Courtesy of C Barros.)

barrier to penetration by foreign materials and provides a very important sealing effect in the skin. Formation of this barrier, which appeared first in reptiles, was one of the important evolutionary events that permitted development of terrestrial life.

Stratum Lucidum

More apparent in thick skin, the stratum lucidum is a translucent, thin layer of extremely flattened eosinophilic epidermal cells (Figure 18–1). The organelles and nuclei are no longer evident, and the cytoplasm consists primarily of densely packed keratin filaments embedded in an electron-dense matrix. Desmosomes are still evident between adjacent cells.

Stratum Corneum

The stratum corneum (Figure 18–1) consists of 15–20 layers of flattened nonnucleated keratinized cells whose cytoplasm is filled with a birefringent filamentous scleroprotein, **keratin.** Keratin contains at least 6 different polypeptides with molecular mass ranging from 40 to 70 kDa. The composition of tonofilaments changes as epidermal cells differentiate. Basal cells contain polypeptides of lower molecular weight, whereas more differentiated cells synthesize higher-molecular-weight polypeptides. Tonofilaments are packed together in a matrix contributed by the keratohyalin granules.

After keratinization, the cells consist of only fibrillar and amorphous proteins and thickened plasma membranes; they are called **horny cells.** During keratinization, lysosomal hydrolytic enzymes play a role in the disappearance of the cytoplasmic organelles. These cells are continuously shed at the surface of the stratum corneum.

This description of the epidermis corresponds to its most complex structure in areas where it is very thick, as on the soles of the feet. In thin skin, the stratum granulosum and the stratum lucidum are often less well developed, and the stratum corneum may be quite thin (Figure 18–2).

MEDICAL APPLICATION

*In **psoriasis,** a common skin disease, there is an increase in the number of proliferating cells in the stratum basale and the stratum spinosum as well as a decrease in the cycle time of these cells. This results in greater epidermal thickness and more rapid renewal of epidermis.*

Melanocytes

The color of the skin is the result of several factors, the most important of which are its content of **melanin** and **carotene,** the number of blood vessels in the dermis, and the color of the blood flowing in them.

Eumelanin is a dark brown pigment produced by the **melanocyte** (Figures 18–6 and 18–7), a specialized cell of the epidermis found beneath or between the cells of the stratum basale and in the hair follicles. The pigment found in red hair is called **pheomelanin** (Gr. *phaios,* dusky, + *melas,* black) and contains **cysteine** as part of its structure. Melanocytes are derived from neural crest cells. They have rounded cell bodies from which long irregular extensions branch into the epidermis, running between the cells of the strata basale and spinosum. Tips of these extensions terminate in invaginations of the cells present in the two layers. The electron microscope reveals a pale-staining cell containing numerous small mitochondria, a well-developed Golgi complex, and short cisternae of rough endoplasmic reticulum. Although melanocytes are not attached to the adjacent keratinocytes by desmosomes, they are bound to the basal lamina by hemidesmosomes (Figure 18–7).

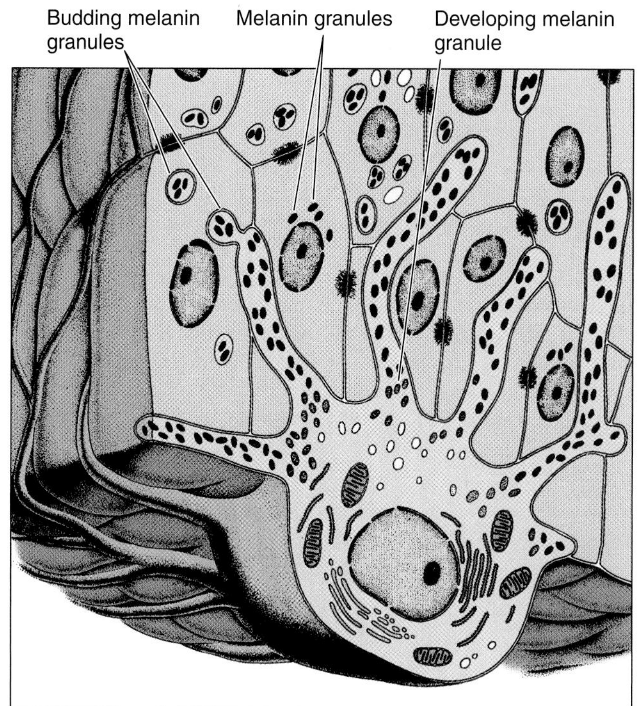

Figure 18–6. Diagram of a melanocyte. Its processes extend into the interstices between keratinocytes. The melanin granules are synthesized in the melanocyte, migrate to its processes, and are transferred into the cytoplasm of keratinocytes.

Melanin is synthesized in the melanocyte, with tyrosinase playing an important role in the process. As a result of tyrosinase activity, tyrosine is transformed first into **3,4-dihydroxyphenylalanine (dopa)** and then into **dopaquinone,** which is converted, after a series of transformations, into melanin. Tyrosinase is synthesized on ribosomes, transported in the lumen of the rough endoplasmic reticulum of melanocytes, and accumulated in vesicles formed in the Golgi complex (Figure 18–8). Four stages can be distinguished in the development of the mature melanin granule:

Stage I

A vesicle is surrounded by a membrane and shows the beginning of tyrosinase activity and formation of fine granular material; at its periphery, electron-dense strands have an orderly arrangement of tyrosinase molecules on a protein matrix.

Stage II

The vesicle (**melanosome**) is ovoid and shows, in its interior, parallel filaments with a periodicity of about 10 nm or cross-striations of about the same periodicity. Melanin is deposited on the protein matrix.

Stage III

Increased melanin formation makes the periodic fine structure less visible.

Stage IV

The mature melanin granule is visible in the light microscope, and melanin completely fills the vesicle. No ultrastructure is visible. The mature granules are ellipsoid, with a length of 1 μm and a diameter of 0.4 μm.

Once formed, melanin granules migrate within cytoplasmic extensions of the melanocyte and are transferred to cells of the

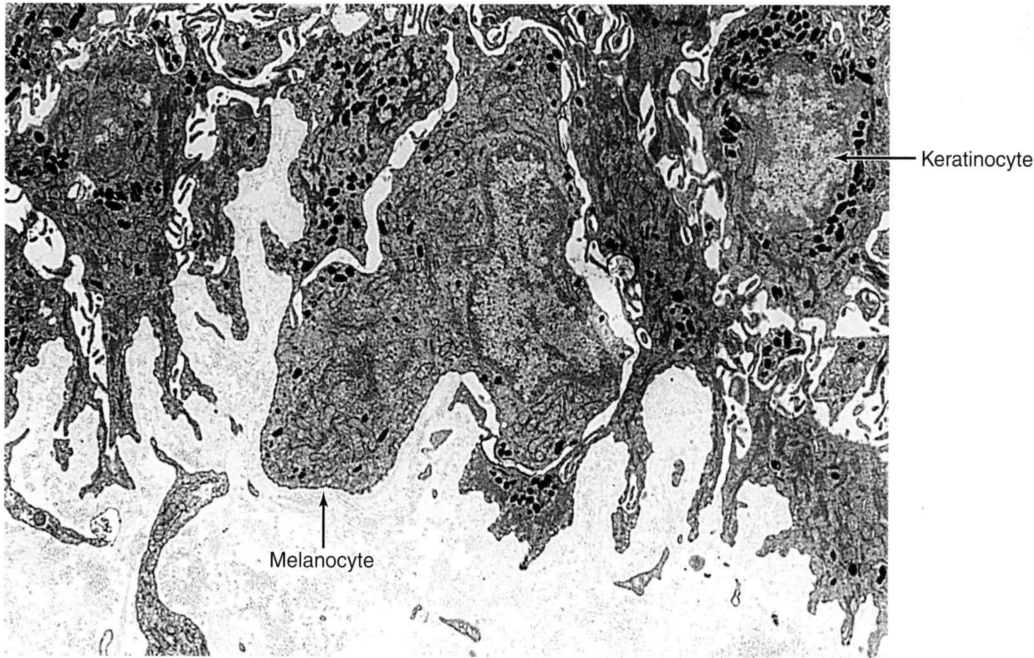

Figure 18–7. Electron micrograph of human skin containing melanocytes and keratinocytes. Note the greater abundance of melanin granules in the keratinocyte at right than in the adjacent melanocyte. The clear material at the bottom is dermal collagen. ×1800.

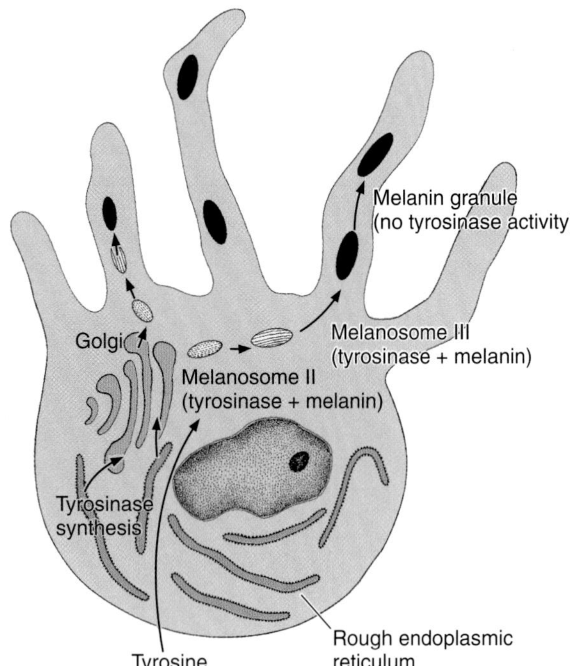

Figure 18–8. Diagram of a melanocyte, illustrating the main features of melanogenesis. Tyrosinase is synthesized in the rough endoplasmic reticulum and accumulated in vesicles of the Golgi complex. The free vesicles are now called melanosomes. Melanin synthesis begins in the stage II melanosomes, where melanin is accumulated and forms stage III melanosomes. Later, this structure loses its tyrosinase activity and becomes a melanin granule. Melanin granules migrate to the tips of the melanocyte's processes and are then transferred to the keratinocytes of the malpighian layer.

strata germinativum and spinosum of the epidermis. This transfer has been directly observed in tissue cultures of skin.

Melanin granules are essentially injected into keratinocytes. Once inside the keratinocyte, melanin granules accumulate in the supranuclear region of the cytoplasm, thus protecting the nuclei from the deleterious effects of solar radiation (Figure 18–9).

Although melanocytes synthesize melanin, epithelial cells act as a depot and contain more of this pigment than do melanocytes. Within the keratinocytes, melanin granules fuse with lysosomes—the reason that melanin disappears in upper epithelial cells. In this interaction between keratinocytes and melanocytes, which creates the pigmentation of the skin, the important factors are the rate of formation of melanin granules within the melanocyte, the transfer of the granules into the keratinocytes, and their ultimate disposition by the keratinocytes. A feedback mechanism may exist between melanocytes and keratinocytes.

Melanocytes can be easily seen by incubating fragments of epidermis in dopa. This compound is converted to dark brown deposits of melanin in melanocytes, a reaction catalyzed by the enzyme tyrosinase. This method makes it possible to count the number of melanocytes per unit area of the epidermis. Such studies show that these cells are not distributed at random among keratinocytes; rather, there is a pattern in their distribution, called the **epidermal-melanin unit.** In humans, the ratio of dopa-positive melanocytes to keratinocytes in the stratum basale is constant within each area of the body but varies from one region to another. For example, there are about 1000 melanocytes/mm^2 in the skin of the thigh and 2000/mm^2 in the skin of the scrotum. Sex or race does not influence the number of melanocytes per unit area; differences in skin color are due mainly to differences in the number of melanin granules in the keratinocytes.

Darkening of the skin (tanning) after exposure to solar radiation (wavelength of 290–320 nm) is the result of a 2-step process. First, a physicochemical reaction darkens the preexisting melanin and releases it rapidly into the keratinocytes. Next, the rate of melanin synthesis in the melanocytes accelerates, increasing the amount of this pigment.

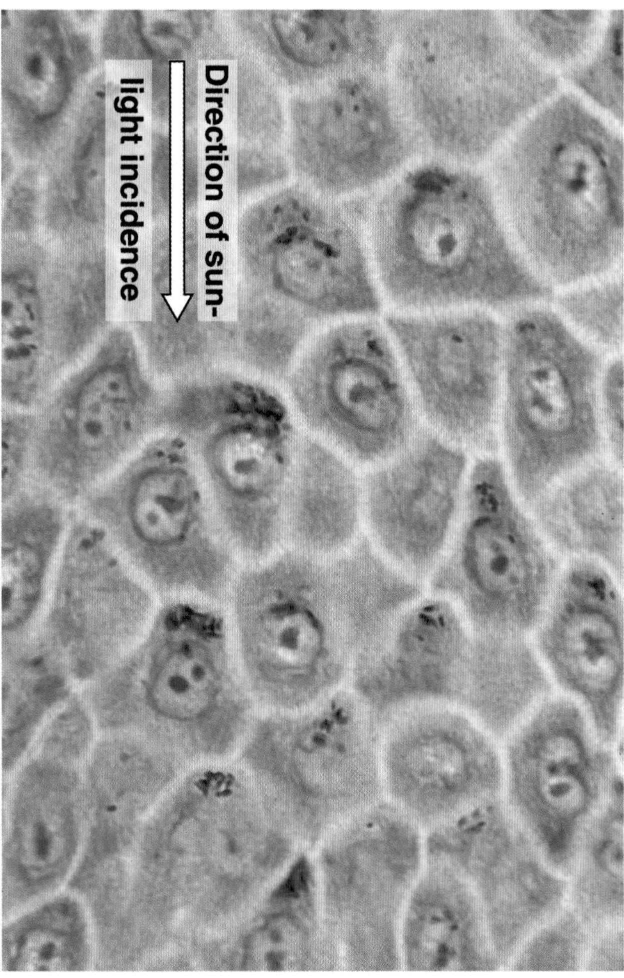

Figure 18–9. Section of the stratum spinosum showing the localized deposits of melanin covering the cell nuclei. Melanin protects the DNA from the UV radiation of the sun. This explains why people with light skin have a higher incidence of skin cancer than do people with dark skin. The highest concentration of melanin occurs in the cells that are more deeply localized; these cells divide more actively. (The DNA of cell populations that multiply more actively is particularly sensitive to harmful agents.)

MEDICAL APPLICATION

*In humans, lack of cortisol from the adrenal cortex causes overproduction of adrenocorticotropic hormone, which increases the pigmentation of the skin. An example of this is **Addison disease,** which is caused by dysfunction of the adrenal glands.*

***Albinism,** a hereditary inability of the melanocytes to synthesize melanin, is caused by the absence of tyrosinase activity or the inability of cells to take up tyrosine. As a result, the skin is not protected from solar radiation by melanin, and there is a greater incidence of basal and squamous cell carcinomas (skin cancers).*

*The degeneration and disappearance of entire melanocytes results in a depigmentation disorder called **vitiligo.***

Langerhans Cells

Langerhans cells, star-shaped cells found mainly in the stratum spinosum of the epidermis, represent 2–8% of the epidermal cells. They are bone marrow–derived, carried to the skin by the blood, and capable of binding, processing, and presenting antigens to T lymphocytes, thus participating in the stimulation of these cells. Consequently, they have a significant role in immunologic skin reactions.

Merkel's Cells

Merkel's cells, generally present in the thick skin of palms and soles, somewhat resemble the epidermal epithelial cells but have small dense granules in their cytoplasm. The composition of these granules is not known. Free nerve endings that form an expanded terminal disk are present at the base of Merkel's cells. These cells may serve as sensory mechanoreceptors, although other evidence suggests that they have functions related to the diffuse neuroendocrine system.

IMMUNOLOGIC ACTIVITY IN THE SKIN

Because of its large size, the skin has an impressive number of lymphocytes and antigen-presenting cells (Langerhans cells), and because of its location it is in close contact with many antigenic molecules. For these reasons, the epidermis has an important role in some types of immune responses. Most lymphocytes found in the skin are "homed" in the epidermis.

DERMIS

The dermis is the connective tissue (Figures 18–2 and 18–10) that supports the epidermis and binds it to the subcutaneous tissue (hypodermis). The thickness of the dermis varies according to the region of the body and reaches its maximum of 4 mm on the back. The surface of the dermis is very irregular and has many projections (dermal papillae) that interdigitate with projections (epidermal pegs or ridges) of the epidermis (Figure 18–1). Dermal papillae are more numerous in skin that is subjected to frequent pressure; they increase and reinforce the dermal-epidermal junction. During embryonic development, the dermis determines the developmental pattern of the overlying epidermis. Dermis obtained from the sole always induces the formation of a heavily keratinized epidermis irrespective of the site of origin of the epithelial cells.

A **basal lamina** is always found between the stratum germinativum and the papillary layer of the dermis and follows the contour of the interdigitations between these layers. Underlying the basal lamina is a delicate net of reticular fibers, the **lamina reticularis.** This composite structure is called the **basement membrane** and can be seen with the light microscope.

MEDICAL APPLICATION

*Abnormalities of the dermal-epidermal junction can lead to one type of blistering disorder (**bullous pemphigoid**). Another type of blistering disorder (**pemphigus**) is caused by the loss of intercellular junctions between keratinocytes.*

The dermis contains 2 layers with rather indistinct boundaries—the outermost papillary layer and the deeper reticular layer. The thin **papillary layer** is composed of loose connective tissue; fibroblasts and other connective tissue cells, such as mast

Figure 18–10. Section of dermal reticular layer, which is made of dense connective tissue containing irregular bundles of thick collagen type I fibers. Picrosirius–polarized light (PSP) stain. Medium magnification.

cells and macrophages, are present. Extravasated leukocytes are also seen. The papillary layer is so called because it constitutes the major part of the dermal papillae. From this layer, special collagen fibrils insert into the basal lamina and extend into the dermis. They bind the dermis to the epidermis and are called **anchoring fibrils.** The **reticular layer** is thicker, composed of irregular dense connective tissue (mainly type I collagen), and therefore has more fibers and fewer cells than does the papillary layer. The principal glycosaminoglycan is dermatan sulfate. The dermis contains a network of fibers of the elastic system (Figures 18–11 and 18–12), with the thicker fibers characteristically found in the reticular layer. From this region emerge fibers that become gradually thinner and end by inserting into the basal lamina. As these fibers progress toward the basal lamina, they gradually lose their amorphous elastin component (see Chapter 5), and only the microfibrillar component inserts into the basal lamina. This elastic network is responsible for the elasticity of the skin.

MEDICAL APPLICATION

*With age, collagen fibers thicken and collagen synthesis decreases. Elastic fibers steadily increase in number and thickness, so the elastin content of human skin increases approximately fivefold from fetal to adult life. In old age, extensive cross-linking of collagen fibers, the loss of elastic fibers, and degeneration of these fibers caused by excessive exposure to the sun (**solar elastosis**) cause the skin to become more fragile, lose its suppleness, and develop wrinkles.*

*In several disorders, such as **cutis laxa** and **Ehlers-Danlos syndromes** (Table 5–4), there is a considerable increase in skin and ligament extensibility caused by defective collagen-fibril processing.*

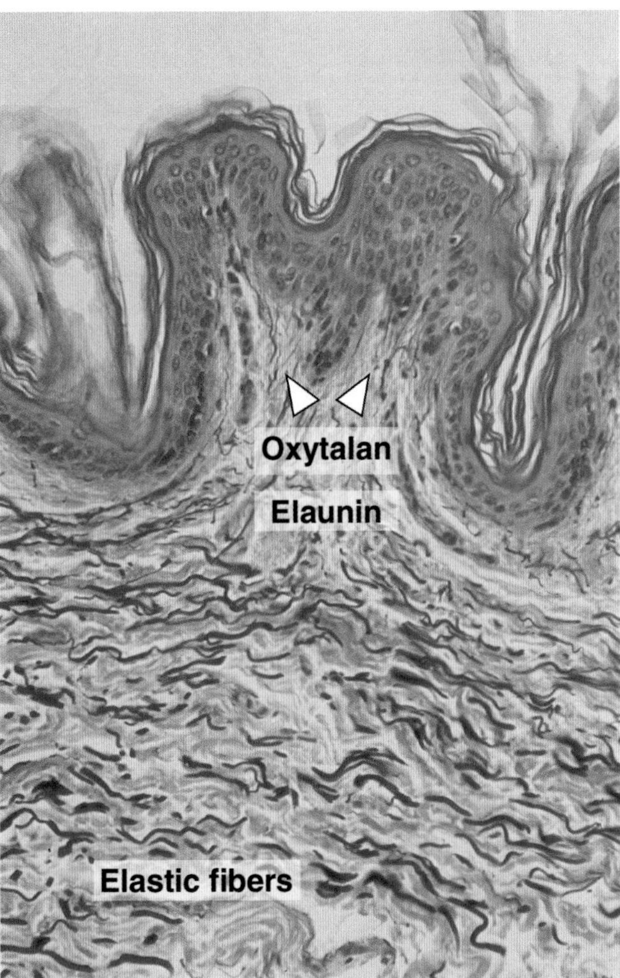

Figure 18–11. Photomicrograph of thin skin stained for fibers of the elastic system. Note the gradual decrease in the diameter of fibers as they approach the epidermis. The thick fibers are elastic fibers. Those with an intermediate diameter are elaunin fibers. The very thin superficial fibers are oxytalan fibers formed by microfibrils that insert into the basement membrane. Weigert's stain. Medium magnification.

The dermis has a rich network of blood and lymph vessels. In certain areas of the skin, blood can pass directly from arteries to veins through arteriovenous anastomoses, or shunts. These shunts play a very important role in temperature regulation.

In addition to these components, the dermis contains such epidermal derivatives as the hair follicles and sweat and sebaceous glands. There is a rich supply of nerves in the dermis, and the effector nerves to the skin are postganglionic fibers of sympathetic ganglia of the paravertebral chain. No parasympathetic innervation is present. The afferent nerve endings form a superficial dermal network with free nerve endings, a hair follicle network, and the innervation of encapsulated sensory organs.

SUBCUTANEOUS TISSUE

The subcutaneous tissue layer consists of loose connective tissue that binds the skin loosely to the subjacent organs, making it possible for the skin to slide over them. The hypodermis often contains fat cells that vary in number according to the area of the body and vary in size according to nutritional state. This layer is also referred to as the superficial fascia and, where thick enough, the panniculus adiposus.

VESSELS & SKIN SENSORIAL RECEPTORS

The connective tissue of the skin contains a rich network of blood and lymphatic vessels. The arterial vessels that nourish the skin form 2 plexuses. One is located between the papillary and reticular layers; the other, between the dermis and the subcutaneous tissue. Thin branches leave these plexuses and vascularize the dermal papillae. Each papilla has only one arterial ascending branch and one venous descending branch. Veins are disposed in 3 plexuses, 2 in the position described for arterial vessels and the third in the middle of the dermis. Arteriovenous anastomoses with glomera (see Chapter 11) are frequent in the skin, participating in the regulation of body temperature. Lymphatic vessels begin as closed sacs in the papillae of the dermis and converge to form 2 plexuses, as described for the arterial vessels.

One of the most important functions of the skin, with its great extension and abundant sensory innervation, is to receive stimuli from the environment. The skin is the most extensive sensory receptor. In addition to numerous free nerve endings in

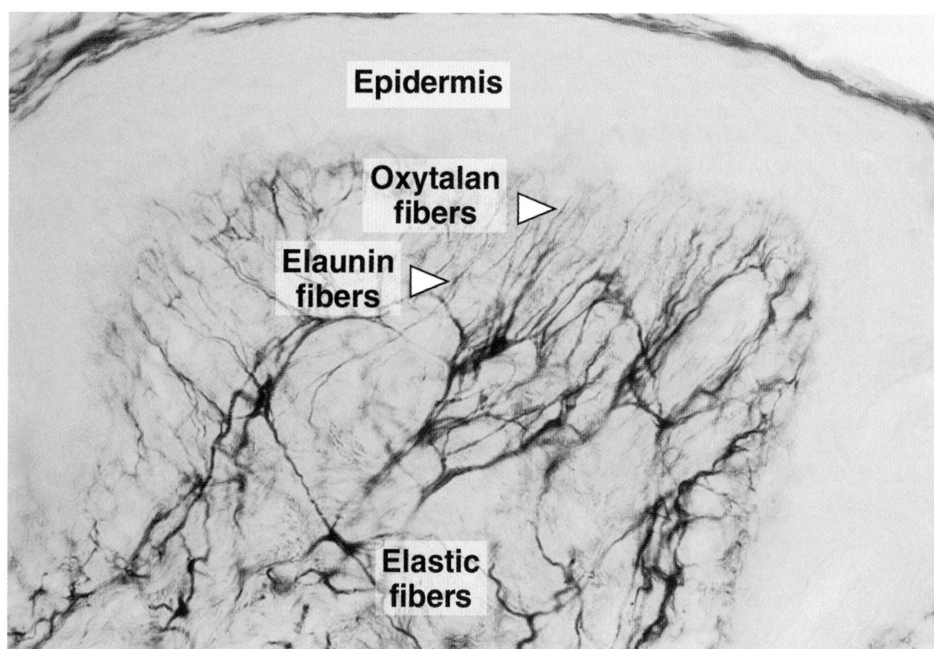

Figure 18–12. Thick section of skin stained for the fibers of the elastic system, photographed with a yellow filter to increase contrast. This procedure increases the visibility of the elastic, elaunin, and oxytalan fibers. Weigert's stain. Medium magnification.

the epidermis, hair follicles, and cutaneous glands, encapsulated and expanded receptors are present in the dermis and subcutaneous tissue; they are more frequently found in the dermal papillae. Free nerve endings are sensitive to touch-pressure (pressure is sustained touch), tactile reception, high and low temperatures, pain, itching, and other sensations. The expanded ending includes the **Ruffini endings,** and the encapsulated ending includes the **Vater-Pacini, Meissner,** and **Krause** corpuscles (Figure 18–13). There is evidence showing that the expanded and encapsulated corpuscles are not necessary for cutaneous sensation. Their distribution is irregular, with many areas of skin containing only free nerve endings. However, when present, the expanded and encapsulated receptors respond to tactile stimuli, functioning as mechanoreceptors. Vater-Pacini corpuscles and Ruffini endings are also found in the connective tissue of organs located deep in the body, where they probably are sensitive to movements of internal organs and to pressure of one organ over another.

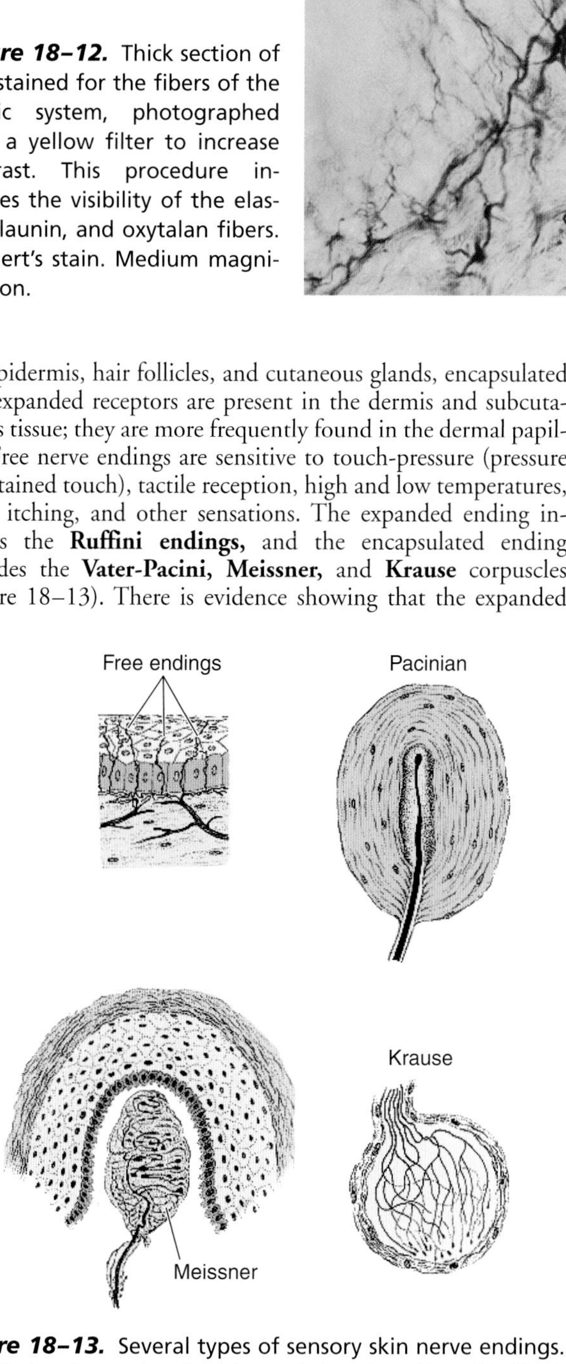

Free endings
Pacinian
Krause
Meissner

Figure 18–13. Several types of sensory skin nerve endings. (Modified and reproduced, with permission, from Ham AV: *Histology,* 6th ed. Lippincott, 1969.)

HAIRS

Hairs (Figure 18–14) are elongated keratinized structures derived from invaginations of epidermal epithelium. Their color, size, and disposition vary according to race, age, sex, and region of the body. Hairs are found everywhere on the body except on the palms, soles, lips, glans penis, clitoris, and labia minora. The face has about 600 hairs/cm^2, and the remainder of the body has about 60/cm^2. Hairs grow discontinuously and have periods of growth followed by periods of rest. This growth does not occur synchronously in all regions of the body or even in the same area; rather, it tends to occur in patches. The duration of the growth and rest periods also varies according to the region of the body. Thus, in the scalp, the growth periods (anagen) may last for several years, whereas the rest periods (catagen and telogen) average 3 months. Hair growth in such regions of the body as the scalp, face, and pubis is strongly influenced not only by sex hormones—especially androgens—but also by adrenal and thyroid hormones.

Each hair arises from an epidermal invagination, the **hair follicle** (Figures 18–14 and 18–15), that during its growth period has a terminal dilatation called a **hair bulb.** At the base of the hair bulb, a **dermal papilla** can be observed. The dermal papilla contains a capillary network that is vital in sustaining the hair follicle. Loss of blood flow or loss of the vitality of the dermal papilla will result in death of the follicle. The epidermal cells covering this dermal papilla form the hair root that produces and is continuous with the hair shaft, which protrudes beyond the skin.

During periods of growth, the epithelial cells that make up the hair bulb are equivalent to those in the stratum germinativum of the skin. They divide constantly and differentiate into specific cell types. In certain types of thick hairs, the cells of the central region of the root at the apex of the dermal papilla produce large, vacuolated, and moderately keratinized cells that form the **medulla** of the hair (Figure 18–14). Root cells multiply and differentiate into heavily keratinized, compactly grouped fusiform cells that form the **hair cortex.**

Farther toward the periphery are the cells that produce the **hair cuticle,** a layer of cells that are cuboidal midway up the bulb, then become tall and columnar. Higher up, they change from horizontal to vertical, at which point they form a layer of flattened, heavily keratinized, shinglelike cells covering the cortex. These cuticle cells are the last cell type in the hair follicle to differentiate.

The outermost cells give rise to the **internal root sheath,** which completely surrounds the initial part of the hair shaft. The internal sheath is a transient structure whose cells degenerate and

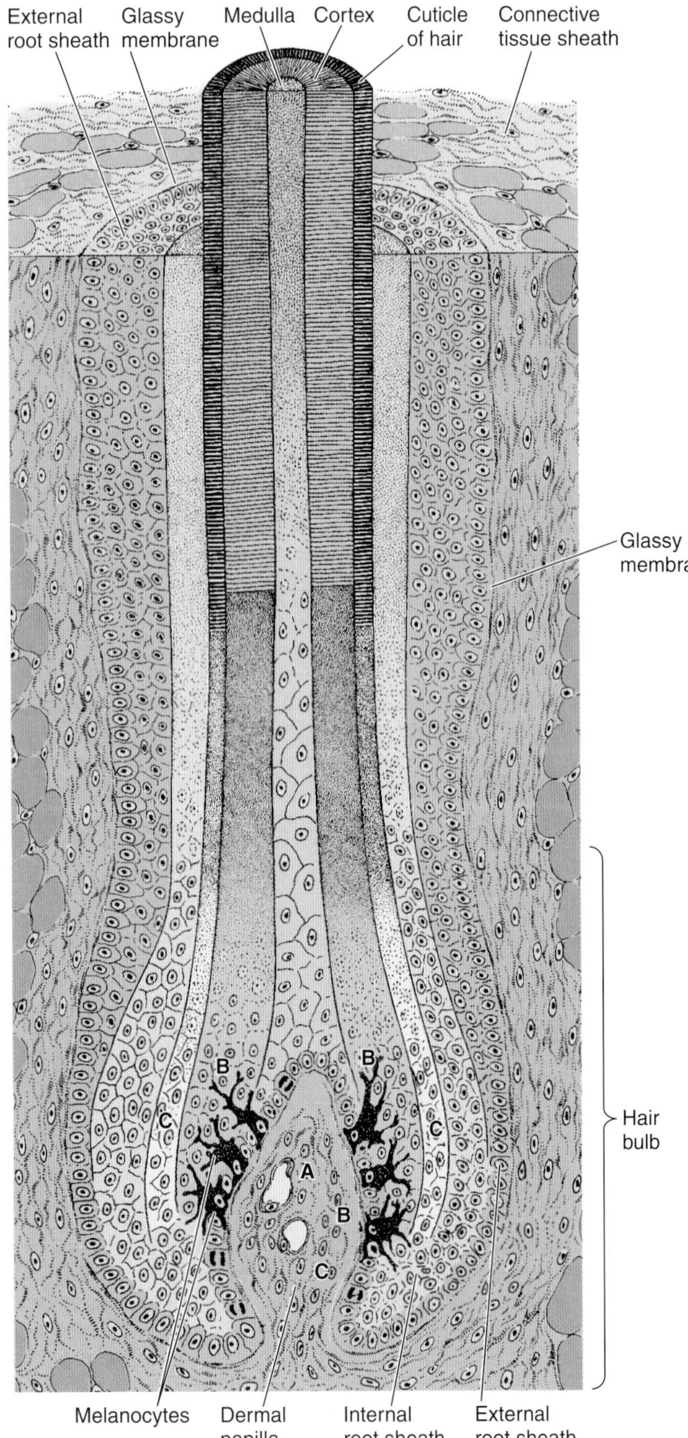

Figure 18–14. Hair follicle. The follicle has a bulbous terminal expansion with a dermal papilla. The papilla contains capillaries and is covered by cells that form the hair root and develop into the hair shaft. The central cells (A) produce large, vacuolated, moderately keratinized cells that form the medulla of the hair. The cells that produce the cortex of the hair are located laterally (B). Cells forming the hair cuticle originate in the next layer (C). The peripheral epithelial cells develop into the internal and external root sheaths. The external root sheath is continuous with the epidermis, whereas the cells of the internal root sheath disappear at the level of the openings of the sebaceous gland ducts (not shown).

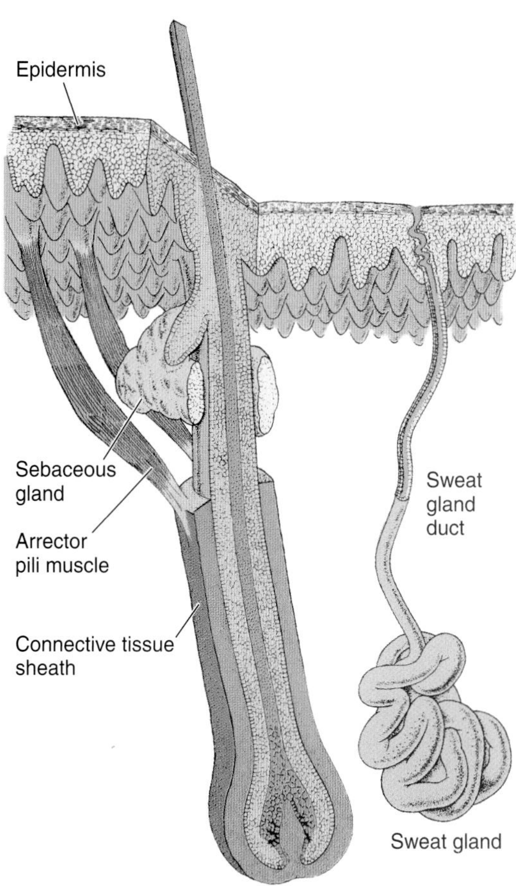

Figure 18–15. Relationships between the skin, hair follicle, arrector pili muscle, and sebaceous and sweat glands. The arrector pili muscle originates in the connective tissue sheath of the hair follicle and inserts into the papillary layer of the dermis, where it ends.

disappear above the level of the sebaceous glands. The **external root sheath** is continuous with epidermal cells and, near the surface, shows all the layers of epidermis. Near the dermal papilla, the external root sheath is thinner and is composed of cells corresponding to the stratum germinativum of the epidermis.

Separating the hair follicle from the dermis is a noncellular hyaline layer, the **glassy membrane** (Figure 18–14), which results from a thickening of the basal lamina. The dermis that surrounds the follicle is denser, forming a sheath of connective tissue. Bound to this sheath and connecting it to the papillary layer of the dermis are bundles of smooth muscle cells, the **arrector pili** muscles (Figure 18–15). They are disposed in an oblique direction, and their contraction results in the erection of the hair shaft to a more upright position. Contraction of arrector pili muscles also causes a depression in the skin where the muscles attach to the dermis. This contraction produces the "gooseflesh" of common parlance.

Hair color is created by the activity of melanocytes located between the papilla and the epithelial cells of the hair root. The epithelial cells produce the pigment found in the medullary and cortical cells of the hair shaft (Figure 18–14). The melanocytes produce and transfer melanin to the epithelial cells by a mechanism similar to that described for the epidermis.

Although the keratinization processes in the epidermis and hair appear to be similar, they differ in several ways:

1. The epidermis produces relatively soft keratinized outer layers of dead cells that adhere slightly to the skin and desquamate continuously. The opposite occurs in the hair, which has a hard and compact keratinized structure.
2. Although keratinization in the epidermis occurs continuously and over the entire surface, it is intermittent in the hair and occurs only in the hair root. The connective tissue of the hair papilla has an inductive action on the covering epithelial cells, promoting their proliferation and differentiation. Injuries to the dermal papillae thus result in the loss of hair.
3. Contrary to what happens in the epidermis, where the differentiation of all cells in the same direction gives rise to the final keratinized layer, cells in the hair root differentiate into various cell types that differ in ultrastructure, histochemical characteristics, and function. Mitotic activity in hair follicles is influenced by androgens.

NAILS

Nails are plates of keratinized epithelial cells on the dorsal surface of each distal phalanx. The proximal part of the nail, hidden in the nail groove, is the **nail root.** The epithelium of the fold of skin covering the nail root consists of the usual layers of cells. The stratum corneum of this epithelium forms the **eponychium,** or **cuticle.** The **nail plate,** which corresponds to the stratum corneum of the skin, rests on a bed of epidermis called the **nail bed.** Only the stratum basale and the stratum spinosum are present in the nail bed. Nail plate epithelium arises from the **nail matrix.** The proximal end of the matrix extends deep to the nail root. Cells of the matrix divide, move distally, and eventually cornify, forming the proximal part of the nail plate. The nail plate then slides forward over the nail bed (which makes no contribution to the formation of the plate). The distal end of the plate becomes free of the nail bed and is worn away or cut off. The nearly transparent nail plate and the thin epithelium of the nail bed provide a useful window on the amount of oxygen in the blood by showing the color of blood in the dermal vessels.

GLANDS OF THE SKIN

Sebaceous Glands

Sebaceous glands are embedded in the dermis over most of the body surface. There are about 100 of these glands per square centimeter over most of the body, but the frequency increases to 400–900/cm^2 in the face, forehead, and scalp. Sebaceous glands, which are not found in the glabrous skin of the palms and soles, are acinar glands that usually have several acini opening into a short duct. This duct usually ends in the upper portion of a hair follicle (Figure 18–15); in certain regions, such as the glans penis, glans clitoridis, and lips, it opens directly onto the epidermal surface. The acini consist of a basal layer of undifferentiated flattened epithelial cells that rest on the basal lamina. These cells proliferate and differentiate, filling the acini with rounded cells containing increasing amounts of fat droplets in their cytoplasm (Figure 18–16). Their nuclei gradually shrink, and the cells simultaneously become filled with fat droplets and burst. The product of this process is **sebum,** the secretion of the sebaceous gland, which is gradually moved to the surface of the skin.

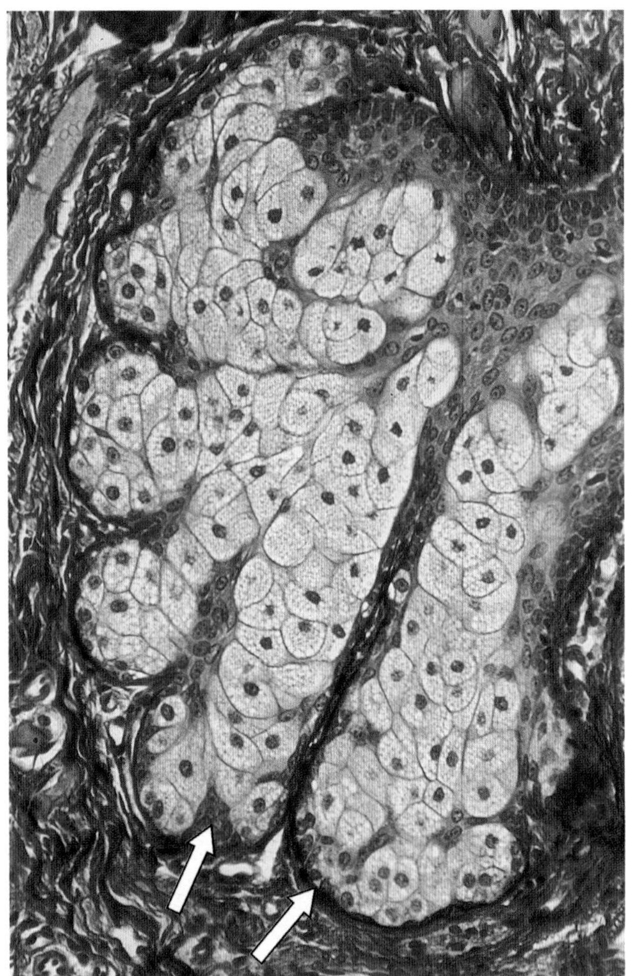

Figure 18–16. Sebaceous gland. This is a holocrine gland, because its product is secreted with the remnants of a dead cell. Stem cells (arrows) in the base of the gland proliferate to replace the lost cells. Collagen fibers are stained in red. PSP stain. Medium magnification.

The sebaceous gland is an example of a **holocrine** gland, because its product of secretion is released with remnants of dead cells. This product comprises a complex mixture of lipids that includes triglycerides, waxes, squalene, and cholesterol and its esters. Sebaceous glands begin to function at puberty. The primary controlling factor of sebaceous gland secretion in men is testosterone; in women it is a combination of ovarian and adrenal androgens.

MEDICAL APPLICATION

The flow of sebum is continuous, and a disturbance in the normal secretion and flow of sebum is one of the reasons for the development of acne, a chronic inflammation of obstructed sebaceous glands. It occurs mainly during puberty.

The functions of sebum in humans are largely unknown. It may have weak antibacterial and antifungal properties. Sebum does not have any importance in preventing water loss.

Sweat Glands

Sweat glands (Figures 18–17 and 18–18) are widely distributed in the skin except for certain regions, such as the glans penis.

The **merocrine** sweat glands are simple, coiled tubular glands whose ducts open at the skin surface (Figure 18–15). Their ducts do not divide, and their diameter is thinner than that of the secretory portion. The secretory part of the gland is embedded in the dermis; it measures approximately 0.4 mm in diameter and is surrounded by myoepithelial cells (described in Chapter 4). Contraction of these cells helps to discharge the secretion. Two types of cells have been described in the secretory portion of sweat glands. **Dark cells** are pyramidal cells that line most of the luminal surface of this portion of the gland. Their basal surface does not touch the basal lamina. Secretory granules containing glycoproteins are abundant in their apical cytoplasm. **Clear cells** are devoid of secretory granules. Their basal plasmalemma has the numerous invaginations characteristic of cells

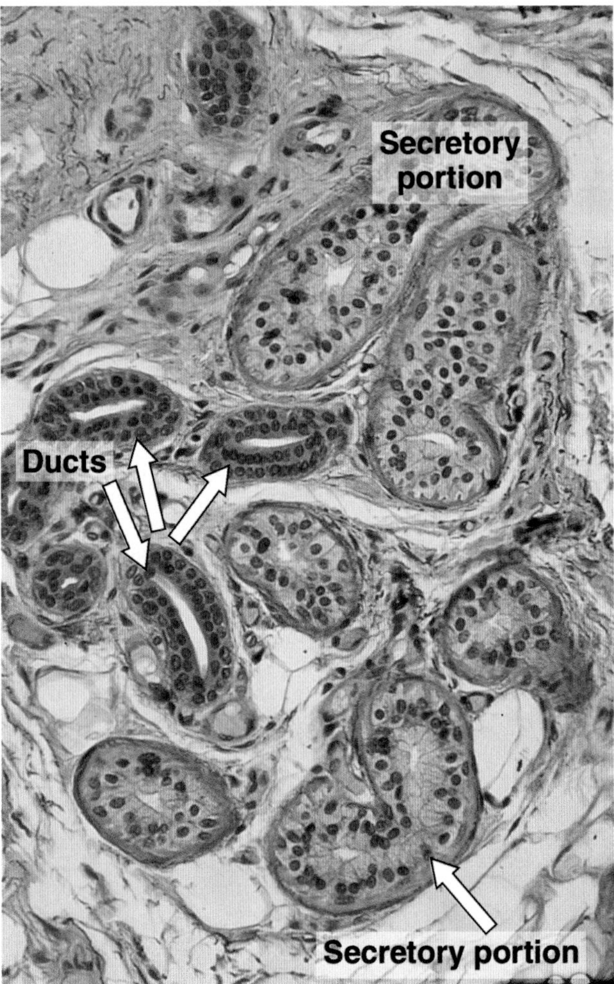

Figure 18–17. Low-magnification photomicrograph of a section of sweat gland. This is a simple coiled tubular gland. H&E stain.

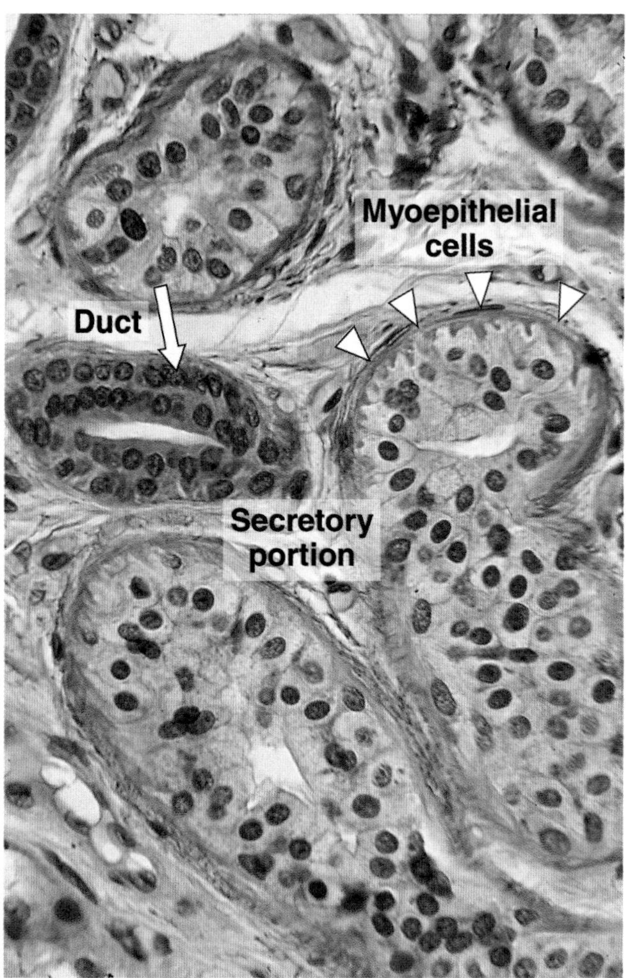

Figure 18–18. Section of sweat gland. Note the duct lined by stratified cuboidal epithelium. The myoepithelial cells, whose contraction helps to discharge the glandular secretion, surround the secretory portion. H&E stain. Medium magnification.

involved in transepithelial salt and fluid transport. The ducts of these glands are lined by stratified cuboidal epithelium (Figures 18–17 and 18–18).

The fluid secreted by sweat glands is not viscous and contains little protein. Its main components are water, sodium chloride, urea, ammonia, and uric acid. Its sodium content of 85 mEq/L is distinctly below that of blood (144 mEq/L), and the cells present in the sweat ducts are responsible for sodium absorption, to prevent excessive loss of this ion. The fluid in the lumen of the secretory portion of the gland is an ultrafiltrate of the blood plasma. This ultrafiltrate is derived from a network of capillaries that intimately envelop the secretory region of each gland. After its release on the surface of the skin, sweat evaporates, cooling the surface. Besides its important cooling role, sweat glands also function as an auxiliary excretory organ, eliminating several substances not necessary for the organism.

In addition to the merocrine sweat glands just described, another type of sweat gland—the **apocrine** gland—is present in the axillary, areolar, and anal regions. Apocrine glands are much larger (3–5 mm in diameter) than merocrine sweat glands. They are embedded in the dermis and hypodermis, and their ducts open into hair follicles. These glands produce a viscous secretion that is initially odorless but may acquire a distinctive odor as a result of bacterial decomposition. Apocrine glands are innervated by adrenergic nerve endings, whereas merocrine glands receive cholinergic fibers. The glands of Moll in the margins of the eyelids and the ceruminous glands of the ear are modified sweat glands.

Tumors of the Skin

MEDICAL APPLICATION

*In adults, one third of all tumors are of the skin. Most of these tumors derive from the basal cells, the squamous cells of the stratum spinosum, and melanocytes. They produce, respectively, basal cell carcinomas, squamous cell carcinomas, and melanomas. The first two types of tumors can be diagnosed and excised early and consequently are rarely lethal. Skin tumors show an increased incidence in fair-skinned individuals residing in regions with high amounts of solar radiation. **Malignant melanoma** is an invasive tumor of melanocytes. Dividing rapidly, malignantly transformed melanocytes penetrate the basal lamina, enter the dermis, and invade the blood and lymphatic vessels to gain wide distribution throughout the body.*

REFERENCES

Edelson RL, Fink JM: The immunologic function of the skin. Sci Am 1985;252:46.

Goldsmith LA (ed): *Biochemistry and Physiology of the Skin.* Vols 1 and 2. Oxford University Press, 1983.

Green H et al: Differentiated structural components of the keratinocyte. Cold Spring Harbor Symp Quant Biol 1982;46(Pt 1):293.

Hentula M et al: Expression profiles of cell-cell and cell-matrix junction proteins in developing human epidermis. Arch Dermatol Res 2001;293:259.

Millington PF, Wilkinson R: *Skin.* Cambridge Univ Press, 1983.

Montagna W: *The Structure and Function of Skin,* 3rd ed. Academic Press, 1974.

Strauss JS et al: The sebaceous glands: twenty-five years of progress. J Invest Dermatol 1976;67:90.

Winkelmann RK: The Merkel cell system and a comparison between it and the neurosecretory or APUD cell system. J Invest Dermatol 1977;69:41.

The Urinary System

19

The urinary system consists of the paired kidneys and ureters and the unpaired bladder and urethra. This system contributes to the maintenance of homeostasis by a complex process that involves **filtration, active absorption, passive absorption,** and **secretion.** The result is the production of urine, in which various metabolic waste products are eliminated. Urine produced in the kidneys passes through the ureters to the bladder, where it is temporarily stored and then released to the exterior through the urethra. The two kidneys produce about 125 mL of filtrate per minute; of this amount, 124 mL is absorbed in the organ, and only 1 mL is released into the ureters as urine. About 1500 mL of urine is formed every 24 hours. The kidneys also regulate the fluid and electrolyte balance of the body and are the site of production of renin, a substance that participates in the regulation of blood pressure. Erythropoietin, a growth factor glycoprotein of 30 kDa that stimulates the production of erythrocytes, is also produced in the kidneys. Erythropoietin also hydroxylates vitamin D_3, a steroid prohormone, to its active form.

KIDNEYS

Each kidney has a concave medial border, the **hilum**—where nerves enter, blood and lymph vessels enter and exit, and the ureter exits—and a convex lateral surface (Figure 19–1). The **renal pelvis,** the expanded upper end of the ureter, is divided into 2 or 3 **major calyces.** Several small branches, the **minor calyces,** arise from each major calyx.

The kidney can be divided into an outer **cortex** and an inner **medulla** (Figures 19–1 and 19–2). In humans, the renal medulla consists of 10–18 conical or pyramidal structures, the **medullary pyramids.** From the base of each medullary pyramid, parallel arrays of tubules, the **medullary rays,** penetrate the cortex (Figure 19–1).

Each kidney is composed of 1–4 million **nephrons** (Gr. *nephros,* kidney). Each nephron consists of a dilated portion, the **renal corpuscle;** the **proximal convoluted tubule;** the **thin** and **thick limbs** of **Henle's loop;** the **distal convoluted tubule** (Figure 19–1); and the **collecting tubules** and **ducts.** Some authors do not consider the collecting tubules and ducts to be part of the nephron. The nephron is the functional unit of the kidney.

Renal Corpuscles & Blood Filtration

Each renal corpuscle is about 200 μm in diameter and consists of a tuft of capillaries, the **glomerulus,** surrounded by a double-walled epithelial capsule called **glomerular (Bowman's) capsule** (Figures 19–1, 19–2, and 19–3). The internal layer (the **visceral layer**) of the capsule envelops the capillaries of the glomerulus. The external layer forms the outer limit of the renal corpuscle and is called the **parietal layer** of Bowman's capsule (Figures 19–2, 19–3, and 19–4). Between the two layers of Bowman's capsule is the **urinary space,** which receives the fluid filtered through the capillary wall and the visceral layer. Each renal corpuscle has a **vascular pole,** where the **afferent arteriole** enters and the efferent arteriole leaves (Figure 19–3), and a **urinary pole,** where the proximal convoluted tubule begins (Figure 19–3). After entering the renal corpuscle, the afferent arteriole usually divides into 2 to 5 primary branches, each subdividing into capillaries and forming the renal glomerulus.

The parietal layer of Bowman's capsule consists of a simple squamous epithelium supported by a basal lamina and a thin layer of reticular fibers. At the urinary pole, the epithelium changes to the simple cuboidal, or low columnar, epithelium characteristic of the proximal tubule (Figure 19–3).

During embryonic development, the epithelium of the parietal layer remains relatively unchanged, whereas the internal, or visceral, layer is greatly modified. The cells of this internal layer, the **podocytes** (Figures 19–3, 19–5, 19–6, and 19–7), have a cell body from which arise several **primary processes.** Each primary process gives rise to numerous **secondary processes,** called **pedicels** (Figures 19–5, 19–6, and 19–7), that embrace the capillaries of the glomerulus. At a periodic distance of 25 nm, the secondary processes are in direct contact with the basal lamina. However, the cell bodies of podocytes and their primary processes do not touch the basement membrane (Figures 19–5 and 19–7).

The secondary processes of podocytes interdigitate, defining elongated spaces about 25 nm wide—the **filtration slits.** Spanning adjacent processes (and thus bridging the filtration slits) is a diaphragm about 6 nm thick. Podocytes have bundles of actin microfilaments in their cytoplasm that give them a contractile capacity (Figures 19–7 and 19–8).

Between the fenestrated endothelial cells of the glomerular capillaries and the podocytes that cover their external surfaces is a thick (~0.1-μm) basement membrane (Figures 19–8 and 19–9). This membrane is believed to be the filtration barrier that separates the urinary space and the blood in the capillaries. The basement membrane is derived from the fusion of capillary- and podocyte-produced basal laminae. With the aid of the electron microscope, one can distinguish a central electron-dense layer (**lamina densa**) and, on each side, a more electron-lucent layer (**lamina rara;** Figure 19–8). The two electron-lucent laminae rarae contain fibronectin, which may serve to bind them to the cells. The lamina densa is a meshwork of type IV collagen and laminin in a matrix containing the negatively charged

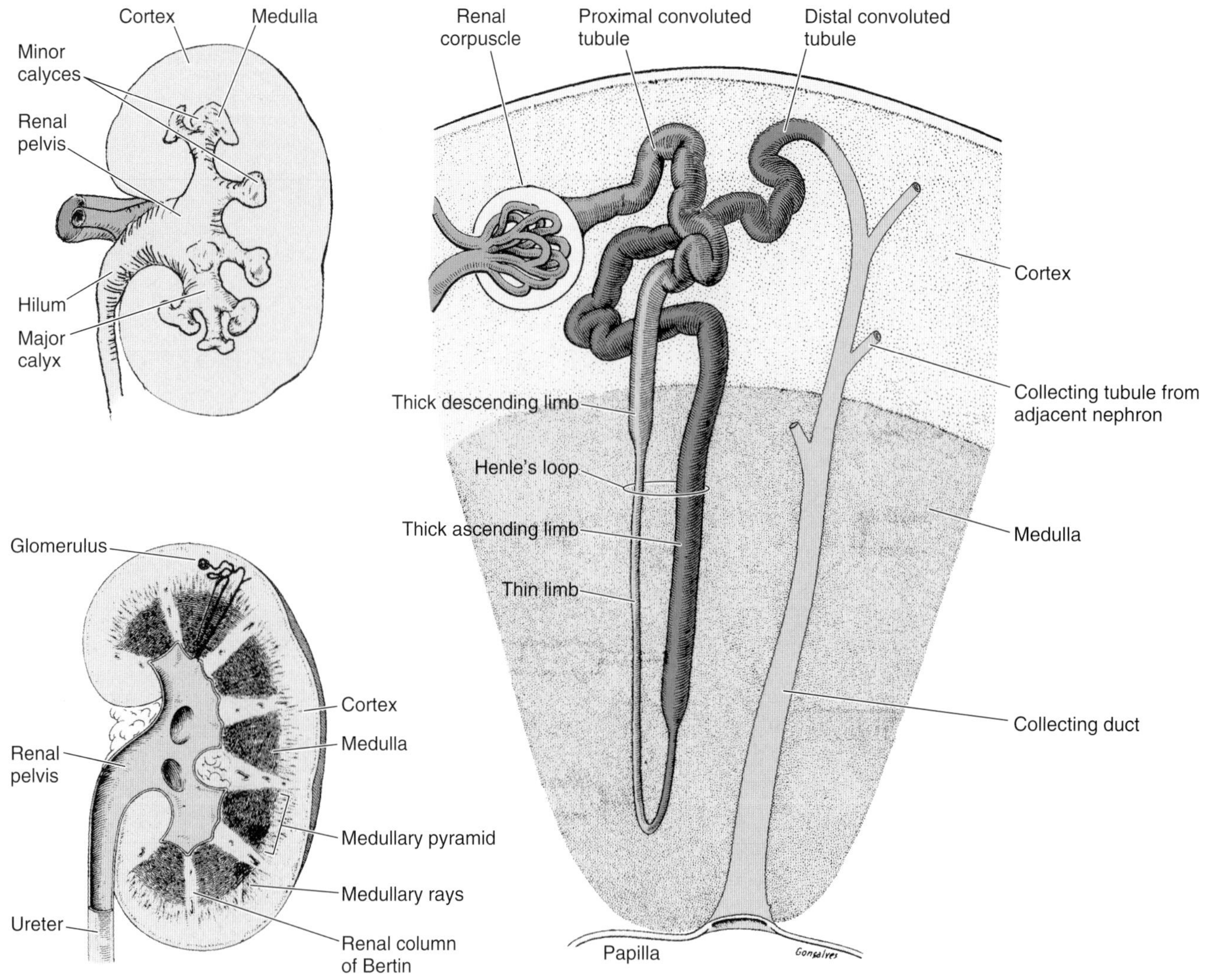

Figure 19–1. **Left:** General organization of the kidney. **Right:** Parts of a juxtamedullary nephron and its collecting duct and tubule.

proteoglycan heparan sulfate that restricts the passage of cationic molecules. Thus, the glomerular basement membrane is a selective macromolecular filter in which the lamina densa acts as a physical filter, whereas the anionic sites in the laminae rarae act as a charge barrier. Particles greater than 10 nm in diameter do not readily cross the basal lamina, and negatively charged proteins with a molecular mass greater than that of albumin (69 kDa) pass across only sparingly.

The blood flow in the two kidneys of an adult amounts to 1.2–1.3 L of blood per minute. This means that all the circulating blood in the body passes through the kidneys every 4–5 minutes. The glomeruli are composed of arterial capillaries in which the hydrostatic pressure—about 45 mm Hg—is higher than that found in other capillaries.

The glomerular filtrate is formed in response to the hydrostatic pressure of blood, which is opposed by the osmotic (oncotic) pressure of plasma colloids (20 mm Hg), and the hydrostatic pressure of the fluids in Bowman's capsule (10 mm Hg). The net filtration pressure at the afferent end of glomerular capillaries is 15 mm Hg.

The glomerular filtrate has a chemical composition similar to that of blood plasma but contains almost no protein, because macromolecules do not readily cross the glomerular filter. The largest protein molecules that succeed in crossing the glomerular filter have a molecular mass of about 70 kDa, and small amounts of plasma albumin appear in the filtrate.

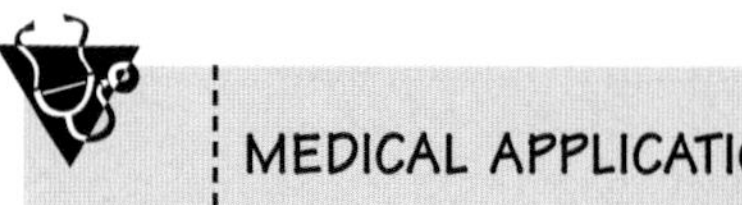

MEDICAL APPLICATION

*In diseases such as diabetes mellitus and glomerulonephritis, the glomerular filter is altered and becomes much more permeable to proteins, with the subsequent release of protein into the urine (**proteinuria**).*

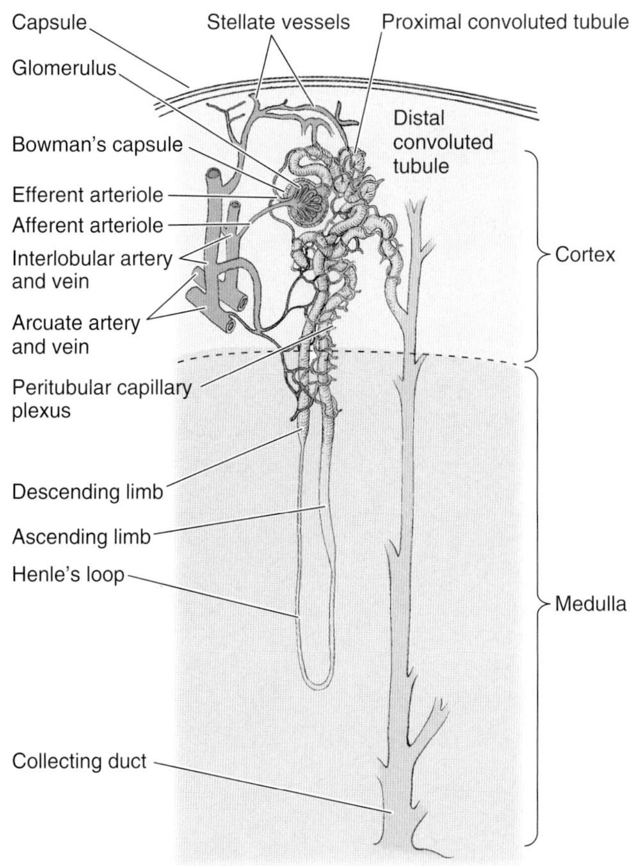

Figure 19–2. Diagram of the vascular supply of a nephron in the outer part of the cortex. Arteries and capillaries are red; veins are blue.

The endothelial cells of glomerular capillaries are of the fenestrated variety, but they lack the thin diaphragm that spans the openings of other fenestrated capillaries (Figure 19–7).

Besides endothelial cells and podocytes, the glomerular capillaries have **mesangial** (Gr. *mesos,* middle, + *angeion,* vessel) cells adhering to their walls (Figures 19–10 and 19–11). Mesangial cells are contractile and have receptors for angiotensin II. When these receptors are activated, the glomerular flow is reduced. Mesangial cells also have receptors for the natriuretic factor produced by cardiac atria cells. This factor is a vasodilator and relaxes the mesangial cells, probably increasing the blood flow and the effective surface area available for filtration. Mesangial cells also have several other functions: they give structural support to the glomerulus, synthesize extracellular matrix, endocytose and dispose of normal and pathologic (immune complex) molecules trapped by the glomerular basement membrane, and probably produce chemical mediators such as cytokines and prostaglandins. In the vascular pole but outside the glomerulus, there are the so-called **extraglomerular mesangial cells** that form part of the juxtaglomerular apparatus (described below).

Proximal Convoluted Tubule

At the urinary pole of the renal corpuscle, the squamous epithelium of the parietal layer of Bowman's capsule is continuous with the cuboidal, or low columnar, epithelium of the proximal convoluted tubule (Figures 19–1, 19–3, and 19–9). This tubule is longer than the distal convoluted tubule and is therefore more frequently seen near renal corpuscles in the renal cortex.

The cells of this cuboidal epithelium have an acidophilic cytoplasm (Figures 19–12, 19–13, and 19–14) because of the presence of numerous elongated mitochondria. The cell apex has abundant microvilli about 1 μm in length, which form a **brush border** (Figures 19–14, 19–15, and 19–16). Because the cells are large, each transverse section of a proximal tubule contains only 3–5 spherical nuclei.

In the living animal, proximal convoluted tubules have a wide lumen and are surrounded by peritubular capillaries. In routine histologic preparations, the brush border is usually disorganized and the peritubular capillary lumens are greatly reduced in size or collapsed.

The apical cytoplasm of these cells has numerous canaliculi between the bases of the microvilli; these canaliculi increase the capacity of the proximal tubule cells to absorb macromolecules. Pinocytotic vesicles are formed by evaginations of the apical membranes and contain macromolecules (mainly proteins with a molecular mass less than 70 kDa) that have passed across the

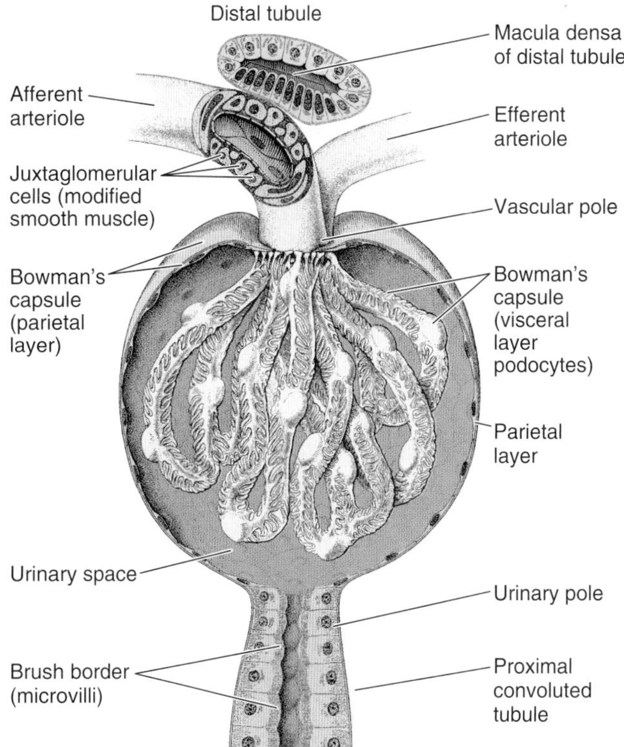

Figure 19–3. The renal corpuscle. The upper part of the drawing shows the vascular pole, with afferent and efferent arterioles and the macula densa. Note the juxtaglomerular cells in the wall of the afferent arteriole. Podocyte processes cover the outer surfaces of the glomerular capillaries; the part of the podocyte containing the nucleus protrudes into the urinary space. Note the flattened cells of the parietal layer of Bowman's capsule. The lower part of the drawing shows the urinary pole and the proximal convoluted tubule.

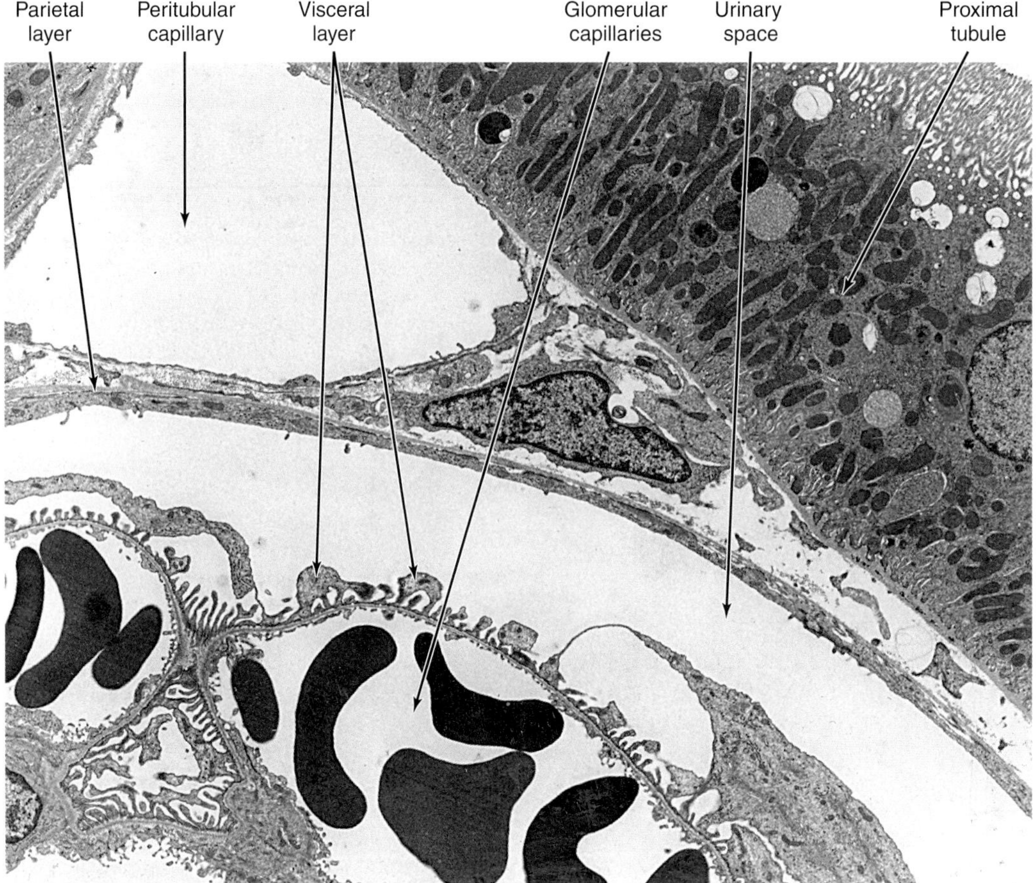

Figure 19–4. Electron micrograph of a rat kidney showing part of a renal corpuscle, including the parietal layer of Bowman's capsule, the urinary space, glomerular capillaries containing erythrocytes, the visceral layer of Bowman's capsule, the peritubular capillary, and the proximal tubule. ×2850. (Courtesy of SL Wissig.)

glomerular filter. The pinocytotic vesicles fuse with lysosomes, where macromolecules are degraded, and monomers are returned to the circulation. The basal portions of these cells have abundant membrane invaginations and lateral interdigitations with neighboring cells. The Na^+/K^+-ATPase (sodium pump) responsible for actively transporting sodium ions out of the cells is localized in these basolateral membranes. Mitochondria are concentrated at the base of the cell (Figure 19–4) and arranged parallel to the long axis of the cell. This mitochondrial location and the increase in the area of the cell membrane at the base of the cell are characteristic of cells engaged in active ion transport (see Chapter 4). Because of the extensive interdigitations of the lateral membranes, no discrete limits can be observed (in the light microscope) between cells of the proximal tubule. The glomerular filtrate formed in the renal corpuscle passes into the proximal convoluted tubule, where the processes of absorption and excretion begin. The proximal convoluted tubule absorbs all the glucose and amino acids and about 85% of the sodium chloride and water contained in the filtrate, in addition to phosphate and calcium. Glucose, amino acids, and sodium are absorbed by these tubular cells through an active process involving Na^+/K^+-ATPase (sodium pump) located in the basolateral cell membranes. Water diffuses passively, following the osmotic gradient. When the amount of glucose in the filtrate exceeds the absorbing capacity of the proximal tubule, urine becomes more abundant and contains glucose.

In addition to these activities, the proximal convoluted tubule secretes creatinine and substances foreign to the organism, such as para-aminohippuric acid and penicillin, from the interstitial plasma into the filtrate. This is an active process referred to as tubular secretion. Study of the rates of secretion of these substances is useful in the clinical evaluation of kidney function.

Henle's Loop

Henle's loop is a U-shaped structure consisting of a **thick descending limb,** a **thin descending limb,** a **thin ascending limb,** and a **thick ascending limb.** The thick limbs are very similar in structure to the distal convoluted tubule (Figure 19–16). In the outer medulla, the thick descending limb, with an outer diameter of about 60 μm, suddenly narrows to about 12 μm and continues as the thin descending limb. The lumen of this segment of the nephron is wide because the wall consists of squamous epithelial cells whose nuclei protrude only slightly into the lumen (Figures 19–16, 19–17, and 19–18).

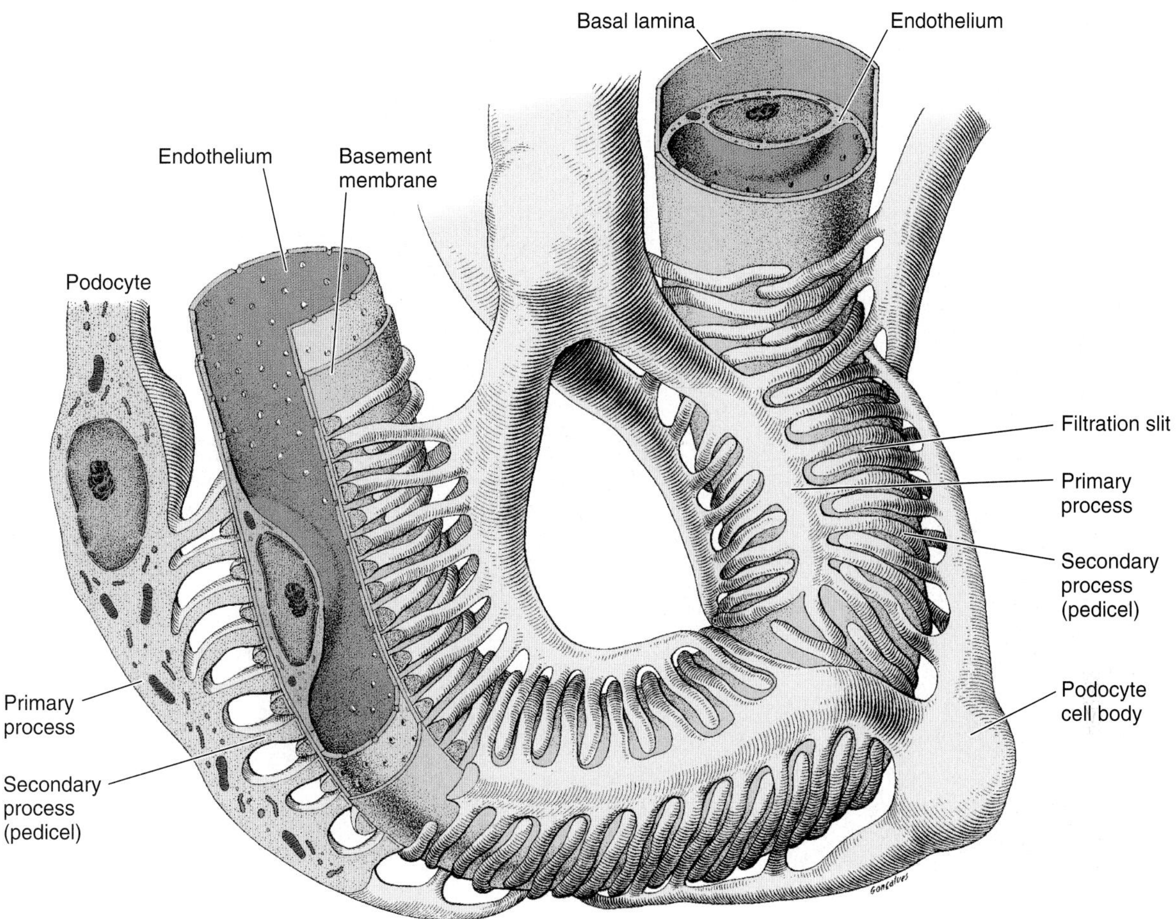

Figure 19–5. Schematic representation of a glomerular capillary with the visceral layer of Bowman's capsule (formed of podocytes). In this capillary, endothelial cells are fenestrated, but the basal lamina on which they rest is continuous. At left is a podocyte shown in partial section. As viewed from the outside, the part of the podocyte that contains the nucleus protrudes into the urinary space. Each podocyte has many primary processes, from which arise an even greater number of secondary processes that are in contact with the basal lamina. (Modified and redrawn from Gordon. Reproduced, with permission, from Ham AW: *Histology,* 6th ed. Lippincott, 1969.)

Approximately one seventh of all nephrons are located near the corticomedullary junction and are therefore called **juxtamedullary nephrons.** The other nephrons are called **cortical nephrons.** All nephrons participate in the processes of filtration, absorption, and secretion. Juxtamedullary nephrons, however, are of prime importance in establishing the gradient of hypertonicity in the medullary interstitium—the basis of the kidneys' ability to produce hypertonic urine. Juxtamedullary nephrons have very long Henle's loops, extending deep into the medulla. These loops consist of a short thick descending limb, long thin descending and ascending limbs, and a thick ascending limb. Cortical nephrons, on the other hand, have very short thin descending limbs and no thin ascending limbs (Figure 19–2).

Henle's loop is involved in water retention; only animals with such loops in their kidneys are capable of producing hypertonic urine and thus maintaining body water. Henle's loop creates a gradient of hypertonicity in the medullary interstitium that influences the concentration of the urine as it flows through the collecting ducts.

Although the thin descending limb of the loop is freely permeable to water, the entire ascending limb is impermeable to water. In the thick ascending limb, sodium chloride is actively transported out of the tubule to establish the gradient of hypertonicity in the medullary interstitium that is necessary for urine concentration. The osmolarity of the interstitium at the tips of the medullary pyramids is about 4 times that of blood.

Distal Convoluted Tubule

The thick ascending limb of Henle's loop penetrates the cortex; after describing a certain trajectory, it becomes tortuous and is called the distal convoluted tubule. This tubule, like the ascending limb, is lined with simple cuboidal epithelium (Figures 19–16, 19–17, and 19–19).

The distal convoluted tubules differ from the proximal convoluted tubules (both found in the cortex) because they have no brush border, no apical canaliculi, and smaller cells. Because distal tubule cells are flatter and smaller than those of the proximal tubule, more nuclei are seen in the distal tubule than in

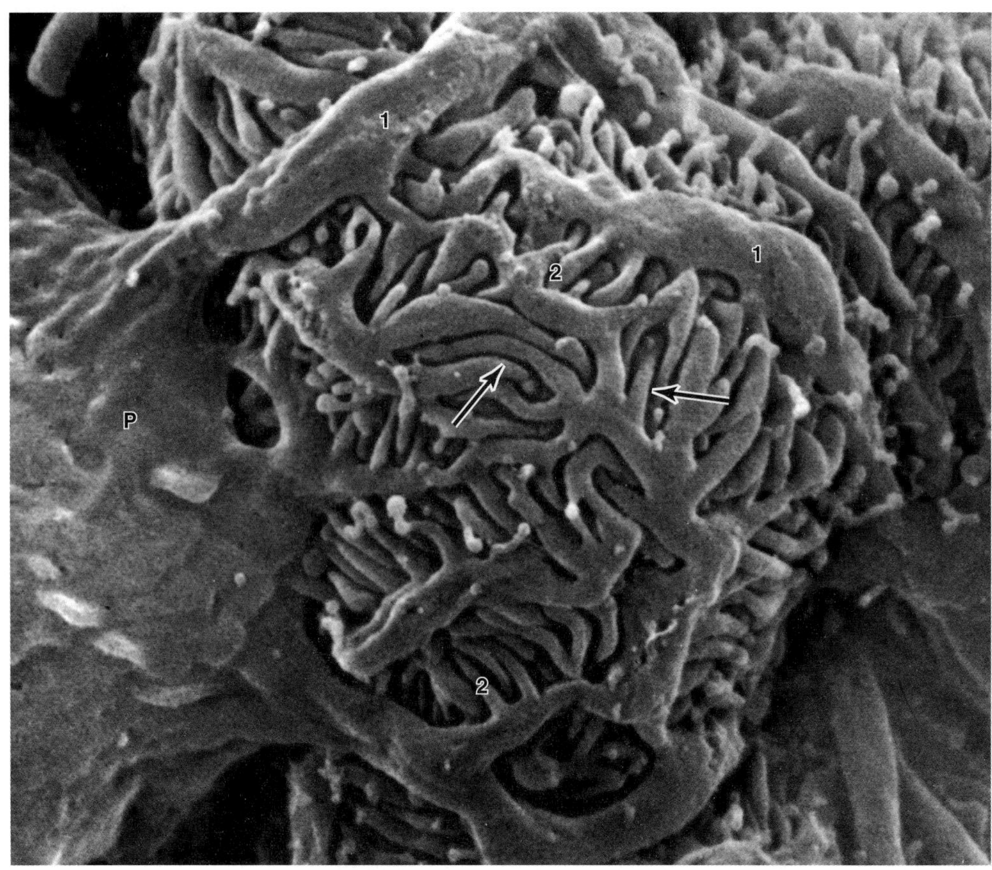

Figure 19–6. Scanning electron micrograph showing Bowman's visceral epithelial cells, or podocytes (P), surrounding capillaries of the renal glomerulus. Two orders of branching of the podocyte processes are apparent: the primary processes (1) and the secondary processes, or pedicels (2). The small spaces between adjacent processes constitute the filtration slits (arrows). ×10,700.

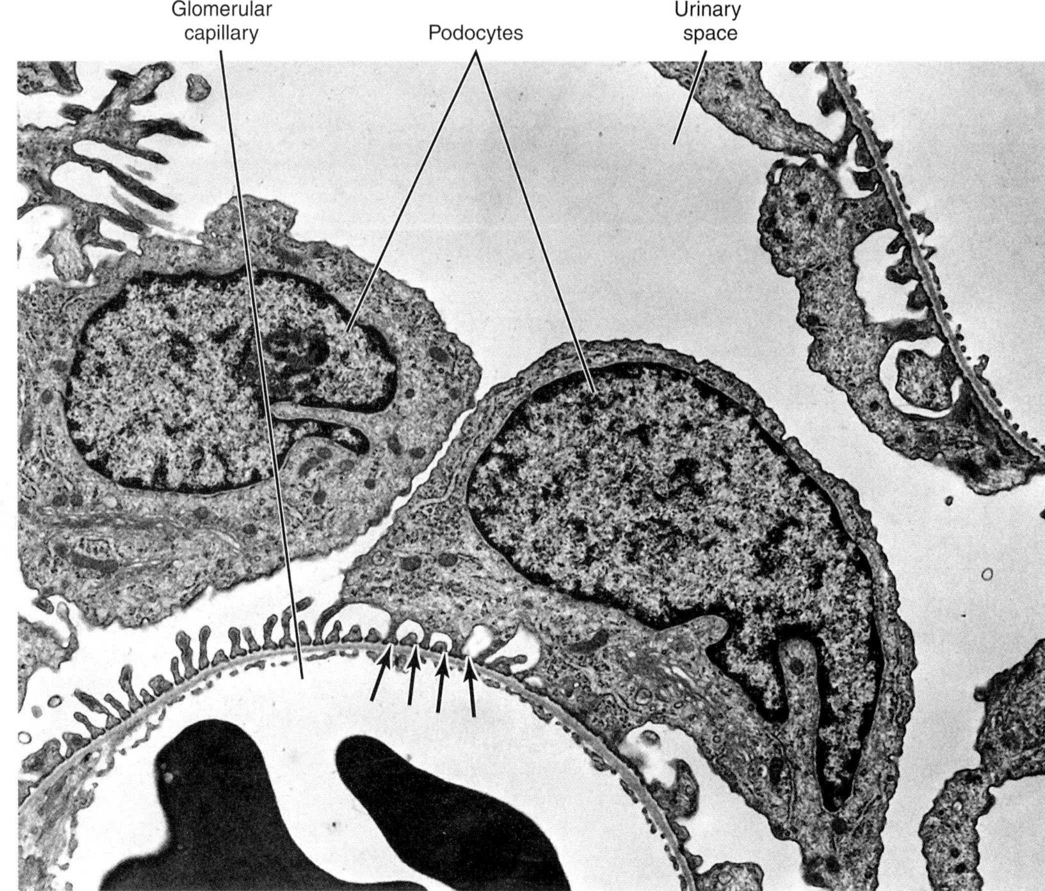

Figure 19–7. Electron micrograph showing the cell bodies of 2 podocytes and the alternation of secondary processes from 2 different cells (arrows). The urinary space and the glomerular capillary are indicated. ×9000. (Courtesy of SL Wissig.)

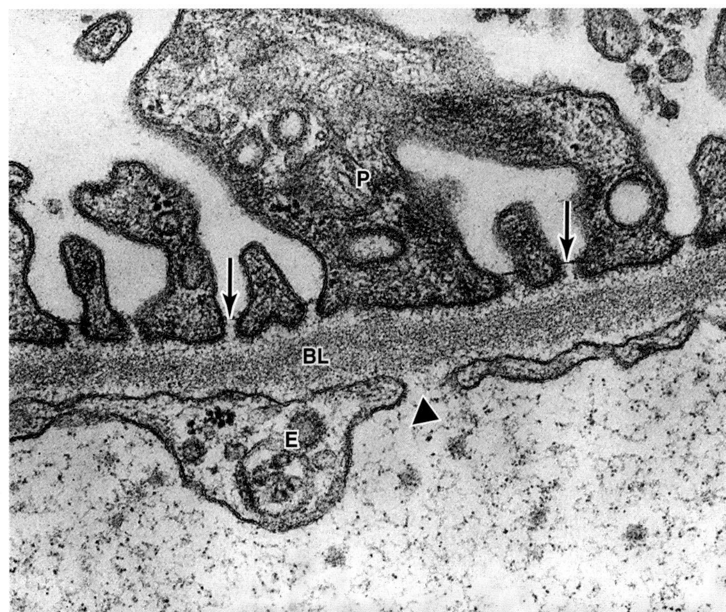

Figure 19–8. Electron micrograph of the filtration barrier in a renal corpuscle. Note the endothelium (E) with open fenestrae (arrowhead), the fused basal laminae (basement membrane) of epithelial and endothelial cells (BL), and the processes of podocytes (P). The basement membrane consists of a central lamina densa bounded on both sides by a light-staining lamina rara. Arrows indicate the thin diaphragms crossing the filtration slits. ×45,750. (Courtesy of SL Wissig.)

the proximal tubule. Cells of the distal convoluted tubule have elaborate basal membrane invaginations and associated mitochondria indicative of their ion-transporting function (Figure 19–16).

The distal convoluted tubule establishes contact with the vascular pole of the renal corpuscle of its parent nephron. At this point of close contact, the distal tubule is modified, as is the afferent arteriole. In this juxtaglomerular region, cells of the distal convoluted tubule usually become columnar, and their nuclei are closely packed together. Most of the cells have a Golgi complex in the basal region. This modified segment of the wall of the distal tubule, which appears darker in microscopic preparations because of the close proximity of its nuclei, is called the **macula densa** (Figures 19–3, 19–20, and 19–21). The cells of the macula densa are sensitive to the ionic content and water volume of the tubular fluid, producing molecular signals that promote the liberation of the enzyme renin in the circulation.

In the distal convoluted tubule, there is ion exchange if aldosterone is present in high enough concentration: Sodium is absorbed, and potassium ions are secreted. This mechanism influences the total salt and water content of the body. The distal tubule also secretes hydrogen and ammonium ions into tubular urine. This activity is essential for maintenance of the acid-base balance in the blood.

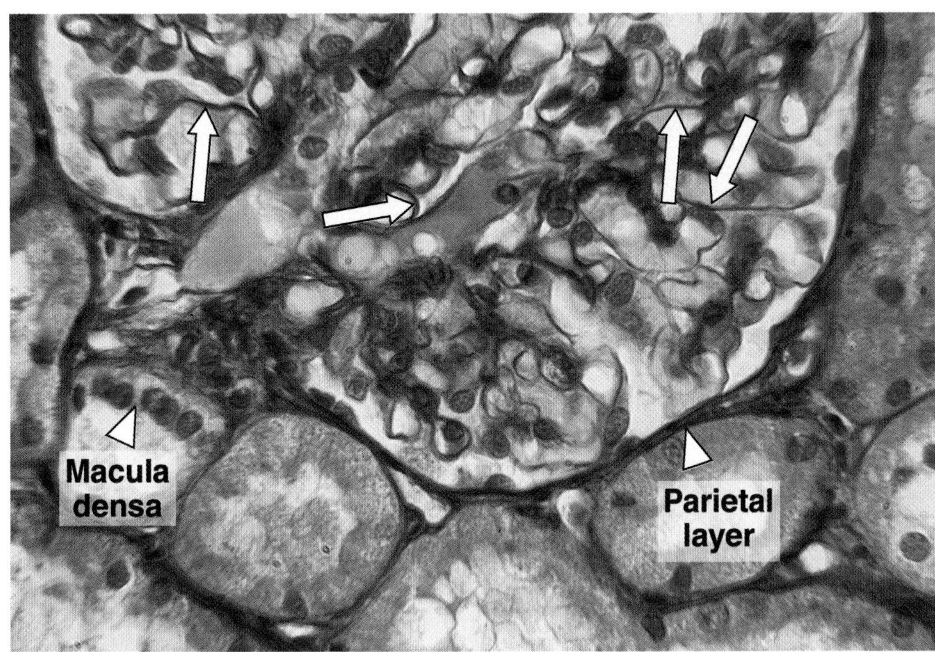

Figure 19–9. Photomicrograph of a renal cortex showing parts of 2 renal corpuscles, macula densa, and distal and proximal convoluted tubules. The collagen type IV of the basement membrane of the glomerular capillaries is clearly visible (arrows). The collagen of the parietal layer of the Bowman's capsule and basal membrane of a distal tubule are shown by the arrowhead. Picrosirius stain. Medium magnification.

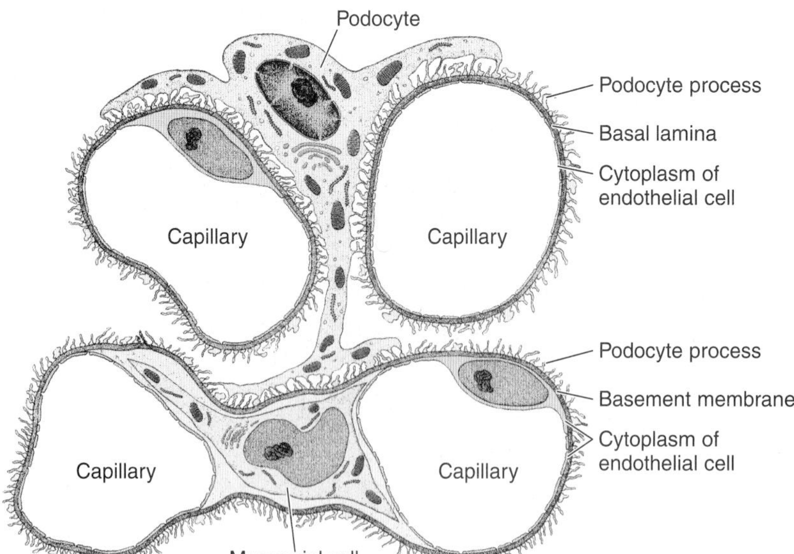

Figure 19–10. Mesangial cell located between capillaries enveloped by the basement membrane.

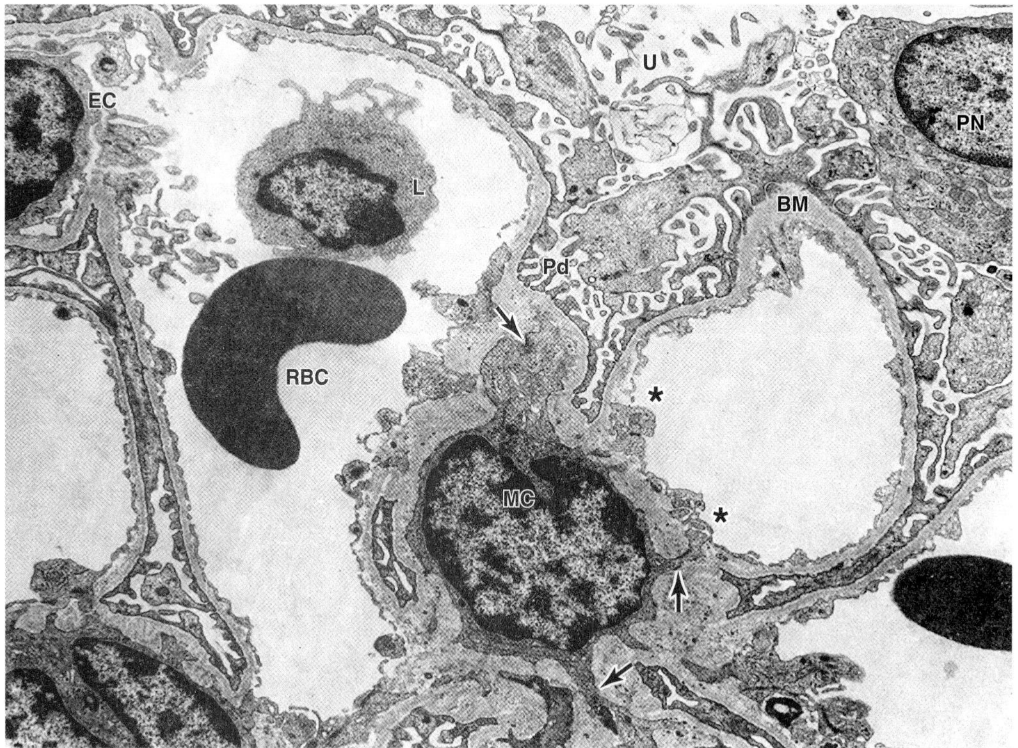

Figure 19–11. Electron micrograph showing a mesangial cell (MC) and the amorphous mesangial matrix surrounding it. The matrix helps to support the capillary loops where a basement membrane is lacking. Some of the mesangial cell's processes (arrows) reach the capillary lumen, passing between endothelial cells (asterisks). The capillary at left contains an erythrocyte (RBC) and a leukocyte (L). BM, basement membrane; EC, endothelial cell; Pd, pedicels; PN, podocyte nucleus; U, urinary space.

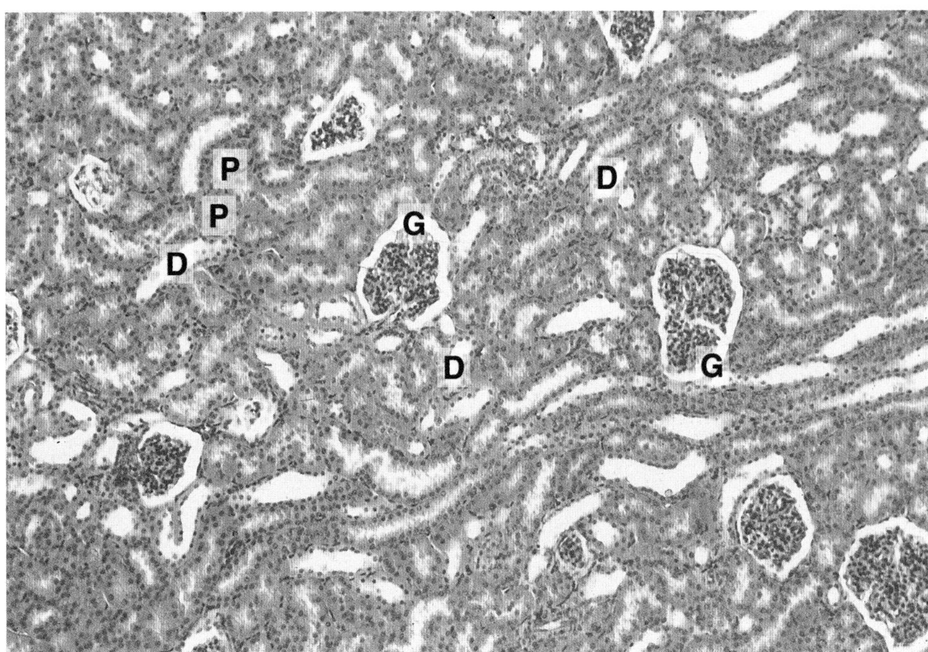

Figure 19–12. Bird's-eye view of the renal cortex, which is composed mainly of proximal (P) and distal (D) convoluted tubules, and renal glomeruli (G). Pararosaniline–toluidine blue (PT) stain. Low magnification.

Collecting Tubules & Ducts

Urine passes from the distal convoluted tubules to collecting tubules that join each other to form larger, straight collecting ducts, which widen gradually as they approach the tips of the medullary pyramids (Figure 19–1).

The smaller collecting tubules are lined with cuboidal epithelium and have a diameter of approximately 40 μm. As they penetrate deeper into the medulla, their cells increase in height until they become columnar. The diameter of the collecting duct reaches 200 μm near the tips of the medullary pyramids.

Along their entire extent, collecting tubules and ducts are composed of cells that stain weakly (Figure 19–22) with the usual stains. They have an electron-lucent cytoplasm with few organelles (Figure 19–23). In collecting tubules and cortical collecting ducts, a dark-staining intercalated cell is also seen; its significance is not understood. The intercellular limits of the collecting tubule and duct cells are clearly visible in the light microscope (Figure 19–22). Cortical collecting ducts are joined at right angles by several generations of smaller collecting tubules that drain each medullary ray. In the medulla, collecting ducts are a major component of the urine-concentrating mechanism.

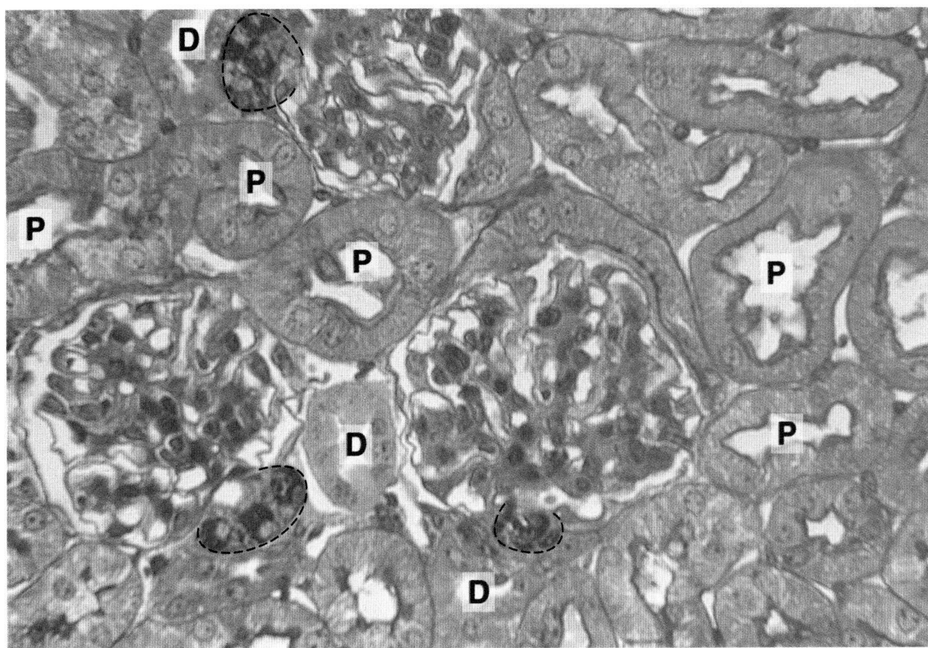

Figure 19–13. Renal cortex showing proximal (P) and distal (D) convoluted tubules. One can see sections through the vascular pole of 3 renal corpuscles where juxtaglomerular renin-secreting cells appear well stained (broken lines). PT stain. Medium magnification.

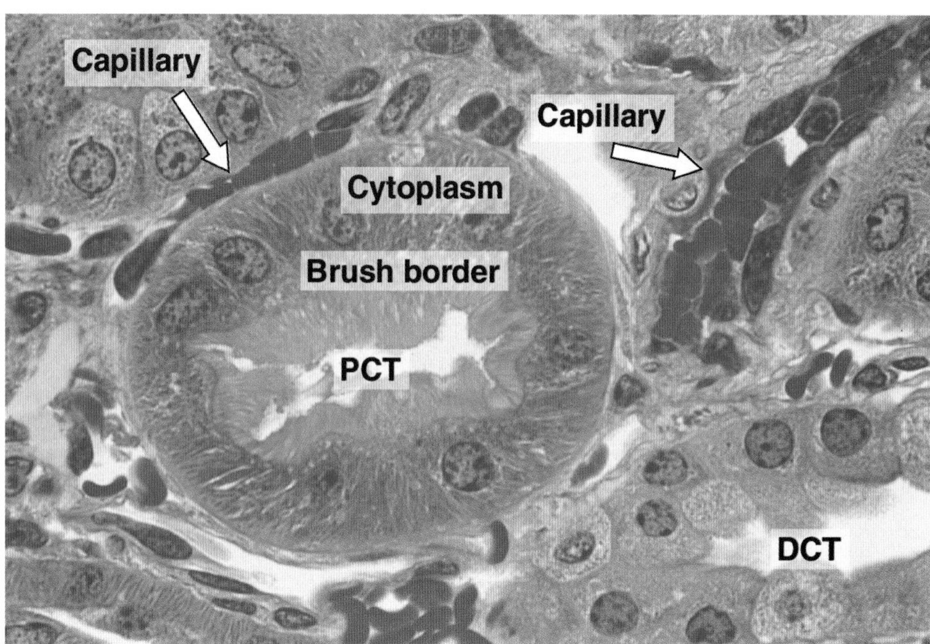

Figure 19–14. Renal cortex section showing a proximal convoluted tubule (PCT) with its large cuboidal cells presenting a brush border formed by numerous microvilli. Distal convoluted tubules (DCT) are also present. PT stain. Medium magnification.

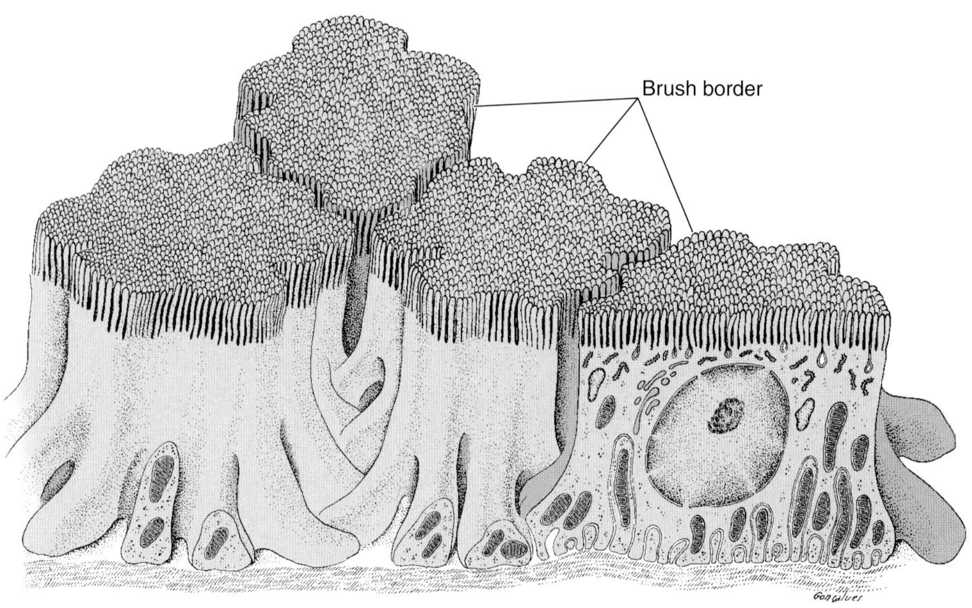

Figure 19–15. Schematic drawing of proximal convoluted tubule cells. The apical surfaces of these cuboidal cells have abundant microvilli constituting a brush border. Note the distribution of mitochondria and associated basilar infoldings of the cell membrane. The latter processes are longer than the former and penetrate deeply among the neighboring cells. Artificial spaces between the cells are shown to make the drawing easier to understand. (Modified from Bulger R: Amer J Anat 1965;116:237.)

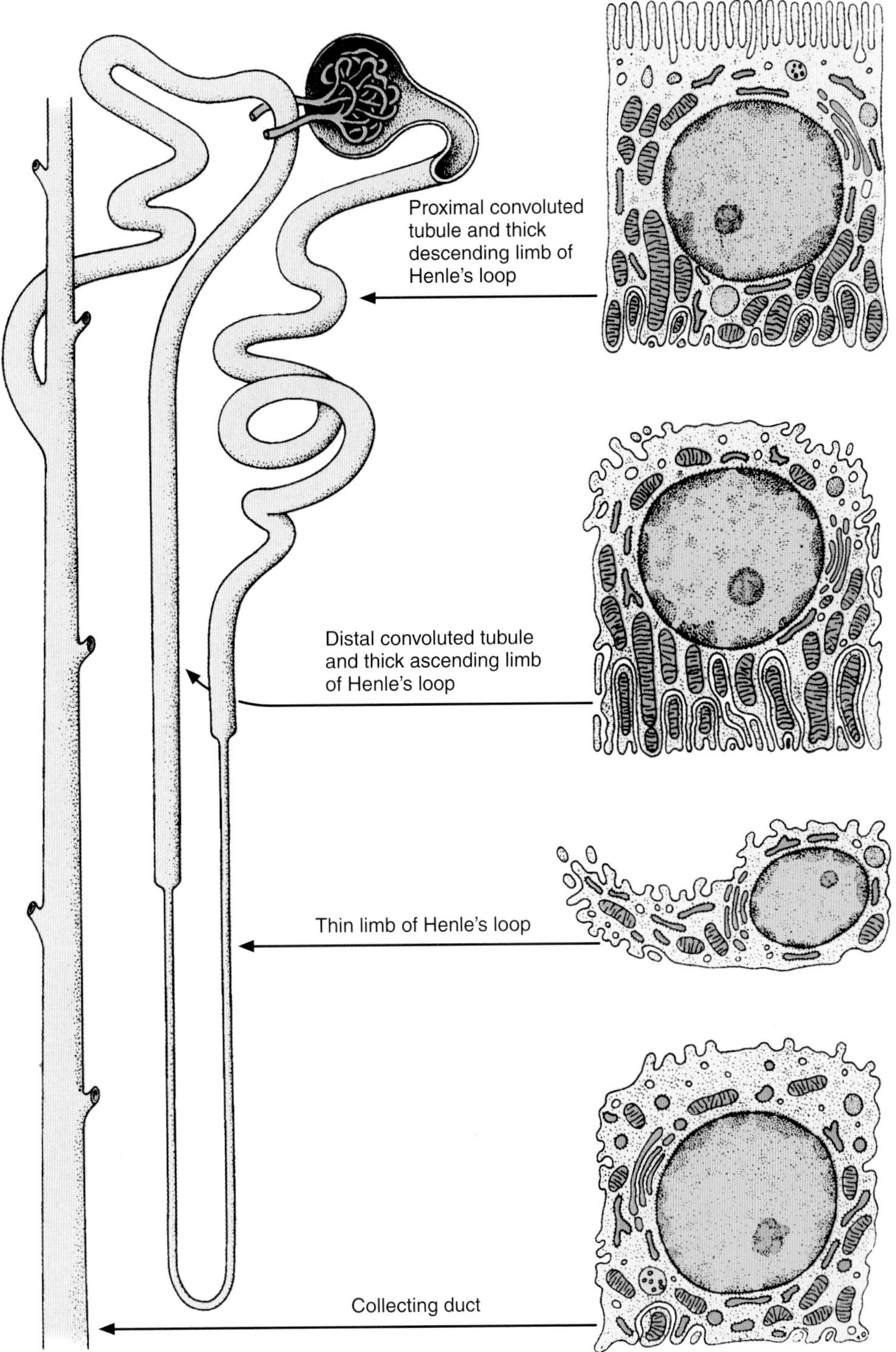

Figure 19–16. Cellular ultrastructure of the nephron, represented schematically. Cells of the thick ascending limb of Henle's loop and the distal tubule are different in their ultrastructures and functions.

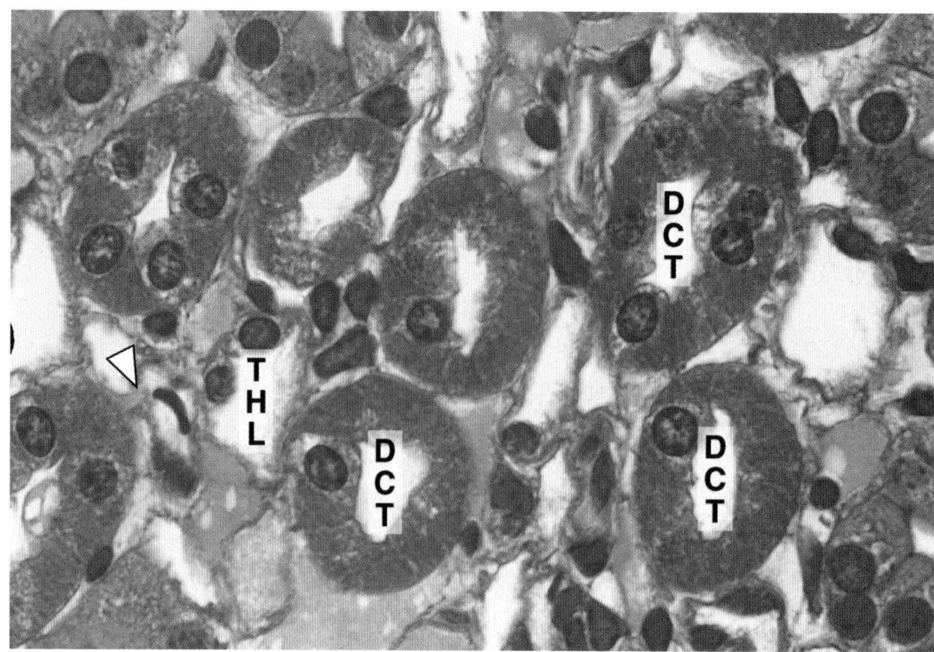

Figure 19–17. Distal convoluted tubules (DCT) characterized by the absence of brush border. Note also a thin portion of Henle's loop (THL) and a blood capillary (arrowhead). PT stain. Medium magnification.

The epithelium of collecting ducts is responsive to arginine vasopressin, or antidiuretic hormone, secreted by the posterior pituitary. If water intake is limited, antidiuretic hormone is secreted and the epithelium of the collecting ducts becomes permeable to water, which is absorbed from the glomerular filtrate, transferred to blood capillaries, and thus retained in the body. In the presence of antidiuretic hormone, intramembrane particles in the luminal membrane aggregate to form what may be channels for water absorption.

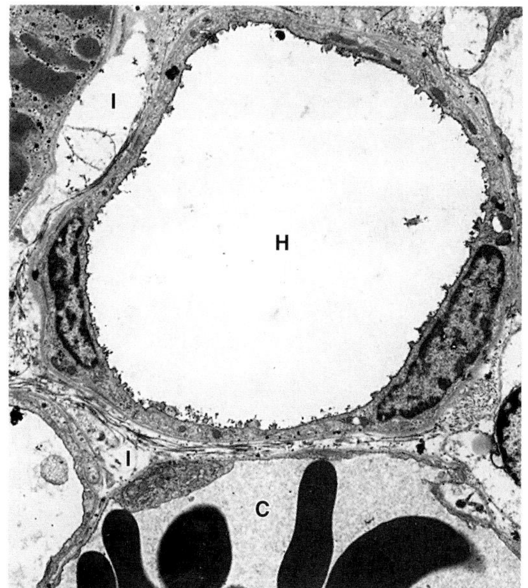

Figure 19–18. Electron micrograph of the thin limb of Henle's loop (H) composed entirely of squamous cells. Note the fenestrated capillaries with erythrocytes (C) and the interstitium (I) with bundles of collagen fibrils. ×3300. (Courtesy of J Rhodin.)

Juxtaglomerular Apparatus

Adjacent to the renal corpuscle, the tunica media of the afferent arteriole has modified smooth muscle cells. These cells are called **juxtaglomerular (JG) cells** (Figures 19–3, Figures 19–13, 19–24, and 19–25) and have a cytoplasm full of secretory granules. Secretions of JG cells play a role in the maintenance of blood pressure. The macula densa of the distal convoluted tubule is usually located near the region of the afferent arteriole that contains the JG cells; together, this portion of the arteriole and the macula densa form the JG apparatus (Figures 19–3 and 19–24). Also a part of the JG apparatus are some light-staining cells whose functions are not well understood. They are called **extraglomerular mesangial cells** or **lacis cells.** The internal elastic membrane of the afferent arteriole disappears in the area of the JG cells.

When examined with the electron microscope, JG cells show characteristics of protein-secreting cells, including an abundant rough endoplasmic reticulum, a highly developed Golgi complex, and secretory granules measuring approximately 10–40 nm in diameter. JG cells produce the enzyme **renin,** which acts on a plasma protein—**angiotensinogen**—to produce an inactive decapeptide, **angiotensin I.** As a result of the action of a converting enzyme present in high concentration in lung endothelial cells, this substance loses 2 amino acids and becomes an active vasopressive octapeptide, **angiotensin II.**

MEDICAL APPLICATION

After a significant hemorrhage (decreased blood volume promotes a decreased blood pressure), there is an increase in renin secretion. Angiotensin II is produced, enhancing blood pressure by both constricting arterioles and stimulating the secretion of the adrenocortical hormone ***aldosterone.*** *Aldosterone acts on cells of the renal tubules (mostly the distal*

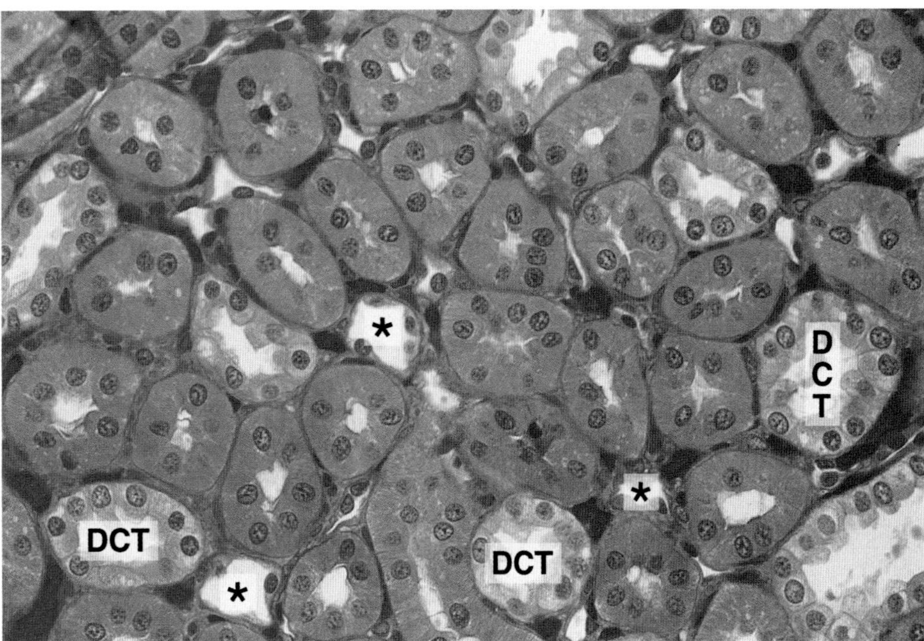

Figure 19–19. Region of the kidney consisting mainly of distal convoluted tubules (DCT) and thin segments of Henle's loop (asterisks). Capillaries filled with blood appear in red. PT stain. Medium magnification.

tubules) to increase the absorption of sodium and chloride ions from the glomerular filtrate. This increase in sodium and chloride ions, in turn, expands the fluid volume (particularly blood plasma volume), leading to an increase in blood pressure due to increased blood volume.

Decreased blood pressure caused by other factors (eg, sodium depletion, dehydration) that decrease blood volume also activates the renin–angiotensin II–aldosterone mechanism that contributes to the maintenance of blood pressure.

Blood Circulation

Each kidney receives blood from its **renal artery,** which usually divides into 2 branches before entering the organ. One branch goes to the anterior part of the kidney, the other to the posterior part. While still in the hilum, these branches give rise to arteries that branch again to form the **interlobar arteries** located between the renal pyramids (Figure 19–26). At the level of the corticomedullary junction, the interlobar arteries form the **arcuate arteries. Interlobular arteries** branch off at right angles from the arcuate arteries and follow a course in the cortex perpendicular to the renal capsule. Interlobular arteries form the boundaries of

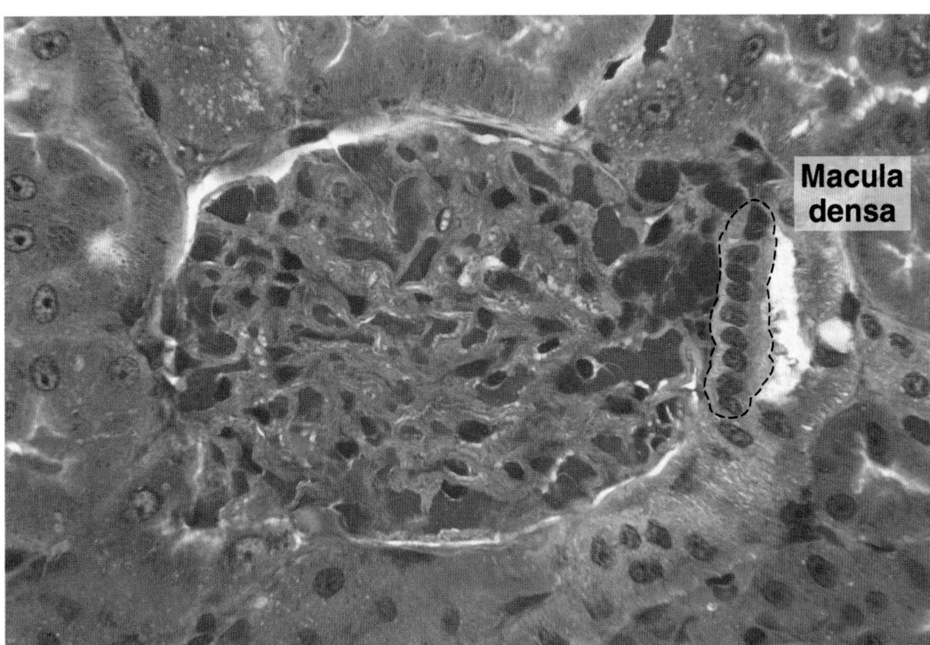

Figure 19–20. Renal cortex showing a distal convoluted tubule with a macula densa formed by closely packed epithelial cells (broken line). This structure is sensitive to the ionic concentration of the filtrate in the distal tubule and is believed to influence glomerular filtration. PT stain. Medium magnification.

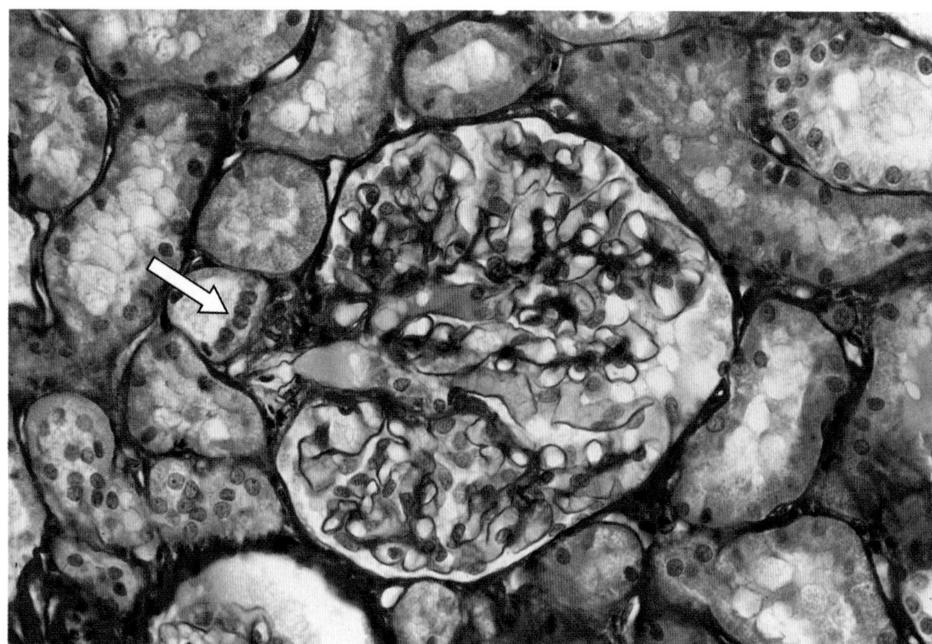

Figure 19–21. Photomicrograph of renal cortex. A macula densa is clearly seen (arrow) at the vascular pole of a renal corpuscle. Picrosirius-hematoxylin (PSH) stain. Medium magnification.

the renal lobules, which consist of a medullary ray and the adjacent cortical labyrinth (Figure 19–26). From the interlobular arteries arise the **afferent arterioles,** which supply blood to the capillaries of the glomeruli. Blood passes from these capillaries into the **efferent arterioles,** which at once branch again to form a **peritubular capillary network** that will nourish the proximal and distal tubules and carry away absorbed ions and low-molecular-weight materials. The efferent arterioles that are associated with juxtamedullary nephrons form long, thin capillary vessels. These vessels, which follow a straight path into the medulla and then loop back toward the corticomedullary boundary, are called **vasa recta** (straight vessels). The descending vessel is a continuous-type capillary, whereas the ascending vessel has a fenestrated endothelium. These vessels, containing blood that has been filtered through the glomeruli, provide nourishment and oxygen to the medulla. Because of their looped structure, they do not carry away the high osmotic gradient set up in the interstitium by Henle's loop.

The capillaries of the outer cortex and the capsule of the kidney converge to form the **stellate veins** (so called because of their configuration when seen from the surface of the kidney), which empty into the interlobular veins.

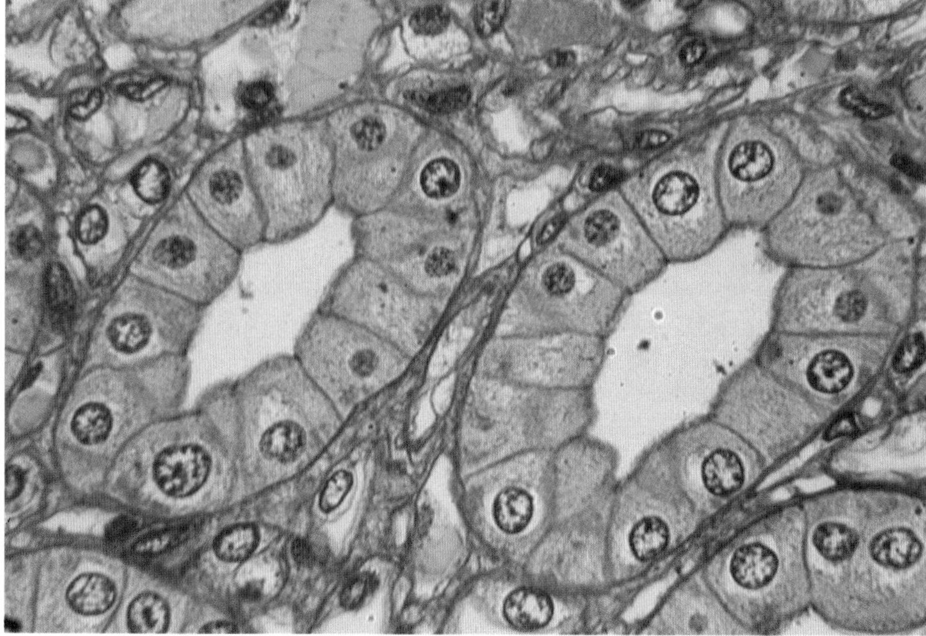

Figure 19–22. Photomicrograph of renal medulla with 2 collecting ducts consisting of cuboidal cells resting on a basement membrane. In this hypertonic region of the kidney, because of the action of the hypophyseal antidiuretic hormone, water is reabsorbed, controlling the water balance of the body. PT stain. Medium magnification.

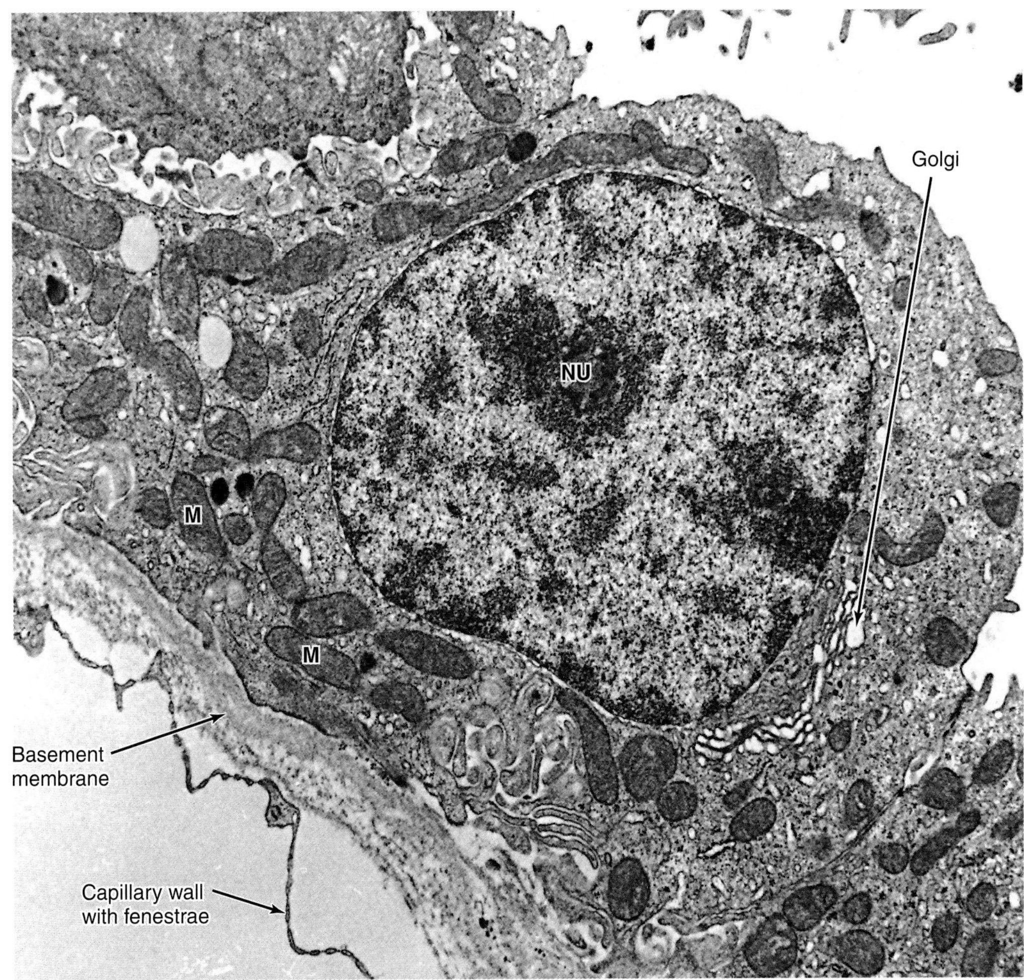

Figure 19–23. Electron micrograph of a collecting tubule wall. M, mitochondria; NU, nucleolus. ×15,000.

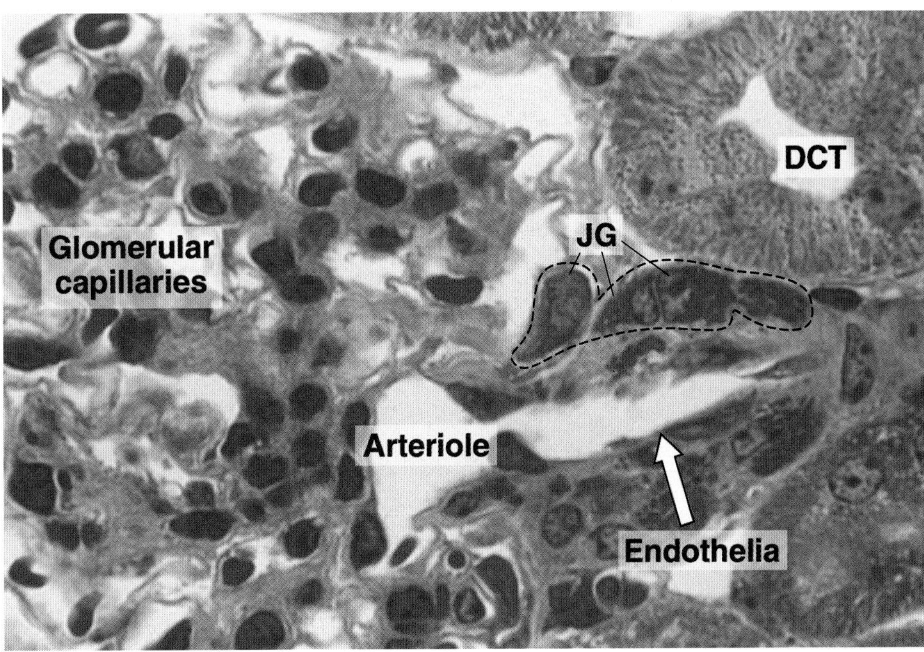

Figure 19–24. Photomicrograph of an afferent arteriole entering a renal corpuscle. The wall of this arteriole shows the renin-producing juxtaglomerular (JG) cells (broken line). At the upper right is a distal convoluted tubule (DCT) with many elongated mitochondria. PT stain. High magnification.

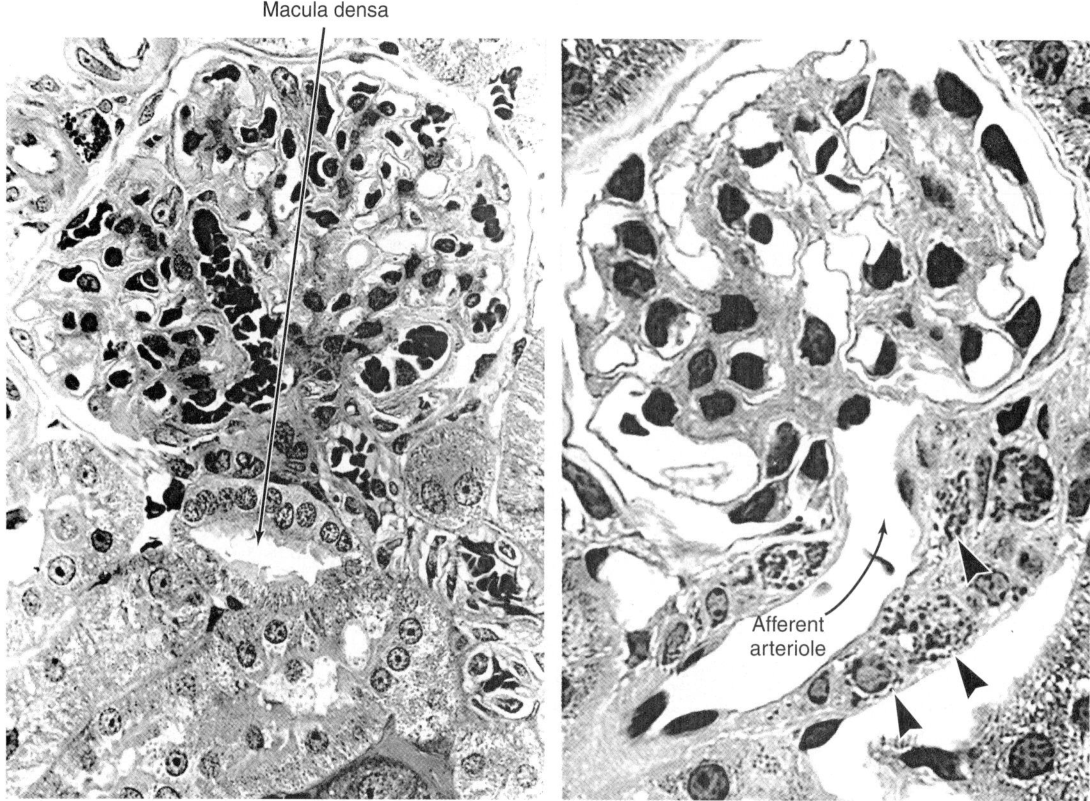

Figure 19–25. Photomicrographs of 2 renal corpuscles. **Left:** A macula densa with the characteristic close proximity of its nuclei. In this location, cells of the distal tubules are smaller. **Right:** A portion of a juxtaglomerular apparatus showing the wall of the afferent arteriole with cells that have secretory granules (arrowheads) containing renin.

Veins follow the same course as arteries (Figure 19–26). Blood from interlobular veins flows into arcuate veins and from there to the interlobar veins. Interlobar veins converge to form the renal vein through which blood leaves the kidney.

Renal Interstitium

The space between uriniferous tubules and blood and lymph vessels is called the **renal interstitium.** It occupies a very small volume in the cortex but increases in the medulla. The renal interstitium contains a small amount of connective tissue with fibroblasts, some collagen fibers and, mainly in the medulla, a highly hydrated ground substance rich in proteoglycan. In the medulla the secreting cells called **interstitial cells** are found. They contain cytoplasmic lipid droplets and are implicated in the synthesis of prostaglandins and prostacyclin.

Effects of Adrenal Steroids

Steroid hormones of the adrenal cortex, mainly **aldosterone,** increase distal tubular absorption of sodium from the filtrate and thus decrease sodium loss in the urine. Aldosterone also facilitates the elimination of potassium and hydrogen ions. This hormone is crucial in maintaining electrolyte balance in the body.

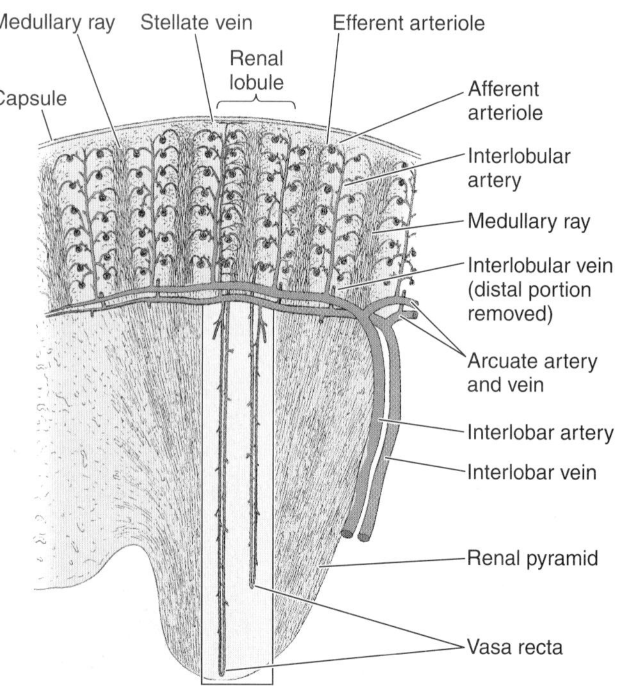

Figure 19–26. Circulation of blood in the kidney. Arcuate arteries are seen in the border between the cortex and the medulla.

MEDICAL APPLICATION

*Aldosterone deficiency in adrenalectomized animals and in humans with **Addison disease** results in an excessive loss of sodium in the urine.*

BLADDER & URINARY PASSAGES

The bladder and the urinary passages store the urine formed in the kidneys and conduct it to the exterior. The calyces, renal pelvis, ureter, and bladder have the same basic histologic structure, with the walls of the ureters becoming gradually thicker as proximity to the bladder increases.

The mucosa of these organs consists of **transitional epithelium** (Figures 19–27 A and B) and a lamina propria of loose-to-dense connective tissue. Surrounding the lamina propria of these organs is a dense woven sheath of smooth muscle.

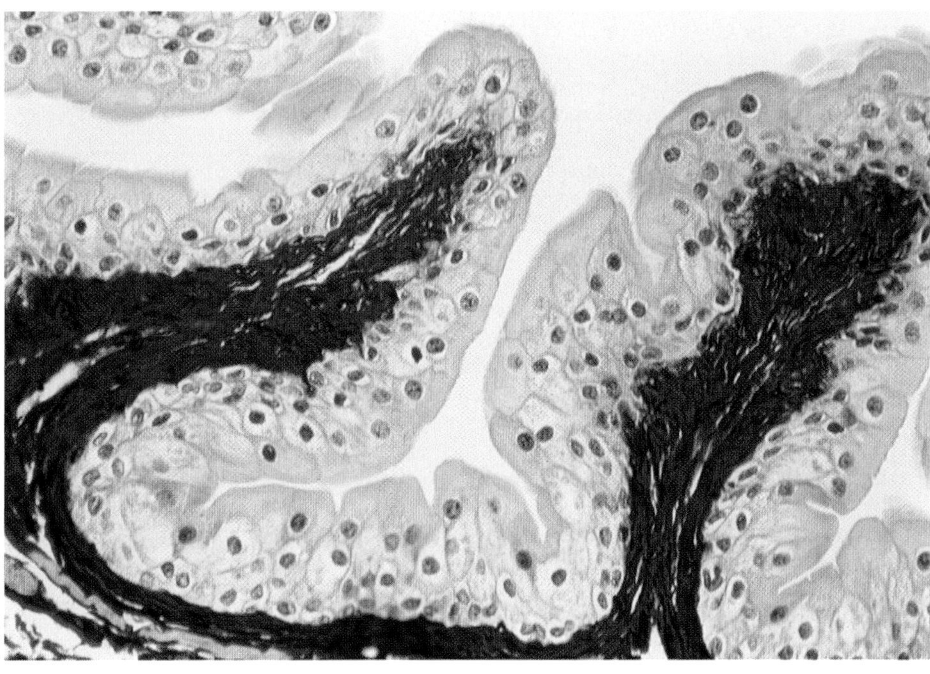

A

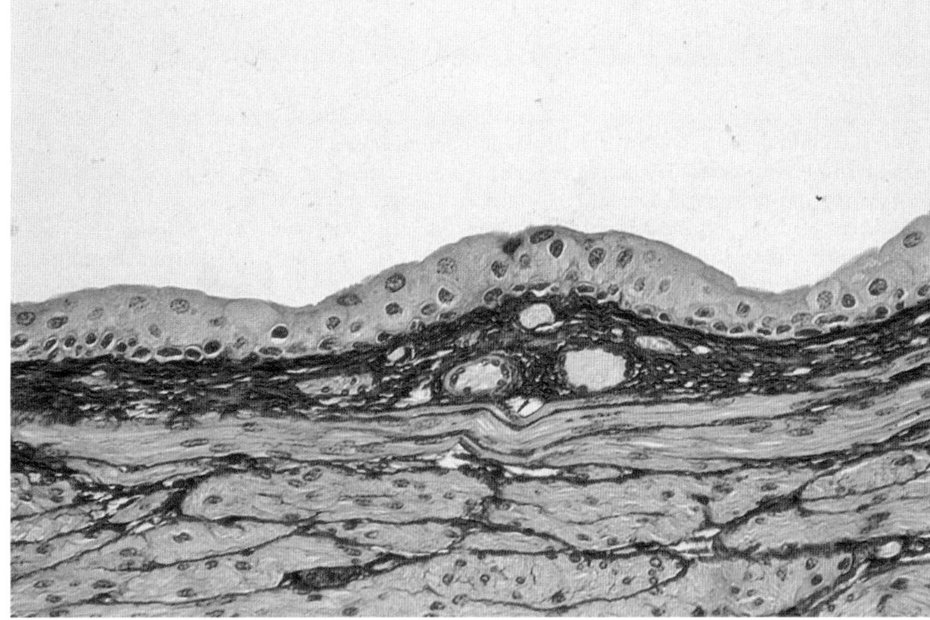

B

Figure 19–27. Compare the structure of the transitional epithelium when the urinary bladder is empty (**A**) or full (**B**). When the bladder is full, the capacity of epithelial cells to slide upon one another reduces the thickness of the epithelium. As a result, the interior surface of the bladder increases. In **B,** note the thin strands of collagen fibers separating bundles of smooth muscle cells. PSH stain. Medium magnification.

The transitional epithelium of the bladder in the undistended state is 5 or 6 cells in thickness; the superficial cells are rounded and bulge into the lumen. These cells are frequently polyploid or binucleate. When the epithelium is stretched, as when the bladder is full of urine, the epithelium is only 3 or 4 cells in thickness, and the superficial cells become squamous.

The superficial cells of the transitional epithelium have a special membrane of thick plates separated by narrow bands of thinner membrane that are responsible for the osmotic barrier between urine and tissue fluids. When the bladder contracts, the membrane folds along the thinner regions, and the thicker plates invaginate to form fusiform cytoplasmic vesicles. These vesicles represent a reservoir of these thick plates that can be stored in the cytoplasm of the cells of the empty bladder and used to cover the increased cell surface in the full bladder. This luminal membrane is assembled in the Golgi complex and has an unusual chemical composition; cerebroside is the major component of the polar lipid fraction.

The muscular layers in the calyces, renal pelvis, and ureters have a helical arrangement. As the ureteral muscle cells reach the bladder, they become longitudinal. The muscle fibers of the bladder run in every direction (without distinct layers) until they approach the bladder neck, where 3 distinct layers can be identified: The internal longitudinal layer, distal to the bladder neck, becomes circular around the prostatic urethra and the prostatic parenchyma in men. It extends to the external meatus in women. Its fibers form the true involuntary urethral sphincter. The middle layer ends at the bladder neck, and the outer longitudinal layer continues to the end of the prostate in men and to the external urethral meatus in women.

The ureters (Figure 19–28) pass through the wall of the bladder obliquely, forming a valve that prevents the backflow of urine. The intravesical ureter has only longitudinal muscle fibers.

The urinary passages are covered externally by an adventitial membrane, except for the upper part of the bladder, which is covered by serous peritoneum.

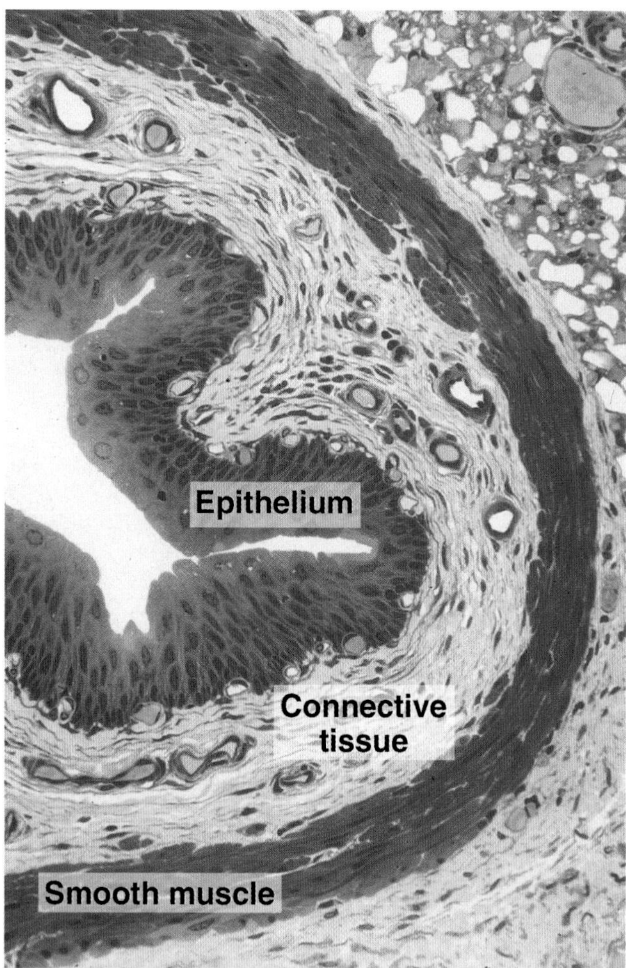

Figure 19–28. Photomicrograph showing the main components of the ureter, which consists of an inner layer of transitional epithelium, a highly vascularized connective tissue, a smooth muscle layer, and an outer layer of connective tissue. PT stain. Low magnification.

Urethra

The urethra is a tube that carries the urine from the bladder to the exterior. In men, sperm also pass through it during ejaculation. In women, the urethra is exclusively a urinary organ.

Male Urethra

The male urethra consists of 4 parts: **prostatic, membranous, bulbous,** and **pendulous.** The initial part of the urethra passes through the prostate (see Chapter 22), which is situated very close to the bladder, and the ducts that transport the secretions of the prostate open into the prostatic urethra.

In the dorsal and distal part of the **prostatic urethra,** there is an elevation, the **verumontanum** (from Latin, meaning mountain ridge), that protrudes into its interior. A closed tube called the prostatic utricle opens into the tip of the verumontanum; this tube has no known function. The ejaculatory ducts open on the sides of the verumontanum. The seminal fluid enters the proximal urethra through these ducts to be stored just before ejaculation. The prostatic urethra is lined with transitional epithelium.

The **membranous urethra** extends for only 1 cm and is lined with stratified or pseudostratified columnar epithelium. Surrounding this part of the urethra is a sphincter of striated muscle, the **external sphincter** of the urethra. The voluntary external striated sphincter adds further closing pressure to that exerted by the involuntary urethral sphincter. The latter is formed by the continuation of the internal longitudinal muscle of the bladder.

The **bulbous** and **pendulous** parts of the urethra are located in the **corpus spongiosum** of the penis. The urethral lumen dilates distally, forming the **fossa navicularis.** The epithelium of this portion of the urethra is mostly pseudostratified and columnar, with stratified and squamous areas.

Littre's glands are mucous glands found along the entire length of the urethra but mostly in the pendulous part. The secretory portions of some of these glands are directly linked to the epithelial lining of the urethra; others have excretory ducts.

Female Urethra

The female urethra is a tube 4–5 cm long, lined with stratified squamous epithelium and areas of pseudostratified columnar epithelium. The mid part of the female urethra is surrounded by an external striated voluntary sphincter.

REFERENCES

Barger AC, Herd JA: The renal circulation. N Engl J Med 1971;284:482.

Bulger RE, Dobyan DC: Recent advances in renal morphology. Annu Rev Physiol 1982;44:147.

Farquhar MG: The glomerular basement membrane: a selective macromolecular filter. In: Hay ED (editor): *Cell Biology of Extracellular Matrix.* Plenum Press, 1981.

Friis UG et al: Exocytosis and endocytosis in juxtaglomerular cells. Acta Physiol Scand 2001;168:95.

Ganong WF: Formation and excretion of urine. In: *Review of Medical Physiology,* 20th ed. McGraw-Hill, 2001.

Hicks RM: The mammalian urinary bladder: an accommodating organ. Biol Rev 1975;50:215.

Levy BJ, Wight TN: The role of proteoglycans in bladder structure and function. Adv Exp Med Biol 1995;385:191.

Maunsbach AB (editor): *Functional Ultrastructure of the Kidney.* Academic Press, 1981.

Staehelin LA et al: Luminal plasma membrane of the urinary bladder. 1. Three-dimensional reconstruction from freeze-etch images. J Cell Biol 1972;53:73.

Hypophysis 20

HORMONES

Hormones are molecules that function in the body as chemical signals. They are liberated by specialized cells that are called endocrine cells because they secrete "inward," as opposed to exocrine cells, which secrete into a body cavity or toward the body surface. Endocrine cells usually aggregate as endocrine glands, where they typically arrange themselves as cords of cells. A notable exception is the thyroid gland, in which the cells are organized as follicles. Besides the glands, there are many isolated endocrine cells in the body, such as the endocrine cells of the digestive tract. The endocrine cells are always very close to blood capillaries, which receive the secreted hormones and distribute them throughout the organism. Thus, many hormones act at a distance from the site of their secretion. Many endocrine cells, however, produce hormones that act at a short distance. This is called **paracrine** secretion. These hormones may reach their site of action through short loops of blood vessels. A good example of paracrine secretion is the hormone gastrin, which is released by the G cells located mainly in the pylorus. Gastrin reaches the fundic glands, stimulating the production of hydrochloric acid. Another method of secretion is the **juxtacrine** secretion, by which a molecule is released into the extracellular tissue, diffuses through the matrix, and acts on cells at a very short distance. In islets of Langerhans, the inhibition of insulin secretion by somatostatin, produced by the same islet, is an example of juxtacrine secretion. In **autocrine** secretion, cells may produce molecules that act on themselves or on cells of the same type. Insulin-like growth factor (IGF) produced by several cell types may act on the same cells that produced it.

The tissues and organs on which the hormones act are called **target tissues** or **target organs.** They react because their cells have receptors that specifically recognize and respond to the hormones. Because of this, the hormones may circulate in the blood and do not indiscriminately influence all cells of the body. Another advantage of receptors is that target cells respond to the respective hormones even when they are present in very small concentrations in the blood, which typically is the case. Endocrine glands are also target organs. In this way, the body is able to control hormone secretion through a mechanism of feedback and to keep blood hormonal levels within strict limits.

HYPOPHYSIS

The **hypophysis** (Gr. *hypo,* under, + *physis,* growth), or **pituitary gland,** weighs about 0.5 g, and its normal dimensions in humans are about 10 × 13 × 6 mm. It lies in a cavity of the sphenoid bone—**the sella turcica—**an important radiologic landmark. During embryogenesis, the hypophysis develops partly from oral ectoderm and partly from nerve tissue. The neural component arises as an evagination from the floor of the diencephalon and grows caudally as a stalk without detaching itself from the brain. The oral component arises as an outpocketing of ectoderm from the roof of the primitive mouth of the embryo and grows cranially, forming a structure called **Rathke's pouch.** Later, a constriction at the base of this pouch separates it from the oral cavity. At the same time, its anterior wall thickens, reducing the lumen of Rathke's pouch to a small fissure (Figure 20–1).

Because of its dual origin, the hypophysis actually consists of 2 glands—the **neurohypophysis** and the **adenohypophysis—** that are united anatomically but that have different functions. The **neurohypophysis,** the part of the hypophysis that develops from nerve tissue, consists of a large portion, the **pars nervosa,** and the smaller **infundibulum,** or **neural stalk** (Figure 20–2). The neural stalk is composed of the stem and median eminence. The part of the hypophysis that arises from oral ectoderm is known as the **adenohypophysis** and is subdivided into 3 portions: a large **pars distalis,** or **anterior lobe;** a cranial part, the **pars tuberalis,** which surrounds the neural stalk; and the **pars intermedia** (Figures 20–1 and 20–2).

Blood Supply

To understand the functioning of the hypophysis, it is important to first study its blood supply. The blood supply of the hypophysis derives from 2 groups of blood vessels that come from the internal carotid artery. From above, the right and left **superior hypophyseal arteries** supply the median eminence and the neural stalk; from below, the right and left **inferior hypophyseal arteries** provide blood mainly for the neurohypophysis, with a small supply to the stalk. The superior hypophyseal arteries form a **primary capillary plexus** of fenestrated capillaries that irrigate the stalk and median eminence. They then rejoin to form veins that develop a **secondary capillary plexus** in the adenohypophysis (Figure 20–2). This **hypophyseal portal system** is of utmost importance because it carries neurohormones that control the function of the cells of the adenohypophysis from the median eminence to the adenohypophysis.

The Hypothalamo-Hypophyseal System

Because of its embryologic origin, the hypophysis is connected to the hypothalamus at the base of the brain, with which it has important anatomic and functional relationships.

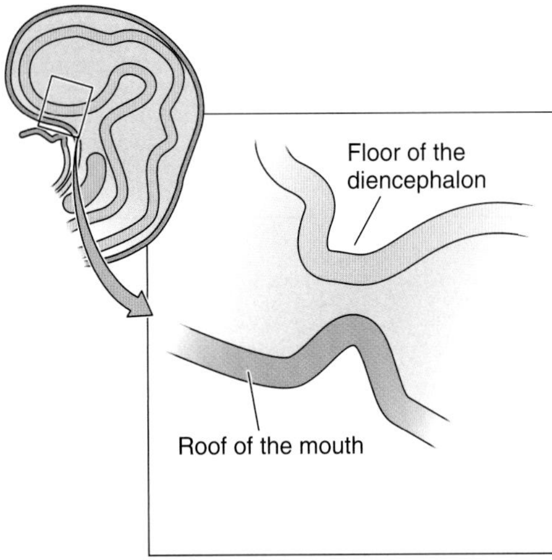

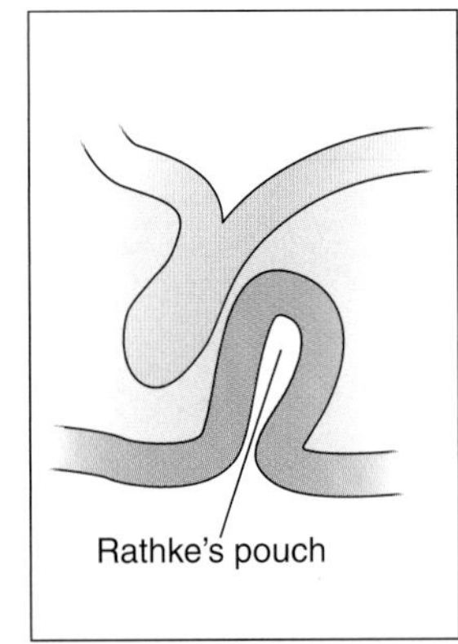

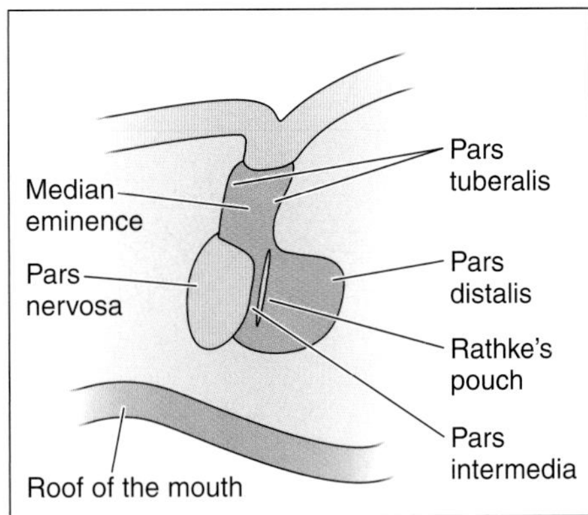

Figure 20–1. Development of the adenohypophysis and the neurohypophysis from the ectoderm of the roof of the mouth and from the floor of the diencephalon.

In the hypothalamo-hypophyseal system there are 3 known sites of production of hormones that liberate 3 groups of hormones:

1. The first group consists of peptides produced by aggregates (nuclei) of secretory neurons in the hypothalamus: the supraoptic and the paraventricular nuclei. The hormones are transported along the axons of these neurons and accumulate in the ends of these axons, which are situated in the neurohypophysis (Figure 20–2).
2. The second group of hormones (peptides) is produced by neurons of the dorsal medial, ventral medial, and infundibular nuclei of the hypothalamus. These hormones are carried along axons until they end in the median eminence where the hormones are stored and liberated. These hormones enter the blood capillaries of the median eminence and are transported to the adenohypophysis through the first stretch of the hypophyseal portal system (Figure 20–2).
3. The third group of hormones consists of proteins and glycoproteins produced by cells of the pars distalis and liberated into blood capillaries of the second stretch of the portal system. These capillaries surround the secretory cells and distribute the hormones to the general circulation (Figure 20–2).

ADENOHYPOPHYSIS

Pars Distalis

The main components of the pars distalis are cords of epithelial cells interspersed with capillaries (Figures 20–3 and 20–4). The hormones produced by these cells are stored as secretory granules (Figure 20–5 and 20–6). The few fibroblasts that are present produce reticular fibers that support the cords of hormone-secreting cells. The pars distalis accounts for 75% of the mass of the hypophysis. Common stains allow the recognition of 3 cell types in the pars distalis: **chromophobes** (Gr. *chroma,* color, + *phobos,* fear) and 2 types of **chromophils** (Gr. *chroma* + *philein,* to love) called basophils and acidophils according to their affinity for basic and acid dyes, respectively (Figure 20–6). The subtypes of basophil and acidophil cells are named for the hormones they produce (Table 20–1). Chromophobes do not stain intensely and, when observed with an electron microscope, show 2 populations of cells. One has few secretory granules, and the other has none. The group with no secretory granules probably contains undifferentiated cells and follicular cells. The long branching processes of follicular cells form a supporting network for the other cells. With the exception of the gonadotropic cell, which produces 2 hormones, the other cells produce only a single hormone. Many dyes have been used in attempts to distinguish the 5 types of hormone-secreting cells, but with little success. Immunocytochemical methods and electron microscopy are currently the only reliable techniques to distinguish these cell types (Figure 20–5). The hormones produced by the hypophysis have widespread physiologic activity (Figure 20–7 and Table 20–1); they regulate almost all other endocrine glands, the secretion of milk, and the metabolism of muscle, bone, and adipose tissue.

Control of the Pars Distalis

The activities of the cells of the pars distalis are controlled by more than one mechanism. The main mechanism uses the peptide hormones (Table 20–2) produced in the hypothalamic aggregates of neurosecretory cells and stored in the median eminence. Most of these hormones are called **hypothalamic releasing hormones;** when liberated, they are transported to the pars distalis through the capillary plexuses (Figure 20–2). Two of these hormones, which act on specific cells of the pars distalis, inhibit the release of hormones (**hypothalamic inhibiting hormones;** see Table 20–1). Because of the strategic position of the hypothalamic neurons and the control they exert on the hypophysis and therefore on many bodily functions, many exterior stimuli as well as stimuli arising in the brain may affect the function of the hypophysis and consequently the function of many organs and tissues.

A second control mechanism is the direct effect of hormones from stimulated endocrine cells on the release of peptides from the median eminence and the pars distalis (Figure 20–7). Figure 20–8 also illustrates these mechanisms, using the thyroid as an example, and shows the complex chain of events that begin with the action of neurons on neurosecretory cells of the hypothalamic nuclei and end on the effector cells with the action of the last

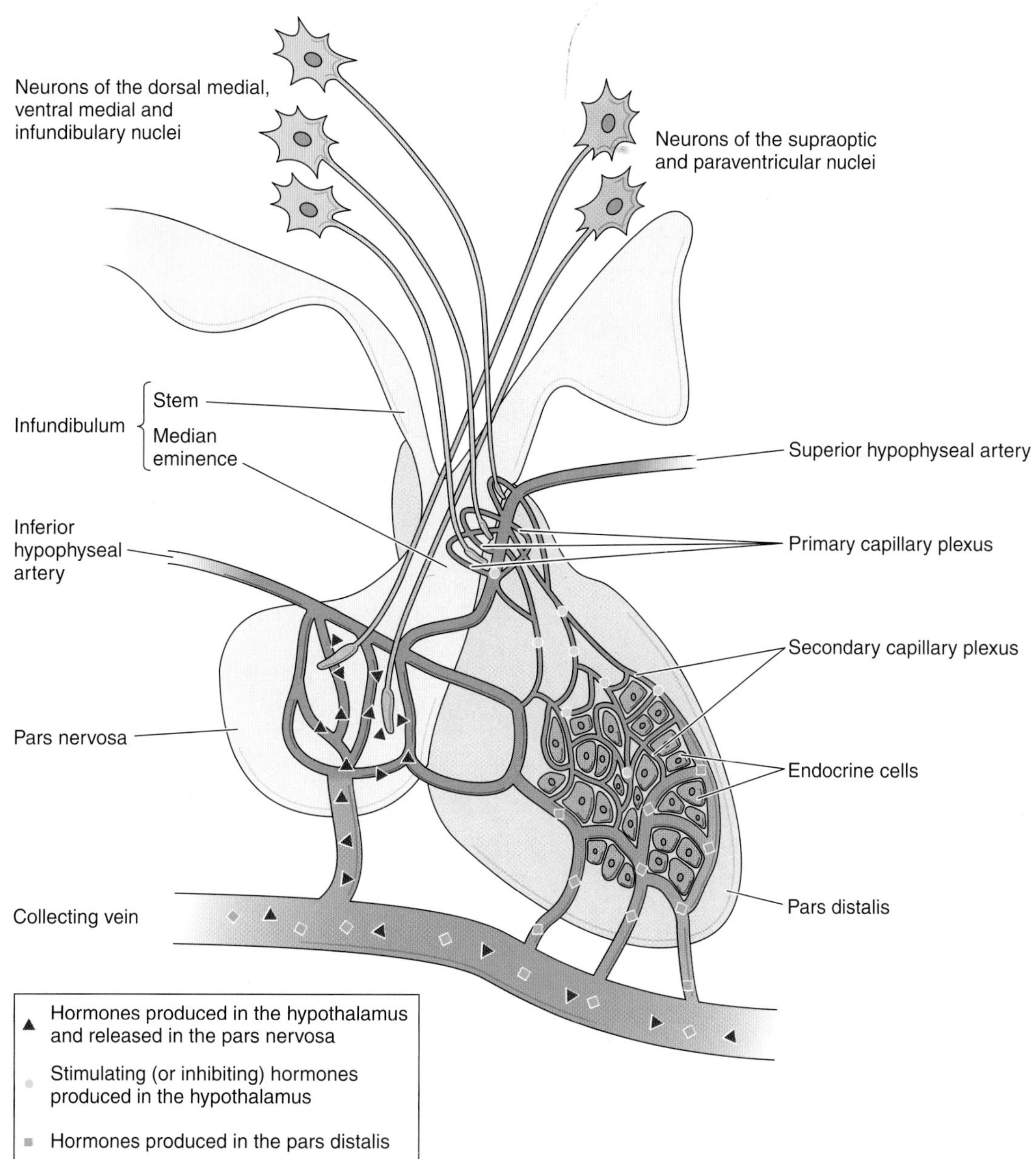

Figure 20–2. The hypothalamo-hypophyseal system, with its vascularization and sites of hormone production, storage, and release.

hormone in the sequence. Nonhormonal molecules, such as inhibin and activin, which are peptides and members of the transforming growth factor β family produced in the gonads, control secretion of follicle-stimulating hormone. All these mechanisms allow the fine-tuning of the secretion of hormones by cells of the pars distalis.

Pars Tuberalis

The pars tuberalis is a funnel-shaped region surrounding the infundibulum of the neurohypophysis (Figure 20–2). Most of the cells of the pars tuberalis secrete gonadotropins (follicle-stimulating hormone and luteinizing hormone) and are arranged in cords alongside the blood vessels.

Pars Intermedia

The pars intermedia, which develops from the dorsal portion of Rathke's pouch (Figures 20–1 and 20–3), is, in humans, a rudimentary region made up of cords and follicles of weakly basophilic cells that contain small secretory granules. The function of these cells is not known.

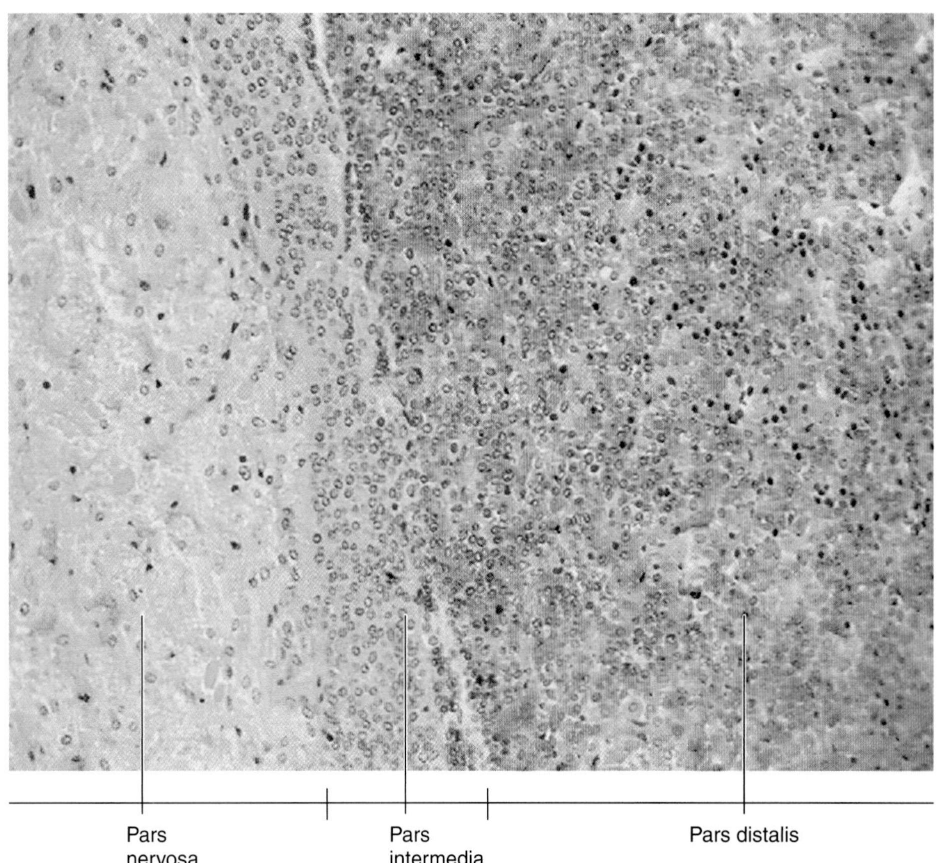

Figure 20–3. Section of a hypophysis showing the pars nervosa, the pars intermedia, and the pars distalis. Yellow-stained erythrocytes show the arrangement of blood vessels. Mallory's trichrome stain. Low magnification.

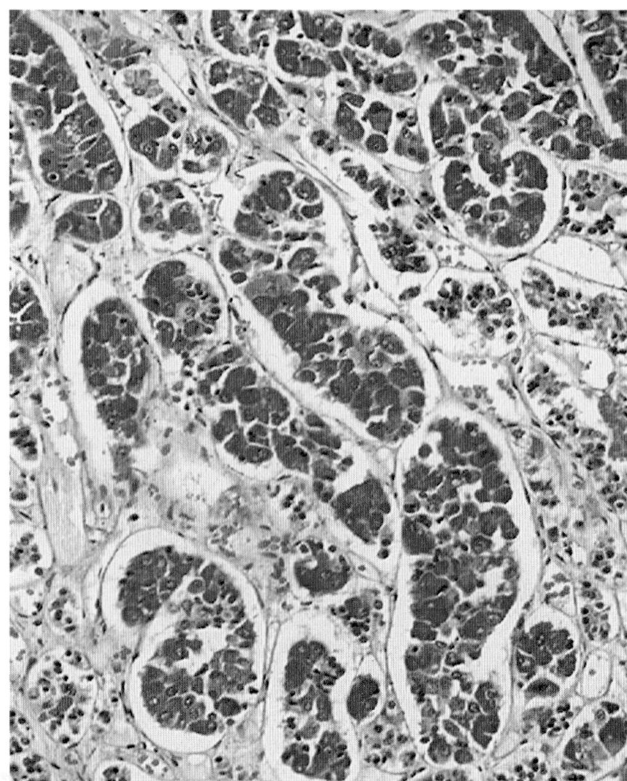

Figure 20–4. In the pars distalis the endocrine cells are organized as cords. Gomori's trichrome stain. Low magnification.

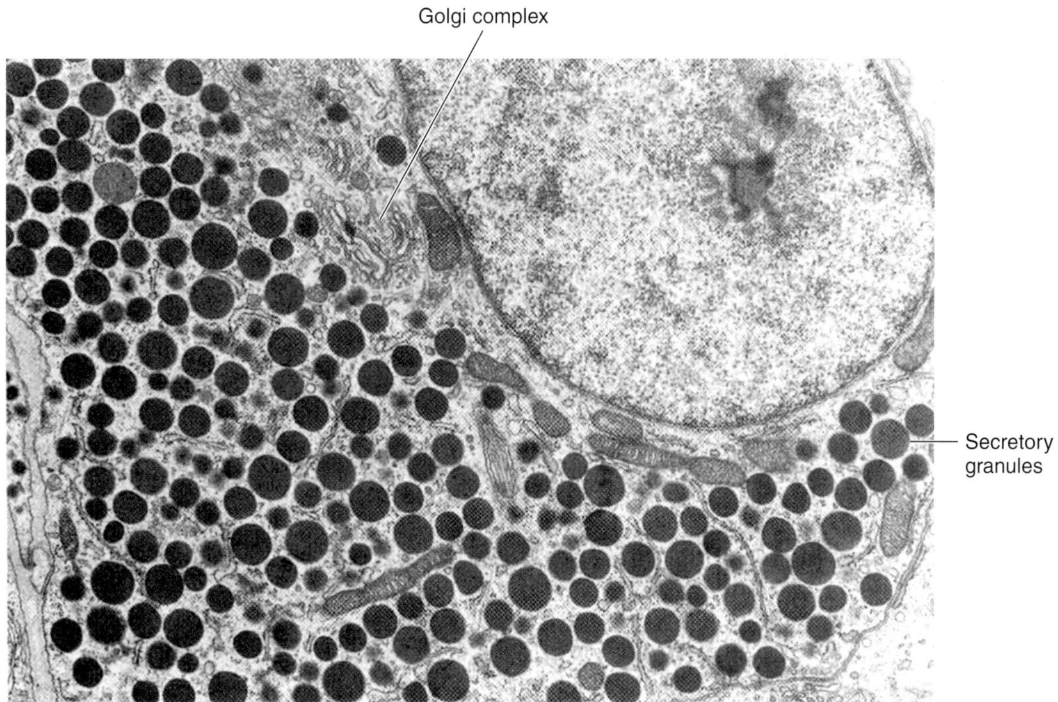

Figure 20–5. Electron micrograph of a somatotroph (growth hormone–secreting cell) of a cat anterior hypophysis. Note the numerous secretory granules, long mitochondria, cisternae of rough endoplasmic reticulum, and Golgi complex. ×10,270.

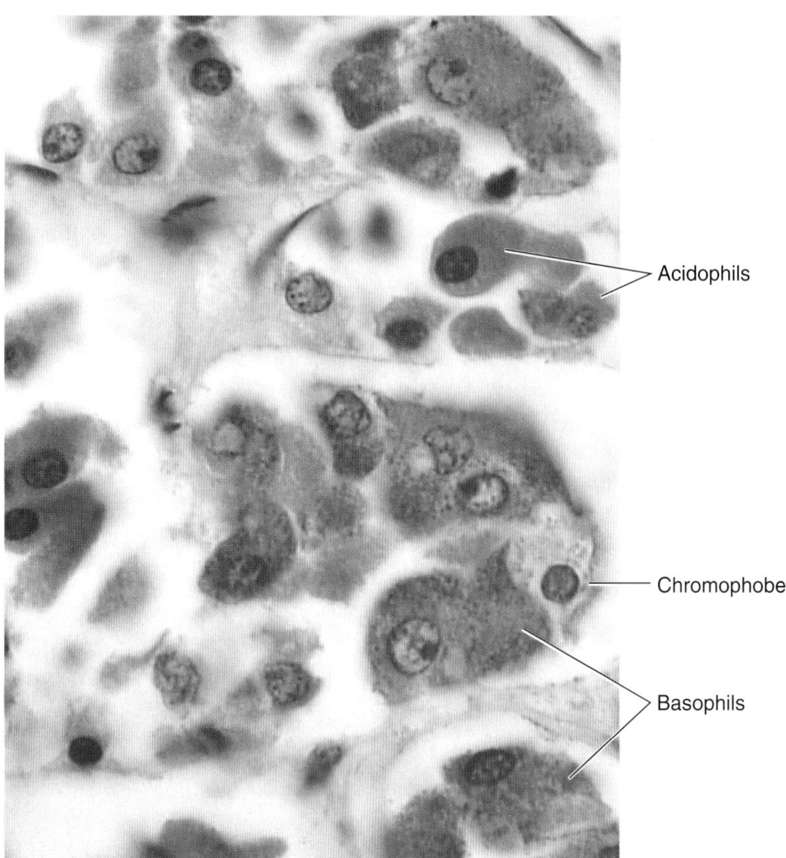

Figure 20–6. Some stains allow the recognition of cell types of the pars distalis: chromophils (acidophils and basophils) and chromophobes. Gomori's trichrome stain. High magnification.

Table 20–1. Secretory cells of the pars distalis.

Cell Type	Stain Affinity	Hormone Produced	Main Physiologic Activity	Secretory Granules in Humans	Hypothalamic Releasing Hormones	Hypothalamic Inhibiting Hormones
Somatotropic cell	Acidophilic	Somatotropin (growth hormone)	Acts on growth of long bones via somatomedins synthesized in liver	Numerous, round or oval; 300 to 400 nm in diameter	Somatotropin-releasing hormone (SRH)	Somatostatin
Mammotropic cell	Acidophilic	Prolactin	Promotes milk secretion	200 nm; increases in size during pregnancy and lactation (600 nm in diameter)	Prolactin-releasing hormone (PRH)	Prolactin-inhibiting hormone (PIH)
Gonadotropic cell	Basophilic	Follicle-stimulating hormone (FSH) and luteinizing hormone (LH) in the same cell type	FSH promotes ovarian follicle development and estrogen secretion in women and stimulates spermatogenesis in men. LH promotes ovarian follicle maturation and progesterone secretion in women and Leydig cell stimulation and androgen secretion in men.	250 to 400 nm diameter	Gonadotropin-releasing hormone (GnRH). May be two releasing hormones: FRH (follicle-releasing) and LRH (lutein-releasing)	
Thyrotropic cell	Basophilic	Thyrotropin (TSH)	Stimulates thyroid hormone synthesis, storage, and liberation	Small granules, 120 to 200 nm in diameter	Thyrotropin-releasing hormone (TRH)	
Corticotropic cell	Basophilic	Corticotropin (ACTH)	Stimulates secretion of adrenal cortex hormones	Large granules, 400 to 550 nm in diameter	Corticotropin-releasing hormone (CRH)	

NEUROHYPOPHYSIS

The neurohypophysis consists of the pars nervosa and the neural stalk. The pars nervosa, unlike the adenohypophysis, does not contain secretory cells. It is composed of some 100,000 unmyelinated axons of secretory neurons situated in the supraoptic and paraventricular nuclei (Figure 20–2). The secretory neurons have all the characteristics of typical neurons, including the ability to conduct an action potential, but have more developed Nissl bodies related to the production of the neurosecretory material. The neurosecretions (which can be studied by specific techniques such as Gomori's chrome hematoxylin stain) are transported along the axons and accumulate at their endings in the pars nervosa. Here they form structures known as **Herring bodies,** that are visible in the light microscope (Figure 20–9). The electron microscope reveals that the Herring bodies contain neurosecretory granules that have a diameter of 100–200 nm and are surrounded by a membrane. The granules are released and enter the fenestrated capillaries that exist in large numbers in the pars nervosa; the hormones are then distributed to the general circulation.

The **neurosecretory material** consists of 2 hormones, both cyclic peptides made up of 9 amino acids. The hormones have a slightly different amino acid composition, which results in very different functions. They are **arginine vasopressin**—also called **antidiuretic hormone**—and **oxytocin.** Each hormone is joined to a binding protein (**neurophysin**). The hormone-neurophysin complex is synthesized as a single long peptide. Proteolysis of the precursor yields the hormone and its specific binding protein. Vasopressin and oxytocin are stored in the neurohypophysis and released into the blood by impulses in the nerve fibers from the hypothalamus. Although there is some overlap, the fibers from supraoptic nuclei are mainly concerned with vasopressin secretion, whereas most of the fibers from the paraventricular nuclei are concerned with oxytocin secretion.

Cells of the Neurohypophysis

Although the neurohypophysis consists mainly of axons from hypothalamic neurons, about 25% of the volume of this structure consists of a specific type of highly branched glial cell called a **pituicyte** (Figure 20–9).

Actions of the Hormones of the Neurohypophysis

Vasopressin or antidiuretic hormone is secreted whenever the osmotic pressure of the blood increases. The blood then acts

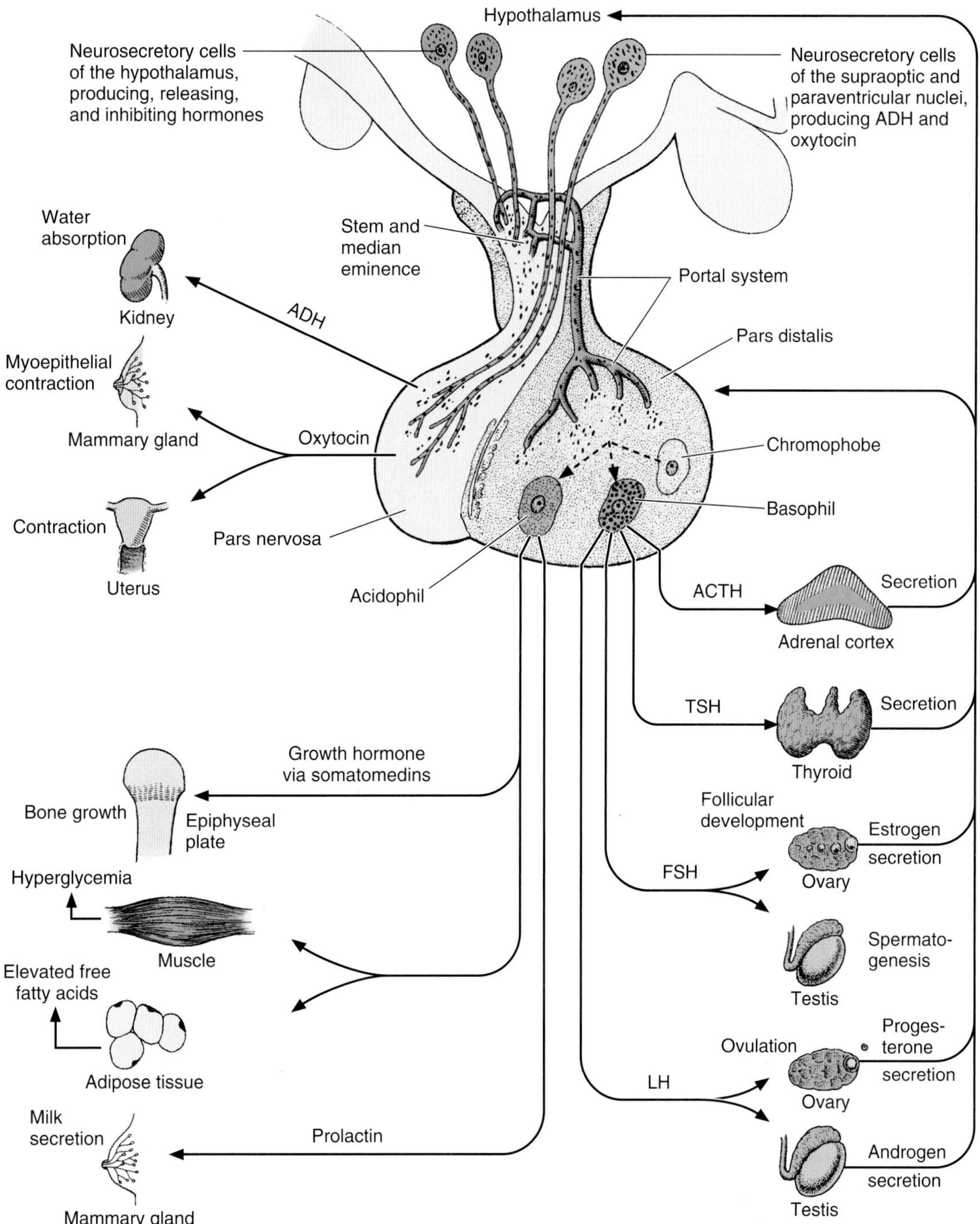

Figure 20–7. The effects of various hypophyseal hormones on target organs and the feedback mechanisms that control their secretion. For definitions of abbreviations, see Tables 20–1 and 20–2.

on osmoreceptor cells in the anterior hypothalamus, stimulating the secretion of the hormone from supraoptic neurons. Its main effect is to increase the permeability to water of the collecting tubules of the kidney. As a result, water is absorbed by these tubules, and urine becomes hypertonic. Thus, vasopressin helps to regulate the osmotic balance of the internal milieu. In large doses, vasopressin promotes the contraction of smooth muscle of blood vessels, raising the blood pressure. It acts mainly on the muscle layers of small arteries and arterioles. It is doubtful if the amount of endogenous vasopressin secreted is sufficient to exert any appreciable effect on blood pressure homeostasis.

Table 20–2. Hormones of the neurohypophysis.

Hypothalamus		Pars Nervosa	
Hormone	**Function**	**Hormone**	**Function**
Thyrotropin-releasing hormone (TRH)	Stimulates release of thyrotropin and prolactin	Vasopressin/ antidiuretic hormone (ADH)	Increases water permeability of kidney collecting ducts and promotes vascular smooth muscle contraction
Gonadotropin-releasing hormone (GnRH)	Stimulates the release of both follicle-stimulating hormone and luteinizing hormone	Oxytocin	Acts on contraction of uterine smooth muscle and the myoepithelial cells of the mammary gland
Somatostatin	Inhibits release of both growth hormone and thyrotropin		
Growth hormone–releasing hormone (GRH)	Stimulates release of growth hormone		
Prolactin-inhibiting hormone (PIH) Dopamine	Inhibits release of prolactin		
Corticotropin-releasing hormone (CRH)	Stimulates release of both B lipotropin and corticotropin		

MEDICAL APPLICATION

Oxytocin stimulates contraction of the smooth muscle of the uterine wall during copulation and childbirth and of the myoepithelial cells that surround the alveoli and ducts of the mammary glands during nursing. The secretion of oxytocin is stimulated by distention of the vagina or of the uterine cervix and by nursing. This occurs via nerve tracts that act on the hypothalamus. The neurohormonal reflex triggered by nursing is called the ***milk-ejection reflex*** *(Figure 20–7).*

Lesions of the hypothalamus, which destroy the neurosecretory cells that produce antidiuretic hormone, cause ***diabetes insipidus,*** *a disease characterized by loss of renal capacity to concentrate urine. As a result, an individual suffering from this disease may excrete up to 20 liters of urine per day (polyuria) and will drink enormous quantities of liquids.*

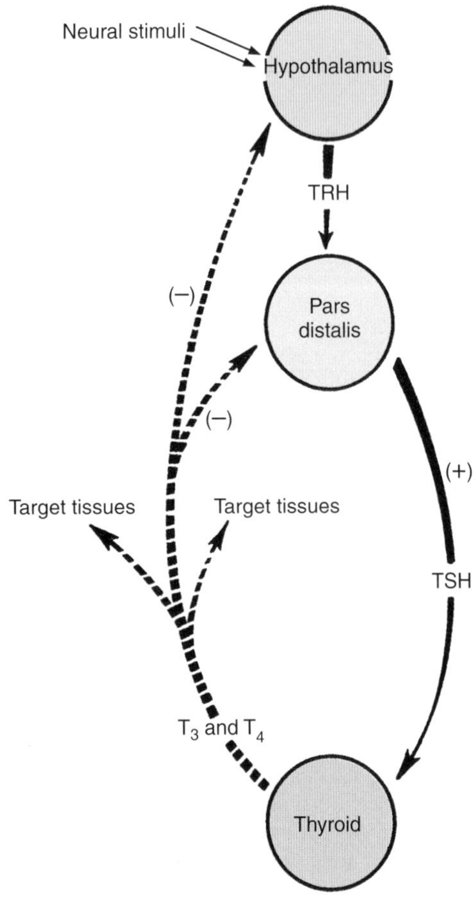

Figure 20–8. Relationship between the hypothalamus, the hypophysis, and the thyroid. Thyrotropin-releasing hormone (TRH) promotes secretion of thyrotropin (TSH), which regulates the synthesis and secretion of the hormones T_3 and T_4. In addition to their effect on target tissues and organs, these hormones regulate TSH and TRH secretion from the pars distalis and the hypothalamus by a negative-feedback. Solid arrows indicate stimulation; dashed arrows, inhibition.

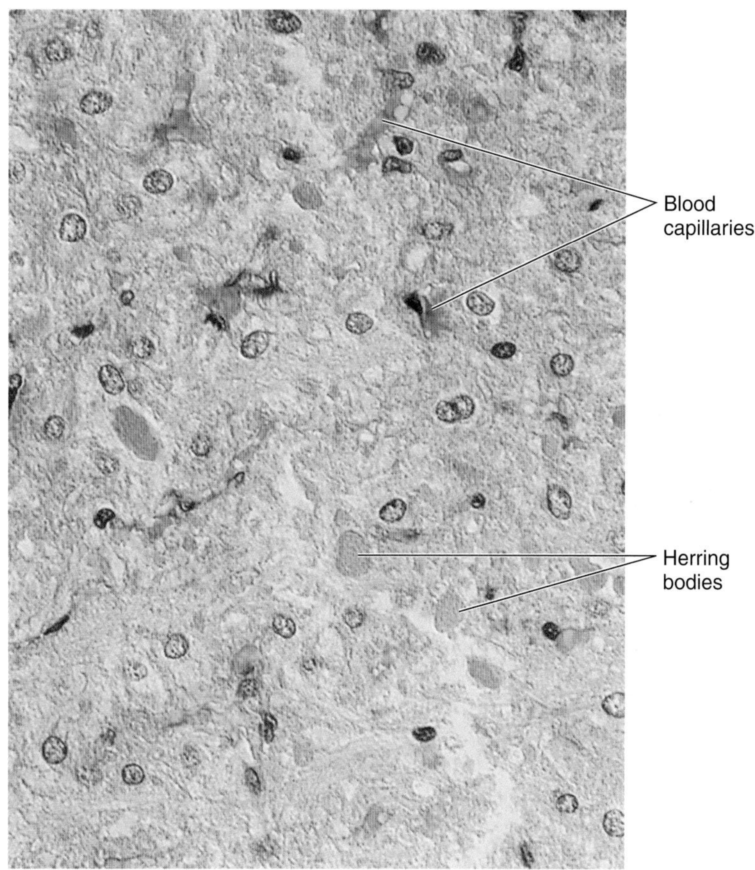

Figure 20–9. Section of the pars nervosa mechanism. Most of the tissue is formed by axons. Herring bodies and nuclei of pituicytes can be seen as well as erythrocytes (in yellow) within blood capillaries. Mallory's trichrome stain. High magnification.

Tumors of the Hypophysis

MEDICAL APPLICATION

Tumors of the hypophysis are usually benign. About two thirds of them produce hormones that cause clinical symptoms. These tumors can produce growth hormone, prolactin, adrenocorticotropin, and, less frequently, thyroid-stimulating hormone. Clinical diagnosis of these tumors can be confirmed by immunocytochemical methods after surgical removal.

REFERENCES

Bhatnagar AS (editor): *The Anterior Pituitary Gland.* Raven Press, 1983.

Braunwald E et al: *Harrison's Principles of Internal Medicine,* 15th ed. McGraw-Hill, 2001.

Brownstein MJ et al: Synthesis, transport, and release of posterior pituitary hormones. Science 1980;207:373.

Cross BA, Leng G (editors): The neurohypophysis; structure, function and control. Prog Brain Res 1982;60:3.

Daniel PM: The blood supply of the hypothalamus and pituitary gland. Br Med Bull 1966;22:202.

Pelletier G et al: Identification of human anterior pituitary cells by immunoelectron microscopy. J Clin Endocrinol Metab 1978;46:534.

Phifer RF et al: Immunohistologic and histologic evidence that follicle-stimulating hormone and luteinizing hormone are present in the same cell type in the human pars distalis. J Clin Endocrinol Metab 1973;36:125.

Phifer RF et al: Specific demonstration of the human hypophyseal cells which produce adrenocorticotropic hormone. J Clin Endocrinol 1970;31:347.

Reichlin S (editor): *The Neurohypophysis: Physiological and Clinical Aspects.* Plenum, 1984.

Adrenals, Islets of Langerhans, Thyroid, Parathyroids, & Pineal Gland

21

ADRENAL (SUPRARENAL) GLANDS

The adrenal glands are paired organs that lie near the superior poles of the kidneys, embedded in adipose tissue (Figure 21–1). They are flattened structures with a half-moon shape; in the human, they are about 4–6 cm long, 1–2 cm wide, and 4–6 mm thick. Together they weigh about 8 g, but their weight and size vary with the age and physiologic condition of the individual. Examination of a fresh section of adrenal gland shows it to be covered by a capsule of dense collagenous connective tissue. The gland consists of 2 concentric layers: a yellow peripheral layer, the **adrenal cortex;** and a reddish-brown central layer, the **adrenal medulla** (Figures 21–1 and 21–2).

The adrenal cortex and the adrenal medulla can be considered two organs with distinct origins, functions, and morphologic characteristics that become united during embryonic development. They arise from different germ layers. The cortex arises from the coelomic intermediate mesoderm; the medulla consists of cells derived from the neural crest, from which sympathetic ganglion cells also originate. The general histologic appearance of the adrenal gland is typical of an endocrine gland in which cells of both cortex and medulla are grouped in cords along capillaries (see Chapter 4).

The collagenous connective tissue capsule that covers the adrenal gland sends thin septa to the interior of the gland as trabeculae. The stroma consists mainly of a rich network of reticular fibers that support the secretory cells.

Blood Supply

The adrenal glands are supplied by several arteries that enter at various points around their periphery (Figure 21–2). The arterial branches form a subcapsular plexus from which arise 3 groups of vessels: arteries of the capsule; arteries of the cortex, which branch repeatedly between the gland cells, forming capillaries that drain into medullary capillaries; and arteries of the medulla, which pass through the cortex before breaking up to form part of the extensive capillary network of the medulla (Figure 21–2).

A dual blood supply thus provides the medulla with both arterial (via **medullary arteries**) and venous (via **cortical arteries**) blood. The capillary endothelium is extremely attenuated and interrupted by small fenestrae that are closed by thin diaphragms. A continuous basal lamina is present beneath the endothelium. Capillaries of the medulla, together with capillaries that supply the cortex, form the medullary veins, which join to constitute the **adrenal** or **suprarenal vein** (Figure 21–2).

Adrenal Cortex

Because of the differences in disposition and appearance of its cells, the adrenal cortex can be subdivided into 3 concentric layers whose limits are usually not sharply defined in humans (Figure 21–3): the **zona glomerulosa,** the **zona fasciculata,** and the **zona reticularis.** These layers occupy 15%, 65%, and 7%, respectively, of the total volume of the adrenal glands.

The layer immediately beneath the connective tissue capsule is the zona glomerulosa, in which columnar or pyramidal cells

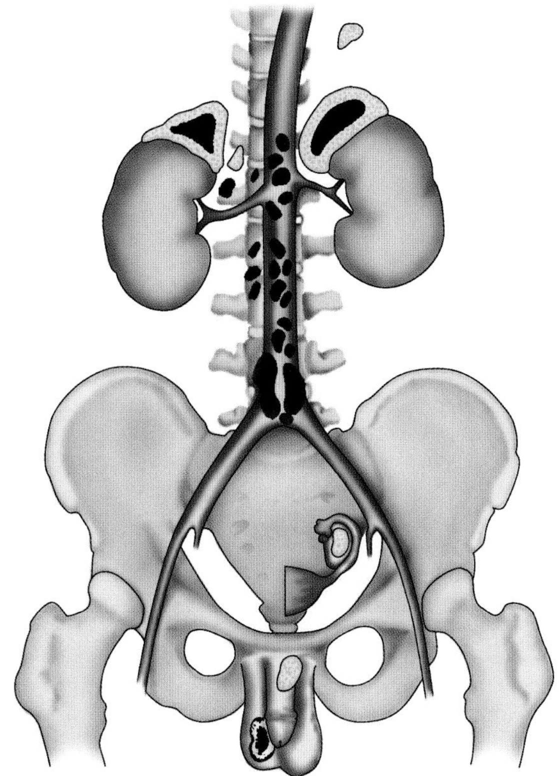

Figure 21–1. Human adrenal glands. Adrenocortical tissue is shown stippled; adrenal medullary tissue is shown black. Note the location of the adrenal glands at the superior pole of each kidney. Also shown are extra-adrenal sites where cortical and medullary tissues are sometimes found. (Reproduced, with permission, from Forsham in: *Textbook of Endocrinology,* 4th ed. Williams RH [editor]. Saunders, 1968.)

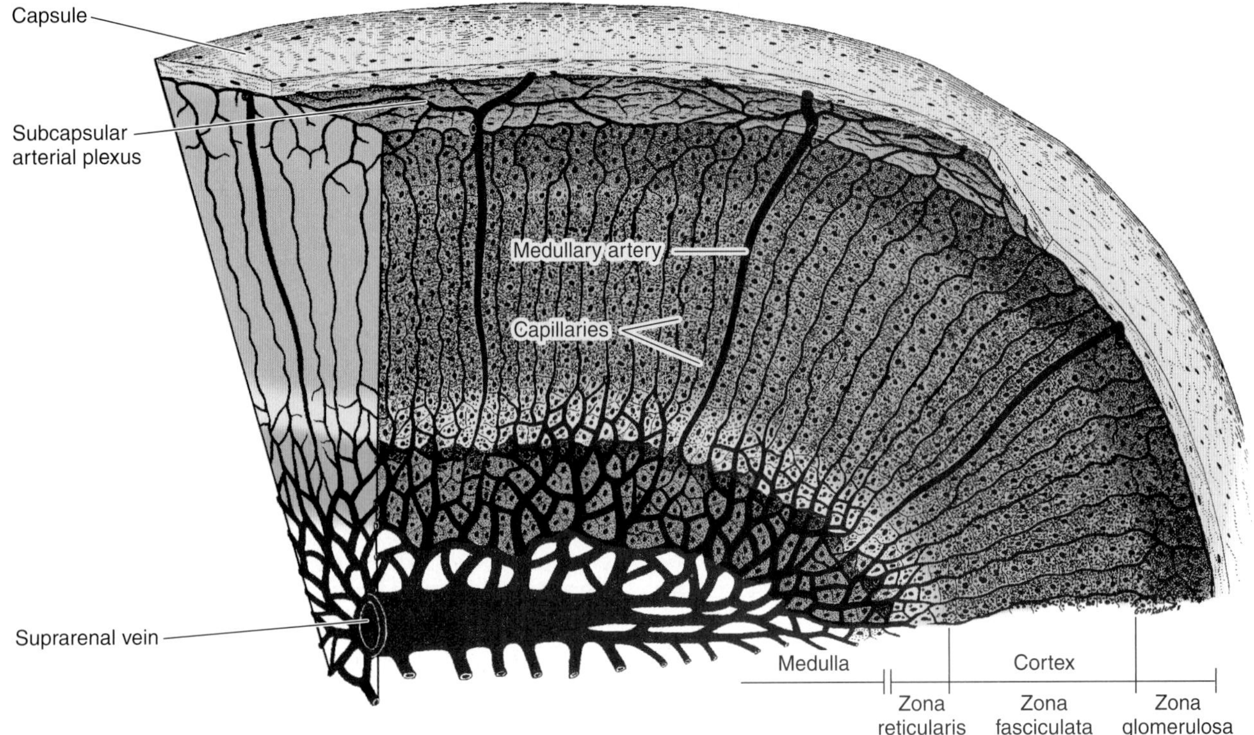

Figure 21–2. General architecture and blood circulation of the adrenal gland.

are arranged in closely packed, rounded, or arched cords surrounded by capillaries (Figure 21–3A and B).

The next layer of cells is known as the zona fasciculata because of the arrangement of the cells in straight cords, one or two cells thick, that run at right angles to the surface of the organ and have capillaries between them (Figure 21–3C). The cells of the zona fasciculata are polyhedral, with a great number of lipid droplets in their cytoplasm. As a result of the dissolution of the lipids during tissue preparation, the fasciculata cells appear vacuolated in common histologic preparations. Because of their vacuolization, the cells of the fasciculate are also called **spongyocytes.**

The zona reticularis (Figure 21–3D), the innermost layer of the cortex, lies between the zona fasciculata and the medulla; it contains cells disposed in irregular cords that form an anastomosing network. These cells are smaller than those of the other two layers. Lipofuscin pigment granules in the cells are large and quite numerous. Irregularly shaped cells with pyknotic nuclei—suggesting cell death—are often found in this layer.

Cells of the adrenal cortex do not store their secretory products in granules; rather, they synthesize and secrete steroid hormones only upon demand. Steroids, being low-molecular-weight lipid-soluble molecules, can freely diffuse through the plasma membrane and do not require the specialized process of exocytosis for their release. Cells of the adrenal cortex (Figure 21–4) have the typical ultrastructure of steroid-secreting cells (see Chapter 4).

Cortical Hormones & Their Actions

The steroids secreted by the cortex can be divided into 3 groups, according to their main physiologic actions: **glucocorticoids, mineralocorticoids,** and **androgens** (Figure 21–5). The zona glomerulosa secretes mineralocorticoids, primarily aldosterone, that maintain electrolyte (eg, sodium and potassium) and water balance. The zona fasciculata and probably the zona reticularis secrete the glucocorticoids cortisone and cortisol or, in some animals, corticosterone; these glucocorticoids regulate carbohydrate, protein, and fat metabolism. These zones also produce androgens (mainly dehydroepiandrosterone) and perhaps estrogens in small amounts.

The synthesis of cholesterol from acetate takes place in smooth endoplasmic reticulum, and the conversion of cholesterol to pregnenolone takes place in the mitochondria. The enzymes associated with the synthesis of progesterone and deoxycorticosterone from pregnenolone are found in smooth endoplasmic reticulum; those enzymes that convert deoxycorticosterone to aldosterone are located in mitochondria—a clear example of collaboration between two cell organelles.

The **glucocorticoids,** mainly cortisol and corticosterone, exert a profound effect on the metabolism of carbohydrates, as well as on that of proteins and lipids. In the liver, glucocorticoids promote the uptake and use of fatty acids (energy source), amino acids (enzyme synthesis), and carbohydrates (glucose synthesis). Glucocorticoids also stimulate the synthesis of glycogen from noncarbohydrate precursors, a process called **glyconeogenesis,** and the assembly of glucose molecules into glycogen, a synthesis named **glycogenesis.** Glucocorticoids can stimulate the synthesis of so much glucose that the resulting high levels in the blood produce a condition similar to diabetes mellitus. Outside the liver, however, glucocorticoids induce an opposite, or catabolic, effect on peripheral organs (eg, skin, muscle, adipose tissue). In these structures, glucocorticoids not only decrease synthetic activity but also promote protein and lipid degradation. The by-products of degradation, amino and fatty acids, are removed from the blood and used by the hepatocytes.

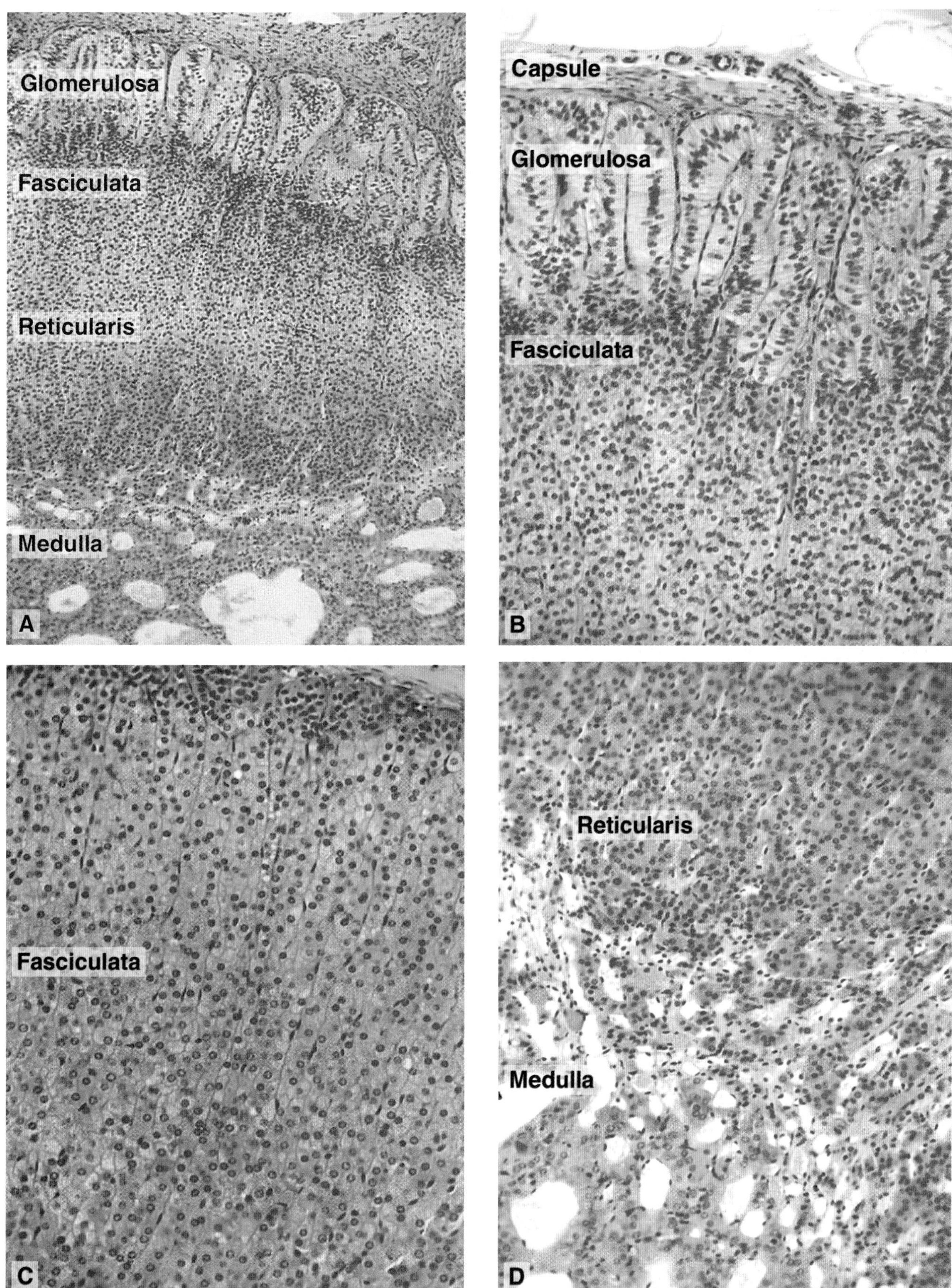

Figure 21–3. Photomicrographs of several regions of the adrenal cortex. **A:** A low-power general view showing the gland's layers. Low magnification. **B:** The capsule, the zona glomerulosa, and the beginning of the zona fasciculata. One of the arcuate cords of this zona is delineated. Medium magnification. **C:** The zona fasciculata, showing parallel cords of cells. Medium magnification. **D:** The zona reticularis and the adrenal medulla. Medium magnification. H&E stain.

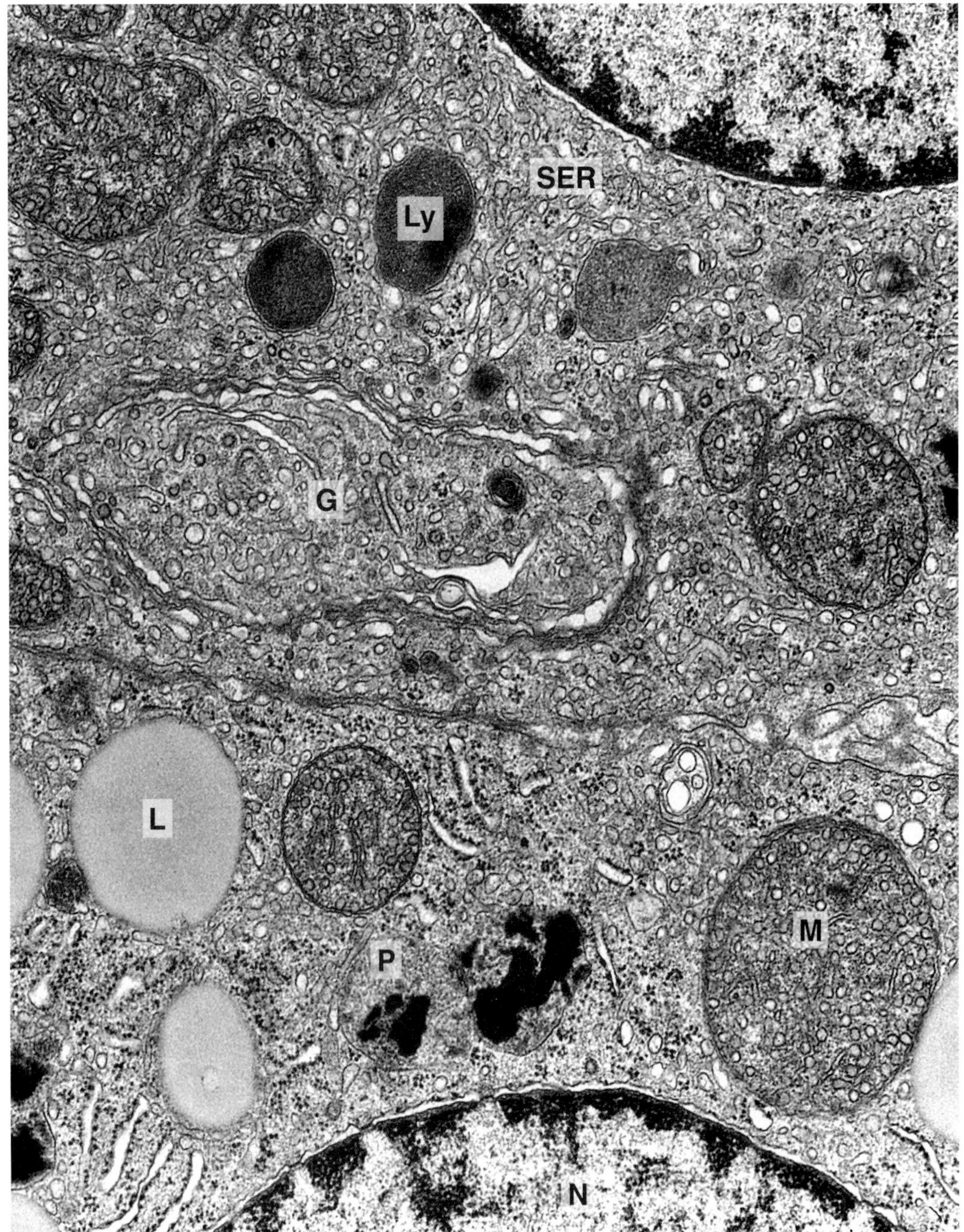

Figure 21–4. Fine structure of 2 steroid-secreting cells from the zona fasciculata of the human adrenal cortex. The lipid droplets (L) contain cholesterol esters. M, mitochondria with characteristic tubular and vesicular cristae; SER, smooth endoplasmic reticulum; N, nucleus; G, Golgi complex; Ly, lysosome; P, lipofuscin pigment granule. ×25,700.

Glucocorticoids also suppress the immune response by destroying circulating lymphocytes and inhibiting mitotic activity in lymphocyte-forming organs.

The **mineralocorticoids** act mainly on the distal renal tubules as well as on the gastric mucosa and the salivary and sweat glands, stimulating the absorption of sodium. They may increase the concentration of potassium and decrease the concentration of sodium in muscle and brain cells.

The separation of steroids produced by the adrenal cortex into glucocorticoids and mineralocorticoids is, however, somewhat arbitrary, because most glucocorticoids also act on ion transport.

The defense system of the body and the adrenal cortex are closely associated because cortisol has anti-inflammatory properties via white blood cells and suppression of cytokines and is also an immunosuppressant.

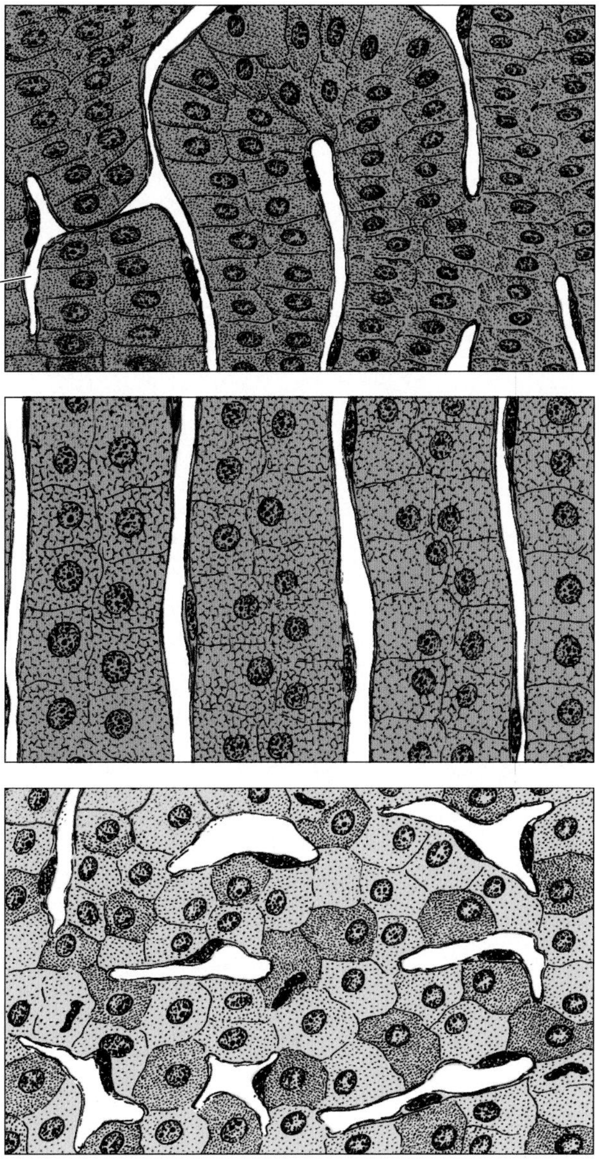

Factors acting on the gland

Hormones secreted

Zona glomerulosa

Angiotensin and corticotropin (ACTH)

Mineralocorticoids (aldosterone)

Capillaries

Zona fasciculata

Corticotropin

Glucocorticoids (cortisol and corticosterone)

Androgens (dihydroepiandrosterone; androstenedione)

Zona reticularis

Corticotropin

Glucocorticoids

Androgens

Figure 21–5. Structure and physiology of the adrenal cortex.

Dehydroepiandrosterone is the only sex hormone that is secreted in significant physiologic quantities by the adrenal cortex. Smaller amounts of other androgens such as androstenedione, 11 beta-hydroxyandrostenedione, and testosterone, are also secreted. Dehydroepiandrosterone and androstenedione are weak androgens and exert their actions after conversion into testosterone in other parts of the body. As in other endocrine glands, the adrenal cortex is controlled initially through the release of its corresponding releasing hormone stored in the median eminence. This is followed by secretion of adrenocorticotropic hormone (ACTH), or corticotropin, by the pars distalis of the hypophysis, which stimulates the synthesis and secretion of cortical hormones (eg, glucocorticoids). Free glucocorticoids may then inhibit ACTH secretion. The degree of pituitary inhibition is proportionate to the concentration of circulating glucocorticoids; inhibition is exerted at both the pituitary and hypothalamic levels (Figures 21–6 and 21–7).

MEDICAL APPLICATION

Because of this mechanism, patients who are treated with corticoids for long periods should never stop taking these hormones suddenly: secretion of ACTH in these patients is inhibited, and thus the cortex will not be induced to produce corticoids, causing severe drops in the levels of sodium and potassium.

Fetal, or Provisional, Cortex

In humans and some other animals, the adrenal gland of the newborn is proportionately larger than that of the adult. At this

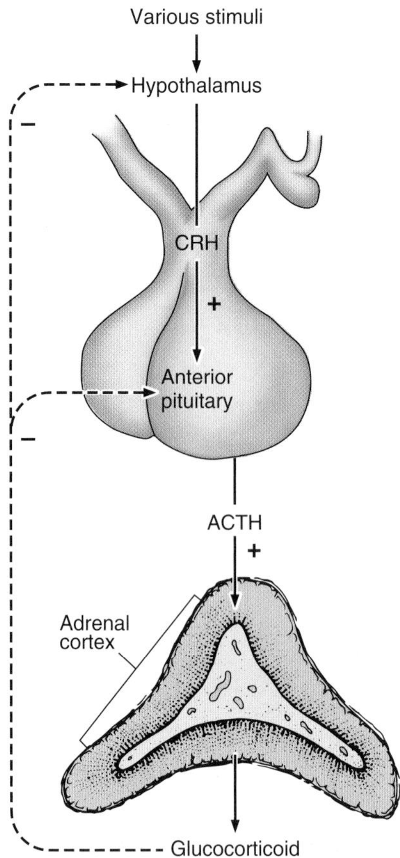

Figure 21–6. Feedback mechanism of ACTH and glucocorticoid secretion. Solid arrows indicate stimulation; dashed arrows, inhibition. CRH, corticotropin-releasing hormone, ACTH, corticotropin.

early age, a layer known as the **fetal,** or **provisional, cortex** is present between the medulla and the thin permanent cortex. This layer is fairly thick, and its cells are disposed in cords. After birth, the provisional cortex undergoes involution while the permanent cortex—the initially thin layer—develops, differentiating into the 3 layers (zones) described above. A major function of the fetal cortex is the secretion of sulfate conjugates of androgens, which are converted in the placenta to active androgens and estrogens that enter the maternal circulation.

Adrenal Medulla

The adrenal medulla is composed of polyhedral parenchymal cells arranged in cords or clumps and supported by a reticular fiber network (Figure 21–3). A profuse capillary supply intervenes between adjacent cords, and there are a few parasympathetic ganglion cells. Medullary parenchymal cells arise from neural crest cells, as do the postganglionic neurons of sympathetic and parasympathetic ganglia. Parenchymal cells of the adrenal medulla can be regarded as modified sympathetic postganglionic neurons that have lost their axons and dendrites during embryonic development and have become secretory cells.

Medullary parenchymal cells have abundant membrane-limited electron-dense secretory granules, 150–350 nm in diameter. These granules contain one or the other of the catecholamines, epinephrine or norepinephrine. The secretory granules also contain ATP, proteins called **chromogranins** (which may serve as binding proteins for catecholamines), dopamine beta-hydroxylase (which converts dopamine to norepinephrine), and opiatelike peptides (enkephalins) (Figure 21–8).

A large body of evidence shows that epinephrine and norepinephrine are secreted by 2 different types of cells in the medulla. Epinephrine-secreting cells have smaller granules that are less electron-dense, and their contents fill the granule. Norepinephrine-secreting cells have larger granules that are more electron-dense; their contents are irregular in shape, and there is an electron-lucent layer beneath the surrounding membrane. About 80% of the catecholamine output of the adrenal vein is epinephrine.

All adrenal medullary cells are innervated by cholinergic endings of preganglionic sympathetic neurons. Unlike the cortex, which does not store steroids, cells of the medulla accumulate and store their hormones in granules.

MEDICAL APPLICATION

Epinephrine and norepinephrine are secreted in large quantities in response to intense emotional reactions (eg, fright). Secretion of these substances is

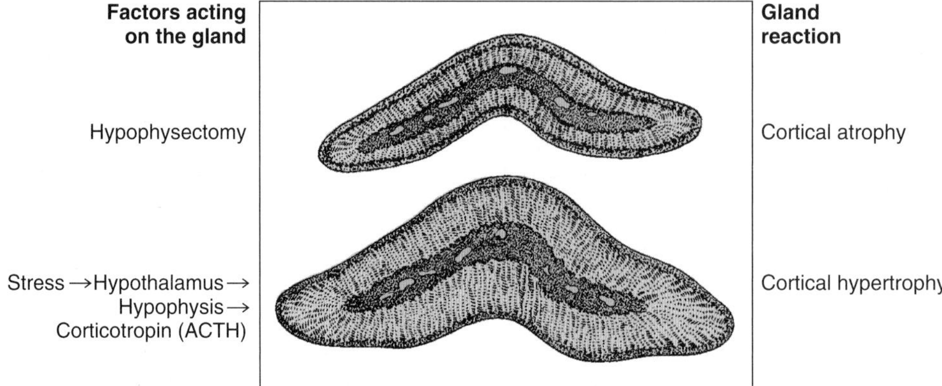

Figure 21–7. Effects of decreased or increased stimulation on the structure of the adrenal cortex.

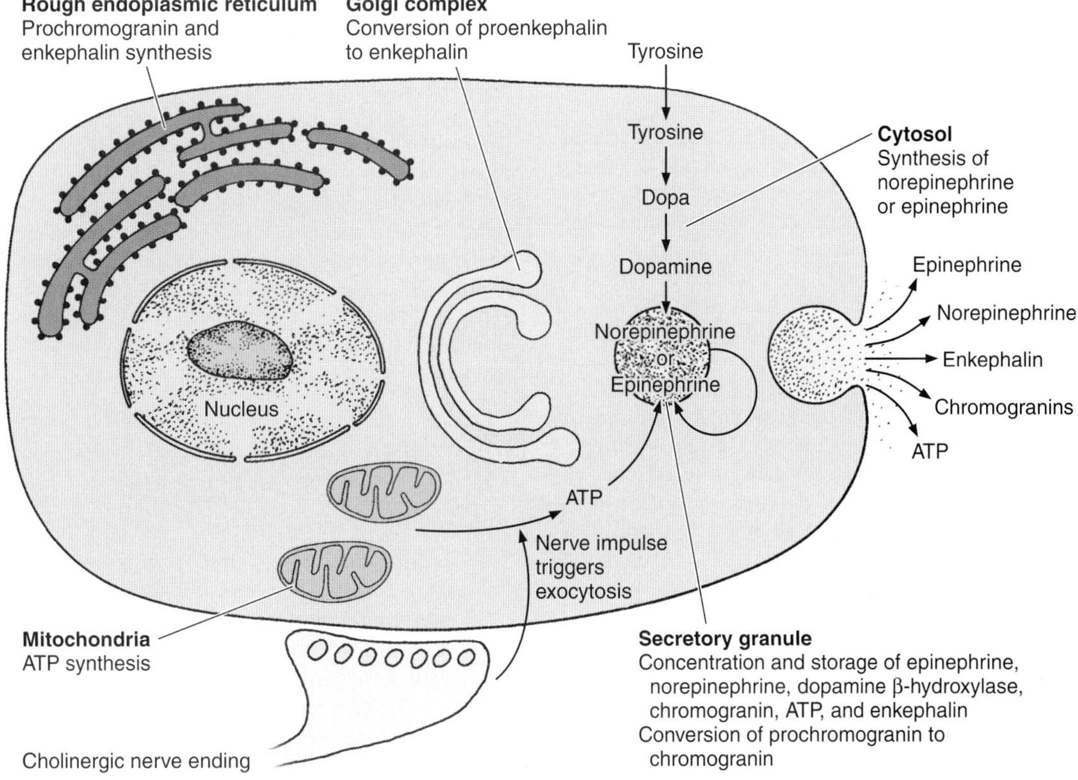

Figure 21–8. Diagram of an adrenal medullary cell showing the role of several organelles in synthesizing the constituents of secretory granules. Synthesis of norepinephrine and conversion to epinephrine take place in the cytosol.

mediated by the preganglionic fibers that innervate medullary cells. Vasoconstriction, hypertension, changes in heart rate, and metabolic effects such as elevated blood glucose result from the secretion and release of catecholamines into the bloodstream. These effects are part of the organism's defense reaction to stress (the fight-or-flight response). During normal activity, the medulla continuously secretes small quantities of these hormones.

Medullary cells are also found in the paraganglia (collections of catecholamine-secreting cells adjacent to the autonomic ganglia) as well as in various viscera. Paraganglia are a diffuse source of catecholamines.

Adrenal Dysfunction

MEDICAL APPLICATION

*One disorder of the adrenal medulla is **pheochromocytoma,** a tumor of its cells that causes hyperglycemia and transient elevations of blood pressure. These tumors can also develop in extramedullary sites (Figure 21–1).*

*Disorders of the adrenal cortex can be classified as **hyperfunctional** or **hypofunctional.** Tumors of the adrenal cortex can result in excessive production of glucocorticoids (**Cushing syndrome**) or aldosterone (**Conn syndrome**). Cushing syndrome is most often (90%) due to a pituitary adenoma that results in excessive production of ACTH; it is rarely caused by adrenal hyperplasia or an adrenal tumor. Excessive production of adrenal androgens has little effect in men. Hirsutism (abnormal hair growth) is seen in women, and precocious puberty (in boys) and virilization (in girls) are encountered in prepubertal children. These adrenogenital syndromes are the result of several enzymatic defects in steroid metabolism that cause increased biosynthesis of androgens by the adrenal cortex.*

*Adrenocortical insufficiency (**Addison disease**) is caused by destruction of the adrenal cortex in some diseases. The signs and symptoms suggest failure of secretion of both glucocorticoids and mineralocorticoids by the adrenal cortex.*

Carcinomas of the adrenal cortex are rare, but most are highly malignant. About 90% of these tumors produce steroids associated with endocrine glands.

ISLETS OF LANGERHANS

The islets of Langerhans are multihormonal endocrine microorgans of the pancreas; they appear as rounded clusters of cells embedded within exocrine pancreatic tissue (Figure 21–9).

Although most islets are 100–200 μm in diameter and contain several hundred cells, small islets of endocrine cells are also found interspersed among the pancreatic exocrine cells. There may be more than 1 million islets in the human pancreas, with a slight tendency for islets to be more abundant in the tail region.

In sections, each islet consists of lightly stained polygonal or rounded cells, arranged in cords separated by a network of blood capillaries (Figures 21–9). In 3-dimensional reconstructions, islets of Langerhans are seen as round, compact masses of secretory epithelial cells pervaded by a labyrinthine network of blood capillaries. Both the endocrine cells and the blood vessels are innervated by autonomic nerve fibers. A fine capsule of reticular fibers surrounds each islet, separating it from the adjacent exocrine pancreatic tissue.

Routine stains or trichrome stains allow the recognition of acidophils (alpha) and basophils (beta) (Figure 21–10). Using immunocytochemical methods four types of cells—A, B, D, and F—have been located in the islets. Figures 21–11 and 21–12 show immunocytochemical demonstrations of A (glucagon-producing) and B (insulin-producing) cells. The ultrastructure of these cells (Figure 21–13) resembles that of cells synthesizing polypeptides (see Chapter 4). The secretory granules of cells of the islets vary according to the species studied. In humans, the A cells have regular granules with a dense core surrounded by a clear region bounded by a membrane. The B (insulin-producing) cells have irregular granules with a core formed of irregular crystals of insulin in complex with zinc. The main steps of insulin synthesis are shown in Figure 21–14.

The relative quantities of the 4 cell types found in islets are not uniform; they vary considerably with the islet's location in the pancreas. Table 21–1 summarizes the types, quantities, and functions of the hormones produced by the islet cells.

Terminations of nerve fibers on islet cells can be observed by light or electron microscopy. Both sympathetic and parasympathetic nerve endings have been found in close association with about 10% of the A, B, and D cells. Gap junctions presumably serve to transfer the ionic changes associated with autonomic discharge to the other cells. These nerves function as part of the insulin and glucagon control system.

MEDICAL APPLICATION

Several tumor types arise from islet cells that produce such hormones as insulin, glucagon, somatostatin, and pancreatic polypeptide. Some pancreatic tumors produce 2 or more of these hormones simultaneously, generating complex clinical symptoms.

One of the principal types of diabetes (type I) is an autoimmune disease in which antibodies against B cells depress the cells' activity.

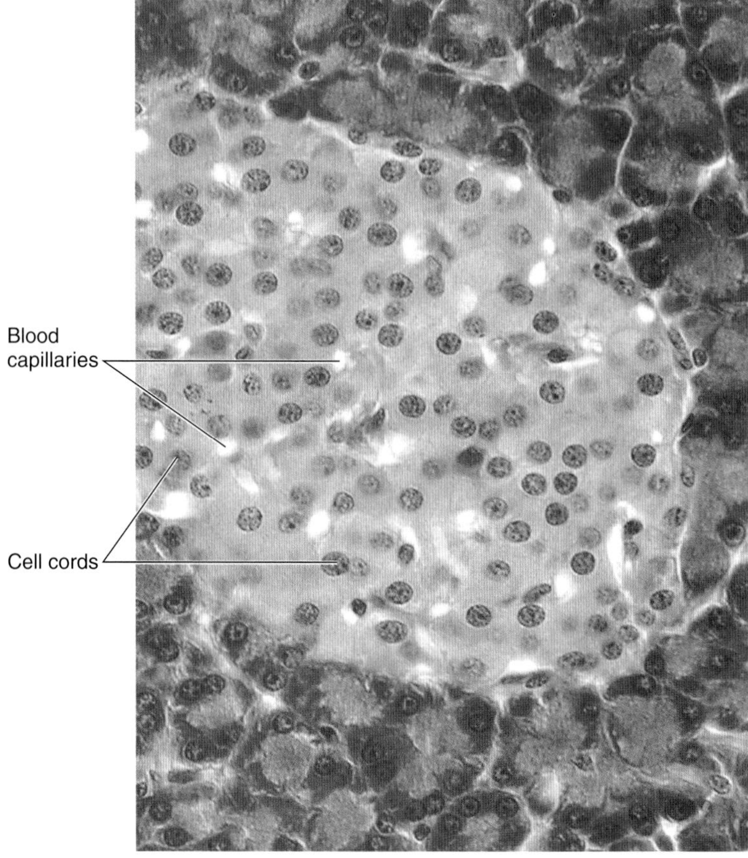

Figure 21–9. Photomicrograph of a section of the pancreas showing an islet of Langerhans surrounded by pancreatic acinar cells. The islet cells form cords separated by blood capillaries, here seen as white spaces. H&E stain. Medium magnification.

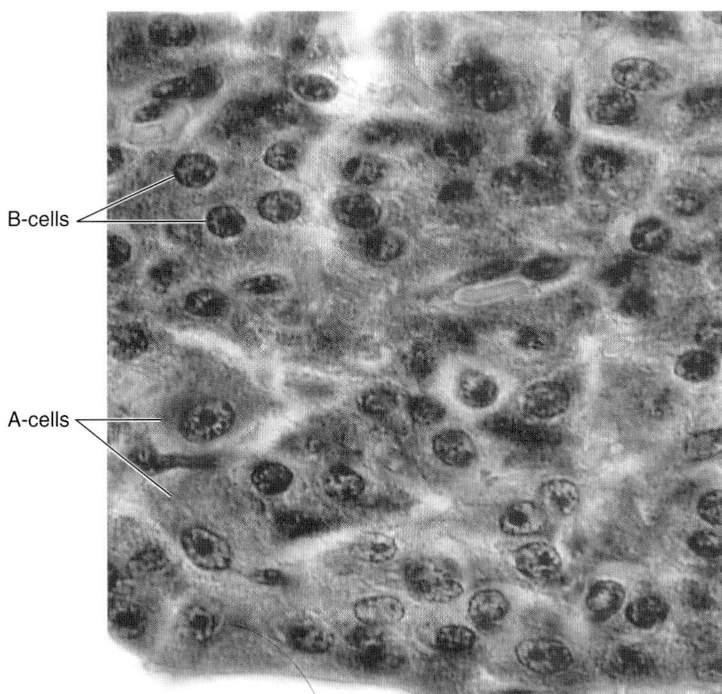

Figure 21–10. Photomicrograph of an islet of Langerhans showing alpha (A) cells and beta (B) cells. Gomori's trichrome stain. High magnification.

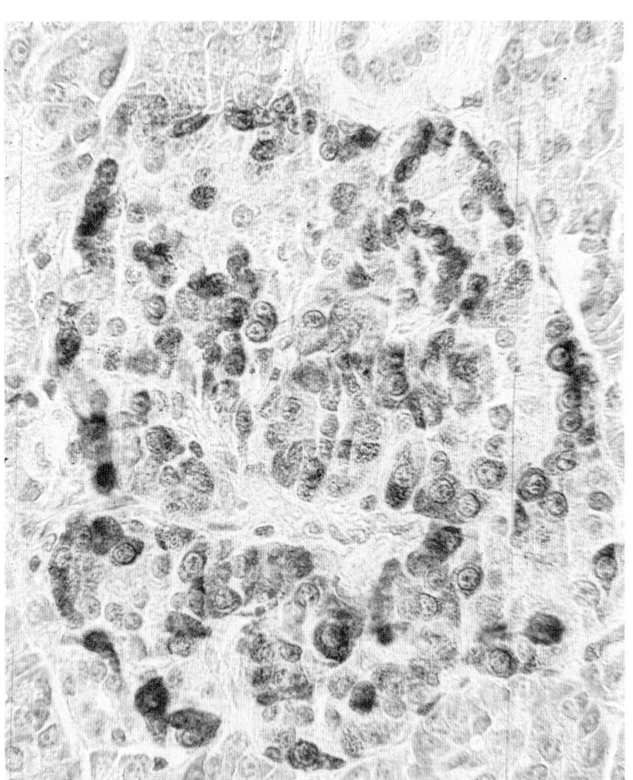

Figure 21–11. Light-microscope immunocytochemical localization of glucagon in A cells (stained in brown). Medium magnification.

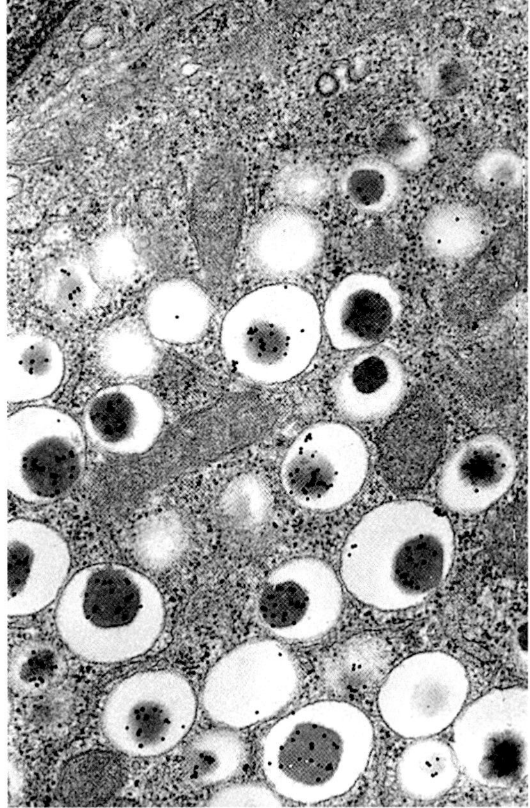

Figure 21–12. Electron-microscope immunocytochemical localization of insulin in a B cell of an islet of Langerhans. The black granules are gold particles used to label anti-insulin. They indicate the sites where this antibody was attached to the insulin in the secretory granules. Note also the clear zone between the secretory material and the granule membrane. (Courtesy of M Bendayan.)

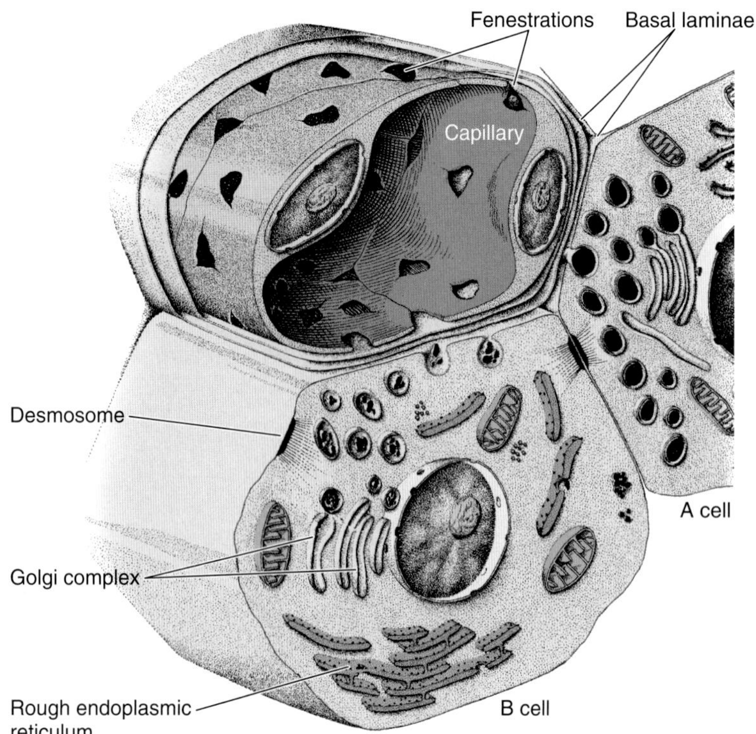

Figure 21–13. Drawing of the A and B cells, showing their main ultrastructural features. The B cell's granules are irregular, whereas the A cell's granules are round and uniform.

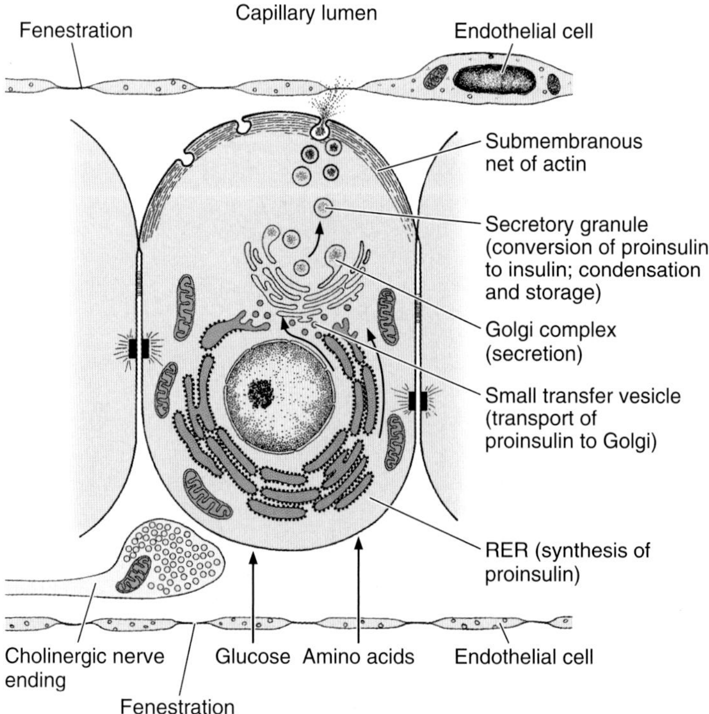

Figure 21–14. The main steps of insulin synthesis and secretion by a B cell in the islets of Langerhans. RER, rough endoplasmic reticulum. (Based on Orci L: A portrait of the pancreatic B cell. Diabetologia 1974;10:163.)

Table 21–1. Cell types in human islets of Langerhans.[a]

Cell Type	Quantity	Position	Hormone Produced	Hormonal Function
A	~20%	Usually in periphery	Glucagon	Acts on several tissues to make energy stored in glycogen and fat available through glycogenolysis and lipolysis; increases blood glucose content
B	~70%	Central region	Insulin	Acts on several tissues to cause entry of glucose into cells and promotes decrease of blood glucose content
D	<5%	Variable	Somatostatin	Inhibits release of other islet cell hormones through local paracrine action
F	Rare	Variable	Pancreatic polypeptide	Not well established

[a]The islets of Langerhans contain several cell types that secrete hormones that increase or decrease blood glucose. This mechanism precisely controls blood glucose concentration, an important factor in body homeostasis.

THYROID

In early embryonic life, the thyroid is derived from the cephalic portion of the alimentary canal endoderm. Its function is to synthesize the hormones thyroxine (T_4) and triiodothyronine (T_3), which stimulate the rate of metabolism in the body.

The thyroid gland, located in the cervical region anterior to the larynx, consists of 2 lobes united by an isthmus (Figure 21–15). Thyroid tissue is composed of thousands of **follicles** that consist of spheres formed by simple epithelium whose lumen contains a gelatinous substance called **colloid** (Figures 21–16, 21–17, and 21–18). In sections, follicular cells range from squamous to columnar and the follicles have an extremely variable diameter. The gland is covered by a loose connective tissue capsule that sends septa into the parenchyma. As these septa gradually become thinner they reach all the follicles, separated from one another by fine, irregular connective tissue composed mainly of reticular fibers. The thyroid is an extremely vascularized organ, with an extensive blood and lymphatic capillary network surrounding the follicles. Endothelial cells of these capillaries are fenestrated, as they are in other endocrine glands. This configuration facilitates the transport of molecules between the gland cells and the blood capillaries.

The major regulator of the anatomic and functional state of the thyroid gland is thyroid-stimulating hormone (thyrotropin), which is secreted by the anterior pituitary.

The morphologic appearance of thyroid follicles varies according to the region of the gland and its functional activity. In the same gland, larger follicles that are full of colloid and have a cuboidal or squamous epithelium are found alongside follicles that are lined by columnar epithelium. Despite this variation, the gland is considered hypoactive when the average composition of these follicles is squamous. Thyrotropin stimulates the synthesis of thyroid hormone, increases the height of the follicular epithelium, and decreases the quantity of the colloid and the size of the follicles. The cell membrane of the basal portion of follicular cells is rich in receptors for thyrotropin.

The thyroid epithelium rests on a basal lamina. The follicular epithelium exhibits all the characteristics of a cell that simultaneously synthesizes, secretes, absorbs, and digests proteins (Figure 21–19). The basal part of these cells is rich in rough endoplasmic reticulum. The nucleus is generally round and situated in the center of the cell. The apical pole has a discrete Golgi complex and small secretory granules with the morphologic characteristics of follicular colloid. Abundant lysosomes, 0.5–0.6 μm in diameter, and some large phagosomes are found in this region. The cell membrane of the apical pole has a moderate number

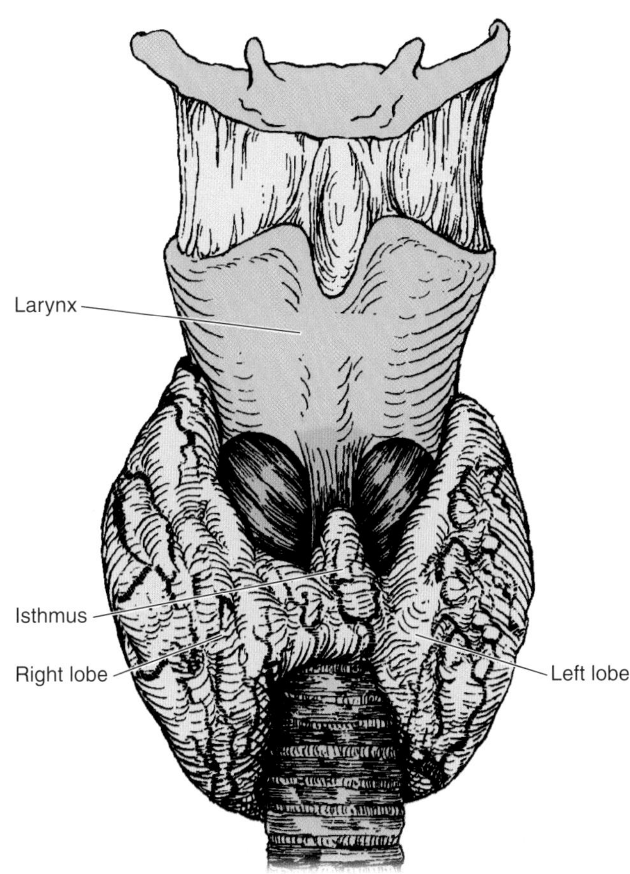

Figure 21–15. Anatomy of the human thyroid. (Reproduced, with permission, from Ganong WF: *Review of Medical Physiology,* 20th ed. McGraw-Hill, 2001.)

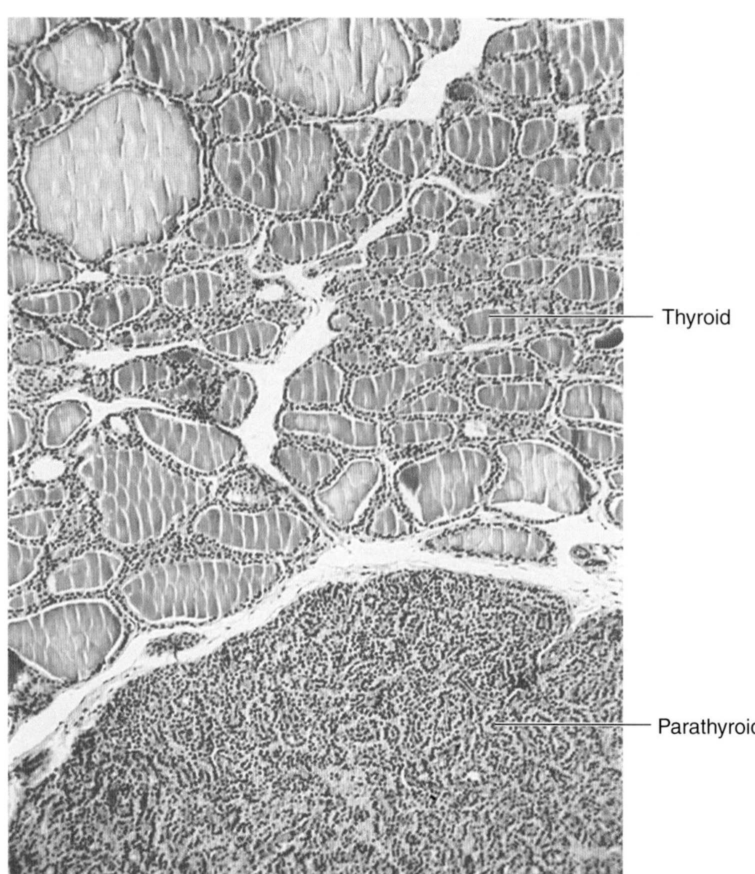

Figure 21–16. Photomicrograph of a section of thyroid and parathyroid. The thyroid is formed by thousands of spheres called thyroid follicles. They are filled with a glycoprotein, the colloid, which appears fragmented here because of an artifact. The parathyroid is separated from the thyroid by a thin connective tissue capsule. H&E stain. Low magnification.

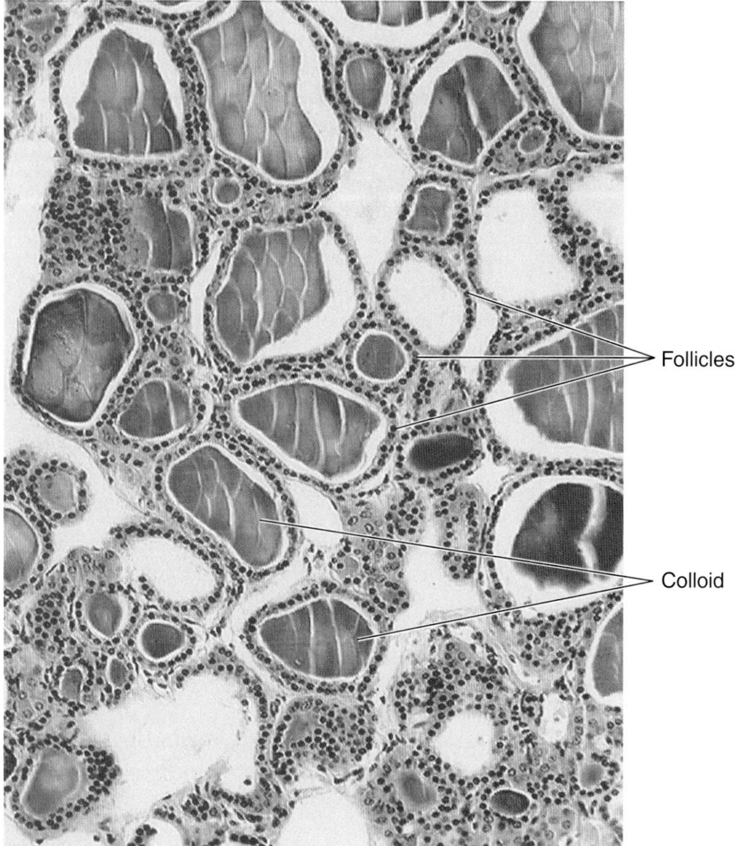

Figure 21–17. Section of a thyroid showing the follicles formed by a simple epithelium, containing colloid. H&E stain. Medium magnification.

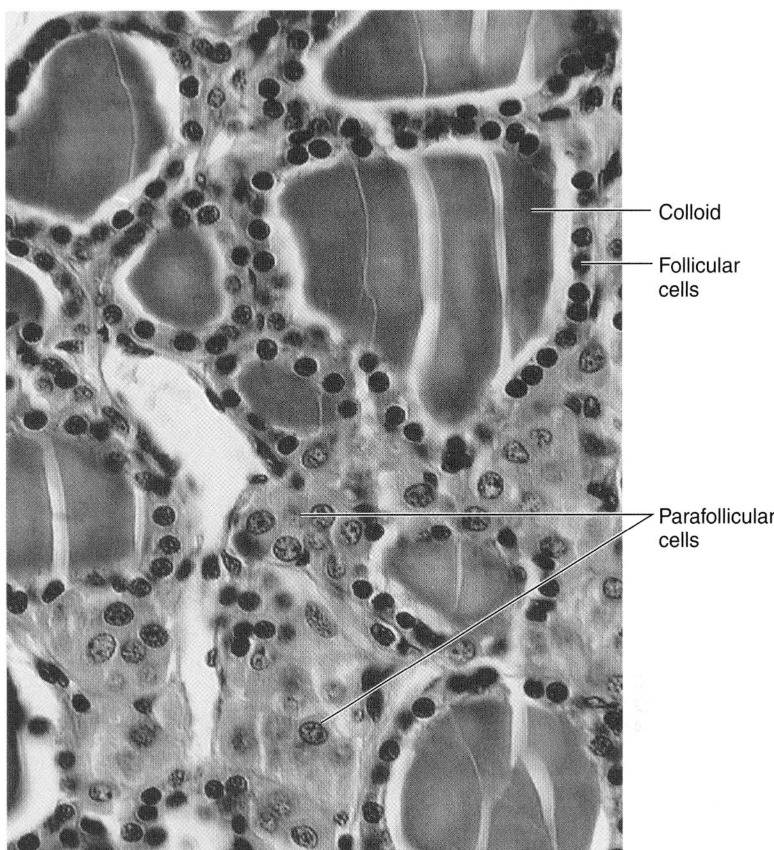

Figure 21–18. High magnification of a section of a thyroid. Calcitonin-producing parafollicular cells can be distinguished from the follicular epithelial cells because they are larger and their nuclei stain lighter. H&E stain. High magnification.

of microvilli. Mitochondria and cisternae of rough endoplasmic reticulum are dispersed throughout the cytoplasm.

Another type of cell, the **parafollicular,** or **C, cell,** is found as part of the follicular epithelium or as isolated clusters between thyroid follicles (Figures 21–18 and 21–20). Parafollicular cells are somewhat larger than thyroid follicular cells and stain less intensely. They have a small amount of rough endoplasmic reticulum, long mitochondria, and a large Golgi complex. The most striking feature of these cells is their numerous small (100–180 nm in diameter) granules containing hormone (Figure 21–21). These cells are responsible for the synthesis and secretion of **calcitonin,** a hormone whose main effect is to lower blood calcium levels by inhibiting bone resorption. Secretion of calcitonin is triggered by an elevation in blood calcium concentration.

Synthesis & Secretion of Thyroid Hormones

The thyroid is the only endocrine gland whose secretory product is stored in great quantity. This accumulation is also unusual in that it occurs in the extracellular colloid. In humans, there is sufficient hormone within the follicles to supply the organism for up to 3 months. Thyroid colloid is composed of a glycoprotein (thyroglobulin) of high molecular mass (660 kDa).

Control of the activity of thyroid follicular cells is summarized in Figure 20–8. This mechanism maintains an adequate quantity of T_4 and T_3 within the organism. Secretion of thyrotropin is also increased by exposure to cold and decreased by heat and stressful stimuli.

Synthesis & Accumulation of Hormones by Follicular Cells

Synthesis and accumulation of hormones take place in 4 stages (Figure 21–19): synthesis of thyroglobulin, uptake of iodide from the blood, activation of iodide, and iodination of the tyrosine residues of thyroglobulin.

1. The **synthesis of thyroglobulin** is similar to that in other protein-exporting cells (described in Chapter 4). Briefly, the secretory pathway consists of the synthesis of protein in the rough endoplasmic reticulum, the addition of carbohydrate in the endoplasmic reticulum and the Golgi complex, and the release of thyroglobulin from formed vesicles at the apical surface of the cell into the lumen of the follicle.
2. The **uptake of circulating iodide** is accomplished in the thyroid follicular cells by a membrane transport protein. This protein, which simultaneously carries 2 molecules, sodium and iodide, is called the Na/I symporter and is located in the basolateral membrane of the follicular cells. Iodine plays an important role in regulating thyroid function because low iodine levels increase the amount of Na/I symporter and thus increase the uptake, compensating for the lower serum concentration.
3. Iodide is **oxidized** by thyroid peroxidase and is transported into the follicle cavity by an anion transporter called pendrin.
4. Within the colloid occurs the **iodination of tyrosine residues** of thyroglobulin, also catalyzed by thyroid peroxidase. In this way, T_3 and T_4 are produced, but they become part of the much larger thyroglobulin molecule.

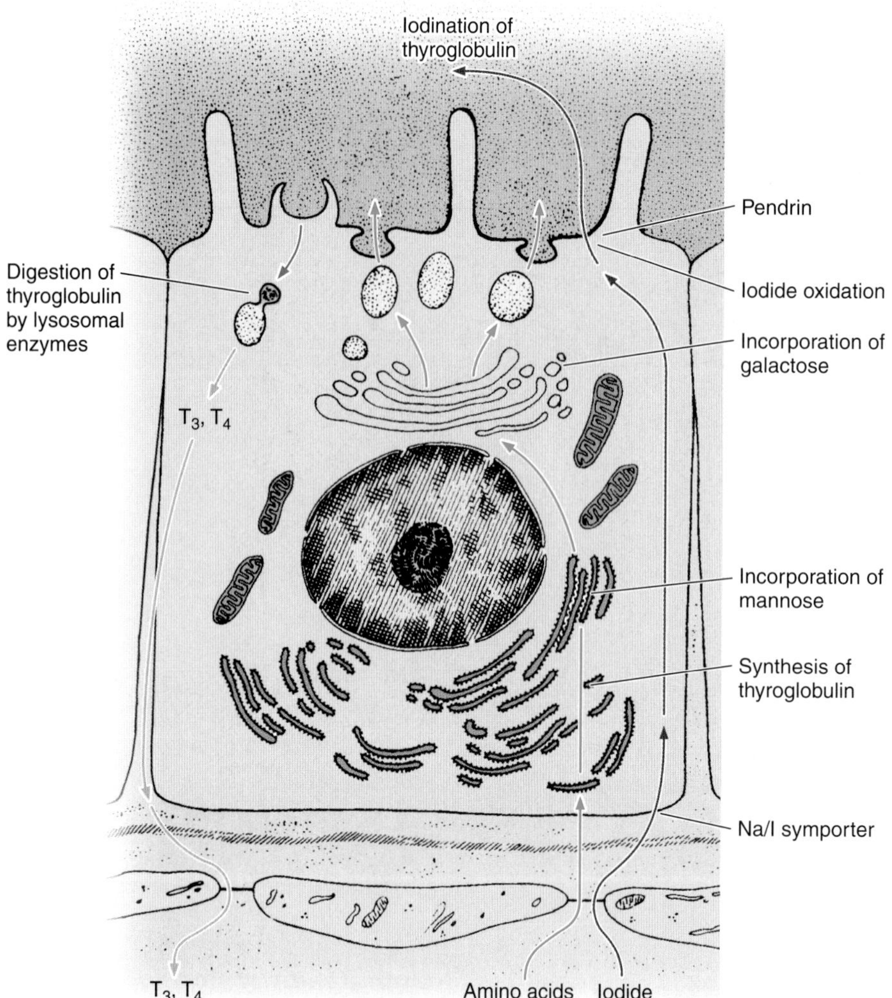

Figure 21–19. The processes of synthesis and iodination of thyroglobulin and its absorption and digestion. These events may occur simultaneously in the same cell.

Liberation of T_3 & T_4

When stimulated by thyrotropin, thyroid follicular cells take up colloid by endocytosis. The colloid within the endocytic vesicles is then digested by lysosomal enzymes. The bonds between the iodinated residues and the thyroglobulin molecule are broken by proteases, and T_4, T_3, diiodotyrosine, and monoiodotyrosine are liberated into the cytoplasm. The free T_4 and T_3 then cross the basolateral cell membrane and are discharged into the capillaries. Monoiodotyrosine and diiodotyrosine are not secreted into the blood, because their iodine is removed as a result of the intracellular action of **iodotyrosine dehalogenase.** The products of this enzymatic reaction, iodine and tyrosine, are reused by the follicular cells. T_4 is the more abundant compound, constituting 90% of the circulating thyroid hormone, although T_3 acts more rapidly and is more potent.

Thyroxine has a gradual effect, stimulating mitochondrial respiration and oxidative phosphorylation; this effect is dependent on mRNA synthesis. T_3 and T_4 increase the numbers of both mitochondria and their cristae. Synthesis of mitochondrial proteins is increased, and degradation of the proteins is decreased.

Most of the effects of thyroid hormones are the result of their action on the basal metabolic rate; they increase the absorption of carbohydrates from the intestine and regulate lipid metabolism. Thyroid hormones also influence body growth and development of the nervous system during fetal life.

Thyroid Disorders

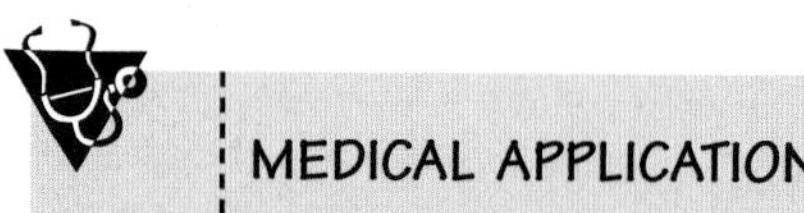

*A diet low in iodine hinders the synthesis of thyroid hormones, causing hypothyroidism. Thyroid hypertrophy as a result of increased thyrotropin secretion causes the disorder known as **iodine deficiency goiter,** which occurs widely in some regions of the world.*

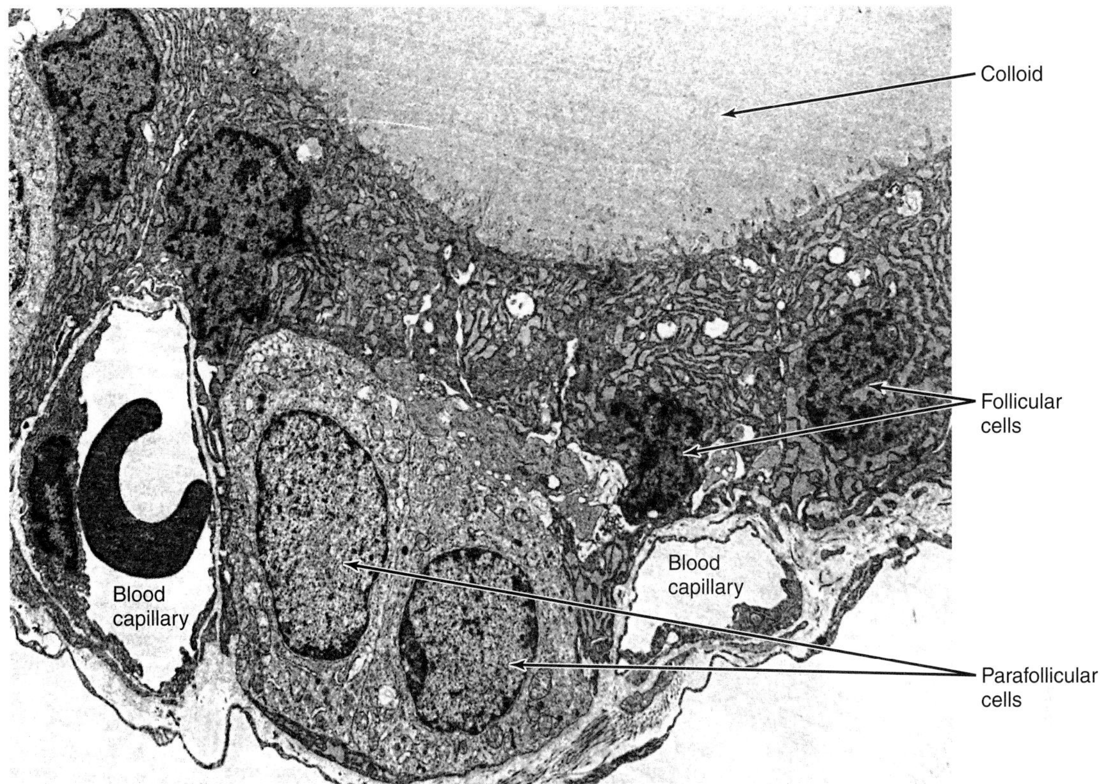

Figure 21–20. Electron micrograph of thyroid showing 2 calcitonin-producing parafollicular cells and part of a thyroid follicle. Note 2 blood capillaries at both sides of the parafollicular cells.

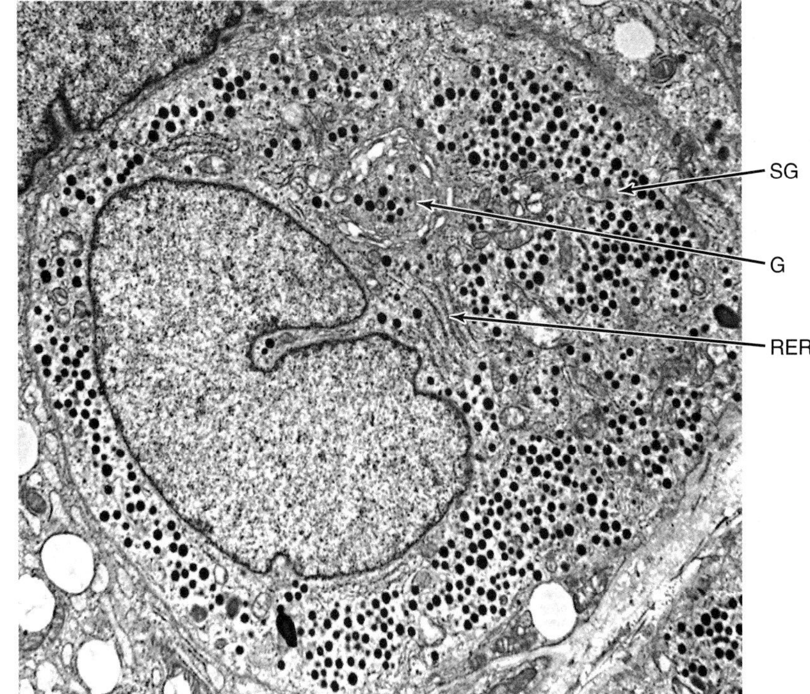

Figure 21–21. Electron micrograph of a calcitonin-producing cell. Note the small secretory granules (SG) and the scarcity of rough endoplasmic reticulum (RER). G, Golgi region. ×5000.

*The syndrome of adult hypothyroidism, **myxedema,** may be the result of a number of diseases of the thyroid gland, or it may be secondary to pituitary or hypothalamic failure. Autoimmune diseases of this gland impair its function, with consequent hypothyroidism. In Hashimoto thyroiditis it is possible to detect antibodies against thyroid tissue in the patient's blood. As with other autoimmune malfunctions, Hashimoto disease is more common in women.*

*Children who are hypothyroid from birth are called **cretins;** cretinism is characterized by arrested physical and mental development.*

*Hyperthyroidism, or thyrotoxicosis, may be caused by a variety of thyroid diseases, of which the most common form is **Graves disease,** or **exophthalmic goiter.** This thyroid hyperfunction is due to an immunologic dysfunction, with production of a circulating immunoglobulin that binds to thyrotropin receptors in thyroid follicular cells, and whose effects resemble those of thyrotropin. Patients with Graves disease exhibit decreased body weight, nervousness, eye protrusion, asthenia, and accelerated heart rate.*

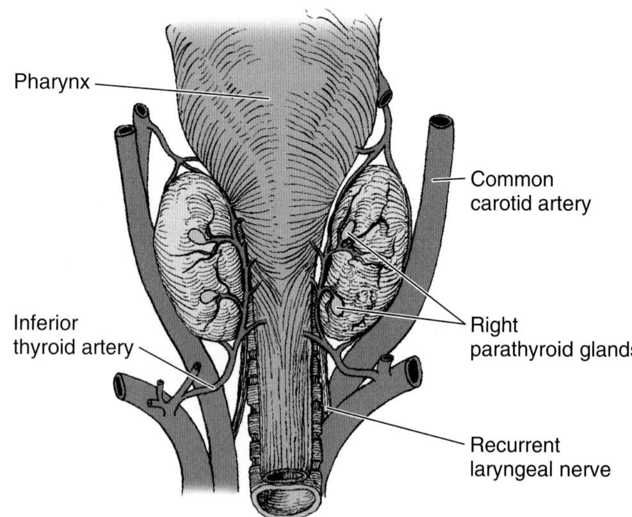

Figure 21–22. **The human parathyroid glands, viewed from behind.** (Redrawn and reproduced, with permission, from Nordland in: Surg Gynecol Obstet 130;51:449; and from *Gray's Anatomy of the Human Body,* 29th ed. Goss CM [editor]. Lea & Febiger, 1973.)

PARATHYROID GLANDS

The parathyroids are 4 small glands—3 × 6 mm—with a total weight of about 0.4 g. They are located behind the thyroid gland, one at each end of the upper and lower poles, usually in the capsule that covers the lobes of the thyroid (Figure 21–22). Sometimes they are embedded in the thyroid gland (Figure 21–16). The parathyroid glands are derived from the pharyngeal pouches—the superior glands from the fourth pouch and the inferior glands from the third pouch. They can also be found in the mediastinum, lying beside the thymus, which originates from the same pharyngeal pouches.

Each parathyroid gland is contained within a connective tissue capsule. These capsules send septa into the gland, where they merge with the reticular fibers that support elongated cordlike clusters of secretory cells.

The endocrine cells of the parathyroid are arranged in cords (Figure 21–23). There are 2 types of cells: the chief, or principal, cells and the oxyphil cells.

The **chief cells** are small polygonal cells with a vesicular nucleus and a pale-staining, slightly acidophilic cytoplasm. Electron microscopy shows irregularly shaped granules (200–400 nm in diameter) in their cytoplasm. They are the secretory granules containing **parathyroid hormone,** which is a polypeptide in its active form. **Oxyphil cells** constitute a smaller population (Figure 21–24). They are larger polygonal cells, and their cytoplasm contains many acidophilic mitochondria with abundant cristae. The function of the oxyphil cells is not known.

With increasing age, secretory cells are replaced with adipocytes. Adipose cells constitute more than 50% of the gland in older people.

Action of Parathyroid Hormone & Its Interrelation with Calcitonin

Parathyroid hormone binds to receptors in osteoblasts. This is a signal for these cells to produce an osteoclast-stimulating factor, which increases the number and activity of osteoclasts and thus promotes the absorption of the calcified bone matrix and the release of Ca^{2+} into the blood. The resulting increase in the concentration of Ca^{2+} in the blood suppresses the production of parathyroid hormone. Calcitonin from the thyroid gland also influences osteoclasts by inhibiting both their resorptive action on bone and the liberation of Ca^{2+}. Calcitonin thus lowers blood Ca^{2+} concentration and increases osteogenesis; its effect is opposite to that of parathyroid hormone. These hormones constitute a dual mechanism to regulate blood levels of Ca^{2+}, an important factor in homeostasis.

In addition to increasing the concentration of Ca^{2+}, parathyroid hormone reduces the concentration of phosphate in the blood. This effect is a result of the activity of parathyroid hormone on kidney tubule cells, diminishing the absorption of phosphate and causing an increase of phosphate excretion in urine. Parathyroid hormone indirectly increases the absorption of Ca^{2+} from the gastrointestinal tract by stimulating the synthesis of vitamin D, which is necessary for this absorption. The secretion of parathyroid cells is regulated by blood Ca^{2+} levels.

MEDICAL APPLICATION

*In **hyperparathyroidism,** concentrations of blood phosphate are decreased and concentrations of blood Ca^{2+} are increased. This condition frequently produces pathologic deposits of calcium in several organs, such as the kidneys and arteries. The bone disease caused by hyperparathyroidism, which is characterized by an increased number of osteoclasts and multiple bone cavities, is known as **osteitis fibrosa***

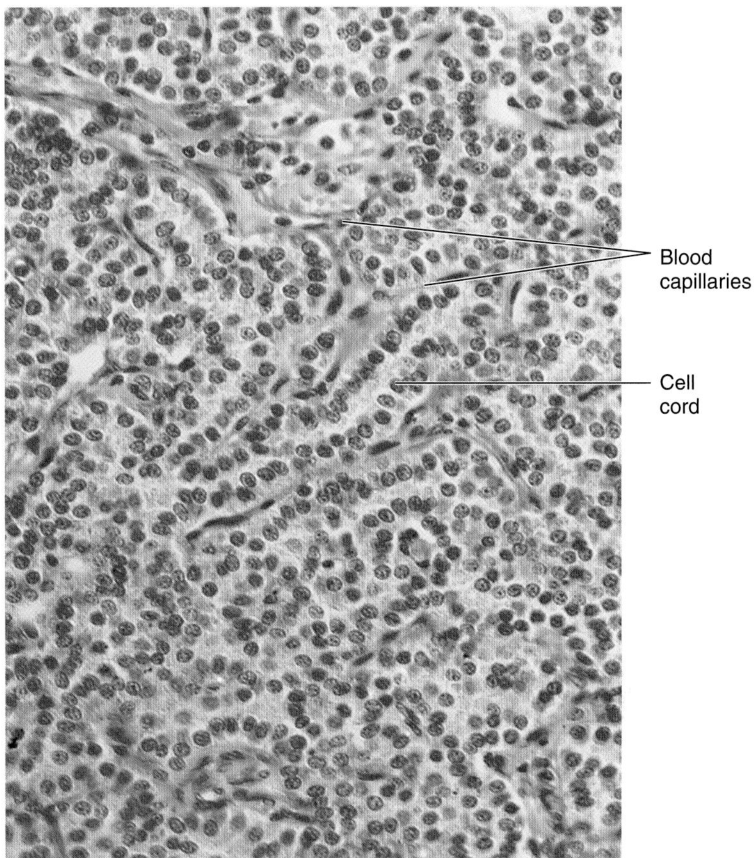

Figure 21–23. Section of a parathyroid gland showing the chief cells of the gland arranged as cords separated by blood capillaries. H&E stain. Medium magnification.

cystica. *Bones from patients with osteitis fibrosa cystica are less resistant and prone to fractures.*

In ***hypoparathyroidism,*** *concentrations of blood phosphate are increased and concentrations of blood* Ca^{2+} *are decreased. The bones become denser and more mineralized. This condition causes spastic contractions of the skeletal muscles and generalized convulsions called* ***tetany.*** *These symptoms are caused by the exaggerated excitability of the nervous system, due to the lack of* Ca^{2+} *in the blood. Patients with hypoparathyroidism are treated with calcium salts and vitamin D.*

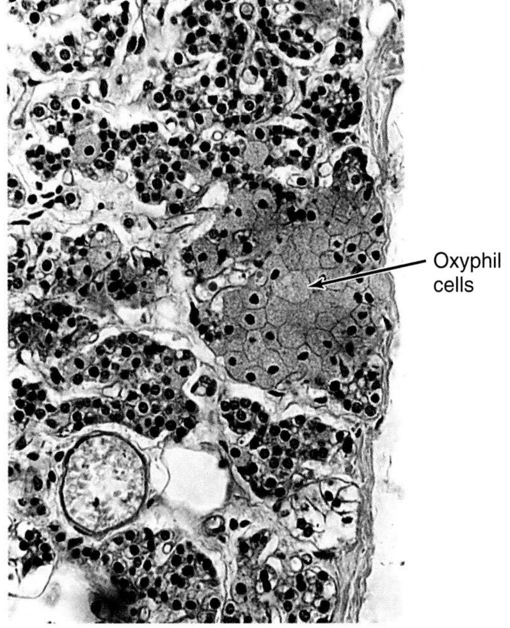

Figure 21–24. Photomicrograph of a section of a parathyroid gland. Note a group of large, acidophilic oxyphil cells at the right. Medium magnification. (Courtesy of J James.)

PINEAL GLAND

The pineal gland is also known as the **epiphysis cerebri,** or **pineal body.** In the adult, it is a flattened conical organ measuring approximately 5–8 mm in length and 3–5 mm at its greatest width and weighing about 120 mg. It is found in the posterior extremity of the third ventricle, above the roof of the diencephalon, to which it is connected by a short stalk.

The pineal gland is covered by pia mater. Connective tissue septa (containing blood vessels and unmyelinated nerve fibers) originate in the pia mater and penetrate the pineal tissue. Along with the capillaries, they surround the cellular cords and follicles, forming irregular lobules.

The pineal gland consists of several types of cells, principally pinealocytes and astrocytes. **Pinealocytes** have a slightly basophilic cytoplasm with large irregular or lobate nuclei and sharply defined nucleoli. When impregnated with silver salts, the pinealocytes appear to have long and tortuous branches reaching out to the vascular connective tissue septa, where they end as flattened dilatations. These cells produce **melatonin** and some ill-defined pineal peptides.

The **astrocytes** of the pineal gland are a specific type of cell characterized by elongated nuclei that stain more heavily than do those of parenchymal cells. They are observed between the cords of pinealocytes and in perivascular areas. These cells have long cytoplasmic processes that contain a large number of intermediate filaments 10 nm in diameter.

Innervation

Nerve fibers lose their myelin sheaths when they penetrate the pineal gland; the unmyelinated axons end among pinealocytes, with some forming synapses. A great number of small vesicles containing norepinephrine are seen in these nerve endings. Serotonin is also present, in both the pinealocytes and the sympathetic nerve terminals.

Role of the Pineal Gland in Controlling Biological Cycles

The pineal gland is involved in both circadian (24-hour) and seasonal biorhythms. It responds to light transmitted to the cerebral cortex and relayed to the pineal gland by secretion of melatonin and several peptides. The number of these molecules liberated into the blood increases greatly during the dark hours of the 24-hour daily cycle. In turn, these secreted molecules promote rhythmic changes in the secretory activities of the gonads and other organs. The pineal gland is therefore a neuroendocrine transducer, converting nerve input into variations in hormone output.

REFERENCES

Adrenal Glands

Braunwald E et al: *Harrison's Principles of Internal Medicine,* 15th ed. McGraw-Hill, 2001.

Christy NP (editor): *The Human Adrenal Cortex.* Harper & Row, 1971.

James VHT (editor): *The Adrenal Gland.* Raven Press, 1979.

Neville AM, O'Hare MJ: *The Human Adrenal Cortex.* Springer-Verlag, 1982.

Islets of Langerhans

Cooperstein SJ, Watkins D (editors): *The Islets of Langerhans.* Academic Press, 1981.

Ganong WF: *Review of Medical Physiology,* 20th ed. McGraw-Hill, 2001.

Gruppuso PA: Familial hyperproinsulinemia due to proposed defect in conversion of proinsulin to insulin. New Engl J Med 1984;629:311.

Orci L et al: The insulin factory. Sci Am 1988;259:85.

Thyroid Gland

Braunwald E et al: *Harrison's Principles of Internal Medicine,* 15th ed. McGraw-Hill, 2001.

Dunn JT, Dunn AD: Update on intrathyroidal iodine metabolism. Thyroid 2001;11:407.

Kohn LD et al: Effects of thyroglobulin and pendrin on iodide flux through the thyrocyte. Trends Endocrinol Metab 2001;12:10.

Nilsson M: Iodide handling by the thyroid epithelial cell. Exp Clin Endocrinol Diabetes 2001;109:13.

Nunez EA, Gershon MD: Cytophysiology of thyroid parafollicular cells. Int Rev Cytol 1978;52:1.

Parathyroid Glands

Gaillard PJ et al (editors): *The Parathyroid Glands.* Univ of Chicago Press, 1965.

Pineal Gland

Sugden D: Melatonin: binding site characteristics and biochemical and cellular responses. Neurochem Int 1994;24:147.

Tapp E, Huxley M: The histological appearance of the human pineal gland from puberty to old age. J Pathol 1972;108:137.

The Male Reproductive System 22

The male reproductive system is composed of the testes, genital ducts, accessory glands, and penis. The dual function of the **testis** is to produce hormones and spermatozoa. Although testosterone is the main hormone produced in the testes, both testosterone and its metabolite, dihydrotestosterone, are necessary for the physiology of men. Testosterone is important for spermatogenesis, sexual differentiation during embryonic and fetal development, and control of gonadotropin secretion. Dihydrotestosterone acts on many organs and tissues of the body during puberty and adulthood (eg, muscle, hair pattern, and hair growth).

The genital ducts and accessory glands produce secretions that, aided by smooth muscle contractions, propel spermatozoa toward the exterior. These secretions also provide nutrients for spermatozoa while they are confined to the male reproductive tract. Spermatozoa and the secretions of the genital ducts and accessory glands make up the **semen** (from Latin, meaning seed), which is introduced into the female reproductive tract through the penis.

TESTES

Each testis is surrounded by a thick capsule of dense connective tissue, the **tunica albuginea.** The tunica albuginea is thickened on the posterior surface of the testis to form the **mediastinum testis,** from which fibrous septa penetrate the gland, dividing it into about 250 pyramidal compartments called the **testicular lobules** (Figure 22–1). These septa are incomplete, and there is frequently intercommunication between the lobules. Each lobule is occupied by 1–4 **seminiferous tubules** enmeshed in a web of loose connective tissue that is rich in blood and lymphatic vessels, nerves, and **interstitial** (**Leydig**) **cells.** Seminiferous tubules produce male reproductive cells, the spermatozoa, whereas interstitial cells secrete testicular androgens.

The testes develop retroperitoneally in the dorsal wall of the abdominal cavity. They migrate during fetal development and eventually are suspended within the scrotum at the ends of the spermatic cords. Because of the migration toward the scrotum, each testis carries with it a serous sac, the **tunica vaginalis** (Fig-

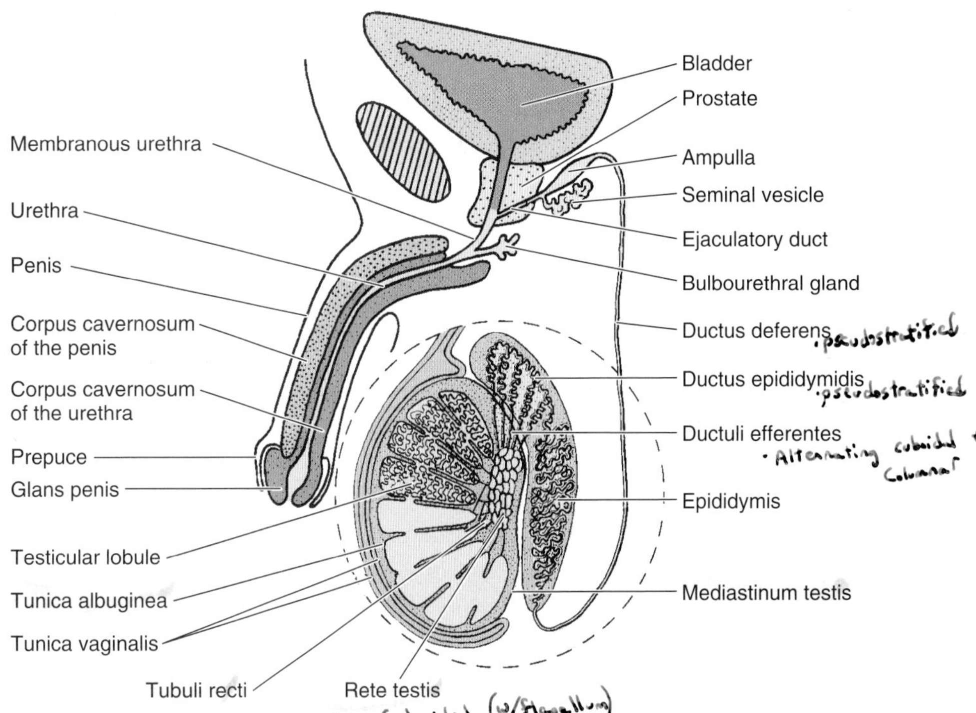

Figure 22–1. The male genital system. The testis and the epididymis are shown in different scales than the other parts of the reproductive system. Note the communication between the testicular lobules.

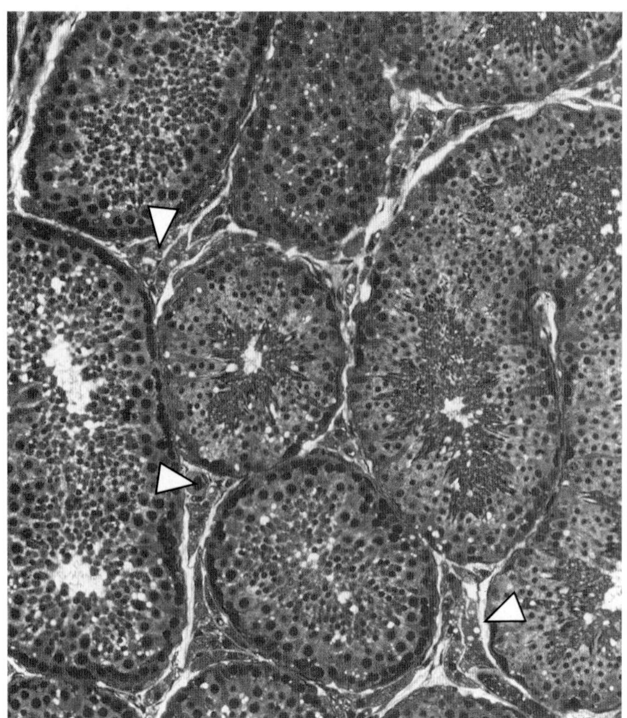

Figure 22–2. Section of a testis showing seminiferous tubules, some of which are outlined, and groups of pale-stained interstitial (Leydig) cells (arrowheads). Pararosaniline–toluidine blue (PT) stain. Medium magnification.

ure 22–1), derived from the peritoneum. The tunic consists of an outer parietal layer and an inner visceral layer, covering the tunica albuginea on the anterior and lateral sides of the testis. The scrotum has an important role in maintaining the testes at a temperature lower than the abdominal temperature.

Seminiferous Tubules

Spermatozoids are produced in the seminiferous tubules. Each testicle has 250–1000 seminiferous tubules. Each seminiferous tubule is lined with a complex stratified epithelium; it is about 150–250 μm in diameter and 30–70 cm long. The combined length of the tubules of one testis is about 250 m. The tubules are convoluted and begin closed-ended. At their termination, the lumen narrows and continues in short segments, known as **straight tubules,** or **tubuli recti,** that connect the seminiferous tubules to an anastomosing labyrinth of epithelium-lined channels, the **rete testis.** About 10–20 **ductuli efferentes** connect the rete testis to the cephalic portion of the **epididymis** (Figure 22–1).

The seminiferous tubules consist of a tunic of fibrous connective tissue, a well-defined basal lamina, and a complex **germinal,** or **seminiferous, epithelium** (Figure 22–2). The fibrous **tunica propria** enveloping the seminiferous tubule consists of several layers of fibroblasts. The innermost layer adhering to the basal lamina consists of flattened **myoid cells** (Figure 22–3),

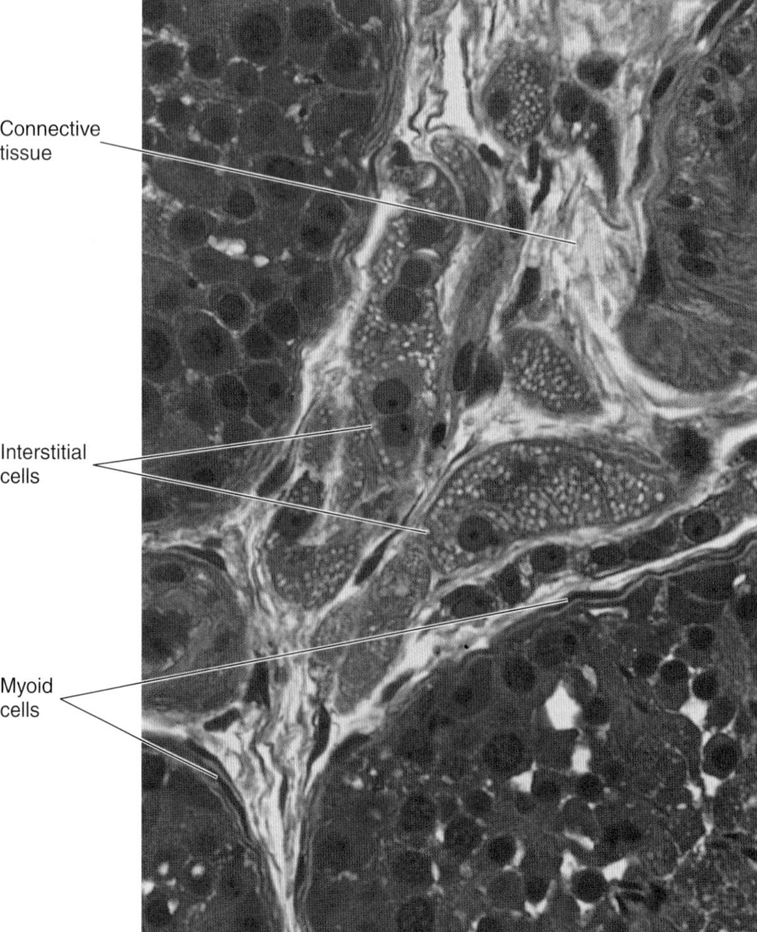

Figure 22–3. Epithelium of seminiferous tubules surrounded by myoid cells. The spaces between the tubules contain connective tissue, blood and lymphatic vessels, and interstitial cells. PT stain. Medium magnification.

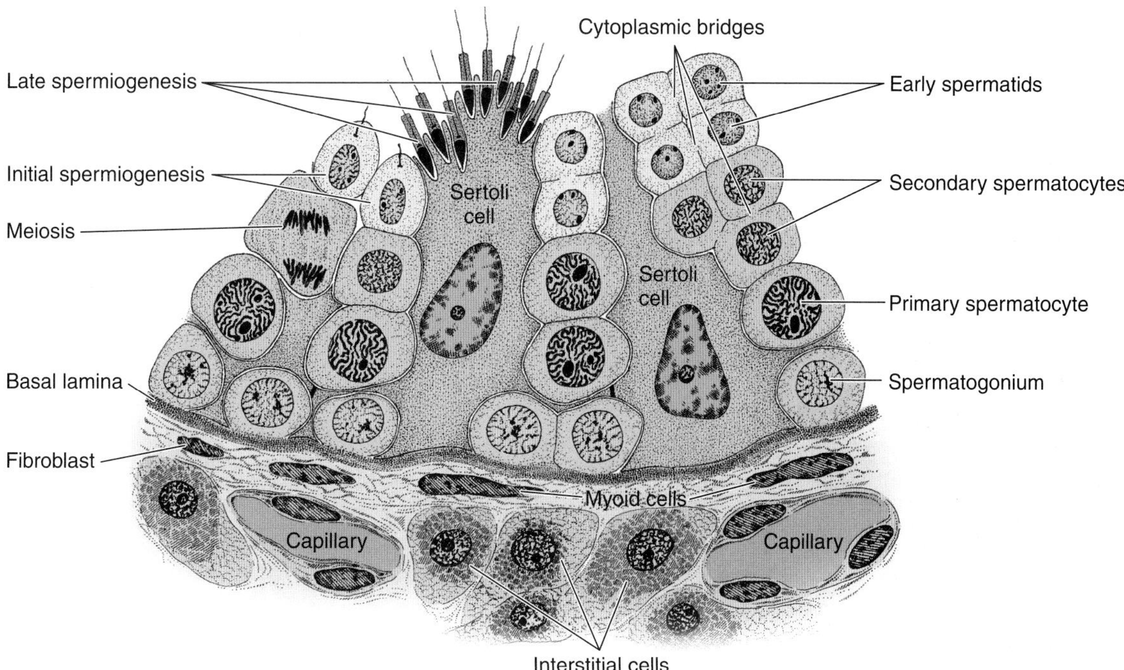

Figure 22–4. Part of a seminiferous tubule with its surrounding tissues. The seminiferous epithelium is formed by 2 cell populations: the cells of the spermatogenic lineage and the supporting or Sertoli cells.

which have characteristics of smooth muscle. Interstitial cells occupy much of the space between the seminiferous tubules (Figures 22–2 and 22–3).

The seminiferous epithelium consists of 2 types of cells: **Sertoli,** or **supporting, cells** and cells that constitute the **spermatogenic lineage** (Figure 22–4). The cells of the spermatogenic lineage are stacked in 4–8 layers; their function is to produce spermatozoa. The production of spermatozoa is called **spermatogenesis,** a process that includes cell division through mitosis and meiosis and the final differentiation of spermatozoids, called **spermiogenesis.**

Spermatogenesis

Spermatogenesis is the process of formation of spermatozoids. The process begins with a primitive germ cell, the **spermatogonium** (Gr. *sperma* + *gone,* generation), which is a relatively small cell, about 12 μm in diameter, situated next to the basal lamina of the epithelium (Figures 22–4, 22–5, and 22–6). At sexual maturity, spermatogonia begin dividing by mitosis, producing successive generations of cells. The newly formed cells can follow one of 2 paths: They can continue dividing as stem cells, also called **type A spermatogonia,** or they can differentiate during progressive mitotic cycles to become **type B spermatogonia** (Figure 22–7). Type B spermatogonia are progenitor cells that will differentiate into **primary spermatocytes** (Figure 22–7). The primary spermatocyte has 46 (44 + XY) chromosomes and 4N of DNA. (N denotes either the haploid set of chromosomes [23 chromosomes in humans] or the amount of DNA in this set.) Soon after their formation, these cells enter the prophase of the first meiotic division. Since the prophase of this division takes about 22 days, the majority of spermatocytes seen in sections will be in this phase. The primary spermatocytes are the largest cells of the spermatogenic lineage and are characterized by the presence of chromosomes in various stages of the coiling process within their nuclei (Figures 22–5, 22–6, and 22–8).

From this first meiotic division arise smaller cells called **secondary spermatocytes** (Figures 22–4 and 22–7) with only 23 chromosomes (22 + X or 22 + Y). This decrease in number (from 46 to 23) is accompanied by a reduction in the amount of DNA per cell (from 4N to 2N). Secondary spermatocytes are difficult to observe in sections of the testis because they are short-lived cells that remain in interphase very briefly and quickly enter into the second meiotic division. Division of each secondary spermatocyte results in 2 cells that contain 23 chromosomes, the **spermatids** (Figure 22–7). Because no S phase (DNA synthesis) occurs between the first and second meiotic divisions of the spermatocytes, the amount of DNA per cell in this second division is reduced by half, forming haploid (1N) cells. The meiotic process therefore results in the formation of cells with a haploid number of chromosomes. With fertilization, they return to the normal diploid number.

Spermiogenesis

Spermiogenesis is the final stage of production of spermatozoids. Spermiogenesis is the process by which spermatids transform into spermatozoa, cells which are highly specialized to deliver male DNA to the ovum. No cell division occurs during this process.

The spermatids can be distinguished by their small size (7–8 μm in diameter) and by nuclei with areas of condensed chromatin. Their position within the seminiferous tubules is close to the lumen (Figures 22–4, 22–5, and 22–6). Spermiogenesis is a complex process that includes formation of the acrosome (Gr. *akron,* extremity, + *soma,* body), condensation and elongation of the nucleus, development of the flagellum, and

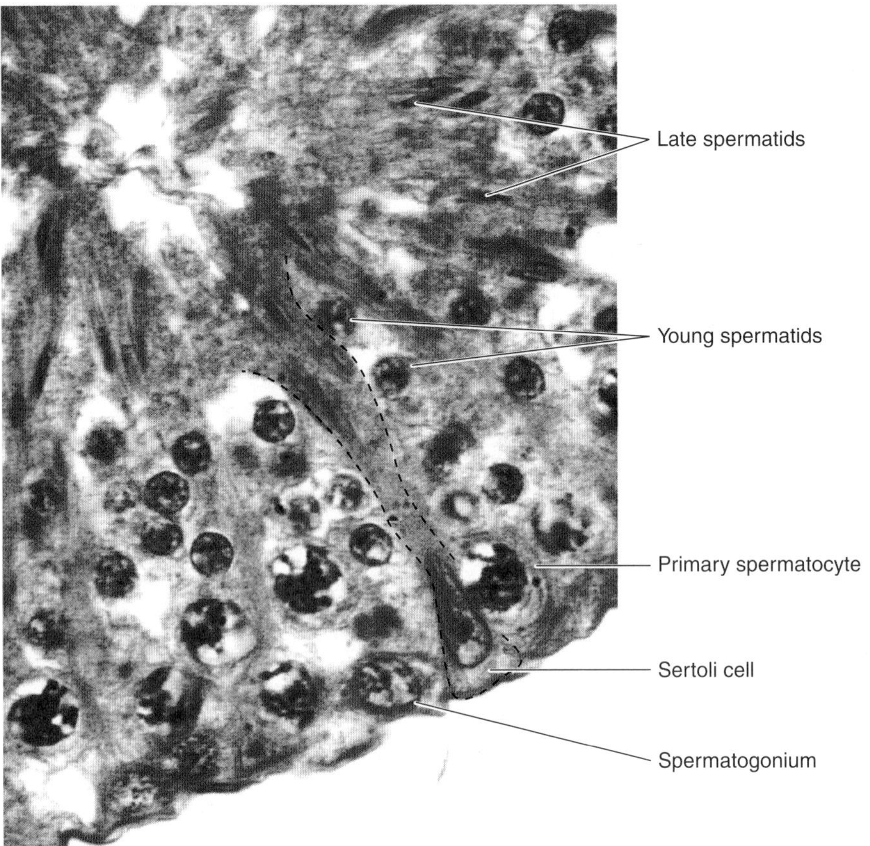

Figure 22–5. Part of the wall of a seminiferous tubule. Several cells of the spermatogenic lineage are present: a spermatogonium, primary spermatocytes, and young and late spermatids. The approximate limits of a Sertoli cell holding several spermatids are delineated. H&E stain. High magnification.

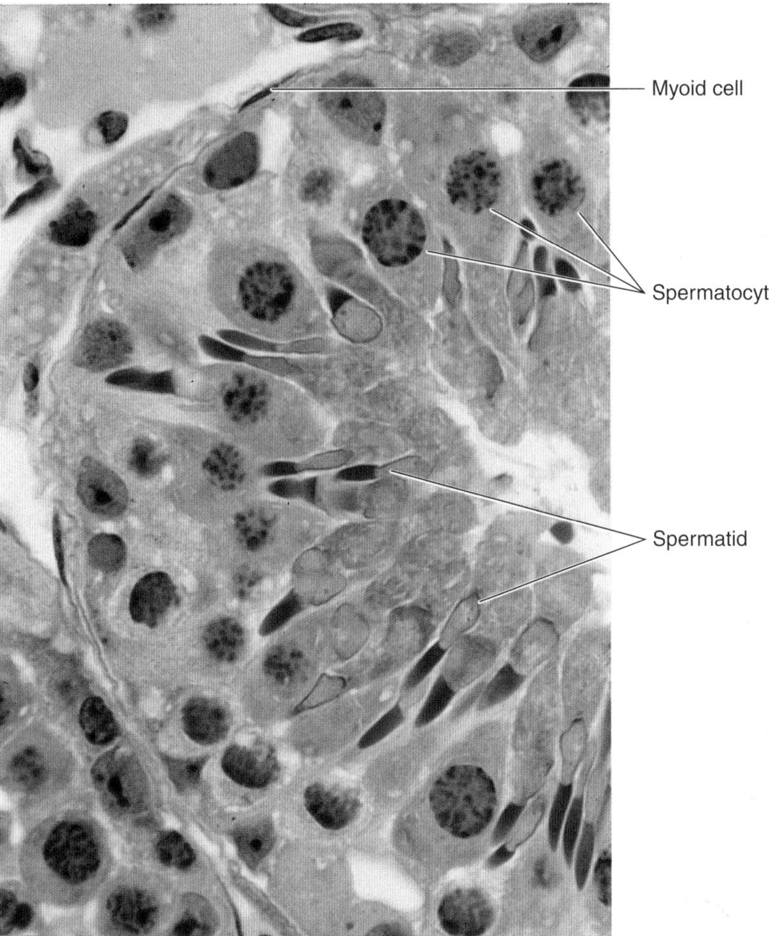

Figure 22–6. Spermatocytes and spermatids in the epithelium of a seminiferous tubule. The tubule is covered by myoid cells. Picrosirius-hematoxylin (PSH) stain. Medium magnification.

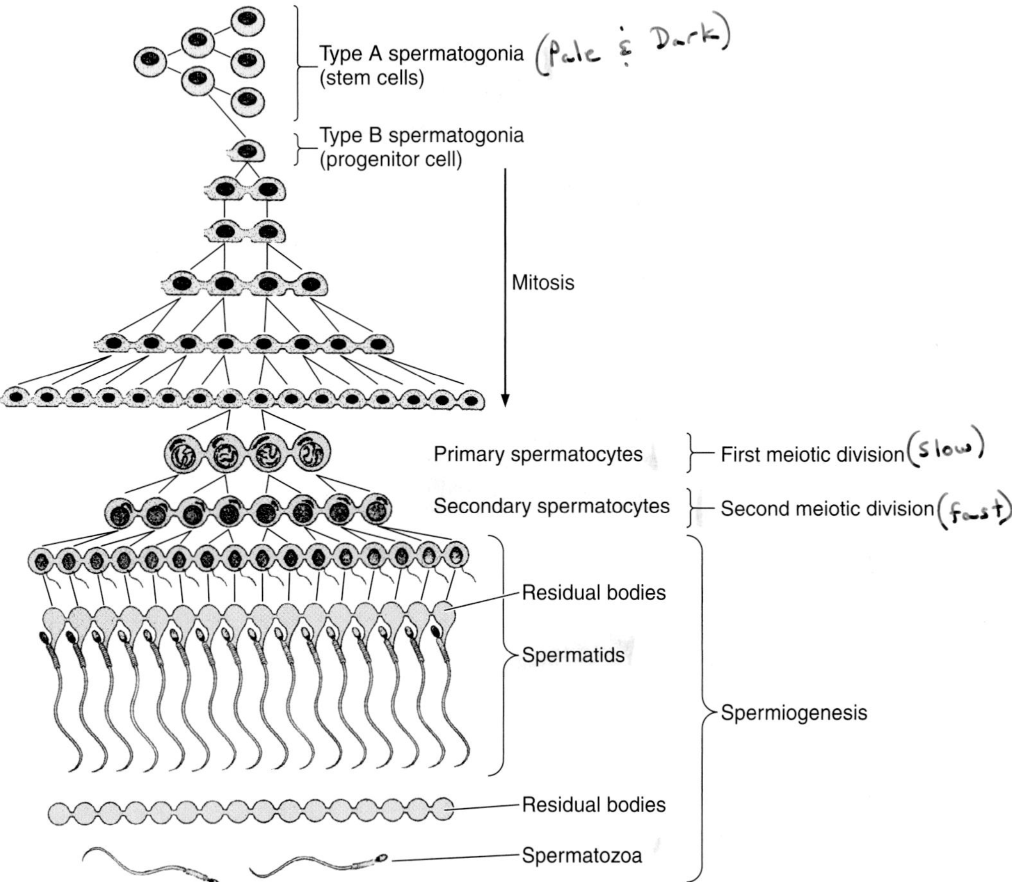

Figure 22–7. **Diagram showing the clonal nature of the germ cells. Only the initial spermatogonia divide and produce separate daughter cells. Once committed to differentiation, the cells of all subsequent divisions stay connected by intercellular cytoplasmic bridges. Only after they are separated from the residual bodies can the spermatozoa be considered isolated cells.** (Modified and reproduced, with permission, from Bloom W, Fawcett DW: *A Textbook of Histology*, 10th ed. Saunders, 1975.)

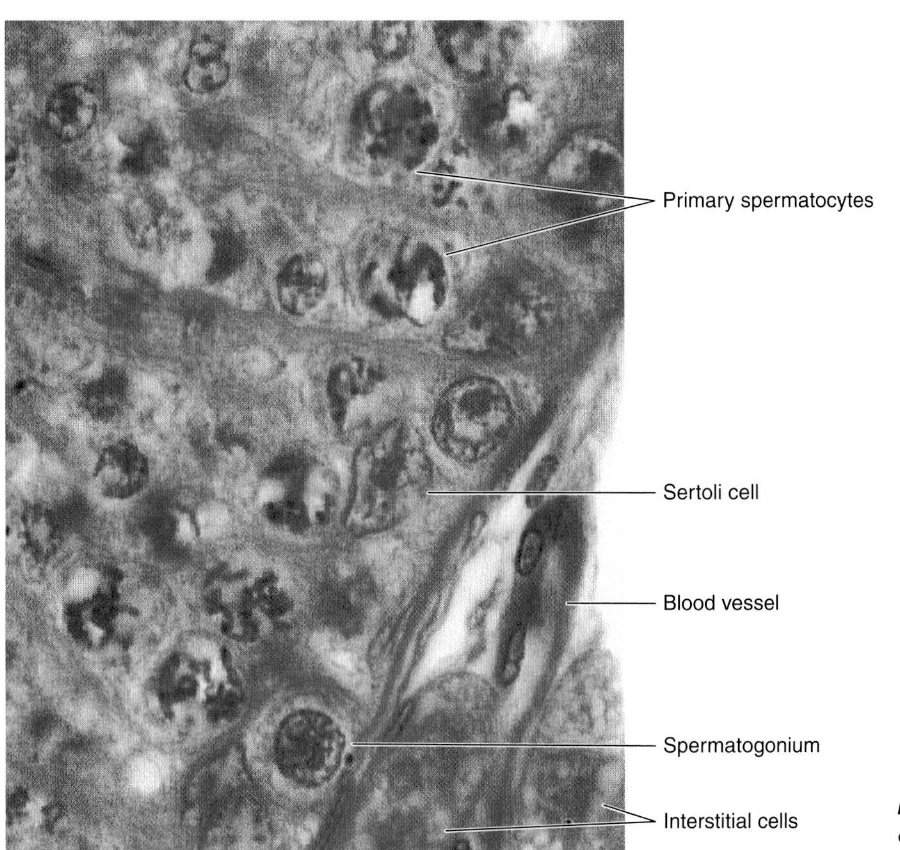

Figure 22–8. Interstitial cells and cells of the seminiferous epithelium. H&E stain. High magnification.

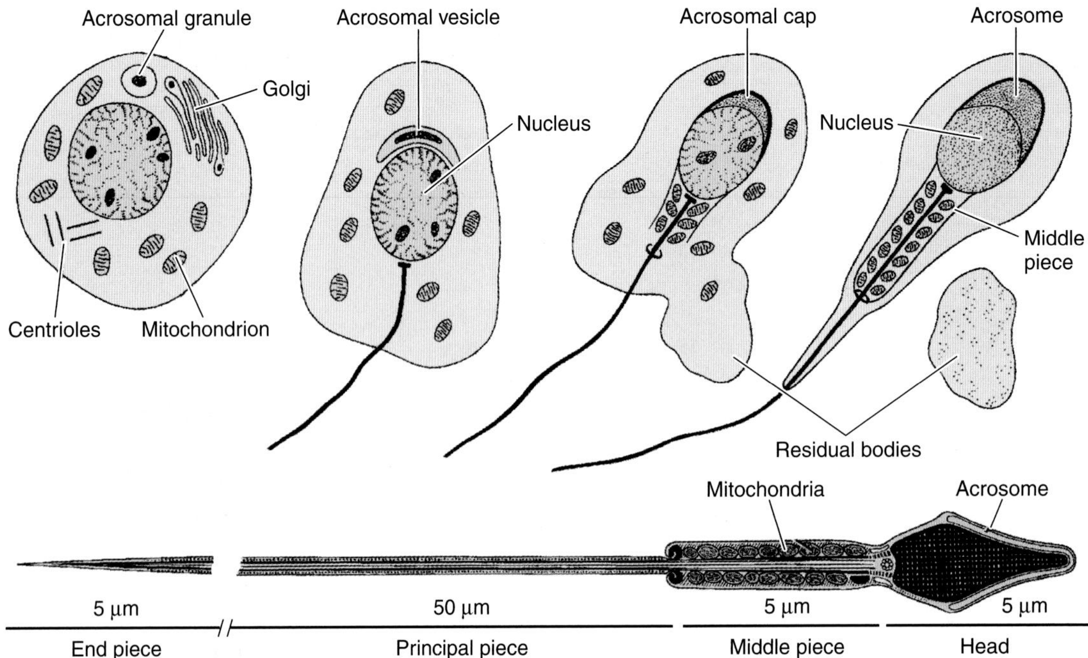

Figure 22–9. **Top:** The principal changes occurring in spermatids during spermiogenesis. The basic structural feature of the spermatozoon is the head, which consists primarily of condensed nuclear chromatin. The reduced volume of the nucleus affords the sperm greater mobility and may protect the genome from damage while in transit to the egg. The rest of the spermatozoon is structurally arranged to promote motility. **Bottom:** The structure of a mature spermatozoon.

the loss of much of the cytoplasm. The end result is the mature spermatozoon, which is then released into the lumen of the seminiferous tubule. Spermiogenesis can be divided into 3 phases:

THE GOLGI PHASE

The cytoplasm of spermatids contains a prominent Golgi complex near the nucleus, mitochondria, a pair of centrioles, free ribosomes, and tubules of smooth endoplasmic reticulum (Figure 22–9). Small PAS-positive granules called **proacrosomal granules** accumulate in the Golgi complex. They subsequently coalesce to form a single **acrosomal granule** within a membrane-limited **acrosomal vesicle** (Figure 22–9). The centrioles migrate to a position near the cell surface and opposite the forming acrosome. The flagellar axoneme begins to form, and the centrioles migrate back toward the nucleus, spinning out the axonemal components as they move.

THE ACROSOMAL PHASE

The acrosomal vesicle and granule spread to cover the anterior half of the condensing nucleus and are then known as the **acrosome** (Figures 22–9 and 22–10). The acrosome contains several hydrolytic enzymes, such as hyaluronidase, neuraminidase, acid phosphatase, and a protease that has trypsinlike activity. The acrosome thus serves as a specialized type of lysosome. These enzymes are known to dissociate cells of the corona radiata and to digest the zona pellucida, structures that surround the oocytes. When spermatozoa encounter an oocyte, the outer membrane of the acrosome fuses with the plasma membrane of a spermatozoon at several sites, liberating the acrosomal enzymes to the extracellular space. This process, the **acrosomal reaction,** is one of the first steps in fertilization.

During this phase of spermiogenesis, the nucleus of the spermatid becomes oriented toward the base of the seminiferous tubule, and the axoneme projects into its lumen. In addition, the nucleus becomes more elongated and condensed. One of the centrioles grows concomitantly, forming the **flagellum.** Mitochondria aggregate around the proximal part of the flagellum, forming a thickened region known as the **middle piece,** the region where the movements of the spermatozoa are generated (Figure 22–9).

This disposition of mitochondria is another example of a concentration of these organelles in sites related to cell movement and high energy consumption. (Flagellar structure and function are described in Chapter 2.) Movement of the flagellum is a result of the interaction among microtubules, ATP, and **dynein,** a protein with ATPase activity.

MEDICAL APPLICATION

Immotile cilia syndrome *is characterized by immotile spermatozoa and consequent infertility. It is due to a lack of dynein or other proteins required for ciliar and flagellar motility in the cells of the diseased person. This disorder usually coincides with chronic respiratory infections because of impaired motility of the ciliary axonemes of respiratory epithelial cells.*

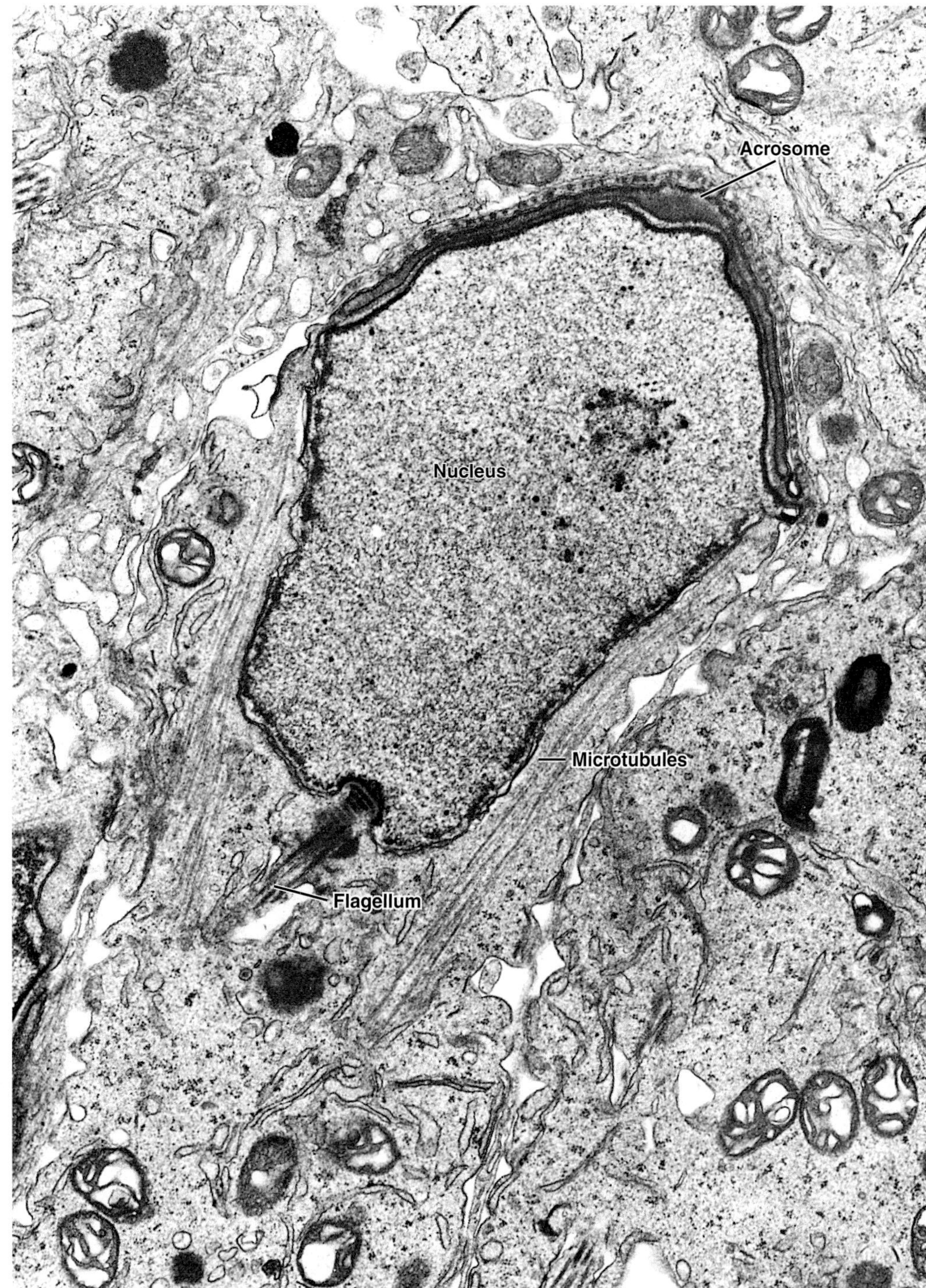

Figure 22–10. Electron micrograph of a mouse spermatid. In the center is the nucleus, covered by the acrosome. The flagellum can be seen emerging in the lower region below the nucleus. A cylindrical bundle of microtubules, the manchette, limits the nucleus laterally. (Courtesy of KR Porter.)

The Maturation Phase

Residual cytoplasm is shed and phagocytized by Sertoli cells, and the spermatozoa are released into the lumen of the tubule. Mature spermatozoa are shown in Figures 22–9 and 22–11.

The Clonal Nature of Germ Cells

The daughter cells that result from the division of type A spermatogonia remain separated until one of these cells becomes committed to transform into a type B spermatogonium. From

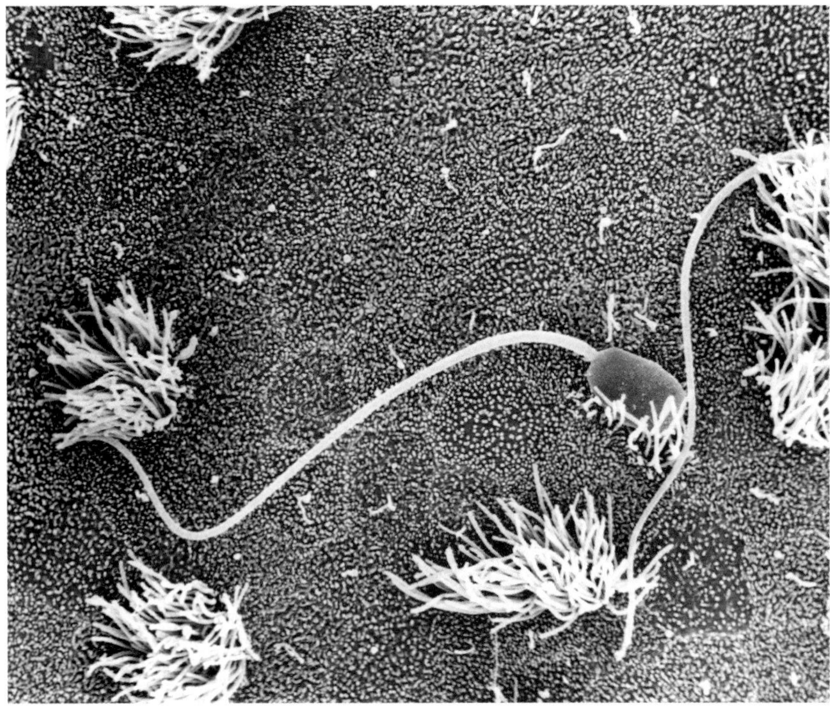

Figure 22–11. **Scanning electron micrograph of a spermatozoon in the uterine cavity of a rodent. The tufts are ciliated epithelial cells. ×2,000.** (Reproduced, with permission, from Motta P et al: *Microanatomy of Cell and Tissue Surfaces: An Atlas of Scanning Electron Microscopy.* Lea & Febiger, 1977. Copyright © Societa Editrice Libraria [Milan].)

this moment on, the cells that result from the division of these cells do not separate completely but remain attached by cytoplasmic bridges (Figure 22–7). The intercellular bridges provide communication between the primary and secondary spermatocytes and spermatids derived from a single spermatogonium. By permitting the interchange of information from cell to cell, these bridges play an important role in coordinating the sequence of events in spermatogenesis. This detail may be of importance in understanding the cycle of the seminiferous epithelium (described below). When the process of spermatogenesis is completed, the sloughing of the cytoplasm and cytoplasmic bridges as residual bodies leads to a separation of the late spermatids.

Spermatozoa are transported to the epididymis in an appropriate medium, **testicular fluid,** produced by the Sertoli cells and rete testis. This fluid contains steroids, proteins, ions, and androgen-binding protein associated with testosterone.

Experimental injection of ^{3}H-thymidine into the testes of volunteers showed that, in humans, the changes that occur between the spermatogonia stage and the formation of the spermatozoa take about 64 days. Aside from the slowness of the process, spermatogenesis occurs neither simultaneously nor synchronously inside each seminiferous tubule and among all the seminiferous tubules. This means that at each small site of the wall of the tubule, spermatogenesis proceeds more or less independently from the neighbor sites. Because of this asynchrony, different areas of the section of one tubule as well as sections of different tubules exhibit different phases of spermatogenesis. It also explains why spermatozoa are encountered in some regions of the seminiferous tubules, whereas only spermatids are found in others. This asynchrony is called the **cycle of the seminiferous epithelium.**

Sertoli Cells

The **Sertoli cells** are important for the function of the testes. These **cells** are elongated pyramidal cells that partially envelop cells of the spermatogenic lineage. The bases of the Sertoli cells adhere to the basal lamina, and their apical ends frequently extend into the lumen of the seminiferous tubule. In the light microscope, the outlines of Sertoli cells appear poorly defined because of the numerous lateral processes that surround spermatogenic cells (Figures 22–4, 22–5, and 22–12). Studies with the electron microscope reveal that these cells contain abundant smooth endoplasmic reticulum, some rough endoplasmic reticulum, a well-developed Golgi complex, and numerous mitochondria and lysosomes. The elongated nucleus, which is often triangular in outline, possesses numerous infoldings and a prominent nucleolus; it exhibits little heterochromatin.

Adjacent Sertoli cells are bound together by occluding junctions at the basolateral part of the cell, forming a **blood-testis barrier.** The spermatogonia lie in a **basal compartment** that is situated below the barrier. During spermatogenesis, some of the cells resulting from division of spermatogonia somehow traverse these junctions and come to lie in the **adluminal compartment** situated above the barrier. Spermatocytes and spermatids lie within deep invaginations of the lateral and apical margins of the Sertoli cells, above the barrier. As the flagellar tails of the spermatids develop, they appear as tufts extending from the apical ends of the Sertoli cells. Sertoli cells are also connected by gap junctions that provide ionic and chemical coupling of the cells; this may be important in coordinating the cycle of the seminiferous epithelium described above.

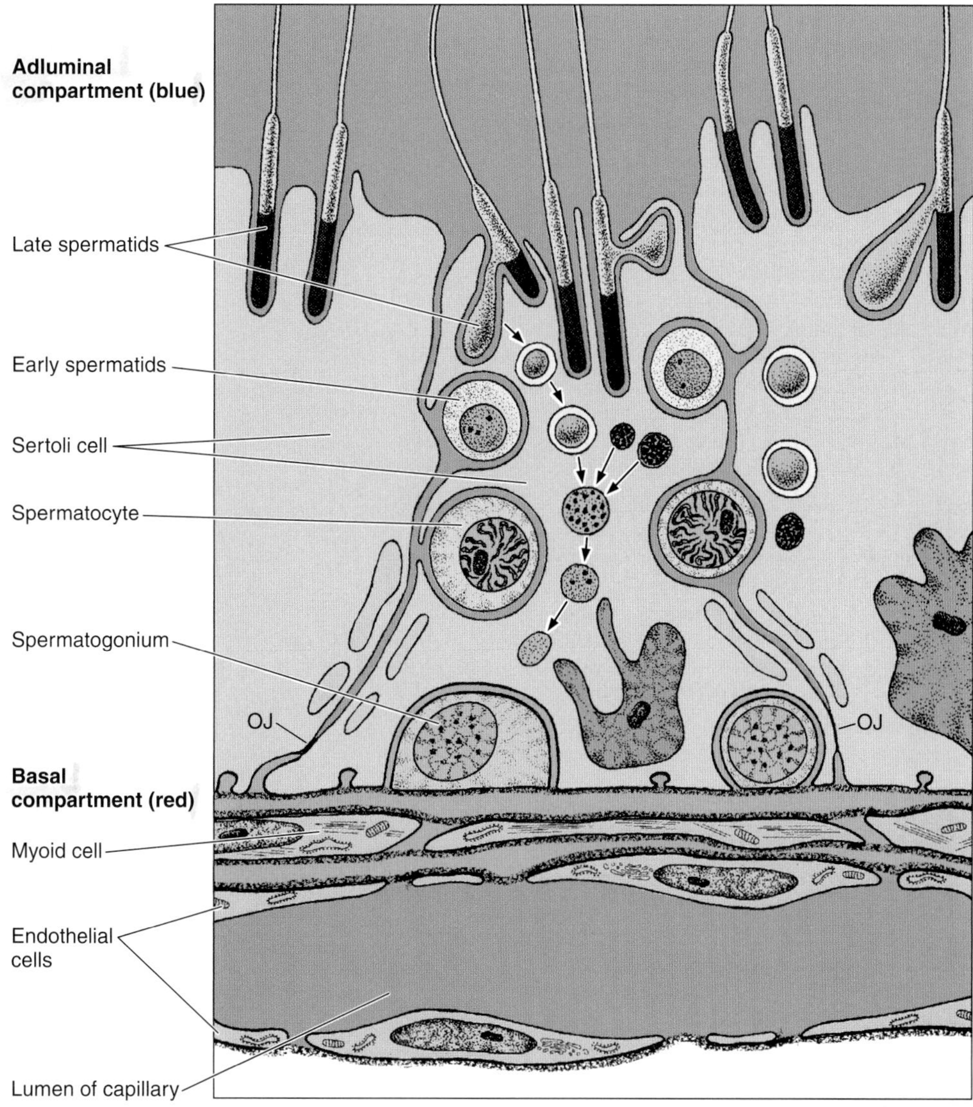

Figure 22–12. The Sertoli cells form the blood-testis barrier. Neighbor Sertoli cells are attached by occluding junctions that divide the seminiferous tubules into 2 compartments and impede the passage of substances between both compartments. The basal compartment comprises the interstitial space and the spaces occupied by the spermatogonia. The adluminal compartment comprises the tubule lumen and the intercellular spaces down to the level of the occluding junctions (OJ). In this compartment are spermatocytes, spermatids, and spermatozoa. Cytoplasmic residual bodies from spermatids undergo phagocytosis by the Sertoli cells and are digested by lysosomal enzymes. The myoid cells surround the seminiferous epithelium.

Sertoli cells have several functions:

- **Support, protection, and nutritional regulation of the developing spermatozoa.** As mentioned above, the cells of the spermatogenic series are interconnected via cytoplasmic bridges. This network of cells is physically supported by extensive cytoplasmic ramifications of the Sertoli cells. Because spermatocytes, spermatids, and spermatozoa are isolated from the blood supply by the blood-testis barrier, these spermatogenic cells depend on the Sertoli cells to mediate the exchange of nutrients and metabolites. The Sertoli cell barrier also protects the developing sperm cells from immunologic attack (discussed below).
- **Phagocytosis.** During spermiogenesis, excess spermatid cytoplasm is shed as residual bodies. These cytoplasmic fragments are phagocytized and digested by Sertoli cell lysosomes.
- **Secretion.** Sertoli cells continuously secrete into the seminiferous tubules a fluid that flows in the direction of the genital ducts and is used for sperm transport. Secretion of an **androgen-binding protein** by Sertoli cells is under the control of follicle-stimulating hormone (FSH) and testosterone

and serves to concentrate testosterone in the seminiferous tubule, where it is necessary for spermatogenesis. Sertoli cells can convert testosterone to estradiol. They also secrete a peptide called **inhibin,** which suppresses synthesis and release of FSH in the anterior pituitary gland.

- **Production of the anti-müllerian hormone.** Anti-müllerian hormone (also called **müllerian-inhibiting hormone**) is a glycoprotein that acts during embryonic development to promote regression of the müllerian (paramesonephric) ducts in the male fetus; testosterone fosters the development of structures derived from the Wolffian (mesonephric) ducts. Sertoli cells in humans and other animals do not divide during the reproductive period. They are extremely resistant to such adverse conditions as infection, malnutrition, and x-irradiation and have a much better rate of survival after these insults than do cells of the spermatogenic lineage. In mammals, spermatozoa are probably released as a result of cellular movements, with the participation of microtubules and microfilaments in the Sertoli cell apex.
- **The blood-testis barrier.** The existence of a barrier between the blood and the interior of the seminiferous tubules accounts for the fact that few substances from the blood are found in the testicular fluid. The testicular capillaries are of the fenestrated type and permit passage of large molecules. Spermatogonia have free access to materials found in blood. However, occluding junctions between the Sertoli cells form a barrier to the transport of large molecules along the space between Sertoli cells. Thus, the more advanced stages of spermatogenesis are protected from blood-borne products protecting male germ cells against blood-borne noxious agents.

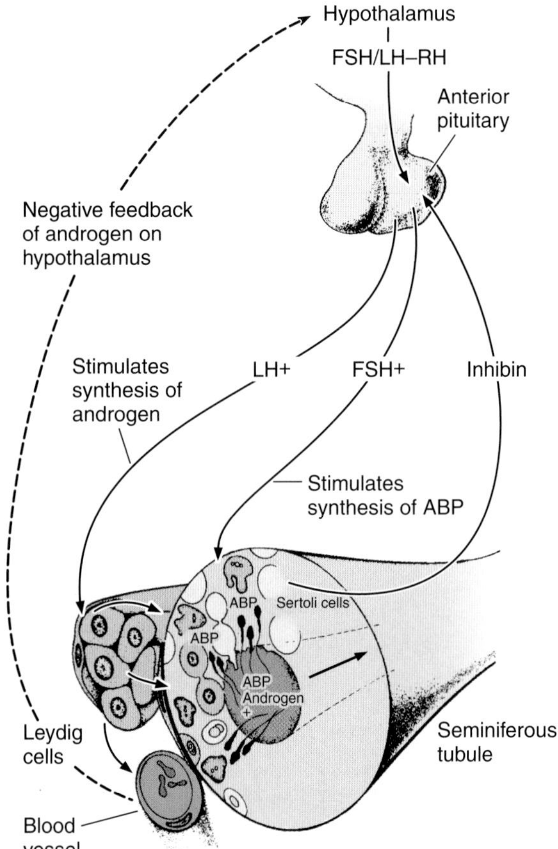

Figure 22–13. Hypophyseal control of male reproduction. Luteinizing hormone (LH) acts on the Leydig cells, and follicle-stimulating hormone (FSH) acts on the seminiferous tubules. A testicular hormone, inhibin, inhibits FSH secretion in the pituitary. ABP, androgen-binding protein. (Modified and reproduced, with permission, from Bloom W, Fawcett DW: *A Textbook of Histology,* 10th ed. Saunders, 1975.)

MEDICAL APPLICATION

Differentiation of spermatogonial cells leads to the appearance of sperm-specific proteins. Since sexual maturity occurs long after the development of immunocompetence, differentiating sperm cells could be recognized as foreign and provoke an immune response that would destroy the germ cells. The blood-testis barrier eliminates any interaction between developing sperm and the immune system. This barrier prevents the passage of immunoglobulins into the seminiferous tubule and accounts for the lack of impaired fertility in men whose serum contains high levels of sperm antibodies. The Sertoli cell barrier thus functions to protect the seminiferous epithelium against an autoimmune reaction.

Factors That Influence Spermatogenesis

Hormones are the factors that have the most important effect on spermatogenesis. Spermatogenesis depends on the action of the FSH and luteinizing hormone (LH) of the hypophysis on the testicular cells. LH acts on the interstitial cells, stimulating the production of testosterone necessary for the normal development of cells of the spermatogenic lineage. FSH is known to act on the Sertoli cells, stimulating adenylate cyclase and consequently increasing the presence of cAMP; it also promotes the synthesis and secretion of **androgen-binding protein.** This protein combines with testosterone and transports it into the lumen of the seminiferous tubules (Figure 22–13). Spermatogenesis is stimulated by testosterone and inhibited by estrogens and progestogens.

Temperature is very important in the regulation of spermatogenesis, which occurs only below the core body temperature of 37 °C. Testicular temperature is about 35 °C and is controlled by several mechanisms. A rich venous plexus (the **pampiniform plexus**) surrounds each testicular artery and forms a countercurrent heat-exchange system that is important in maintaining the testicular temperature. Other factors are evaporation of sweat from the scrotum, which contributes to heat loss, and contraction of cremaster muscles of the spermatic cords, which pull the testes into the inguinal canals, where their temperature can be increased.

MEDICAL APPLICATION

*Failure of descent of the testes (**cryptorchidism** [Gr. kryptos, hidden, + orchis, testis]) maintains the testes at the core body temperature of 37 °C, which inhibits spermatogenesis. In cases that are not too far advanced, spermatogenesis can occur normally if the testes are moved surgically to the scrotum. For this reason, it is important to examine male newborns to check if the testicles are present in the scrotum. Although germ cell proliferation is inhibited by abdominal temperature, testosterone synthesis is not. This explains why men with cryptorchidism can be sterile but still develop secondary male characteristics and achieve erection.*

Malnutrition, alcoholism, and the action of certain drugs lead to alterations in spermatogonia, with a resulting decrease in production of spermatozoa. X-irradiation and cadmium salts are quite toxic to cells of the spermatogenic lineage, causing the death of those cells and sterility in animals. The drug busulfan acts on the germinal cells; when administered to pregnant female rats, it promotes the death of the germinal cells of their offspring. The offspring are therefore sterile, and their seminiferous tubules contain only Sertoli cells. Androgen-producing interstitial cell tumors can cause precocious puberty in males.

Interstitial Tissue

The interstitial tissue of the testis is an important site of production of androgens. The spaces between the seminiferous tubules in the testis are filled with connective tissue, nerves, blood, and lymphatic vessels. Testicular capillaries are fenestrated and permit the free passage of macromolecules such as blood proteins. The extensive network of lymphatic vessels in the interstitial space explains the similarity of composition between the interstitial fluid and lymph collected from this organ. The connective tissue consists of various cell types, including fibroblasts, undifferentiated connective cells, mast cells, and macrophages. During puberty, an additional cell type becomes apparent; it is either rounded or polygonal in shape and has a central nucleus and an eosinophilic cytoplasm rich in small lipid droplets (Figures 22–3 and 22–14). These are the **interstitial,** or **Leydig,** cells of the testis, and they have the characteristics of steroid-secreting cells (described in Chapter 4). These cells produce the male hormone **testosterone,** which is responsible for the development of the secondary male sex characteristics. Testosterone is synthesized by enzymes present in mitochondria and the smooth endoplasmic reticulum, an example of cooperation between organelles.

Both the activity and the number of the interstitial cells depend on hormonal stimuli. During human pregnancy, placental gonadotropic hormone passes from the maternal blood to the male fetus, stimulating the abundant fetal testicular interstitial cells that produce androgenic hormones. The presence of these hormones is required for the embryonic differentiation of the male genitalia. The embryonic interstitial cells remain fully differentiated for up to 4 months of gestation; they then regress, with an associated decrease in testosterone synthesis. They remain quiescent throughout the rest of the pregnancy and up to

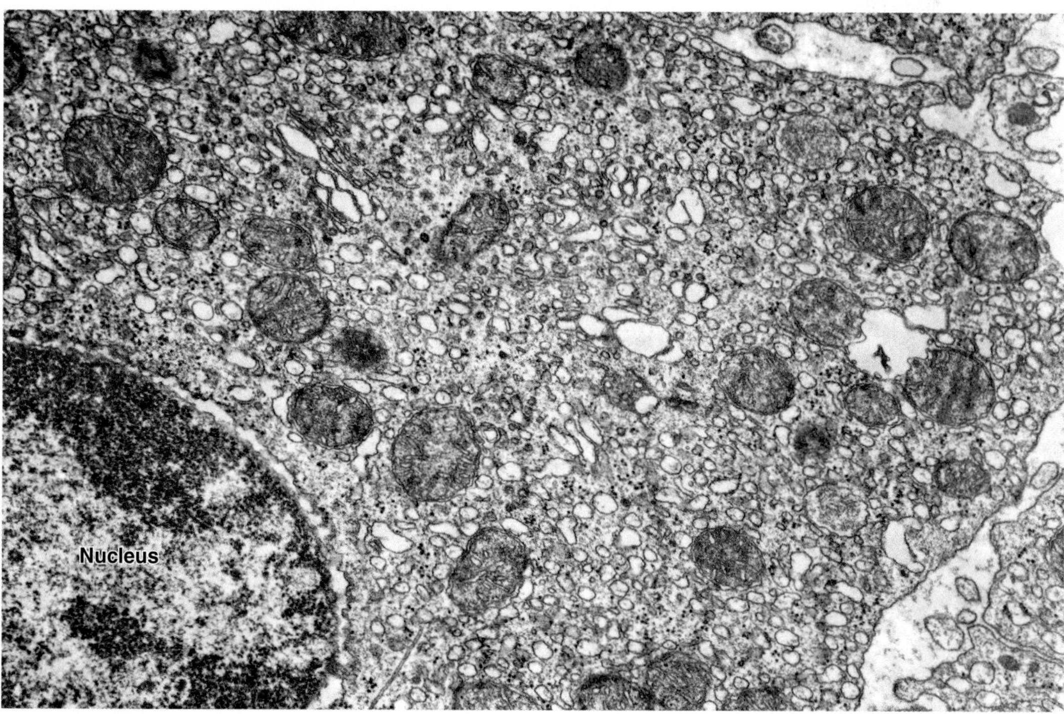

Figure 22–14. Electron micrograph of a section of an interstitial cell. There is abundant smooth endoplasmic reticulum as well as mitochondria. Medium magnification.

the prepubertal period, when they resume testosterone synthesis in response to the stimulus of LH from the hypophysis.

INTRATESTICULAR GENITAL DUCTS

The intratesticular genital ducts are the **tubuli recti** (straight tubules), the **rete testis,** and the **ductuli efferentes** (Figure 22–1). These ducts carry spermatozoa and liquid from the seminiferous tubules to the ductus epididymidis.

Most seminiferous tubules are in the form of loops, both ends of which join the rete testis by structures known as **tubuli recti.** These tubules are recognized by the gradual loss of spermatogenic cells, with an initial segment in which only Sertoli cells remain to form their walls, followed by a main segment consisting of cuboidal epithelium supported by a dense connective tissue sheath.

Tubuli recti empty into the **rete testis,** contained within the mediastinum, a thickening of the tunica albuginea. The rete testis is a highly anastomotic network of channels lined with cuboidal epithelium.

From the rete testis extend 10–20 **ductuli efferentes** (Figure 22–1). They have an epithelium composed of groups of nonciliated cuboidal cells alternating with ciliated cells that beat in the direction of the epididymis. This gives the epithelium a characteristic scalloped appearance. The nonciliated cells absorb much of the fluid secreted by the seminiferous tubules. The activity of ciliated cells and fluid absorption create a fluid flow that sweeps spermatozoa toward the epididymis. A thin layer of circularly oriented smooth muscle cells is seen outside the basal lamina of the epithelium. The ductuli efferentes gradually fuse to form the ductus epididymidis of the epididymis.

EXCRETORY GENITAL DUCTS

Excretory genital ducts transport the spermatozoa produced in the testis toward the penile meatus. These ducts are the **ductus epididymidis,** the **ductus (vas) deferens,** and the **urethra.**

The **ductus epididymidis** is a single highly coiled tube (Figure 22–1) about 4–6 m in length. Together with surrounding connective tissue and blood vessels, this long canal forms the

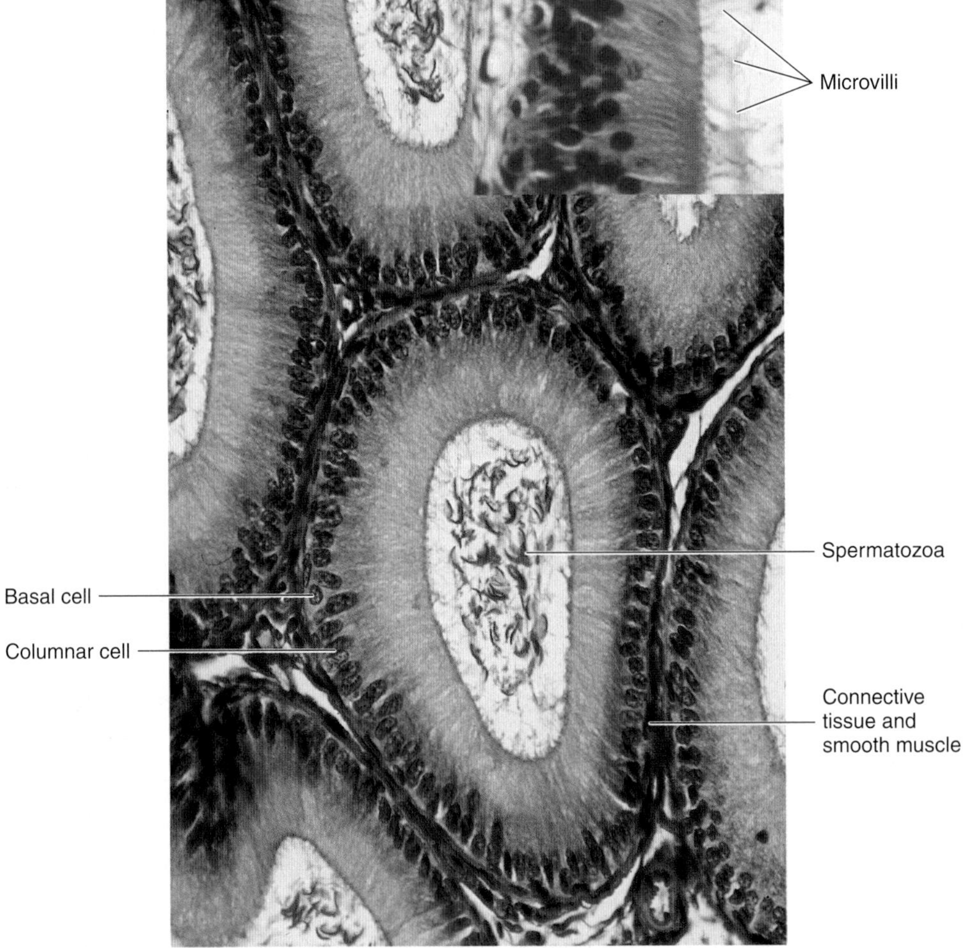

Figure 22–15. The highly coiled ductus **epididymidis,** sectioned several times. Its wall is made of a pseudostratified columnar epithelium surrounded by connective tissue and smooth muscle. PSH stain. Medium magnification. Inset: Higher magnification of the epithelial cells with their long microvilli (stereocilia).

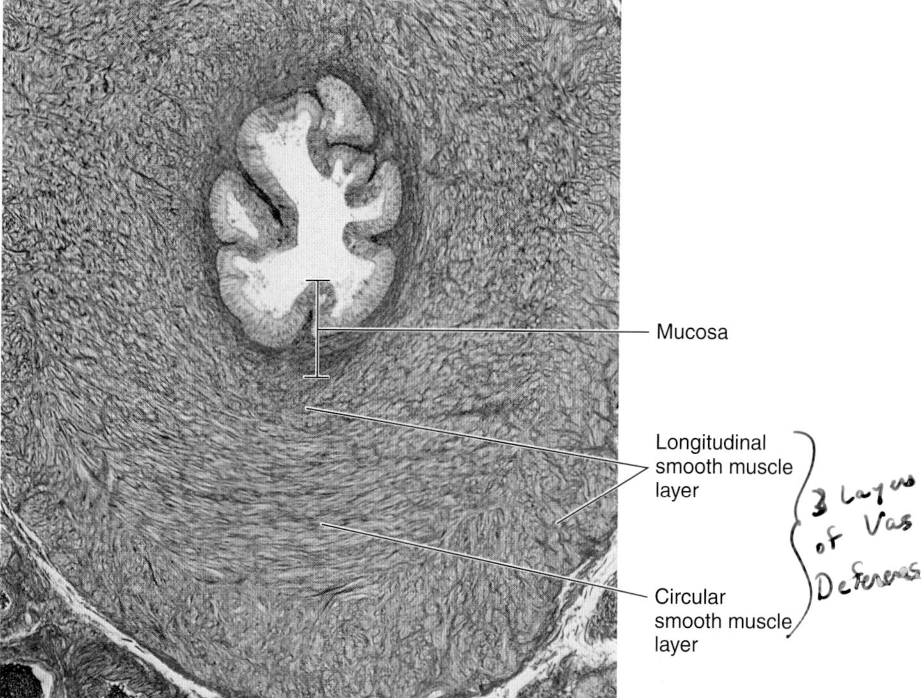

Figure 22–16. Section of the ductus deferens showing the mucosa formed by pseudostratified columnar epithelium with stereocilia and a lamina propria. The thick outer wall is formed of smooth muscle (brown) and collagen fibers (blue). Trichrome stain. Low magnification.

body and tail of the **epididymis.** It is lined with pseudostratified columnar epithelium composed of rounded basal cells and columnar cells (Figure 22–15). These cells are supported on a basal lamina surrounded by smooth muscle cells, whose peristaltic contractions help to move the sperm along the duct, and by loose connective tissue rich in blood capillaries. Their surface is covered by long, branched, irregular microvilli called **stereocilia.** The epithelium of the ductus epididymidis participates in the uptake and digestion of residual bodies that are eliminated during spermatogenesis.

From the epididymis the **ductus** (**vas**) **deferens,** a straight tube with a thick, muscular wall, continues toward the prostatic urethra and empties into it (Figure 22–1). It is characterized by a narrow lumen and a thick layer of smooth muscle (Figure 22–16). Its mucosa forms longitudinal folds and is covered along most of its extent by pseudostratified columnar epithelium with stereocilia. The lamina propria is a layer of connective tissue rich in elastic fibers, and the thick muscular layer consists of longitudinal inner and outer layers separated by a circular layer. The abundant smooth muscle produces strong peristaltic contractions that participate in the expulsion of the spermatozoa during ejaculation.

The ductus deferens forms part of the spermatic cord, which includes the testicular artery, the pampiniform plexus, and nerves. Before it enters the prostate, the ductus deferens dilates, forming a region called the **ampulla** (Figure 22–1). In this area, the epithelium becomes thicker and extensively folded. At the final portion of the ampulla, the seminal vesicles join the duct. From there on, the ductus deferens enters the prostate, opening into the prostatic **urethra.** The segment entering the prostate is called the **ejaculatory duct.** The mucous layer of the ductus deferens continues through the ampulla into the ejaculatory duct, but the muscle layer ends after the ampulla.

ACCESSORY GENITAL GLANDS

The accessory genital glands produce secretions that are essential for the reproductive function in men. The accessory genital glands are the **seminal vesicles,** the **prostate,** and the **bulbourethral glands.**

The **seminal vesicles** consist of 2 highly tortuous tubes about 15 cm in length. When the organ is sectioned, the same tube is observed in different orientations. It has a folded mucosa that is lined with cuboidal or pseudostratified columnar epithelium rich in secretory granules. These granules have ultrastructural characteristics similar to those found in protein-synthesizing cells (see Chapter 4). The lamina propria of the seminal vesicles is rich in elastic fibers and surrounded by a thin layer of smooth muscle (Figure 22–17). The seminal vesicles are not reservoirs for spermatozoa. They are glands that produce a viscid, yellowish secretion that contains spermatozoa-activating substances such as fructose, citrate, inositol, prostaglandins, and several proteins. Carbohydrates produced by the glands associated with the male reproductive system and secreted in the seminal fluid are the source of energy for sperm motility. The monosaccharide **fructose** is the most abundant of these carbohydrates. Seventy percent of human ejaculate originates in the seminal vesicles. The height of the epithelial cells of the seminal vesicles and the degree of activity of the secretory processes are dependent on testosterone levels.

The **prostate** is a collection of 30–50 branched tubuloalveolar glands. Their ducts empty into the prostatic urethra, which

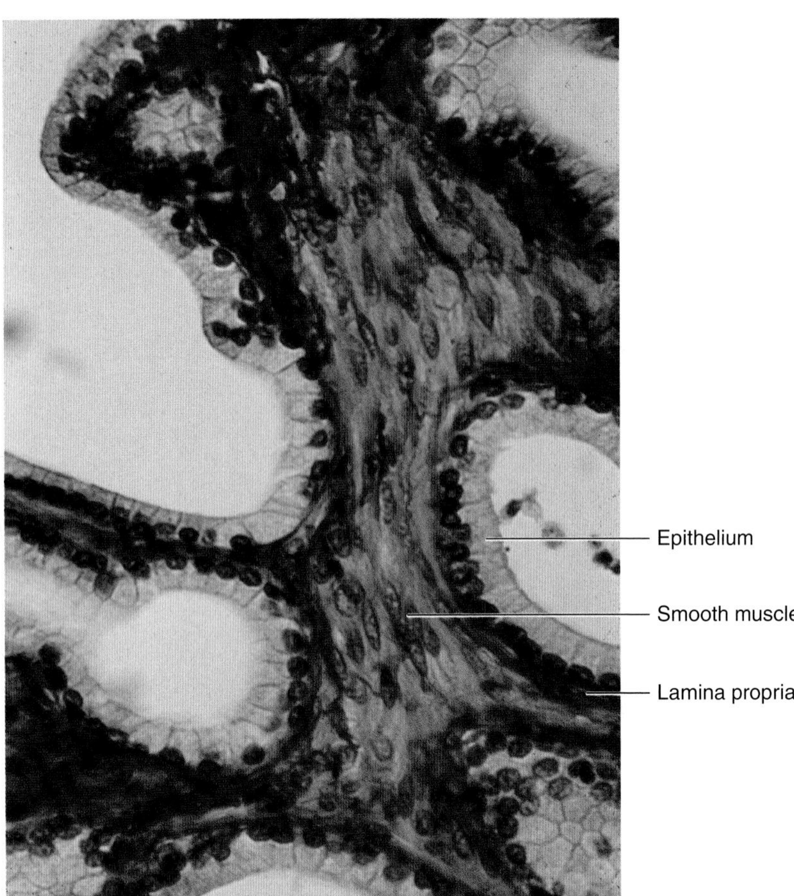

Figure 22–17. Seminal vesicle. A section of this tortuous tubular gland with a much-folded mucosa gives the impression that the gland consists of many tubules. PSH stain. Medium magnification.

crosses the prostate (Figures 22–1, 22–18, and 22–19). The prostate has 3 distinct zones: The **central zone** occupies 25% of the gland's volume. Seventy percent of the gland is formed by the **peripheral zone,** which is the major site of prostatic cancer. The **transition zone** is of medical importance because it is the site where most benign prostatic hyperplasia originates.

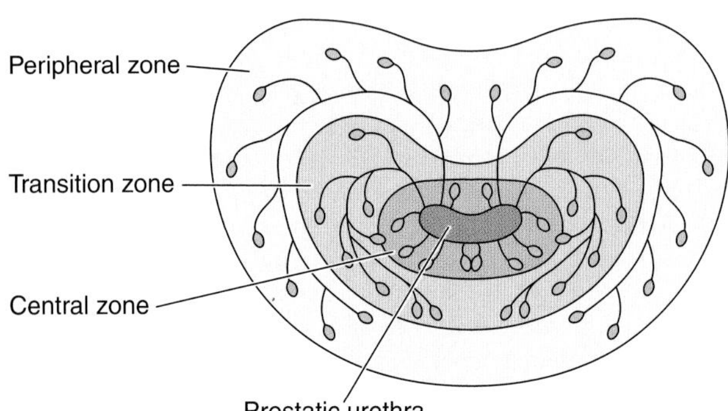

Figure 22–18. Section of prostate showing the distribution of its glands in 3 zones. The gland ducts open into the urethra.

The tubuloalveolar glands of the prostate are formed by a cuboidal or a columnar pseudostratified epithelium. An exceptionally rich fibromuscular stroma surrounds the glands (Figure 22–20). The prostate is surrounded by a fibroelastic capsule rich in smooth muscle. Septa from this capsule penetrate the gland and divide it into lobes that are indistinct in adult men.

The glands produce prostatic fluid and store it for expulsion during ejaculation. As with the seminal vesicle, the structure and function of the prostate depends on the level of testosterone.

MEDICAL APPLICATION

Benign prostatic hypertrophy *is present in 50% of men more than 50 years of age and in 95% of men more than 70 years of age. It leads to obstruction of the urethra with clinical symptoms in only 5–10% of cases.*

Malignant prostatic tumor *is the second most common form of cancer in men and the third leading cause of cancer deaths. One of the products of the prostate, the prostate specific antigen, is se-*

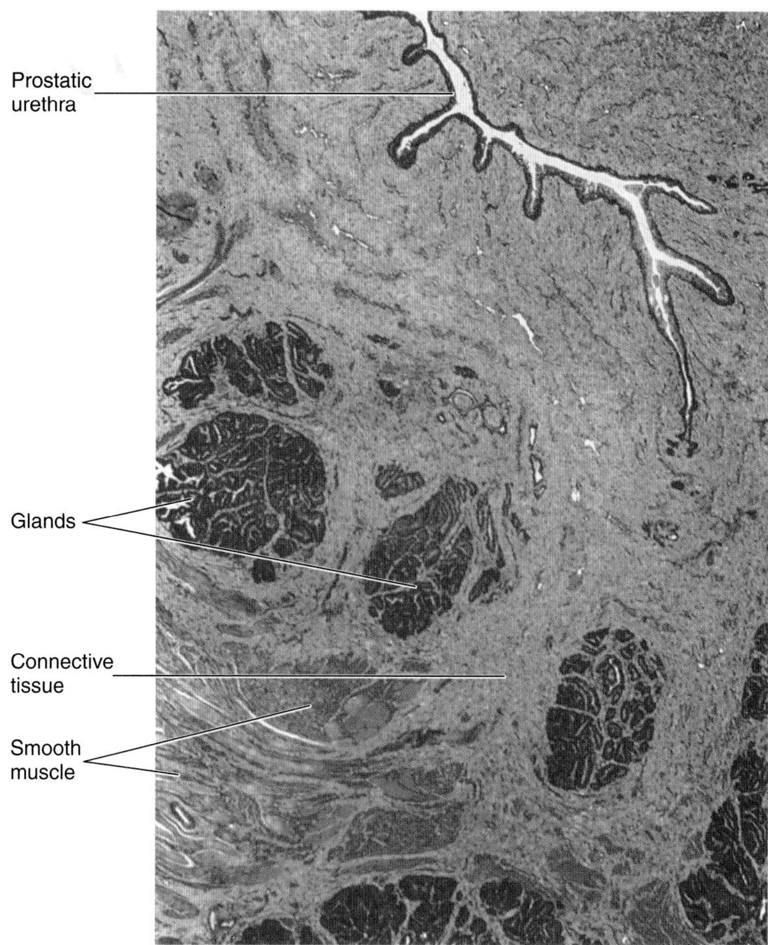

Figure 22–19. Section of the central region of the prostate showing the prostatic urethra and tubuloalveolar glands surrounded by connective tissue and smooth muscle. PT stain. Low magnification.

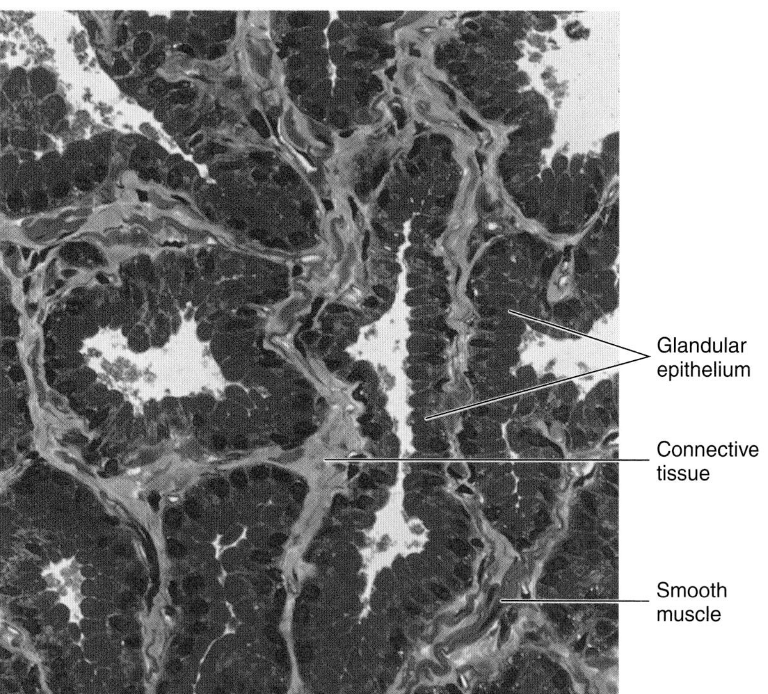

Figure 22–20. Glands of the prostate surrounded by connective tissue and smooth muscle. PT stain. Medium magnification.

creted into the blood. Because its concentration in the serum often increases during malignancy, it is useful for diagnosis and control of treatment of the tumor.

Small spherical bodies of glycoproteins, 0.2–2 mm in diameter and often calcified, are frequently observed in the lumen of prostatic glands. They are called **prostatic concretions,** or **corpora amylacea.** Their significance is not understood, but their number increases with age.

The **bulbourethral glands (Cowper's glands)**, 3–5 mm in diameter, are proximal to the membranous portion of the urethra and empty into it (Figure 22–1). They are tubuloalveolar glands lined with mucus-secreting simple cuboidal epithelium. Skeletal and smooth muscle cells are present in the septa that divide each gland into lobes. The secreted mucus is clear and acts as a lubricant.

PENIS

The main components of the penis are 3 cylindrical masses of erectile tissue, plus the urethra, surrounded by skin. Two of these cylinders—the **corpora cavernosa of the penis**—are placed dorsally. The other—the **corpus cavernosum of the urethra,** or **corpus spongiosum**—is ventrally located and surrounds the urethra. At its end it dilates, forming the **glans penis** (Figure 22–1). Most of the penile urethra is lined with pseudostratified columnar epithelium; in the glans penis, it becomes stratified squamous epithelium. Mucus-secreting **glands of Littre** are found throughout the length of the penile urethra.

The prepuce is a retractile fold of skin that contains connective tissue with smooth muscle in its interior. Sebaceous glands are present in the internal fold and in the skin that covers the glans.

The corpora cavernosa are covered by a resistant layer of dense connective tissue, the **tunica albuginea** (Figure 22–21). The corpora cavernosa of the penis and the corpus cavernosum of the urethra are composed of erectile tissue. This is a tissue with a large number of venous spaces lined with endothelial cells and separated by trabeculae of connective tissue fibers and smooth muscle cells.

The arterial supply of the penis derives from the internal pudendal arteries, which give rise to the deep arteries and the dorsal arteries of the penis. Deep arteries branch to form nutritive and helicine arteries. Nutritive arteries supply oxygen and nutrients to the trabeculae, and helicine arteries empty directly into the cavernous spaces (erectile tissue). There are arteriovenous shunts between the helicine arteries and the deep dorsal vein.

Penile erection is a hemodynamic event that is controlled by neural input to both arterial muscle and smooth muscle in the walls of the vascular spaces in the penis; in the flaccid state, there is minimal blood flow in the penis. The nonerect state is maintained by both the intrinsic tone of penile smooth muscle and the tone induced by continuous sympathetic input. Erection occurs when vasodilator impulses of parasympathetic origin cause relaxation of the penile vessels and cavernous smooth muscle. Vasodilatation also involves the concomitant inhibition of sympathetic vasoconstrictor impulses to penile tissues. Opening of the penile arteries and cavernous spaces accounts for the increase in blood flow, the filling of the cavernous spaces, and the resulting rigidity of the penis. Contraction and relaxation of corpora cavernosa depend on intracellular calcium, which in turn is modulated by guanosine monophosphate.

After ejaculation and orgasm, parasympathetic activity declines, and the penis returns to its flaccid state.

MEDICAL APPLICATION

The new drugs developed for treatment of penile erectile dysfunction act on an enzyme, a phosphodiesterase, present in the corpus cavernosum, which regulates cyclic nucleotides such as guanosine monophosphate.

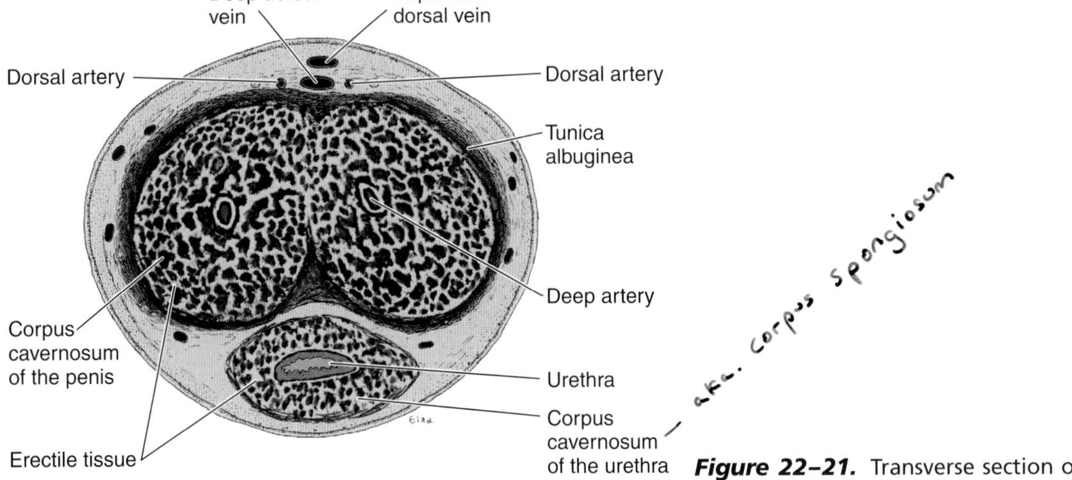

Figure 22–21. Transverse section of the penis.

REFERENCES

Afzelius BA et al: Lack of dynein arms in immotile human spermatozoa. J Cell Biol 1975;66:225.

Bonkhoff H, Remberger K: Morphogenetic aspects of normal and abnormal prostatic growth. Pathol Res Pract 1995;191:833.

Braunwald E et al: *Harrison's Principles of Internal Medicine,* 15th ed. McGraw-Hill, 2001.

Dail WG: Autonomic control of penile erectile tissue. In: *Experimental Brain Research.* Series 16. Springer-Verlag, 1987.

Fawcett DW: The mammalian spermatozoon. Dev Biol 1975;44:394.

Hafez ESE, Spring-Mills E (editors): *Accessory Glands of the Male Reproductive Tract.* Ann Arbor Science Publishers, 1979.

Johnson AD, Gomes WR (editors): *The Testis.* Vols 1–4. Academic Press, 1970–1977.

McNeal JE: Normal histology of the prostate. Am J Surg Pathol 1988;12:619.

Tindall DJ et al: Structure and biochemistry of the Sertoli cell. Int Rev Cytol 1985;94:127.

Trainer TD: Histology of the normal testis. Am J Surg Pathol 1987;11:797.

The Female Reproductive System 23

The female reproductive system consists of 2 ovaries, 2 oviducts (uterine tubes), the uterus, the vagina, and the external genitalia (Figure 23–1). Its functions are to produce female gametes (**oocytes**) and to hold a fertilized oocyte during its complete development through embryonic and fetal stages until birth. The system also produces sexual hormones that control organs of the reproductive system and have influence on other organs of the body. Beginning at **menarche,** when the first menses occurs, the reproductive system undergoes cyclic changes in structure and functional activity. These modifications are controlled by neurohumoral mechanisms. **Menopause** is a variable period during which the cyclic changes become irregular and eventually disappear. In the postmenopausal period there is a slow involution of the reproductive system. Although the mammary glands do not belong to the genital system, they are studied here because they undergo changes directly connected to the functional state of the reproductive system.

OVARIES

Ovaries are almond-shaped bodies approximately 3 cm long, 1.5 cm wide, and 1 cm thick. Their surface is covered by a simple squamous or cuboidal epithelium, the **germinal epithelium.** Under the germinal epithelium is a layer of dense connective tissue, the **tunica albuginea,** which is responsible for the whitish color of the ovary. Underneath the tunica albuginea is the **cortical region,** where ovarian follicles containing the oocytes predominate. The follicles are embedded in the connective tissue (**stroma**) of the cortical region. This stroma is composed of characteristic spindle-shaped fibroblasts that respond in a different way to hormonal stimuli than do fibroblasts of other organs. The most internal part of the ovary is the **medullary region,** containing a rich vascular bed within a loose connective tissue. There are no sharp limits between the cortical and medullary regions (Figures 23–2 and 23–3).

Early Development of the Ovary

Around the first month of embryonic life, a small population of **primordial germ cells** migrates from the yolk sac to the gonadal primordia. In the gonads these cells divide and transform into **oogonia.** Division is so intense that in the second month of intrauterine life there are around 600,000 oogonia, and around the fifth month more than 7 million. Beginning in the third month, oogonia begin to enter the prophase of the first meiotic division but stop at the diplotene stage and do not progress to other stages of meiosis. These cells are the **primary oocytes** (Gr. *oon,* egg, + *kytos,* cell), and they become surrounded by flattened cells called **follicular cells.** By the seventh month of pregnancy, most oogonia have transformed into primary oocytes. Many primary oocytes, however, are lost through a degenerative process called **atresia.** As a result, around puberty the ovaries contain about 300,000 oocytes. Atresia continues over the entire span of the woman's reproductive life so that at 40–45 years

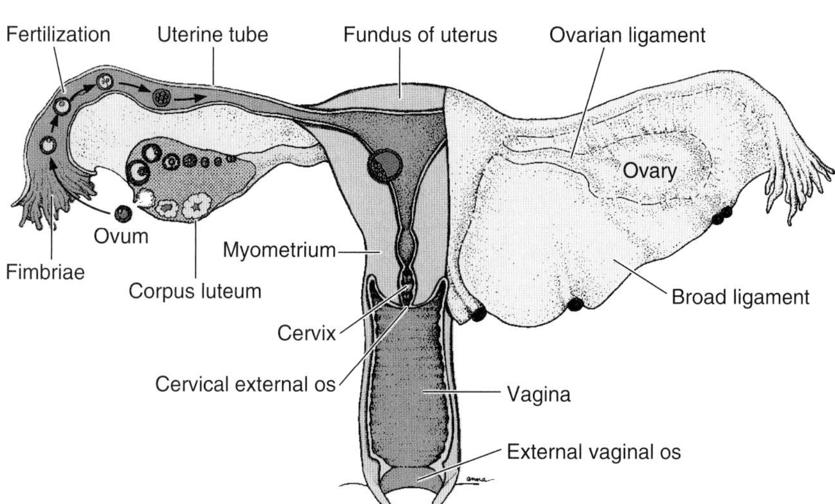

Figure 23–1. Internal organs of the female reproductive system.

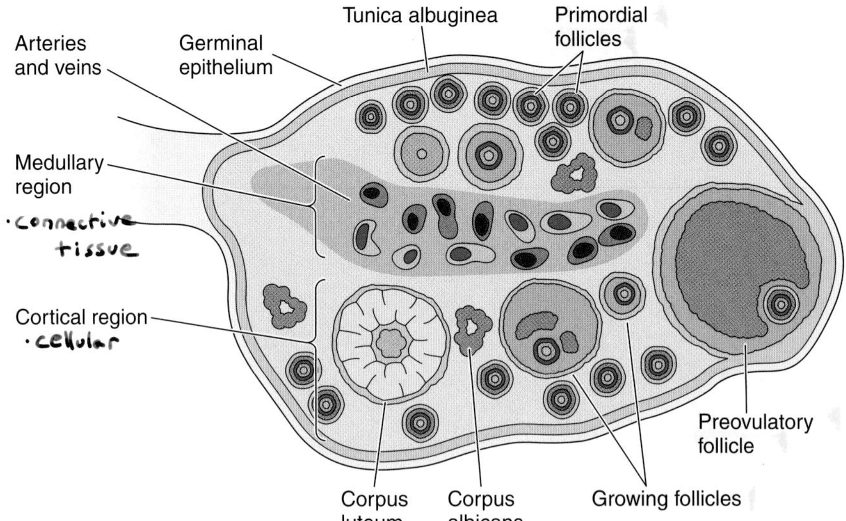

Figure 23–2. Ovary of a woman of reproductive age showing its main components: germinal epithelium, tunica albuginea, cortical region, and medullary region.

there are about 8,000 oocytes left. Since generally only one oocyte is liberated by the ovaries in each menstrual cycle (average duration, 28 days) and the reproductive life of a woman lasts about 30–40 years, only about 450 oocytes are liberated. All others degenerate through atresia.

Ovarian Follicles

An ovarian follicle consists of an oocyte surrounded by one or more layers of **follicular** cells, or **granulosa cells.** The follicles that are formed during fetal life—**primordial follicles**—consist of a primary oocyte enveloped by a single layer of flattened follicular cells (Figures 23–4, 23–5, and 23–6). These follicles are found in the superficial layer of the cortical region. The oocyte in the primordial follicle is a spherical cell about 25 μm in diameter. Its nucleus is large and has a large nucleolus. These cells are in the first prophase of meiosis. The chromosomes are mostly uncoiled and do not stain intensely. The organelles in the cytoplasm tend to form a clump adjacent to the nucleus. There are numerous mitochondria, several Golgi complexes, and cisternae of endoplasmic reticulum. A basal lamina underlies the follicular cells and marks the boundary between the follicle and the surrounding stroma.

Beginning in puberty, a small group of primordial follicles begins a daily process called follicular growth. This consists of modifications of the oocyte, of the granulosa cells, and of the stromal fibroblasts that surround these follicles. It is not known how the particular follicles that enter the growth stage are selected from the large population of primordial follicles.

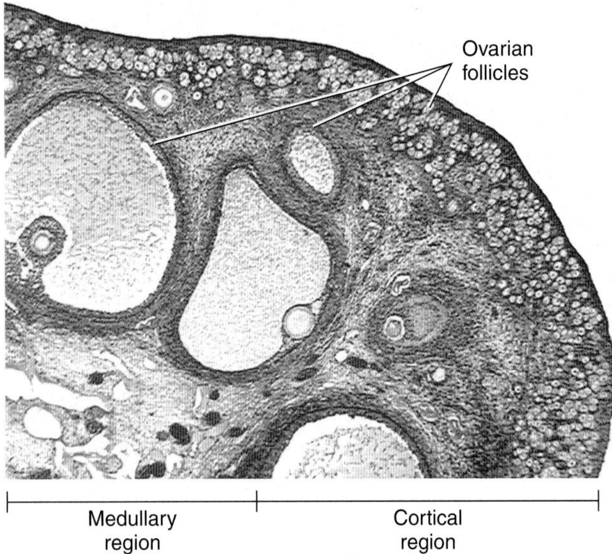

Figure 23–3. Photomicrograph of part of an ovary showing the cortical and medullary regions. H&E stain. Low magnification.

Follicular Growth

Follicular growth is stimulated by follicle-stimulating hormone, secreted by the hypophysis. Oocyte growth is most rapid during the first part of follicular growth, with the oocyte reaching a maximum diameter of about 120 μm. The nucleus enlarges; the mitochondria increase in number and become uniformly distributed throughout the cytoplasm; the endoplasmic reticulum hypertrophies, and the Golgi complexes migrate to just beneath the cell surface. Follicular cells divide by mitosis and form a single layer of cuboidal cells; the follicle is then called a **unilaminar primary follicle** (Figure 23–4 and 23–6). The follicular cells continue to proliferate and form a stratified follicular epithelium, or **granulosa layer,** whose cells communicate through gap junctions. The follicle is then called a **multilaminar primary** or **preantral follicle** (Figures 23–4 and 23–7). A thick amorphous layer, the **zona pellucida,** composed of at least 3 glycoproteins, is secreted and surrounds the oocyte (Figure 23–8). Both the oocyte and follicular cells are believed to contribute to the synthesis of the zona pellucida. Filopodia of follicular cells and microvilli of the oocyte penetrate the zona pellucida and make contact with one another via gap junctions.

As the follicles grow—owing mainly to the increase in size and number of granulosa cells—they move to deeper areas of the cortical region. Liquid (**liquor folliculi**) begins to accumulate between the follicular cells. The small spaces that contain

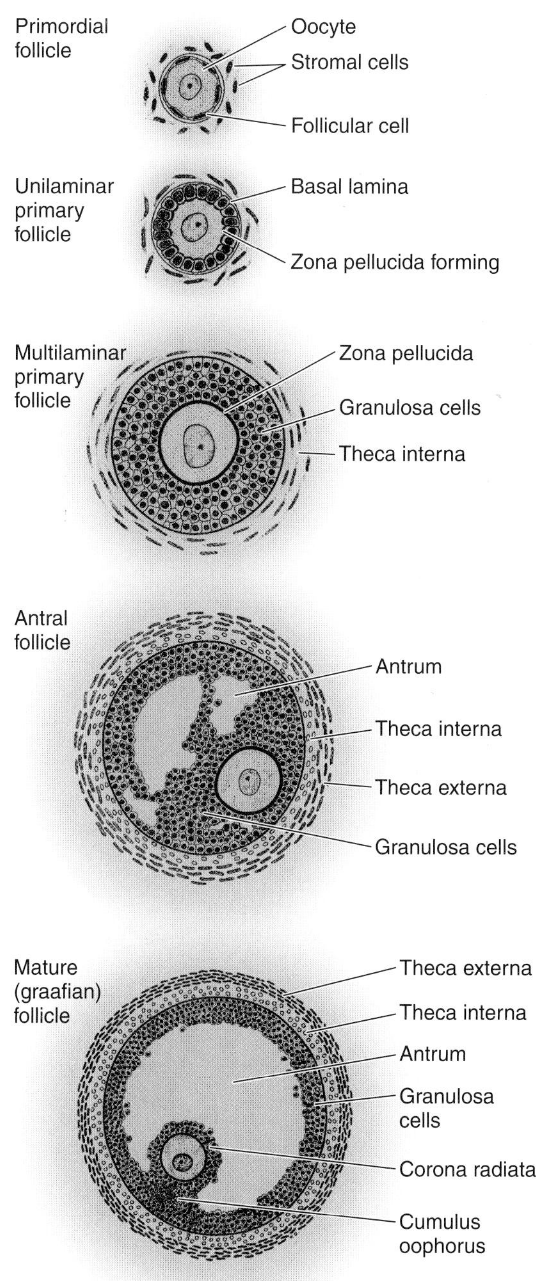

Figure 23–4. Types of ovarian follicles, from primordial to mature. The relative proportions of the follicles are not maintained in this drawing.

this fluid coalesce, and the granulosa cells reorganize themselves to form a larger cavity, the **antrum** (Figures 23–4 and 23–9). The follicles are then called **secondary** or **antral follicles.** Follicular fluid contains components of the plasma and products secreted by follicular cells. Glycosaminoglycans, several proteins (including steroid-binding proteins), and high concentrations of steroids (progesterone, androgens, and estrogens) are present.

During the reorganization of the granulosa cells to form the antrum, some cells of this layer concentrate at a certain point on the follicular wall. This group forms a small hillock of cells, the **cumulus oophorus,** which protrudes toward the interior of the antrum and contains the oocyte (Figure 23–10). A group of granulosa cells concentrate around the oocyte and form the **corona radiata.** These granulosa cells accompany the oocyte when it leaves the ovary (Figure 23–10).

While modifications are taking place in the oocyte and granulosa layer, the fibroblasts of the stroma immediately around the follicle differentiate to form the **theca folliculi** (theca from Greek, meaning box). This layer subsequently differentiates into the **theca interna** and the **theca externa** (Figures 23–4, 23–10, and 23–11). The cells of the theca interna, when completely differentiated, have the same ultrastructural characteristics as cells that produce steroids. These characteristics include abundant profiles of smooth endoplasmic reticulum, mitochondria with tubular cristae, and numerous lipid droplets. These cells are known to synthesize a steroid hormone—**androstenedione**—that is transported to the granulosa layer. The cells of the gran-

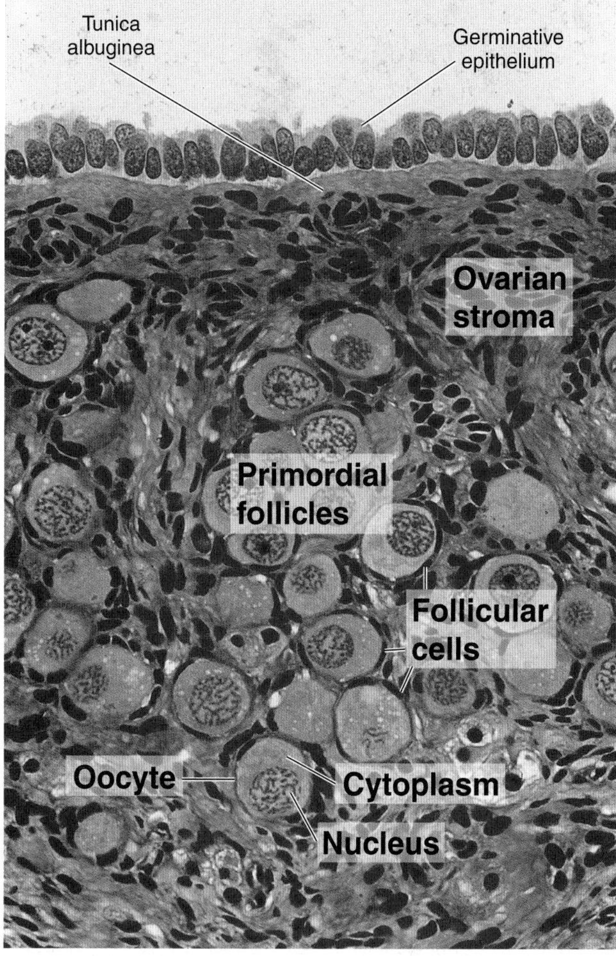

Figure 23–5. Photomicrograph of the cortical region of an ovary. The ovary is surrounded by the germinal epithelium and by the tunica albuginea. Groups of primordial follicles, each formed by an oocyte surrounded by a layer of flat follicular cells, are present in the ovarian connective tissue (stroma). Giemsa stain. Low magnification.

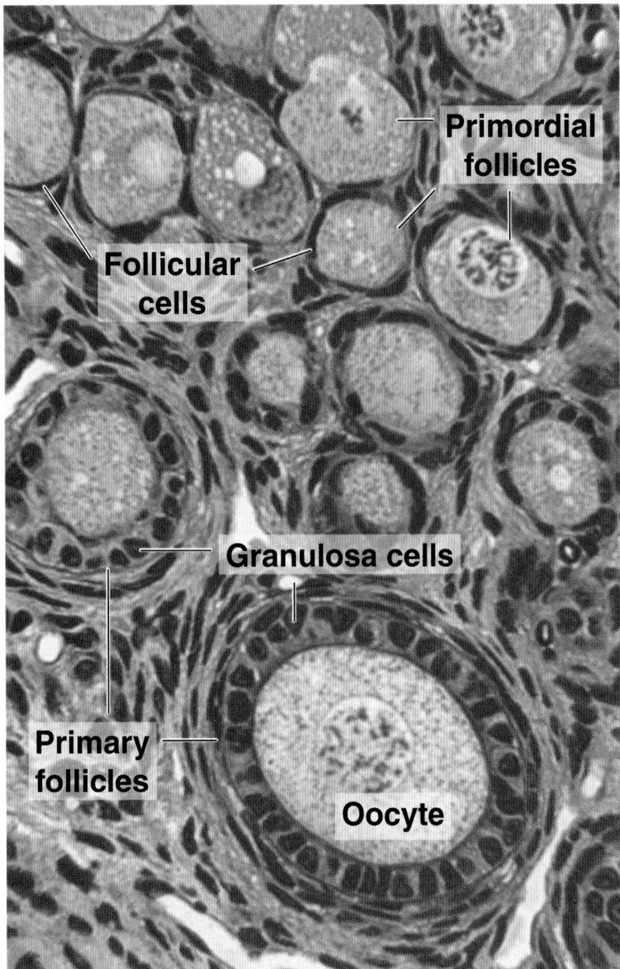

Figure 23–6. Cortical region of an ovary. Besides primordial follicles formed by an oocyte and flat follicular cells, a few follicles at the initial stage of growth (unilaminar primary follicles) are present. These are formed by an oocyte and one layer of cuboidal granulosa cells. Pararosaniline–toluidine blue (PT) stain. Low magnification.

ulosa, under influence of follicle-stimulating hormone, synthesize an enzyme, aromatase, that transforms androstenedione into estrogen. Estrogen returns to the stroma surrounding the follicle, enters the blood vessels, and is distributed throughout the body. The theca externa, on the other hand, consists mainly of organized layers of fibroblasts that surround the theca interna. The boundary between the two thecas is not clear; neither is there a clear boundary between the theca externa and the ovarian stroma. On the other hand, the boundary between the theca interna and the granulosa layer is well defined, since their cells are morphologically different and there is a thick basement membrane between them (Figure 23–11).

Small blood vessels penetrate the theca interna and supply a rich capillary plexus around the secretory cells of this region, which, like all organs of endocrine function, is richly vascularized. There are no blood vessels in the granulosa cell layer during the stage of follicular growth.

During each menstrual cycle, usually one follicle grows much more than the others and becomes the dominant follicle. The other follicles enter atresia. The dominant follicle may reach the most developed stage of follicular growth and may ovulate. This follicle, the **mature, preovulatory,** or **graafian follicle,** is so large (about 2.5 cm in diameter) that it protrudes from the surface of the ovary and can be detected with ultrasound. As a result of the accumulation of liquid, the follicular cavity increases in size, and the oocyte adheres to the wall of the follicle through a pedicle formed by granulosa cells. Since the granulosa cells of the follicle wall do not multiply in proportion to the growth of the follicle, the granulosa layer becomes thinner. These follicles have a very thick theca layer.

The whole process of growth from primordial to mature follicle lasts about 90 days.

Follicular Atresia

Most ovarian follicles undergo atresia, in which follicular cells and oocytes die and are disposed of by phagocytic cells. Follicles at any stage of development (primordial, primary, preantral, and antral) may undergo atresia (Figure 23–12). This process is characterized by cessation of mitosis in the granulosa cells, detachment of granulosa cells from the basal lamina, and death of the oocyte. After cell death, macrophages invade the follicle to

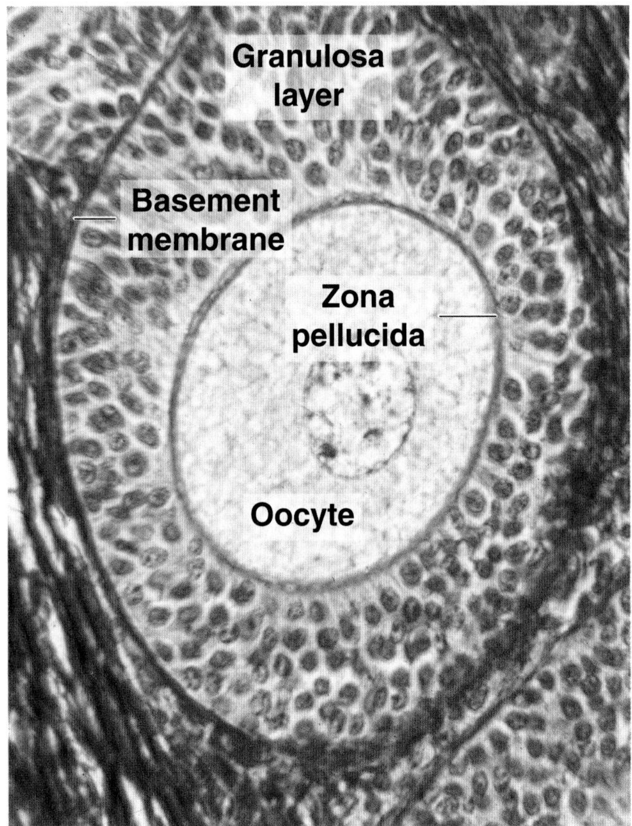

Figure 23–7. Photomicrograph of a preantral ovarian follicle formed by an oocyte and several layers of granulosa cells. The oocyte is surrounded by the zona pellucida. Picrosirius-hematoxylin (PSH) stain. Medium magnification.

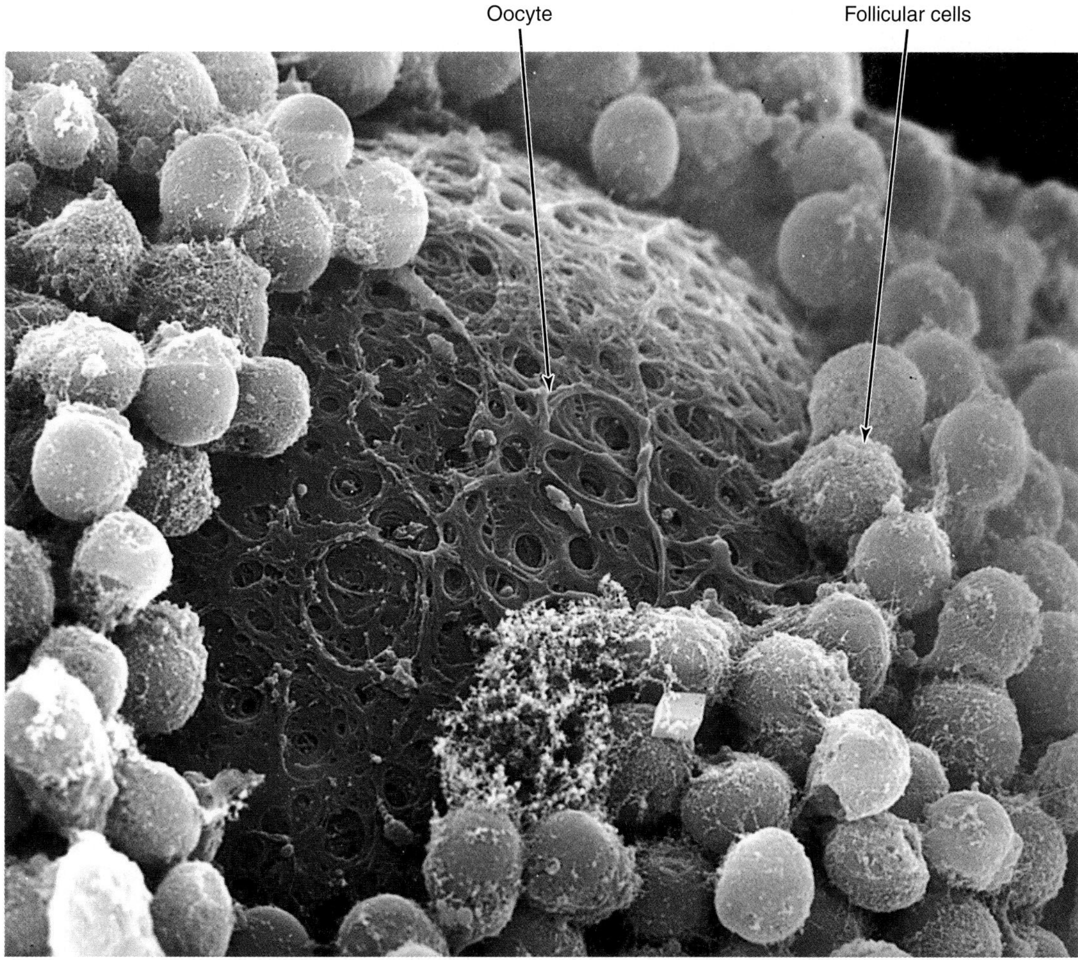

Figure 23–8. Scanning electron micrograph of an ovary, showing an oocyte surrounded by follicular cells. The structure covering the oocyte is the zona pellucida, which appears as an irregular meshwork. ×2950. (Courtesy of C Barros.)

phagocytose the debris. At a later stage, fibroblasts occupy the area of the follicle and produce a scar of collagen that may persist for a long time. Although follicular atresia takes place from before birth until a few years after menopause, there are times at which it is particularly intense. Atresia is greatly accentuated just after birth, when the effect of maternal hormones ceases, and during puberty and pregnancy, when marked qualitative and quantitative hormonal modifications take place.

Ovulation

Ovulation consists of the rupture of part of the wall of the mature follicle and liberation of the oocyte, which is caught by the dilated extremity of the oviduct. It takes place in approximately the middle of the menstrual cycle, ie, around the fourteenth day of a 28-day cycle. In the human, usually only one oocyte is liberated by the ovary during each cycle, but sometimes no oocyte is ovulated at all (anovulatory cycle). Sometimes 2 or more oocytes can be expelled at the same time, and if they are fertilized, there may be 2 or more fetuses.

The stimulus for ovulation is a surge of luteinizing hormone (LH) secreted by the anterior pituitary gland in response to high levels of circulating estrogen produced by the growing follicles. Within minutes after the increase in blood LH, there is an increase in blood flow through the ovary, and plasma proteins leak through capillaries and postcapillary venules, resulting in edema. There is a local release of prostaglandins, histamine, vasopressin, and collagenase. The granulosa cells produce more hyaluronic acid and become loose. A small area of the wall of the follicle becomes weak because of collagen degradation of the tunica albuginea, ischemia, and the death of some cells. This weakness, combined with an increased pressure of the follicular fluid and possibly the contraction of smooth muscle cells, leads to the rupture of the outer follicular wall and ovulation. An indication of impending ovulation is the appearance on the surface of the follicle of the **stigma,** in which the flow of blood ceases, resulting in a local change in color and translucence of the follicular wall.

The first meiotic division is completed just before ovulation (until this moment the oocyte was in prophase I of meiosis, initiated during fetal life). The chromosomes are equally divided between the daughter cells, but one of the secondary oocytes retains almost all of the cytoplasm. The other becomes the **first polar body,** a very small cell containing a small nucleus and a minimal amount of cytoplasm. Immediately after expulsion of

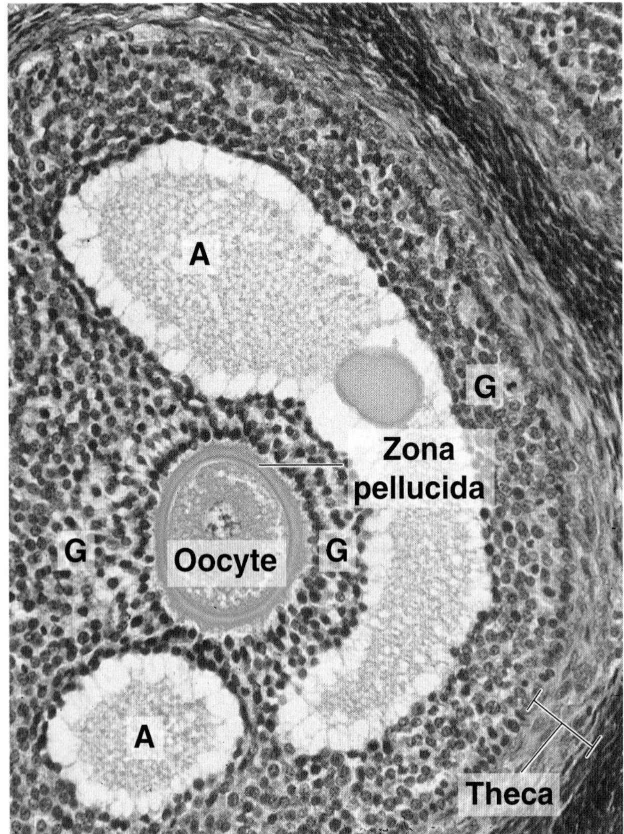

Figure 23–9. Photomicrograph of part of an antral follicle. Cavities (A) that appear in the granulosa layer will fuse and form one large cavity, the antrum. The oocyte is surrounded by the zona pellucida. Granulosa cells (G) surround the oocyte and cover the wall of the follicle. A theca can be seen around the follicle. H&E. Medium magnification.

the first polar body, the nucleus of the oocyte starts the second meiotic division, which stops in metaphase.

Because of the rupture of the follicular wall, the oocyte and the first polar body, enclosed by the zona pellucida, the corona radiata, and some follicular fluid, leave the ovary and enter the open extremity of the uterine tube where the oocyte may be fertilized. If this does not happen within the first 24 hours after ovulation, it degenerates and is phagocytosed.

Corpus Luteum

After ovulation, the granulosa cells and the cells of the theca interna of the ovulated follicle reorganize to form a temporary endocrine gland called the corpus luteum, which becomes embedded within the cortical region.

Release of the follicular fluid results in collapse of the follicle's wall so that it becomes folded (Figure 23–13). Some blood flows into the follicular cavity, where it coagulates and is later invaded by connective tissue. This connective tissue, with remnants of blood clots that are gradually removed, remains as the most central part of the corpus luteum.

Although the granulosa cells do not divide after ovulation, they increase greatly in size (20–35 μm in diameter). They make up about 80% of the parenchyma of the corpus luteum and are then called **granulosa lutein cells** (Figure 23–14), with the characteristics of steroid-secreting cells. This is in contrast to their structure in the preovulatory follicle, where they appear to be protein-secreting cells.

Cells of the theca interna also contribute to the formation of the corpus luteum by giving rise to **theca lutein cells** (Figure 23–13). These cells are similar in structure to granulosa lutein cells but are smaller (about 15 μm in diameter) and stain more intensely. They are located in the folds of the wall of the corpus luteum.

The blood capillaries and lymphatics that were restricted to the theca interna now grow into the interior of the corpus luteum and form the rich vascular network of this structure.

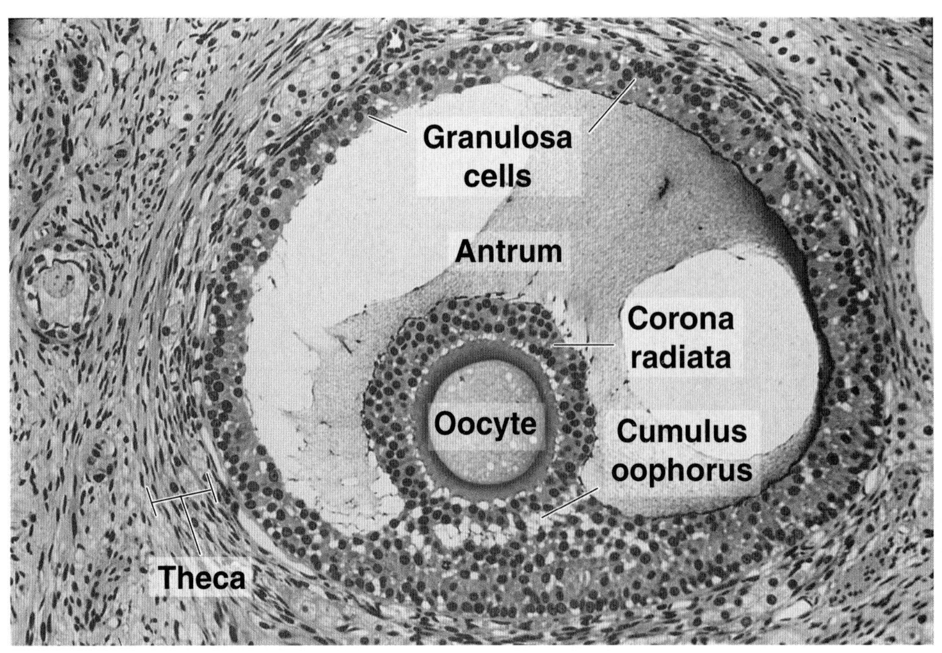

Figure 23–10. Photomicrograph of an antral follicle showing the oocyte surrounded by the granulosa cells of the corona radiata and supported by the cells of the cumulus oophorus. The remaining granulosa cells form the wall of the follicle and surround the large antrum. A theca surrounds the whole follicle. PT stain. Medium magnification.

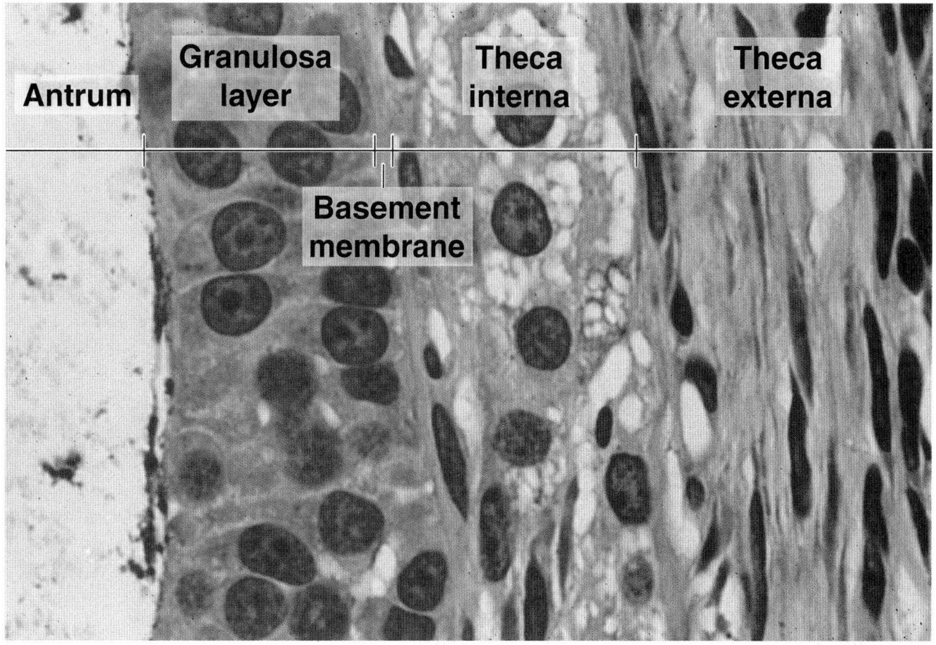

Figure 23–11. Photomicrograph of a small part of the wall of an antral follicle, showing the antrum, the layer of granulosa cells, and the thecas. The theca interna surrounds the follicle, and its cells appear lightly stained because their cytoplasm contains lipid droplets, a characteristic of steroid-producing cells. The theca interna is surrounded by the theca externa, which merges with the stroma of the ovary. A basement membrane separates the granulosa layer from the theca interna. PT stain. High magnification.

The reorganization of the ovulated follicle and the development of the corpus luteum result from the stimulus by LH released before ovulation (Figure 23–15). Also under stimulus by LH, the cells of the corpus luteum change their sets of enzymes and begin secreting progesterone and estrogens.

The fate of the corpus luteum depends on whether pregnancy is established. Following the stimulus by LH, the corpus luteum is programmed to secrete for 10–12 days. If no further hormonal stimulation takes place and pregnancy does not occur, the cells of the corpus luteum degenerate by apoptosis. One of the consequences of the decreasing secretion of progesterone is menstruation, which constitutes the shedding of part of the uterine mucosa. Estrogen inhibits the liberation of follicle-stimulating hormone from the hypophysis. After the corpus luteum degenerates, the concentration of blood steroids decreases and follicle-stimulating hormone is liberated, stimulating the growth of another group of follicles, beginning the next menstrual cycle. The corpus luteum that lasts only part of a menstrual cycle is called the **corpus luteum of menstruation.** Its cellular remnants are phagocytosed by macrophages. Neighboring fibroblasts in-

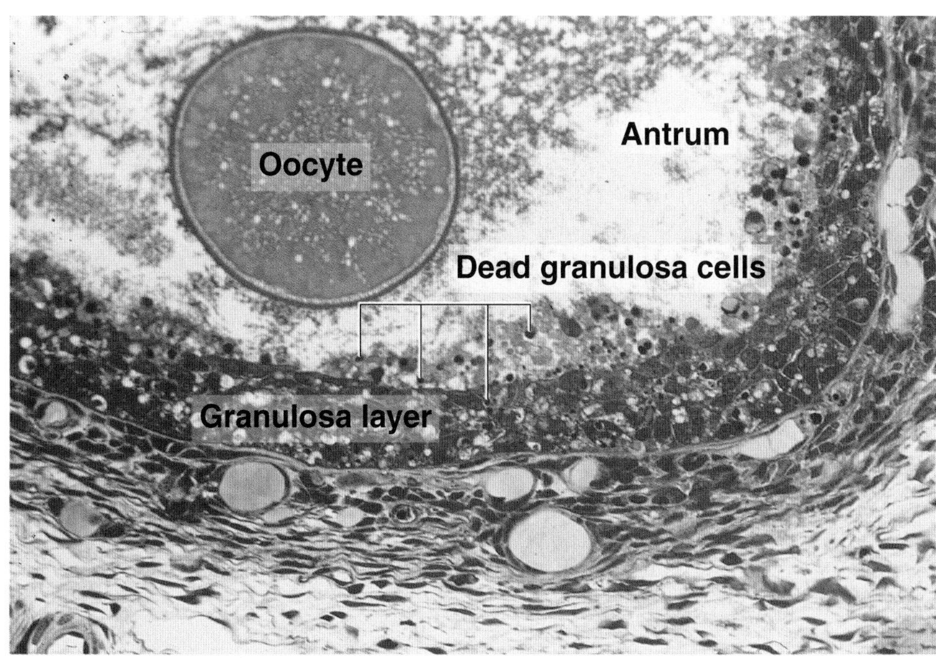

Figure 23–12. Photomicrograph of the atresia of a follicle characterized by: (1) the death of granulosa cells, many of which are seen loose in the antrum; (2) loss of the cells of the corona radiata; and (3) the oocyte floating free within the antrum. PT stain. Medium magnification.

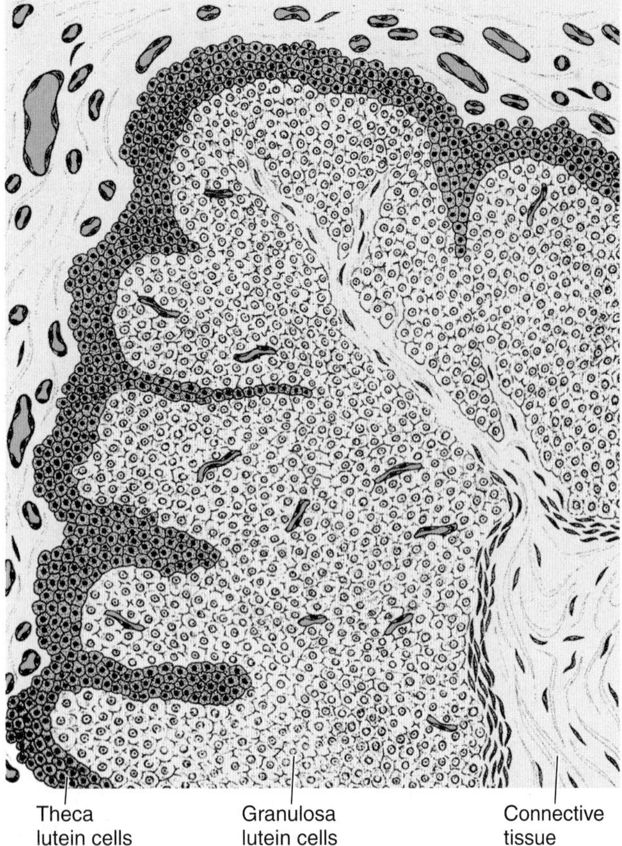

Figure 23–13. Part of a corpus luteum. Granulosa lutein cells, which constitute the majority of the cells, derive from the granulosa layer. They are larger and stain more lightly than the theca lutein cells, which originate from the theca interna.

vade the area and produce a scar of dense connective tissue called **corpus albicans** ("white body," because of the large amount of collagen) (Figure 23–16).

If pregnancy occurs, the uterine mucosa cannot be allowed to shed. If this happens, the implanting embryo dies and the pregnancy results in an abortion. A signal to the corpus luteum is given by the implanting embryo, whose trophoblastic cells synthesize a hormone called **human chorionic gonadotropin (HCG)**. The action of HCG is similar to that of LH. Thus, HCG rescues the corpus luteum from degeneration, causes further growth of this endocrine gland, and stimulates secretion of progesterone (which will maintain the uterine mucosa throughout pregnancy). Besides maintaining the uterine mucosa, progesterone also stimulates secretion of the uterine glands, which is thought to be important for the nutrition of the embryo before the placenta is functional. This is the **corpus luteum of pregnancy.** It persists for 4–5 months and then degenerates and is replaced by a corpus albicans that is much larger than the corpus albicans of menstruation.

Interstitial Cells

Although granulosa cells and the oocytes undergo degeneration during follicular atresia, the theca interna cells frequently persist in isolation or in small groups throughout the cortical stroma and are called **interstitial cells.** Present from childhood through menopause, interstitial cells are active steroid secretors, stimulated by LH.

OVIDUCTS

The oviducts are 2 muscular tubes (Figure 23–1) of great mobility, each measuring about 12 cm in length. One of its extremities, the infundibulum, opens into the peritoneal cavity next to the ovary and has a fringe of fingerlike extensions called **fimbriae;** the other extremity, the intramural portion, passes through the wall of the uterus and opens into the interior of this organ.

The wall of the oviduct is composed of 3 layers: (1) a mucosa, (2) a thick muscularis composed of smooth muscle disposed as an inner circular or spiral layer and an outer longitudinal layer, (3) and a serosa composed of visceral peritoneum.

The mucosa has longitudinal folds that are most numerous in the ampulla. In cross sections, the lumen of the ampulla resembles a labyrinth (Figure 23–17). These folds become smaller in the segments of the tube that are closer to the uterus. In the intramural portion, the folds are reduced to small bulges in the lumen, so its internal surface is almost smooth.

The mucosa is composed of a simple columnar epithelium and of a lamina propria composed of loose connective tissue. The epithelium contains 2 types of cells: one has cilia, the other

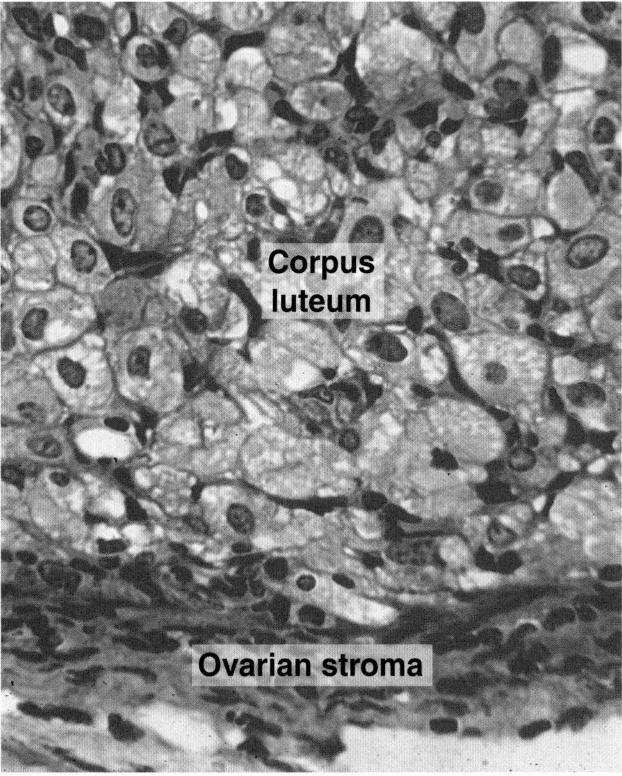

Figure 23–14. Photomicrograph of a small portion of a corpus luteum. Most cells present in the figure are granulosa lutein cells. PT stain. High magnification.

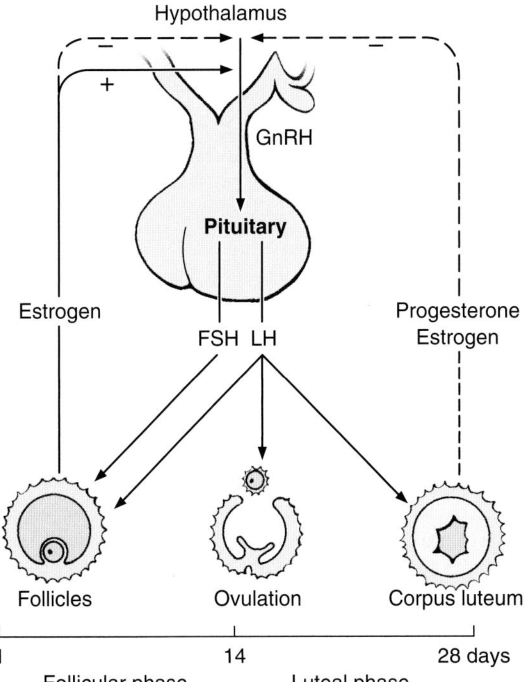

Figure 23–15. Pituitary hormones control most ovarian functions. Follicle-stimulating hormone (FSH) stimulates follicular growth and synthesis of estrogen by the granulosa cells. Luteinizing hormone (LH) induces ovulation and transforms the granulosa layer and the theca interna into an actively secreting gland, the corpus luteum. Estrogen and progesterone produced in the ovary act on the hypothalamus, stimulating or inhibiting the liberation of gonadotropin-releasing hormone (GnRH).

is secretory (Figures 23–18 and 23–19). The cilia beat toward the uterus, causing movement of the viscous liquid film that covers its surface. This liquid consists mainly of products of the secretory cells interspersed between ciliated cells.

At the moment of ovulation, the oviduct exhibits active movement. The funnel-shaped extremity (fringed with numerous fimbriae) comes very close to the surface of the ovary. This favors the transport of the ovulated oocyte into the tube. Promoted by muscle contraction and the activity of ciliated cells, the oocyte enters the infundibulum of the oviduct. The secretion of the tube epithelium has nutrient and protective functions for the oocyte. Unless it is fertilized, the oocyte remains viable for a maximum of about 24 hours. The secretion also promotes activation (**capacitation**) of spermatozoa.

Fertilization usually occurs in the ampulla and reconstitutes the diploid number of chromosomes typical of the species. It also serves as a stimulus for the oocyte to complete the second meiotic division. Only at this moment does the primary oocyte transform into a secondary oocyte. The corona radiata is usually still present when the spermatozoon fertilizes the oocyte; it is retained for some time during the passage of the oocyte through the oviduct. When fertilization does not take place, the oocyte undergoes autolysis in the oviduct without completing the second **meiotic** division.

Once fertilized, the oocyte, now called a zygote (Gr. *zygotos,* yolked), begins cell division and is transported to the uterus, a process that lasts about 5 days. Movement of the film that covers the mucosa of the tube, in conjunction with contractions of the muscle layer, helps to transport the oocyte or the conceptus toward the uterus. This movement also hampers the passage of microorganisms from the uterus to the peritoneal cavity. Transport of the oocyte or conceptus to the uterus, however, is normal in females with **immotile cilia syndrome,** showing that ciliary activity is not essential for transport.

MEDICAL APPLICATION

*In cases of abnormal nidation, the embryo may implant itself in the oviduct (**ectopic pregnancy**). In this case, the lamina propria reacts like the endometrium, forming numerous decidual cells. Because of its small diameter, the oviduct cannot contain the growing embryo and bursts, causing extensive hemorrhage that can be fatal if not treated immediately.*

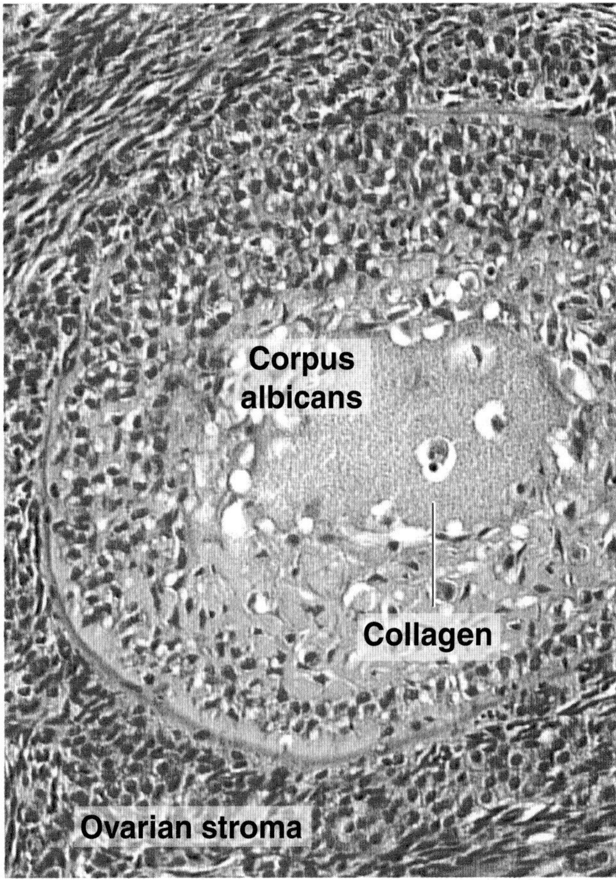

Figure 23–16. Corpus albicans, the scar of connective tissue that replaces a corpus luteum after its involution.

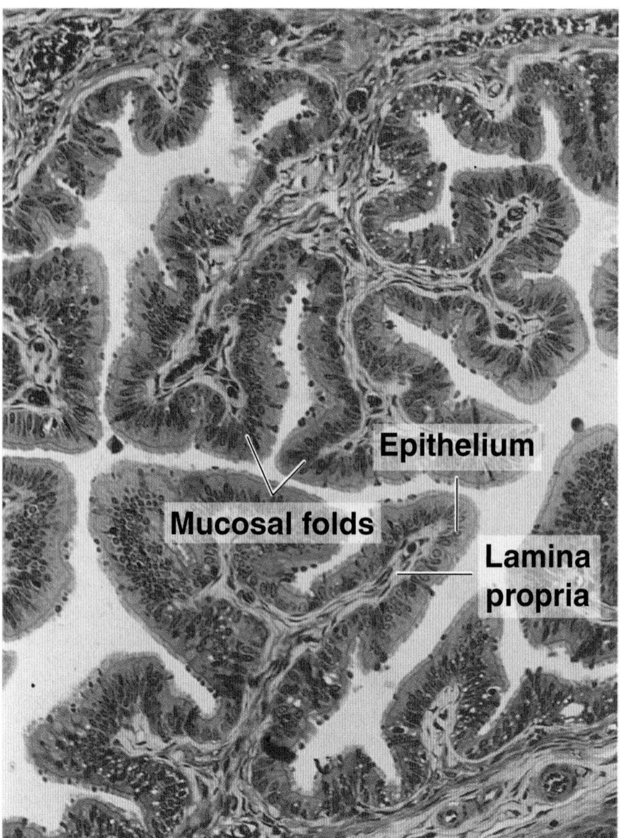

Figure 23–17. Photomicrograph of part of the wall of an oviduct. The highly folded mucosa indicates that this region is close to the ovary. PT stain. Low magnification.

UTERUS

The uterus is a pear-shaped organ that consists of a **body** (**corpus**), which lies above a narrowing of the uterine cavity (**the internal os**), and a lower cylindrical structure, the **cervix,** which lies below the internal os. The dome-shaped part of the body of the uterus is called the **fundus** (Figure 23–1).

The wall of the uterus is relatively thick and is formed of 3 layers. Depending on the part of the uterus, there is either an outer **serosa** (connective tissue and mesothelium) or **adventitia** (connective tissue). The other uterine layers are the **myometrium,** a thick tunic of smooth muscle, and the **endometrium,** or mucosa of the uterus.

Myometrium

The myometrium (Gr. *mys,* muscle, + *metra,* uterus), the thickest tunic of the uterus, is composed of bundles of smooth muscle fibers separated by connective tissue. The bundles of smooth muscle form 4 poorly defined layers. The first and fourth layers are composed mainly of fibers disposed longitudinally, ie, parallel to the long axis of the organ. The middle layers contain the larger blood vessels.

During pregnancy, the myometrium goes through a period of great growth as a result of both **hyperplasia** (an increase in the number of smooth muscle cells) and **hypertrophy** (an increase in cell size). During pregnancy, many smooth muscle cells have ultrastructural characteristics of protein-secreting cells and actively synthesize collagen, promoting a significant increase in uterine collagen content.

After pregnancy, there is destruction of some smooth muscle cells, reduction in the size of others, and enzymatic degradation of the collagen. The uterus is reduced in size almost to its prepregnancy dimensions.

Endometrium

The endometrium consists of epithelium and a lamina propria containing simple tubular glands that sometimes branch in their deeper portions (near the myometrium). Its covering epithelial cells are a mixture of ciliated and secretory simple columnar cells. The epithelium of the uterine glands is similar to the superficial epithelium, but ciliated cells are rare within the glands.

The connective tissue of the lamina propria is rich in fibroblasts and contains abundant ground substance. Connective tissue fibers are mostly made of collagen type III.

The endometrial layer can be subdivided into 2 zones: (1) The **basalis** is the deepest one, adjacent to the myometrium; it contains lamina propria and the beginning of the uterine glands. (2) The **functionalis** contains the remainder of the lamina propria and of the glands, as well as the surface epithelium. While the functionalis undergoes profound changes during the menstrual cycles, the basalis remains mostly unchanged.

The blood vessels supplying the endometrium are of special significance in the periodic sloughing of most of this layer. **Arcuate arteries** are circumferentially oriented in the middle layers of the myometrium. From these vessels, 2 sets of arteries arise to supply blood to the endometrium: **straight arteries,** which supply the basalis, and **spiral arteries,** which bring blood to the functionalis.

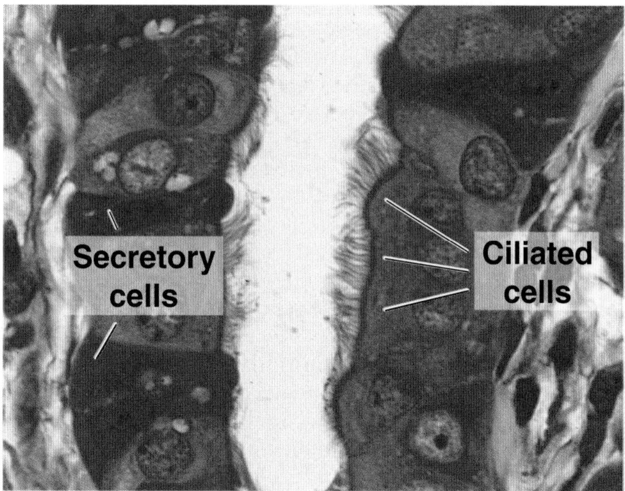

Figure 23–18. Photomicrograph of the epithelial lining of an oviduct. The epithelial lining is formed by ciliated and more darkly staining nonciliated secretory cells. Ciliated cells contribute to the movement of the oocyte or conceptus to the uterus. PT stain. High magnification.

Figure 23–19. Scanning electron micrograph of the lining of an oviduct. Note the abundant cilia. In the center is the apex of a secretory cell covered by short microvilli. ×8000. (Courtesy of KR Porter.)

The Menstrual Cycle

Estrogens and progesterone control the organs of the female reproductive system. The proliferation and the differentiation of epithelial cells and the associated connective tissues depend on these hormones. Even before birth, these organs are influenced by estrogen and progesterone that circulate in the maternal blood and reach the fetus through the placenta (Figure 23–20). After menopause, the diminished synthesis of these hormones causes a general involution of the reproductive organs.

After puberty, the ovarian hormones, under the stimulus of the anterior lobe of the pituitary, cause the endometrium to undergo cyclic structural modifications during the menstrual cycle. The duration of the menstrual cycle is variable but averages 28 days.

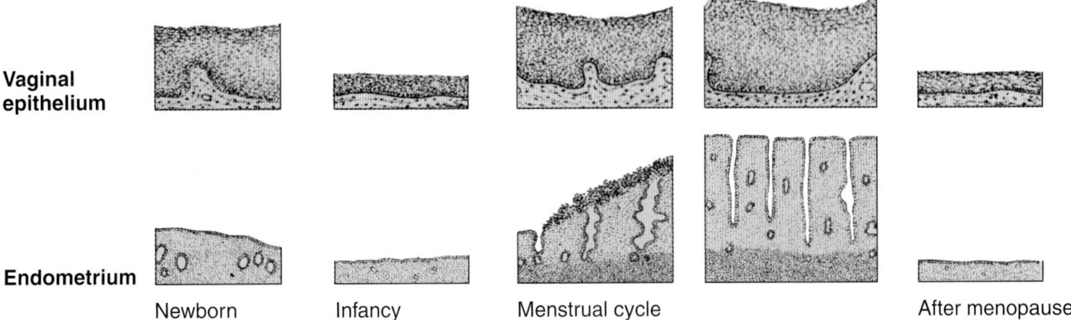

Figure 23–20. During the entire life of a woman, the structure and functions of the vaginal epithelium and of the endometrium depend on ovarian hormones.

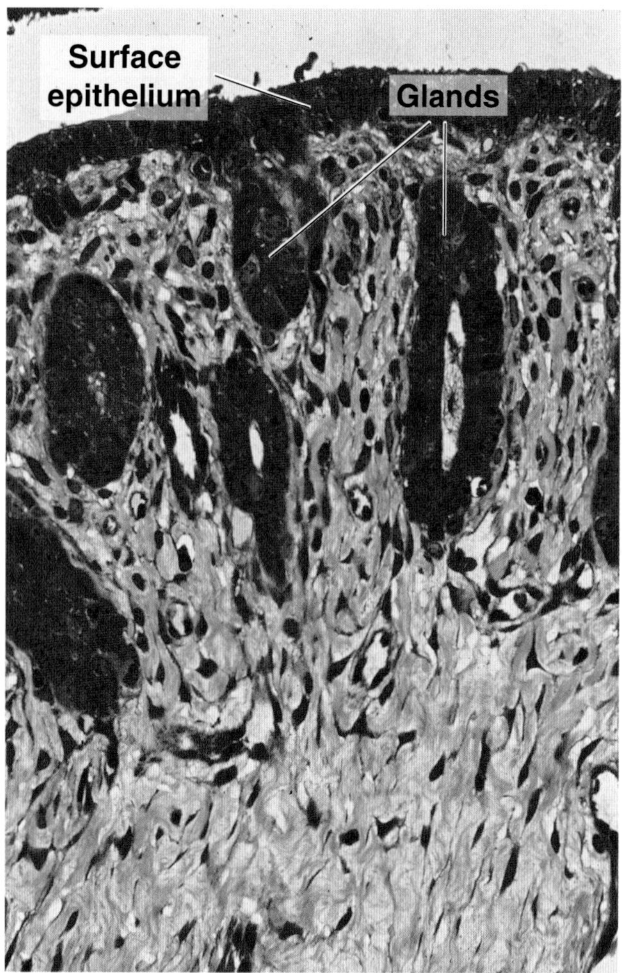

Figure 23–21. Photomicrograph of the superficial layer of the endometrium during the proliferative phase. The surface epithelium and the uterine glands are embedded in a lamina propria made of very loose connective tissue. PT stain. Medium magnification.

Menstrual cycles usually start between 12 and 15 years of age and continue until about age 45–50. Since menstrual cycles are a consequence of ovarian modifications related to the production of oocytes, the female is fertile only during the years when she is having menstrual cycles. This does not mean that sexual activity is terminated by menopause—only that fertility ceases.

For practical purposes, the beginning of the menstrual cycle is taken as the day when menstrual bleeding appears. The menstrual discharge consists of degenerating endometrium mixed with blood from the ruptured blood vessels. The **menstrual phase** lasts 3–4 days on average. The next phases of the menstrual cycle are called the **proliferative** and **secretory** (or **luteal**) **phases.** The secretory phase begins at ovulation and lasts about 14 days. The duration of the proliferative phase is variable, 10 days on average. The structural changes that occur during the cycle are gradual, and the clear division of the phases implied here is mainly for teaching value.

The Proliferative, Follicular, or Estrogenic Phase

After the menstrual phase, the uterine mucosa is relatively thin (about 0.5 mm). The beginning of the proliferative phase coincides with the rapid growth of a small group of ovarian follicles that, when the cycle began, were probably at the transition from preantral to antral follicles. When their theca interna develops, these follicles begin to actively secrete estrogens, whose plasma concentrations increase gradually.

Estrogens act on the endometrium, inducing cell proliferation and reconstituting the endometrium lost during menstruation. (Estrogen also acts on other parts of the reproductive system, eg, inducing the production of cilia by epithelial cells of the oviduct.)

During the proliferative phase, the endometrium is covered by a simple columnar epithelium (Figure 23–21). The glands, formed by simple columnar epithelial cells, are straight tubules with narrow lumens (Figure 23–22). These cells gradually accumulate more cisternae of rough endoplasmic reticulum, and the Golgi complex increases in size in preparation for secretory activity. At the end of the proliferative phase, the endometrium is 2–3 mm thick.

The Secretory, or Luteal, Phase

The secretory phase starts after ovulation and results from the action of progesterone secreted by the corpus luteum. Acting on

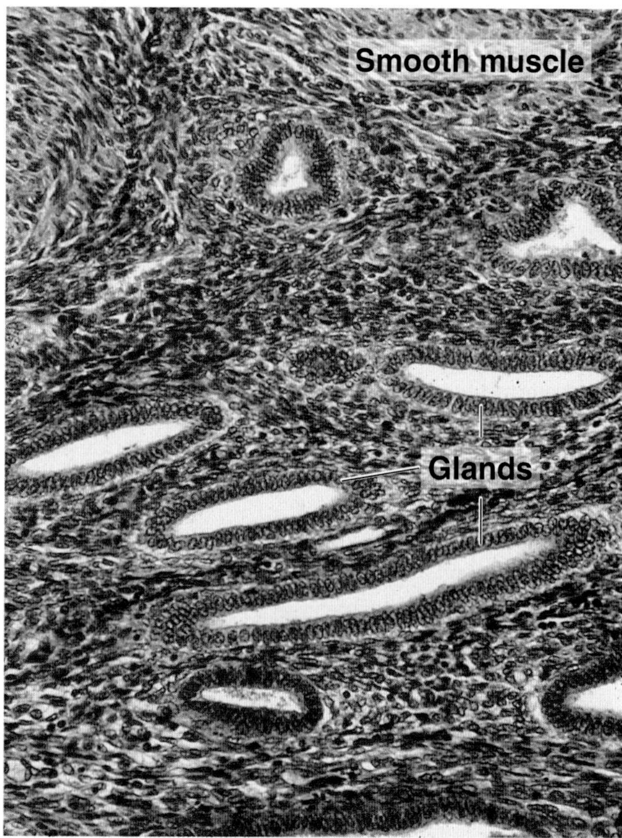

Figure 23–22. Photomicrograph of straight uterine glands in the deep endometrium during the proliferative phase. Smooth muscle of the myometrium is also seen. H&E stain. Medium magnification.

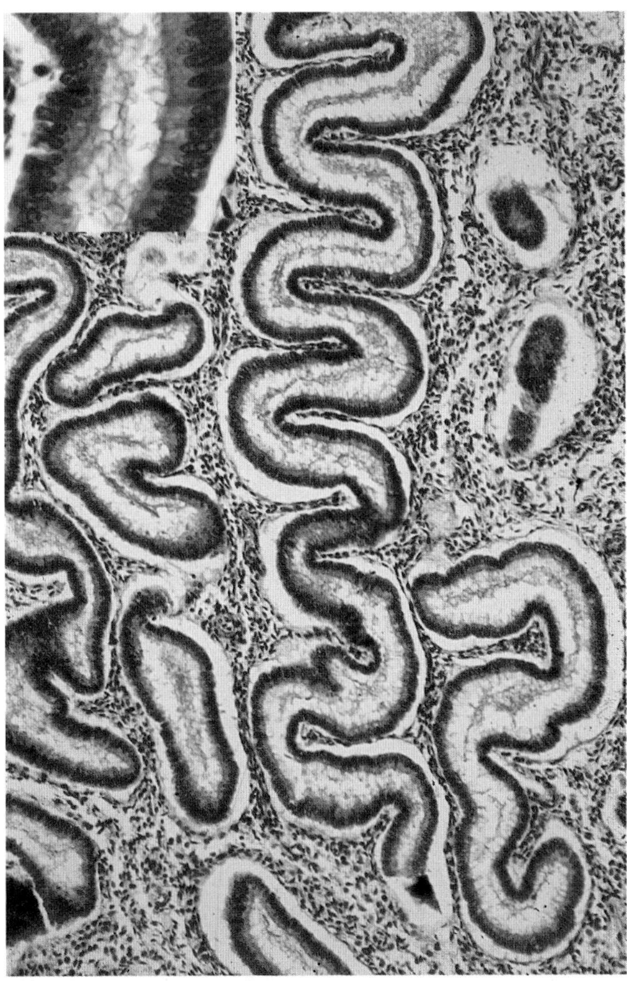

Figure 23–23. Photomicrograph of uterine glands. During the luteal phase, the uterine glands become tortuous and their lumen is filled with secretions. Some edema is present in the connective tissue. H&E stain. Medium magnification. **Inset:** High magnification.

glands already developed by the action of estrogen, progesterone further stimulates the gland cells. The epithelial cells begin to accumulate glycogen below their nuclei. Later, the amount of glycogen diminishes, and glycoprotein secretory products dilate the lumens of the glands. One important feature of this phase is that the glands become highly coiled (Figures 23–23 and 23–24). In this phase, the endometrium reaches its maximum thickness (5 mm) as a result of the accumulation of secretions and of edema in the stroma. Mitoses are rare during the secretory phase.

If fertilization has taken place, the embryo has been transported to the uterus and attaches to the uterine epithelium during the secretory stage, around 7 or 8 days after ovulation. It is thought that the secretion of the glands is the major source of embryonic nutrition before embryo implantation.

Progesterone inhibits the contractions of smooth muscle cells of the myometrium that might otherwise interfere with the implantation of the embryo.

The Menstrual Phase

When fertilization of the oocyte and embryo implantation do not occur, the corpus luteum ceases functioning 10–12 days after ovulation. As a result, the levels of progesterone and estrogens in the blood decrease rapidly. This causes several cycles of contraction of the spiral arteries, closing off the blood flow and producing ischemia, which leads to death (necrosis) of their walls and of part of the functionalis layer of the endometrium. Blood vessels rupture above the constrictions, and bleeding begins. A

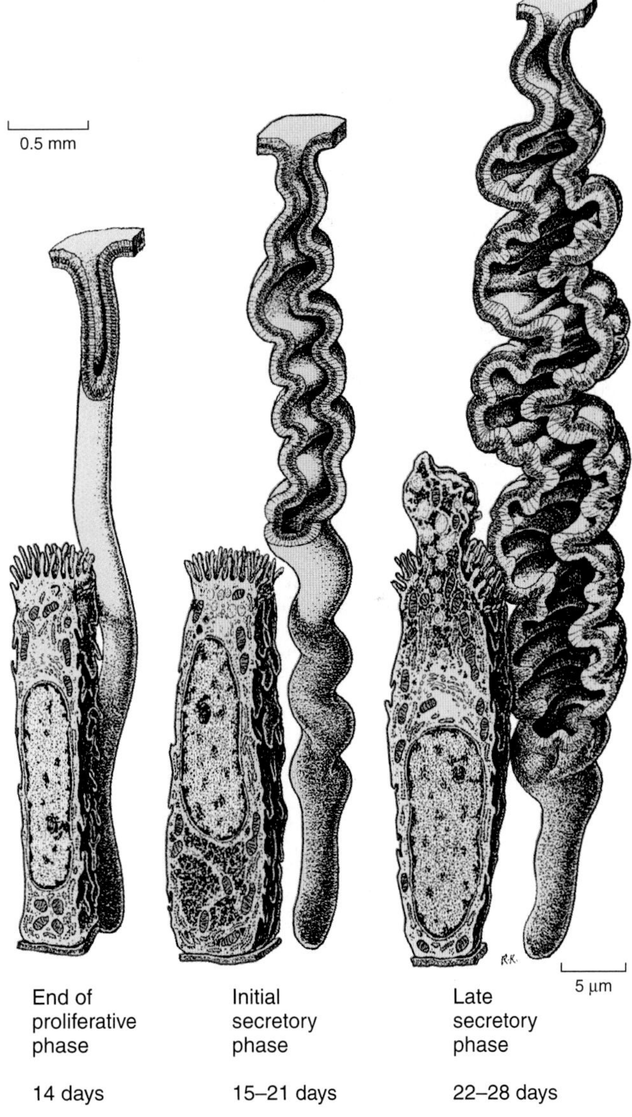

Figure 23–24. Changes in the uterine glands and in the gland cells during the menstrual cycle. In the proliferative stage the glands are straight tubules, and their cells show no secretory activity. In the initial secretory phase the glands begin to coil, and their cells accumulate glycogen in the basal region. In the late secretory phase the glands are highly coiled, and their cells present secretory activity at their apical portion. (Reproduced, with permission, from Krstíc RV: *Human Microscopic Anatomy*, Springer, 1991.)

portion of the functional layer of the endometrium becomes detached, and the rest of the endometrium shrinks owing to loss of interstitial fluid. The amount of endometrium and blood lost varies between women and even in the same woman at different times.

At the end of the menstrual phase, the endometrium is usually reduced to a thin layer. The endometrium is thus ready to begin a new cycle as its cells begin dividing to reconstitute the mucosa. Table 23–1 summarizes of the main events of the menstrual cycle.

PREGNANT ENDOMETRIUM

If implantation occurs, embryonic trophoblast cells produce HCG, which stimulates the corpus luteum to continue secreting progesterone. As pregnancy is established, menstruation does not occur, and the menstrual cycle is deferred during the whole duration of pregnancy. Progesterone makes the uterine glands wider, more tortuous, and able to contain more secretions than during the secretory stage.

Implantation, Decidua, & Placenta

The human oocyte is fertilized in the lateral third of the uterine tube, and the zygote is cleaved as it is moved passively toward the uterus. Through successive mitoses, a compact collection of cells, the **morula,** is formed. The morula, covered by the zona pellucida, is about the same size as the fertilized oocyte. The cells that result from segmentation of the zygote are called **blastomeres** (Gr. *blastos,* germ, + *meros,* part). Because the zygote does not grow in size, at each division the blastomeres become smaller.

At the center of the morula a liquid-filled cavity develops; this is the **blastocyst,** which is the stage at which the embryo arrives in the uterus. The blastomeres arrange themselves in a peripheral layer (**trophoblast**) while a few blastomeres collect inside the cavity (**inner cell mass**). This stage of development corresponds approximately to the fourth or fifth day after ovulation. The blastocyst remains in the lumen of the uterus for 2 or 3 days and comes into contact with the surface of the endometrium, immersed in the secretion of the endometrial glands. The zona pellucida is dissolved, allowing cells of the trophoblast to interact with cells of the uterine surface epithelium.

Implantation, or nidation, involves the attachment of the embryo to the epithelial cells and its penetration through the uterine epithelium. This type of implantation is called **interstitial** and occurs in humans and a few other mammals. The process starts around the seventh day; on about the ninth day after ovulation, the embryo is totally submerged in the endometrium, from which it will receive protection and nourishment during pregnancy.

After implantation of the embryo, the endometrial connective tissue goes through profound changes. The fibroblasts of the lamina propria become enlarged and polygonal and exhibit the characteristics of protein-synthesizing cells. They are now called decidual cells, and the whole endometrium is called the **decidua.** The decidua can be divided into the **decidua basalis,** situated between the embryo and the myometrium; **decidua capsularis,** between the embryo and the lumen of the uterus; and **decidua parietalis,** the remainder of the decidua (Figure 23–25).

Placenta

The placenta is a temporary organ and is the site of physiologic exchanges between the mother and the fetus. It consists of a fetal part (**chorion**) and a maternal part (decidua basalis). Thus the placenta is composed of cells derived from 2 genetically distinct individuals.

The decidua basalis supplies maternal arterial blood to, and receives venous blood from, spaces that exist inside the placenta. The placenta is also an endocrine organ, producing such hormones as HCG, chorionic thyrotropin, chorionic corticotropin, estrogens, and progesterone. It also secretes a protein hormone called human chorionic somatomammotropin, which has lactogenic and growth-stimulating activity.

More detailed information on the developing embryo and on the formation and structure of the placenta should be sought in embryology textbooks.

Table 23–1. Summary of events of the menstrual cycle.

	Stage of Cycle				
	Proliferative	**Secretory or Luteal**			**Menstrual**
Main actions of pituitary hormones	Follicle-stimulating hormone stimulates rapid growth of ovarian follicles.	Peak of luteinizing hormone at the beginning of secretory stage, secreted by stimulation of estrogen, induces ovulation and development of the corpus luteum.			
Main events in the ovary	Growth of ovarian follicles; dominant follicle reaches preovulatory stage.	Ovulation.	Development of the corpus luteum.	Degeneration of the corpus luteum.	
Dominant ovarian hormone	Estrogens, produced by the growing follicles, act on vagina, tubes, and uterus.	Progesterone, produced by the corpus luteum, acts mainly on the uterus.		Progesterone production ceases.	
Main events in the endometrium	Growth of the mucosa after menstruation.	Further growth of the mucosa, coiling of glands, secretion.			Shedding of part of the mucosa about 14 days after ovulation.

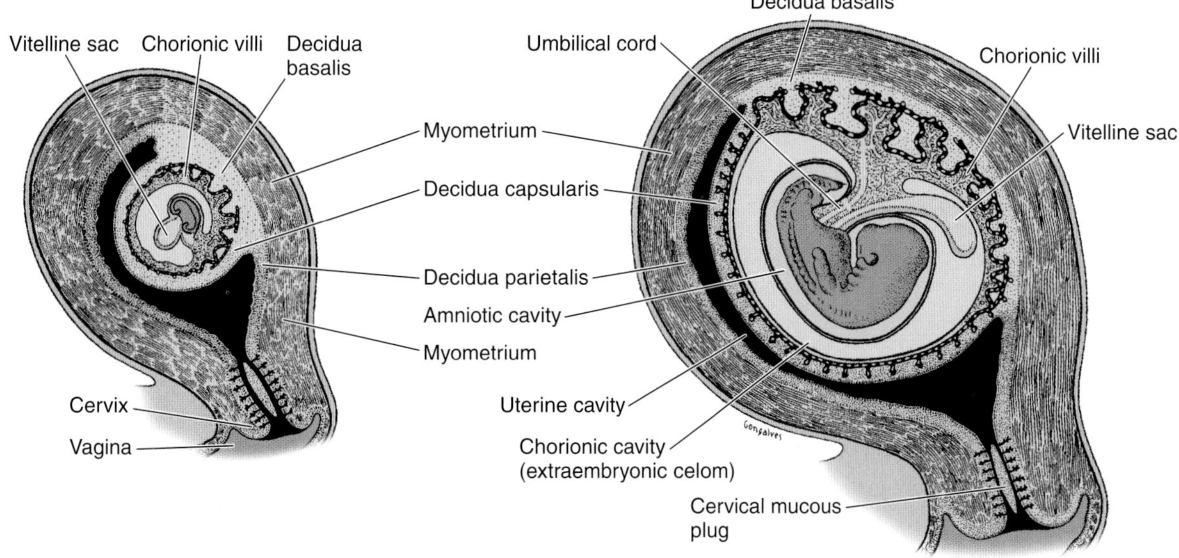

Figure 23–25. During pregnancy the endometrial connective cells transform into decidual cells. The endometrium is then called decidua, and 3 regions of mucosa can be recognized: decidua basalis, capsularis, and parietalis.

MEDICAL APPLICATION

The initial attachment of the embryo usually occurs on the ventral or dorsal walls of the corpus of the uterus. Sometimes the embryo attaches close to the internal os. In this case the placenta will be interposed between the fetus and the vagina, obstructing the passage of the fetus at parturition. This situation, called placenta previa, must be recognized by the physician, and the fetus must be delivered by cesarean section; otherwise, the fetus may die. Sometimes, as already mentioned, the embryo attaches to the epithelium of the uterine tube. Very rarely the zygote may enter the abdominal cavity, attach to the peritoneum, and develop there.

Uterine Cervix

The **cervix** is the lower, cylindrical part of the uterus (Figure 23–1), and it differs in histologic structure from the rest of the uterus. The lining consists of a mucus-secreting simple columnar epithelium. The cervix has few smooth muscle fibers and consists mainly (85%) of dense connective tissue. The external aspect of the cervix that bulges into the lumen of the vagina is covered with stratified squamous epithelium.

The mucosa of the cervix contains the mucous **cervical glands,** which are extensively branched. This mucosa does not undergo remarkable changes during the menstrual cycle and does not desquamate during menstruation. During pregnancy, the cervical mucous glands proliferate and secrete a more viscous and abundant mucus.

Cervical secretions play a significant role in fertilization of the oocyte. At the time of ovulation, the mucous secretions are watery and allow penetration of the uterus by sperm. In the luteal phase or in pregnancy, the progesterone levels alter the mucous secretions so that they become more viscous and prevent the passage of sperm, as well as microorganisms, into the body of the uterus. The dilation of the cervix that precedes parturition is due to intense collagenolysis, which promotes its softening.

MEDICAL APPLICATION

*Cancer of the cervix (**cervical carcinoma**) is derived from its stratified squamous epithelium. Although it is frequently observed, the mortality rate is low (8 per 100,000). This low rate is due to the usual discovery of the carcinoma in its early stages, made possible by yearly physical observation of the cervix and by cytologic analysis of smears of the cervical epithelium (Papanicolaou test).*

VAGINA

The wall of the vagina (from Latin, meaning sheath) is devoid of glands and consists of 3 layers: a **mucosa,** a **muscular layer,** and an **adventitia.** The mucus found in the lumen of the vagina comes from the glands of the uterine cervix.

The epithelium of the vaginal mucosa of an adult woman is stratified squamous and has a thickness of 150–200 μm. Its cells may contain a small amount of keratohyalin. Intense keratinization, however, with the cells changing into keratin plates, as in typical keratinized epithelia, does not occur (Figure 23–26). Under the stimulus of estrogen, the vaginal epithelium synthesizes and accumulates a large quantity of glycogen, which is deposited in the lumen of the vagina when the vaginal cells desquamate. Bacteria in the vagina metabolize glycogen and form lactic acid, which is responsible for the usually low pH of the vagina. The acidic vaginal environment provides a protective action against some pathogenic microorganisms.

The lamina propria of the vaginal mucosa is composed of loose connective tissue that is very rich in elastic fibers. Among the cells present are lymphocytes and neutrophils in relatively large quantities. During certain phases of the menstrual cycle, these 2 types of leukocytes invade the epithelium and pass into the lumen of the vagina. The vaginal mucosa is virtually devoid of sensory nerve endings, and the few naked nerve endings that do exist are probably pain fibers.

The muscular layer of the vagina is composed mainly of longitudinal bundles of smooth muscle fibers. There are some circular bundles, especially in the innermost part (next to the mucosa).

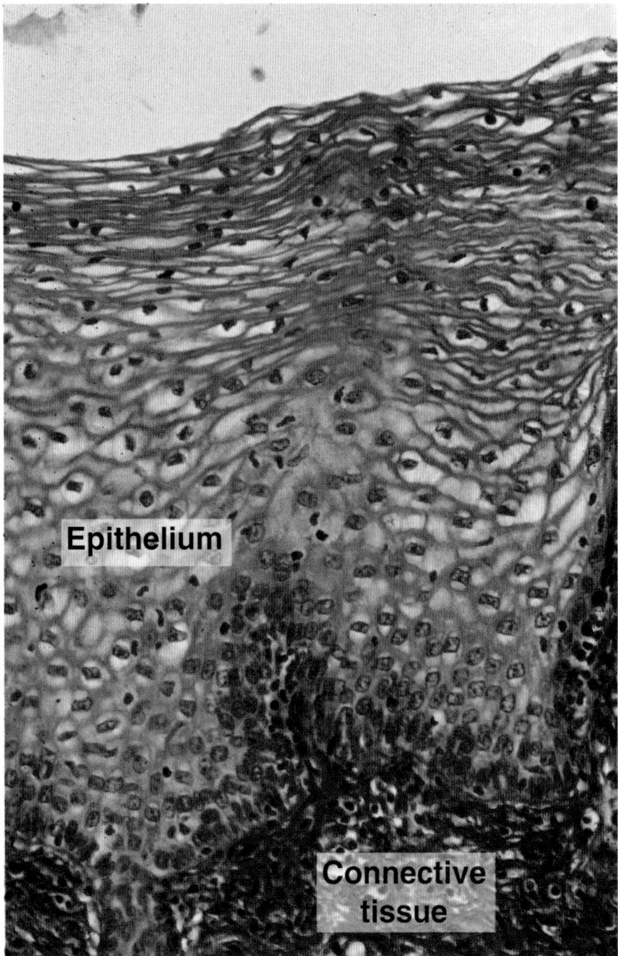

Figure 23–26. Photomicrograph of stratified squamous epithelium of the vagina supported by a dense connective tissue. The cytoplasm of these epithelial cells is clear because of accumulated glycogen. PSH stain. Medium magnification.

Outside the muscular layer, a coat of dense connective tissue, the adventitia, rich in thick elastic fibers, unites the vagina with the surrounding tissues. The great elasticity of the vagina is related to the large number of elastic fibers in the connective tissues of its wall. In this connective tissue are an extensive venous plexus, nerve bundles, and groups of nerve cells.

EXFOLIATIVE CYTOLOGY

MEDICAL APPLICATION

Exfoliative cytology is the study of the characteristics of cells that normally desquamate from various surfaces of the body. Cytologic examination of cells collected from the vagina gives information of clinical importance.

In fully mature vaginal mucosa, 5 types of cells are easily identified: cells of the internal portion of the basal layer (basal cells), cells of the external portion of the basal layer (parabasal cells), cells of the intermediate layers, precornified cells, and cornified cells. Based on the numbers of cell types that appear in a vaginal smear, valuable information can be obtained on the hormonal status of the patient (action of estrogen and progesterone). The vaginal smear is also useful in the early detection of cervical cancer.

EXTERNAL GENITALIA

The female external genitalia, or vulva, consist of the **clitoris, labia minora, labia majora,** and some glands that open into the vestibulum, a space enclosed by the labia minora.

The urethra and the ducts of the vestibular glands open into the vestibulum. The two **glandulae vestibulares majores,** or **glands of Bartholin,** are situated on either side of the vestibulum. These glands are homologous to the bulbourethral glands in the male. Women frequently experience inflammation of these glands and the formation of very painful cysts. The more numerous **glandulae vestibulares minores** are scattered, found with greater frequency around the urethra and clitoris. All the glandulae vestibulares secrete mucus.

The clitoris and the penis are homologous in embryonic origin and histologic structure. The clitoris is formed by 2 erectile bodies ending in a rudimentary **glans clitoridis** and a prepuce. The clitoris is covered with stratified squamous epithelium.

The labia minora are folds of skin with a core of spongy connective tissue permeated by elastic fibers. The stratified squamous epithelium that covers them has a thin layer of keratinized cells on the surface. Sebaceous and sweat glands are present on the inner and outer surfaces of the labia minora.

The labia majora are folds of skin that contain a large quantity of adipose tissue and a thin layer of smooth muscle. Their inner surface has a histologic structure similar to that of the labia minora. The external surface is covered by skin and coarse, curly hair. Sebaceous and sweat glands are numerous on both surfaces.

The external genitalia are abundantly supplied with sensory tactile nerve endings, including Meissner's and Pacinian corpuscles, which contribute to the physiology of sexual arousal.

MAMMARY GLANDS

Each mammary gland consists of 15–25 **lobes** of the compound tubuloalveolar type whose function is to secrete milk to nourish newborns. Each lobe, separated from the others by dense connective tissue and much adipose tissue, is really a gland in itself with its own **excretory lactiferous duct** (Figure 23–27). These ducts, 2–4.5 cm long, emerge independently in the **nipple,** which has 15–25 openings, each about 0.5 mm in diameter. The histologic structure of the mammary glands varies according to sex, age, and physiologic status.

Breast Development

Before puberty, the mammary glands are composed of **lactiferous sinuses** and several branches of these sinuses, the **lactiferous ducts** (Figure 23–27).

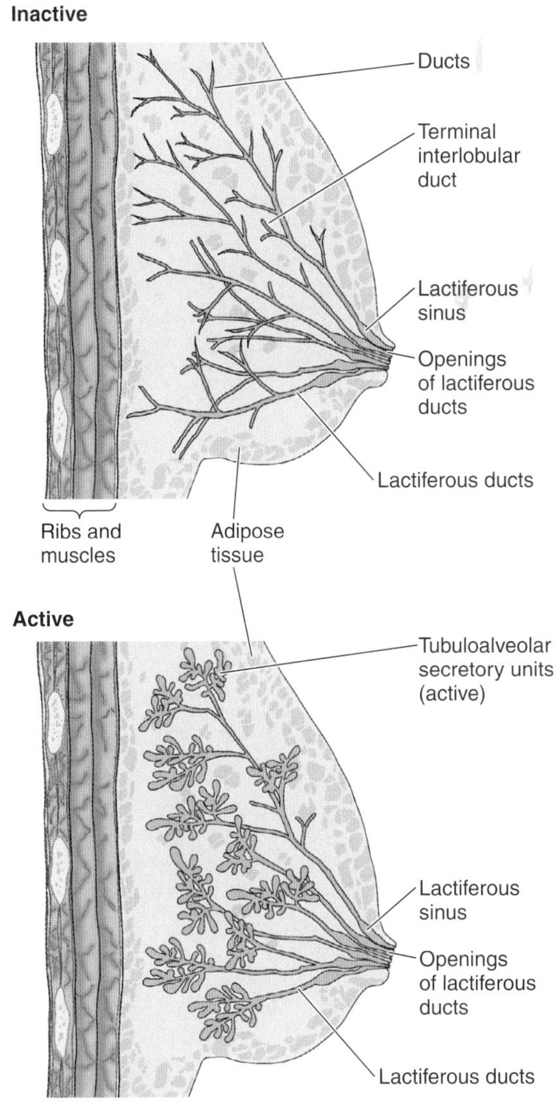

Figure 23–27. Schematic drawing of the female breast showing inactive and active mammary glands. Each lactiferous duct with its accompanying smaller ducts is a gland in itself and constitutes the lobes of the gland.

In girls during puberty the breasts increase in size and develop a prominent nipple. In boys, the breasts remain flattened.

Breast enlargement during puberty is the result of the accumulation of adipose tissue and connective tissue, with increased growth and branching of lactiferous ducts due to an increase in the amount of ovarian estrogens.

The characteristic structure of the gland—the **lobe**—in the adult woman is developed at the tips of the smallest ducts (Figure 23–27). A lobe consists of several ducts that empty into one terminal duct. Each lobe is embedded in loose connective tissue. A denser, less cellular connective tissue separates the lobes.

Near the opening of the nipple, the lactiferous ducts dilate to form the lactiferous sinuses (Figure 23–27). The lactiferous sinuses are lined with stratified squamous epithelium at their external openings. This epithelium very quickly changes to stratified columnar or cuboidal epithelium. The lining of the lactiferous ducts and terminal ducts is formed of simple cuboidal epithelium covered by closely packed myoepithelial cells.

The connective tissue surrounding the alveoli contains many lymphocytes and plasma cells. The plasma cell population increases significantly toward the end of pregnancy; it is responsible for the secretion of immunoglobulins (secretory IgA) that confer passive immunity on the newborn.

The histologic structure of these glands undergoes small alterations during the menstrual cycle, eg, proliferation of cells of the ducts at about the time of ovulation. These changes coincide with the time at which circulating estrogen is at its peak. Greater hydration of connective tissue in the premenstrual phase produces breast enlargement.

The **nipple** has a conical shape and may be pink, light brown, or dark brown. Externally, it is covered by keratinized stratified squamous epithelium continuous with that of the adjacent skin. The skin around the nipple constitutes the **areola.** The color of the areola darkens during pregnancy, as a result of the local accumulation of melanin. After delivery, the areola may become lighter in color but rarely returns to its original shade. The epithelium of the nipple rests on a layer of connective tissue rich in smooth muscle fibers. These fibers are disposed in circles around the deeper lactiferous ducts and parallel to them where they enter the nipple. The nipple is abundantly supplied with sensory nerve endings.

The Breasts During Pregnancy & Lactation

The mammary glands undergo intense growth during pregnancy as a result of the synergistic action of several hormones, mainly estrogen, progesterone, prolactin, and human placental lactogen. One of the actions of these hormones is the proliferation of **alveoli** at the ends of the terminal ducts. Alveoli are spherical collections of epithelial cells that become the active milk-secreting structures in lactation (Figures 23–28 and 23–29). A few fat droplets and membrane-limited secretory vacuoles containing from one to several dense aggregates of milk proteins can be seen in the apical cytoplasm of alveolar cells. The number of secretory vacuoles and fat droplets greatly increases in lactation (see below). Stellate myoepithelial cells are found between the alveolar epithelial cells and the basal lamina. The amounts of connective tissue and adipose tissue, relative to the parenchyma, decrease considerably during lactation.

During lactation, milk is produced by the epithelial cells of the alveoli (Figure 23–29) and accumulates in their lumens and inside the lactiferous ducts. The secretory cells become small and low cuboidal, and their cytoplasm contains spherical droplets of

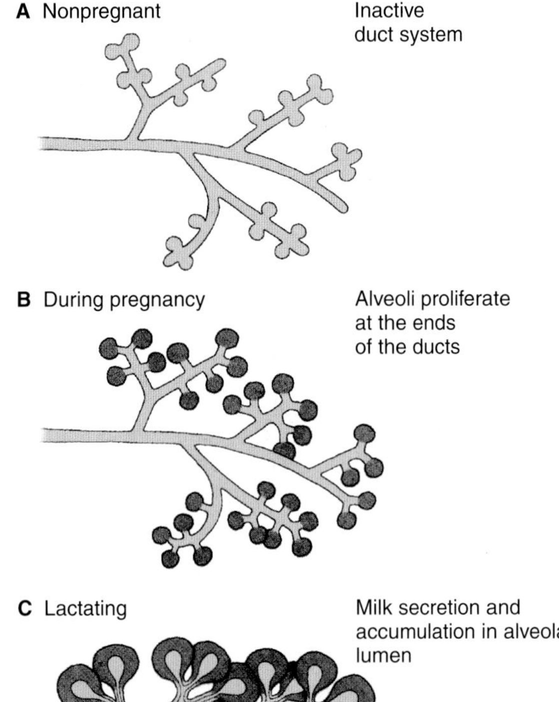

Figure 23–28. Changes in the mammary gland. **A:** In nonpregnant women, the gland is quiescent and undifferentiated, and its duct system is inactive. **B:** During pregnancy, alveoli proliferate at the ends of the ducts and prepare for the secretion of milk. **C:** During lactation, alveoli are fully differentiated, and milk secretion is abundant. Once lactation is completed, the gland reverts to the nonpregnant condition.

various sizes containing mainly neutral triglycerides. These lipid droplets pass out of the cells into the lumen and in the process are enveloped with a portion of the apical cell membrane. Lipids constitute about 4% of human milk.

In addition to the lipid droplets, there are a large number of membrane-limited vacuoles that contain granules composed of caseins and other milk proteins (Figure 23–30). Milk proteins include several caseins, α-lactalbumin, and plasmocyte-produced IgA. Proteins constitute approximately 1.5% of human milk. Lactose, the sugar of milk, is synthesized from glucose and galactose and constitutes about 7% of human milk.

MEDICAL APPLICATION

The first secretion from the mammary glands to appear after birth is called ***colostrum.*** *It contains less fat and more protein than regular milk and is rich in antibodies (predominantly secretory IgA) that provide some degree of passive immunity to the newborn, especially within the gut lumen.*

When a woman is breast-feeding, the nursing action of the child stimulates tactile receptors in the nipple, resulting in liberation of the posterior pituitary hormone ***oxytocin.*** *This hormone causes contraction of myoepithelial cells in alveoli and ducts, resulting in ejection of milk (****milk-ejection reflex****). Negative emotional stimuli, such as frustration, anxiety, or anger, can inhibit the liberation of oxytocin and thus prevent the reflex.*

Postlactational Regression of the Breasts

With cessation of breast-feeding (weaning), most alveoli that develop during pregnancy undergo degeneration through apoptosis (see Figure 3–24). This includes sloughing of whole cells as

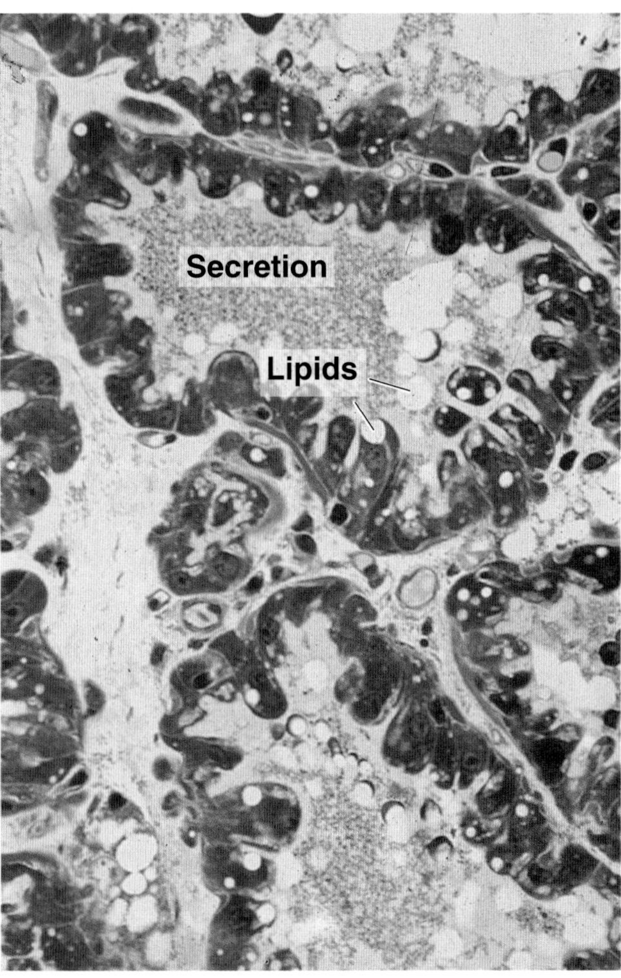

Figure 23–29. Photomicrograph of lactating mammary gland. Several alveoli are filled with milk, visible as granular material. The vacuoles in the lumen and in the alveolar cell cytoplasm represent the lipid portion of milk. PT stain. Medium magnification.

well as autophagic absorption of cellular components. Dead cells and debris are removed by macrophages.

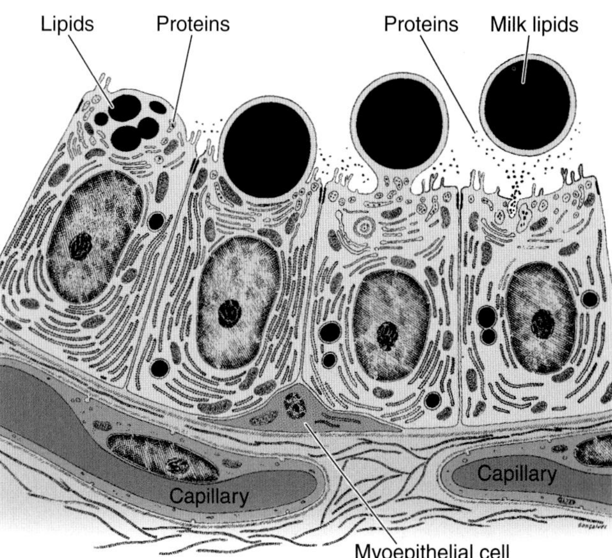

Figure 23–30. Secreting cells from the mammary gland. From left to right, note the accumulation and extrusion of lipids and proteins. The proteins are released through exocytosis.

Senile Involution of the Breasts

After menopause, involution of the mammary glands is characterized by a reduction in size and the atrophy of their secretory portions and, to a certain extent, the ducts. Atrophic changes also take place in the connective tissue.

Cancer of the Breast

MEDICAL APPLICATION

About 9% of all women born in the United States will develop breast cancer at some time during their lives. Most of these cancers (carcinomas) arise from epithelial cells of the lactiferous ducts. If these cells metastasize to the lungs, brain, or bone, breast carcinoma becomes a major cause of death. Early detection (eg, through self-examination, mammography, ultrasound, and other techniques) and consequent early treatment have significantly reduced the mortality rate from breast cancer.

REFERENCES

Brenner RM, Slayden OD: Cyclic changes in the primate oviduct and endometrium. In Knobil E et al (editors): *The Physiology of Reproduction.* Raven Press, 1994.

Gosden RG: Ovulation 1: oocyte development throughout life. In Gudzinskas JG, Yovich JL (editors): *Gametes—The Oocyte.* Cambridge University Press, 1995.

Hillier SG: Hormonal control of folliculogenesis and luteinization. In Findlay JK (editor): *Molecular Biology of the Female Reproductive System.* Academic Press, 1994.

Kenisgsberg D et al: Ovarian follicular maturation, ovulation, and ovulation induction. In DeGroot LJ et al (editors): *Endocrinology.* Saunders, 1995.

Ledger WL, Baird DT: Ovulation 3: endocrinology of ovulation. In Gudzinskas JG, Yovich JL (editors): *Gametes—The Oocyte.* Cambridge University Press, 1995.

Mishell DR Jr: Contraception. In DeGroot LJ et al (editors): *Endocrinology.* Saunders, 1995.

Peters H, McNatty KP: *The Ovary: A Correlation of Structure and Function in Mammals.* Granada Publishing, 1980.

Pitelka DR, Hamamoto ST: Ultrastructure of the mammary secretory cell. In Mepham TB (editor): *Biochemistry of Lactation.* Elsevier, 1983.

Tsafiri A, Dekel N: Molecular mechanisms in ovulation. In Findlay JK (editor): *Molecular Biology of the Female Reproductive System.* Academic Press, 1994.

Vorherr H: *The Breast: Morphology, Physiology and Lactation.* Academic Press, 1974.

Wynn RM (editor): *Biology of the Uterus.* Plenum Press, 1977.

Zuckerman S, Weir BJ (editors): *The Ovary,* 2nd ed. Vol. 1. *General Aspects.* Academic Press, 1977.

Photoreceptor and Audioreceptor Systems

24

Information about the external world is conveyed to the central nervous system by sensory units called **receptors.** This chapter examines the systems responsible for the reception of light and sound waves.

VISION: THE PHOTORECEPTOR SYSTEM

The Eye

The eye (Figure 24–1) is a complex and highly developed photosensitive organ that permits an accurate analysis of the form, light intensity, and color reflected from objects. The eyes are located in protective bony structures of the skull, the **orbits.** Each eye includes a tough, fibrous globe to maintain its shape, a lens system to focus the image, a layer of photosensitive cells, and a system of cells and nerves whose function it is to collect, process, and transmit visual information to the brain. Each eye (Figure 24–2) is composed of 3 concentric layers: an external layer that consists of the **sclera** and the **cornea;** a middle layer—also called the **vascular layer**—consisting of the **choroid, ciliary body,** and **iris;** and an inner layer of nerve tissue, the **retina,** which consists of an outer pigment epithelium and an inner retina proper. The photosensitive retina proper communicates with the cerebrum through the **optic nerve** (Figures 24–1 and 24–2) and extends forward to the **ora serrata.** The optic nerve arises in the embryo as an evagination of the prosencephalon. Consequently, it is not considered a true peripheral nerve like the other cranial nerves. Since it is a tract of the central nervous system, the myelin of its nerve fibers is produced by oligodendrocytes, not by Schwann cells. This may explain the visual dysfunction often associated with multiple sclerosis, a demyelinating disorder of the central nervous system.

The **lens** of the eye is a biconvex transparent structure held in place by a circular system of fibers, the **zonule,** which extends from the lens into a thickening of the middle layer, the **ciliary body,** and, by close apposition, to the vitreous body on its posterior side (Figures 24–1 and 24–2). Partly covering the anterior surface of the lens is an opaque pigmented expansion of the middle layer called the **iris.** The round hole in the middle of the iris is the **pupil** (Figure 24–1).

The eye contains 3 compartments: the **anterior chamber,** which occupies the space between the cornea and the iris and lens; the **posterior chamber,** between the iris, ciliary process, zonular attachments, and lens; and the **vitreous space,** which lies behind the lens and zonular attachments and is surrounded by the retina (Figures 24–1 and 24–2). Both the anterior and posterior chambers contain a protein-poor fluid called **aqueous humor.** The vitreous space is filled with a gelatinous substance called the **vitreous body.**

Note that the terms **outer** (**external**) and **inner** (**internal**) refer to the gross structure of the eye. Inner denotes a structure closer to the center of the globe, while outer means closer to the surface of the eyeball.

External Layer, or Tunica Fibrosa

The opaque white posterior five sixths of the external layer of the eye is the **sclera** (Figure 24–3); in the human, this forms a segment of a sphere approximately 22 mm in diameter (Figures 24–1 and 24–2). The sclera consists of tough, dense connective tissue made up mainly of flat collagen bundles intersecting in various directions while remaining parallel to the surface of the organ, a moderate amount of ground substance, and a few fibroblasts. The external surface of the sclera—the **episclera**—is connected by a loose system of thin collagen fibers to a dense layer of connective tissue called **Tenon's capsule.** Tenon's capsule comes into contact with the loose conjunctival stroma at the junction of the cornea with the sclera. Between Tenon's capsule and the sclera is **Tenon's space.** Because of this loose space, the eyeball can make rotating movements. Between the sclera and the choroid is the **suprachoroidal lamina,** a thin layer of loose connective tissue rich in melanocytes, fibroblasts, and elastic fibers. The sclera is relatively avascular.

In contrast to the posterior five sixths of the eye, the anterior one sixth—the **cornea**—is colorless and transparent (Figures 24–1 and 24–2). A transverse section of the cornea shows that it consists of 5 layers: epithelium, Bowman's membrane, stroma, Descemet's membrane, and endothelium. The corneal epithelium is stratified, squamous, and nonkeratinized and consists of 5 or 6 layers of cells (Figure 24–4). In the basal part of the epithelium are numerous mitotic figures that are responsible for the cornea's remarkable regenerative capacity: The turnover time for these cells is approximately 7 days. The surface corneal cells show microvilli protruding into the space filled by the precorneal tear film. This epithelial tissue is covered by a protective layer of lipid and glycoprotein, about 7 μm thick. The cornea has one of the richest sensory nerve supplies of any eye tissue.

Beneath the corneal epithelium lies a thick homogeneous layer 7–12 μm thick. This layer, **Bowman's membrane,** consists of collagen fibers crossing at random, a condensation of the intercellular substance, and no cells (Figure 24–4). Bowman's membrane contributes greatly to the stability and strength of the cornea.

The **stroma** is formed of many layers of parallel collagen bundles that cross at approximately right angles to each other. The

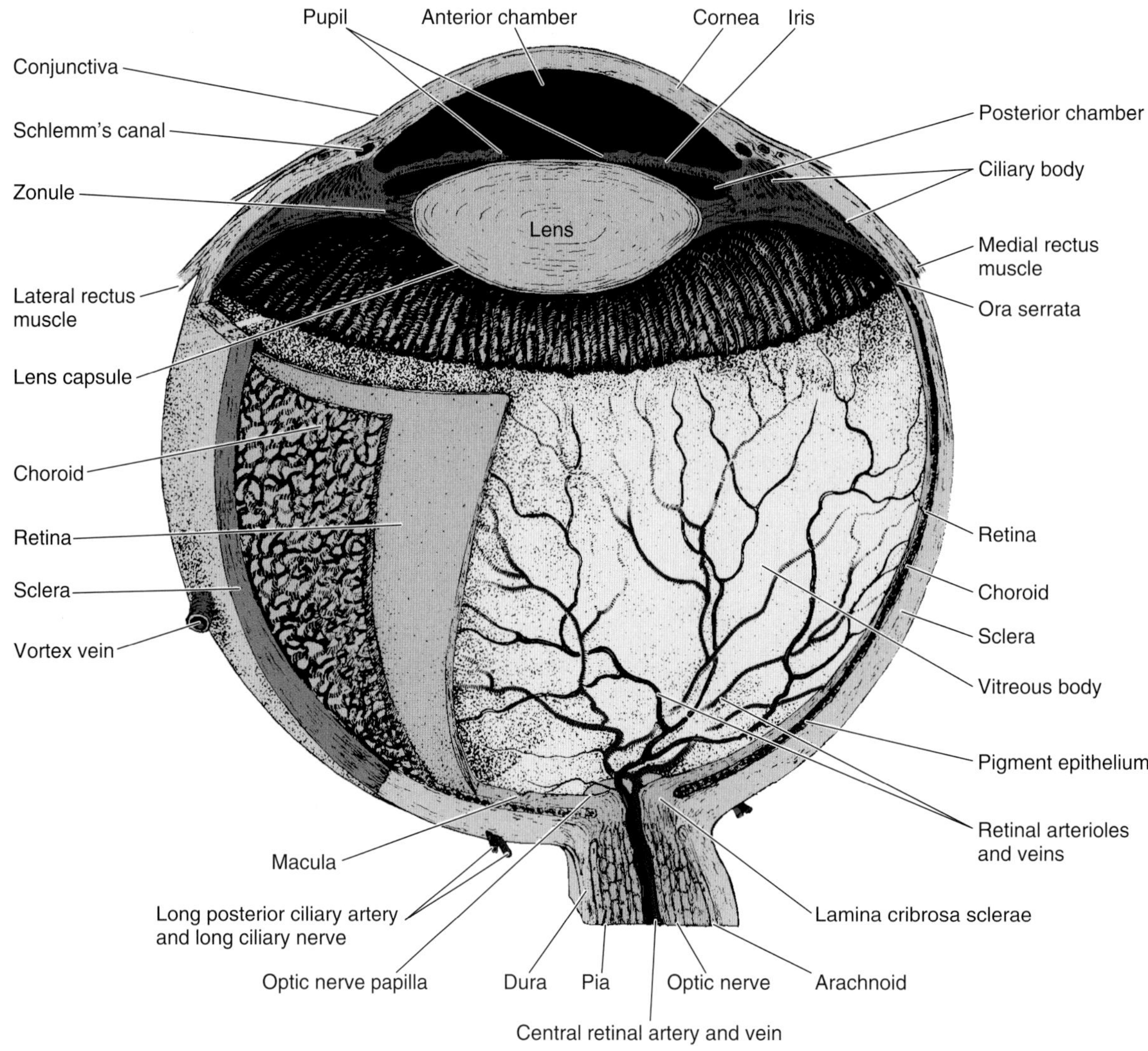

Figure 24–1. Internal structures of the human eye. (Redrawn from an original drawing by Paul Peck and reproduced, with permission, from *The Anatomy of the Eye* [Courtesy of Lederle Laboratories].)

collagen fibrils within each lamella are parallel to each other and run the full width of the cornea. Between the several layers, the cytoplasmic extensions of fibroblasts are flattened like the wings of a butterfly. Both cells and fibers of the stroma are immersed in a substance rich in glycoproteins and chondroitin sulfate. Although the stroma is avascular, migrating lymphoid cells are normally present in the cornea.

Descemet's membrane is a thick (5–10 μm) homogeneous structure composed of fine collagenous filaments organized in a 3-dimensional network.

The **endothelium** of the cornea is a simple squamous epithelium. These cells possess organelles for secretion that are characteristic of cells engaged in active transport and protein synthesis and that may be related to the synthesis and maintenance of Descemet's membrane. The corneal endothelium and epithelium are responsible for maintaining the transparency of the cornea. Both layers are capable of transporting sodium ions toward their apical surfaces. Chloride ions and water follow passively, maintaining the corneal stroma in a relatively dehydrated state. This state, along with the regular orientation of the very thin collagen fibrils of the stroma, accounts for the transparency of the cornea.

The **corneoscleral junction,** or **limbus,** is an area of transition from the transparent collagen bundles of the cornea to the white opaque fibers of the sclera. It is highly vascularized, and its blood vessels assume an important role in corneal inflammatory processes. The cornea, an avascular structure, receives its metabolites by diffusion from adjacent vessels and from the fluid of the anterior chamber of the eye. In the region of the limbus in the stromal layer, irregular endothelium-lined channels, the trabecular meshwork, merge to form **Schlemm's canal** (Figures 24–1 and 24–2), which drains fluid from the anterior chamber of the eye. Schlemm's canal communicates externally with the venous system.

Middle, or Vascular, Layer

The middle (vascular) layer of the eye consists of 3 parts: choroid, ciliary body, and iris (Figure 24–1), known collectively as the uveal tract.

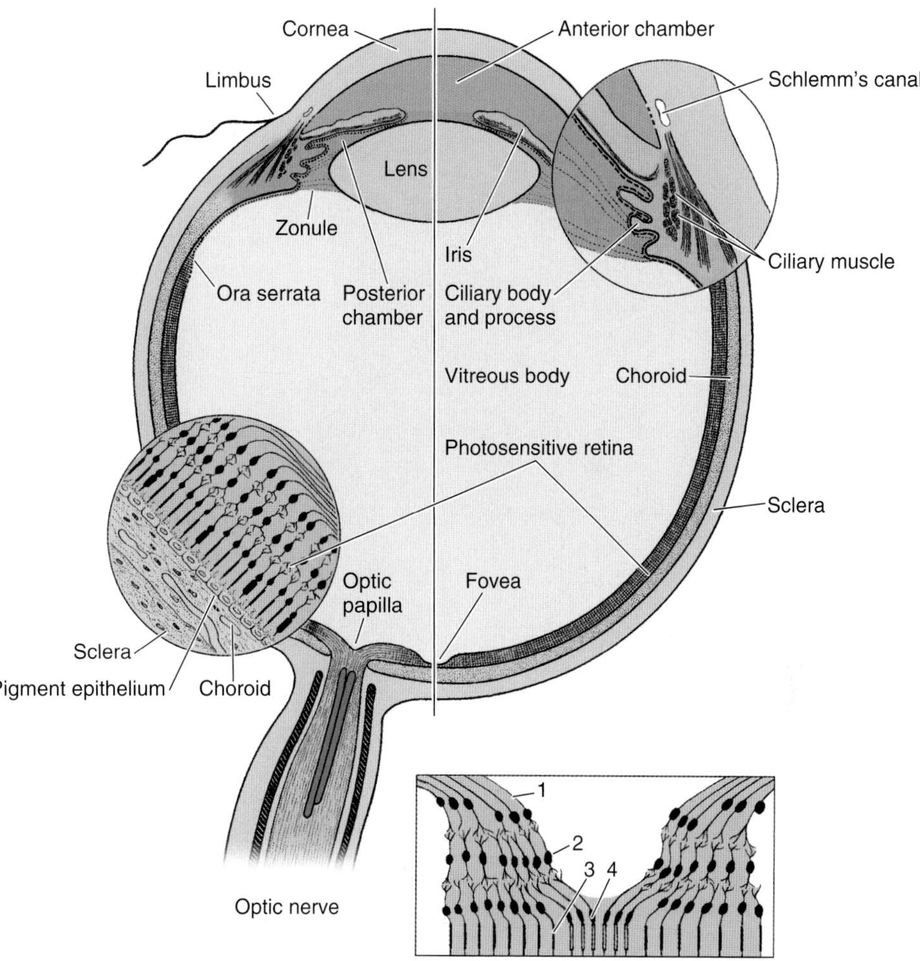

Figure 24–2. Diagram of the right eye, seen from above, showing the structure of the eye, retina, fovea, and ciliary body. An enlarged diagram of the fovea is shown at lower right: (**1**) axons of ganglion cells; (**2**) bipolar cells; (**3**) rods; (**4**) cones. Enlarged diagrams of the ciliary body (upper right) and retina (lower left) are also shown. (Modified and reproduced, with permission, from Ham AW: *Histology*, 6th ed. Lippincott, 1969.)

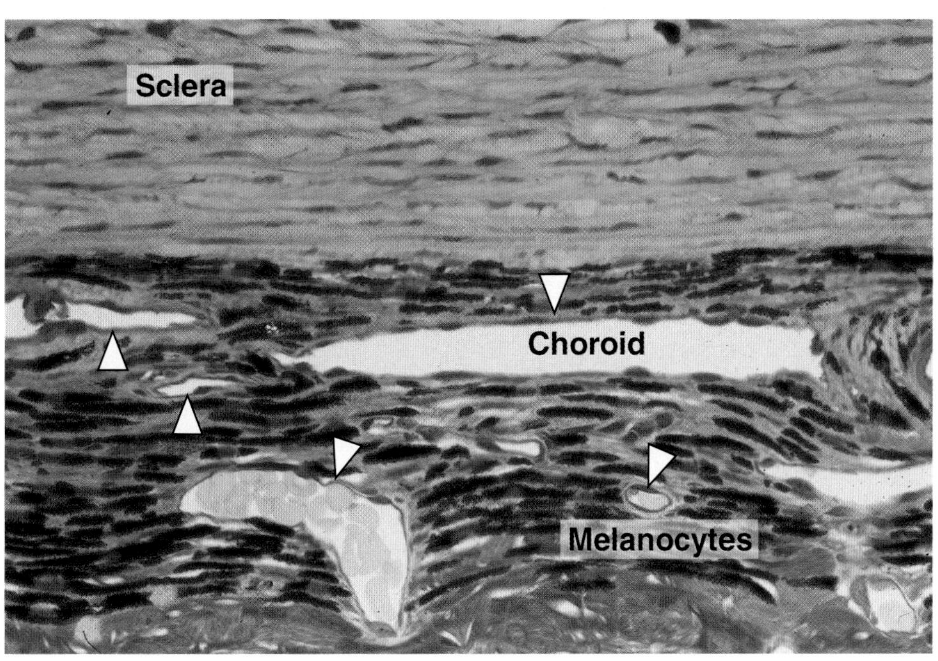

Figure 24–3. Section of choroid and sclera. The choroid is a highly vascular layer (arrowheads) of connective tissue containing melanocytes that prevent the reflection of incident light. Many of the nutrients for the retina come from choroid blood vessels. The sclera is a dense layer of connective tissue rich in fibers of collagen type I, arranged in parallel bundles. Pararosaniline–toluidine blue (PT) stain. Medium magnification.

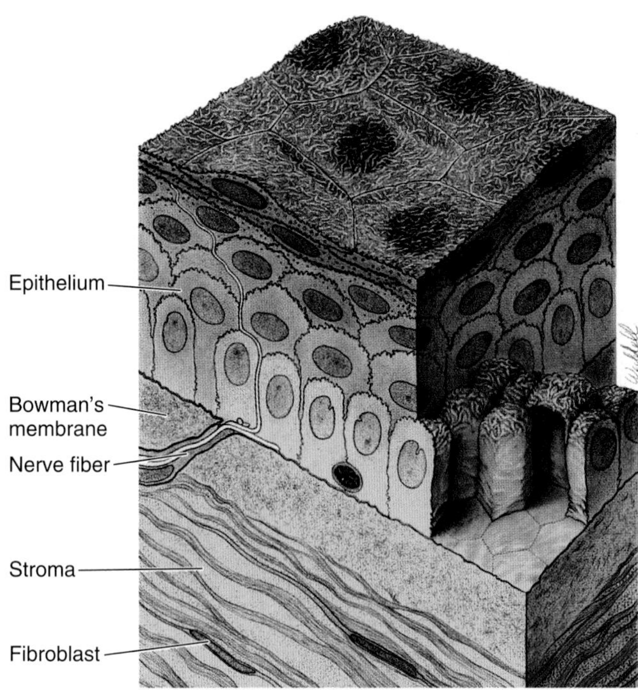

Figure 24–4. Three-dimensional drawing of the cornea. (Reproduced, with permission, from Hogan MJ et al: *Histology of the Human Eye.* Saunders, 1971.)

CHOROID

The choroid (Figure 24–3) is a highly vascularized coat, with loose connective tissue between its blood vessels that is rich in fibroblasts, macrophages, lymphocytes, mast cells, plasma cells, collagen fibers, and elastic fibers. Melanocytes are abundant in this layer and give it its characteristic black color. The inner layer of the choroid is richer than the outer layer in small vessels and is called the **choriocapillary layer.** It has an important function in nutrition of the retina, and damage to this tissue causes serious damage to the retina. A thin (3–4 μm) hyaline membrane separates the choriocapillary layer from the retina. This is known as **Bruch's membrane** and extends from the optic papilla to the ora serrata. The **optic papilla** is the region where the optic nerve enters the eyeball (Figure 24–2).

Bruch's membrane is formed of 5 layers. The central layer is composed of a network of elastic fibers. This network is lined on its two surfaces with layers of collagen fibers that are covered by the basal lamina of the capillaries of the choriocapillary layer on one side and the basal lamina of the pigment epithelium on the other side. (See Retina, below, for a description of the pigment epithelium.) The choroid is bound to the sclera by the **suprachoroidal lamina,** a loose layer of connective tissue rich in melanocytes.

CILIARY BODY

The ciliary body, an anterior expansion of the choroid at the level of the lens (Figures 24–1 and 24–2), is a continuous thickened ring that lies at the inner surface of the anterior portion of the sclera; in transverse section, it forms a triangle. One of its faces is in contact with the vitreous body, one with the sclera, and the third with the lens and the posterior chamber of the eye. The histologic structure of the ciliary body is basically loose connective tissue (rich in elastic fibers, vessels, and melanocytes) surrounding the **ciliary muscle** (Figure 24–2). This structure consists of 2 bundles of smooth muscle fibers that insert on the sclera anteriorly and on different regions of the ciliary body pos-

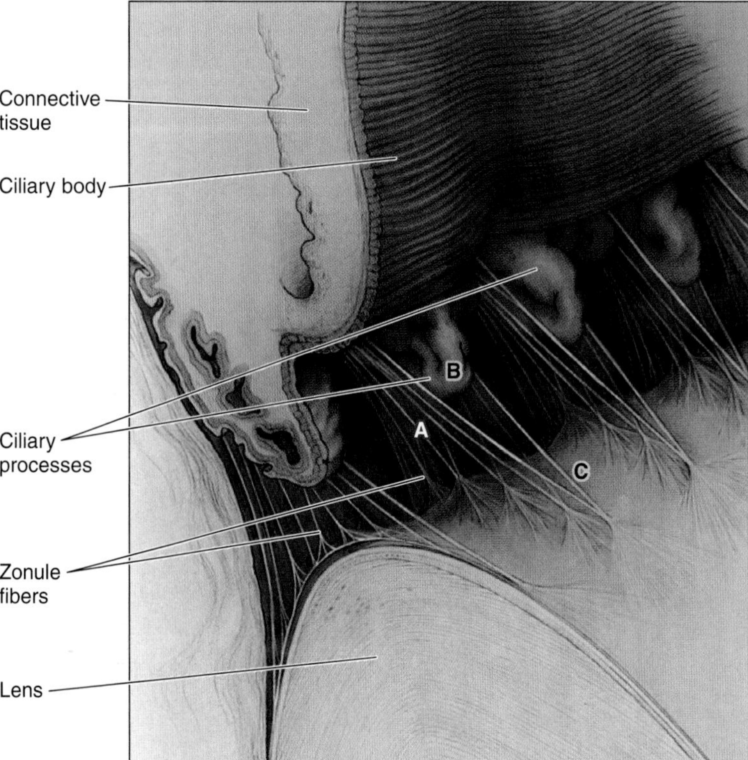

Figure 24–5. Anterior view of the ciliary processes showing the zonules attaching to the lens. Zonule fibers are bundles of microfilaments (oxytalan fibers) from the elastic fiber system. The zonules form columns **(A)** on either side of the ciliary processes **(B)**, which meet on a single site **(C)** as they attach to the lens. (Reproduced, with permission, from Hogan MJ et al: *Histology of the Human Eye.* Saunders, 1971.)

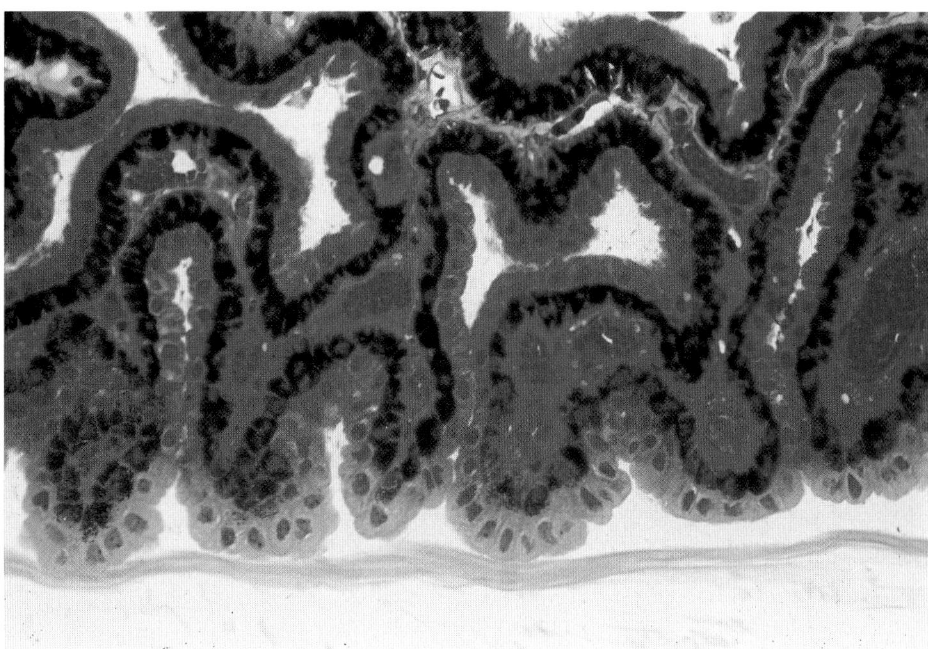

Figure 24–6. Section of ciliary processes showing their double layer of pigmented and nonpigmented epithelial cells. Note also the core of connective tissue. PT stain. Medium magnification.

teriorly. One of these bundles has the function of stretching the choroid; another bundle, when contracted, relaxes the tension on the lens. These muscular movements are important in visual accommodation (see Lens, below). The surfaces of the ciliary body that face the vitreous body, posterior chamber, and lens are covered by the anterior extension of the retina (Figure 24–2). In this region, the retina consists of only 2 cell layers. The layer directly adjacent to the ciliary body consists of simple columnar cells rich in melanin and corresponds to the forward projection of the pigment layer of the retina. The second layer, which covers the first, is derived from the sensory layer of the retina and consists of simple nonpigmented columnar epithelium.

Ciliary Processes

The ciliary processes are ridgelike extensions of the ciliary body (Figure 24–5). They have a loose connective tissue core and numerous fenestrated capillaries (see Chapter 11) and are covered by the two simple epithelial layers described above (Figures 24–6 and 24–7). From the ciliary processes emerge oxytalan fibers (**zonule fibers**) that insert into the capsule of the lens and anchor

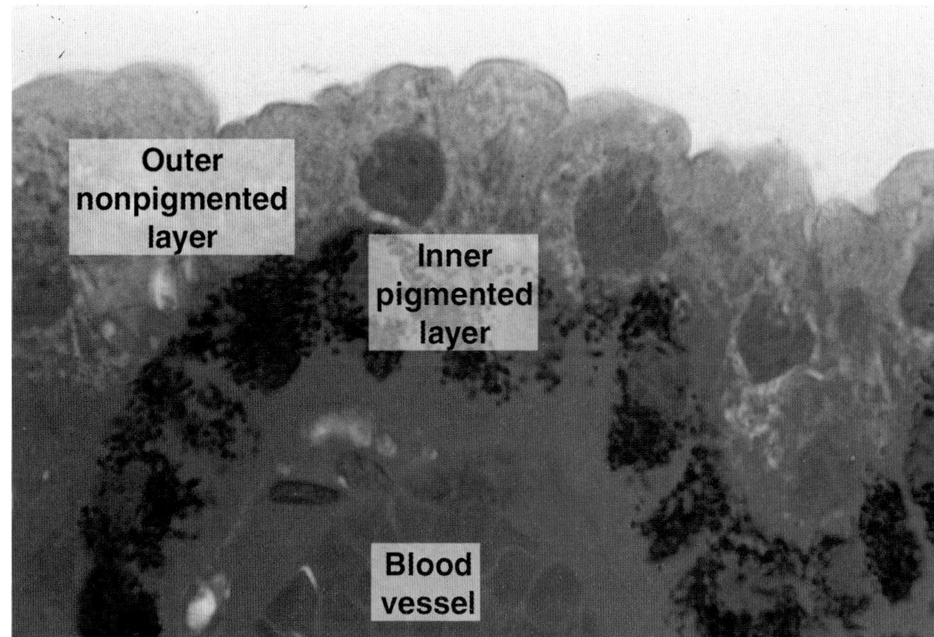

Figure 24–7. Section of a ciliary process. Note the dark granules of melanin located in the cytoplasm of the inner epithelial cells. The outer epithelium is devoid of melanin. PT stain. High magnification.

it in place (oxytalan fibers are described in Chapter 5). The apical ends of the epithelial cells are found at the junction between pigmented and nonpigmented cells, and the cells thus meet each other head-to-head. The zonular fibers have their origin in the basement membrane of the inner cells. The apical ends of the epithelial cells are joined by desmosomes, and elaborate tight junctions are found around the apical surfaces of epithelial cells of both layers. The nonpigmented inner layer of cells has extensive basal infoldings and interdigitations characteristic of ion-transporting cells (see Chapter 4). These cells actively transport certain constituents of plasma into the posterior chamber, thus forming the **aqueous humor.** This fluid has an inorganic ion composition similar to that of plasma but contains less than 0.1% protein (plasma has about 7% protein). Aqueous humor flows toward the lens and passes between it and the iris, reaching the anterior chamber of the eye (Figure 24–2). Once in the anterior chamber, the humor proceeds to the angle formed by the cornea with the basal part of the iris. It penetrates the tissue of the limbus in a series of labyrinthine spaces (the trabecular meshwork) and finally reaches the irregular Schlemm's canal, lined with endothelial cells (Figures 24–1 and 24–2). This structure communicates with small veins of the sclera, through which the aqueous humor escapes.

MEDICAL APPLICATION

Any impediment to the drainage of aqueous humor caused by an obstruction in the outflow channels results in an increase in intraocular pressure, causing ***glaucoma.***

Iris

The iris is an extension of the choroid that partially covers the lens, leaving a round opening in the center called the **pupil** (Figure 24–1). The anterior surface of the iris is irregular and rough, with grooves and ridges. It is formed of a discontinuous layer of pigment cells and fibroblasts. Beneath this layer is a poorly vascularized connective tissue with few fibers and many fibroblasts and melanocytes. The next layer is rich in blood vessels embedded in loose connective tissue (Figure 24–8). The smooth posterior surface of the iris is covered by 2 layers of epithelium, which also cover the ciliary body and its processes. The inner epithelium, in contact with the posterior chamber, is heavily pigmented with melanin granules. The outer epithelial cells have radially directed tonguelike extensions of their basal region; they are filled with overlapping myofilaments, creating the **dilator pupillae muscle** of the iris. The heavy pigmentation prevents the passage of light into the interior of the eye except through the pupil.

The function of the abundant melanocytes or pigment cells containing melanin in several regions of the eye is to keep stray light rays from interfering with image formation. The melanocytes of the stroma of the iris are responsible for the color of the eyes. If the layer of pigment in the interior region of the iris consists of only a few cells, the light reflected from the black pigment epithelium in the posterior surface of the iris will be blue. As the amount of pigment increases, the iris assumes various shades of greenish-blue, gray, and finally brown. Albinos have almost no pigment, and the pink color of their irises is due to the reflection of incident light from the blood vessels of the iris.

The iris contains smooth muscle bundles disposed in circles concentric with the pupillary margin, forming the **sphincter pupillae muscle** of the iris. The dilator and sphincter muscles have sympathetic and parasympathetic innervation, respectively.

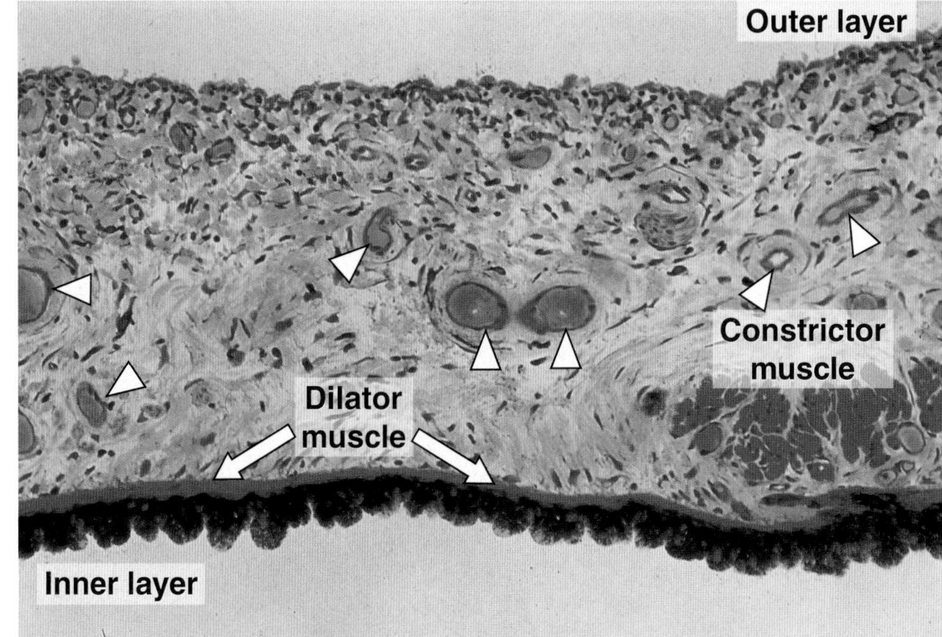

Figure 24–8. Section of iris, a structure consisting of a core of connective tissue highly vascularized in certain regions (arrowheads). The outer layer contains fibroblasts and very few pigmented cells (not seen in this photomicrograph). In contrast, the inner covering layer is heavily pigmented to protect the eye's interior from stray light. Dilator and constrictor (sphincter) pupillary muscles control the diameter of the pupil. PT stain. Medium magnification.

Lens

The lens is a biconvex structure characterized by great elasticity, a feature that is lost with age as the lens hardens. The lens has 3 principal components:

Lens Capsule

The lens is enveloped by a thick (10–20 μm), homogeneous, refractile, carbohydrate-rich capsule (Figure 24–9) coating the outer surface of the epithelial cells. It is a very thick basement membrane and consists mainly of collagen type IV and glycoprotein.

Subcapsular Epithelium

Subcapsular epithelium consists of a single layer of cuboidal epithelial cells that are present only on the anterior surface of the lens. The lens increases in size and grows throughout life as new lens fibers develop from cells located at the equator of the lens. The cells of this epithelium exhibit many interdigitations with the lens fibers.

Lens Fibers

Lens fibers are elongated and appear as thin, flattened structures (Figure 24–9). They are highly differentiated cells derived from cells of the subcapsular epithelium. Lens fibers eventually lose their nuclei and other organelles and become greatly elongated, attaining dimensions of 7–10 mm in length, 8–10 μm in width, and 2 μm in thickness. These cells are filled with a group of proteins called **crystallins.** Lens fibers are produced throughout life, at an ever-decreasing rate.

The lens is held in place by a radially oriented group of fibers, the **zonule,** that inserts on one side on the lens capsule and on the other on the ciliary body (Figure 24–5). Zonular fibers are similar to the microfibrils of elastic fibers. This system is important in the process known as **accommodation,** which permits focusing on near and far objects by changing the curvature of the lens. When the eye is at rest or gazing at distant objects, the lens is kept stretched by the zonule in a plane perpendicular to the optical axis. To focus on a near object, the ciliary muscles contract, causing forward displacement of the choroid and ciliary body. The tension exerted by the zonule is relieved, and the lens becomes thicker, keeping the object in focus.

MEDICAL APPLICATION

*Advancing age reduces the elasticity of the lens, making accommodation for near objects difficult. This is a normal aging process (**presbyopia**), which can be corrected by wearing glasses with convex lenses. In older individuals, a brownish pigment accumulates in lens fibers, making them less transparent. When the lens becomes opaque, the condition is termed **cataract,** which may also be caused by excessive exposure to ultraviolet radiation. In diabetes mellitus, the high levels of glucose are believed to produce cataract.*

Vitreous Body

The vitreous body occupies the region of the eye behind the lens. It is a transparent gel that consists of water (about 99%), a small amount of collagen, and heavily hydrated hyaluronic acid molecules. The vitreous body contains very few cells, which synthesize collagen and hyaluronic acid.

Retina

The retina, the inner layer of the globe, consists of 2 portions. The posterior portion is photosensitive; the anterior part, which is not photosensitive, constitutes the inner lining of the ciliary

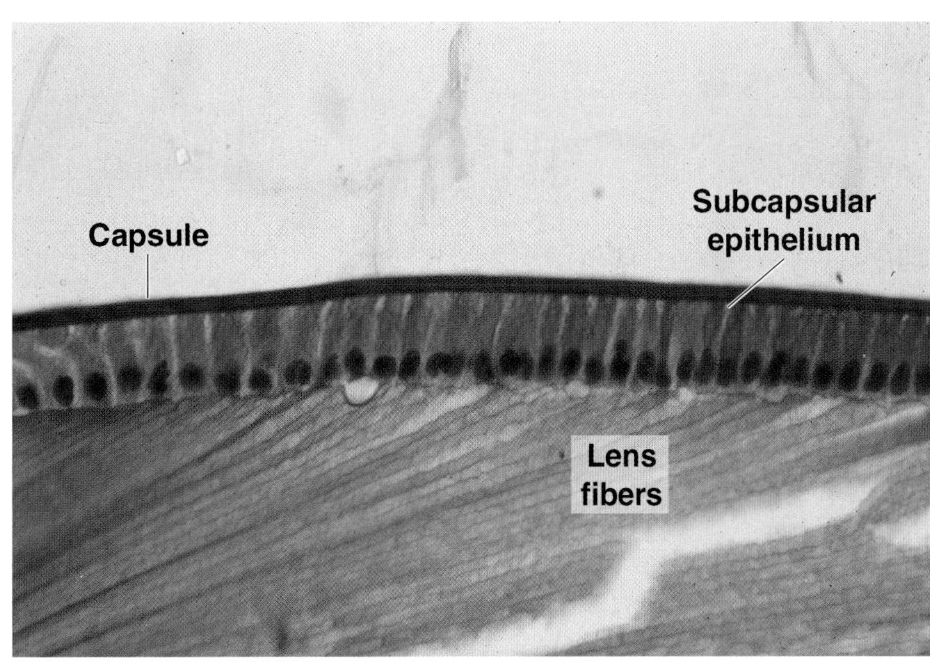

Figure 24–9. Section of the anterior portion of the lens. The subcapsular epithelium secretes the lens capsule, which appears stained in red. The lens capsule is a thick basement membrane containing collagen type IV and laminin. Below the subcapsular epithelium, note the lens fibers, which are cells that have lost their nuclei and organelles, becoming thin, elongated, transparent structures. Picrosirius-hematoxylin. Medium magnification.

body and the posterior part of the iris (Figures 24–2). The retina derives from an evagination of the anterior cephalic vesicle, or prosencephalon. As this **optic vesicle** comes into contact with the surface ectoderm, it gradually invaginates in its central region, forming a double-walled **optic cup.** In adults, the outer wall gives rise to a thin membrane called the **pigment epithelium;** the optical or functioning part of the retina—the **neural retina**—is derived from the inner layer.

The pigment epithelium consists of columnar cells with a basal nucleus. The basal regions of the cells adhere firmly to Bruch's membrane, and the cell membranes have numerous basal invaginations. Mitochondria are more abundant in the region of the cytoplasm near these invaginations. These characteristics suggest an ion-transporting activity for this region.

The lateral cell membranes show cell junctions with conspicuous zonulae occludentes and zonulae adherentes at their apexes; there are also desmosomes and gap junctions. These morphologic characteristics indicate that the apical and basal regions of this epithelial sheet are sealed off and that there is intercellular communication. These junctional specializations account for the electrical potential difference that results from ion transport between the two surfaces of this epithelium.

The cell apex has abundant extensions of 2 types: slender microvilli and cylindrical sheaths that envelop the tips of the photoreceptors.

MEDICAL APPLICATION

Because neither type of extension is anatomically joined to the photoreceptors, these regions can become separated, as in ***detachment of the retina.*** *This common and serious disorder in humans can be treated effectively with laser surgery.*

The cytoplasm of pigment epithelial cells has abundant smooth endoplasmic reticulum, believed to be a site of vitamin A esterification and transport to the photoreceptors. Melanin granules are numerous in the apical cytoplasm and microvilli. Melanin is synthesized in these cells by a mechanism similar to that described for the melanocytes in the skin (see Chapter 18). This dark pigment has the function of absorbing light after the photoreceptors have been stimulated.

The cell apex has numerous dense vesicles of variable shape that represent various stages in the phagocytosis and digestion of the tips of photoreceptor outer segments. Pigment cell structure and functions are shown in Figure 24–16.

The optical part of the retina—the posterior, or photosensitive, part—is a complex structure containing at least 15 types of neurons, and these cells form at least 38 distinct kinds of synapses with one another. The optical retina consists of an outer layer of photosensitive cells, the **rods** and **cones** (Figures 24–2, 24–10, and 24–11); an intermediate layer of **bipolar neurons,** which connect the rods and cones to the **ganglion cells;** and an internal layer of ganglion cells, which establish contact with the bipolar cells through their dendrites and send axons to the brain. These axons converge at the optic papilla, forming the **optic nerve.**

Between the layer of rods and cones and the bipolar cells is a region called the **external plexiform,** or **synaptic, layer,** where synapses between these two types of cells occur. The region where the synapses between the bipolar and ganglion cells are established is called the **internal plexiform layer** (Figures 24–11 and 24–12). The retina has an inverted structure, for the light will first cross the ganglion layer and then the bipolar layer to reach the rods and cones. The structure of the retina will now be examined in greater detail.

The rods and cones, named for the forms they assume, are polarized neurons; at one pole is a single photosensitive dendrite, and at the other are synapses with cells of the bipolar layer. The rod and cone cells (Figures 24–13, 24–14, and 24–15) can be divided into outer and inner segments, a nuclear region, and a synaptic region. The outer segments are modified cilia and con-

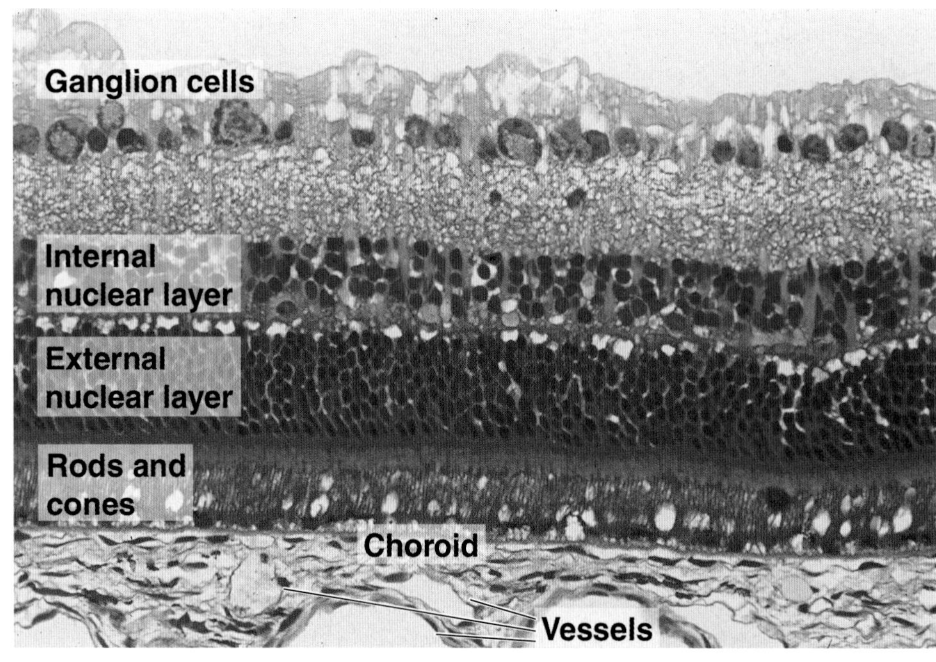

Figure 24–10. Section of retina showing most of its components. PT stain. Low magnification.

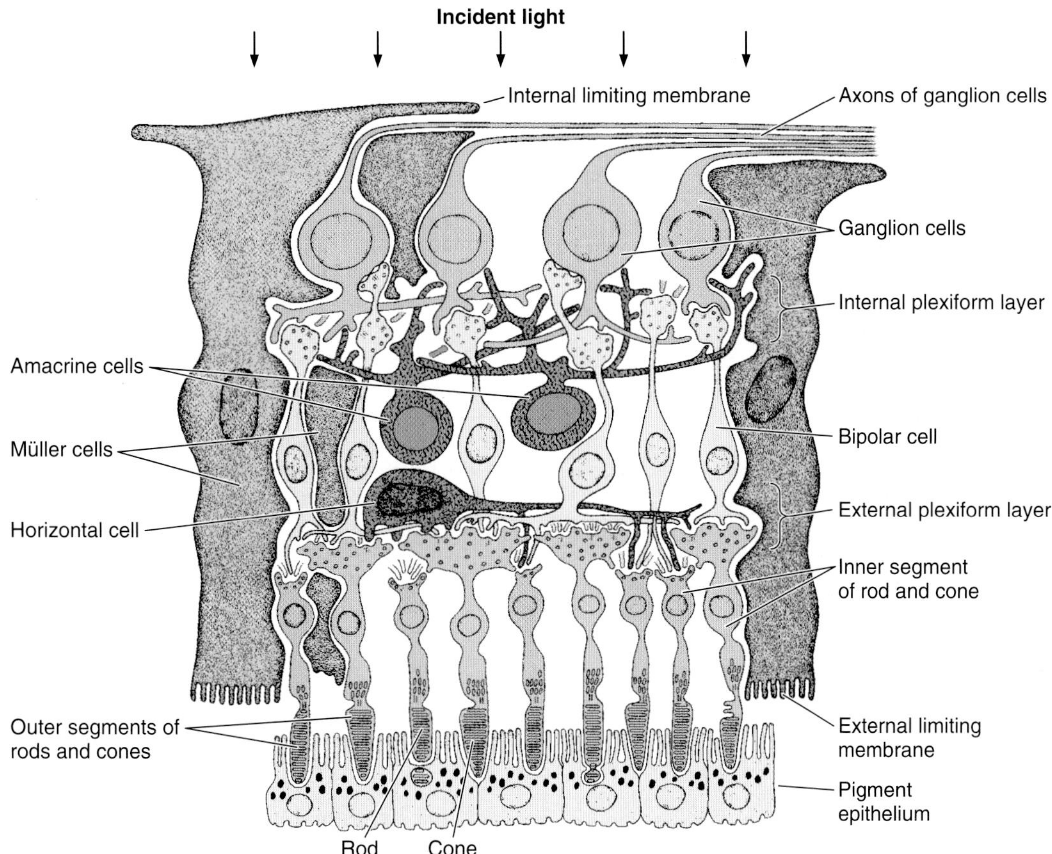

Figure 24–11. The three layers of retinal neurons. The arrows indicate the direction of the light path. The stimulation generated by the incident light on rods and cones proceeds in the opposite direction. (Redrawn and reproduced, with permission, from Boycott and Dowling: Proc R Soc Lond [Biol] 1966;166:80.)

tain stacks of membrane-limited saccules with a flattened, disklike shape. The photosensitive pigment of the retina is in the membranes of these saccules. Both rod and cone cells pass through a thin layer, the **external limiting membrane,** that is a series of junctional complexes between the photoreceptors and glial cells of the retina (Müller cells). The nuclei of the cones are generally disposed near the limiting membrane, whereas the nuclei of the rods lie near the center of the inner segment.

Rod Cells

Rod cells are thin, elongated cells (50 × 3 μm) composed of 2 portions (Figures 24–11 and 24–13). The external photosensitive rod-shaped portion is composed mainly of numerous (600–1000) flattened membranous disks stacked up like coins. The disks in rods are not continuous with the plasma membrane; the **outer segment** is separated from the **inner segment** by a constriction. Just below this constriction is a basal segment from which a cilium arises and passes to the outer segment. The inner segment is rich in glycogen and has a remarkable accumulation of mitochondria, most of which lie near the constriction (Figures 24–13 and 24–14). This local accumulation of mitochondria is related to the production of energy necessary for the visual process and protein synthesis. Polyribosomes, present in large numbers below the mitochondrial region of the inner segment, are involved in protein synthesis. Some of these proteins migrate to the outer segment of the rod cells, where they are incorporated into membranous disks. The flattened disks of the rod cells contain the pigment **visual purple,** or **rhodopsin,** which is bleached by light and initiates the visual stimulus. This substance is globular and is located in the outer surface of the lipid bilayer of the flattened membranous disks.

The human retina has approximately 120 million rods. They are extremely sensitive to light and are considered the receptors used when low levels of light are encountered, such as at dusk or nighttime. The outer segment is the site of photosensitivity; the inner segment contains the metabolic machinery necessary for the biosynthetic and energy-producing processes of these cells.

Autoradiographic studies show that proteins of the rod vesicles are synthesized in the polyribosome-rich inner segments of these cells. From there, they migrate to the outer segment and aggregate at its basal region, where they are incorporated into membranes formed by a double layer of phospholipids, producing flattened disks (Figures 24–13 and 24–14). These structures gradually migrate to the cell apex, where they are shed, phagocytized, and digested by the cells of the pigment epithelium (Figures 24–15 and 24–16). It has been calculated that, in the monkey, approximately 90 vesicles per cell are produced daily. The whole process of migration, from assembly at the basal cell region to apical shedding, takes from 9 to 13 days.

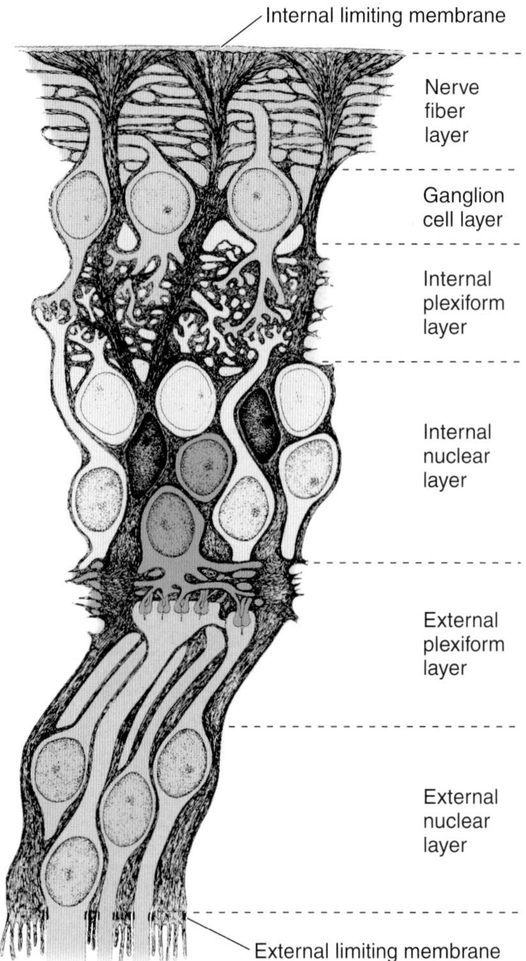

Figure 24–12. The close association of Müller cells with neural elements in the sensory retina. Müller cells (dark fibrous cells) appear to be structurally and functionally equivalent to the astrocytes of the central nervous system, in that they envelop and support the neurons and nerve processes of the retina. (Reproduced, with permission, from Hogan MJ et al: *Histology of the Human Eye.* Saunders, 1971.)

Cone Cells

Cone cells (Figure 24–13) are also elongated (60 × 1.5 μm) neurons. Each human retina has about 6 million cone cells. The structure is similar to that of rods, with outer and inner segments, a basal body with cilium, and an accumulation of mitochondria and polyribosomes. Cones differ from rods in their form (conical) and the structure of their outer segments. As in rods, this region is composed of stacked membranous disks; however, they are not independent of the outer plasma membrane but arise as invaginations of this structure (Figure 24–13). In cones, newly synthesized protein is not concentrated in recently assembled disks, as it is in rods, but is distributed uniformly throughout the outer segment.

There are at least 3 functional types of cones that cannot be distinguished by their morphologic characteristics. Each type contains a variety of the cone photopigment called **iodopsin,** and its maximum sensitivity is in the red, green, or blue region of the visible spectrum. Cones, sensitive only to light of a higher intensity than that required to stimulate rods, are believed to permit better visual acuity than do rods.

Other Cells

The layer of bipolar cells consists of 2 types of cells (Figure 24–11): **diffuse bipolar cells,** which have synapses with 2 or more photoreceptors; and **monosynaptic bipolar cells,** which establish contact with the axon of only one cone photoreceptor and only one ganglion cell. A certain number of cones therefore transmit their impulses directly to the brain.

In addition to establishing contact with the bipolar cells, the cells of the ganglion layer project their axons to a specific region of the retina, where they come together to form the **optic nerve** (Figure 24–11). This region, which is devoid of receptors, is known as the **blind spot** of the retina, the **papilla of the optic nerve,** or the **optic nerve head** (Figure 24–2). The **ganglion cells** are typical nerve cells, containing a large euchromatic nucleus and basophilic Nissl bodies. These cells, like the bipolar cells, are classified as diffuse or monosynaptic in their connections with other cells.

In addition to these 3 main types of cells (photoreceptor, bipolar, and ganglion cells), there are other types of cells that are distributed more diffusely in the layers of the retina.

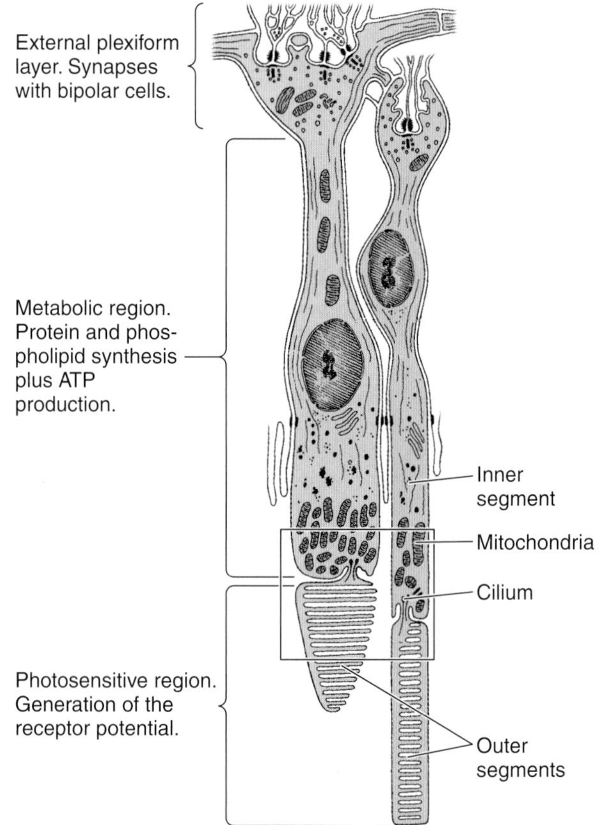

Figure 24–13. Ultrastructure of rods (at right) and cones (at left). The rectangular outlined region is shown in the electron micrograph in Figure 24–14. (Redrawn and reproduced, with permission, from Chevremont M: *Notions de Cytologie et Histologie.* S.A. Desoer Editions [Liege], 1966.)

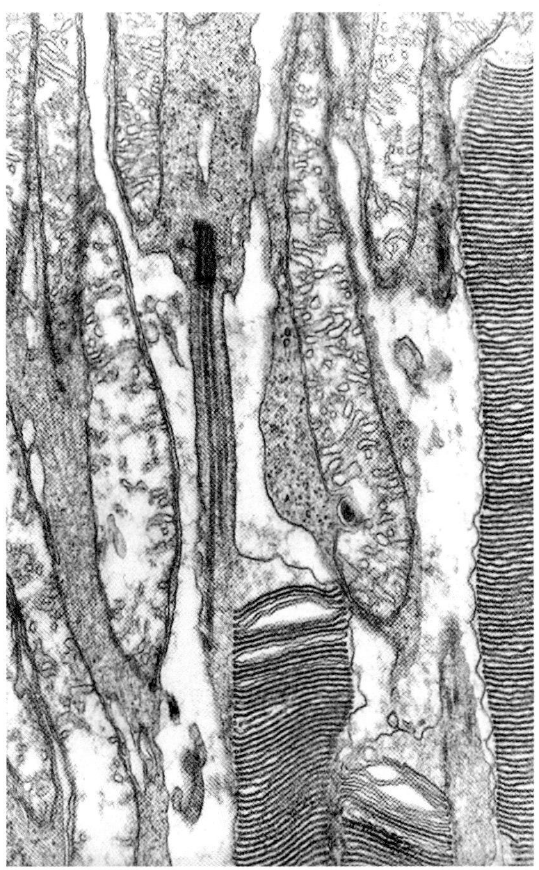

Figure 24–14. Electron micrograph of a section of the retina. In the upper part of the figure are the inner segments. This photosensitive region consists of parallel membranous flat disks. Accumulation of mitochondria takes place in the inner segment (see Figure 24–13). In the middle of the figure is a basal body giving rise to a cilium that is further modified into an outer segment.

Horizontal cells (Figure 24–11) establish contact between different photoreceptors. Their exact function is not known, but they may act to integrate stimuli.

Amacrine cells (Figure 24–11) are various types of neurons that establish contact between the ganglion cells. Their function is also obscure.

Supporting cells are neuroglia that possess, in addition to the astrocyte and microglial cell types, some large, extensively ramified cells (**Müller cells**). The processes of these cells bind the neural cells of the retina and extend from the internal to the external limiting membranes of the retina (Figure 24–11). The external limiting membrane is a zone of adhesion (tight junctions) between photoreceptors and Müller cells. Müller cells are functionally analogous to neuroglia in that they support, nourish, and insulate the retinal neurons and fibers.

Retinal Histophysiology

Light passes through the layers of the retina to the rods and cones, where it is absorbed, initiating a series of reactions that result in vision—an extraordinarily sensitive process. Experimental evidence suggests that a single photon is enough to trigger the production of a receptor potential in a rod. Light acts to bleach the visual pigments, a photochemical process amplified by mechanisms that cause the local production of responses that are subsequently transmitted to the brain.

The visual pigment of rods, **rhodopsin,** is composed of an aldehyde of vitamin A (retinaldehyde) bound to specific proteins called **opsins.** Because rods have a low resolution, they form images without clear details; they are not sensitive to colors. Cones, on the other hand, have a higher threshold and are responsible for sharp images and color vision. In humans, they contain 3 incompletely characterized pigments (**iodopsins**), which may provide a chemical basis for the classic tricolor theory of color vision.

When light strikes rhodopsin molecules, retinaldehyde undergoes isomerization from the all-*cis* form to the all-*trans* form. This change results in the dissociation of retinaldehyde from opsin, a reaction called **bleaching.** Bleaching of the visual pigment incorporated in the membrane disks increases the calcium conductance of the disk membranes and promotes the diffusion of calcium to the intracellular space of the outer segment of the photoreceptor. Calcium acts on the cell membrane, reducing its permeability to sodium ions and promoting cell hyperpolarization. The electrical signals produced by closing these sodium channels spread to the inner segment and through gap junctions to neighboring cells.

In a second step, the visual pigment is reassembled, and the calcium ions are transported back into the disks in an energy-consuming process. The high energy requirement would seem to account for the abundance of mitochondria near the photosensitive site of rods and cones. Contrary to what happens in other receptors where action potentials are generated through cell depolarization, the rods and cones are hyperpolarized by light. This signal is transmitted to the bipolar, amacrine, and horizontal cells and then to the ganglion cells. Only the ganglion cells generate action potentials along their axons, which relay the information to the brain.

MEDICAL APPLICATION

The clinical observation that the retina is damaged when it becomes detached suggests that the photosensitive cells derive their metabolites from the choriocapillary layer. The superficial localization of the vessels of the retina provides for their easy observation with an ophthalmoscope. This examination is of great value in the diagnosis and evaluation of disorders that affect blood vessels, such as diabetes mellitus and hypertension.

At the posterior pole of the optical axis lies the **fovea,** a shallow depression with very thin retina in the center. This is because the bipolar and ganglion cells accumulate in the periphery of this depression, so that its center consists only of cone cells

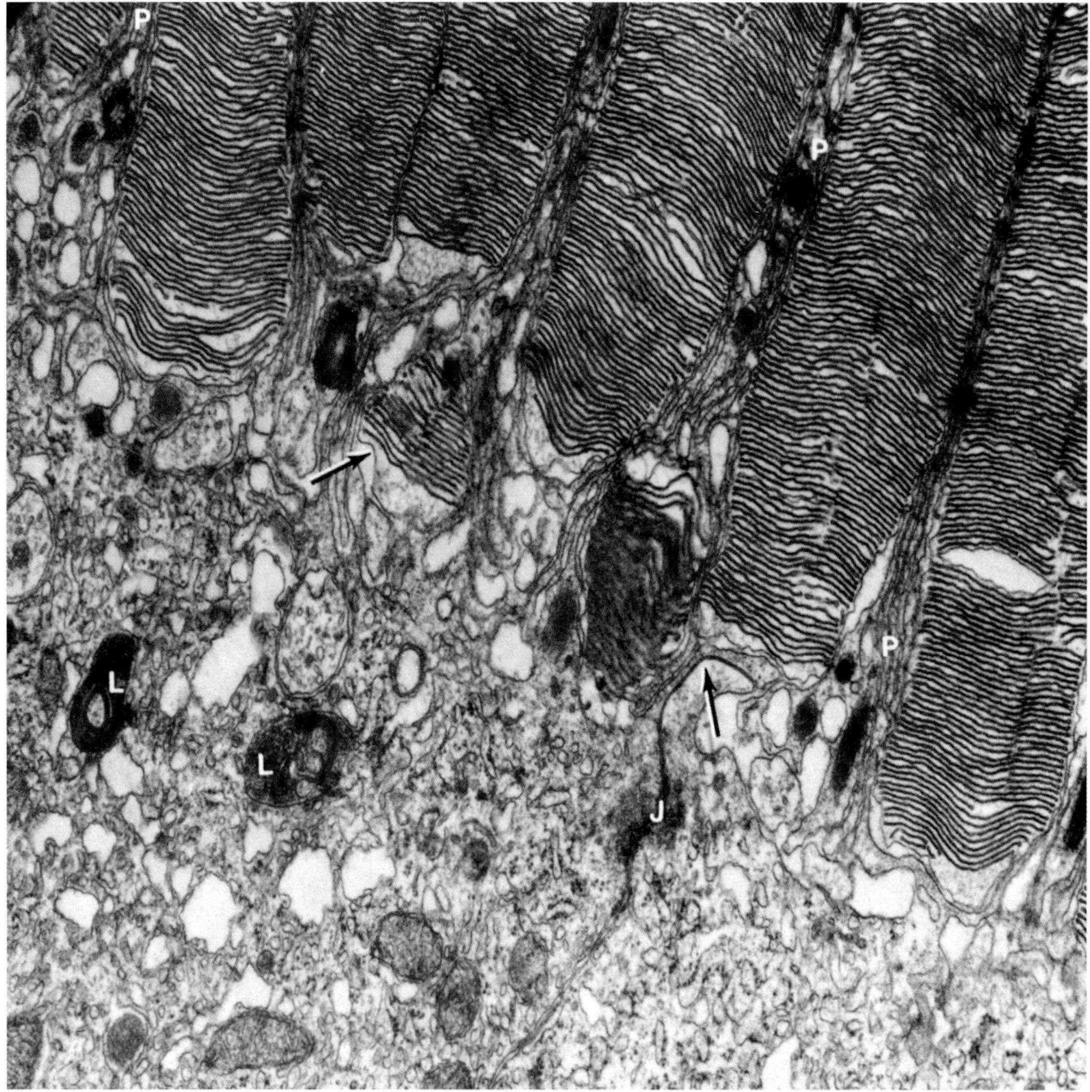

Figure 24–15. Electron micrograph of the interface between the photosensitive and pigmented layers in the retina. In the lower portion are parts of 2 pigment epithelium cells, revealing specialized junctions (J) between their lateral plasmalemmas. Above the pigment cells are the tips of several outer segments of rod cells that interdigitate with apical processes of the pigment epithelium (P). The large vacuoles containing flattened membranes (arrows) have been shed from the tips of the rods. L, lysosomal vesicles.

(Figure 24–2). Cone cells in the fovea are long and narrow, resembling rod cells. This is an adaptation to permit closer packing of cones and thereby increase visual acuity. In this area, blood vessels do not cross over the photosensitive cells. Light falls directly on the cones in the central part of the fovea, which helps account for the extremely precise visual acuity of this region.

Accessory Structures of the Eye

Conjunctiva

The conjunctiva is a thin, transparent mucous membrane that covers the anterior portion of the eye up to the cornea and the internal surface of the eyelids. It has a stratified columnar epithelium with numerous goblet cells, and its lamina propria is composed of loose connective tissue.

Eyelids

Eyelids (Figure 24–17) are movable folds of tissue that protect the eye. The skin of the lids is loose and elastic, permitting extreme swelling and subsequent return to normal shape and size.

The 3 types of glands in the lid are the Meibomian glands and the glands of Moll and Zeis. The Meibomian glands are long sebaceous glands in the tarsal plate. They do not communicate with the hair follicles. The Meibomian glands produce a sebaceous substance that creates an oily layer on the surface of the tear film, helping to prevent rapid evaporation of the

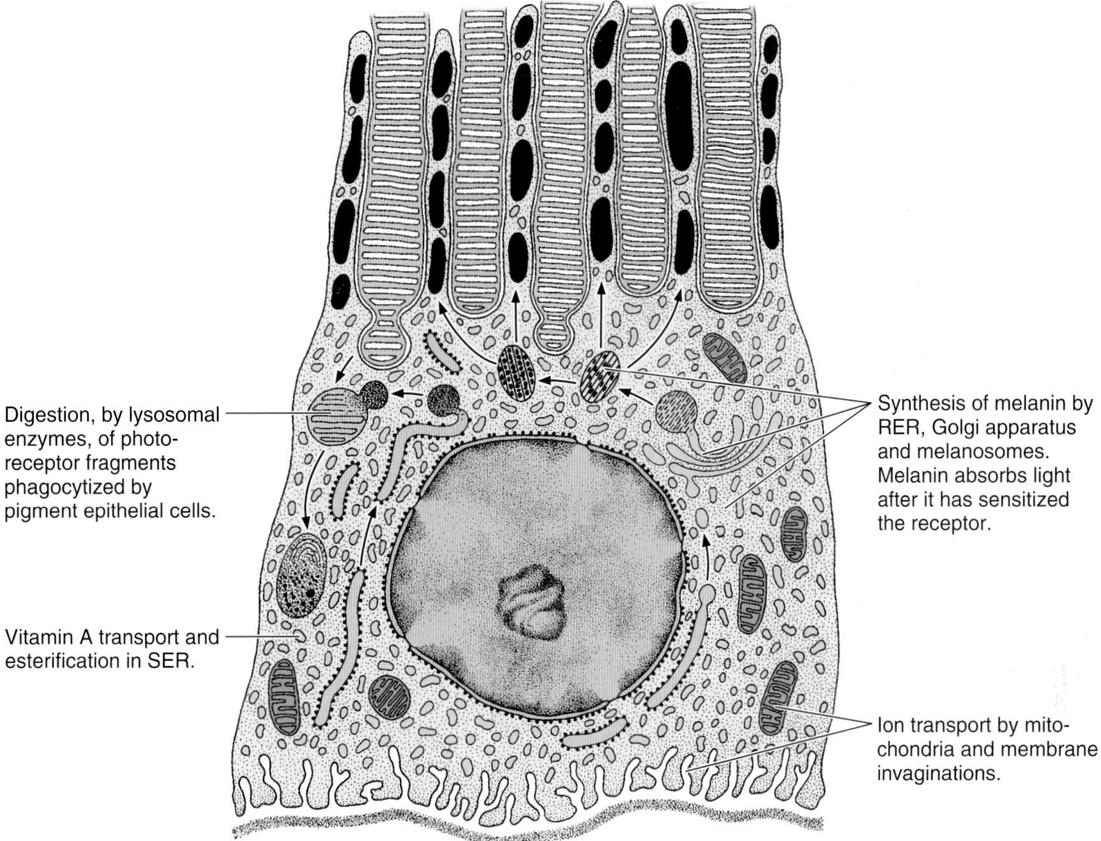

Figure 24–16. Functions of a retinal pigment epithelial cell. Note that the apical portion has abundant cell processes that fill the spaces between the outer segments of the photosensitive cells, and the membrane of the basal region has invaginations into the cytoplasm. This is a cell type with several functions, including the synthesis of melanin granules (by a process described in Chapter 18) that absorb stray light in the eye chamber. This is depicted on the right side of the figure, which shows the organelles that participate in melanin synthesis. On the left side of the figure, lysosomes containing enzymes synthesized in the rough endoplasmic reticulum (RER) coalesce with the phagocytized apical parts of the photoreceptor, digesting them. In addition to these activities, pigment cells are probably active in ion transport, since they maintain an electrical potential between the two surfaces of the epithelium membrane. The relatively well-developed smooth endoplasmic reticulum (SER) participates in the processes of vitamin A esterification.

normal tear layer. The glands of Zeis are smaller, modified sebaceous glands connected with the follicles of the eyelashes. The sweat glands of Moll are unbranched sinuous tubules that begin in a simple spiral and not in a glomerulus like ordinary sweat glands. They empty their secretion into the follicles of the eyelashes.

Lacrimal Apparatus

The lacrimal apparatus consists of the lacrimal gland, canaliculi, lacrimal sac, and nasolacrimal duct. The **lacrimal gland** (Figures 24–18 and 24–19) is a tear-secreting gland located in the anterior superior temporal portion of the orbit. It consists of several separate glandular lobes with 6–12 excretory ducts that connect the gland to the superior conjunctival fornix. (Fornices are the conjunctiva-lined recesses between the lids and the eyeball.) The lacrimal gland is a tubuloalveolar gland that usually has distended lumens and is composed of column-shaped cells of the serous type, resembling the parotid acinar cells. These cells show lightly stained secretory granules, and a basal lamina separates them from the surrounding connective tissue.

Well-developed myoepithelial cells surround the secretory portions of the lacrimal gland. The secretion of the gland passes down over the cornea and the bulbar and palpebral conjunctiva, moistening the surfaces of these structures. It drains into the **lacrimal canaliculi** through the **lacrimal puncta,** which are round apertures about 0.5 mm in diameter on the medial aspect of both the upper and lower lid margins. The canaliculi, which are about 1 mm in diameter and 8 mm long and join to form a common canaliculus just before opening into the lacrimal sac, are lined with a thick stratified squamous epithelium. Diverticuli of the common canaliculus, which may be part of the normal structure, are frequently susceptible to infections.

The lacrimal glands secrete a fluid rich in lysozyme, an enzyme that hydrolyzes the cell walls of certain species of bacteria, facilitating their destruction.

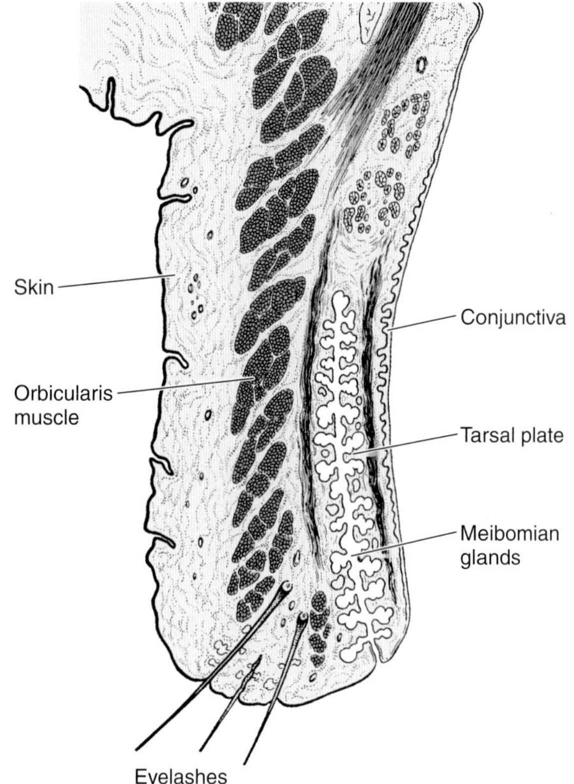

Figure 24–17. Structure of the eyelid.

HEARING: THE AUDIORECEPTOR SYSTEM

The Ear (Vestibulocochlear Apparatus)

The functions of the vestibulocochlear apparatus (Figure 24–20) are related to equilibrium and hearing. The organ consists of 3 parts: the **external ear,** which receives sound waves; the **middle ear,** where sound waves are transmitted from air to bone and by bone to the internal ear; and the **internal ear,** where these vibrations are transduced to specific nerve impulses that pass via the acoustic nerve to the central nervous system. The internal ear also contains the vestibular organ, which maintains equilibrium.

External Ear

The **auricle** (**pinna**) consists of an irregularly shaped plate of elastic cartilage covered by tightly adherent skin on all sides.

The **external auditory meatus** is a somewhat flattened canal extending from the surface into the temporal bone. Its internal limit is the tympanic membrane. A stratified squamous epithelium continuous with the skin lines the canal. Hair follicles, sebaceous glands, and the **ceruminous glands** (a type of modified sweat gland) are found in the submucosa. Ceruminous glands are coiled tubular glands that produce the cerumen—or earwax—a brownish, semisolid mixture of fats and waxes. Hairs and cerumen probably have a protective function. The wall of the external auditory meatus is supported by elastic cartilage in its outer third, whereas the temporal bone provides support for the inner part of the canal.

Across the deep end of the external auditory meatus lies an oval membrane, the **tympanic membrane** (eardrum). Its external surface is covered with a thin layer of epidermis, and its inner surface is covered with simple cuboidal epithelium continuous with the lining of the tympanic cavity (see below). Between the two epithelial coverings is a tough connective tissue layer composed of collagen and elastic fibers and fibroblasts. The tympanic membrane is the structure that transmits sound waves to the ossicles of the middle ear (Figure 24–20).

Middle Ear

The middle ear, or tympanic cavity, is an irregular space that lies in the interior of the temporal bone between the tympanic mem-

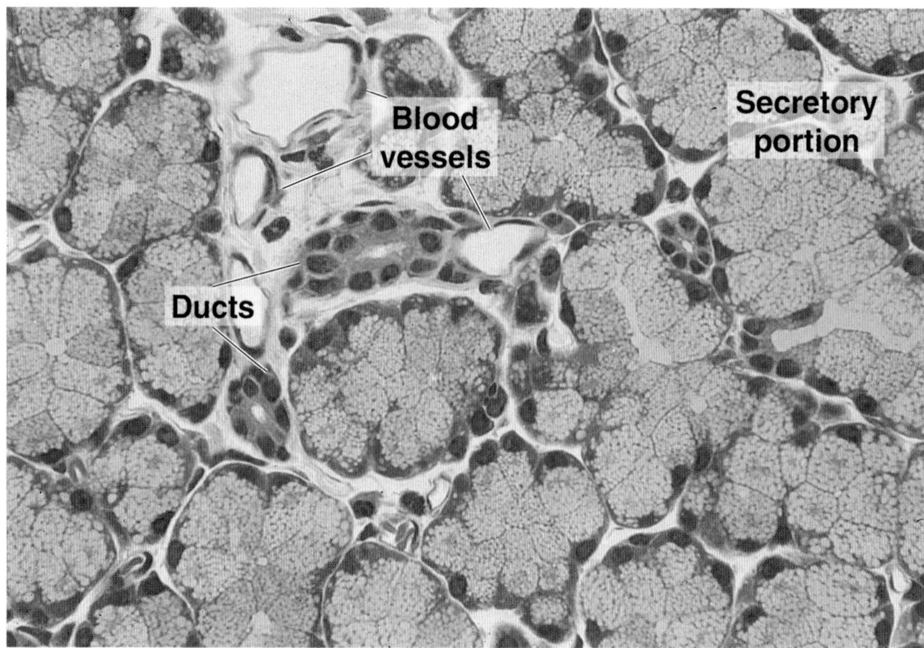

Figure 24–18. Photomicrograph of a section of a lacrimal gland. Note the sections of the tubuloalveolar secretory portions, excretory ducts, and blood vessels. The secretory cells contain only small amounts of RNA (basophilia). They produce secretory material poor in proteins. H&E stain. Medium magnification.

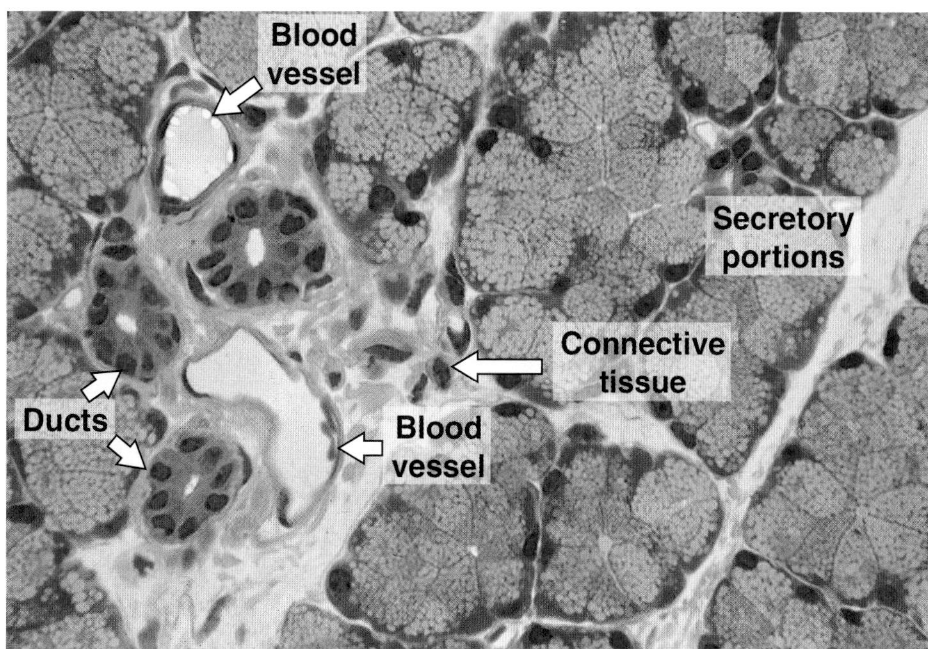

Figure 24–19. Photomicrograph of a section of a lacrimal gland. Three small excretory ducts, 2 blood vessels, and numerous tubuloalveolar secretory units can be seen. PT stain. Medium magnification.

brane and the bony surface of the internal ear. It communicates anteriorly with the pharynx via the **auditory tube** (**eustachian tube**) and posteriorly with the air-filled cavities of the mastoid process of the temporal bone. The middle ear is lined with simple squamous epithelium resting on a thin lamina propria that is strongly adherent to the subjacent periosteum. Near the auditory tube and in its interior, the simple epithelium that lines the middle ear is gradually transformed into ciliated pseudostratified columnar epithelium. Although the walls of the tube are usually collapsed, the tube opens during the process of swallowing, balancing the pressure of the air in the middle ear with atmospheric pressure. In the medial bony wall of the middle ear are 2 membrane-covered oblong regions devoid of bone; these are the **oval** and **round windows** (Figure 24–20).

The tympanic membrane is connected to the oval window by a series of 3 small bones, the **auditory ossicles**—the **malleus, incus,** and **stapes** (Figure 24–20)—which transmit the mechanical vibrations generated in the tympanic membrane to the

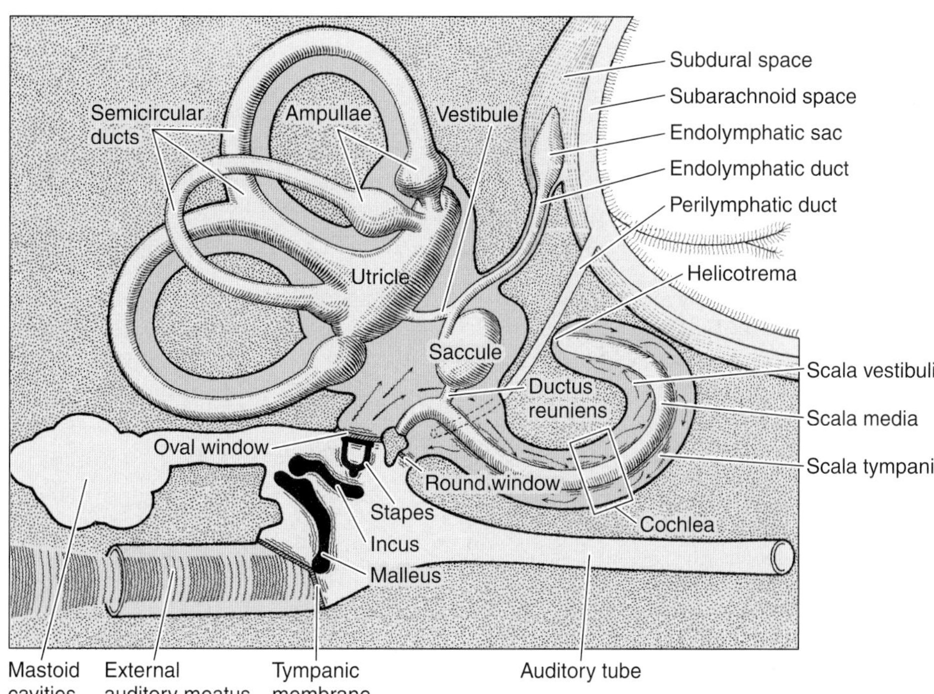

Figure 24–20. The vestibulocochlear organ and the path of sound waves in the external, middle, and internal ear. Components of the internal ear are shown in color. A cross section of the cochlea, as in the rectangular outline, is illustrated in Figure 24–24. (Redrawn and reproduced, with permission, from Best CH, Taylor NB: *The Physiological Basis of Medical Practice,* 8th ed. Williams & Wilkins, 1966.)

internal ear. The malleus inserts itself into the tympanic membrane and the stapes into the membrane of the oval window. These bones are articulated by synovial joints and, like all structures of this cavity, are covered with simple squamous epithelium. In the middle ear are 2 small muscles that insert themselves into the malleus and stapes. They have a function in regulating sound conduction.

Internal Ear

The internal ear is composed of 2 **labyrinths.** The **bony labyrinth** consists of a series of spaces within the petrous portion of the temporal bone that house the **membranous labyrinth** (Figure 24–20). The membranous labyrinth is a continuous epithelium-lined series of cavities of ectodermal origin. It derives from the auditory vesicle that is developed from the ectoderm of the lateral part of the embryo's head. During embryonic development, this vesicle invaginates into the subjacent connective tissue, loses contact with the cephalic ectoderm, and moves deeply into the rudiments of the future temporal bone. During this process, it undergoes a complex series of changes in form, giving rise to 2 specialized regions of the membranous labyrinth: the **utricle** and the **saccule.** The **semicircular ducts** take their origin from the utricle, whereas the elaborate **cochlear duct** is formed from the saccule. In each of these areas, the epithelial lining becomes specialized to form such sensory structures as the **maculae** of the utricle and saccule, the **cristae** of the semicircular ducts, and the **organ of Corti** of the cochlear duct.

The **bony labyrinth** consists of spaces in the temporal bone. There is an irregular central cavity, the **vestibule,** housing the saccule and the utricle. Behind this, 3 **semicircular canals** enclose the semicircular ducts; the anterolateral **cochlea** contains the cochlear duct (Figure 24–20).

The cochlea, about 35 mm in total length, makes two-and-one-half turns around a bony core known as the **modiolus.** The modiolus has spaces containing blood vessels and the cell bodies and processes of the acoustic branch of the eighth cranial nerve (spiral ganglion). Extending laterally from the modiolus is a thin bony ridge, the **osseous spiral lamina.** This structure extends farther across the cochlea in the basal region than it does at the apex (see Figure 24–24).

The bony labyrinth is filled with **perilymph,** which is similar in ionic composition to extracellular fluids elsewhere but has a very low protein content. The membranous labyrinth contains **endolymph,** which is characterized by its low sodium and high potassium content. The protein concentration in endolymph is low.

The Membranous Labyrinth

Saccule and Utricle

The saccule and the utricle are composed of a thin sheath of connective tissue lined with simple squamous epithelium. The membranous labyrinth is bound to the periosteum of the osseous labyrinth by thin strands of connective tissue that also contain blood vessels supplying the epithelia of the membranous labyrinth. In the wall of the saccule and utricle, one can observe small regions, called **maculae,** of differentiated neuroepithelial cells that are innervated by branches of the vestibular nerve (Figure 24–21). The macula of the saccule lies in its floor, whereas the macula of the utricle occupies the lateral wall so the maculae are perpendicular to one another. Maculae in both locations have the same basic histologic structure. They consist of a thickening of the wall and possess 2 types of receptor cells, some supporting cells, and the afferent and efferent nerve endings.

Receptor cells (**hair cells**) are characterized by the presence of 40–80 long, rigid stereocilia, which are actually highly specialized microvilli, and one cilium (Figure 24–21). Stereocilia are arranged in rows of increasing length, with the longest—about 100 μm—located adjacent to a cilium. The cilium has a basal body and the usual 9 + 2 arrangement of microtubules in its proximal portion, but the two central microtubules soon disappear. This cilium is usually called a kinocilium, but it probably is immotile. There are 2 types of hair cells, distinguished by the form of their afferent innervation. Type I cells have a large, cup-shaped ending surrounding most of the base of the cell, whereas type II cells have many afferent endings. Both cell types have efferent nerve endings that are probably inhibitory.

The supporting cells disposed between the hair cells are columnar in shape, with microvilli on the apical surface (Figure 24–21). Covering this neuroepithelium is a thick, gelatinous glycoprotein layer, probably secreted by the supporting cells, with surface deposits of crystals composed mainly of calcium carbonate and called **otoliths,** or **otoconia** (Figures 24–21 and 24–22).

Semicircular Ducts

Semicircular ducts have the same general form as the corresponding parts of the bony labyrinth. The receptor areas in their **ampullae** (Figure 24–20) have an elongated ridgelike form and are called **cristae ampullares.** The ridge is perpendicular to the long axis of the duct. Cristae are structurally similar to maculae, but their glycoprotein layer is thicker; this layer has a conical form called a **cupula** and is not covered with otoliths. The cupula extends across the ampullae, establishing contact with its opposite wall (Figure 24–23).

Endolymphatic Duct and Sac

The endolymphatic duct initially has a simple squamous epithelial lining. As it nears the endolymphatic sac, it gradually changes to tall columnar epithelium composed of 2 cell types; one of these cell types has microvilli on its apical surface and

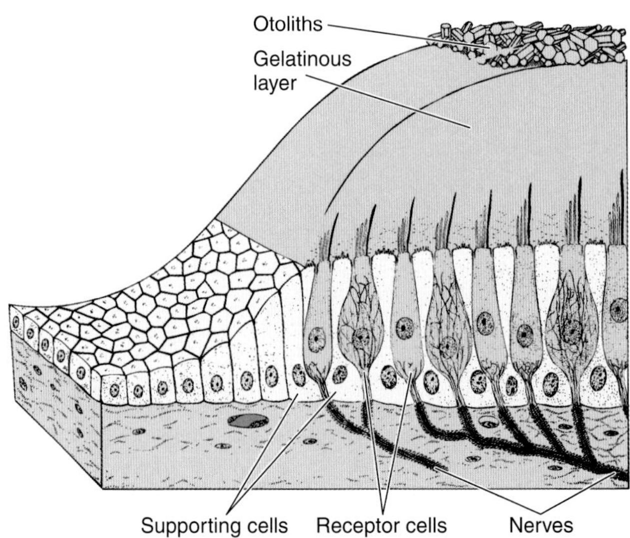

Figure 24–21. Structure of maculae.

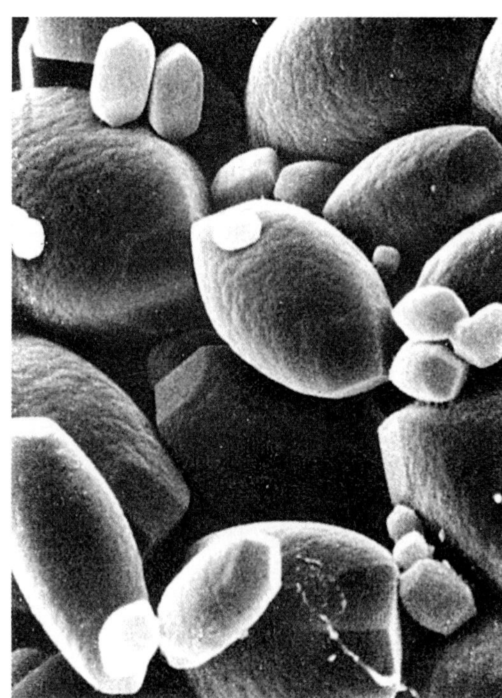

Figure 24–22. **Scanning electron micrograph of the surface of a pigeon's macula showing the otoliths.** (Courtesy of DJ Lim.)

abundant pinocytotic vesicles and vacuoles. It has been suggested that these cells are responsible for the absorption of endolymph and for the endocytosis of foreign material and cellular remnants that may be present in endolymph.

Cochlear Duct

The cochlear duct, a diverticulum of the saccule, is highly specialized as a sound receptor. It is about 35 mm long and is surrounded by specialized perilymphatic spaces. When observed in histologic sections, the cochlea (in the bony labyrinth) appears to be divided into 3 spaces: the **scala vestibuli** (above), the **scala media** (cochlear duct) in the middle, and the **scala tympani** (Figure 24–24). The cochlear duct, which contains endolymph, ends at the apex of the cochlea. The other two scalae contain perilymph and are in reality one long tube, beginning at the **oval window** and terminating at the **round window** (Figure 24–20). They communicate at the apex of the cochlea via an opening known as the **helicotrema.**

The cochlear duct has the following histologic structure (Figure 24–24). The **vestibular (Reissner's) membrane** consists of 2 layers of squamous epithelium, one derived from the scala media and the other from the lining of the scala vestibuli. Cells of both layers are joined by means of extensive tight junctions that help preserve the very high ionic gradients across this membrane. The **stria vascularis** is an unusual vascularized epithelium located in the lateral wall of the cochlear duct. It consists of cells that have many deep infoldings of their basal plasma membranes, where numerous mitochondria are located. These characteristics indicate that they are ion- and water-transporting cells, and it is generally believed that they are responsible for the characteristic ionic composition of endolymph.

The structure of the internal ear that contains special auditory receptors is called the **organ of Corti;** it contains hair cells that respond to different sound frequencies. It rests on a thick layer of ground substance—the **basilar membrane.** Supporting cells and 2 types of hair cells can be distinguished. Three to 5 rows of **outer hair cells** can be seen, depending on the distance from the base of the organ, and there is a single row of **inner hair cells.** The most characteristic feature of these cells is the W-shaped (outer hair cells) or linear (inner hair cells) array of stereocilia (Figure 24–25). A basal body is found in the cytoplasm adjacent to the tallest stereocilia. In contrast to vestibular receptors, no kinocilium is present. This absence of a kinocilium imparts symmetry to the hair cell that is important in sensory transduction.

The tips of the tallest stereocilia of the outer hair cells are embedded in the **tectorial membrane,** a glycoprotein-rich secretion of certain cells of the spiral limbus (Figure 24–24).

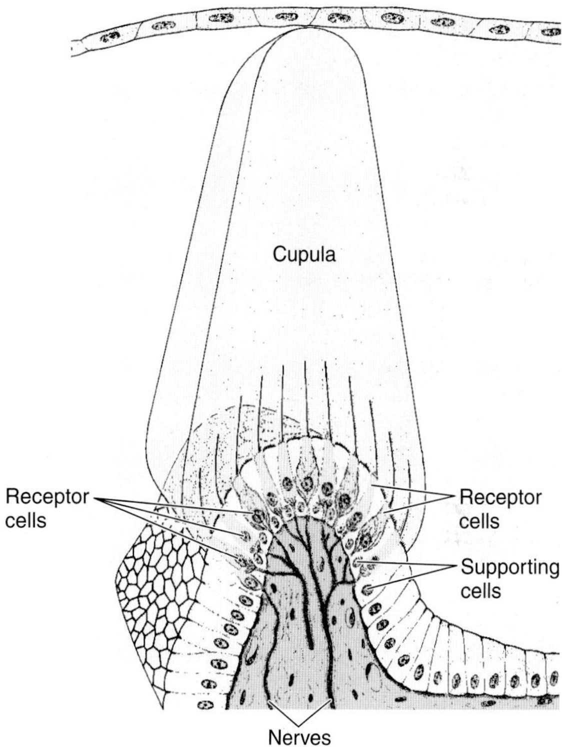

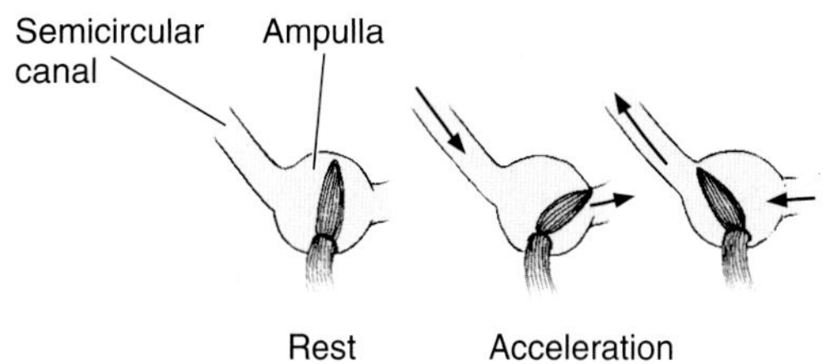

Figure 24–23. **Crista ampullaris. Top: Structure of the crista ampullaris. Bottom: Movements of the cupula in a crista ampullaris during rotational acceleration. Arrows indicate the direction of fluid movement.** (Redrawn and reproduced, with permission, from Wersall J: Studies of the structure and innervation of the sensory epithelium of the cristae ampullares in the guinea pig. Acta Otolaryngol [Stockh] Suppl 1956;126:1.)

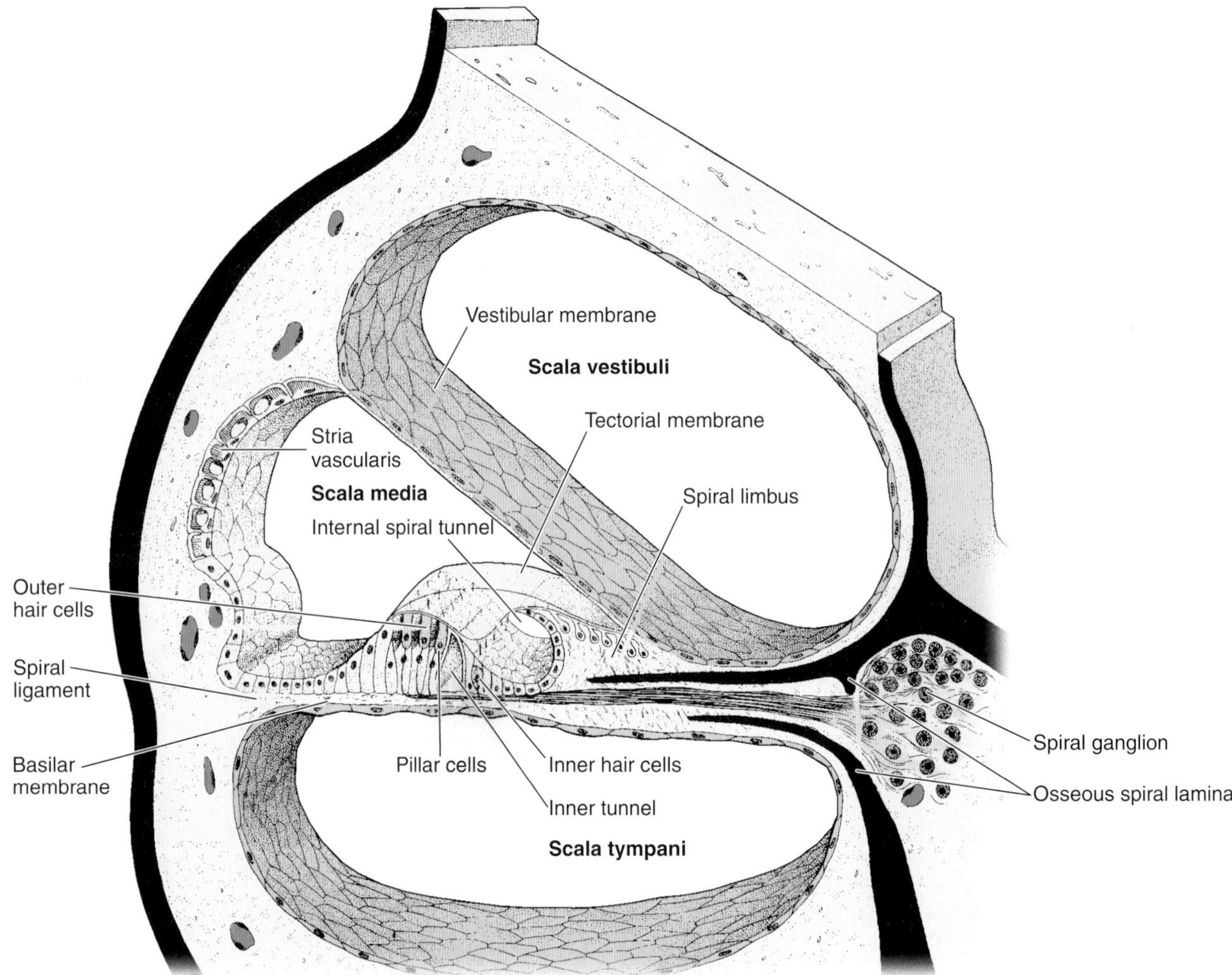

Figure 24–24. Structure of the cochlea. (Redrawn and reproduced, with permission, from Bloom W, Fawcett DW: *A Textbook of Histology,* 9th ed. Saunders, 1968.)

Of the supporting cells, the **pillar cells** should be singled out for special mention. Pillar cells contain a large number of microtubules that seem to impart stiffness to these cells. They outline a triangular space between the outer and inner hair cells—the **inner tunnel** (Figure 24–24). This structure is of importance in sound transduction.

Both outer and inner hair cells have afferent and efferent nerve endings. Although the inner hair cells have by far the greater afferent innervation, the functional significance of this difference is not understood. The cell bodies of the bipolar afferent neurons of the organ of Corti are located in a bony core in the modiolus and constitute the spiral ganglion (Figure 24–24).

Histophysiology of the Internal Ear

Vestibular Functions

An increase or decrease in the velocity of circular movement—angular acceleration or deceleration—results in a flow of fluid in the semicircular ducts as a consequence of the inertia of the endolymph. This induces a corresponding movement of the cupula over the crista ampullaris and results in the bending of the stereocilia on the sensory cells. Measurement of electrical impulses along vestibular nerve fibers indicates that movement of the cupula in the direction of the kinocilium results in excitation of the receptors, accompanied by the action potentials in the vestibular nerve fibers. Movement in the opposite direction inhibits neuronal activity. When uniform movement returns, acceleration ceases; the cupula returns to its normal position; and excitation or inhibition of the receptors no longer occurs (Figure 24–23).

The semicircular ducts respond to fluid displacement and therefore body position after angular acceleration. The maculae of the saccules and the utricles respond to linear acceleration. Because of their greater density, the otoliths are displaced when there is a change in the position of the head. This displacement is transferred to the underlying hair cells via the gelatinous otolithic membrane. Deformation of stereocilia of the hair cells results in action potentials that are carried to the central nervous system by the vestibular branch of the eighth cranial nerve. Maculae are thus sensitive to the force of gravity on the otoliths. The vestibular apparatus is important for perceiving movement and orientation in space and for maintaining equilibrium or balance.

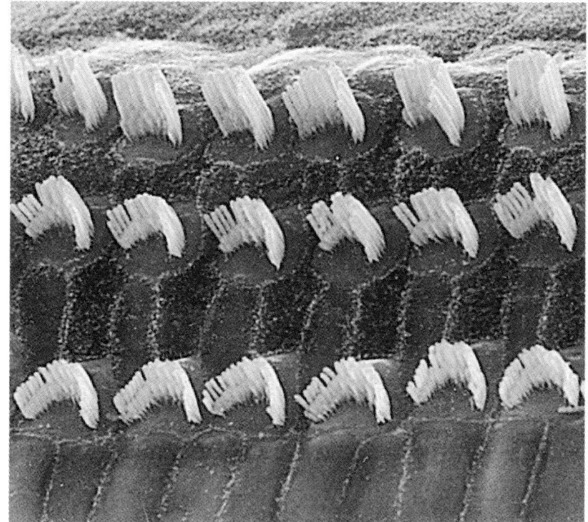

A

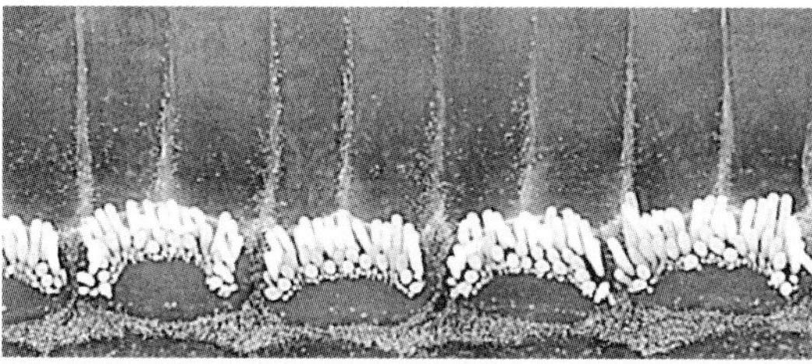

B

Figure 24–25. Scanning electron micrograph of 3 rows of outer hair cells (**A**) and a single row of inner hair cells (**B**) in the middle turn of a cat cochlear duct. ×2700. (Courtesy of P Leake.)

Auditory Functions

Sound waves impinging on the tympanic membrane set the auditory ossicles into motion. The large difference in area of the tympanic membrane and the footplate of the stapes ensures the efficient transmission of mechanical motion from air to the fluids of the internal ear. Two striated skeletal muscles are found in the middle ear—the **tensor tympani** muscle (attached to the malleus) and the **stapedius** muscle (attached to the stapes). Loud sounds cause reflex contractions of these muscles, which limit excursions of the tympanic membrane and the stapes; this helps prevent damage to the internal ear. These reflexes are too slow, however, to guard against sudden loud sounds, such as gunshots.

The following is a step-by-step explanation of how sound waves are converted to electrical impulses in the internal ear (Figure 24–20). Sound waves are longitudinal waves, with **compression** and **rarefaction** phases. The compression phase causes the stapes to move inward. Since the fluids of the internal ear are almost incompressible, the pressure change is transmitted across the vestibular membrane and the basilar membrane, causing them to be deflected downward toward the scala tympani. This pressure change also causes the covering of the round window to bulge outward, thereby relieving the pressure. Because the tips of the pillar cells form a pivot, downward deflection of the basilar membrane is converted into lateral shearing of the stereocilia of hair cells against the tectorial membrane. The tips of the stereocilia are deflected toward the modiolus and away from the basal body.

During the rarefaction phase of the sound wave, everything is reversed: The stapes move outward, the basilar membrane moves upward toward the scala vestibuli, and the stereocilia of the hair cells bend toward the stria vascularis and the basal body. Deflection in this direction sets up depolarizing generator potentials in the hair cells, resulting in the release of a neurotransmitter (whose chemical nature is unknown) that causes the production of action potentials in bipolar neurons of the spiral ganglion (**excitation**).

Discrimination between sound frequencies is based on the response of the basilar membrane. The membrane responds to the frequency of sound with different displacement at different points along its length. High frequencies are detected at the basal end of the membrane, whereas low frequencies are detected in the apex of the organ of Corti. This **tonotopic** localization can be correlated with the width and stiffness of the basilar membrane: The narrow basilar membrane, with greater stiffness at the base, responds best to high-frequency sounds.

REFERENCES

The Eye

Bok D, Hall MO: The role of the retinal pigment epithelium in the etiology of inherited retinal dystrophy in the rat. J Cell Biol 1971;49:664.

Botelho SY: Tears and the lacrimal gland. Sci Am 1964;211:78.

Dowling JE: Organization of vertebrate retinas. Invest Ophthalmol 1970;9:665.

Hogan MJ et al: *Histology of the Human Eye.* Saunders, 1971.

McDevitt D (editor): *Cell Biology of the Eye.* Academic Press, 1982.

Nguyen LJ, Hicks D: Renewal of photoreceptor outer segments and their phagocytosis by the retinal pigment epithelium. Int Rev Cytol 2001;196:245.

Young RW: Visual cells and the concept of renewal. Invest Ophthalmol 1976;15:700.

The Ear

Dallos P: The active cochlea. J Neurosci 1992;12:4575.

Hudspeth AJ: The hair cells of the inner ear. Sci Am 1983;248:54.

Kimura RS: The ultrastructure of the organ of Corti. Int Rev Cytol 1975;42:173.

Lim DJ: Functional structure of the organ of Corti: a review. Hear Res 1986;22:117.

Index

Index

NOTE: Page numbers in **boldface** type indicate a major discussion. A *t* following a page number indicates tabular material, an *f* following a page number indicates a figure, and a *b* following a page number indicates a medical application box.